Heavy-Duty
Truck Systems
Third Edition

Ian Andrew Norman
John A. Corinchock
Sean Bennett

DELMAR

™

THOMSON LEARNING

Australia   Canada   Mexico   Singapore   Spain   United Kingdom   United States

DELMAR

™

THOMSON LEARNING

Heavy-Duty Truck Systems Third Edition

Ian Andrew Norman

John A. Corinchock

Sean Bennett

Business Unit Director:
Alar Elken

Executive Marketing Manager:
Maura Theriault

Production Editor:
Barbara L. Diaz

Developmental Editor:
Christopher Shortt

Channel Manager:
Mona Caron

Art/Design Coordinator:
Cheri Plasse

Editorial Assistant:
Matthew Seeley

Executive Production Manager:
Mary Ellen Black

For permission to use material from this text or product, contact us by
Tel   (800) 730-2214
Fax   (800) 730-2215
www.thomsonrights.com

Library of Congress Cataloging-in-Publication Data
Norman, Andrew.
   Heavy-duty truck systems/Ian Andrew Norman, Sean Bennett, John A. Corinchock—3rd ed.
      p. cm.
   Includes index.
   ISBN 0-7668-1340-1
   1. Trucks—Maintenance and repair. I. Bennett, Sean.
   II. Corinchock, John A. III. Title.
TL230.2.N67 N67 2000
629.28'74—dc21
                                          00-040466

NOTICE TO THE READER

Publisher does not warrant or guarantee any of the products described herein or perform any independent analysis in connection with any of the product information contained herein. Publisher does not assume, and expressly disclaims, any obligation to obtain and include information other than that provided to it by the manufacturer.

The reader is expressly warned to consider and adopt all safety precautions that might be indicated by the activities herein and to avoid all potential hazards. By following the instructions contained herein, the reader willingly assumes all risks in connection with such instructions.

The publisher makes no representation or warranties of any kind, including but not limited to, the warranties of fitness for particular purpose or merchantability, nor are any such representations implied with respect to the material set forth herein, and the publisher takes no responsibility with respect to such material. The publisher shall not be liable for any special, consequential, or exemplary damages resulting, in whole or part, from the readers' use of, or reliance upon, this material.

Contents

Photo Sequences

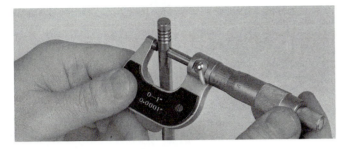

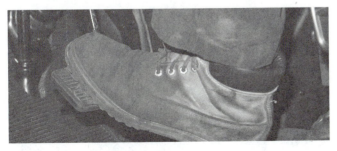

11 Wheel-End Procedure: TMC Method of Bearing Adjustment 774

12 Installing a Wheel Hub Seal 779

13 Wheel Speed Sensor Testing 867

14 Rebuilding an Air Dryer with an Integral Cartridge 904

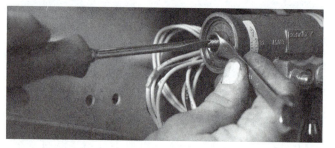

15 Governor Adjustment 906

16 Check Free-Stroke on S-Cam Foundation Brakes 928

17 Wheel-Down Brake Adjustment Procedure on a Tractor/Trailer 929

18 Wheel-Up Brake Adjustment Procedure on a Tractor/Trailer 930

19 Measuring a Brake Rotor 936

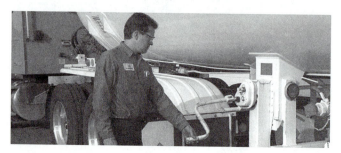

20 Trailer Coupling 994

Preface

ABOUT THE BOOK

I was flattered when I was asked to head up a third revision of *Heavy Duty Truck Systems.* The success of this textbook has been due to its clear-cut delivery of information in language that entry-level technicians can understand. *Heavy Duty Truck Systems* has established itself as the textbook of choice in most college programs providing specialty truck technician training and as such has become the backbone of technical understanding for tens of thousands of technicians in North America. In undertaking this third edition, I have attempted to adhere to the spirit of the previous two editions while adding the new content essential for the truck technician of the 21st century.

CHANGES IN TECHNOLOGY

During the past decade and a half, truck technology has probably been subject to more change than in the 40 years previous to 1985. During the past 15 years, we have gone from almost no use of computer technology in truck components and systems to an acceptance that microprocessors will play a role in most of the chassis systems on a truck. This new reliance on computer technology has changed the type of technician needed by the trucking industry. The technician of the new millennium will be required to have all of the mechanical skills of the truck technician of 20 years ago, but additionally will be required to understand and repair computer-managed systems and be fully computer literate in non-vehicle computer systems. Even for a technician whose specialty requires little contact with chassis computer systems, computer literacy is mandated today because it is the means by which we organize our business life, communicate with each other, download service repair information, track time, dispatch trucks, and network with data hubs. A clear objective of this third revision of *Heavy Duty Truck Systems* has been to prepare technicians for the real world of the contemporary truck service facility.

NEW TO THIS EDITION

The original sequence of chapters has been more or less adhered to. A number of reviewers felt that the chapter on preventive maintenance (PM) should be located close to the front of the book, as many of the procedures are practiced by entry-level technicians. This is true, and because this has become the fourth chapter of the book, the reader is informed that in order to properly understand some of the PM content it is necessary to consult information in later chapters. The clustering of chapters by subject used in the second edition has otherwise not changed. Most formal programs of study will require the user to jump around the chapters rather than strictly follow the chapter sequence used in the book. It should be noted, however, that the electricity and electronics chapters appear toward the beginning of the book for a reason. A foundation in electricity and electronics is required before attempting to progress to any chapter that deals with computer controls.

This edition uses a more detailed approach to the subject matter of electricity and electronics. The original two chapters dedicated to this subject matter have been expanded to four, two devoted to electrical and electronics theory and two devoted to the practical application of the theory. A complete introduction to electronic service tools (ESTs) and how to use them is also provided. Commonly used ESTs such as digital multimeters, ProLink, and PCs are covered from the entry-level user perspective.

The four chapters on transmissions in the second edition have been expanded by three chapters. The three new chapters address new technology transmissions that have been identified as electronically automated standard transmissions, partial-authority electronically controlled automatics, and full-authority electronically controlled automatics. Most automatics used in truck and bus chassis today are electronically controlled, and the textbook addresses these technologies at a beginner level. An especially important innovation in recent years is the electronically automated standard transmission. This essentially adapts an established standard transmission to computer controls, greatly reducing driver fatigue and the driver skill required to operate it.

The four chapters addressing truck brake systems have also been reorganized. The reorganization addresses some of the recent federal requirements of truck brake systems and clearly separates air brake theory, hydraulic brakes, ABS, and diagnosis and repair of air brake systems. Here the focus is primarily on the air brake systems used by most medium- and heavy-duty highway trucks but also includes the hydraulic brakes used on older and light-duty trucks.

Toward the end of the book, both the chapters on fifth wheels and air conditioning systems have been reworked. The fifth wheel chapter has been expanded to explain the operating principles of a wider range of current fifth wheels. The air conditioning chapter now has a more thorough approach to basic thermodynamics and more detail on the current HFC refrigerant-type A/C systems.

STUDENTS IN THE TEACHING AND LEARNING PROCESS

Teachers who are experienced technicians will often confess that they never truly understand a technology until they have to teach it. The classroom and lab have a way of putting technology under a microscope, and this aspect of teaching and learning depends on the students. So, finally, I would like to thank my many students at Centennial, Mack Trucks Corporate, and Freightliner Training and Apprenticeship who have asked the thousands of questions that have ensured that I remain a perpetual learner.

Sean Bennett, Centennial College, Toronto, April 2000

Features of the Text

Learning how to maintain and repair heavy-duty truck systems can be a daunting endeavor. To guide the readers through this complex material, we have built in a series of features that will ease the teaching and learning processes.

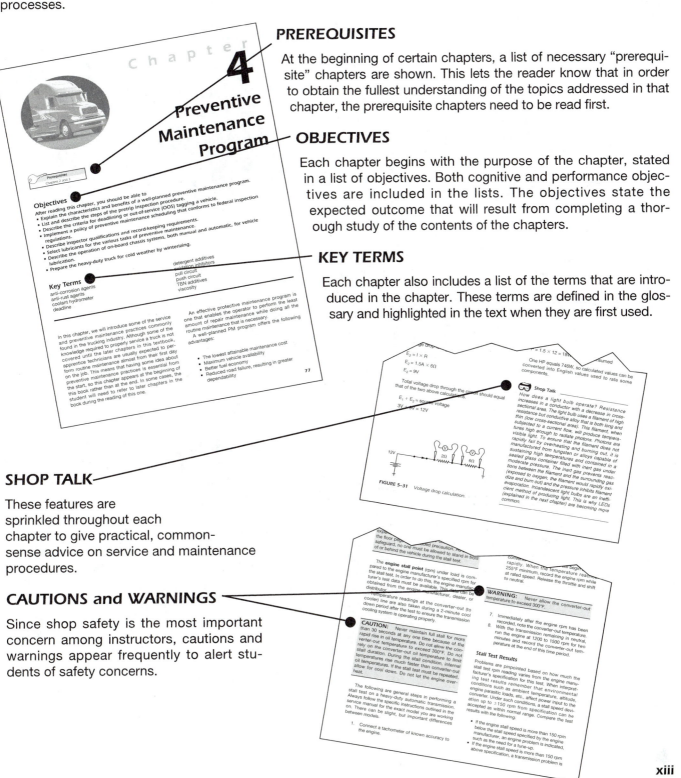

PREREQUISITES

At the beginning of certain chapters, a list of necessary "prerequisite" chapters are shown. This lets the reader know that in order to obtain the fullest understanding of the topics addressed in that chapter, the prerequisite chapters need to be read first.

OBJECTIVES

Each chapter begins with the purpose of the chapter, stated in a list of objectives. Both cognitive and performance objectives are included in the lists. The objectives state the expected outcome that will result from completing a thorough study of the contents of the chapters.

KEY TERMS

Each chapter also includes a list of the terms that are introduced in the chapter. These terms are defined in the glossary and highlighted in the text when they are first used.

SHOP TALK

These features are sprinkled throughout each chapter to give practical, common-sense advice on service and maintenance procedures.

CAUTIONS and WARNINGS

Since shop safety is the most important concern among instructors, cautions and warnings appear frequently to alert students of safety concerns.

PHOTO SEQUENCES

Step-by-step photo sequences illustrate practical shop techniques. The photo sequences focus on techniques that are common, need-to-know service and maintenance procedures. These photo sequences give students a clean, detailed image of what to look for when they perform these procedures.

PHOTO SEQUENCE 2
REPAIRING DAMAGED THREADS

Tools and Fasteners   69

P2–1   Using a threaded pitch gauge, determine the thread size of the fastener that should fit into the damaged internal threads.

P2–2   Select the correct size and type of tap for the threads and bore to be repaired.

P2–3   Install the tap into a tap wrench.

P2–4   Start the tap squarely in the threaded hole using a machinist square as a guide.

P2–5   Rotate the tap clockwise into the bore until the tap has been run through the entire length of the threads.

P2–6   Drive the tap back out of the hole by turning it counterclockwise.

P2–7   Clean the metal chips left by the tap out of the hole.

P2–8   Inspect the threads left by the tap to be sure they are acceptable.

P2–9   Test the threads by threading the correct fastener into the threaded hole.

SUMMARY

Highlights and key bits of information from the chapter are listed at the end of each chapter. This listing is designed to serve as a refresher for the reader.

REVIEW QUESTIONS

A combination of short-answer essay, fill-in-the-blank, multiple-choice, and ASE-style questions make up the end-of-chapter questions. Different question types are used to challenge the reader's understanding of the chapter's contents. The chapter objectives are used as the basis for the review questions.

Supplements

INSTRUCTOR'S ANSWER KEY

Contains the complete list of objectives for each chapter that correspond with the text, answers to the text review questions, and answers to the practice questions from the student workbook. *The Instructor's Answer Key* also includes ten additional test questions and answers for each chapter.

WORKBOOK

This student workbook gives hands-on, practical experience. It contains a list of objectives and practice questions for each chapter as well as interactive job sheets with procedures. Each job sheet includes performance objectives, tools, materials, and a list of protective clothing needed to complete the procedures.

ACKNOWLEDGMENTS

In thanking those who have assisted with the development of this third edition I would like to begin with the original authors, Andy Norman, Robert Scharff, and John Corinchock, whose efforts still provide the backbone of *Heavy Duty Truck Systems*. In addition, the feedback from my reviewers identified in the credits also proved to be invaluable. My colleagues at Centennial College helped me with comments on the early drafts of many of the new chapters, and in this respect I am particularly indebted to John Murphy. John's incisive critiquing of most of the new content for the third edition was key to making this a better book. I would also like to thank the many technicians who took the trouble to teach me the importance of fundamentals in my early days with a Freightliner dealership and later at Mack Trucks Incorporated.

REVIEWERS

We would like to acknowledge and thank the following dedicated and knowledgeable educators for their comments, criticisms, and suggestions during the review process:

Leland Redding, Texas State Technical Institute
Alan B. Clark, Layne Community College
Michael Henich, Linn Benton Community College
Rick Higinbotham, Ohio Auto Diesel Technical College
Winston Ingraham, University College of Cape Breton
Kenneth W. Kephart, Central Texas College
Walter Brueggeman, Tidewater Community College
Douglas Bradley, Utah Valley State College
John Stone, Washington State Community College
Robert D. Ressler, Lebannon Co. Career and Technology Center
John Murphy, Centennial College

Ted Hrdlicka, Denver Automotive and Diesel College
T. Grant Ralston, Fairview College
Terryl Lindsey, Oklahoma State University—Okmulgee
Dexter Rammage, Santa Rosa Junior College
Mark Koslan, Texas State Technical College
Dennis Chapin, Rogue Community College
Robert Van Dyke, Denver Automotive and Diesel College
Steve Musser, Wyoming Technical Institute
Doug Anderson
Wayne Lehnert

INDIVIDUALS

Ken Attwood, Centennial College
Jim Bardeau, Centennial College
Dan Bloomer, Centennial College
Dave Coffey, Toromont Caterpillar
Dara Cruickshank-Bennett, Angharan Consulting
John Dixon, Centennial College
Dave Drummond, Mack Trucks Inc.
Owen Duffy, Centennial College
Dave Emburey, Eaton Dana Corporation
Lou Gilbert, Allison Transmission
Terry Harkness, Toromont Caterpillar

John Holloway, Freightliner Corp.
Helmut Hryciuk, Centennial College
Serge Joncas, Volvo Training
George Liidemann, Centennial College
Dave Morgan, Centennial College
John Murphy, Centennial College
Fred Pedler, Haldex Midland
Wayne Scott, Freightliner Corp.
Angelo Spano, Centennial College
Gino Tamburro, Centennial College
Gus Wright, Centennial College

CONTRIBUTING COMPANIES

We would like to thank the following companies who provided technical information and art for this edition:

AC Delco
Accuride Corporation
Aidco International, Inc.
Air-O-Matic Power Steering, Div. of Sycon Corp.
Alcoa Wheel Products International
Allison Transmission Division of GM
American Steel Foundries
American Trailer Industries Inc.
APA Engineered Solutions (Power-Packer)
ArvinMeritor Inc.
Auto Electric Supplier and Equipment Co.
Automotive Educational Products, Inc.
Battery Council International
BeeLine Co., Bettendorf, IA
Bendix Commercial Vehicle Systems
Bostrom Seating, Inc.
Bridgestone/Firestone, Inc.
Carrier Transicold
Central Tools, Inc.
Chicago Rawhide
Commercial Carrier Journal
CR Industries
Dalloz Safety
Dayton-Walther Corporation
Deere & Company
Delco-Remy Co.
Dorsey Trailers, Inc.
DuPont Co.
Eaton Corp.—Eaton Clutch Div.
Eaton Corporation
Firestone Industrial Products Co.
Fontaine Fifth Wheel®
Freightliner Trucks
Goodson Shop Supplies
Haldex Brake Systems
Heavy Duty Trucking
Heavy Duty Trucking/Andrew Ryder
Hendrickson International, Truck Suspension
   Systems

Hennessy Industries, Inc.
Holland Hitch Co.
Hunter Engineering Company
International Business Machines
International Truck and Engine Corp.
International Truck and Engine Transportation
   Corporation
International Truck and Freight Corp.
Jacobs Vehicle Systems
James Winsor, Newport Communications
Mack Trucks, Inc.
MPSI (Micro Processor Systems, Inc.)
Penske Truck Leasing Corp.
Peterbilt Motors Co.
Phillips Temro, Inc.
Prestolite Electric Corp.
R.H. Sheppard Co., Inc.
Roadranger Marketing
Robinair, SPX Corporation
RTI Technologies, Inc.
Snap-On Tools Company
Spicer Universal Joint Division/Dana Corporation
Stanley Tools, a Product Group of The Stanley Works,
   New Britain, CT
The Budd Company
The Hearst Corporation
The Holland Group Inc.—Anchorlok Div.
The Maintenance Council of the American Trucking
   Associations, Inc.
The Protectoseal Company
Trailmobile, Inc.
TRW Chassis Systems
TTMA
Utah Technical College
Voith Transmissions Inc.
Volvo Trucks North America, Inc.
Vulcan Materials Company
Wagner Brake, div. of Federal-Mogul Corp.
White Industries, Div. of K-Whit Tools, Inc.

Introduction to Servicing Heavy-Duty Trucks

Objectives

After reading this chapter, you should be able to
- Explain the basic truck classifications.
- Define Gross Vehicle Weight (GVW).
- Classify a truck by the number of axles it has.
- Identify an on-highway truck's major systems and its related components.
- Identify various career opportunities in the heavy-duty trucking industry.
- Understand and explain the basic job classifications the heavy-duty truck industry offers to trained technicians with a sound understanding of truck systems.
- Understand the ASE certification program and how it benefits technicians.

Key Term List

air brakes
axle
dedicated contract carriage
differential
drive shaft
end yoke
fifth wheel
franchised dealership
gross vehicle weight (GVW)
heavy-duty truck
kingpin
LTL carriers

National Automotive Technicians Education Foundation (NATEF)
National Institute for Automotive Service Excellence (ASE)
National Highway Traffic Safety Administration (NHTSA)
nose
original equipment manufacturer (OEM)
power train
semitrailer landing gear assembly
specialty service shops
splined yokes
suspension
torque
universal joints

Well over 200 million vehicles are registered for use on North American highways, more than 15 percent of which are commercially used trucks. For most of the past twenty years, through periods of boom and recession, a shortage of truck technicians has existed throughout the continent. The job opportunities and rates of pay in the trucking industry have never been better. Although the modern highway truck requires much less service work to keep it in top mechanical condition, most service work per- formed by a truck technician today requires a higher level of skills.

Good truck technicians are in high demand. A good truck technician today is required not just to diagnose and repair trucks but also to be computer-literate, regularly update technical knowledge, and practice customer service skills. Perhaps more than any other, the skill required of the truck technician is that of being a lifelong learner to keep abreast of the fast-changing technology of this industry.

1.1 TRUCK CLASSIFICATIONS

Trucks are classified by their **gross vehicle weight (GVW)** or weight of the vehicle and the weight of the load it can safely carry. There are three classes of "light-duty" trucks, three classes of "medium-duty" trucks, and two classes of "heavy-duty" trucks (**Table 1–1**). A **heavy-duty truck** has a GVW of 26,001 pounds or more. **Table 1–2** lists the major truck manufacturers of Class 7 and Class 8 heavy-duty trucks.

TABLE 1–1:	TRUCK WEIGHT CLASSIFICATIONS	
Light-duty	Class 1	up to 6000 GVW*
	Class 2	6001–10,000 GVW*
	Class 3	10,001–14,000 GVW*
Medium-duty	Class 4	14,001–16,000 GVW*
	Class 5	16,001–19,500 GVW*
	Class 6	19,501–26,000 GVW*
Heavy-duty	Class 7	26,001–33,000 GVW*
	Class 8	33,001 GVW* and over

*Gross Vehicle Weight in Pounds

TABLE 1–2:	MAJOR TRUCK MANUFACTURERS
Class 7 and 8	
Manufacturer	
Freightliner	Sterling
Navistar	Peterbilt
Kenworth	Volvo Heavy Truck
Mack	Western Star
Class 7	
GMC	Mercedes-Benz
Chevrolet	Mitsubishi Fuse
Hino	Nissan Diesel (UD Trucks)
Isuzu	Mack (Renault)

A truck is also classified by the number of axles it has. For example, a tractor with a tandem (close-coupled pairs) rear axle will be either a 6 × 2 or a 6 × 4 (**Figure 1–1A & B**). The first number refers to the total number of wheels (or sets of wheels in the case of dual wheels) and the second number indicates the number of wheels that are driven by the **power train** (which consists of a drive shaft, coupling, clutch, and transmission-differential). A tractor with a tandem rear axle where only the forward rear axle is driven

A

B

FIGURE 1–1   (A) 4 × 2 Class 7 truck; and (B) 6 × 4 Class 8 truck.

TABLE 1–3:	TRUCK CLASSIFICATION BY WHEEL NUMBER			
Motor Vehicle	Total Wheels	Driven Wheels	Total Axles	Drive Axles
4 × 2	4	2	2	1
4 × 4	4	4	2	2
6 × 2	6	2	3	1
6 × 4	6	4	3	2
6 × 6	6	6	3	3
8 × 4	8	4	4	2
8 × 8	8	8	4	4

would be a 6 × 2; it has six wheels, but only two wheels drive the vehicle. **Table 1–3** lists the common axle wheel types and their driven wheels and axles. As shown in **Figure 1–2A-D** and described in Chapter 29, there are many trailer designs, sizes, and applications of semi- and twin-trailers that are hauled by truck tractors. Some semitrailers are used to carry a wide variety of cargo, while others such as refrigerated (reefers), cement carriers, and tank trailers are designed to haul specific products.

Tractor semitrailer dimensions are provided in **Figure 1–3**. An important consideration in determining tractor/semitrailer dimensions is the distance between the two vehicles when hooked together.

FIGURE 1–2   Typical trailers: (A) van; (B) platform; (C) reefer; and (D) cement carrier. (Courtesy of Mack Trucks Inc.)

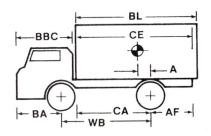

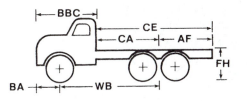

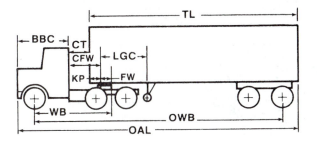

A	Distance from centerline of rear axle to centerilne of body and/or payload. Centerline of body (as 1/2 body length)
AF	Center or rear axle to end of frame
BA	Bumper to centerline of front axle
BBC	Bumper to back of cab
BL	Body length
CA	Back of cab to centerline of rear axle or tandem suspension
CE	Back of cab to end of frame
CFW	Back of cab to centerpoint of kingpin hole in fifth wheel
CT	Back of cab to front of semitrailer in straight-ahead relationship
FH	Frame height
FW	Centerline of rear axle or tandem to centerpoint of fifth wheel
KP	Kingpin setting—front of semitrailer to centerpoint of kingpin on semitrailer
LGC	Landing gear clearance—center point of kingpin to nearest interface point of landing gear assembly
OAL	Overall length
OWB	Overall wheel base
TL	Semitrailer length
WB	Wheel base—distance between centerline of front and rear axle or tandem suspension

TERMS
Chassis: Basic vehicle-cab, frame, and running gear
Body: Container in which the load is carried
Payload: Commodity to be carried
Curb Weight: Weight of chassis only
Body Weight: Weight of complete body to be installed on chassis
Payload Weight: Weight of commodity to be carried
Gross Vehicle Weight (GVW): Total or curb, body, and payload weight

FIGURE 1–3   Important truck/trailer dimensions and terms. (Courtesy of Heavy Duty Trucking)

There must be enough space between the tractor cab and semitrailer front (**nose**), as well as between the rear of the tractor and **semitrailer landing gear assembly**, to allow for sharp turns and the effect of grade changes.

1.2 HEAVY-DUTY TRUCKS

Heavy-duty truck technicians need to know about the systems and components that power or move, slow and stop, control, direct, support, and stabilize a tractor/trailer. **Figure 1–4** shows some of the components that will be discussed in this book. The following major systems are found in on-highway trucks.

More information on the classification of trailers can be found in Chapter 29.

ENGINES

Heavy-duty trucks are powered almost always by diesel engines (**Figure 1–5**), and are usually serviced by engine technicians. Some vehicles are powered by gasoline engines. The diesel costs less to operate, is more dependable, requires less downtime for repairs, and is capable of generating more power (140 to 600 horsepower) and **torque** (180 to 2000 lb.-ft.) than a gasoline engine. Power is the engine's ability to move, or propel the vehicle, and is measured in horsepower; torque is the potential ability of a rotating element (gear or shaft) to overcome turning resistance and is measured in pounds per square foot.

Most current highway diesel engines are electronically controlled to maximize power output while minimizing emissions. In addition, various engine support systems such as air intake, fuel injection, exhaust, lubrication, and cooling work together to keep the engine running properly.

This textbook covers heavy-duty chassis systems but does not include information on diesel engine systems. Variations between engine manufacturers make this subject too broad and complex to be properly presented in one or even several chapters.

ELECTRICAL SYSTEMS

The batteries, alternator, and starter must be sized to match the operating requirements of the engine and truck electrical systems (see Chapters 5–8). Most

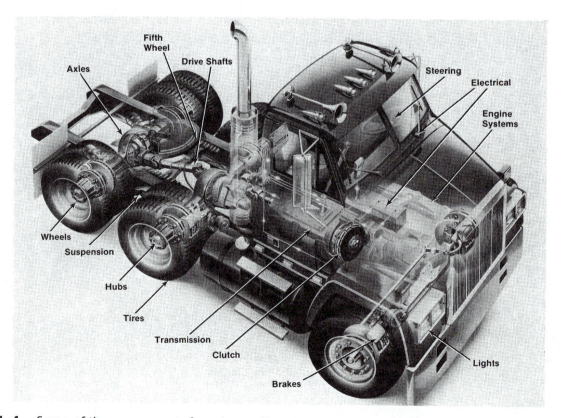

FIGURE 1–4　*Some of the components found on a Class 8 heavy-duty truck. (Courtesy of ArvinMeritor Inc.)*

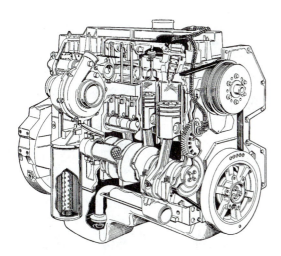

FIGURE 1–5 *Heavy-duty Class 8 trucks are powered by diesel engines. (Courtesy of International Truck and Engine Corp.)*

heavy-duty trucks have two to four batteries to supply current for the starter motor. Some trucks have 24-volt starters that require a series/parallel switch in the system. In addition, the electrical systems of heavy-duty trucks provide power to operate such safety components as lighting, windshield wiper motors, gauges, plus the operation of the accessories and tractor cab amenities.

CLUTCHES

A heavy-duty truck will use either a push- or pull-type clutch. Trucks equipped with high torque engines and designed to haul heavy payloads use a two-plate clutch. The additional contact surface area is necessary to transmit the high torque to the transmission without slippage.

TRANSMISSIONS

Heavy-duty truck transmissions (**Figure 1–6**) are complex pieces of machinery. It is a tribute to materials, engineers, production workers, and truck technicians that transmissions can last as long as they do. Their gears, shafts, bearings, forks, and other components last for thousands of hours, hundreds of thousands of miles, year after year. It is not unusual for a Class 8 transmission to run for a half-million, even a million miles, with little maintenance other than checking lubrication levels and observing drain intervals.

The majority of heavy-duty trucks are equipped with standard transmissions (see Chapter 10). Depending on the engine output, the intended application of the vehicle, and the terrain over which it must operate, the transmission might have from six

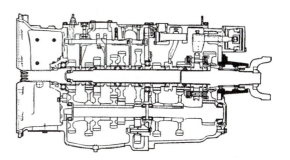

FIGURE 1-6 *Typical conventional transmission. (Courtesy of Mack Trucks Inc.)*

to twenty forward gears. These transmissions have two or three countershafts that transmit the engine torque from the input shaft to the output shaft (main shaft). This splits the torque two or three ways so that there is less strain on individual gears, prolonging the service life of the transmission.

As described in later chapters, heavy-duty trucks are also available with automatic transmissions.

DRIVE SHAFTS

A flange or **end yoke** splined to the output shaft of the transmission transfers engine torque to the drive shaft (see Chapter 18). The **drive shaft** is a hollow tube with end yokes welded or splined to each end. **Splined yokes** allow the drive shaft to increase in length to accommodate movements of the drive axles. Sections of the drive shaft are connected to each other and to the transmission and differentials with **universal joints** ("U-joints"). The U-joints allow torque to be transmitted to components that are operating on different planes.

AXLES

Axles provide a mounting point for the suspension system components, wheels, and steering components. The drive axles also carry the **differential** and axle shafts (**Figure 1–7**). The differential transfers the motion

FIGURE 1–7 *Typical axle and drive shaft components. (Courtesy of ArvinMeritor Inc.)*

of the drive shaft, which is turning perpendicular (at a right angle to) to the rotation of the axle shaft, into motion that is the same as the direction the vehicle is moving. A differential also provides a gear reduction, increasing the torque delivered to the drive wheels. A differential also permits torque to be evenly divided between the left and right wheels. Further information on axles and their servicing can be found in Chapter 19.

STEERING

Both manual and power steering systems are installed on heavy-duty trucks (see Chapter 21). Two types of manual steering gears are used: worm roller and recirculating ball. Power steering systems use a hydraulic pump to provide steering assist. Steering assist can also be provided by an air-powered cylinder installed in the steering linkage.

SUSPENSION SYSTEMS

Smooth riding axle air **suspensions** are popular for reasons of comfort and cargo protection (**Figure 1–8**). But while truckers and the cargo benefit, so does the vehicle itself. Trucks and trailers equipped with air suspensions, which absorb more road shock than conventional spring suspensions, require less maintenance.

Another advantage of an air suspension is better axle control. Ride and handling are much improved when running empty, as an air suspension maintains the same ride height regardless of gross vehicle weight. Like axle air suspensions, cab air suspension systems are getting more popular and for some of the same reasons. By minimizing vibration transmitted to the cab, they reduce repairs to cab hardware and electrical components. They also reduce driver fatigue and can add to resale value.

FIGURE 1–9 Disc wheels, which originated on the West Coast, have now caught on nationwide. (Courtesy of Alcoa Wheel Product International)

WHEELS AND TIRES

There are four basic types of wheel systems: cast spoke, steel disc, aluminum disc, and wide base disc. Cast spoke wheels, used almost universally 20 years ago, are less popular today than disc wheels. Disc wheels (both steel and aluminum) originated on the West Coast and have now caught on nationwide (**Figure 1–9**). Wide base discs have been principally used on front axles of some extra heavy-duty applications such as construction trucks. Interest has been increasing, though, and they are being used on other axles and in other applications. Truck tires are studied in Chapter 23 and are available in a variety of tread patterns to suit different driving conditions.

BRAKES

Service brakes are one of the most important systems on a vehicle. If they do not function properly, a serious accident could occur. Even if they perform well, brake systems can lead to major repair costs if they wear out prematurely. For this reason, it is important that truck technicians observe the information provided in Chapters 24 to 27.

Heavy-duty trucks use **air brakes** exclusively. An extensive arrangement of pneumatic lines, valves, and cylinders controls the delivery of compressed air to the brakes. Trucks are equipped with drum brakes and/or disc brakes. Some Class 7 trucks are equipped with air-over-hydraulic brake systems.

The **National Highway Traffic Safety Administration (NHTSA)** regulations require that all current tractors and trailers be equipped with antilock brakes (ABS). Chapter 26 describes the operating principles of truck ABS.

FIGURE 1–8 Typical air suspension system. (Courtesy of Mack Trucks, Inc.)

VEHICLE RETARDERS

There are other ways to slow down a moving tractor/ trailer without engaging the service brakes. The most popular vehicle retarder is the internal engine brake; it turns the engine into a power absorbing compressor to slow the vehicle. An exhaust brake creates a restriction in the exhaust system to slow the engine and truck. Some vehicles, particularly those equipped with an automatic transmission, use a hydraulic retarder to reduce vehicle speed. A fourth type of retarder is the electrical retarder, which uses magnetism to resist the rotation of the power train.

CHASSIS FRAME

A truck is nothing more than a mechanical beast of burden. And like the backbone of a horse or ox, the truck's mainframe must be strong enough to bear a heavy load, and in many cases, flexible enough to shrug off the flexing generated by its passage over roads and trails.

Most tractor frames are shaped like a ladder. Although a ladder's function is far different, its two main components—rails and steps—can be compared to the truck frame's rails and crossmembers. The cross-sections of the rails resemble a "C" or "I," and as they are increased in size, the "duty rating" of the ladder or truck frame is upgraded. Steps and crossmembers also get stronger as the anticipated workload increases. As outlined in Chapter 28, truck building is a far more complex business than ladder making. But as with ladders, there are many ways of designing and building the frame, and these become visually apparent on close observation. These differences show up within one truck manufacturer's model lineup and to a greater extent among the different **original equipment manufacturers (OEMs)**.

FIFTH WHEEL

A **fifth wheel** is used on a tractor/trailer combination to couple the tractor to the trailer's **kingpin (Figure 1–10)**. The weight of the trailer rests on the fifth wheel plate, and the truck pivots around the kingpin when turning corners or rounding curves. Various locking mechanisms are used to fasten the kingpin to the fifth wheel, and most fifth wheels can be slid backward or forward to properly position the weight of the trailer on the tractor (see Chapter 30).

HEATING/AIR CONDITIONING

Most, if not all, tractor cabs are equipped with a heater/air conditioner. And these cab climate control

FIGURE 1–10 *Typical fifth wheel connection. (Courtesy of Heavy Duty Trucking/Andrew Ryder)*

devices extend to the sleeping quarters. There are also auxiliary heaters used to prevent engine freeze-up. More information on cab and sleeper heating/air-conditioning systems can be found in Chapter 31.

ELECTRONIC CONTROLS

Almost every truck on the roads today has at least one of its major systems managed by computer. In a short period of time, truck technicians have had to adapt from the mainly hydromechanical controls used up to 1990 to the electronic management systems used today. Today, almost all engines and brake systems use some form of electronic controls and other systems such as transmissions, climate control, suspension, and safety radar can use electronics. Understanding "smart" or electronic systems is required knowledge for the technician, and this textbook will describe the basics of computer control systems.

ACCESSORIES

Several safety and driver *comfort accessories* can be found on a heavy-duty truck. The servicing of some of them is not always given in great detail in this book. If the truck technician is responsible for their servicing, the necessary information can be found in

1.3 CAREER OPPORTUNITIES FOR HEAVY-DUTY TRUCK TECHNICIANS

the manufacturer's service manual.

Qualified heavy-duty service technicians are needed in the different branches of the heavy-duty trucking industry. Trucking accounts for nearly 80 percent of all domestic freight revenues. Percentage breakdown for other domestic revenues are as follows: less than 10 percent by rail; 5 percent by air; 3 percent by pipeline; and 7 percent by water, according to the U.S. Department of Commerce.

FLEET OPERATIONS

The trucking industry was deregulated in the United States by the Motor Carrier Act of 1980, and this changed the structure of the trucking industry. Truck haulage used to be an industry of small businesses, but this has changed due to competition, which makes it difficult for small operations to compete. The owner-operator today still represents a significant portion of the market, but to survive the owner-operator must be affiliated with one of the large operators. As each year goes by, the top 100 carriers increase their percentage of the total haulage market revenues, and competition for those revenues is fierce. There are more than 2 million Class 8 trucks registered on our highways, and the major carriers (those that own more than 500 units) operate more than half of them (**Figure 1-11**). Fleets operate in different ways; some have specialties of some kind, whereas the largest keep as much of the operation in-house as possible. The following section explains some of the terminology used to describe fleet operations.

- **Less-than-truckload. LTL carriers** are networks of consolidation centers and satellite terminals. The average haul for national LTL carriers is about 650 miles, whereas the average haul for a regional carrier is approximately 250 miles.
- **Truckload.** This is trucking's largest for-hire segment and its most diverse. These are typically nonunion operations that often use driving teams to increase vehicle productivity or owner/operators to minimize costs. They concentrate on high-density traffic corridors and balanced freight flows to ensure high vehicle use and low costs.
- **Dedicated Contract Carriage.** Trucking operations set up and run according to a specific shipper's needs are a **dedicated contract carriage**. In addition to transportation, they often provide other services such as warehousing and logistics planning.
- **Private Fleets.** Private carriage accounts for about 40 percent of all truck movement and load volume. Private carriage is mostly medium to short haul.
- **Household Goods Carriers.** Sometimes called "van lines," these for-hire trucking companies are set up specifically to move household goods, office equipment, trade show displays, and high-value shipments such as museum displays, which require special packing and handling.
- **Owner/Operators.** The term traditionally applies to a driver who operates his or her own truck. In recent years, however, the owner/operator or "independent trucker" segment has come to signify both single vehicle operations and small fleets that operate under another carrier's ICC authority. Owner/operators account for about 8 percent of Class 8 trucks in service today.
- **Renting and Leasing.** Not all trucking operations own their equipment. Many fleets—particularly private fleets—are set up through full-service lease arrangements under which the lessor provides equipment, maintenance, and maybe even fuel, fuel tax reporting, and other services. Rental trucks and trailers are also used extensively to augment fleets in peak business cycles and to avoid long-term commitments in times of economic uncertainty.

Shop Talk

Figure 1–12 shows Class 8 and Class 7 ownership by vocations based on a total 1,506,500 trucks in Class 8 and 841,500 in Class 7. U.S. retail truck and trailer sales from 1975 to 1993 (Table 1–4) indicate that there are increasing numbers of new vehicles out in the field to be serviced.

There are different forms of shop operations that handle the servicing of the various types of trucking services.

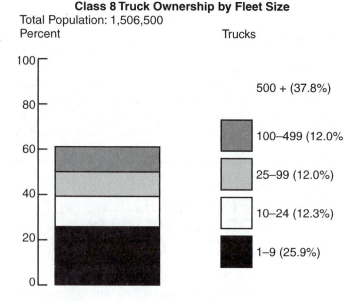

Class 8 Truck Ownership by Fleet Size
Total Population: 1,506,500
Percent   Trucks

500 + (37.8%)
100–499 (12.0%)
25–99 (12.0%)
10–24 (12.3%)
1–9 (25.9%)

FIGURE 1–11   Class 8 truck ownership by fleet size. (Courtesy of Heavy Duty Trucking)

Truck Ownership by Vocation

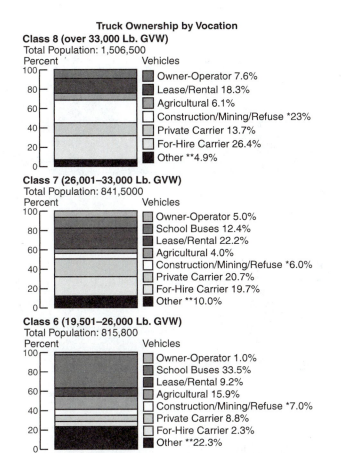

Class 8 (over 33,000 Lb. GVW)
Total Population: 1,506,500

Vehicles
- Owner-Operator 7.6%
- Lease/Rental 18.3%
- Agricultural 6.1%
- Construction/Mining/Refuse *23%
- Private Carrier 13.7%
- For-Hire Carrier 26.4%
- Other **4.9%

Class 7 (26,001–33,000 Lb. GVW)
Total Population: 841,5000

Vehicles
- Owner-Operator 5.0%
- School Buses 12.4%
- Lease/Rental 22.2%
- Agricultural 4.0%
- Construction/Mining/Refuse *6.0%
- Private Carrier 20.7%
- For-Hire Carrier 19.7%
- Other **10.0%

Class 6 (19,501–26,000 Lb. GVW)
Total Population: 815,800

Vehicles
- Owner-Operator 1.0%
- School Buses 33.5%
- Lease/Rental 9.2%
- Agricultural 15.9%
- Construction/Mining/Refuse *7.0%
- Private Carrier 8.8%
- For-Hire Carrier 2.3%
- Other **22.3%

* On-highway only
**Includes utilities, government, logging

FIGURE 1–12 Heavy-duty truck ownership by vocation. (Courtesy of Heavy Duty Trucking)

TABLE 1–4: U.S. RETAIL TRUCK AND TRAILER SALES: 1975–1993			
Year	Class 7	Class 8	Trailers
1975	22,993	83,148	78,300
1976	22,282	97,286	105,400
1977	28,491	140,643	160,600
1978	41,032	161,608	194,900
1979	49,623	173,543	209,500
1980	58,436	117,270	136,700
1981	51,402	100,334	122,500
1982	62,488	75,777	103,900
1983	59,383	81,647	117,700
1984	78,479	137,693	213,900
1985	96,973	133,581	179,800
1986	100,713	112,871	167,300
1987	102,583	131,156	180,100
1988	103,042	148,361	186,500
1989	93,446	145,086	181,500
1990	85,345	121,324	149,100
1991	72,598	98,730	126,200
1992	73,229	119,057	175,300
1993	80,793	157,886	188,100

Courtesy of *Heavy Duty Trucking*

FLEET SHOPS

Any company that operates more than several vehicles faces on-going vehicle service and preventive maintenance requirements. Whereas small fleets often employ the services of a dealership or an independent shop or freelance service technician to do this work, larger companies usually have their own preventive maintenance and repair facility.

Employment in a medium to large fleet is an excellent way to start your career as a heavy-duty truck technician. They offer good advancement possibilities. Most offer some type of training and are usually equipped with the latest equipment. Large fleets with 500 or more vehicles will probably have several shops along their routes (**Figure 1–13**).

DEALERSHIP SHOPS

Heavy-duty truck **franchised dealership** shops are becoming the major employers of technicians and related personnel. The truck dealership is the major link between the various segments of the trucking industry shown in **Figure 1–14**.

Dealerships are privately owned businesses. A franchised dealership is one that has signed a contract with a particular manufacturer to sell and service a particular line of vehicles. A few dealerships now sign

FIGURE 1–13 Typical large fleet shop. (Courtesy of Penske Truck Leasing Corp.)

FIGURE 1–14 Typical franchised dealership shop. A dealership may have a franchise to service more than one line of trucks.

contracts with more than one manufacturer—often one domestic and one import line. A dealership may also handle a line of trailers under an arrangement similar to one with a truck manufacturer.

The sales and service policies of the dealership are usually set by the manufacturer. Service performed while the vehicle is under warranty is usually performed by dealerships or authorized service centers. Truck manufacturers have been taking an increased role in securing service business for their dealerships. Extended warranties and service plans are designed to channel repair and maintenance work to the dealership shop(s). Manufacturers provide special diagnostic equipment designed specifically for their vehicles. They stress the compatibility of their replacement parts and actively promote their service personnel as the most qualified to work on their products.

Working for a dealership backed by a major truck maker can have many advantages. Technical support, equipment, and the opportunity for on-going training are usually excellent. When working for an auto dealership, the service technician's scope of service expertise may be limited to one or two particular model lines. This is not true for the heavy-duty tractors and trailers, as they are more or less custom built or "spec'ed" for customers. In other words, the heavy-duty truck dealership technician requires expertise on various original equipment manufacturer (OEM) systems.

INDEPENDENT TRUCK SERVICE SHOPS

As the name states, an independent heavy-duty service shop is not associated with any particular manufacturer or trucking fleet, although they may service both segments of the industry. Some shops are authorized under agreement with the manufacturer to make warranty repair and replacements. Today many small and mid-size fleets that at one time did their own servicing now depend on independent shops (**Figure 1–15**).

SPECIALTY SERVICE SHOPS

Specialty service shops are shops that specialize in areas such as engine rebuilding, transmission/axle overhauling, brake, air-conditioning/heating repairs, and electrical/electronic work (**Figure 1–16**). The number of specialty service shops that maintain and repair only one or two systems of the truck has steadily increased over the past 10 to 20 years. Service technicians employed by such shops have the opportunity to become very good in one particular area of vehicle service and repair.

FIGURE 1–15   Typical small independent truck service shop.

FIGURE 1–16   Typical specialty service shop.

FIGURE 1–17   Typical truck leasing/rental shop. (Courtesy of Penske Truck Leasing Corp.)

OTHER TRUCK SHOPS

Truck leasing/rental companies, construction/mining/refuse haulers, van truckers, buses, agriculture haulers, and private and for-hire carriers may operate their own service shops (**Figure 1–17**). They may hire out their service work, depending on their size of operation. Many career opportunities are available in this segment of the trucking industry.

1.4   JOB CLASSIFICATIONS

The heavy-duty truck industry offers numerous types of employment for people with a sound understanding of truck systems. Not all jobs involve hands-on service and diagnostic testing.

SERVICE TECHNICIANS

The most important and popular career choice in the trucking industry is the service technician (**Figure 1–18**). The service technician assesses vehicle problems, performs necessary diagnostic tests, and competently repairs or replaces defective components. The skills to do this job are based on a sound understanding of truck technology.

Skilled service personnel are now called technicians. There is a good reason for this. The word "mechanic" emphasizes the ability to repair and service mechanical systems. Although this skill is still absolutely needed, it is only part of the service technician's overall job. Today's heavy-duty vehicles require mechanical knowledge plus an understanding of electronics, hydraulics, and pneumatics.

Service technicians must be skilled in all areas of heavy-duty truck maintenance and repair. The **National Institute for Automotive Service Excellence (ASE)** has established a certification program for automotive, heavy-duty truck, auto body repair, engine machine shop technicians, and parts specialists. This certification system combines voluntary testing with on-the-job experience to verify that technicians have the skills needed to work on today's more complex heavy-duty vehicles. ASE recognizes two distinct levels of service capability—the medium heavy-duty truck technician and the master heavy-duty technician. The master technician is certified by ASE in all major heavy-duty truck systems, including engine repair. The technician may have certification in one or more areas. More information

FIGURE 1–19 *A tire/wheel specialty technician at work inspecting a tire.*

on ASE and the certification process is given in **Table 1–5** and **Table 1–6**. There are eight heavy-duty truck ASE certification tests. Master technician status is achieved by certification in five test areas.

SPECIALTY TECHNICIAN

The heavy-duty truck specialty technician concentrates on servicing a single system of a vehicle, such as electrical (and/or electronic), brakes, transmission, drive train, suspension/steering, trailers, heating/air conditioning, or tire/wheel (**Figure 1–19**). These specialties require advanced and continuous training in that particular field. The number of specialty technicians and the area covered usually depends on the size of the shop.

SERVICE WRITER

The person who handles the drivers' problems in a fleet operation or customer concerns in independent shops is the service writer or service advisor. The service writer must have a sound understanding of truck technology. A friendly attitude and an ability to deal with people are important. The driver's (or the driver's inspection sheet) or customer's problems and needs are discussed with the service writer, who in turn consults the service technician or specialist on vehicle diagnosis. In many shops, the service writer prepares the work order (see Chapter 3).

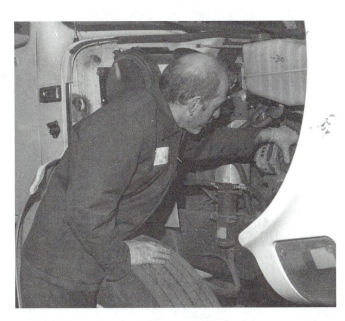

FIGURE 1–18 *An ASE master technician at work.*

TABLE 1–5: SPECIFICATIONS FOR HEAVY-DUTY TRUCK TESTS

Gasoline Engines (T1)

Content Area	Number of Questions in Test
A. General Engine Diagnosis	15
B. Cylinder Head and Valve Train Diagnosis and Repair	12
C. Engine Block Diagnosis and Repair	12
D. Lubrication and Cooling Systems Diagnosis and Repair	9
E. Ignition System Diagnosis and Repair	10
F. Fuel and Exhaust Systems Diagnosis and Repair	10
G. Battery and Starting Systems Diagnosis and Repair	8
H. Emissions Control Systems Diagnosis and Repair	4
Total:	80

Diesel Engines (T2)

Content Area	Number of Questions in Test
A. General Engine Diagnosis	15
B. Cylinder Head and Valve Train Diagnosis and Repair	11
C. Engine Block Diagnosis and Repair	11
D. Lubrication and Cooling Systems Diagnosis and Repair	12
E. Air Induction and Exhaust Systems Diagnosis and Repair	11
F. Fuel System Diagnosis and Repair	12
G. Starting System Diagnosis and Repair	6
H. Miscellaneous	2
Total:	80

Drivetrain (T3)

Content Area	Number of Questions in Test
A. Clutch Diagnosis and Repair	11
B. Transmission Diagnosis and Repair	20
C. Drive Shaft and Universal Joint Diagnosis and Repair	9
D. Drive Axle Diagnosis and Repair	20
Total:	80

Brakes (T4)

Content Area	Number of Questions in Test
A. Air Brake Diagnosis and Repair	36
1. Air Supply and Service Systems	(18)
2. Mechanical/Foundation	(11)
3. Parking Brakes	(7)
B. Hydraulic Brakes Diagnosis and Repair	19
C. 1. Hydraulic System	(8)
2. Mechanical System	(6)
3. Power Assist Units and Miscellaneous	(5)
D. Wheel Bearings Diagnosis and Repair	5
Total:	60

Suspension and Steering (T5)

Content Area	Number of Questions in Test
A. Steering Systems Diagnosis and Repair	19
1. Steering Column and Manual Steering Gear	(5)
2. Power Steering Units	(9)
3. Steering Linkage	(5)
B. Suspension System Diagnosis and Repair	15
C. Wheel Alignment Diagnosis, Adjustment, and Repair	14
D. Wheels and Tires Diagnosis and Repair	7
E. Miscellaneous	5
Total:	60

Electrical Systems (T6)

Content Area	Number of Questions in Test
A. General Electrical Diagnosis	6
B. Battery Diagnosis and Repair	5
C. Starting System Diagnosis and Repair	7
D. Charging System Diagnosis and Repair	8
E. Lighting System Diagnosis and Repair	7
1. Headlights, Parking, Clearance, Tail Cab, and Dash Lights	(3–4)
2. Stoplights, Turn Signals, Hazard Lights, and Back-up Lights	(3–4)
F. Gauges and Warning Devices Diagnosis and Repair	4
G. Miscellaneous	3
Total:	40

TABLE 1–6: SPECIFICATIONS FOR MEDIUM/HEAVY-DUTY TRUCK PARTS SPECIALIST TEST (PI)

Content Area	Number of Questions in Test
A. Communications Skills	10
B. Sales Skills	12
C. Product Knowledge	38
1. Brakes	(7)
2. Electrical Systems	(6)
3. Drive Train	(8)
4. Suspension and Steering	(6)
5. Cab Heating and Air Conditioning	(3)
6. Engines	(8)
a. General/Major Components	(3)
b. Fuel System	(2)
c. Cooling System	(1)
d. Lubrication System	(1)
e. Air Induction and Exhaust Systems	(1)
Total:	60

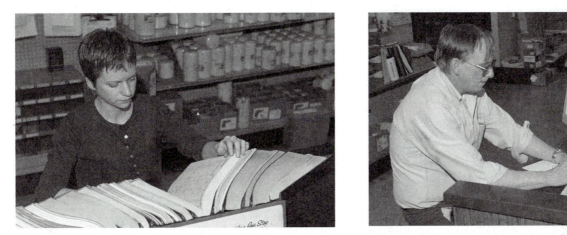

FIGURE 1–20 Today's parts manager, or specialist, uses both catalogs (left) and the computer (right) to check and order parts.

PARTS MANAGER

The parts manager (**Figure 1–20**) is in charge of ordering all replacement parts for the repairs the shop performs. The ordering and timely delivery of parts is extremely important if the shop is to operate smoothly and on schedule. Delays in obtaining parts or omitting a small but crucial part from the initial parts order can cause frustrating holdups for both the service technicians and customers or fleet operators.

Most fleets and large independent service shops maintain a set inventory of commonly-used parts such as filters, belts, hoses, and gaskets. The parts manager and specialists are responsible for maintaining this inventory. ASE certifies medium-/heavy-duty parts specialists.

SHOP SUPERVISOR

The shop supervisor or foreperson is directly in charge of the service technicians, including directing, routing, and scheduling service and repair work (**Figure 1–21**). The supervisor often helps hire, transfer, promote, and discharge technicians to meet the needs of the service department. The supervisor also instructs and oversees the technicians in their work procedures, inspects completed repairs, and is responsible for quality service and satisfactory shop operation. In small shops most of the duties of shop supervisor or foreperson are performed by the service manager.

SERVICE MANAGER

The service manager oversees the entire service operation of a large dealership, fleet, or independent shop. Driver, customer concerns, and complaints are usually handled through the service manager. One

FIGURE 1–21 The shop supervisor checking his paperwork.

must have good communications skills as well as a sound technical background.

In a franchised or company dealership, the service manager makes certain the manufacturer's policies concerning warranties, service procedures, and customer relations are carried out. The service manager normally coordinates on-going training programs and keeps all other shop personnel informed and working together.

The trucking industry is extremely diverse and capable of supporting tens of thousands of interesting careers with unlimited opportunity for advancement. Many people working in related trucking fields such as sales and parts distribution began their careers as service technicians. In this text, we will concentrate on this area of the service industry.

1.5 ADVANCEMENT IN THE PROFESSION

The three most common sources of training are

- vocational/technical schools;
- fleet training programs; and
- manufacturer training programs (**Figure 1–22**).

Heavy-duty truck service courses are given at various training levels—secondary, postsecondary, vocational/technical, or community colleges, both private and public. To help these schools keep pace with the rapidly changing technology and equipment needed for training and maintaining a curriculum that meets the service industry's needs, many truck manufacturers and fleets now work with these schools through their own cooperative programs. A sister organization to ASE, the **National Automotive Technicians Education Foundation (NATEF)** has started a program of certifying secondary and post-secondary heavy-duty truck training programs.

Apprenticeship programs offered by some large dealerships and fleets are another good way to receive training. In such a program, the new worker receives all job training under full supervision. For information on apprenticeship programs available, write to U.S. Department of Labor, Bureau of Apprenticeships and Training, Washington, D.C. 20006.

The professional heavy-duty truck technician must constantly learn. Truck manufacturers, aftermarket parts manufacturers, and independent publishers are always producing new training materials to keep technicians informed on how to service the next generation of trucks.

In addition, technical clinics are often sponsored by truck manufacturers, aftermarket parts manufacturers, and parts dealers. Reading trade magazines and publications is also an excellent way to stay informed and up to date. A competent technician takes advantage of every opportunity to acquire updated information concerning the latest in technology.

SUMMARY

- Although the number of trucks and automobiles in America is increasing, the number of technicians available to service and maintain them is decreasing.
- Trucks are classified by their gross vehicle weight (GVW), the weight of the vehicle and maximum load, and by the number of axles they have. Heavy-duty trucks have a GVW of 26,001 pounds or more.
- The major systems in on-highway trucks are engines, electrical systems, clutches, transmissions, drive shafts, axles, steering, suspension systems, wheels and tires, brakes, vehicle retarders, chassis frame, fifth wheel, heating and air conditioning, electronic controls, and accessories.
- Heavy-duty truck technicians are employed by fleet operations, fleet shops, dealership shops, independent truck service shops, specialty service shops, and other types of truck shops such as truck leasing and refuse haulers.
- Job classifications in the heavy-duty truck industry include the service technician, who maintains and repairs all systems; the specialty technician, who maintains and repairs a single system; the service writer, who deals directly with drivers and communicates truck problems to the service technician; the parts manager, who maintains the inventory of parts needed for maintenance and repair; the shop supervisor, who is in charge of the service technicians; and the service manager, who oversees the entire service operation of a large dealership, fleet, or independent shop.
- A successful heavy-duty truck technician must be able to maintain good customer relations and working relations, use effective communication skills, maintain a safe work environment, perform preventive maintenance, use tools and equipment properly, troubleshoot, correct problems by repairing or replacing, and upgrade skills and knowledge continuously.
- Training for heavy-duty truck technicians is offered by vocational/technical schools, fleet training programs, and manufacturer training programs.

FIGURE 1–22   *Typical manufacturer's training program in session. (Courtesy of Heavy Duty Trucking/ Andrew Ryder)*

■ Heavy-duty truck technicians may obtain certification by the National Institute for Automotive Service Excellence (ASE) through passing written exams. They must meet formal training requirements for government certification to work on brake and air-conditioning systems.

REVIEW QUESTIONS

1. Gross vehicle weight is _____
   a. the weight of the vehicle
   b. the weight of the load the vehicle can safely carry
   c. the weight of the vehicle plus the weight of the load it can safely carry
   d. the weight of a vehicle's chassis frame and engine

2. What is the minimum gross vehicle weight of a heavy-duty truck, and what are the two classes of heavy-duty trucks? _____ and
   _____

3. A truck with 6 wheels, of which 4 are driven by the power train, has what classification number?
   a. 4 × 6
   b. 2 × 8
   c. 8 × 2
   d. 6 × 4

4. Almost all heavy-duty trucks are powered by _____ engines.

5. The number of forward gears in a heavy-duty standard transmission ranges between
   a. 3 and 5
   b. 4 and 8
   c. 7 and 13
   d. 6 and 20

6. Which of the following systems propel heavy-duty trucks?
   a. suspension
   b. fifth wheel
   c. engine
   d. steering

7. Which of the following is NOT a component of truck steering systems?
   a. differential
   b. worm roller gears
   c. drag link and pitman arm
   d. tie-rod

8. Air suspension systems are popular for what reason?
   a. improved fuel economy
   b. better cargo protection
   c. low initial cost
   d. no maintenance required

9. What is the most important safety system on the heavy-duty truck?
   a. transmission
   b. suspension
   c. engine
   d. brakes

10. Most tractor mainframes are shaped like _____
   a. a lattice
   b. a ladder
   c. an "I" beam
   d. the letter "C"

11. What heavy-duty truck system has changed most in the last few years?
   a. wheels and tires
   b. heating/air conditioning
   c. electronic controls
   d. transmissions

12. What percentage of all cargo transportation revenues comes from trucking in this country?
   a. 80
   b. 60
   c. 20
   d. 7

13. Most U.S. trucking fleets of Class 8 trucks have
   _____
   a. more than 500 vehicles
   b. between 100 and 500 vehicles
   c. fewer than 100 vehicles
   d. fewer than 10 vehicles

14. Trucking operations set up and run according to a specific shipper's needs are called _____
   a. LTL carriers
   b. dedicated contract carriage
   c. van lines
   d. rental and leasing operations

15. How many Class 7 and Class 8 trucks are owned and operated in this country today?
   a. almost a million
   b. a million and a half
   c. almost two million
   d. over two million

16. Which job classification requires good communication skills as well as solid technical knowledge?
   a. service technician
   b. service writer
   c. service manager
   d. all of the above

17. Why is "service technician" better than the term "mechanic" in describing the personnel who maintain and repair heavy-duty trucks?

18. "Troubleshooting" might include which of the following tasks?
   a. organizing tools
   b. using diagnostic equipment

c. cleaning components
d. attending training programs

19. The ASE certification program benefits service technicians by _____
    a. raising their professional status
    b. providing specialty education
    c. offering lifetime certification without retesting
    d. ensuring government certification

20. How many heavy-duty truck ASE Certification tests are there?
    a. four
    b. five
    c. eight
    d. ten

Shop Safety and Operations

Objectives

After reading this chapter, you should be able to

- Explain the special notations in the text labeled **SHOP TALK, CAUTION,** and **WARNING.**
- Identify the basic procedures for lifting and carrying heavy objects and materials.
- Explain the role of personal protective equipment in providing for the technician's personal safety.
- Describe other personal safety warnings as they relate to work area safety.
- Identify the different classifications of fires and the proper procedures for extinguishing each.
- Operate the various types of fire extinguisher based on the type of extinguishing agent each uses.
- Identify the four types of hazardous wastes and their respective hazards to people's health and to the environment.
- Explain laws regulating hazardous materials, including both the "Right-To-Know" and employee/employer obligations.
- Identify which types of records are required by law to be kept on trucks involved in interstate shipping.
- Discuss the ever-increasing role of computers in the administration, accounting, and record-keeping functions of heavy-duty truck operations.

Key Term List

corrosive
dispatch sheet
Federal Motor Vehicle Safety Standard (FMVSS)
flammable
hazardous materials
Occupational Safety and Health Administration
 (OSHA)

parts requisition
Resource Conservation and Recovery Act (RCRA)
reactive
Right-To-Know Law
solvents
spontaneous combustion
toxic
Vehicle Identification Number (VIN)

Shop safety should be the utmost concern for all technicians, forepersons, managers, and shop and fleet owners. Safety rules and regulations must be followed to prevent injuries to yourself, to fellow employees, and to the public. Carelessness and the lack of good safety habits cause accidents. Accidents have a far-reaching effect, not only on the victim, but also on the victim's family and society in general. More important, accidents can cause serious injury, temporary or permanent, or even death. Therefore, it is the obligation of all shop employees and the employer to develop a safety program to protect the health and welfare of those involved.

Throughout this book, the text contains special notations labeled **SHOP TALK, CAUTION,** and **WARNING.** Each one has a specific purpose. **SHOP**

TALK gives added information that will help the technician to complete a particular procedure or make a task easier. **CAUTION** is given to prevent the technician from making an error that could damage the vehicle. **WARNING** reminds the technician to be especially careful of those areas where carelessness can cause personal injury. The following text contains some general warnings that should be followed when working in a truck service facility.

2.1 PERSONAL SAFETY

Personal safety refers to the steps you take to protect yourself from injury. It involves wearing protective gear, dressing for safety, and handling tools and equipment correctly.

EYE PROTECTION

The eyes are sensitive to dust, vapors, metal shavings, and liquids. Grinding and machining generates tiny particles that are thrown off at high speeds. Gases and liquids escaping a broken hose or fuel line fitting can be sprayed great distances under great force. Dirt and sharp bits of corroded metal can easily fall into your eyes when working under a vehicle.

Eye protection should be worn whenever you are exposed to these risks. It is good practice to wear safety glasses at all times in the shop. There are many types of eye protection available (**Figure 2–1**). The lenses must be made of safety glass and must offer some sort of side protection. Regular prescription glasses will not suffice. Select safety glasses that fit well and feel comfortable. Make a habit of putting eye protection on and leaving it on.

If chemicals such as battery acid, fuel, or solvents enter your eyes, flush them continuously with clean water until you can get medical help.

CLOTHING

Clothing should be durable, comfortable, and well-fitted. Loose, baggy clothing can get caught on moving parts and machinery. Neck ties should not be worn. Many service technicians prefer to wear coveralls or shop coats to protect their personal clothing (**Figure 2–2**). Cut-offs and short pants are not satisfactory for shop work.

SHOES

Service work involves the handling of many heavy objects that could be accidentally dropped onto feet and toes. Always wear steel-toe safety shoes with

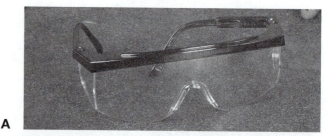
A

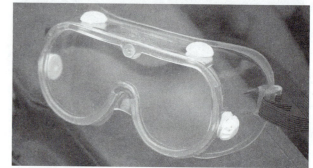
B

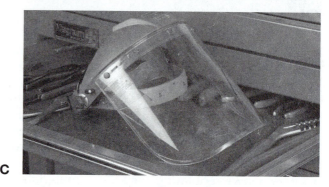
C

FIGURE 2–1   (A) Safety glasses; (B) splash goggles; and (C) face shield. (Courtesy of Goodson Shop Supplies)

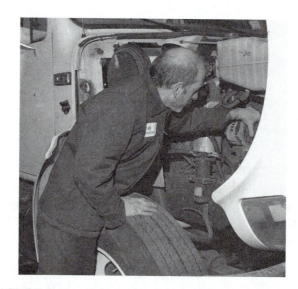

FIGURE 2–2   Shop coats can be worn to protect clothing.

nonslip soles. Athletic shoes, street shoes, and sandals are inappropriate in the shop.

GLOVES

Good hand protection is often overlooked. A scrape, cut, or burn can seriously impair your ability to work for many days. A well-fitted pair of heavy work gloves should be worn during operations such as grinding and welding, or when handling caustic chemicals or high-temperature components.

EAR PROTECTION

Exposure to very high noise levels for extended periods of time can lead to ear damage and hearing loss. Air wrenches, engines run on dynamometers, and vehicles running in enclosed areas can all generate annoying and harmful levels of noise. Simple ear plugs or earphone-type protectors (**Figure 2–3**) should be worn in constantly noisy environments.

HAIR AND JEWELRY

Long hair and hanging jewelry can create the same type of hazard as loose-fitting clothing. They can become caught on moving engine parts and machinery. Tie up long hair securely behind your head or cover it with a cap. Bump caps (similar to construction helmets) are recommended when working in pits or under overhead hoists.

Remove all rings, watches, bracelets, and neck chains. These items can easily be caught on moving parts, causing serious injury.

LIFTING AND CARRYING

Knowing the proper way to lift heavy materials is important. You should always lift and work within your ability and seek help from others when you are not sure you can handle the size or weight of the material or object. Even small, compact auto parts can be surprisingly heavy or unbalanced. Always examine the lifting task before beginning. When lifting any object, follow these steps.

1. Place your feet close to the load and properly positioned for balance.
2. Keep your back and elbows as straight as possible. Bend your knees until your hands reach the best place for getting a strong grip on the load (**Figure 2–4**).
3. If the part or component is stored in a cardboard box, be certain the box is in good condition. Old, damp, or poorly-sealed boxes will tear or otherwise fail. A heavy object could tear through the side or bottom of the container, causing injury or damage.
4. Grasp the object close to your body and lift by straightening your legs. Use your leg muscles, not back muscles.
5. When changing direction of travel, do not twist your body. Turn your whole body, including your feet.
6. When placing the object on a shelf or counter, do not bend forward. Place the edge of the load on the surface and slide it forward. Be careful not to pinch your fingers.
7. When placing a load down, bend your knees and keep your back straight. Do not bend forward—this strains the back muscles.
8. Use blocks to protect your fingers when picking up or lowering heavy objects to the floor.

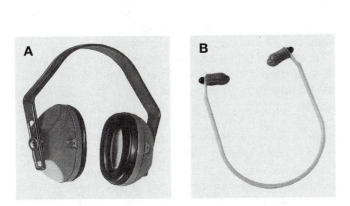

FIGURE 2–3 Typical (A) ear muffs and (B) ear plugs. (Courtesy of Dalloz Safety)

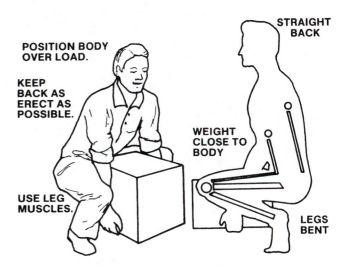

POSITION BODY OVER LOAD.

KEEP BACK AS ERECT AS POSSIBLE.

USE LEG MUSCLES.

STRAIGHT BACK

WEIGHT CLOSE TO BODY

LEGS BENT

FIGURE 2–4 Use your leg muscles, never your back, when lifting any heavy load.

OTHER PERSONAL SAFETY WARNINGS

Never smoke while working on any vehicle or machine in the shop. Tilt the cab with care (**Figure 2–5**).

Proper conduct can help prevent accidents. Horseplay is not fun when it sends someone to the hospital. Such things as air nozzle fights, creeper races, or practical jokes do not have any place in the shop.

A welding helmet or welding goggles with the proper shade lens must be worn when welding (**Figure 2–6**). These will protect the eyes and face from flying molten pieces of steel and harmful light rays. Never use welding equipment unless thoroughly instructed in its use.

To prevent serious burns, avoid contact with hot metal parts such as the radiator, exhaust manifold, tail pipe, and muffler.

When working with a hydraulic press, make sure that hydraulic pressure is applied in a safe manner. It is generally wise to stand to the side when operating the press. Always wear safety glasses.

Store all parts and tools properly by putting them away neatly where people will not trip over them. This practice not only cuts down on injuries, it also reduces time wasted looking for a misplaced part or tool.

B

FIGURE 2–6 (A) Welding helmet and (B) welding goggles.

FIGURE 2–5 The tilting of a hood must be done with care. Make sure there is adequate clearance in front of the vehicle and that the area is free of people and all objects. Do not tilt a cab with the engine running. Tilting the cab could engage the transmission. If the engine is running, the vehicle could move, causing an accident that could result in personal injury or property damage.

2.2 WORK AREA SAFETY

It is important that the work area be kept safe. All surfaces should be kept clean, dry, and orderly. Any oil, coolant, or grease on the floor can cause slips that could result in serious injuries. To clean up oil, be sure to use a commercial oil absorbent. Keep all water off the floor. Remember, water is a conductor of electricity. Aisles and walkways should be kept clean and wide enough for safe clearance. Provide for adequate work space around any machines. Also, keep workbenches clean and orderly.

Another important safety requirement for any work area is proper circulation of air, or ventilation. While diesel engines produce less carbon monoxide (CO) fumes than gasoline engines, their exhaust pipes must be connected to an operating shop exhaust system because of the harmful particulates they produce. Space heaters used in some shops can also be a serious source of deadly CO and therefore must be periodically inspected to make sure they are adequately vented and do not become blocked. Proper ventilation is very important in areas where flammable chemicals are used.

Keep a list of up-to-date emergency telephone numbers clearly posted next to the telephone. These numbers should include a doctor, hospital, and fire and police departments. Also, the work area should have a first-aid kit for treating minor injuries. Facilities for flushing the eyes should also be in or near the shop area.

Gasoline is a highly flammable liquid. **Flammable** liquids are very easily ignited and burn very quickly. For this reason, always keep gasoline or diesel fuel in an approved safety can and never use it to wash hands or tools. Oily rags should also be stored in an approved metal container. When these oily, greasy, or paint-soaked rags are left lying about or are stored improperly, they are prime candidates for **spontaneous combustion;** that is, fires that start by themselves.

Check to be sure that all drain covers are snugly in place. Open drains can cause toe, ankle, and leg injuries.

Handle all **solvents** (solvents are substances that dissolve other substances) with care to avoid spillage. Keep all solvent containers closed, except when pouring. Extra caution should also be used when moving flammable materials from bulk storage (**Figure 2–7**). Otherwise, static electricity can build up enough to create a spark that could cause an

FIGURE 2–8   *Store combustible materials in approved safety cabinets. (Courtesy of the Protectoseal Company)*

explosion. Discard or clean all empty solvent containers. Solvent fumes in the bottom of these containers can start a fire or explosion. Do not light matches or smoke near flammable solvents and chemicals, including battery acids. Solvents and other combustible materials must be stored in approved and designated storage cabinets or rooms (**Figure 2–8**). Storage rooms should have adequate ventilation.

Storage of flammable material and combustible liquids should never be near exits, stairways, or areas normally used for the safe movement of people.

FIRE SAFETY

Familiarize yourself with the location and operation of the fire-fighting equipment in the work area. Keep in mind that all fires are classified in the following manner:

- **Class A**—Fires in which the burning materials are ordinary combustibles, such as paper, wood, cloth, or trash. Putting out this type of fire requires drowning with water or foam solutions containing a high percentage of water, or a multipurpose dry chemical extinguisher.
- **Class B**—Fires in which the burning material is a liquid, such as gasoline, diesel fuel, oil, grease, or solvents. To extinguish this type of fire requires a smothering action from foam, carbon dioxide, or dry chemical type extinguisher. *Do not use water on this type of fire. It will cause the fire to spread.*
- **Class C**—Fires in which the burning material is "live" electrical equipment: motors, switches, generators, transformers, or general wiring. To extinguish this type of fire requires a nonconductive

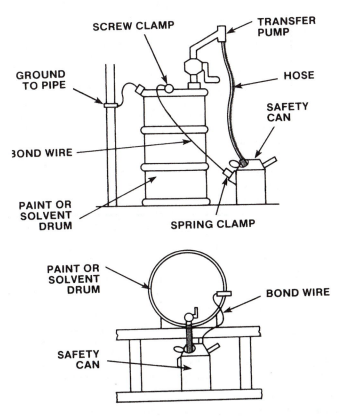

FIGURE 2–7   *Safe methods of transferring flammable materials from bulk storage. (Courtesy of DuPont Co.)*

smothering action, such as carbon dioxide or dry chemical extinguisher. Do not use water on this type of fire.

- **Class D**—Fires in which the burning materials are combustible metals. Special extinguishing agents are required to put out this type of fire.

Following are some general tips for operating the various types of portable extinguishers based on the type of extinguishing agent they use (**Figure 2–9**):

- Foam—Don't spray the stream into the burning liquid. Allow the foam to fall lightly on the fire.
- Carbon dioxide—Direct discharge as close to fire as possible, first at the edge of the flames and gradually forward and upward.
- Soda-acid, gas cartridge—Direct stream at the base of the flame.
- Pump tank—Place foot on footrest and direct the stream at the base of the flame.
- Dry chemical—Direct at the base of the flames. In the case of Class A fires, follow up by direct-

	Class of Fire	Typical Fuel Involved	Type of Extinguisher
Class **A** Fires (green)	**For Ordinary Combustibles** Put out a class A fire by lowering its temperature or by coating the burning combustibles.	Wood Paper Cloth Rubber Plastics Rubbish Upholstery	Water*[1] Foam* Multipurpose dry chemical[4]
Class **B** Fires (red)	**For Flammable Liquids** Put out a class B fire by smothering it. Use an extinguisher that gives a blanketing, flame-interrupting effect; cover whole flaming liquid surface.	Gasoline Oil Grease Paint Lighter fluid	Foam* Carbon dioxide[5] Halogenated agent[6] Standard dry chemical[2] Purple K dry chemical[3] Multipurpose dry chemical[4]
Class **C** Fires (blue)	**For Electrical Equipment** Put out a class C fire by shutting off power as quickly as possible and by always using a nonconducting extinguishing agent to prevent electric shock.	Motors Appliances Wiring Fuse boxes Switchboards	Carbon dioxide[5] Halogenated agent[6] Standard dry chemical[2] Purple K dry chemical[3] Multipurpose dry chemical[4]
Class **D** Fires (yellow)	**For Combustible Metals** Put out a class D fire of metal chips, turnings, or shavings by smothering or coating with a specially designed extinguishing agent.	Aluminum Magnesium Potassium Sodium Titanium Zirconium	Dry powder extinguishers and agents only

*Cartridge-operated water, foam, and soda-acid types of extinguishers are no longer manufactured. These extinguishers should be removed from service when they become due for their next hydrostatic pressure test.

Notes:

(1) Freezes in low temperatures unless treated with antifreeze solution, usually weighs over 20 pounds (9 kg), and is heavier than any other extinguisher mentioned.

(2) Also called ordinary or regular dry chemical. (sodium bicarbonate)

(3) Has the greatest initial fire-stopping power of the extinguishers mentioned for class B fires. Be sure to clean residue immediately after using the extinguisher so sprayed surfaces will not be damaged. (potassium bicarbonate)

(4) The only extinguishers that fight A, B, and C classes of fires. However, they should not be used on fires in liquefied fat or oil of appreciable depth. Be sure to clean residue immediately after using the extinguisher so sprayed surfaces will not be damaged. (ammonium phosphates)

(5) Use with caution in unventilated, confined spaces.

(6) May cause injury to the operator if the extinguishing agent (a gas) or the gases produced when the agent is applied to a fire is inhaled.

FIGURE 2–9   Guide to fire extinguisher selection.

ing the dry chemicals at the remaining material that is burning.

If a fire extinguisher is used, report it to your instructor or service manager so that it can be immediately recharged.

2.3 SHOP TOOL SAFETY

Understanding the proper use of nonpower–driven hand tools, portable electric power tools, pneumatic power tools, and stationary equipment will help eliminate many accidents. Observe the following:

- Select the proper size and type of tool for the job.
- Use tools only for the purpose for which they are designed.
- Keep tools in safe working condition.
- Store tools safely when not in use.
- Report any breakage or malfunctions to your instructor or service manager.
- Make sure that cutting tools are properly sharpened and in good condition.
- Do not use tools with loose or cracked handles.
- Never use tools unless you know how to operate them.

Shop tool safety depends mainly on the person who uses the tool. Knowing what a tool is designed to do and how to use it correctly is the key. Because of the importance of this, Chapter 3 is devoted to safe use of the tools used by the heavy-duty truck service technician.

2.4 HAZARDOUS MATERIALS

Heavy-duty truck repair work involves use of many materials classified as hazardous by both state and federal governments. These materials include such items as solvent and cleaners, paint and body repair products, adhesives, acids, coolants, and refrigerant products.

Hazardous materials are those that could cause harm to a person's well-being. Hazardous materials can also damage and pollute land, air, or water. There are four types of hazardous waste:

- **Flammable**—Materials that will easily catch fire or explode.
- **Corrosive**—**Corrosive** materials are so caustic that they can dissolve metals and burn skin and eyes.

- **Reactive**—**Reactive** materials will become unstable (burn, explode, or give off **toxic** vapors) if mixed with air, water, heat, or other materials.
- **Toxic**—Materials that can cause illness or death after being inhaled or contacting the skin.

LAWS REGULATING HAZARDOUS MATERIALS

The Hazard Communication Regulation—commonly called the **Right-To-Know Law**—was passed by the federal government and is administered by **Occupational Safety and Health Administration (OSHA)**. This law mandates that any company that uses or produces hazardous chemicals or substances must inform its employees, customers, and vendors of any potential hazards that may exist in the workplace as a result of using the products.

Most important is that you keep yourself informed. You are the only person who can keep yourself and those with whom you work protected from the dangers of hazardous materials. These are some of the highlights of the Right-To-Know Law:

- You have a right to know what hazards you may face on the job.
- You have a right to learn about these materials, and how to protect yourself from them.
- You cannot be fired or discriminated against for requesting information and training on how to handle hazardous materials.
- You have the right for your doctor to receive the same hazardous material information that you receive.

Employee/Employer Obligations

An employer or school that uses hazardous materials must

- Provide a safe workplace.
- Educate employees about the hazardous materials they will encounter while on the job (**Figure 2–10**).
- Recognize, understand, and use warning labels and Material Safety Data Sheets (MSDSs) (Workplace Hazardous Material Information Systems—WHMISs).
- Provide personal protective clothing and equipment and train employees to use them properly.

You, the employee or student, must

- Read the warning labels on the materials.
- Follow the instructions and warnings on the MSDS or WHMIS.
- Take the time to learn to use protective equipment and clothing.

MATERIAL SAFETY DATA SHEET

24 Hour Emergency Phone (316) 524-5751

Division of Vulcan Materials Company / P. O. Box 530390 • Birmingham, AL 35253-0390

I – IDENTIFICATION

CHEMICAL NAME	CHEMICAL FORMULA	MOLECULAR WEIGHT
Sodium Hydroxide Solution	NaOH	40.00

TRADE NAME
Caustic Soda, 73%, 50% and Weaker Solutions

SYNONYMS	DOT IDENTIFICATION NO.
Liquid Caustic, Lye Solution, Caustic, Lye, Soda Lye	UN 1824

II – PRODUCT AND COMPONENT DATA

COMPONENT(S) CHEMICAL NAME	CAS REGISTRY NO.	% (wt.) Approx.	OSHA PEL
Sodium Hydroxide	1310-73-2	73, 50 and less	2 mg/m³ Ceiling

Note: This Material Safety Data Sheet is also valid
for caustic soda solutions weaker than 50%. The
boiling point, vapor pressure, and specific gravity
will be different from those listed.

* Denotes chemical subject to reporting requirements of Section 313 of Title III of the 1986
Superfund Amendments and Reauthorization Act (SARA) and 40 CFR Part 372

III – PHYSICAL DATA

APPEARANCE AND ODOR	SPECIFIC GRAVITY
Colorless or slightly colored, clear or opaque; odorless	50% Solution: 1.53 @ 60°F/60°F 73% Solution: 1.72 @ 140°F/4°F

BOILING POINT	VAPOR DENSITY IN AIR (Air = 1)
50% Solution: 293°F (145°C) 73% Solution: 379°F (192.8°C)	N/A

VAPOR PRESSURE	% VOLATILE, BY VOLUME
50% = 6.3 mm Hg @ 104°F 73% = 6.0 mm Hg @ 158°F	0

EVAPORATION RATE	SOLUBILITY IN WATER
0	100%

IV – REACTIVITY DATA

STABILITY	CONDITIONS TO AVOID Mixture with water, acid or incompatible materials can cause splattering and release of large amounts of heat (Refer to Section VIII). Will react with some metals forming flammable hydrogen gas.
Stable	

INCOMPATIBILITY (Materials to avoid)
Chlorinated and fluorinated hydrocarbons (i.e. chloroform, difluoroethane), acetaldehyde, acrolein, aluminum,
chlorine trifluoride, hydroquinone, maleic anhydride, phosphorous pentoxide and tetrahydrofuran.

HAZARDOUS DECOMPOSITION PRODUCTS
Will not decompose

HAZARDOUS POLYMERIZATION
Will not occur

FIGURE 2–10 Typical Material Safety Data Sheet. (Courtesy of Vulcan Materials Company)

V – FIRE AND EXPLOSION HAZARD DATA

FLASHPOINT (Method used)	FLAMMABLE LIMITS IN AIR
None	None

EXTINGUISHING AGENTS	
N/A	NFPA Hazard Ratings: Health 3; Flammability 0; Reactivity 1

UNUSUAL FIRE AND EXPLOSION HAZARDS

Firefighters should wear self-contained positive pressure breathing apparatus, and avoid skin contact. Refer to Reactivity Data, Section IV.

VI – TOXICITY AND FIRST AID

EXPOSURE LIMITS (When exposure to this product and other chemicals is concurrent, the exposure limit must be defined in the workplace.)

ACGIH: 2 mg/m^3 Ceiling

OSHA   2 mg/m^3 Ceiling

IDLH: 250 mg/m^3

Effects described in this section are believed not to occur if exposures are maintained at or below appropriate TLVs.
Because of the wide variation in individual susceptibility, these exposure limits may not be applicable to all persons and those with medical conditions listed below.

MEDICAL CONDITIONS AGGRAVATED BY EXPOSURE

May aggravate existing skin and/or eye conditions on contact.

ACUTE TOXICITY   Primary route(s) of exposure: ☒ Inhalation   ☒ Skin Absorption   ☐ Ingestion

Inhalation: Inhalation of solution mist can cause mild irritation at 2 mg/m^3. More severe burns and tissue damage at the upper respiratory tract, can occur at higher concentrations. Pneumonitis can result from severe exposures.

Skin: Major potential hazard - contact with the skin can cause severe burns with deep ulcerations. Contact with solution or mist can cause multiple burns with temporary loss of hair at burn site. Solutions of 4% may not cause irritation and burning for several hours, while 25 to 50% solutions can cause these effects in less than 3 minutes.

Eyes: Major potential hazard - Liquid in the eye can cause severe destruction and blindness. These effects can occur rapidly effecting all parts of the eye. Mist or dust can cause irritation with high concentrations causing destructive burns.

Ingestion: Ingestion of sodium hydroxide can cause severe burning and pain in lips, mouth, tongue, throat and stomach. Severe scarring of the throat can occur after swallowing. Death can result from ingestion.

FIRST AID

Inhalation: Move person to fresh air. If breathing stops, administer artificial respiration. Get medical attention immediately.

Skin: Remove contaminated clothing immediately and wash skin thoroughly for a minimum of 15 minutes with large quantities of water (preferably a safety shower). Get medical attention immediately.

Eyes: Wash eyes immediately with large amounts of water (preferably eye wash fountain), lifting the upper and lower eyelids and rotating eyeball. Continue washing for a minimum of 15 minutes. Get medical attention immediately.

Ingestion: If person is conscious, give large quantities of water to dilute caustic. Do not induce vomiting. Get medical attention immediately. Do not give anything by mouth to an unconscious person.

CHRONIC TOXICITY

No known chronic effects

Carcinogenicity: No studies were identified relative to sodium hydroxide and carcinogenicity. Sodium hydroxide is not listed on the IARC, NTP or OSHA carcinogen lists.

Reproductive Toxicity: No studies were identified relative to sodium hydroxide and reproductive toxicity.

VII – PERSONAL PROTECTION AND CONTROLS

RESPIRATORY PROTECTION

Where concentrations exceed or are likely to exceed $2mg/m^3$ use a NIOSH/MSHA approved high-efficiency particulate filter with full facepiece or self-contained breathing apparatus. Follow any applicable respirator use standards and regulations.

VENTILATION

As necessary to maintain concentration in air below $2 mg/m^3$ at all times.

SKIN PROTECTION

Wear neoprene, PVC, or rubber gloves; PVC rain suit; rubber boots with pant legs over boots.

EYE PROTECTION

Chemical goggles which are splashproof and faceshield.

HYGIENE

Avoid contact with skin and avoid breathing mist. Do not eat, drink, or smoke in work area. Wash hands prior to eating, drinking, or using restroom. Any protective clothing or shoes which become contaminated with caustic should be removed immediately and thoroughly laundered before wearing again.

OTHER CONTROL MEASURES

Safety shower and eyewash station must be located in immediate work area. To determine the exposure level(s), monitoring should be performed regularly.
NOTE: Protective equipment and clothing should be selected, used, and maintained according to applicable standards and regulations. For further information, contact the clothing or equipment manufacturer or the Vulcan Chemicals Technical Service Department.

VIII – STORAGE AND HANDLING PRECAUTIONS

Follow protective controls set forth in Section VII when handling this product.

Store in closed, properly labeled tanks or containers. Do not remove or deface labels or tags.

When diluting with water, slowly add caustic solution to the water. Heat will be produced during dilution. Full protective clothing, goggles and faceshield should be worn. Do not add water to caustic because excessive heat formation will cause boiling and spattering.

Contact of caustic soda cleaning solutions with food and beverage products (in enclosed vessels or spaces) can produce lethal concentrations of carbon monoxide gas. Do not enter confined spaces such as tanks or pits without following proper entry procedures as required by 29 CFR 1910.146.

SARA Title III Hazard Categories: Immediate Health.

IX – SPILL, LEAK AND DISPOSAL PRACTICES

STEPS TO BE TAKEN IN CASE MATERIAL IS RELEASED OR SPILLED

Cleanup personnel must wear proper protective equipment (refer to Section VII). Completely contain spilled material with dikes, sandbags, etc., and prevent run-off into ground or surface waters or sewers. Recover as much material as possible into containers for disposal. Remaining material may be diluted with water and neutralized with dilute hydrochloric acid. Neutralization products, both liquid and solid, must be recovered for disposal. Reportable Quantity (RQ) is 1000 lbs. Notify National Response Center (800/424-8802) of uncontained releases to the environment in excess of the RQ.

WASTE DISPOSAL METHOD

Recovered solids or liquids may be sent to a licensed reclaimer or disposed of in a permitted waste management facility. Consult federal, state, or local disposal authorities for approved procedures.

X – TRANSPORTATION

DOT HAZARD CLASSIFICATION

Sodium Hydroxide Solution, 8, UN 1824, PG II, RQ

PLACARD REQUIRED

Corrosive, 1824, Class 8

LABEL REQUIRED

Corrosive, Class 8. Label as required by OSHA Hazard Communication Standard, and any applicable state and local regulations.

Medical Emergencies	**For any other information contact:**
Call collect 24 hours a day for emergency toxicological information 415/821-5338	**Vulcan Chemicals Technical Service Department P.O. Box 530390 Birmingham, AL 35253-0390 800/873-4898 8 AM to 5 PM Central Time Monday Through Friday**
Other Emergency information	
Call 316/524-5751 (24 hours)	

DATE OF PREPARATION: November 1, 1993

- Use common sense when working with hazardous materials.
- Ask the service manager if you have any questions about a hazardous material.

Personal Protection

One of the greatest concerns to personal safety is the effect of long-term exposure to hazardous materials, particularly solvents, cleaning agents, and paint products. The importance of this is easily seen in concerns over asbestos. When first introduced on the market, asbestos was widely used in brake pads, shoes, clutches, and other automotive applications. We now know that asbestos fibers pose a health risk and that long-term exposure to small amounts of asbestos can cause health problems. For this reason, asbestos has gradually been phased out of the automotive and truck parts market in favor of safer products.

To handle hazardous materials properly, you need to

- Know what the material is.
- Know the material is dangerous.
- Know the correct safety equipment needed for working with that material.
- Know how to use the safety equipment properly (**Figure 2–11**).
- Make sure that the safety equipment fits properly and is in working order.

Good personal hygiene practices are important in minimizing exposure to asbestos dust and other hazardous wastes.

- Do not smoke.
- Wash before eating.
- Shower after work.
- Change to work clothes upon arrival at work, and change from work clothes after work. Work clothing should not be taken home.

2.5   HANDLING AND DISPOSAL OF HAZARDOUS WASTE

Specific laws govern the disposal of hazardous wastes. You and your shop must be aware of how these laws affect shop operation.

These laws include the **Resource Conservation and Recovery Act (RCRA).** This law states that after you have used hazardous materials, they must be properly stored until an approved hazardous waste hauler (**Figure 2–12**) arrives to take them to the disposal site. In addition, your responsibility continues until the materials arrive at an approved disposal site and are processed in accordance with the law.

When dealing with hazardous wastes

1. Consult the MSDS or WHMIS under the "Waste Disposal Method" category.
2. Check with your instructor or service manager for the exact method for correct storage and disposal.
3. Follow their recommendations exactly.

FIGURE 2–11   Wear proper safety equipment when handling hazardous waste. (Courtesy of DuPont Co.)

FIGURE 2–12   Many automotive shops hire full-service haulers for hazardous waste removal. (Courtesy of DuPont Co.)

WARNING: The shop is ultimately responsible for the safe disposal of hazardous wastes, even after the waste leaves the shop. In the event of an emergency hazardous waste spill, contact the National Response Center (1-800-424-8802) immediately. Failure to do so can result in a $10,000 fine, a year in jail, or both.

Never, under any circumstance

- Throw hazardous materials into a dumpster.
- Dump waste anywhere but into a collection site of a licensed facility.
- Pour waste down drains, toilets, sinks, or floor drains.
- Use hazardous waste to kill weeds or to keep dust down in gravel.

2.6   SHOP RECORDS

It is the "law" that certain records must be kept by the shop's technician if the trucks are involved in interstate shipping. These records include the following:

- Identification of each vehicle including the company unit number, model, serial number, year, and tire size.
- A schedule showing the nature and due date of the various inspections and maintenance to be performed.
- A record of the nature and date of inspections, maintenance, and repairs made.
- Lubrication record.

Even if the trucks are not involved in interstate commerce or local heavy-duty truck servicing operations, these records are important because

- They are required by the Department of Transportation (DOT).
- A study of parts or component failures can alert the manager to problems or highlight components that have performed well.
- A service manager cannot develop a good preventive maintenance (PM) program without the use of these records (see Chapter 4).
- In the event of a serious accident, a vehicle maintenance file is usually beneficial in the defense of a lawsuit.

Shop Talk

Federal Motor Vehicle Safety Standard (FMVSS) specifies that all vehicles in the United States be assigned a **Vehicle Identification Number (VIN)**. The VIN is located on the left frame rail over the front axle and on the Vehicle Specification Decal (see the drivers manual for the location of the decal). All heavy-duty trucks are assigned a 17-character VIN (**Figure 2–13**). Using a combination of letters and numerals, the VIN codes the vehicle make, series or type, application, chassis, cab, axle configuration, gross vehicle weight rating (GVWR), engine type, model year, manufacturing plant location, and production serial number. A check digit (ninth position) is determined by assignment of weighted values to the other 16 characters. These weighted values are processed through a series of equations designed to check the validity of the VIN and to detect VIN alteration.*

WORK OR REPAIR ORDER

In most shops, the work or repair is the result of collaboration between truck driver, service manager, and technician. The need for a vehicle's service is noted in the driver's pre- and post-trip report. After studying the problem, observations are passed on to the service manager. The service manager then turns it over to the technician in the form of a work (or repair) order to correct the situation (**Figure 2–14**).

Another form that is used in some shops that concerns the technician is the **parts requisition.** To order new parts, the technician writes the names of what part(s) are needed along with the vehicle's VIN or company's identification folder.

Parts managers or specialists (see Chapter 1) must fill out forms to keep enough parts in stock. When the stock gets low, new parts are ordered. Occasionally it is necessary to order special parts.

Description	Typical Identification Number							
	1FU	P	D	CY	B	2	J	P	345678
Decoding Table Number*	1	2	3	4	5		6	7	
Manufacturer, Make, Type of Vehicle								
Chassis, Front Axle Position, Brakes								
Model Series, Cab								
Engine Type								
Gross Vehicle Weight Rating (GVWR)								
Check Digit								
Vehicle Model Year								
Plant of Manufacture								
Production Number								

FIGURE 2–13   Typical VIN.

REPAIR ORDER INFORMATION

Portland Freightliner, Inc.

REPAIR ORDER NO. _____

WRITTEN BY	DELIVERY DATE
SERIAL NO.	DELIVERY MILES
CUSTOMER ACCOUNT NO.	YR/MAKE/MODEL
NAME	UNIT NO.
ADDRESS	UNIT DATA
CITY	ENGINE MODEL
PHONE	ENGINE SERIAL NO.
P.O. NO.	TRANS. MODEL
ESTIMATE	TRANS. SERIAL NO.
MILEAGE	R. AXLE MODEL
MEMO - 1	R. AXLE SERIAL NO.
MEMO - 2	R. AXLE MODEL
	R. AXLE SERIAL NO.

1 CONDITION / TYPE
2 CONDITION / TYPE
3 CONDITION / TYPE
4 CONDITION / TYPE
5 CONDITION / TYPE
6 CONDITION / TYPE
7 CONDITION / TYPE
8 CONDITION / TYPE
9 CONDITION / TYPE
10 CONDITION / TYPE

TERMS: STRICTLY CASH, APPROVED ACCOUNT OR CREDIT CARD

I the undersigned authorize you to perform the repairs and furnish the necessary materials. I understand any costs verbally quoted are an estimate only and not binding. Your employees may operate vehicle for inspecting, testing, and delivery at my risk. You will not be responsible for loss or damage to vehicle or articles left in it. I agree to pay reasonable storage on vehicle left more than 48 hours after notification that repairs are completed.

I AGREE THAT YOU HAVE AN EXPRESS LIEN ON THE DESCRIBED VEHICLE FOR THE CHARGES FOR PARTS AND LABOR FURNISHED UNDER THIS REPAIR ORDER INCLUDING THOSE FROM ANY PRIOR REPAIR ORDERS ON THE VEHICLE. IF I FAIL TO PAY SUCH CHARGES, I AGREE THAT THE VEHICLE MAY BE HELD UNTIL CHARGES ARE PAID IN FULL. IN THE EVENT OF LEGAL ACTION TO COLLECT ANY SUMS DUE, I AGREE TO PAY COSTS OF COLLECTION AND FEES INCLUDING REASONABLE ATTORNEY FEES.

If charges are handled on an approved open account, they are due on the 10th of the month following the purchase. A FINANCE CHARGE of 1½% per month (18% per year) will be added to all balances 30 days past due.

ABOVE REPAIRS AUTHORIZED BY: _____

FIGURE 2–14 Typical work or repair order. (Courtesy of Freightliner Trucks)

In many shops, a **dispatch sheet,** or work schedule, keeps track of dates when the work is to be completed. Some dispatch sheets follow the job through each step of the servicing process. This information helps the truck dispatcher know the whereabouts of vehicles in the shop and when it will be ready for service use again.

Most fleets, large or small, keep a file on each individual vehicle, whether it be a tractor or a trailer. This file includes all the vehicle's maintenance and repair records, schedule, PM inspection results, and any other data about that vehicle. All work orders performed on that particular vehicle are generally kept in its history folder.

If a shop wishes to maintain a history of individual component life or performance by a vehicle, a major component history can be a part of the vehicle file also. It is a simple matter of updating this record as each repair order is placed in the file folder. A copy of invoices covering outside repairs should be placed in the vehicle file folder. If major components are involved, the component history records should be updated.

The Federal Motor Carrier Safety Regulations require the mandatory records of a vehicle file to be retained where the vehicle is housed or maintained for a period of one year. They also require these records to be kept on hand for six months after the vehicle leaves the carrier's control. Experience has shown that it is a good idea to keep work orders in the vehicle file folder as long as the vehicle is in service within a given operation.

The same regulations require that the driver's vehicle condition report be retained for at least three months from the date of the report.

COMPUTERS IN THE SHOP

As can be seen, the heavy-duty truck technician's responsibilities go beyond just inspecting, maintaining, and servicing the vehicles. He/she must keep neat and accurate records. Shop computers (**Figure 2–15**) equipped with bar-code wands can relieve technicians of the burden of writing legible repair orders, help shop managers improve the scheduling, and keep mistakes out of parts department transactions. With the proper software, a computer can turn the repair/maintenance process into a "paperless" operation. (Shop software can best be described as the programs a computer uses to perform maintenance shop tasks—a job which used to be accomplished with a sharp pencil and someone with a good head for figures.) Software packages are available to track every aspect of a vehicle. Here are some examples:

FIGURE 2–15 A computer makes shop record keeping a great deal easier.

- **Vehicle Maintenance Reporting Standards (VRMS).** Developed by the American Trucking Association, this is a coding system that can be used at an individual part level up to total operating systems. Used within a software package, it can provide information on items such as part inventory, part usage, part wear rates, warranty information, and overall cost control.
- **Parts Inventory Control.** A basic inventory control system should maintain at least the following information on each part: part number, description, current quantity, and minimum (reorder) quantity. The goal is to balance maximum part availability against minimum investment in inventory.
- **Bar Coding.** One of the weaknesses of a computer inventory system is human input error. Bar coding systems reduce the error rate from one in a 100 to one in a 100 million. Bar coding used in maintenance shops is similar to that used in grocery stores, except it uses the 3-of-9 code instead of the universal product code (UPC).
- **Replace/Repair Analysis.** The ability to keep track of which parts are being replaced and repaired.
- **Preventive Maintenance.** A schedule set up in the computer tells you which vehicle needs servicing at what time.
- **Electronic Data Interchange (EDI).** A rapidly growing technology that allows computers of any type to talk to each other and exchange data about business transactions that formerly had to be transmitted on paper or by telephone.

- **Vehicle History.** A database that can provide a detailed profile on any vehicle in the fleet.
- **Work Order Generating.** The ability to automatically generate a work order when a vehicle needs servicing.
- **Cost Tracking.** Using databases on parts, vehicle history, PM scheduling, and warranty information to keep track of overall costs.
- **Warranty Information.** A database linked with vehicle history that provides specific information on parts warranty. Helpful in obtaining prompt warranty payments from vendors.
- **Vehicle and Driver Performance Analysis.** Trip data can be wirelessly downloaded during or after each trip.

SUMMARY

- Personal safety on the job may require eye or ear protection, or both, and protective clothing and shoes. Long hair and loose jewelry are hazards.
- Lifting and carrying heavy materials the correct way will protect against injury.
- Tilt hoods and cabs with care.
- Do not smoke or engage in horseplay in the shop.
- Take care when welding or working with a hydraulic press. Use protective eyewear. Avoid contact with hot metal parts.
- The work area should be kept clean, dry, and neat, and flammable liquids and solvents should be handled and stored carefully.
- Emergency telephone numbers and a first-aid kit should be handy.
- Use fire-fighting equipment appropriately: water or foam on ordinary combustibles; foam, carbon dioxide, or dry chemicals on burning liquids; carbon dioxide or dry chemicals on burning "live" electrical equipment; and special extinguishing agents on burning metals.
- Select, store, use, and maintain shop tools properly.
- Hazardous materials used in heavy-duty truck repair include flammable, corrosive, reactive, and toxic materials. Your employer is obligated to inform you of potential hazards in your workplace, and you have a right to protect yourself from them.
- Specific laws govern the disposal of hazardous wastes, including oil, antifreeze/coolants, refrigerants, batteries, battery acids, acids and solvents used for cleaning, and paint and body repair product wastes. Hazardous wastes may be recycled in the shop or removed by a licensed disposal hauler.

- By law, records must be kept by each shop of the repair and maintenance of trucks involved in interstate shipping.
- Computers can be useful in the shop for such tasks as parts inventory control, work order generating, and cost tracking.

REVIEW QUESTIONS

1. In this book, the sections that contain cautions about situations that might result in personal injury are labeled with what special notation?
   a. SHOP TALK
   b. SAFETY RULES
   c. CAUTION
   d. WARNING

2. The best way to prevent eye injury is to _____
   a. take care during grinding or other operations that throw off particles
   b. always wear safety glasses
   c. always wear a bump cap
   d. make sure a source of running water is available to flush foreign matter out of the eyes

3. Which of the following describes a safe lifting and carrying practice?
   a. Twist your body when changing your direction of travel while carrying a heavy object.
   b. Bend forward to place a heavy object on a shelf or counter.
   c. Lift by bending and then straightening your legs, rather than by using your back.
   d. Position your feet as far as possible from the load when you begin to lift.

4. Oil on the floor of the work area can be cleaned up using a _____.

5. The exhaust pipe of a diesel engine must be connected to the shop exhaust system to protect against _____.
   a. particulates
   b. carbon monoxide
   c. carbon dioxide
   d. fire

6. Do NOT attempt to put out a Class B fire using
   _____
   a. foam
   b. carbon dioxide
   c. a dry chemical type extinguisher
   d. water

7. A Class C fire involves _____
   a. ordinary combustibles, such as paper or cloth
   b. a flammable liquid
   c. live electrical equipment
   d. combustible metals

8. To use carbon dioxide to extinguish a fire, direct the discharge _____
   a. at the top of the flames
   b. first at the edge of the fire, then forward and upward
   c. at the base of the flames
   d. several feet over the top of the flames

9. What are the four types of hazardous wastes?

10. The "Right-To-Know" law was passed by the government to _____
    a. require any company that disposes of hazardous materials to inform their community
    b. protect employees, customers, or vendors from hazards in the workplace caused by hazardous chemicals
    c. require industries to compensate employees injured by contact with hazardous materials
    d. require chemical industries to reveal complete information about the chemicals they produce

11. Which of the following is covered under the Resource Conservation and Recovery Act?
    a. waste water
    b. waste oil
    c. cleaning solvents
    d. all of the above

12. Which of the following is an approved way of disposing of hazardous wastes?
    a. washing them down the drain with plenty of water
    b. using them as weed killer
    c. recycling them by reusing them in the shop
    d. placing them in leak-proof containers and disposing of them in an RCRA approved method.

13. What record must be kept by the shop on trucks involved in interstate shipping?
    a. out-of-service times
    b. names of all drivers
    c. names of all service technicians
    d. nature and date of inspections

14. What information is provided by the first digit of the VIN of a heavy-duty truck?
    a. model year
    b. axle configuration
    c. manufacturer
    d. gross weight rating

15. Bar coding, along with a computer system, can be used effectively to _____
    a. improve parts inventory control
    b. track preventive maintenance
    c. provide a detailed profile on any fleet vehicle
    d. generate work orders efficiently

Tools and Fasteners

Objectives

After reading this chapter, you should be able to
- List some of the common hand tools used in heavy-duty truck repair.
- Describe the use of common pneumatic, electrical, and hydraulic power tools used in heavy-duty truck repair.
- List the mechanical measuring tools used in the heavy-duty truck shop.
- Describe the proper procedure for measuring with a micrometer.
- List the major manufacturer service publications used in heavy-duty truck repair and describe the type of information each provides.
- Explain the principles and precautions of working with various heavy-duty truck fasteners.

Key Terms

adjustable pliers
adjustable wrench
air ratchet wrench
Allen wrench
bench grinder
block diagnosis chart
blowgun
box-end wrench
chisel
combination pliers
combination wrench
deburring
diagonal cutting pliers
dial caliper
driver's manual

extractor
flaring
frame machine
hacksaw
hand tap
hand-threading die
Huck fastener
impact sockets
impact wrench
locking pliers
machinist's rule
maintenance manual
micrometer
needlenose pliers
open-end wrench
Phillips screwdriver
pliers

Posi-Drive™ screwdriver
press
punch
recall bulletin
rivet
screw pitch gauge
service bulletins
service manual
socket wrench
stud remover
swaging
thickness gauge
time guide
torque wrench
tree diagnosis chart
validity list
wrench

There is no substitute for getting the job done right the first time. For parts personnel, that means being able to provide both the correct part(s) and helpful information. For the technician, it means doing the job as quickly and safely as possible.

Many of the tools the heavy-duty truck technician uses every day are general purpose hand and power tools. For instance, a complete collection of wrenches is indispensable. A variety of truck parts, accessories, and related parts, not to mention shop equipment, uses common bolt and nut fasteners as well as special hex screws and fasteners. Depending on the make and model of the vehicle, the fasteners can be standard SAE or metric size fasteners. A well-equipped technician will have both metric and SAE wrenches in a variety of sizes and styles. The proper use of the appropriate hand and power tools by the technician is very important for performing quality heavy-duty truck service.

3.1 | HAND TOOLS

There are many repair tasks that do not lend themselves to the use of power tools. For those jobs, the heavy-duty truck technician must have suitable hand tools. Most service departments and garages require their technicians to buy their own hand tools. A set of master technician's hand tools and tool chest are shown in **Figure 3–1**.

HAMMERS

Hammers are identified by the material and weight of the head. There are two groups of hammer heads:

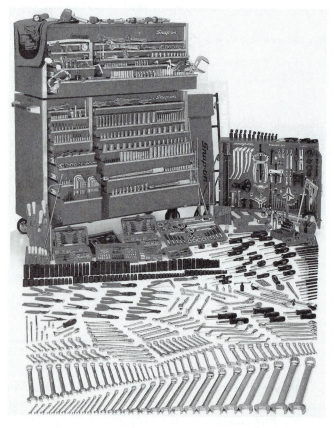

FIGURE 3–1   *Professional set of heavy-duty truck hand tools. (Courtesy of Snap-on Tools Company)*

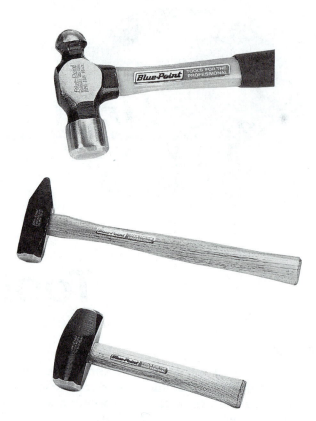

FIGURE 3–2   *Steel-face hammers.*

steel and soft face. The heads of steel-face hammers (**Figure 3–2**) are made from high-grade alloy steel. The steel is deep forged and heat treated to a suitable degree of hardness. Soft-face hammers (**Figure 3–3**) have a surface that yields when it strikes an object. Soft-face hammers are preferred when machined surfaces and precision are involved or when marring a finish is undesirable. For example, a brass hammer is used to drive in gears or shafts.

HAMMER SAFETY

- Wear eye protection. Always wear eye protection when striking hardened tools and hardened metal surfaces. This will protect your eyes from flying chips. Whenever possible, use soft-faced hammers (plastic, wood, or rawhide) when striking hardened surfaces.
- Never strike one hammer against another. A hammer can chip if struck against another hammer or hardened surface, resulting not only in damage to the hammer but possibly in bodily injury.
- Check the fit and condition of the handle. Keep handles tightly wedged in hammer heads to prevent injury to yourself and others nearby.
- Replace cracked or splintered handles and do not use the handle for prying or bumping. Handles are easily damaged and broken this way.

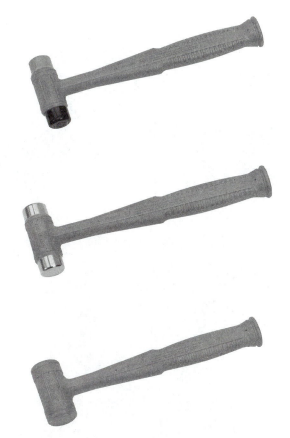

FIGURE 3–3   Soft-face hammers.

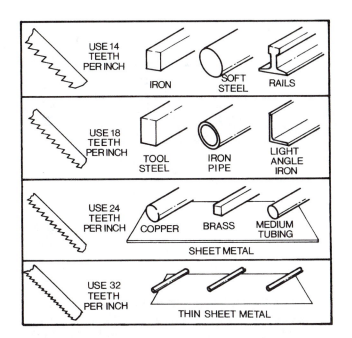

FIGURE 3–4   Typical type of work to be cut determines the best blade tooth per inch selection.

- Select the right size for the job. A light hammer bounces off the work. One that is too heavy is hard to control.
- Grip the handle close to the end. This increases leverage for harder, less tiresome blows. It also reduces the possibility of crushing your fingers between the handle and the projecting parts and edges of the work piece if you should miss.
- Prevent injuries to others. Swing in a direction that will not let your hammer strike someone if it slips from your hand. Keep the handle dry and free of grease and oil.
- Keep the hammer face parallel with your work. Force is then distributed over the entire hammer face, reducing the tendency of the edges of the hammerhead to chip or slip off the object.

SAWS AND KNIVES

The **hacksaw** is a much-used tool by truck technicians. It is excellent for cutting bolts, angle iron, tubing, and so on.

Hacksaw blades are generally available with 14, 18, 24, or 32 cutting teeth per inch. As shown in

Figure 3–4, the 14-tooth blade is used for fairly thick metal, while the 18-tooth one is employed to cutting medium thick material. The 24-tooth blade is usually used on heavy sheet metal, copper, brass, and medium tubing. For thin sheet metal and thinwall tubing, use a 32-tooth blade. While the blades may be made from a variety of material, high speed and tungsten steels are best for cutting alloy steels.

When hacksawing, be sure the blade is held tightly in the saw frame with the cutting teeth edges facing away from the handle. Securing the work firmly to prevent slippage, grasp the top of the frame in one hand while holding the handle in the other (**Figure 3–5**). Apply enough pressure on the forward stroke to ensure cutting, and then raise the blade slightly on the return stroke. Avoid starting the cut on a sharp edge; this may chip the saw.

Saw and Knife Safety

- Keep blades sharp. The greater the force you have to apply, the less control you have over the cutting action of the knife. The safest knife usually has the sharpest edge.
- Cut away from the body. The hands and fingers should always be behind the cutting edge. Keep knife handles clean and dry to keep the hand from slipping onto the blade.
- Never pry with a knife. Blades are hardened and can break with a snap.
- Store knives safely. An unguarded blade could cut you severely. Sharp pointed tools and knives should be kept sheathed while not in use.

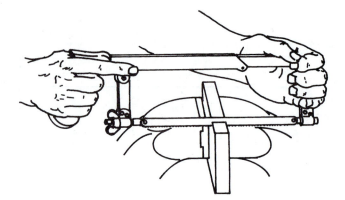

FIGURE 3–5   Proper way to hold a hacksaw when cutting.

- Before completing a saw cut, slow down to avoid injury when the saw cuts through the material. Cut with the saw in one direction only to prevent dulling the blade.
- Never use a damaged blade (cracked, kinked, missing teeth).
- Always wear safety glasses when using a hacksaw.
- Do not use the thumb to aid in starting a hacksaw. If starting is a problem, use a file to make a starting notch in the work.
- Use full strokes to get maximum life from the blade.

Tube Cutting Tools

Tubing made from steel, copper, aluminum, and plastic is used frequently on trucks. During service work, tubing often needs to be repaired or replaced. Tubing tools are made for such tasks as cutting, **deburring** (remove sharp edges from a cut), **flaring** (spreading gradually outward), **swaging** (reducing or tapering), bending, and removing fittings. **Figure 3–6** illustrates some of these tools.

CHISELS AND PUNCHES

Chisels (Figure 3–7) are used to cut metal by driving them with a hammer. Technicians use a variety of chisels for cutting sheet metal, shearing off rivet and bolt heads, splitting rusted nuts, and chipping metal.

When cutting with a chisel, the blade should be at least as large as the cut being made. Hold the chisel firmly enough to guide it, but lightly enough to ease the shock of the hammer blows. Hold the chisel just below the head to prevent pinching the hand in the event that the hammer misses striking the chisel. Grip the end of the hammer handle and strike with enough force to cut rivet head. Strike the chisel head squarely with the hammer face. Check your work

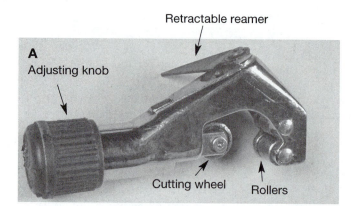

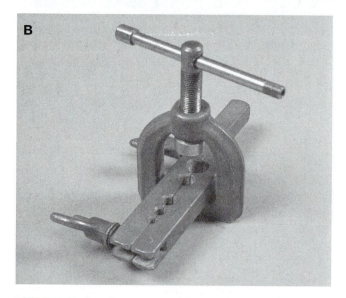

FIGURE 3–6   Common tubing tools: (A) tubing cutters; and (B) single loop flaring tool.

FIGURE 3–7   Chisels and their correct cutting edges.

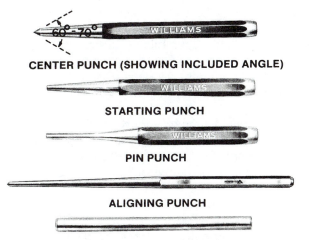

FIGURE 3–8   *Punches are designated by point diameter and punch shape.*

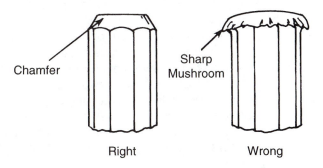

FIGURE 3–9   *Proper method for dressing a punch.*

every two or three blows. Correct your chisel angle as needed until the rivet is cut off.

Punches (**Figure 3–8**) are used for driving out pins, rivets, or shafts; aligning holes in parts during assembly; and marking the starting point for drilling a hole. Punches are designated by their point diameter and punch shape.

Chisel and Punch Safety

- Wear eye protection when cutting with a chisel or using a punch.
- Never use a punch or chisel on hardened metal.
- Remove any mushroom heads (**Figure 3–9**) before using a chisel or punch. This will help prevent cutting your hand and will keep chips from flying into your eyes.
- Do not drive a punch too deep into the bore or it may become wedged due to its taper.

SCREWDRIVERS

A variety of threaded fasteners used in the trucking industry are driven by a screwdriver. Each fastener

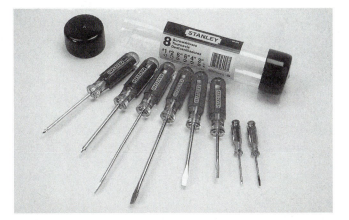

FIGURE 3–10   *The prepared technician keeps several sizes of screwdrivers on hand. (©1998 Stanley Tools, a Product Group of The Stanley Works, New Britain, CT)*

requires a specific kind of screwdriver, and the well-equipped technician will have several sizes of each (**Figure 3–10**).

All screwdrivers, regardless of the type of fastener they were designed for, have several things in common. The size of the screwdriver is determined by the length of the shank or blade. The size of the handle is important, too. The larger the handle diameter, the better grip it has and the more torque it will generate when turned. Screwdriver handles should also be insulated from the blade and made with a material that does not conduct electricity.

Standard Tip Screwdrivers

A slotted screw accepts a screwdriver with a standard tip. The standard tip screwdriver is probably the most common type (**Figure 3–11**). It is used for turning carriage bolts, machine screws, and sheet metal screws.

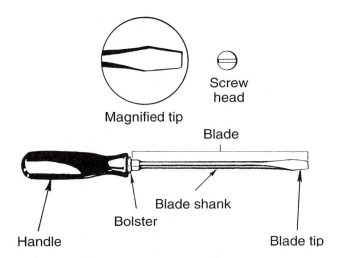

FIGURE 3–11   *The standard tip screwdriver fits slotted head screws.*

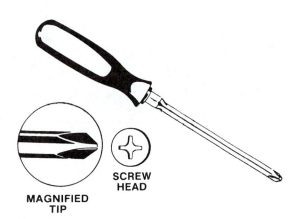

MAGNIFIED TIP

SCREW HEAD

FIGURE 3–12 The tip of a Phillips screwdriver has four prongs that help prevent slippage of the fastener.

Phillips Screwdrivers

The tip of a **Phillips screwdriver** has four prongs that fit the four slots in a Phillips head screw (**Figure 3–12**). This type of fastener is used very often in the automotive field. Not only does it look nicer than the slot head screw, it is also easier to install. The four surfaces enclose the screwdriver tip so that there is less likelihood that the screwdriver will slip off the fastener.

Specialty Screwdrivers

A number of specialty fasteners have been replacing the slot and Phillips head screws. This new breed of fastener is designed to improve transfer of torque from the screwdriver to the fastener, to slip less, and to offer tamper resistance.

The **Posi-Drive™ screwdriver** is like a Phillips but with a tip that is flatter and blunter (**Figure 3–13**). The

squared tip grips the screw's head and slips less than a Phillips screwdriver. The Torx® fastener is used to secure headlight assemblies, mirrors, and various racks. Not only does the six-prong tip provide greater turning power and less slippage (**Figure 3–14**), but it also provides a measure of tamper resistance.

- Use screwdrivers only for driving screws. Using them as punches or prybars breaks handles, bends shanks, and dulls and twists the tips. This makes them unfit to tighten or loosen screws safely.
- A slotted screwdriver tip can easily be dressed to its original shape. Never grind the tip, as excess heat will destroy the metal's temper. Always file by hand; you will have more control over the shape of the tip.
- If the screwdriver blade fits the screw slot properly, you will get maximum turning power with minimum pressure. A blade tip that does not fit properly will not only damage the screw slot but perhaps the tip itself.
- It is a good rule to keep your other hand clear when putting force on any type of screwdriver.
- Always have the screwdriver and the screw correctly lined up. You cannot get a good grip in the slot if the tip is held at an angle.
- Screwdrivers designed for use with wrenches have either a square shank or a special bolster at the handle to withstand the application of extra force.
- Do not hold parts in your hand. Put the work on a bench or in a vise to avoid the possibility of piercing your hand with the screwdriver tip.
- When working around anything electrical, use a screwdriver with an insulated or molded plastic handle to avoid shock (wooden handles with set screws are not acceptable).

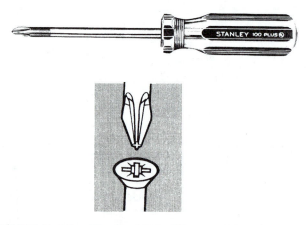

FIGURE 3–13 The Posi-Drive™ screwdriver is similar to a Phillips. (©1998 Stanley Tools, a Product Group of The Stanley Works, New Britain, CT)

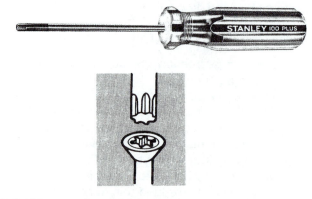

FIGURE 3–14 Torx® screwdrivers feature a six-prong tip. (©1998 Stanley Tools, a Product Group of The Stanley Works, New Britain, CT; Torx® is a registered trademark of Textron, Inc.)

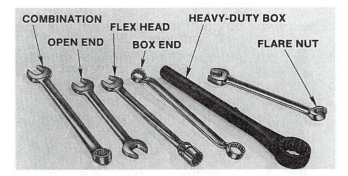

FIGURE 3–15   Wrench assortment. (Courtesy of Snap-on Tools Company)

WRENCHES

A complete collection of wrenches is indispensable for a truck technician. A well-equipped technician will have both metric and SAE wrenches in a variety of sizes and styles (**Figure 3–15**). But, remember that metric and SAE wrenches are not interchangeable. For example, a $9/16$-inch wrench is 0.02 inch larger than a 14-millimeter nut. If the $9/16$-inch wrench is used to turn or hold a 14-millimeter nut, the wrench will probably slip. This may cause rounding of the points of the nut and possibly skinned knuckles as well.

The word **wrench** means twist. A wrench is a tool for twisting and/or holding bolt heads or nuts. The width of the jaw opening determines its size. For example, a $1/2$-inch wrench has a jaw opening (from face to face) of $1/2$ inch. The size is actually slightly larger than its nominal size so that the wrench fits around a nut or bolt head of equal size.

Open-End Wrenches

The jaws of the **open-end wrench** (**Figure 3–16**) allow the wrench to slide around bolts or nuts where

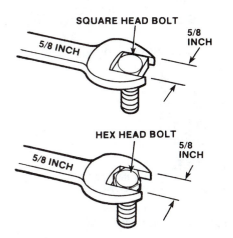

FIGURE 3–16   The open-end wrench grips only two faces of a fastener.

FIGURE 3–17   Box-end wrenches are closed to provide better holding power. (Courtesy of Snap-on Tools Company)

there might be insufficient clearance above or on one side of the nut to accept a box wrench.

Box-End Wrenches

The end of the **box-end wrench** is boxed or closed rather than open (**Figure 3–17**). The jaws of the wrench fit completely around a bolt or nut, gripping each point on the fastener. The box-end wrench is not likely to slip off a nut or bolt. It is safer than an open-end wrench.

Combination Wrenches

The **combination wrench** has an open-end jaw on one end and a box-end on the other. Both ends are the same size. Every auto technician should have two sets of wrenches, one for holding and one for turning. The combination wrench is probably the best choice for the second set. It complements either open-end or box-end wrench sets.

Adjustable Wrenches

An **adjustable wrench** has one fixed jaw and one movable jaw. The wrench opening can be adjusted by rotating a helical adjusting screw that is mated to teeth in the lower jaw. Because this type of wrench does not firmly grip a bolt's head, it is likely to slip. Adjustable wrenches should be carefully used and only when necessary. Be sure to apply all of the turning pressure on the fixed jaw (**Figure 3–18**).

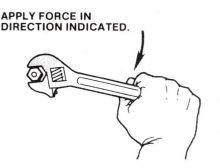

FIGURE 3–18   Pull the adjustable wrench so that the force bears against the fixed jaw.

Allen Wrenches

Set screws are used to fasten door handles, instrument panel knobs, and even brake calipers. A set of hex head wrenches, or **Allen wrenches** (**Figure 3–19**), should be in every technician's tool box.

Socket Wrenches

In many situations, a socket wrench is safer, faster, and easier to use than an open-end or box-end wrench. Some applications require the use of a socket.

The basic **socket wrench** set consists of a ratchet handle and several barrel-shaped sockets. The socket fits over and around a given size bolt, nut, or wrench (**Figure 3–20**). Inside, it is shaped like

a box-end wrench. Sockets are available in 6, 8, or 12 points. The top side of a socket has a square hole that accepts a square lug on the socket handle (**Figure 3–21**). One handle fits all the sockets in a set. The size of the lug ($3/8$ inch, $1/2$ inch, and so on) indicates the drive size of the socket wrench. On better quality handles, a spring-loaded ball in the square lug fits into a depression in the socket. This ball holds the socket to the handle.

Sockets are available in various sizes, lengths, and bore depths. Both standard SAE and metric socket wrench sets are necessary for automotive service. Normally, the larger the socket size, the deeper the well. Deep-well sockets (**Figure 3–22**) are

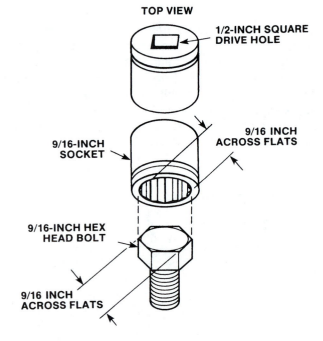

FIGURE 3–19 Typical Allen wrench set. (Courtesy of Snap-on Tools Company)

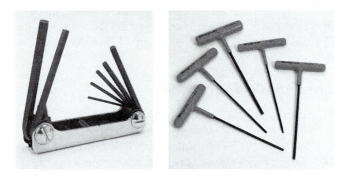

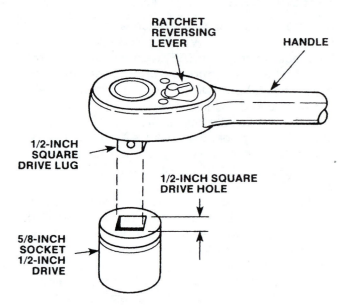

FIGURE 3–21 The socket drive size is equal to the diameter of the handle lug.

FIGURE 3–20 The size of a socket is the same as the bolt or nut size it fits.

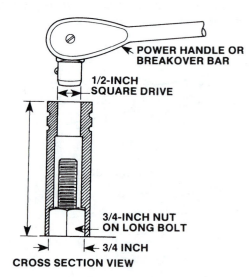

FIGURE 3–22 Deep-well sockets fit over bolt ends and studs.

made extra long to fit over bolt ends or studs. They are also good for reaching nuts or bolts in limited access areas. Deep-well sockets should not be used when a regular size socket will do the job. The longer socket develops more twist torque and tends to slip off the fastener.

Heavier walled sockets made of softer steel are designed for use with an **impact wrench** and are called **impact sockets.**

Figure 3–23 shows a number of socket wrench set accessories. Accessories multiply the usefulness of a socket wrench. A good socket wrench set has a variety of accessories.

Screwdriver attachments are also available for use with a socket wrench. **Figure 3–24** shows a typical

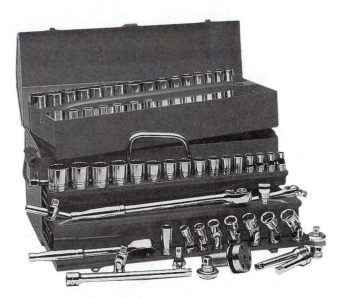

FIGURE 3–23   Typical $^3/_8$-inch and $^1/_2$-inch socket wrench sets. (Courtesy of Snap-on Tools Company)

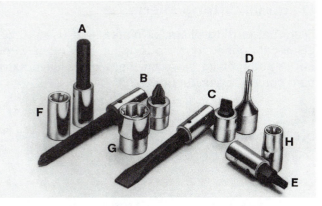

FIGURE 3–24   Typical screwdriver attachment set, including (A) hex driver, (B) Phillips driver, (C) flat tip driver, (D) clutch head driver, (E) Torx® driver, (F) three wing socket, (G) double square socket, and (H) Torx® socket. (Courtesy of Snap-on Tools Company)

set of screwdriver attachments and three specialty sockets. These socket wrench attachments are very handy when a fastener cannot be loosened with a regular screwdriver. The leverage given by the ratchet handle is often just what it takes to break a stubborn screw loose.

Wrench Safety

- Use wrenches that fit. Wrenches that slip damage bolt heads and nuts, skin knuckles, and lead to falls. Do not try to make wrenches fit by using shims.
- Use the proper wrench to get the job done—the one that gives you the surest grip and a straight clean pull. Cocking a wrench puts concentrated stress on the points of contact, a frequent cause of tool failure under pressure. (Other types of wrenches, such as the angle head, offset, and socket type, give you the ability to work in difficult places and still get a straight, clean pull.)
- Do not extend the length of a wrench. Do not use a pipe to increase the leverage of the wrench. The handle was made long enough for the maximum safe force to be applied. Excessive force may break the wrench or bolt unexpectedly, or the wrench may slip, rounding off the corners of the bolt head or nut; skinned knuckles, a fall, or a broken wrench may result.
- Do not use a hammer on wrenches unless they are designed for that type of use.
- Pull on the wrench. This is not always possible, but if you push, you take the risk that if the wrench slips or if the nut suddenly breaks loose, you may skin your knuckles or cut yourself on a sharp edge. Use the open palm of your hand to push on a wrench when you cannot pull it toward you.
- Replace damaged wrenches. Straightening a bent wrench weakens it. Cracked and worn wrenches are dangerous to use, because they could break or slip at any time.
- The adjustable wrench is a very convenient tool, but it should never be used if a properly fitting solid wrench is at hand.

PLIERS

Pliers are an all-around gripping tool used for working with wires, clips, and pins. The auto technician must own several types: standard pliers for common parts and wires, needlenose for really small parts, and large adjustable pliers for large items and heavy-duty work.

Combination Pliers

Combination pliers (Figure 3–25) are the most common type of pliers and are frequently used in many

FIGURE 3–25 *Combination, or slip-joint, pliers.* (Courtesy of Snap-on Tools Company)

kinds of automotive repair. The jaws have both flat and curved surfaces for holding flat or round objects. Also called slip-joint pliers, the combination pliers have many jaw-opening sizes. One jaw can be moved up or down on a pin attached to the other jaw to change the size of the opening.

Adjustable Pliers

Adjustable pliers, commonly called *channel locks* (**Figure 3–26**), have a multiposition slip joint that allows for many jaw-opening sizes.

Needlenose Pliers

Needlenose pliers have long tapered jaws (**Figure 3–27**). They are indispensable for grasping small parts or for reaching into tight spots. Many needlenose pliers also have wire cutting edges and a wire stripper.

Locking Pliers

Locking pliers, or vise grips, are similar to standard pliers, except that they can be locked closed with a very tight grip (**Figure 3–28**). They are extremely useful for holding parts together. They are also useful for

FIGURE 3–26 Adjustable pliers. (Courtesy of Snap-on Tools Company)

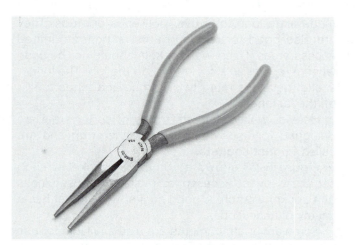

FIGURE 3–27 Needlenose pliers. (Courtesy of Snap-on Tools Company)

FIGURE 3–28 Locking pliers, or vise grips, can be locked closed. (Courtesy of Snap-on Tools Company)

getting a firm grip on a badly rounded fastener on which wrenches and sockets are no longer effective. Locking pliers come in several sizes and jaw configurations for use in many auto repair jobs.

Diagonal Cutting Pliers

Diagonal cutting pliers, or cutters, are used to cut electrical connections, cotter pins, and other wires on an automobile. Jaws on these pliers have extra hard cutting edges (**Figure 3–29**).

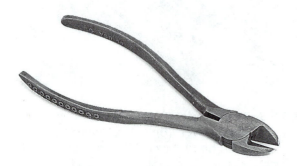

FIGURE 3–29 Diagonal cutting pliers.

Plier and Cutter Safety

- Do not use pliers as a wrench. They do not hold the work securely and can damage bolt heads and nuts.
- Guard against eye injuries when cutting with pliers or cutters. Short and long ends of wire often fly or whip through the air when cut. Wear eye protection and cup your hand over the pliers to guard yourself.
- Observe the following precautions:
  — Select a cutter big enough for the job.
  — Keep the blades at right angles to the stock.
  — Do not rock the cutter to get a faster cut.
  — Adjust the cutters to maintain a small clearance between the blades. This prevents the hardened blades from striking each other when the handles are closed.
- Pliers are made for holding, pinching, squeezing, and cutting—not for turning.

FILES

Files are designed to remove small amounts of metal, for smoothing, or sharpening parts. They are classified as single cut, double cut, fine, and coarse. These are easily identified by the file tooth pattern. Files come in a variety of shapes and sizes. **Figure 3–30** shows various file cuts and shapes.

File teeth cut in only one direction—on the forward stroke. Choose a file that is correct for the job at hand. Use a coarse cut file on soft material to prevent clogging. A slight drag on the return stroke, when filing soft material, will help keep the cutting teeth clear of filings.

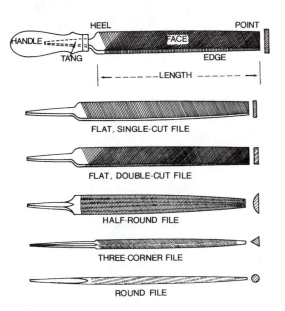

FIGURE 3–30 Various types of files.

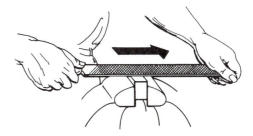

FIGURE 3–31 Correct way to hold a file.

When filing, grasp the file in one hand, with the thumb on top. Hold the point of the file with the thumb and first two fingers of the other hand (**Figure 3–31**). File only on the push strokes. Do not drag the file back over the work or cut on the return stroke. Cross the stroke at short intervals.

File Safety

- Wear eye protection while performing any filing activities.
- Never hit the file with a hammer. The file may shatter, resulting in serious injury.
- Always cut away from the body.
- Never use a file without a securely attached handle.
- Do not use worn (dull) files; replace them.

SPECIAL TOOLS

There are several other tools that can be found in a well-equipped truck technician's tool chest. Among them are the following:

Taps and Dies

The **hand tap** is a small tool used for hand cutting internal threads (**Figure 3–32**). An internal thread is cut on the inside of a part, such as a thread on the inside of a nut. This tap is also used for cleaning and restoring threads previously cut.

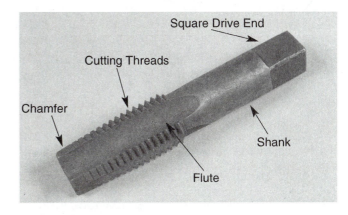

FIGURE 3–32 Parts of a hand tap.

Hand-threading dies (Figure 3–33) are the opposite of taps because they cut external (outside) threads on bolts, rods, and pipes rather than internal threads. Dies are made in various sizes and shapes, depending on the particular work for which they are intended. Dies may be solid (fixed size), split on one side to permit adjustment, or have two halves held together in a collet that provides for individual adjustments. Dies fit into holders called die stocks.

Gear and Bearing Pullers

Many precision gears and bearings have a slight interference fit (press-fit) when installed on a shaft or a housing. The press-fit allows no motion between parts and, therefore, prevents wear. The removal of gears and bearings must be done carefully. Prying or hammering can break or bind the parts. A puller with the proper jaws and adapters should be used when applying force to remove gears and bearings. Using proper tools, the force can be applied with a slight and steady motion. Various gears and bearing puller styles and sizes are shown in **Figure 3–34**.

Other hand tools that come in handy when servicing trucks include pry bars, wire brushes, scrapers, strap wrenches, propane torch, tire gauges, grease guns, tire irons, and trouble lights. Other special hand tools used are mentioned in the appropriate chapters in this book. There are also manufacturer's special tools (factory shop tools) that are necessary to make repairs on their equipment. These are usually assigned a part number by the manufacturer **(Figure 3–35)**.

Adjustable Round   Solid Square

FIGURE 3–33   *Common die shapes.*

FIGURE 3–34   *Various gear and bearing pullers.*

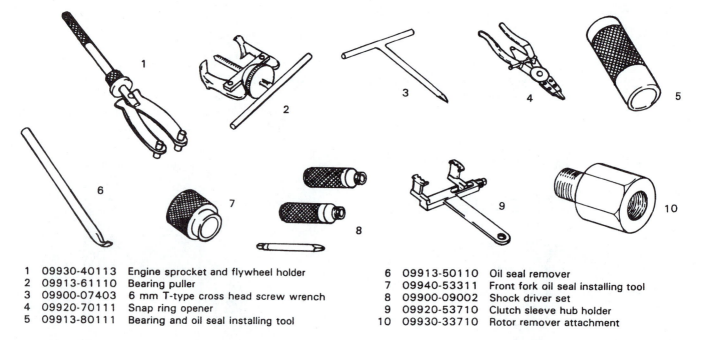

1	09930-40113	Engine sprocket and flywheel holder
2	09913-61110	Bearing puller
3	09900-07403	6 mm T-type cross head screw wrench
4	09920-70111	Snap ring opener
5	09913-80111	Bearing and oil seal installing tool
6	09913-50110	Oil seal remover
7	09940-53311	Front fork oil seal installing tool
8	09900-09002	Shock driver set
9	09920-53710	Clutch sleeve hub holder
10	09930-33710	Rotor remover attachment

FIGURE 3–35   *Manufacturer's special tools make jobs easier and are assigned a part number that can be found in the service manual.*

3.2 POWER TOOLS

Power tools make a technician's job easier. They operate faster and with more torque than hand tools. However, power tools require greater safety measures. Power tools do not stop unless they are turned off. Power is supplied by either air (pneumatic), electricity, or hydraulic fluid.

Although electric drills, wrenches, grinders, chisels, drill presses, and various other tools are found in shops, pneumatic (air) tools are used more frequently. Pneumatic tools have four major advantages over electrically powered equipment in an engine rebuilding shop:

- **Flexibility.** Air tools run cooler and have the advantage of variable speed and torque; damage from overload or stalling is eliminated. They can fit in tight spaces.
- **Lightweight.** The air tool is lighter in weight and lends itself to a higher rate of production with less fatigue (**Figure 3–36**).
- **Safety.** Air equipment reduces the danger of fire and shock hazards in some environments where the sparking of electric power tools can be a problem.
- **Low-Cost Operation and Maintenance.** Due to fewer parts, air tools require fewer repairs and less preventive maintenance. Also, the original cost of air-driven tools is usually less than the equivalent electric type.

The automotive industry was one of the first industries to recognize the advantages of air powered tools. Today they are known as tools of the professional truck technician. However, one major disadvantage of air tools is noise.

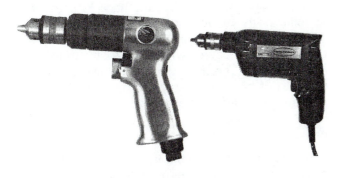

FIGURE 3–36 (Left) $3/8$-inch air drill and (right) $3/8$-inch electric drill. The air drill weighs $2^{1}/_{2}$ pounds; the electric drill weighs $4^{1}/_{2}$ pounds.

POWER TOOL SAFETY

Safety is critical when using power tools, regardless how they are powered. Carelessness or mishandling of power tools can cause serious injury. Here are some general power tool safety rules that must always be followed:

- Return all equipment to its proper place when finished.
- Wear eye protection.
- Noise may be a hazard with some portable power tools, especially pneumatic tools. Wear ear protection whenever noise is excessive.
- Wear gloves when operating air chisels or air hammers.
- All electrical equipment should be grounded, unless it is the double insulated type.
- Never make adjustments, lubricate, or clean a machine while it is running.
- Do not clean yourself or anyone else with compressed air.
- Report any suspicious or malfunctioning machinery to the instructor or service manager.
- Know your power tool. Read the operator's manual carefully. Learn its applications and limitations as well as the specific potential hazards peculiar to this tool. In other words, do not operate any power tools in damp or wet locations. Keep your work area well lighted.
- Do not abuse the electric power cord. Never yank it to disconnect it from a receptacle.
- No machine should be started unless the guards are in place and in good condition. Defective or missing guards should be reported to the instructor or service manager immediately.
- Check and make all adjustments before applying power.
- Give the machine your undivided attention while you are using it. Do not look away or talk to others.
- Inspect all equipment for safety and for apparent defects before using.
- Whenever safeguards or devices are removed to make repairs or adjustments, equipment should be turned off and the main switch locked and tagged.
- Start and stop your own machine and remain with it until it has come to a complete stop.
- Always allow any machine to reach full operating speed before applying work.
- No attempt shall be made to retard rotation of the tool or work.
- Do not try to strip broken belts or other debris from the pulley while in motion or reach between belts and pulleys.

- Do not use loose rags around operating machinery.
- Use the right tool. Do not force a small tool or attachment to do the job of a heavy-duty tool.
- Do not force the tool. It will do the job better and safer at the rate for which it was designed. That is, do not attempt to use a machine or tool beyond its stated capacity or for operations requiring more than the rated horsepower of the tool.
- Maintain tools with care. Keep tools sharp and clean at all times, for best and safest performance. Follow instructions for lubricating and changing accessories.
- Remove adjusting keys and wrenches. Form the habit of checking to see that keys and adjusting wrenches are removed from the tool before turning it on.
- Do not over-reach. Keep your proper footing and balance at all times. Maintain a balanced stance to avoid slipping.
- Disconnect tools when not in use, before servicing, or when changing attachments, blades, bits, cutters, and so on. Before plugging in any electric tool or machine, make sure the switch is off to prevent serious injury. When the task is completed, turn it off and unplug it.
- Remove all sharp edges and burrs before completing any job.
- Keep all guards in place and in working order.
- Use a power tool only for its intended purpose.
- Never leave a power tool running unattended. Before leaving, turn off the machine.
- When using power equipment on small parts, never hold the part in your hand. Always mount the part in a bench vise or use vise grip pliers.

Impact Wrenches

An impact wrench (**Figure 3–37**) is a portable hand-held reversible wrench. A medium-duty model can deliver up to 200 lb.ft. of torque. When triggered, the output shaft, onto which the impact socket is fastened, spins freely at 2,000 to 14,000 rpm, depending on the wrench's make and model. When the impact wrench meets resistance, a small spring loaded hammer, which is situated near the end of the tool, strikes an anvil attached to the drive shaft onto which the socket is mounted. Each impact moves the socket around a little until torque equilibrium is reached, the fastener breaks, or the trigger is released.

When using an impact wrench, keep the following safety rules in mind:

- Sockets designed for hand tool use should not be used on impact wrenches. They can break,

FIGURE 3–37 Impact wrench in use removing tire nuts.

causing damage or injury. Use impact sockets only.
- Make sure the socket is securely snapped in place before using the wrench.
- Hold the wrench firmly with both hands.
- Keep hands clear of moving parts.
- Ear protection is recommended for use of more than 15 minutes.
- Keep face away from work when using an impact wrench.
- Wear eye protection.

Air Ratchet Wrenches

An **air ratchet wrench,** like the hand ratchet, has a special ability to work in hard to reach places. Its angle drive reaches in and loosens or tightens where other hand or power wrenches just cannot work (**Figure 3–38**). The air wrench looks like an ordinary ratchet but has a fat handgrip that contains the air vane motor and drive mechanism.

Air Drills

They are usually available in $1/4$-, $3/8$-, and $1/2$-inch chuck sizes and operate in much the same manner as an electric drill, but they are smaller and lighter. This compactness makes them a great deal easier to use for drilling operations in truck work.

FIGURE 3-38   Typical air ratchet wrench.

Air Chisels and Hammers

Of all truck tools, the air chisel or hammer (**Figure 3-39**) is one of the most useful. Used with the accessories illustrated in **Figure 3-40**, this tool will perform many different operations.

- **Universal joint and tie-rod tool.** This tool helps to shake loose stubborn universal joints and tie-rod ends.
- **Ball joint separator.** The wedge action breaks apart frozen ball joints.
- **Shock absorber chisel.** With this tool, quick work is made of the roughest jobs without the usual bruised knuckles and lost time. It easily cracks frozen shock absorber nuts.
- **Exhaust pipe cutter.** The cutter slices through mufflers and exhaust pipes.
- **Tapered punch.** Driving frozen bolts, installing pins, and punching or aligning holes are some of the many uses for this accessory.
- **Rubber bushing splitter.** Old bushings can be opened up for easy removal.
- **Bushing removal.** This accessory is designed to remove all types of bushings. The blunt edge pushes but does not cut.
- **Bushing installer.** The installer drives all types of bushings to the correct depth. A pilot prevents the tool from sliding.

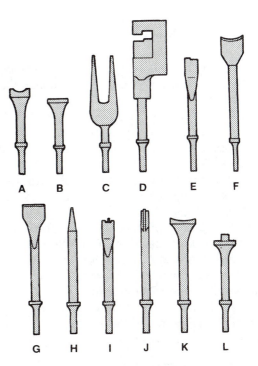

FIGURE 3-40   Air chisel accessories: (A) universal joint and tie-rod tool; (B) smoothing hammer; (C) ball joint separator; (D) panel crimper; (E) shock absorber chisel; (F) tail pipe cutter; (G) scraper, (H) tapered punch; (I) edging tool; (J) rubber bushing splitter; (K) bushing remover; and (L) bushing installer.

Blowgun

One way to use compressed air from a pneumatic hose is with a **blowgun** (**Figure 3-41**). A blowgun snaps into one end of the hose and directs airflow when a button is pressed. Before using a blowgun, be sure it has not been modified to eliminate air-bleed holes on the side. Blowguns are used for blowing off parts during cleaning. Never point a blowgun at yourself or someone else.

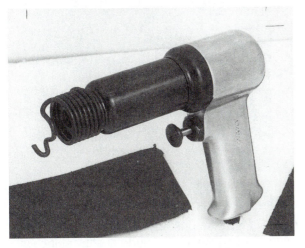

FIGURE 3-39   Typical air hammer or chisel.

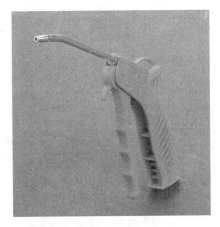

FIGURE 3-41   Typical blowgun.

Other safety rules to remember when using pneumatic powered tools are

- A tool retainer must be installed on each pipe of equipment that, without such a retainer, may eject the tool.
- Hose and hose connections used for conducting compressed air to equipment must be used only for the pressure and service for which they are designed.
- When connecting pneumatic tools and lines, check to make sure they are attached securely and properly.
- Do not use compressed air for cleaning, unless it is reduced to less than 30 psi (pounds per square inch) and then only with personal protection equipment. (More than 30 psi may be used to power pneumatic tools.)
- Ear protection is required when using pneumatic tools for more than 15 minutes, or whenever noise is excessive.

OTHER POWER TOOLS

There are several other power tools the truck technician may use, including the following.

Bench Grinders

These electric power tools are generally bolted to a workbench. A **bench grinder** is classified by wheel size. Six- to ten-inch wheels are the most common in auto repair shops. Three types of wheels are available with this bench tool.

- **Grinding wheel.** For a wide variety of grinding jobs from sharpening cutting tools to deburring.
- **Wire wheel brush.** Used for general cleaning and buffing, removing rust and scale, paint removal, deburring, and so forth.
- **Buffing wheel.** For general purpose buffing, polishing, light cutting, and finish coloring operations.

Lift and Hoist Safety

Raising a heavy-duty truck trailer on a lift or **frame machine (Figure 3-42)** requires special care. Adapters and hoist plates must be positioned correctly on twin post and rail type lifts to prevent damage to the underbody of the vehicle. There are specific lift points to use where the weight of the vehicle is evenly supported by the adapters or hoist plates. The correct lift points can be found in the vehicle's service manual. Before operating any lift or alignment machine, carefully read the manufacturer's manual and understand all the operating and maintenance instructions.

FIGURE 3-42 *Typical frame straightening machine. (Courtesy BeeLine Co., Bettendorf, IA)*

Heavy parts of a heavy-duty truck, such as engines and transmissions, are removed by using chain hoists **(Figure 3-43)** or cranes. To prevent serious injury, chain hoists and cranes must be properly attached to the parts being lifted. Always use bolts of sufficient strength to attach the hoist to the object being lifted. Attach the lifting chain or cable to the system that is to be removed. Before lifting, secure the chain or cable.

The following are some general rules for jacks, lifts, frame machines, and hoists:

- Do not let anyone remain in the vehicle when it is being raised.
- Make certain you know how to operate the equipment and know its limitations.
- Never overload a lift, hoist, or jack.
- Chain hoists and cranes must be properly attached to the parts being lifted. Always use

FIGURE 3-43 *A pulling sling helps to lift an engine out of a vehicle. (Courtesy of BeeLine Co., Bettendorf, IA)*

bolts of sufficient strength to attach the hoist to the object being lifted. Attach the lifting chain or cable to the system that is to be removed.

- Do not use any lift, hoist, or jack that you believe to be defective or not operating properly. Report it to your instructor or service manager immediately.
- Make sure all persons and obstructions are clear before raising or lowering an engine or car.
- Avoid working, walking, or standing under suspended objects; lower them to the floor as soon as possible.

Presses

Many truck repair jobs require the use of considerable force to assemble or disassemble parts that are press fit together. Servicing rear axle bearing, pressing brake drum and rotor studs, transmission assembly work, and frame work are just a few examples. **Presses** can be hydraulic, electric, air, or hand driven. Capacities range up to 150 tons of pressing force, depending on the size and design of the press. Smaller arbor and C-frame presses can be bench or pedestal mounted (**Figure 3–44**), while high capacity units are free standing or floor mounted (**Figure 3–45**).

FIGURE 3–44 _Small bench-mounted, manually operated arbor press. (Courtesy of Auto Electric Supplier and Equipment Co.)_

FIGURE 3–45 _Typical heavy-duty shop press. (Courtesy of Snap-On Tools Company)_

3.3 MEASURING TOOLS

Compared to the number used by an engine technician, the so-called precision measuring tools used by the truck technician are few. However, it is most important that he/she know how to use them.

MACHINIST'S RULE

The **machinist's rule** looks like an ordinary ruler (**Figure 3–46**). However, unlike the common ruler, it is precisely divided into small ($1/64$ inch) increments. A typical machinist's rule is marked on both sides. One side is marked off at $1/16$-, $1/16$-, $1/4$-, $1/2$- and 1-inch intervals. The other side is marked at $1/32$- and $1/64$-inch intervals.

Machinist's rules are also available with metric or decimal graduations. Metric rules are usually divided into 0.5-mm and 1-mm increments. Decimal rules are typically divided into $1/10$-, $1/50$-, and $1/100$-inch (0.10-, 0.50-, and 0.01-inch) increments. Decimal machinist's rules are convenient when measuring component dimensions that are specified in decimals.

DIAL CALIPERS

The **dial caliper** (**Figure 3–47**) is a versatile measuring instrument. It is capable of taking inside, outside,

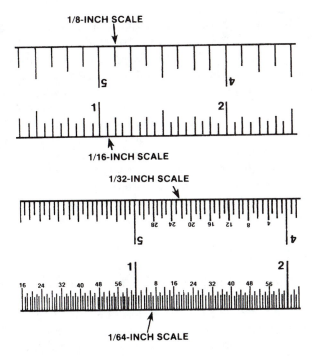

FIGURE 3–46 _Typical machinist's rule graduations._

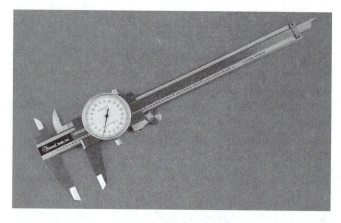

FIGURE 3–47 *Typical dial caliper. (Courtesy of Central Tools, Inc.)*

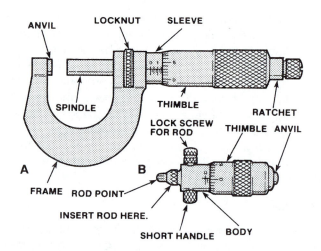

FIGURE 3–48 *Major components of (A) an outside and (B) an inside micrometer.*

depth, and step measurements. It can measure these dimensions from zero to 150 mm in increments of 0.02 mm.

The dial caliper features a depth scale, bar scale, dial indicator, inside measurement jaws, and outside measurement jaws. The bar scale is divided into one-tenth (0.10) of an inch graduations. The dial indicator is divided into one thousandth (0.001) of an inch graduations. Therefore, one revolution of the dial indicator needed equals one-tenth of an inch on the bar scale (one-hundred thousandth of an inch equals one-tenth of an inch).

The metric dial caliper is similar in appearance to the standard (inch-reading) model. However, the bar scale is divided into 2-mm increments. Additionally, on the metric dial caliper, one revolution of the dial indicator needle equals 2 mm.

Both inch reading and metric dial calipers use a thumb-operated roll knob for fine adjustment. When you use a dial caliper, always move the measuring jaws backward and forward to center the jaws in the work. Always make sure that the caliper jaws lay flat on the work. If the jaws or the work are tilted in any way, you will not obtain an accurate measurement. However, although dial calipers are precision measuring instruments, they are only accurate to within ±0.002 inch. The factors that limit dial caliper accuracy include jaw flatness and "feel." Micrometers are better suited to measuring tasks that require extreme precision.

MICROMETERS

Measurements required in some truck servicing jobs are either the size of the outside or inside diameter or the diameter of a shaft and the bore of a hole. The **micrometer** is the common instrument for taking these measurements. Both outside and inside micrometers are calibrated and read in the same

manner and are both operated so that the measuring points exactly contact the surfaces being measured.

The major components and markings of the micrometer include the frame, anvil, spindle, locknut, sleeve, sleeve numbers, sleeve long line, thimble marks, thimble, and ratchet (**Figure 3–48**). Micrometers are calibrated in either inch or metric graduations.

On both the outside and inside micrometers, the thimble is revolved between the thumb and forefinger. Very light pressure is required when bringing the measuring points into contact with the surfaces being measured. It is *important* to remember that the micrometer is a delicate instrument and that even slight excessive pressure will result in an incorrect reading.

Reading an Inch-Graduated Outside Micrometer

These micrometers are made so that each turn of the thimble moves the spindle 0.025 inch (twenty-five thousandths of an inch). This is accomplished by using forty threads per inch on the thimble. The sleeve long line is marked with sleeve numbers 1, 2, 3, and so on. These sleeve numbers represent 0.100 inch, 0.200 inch, 0.300 inch, and so on. The sleeve on the micrometer contains sleeve marks that represent 1 inch in 0.025-inch (twenty-five thousandths of an inch) increments. Each of the thimble marks represents 0.001 inch (one thousandth of an inch). In one complete turn, the spindle will move 25 marks or 0.025 inch (twenty-five thousandths of an inch). Inch-graduated micrometers come in a range of sizes—zero to 1 inch, 1 inch to 2, 2 inches to 3, 3 inches to 4, etc. The most commonly used micrometers are calibrated in one-thousandths of an inch increments.

To read a micrometer, first read the last whole sleeve number visible on the sleeve long line. Next, count the number full sleeve marks past the number. Finally, count the number of thimble marks past the sleeve marks. Add these measurements together for the measurement. These three readings indicate tenths, hundredths, and thousandths of an inch, respectively. For example, a 2- to 3-inch micrometer that has taken a measurement is described as follows (**Figure 3–49**):

1. The largest sleeve number visible is 4, indicating 0.400 inch (four-tenths of an inch).
2. The thimble is three full sleeve marks past the sleeve number. Each sleeve mark indicates 0.025 inch, so this indicates 0.075 inch (seventy-five hundredths of an inch).

3. The number 12 thimble mark is lined up with the sleeve long line. This indicates 0.012 inch (twelve thousandths of an inch).
4. Add the readings from Steps 1, 2, and 3. The total of the three is the correct reading. In our example

Sleeve	0.400 inch
Sleeve marks	0.075 inch
Thimble marks	0.012 inch
Total =	0.487 inch

5. Now add 2 inches to the measurement, since this is a 2- to 3-inch micrometer. The final reading is 2.487 inches.

Reading an Outside Micrometer with a Vernier Scale

In cases where a measurement must be within 0.0001 inch (one ten-thousandth of an inch), a micrometer with a vernier scale should be used. This micrometer is read in the same way as a standard micrometer. However, in addition to the three scales found on the typical micrometer, this type has a vernier scale on the sleeve. When taking measurements with this micrometer (sometimes called a mike), read it in the same way as you would a standard mike. Then, find the thimble mark that lines up precisely with one of the vernier scale lines (**Figure 3–50**). Only one of these lines will match up correctly. All other lines will be mismatched. The vernier scale number that matches up with a thimble mark is the 0.0001-inch (one ten-thousandth of an inch) measurement.

Reading a Metric Outside Micrometer

The metric micrometer is read in the same manner as the inch-graduated micrometer except that the

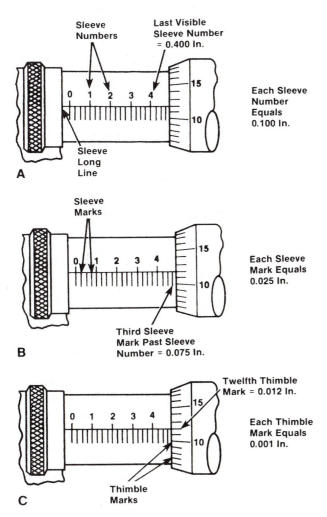

FIGURE 3–49 The three steps in reading a micrometer: (A) measuring tenths of an inch; (B) measuring hundredths of an inch; (C) measuring thousandths of an inch.

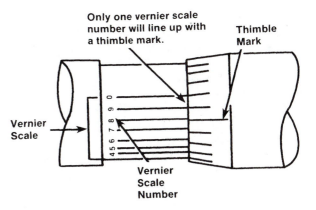

FIGURE 3–50 Measuring ten thousandths of an inch using a micrometer with vernier scale.

graduations are in the metric system of measurement. Readings are obtained as follows:

- Each number on the sleeve of the micrometer represents 5 millimeters or $5/1000$ of a meter (**Figure 3–51A**).
- Each of the ten equal spaces between each number, with index lines alternating above and below the horizontal line, represents 0.5 millimeter or five tenths of a millimeter. One revolution of the thimble changes the reading one space on the sleeve scale or 0.5 mm (**Figure 3–51B**).
- The beveled edge of the thimble is divided into 50 equal divisions with every fifth line numbered 0, 5, 10 . . . 45. Since one complete revolution of the thimble advances the spindle 0.5 mm, each graduation on the thimble advances the spindle 0.5, each graduation on the thimble is equal to $1/50$ of 0.5 mm or one hundredth of a millimeter (**Figure 3–51C**).

As with the inch-graduated micrometer, the three separate readings are added together to obtain the total reading (**Figure 3–52**).

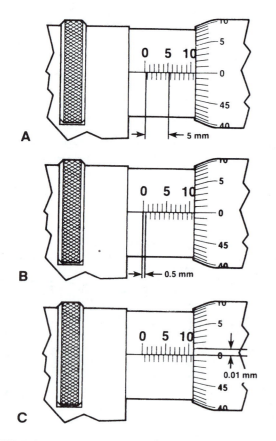

FIGURE 3–51 Reading a metric micrometer: (A) 5 mm; (B) 0.5 mm; (C) 0.01 mm.

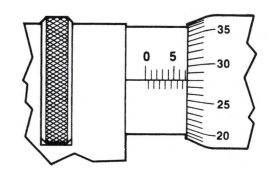

FIGURE 3–52 The total reading on this micrometer is 7.28 mm.

1. Read the largest number on the sleeve that has been uncovered by the thimble. In the illustration it is 5, which means the first number in the series is 5 mm.
2. Count the number of lines past the number 5 that the thimble has uncovered. In the example, this is 4, and since each space is equal to 0.5 mm, 4 spaces equal 4 × 0.05 or 2 mm. This added to the figure obtained in Step 1 gives 7 mm.
3. Read the graduation line on the thimble that coincides with the horizontal line of the sleeve scale and add this to the total obtained in Step 2. In the example, the thimble scale reads 28 or 0.28 mm. This, added to the 7 mm from Step 2 gives a total reading of 7.28 mm.

Using an Outside Micrometer

To measure small objects using an outside micrometer, grasp the micrometer with the right hand and slip the object to be measured between the spindle and anvil. While holding the object against the anvil, turn the thimble using the thumb and forefinger until the spindle contacts the object. Never clamp the micrometer tightly. Use only enough pressure on the thimble to allow the work to just fit between the anvil and spindle. If the micrometer is equipped with a ratchet screw, use it to tighten the micrometer around the object for final adjustment. For a correct measurement, the object must slip while adjusting the thimble (**Figure 3–53**). It is important to slip the mike back and forth over the work until you feel a very light resistance, while at the same time rocking the mike from side to side to make certain the spindle cannot be closed any further (**Figure 3–54**). These steps should be taken with any precision measuring device to ensure correct measurements.

Measurements will be reliable if the mike is calibrated correctly. To calibrate a micrometer, close the mike over a micrometer standard. If the reading dif-

FIGURE 3–53 Using outside micrometer for measuring small objects.

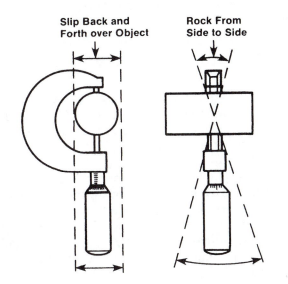

FIGURE 3–54 Slip the micrometer.

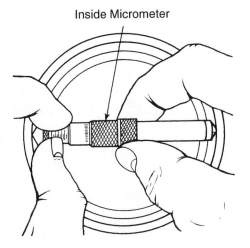

Inside Micrometer

FIGURE 3–55 Obtaining a precise measurement with an inside micrometer.

After taking the measurement with the inside mike, use an outside mike to take a comparison measurement. This reduces the chance of errors and ensures an accurate measurement.

Shop Talk

Follow these tips for taking care of a micrometer:

- *Always clean the micrometer before using it.*
- *Do not touch measuring surfaces.*
- *Store the micrometer properly. The spindle face should not touch the anvil face, or a change in temperature might spring the micrometer.*
- *Clean the mike after use. Wipe it clean of any oil, dirt, or dust using a lint-free cloth.*
- *Do not drop the mike. It is a sensitive instrument and must be handled with care.*
- *Check the calibration weekly. If it drops at any time, check it immediately.*

OTHER MEASURING GAUGES

There are several measuring gauges that truck technicians may use. They are

Thickness Gauges

The **thickness gauge** is a strip of metal of a known and closely controlled thickness. Several of these metal strips are often combined into a multiple measuring instrument that pivots in a manner similar to a pocket knife (**Figure 3–56**). The desired thickness gauge can be pivoted away from others for convenient use. A steel thickness pack usually will contain

fers from that of the known micrometer standard, then the mike will need adjustment.

Reading an Inside Micrometer

Inside mikes (**Figure 3–55**) are used to measure bore sizes. They are frequently used with outside mikes to reduce the chance of error.

To use an inside mike, place it inside the bore or hole and extend the measuring surfaces until each end touches the bore's surface. If the bore is large, it might be necessary to use an extension rod to increase the mike's range. These extension rods come in various lengths. The inside micrometer is read in the same manner as an outside micrometer.

To obtain a precise measurement in either inch or metric graduations, keep the anvil firmly against one side of the bore and rock the inside mike back and forth and side to side. This ensures that the mike fits in the center of the work with the correct amount of resistance. As with the outside micrometer, this procedure will require a little practice until you get the feel for the correct resistance and fit of the mike.

FIGURE 3-56 Typical thickness gauge pack.

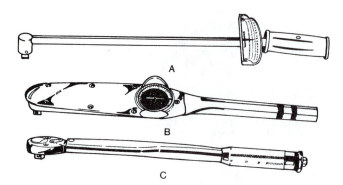

FIGURE 3–58 Three styles of torque-indicating wrenches: (A) flex bar; (B) dial indicator; and (C) sound-indicating. (Courtesy of Snap-on Tools Company)

leaves of 0.002-to 0.010-inch thickness (in steps of 0.001 inch) and leaves of 0.012- to 0.024-inch thickness (in steps of 0.002 inch).

Screw Pitch Gauges

Sometimes it is necessary to determine the pitch (threads per inch) of a fastener. The use of a **screw pitch gauge** (**Figure 3–57**), provides a quick and accurate method of checking the threads per inch. The leaves of this measuring tool are marked with various pitches. Just match the teeth of the gauge with the threads of the fastener and the correct pitch can be read directly from the leaf.

Screw pitch gauges are available for the various types of fastener threads used in the truck industry: American National coarse and fine threads, metric threads, International Standard threads, and Whitworth threads.

Torque-Indicating Wrenches

A **torque wrench** measures the amount of turning force being applied to a fastener (bolt or nut). Conventional torque wrench scales are usually read in foot-pounds (lb.ft.) and metric scales in newton-meters (N-m). The fact that practically every truck manufacturer publishes a list of torque recommendations is ample proof of the importance of using the proper amounts of torque when tightening nuts or cap screws.

The three general types of torque wrenches are the flex bar, dial indicator, and sound-indicating types. These are shown in **Figure 3–58**. The flex bar torque wrench is inexpensive and accurate. The dial indicator torque wrench is very accurate, but can be hard to use in tight quarters. The sound-indicating type torque wrench is very fast and easy to use. It makes a "pop" or "click" sound when a preset torque value is reached. With this type of torque wrench it is not necessary to watch an indicating needle while torquing.

As described later in this chapter, a torque wrench is a very safe tool because it makes it possible to apply the right amount of torque without overstressing either the tool or the fastener. Do not use a torque wrench, however, to break nuts and bolts loose. A torque wrench should be used for tightening purposes only.

3.4 MANUFACTURERS' SERVICE PUBLICATIONS

Information is a very important servicing tool in the modern trucking industry. Rapid change makes up-to-date information extremely valuable to the service technician. Because of this, truck manufacturers provide all or some of the following major type service publications:

Service Manuals

The manufacturer's (vehicle) **service manual** contains service and repair information for all vehicle

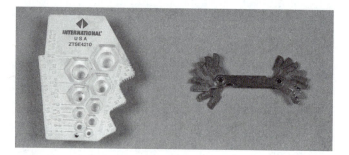

FIGURE 3–57 Typical screw pitch gauge.

PHOTO SEQUENCE 1
USING A MICROMETER

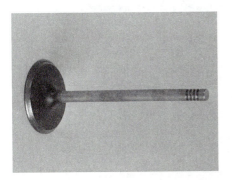

P1–1 Micrometers can be used to measure the diameter of many different objects. By measuring the diameter of a valve stem in two places, the wear of the stem can be determined.

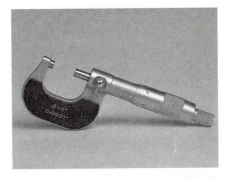

P1–2 Because the diameter of a valve stem is less than one inch, a 0-to-1-inch (0-to-25-mm) outside micrometer is used.

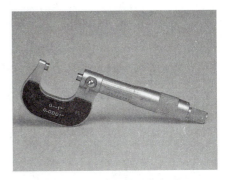

P1–3 The graduations on the sleeve each represent 0.025 inch. To read a measurement on a micrometer, begin by counting the visible lines on the sleeve and multiplying them by 0.025.

P1–4 The graduations on the thimble assembly define the area between the lines on the sleeve. The number indicated on the thimble is added to the measurement shown on the sleeve.

P1–5 A micrometer reading of 0.500 inch (12.70 mm).

P1–6 A micrometer reading of 0.375 inch (9.53 mm).

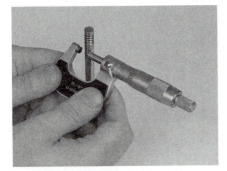

P1–7 Normally, little stem wear is evident directly below the keeper grooves. To measure the diameter of the stem at that point, close the micrometer around the stem.

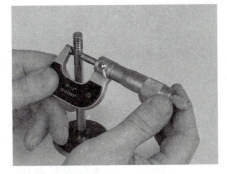

P1–8 To get an accurate reading, slowly close the micrometer until a slight drag is felt while passing the valve in and out of the micrometer.

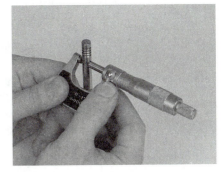

P1–9 To prevent the reading from changing while you move the micrometer away from the stem, use your thumb to activate the lock lever.

PHOTO SEQUENCE 1 (CONTINUED)
USING A MICROMETER

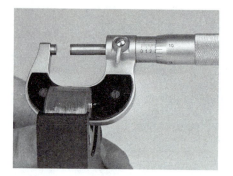

P1–10   This reading (0.311 inch [7.89 mm]) represents the diameter of the valve stem at the top of the wear area.

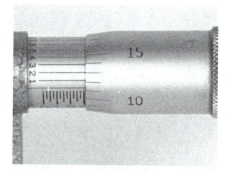

P1–11   Some micrometers are able to measure in 0.0001 (ten-thousandths) of an inch. Use this type of micrometer if the specifications call for this much accuracy. Note that the exact diameter of the valve stem is 0.3112 inch (7.90 mm).

P1–12   Most valve stem wear occurs above the valve head. The diameter here should also be measured. The difference between the diameter of the valve stem just below the keepers and just above the valve head represents the amount of valve stem wear.

systems and components. As shown in **Figure 3–59**, the service manuals can be made up of a separate manual on an individual system, or it can encompass several volumes in which truck servicing areas are broken down in component groups or sections (**Figure 3–60**). A typical page layout of each type of manual is shown in **Figure 3–61**.

Regardless of the system used, the service manual (sometimes called a "shop manual") is divided into subjects that usually include general information, principles of operation, removal, disassembly, assembly, installation, specifications, and troubleshooting. Many truck OEMs produce service manuals on CD or on-line: This makes them easier to use and update. If the source of the problem is hard to find, a troubleshooting chart can be used. It will guide you to the most common causes for specific problems. There are usually two basic types of troubleshooting charts:

- A **tree diagnosis chart** (**Figure 3–62**) provides a logical sequence for what should be inspected or tested when trying to solve a repair problem.
- A **block diagnosis chart** (**Figure 3–63**) lists conditions (problem symptoms), causes (problem sources) and remedy (needed repairs) in columns.

Maintenance Manuals

Maintenance manuals contain routine maintenance procedures and intervals for vehicle components and

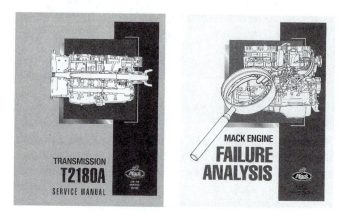

FIGURE 3–59   Separate types of service manuals.

FIGURE 3–60   Service manual divided by sections or groups.

systems. They are important tools for PM service technicians since they have information such as lubrication procedures and charts, replacement of fluids,

Removal

1. Review the information under "Safety Precautions, 100."

WARNING: Failure to review the safety precautions, and to be aware of the danger involved when working with refrigerant, could result in serious personal injury.

2. Depending on the type of temperature control, open the water shutoff valve.

   If equipped with a toggle switch control, set the heater water-temperature toggle switch to HOT (**Fig. 1**). When in this position, there is no air pressure to the water shutoff valve.

   If equipped with a cable control, pull the temperature control cable up (**Fig. 2**).

   If equipped with C.T.C., turn the temperature control knob clockwise to HEAT (**Fig. 3**).

3. Disconnect the batteries.

Heater and Air Conditioner Removal and Installation

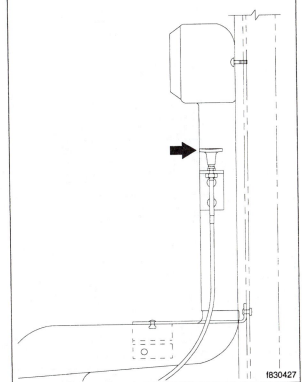

Fig. 2, Manual Temperature Control Cable Handle

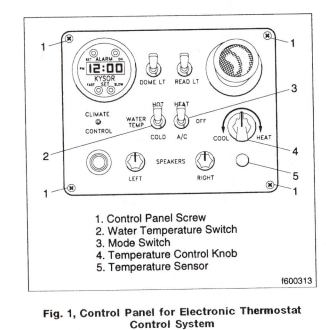

1. Control Panel Screw
2. Water Temperature Switch
3. Mode Switch
4. Temperature Control Knob
5. Temperature Sensor

Fig. 1, Control Panel for Electronic Thermostat Control System

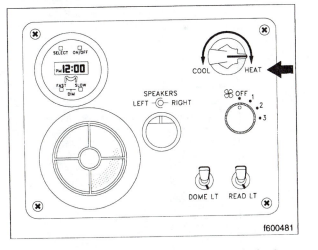

Fig. 3, Temperature Control Knob for C.T.C. System

FIGURE 3–61 Typical service manual page. (Courtesy of Freightliner Trucks)

fluid capacities, specifications, procedures for adjustments and for checking the tightness of fasteners, and torque value charts. Maintenance manuals generally do not contain detailed repair or service information.

Driver's Manuals

Driver's manuals contain information needed by the driver to understand, operate, and care for the vehicle

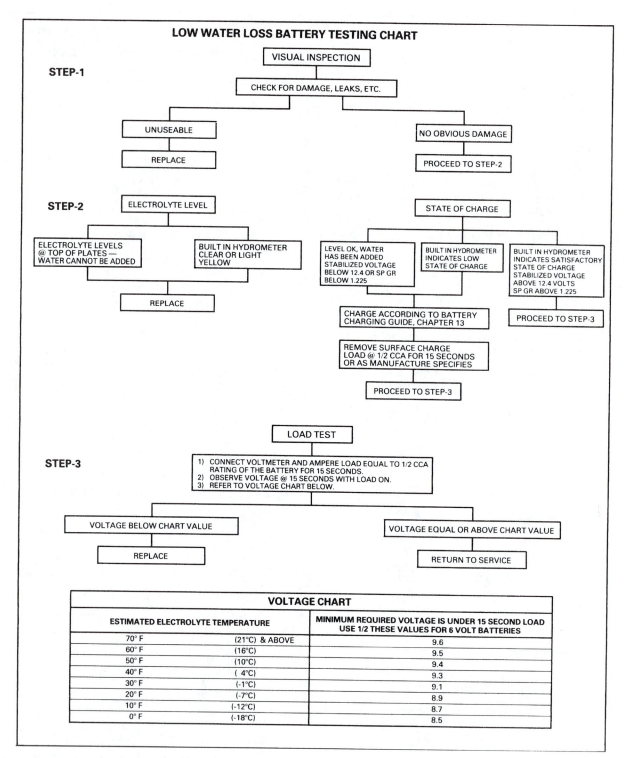

LOW WATER LOSS BATTERY TESTING CHART

STEP-1

VISUAL INSPECTION

CHECK FOR DAMAGE, LEAKS, ETC.

UNUSEABLE → REPLACE

NO OBVIOUS DAMAGE → PROCEED TO STEP-2

STEP-2

ELECTROLYTE LEVEL

ELECTROLYTE LEVELS @ TOP OF PLATES — WATER CANNOT BE ADDED

BUILT IN HYDROMETER CLEAR OR LIGHT YELLOW

REPLACE

STATE OF CHARGE

LEVEL OK, WATER HAS BEEN ADDED STABILIZED VOLTAGE BELOW 12.4 OR SP GR BELOW 1.225

BUILT IN HYDROMETER INDICATES LOW STATE OF CHARGE

BUILT IN HYDROMETER INDICATES SATISFACTORY STATE OF CHARGE STABILIZED VOLTAGE ABOVE 12.4 VOLTS SP GR ABOVE 1.225

CHARGE ACCORDING TO BATTERY CHARGING GUIDE, CHAPTER 13

PROCEED TO STEP-3

REMOVE SURFACE CHARGE LOAD @ 1/2 CCA FOR 15 SECONDS OR AS MANUFACTURE SPECIFIES

PROCEED TO STEP-3

STEP-3

LOAD TEST

1) CONNECT VOLTMETER AND AMPERE LOAD EQUAL TO 1/2 CCA RATING OF THE BATTERY FOR 15 SECONDS.
2) OBSERVE VOLTAGE @ 15 SECONDS WITH LOAD ON.
3) REFER TO VOLTAGE CHART BELOW.

VOLTAGE BELOW CHART VALUE → REPLACE

VOLTAGE EQUAL OR ABOVE CHART VALUE → RETURN TO SERVICE

VOLTAGE CHART		
ESTIMATED ELECTROLYTE TEMPERATURE		**MINIMUM REQUIRED VOLTAGE IS UNDER 15 SECOND LOAD USE 1/2 THESE VALUES FOR 6 VOLT BATTERIES**
70° F	(21°C) & ABOVE	9.6
60° F	(16°C)	9.5
50° F	(10°C)	9.4
40° F	(4°C)	9.3
30° F	(-1°C)	9.1
20° F	(-7°C)	8.9
10° F	(-12°C)	8.7
0° F	(-18°C)	8.5

FIGURE 3–62   Tree diagnosis chart layout. (Courtesy of Battery Council International)

and its components. Each manual contains a chapter that covers pretrip inspection and daily maintenance of vehicle components. Driver's manuals do not contain detailed repair or service information.

Parts Books

These manuals contain illustrations and part numbers, and are designed to aid in the identification of all ser-

Problem— Little or No Airflow

Possible Cause	Remedy
The blower is not operating.	Check for an open circuit breaker. An open circuit indicates a short in the electrical system, which must be located and repaired.
	Check the air conditioner relays for operation. Replace, as necessary.
	Make sure the blower motor switch is working. Replace, if necessary.
	Check the wiring to the blower motor. If any connections are loose, securely tighten them. Make sure the wiring conforms to the applicable diagram under **"Specifications, 400."**
	Check the blower motor for operation. Replace if sticking or otherwise inoperative.
	Check the resistor block. Replace, if necessary. **CAUTION:** Never try to by pass the fuse in the resistor block. To do so could cause the blower motor to overheat, resulting in serious damage to the heater/air-conditioning system.
There are restrictions or leaks in the air ducts.	Examine all air ducts and remove any blockages. Stop any leaks or replace any portion where the leaks cannot be stopped.
Ice has formed on the evaporator coil.	Defrost the evaporator coil before resuming operation of the air conditioner. Review "Performance Tests" in this subject for possible causes and corrective action.

Problem— Warm Airflow When the Air Conditioner Is On

Possible Cause	Remedy
There is no refrigerant charge in the system.	Perform a leak test. Repair any leaks, purge the system, replace the receiver-drier, and add a full charge of refrigerant.
Moisture in the system.	If moisture is in the system, ice crystals may form at the expansion valve, blocking the flow of refrigerant (off and on). Discharge the refrigerant charge, purge the system, replace the receiver-drier, and add a full charge of refrigerant.
The refrigerant compressor is not operating.	If the refrigerant charge is low, charge and leak test the system. Repair any leaks.
	The refrigerant compressor clutch or drive belt needs repair or replacement. For instructions, refer to the refrigerant compressor section elsewhere in this group.

FIGURE 3–63 *Block diagnosis chart layout. (Courtesy of Freightliner Trucks)*

viceable replacement parts for heavy-duty vehicles. They are organized by service group number. Parts books are a must for the fleet or shop parts specialists.

Service Bulletins

Service bulletins provide the latest service tips, field repairs, product improvements, and related informa-

tion of benefit to service personnel (**Figure 3–64**). Some service bulletins are updates to information in the service manual; these take precedence over service manual information, until the latter is updated. At that time, the bulletin is usually canceled. The service bulletins manual is usually available only to dealers who in turn supply the owners of the vehicle. When doing service work on a vehicle system or

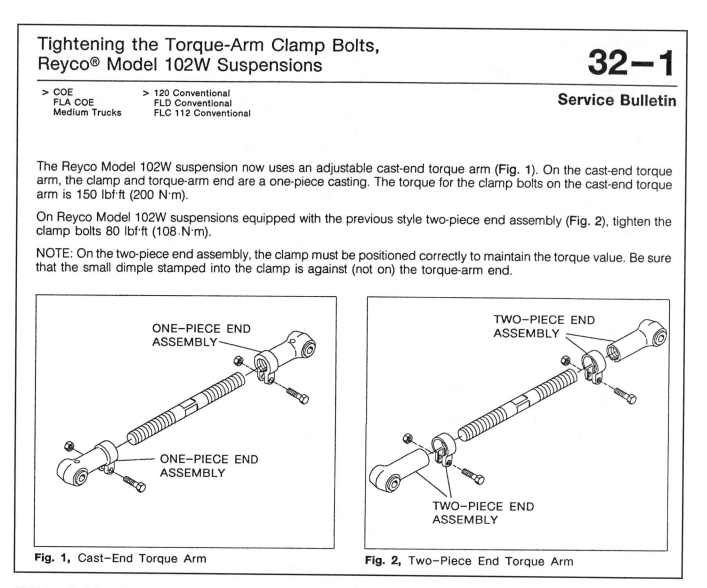

Tightening the Torque-Arm Clamp Bolts, Reyco® Model 102W Suspensions

32–1

Service Bulletin

> COE   > 120 Conventional
> FLA COE   FLD Conventional
> Medium Trucks   FLC 112 Conventional

The Reyco Model 102W suspension now uses an adjustable cast-end torque arm **(Fig. 1)**. On the cast-end torque arm, the clamp and torque-arm end are a one-piece casting. The torque for the clamp bolts on the cast-end torque arm is 150 lbf·ft (200 N·m).

On Reyco Model 102W suspensions equipped with the previous style two-piece end assembly **(Fig. 2)**, tighten the clamp bolts 80 lbf·ft (108·N·m).

NOTE: On the two-piece end assembly, the clamp must be positioned correctly to maintain the torque value. Be sure that the small dimple stamped into the clamp is against (not on) the torque-arm end.

ONE–PIECE END ASSEMBLY

ONE–PIECE END ASSEMBLY

Fig. 1, Cast–End Torque Arm

TWO–PIECE END ASSEMBLY

TWO–PIECE END ASSEMBLY

Fig. 2, Two–Piece End Torque Arm

FIGURE 3–64   *Typical service bulletin. (Courtesy of Freightliner Trucks)*

part, check for a valid service bulletin for the latest information on the subject.

Shop Talk

Before using a particular service bulletin, check the current **validity list** *(supplied by the manufacturer) to be sure the bulletin is valid.*

Recall Bulletins

Recall bulletins pertain to special situations that involve service work or replacement of parts in connection with a recall notice. Recall bulletins pertain to matters of vehicle safety. All bulletins are distributed to dealers; customers receive notices that apply to their vehicles.

Field Service Modifications

The Field Service Modification publication is concerned with nonsafety–related service work or replacement of parts. All field service modifications are distributed to dealers; customers receive notices that apply to their vehicles.

Time Guides

The **time guide** is used for computing compensation payable by the truck manufacturer for repairs or service work to vehicles under warranty, or for other special conditions authorized by the company. The time guide usually covers all vehicle models of the manufacturer. This software contains operation numbers and time allowances for various procedures on vehicle systems and parts. The time guide software is available only to dealers.

OTHER SERVICE PUBLICATIONS

There are other sources of servicing information available to truck technicians, such as

Supplier Manufacturers' Guides and Catalogs

Many of the larger parts manufacturers have excellent guides on the various parts that they manufacture or supply for the truck builder (**Figure 3–65**). They also provide updated service bulletins on their products.

General and Specialty Repair Manuals

These are published by independent publishers rather than the manufacturers. However, they pay for and get most of their information from the manufacturer. They contain component information, diagnostic steps, repair procedures, and specifications for several types of trucks in one book. Information is usually condensed and is more general in nature, depending on which manual is used.

The same information available in the service manuals and bulletins is also sometimes available on compact disc/read only memory systems. A single compact disc can hold a quarter million pages of text. One great advantage of this system is that it makes accessing the right information much easier and quicker. The manufacturer sends monthly update discs that not only contain the most recent service bulletins but also engineering and field ser-

FIGURE 3–66   Leading heavy-duty truck publications.

vice fixes. Other sources of up-to-date technical information are trade magazines (**Figure 3–66**) and trade associations such as The Maintenance Council (TMC) of the American Trucking Association (ATA).

3.5   FASTENERS

Trucks and trailers on our roads today use both English and Metric threads: metric fasteners become more common in late model applications. See **Figure 3–67**. The engine and major components use metric fasteners almost exclusively (diameter and pitch are measured in millimeters).

Most fasteners used on the vehicle that are $1/2$-inch diameter or larger are plain hex type fasteners (nonflanged); all metric fasteners are nonflanged. Hardened flatwashers are used under the bolt head, between the part being clamped and the hexnut, to distribute the load: this prevents localized stress. The washers are cadmium- or zinc-plated and have a hardness rating of 38 to 45 Rockwell "C" hardness (HRC).

Some fasteners smaller than $1/2$-inch diameter are flanged fasteners, which have integral flanges that fit against the parts being clamped. The flanges eliminate the need for washers.

FASTENER GRADES AND CLASSES

Fasteners are divided into grades established by the Society of Automotive Engineers (S.A.E.) or the International Fastener Institute (I.F.I.). The fastener grades indicate the tensile strength of the fastener;

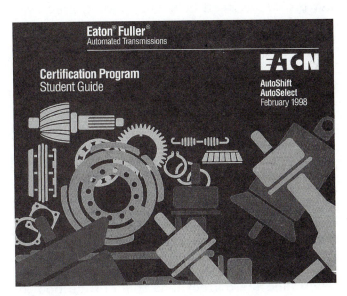

FIGURE 3–65   Typical aftermarket publications.

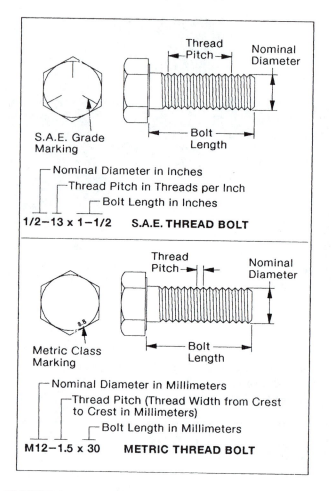

FIGURE 3–67 SAE and metric thread bolts. (Courtesy of Freightliner Trucks)

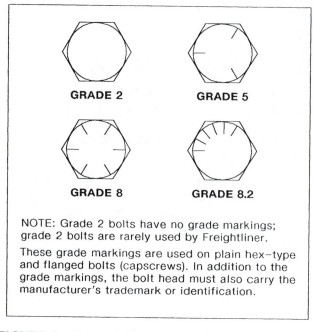

NOTE: Grade 2 bolts have no grade markings; grade 2 bolts are rarely used by Freightliner.

These grade markings are used on plain hex–type and flanged bolts (capscrews). In addition to the grade markings, the bolt head must also carry the manufacturer's trademark or identification.

FIGURE 3–68 Bolt (capscrew) indication. (Courtesy of Freightliner Trucks)

the higher the number (or letter), the stronger the fastener. Bolt (capscrew) grades can be identified by the number and pattern of radial lines forged on the bolt head (**Figure 3–68**).

Hex nut (and lock nut) grades can also be identified by the number and pattern of lines and dots on various surfaces of the nut (**Figure 3–69**). Nearly all of the bolts used on the heavy-duty vehicle are grades 5, 8, and 8.2. Matching grades of hex nuts are used with grade 5 bolts; grade 8, grade C, or grade G (flanged) hex nuts are used with grade 8 or 8.2 bolts.

Fasteners with metric threads are divided into classes adopted by the American National Standards Institute (ANSI). The higher the class number, the stronger the fastener. Bolt classes can be identified by the numbers forged on the head of the bolt (**Figure 3–70**). Hex nut (and lock nut) classes can be identified by the marks or numbers on various surfaces of the nut (**Figure 3–71**). Class 8 hex nuts are always used with class 8.8 bolts; class 10 hex nuts

with class 10.9 bolts. Threads can be measured with a screw pitch gauge.

Frame Fasteners

For most components attached to the frame by the threaded fasteners, grade 8 and 8.2 phosphate- and oil-coated hex head bolts and grade C cadmium-plated and wax-coated prevailing torque lock nuts are used. The prevailing torque lock nuts have distorted sections of threads to provide torque retention. For attachments where clearance is minimal, low-profile hex head bolts and grade C prevailing torque lock nuts are used (**Figure 3–72**).

Many trucks today are manufactured with constant clamping force fasteners such as **Huck fasteners**. These fasteners can be installed by automatic assembly processes but are usually destroyed on removal. Replace with either Huck fasteners or a similar grade of bolt.

Tightening Fasteners

When a capscrew or bolt is tightened to its torque value in a threaded hole, or a nut is tightened to its torque value on a bolt, the shank of the capscrew or bolt is stretched slightly. This stretching (tensioning) results in a preload. The torque values given in the tables in Appendix C have been calculated to provide enough clamping force on the parts being fastened, and the correct tensioning of the bolt to maintain the clamping force.

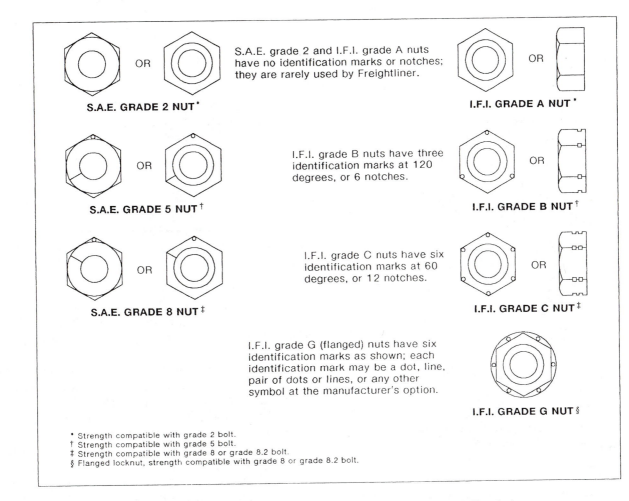

S.A.E. grade 2 and I.F.I. grade A nuts have no identification marks or notches; they are rarely used by Freightliner.

S.A.E. GRADE 2 NUT *

I.F.I. GRADE A NUT *

I.F.I. grade B nuts have three identification marks at 120 degrees, or 6 notches.

S.A.E. GRADE 5 NUT †

I.F.I. GRADE B NUT †

I.F.I. grade C nuts have six identification marks at 60 degrees, or 12 notches.

S.A.E. GRADE 8 NUT ‡

I.F.I. GRADE C NUT ‡

I.F.I. grade G (flanged) nuts have six identification marks as shown; each identification mark may be a dot, line, pair of dots or lines, or any other symbol at the manufacturer's option.

I.F.I. GRADE G NUT §

* Strength compatible with grade 2 bolt.
† Strength compatible with grade 5 bolt.
‡ Strength compatible with grade 8 or grade 8.2 bolt.
§ Flanged locknut, strength compatible with grade 8 or grade 8.2 bolt.

FIGURE 3–69 Hex nut (and lock washer) indication. (Courtesy of Freightliner Trucks)

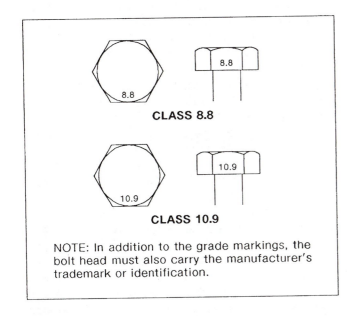

CLASS 8.8

CLASS 10.9

NOTE: In addition to the grade markings, the bolt head must also carry the manufacturer's trademark or identification.

FIGURE 3–70 Bolt classes can be identified by the numbers forged on the head of the bolt. (Courtesy of Freightliner Trucks)

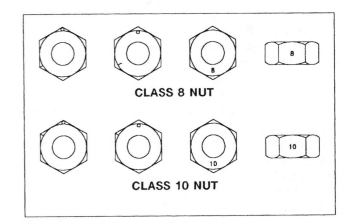

CLASS 8 NUT

CLASS 10 NUT

FIGURE 3–71 Markings on Class 8 and Class 10 nuts. (Courtesy of Freightliner Trucks)

Use of a torque wrench to tighten fasteners will help prevent overtensioning them. Overtensioning causes permanent stretching of the fasteners, which can result in breakage of the parts or fasteners.

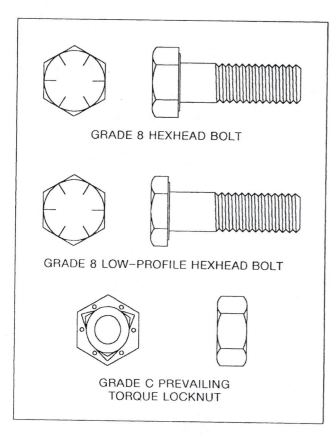

FIGURE 3-72 Grade C prevailing torque lock nut. (Courtesy of Freightliner Trucks)

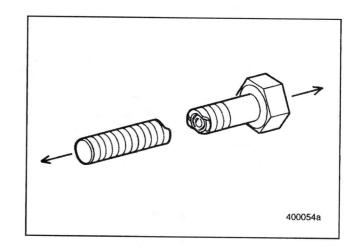

FIGURE 3-73 This bolt has been torqued past its yield point, resulting in shear.

When torquing a fastener, typically 80 to 90 percent of the turning force is used to overcome thread, cap, and nutface friction; only 10 to 20 percent results in capscrew or bolt clamping force. About 40 to 50 percent of the turning force is needed to overcome the friction between the underside of the capscrew head or nut and the washer. Another 30 to 40 percent is needed to overcome the friction between the threads of the capscrew and the threaded hole, or the friction between the threads of the nut and bolt.

All metals are elastic to some extent, which means they can be stretched and compressed to a certain point. This elastic, spring-like property is what provides the clamping force when a bolt is threaded into a tapped hole or when a nut is tightened. As the bolt is stretched, clamping force or holding power is created due to the bolt's tension.

Like a spring, the more a bolt is stretched, the tighter it becomes. However, a bolt can be stretched too far which will result in shear. At this point, the bolt can no longer safely carry the load it was designed to support.

Elasticity means that a bolt can be stretched a certain amount, and each time the stretching load is reduced, the bolt will return to its original, normal size. In other words, it is reusable. However, if the bolt is stretched into "yield," it deforms and never returns to normal (**Figure 3-73**); the bolt will continue to stretch more each time it is used, just like a piece of taffy that is stretched until it breaks.

Proper use of torque will avoid this yield condition. Torque values are calculated with a 25 percent safety factor below the yield point. There are some fasteners, however, that are torqued intentionally just barely into a yield condition, although not far enough to create the classic coke bottle shape of a necked out bolt. This type of fastener, known as a torque-to-yield (TTY) bolt, will produce 100 percent of its intended strength, compared to 75 percent when torqued to normal values. These fasteners, however, should not be reused, unless otherwise specified.

The amount of torque required to tighten a fastener is reduced when the amount of friction is reduced. If a fastener is dry (unlubricated) and plain (unplated), the amount of friction is high. If a fastener is wax-coated or oiled, or has a zinc phosphate coating or cadmium plating, the amount of friction is reduced. Each of these coatings and combinations of coatings has a different effect. Using zinc plated hardened flat washers under the bolt (capscrew) head and nut reduces the amount of friction. Dirt or other foreign material on the threads or clamping surfaces of the fastener or clamped part increases the amount of friction.

Even though different conditions affect the amount of friction, a different torque value cannot be given for each. To ensure they are always torqued accurately, most OEMs recommend that all fasteners be lubricated with oil (unless specifically instructed to install them dry), and then torqued to the values for lubricated- and plated-thread fasteners. When locking compound or antiseize compound is recommended for a fastener, the compound acts as a lubricant, and oil is not needed.

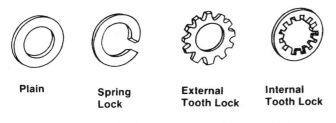

Plain   Spring   External   Internal
          Lock   Tooth Lock   Tooth Lock

FIGURE 3–74 *Washers are sometimes used to lock the bolts to the structure to keep them from coming loose and to prevent damage to soft metal parts.*

Be wary of hexagon (hex) nut thread stripping by power wrenches. It is deceptively easy to place a bolt into a yield condition within seconds. Impact wrenches are the worst offenders. Friction is needed to prevent the nut from spinning. If the nut is lubricated, there is no friction left to stop the impact wrench from hammering the nut past the bolt's yield point and/or stripping the threads.

Do not run the nut full speed onto the bolt. Instead, run it up slowly until it contacts the work, and then mark the socket and watch how far it turns. Smaller air powered speed wrenches do not produce the severe force of impact wrenches and are much safer to use. Follow this procedure with a torque wrench as well.

A rule of thumb about lock washers is that if the connection did not come with one, do not add one. Lock washers are extremely hard and tend to break under severe pressure. Use lock nuts and hard, flat washers. Properly torqued, this type of fastener will never come loose—even if lubricated (**Figure 3–74**).

FASTENER REPLACEMENT

When selecting and installing replacement fasteners, keep the following points in mind:

- When replacing fasteners, use only identical bolts, washers, and nuts; they must be the same size, strength, and finish as originally specified.
- When replacing graded (or metric class) bolts and capscrews, use only fasteners that have the manufacturer's trademark or identification on the bolt head; do not use substandard bolts.
- When using nuts with bolts, use a grade (or class) of nut that matches the bolt.
- When installing nonflanged fasteners, use hardened steel flat washers under the bolt (capscrew) head and under the hex nut or lock nut.
- For bolts 4 inches (100 mm) or less in length, make sure that at least 1 1/2 threads and no more than 5/8 inch (16 mm) bolt length extends

through the nut after it has been tightened. For bolts longer than 4 inches (100 mm), allow a minimum of 1 1/2 threads and a maximum of 3/4 inch (19-mm) bolt length protrusion.
- Never hammer or screw bolts into place. Align the holes of the parts being attached, so that the nut and bolt surfaces are flush against the washers, and the washers are flush against the parts.
- When installing fasteners in aluminum or plastic parts with threaded holes, start the fasteners by hand, to ensure straight starting and to prevent damaged threads.
- Do not use lock washers (split or toothed) next to aluminum surfaces.
- When installing studs that do not have an interference fit, install them with thread locking compound.
- When installing parts that are mounted on studs, use free-spinning (nonlocking) nuts and helical-spring (split) lock washers or internal-tooth lock washers. Do not use lock nuts because they tend to loosen the studs during removal. Do not use plain washers (flat washers).
- Do not use lock washers and flat washers in combination (against each other); each defeats the other's purpose.
- Use stainless steel fasteners against chrome plating, unpainted aluminum, or stainless steel.

👓 Shop Talk

If a torque-to-yield bolt is replaced with a new bolt of identical grade but torqued to a value found in a regular torque chart, the clamping force produced will be at least 25 percent less.

Fastener Tightening

When tightening fasteners, remember the following procedures:

- Clean all fasteners (and parts) threads, and all surfaces before installing fasteners.
- To ensure they are always torqued accurately, all fasteners should be lubricated with oil (unless specifically instructed to install them dry), and then torqued to the values for lubricated- and plated-thread fasteners. When locking compound or antiseize compound is recommended for a fastener, the compound acts as a lubricant, and oil is not needed.
- Bring parts and fasteners into contact, with no gaps between them, before using a torque wrench to tighten fasteners to their final torque values.

- Tighten the nut, not the bolt head. This will give a truer torque reading by eliminating bolt body friction.
- Always use a torque wrench to tighten fasteners, and use a slow, smooth, even pull on the wrench. Do not use a short, jerky motion or inaccurate readings can result.
- When reading a torque wrench, look straight down at the scale. Viewing from an angle can give a false reading.
- Only pull on the handle of the torque wrench. Do not allow the beam of the wrench to touch anything.
- Tighten bolts and nuts in at least four steps: to one-half recommended torque, to three-fourth torque, to full torque, and to full torque a second time.
- Do not overtorque fasteners; as already mentioned, overtightening causes permanent stretching of the fasteners, which can result in breakage of the parts or fasteners.
- If specific torque values are not given for countersunk bolts, use the torque value for the corresponding size and grade of regular bolt.
- Always follow the torque sequence or torque interval when provided, to ensure that clamping forces are even, and parts and fasteners are not distorted.
- Tighten frame fasteners periodically to offset the effects of "bedding in" (seating). Continued vehicle operation with loose frame fasteners could result in component, bracket, and frame damage. Refer to the Maintenance Schedule and the frame section in the vehicle maintenance manual for intervals.

Thread Repair

A common fastening problem is stripped threads. This problem is usually caused by torque that is too high or by cross-threading. Rather than replacing the block or cylinder head, the threads can be replaced by the use of threaded inserts. Several types of threaded inserts are available—the helically coiled insert is the most popular (**Figure 3–75**). To install this and similar thread reconditioning inserts, proceed as follows:

1. Establish the size, pitch, and length of the thread required. Refer to the insert manufacturer's instructions for proper size drill to use for the thread to be repaired.
2. Drill out the damaged threads with the specified drill. Clean out all metal chips from the hole.
3. Tap new threads in the hole using the specified tap. Lubricate the tap while threading the hole. Back out the tap every turn or two.

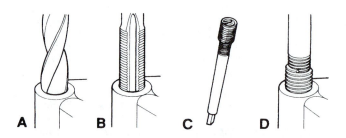

FIGURE 3–75   Steps in the installation of a helical screw repair coil: (A) Drill the damaged threads using the correct size drill bit. Clean all metal chips out of the hole. (B) Tap new threads in the hole, using the specified tap. The thread depth should exceed the length of the bolt. (C) Install the proper size coil insert on the mandrel provided in the installation kit. Bottom it against the tang. (D) Lubricate the insert with oil and thread it into the hole until flush with the surface. Use a punch or side cutter to break off the tang.

When the hole is threaded to the proper depth, remove the tap and all metal chips from the hole.

4. Select the proper size insert and thread it onto the special installing mandrel or tool. Make sure the tool is engaged with the tang of the insert. Screw the insert in the hole by turning the installing tool clockwise. Lubricate the thread insert with motor oil if it is installed in cast iron (do not lubricate if installing in aluminum). Install the thread into the hole until it is flush with the surface or one turn below. Remove the installer.

Screw/Stud Removers and Extractors

Several **stud removers** are shown in **Figure 3–76**. These tools are also used to install studs. Stud removers have a hardened, knurled, or grooved eccentric roller or jaws that grip the stud tightly when operated. Stud removers/installers may be turned by a socket wrench drive handle.

FIGURE 3–76   Various stud removers.

PHOTO SEQUENCE 2
REPAIRING DAMAGED THREADS

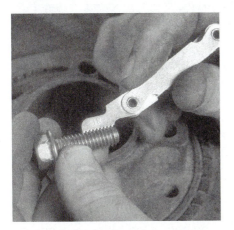

P2–1   Using a threaded pitch gauge, determine the thread size of the fastener that should fit into the damaged internal threads.

P2–2   Select the correct size and type of tap for the threads and bore to be repaired.

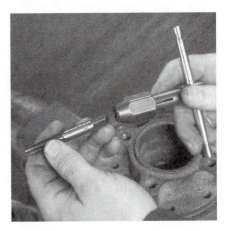

P2–3   Install the tap into a tap wrench.

P2–4   Start the tap squarely in the threaded hole using a machinist square as a guide.

P2–5   Rotate the tap clockwise into the bore until the tap has been run through the entire length of the threads.

P2–6   Drive the tap back out of the hole by turning it counterclockwise.

P2–7   Clean the metal chips left by the tap out of the hole.

P2–8   Inspect the threads left by the tap to be sure they are acceptable.

P2–9   Test the threads by threading the correct fastener into the threaded hole.

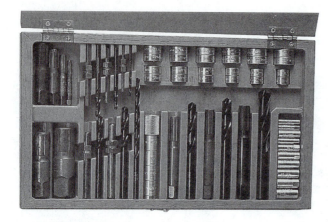

FIGURE 3–77   Screw extractor set. (Courtesy of Snap-On Tools Company)

Extractors are used on screws and bolts that are broken off below the surface. Twist drills, fluted extractors, and hex nuts are included in a screw extractor set (**Figure 3–77**). This type of extractor lessens the tendency to expand the screw or stud that has been drilled out by providing gripping power along the full length of the stud.

Thread Locking Compound Application

When applying a thread locking compound, follow the safety precautions given on the locking compound container. Then proceed as follows:

1. Clean the male and female threads of the fasteners, removing all dirt, oil, and other foreign material. If parts are contaminated, use solvent for cleaning, and then allow the fasteners to air dry for 10 minutes. Be sure solvent is completely gone before applying adhesive.
2. Transfer a small amount of the locking compound from the container to a paper cup or small nonmetal dish.
3. Use a plastic brush to apply because a metal brush will contaminate the compound. Apply a small amount of compound to the entire circumference of three or four of the male threads that will be covered by the nut after it has been tightened. Be sure enough compound is applied to fill the inside of the nut threads, with a slight excess.
4. Install and torque the nut. Readjustment of the nut position is not possible after installation is complete without destroying the locking effect.

To disassemble fasteners that have been held together with a thread locking compound, it may be necessary to heat the bond line to 400 degrees Fahrenheit before removing the nut. Every time the fasteners are disassembled, replace them. If any parts are damaged by overheating, replace the parts.

RIVETING

The overlapping skin panels on the tractor cab and some trailer parts are fastened together with aluminum alloy or mild steel rivets. A **rivet** is a fastening device and as such, it must have two ends (or heads) to hold the material together. One of these heads is preformed by the manufacturer and is referred to as the "manufactured head." The other end of the rivet is formed after the rivet has been driven through the material that is to be clamped. This head is referred to as the "bucked head" and is formed (or shaped) when driven against a bucking bar.

A rivet set connects to an air driven rivet gun, and when the proper size rivet set is positioned on the rivet, the rivet gun delivers hammer-like blows that form the bucked head when it is backed up against a bucking bar (**Figure 3–78**). The size of the rivet set must be the same size as the rivet. An incorrect size will damage either the rivet or the skin panels. Too large a rivet set will flatten the rivet head and mar the skin panels.

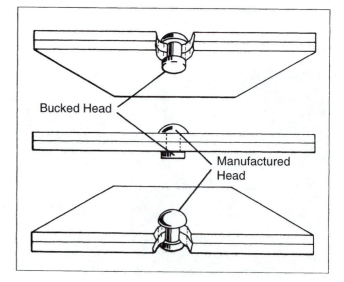

FIGURE 3–78   How a rivet holds. (Courtesy of Freightliner Trucks)

Types of Rivets

There are two methods of clamping trailer and cab panels depending on the type of rivet: brazier head and flush type (**Figure 3–79**). The rivet set used to drive the brazier head rivet is cupped to fit the round head of the rivet. The flush type derives its name from the angle between the shoulder of the rivet. The rivet requires the use of a "mushroom" flush-type rivet set. The rivet set is the link that transmits the power of the blow from the rivet gun to the rivet head. The rivet set end must fit the manufactured head of the rivet at all times (**Figure 3–80**). If the size of the rivet and the rivet set are not identical to each other, damage to the rivet or the material being fastened can result. The rivet set has the size of the rivet that is to be used for stamped on the rivet set.

Rivet Gun

The air driven or pneumatic rivet gun (**Figure 3–81**) provides the force needed for riveting. Rivet guns are

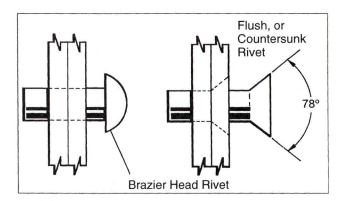

FIGURE 3–79 *Two common types of rivets.* (Courtesy of Freightliner Trucks)

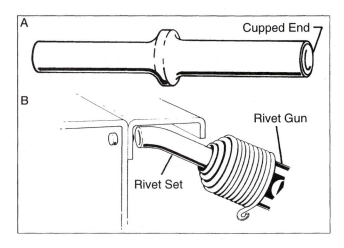

FIGURE 3–80 *(A) Straight rivet set; and (B) offset rivet set.* (Courtesy of Freightliner Trucks)

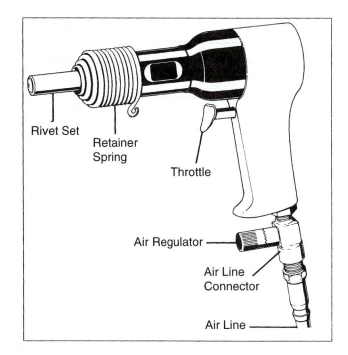

FIGURE 3–81 *Typical pneumatic rivet gun.* (Courtesy of Freightliner Trucks)

made in varying sizes. Each rivet gun size has a definite range of rivet sizes that it can drive. The gun sizes are identified as follows; 2X, 3X, 4X, 5X, 9X, and so on. The larger the number, the larger the size rivet it will handle. In general, rivets $3/16$-inch diameter or less can be driven with any one of the following gun sizes: 2X, 3X, or 4X.

The air connector attaches the air line to the rivet gun. The air regulator adjusts the volume of air pressure to the desired impact power of the rivet gun. The retainer spring on the rivet gun is a safety feature that prevents the rivet set from flying out of the gun during operation.

Bucking Bars

Bucking bars are made of steel and vary in size and shape. (Various shapes are required for "hard to get" rivets.) The proper weight of a bucking bar depends on the size, strength, and position of the rivet being used. Bucking bars vary in weight from approximately 1 to 15 pounds.

The polished surface (or end) of the bucking bar is called the "face." The "face" of the bucking bar is positioned against the rivet shank and parallel to the panels being fastened when the rivet is being "shot." This forms the bucked head of the rivet. When using the rivet gun, be sure to follow the makers instructions to the letter.

GUIDELINES FOR PROPER RIVETING

All rivets must be driven and bucked properly. Failure to do so can result in damage to the cab or trailer skin panels or rivet(s). Failure of one rivet in a row of rivets will place additional clamping stresses on the remaining rivets, reducing rivet strength and causing possible rivet failure.

To performing the riveting, proceed as follows:

1. Connect the air line to the rivet gun. Adjust the air regulator to obtain the desired impact power as directed by the rivet gun manufacturer. Check the impact power by pressing the rivet set against a block of wood and opening the throttle of the gun. Adjust to the desired impact power.
2. Drill the rivet holes to specified size. Place the rivet in the rivet holes of the panels to be fastened. As a general rule, the projection of the rivet shank, before driving, should be about $1\frac{1}{2}$ "D" wide, with "D" being the original diameter of the rivet shank (**Figure 3–82**).
3. Align the rivet and rivet gun, rivet, and bucking bar as shown in **Figure 3–83**. Be sure to keep the line of pressure in a straight line with the rivet head. Be sure that the "face" of the bucking bar is kept parallel to the skin panels when the rivet is being "shot," as this will form the bucked head of the rivet properly. If the "face" of the bucking bar is not parallel to the skin panels, damage to the skin panels and an improperly formed bucked head can result.
4. Drive the rivet by opening the throttle of the rivet gun. The manufactured head of the rivets must be driven carefully to avoid dimpling of the skin panels and "eye-browing" of the heads. These conditions will blemish the

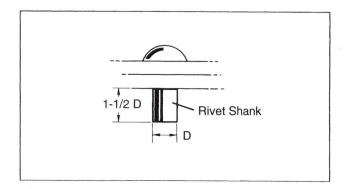

FIGURE 3–82   Correct rivet shank length. (Courtesy of Freightliner Trucks)

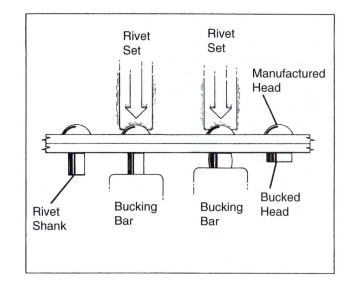

FIGURE 3–83   Riveting operation. (Courtesy of Freightliner Trucks)

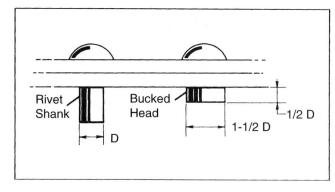

FIGURE 3–84   Correct bucked head sizes. (Courtesy of Freightliner Trucks)

overall appearance of the vehicle, as they are highly visible after painting.

5. Inspect the rivet. The bucked head of the rivet should be $1\frac{1}{2}$ "D" wide, and $\frac{1}{2}$ "D" thick, with "D" being the original diameter of the rivet shank (**Figure 3–84**).

Removing Rivets

As shown in **Figure 3–85**, rivets can be removed in one of the following three ways:

1. Using a cold chisel, cut off the manufactured head of the rivet. See **Figure 3–85A**. Take care not to damage the surrounding material.
2. Using a drill bit that has the same diameter as the rivet shank to be removed, drill into the exact center of the manufactured head to the depth shown in **Figure 3–85B**.

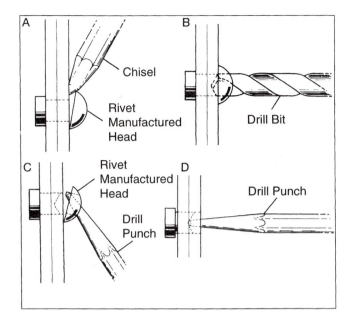

FIGURE 3–85   Removing a rivet. (Courtesy of Freightliner Trucks)

3. Using a drill punch, pry off the rivet head. See **Figure 3–85C**. Then drive out the shank with the drill punch (**Figure 3–85D**).

ADHESIVES AND CHEMICAL SEALING MATERIALS

Chemical adhesives and sealants give added holding power and sealing ability where two parts are joined. Sealants are added to threads where fluid contact is frequent. Chemical thread retainers are either aerobic (cures in the presence of air) or anaerobic (cures in the absence of air).

When using a chemical adhesive or sealant, always follow the manufacturer's instructions on the container to the letter.

SUMMARY

- Hand tools are used in many repair tasks. Proper use of the appropriate hand tools by the technician is important for performing quality heavy-duty truck service.
- Power tools can make a technician's job easier. Pneumatic (air) tools have four advantages over electrically powered equipment: flexibility, light weight, safety, and low cost operation and maintenance. The disadvantage is that they are noisy. Proper power tool safety rules must always be followed.

- The power tools used in heavy-duty truck repair include impact wrenches, air ratchet wrenches, air drills, air chisels and hammers, blowguns, bench grinders, grinding wheels, wire wheel brushes, buffing wheels, presses, and lifts and hoists.
- Using jacks, lifts, frame machines, and hoists to lift truck trailers or heavy parts of a heavy-duty truck requires careful adherence to safety rules.
- The machinist's rule looks like an ordinary ruler, but it is precisely divided into small increments, either in metric or decimal graduations.
- The dial caliper is used for taking inside, outside, depth, and step measurements, and is available in either metric or inch reading models.
- The micrometer is used for measuring the inside or outside diameter of a shaft or the bore of a hole, either in metric or inch units.
- Gauges are used to measure thickness and screw pitch. The torque wrench is used to measure the amount of force that is being applied in turning a fastener.
- The main source of repair and specification information for any heavy-duty truck are the OEM service procedures in a manual, on CD, or on-line.
- Troubleshooting charts are supplied in manuals to provide a way to track problems and identify the type of repair needed.
- Fasteners used on heavy-duty trucks come in many grades and classes. Proper use of torque is necessary for the installation of threaded fasteners. Care must be taken to make the correct choice of fastener when fasteners are replaced. In some cases, the thread must be repaired to stop slippage of a fastener.
- Rivets are used to fasten overlapping skin panels on a truck cab and some trailer parts. A rivet gun and bucking bar must be used to drive and buck rivets.

REVIEW QUESTIONS

1. Which of the following is NOT a requirement for using a hammer safely?
    a. Always wear eye protection.
    b. Grip the hammer close to the head.
    c. Select the right size hammer for the job.
    d. Keep the hammer face parallel with the surface being hammered.

2. You should slow down before completing a saw cut for what reason?
    a. to keep from dulling the blade
    b. to prevent the blade from breaking
    c. to prevent the surface being sawed from slipping
    d. to prevent injury

3. You should never use a punch or chisel on what material? _____

4. Which of the following is a specialty fastener used to secure mirrors, headlight assemblies, and often racks?
   a. Torx®
   b. Posi-Drive™
   c. Phillips
   d. Robertson

5. Which of the following is a requirement for the proper use and care of a screwdriver?
   a. Use screwdrivers only for driving screws, punching holes, or as prybars.
   b. File by hand to dress a screwdriver tip.
   c. Hold parts in your hand to work on them whenever possible.
   d. Use a screwdriver blade that is smaller than the screw slot in order to preserve the screwdriver tip.

6. Which type of wrench is best for turning, as opposed to holding?
   a. combination wrench
   b. box-end wrench
   c. open-end wrench
   d. socket wrench

7. Which is NOT a safety rule for using wrenches?
   a. Use wrenches that fit firmly.
   b. Do not extend the length of a wrench.
   c. Do not hammer on wrenches.
   d. Carefully straighten a bent wrench before using it again.

8. Which kind of pliers is best for grasping small components?
   a. combination pliers
   b. needlenose pliers
   c. locking pliers
   d. diagonal cutting pliers

9. Using a coarse cut file on soft material will prevent _____
   a. dragging
   b. clogging
   c. injury
   d. damage to the file

10. How do hand-threading dies differ from hand taps?
    a. They remove, rather than form, threads.
    b. They widen threads, but taps narrow them.
    c. They cut external threads, but hand taps cut internal threads.
    d. They cut threads in the opposite direction to those cut by hand taps.

11. Precision bearings and gears must be removed with
    a. a hammer
    b. a prybar
    c. a puller designed specifically for the task
    d. gear and bearing wrenches

12. Which of the following is NOT true of pneumatic tools compared to electrically powered equipment?
    a. They run cooler and speed and torque can be varied.
    b. They are lighter in weight than electrical tools that do the same job.
    c. There is less risk of fire and shock.
    d. They are not as noisy.

13. Which of the following accessories, used with an air chisel or hammer, can be used to install pins and drive frozen bolts?
    a. universal joint and tie-rod tool
    b. ball joint separator
    c. exhaust pipe cutter
    d. tapered punch

14. What device must always be used with a blowgun to prevent the attached tool from being ejected?_____

15. Which of the following is NOT a safe practice when using lifts, frame machines, and hoists?
    a. letting someone remain in the vehicle while it is being raised
    b. finding the specific lift points to use by consulting the vehicle's service manual
    c. using cables or chains to attach the object being lifted
    d. checking to see that attachments are secure before lifting

16. Which measuring tool is best suited to measuring tasks that require extreme precision?
    a. a metric machinist's rule
    b. a decimal machinist's rule
    c. a dial caliper
    d. a micrometer

17. What part of the micrometer is revolved between the thumb and forefinger to bring the measuring points into contact with surfaces being measured?
    a. thimble
    b. spindle
    c. sleeve
    d. frame

18. When should the calibration of a micrometer be checked?

19. Which of the following would a screw pitch gauge be used for?
    a. to determine the number of threads per inch
    b. to measure thickness
    c. to measure turning force
    d. to measure air pressure

20. Technician A uses the tree diagnosis chart in a heavy-duty truck's service manual to troubleshoot; technician B uses the block diagnosis chart for the same purpose. Who is correct?
    a. Technician A
    b. Technician B
    c. both of them
    d. neither of them

21. Hardened flatwashers are used with _____
    a. flanged fasteners
    b. hex type fasteners
    c. all fasteners with metric threads
    d. rivets

22. Tightening a capscrew or bolt beyond its torque value may result in _____
    a. reducing the amount of friction
    b. reducing the elasticity of the metal
    c. permanent stretching or breakage
    d. increasing the elasticity of the metal

23. Which of the following should NOT be done when replacing fasteners?
    a. Match the grade (or class) of the nut and bolt when using nuts and bolts.
    b. Use locknuts when installing parts that are mounted on studs.
    c. Use stainless steel fasteners against chrome plating, unplated aluminum, or stainless steel.
    d. Use hardened steel flat washers under the bolt when installing nonflanged fasteners.

24. Threads can be repaired using _____
    a. inserts
    b. studs
    c. extractors
    d. thread-locking compound

25. Which size rivet gun can drive the largest rivets?
    a. 2X
    b. 4X
    c. 5X
    d. 9X

Preventive Maintenance Program

🔑 **Prerequisites**
Chapters 2 and 3

Objectives

After reading this chapter, you should be able to

- Explain the characteristics and benefits of a well-planned preventive maintenance program.
- List and describe the steps of the pretrip inspection procedure.
- Describe the criteria for deadlining or out-of-service (OOS) tagging a vehicle.
- Implement a policy of preventive maintenance scheduling that conforms to federal inspection regulations.
- Describe inspector qualifications and record-keeping requirements.
- Select lubricants for the various tasks of preventive maintenance.
- Describe the operation of on-board chassis systems, both manual and automatic, for vehicle lubrication.
- Prepare the heavy-duty truck for cold weather by winterizing.

Key Terms

anti-corrosion agents
anti-rust agents
coolant hydrometer
deadline

detergent additives
oxidation inhibitors
pull circuit
push circuit
TBN additives
viscosity

In this chapter, we will introduce some of the service and preventive maintenance practices commonly found in the trucking industry. Although some of the knowledge required to properly service a truck is not covered until the later chapters in this textbook, apprentice technicians are usually expected to perform routine maintenance almost from their first day on the job. This means that having some idea about preventive maintenance practices is essential from the start, so this chapter appears at the beginning of this book rather than at the end. In some cases, the student will need to refer to later chapters in the book during the reading of this one.

An effective protective maintenance program is one that enables the operator to perform the least amount of repair maintenance while doing all the routine maintenance that is necessary. A well-planned PM program offers the following advantages:

- The lowest attainable maintenance cost
- Maximum vehicle availability
- Better fuel economy
- Reduced road failure, resulting in greater dependability

- Increased customer confidence, resulting in better public relations
- Reduced possibilities of accidents due to defective equipment
- Fewer driver complaints

Many factors are involved in the success of a PM program. The program's success depends on careful planning. Although this is primarily the responsibility of the maintenance manager, it is a task that cannot be accomplished unless all personnel involved are performing their responsibilities. Unless the inspections are performed properly and as scheduled and accurate records are kept, the manager cannot adjust the program for maximum effectiveness.

Even though careful planning is important, that alone is not enough. Unless the maintenance manager and technicians implement the preplanned PM program through disciplined scheduling of the vehicles for inspection and maintenance, the best plans become useless.

Both technicians and drivers should conduct and coordinate their inspections. Since the purpose of these inspections is to detect any impending failures (**Figure 4–1**), their importance to a preventive maintenance program is obvious. Drivers must perform pre- and post-trip inspections. The nature and extent of these inspections are largely determined by laws with a focus on safety. The inspections made by technicians are designed by the maintenance manager, and they are performed at specified intervals. Both drivers and technicians should understand what is involved in each inspection. They must know how to execute the details of each step, and they must perform them properly and honestly. A major cause of PM program failure is a "pencil inspection." In other words, the driver or technician assigned the inspection checks each operation as being "OK," without making the actual inspection. This type of inspection is worse than useless; it wastes time and resources, and it also deceives and can endanger others.

The technician's responsibilities go beyond inspecting the vehicles; he or she must also maintain and service them. Maintenance includes cleaning, lubricating, adjusting, and tightening. Servicing is done in response to any problems noted on the driver inspection report or on the PM form.

Neat and accurate records are another important aspect of the successful PM program. Drivers must fill out driver inspection reports. Technicians work with PM forms, repair orders, vehicle files, and major component histories. Computer data records are making these tasks easier for both the driver and technician.

Records are important for several reasons:

- In many operations they are required by the Department of Transportation (DOT).
- Permanent records are invaluable from a performance standpoint. A study of failures of individual parts or components can alert the manager of the need to take corrective action. It also highlights components that have performed satisfactorily.
- A maintenance manager cannot develop a systematic, preplanned preventive maintenance program without the use of these records.
- In the event of a serious accident, a vehicle file, which contains written proof that maintenance inspections were performed, and repairs were made as required, is usually beneficial in the defense of a lawsuit.

4.1 SETTING UP A PM PROGRAM

No one maintenance program applies to all operations. There are certain factors, such as the nature of each fleet operation and the age and type of equipment, that have a bearing on the best shape of a PM program for any particular operation. A basic PM program is not difficult to set up. However, the program that is most effective is built around the needs and experiences of each individual operation. Following the service recommendations outlined in each major OEM's preventive maintenance checklist is a safe foundation for any program. Another source

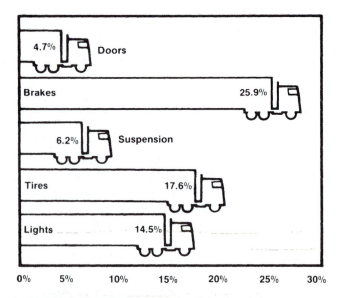

FIGURE 4–1 *Brakes top the list of maintenance concerns that fleets cite when asked to name the top five trailer maintenance problems. (Courtesy of Heavy Duty Trucking)*

of help is government information. The Department of Transportation for years has influenced the maintenance practices and record-keeping procedures of those involved in interstate commerce. Rules and guidelines are covered in a manual called *Federal Motor Carrier Safety Regulations.* Every truck operation may or may not be governed by the regulations. However, whether or not compliance is required, the guidelines and requirements are the basis for a planned, controlled maintenance program on any type of equipment.

A key to any successful PM program is the preventive maintenance form (**Figure 4–2**). This form provides the technician with an orderly list of items to be checked or inspected and instructions on those items that should be cleaned, lubricated, tightened, adjusted, and replaced. It is the technician's responsibility to know specifically what to do at each step. The following is a list of the basic types of inspections in A, B, C, D order:

Schedule "A" is a light inspection.
Schedule "B" is a more detailed check.
Schedule "C" is a detailed inspection, service, and adjustment.
Schedule "D" is a comprehensive inspection and adjustment.
Schedule "L" is a chassis lubrication.

A road test or driver's inspection report is conducted prior to performing any scheduled inspection.

Sometimes the instructions on a PM form are too general. For instance, what should the technician do in response to the instruction "Inspect the cooling system"? Perhaps if the instructions were written more specifically, the technician would know that he or she should pressure test the system to 15 psi; let it sit for 10 minutes while checking the cab heater, hoses, water pump, tanks, radiator core, and head gasket for leaks; then OK the system if it lost less than 1 psi. If the technician is unsure of the exact requirements of any step, the maintenance manager should be consulted. Once the specifics of each step are known, the technician should perform each step conscientiously. It is important to keep in mind the goal of eliminating any potential problem through careful inspection and proper maintenance.

DRIVER INSPECTION

It is the driver who first identifies the need for a repair when he or she performs both a pre-trip and a post-trip inspection. A form that is vital to a well-managed maintenance program is the "Driver Inspection Report." It is not only helpful to the repair technician or maintenance management in keeping the vehicle in a safe and operable condition, it is required by federal law.

A driver's post-trip responsibilities, as spelled out in the Department of Transportation's Federal Motor Carrier Safety Regulations, require preparation of a vehicle inspection report after each day's work and each vehicle driven. The form is almost always printed on the back of the driver hours-of-service log and completed before drivers turn in their daily logs. **Figure 4–3** is a copy of the driver's report form that is issued by the American Trucking Association. Items the rules say must be addressed include

- Service brakes
- Frames
- Parking brakes
- Sliding subframes
- Brake drums
- Tire and wheel clearance
- Brake hoses and tubing
- Tires
- Low air pressure warning
- Wheels and rims
- Tractor protection valve
- Windshield glazing
- Air compressor
- Wipers
- Hydraulic brake systems
- Vacuum brake systems
- Fifth wheels
- Exhaust system
- Fuel system
- Lighting
- Cargo securement
- Steering componentry
- Suspensions

In addition, drivers are required to note any defects or problems that could affect safe operation, and then sign the report. The use of this inspection report makes the driver a part of the maintenance program, and places direct responsibility on the driver to report malfunctions that might arise on a daily basis. When properly used, there should be no excuse for a defective vehicle being in service.

Pre-trip inspection rules are less specific. In addition to ensuring that the vehicle is safe, drivers are required to review the last vehicle inspection report.

If any defect(s) noted by the previous driver have not been repaired, both the technician and the new driver must sign off on the work before the vehicle can be dispatched. If the problem noted did not require repair, that too should be noted. In addition, a copy of the latest inspection report must remain with the vehicle. (Carriers are required to keep the original copy of each report and the certification of repairs for at least three months.)

PM INSPECTION LINE-HAUL TRACTOR

Unit # _____   ☐ "A" PM   ☐ "B" PM   ☐ "C" PM

Date _____   Check-High Oil/Fuel Checklist ☐

Mechanic _____
Supervisor _____
Mileage _____

On D/L, Check and Inspect

	Ck	Remarks
Verify PM on History Card		
Start Engine		
Gauges, Warning Lights		
Low Air Buzzer & Light on at 60#		
Air Buildup & Cutoff		
Air Dryer Cutoff		
Air Loss Test, 3# Per Min.		
Heater/Defroster Operation		
A/C Operation		
Horns, Air & Electric		

Moving to Shop, Check and Inspect

	Ck	Remarks
Park Brake Application		
Clutch Operation, Free Travel		
Transmission Shift, Hi-Lo		
Steering Free Travel, Bind, Pull		
Speedometer Operation		
Any Unusual Noise or Vibration		
Windshield Wiper/Washer Operation		
Foot Brake Application		
Headlights & Driving Lights		

In Shop, Check and Inspect

	Ck	Remarks
Turn Signals, Marker Lights		
Brake Lights–Reflectors		
Dome Light (Rear Light)		
Floor Mats, Boots, Coat Hook, DH Seal		
Sun Visors, Dash Screws		
All Glass and Mirrors		
Door Locks, Regulators		
Safety Equipment		
Seat Belts, Retractors		
(Sleeper Compartment Items)		
Trailer Cord Test		
Air Hoses and Hangers		
Lube: Door Latches, Hinges		
Seat Rails, Pivot Points		
Brake/Clutch Peddle Pivot		
Accelerator Peddle Pivots		
Heater/Defroster Cables		
Clean & Lube Floor Mtd. Foot Valve		

Engine Compartment, Check and Inspect

	Ck	Remarks
Hood Condition, Cracks, Damage		
Hinges, Bug Screen, Brackets, Wiring		
(Cab Over, Ck Jack Lift Cyl and Lines)		
Air Cleaner Ducts, Hoses, Brackets		
Air Restriction, Repl. Element @ + 20 in.		
All Belts, Tension-Condition		
A/C Compressor, Condenser,		
Receiver-Dryer, Wiring, Pressure Lines		
Pressure Test Cooling System		
Radiator, Leaks, Mounts, & Brackets		
Cooling Hoses, Any Leaks		
Water Pump Leaks		
Antifreeze, –20°		
DCA Check, Add Nalcool if Required		
Change Water Filter if Equipped		
Alternator Mounting, Wiring		
Starter Mounting, Wiring		
Spray Protectant on All Wiring Term.		
Fuel System, Leaks, Lines Rubbing		
Air Comp. Lines, Mtgs., Air Gov. Screws		
Lube Linkage (Must Have 2 Return Springs)		
Drain Fuel Tank Sumps of Water		
Drain Fuel Heater of Water		
Exhaust System, Clamps, Leaks		
Inspect Fan Hub		
Engine and Transmission Mounts		
Cab Mounts		
Drain Air Tanks of Water		

Under Vehicle

	Ck	Remarks
Change Oil, Oil Filters and Fuel Filters		
Check Trans Lube ("C" PM Change)		
Check Diff(s) Lube (C' PM Change)		
Inspect Axle Breather		
Inspect Pinion Seals, Wheel Seals		
Inspect Dr. Shaft, U-Joints Yokes, Lube		
Inspect Clutch Linkage, Lube as Req.		
Insp. Exh. Pipe, Muffler, Hangers, Leaks		
Insp. Fuel Tanks, Lines, Hangers, Leaks		
Insp. Air Tanks, Hangers, Lines Rubbing		
Record Oil Pressure Hi-Lo		
Check Dipstick Full Mark		
Lube All Required Points		

Used by a major carrier, this PM form: provides space for remarks and special instructions; is clearly sequenced; notes specific requirements.

FIGURE 4–2   An example of a PM form. (Courtesy of Commercial Carrier Journal)

While the regulations seem to place greater emphasis on the post-trip, most operators agree the pre-trip is more important. Since the driver doing the inspection will soon be operating that particular vehicle, the incentive to ensure that the vehicle is safe is probably greater than if he or she had just returned from a long day on the road. Also repairs are cheaper and less time-consuming if the driver finds them on the pre-trip before inspectors levy fines or impose penalties.

PM INSPECTION LINE-HAUL TRACTOR

Unit # _____

Date _____

☐ "A" PM   ☐ "B" PM   ☐ "C" PM

Check-High Oil/Fuel Checklist ☐

Mechanic _____

Supervisor _____

Mileage _____

Axles and Chassis

	Ck	Remarks
Vehicle Jacked Up		
Insp. Frt. Spgs., Pins, Hanger, U-Bolts		
Ins. Frt. Brake Chamber, Slacks,		
Lining, Air Lines, Brake Adjustment		
Insp. Frt. End. Tie-Rod Ends, Drag Link		
Pitman Arm, Kingpin, Steering		
Box, Steering Shaft, Hub Lube		
Level, Lube All Points.		
Lower Vehicle		
Insp. Rear Spgs., Pins, Hangers, U-Bolts		
Torque Rods, Rayco Equalizer Wear		
Insp. Rear Brake Chambers, Slacks,		
Lining, Air Lines, Brake Adjustment		
Insp. Frame and Cross Members		
Insp. 5th Wheel, Legs, Brackets, Ground		
Strap, Lube Slider Assembly		
Insp. Battery Box Mounts		
Clean Battery and Post		
Check Battery		
Insp. Battery Cables, Spray Prot		
Insp. All Chassis Wiring, Spray Prot. on All Terminals		
Check Lisc., Regis, Permits, State Insp.		
Clean Glass & Cab Interior		

Add for 'C' PM

	Ck	Remarks
See Foreman for Compuchek		
Replace Gladhand Rubbers		
Check Air Shield Angle		
Check Toe-in		
Check Tandem Alignment		
Change Trans and Diff Lube		
Change Rear Center Pump Filter/Screen		
Check U-Bolt Torque		
Change Steering Box Oil, As Required		
(Delvac SHC Synthetic)		
Clean Heater Core & A/C Evap.		
Clean Trans. Air Valve Filter		
Air System Check		
Clean & Check 7 Way Cord		
Use UYK Comp. Reverse Ends		

"A" PM ONLY

On D/L & Moving to Shop, Check & Insp.

	Ck	Remarks
Verify PM on History Card		
Start Eng.: Ck Gauges, Warning Lights		
Low Air Buzzer and Light On @ 60#		
Air Build Up & Cut Off		
Air Dryer Cut Off		
Heater, Defroster and A/C Operation		
Horns: Air & Electric		
Clutch Operation, Free Travel		
Steering Free Travel, Bind, Pull		
Foot & Park Brake Application		

In Shop, Check and Inspection

	Ck	Remarks
All Lights		
Floor Mats, Boots, Coat Hook		
Safety Equipment		
All Glass and Mirrors		
Trailer Cord Test		
Check Engine Oil, Coolant Level		
Change Fuel Filter		
Check Front End, Lube All Points		
Check Trans and Diff Lube Levels		
Lube Drive Shaft U-Joints		
Lube 5th Wheel and Sliders		
Drain Air Tanks		
Check Batteries		
Clean All Glass, Inside Cab		

Wheels and Tires, Check and Inspect

Air Pressure	LF	RF	RFI	RFO	RRI	RRO	LRI	LRO	LFI	LFO
Tread Depth										
Loose Lugs										
Cracked Rims										
Valve Caps										

Used by a major carrier, this PM form: provides space for remarks and special instructions; is clearly sequenced; notes specific requirements.

FIGURE 4–2 Continued.

INSPECTION PROCEDURE

The driver (and technician) should perform a pre-trip inspection in the same way every time to learn all the steps and to reduce the possibility of forgetting something (**Figure 4–4**). The following procedure is an adaptation of the Commercial Vehicle Safety Alliance (CVSA) standard inspection and should be a useful guide.

Step 1 (Vehicle Overview). Notice the general condition of the vehicle. If the unit is leaning to one side, this might indicate a broken spring, poor load

DRIVER'S INSPECTION REPORT

(SEE INSTRUCTIONS ON REVERSE SIDE)

CHECK DEFECTS ONLY Explain under REMARKS

COMPLETION OF THIS REPORT REQUIRED BY FEDERAL LAW, 49CFR 396.11 & 396.13.

Truck or
Tractor No. _____   Mileage (No Tenths) | | | | | | | Trailer No. _____

Dolly No. _____   Trailer No. _____   Location: _____

ATA/VMRS System Code Numbers for Shop Use Only   © 1980 American Trucking Associations, Inc.

POWER UNIT

GENERAL CONDITION
- ☐ 02 Cab/Doors/Windows
- ☐ 02 Body/Doors
- ☐ ___ Oil Leak _____
- ☐ ___ Grease Leak _____
- ☐ 42 Coolant Leak
- ☐ 44 Fuel Leak
- ☐ ___ Other _____

_____
(IDENTIFY)

ENGINE COMPARTMENT
- ☐ 45 Oil Level
- ☐ 42 Coolant Level
- ☐ ___ Belts _____

_____
(IDENTIFY)

IN-CAB
- ☐ 03 Gauges/Warning Indicators
- ☐ 02 Windshield Wiper/Washers
- ☐ 54 Horn(s)
- ☐ 01 Heater/Defroster
- ☐ 02 Mirrors
- ☐ 15 Steering
- ☐ 23 Clutch
- ☐ 13 Service Brakes
- ☐ 13 Parking Brake
- ☐ 13 Emergency Brakes
- ☐ 53 Triangles
- ☐ 53 Fire Extinguisher
- ☐ 53 Other Safety Equipment
- ☐ 34 Spare Fuses
- ☐ 02 Seat Belts
- ☐ ___ Other _____

_____
(IDENTIFY)

EXTERIOR
- ☐ 34 Lights
- ☐ 34 Reflectors
- ☐ 16 Suspension
- ☐ 17 Tires
- ☐ 18 Wheels/Rims/Lugs
- ☐ 32 Battery
- ☐ 43 Exhaust
- ☐ 13 Brakes
- ☐ 13 Air Lines
- ☐ 34 Light Line
- ☐ 49 Fifth-Wheel
- ☐ 49 Other Coupling
- ☐ 71 Tie-Downs
- ☐ 14 Rear-End Protection
- ☐ ___ Other _____

_____
(IDENTIFY)

☐ NO DEFECTS

TOWED UNIT(S)

- ☐ 71 Body/Doors
- ☐ 71 Tie-Downs
- ☐ 34 Lights
- ☐ 34 Reflectors

- ☐ 16 Suspension
- ☐ 17 Tires
- ☐ 18 Wheels/Rims/Lugs
- ☐ 13 Brakes

- ☐ 77 Landing Gear
- ☐ 59 Kingpin/Upper Plate
- ☐ 59 Fifth-Wheel (Dolly)
- ☐ 59 Other Coupling Devices

- ☐ 79 Rear-End Protection
- ☐ ___ Other _____

_____
(IDENTIFY)

☐ NO DEFECTS

REMARKS: _____
_____
_____
_____
_____

REPORTING DRIVER: Date_____
Name _____   Emp. No. _____

REVIEWING DRIVER: Date_____
Name _____   Emp. No. _____

MAINTENANCE ACTION: Date_____
Repairs Made ☐   No Repairs Needed ☐
R.O. –s _____
Certified By: _____
Location: _____

SHOP REMARKS: _____
_____

FIGURE 4–3 Driver's report form as issued by the American Trucking Association. (Courtesy of Volvo Trucks North America Inc.)

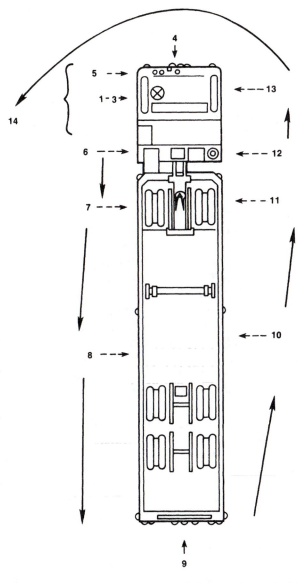

FIGURE 4–4   *Make a thorough inspection a habit by following the same logical sequence (numbers correspond with the text). (Courtesy of Heavy Duty Trucking)*

distribution, or a flat tire. Look under the vehicle for signs of fuel, oil, grease, or coolant leaks. Check the area around the vehicle for hazards to vehicle movement (people, other vehicles, objects, low-hanging wires or tree limbs, and so forth).

Review the last vehicle inspection report. Note any defects reported by the previous driver. Inspect the vehicle and confirm that any necessary repairs were made.

Step 2 (Engine Compartment). Check that the parking brakes are applied and/or the wheels are chocked. Raise the hood, tilt the cab (secure loose items so they do not fall and break something), or

open the engine compartment door, and check the following:

- Coolant level in radiator; condition of hoses
- Power steering fluid level; hose condition (if so equipped)
- Windshield washer fluid level
- Battery fluid level, connections, and tiedowns (battery may be located elsewhere)
- Automatic transmission fluid level (may require engine to be running)
- Check belts for tightness and excessive wear (alternator, water pump, air compressor)
- Leaks in the engine compartment (fuel, coolant, oil, power steering fluid, hydraulic fluid, battery fluid)
- Cracked, worn electrical wiring insulation

Lower and secure the hood, cab, or engine compartment door.

Step 3 (Inside the Cab). Make sure the parking brake is applied and put the gearshift in neutral (or park if automatic). Then start the engine and listen for unusual noises. Check the gauges for the following readings:

- Oil pressure should come up to normal within seconds after the engine is started.
- Ammeter and/or voltmeter should be in the normal range(s).
- Coolant temperature should begin a gradual rise to the normal operating range.
- Engine oil temperature should begin a gradual rise to the normal operating range.
- Warning lights and buzzers for the oil, coolant, and charging circuit should go out right away.

Check the seats and the seat belts. Inspect the windshield and mirrors for cracks, dirt, illegal stickers, or other obstructions to vision. Clean and adjust as necessary. Check for excessive play in the steering wheel. Manual steering play should not exceed about 2 inches, and power steering play should not exceed about $2\frac{1}{2}$ inches. Check all of the following for looseness, sticking, damage, or improper setting:

- Steering wheel
- Clutch
- Accelerator (gas pedal)
- Brake controls
- Foot brake
- Trailer brake
- Parking brake
- Retarder controls (if vehicle has them)
- Transmission controls
- Interaxle differential lock (if vehicle has one)

- Horn(s)
- Windshield wiper/washer
- Lights
  - Headlights
  - Dimmer switch
  - Turn signal
  - Four-way flashers
  - Clearance, identification, marker light switch(s)

Shop Talk

The National Highway Traffic Safety Administration (NHTSA) issued a rule that standardizes the buckle release mechanism that requires the belts to have either an emergency locking retractor or an automatic locking retractor. The rule also requires that retractors be attached to the seat so that they will move along with the air suspension system.

Check to see that the truck is equipped with all the necessary emergency equipment. There should be spare fuses (unless the vehicle has circuit breakers), three red reflective triangles, and a properly charged and rated fire extinguisher within arm's reach of the driver's seat. Flares, lanterns, and flags are optional. Other options include tire chains (where winter conditions require them), tire changing equipment, a list of emergency phone numbers, and an accident reporting kit.

Step 4 (Front of Cab). Check the steering system for defects. Look for missing nuts, bolts, cotter pins, or other parts. Check for bent, loose, or broken parts such as the steering column, steering gearbox, or tie-rods.

Inspect the headlights, turn signals, and emergency flashers for proper color and operation.

Check for the following on both sides of the suspension system:

- Spring hangers that allow movement of axle from proper position
- Cracked or broken spring hangers
- Missing or broken leaves in any leaf spring. If one-fourth or more are missing, it will put the vehicle out of service, but any defect could be dangerous.
- Broken leaves in a multileaf spring or leaves that have shifted so they might hit a tire or other part
- Leaking shock absorbers
- Torque rod or arm, U-bolts, spring hangers, or other axle positioning parts that are cracked, damaged, loose, or missing

- Air suspension systems that are damaged and/or leaking
- Any loose, cracked, broken, or missing frame members

Check the front brakes on both sides to make sure that all components are attached and operational. Check brake lines for leaks or damage and chambers (if visible) for cracks or insecure mounting. Check brake linings. They should be free of large cracks or missing pieces. No grease or oil should be on the linings or drums. Make certain the pushrod and slack adjuster are mechanically operational. Check for audible air leaks. If possible, ask a helper to apply the brakes, hold them, then release them when signaled. Check for excessive slack adjuster travel. If visible, check brake drums for external cracks that open upon application.

Step 5 (Left Side of Cab). Inspect the left front wheel. Check for defective welds, cracks, or breaks (especially between handholds or stud holes); unseated locking rings; broken, missing, or loose lugs, studs, or clamps; bent or cracked rims. Check for "bleeding" rust stains, defective nuts, or elongated stud holes. Spoke wheels should be checked for cracks across the spokes. Inspect for scrubbed or polished areas on either side of the lug, which indicates a slipped rim. Rims should also be checked for cracks or bends. The valve stem should be straight and equidistant from the wheel spokes.

Check the left front tire for bulges, leaks, sidewall separation, cuts, exposed fabric, and worn spots. Check for proper inflation. Measure the tread depth. Check for tire contact with any part of the vehicle.

Inspect the frame for cracked, sagging rails. Check for broken or loose bolts or brackets.

Step 6 (Left Fuel Tank Area). Check the fuel level. Check for unsecured mounting, leaks, or other damage. Check for an unsecured cap or loose connections. Verify that the fuel crossover line is secure.

Check the air and electrical lines between the tractor and the trailer for tangles, crimps, chafing, or dragging. Check the connections. Listen for air leaks.

Check the mounting of the hose couplers (gladhands). Look for leaks or other damage.

Inspect the frame for cracked, sagging rails. Check for broken or loose bolts or brackets.

Step 7 (Left Rear Tractor Area). Inspect the wheels, rims, and tires as described in step 5. Examine the inside tire and make sure that both tires are the same height. Check between the tires for debris or contact.

Inspect the suspension and brakes as described in step 4.

Check for cracks along the fifth wheel plate and mounting area. Make sure the locking jaws are properly engaged. Check for loose or missing nuts and bolts. For sliding fifth wheels, make sure the slider is locked.

Check to see that the operating handle is closed and latched. Check the tractor stoplights and turn signals for proper color and operation. Inspect the frame for cracked, sagging rails and check for broken or loose bolts or brackets.

Step 8 (Left Side of Trailer). Inspect the wheels, rims, and tires as described in steps 5 and 7. Inspect the visible suspension and brake components as described in step 4. Check for burned out or missing marker lights or reflectors.

For flatbeds, check the header board for proper type and mounting. Check the blocking and bracing, chains, straps, and side posts. Check for shifted cargo. Inspect the tarp.

Step 9 (Rear of Trailer). Check for proper operation, color, and cleanliness of stoplights, taillights, turn signals, emergency flashers, reflectors, and clearance and marker lights.

Check the suspension and brakes as described in step 4. Inspect the wheels, tires, and rims as described in steps 5 and 7. Check the rear bumper for damage or missing pieces. Verify that the doors are locked/latched. For flatbeds, inspect as described in step 8. Check for proper placarding. Make sure that the license plate is visible and the license plate light is operable. Inspect the frame for cracked, sagging rails and check for broken or loose bolts or brackets.

Step 10 (Right Side of Trailer). Check the same items that were checked on the left side. In addition, check the spare tire for secure mounting and proper inflation. Make sure the landing gear or dollies are fully raised and the crank handle is secured. Check for missing, bent, or damaged parts.

Step 11 (Right Rear Tractor Area). Check the same items that were checked on the left side.

Step 12 (Right Fuel Tank Area). Check the same items that were checked on the left side. In addition, inspect the exhaust system. Check it for secure mounting and leaks (under the cab). Make sure the exhaust system is not contacted by fuel or air lines or electrical wires. Look for carbon deposits around seams and clamps indicating exhaust leaks.

Step 13 (Right Side of Cab). Check the same items that were checked on the left side.

Step 14 (Cab). Pump the air system and check the air pressure gauge. Check the low air pressure warning device by depleting the air supply (pump the foot brake valve). The warning light/buzzer should activate at about 55 psi or above.

With the seat belt fastened, release the brakes. As the vehicle begins to move, activate the parking brakes to check their operation.

At about 5 mph, apply the service brakes. Note any unusual pulling, delay, or play in the brake pedal.

If any problem is noted during the pre-trip inspection, the technician should record it on the pre-trip form. This allows the technician to make a judgment call on whether a questionable item is to be repaired or allowed to be placed back in service. The technician is usually responsible for verifying that the vehicle is safe.

4.2 OUT-OF-SERVICE OR DEADLINING A VEHICLE

The decision to take a vehicle out of service can be a tough one. It is necessary to balance the possibility that the mechanical problem will cause an accident or breakdown against the need to get the load moved. So how does the dispatcher—or anyone else—determine when to **deadline** a vehicle?

The Federal Motor Carrier Safety Regulations, Part 393, say that anything mechanical that can either cause or prevent an accident is a safety item. All alleged safety problems should be addressed, no matter how minor they appear. For instance, drivers can be injured by sticky trailer doors and slippery tractor steps. Ragged sheet metal can injure dock workers and pedestrians.

The regulations also forbid dispatching a vehicle that is likely to break down. Regardless of whether any fine might be levied for that, dispatching a truck that is on the brink of breaking down just is not good business—nor is it safe. A dead or sputtering vehicle on the highway is a traffic hazard. Even if the driver makes it to the shoulder, it is a high-risk situation. Adding up the costs of idle driver time, unhappy freight customers, and a towing and/or road repair bill quickly verifies that sending out such a vehicle is a mistake. Management can help minimize deadline situations by listening to the driver and the technician (**Figure 4–5**).

FIGURE 4–5 *Vehicles with serious defects should be tagged immediately. (Courtesy of Heavy Duty Trucking)*

It is a must that all drivers and technicians familiarize themselves with the CVSA out-of-service criteria (**Table 4–1**). They represent the guidelines being used by law enforcement officials in virtually every state and Canadian province in increasingly frequent truck inspections.

A vehicle considered likely to cause an accident or breakdown because of mechanical conditions or improper loading will be taken out of service. Tagged with an out-of-service (OOS) sticker, the vehicle cannot be moved under its own power until all required repairs are completed. A "restricted service" designation (indicated by an asterisk (*) in **Table 4–1**) makes the provision for a vehicle to travel up to 25 miles to a repair shop—if the inspector thinks allowing the truck to continue is less hazardous than requiring it to remain at the inspection site.

When the decision to deadline a vehicle is made, the next step is to have a technician repair the problem. In many cases a safety defect can be corrected in less than an hour (brake adjustment, kinked hoses, dead headlights, gladhand replacement). If the problem is more serious, the manager may have to move to a contingency plan and rent a vehicle, bring in a spare, or distribute the freight among others that can make deliveries. After the repair is made on the down vehicle, the technician is required to certify in writing that the defect has been fixed.

Shop Talk

Remember that if the technician misses something or fails to fix a defect before an inspection, the inspectors may take the vehicle out of service. That can be costly—in both downtime and fines. It can also hurt a fleet's Department of Transportation safety rating. Worse yet, the defect could cause an accident that could put the carrier out of business by the time the attorneys are through.

TABLE 4–1: CVSA VEHICLE OUT-OF-SERVICE CRITERIA

Inspected Item	Out-of-service if
BRAKE SYSTEM Defective Brakes	20 percent or more of the brakes on a vehicle are defective. A brake is defective if:

1. Brakes are out of adjustment. (Measured with engine off and reservoir pressure at 80–90 psi with brakes fully applied.)

CLAMP-TYPE BRAKE CHAMBER			ROTOCHAMBER		
Type	Outside Dia.	Max. Stroke	Type	Outside Dia.	Max. Stroke
6	$4^1/_2$"	$1^1/_4$"	9	$4^9/_{32}$"	$1^1/_2$"
9	$5^1/_4$"	$1^3/_8$"	12	$4^{13}/_{16}$"	$1^1/_2$"
12	$5^{11}/_{16}$"	$1^3/_8$"	16	$5^{13}/_{32}$"	2"
16	$6^3/_8$"	$1^3/_4$"	20	$5^{15}/_{16}$"	2"
20	$6^{-25}/_{32}$"	$1^3/_4$"	24	$6^{13}/_{32}$"	2"
24	$7^7/_{32}$"	$1^3/_4$"	30	$7^1/_{16}$"	$2^1/_4$"
		(2" for long stroke)			
30	$8^3/_{32}$"	2"	36	$7^5/_8$"	$2^3/_4$"
36	9"	$2^1/_4$"	50	$8^7/_8$"	3"

BOLT-TYPE BRAKE CHAMBER				WEDGE BRAKE
Type	Effective Area	Outside Dia.	Max. Stroke	
A	12 sq. in.	$6^{15}/_{16}$"	$1^3/_8$"	Movement of the scribe mark on the lining shall not exceed $1/_{16}$".
B	24 sq. in.	$9^3/_{16}$"	$1^3/_4$"	
C	16 sq. in.	$8^1/_{16}$"	$1^3/_4$"	
D	6 sq. in.	$5^1/_4$"	$1^1/_4$"	
E	9 sq. in.	$6^3/_{16}$"	$1^3/_8$"	
F	36 sq. in.	11"	$2^1/_4$"	
G	30 sq. in.	$9^7/_8$"	2"	

2. On application of service brakes, there is no braking action (such as brake shoe(s) failing to move on a wedge, S-cam, cam, or disc brake).
3. Mechanical components are missing, broken, or loose (such as shoes, linings, pads, springs, anchor pins, spiders, cam rollers, pushrods, and air chamber mounting bolts).
4. Audible air leak at brake chamber is present.

TABLE 4–1:   CVSA VEHICLE OUT-OF-SERVICE CRITERIA (CONTINUED)

Inspected Item	Out-of-service if
	5. Brake linings or pads are not firmly attached or are saturated with oil, grease, or brake fluid.
	6. Linings show excessive wear.
	7. Required brake(s) are missing.
Steering Axle Brakes	1. On vehicles required to have steering axle brakes, there is no braking action on application (includes the dolly and front axle of a full trailer).
	2. Air chamber sizes or slack adjuster lengths are mismatched on tractor steering axles.
	3. Brake linings or pads on tractor steering axles are not firmly attached to the shoe; are saturated with oil, grease, or brake fluid; or lining thickness is insufficient.
Parking Brakes	Upon actuation, no brakes are applied, including driveline hand-controlled parking brake.*
Brake Drums or Rotors	1. External crack(s) on drums open on brake application.
	2. A portion of the drum or rotor is missing or in danger of falling away.
Brake Hose	1. Hose damage extends through the outer reinforcement ply.
	2. Hose bulges/swells when air pressure is applied or has audible leak at other than proper connection.
	3. Two hoses are improperly joined and can be moved or separated by hand at splice.
	4. Hoses are improperly joined but cannot be moved or separated by hand.*
	5. Air hose is cracked, broken, or crimped and airflow is restricted.
Brake Tubing	Tubing has audible leak at other than proper connection or is cracked, heat-damaged, broken, or crimped.
Low-Pressure Warning Device	Device is missing, inoperative, or does not operate at 55 psi and below, or half the governor cutout pressure, whichever is less.*
Air Loss Rate	If air leak is discovered and the reservoir pressure is not maintained when governor is cut in, pressure is between 80 and 90 psi, engine is at idle, and service brakes are fully applied.
Tractor Protection	Protection valve on tractor is missing or inoperable.
Air Reservoir	Mounting bolts or brackets are broken, missing, or loose.*
Air Compressor	1. Drive belt condition indicates impending or probable failure.*
	2. Mounting bolts are loose. Pulley is cracked, broken, or loose. Mounting brackets, braces, and adapters are cracked or broken.
Electric Brakes	1. Absence of braking on 20 percent or more of the braked wheels.
	2. Breakaway braking device is missing or inoperable.
Hydraulic Brakes	1. No pedal reserve with engine running, except by pumping the pedal.
	2. Master cylinder is less than one-quarter full.
	3. Power-assist unit fails.
	4. Brake hoses seep or swell under application.
	5. Hydraulic fluid is visibly leaking from brake system and master cylinder is less than one-quarter full.*
	6. Check valve is missing or inoperative.
	7. Hydraulic fluid is visibly leaking from brake system.
	8. Hydraulic hoses are worn through outer cover-to-fabric layer.
	9. Fluid lines or connections are restricted, crimped, cracked, or broken.
	10. Brake failure light/low fluid warning light is on and/or inoperative.*

Fm boss 185 (handwritten)

TABLE 4–1: CVSA VEHICLE OUT-OF-SERVICE CRITERIA (CONTINUED)	
Inspected Item	**Out-of-service if**
Vacuum System	1. Vacuum reserve is insufficient to permit one full brake application after engine is shut off.
	2. Hoses or lines are restricted, excessively worn, crimped, cracked, broken, or collapsed under pressure.
	3. Low-vacuum warning light is missing or inoperative.*
COUPLING DEVICES Fifth Wheels	1. Mounting to frame—more than 20 percent of frame mounting fasteners are missing or ineffective.
	2. Movement between mounting components is observed. Mounting angle iron is cracked or broken.
	3. Mounting plates and pivot brackets, more than 20 percent of fasteners on either side are missing or ineffective. Movement between pivot bracket pin and bracket exceeds $3/8$". Pivot bracket pin is missing or not secured.
	4. Cracks in any weld(s) or parent metal are observed on mounting plates or pivot brackets.*
	5. Sliders—more than 25 percent of latching fasteners, per side, are ineffective. Any fore or aft stop is missing or not secured. Movement between slider bracket and base exceeds $3/8$".
	6. Cracks are observed in any slider component parent metal or weld.*
	7. Lower coupler—horizontal movement between the upper and lower fifth wheel halves exceeds $1/2$". Operating handle is not closed or locked in position. Kingpin is not properly engaged. Cracks are observed in fifth wheel plate. Locking mechanism parts are missing, broken, or deformed to the extent the kingpin is not securely held.
	8. Space between upper and lower coupler allows light to show through from side to side.*
Pintle Hooks	1. Mounting to frame—fasteners are missing or ineffective. Frame is cracked at mounting bolt holes. Loose mounting. Frame cross member providing pintle hook attachment is cracked.
	2. Integrity—cracks are anywhere in the pintle hook assembly. Section reduction is visible when coupled. Latch is insecure.
	3. Any welded repairs to the pintle hook are visible.*
Drawbar Eye	1. Mounting—any cracks are visible in attachment welds. Fasteners are missing or ineffective.
	2. Integrity—any cracks are visible. Section reduction is visible when coupled. Note: No part of the eye should have any section reduced by more than 20 percent.
Drawbar/Tongue	1. Slider (power/manual)—latching mechanism is ineffective. Stop is missing or ineffective. Movement of more than $1/4$" between slider and housing. Any leaking air or hydraulic cylinders, hoses, or chambers.
	2. Integrity—any cracks. Movement of $1/4$" between subframe and drawbar at point of attachment.
Safety Devices	1. Missing or unattached, or incapable of secure attachment.
	2. Chains and hooks worn to the extent of a measurable reduction in link cross section.*
	3. Improper repairs to chains and hooks including welding, wire, small bolts, rope, and tape.*
	4. Kinks or broken cable strands. Improper clamps or clamping on cables.*
Saddlemounts (method of attachment)	1. Any missing or ineffective fasteners, loose mountings, or cracks in a stress or load-bearing member.
	2. Horizontal movement between upper and saddlemount halves exceeds $1/4$".

TABLE 4–1:   CVSA VEHICLE OUT-OF-SERVICE CRITERIA (CONTINUED)

Inspected Item	Out-of-service if
EXHAUST SYSTEM	1. Any exhaust system leaking in front of or below the driver/sleeper compartment and the floor pan permits entry of exhaust fumes.* 2. Location of exhaust system is likely to result in burning, charring, or damaging wiring, fuel supply, or combustible parts.
FUEL SYSTEM	1. System with visible leak at any point. 2. A fuel tank filter cap missing.* 3. Fuel tank not securely attached due to loose, broken, or missing mounting bolts or brackets.*
LIGHTING DEVICES When Lights are Required	1. Single vehicle without at least one headlight operative on low beam and without a stoplight on the rearmost vehicle. 2. Vehicle that does not have at least one steady burning red light on the rear of the rearmost vehicle (visible at 500 feet). 3. Projecting loads without at least one operative red or amber light on the rear of loads projecting more than 4 feet beyond vehicle body and visible from 500 feet.
At All Times	1. No stop light on the rearmost vehicle is operative. 2. Rearmost turn signal(s) do not work.*
SAFE LOADING	1. Spare tire or part of the load is in danger of falling onto the roadway. 2. Protection against shifting cargo—any vehicle without front-end or equivalent structure as required.*
STEERING MECHANISM Steering Wheel Free Play	When any of the following values are met or exceeded, the vehicle will be taken out of service.

	Steering Wheel Diameter	Manual System Movement	Power System Movement
		30 Degrees or	45 Degrees or
	16"	$4^1/_2$" (or more)	$6^3/_4$" (or more)
	18"	$4^3/_4$" (or more)	$7^1/_8$" (or more)
	20"	$5^1/_4$" (or more)	$7^7/_8$" (or more)
	21"	$5^1/_2$" (or more)	$8^1/_4$" (or more)
	22"	$5^3/_4$" (or more)	$8^5/_8$" (or more)

Inspected Item	Out-of-service if
Steering Columns	1. Any absence or looseness of U-bolts or positioning parts. 2. Worn, faulty, or obviously repair-welded universal joints. 3. Steering wheel not properly secured.
Front Axle Beam	1. Includes all steering components other than steering column, including hub. 2. Any cracks. 3. Any obvious welded repair(s).
Steering Gear Box	1. Any mounting bolt(s) loose or missing. 2. Any cracks in gear box or mounting brackets.
Pitman Arm	Any looseness of the arm or steering gear output shaft.
Power Steering	Auxiliary power-assist cylinder loose.
Ball and Socket Joints	1. Movement under steering load of a stud nut. 2. Any motion (other than rotational) between linkage member and its attachment point over $1/_4$".
Tie-Rods and Drag Links	1. Loose clamp(s) or clamp bolt(s) on tie-rods or drag links. 2. Any looseness in any threaded joint.

TABLE 4–1: CVSA VEHICLE OUT-OF-SERVICE CRITERIA (CONTINUED)	
Inspected Item	**Out-of-service if**
Nuts	Loose or missing nuts in tie-rods, pitman arm, drag link, steering arm, or tie-rod arm.
Steering System	Any modification or other condition that interferes with free movement of any steering component.
SUSPENSION Axle Parts/Members	Any U-bolt(s), spring hanger(s), or other axle positioning part(s) cracked, broken, loose, or missing resulting in shifting of an axle from its normal position.
Spring Assembly	1. One-fourth of the leaves in any spring assembly broken or missing. 2. Any broken main leaf in a leaf spring. 3. Coil spring broken. 4. Rubber spring missing. 5. Leaf displacement that could result in contact with a tire, rim, brake drum, or frame. 6. Broken torsion bar spring in torsion bar suspension. 7. Deflated air suspension.
Torque, Radius, or Tracking Components	Any part of a torque, radius, or tracking component assembly that is cracked, loose, broken, or missing.
FRAME Frame Members	1. Any cracked, loose, or sagging frame member permitting shifting of the body onto moving parts or other condition indicating an imminent collapse of the frame. 2. Any cracked, loose, or broken frame member adversely affecting support of functional components such as steering gear, fifth wheel, engine, transmission, or suspension. 3. Any crack $1\frac{1}{2}$" or longer in the frame web that is directed toward the bottom flange. 4. Any crack extending from the frame web around the radius and into the bottom flange. 5. Any crack 1" or longer in bottom flange.
Tire and Wheel Clearance	Any condition, including loading, causes the body or frame to be in contact with a tire or any part of the wheel assemblies at the time of inspection.
Adjustable Axle (sliding subframe)	1. Adjustable axle assembly with more than one-quarter of the locking pins missing or not engaged. 2. Locking bar not closed or not in the locked position.
TIRES Any Tire on Any Steering Axle of a Power Unit	1. With less than $2/32$" tread when measured in any two adjacent major tread grooves at any location on the tire. 2. Any part of the breaker strip or casing ply is showing in the tread. 3. The sidewall is cut, worn, or damaged so that ply cord is exposed. 4. Labeled "Not for Highway Use" or carrying markings that would exclude use on steering axle. 5. A tube-type radial tire without the stem markings. These include a red band around the tube stem, the word "radial" embossed in metal stems, or the word "radial" molded in rubber stems.* 6. Mixing bias and radial tires on the same axle.* 7. Tire flap protrudes through valve slot rim and touches stem.* 8. Regrooved tire except on motor vehicles used solely in urban or suburban service.* 9. Visually observable bump, bulge, or knot related to tread or sidewall separation. 10. Boot, blowout patch, or other ply repair.* 11. Weight carried exceeds tire load limit. This includes overloaded tire resulting from low air pressure.* 12. Tire is flat or has a noticeable leak. 13. So mounted or inflated that it comes in contact with any part of the vehicle.

TABLE 4–1: CVSA VEHICLE OUT-OF-SERVICE CRITERIA (CONTINUED)

Inspected Item	Out-of-service if
All Tires Other Than Those Found on the Steering Axle of a Powered Vehicle	1. Weight carried exceeds tire load limit. This includes overloaded tires resulting from low air pressure.* 2. Tire is flat or has noticeable leak (that is, one that can be heard or felt). 3. Bias ply tire—when more than one ply is exposed in the tread area or sidewall, or when the exposed area of the top ply exceeds 2 sq. in. Note: On duals, both tires must meet this condition. 4. Bias ply tire—when more than one ply is exposed in the tread area or sidewall, or when the exposed area of the top ply exceeds 2 sq. in.* 5. Radial ply tire—when two or more plies are exposed in the tread area or damaged cords are evident in the sidewall, or when the exposed area exceeds 2 sq. in., tread or sidewall. Note: On dual wheels, both tires must meet this condition. 6. Radial ply tire—when two or more plies are exposed in the tread area or damaged cords are evident in the sidewall, or when the exposed area exceeds 2 sq. in., tread or sidewall.* 7. Any tire with visually observable bump or knot apparently related to tread or sidewall separation. 8. So mounted or inflated that it comes in contact with any part of the vehicle. (This includes any tire contacting its mate in a dual set.) 9. Is marked "Not for Highway Use" or otherwise marked and having similar meaning.* 10. So worn that less than $^1/_{32}$" tread remains when measured in any two adjacent major tread grooves at any location on the tire. Exception: On duals, both tires must be so worn.*
WHEELS AND RIMS	1. Lock or side ring is bent, broken, cracked, improperly sealed, sprung, or mismatched. 2. Rim cracks—any circumferential crack except at valve hole. 3. Disc wheel cracks—extending between any two holes. 4. Stud holes (disc wheels)—50 percent or more elongated stud holes (fasteners tight). 5. Spoke wheel cracks—two or more cracks more than 1" long across a spoke or hub section. Two or more web areas with cracks. 6. Tubeless demountable adapter cracks—cracks at three or more spokes. 7. Fasteners—loose, missing, broken, cracked, or stripped (both spoke and disc wheels) ineffective as follows: For 10 fastener positions: 3 anywhere, 2 adjacent. For 8 fastener positions or less (including spoke wheel and hub bolts): 2 anywhere. 8. Welds—any cracks in welds attaching disc wheel disc to rim. Any crack in welds attaching tubeless demountable rim to adapter. Any welded repair on aluminum wheel(s) on a steering axle. Any welded repair other than disc to rim attachment on steel disc wheel(s) mounted on the steering axle.
WINDSHIELD GLAZING	Any crack over $^1/_4$" wide, intersecting cracks, discoloration not applied in manufacture, or other vision-distorting matter in the sweep of the wiper on the driver's side.*
WINDSHIELD WIPERS	Any power unit that has inoperable parts or missing wipers that render the system ineffective on the driver's side.

4.3 PM SCHEDULING

An important phase of any maintenance program is *scheduled* periodic preventive maintenance, accomplished by time-tabled servicing of equipment, systematic inspection, detection, and correction of potential failures before major defects develop. Implementing this policy requires good scheduling and controls to ensure that all vehicles are serviced on their set cycle whether it be miles, hours, gallons of fuel consumed, or monthly (**Figure 4–6**).

🥽 **Shop Talk**

It should be noted that off-highway vehicles rarely have odometers. Most use engine service or the vehicle's hour meter to measure PM intervals, as opposed to distance traveled. One current trend bases PM time-tabling on the amount of fuel consumed.

As shown on the PM form illustrated in **Figure 4–2**, there are three categories of maintenance: "A" PM, "B" PM, and "C" PM. The "A" PM is the simplest to perform, and it is performed at the shortest interval. The "B" PM is more involved and performed less often. The "C" PM adds more steps to the "B" PM. It is performed the least often of the three inspections. A recent TMC report indicated that

- "A" PM services averaged 2.4 hours but ranged from 45 minutes to 5½ hours.
- "B" PM services averaged 5 hours but ranged from 2 to 8 hours. "C" PM (major inspection) averaged about 1 day, but may be longer, depending on the servicing required.
- Miles between unscheduled repairs averaged about 20,000. Some PM supervisors have created a "Mini-A" at 6,000 miles specifically to

look for items that cause unplanned shop visits. A mini check takes about half an hour and can eliminate unscheduled shop visits.

In most fleet operations, vehicle usage generally determines the basis of PM scheduling. The number of different inspections and the intervals at which they are usually conducted must be determined by the PM manager according to the specific needs of the trucking operation.

TRAILER PREVENTIVE MAINTENANCE (PM)

Like the tractor, trailer components require preventive inspection and servicing at periodic intervals. Maintenance will help ensure maximum service life from the trailer. **Table 4–2** is a typical trailer inspection guide.

VEHICLE MAINTENANCE RECORD & SCHEDULE

TRUCK #_____   MAKE _____   MODEL _____   SERIAL# _____

MONTH OR MILEAGE	DUE DATE	SERVICE DUE	MILEAGE SERVICED	DATE SERVICED	REPAIRS PERFORMED	REPAIR ORDER NO.
JANUARY		(1) O&F				
FEBRUARY		(1)				
MARCH		(1) O&F				
APRIL		(1) WF				
MAY		(1) O&F				
JUNE		(1)				
JULY		(1) O&F				
AUGUST		(1) WF				
SEPTEMBER		(1&2) O&F				
OCTOBER		(1)				
NOVEMBER		(1) O&F				
DECEMBER		(1) WF				

O = OIL
F = OIL FILTER
WF = WATER FILTER

FIGURE 4–6 *Typical monthly PM schedules. (Courtesy of Volvo Trucks North America Inc.)*

TABLE 4–2: PM TRAILER INSPECTION GUIDE	
Inspection after the first 500 miles	Conduct pre-trip inspection. Check axle alignment. Tighten axle U-bolts. Adjust brakes. Retorque inner and outer cap nuts on disc wheels or wheel nuts if trailer is equipped with spoke wheels.
Monthly inspection	Conduct pre-trip inspection. Inspect brakes to ensure lining is not worn excessively. When lining is replaced, check drums for excessive wear, heat cracks and grooving. Check travel of brake chamber pushrods and adjust brakes as necessary. Check and retorque wheel nuts. Inspect tires for uneven wear. Check axle alignment. Check for loose or missing fasteners, and replace or tighten as necessary. Check and repair any cracked welds. Inspect all areas where sealant is used. Reseal as necessary. Check under-construction finish, particularly in areas above trailer tires and tractor tires. Replace as necessary. Check optional items (including refrigeration and heating equipment) according to manufacturer's inspection guide. Check other items or special type trailers or construction (FRP trailers, dump trailers, etc.) Lubricate trailer according to the service manual.
Semi-annual inspection	Be sure to conduct 500-mile and monthly inspections. Inspect brake linkage, cams and shoes. Inspect and clean wheel bearings. Refill to proper oil level. Inspect body parts, doors, ventilators, roof, floor, sides and all body hardware. Replace or repair unserviceable parts. Inspect suspension parts, springs, U-bolts, landing gear and tire carrier. Repair or replace worn or damaged parts as needed. Inspect kingpin for excessive wear. Inspect upper fifth wheel plate for excessive wear and cracks. Inspect interior and exterior finishes. Replace as necessary. Lubricate according to service manual recommendations.

NOTE: If the trailer was delivered directly from the manufacturer, check axle alignment, tighten axle U-bolts, and retorque inner and outer cap nuts or wheel nuts on arrival.

FEDERAL INSPECTION REGULATIONS

The Federal Highway Administration has set up a periodic minimum inspection standards program, under which the following vehicles must be inspected:

- Any vehicle involved in interstate commerce with a gross vehicle weight over 10,000 pounds with or without power, including trucks, buses, tractor/trailers, full trailers, semitrailers, converter dollies, container chassis, booster axles, and jeep axles
- Any vehicle, regardless of weight, which is designed to carry more than fifteen passengers, including the driver
- Any vehicle, regardless of weight, carrying hazardous materials in a quantity requiring placarding

Each vehicle must carry proof that an inspection was completed. Proof can be either a copy of the inspection form kept on the vehicle, or a decal. If using a decal, a copy of the inspection form must be kept on file and indicate where an inspector can call or write to get a copy of it.

Size and shape of the decal is not specified. It may be purchased from a supplier or even be made by the PM shop. The only requirement is that they remain legible. Each "vehicle" must be inspected separately. This means a tractor/trailer is two vehicles, each requiring a decal; a converter dolly is a separate vehicle. The decal (**Figure 4–7**) must show the following information:

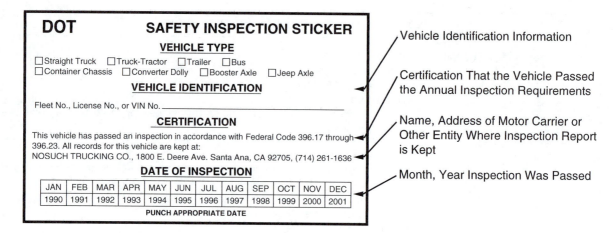

FIGURE 4–7 Make your own inspection sticker, but be sure it meets the Federal requirements shown on the right.

- Date (month and year) the vehicle passed the inspection.
- Name and address to contact about the inspection record. The record can be stored anywhere as long as the decal states where to contact the fleet.
- Certification that the vehicle has passed the inspection
- Vehicle identification information (fleet unit number or serial number) sufficient to identify the vehicle

According to the Office of Motor Carriers, there are three acceptable possibilities:

1. If you have vehicles registered in states with inspection programs meeting federal requirements, those states' inspection programs will suffice.
2. If a vehicle has passed a Commercial Vehicle Safety Alliance (CVSA) roadside inspection, it does not need to be reinspected to meet the federal rules. However, a copy of the inspection report should be retained as proof; a CVSA sticker may not be accepted if it does not have the information required for the annual inspection sticker.
3. If neither of the above applies to the owner's fleet, he/she can self-inspect, or have a commercial repair shop or truck or trailer dealer do the inspections—if they have qualified inspectors. Inspector qualification details are in a later section.

The rules say that the truck owner is responsible for ensuring that the inspector is qualified. This means the fleet must keep records on all inspectors as long as they are doing inspections and for 1 year thereafter.

This requirement includes personnel in outside shops who inspect the fleet's vehicles. The only exception is for those inspections done as part of state inspection programs or at random roadside safety checks.

Inspector Qualifications

Section 396.19 of the Federal Motor Carrier Safety Regulations says it is the fleet's responsibility to ensure that all personnel doing annual inspections are qualified as follows:

- They understand all the inspection criteria and can identify defective parts and components.
- They know how to use the proper tools and equipment to perform inspections.
- They know how to do inspections based on experience and/or training.

Training means completion of a state or federally sponsored training program, or at least 1 year in a truck manufacturer-sponsored training program or comparable commercial training program.

Experience means at least one year working as a technician or inspector in a fleet maintenance program, commercial garage, leasing company or similar facility; or working as a commercial vehicle inspector for a state or the federal government.

Evidence of inspectors' qualifications must be kept on file by the employer while they are employed and for 1 year after leaving. There is an exception for inspectors who are doing state-administered inspection programs or roadside checks: They must be certified by the state.

Some fleets are designing forms to record the date and location of in-shop inspector training, schools attended, etc. All inspectors will be asked to certify that the information is correct and that they are familiar with the inspection requirements spelled out in Appendix G to subchapter B 396, p. 17–23. In addition, many fleets are expected to have a foreman or PM director certify that each person is a qualified inspector.

Record-Keeping Requirements

Federal requirements state that the inspector (fleet or an outside shop) must make a written report which identifies:

- The inspector and the carrier operating the vehicle
- The vehicle inspected and the date

The report must certify that the inspection was complete and complied with all requirements. The original or a copy of the inspection record must be kept by the truck owner (fleet, owner driver, or leasing company for 14 months from the date of the inspection and must be available on demand by an authorized federal, state, or local official. Even when subsequent inspections are performed, the old records may not be discarded until 14 months have passed. In cases where the last annual inspection was performed by some other shop/garage/dealer, the company operating the vehicle is responsible for getting a copy of the earlier inspection record when requested to do so by an official. Some other points:

- The inspection decal can be placed anywhere on the vehicle as long as the driver knows where it is and the information is legible and current.
- New tractor/trailers are not exempt. Every new vehicle must be inspected before it is put on the road and every 12 months thereafter.

Many fleets are modifying their "B" preventive maintenance forms to include all the federal safety checks. This way whenever a "B" service is performed and the vehicle also passes the safety inspection, a new decal can be affixed. This procedure simplifies the task of tracking down each vehicle to perform the annual inspection. Most fleets will want to stagger inspections through the year to prevent shop scheduling congestion.

Experience has shown that it is a good idea to keep work orders in the vehicle file folder as long as

the vehicle is in service within a given operation.

The same regulations require that the driver's vehicle condition report be retained for at least three months from the date of the report.

As mentioned in Chapter 2, computers are beginning to play an important part in both PM programs and record keeping. Today's low-priced desktop computers have increasingly sophisticated maintenance software with automated data entry. Radio-frequency receivers at fuel islands that read tires, and trip computers and diagnostic recorders help the management of driver, vehicle, and trip information. Maintenance managers are increasingly being expected to be computer literate, which means they now have better knowledge of equipment condition than ever before. And that knowledge is important when determining warranty claims, spec'ing decisions, and cost-per-mile reduction.

 Shop Talk

When making some PM checks, as regular servicing procedure, it may be necessary to raise the cab. Check the service manual for the exact procedure, but be sure to follow the safety tips given in Chapter 2. With most hydraulic cab lift systems, there are two circuits: the **push circuit** *that raises the cab from the lowered position to the desired tilt position, and the* **pull circuit** *that brings the cab from a fully tilted position up and over the center (***Figure 4–8***). Remember that in most systems, whenever raising or lowering the cab, stop working the hydraulic pump once the cab goes over center. The cab falls at a controlled rate and continued pumping could lock up the tilt cylinders.*

4.4 LUBRICANTS

Proper lubrication is important in reducing wear and preventing premature failure of expensive truck components. By forming a lubricating film between moving

FIGURE 4–8 *Cab lift positions: (A) lowered or in operating position; (B) 45-degree tilt position; and (C) full-tilt position (80 degrees). (Courtesy of Freightliner Trucks)*

parts, lubricants reduce friction, prevent excessive heat and hold dirt and particles in suspension. Lubricants also contain additives that inhibit corrosion and aid in reducing component wear. Heat and use alter the chemical properties of lubricants, making regular lubricant change intervals necessary.

In later chapters of this book, lubricants and lubrication will be discussed in the context of specific systems. Actually, the proper use of lubricants is the backbone of most good fleet preventive maintenance programs. Of course, satisfying the lubrication requirements of a vehicle is simple. Just follow the manufacturer's recommendations given in the service manual. However, there are other factors to consider—especially in fleets with mixed component brands.

In some cases, manufacturers approve several different types of lubricants (such as engine oil or gear lubricants in high-capacity transmissions). Some still recommend a straight mineral oil (GL-1), while others demand a high-additive lubricant meeting GL-5 specs. There is also the matter of lube **viscosity**: When should a 90 weight, 140 or multi-vis lubricant (such as 80W-90) be used? And how does gear lube viscosity compare with engine oil viscosity? To perform a proper lubrication job, the service technician must have a basic understanding of lubricants and the lubrication requirements of chassis components.

ENGINE OILS

To make the correct lubricating oil selection, there are three basics to understanding diesel engine oil— the oil's rating, its viscosity, and its total base number (TBN). Always refer to the engine manufacturer's service manual for guidance in making the proper selection.

Tables 4–3 and **4–4** outline the American Petroleum Institute (API) classification of engine oils based on the type of engine service it is suitable for. There are two basic classifications—a Standard or "S" class for gasoline driven engines and Commercial or "C" class for diesel driven engines. Actually, the letters refer to the type of ignition that each engine utilizes. "S" is for gasoline engines that employ Spark ignition; "C" is for the Compression ignition system utilized by diesels.

The most important property an oil must have is viscosity, or the ability of the oil to maintain proper lubricating quality under various conditions of oper-

TABLE 4–3: GASOLINE ENGINE OIL CLASSIFICATIONS	
SF	Service typical of gasoline engines in automobiles and some trucks beginning with 1980. SF oils provide increased oxidation stability and improved antiwear performance over oils that meet API designation SE. They also provide protection against engine deposits, rust, and corrosion. SF oils can be used in engines requiring SC, SD, or SE oils.
SG	Service typical of gasoline automobiles and light-duty trucks, plus CC classification diesel engines beginning in the late 1980s. SG oils provide the best protection against engine wear, oxidation, engine deposits, rust, and corrosion. It can be used in engines requiring SC, SD, SE, or SF oils.
SH	Introduced in 1993 to replace the SG category.

TABLE 4–4: DIESEL ENGINE OIL CLASSIFICATIONS	
CE	Used for heavy-duty diesel engine service. Used with turbocharged or supercharged engines manufactured after 1983, but is primarily formulated for 1991 low-emission turbocharged engines. Includes engines operating under both low-speed, high-load and high-speed, high-load conditions.
CF-2	Used in 2-stroke-cycle diesel engines. Used with aspirated, turbocharged, or supercharged diesel engines in moderate to heavy duty applications and with some heavy-duty gasoline engines. These oils are designed to protect against high temperature deposits and bearing corrosion in diesel engines as well as rust, corrosion, and low temperature deposits in gasoline engines. The sulfated ash content should be below 1 percent by weight.
CF-4	Used in heavy-duty diesel engines. Used with turbocharged diesel engines where more effective control of wear and deposits is needed. Also used when there is a wide range in the quality of fuels used, including high sulfur fuels. These oils are designed to protect against bearing corrosion and high temperature deposits in diesel engines. Their sulfated ash content will be greater than 1 percent by weight.

ating speed, temperature, and pressure. Viscosity describes oil thickness or resistance to flow. Viscosity is important because it is directly related to how well the oil lubricates and protects from metal to metal contact. Since oil thins as temperature increases and thickens as temperature drops, viscosity must be matched to ambient air temperature.

The Society of Automotive Engineers (SAE) has established an oil viscosity classification system accepted throughout the trucking industry. In this system the heavier weight oils receive the higher number or rating. For example, an oil with an SAE rating of 50-weight is heavier and flows slower than an SAE 10-weight oil. The heavier weight oils are normally used in warmer climates, while the lighter weight oils are used in low temperature areas.

Multiviscosity oils have been developed to cover a wider range of operational temperatures than the standard weights. These are classified, for example, as 10W/30, 15W/40 and so on (**Table 4–5**). A 10W/30 oil, for example has the same viscosity characteristics as a 10-weight oil at 0°F, but will duplicate flow properties of a SAE 30-weight oil at 210°F. This oil provides both easy starting in cold weather and good protection at all operating temperatures.

The TBN of the oil is important to understand. Most diesel fuel contains some degree of sulfur. A byproduct of sulfur, sulfuric acid, acts as a corrosive to engine components. As mentioned earlier, one of the functions of lubricating oil is to neutralize this sulfuric acid using alkalinity additives. The measure of reserve alkalinity in an oil is called its TBN. The TBN rule is to use a CF or CG oil with a TBN twenty times the fuel sulfur level. The higher the oil's TBN, the greater its neutralizing ability. When an oil's TBN drops below a certain level, it can no longer adequately protect the engine and it should be changed.

The total acid number (TAN) is a companion to the TBN. It measures the acidity of an oil. The higher an oil's TAN number, the more acidic it is. The effects of high TAN values in oil are corrosion, oil thickening, formation of deposits, and accelerated wear. In some applications, a high TAN can be caused by acid contamination from the environment. Using the wrong lubricant or the use of high sulfur fuel may also result in high TAN values.

Resistance to oxidation is an important characteristic governing the service life of lubricating oil. Oxidation occurs when the hydrocarbons in the oil chemically combine with oxygen. Although most oils contain anti-oxidants, they are gradually used up through service. Heat, pressure, and aeration speed up the oxidation process. Other common causes of oxidation include extended oil drain intervals, the wrong oil for the application, excessive combustion gas blowby, and high sulfur fuels. As oil oxidizes, corrosive acids can form and deposits may accumulate on critical engine parts, inhibiting operation and accelerating wear.

Shop Talk

Both SAE and API viscosity ratings can be found on the label of the oil container.

The introduction of synthetic motor oil dates back to World War II: It was often described as the "oil of the future." Synthetic oil is manufactured in a laboratory rather than pumped out of the ground and refined. It offers a variety of advantages over natural oils including the following:

- Better lubrication and easier shifting in extreme cold weather
- Longer life for transmission and axle gears, bearings and seals
- Ability to withstand extremely high temperatures that would break down mineral oils
- Extended warranty coverage, in some cases (purchasable) up to seven years or 750,000 miles
- Less drain oil to dispose of during the vehicle's life
- Fuel savings averaging 1.5 percent over all seasons

	TABLE 4–5: SAE GRADES OF ENGINE OIL	
Lowest Atmospheric Temperature Expected	Single-Grade Oils	Multi-Grade Oils
32°F (0°C)	20, 20W, 30	10W/30, 10W/40, 15W/40, 20W/40, 20W/50
0°F (−18°C)	10W	5W/30, 10W/30, 10W/40, 15W/40
−15°F (−26°C)	10W	10W/30, 10W/40, 5W/30
Below −15°F	5W	5W/20, 5W/30

Synthetic oils have some disadvantages. Synthetic engine oil is not recommended by the Automotive Engine Rebuilder's Association (AERA) for an engine break-in period. Its ability to reduce wear by virtually eliminating friction between moving components, while normally desirable, is not ideal for a break-in engine oil. Certain predictable amounts of friction are required for the proper break-in of pistons and piston rings. This is why AERA does not recommend the use of synthetic motor oils for the first 5,000 miles of service on a rebuilt engine.

To help a lubricating oil achieve maximum performance, additives are generally recommended by lubricating oil manufacturers. They are added to the oil for many reasons—all based on the type of service expected from the oil. The following are some of the more common additives used with lubricating oils.

- **Oxidation inhibitors** keep oil from oxidizing even at very high temperatures. Because they prevent the oil from absorbing oxygen they also help to prevent varnish and sludge formations.
- **Anti-rust agents** prevent rusting of metal parts when the engine is not in use. They form a protective coating that can repel water droplets and neutralize harmful acids.
- **Detergent additives** help keep metal surfaces clean and prevent deposits. These additives suspend particles of carbon and oxidized oil in the oil. These contaminants are then removed from the system when the oil is drained so they cannot accumulate as sludge.
- **Anti-corrosion agents** protect metal surfaces from corrosion. These agents will work with oxidation inhibitors.
- **TBN additives** are acid molecule neutralizers that are blended into the oil to reduce the harmful effects of acid compounds that are produced as a byproduct of the combustion process.

Remember, however, that additives will gradually lose their effectiveness. Drain the oil before the additives are totally depleted. Remember to service filters at regular intervals. Engine oils are blended to a precise chemical formula, and the addition of aftermarket additives can upset the balance of the additives and possibly have an adverse effect on engine performance. Be sure to follow manufacturer's recommendations when using any additive with a lubricating oil.

 Shop Talk

Using the incorrect grade or type of oil in a rebuilt engine can cause a variety of problems. For example, the wrong viscosity oil can cause either an oil consumption problem or a low oil pressure problem, depending upon the weather and driving conditions. Using an oil with the incorrect service rating can result in inadequate protection for engine bearings and other moving engine parts. Always consult the engine manufacturer's recommendations to make certain that the engine is using the correct grade and type of oil.

The frequency of oil changes, as well as oil, fuel, and coolant filters depends on the PM schedule. Most PM programs follow the recommendations given in the service manual.

Used Oil

Most states and provinces consider used oil as hazardous waste; others do not. For this reason, it is wise to check environmental agencies for the regulations concerning the disposal of used oils. Federal regulations state that used oil cannot be stored in unlined surface containers. Tanks and storage containers must be properly maintained and must have a "used oil" label. Spills must be cleaned up.

If you have used oil recycled, make sure both the transporter and recycler are in compliance with federal and state regulations.

If not drained of free flowing oil, used oil filters are regulated as used oil. If oil is drained, state or local recycling rules may apply. Transporters and recyclers who specialize in used oil often also take filters. Crushing may be required.

TRANSMISSION AND AXLE LUBRICANTS

The lubricants and lubrication of transmissions and drive axles will be covered in Chapters 10 through 17. Unfortunately many fleet maintenance personnel do not pay enough attention to axle and transmission lubes because, unlike engine crankcase oils, these oils are in an isolated system. There is no potential for contamination from antifreeze or fuel leaking into the lube. Unfortunately, when there is a failure it is often traced to incorrect or worn-out lubricants. So the "out-of-sight, out-of-mind" approach does not work.

Cold-temperature operation can cause transmission and axle lubrication problems known as *channeling*. High-viscosity oils typically used in transmissions and axles thicken and may not flow at all when the vehicle is first started. Rotating gears can push lubricant aside leaving voids or channels where no lubricant is actually touching the gears. It is not until heat generated from the under-lubricated gears melts the stiffened lube that it starts to flow and do its job. In severe situations, especially with hypoid axle gearing, scoring can start almost immediately. With transmission lubes, a 50-weight oil has a channeling temperature of 0°F. In fleet operations where trucks are parked in cold temperatures, this could cause transmission gears to lack proper lubrication for the first couple of miles of operation.

Extreme pressure (EP) additives were developed to meet the increased stresses of hypoid axle gearing. However, some earlier EP lubes broke down under high temperatures often found in manual transmissions. This is one reason that many manual transmission manufacturers specify a 90-weight straight mineral oil or SAE 50 engine oil.

Axle and transmission manufacturers, in general, prefer the use of a single-viscosity lubricant consistent with high ambient temperature operating environment. In all cases, consult specific recommendations before making any changes. As with engine oils, gear lubricants (GL) have their own classification system starting with GL-1 and progressing through GL-5 (**Table 4–6**).

Synthetic lubricants for transmissions and axles introduced in recent years are solving many of the problems just discussed, and offer greatly extended oil life. Synthetics do not contain the readily oxidized ingredients found in mineral-based lubes, so they have much better oxidation stability, can handle higher operating temperatures, and can double or even triple oil change intervals.

Channeling temperatures are as low as −65°F. They can be used in both manual transmissions and axles and are generally compatible with mineral oil-based lubes and with seals. In addition, they help to improve fuel economy (1 to 3 percent) by reducing gear drag in colder temperatures. Most fleets using synthetics in transmissions report easier gear shifting in cold weather, too. Synthetic gear lubes cost three times as much or more than petroleum-based products.

Automatic transmission fluids (ATF) have various specs that comply with automatic transmission manufacturer's requirements. ATFs are also available as synthetic gear lubricants with longer service intervals and superior cold/hot temperature operations. Most synthetics are compatible with conventional ATFs (see Chapter 14).

CHASSIS LUBRICANTS

Chassis lubricants are usually greases, although several on-board automatic lube systems require gear oil or liquified greases. These systems are covered later in this chapter.

TABLE 4–6:	CLASSIFICATION OF LUBRICANTS
Classification	Description
GL-1	A straight mineral oil not containing EP (extreme pressure) additives is used today in some transmission applications.
GL-2	Semi-obsolete today, it was originally developed for worm gear differentials and contained mild EP agents. Today's EP additives contain borates, molybdenums, and graphites that furnish three to five times thicker surface film than conventional lubes.
GL-3	Originally developed for spiral bevel gear lubricant requirements, it is not for use with hypoid gearing.
GL-4	An early lube spec for hypoid gears, but is now obsolete.
GL-5	Hypoid gearing, severe service. This is the top gear lube spec today. MIL-L-2105D (a common military spec) is closely related to GL-5.
GL-6	No longer in use.

In 1988, the SAE identified specific needs for two additional gear lubricant categories as follows:

PG-1	Lubricant spec for truck manual transmission with high oil temperature (over 300°F). It must also have synchronizer and oil seal compatibility.
PG-2	Heavy-duty gear lubricant with specifications to meet improved gear spalling resistance standards, better oil seal compatibility, and improved thermal stability/component cleanliness. It must also have improved corrosion protection with copper alloys under both wet and dry conditions and it must have improved gear score protection over MIL-L-2105D spec.

Grease is formulated to lubricate just as oil would. But since grease remains in place, it is not expected to perform the cooling and cleansing functions normally associated with oil. Chemists call the oil-holding gel component a "soap," and it can be made from calcium, sodium, lithium or other elements and compounds. Each soap provides its particular characteristics to the grease product. Qualities like appearance, stability, pumpability, and heat and water resistance depend largely on the type of soap base used in a grease.

In addition to soap and liquid lubricant, nearly all greases have additive packages that bring additional properties such as corrosion resistance, adhesiveness, and extreme pressure (EP) capability. Sulfur, phosphorus, zinc, molybdenum disulfide, and graphite are some common additives. The most frequently measured property of grease is consistency. This can be compared to viscosity in engine oil.

The consistency of a grease, a measure of its relative hardness, is commonly expressed in terms of the ASTM penetration or NLGI (National Lubricating Grease Institute) consistency number (**Table 4–7**). The penetration is the depth of penetration of a grease sample, in tenths of a millimeter, by a standard test done under stated conditions (ASTM D 217). The greater that penetration, the softer the grease. The corresponding NLGI grades range from a semifluid 000 ("triple aught") to a very hard 6. The latter grade has been described as "almost hard enough to walk on." While some of the tests and ratings are of little interest to most truck technicians, some technicians might be interested in how the ingredients of a grease perform in service (**Table 4–8**).

Of importance to the technician, however, is the cleanliness and good housekeeping. Dust and dirt will not settle to the bottom of a grease reservoir, as they would in oil. If airborne contaminants are allowed to enter the grease, they will remain in suspension, acting like a lapping compound on metal surfaces or possibly blocking a centralized-system pump or injector.

Greases are not all the same, so avoid mixing them if at all possible. Refill a reservoir only with a grease using the same thickener, or soap, such as calcium, lithium, or aluminum. Sodium-soap greases are water soluble and are not generally recommended for heavy-duty truck applications.

ON-BOARD CHASSIS LUBRICATING SYSTEM

One of the greatest boons to PM chassis lubricating is the on-board system. These systems are manually or automatically operated. Manual systems are used on vehicles such as trailers that have no electrical power to operate the lubricant pump used on automatic systems.

Manual Systems

A manual system consists of a manifold, or distribution block, with twelve to twenty-four grease lines connected to critical lube points. To lube the chassis, a grease gun is connected to one central zerk fitting mounted on the block, through which grease is routed to all connected points.

The disadvantage of the manual system, compared to automatics, is that a technician and a grease gun are still needed. In some installations, a manual unit can be coupled to an automatic unit on the tractor using a quick-connect coupling. Thus, the manual system becomes an extension of the automatic system.

When performing any manual lubrication job, the following must be done:

- All bearings, pins, and bushings must be lubed. Overlooking a zerk fitting because it is hard to reach, or for any other reason, will cause damage and wear.
- Clean zerk nipples before greasing to avoid contamination.
- Wipe off excess grease to avoid attracting dust and grit in and around the lubrication point.
- Over-greasing can be as damaging as under-greasing. Some maintenance personnel judge adequate lubrication by the amount of lubricant

TABLE 4–7:	NLGI GRADES AND METHOD OF APPLICATION
NLGI Grade No.	**Penetration Worked, 77°F. Typical Methods of Application**
000	445–475 Semifluid; used in centralized systems.
00	400–430 Semifluid; used in centralized systems.
0	355–385 Used in centralized systems.
1	310–340 Used in grease guns, centralized systems.
2	265–295 Used in grease guns, centralized systems.
3	220–250 Used in grease guns.
4	175–205 Used in grease guns.
5	130–160 Used in grease guns.
6	85–115 Block grease, very hard; for open grease cellars.

Type Grease Base	Usual Appearance	Mechanical Stability	Cold Weather Pumpability	Heat Resistance	High Temp Life	Water Resistance	Compatibility with Other Greases
Calcium (lime soap)	Buttery	Good	Fair	Fair	NA	Excellent	Excellent
Sodium (soda soap)	Fibrous or spongy	Fair	Poor	Good to excellent	Fair	Poor	Good
Barium	Fibrous	Good	Poor	Excellent	Fair	Excellent	Fair
Lithium (12 Hydroxy)	Buttery	Excellent	Good to excellent	Good to excellent	Good	Excellent	Excellent
Lithium Complex	Buttery	Excellent	Good to excellent	Excellent	Excellent	Excellent	Excellent
Calcium Complex	Buttery to grainy	Good	Fair	Good	Fair	Good to excellent	Fair
Aluminum Complex	Buttery to grainy	Good to excellent	Good	Excellent	Fair	Excellent	Poor
Nonsoap (Bentone)	Buttery	Good	Excellent	Excellent	Good	Good	Poor
Urea	Buttery	Good	Good	Excellent	Excellent	Excellent	Excellent

TABLE 4–8: PROPERTIES OF GREASES

Properties of greases with various soap bases. (Courtesy of Society of Triboloists and Lubrication Engineers)

oozing from around the bearing. However, grease guns exert tremendous pressure and oozing may mean that a bearing seal, if used, has blown out. When this happens, the lubricant will be contaminated and the part will be destroyed.

- Complete a lubrication checklist to help ensure that no lubrication points are missed.

Automatic Chassis Lube Systems (ACLS)

An ACLS mounted on a moving vehicle can deliver a periodic, uniform shot of grease to critical points as often as every few minutes. That helps offset incomplete or inadequate lubrication during PMs, and in some fleets, has reportedly helped kingpins and other components last through the vehicle's trade cycle.

During the typical stationary manual lube job performed as part of PM, roughly 70 to 80 percent of the lubricant applied is essentially wasted because it is not being placed where the load is when the truck is in motion. This is no fault of the technician, but simply a fact of physics.

Another advantage to the mobile system versus stationary lubing is that the continuous addition of fresh lubricant tends to push or flush out the old lubricant in which the oil or other lubricating substance has essentially come out.

To distribute grease, on-board systems use the same basic components:

- A reservoir holds the supply of grease to be distributed (**Figure 4–9**).
- A pump delivers the lubricant through a network of grease lines. Metering valves dispense the grease.
- An automatic timer mechanism tells the pump-and-valve system when to pump grease.
- An electrical motor or other power source (the truck's air system is the most widely used)

In air-driven systems (**Figure 4–10**), progressive feeders, piston distributors, or metering valves at the end of the dispensing lines are strategically located at the chassis' multiple grease points; major suspension points; shackle pins; brake camshafts; slack adjusters; clutch releases; steering fittings (tie-rod ends, drag links); kingpins, and the fifth wheel (pivot points, fifth wheel plate, slider rails). The pump in air-driven systems is activated by a solenoid valve.

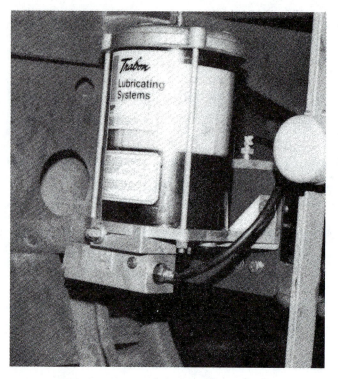

FIGURE 4–9   *Grease reservoirs are usually mounted conveniently to the frame where they may be viewed and filled at PM intervals.*

Air or electric motor-driven systems use a multi-outlet pump and no secondary distribution parts. The grease is metered directly at the pump. A third type of system, using a single outlet, electric-driven gear pump, must use a secondary system of metering valves, piston distributors or progressive feeders, like the truck air system-driven setup. Most U-joints must still be lubed manually. This creates a convenient maintenance interval for the system itself. While technicians are manually lubing the U-joints, they can fill the automatic system's reservoir, which should last for 2 to 4 months, and at the same time check system lines to ensure they are in place and functioning properly.

One system can be configured to lubricate more than 32 points, including steering kingpins, spring pins, brake "S"-cams, brake slack adjusters, the clutch, shift shaft linkage, and fifth wheel plates. The unit's integrated pump, powered by the truck's air supply, feeds grease through a main line to modules that serve various groups of lube points (left and right front, rear axles, fifth wheel, suspension). The modules are fitted with different-sized meters that feed varying amounts of grease to individual points through distribution lines. (Non-module systems require a separate line for each individual lube point.)

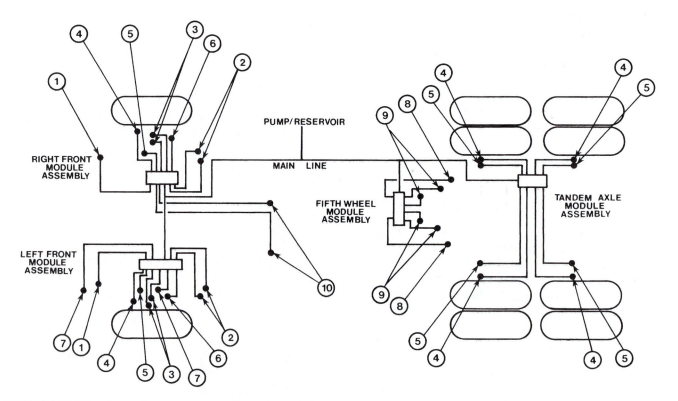

FIGURE 4–10   *Layout of an automatic on-board chassis lubrication system. Key to number for Class 8 vehicle lube points: 1. Front Spring Pin; 2. Front Spring Shackles (upper and lower); 3. Kingpin (upper and lower); 4. Brake cam shaft; 5. Brake slack adjuster; 6. Tie rod; 7. Drag link; 8. Fifth wheel pivot; 9. Fifth wheel plate; 10. Clutch cross shaft. (Courtesy of Lubriquip, Inc.)*

A 12- or 24-volt DC solenoid controller on this particular system, working in conjunction with the pump, initiates lube cycles at regular intervals. Adjustment of a simple knob on the face of a small, solid-state timer programs the controller to trigger a lube cycle anywhere from once every 6 hours to as often as every half-hour. The timer's built-in memory retains elapsed cycle time when the ignition is switched off, and resumes when it is switched back on. The grease reservoir holds 10 pounds of lubricant, or enough to last a truck in an average linehaul operation from 14 to 16 weeks.

The routing and securing of the grease lines from the pumping system to the individual lube points are critical (**Figure 4–11**). Check that the lines are not loose and flopping around; route them away from areas where the line could be pinched or cut. Also include the system in PM checks. Technicians should cycle the automatic lube system each time the tractor is brought to the shop. The objective is to verify that the solenoids are working and to check if the proper amount of grease is being pumped into a component. If not, the controller time cycle can be adjusted. Being able to adjust individual grease amounts is important because some components require more than others.

Both manual and automatic chassis lubricating systems are, in simplest terms, hydraulic systems for pumping grease. Some automatic systems use standard shop greases rated #1 or #2 by NLGI; others require liquid greases with NLGI ratings of 000 and 00. These greases tend to stay fluid over a wide range of operating temperatures. One manufacturer reports that, using these two ratings of grease, its systems are dependable in temperatures ranging from 0 to 70 degrees Fahrenheit.

Automatic chassis lube systems can be added to older trucks that do not have it. When retrofitting a system, there are several installation tips to keep in mind:

- In tilt cab vehicles, do not tilt the cab any further than the truck builder recommends.
- Metering valves, which distribute lubricant, should be placed alongside the vehicle, and if on the ground, away from any dirt or other contamination. All lines or tubing should be fixed a sufficient distance from any heat sources (exhaust components, for example).
- Clip/trim all tubing well to avoid chafing and abrasion.
- Take care when drilling to avoid damage to the compressed air or electrical lines.
- It is important to make the system a part of the truck, but separate from the air and brake lines, and chassis wiring.

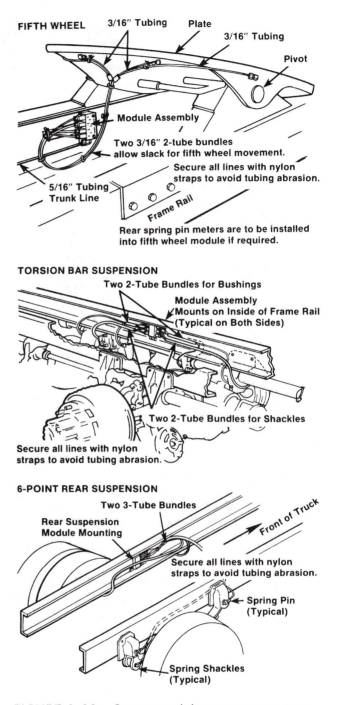

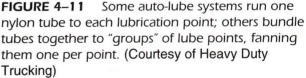

FIGURE 4–11 _Some auto-lube systems run one nylon tube to each lubrication point; others bundle tubes together to "groups" of lube points, fanning them one per point. (Courtesy of Heavy Duty Trucking)_

TRAILER LUBRICATION

Table 4–9 is a typical PM trailer lubrication chart recommended by most trailer manufacturers.

	TABLE 4-9: TRAILER LUBRICATION CHART		
Item	Check or Service	Frequency	Lubricant or Equivalent
Brake camshaft support bearings	Lubricate	3 months	NLGI No. 2 EP grease
Brake slack adjuster	Lubricate	3 months	NLGI No. 2 EP grease
Electrical connections	Add to void areas	3 months	Dielectric grease
Fifth wheel base and ramps	Lubricate	Weekly	Waterproof general purpose grease
Landing gear with grease fittings	Lubricate	6 months	NLGI No. 2 EP grease
Oil type wheel bearings	Fill to full level	Daily	SAE 90 weight oil (API-GL-5)
Grease type wheel bearings	Clean and repack	6 months	NLGI No. 2 EP wheel bearing grease
Sliding suspension: linkages of locking mechanism	Lubricate	3 months	SAE 20 weight oil
Door and ventilator hinge pins	Lubricate	3 months	SAE 20 weight oil
Door handle linkage	Lubricate	3 months	SAE 20 weight oil
Roll up door rollers	Lubricate	3 months	SAE 20 weight oil
Door lock rod retaining points with grease fittings	Lubricate	3 months	NLGI No. 2 EP grease
Miscellaneous moving parts and linkages	Lubricate	3 months or less, check to be sure	SAE 20 weight oil

4.5 WINTERIZING

Winterizing a heavy-duty truck is a very important part of any PM program. The following cold weather (+40° to −40°F) operation maintenance checklist must be integrated into the PM schedule.

ENGINE COOLANT

The cooling system is one of the most important parts of the engine that must be properly winterized to support cold weather operations. Maintenance must be scheduled in mid-fall prior to the onset of cold weather. The outline of the maintenance schedule given here and usually performed by the truck technician should proceed in sequential order.

- **Inspection.** A visual check of the entire cooling system (for example, radiator, water pump, hoses, and so on). Look for leaks or dried inhibitor deposits (gel or crystals)—all heater and water hoses should be sound, soft and pliable. If they are hard, cracked or brittle, replace them. Check all clamps to be sure that they have not lost their hold on the hoses. Do any other necessary repairs or adjustments. Pressure-test the cooling system at its specified system pressure after turning on heat control valves. Repair any leaks. Also pressure-test the radiator cap at its rated pressure. Best

results from these tests will be obtained when the engine is cold.

- **Clean the Cooling System.** If inspection of the coolant reveals rust or cloudy appearance or if deposits have formed on the top header of the radiator, clean the cooling system. Also clean the cooling system if truck history indicates a chronic history of cooling system complaints. Follow the recommended cleaning procedure to remove rust, corrosion by-products, solder bloom, scale, and other deposits. Cleaning opens coolant passages and removes insulating scale and deposits that will cause overheating and possible engine block damage.

- **Antifreeze.** The fluid used as coolant in heavy-duty vehicles is a mixture of water and usually ethylene (EG)- or propylene-glycol (PG)-based antifreeze coolant. The normal mixture used is a 50/50 solution of water and antifreeze/coolant, which protects against freezing down to −34°F. A 60 percent mixture can be used for additional protection, but never higher. Once the level of EG antifreeze exceeds 68 percent, the freeze point quickly rises again. In fact, a 100 percent solution protects only to zero. Running at 80 percent antifreeze can reduce heat transfer causing ring wear, bearing problems, and lube oil breakdown. Over 60 percent antifreeze solutions also encourage formation of silicate drop-out. Ensure antifreeze is low silicate and meets engine manufacturer's specifications. High silicate antifreezes used with a supplemental coolant conditioner can cause a buildup of solids over a period of time, which can cause plugging, loss of heat transfer, and water pump seal damage.

The quality of water used in premixing is of utmost importance. Water that carries a high level of grit or sand should not be used. Also, do not use water that is hard or high in mineral content. Hardness should not exceed 300 parts per million (ppm) nor should chloride and sulfate levels exceed 100 ppm. Most engine manufacturers recommend distilled or deionized water. To be on the safe side, have the water tested prior to premixing.

Checking Condition of Coolant

There are several ways to check the condition of the coolant. The most common ones are as follows:

- **Hydrometers.** A **coolant hydrometer** has one purpose—indicating the freeze point. It does this by measuring the specific gravity of a sample—that is, comparing its weight with the weight of water. A traditional hydrometer contains a single graduated float. Since antifreeze is heavier than water, a high concentration will cause the float to ride high in the sample. With a "floating ball" hydrometer, anywhere from none to all of the balls may float. To find the freeze point, read the temperature graduation at the single float's "water line."

- **Refractometers.** Refractometers test the refractive index of coolant. They do this by measuring the angle at which light bends (refracts) when passing through a liquid droplet. The refractometer compares a known angle for pure water with the angle produced as light passes through a sample. Freeze point is noted on a scale viewed through an eyepiece.

- **Litmus/Chemical Tests.** The basis for litmus tests are chemical reactions affecting the color of a strip of test paper. Often, the sample must be mixed with plain water and/or a chemical reactant before the litmus strip is inserted into it. After the paper has been moistened, comparing its color with a chart indicates the acidity of the solution. Litmus tests measure nitrate concentration and pH. Coolant conditioner protects metal parts by coating them with a chemical barrier against corrosion and cavitation. Nitrates are a major component of that barrier. Measuring nitrate concentration, therefore, gives an indication of the conditioner charge level.

- **Lab Sample Analysis.** Laboratory analysis provides the most information about coolant, at a moderate cost. For example, coolant may be analyzed for the freeze point, silicate concentration, nitrate level, and reserve alkalinity. If these tests indicate a need, or if it is requested, the sample also may be tested for traces of metals such as iron, copper, aluminum and solder (lead and tin). Lab testing gives a thorough picture of the cooling system condition.

- **Supplemental Coolant Additive (SCA).** Use of a SCA is important to proper engine and cooling system protection. If the system has been cleaned, a precharge of SCA may have to be added per manufacturer's recommendations and coolant volume. Otherwise, test coolant for proper inhibition either by test strip or titration method. Do the necessary adjustments to coolant if inhibitor levels are under/over proper concentrations. Follow manufacturer's recommendations. Reinhibit the cooling system with a maintenance charge following the manufacturer's recommendations based on mileage or hours of operation.

Disposal of Antifreeze

The main ingredient in most antifreeze is ethylene glycol, a chemical that is poisonous if taken internally.

Antifreeze is biodegradable and is classified by the EPA as a substance of "unknown" toxicity. The EPA has classified used antifreeze as "toxic waste" due to the amount of heavy metals it can absorb from the engine's cooling system. Lead from soldered radiator cores is the metal that poses the most danger to public drinking water and the environment. Laws have been passed in many areas regulating the disposal of used antifreeze. To meet this environmental challenge, a number of aftermarket companies have introduced antifreeze recycling equipment that allows old antifreeze to be cleaned and replenished, minimizing the disposal problem. The only other alternative in regulated areas is to pay someone to dispose of the antifreeze.

AIR SYSTEM

The air system must receive special attention during cold weather operation. Contaminants such as road dirt, salt, oil and so on must be removed. They must be protected against freezing. Drain the air tanks daily and inspect the system carefully to determine if any repairs are necessary. The inspection should include the following:

1. Remove and clean the air dryer purge line and air governor signal line.
2. Check for the proper vent on the exhaust port if there is an air governor.
3. Check the air dryer heater operation; a 4 to 5 amp draw with the element cold is normal.

There are several methods offering protection to the air brake system: alcohol injectors, which release vaporized methyl alcohol into the system to prevent freezing; both automatic and manual moisture ejectors mounted at the reservoir; and air dryers mounted between the compressor and air reservoir.

STARTING AIDS

There are severaldifferent types of starting aids available. But to be of any value, the starting aids must be performing perfectly.

1. Check the condition of the block heater with an ohmmeter. A 10-to 15-ohm element resistance is normal for 1000- to 1500-watt heaters. Check the condition of the electrical cord and plug. Check for coolant leakage at the mounting points.
2. Check the fuel heater mounting, condition of hoses, and hose routing. Heaters with manual shut-off valves should be turned on before cold weather to ensure proper operation.
3. Check the water separator mounting, condition of hoses, and hose routing. Drain any collected water.
4. Check the ether starting system by removing the cylinder and activating the valve. Weigh the cylinder for contents. Replace it if it is low.
5. For oil pan heaters and battery warmers, check the element resistance with an ohmmeter. (Oil pan heaters have a normal resistance of about 50 ohms, battery warmers 75 to 125 ohms.) Check the condition of the electrical cord and plug.
6. Fuel-fired heaters normally are ready to fire up. Run the unit through at least two thermostatically controlled cycles. Check the hoses for proper routing. Replace the glow plug or ignitor as required. Consult the manufacturer's recommendations for any other tests.

ETHER STARTING SYSTEMS

Diesel fuel will not ignite in an engine until it reaches its fire-point temperature. The heat generated by compressing air in the cylinder of a diesel engine does not reach this temperature if the ambient temperature is below about 40°F because the temperature is directly proportional to the inlet air temperature. Thus, the use of ether-based starting fluid aids in getting the engine started since it will ignite at about 350°F, creating a temperature high enough to ignite the diesel fuel when it is injected.

Typically, starting fluid systems vary from the hand-held aerosol can (total operator control) to the "automatic" and electronic systems that are wired in with the engine cranking or electronic engine control circuits (no operator control).

A spray aerosol can dispenses about 12 cubic centimeters of starting fluid per second. An operator normally will put in at least 2 to 3 seconds of spray or 35 cubic centimeters—six times the amount of ether required. This is the reason starting fluid or "ether" causes damage—improper or excessive usage that causes "pinging," ether-lock, and ultimately, engine damage. The measured shot or metering flow valve limits the amount of ether per activation that can be injected into the engine and is the type of system that is recommended by most diesel engine manufacturers. Measured shot valves are made in three or four shot sizes to match engine displacements. This valve can be remotely operated by a switch on the dashboard. A thermal switch can be wired in series with the ether valve to prevent ether from being injected into a hot engine.

In operation, when the valve is energized (pushing the button), the starting fluid is forced into the measured-shot chamber. De-energizing the valve (releasing the button) dispenses the measured shot of ether into the engine through the atomizer. Push-pull cable operating systems that operate in the same manner are available; pulling the cable charges the shot chamber and pushing the cable dispenses the ether to the engine.

The "automatic" ether systems currently available are wired into the cranking circuit and dispense starting fluid as long as the cranking circuit is energized (**Figure 4–12**). Flow of the ether is controlled by a special valve orifice or rapid cycling of the valve, resulting in a flow from the atomizer that is basically continuous. An engine temperature switch must be used to prevent operation when the engine is warm.

The atomizer or nozzle is the device that directs the flow of starting fluid into the airstream and determines the rate at which it is injected into the airstream. The recommended location for the atomizer is in the intake air system, upstream of, and as close to the inlet of the intake manifold as possible. The direction of spray should be towards the incoming airstream in order to obtain maximum mixing with the air. The valves and atomizers generally are connected with nylon or metal tubing of $1/8$-inch diameter.

The electronic system operates in basically the same manner as the automatic, except the ether is totally controlled, measured, and metered by the electronic black box (**Figure 4–13**). The electronic unit collects engine temperature data and if the engine starter is engaged, it supplies ether only when

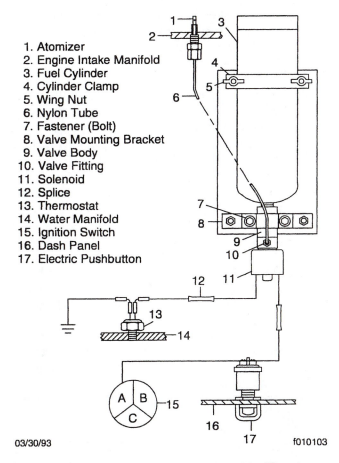

1. Atomizer
2. Engine Intake Manifold
3. Fuel Cylinder
4. Cylinder Clamp
5. Wing Nut
6. Nylon Tube
7. Fastener (Bolt)
8. Valve Mounting Bracket
9. Valve Body
10. Valve Fitting
11. Solenoid
12. Splice
13. Thermostat
14. Water Manifold
15. Ignition Switch
16. Dash Panel
17. Electric Pushbutton
A
B
C
03/30/93
f010103

FIGURE 4–12 Typical ether starting kit. (Courtesy of Freightliner Trucks)

Ether Cylinder
Ether Control Module
J1922
J1939
Engine Control Module
Engine Cylinders
Metering Valve
Air to Air Charge Air Cooler
Turbocharger

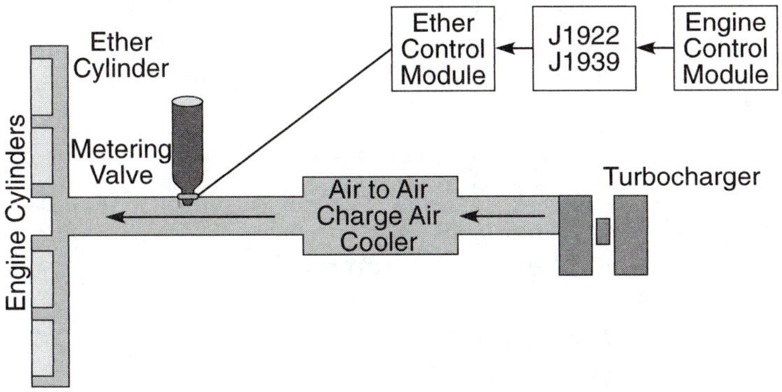

FIGURE 4-13 Black box of the electronic ether system. (Courtesy of Phillips Temro, Inc.)

needed. Most controls have a memory function that immediately alerts the operator when the fluid is low. Retrofit kits are available to change the automatic system to electronic.

WARNING: Never use ether with glow plug-equipped engines. On engines with an air pre-heater, consult the engine owner's manual.

With automatic and electronic systems, the atomizer container must be checked by the technician on a regular basis to be sure that there is always ether in it.

CAB COMFORTS

To keep the driver happy, check the following:

1. Clean out the heater screens and check for any heater core obstructions.
2. Check the heater fan operation using an ammeter to check for current draw. It should be no more than the rated value (see service manual).
3. Check the bunk heater operation and switch condition.
4. Check the heater controls for proper function. Operate in the defrost and floor heat modes. Cab heaters with manual controls should be turned on before cold weather to ensure proper operation.
5. Check the cab floor grommets for any holes and for sealing capability (especially around the clutch throttle, wiring, hoses, and so on).
6. On COE vehicles, make sure the transmission tower seals with the cab down. Replace if it is not sealing properly against the outside air.
7. Check the weatherstripping around the doors and for doors not adjusted properly.

BATTERIES AND ELECTRIC SYSTEM

Night driving during winter months can overload a heavy-duty truck's batteries and electric system (**Table 4–10**). The reserve capacity rating (RCR) is important if long term parking with electrical loads are common to your operation. A battery at 80°F has 100 percent of its power. That same battery at 14°F has only 50 percent power available.

An engine at −20°F is over three times more difficult to crank as an engine at 80°F.

Battery heaters can help solve the problem. The most common battery heaters are the blanket type and the plate type. The plate-type battery heater usually is placed under the battery, but it can also be placed between batteries. The blanket type of heater

TABLE 4–10:   TRACTOR/TRAILER POWER USAGE*

Device	Amps	Hours	Amp Hours
Key on	2.0	10.0	20.0
Starter	1,500.0	0.01	15.0
Wipers	7.5	10.0	75.0
Headlights	8.0	10.0	80.0
Panel lights	3.0	10.0	30.0
Taillights	1.5	10.0	15.0
Marker lights	12.0	10.0	120.0
Brake lights	8.0	0.25	2.0
Turn signals	5.0	0.15	.75
Engine electronics	6.25	10.0	62.5
Heater fan	15.0	10.0	150.0
Heated mirrors	5.0	10.0	150.0
CB radio	1.5	10.0	15.0
AM/FM radio	1.0	10.0	10.0
Radar detector	0.5	10.0	5.0
Subtotal			650.25
+ 25 percent Safety factor			162.56
TOTAL			812.81

*Calculated for 10-hour shift on a winter night

comes in a variety of lengths and wattages and incorporates its own insulation. It is best to heat a battery slowly. A well insulated battery box helps during the heating of the battery and also helps maintain heat.

To keep problems to a minimum in battery and electric systems, prepare them for winter as follows:

1. Look for the accumulation of dirt or corrosion on top of the battery, corroded terminals and cables, broken or loose terminal posts, and containers or covers that are broken or cracked. All affect the battery's ability to perform. Also check the condition and tension of the alternator belts. Inspect all wiring and cables for frays and breaks in insulation.

2. Avoid damaging a battery when installing or removing it. One of the top 10 carriers revealed that 35 percent of its scrapped batteries were discarded because mishandling by shop personnel damaged the case.

3. Check the voltage regulator setting at every PM inspection. Overcharging is an important cause of battery failure. For trucks operating line-haul, the voltage setting should be 13.6 to 13.8 volts. For vehicles involved in pickup and delivery, it should be no more than 14 volts. A side benefit besides longer battery life is increased lamp life. For example, a 23 percent increase in voltage can shorten bulb life by 50 percent. One of the top 10 common carriers claims to have increased battery life by one year on average by dropping voltage output from 14.4 to 13.6.

4. On conventional batteries, maintain the electrolyte level above the plates and below the bottom of the vent well. Never allow the electrolyte level to drop below the top of the plates. A low electrolyte level can cause reduced battery capacity. If water must be added, use distilled water. Overfilling can cause acid spewing and corrosion.

5. Ensure that battery hold-downs are properly attached to keep the batteries from bouncing. Remember, vibration is the biggest killer of batteries. Vibration also breaks alternators and brackets, and elongates mounting holes. Alternators should be mounted using hardened grade 5 or grade 8 (preferred) fasteners and flatwashers. Do not use lock washers.

6. Keep all terminals and connections bright and tight. Use a water-resistant grease or lubricant such as petroleum jelly to seal out moisture on connections that are not totally encapsulated.

7. Keep batteries from freezing by maintaining a full charge. This includes vehicles parked and batteries on the shelf. A completely discharged battery will freeze at 18°F.

8. When winterizing vehicles, start/charge systems should be tested in accordance with The Maintenance Council (TMC) diagnostic procedures. The following diagnostic equipment is required: a variable carbon pile load tester (500-amp minimum with ammeter); a separate digital voltmeter that is calibrated frequently; and an inductive (clamp-on) ammeter. Test areas are

- Batteries
- Cranking motor replacement test
- Cranking circuitry and battery cable test
- Solenoid circuit test
- Magnetic switch control circuit test
- Alternator wiring test
- Alternator output test

9. If using threaded posts, use an adapter on the post when charging.

10. Chassis ground wiring systems on trailers should also be bright and tight. Use an electrical grade lubricant on pigtail connections, lamp sockets, connections, battery terminals and splices. Do not use a sodium-based grease since it will emulsify, allowing corrosion to enter. For longer bulb life, clean lenses and housings frequently and keep the voltage regulated below 14.0 volts.

11. All light bulbs should burn brightly; if not, find out why. Is the alternator producing all the amps it should? How old is it, and how much life does it have left? If the vehicle history folder reveals that it is within a few thousand miles of possible failure, replace it before winter sets in.

DIESEL FUEL

In preparing for cold weather operations, fuel plays an important role. There are three areas relating to fuel that should be examined. They are fuel storage, tank configuration, and chemical properties of the fuel. The following are suggested guidelines and related terms used when dealing with the winterization of fuel and fuel systems.

- **Cloud Point.** This is the temperature at which a cloud or haze of wax crystals is formed. Cloud point is more significant than pour point when determining whether a fuel can be pumped through filters in cold weather. Cloud point has not been shown to be significantly lowered by additives.
- **Pour Point.** This is the lowest temperature at which a diesel fuel will flow when cooled. This temperature is lower than the cloud point and is easier to modify or lower.
- **Fuel Storage.** Storage tanks can form condensation. Checks for water need to be diligent and any water found in the system must be removed. Positive water removal filters need to be installed on fuel dispensing pumps. Microbiological contamination can also be a problem. These living organisms produce slime-like metabolic waste that can plug fuel filters. Fill areas should be raised to eliminate the possibility of water, snow, or debris entering through these areas. Vents should be checked to ensure only air can enter them.
- **Tank Configuration.** Vehicle fuel tanks should be checked for water contamination and any water or debris removed if found. Crossover lines should be checked for both correct size and low spots that can collect water. Any exposed fuel lines can cause freeze-ups during severely cold temperatures.
- **Fuel.** While winter diesel fuel is prepared by refineries on a regional basis, it should be recognized that all fuel is not the same. ASTM specifications for winter fuel are based on a 10 percent mean low temperature for that month only. The fuel refiners therefore adjust the ignition temperature of fuel (Cetane Number/CN) seasonally. The use of aftermarket fuel additives by the user is not generally recommended.

Some operators use fuel heaters. The two most practical heat sources for fuel heating during highway operation are electricity and engine coolant.

The problems caused when wax or ice crystals form at the fuel tank top or bottom suction elbow can be solved by an in-tank heater capable of bringing the fuel's temperature above its cloud point. The fuel temperature should be regulated below 80°F and never exceed 100°F.

FUEL-WATER SEPARATORS

There are many types and sizes of fuel/water separators available on the market. To select the proper unit, it is important to know the requirements in terms of fuel flow rate, usually expressed in gallons per hour, that will be flowing through the fuel/water separator. There are three basic types to choose from: regular fine mesh filters, coalescer screens, and centrifugal fuel/water separators.

When using fuel/water separators during cold weather operation, the following must be considered:

1. Mount the fuel/water separator in a protected area. Some fleets will wrap them with insulation for cold weather operation.
2. If a diesel fuel warmer is used in conjunction with a fuel/water separator, mount the separator downstream from the fuel warmer so it receives warm fuel.
3. Do not use 90-degree elbow fittings; these can act as freeze points in cold weather.
4. Drain the fuel/water separator frequently so there is minimum water in the separator bowl. This water can freeze in operation, especially when exposed to ram air wind chills. Ice crystals can plug a fuel filter just like wax crystals.

GENERAL WINTERIZING TIPS

The following are some other checks that must be looked at while winterizing a vehicle:

1. Carefully check all engine and chassis hoses (fuel, oil, and air) for wear and deterioration. Replace as necessary.
2. Check the operation of the windshield washers and ensure that the wipers remove solution without streaking.
3. Check the windshield washer reservoir for proper type and amount of solution.
4. Check the exhaust system for leakage, for example, soot around connections.

5. Clean and grease the electrical connections to the headlights, taillights, light cord receptacle, and clearance lights.

6. Check the fan blade and fan belt condition and tension.

7. Check the fan clutch for play with the fan clutch released. Play should not be more than $3/16$ inch at fan blade tip, maximum.

8. Lubricate the door latches with a Teflon™ lubricant.

9. Lubricate the door locks with dry graphite or silicone lubricant.

10. Be sure that the steering gear is sound and tight. Make certain the front end and rear axles are aligned. Check for any leaks from power steering components. In the case of manual steering, check boxes for proper lubricant and level. Loosen the bottom plate and strain out any water.

11. Inspect, lube, and adjust the clutch. Check the transmission and adjust or repair it if required. Inspect, lube, and make any needed adjustments in the driveshaft and axle. Check fluid levels in transmission and differentials. If a multiviscosity oil and grease is not used, a switch to a lower-weight lube (or a synthetic lube) may be required.

12. Check the fifth wheel adjustment. Also check to be sure the fifth wheel jaws open and close easily, the sliders work smoothly, that a winter grade lube is in the landing gearbox, and that the trailer sliders operate smoothly.

13. Inspect the condition of the tires and wheels. Replace any damaged or worn tires. It may be shop policy to install lug-type treads on drive wheels to enhance traction on snow and ice. Air up all tires to proper pressures. Look closely at the wheels and rims for bends and cracks, and at lug nuts for signs of looseness or breakage. Repair any problems.

14. Make sure the foundation brakes and the brake cams and slack adjusters are properly lubed. Check brake shoes, linings and drums for wear, cracks, and breakage. Are all hoses and valves in good working condition? Up behind the cab and at the trailer nose, check the gladhands and grommets. Make sure air tanks are securely mounted and that air compressors and air dryers are doing their jobs. All necessary repairs should be made before winter sets in.

15. Be sure the fire extinguishers are charged, safety triangles are available, and tire chains are in good condition.

16. Run the truck outside until it reaches operating temperature. Check the operation of the cooling fan and check the temperature of air blowing out of the heater ducts. It should be 110°F or more.

17. Road test the vehicle and check for leaks (coolant, oil, and air).

SUMMARY

- A preventive maintenance program involves the inspection and servicing of the vehicle as a whole. The program's success depends on careful planning.
- No one maintenance program applies to all operations.
- The program that is most effective is tailored around the needs and experiences of each individual operation.
- It is the driver who identifies many repairs when he or she performs both a pre-trip and a post-trip inspection.
- It is necessary to balance the possibility that the mechanical problem will cause an accident or breakdown against the need to get the load moved.
- A vehicle considered likely to cause an accident or breakdown because of mechanical conditions or improper loading should be taken out of service.
- It is the fleet's responsibility to ensure that all personnel doing annual inspections are qualified.
- Maintenance managers and truck technicians are increasingly being expected to be computer literate.
- Proper lubrication is important in reducing wear and preventing premature failure of expensive truck components.
- Winterizing a heavy-duty truck is an important part of any PM program.

REVIEW QUESTIONS

1. The pretrip and post-trip inspections are a primary responsibility of the
   a. technician
   b. maintenance manager
   c. driver
   d. vehicle inspector

2. In carrying out a pre-trip inspection, which of the following is part of the inspection procedure?
   a. checking under the vehicle for signs of leaks
   b. starting the engine and listening for unusual noises

c. checking to make sure all necessary emergency equipment is on board

d. all of the above

3. A deadlined vehicle is best described as one that
   a. has broken down on the highway
   b. has been taken out of service because it is likely to cause an accident or break down
   c. has not been properly inspected
   d. has been allowed by the inspector to continue on its way to a repair shop within a 25-mile radius of the inspection site

4. Which type of PM service requires the most hours to perform?
   a. category A
   b. category B
   c. category C
   d. If properly performed, A, B, and C preventive maintenance inspections take about a day.

5. When carrying out an inspection of the trailer after the first 500 miles, Technician A adjusts the brakes. When carrying out this inspection, Technician B checks all areas where sealant is used and reseals as necessary. Who is correct?
   a. Technician A
   b. Technician B
   c. both A and B
   d. neither A nor B

6. How many separate inspections are required for a tractor-double combination with a converter dolly?
   a. two   c. four
   b. three   d. five

7. Which of the following vehicles is exempted from record-keeping requirements regarding the inspection schedule?
   a. state-owned tractor/trailers
   b. brand-new vehicles
   c. buses
   d. None of the above

8. A CF-2 engine oil is used in
   a. heavy-duty diesel engines
   b. moderate-duty diesel and gasoline engines
   c. light-duty trucks with gasoline engines
   d. two-stroke-cycle engines

9. Which of the following is affected by an oil's viscosity?
   a. resistance to flow
   b. average temperature in normal service
   c. ability to adjust to varying temperatures
   d. resistance to sulfur contamination

10. High TAN values in an oil can be caused by

a. formation of deposits
b. corrosion
c. burning high sulfur fuel
d. thickening

11. Technician A says synthetic engine oil helps transmission and axle gears, bearings, and seals to last longer. Technician B says it can be used throughout the service life of the vehicle. Who is correct?
    a. Technician A
    b. Technician B
    c. both A and B
    d. neither A nor B

12. Which kind of oil additive neutralizes harmful acids?
    a. oxidation inhibitors
    b. anti-rust agents
    c. detergent additives
    d. anti-corrosion agents

13. Which of the following grades of grease can be used in a grease gun?
    a. 000
    b. 00
    c. 0
    d. None of the above

14. What damage is caused by over-greasing?
    a. channeling
    b. blowing out seals
    c. scoring
    d. grit contamination

15. Which of the following is NOT part of an automatic chassis lube system?
    a. manifold, or distribution block
    b. reservoir
    c. pump
    d. grease lines

16. What is true of both manual and automatic chassis lubricating systems?
    a. They waste 70–80 percent of the lubricant.
    b. They require a technician and a grease gun.
    c. They eliminate A inspections.
    d. They continually add fresh lubricant.

17. SAE 20 weight oil is used to lubricate what part of a trailer every 3 months?
    a. brake camshaft support bearings
    b. fifth wheel base and ramps
    c. brake slack adjuster
    d. roll up door rollers

18. In winterizing a truck, which procedure of the following comes first?
    a. checking the pH of the coolant
    b. visually inspecting the cooling system
    c. adding antifreeze
    d. testing the antifreeze protection of the coolant

19. Power loss of batteries during winter driving is helped by
    a. battery heaters
    b. overcharging
    c. slight overfilling with electrolyte
    d. not maintaining a full charge

20. In winterizing a vehicle, Technician A uses a Teflon™ lubricant on the door latches and a dry graphite lubricant on the door locks and Technician B mounts the fuel/water separator downstream from the fuel warmer. Who is correct?
    a. Technician A
    b. Technician B
    c. both A and B
    d. neither A nor B

5

Fundamentals of Electricity

Objectives

After studying this chapter, you should be able to

- Define the terms *electricity* and *electronics*.
- Describe atomic structure.
- Outline how some of the chemical and electrical properties of atoms are defined by the number of electrons in their outer shells.
- Outline the properties of conductors, insulators, and semiconductors.
- Describe the characteristics of *static electricity*.
- Define what is meant by the *conventional* and *electron theories* of current flow.
- Describe the characteristics of magnetism and the relationship between electricity and magnetism.
- Describe how electromagnetic field strength is measured in common electromagnetic devices.
- Define what is meant by an electrical circuit and the terms *voltage, resistance,* and *current flow*.
- Outline the components required to construct a typical electrical circuit.
- Perform electrical circuit calculations using Ohm's Law.
- Identify the characteristics of DC and AC.
- Describe some methods of generating a current flow in an electrical circuit.
- Describe and apply Kirchhoff's First and Second Laws.
- Define the role of a battery in a vehicle electrical system.
- Outline the construction of standard, maintenance-free, and gelled electrolyte batteries.
- Describe the chemical action within the battery during the charging and discharging cycles.

Key Term List

ampere
ampere-hour rating
anode
Battery Council International (BCI)
cathode
circuit breakers
closed circuit
cold cranking amperes (CCA)
conductor
conventional theory
current flow
deep cycling
electricity
electromotive force (EMF)

electronics
electrons
electron theory
gel cell
ground
hydrometer
insulator
ion
lead-acid battery
ohms
Ohm's Law
open circuit
parallel circuit
recombinant battery
refractometer

reserve capacity rating
resistance
semiconductors
series circuit
series-parallel circuits
short circuit
solenoids
specific gravity

static electricity
step-up transformer
step-down transformers
sulfation
transformer
valence
voltage
watt

This chapter will introduce basic electrical principles. These principles will be applied in later chapters that deal with chassis electrical components. Any vehicle technician must have a good understanding of basic electricity, as almost every sub-component on a modern truck is managed, monitored, or both managed and monitored by electrical and electronic devices.

By definition, electricity is a form of energy that results from charged particles such as **electrons** and protons. These electrically charged particles can be static (at rest), when they would be described as accumulated charge. They may also be dynamic or moving, such as **current flow** in an electrical circuit.

All matter (anything that has mass and occupies space) has some electrical properties. However, only in the last couple of hundred years have humans been able to make electricity work for them. The first type of electricity to be identified was **static electricity.** Over 2,000 years ago, the Greeks observed that amber rubbed with fur would attract lightweight objects such as feathers. The Greek word for amber (a translucent, yellowish resin that comes from fossilized trees) is *electron,* from which we get the word **electricity.** Probably the next significant step forward occurred towards the end of the sixteenth century when the English physicist William Gilbert (1544–1603) made the connection between electricity and magnetism. More than one hundred years later, Benjamin Franklin (1706–1790) proved the electrical nature of thunderstorms in his famous kite experiment. He coined the terms *positive* and *negative* and established the conventional theory of current flow in a circuit. The **conventional theory** of current flow says that current flows from positive to negative in an electrical circuit. From this point in time, progress accelerated. In 1767, Joseph Priestly established that electrical charges attract each other. He saw that this attraction of opposites decreased as the distance between the forces increased. Around 1800, Alessandro Volta invented the first battery. A battery converts chemical energy into electrical potential. Electrical potential means electrical energy. Volta's battery was the forerunner of the modern vehicle battery.

Michael Faraday (1791–1867) opened the doors of the science we now know as electromagnetism. His law of induction states that a magnetic field induces (produces) an **electromotive force** (EMF, electrical potential or **voltage**) in a moving **conductor**. A conductor is something that conducts or transmits heat or electricity. Vehicle alternators and cranking motors each use electromagnetic induction principles. Thomas Edison (1847–1931) invented the incandescent lamp (the electric light bulb) in 1879, but perhaps even more importantly he built the first central power station and distribution system in New York City in 1881. New York City's power grid provided a means of introducing electrical power into industry and the home.

Electronics is a branch of electricity. Electronics deals with the flow of electrons through solids, liquids, gases, and across vacuums. An introduction to electronics theory is provided in Chapter 6. Many electrical components in modern vehicles, such as alternators, have electronic subcircuits. Increasingly, chassis systems are managed by electronic components. It should be noted that terms like *electronics* and *electronic engine management* are often used in vehicle technology to describe any system that is managed by a computer. Many types of vehicle computers exist. Managing a function such as the pulse wiper mode on windshield wipers would be a simple use of a vehicle computer. A more complex application of a vehicle computer is monitoring and controlling the multiple subsystems of engines and transmissions. Because of the ever-increasing use of computers on modern trucks, developing an understanding of electricity and electronics is essential.

This understanding begins with learning about electrical and electronics theory, which will form the building blocks of mastering practical skills. The practical skills are required to diagnose and repair electrical and electronic systems and circuits. Understanding the basic theory of electricity and electronics before advancing to the chapters that address applications of this theory is important.

The days when many truck technicians could avoid working on an electrical circuit through an entire career are long past. A good course of electrical study will always begin with the study of atomic structure. Understanding a little about the behavior of atoms and electrons can help students visualize some of the elements of "invisible" electricity.

<table><tr><td>**5.1**</td><td></td></tr></table>

5.1 ATOMIC STRUCTURE AND ELECTRON MOVEMENT

An atom is the smallest particle of a chemical element that can take part in a chemical reaction. An atom is usually made up of protons, neutrons, and electrons. Protons and neutrons form the center or nucleus of each atom, whereas the electrons orbit the nucleus in a manner similar to that of planets orbiting the sun in our solar system. The simplest atom is hydrogen, which has a single electron orbiting a nucleus consisting of a single proton; it has no neutron. **Figure 5–1** shows how an atom of hydrogen is represented in diagram form.

Electrons orbit the nucleus, the center point of an atom, in a concentric ring known as a shell. **Figure 5–1** is not unlike a picture of a solar system, with the nucleus representing the sun and the orbiting electron representing a single planet. Electrons have a negative electrical charge. Protons and neutrons are clustered into and form the nuclei of all atoms. The electrons orbit around the nucleus in their shells. Protons have a positive electrical charge, whereas neutrons have no electrical charge. The nucleus of an atom makes up 99.9 percent of its mass. This means that most of the weight of any atom is concentrated in its nucleus. The number of protons in the nucleus is the atomic number of any given element. The sum of the neutrons and protons is the atomic mass number. In a balanced or electrically neutral atom, the nucleus is surrounded by as many electrons as there are protons.

All electrons are alike. All protons are alike. The number of protons associated with the nucleus of an atom identifies it as a specific element. Electrons have a tiny fraction of the mass of a proton. Under normal conditions, electrons are bound, that is, held in their orbital shells, to the positively charged nuclei of atoms by the attraction between opposite electrical charges.

Any atom may possess more or fewer electrons than protons. An atom with an excess or deficit of electrons retains the character of the element. Such an atom is described as negatively (an excess of electrons) or positively (a net deficit of electrons) charged and is known as an **ion**. For instance, a copper atom with a shortage or deficit of electrons could be called a positive copper ion, meaning that it would be inclined to steal electrons from other substances.

Each shell within the structure of an atom is an orbital path. The concentric shells of an atom proceed outward from the nucleus of an atom. The electrons in the shells closest to the nucleus of an atom are held most tightly. Those in the outermost shell are held more loosely and tend to be more inclined to move. As we have seen, the simplest element, *hydrogen,* has a single shell containing one electron. The most complex atoms may have seven shells. The maximum number of electrons that can occupy shells one through seven are in sequence, progressing away from the nucleus: 2 (closest to nucleus), 8, 18, 32, 50, 72, 98 (furthest from nucleus). The heaviest elements in their normal states have only the first four shells fully occupied with electrons. The outer three shells are only partially occupied. The outermost shell in any atom is known as its **valence** shell. The number of electrons in the valence shell will dictate some basic electrical and physical characteristics of an element.

The chemical properties of all the elements are defined by how their atoms' shells are occupied with electrons. Atomic elements with similar orbital structures will probably have other similarities in the way they behave. For instance, an atom of the element helium with an atomic number of 2 has a full inner shell, which also happens to be its valence or outermost shell. An atom of the element neon with an atomic number of 10 has both a full first and second shell (2 and 8). Its second shell is its valence. **Figure 5–2** shows a neon atom. Both helium and neon are relatively inert elements. Inert means that these elements are unlikely to participate in chemical reactions. Similarly, other more complex atoms that have eight electrons in their outermost shell (although this shell might not be full) will resemble neon in terms of their chemical inertness. In other words, we can say that atoms with a full valence shell are likely to have inert characteristics.

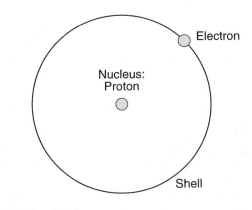

FIGURE 5–1 Hydrogen atom.

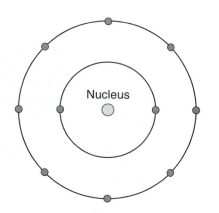

FIGURE 5–2   Structure of a neon atom: Valence shell is full.

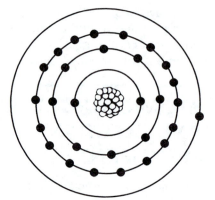

FIGURE 5–3   Atomic structure of a copper atom: The valence shell has 1 electron.

The atomic structure of an atom will always define the chemical characteristics of an element. For instance, a copper atom has 29 protons and 29 electrons. This means that only one electron orbits the outer or valence shell (see **Figure 5–3**). Unlike helium and neon, copper is reactive. It oxidizes easily and is also an excellent conductor of electricity. In fact, copper is often used as material for electrical wiring. Its hold on the electron in its valence shell is a light one; it both readily gives it up and acquires additional electrons.

The atomic structure of the element nickel is very similar in appearance to copper when represented in diagram form. It has an atomic structure with 28 protons and 28 electrons (see **Figure 5–4**) and is just one proton short of an atom of the element copper. Nickel has a full valence or outer shell. It therefore has characteristics that are quite different from those of copper. Nickel is a poor conductor and does not readily react with oxygen. In industry, nickel is an alloyed ingredient of stainless steel that reduces the reactivity of steel, especially its tendency to rust (oxidize).

An electrically balanced atom is one in which the number of electrons equals the number of protons. **Figure 5–5** shows an electrically balanced atom. When atoms of the chemical elements are diagrammed they are almost always shown in a balanced state. Remember that an *ion* is defined as any atom with either a surplus or deficit of electrons. Free electrons can rest on the surface of a material. Electrons resting on a surface will cause that surface to be negatively charged. Because the electrons are not moving, that surface can be described as having a negative static electrical charge. The extent of the charge is measured in voltage or charge differential.

Electrons are also capable of transferring through matter (or a vacuum) at close to the speed of light. A stream of moving electrons is what is known as an

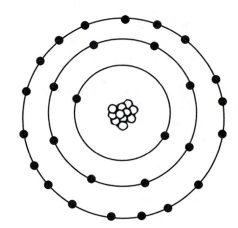

FIGURE 5–4   Atomic structure of a nickel atom: The valence shell is full.

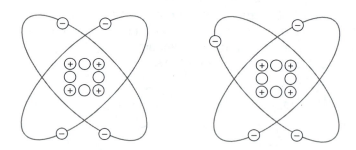

FIGURES 5–5   Balanced atom: The number of protons and electrons are equal.

electrical current. For instance, if a group of positive ions passes in close proximity to electrons resting on a surface, they will attract the electrons (negatively charged) by causing them to fill the "holes" left by the missing electrons in the positive ions. The greater the number of electrons on the move, the greater the current flow. Electron transfer through a conductor is

Conductor

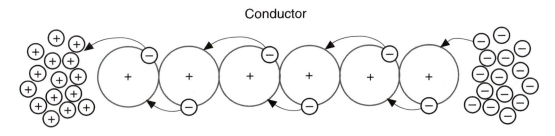

FIGURE 5–6 Electron flow through a conductor.

Conductor

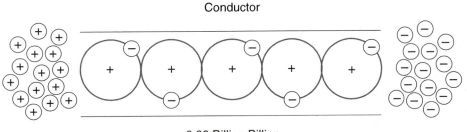

6.28 Billion Billion
electrons per second = one ampere

FIGURE 5–7 The rate of electron flow is measured in amperes.

shown in **Figure 5–6**. The unit of measurement of current flow is the **ampere** (amp). An ampere therefore describes the number of electrons passing through a circuit over a specified time period. Specifically, one ampere is 6.28 billion billion (6.28 × 10^{18}) electrons passing a given point per second. This is shown in **Figure 5–7**. The ampere is the quantitative measurement of current. One ampere or 6.28 billion billion electrons passing a given point in a circuit in one second can be described as one coulomb. It is not especially important to remember these values other than to acknowledge that it is a large number.

A number of factors such as friction, heat, light, and chemical reactions can "steal" electrons from the surface of a material. When this happens, the surface becomes positively charged, meaning that it has a deficit of electrons. Atoms with a deficit of electrons are known as positive ions. Providing positive ions remain at rest, the surface will have a positive static electrical charge differential. Let's take a look at how this works in an everyday example. Every time someone walks across a carpet, electrons are "stolen" from the carpet surface. This has an electrifying effect (*electrification*) on both the substance from which electrons are stolen (the carpet) and the moving body (the person) that does the stealing of the electrons. When this moving body has accumulated a sufficient charge differential (measured in voltage), the excess electrons will be discharged through an arc. This arc, seen as a spark, balances the charge. When the charge has been balanced, an electrically neutral state has been reestablished.

Electrification results in both attractive and repulsive forces. In electricity, like charges repel and unlike charges attract. When a plastic comb is run through hair electrons will be stolen by the comb, giving it a negative charge. This means that the comb can now attract small pieces of paper, as shown in **Figure 5–8**. The attraction will continue until the transfer of electrons results in balancing the charge differential. This experiment will always work better on a dry day. Electrons can travel much more easily through humid air, and any accumulated charge will dissipate (be lost) rapidly. Two balloons rubbed on a woolen fiber will both acquire a negative charge. This enables the two balloons to "stick" to a wall, but at the same time they will tend to repel each other. The attraction of

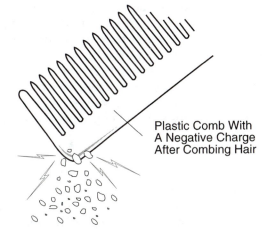

Plastic Comb With
A Negative Charge
After Combing Hair

Small Pieces of Paper

FIGURE 5–8 Unlike electrical charges attract.

unlike charges and repulsion of like charges holds true in both electricity and magnetism.

An atom is held together because of the electrical tendency of unlike charges attracting and like charges repelling each other. Positively charged protons hold the negatively charged electrons in their orbital shells. Also, because like electrical charges repel each other, the electrons do not collide with each other. All matter is composed of atoms. Electrical charge is a component of all atoms. When an atom is balanced (the number of protons match the number of electrons; see **Figure 5–5**), the atom can be described as being in an electrically neutral state. Given that atoms are the building blocks of all matter, it can be said that all matter is electrical in essence.

Atoms seldom just float around by themselves. More often, they join with other atoms; that is, they bond chemically. Chemical bonding of atoms occurs when electrons in the valence shell of atoms are transferred or shared by a companion atom or atoms. When two or more atoms are held together by chemical bonding, they form a molecule. When the atoms are from different elements a compound is formed. A good example of a compound is water, expressed chemically as H_2O. The compound water is formed when an oxygen atom joins with two hydrogen atoms as shown in **Figure 5–9**.

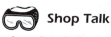 **Shop Talk** _____

A molecule is two or more atoms joined by chemical bonds. The atoms joined may be from the same element or from different elements. The oxygen we breathe is made up of pairs of chemically bonded oxygen atoms. Each pair can be described as a molecule. A single water molecule is made up of two hydrogen atoms chemically bonded to an oxygen atom to form a compound.

A molecule does not necessarily have to be made up of atoms of separate elements. For example, ground-level oxygen is oxygen in bonded in pairs known as O_2, whereas oxygen in the upper stratosphere is ozone, chemically bonded in threes and known as O_3. Although a few more than 100 elements exist, millions of different compounds can be formed from these. Many occur naturally and many are manmade.

 Shop Talk _____

Atoms usually do not float around by themselves. They tend to join with other atoms. Chemical bonds are formed when the electrons in the valence (outermost) shell are shared by other atoms.

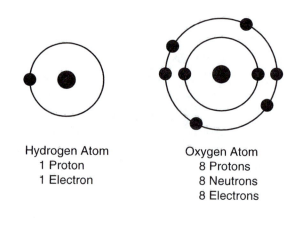

Hydrogen Atom
1 Proton
1 Electron

Oxygen Atom
8 Protons
8 Neutrons
8 Electrons

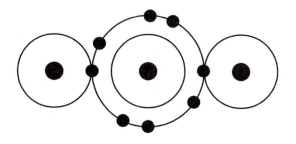

Molecule of Water H_2O
Covalent bonding of 2 hydrogen atoms
to one oxygen.

FIGURE 5–9 *A molecule of the compound water.*

What we describe as *electricity* concerns the behavior of atoms that have become, for whatever reason, unbalanced or electrified. Obviously, there is more to atomic structure than what has been outlined so far. However, understanding a little about atomic structure and behavior can help the technician "see" electricity in a different way when it comes to figuring out failures in electrical circuits. Here is a summary of what we have said about atomic structure so far:

- In the center of every atom is a nucleus.
- The nucleus is made up of positively charged matter called protons.
- The nucleus contains matter with no charge, called neutrons.
- Negatively charged particles called electrons are orbiting each atomic nucleus.
- Electrons orbit the nucleus in concentric paths called shells.
- All electrons are alike.

- All protons are alike.
- Every chemical element has a distinct identity and is made up of distinct atoms. That is, each has a different *number* of protons and electrons.
- In an electrically balanced atom, the number of protons equals the number of electrons. This means that the atom is in what is described as a neutral state of electrical charge.
- An atom with either a deficit or excess of electrons is known as an ion.
- Charge can move from one point to another.
- Like charges repel.
- Unlike charges attract.
- Electrons (negative charge) are held in their orbital shells by the nucleus (positive charge) of an atom.
- Electrons are prevented from colliding with each other because they all have similar negative charges that tend to repel each other.
- A molecule is a chemically bonded union of 2 or more atoms.
- A compound is a chemically bonded union of atoms of 2 or more dissimilar elements.

As we continue to study electricity and magnetism throughout this chapter, the above statements will be reinforced. In fact, they will be repeated many times, although in slightly different formats.

5.2 CONDUCTORS AND INSULATORS

The ease with which an electron moves from one atom to another determines the conductivity of a material. Conductance is the ability of a material to carry electric current. To produce current flow, electrons must move from atom to atom as shown in **Figure 5–6**. Materials that easily permit this flow of electrons from atom to atom are classified as conductors. Some examples of good conductors are copper, aluminum, gold, silver, iron, and platinum. Materials which do not easily permit, or perhaps prevent, a flow of electrons are classified as insulators. Some examples of good insulators are glass, mica, rubber, and plastic.

- A *conductor* is generally a metallic element that contains fewer than 4 electrons in its outer shell or valence. The electrons in atoms of elements with fewer than 4 electrons in the valence shell are not so tightly bound by the nucleus.
- An **insulator** is a nonmetallic substance that contains more than 4 electrons in its outer shell

or valence. When an atom has more than 4 electrons in the valence they are more tightly bound to the nucleus, meaning that they are less likely to be given up.
- **Semiconductors** are a group of materials that cannot be classified either as conductors or insulators. They have exactly 4 electrons in their outer shell. Silicon is an example of a semiconductor.

Any material that can conduct electricity, even when in an electrically neutral state, contains vast numbers of moving electrons that move from atom to atom at random. When a battery is placed at either end of a conductor such as copper wire and a complete circuit is formed, electrons are pumped from the more negative terminal to the more positive. The battery provides the charge differential or potential difference, and the conductive wire provides a path for the flow of electrons. The transfer of electrons continues until either the charge differential ceases to exist or the circuit path is opened. The number of electrons does not change. In order to have a continuous transfer of charge (that is, have a continuous flow of current) between two points in an electrical circuit, it is necessary to produce a new supply of electrons at a negative point in the circuit as fast as this supply is consumed at a positive point in the circuit. A variety of methods of achieving this exist, and they are covered in a later section in this chapter under the heading "Sources of Electricity."

Copper is a commonly used conductor in vehicle wiring systems. Copper wiring may consist of a single extruded wire coated with plastic insulation or multiple thin strands of braided and bound wire coated with insulation. Braided wire has greater flexibility. This greater flexibility is a desirable characteristic in vehicle electrical systems. The **resistance** to current flow in a section of copper wire will depend on its sectional area (wire gauge), temperature, and overall length. The greater the sectional area of copper wire, the less resistance to current flow. Quite simply, thicker electrical wire provides a greater area for flow and is therefore less restrictive. This is because more free electrons are available in heavy-duty wire gauge than in a thin gauge. For instance, 8-gauge wire is thicker than 16-gauge wire. So, 8-gauge wire is capable of permitting greater current flow than 16-gauge wire. The resistance of materials also depends on temperature. The higher the temperature, the greater the resistance to current flow. In most vehicle applications, wiring is designed to be as thin as possible while being capable of handling the electrical load it is designed for. Wire gauge is usually OEM-specified to handle the intended electrical load plus a small safety factor.

When electrons move through a conductor, they often find the atoms of the substance limit their motion to some extent. Some electrical energy is lost in overcoming this resistance to motion in the conductor. If the temperature of the conductor rises, the atoms of the conductor vibrate more vigorously and will further inhibit the movement of electrons through the conductor. In other words, the resistance of the conductor increases with temperature rise.

Insulators are substances with atomic structures that tightly hold their electrons in the valence shell. It is possible to flow very little current through such substances, even when they are subjected to massive charge differentials (very high voltages). Rubber, glass, and some plastics have extremely high resistivity. Both are commonly used as insulators. For instance, plastic is used to protect copper wiring in most vehicle electrical systems, as shown in **Figure 5–10**.

Between low- and high-resistance conductors and insulators is a group of materials known as semiconductors. All semiconductors have 4 electrons in their valence shell. The most commonly used semiconductors are silicon and germanium. These will be studied when electronics theory is introduced in the next chapter.

Electron Movement

Electrons do not actually move through an electrical circuit at high velocity, although the force that moves them does. If a row of pool balls is lined up so that each kiss-contacts the other and the row is struck by a rapidly moving ball, the moving ball stops, imparting a force to the stationary balls that is transmitted through them at high speed but does not move any but the final ball in the row. Most of the force of the moving ball is transferred to the final ball in the row, which separates and moves off from the row. Electron movement through a conductor is somewhat similar. Although actual electron movement through a conductor is not rapid, the electromotive force travels through the conductor at the speed of light.

5.3   CURRENT FLOW

Current flow will occur only when there is a path and a difference in electrical potential. This difference is known as *charge differential* or *potential difference* and is measured in *voltage.* Charge differential exists when the electrical source has a deficit of electrons and therefore is positively charged. Because electrons are negatively charged and unlike charges attract, electrons will flow towards a positive source when provided with a path. Charge differential is electrical pressure (see **Figure 5–11**), and it is measured in voltage.

Initially many scientists thought that current flow in an electrical circuit had one direction of flow, that is, from positive to negative. This is known as the *conventional theory* of current flow. It originated from Ben Franklin's observations and conclusions from his kite in the electrical storm experiment. When the electron was discovered, scientists revised the theory of current flow and called it **electron theory**. In studying electricity, the technician should be acquainted with both conventional and electron theories of current flow. However, vehicle schematics use conventional theory almost exclusively. Conventional and electron theories of electrical circuit flow are shown in **Figure 5–12**.

A conductor such as a piece of copper wire contains billions of neutral atoms whose electrons move randomly from atom to atom vibrating at high frequencies. When an external power source such as a battery is connected to the conductor a deficit (shortage) of electrons occurs at one end of the conductor and an excess of electrons occurs at the other end. The negative terminal will have the effect of repelling free electrons from the conductor's atoms, whereas the positive terminal will attract free electrons. This results in a flow of electrons through the conductor from the negative charge point to the positive charge point in a circuit. The rate of flow will depend on the charge differential (or potential difference/voltage). The greater the voltage (pressure) or charge differential, the greater the rate of flow.

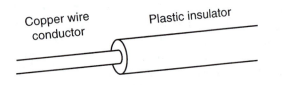

FIGURE 5–10   *A copper wire conductor protected by a plastic insulator.*

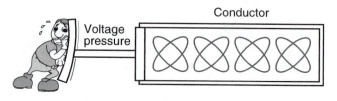

FIGURE 5–11   *Voltage or charge differential is the pressure that causes electrons to move.*

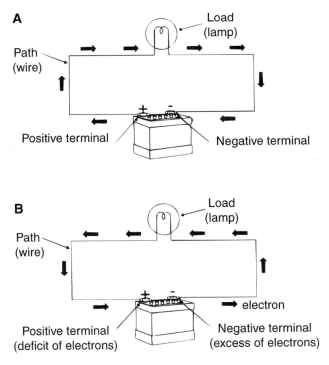

FIGURE 5–12 Electrical circuit flow: (A) Conventional theory; (B) Electron theory.

Voltage is the force that pushes the current through a circuit.

The charge differential or voltage is a measure of electrical *pressure.* The role of a battery, for instance is to act as a sort of electron pump. A 12V battery will pump electrons through a circuit faster than a 6V battery of similar amperage capability. In a closed electrical circuit, electrons move through a conductor, producing a displacement effect close to the speed of light. The higher the voltage, the higher the force potential or pressure available, so the more electrons get pumped through the circuit. Charge differential can also be expressed as *electromotive force (EMF)* or *potential difference (PD).* In fact, because all of the following terms can be used to describe electrical pressure, becoming familiar with them is a good idea:

- charge differential
- voltage
- electromotive force (EMF)
- potential difference (PD)

As mentioned a little earlier, the physical dimensions of a conductor are also a factor. The larger the sectional area (measured by wire gauge size), the more atoms there are over a given sectional area, so there are more free electrons available. Therefore, as wire size increases, so does the ability to flow more

electrical current through the wire. Wire size is specified to the amount of current it is expected to carry. The higher the expected current requirement, the thicker the wire must be; therefore, wire gauge size is matched to the expected current load plus a small safety margin in vehicles. The rate of electron flow is called current and it is measured in amperes. A current flow of 6.28 billion billion electrons per second is equal to one ampere, as shown in **Figure 5–7.**

Terminals

The terminal from which electrons exit in an electrical device connected in a circuit is known as the **anode** or positive terminal. The terminal through which electrons enter an electrical component is known as the **cathode** or negative terminal. Understanding something about the nature of current flow and exactly what is meant by the terms *conventional* and *electron theories* of current flow is crucial because both are used in vehicle technology.

Conventional and Electron Theories

In any electrical circuit, electrons will flow from a more negative point in a circuit toward a more positive point. That is, they will flow from a point in the circuit with an excess of electrons toward a point with a deficit of electrons. So the actual direction of current flow is always from negative to positive. However, many electrical diagrams and schematics use conventional current flow theory. Conventional current flow theory originated from Ben Franklin's observations and mistaken conclusions in his kite experiment. Although the conventional theory of current flow is technically incorrect, it is the basis of all standard vehicle wiring schematics. The conventional theory must be understood so an electrical wiring schematic can be accurately interpreted. The electron theory must be understood so the technician knows electrically what is happening in a circuit for purposes of troubleshooting it.

5.4 MAGNETISM

Magnetism was first observed in lodestone (magnetite) and the way ferrous (iron-based) metals reacted to it. When a bar of lodestone was suspended by string, the same end would always rotate to point toward the earth's north pole. Even today, we do not fully understand magnetism. By observation, we can certainly say quite a bit about magnetism. The molecular theory of magnetism tends to be the most widely accepted. In most materials, the magnetic poles of the composite molecules are arranged randomly so no evident magnetic force exists. However, in certain metals such as iron, nickel, and cobalt, the molecules can be aligned so their north or N poles all point in one direction and their south or S poles point in the opposite direction. In one known material, lodestone, the molecules align themselves in this manner naturally.

Some materials have better magnetic retention than others. This means that when they are magnetized by having their north and south poles aligned, they are able to hold that molecular alignment longer. Other materials are only capable of maintaining their molecular alignment when positioned within a magnetic field. When the field is removed, the molecules disarrange themselves randomly and the substance's magnetic properties are lost. All magnetism is essentially electromagnetism in that it results from the kinetic energy (energy of movement) of electrons. Whenever an electric current is flowed through a conductor, a magnetic field is created. When a bar-shaped permanent magnet is cut in two, each piece will assume the magnetic properties of the parent magnet, with individual north and south poles. **Figure 5–13** shows some basic magnetic principles.

The term *reluctance* is used to describe resistance to the movement of magnetic lines of force. Reluctance can be reduced by using permeable (susceptible to penetration) materials within magnetic fields. The permeability of matter is rated by giving a rating of 1 for air. Air is generally considered to be a poor conductor of magnetic lines of force. In contrast, iron would be given a permeability factor of 2,000. This means that in comparison with air, iron has 2,000 times better permeability. In other words, air has high reluctance and iron has low reluctance. Certain ferrous (iron-based) alloys may have permeability ratings exceeding 50,000.

The force field existing in the space around a magnet can be demonstrated when a piece of thin cardboard is placed over a magnet and iron filings are sprinkled on top of the cardboard. The pattern produced as the filings arrange themselves on the cardboard is referred to as *flux lines.* By observing the behavior of magnetic lines of flux, we can say the following:

- Flux lines flow in one direction.
- Flux lines exit from the magnet's north pole and enter through the south pole.
- The flux density (concentration) determines the magnetic force. A powerful magnetic field has a dense flux field, whereas a weak magnetic field has a low density flux field.
- The flux density is always greatest at the poles of a magnet.
- Flux lines do not cross each other in a permanent magnet.
- Flux lines facing the same direction attract.
- Flux lines facing opposite directions repel.

Atomic Theory and Magnetism

In an atom, all of the electrons in their orbital shells also spin on their own axes. This is much the same as the planets in our solar system orbiting the sun. Each rotates axially and each produces magnetic

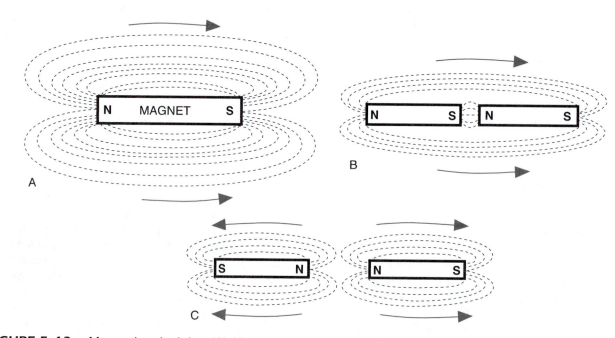

FIGURE 5–13 *Magnetic principles: (A) All magnets have poles; (B) Unlike poles attract; (C) Like poles repel.*

fields. Because of the orbiting electrons' axial rotation, each electron can be regarded as a minute permanent magnet. In most atoms, pairs of electrons spinning in opposite directions will produce magnetic fields that cancel each other out. However, an atom of iron has 26 electrons, only 22 of which are paired. This means that in the second from the outermost shell, four of the eight electrons are not paired, with the result that they rotate in the same direction and do not cancel each other out. This fact accounts for the magnetic character of iron.

A large number of vehicle electrical components use magnetic and electromagnetic principles in some way. Uses for permanent magnets include the AC pulse generators used as shaft speed sensors (rpm sensors would be a good example) and some electric motors. Electromagnetic principles are used extensively in such devices as motors, generators, solenoids, magnetic switches, and coils.

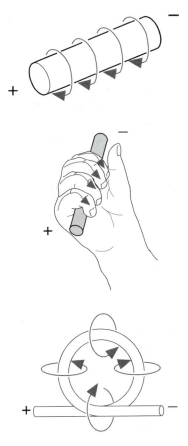

FIGURE 5–14 Electromagnetic field characteristics.

5.5	# ELECTROMAGNETISM

Current flow through any conductor creates a magnetic field. Whenever electrical current is flowed through copper wire, a magnetic field is created surrounding the wire, as shown in **Figure 5–14**. This effect can be observed by passing a copper wire through which current is flowing lengthwise over a compass needle. The compass needle will deflect from its north-south alignment when this occurs. Any magnetic field created by electrical current flow is known as *electromagnetism*. Study of the behavior of electromagnetic fields has proven the following:

- Magnetic lines of force do not move when the current flowed through a conductor remains at a constant. When current flowed through the conductor increases, the magnetic lines of force will extend further away from the conductor.
- The intensity and strength of magnetic lines of force increase proportionally with an increase in current flow through a conductor. Similarly, they diminish proportionally with a decrease in current flow through the conductor.
- A rule called the right-hand rule is used to indicate the direction of the magnetic lines of force: The right hand should enclose the wire with the thumb pointing in the direction of conventional current flow (positive to negative), and the fingertips will then point in the direction of the magnetic lines of force, as shown in **Figure 5–15**.

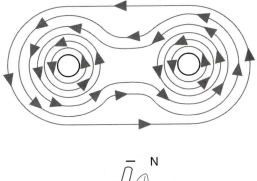

FIGURE 5–15 Magnetic lines of force join together and attract each other and the right-hand rule.

USING ELECTROMAGNETISM

We have already said that when an electric current is flowed through a straight wire, a magnetic field is produced around the wire. When the wire is coiled and electric current is flowed through it, the magnetic field combines to form a larger and more intense magnetic field. Just as in a straight wire, this is identified by north and south poles. This effect can be further amplified by placing an iron core (high permeability) through the center of the coil (see **Figure 5–16**), which reduces the reluctance of the magnetic field.

The polarity of the electromagnet created can be determined by the right-hand rule for coils. The coiled wire should be held with the fingers pointed in the direction of conventional current flow (positive to negative), and the thumb will point to the north pole of the coil. This concept is demonstrated in **Figure 5–15**. The strength of an electromagnet is known as its electromagnetic field force. Electromagnetic field force is often quantified or measured as magnetomotive force (mmf).

Magnetomotive force (mmf) is determined by two factors:

- The amount of current flowed through the conductor
- The number of turns of wire in a coil

Magnetomotive force is measured in ampere-turns (At). Ampere-turn factors are the number of

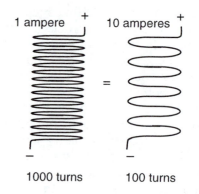

FIGURE 5–17   *Magnetic field strength is determined by the amount of amperage and the number of coils.*

windings (complete turns of a wire conductor) and the quantity of current flowed (measured in amperes). For instance, if a coil with 100 windings has 1 ampere of current flowed through it, the result will be a magnetic field strength rated at 100 At. An identical magnetic field strength rating could be produced by a coil with 10 windings with a current flow by 10 amperes. The actual field strength must factor in reluctance. In other words, the actual field strength of both the above coils would be increased if the coil windings were to be wrapped around an iron core. **Figure 5–17** shows a pair of coils of equal field strength, but because of its larger number of windings, the one on the left requires much less current flow.

A common use of an electromagnet would be that used in an automobile salvage yard crane. By switching the current to the lift magnet on and off, the operator can lift and release scrap cars. Electromagnetic principles are used extensively in vehicle electrical systems. They are the basis of every solenoid, relay, coil, generator and electric motor.

5.6 ELECTRICAL CURRENT CHARACTERISTICS AND SOURCES OF ELECTRICITY

Electrical current is classified as either direct current or alternating current. The technician should understand the basic characteristics of both. Electricity can be produced by any number of means, some of which have been discussed already. This section will discuss the types of current flow and how electricity can be produced.

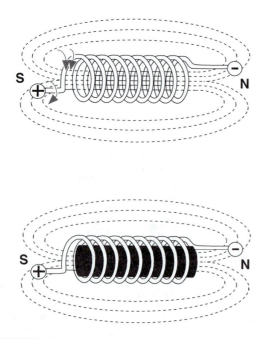

FIGURE 5–16   *Magnetic field characteristics of a coiled conductor.*

DIRECT CURRENT (DC)

Like water through a pipe, electrical current can be made to flow in two directions through a conductor. If the current flows in one direction only, it is known as direct current, usually abbreviated as DC. The current flow may be steady (continuous) or have a pulse characteristic (controlled variable flow). DC can be produced in a number of ways outlined later in this section. Direct current (DC) has many applications and is extensively if not quite exclusively used in highway vehicles throughout most of the chassis electrical circuits.

ALTERNATING CURRENT (AC)

Alternating current or AC describes a flow of electrical charge that cyclically reverses at high speed, due to a reversal in polarity at the voltage source. AC is often shown in graph form (**Figure 5–18**). It is usually produced by rotating a coil in a magnetic field. AC is used in vehicle AC generators (perhaps more often called *alternators*) and in certain sensors on modern vehicles. **Figure 5–19** shows a typical shaft speed

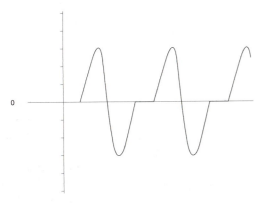

FIGURE 5–18 Graphed AC voltage signal typical of that produced by an inductive pulse generator sensor.

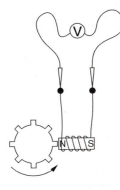

FIGURE 5–19 Principle of an AC inductive pulse generator.

sensor, also called an inductive pulse generator. AC is better suited than DC for transmission through power lines. The frequency at which the current alternates is measured in cycles. A *cycle* is one complete reversal of current from zero though positive to negative and back to the zero point. Frequency is usually measured in hertz. One hertz is one complete cycle per second. For instance, the frequency of household electrical power supply in North America is 60 Hz or 60 cycles per second.

5.7 SOURCES OF ELECTRICITY

Electricity is an energy form. As stated previously, electrical energy may be produced by a variety of methods. The production of electricity involves changing one energy form, such as mechanical or chemical energy, into electrical energy. Some of these methods are outlined in this section.

Chemical

Batteries are a means of producing DC from a chemical reaction. In the lead-acid battery, a *potential difference* is created by the chemical interaction of lead and lead peroxide submerged in sulfuric acid electrolyte. When a circuit is connected to a charged battery (one in which there is a charge differential), the battery will pump electrons through the circuit from the negative terminal, through the circuit load to the positive terminal. This electrochemical process will continue until there is no difference in potential between the two posts of the battery. If a voltmeter were to be placed across the battery terminals, the reading would be zero. When the charge differential ceases to exist, no difference exists between the potential at either terminal and the battery is discharged. The operating principles of lead-acid batteries are explored in much greater depth later in this chapter. A typical lead-acid battery is shown in **Figure 5–20**.

Static Electricity

The term *static electricity* is somewhat misleading because it implies that it is unmoving. Perhaps it is more accurately expressed as *frictional electricity* because it results from the contact of two surfaces. Chemical bonds are formed when any surfaces contact and the atoms on one surface tend to hold electrons more tightly. The result is the "theft" of electrons. Such contact produces a charge imbalance by pulling electrons from one surface to another. As electrons are pulled away from a surface, the result is an excess of electrons (the result is a negative charge) and a

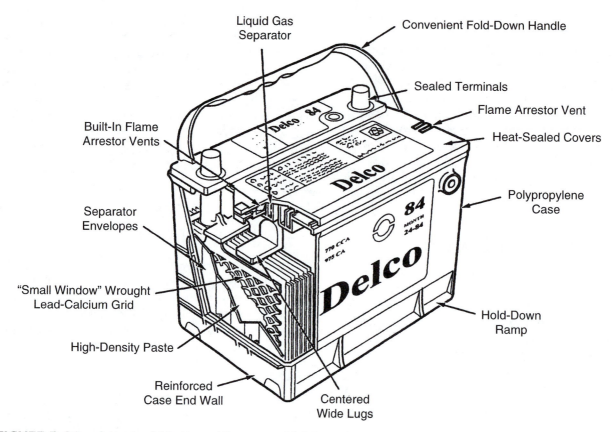

FIGURE 5–20 A lead-acid battery. (Courtesy of AC Delco)

deficit in the other (the result is a positive charge). The extent of the charge differential is of course measured in voltage. While the surfaces with opposite charges remain separate, the charge differential will exist. When the two polarities of charge are united, the charge imbalance will be canceled. Static electricity is an everyday phenomenon as described in examples in the opening to this chapter. It usually involves voltages of over 1,000V and perhaps rising to as much as 50,000V. A fuel tanker trailer towed by a highway tractor steals electrons from the air as it is hauled down the highway (as does any moving vehicle) and can accumulate a significant and potentially dangerous charge differential. This charge differential, which can be as high as 40,000V, must be neutralized by grounding before any attempt is made to load or unload fuel. Failure to ground a fuel tanker before loading or unloading could result in an explosion ignited by an electrostatic arc.

Electromagnetic Induction

Current flow can be created in any conductor that is moved through a magnetic field or alternatively by a mobile magnetic field and a stationary conductor. The voltage induced increases both with speed of movement and the number of conductors. Densely wound conductors will tend to produce higher voltage values. Generators, alternators, and cranking motors all use the principle of electromagnetic induction to either produce or use electricity. Electromagnetic induction is a means of converting mechanical energy into electrical energy—it is much used in the production of electricity. **Figure 5–21** and **Figure 5–22** demonstrate the principle of electromagnetic induction.

Thermoelectric

Electron flow can be created by applying heat to the connection point of two dissimilar metals. A thermocouple used to measure high temperatures consists of two dissimilar metals (typically, iron and constantin, a copper-tin alloy) joined at the "hot" end and connected to a sensitive voltmeter at the gauge end. As temperature increases at the hot end, a potential difference is created at the gauge end. The greater the potential difference, the greater the reading on the display gauge (a millivolt meter). In this way, heat energy is converted into electrical energy. Pyrometers used in truck diesel engine exhaust systems use this thermocouple principle.

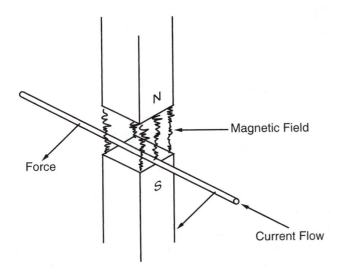

FIGURE 5–21 *Moving a conductor through a magnetic field induces an electric current flow.*

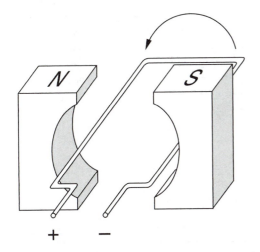

FIGURE 5–22 *Rotating a looped conductor in a magnetic field induces current flow.*

Photoelectric

When light contacts certain materials such as selenium and cesium, electron flow is stimulated. In this way, light energy or photons are converted into electrical energy. Photoelectric cells are used as sensors that can control headlight beams and automatic daylight/night mirrors.

Piezoelectric

Some crystals, notably quartz, become electrified when subjected to direct pressure, the potential difference or voltage increasing with pressure increase. Piezoelectric sensors are used as detonation (knock) sensors used on spark ignited (SI) electronically controlled engines.

5.8 ELECTRICAL CIRCUITS AND OHM'S LAW

The German physicist Georg Simon Ohm (1787–1854) proved the relationship between electrical potential (pressure or voltage), electrical current flow (measured in amperes), and the resistance to the current flow (measured in **ohms**, the symbol of which is Ω). Ohm proved the direct relationship between potential difference circuit resistance and current intensity in mathematical terms. In order to understand the behavior of electricity in an electrical circuit, it is necessary to understand the terminology used to describe its characteristics. When introducing electricity, comparisons are often made between electrical circuits and hydraulic circuits. These comparisons will be used in the following explanations, but it is probably not advisable to think of electricity in terms of hydraulic circuits after having mastered the basics.

A simple electric circuit must have three basic elements:

- **Power source.** This provides the charge differential that will drive the flow of electrons through the circuit. A battery or a generator would be typical power source devices.
- **Path.** A feed path for electron flow to the load and a return path from the load. The wires that connect the power source to the load would provide the circuit path.
- **Load.** The load in an electric circuit acts as a restriction and as a means of converting electrical energy into another energy form. For instance, a light bulb as a load in an electric circuit converts electrical energy into light energy: an electric motor converts electrical energy into kinetic energy (the energy of movement).

Figure 5–23 shows a simple electrical circuit. The power source is a battery and the load is a light bulb. The switch is closed, indicating that there is current flow.

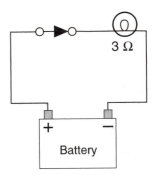

FIGURE 5–23 *Simple series light bulb circuit.*

Next, we should take a look at some terms we have already defined in the context of an actual electric circuit.

Voltage

We said previously that voltage is a measure of charge differential. Charge differential can also be defined as electrical pressure or potential difference. Using a hydraulic comparison, voltage in an electrical circuit can be compared to fluid pressure. Just as in the hydraulic circuit, voltage may be present in an electrical circuit without any current flow taking place. The higher the voltage, the faster the current flow. Voltage is measured in parallel with a circuit. A voltmeter can be used to check circuit voltage or to check one location in a circuit for higher potential than another. A couple of simple applications of a voltmeter are shown in **Figure 5–24**.

Current

Current intensity or the flow of electrons is measured in amperes. One ampere is equal to 6.28 billion billion electrons passing a given point in an electrical circuit in one second—or one coulomb. If electrical current is compared to what happens in a hydraulic circuit, amps can be compared to gpm (gallons per minute). If current flow is to be measured in an electric circuit, the circuit must be electrically active or closed—that is, actively flowing current. Current flow is usually measured with an ammeter as shown in **Figure 5–24**.

Resistance

Resistance to the flow of electrons through a circuit is measured in *ohms*. The resistance to the flow of free electrons through a conductor results from the innumerable collisions that occur. Generally, the greater the sectional area of the conductor (wire gauge size), the less resistance to current flow, simply because more free electrons are available. Once again, using the hydraulic comparison, the resistance to fluid flow through a circuit would be defined by the pipe internal diameter or flow area. In an electrical circuit, resistance generally increases with temperature because of collisions between free electrons and vibrating atoms. As the temperature of a conductor increases, the tendency of the atoms to vibrate also increases, as does the incidence of colliding free electrons. The result is higher resistance. An example of this can be observed when cranking a diesel engine: As the cranking circuit heats up, the less efficient electrical energy is converted into the mechanical energy required to rotate the engine. The resistance in an electrical circuit must be checked with the circuit open; that is, not energized. Resistance checks are often performed on components with an ohmmeter after they have been isolated from their electrical circuit, as shown in **Figure 5–24**.

Typical Vehicle Electrical System Values

Now that we have discussed some of the basic electrical system values, let's apply them to some typical truck chassis components and systems.

Voltage	Current	Resistance
Battery voltage 12.6V	Headlights 10 A	Headlamp 2Ω
Charging voltage 14V	Alternator 80A	Inductive pick-up 150Ω
Spark plug 15,000V	Cranking motor 400A	High tension wire 10,000Ω
V-Ref (ECM) 5V	Clearance light 2A	Thermistor 300Ω

CIRCUIT COMPONENTS

We said before that a complete electrical circuit is an arrangement of a power source, path, and load that permits electrical current to flow. Such a circuit can be described as **closed** whenever current is flowing. The same circuit would be described as **open** when no current is flowing. **Figure 5–25** shows a simple series light bulb circuit in both a closed and open state. A switch is used to open and close a circuit in much the same way a valve acts in a hydraulic cir-

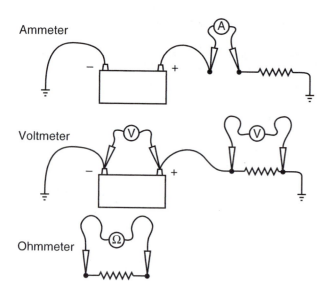

Ammeter

Voltmeter

Ohmmeter

FIGURE 5–24　*Connecting meters into circuits.*

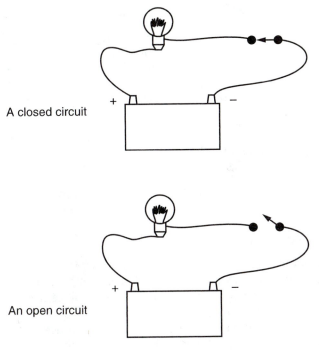

A closed circuit

An open circuit

FIGURE 5–25 *Closed and open circuits.*

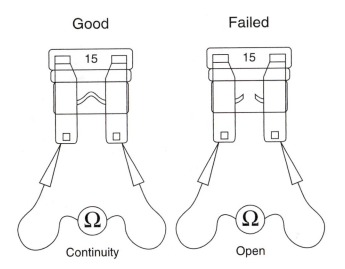

Good   Failed

15   15

Ω   Ω

Continuity   Open

FIGURE 5–26 *A blade-type fuse in good and failed states.*

cuit. **Figure 5–23** shows an electrical circuit consisting of a power source, switch, a load, and wiring to complete the circuit. In most vehicle electrical circuits, a circuit protection device such as a fuse or circuit breaker would also be used.

Vehicle circuit components:

- **Power source.** The energy source of the circuit. The means used to pump electrons through the circuit when it is closed. A battery or a generator are the usual power sources in a vehicle electrical system.
- **Conductors.** The means used to provide a path for current flow in an electrical circuit. This would include electrical wiring and the vehicle chassis in a typical common ground system.
- **Switches.** The means used to open and close an electrical circuit or subcircuits within the main circuit. The ignition switch and dash toggle switches are examples. Note how a switch appears in open and closed states in a circuit diagram by referencing **Figure 5–25**.
- **Circuit protection devices. Circuit breakers** and fuses protect electrical circuits and subcircuits from current overloads by opening the circuit. A fuse permanently fails when overloaded with current and it must be replaced. **Figure 5–26** shows a common blade type fuse in good and failed states. Circuit breakers are widely used in truck chassis in preference to fuses. When a circuit breaker trips (opens) due to a

circuit overload, it may reset automatically (cycling) or the circuit may have to be switched open (noncycling), before current flow can resume.

- **Load.** The objective of the electrical circuit. The function of the load is to convert electrical energy into another energy form, such as light or mechanical energy.

SERIES CIRCUITS

A **series circuit** may have several components such as switches, resistors, and lamps but connected in such a manner that only one path for current flow through the circuit exists, as shown in **Figure 5–23**. In other words, the loads in the circuit are connected one after another. In a series circuit such as the one in **Figure 5–23**, the circuit would open if the element in the light bulb were to fail. Because the path for current flow includes the light bulb element, no current would flow. Series circuits have the following characteristics:

- The resistances placed in the circuit are always added. Total resistance in the circuit is always the sum of all resistances in the circuit because each added resistance represents an added hindrance to current flow.
- Current flow is the same at any point in the circuit when it is closed. That means that the amperage through each resistor in the circuit will be identical.
- The voltage drop through each resistor in the circuit will mathematically relate to its actual

resistance value. This can be calculated using **Ohm's Law**, which will be covered a little later in this chapter.

- The sum of the voltage drops through each resistor in the circuit will equal the source voltage.

Shop Talk

A series circuit is one in which all the components are connected through a single path. For instance, if ten light bulbs were wired in series and the filament in one bulb failed, no current could flow because each load in the circuit represents a portion of the current path.

PARALLEL CIRCUITS

A **parallel circuit** is one with multiple paths for current flow, meaning that the components in the circuit are connected so current flow can flow through a component without having first flowed through other components in the circuit, **Figure 5–27**. In other words, the circuit elements are connected side by side. Parallel circuits have the following characteristics:

- Total circuit resistance will always be less than the lowest value resistor in the circuit. As resistors are added in parallel, they provide additional paths for current flow; therefore, total circuit resistance must drop.
- The current flow through each resistor will depend on the actual resistance value of each resistor. This may be calculated using Ohm's Law.
- Applied voltage is the same on each branch of a parallel circuit.
- The voltage drop through each path in a parallel circuit is the same and equals the source voltage.
- The sum of the current flows measured in amperage in the separate branches of a parallel circuit must equal the total amperage flowed through the circuit.

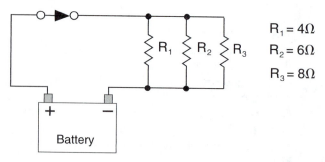

FIGURE 5–27 A parallel circuit with different resistances in each branch.

SERIES-PARALLEL CIRCUIT

Many circuits are constructed using the principles of both series and parallel circuits. These are known as **series-parallel circuits.** The principles used to calculate circuit values in series and parallel circuits are used to calculate circuit behavior in series-parallel circuits.

ELECTRICAL CIRCUIT TERMINOLOGY

The following terminology is used to describe both normal and abnormal behavior in an electrical circuit.

Short Circuit

A **short circuit** describes what occurs in an electrical circuit when a conductor is placed across the connections of a component and some or all of the circuit current flow takes a shortcut. Short circuits are generally undesirable and can quickly overheat electrical circuits. Electricity will generally choose to flow through the shortest possible path—that is, the path of least resistance—in order to complete a circuit. Short circuits result in excessive current flow, which can rapidly overheat wiring harnesses and cause vehicle fires.

Open Circuit

The term open circuit describes any electrical circuit through which no current flow occurs. A switch is used in electrical circuits to intentionally open them. An electrical circuit may also be opened unintentionally. This might occur when a fuse fails, a breaker opens, a wire breaks, or connections corrode.

Grounds

The ground represents the point of a circuit with the lowest voltage potential. In vehicles, **ground** or *chassis ground* is integral. This means that the chassis forms the electrical path that acts to supply electrons for all components in the circuit. It has an excess of electrons because it is directly connected to the negative terminal of the battery. For instance, a clearance light has two terminals. One terminal is connected to the chassis ground, which has a negative potential (an excess of electrons). The second terminal is connected to a switch. When the switch is open, there is no path for current flow, so none takes place. The moment the switch is closed, a path for electrical current flow is provided. Electrons can now flow from the chassis ground (negative potential) through the light bulb filament, exiting at the positive terminal, passing through the closed switch, and returning to the positive terminal of the power source; that is, the battery. This light bulb circuit can now be described as *energized*. Technicians used to working on vehicles always use the term *ground* to mean *chassis ground*.

 Shop Talk _____

In vehicles, one battery terminal is connected to ground; that is, the chassis. In today's vehicles, this is usually the negative terminal. So, to complete a circuit, a load must be connected by a current path only from the nongrounded terminal. This greatly reduces the amount of wiring required.

Short to Ground

The term *grounded circuit* is sometimes used to describe a short to ground. This means that when a circuit is closed, the electron flow bypasses the intended load by taking a shorter route to the positive point or terminal of the power source. Shorts to ground have almost no resistance. The result is excessive current flow and overheating.

High Resistance Circuits

A high resistance circuit is caused by loose or corroded terminals. In a high resistance circuit, a path for current flow exists, but the path is too small for the number of electrons that are being pumped through it. The result is overheating.

OHM'S LAW

Ohm's Law says that an electrical pressure of 1V (volt) is required to move 1A (ampere) of current through a resistance of 1Ω. It is a mathematical formula that the technician must know. In fact, just by playing around with a multimeter, you can prove it. This can be more fun for the beginner than struggling with the math, but doing both is probably best. It is simple to work with, so let's take a look at it:

I = "intensity" = current in amperes
E = EMF (electromotive force) = pressure in
    voltage
R = resistance = resistance to current flow in Ω

$$E = I \times R \qquad R = \frac{E}{I} \qquad I = \frac{E}{R}$$

Ohm's Law can also be expressed in the units of measurement used in the above formulas:

$$V = A \times \Omega \qquad \Omega = \frac{V}{A} \qquad A = \frac{V}{\Omega}$$

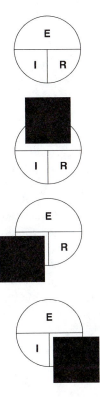

FIGURE 5–28 Ohm's Law graphic.

Or, reference **Figure 5–28**, showing the application of Ohm's Law formulae. Better still, memorize it—it will prove to be invaluable later when we perform electronic troubleshooting.

We can solve Ohm's Law with the circular diagram in **Figure 5–28**, by simply covering the value we wish to find and using simple algebra to calculate it. For instance, to calculate the amount of voltage required to push 4A through a resistance of 3Ω, we would do the following: The unknown value is voltage represented by the letter *E* in the Ohm's Law formula. Cover the E. The result is that the I and R values are side by side. Construct an equation as follows:

$$E = I \times \grave{} \qquad E = 4 \text{ (amps)} \times 3 \text{ (ohms)} = 12 \text{ (volts)}$$

Now we will use another set of values and calculate current flow. To determine how much current will flow in a circuit in which there is a charge differential of 12V acting on a resistance of 4Ω, first we will cover the I, which represents current in the formula. Now we see E over R, so an equation can be constructed as follows:

$$I = \frac{E}{R} \qquad I = \frac{12V}{3\Omega} \qquad = 3A$$

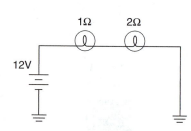

FIGURE 5–29   *Series circuit calculation.*

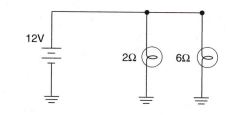

FIGURE 5–30   *Parallel circuit calculation.*

Ohm's Law Applied to Series Circuits

Now we will use some of this theory in some actual circuits. In a series circuit, all of the current flows through all of the resistances in that circuit. The sum of the resistances in the circuit would define the total circuit resistance. If a series circuit were to be constructed with 1Ω and 2Ω resistances, as shown in **Figure 5–29**, total circuit resistance (Rt) would be calculated as follows:

$$Rt = R_1 + R_2$$
$$Rt = 1\Omega + 2\Omega$$
$$Rt = 3\Omega$$

Notice that to construct the formula, each resistance has been identified with a number. Also, the letter *t* is often used to mean total when constructing formulae.

If the power source in the above series circuit is a 12V battery, when the circuit is closed, current flow can be calculated using Ohm's Law as follows:

$$I = \frac{E}{Rt}$$

$$I = \frac{12V}{3\Omega}$$

$$I = 4 \text{ amps}$$

Ohm's Law Applied to Parallel Circuits

According to Kirchhoff's Law of Current (see the next section), current flowed through a parallel circuit divides into each path in the circuit. When the current flow in each path is added, the total current will equal the current flow leaving the power source. When calculating the current flow in parallel circuits, each current flow path must be treated as a series circuit, or the total resistance of the circuit must be calculated before calculating total current. When performing calculation on a parallel circuit, remember that more current will always flow through the path with the lowest resistance. This confirms what we said earlier in this chapter about electricity always

choosing to flow through the path of least resistance. If a parallel circuit is constructed with 2Ω and 6Ω resistors in separate paths and supplied by a 12V power source as shown in **Figure 5–30**, total current flow can be calculated by treating each current flow path separately, as follows:

$$I_1 = \frac{12V}{2\Omega} = 6 \text{ amps}$$

$$I_2 = \frac{12V}{6\Omega} = 2 \text{ amps}$$

$$I_t = 8 \text{ amps}$$

Note how more current flows through the path with the least resistance. An alternative method would be to calculate Rt using the following formula:

$$Rt = \frac{R_1 \times R_2}{R_1 + R_2}$$

$$Rt = \frac{6\Omega \times 2\Omega}{6\Omega + 2\Omega} = \frac{12\Omega}{8} = 1.5\Omega$$

$$I = \frac{E}{R}$$

$$I = \frac{12V}{1.5\Omega} = 8 \text{ amps}$$

Some vehicle electrical circuits are of the series-parallel type; that is, they combine the characteristics of both the series circuit and the parallel circuit. When performing circuit analysis and calculation, visualizing the circuit in terms of paths for current flow helps. When you learn how to visualize electricity in these terms, it becomes common sense rather than a mystery.

Kirchhoff's Law of Current (Kirchhoff's First Law)

Kirchhoff's Law of Current states that the current flowing into a junction or point in an electrical circuit must equal the current flowing out. If you use a hydraulic comparison and visualize a water wheel being rotated by a flow of water through its paddles, this makes some sense. In a parallel circuit, the current flow through each path of the circuit adds up to the total current flow. Basically, what flows into a circuit must exit it. The calculations that accompany **Figure 5–29** and **Figure 5–30** prove this.

Kirchhoff's Law of Voltage Drops (Kirchhoff's Second Law)

Kirchhoff's Law of Voltage Drops states that voltage will drop in exact proportion to the resistance, and that the sum of the voltage drops must equal the voltage applied to the circuit. This is a critical law that has an everyday application for the technician working on electrical circuits. Calculation of voltage drop is frequently performed by technicians troubleshooting electrical circuits. Let's take a closer look. If a series circuit is constructed with a 12V power source and 2Ω and 6Ω resistors, as shown in **Figure 5–31**, the voltage drop across each resistor when the circuit is closed and subject to a current flow of 1.5A can be calculated:

Voltage drop for R_1:

$$E_1 = I \times R_1$$
$$E_1 = 1.5A \times 2\Omega$$
$$E_1 = 3V$$

Voltage drop for R_2:

$$E_2 = I \times R$$
$$E_2 = 1.5A \times 6\Omega$$
$$E_2 = 9V$$

Total voltage drop through the circuit should equal that of the two above calculations.

$$E_1 + E_2 = \text{source voltage}$$
$$3V + 9V = 12V$$

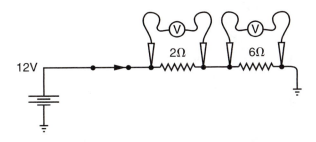

FIGURE 5–31 *Voltage drop calculation.*

In every case, the sum of the voltage drops in a circuit equals the source voltage. Voltage drop testing is an important diagnostic tool for the technician because it is an easy way to locate unwanted resistances caused by corroded connections, damaged wiring, and failed terminals. It must always be performed on a closed or energized circuit.

POWER

Just as in the internal combustion engine, the unit for measuring power is the **watt**, named after James Watt (1736–1819), usually represented by the letter *P*. In engine technology, power is defined as the *rate of accomplishing work* and therefore is always factored by time. This is also true in an electrical circuit. Remember, the definition of an ampere is 6.28 billion billion electrons (1 coulomb) passing a point in a circuit per *second*. The formula for calculating electrical power is

$$P = I \times E \text{ (spells "pie")}$$

Using the data from the previous formula in which the circuit voltage was 12V and the current flow was 1.5A:

$$P = 1.5 \times 12 = 18W = \text{power consumed}$$

One HP equals 746W, so calculated values can be converted into English values used to rate some components.

 Shop Talk

How does a light bulb operate? Resistance increases in a conductor with a decrease in cross-sectional area. The light bulb uses a filament of high resistance but conductive alloy that is both long and thin (low cross-sectional area). This filament, when subjected to a current flow, will produce temperatures high enough to radiate photons. Photons are visible light. To ensure that the filament does not rapidly fail by overheating and burning out, it is manufactured from tungsten or alloys capable of sustaining high temperatures and contained in a sealed glass container filled with inert gas under moderate pressure. The inert gas prevents reactions between the filament and the surrounding gas (exposed to oxygen, the filament would rapidly oxidize and burn out) and the pressure inhibits filament evaporation. Incandescent light bulbs are an inefficient method of producing light. This is why LEDs (explained in the next chapter) are becoming more common.

5.9 | ELECTRIC MOTOR AND GENERATOR PRINCIPLE

Electric motors are used extensively in vehicles. They are required to drive blowers, power windows, wipers, and to crank the engine. After the engine is started, an alternator, a type of generator, is required to keep the battery charged and provide electricity for the entire rig. In this section, we will take a look at the operating principles of both the electric motor and the generator. They will be looked at together because the process by which an electric motor converts electrical energy into mechanical energy can be reversed. The function of a generator is to convert mechanical energy into electrical energy, and the alternator is the type of generator used in all current vehicles.

Both the electric motor and the generator put the principle of electromagnetic induction (discussed earlier in this chapter) to work, so they both require a conductor and a magnetic field, as shown in **Figure 5–32**. One of these two components must rotate while the other is held stationary. In the case of the motor, electric current is input to the conductor and the outcome is motion. The reverse occurs in the generator. Motion is the input and the outcome is current flow.

In electricity, the principle used in a generator and reversed in the electric motor is known as Faraday's Law of Induction. Faraday built the first electric motor after determining that the relative motion between a conductor and a magnetic field induced a current flow in the conductor. The construction of an electric motor is generally similar to that of an electric generator.

DC MOTORS

The objective of an electric motor is to convert electrical energy into mechanical motion. The current carrying conductors on a DC motor are loops of wire arranged on an armature. The armature is the rotating member of the motor. It is supported on bearings, usually bronze bushings. The loops of wire begin and end at copper terminals called *segments* on the armature shank. These segments on the armature rotate between stationary brushes. The brushes are used to create current flow through the loops of wire. Surrounding the armature are stationary magnets called *fields*. The magnets can be either permanent or electromagnets. When an electrical current is flowed through the loops of wire on the armature, a force is exerted on the wire. Because the current is flowing in opposite directions on either side of the loop (it passes up on one side, down on the other), this creates torque, causing the armature to rotate. This principle is shown in **Figure 5–32** and **Figure 5–33**.

Electric motors can produce high torque at low speeds. In fact, some types of DC motor are designed to produce peak torque at zero rpm. This type of DC motor is used as the cranking motor for internal combustion engines, which are unable to produce high torque at low rpms and need considerable help to get them rotating at a speed at which they can work efficiently.

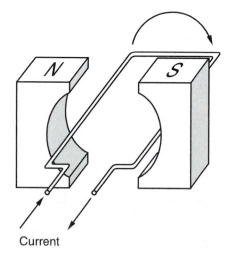

FIGURE 5–32 DC motor principle. When current is flowed through the wire loop, torque (twisting force) is created.

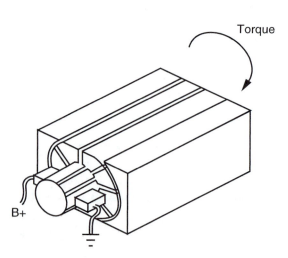

FIGURE 5–33 Construction of a simple electric motor.

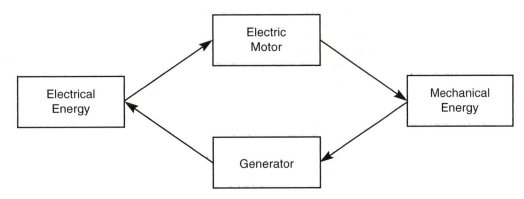

FIGURE 5–34 *Electric motors and generators convert one energy form to another.*

GENERATORS

Until the 1960s, DC generators were used to meet the electrical needs of vehicles. As we have already discovered, the DC generator functions exactly as a DC motor except in reverse. Mechanical energy, usually from the vehicle engine transmitted by a pulley and drive belts, drives the armature through the stationary fields. This results in current flow in the armature wire loops, which flows through the brushes to the vehicle electrical circuit. Because a generator is simply an electric motor with its function reversed, it can be made to function as a motor simply by connecting it to a power source. This concept is shown in **Figure 5–34**. A disadvantage of the DC generator is its inability to function efficiently at low rpms. This means that under conditions of low rpm operation such as city driving, they are incapable of meeting the electrical requirements of the vehicle.

An alternator is an AC generator. Because an alternator produces AC, this must be rectified to DC so it can be used by the chassis electrical system. In an alternator, the rotor is an electromagnet. A small current is flowed into the rotor field windings by means of brushes and copper slip rings. This produces a magnetic field. As the rotor turns, it induces a current flow in stationary conductors called the *stator windings.* An alternator produces AC because the current is induced in opposite directions when opposite magnetic poles are driven through the stator conductor loops. Diodes (electrical one- way valves, described in Chapter 6) are used to change or rectify the AC to DC so it can be used by the vehicle electrical system.

Reluctor-Type Generators

A common type of shaft speed sensor used in today's vehicles is really a miniature AC generator. It consists of a permanent magnet and a coil of wire mounted to toothed reluctor. The reluctor is located on the rotating shaft. As the teeth in the reluctor cut through the magnetic field, voltage is induced in the pick-up coil. This voltage increases proportionally with shaft speed increase. This type of sensor is commonly used to signal wheel speed, transmission tailshaft speed, engine speed, and so on, to ECMs (electronic control modules).

CAPACITANCE

The term *capacitance* is used to describe the electron storage capability of a commonly used electrical component known as a capacitor. *Capacitors,* which are also called condensers, all do the same thing; that is, they store electrons. The simplest type of capacitor consists of two conductors separated by some insulating material called *dielectric.* The conductor plates could be aluminum and the dielectric may be mica (silicate mineral). The greater the dielectric (insulating) properties of the material, the greater the resistance to voltage leakage. When a capacitor is connected to an electrical power source, it becomes capable of storing electrons from that power source. When the capacitor's charge capability is reached, it will cease to accept electrons from the power source. The charge is retained in the capacitor until the plates are connected to a lower voltage electrical circuit. At this point, the stored electrons are discharged from the capacitor into the lower potential (voltage) electrical circuit.

As a capacitor is electrified, for every electron removed from one plate, one is loaded onto the other plate. The total number of electrons in the capacitor is identical when it is in either the electrified or completely neutral states. What changes is the location of the electrons. The electrons in a fully charged capacitor will in time "leak" through the dielectric until both conductor plates have an equal charge. At this point, the capacitor is in a discharged condition. The ability to store electrons is known as capacitance. Capacitance is measured in farads (named after Michael Faraday, the discoverer of the principle). One farad is the ability to store 6.28 billion billion electrons at a 1-volt charge differential. Most capacitors have much less capacitance than this, so they are rated in picofarads (trillionths of a farad) and microfarads (millionths of a farad).

1 farad = 1F

1 microfarad = 1μF = 0.000001F

1 picofarad = 1pF = 0.000000000001F

Types of Capacitors

Many capacitors used in electric and electronic circuits are fixed value capacitors. Fixed value capacitors are coded by capacitance and voltage rating. Some capacitors have a variable capacity. Variable capacitors have a combination of fixed and moving conductor plates and the capacitance is varied by a shaft that rotates the moving plates. In some variable capacitance-type capacitors, the dielectric may be air. One truck engine manufacturer uses a variable capacitance-type sensor to digitally signal throttle position. The advantage of this compared to the more commonly used potentiometer-type throttle position sensor is an accurate digital output signal. Electrolytic type capacitors tend to have much higher capacitance ratings than nonelectrolytic types. They are polarized, so they must be connected accordingly in a circuit. The dielectric in this type of capacitor is the oxide formed on the aluminum conductor plate, but the operating principle is identical.

Figure 5–35 and **Figure 5–36** show how a capacitor functions in an electrical circuit. In **Figure 5–35**, a capacitor is located parallel to the circuit load to smooth voltage changes in the circuit. When the circuit switch is open, no electron flow will occur. When the circuit has been closed by the switch, the battery positive terminal pulls electrons from the capacitor, leaving its positive plate with a deficit of electrons. This will attract electrons to its negative plate: The dielectric keeps the electrons from being pulled through to the capacitor positive plate. In **Figure 5–36**, the capacitor is shown with a stored charge differential that can be unloaded when the circuit switch is opened.

When working on electrical circuits, note that capacitors can retain a charge for a considerable time after the circuit has been opened and current flow has ceased. Accidental discharge can damage circuit components and may represent an electric shock hazard.

Capacitors are used extensively in electrical and electronic circuits performing the following roles:

- **Power supply filter.** Smooths a pulsating voltage supply into a steady DC voltage form

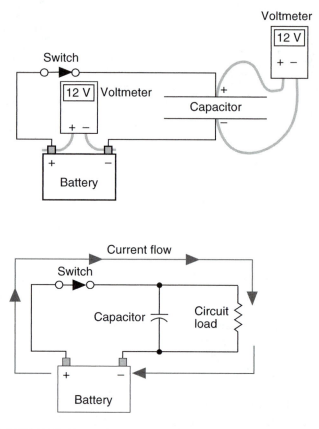

FIGURE 5–35 Operating principles of a capacitor.

- **Spike suppressant.** When digital circuiting is switched at high speed, transient (very brief) voltage reductions can occur. Capacitors can eliminate these spikes or glitches by compensating for them.
- **Resistor-capacitor circuits (R-C circuits).** R-C circuits incorporate a resistor and a capacitor. These are used to reshape a voltage wave or pulse pattern from square wave to sawtooth shaping or modify a wave to an alternating pattern. These concepts are discussed in more detail in the next chapter.

5.10 COILS, TRANSFORMERS, AND SOLENOIDS

When electrons are moved through a conductor, an electromagnetic field is created surrounding the conductor. When the conductor is a wire wound into a coil, the electromagnetic field created is stronger. As discussed previously, such coils are the basis of electric **solenoids** and motors. Coils are used in electronic circuits to shape voltage waves because they tend to resist rapid fluctuations in current flow. Like capacitors, coils can be used in electrical circuits to reshape

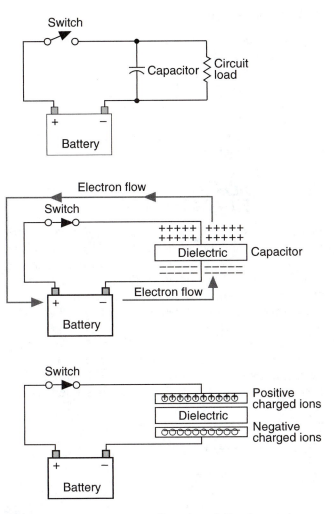

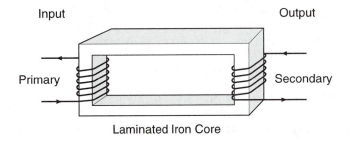

FIGURE: 5–37 *Transformer operating principle.*

FIGURE 5–36 *Current flow in a fully charged capacitor.*

voltage waves. Also, the energy in the electromagnetic field surrounding a coil can be induced in any nearby conductors. If the nearby conductor happens to be a second coil, a current flow can be induced in it. The principle of a transformer is essentially that of flowing current through a primary coil and inducing a current flow in a secondary or output coil. Variations on this principle are coils that are constructed with a movable core that permits their inductance to be varied, thereby altering frequency (remember the variable capacitance throttle position sensor mentioned previously). This type of position sensor uses a movable coil, so mechanical position relates to frequency. The frequency value is then output from the device and used as an input signal to a computer.

TRANSFORMERS

The basic **transformer** functions by having two coils located beside each other. When an electric current

is flowed through either of the coils, the resulting electromagnetic field induces a current flow in the other coil. This principle is shown in **Figure 5–37**.

The coil through which current is flowed is known as the *primary coil.* The coil through which current flow is induced by the electromagnetic effect of the primary coil is known as the *secondary coil.* A transformer requires a changing current characteristic to operate. AC current constantly changes, so when it is flowed through the primary coil, it creates a changing magnetic field required to induce current flow in the secondary coil. The operating principle of a transformer is often described as *mutual inductance.* Transformers of various types are used in vehicle electrical circuits, but they all fall into three categories of function: stepping up voltage, stepping down voltage, and isolating portions of a circuit. Transformers have a number of functions in an electrical circuit. They can be used to step-up voltage, step-down voltage, and isolate portions of a circuit.

Isolation Transformers

In an isolation transformer, the primary and secondary coils have the same number of windings, producing a 1:1 input to output ratio. Their objective is to "isolate" one portion of an electrical circuit from another. Secondary voltage and current equal that of the primary coil. Isolation transformers are used in electronic circuits to isolate or protect subcircuits against voltage surges.

Step-Up Transformer

The objective of a **step-up transformer** is to multiply the primary coil voltage by the winding ratio of the primary coil versus that of the secondary coil. For instance, if the primary to secondary coil winding ratio is 1:10, 12V through the primary coil will produce 120V through the secondary coil with a similarly proportional drop in current flow. With electricity, you never get something for nothing. When voltage is stepped-up by a transformer, current drops by a corresponding amount. Examples of step-up transformers are automotive ignition coils and some types of injector driver units used on diesel electronic unit injection (EUI) systems.

Step-Down Transformer

Step-down transformers function in an opposite way from step-up transformers. The primary coil winding ratio exceeds that of the secondary coil, resulting in a diminished output voltage and increased output current. This results in reduced voltage and amplified current. A common shop use of a step-down transformer is an AC welding station. Electric arc welding requires high current at a relatively low voltage and the AC welding unit uses transformers to achieve this.

Ignition coils

Spark ignited (SI) engines require high voltage, up to 50,000V, to fire spark plugs. The available voltage in a vehicle ignition system is somewhere between 12–14V DC. Direct current will not operate in a transformer because it will not produce a changing electromagnetic field. To make an ignition coil operate, the low voltage primary coil must be pulsed; that is, switched on and off at high speed. The ignition coil uses a high-current, low-voltage primary coil with a low number of windings to induce a low-current, high-voltage output in a secondary coil with a large number of windings. Each time the primary current is pulsed, a magnetic field is created and then collapsed, inducing the current in the secondary coil. The high speed switching is performed by a transistor located in and controlled by the ignition module. **Figure 5–38** shows the coil's role in firing a spark plug.

Transformer Summary

- Two coils are arranged so one is subject to a magnetic field created in the other.
- The input coil is the primary coil.
- The output coil is the secondary coil.
- Step-up transformers have secondary coils with a greater number of windings.
- Step-down transformers have secondary coils with a lower number of windings.

SOLENOIDS AND MAGNETIC SWITCHES

Electromagnetic switches are used in vehicle electrical circuits so a low current circuit can control a high current circuit. A typical example would be in the cranking circuit, as shown in **Figure 5–39**. An electromagnetic switch opens or closes an electrical subcircuit by using an electromagnet to pull on movable contacts when energized.

A solenoid is not much different in its operating principle from a magnetic switch except that it functions specifically to convert electrical energy into mechanical movement. Solenoids are used exten-

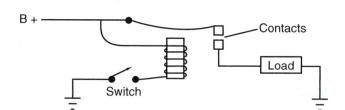

FIGURE 5–39   Magnetic switch shown open.

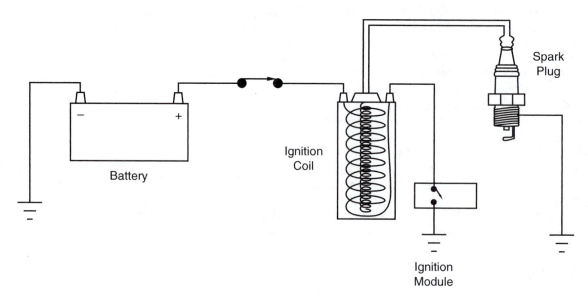

FIGURE 5–38   Operating principle of an ignition coil.

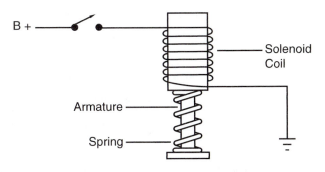

FIGURE 5–40 A solenoid when not energized.

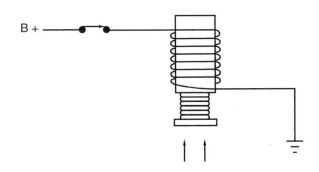

FIGURE 5–41 An energized solenoid.

sively on vehicles in such applications as starter engage mechanisms, diesel electronic unit injector control, gasoline injector switching, pilot switches, automatic transmission clutch controls, and so on. They may be small or large, but all operate on the same principle of converting an electrical signal into mechanical movement. A solenoid consists of two components: a coil and an armature. The coil is generally stationary. When the coil is energized, the armature moves either into or out of the coil, depending on polarity. The action of a typical solenoid is shown in **Figure 5–40** and **Figure 5–41**.

5.11 BATTERY OPERATING PRINCIPLES

The battery principles of operation will be discussed in this section, but their application in truck electrical circuits will be discussed in Chapter 7. A battery is a galvanic device; that is, a device that produces a charge differential by chemical action. The word *galvanic* comes from Luigi Galvani (Italy, 1737–1798), who invented the voltaic pile, an early type of battery. The battery stores electrical energy not as electrons but as chemical energy. It is probably best described as an electron pump. During the charge cycle, electrical energy is converted into chemical energy. During the discharge cycle, chemical energy is converted into electrical energy.

The **lead-acid battery** is the electrical energy storage device used on most vehicles. Its function is to store the electrical energy generated by the alternator. It performs a secondary role as a stabilizer for the voltage in the vehicle electrical system. The battery or batteries on a vehicle must be capable of supplying a high current load to the vehicle cranking circuit and support the operation of other vehicle electrical system loads, especially when the alternator is operating at lower efficiency, such as when idling. Most truck electrical systems use multiple lead-acid battery units connected in series and series-parallel arrangements that are the basis of a 12V electrical system. In some cases, a 12V chassis electrical system may have a 24V cranking circuit, in which case multiple batteries are arranged so they can be switched between the 12V chassis electrical circuit and the 24V cranking circuit.

Battery Construction

A current 12V battery consists of 6 series-connected cells arranged in a polypropylene casing, as shown in **Figure 5–42**. Each cell has positive and negative plates, or anode and cathode plates. Each plate is constructed of a lead grid. Due to the relative softness of lead, the grids are toughened by adding antimony (a metal alloying element) or calcium. The plates must then be coated or "pasted" with chemically active matter. The positive plates are pasted with lead peroxide (lead dioxide), whereas the negative plates are pasted with sponge lead, which is a pure, porous form of lead.

Within each cell in the battery housing, the plates are arranged so positive and negative plates are located next to each other and alternate as shown in **Figure 5–43**. They must not contact each other, so separators are used to prevent short circuits. The separators (**Figure 5–43**) are manufactured from nonconductive materials such as plastics, fiberglass, and resin-coated paper. These separators are porous to permit the chemical interaction of the liquid battery electrolyte with plates. Below the plates in each cell in the battery housing are a series of ribs that help support and separate the plates, but also provide a space to store spent sediment from the pasting on the plates. This spent sediment is capable of causing a short circuit between the plates. Some batteries prevent the spent sediment from settling in the bottom of the battery case by enclosing each plate in a microporous plastic envelope, as shown in **Figure 5–44**. This method prevents the sediment from getting to the bottom of the battery case.

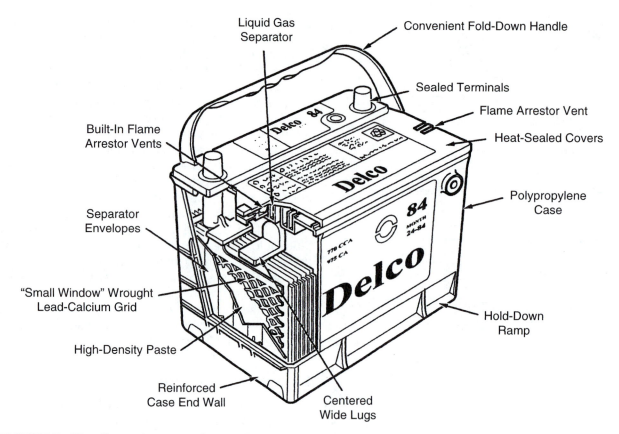

Liquid Gas
Separator

Convenient Fold-Down Handle

Sealed Terminals

Flame Arrestor Vent

Heat-Sealed Covers

Built-In Flame
Arrestor Vents

Polypropylene
Case

Separator
Envelopes

"Small Window" Wrought
Lead-Calcium Grid

High-Density Paste

Hold-Down
Ramp

Reinforced
Case End Wall

Centered
Wide Lugs

FIGURE 5–42 Battery construction and terminology. (Courtesy of AC Delco)

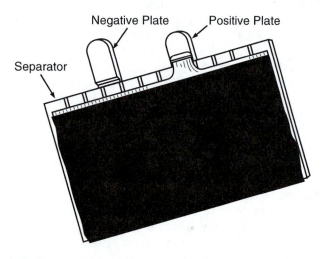

Negative Plate

Positive Plate

Separator

FIGURE 5–43 Positive and negative plates with separator. (Courtesy of AC Delco)

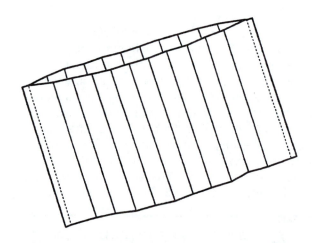

FIGURE 5–44 Envelope separator. (Courtesy of AC Delco)

Each cell in the battery is separated from the other cells by partitions as shown in **Figure 5–45**. The partitions isolate the chemical interactivity of each cell, which means they are only connected electrically. The connectors used are lead and designed to minimize resistance.

👓 **Shop Talk**

A charged battery stores chemical energy. During discharge, a chemical reaction releases this chemical energy as electrical energy. During charging, electrical energy is converted to chemical energy in the battery.

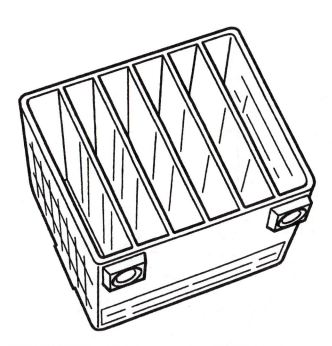

FIGURE 5–45 *Maintenance-free battery case showing partitions. (Courtesy of AC Delco)*

Electrolyte

By definition, electrolyte is any substance that conducts electricity. In the lead-acid battery it is the liquid solution that enables the galvanic or chemical action of the battery. In both standard and maintenance-free batteries, the electrolyte used is a solution (mixture) of sulfuric acid (H_2SO_4) and pure water (H_2O). The solution proportions should be 36 percent sulfuric acid and 64 percent distilled water. This solution will produce a **specific gravity** (the weight of a liquid or solid versus that of the same volume of water) of 1.265 at 80°F. The acid to water proportions should never be tampered with. When a new battery is activated, premixed electrolyte is introduced to the cells. The electrolyte solution in a lead-acid is both conductive (capable of conducting electricity) and reactive (able to participate in chemical reactions).

In batteries that permit addition to the electrolyte when reduced by gassing, only distilled water should be used. Pure water or distilled water does not conduct electricity; in fact, it can be used as an insulator. The conductivity of water depends on its content of total dissolved solids (TDS) or minerals. The minerals are conductive. Using tap water to replenish battery electrolyte can, in time, cause battery cells to short out. The chemical action of a battery during discharge reduces the proportional ratio of sulfuric acid to that of water. This reduces the density (specific gravity) of the solution. This can be measured using a hydrometer or refractometer. A **hydrometer** is a device that directly measures specific gravity by floating a sealed bulb in the solution. A **refractometer** measures the refractive index of a solution. The refractive index is essentially the amount a light ray is bent when it passes through a solution in comparison with pure water. The higher the density of the solution, the more the ray is slowed.

CAUTION: Great care should be exercised when testing battery electrolyte with a refractometer. The process involves placing a drop of electrolyte on the read window close to the eye. Battery electrolyte is highly corrosive and is capable of causing permanent eye damage. Always execute the test instructions precisely as instructed.

Specific Gravity and Temperature

Temperature directly affects the specific gravity of a substance. Specific gravity specifications are based on a temperature of 80°F. When a hydrometer is used to test specific gravity, the reading must always be temperature-adjusted. The usual method of adjusting specific gravity to temperature is to add 0.004 for every 10°F above 80° and to subtract 0.004 for every 10°F below 80°. As we have said, the sulfuric acid concentration in the electrolyte may also be tested using a refractometer—these readings do not require temperature correction. However, a large percentage of batteries used in truck applications today are maintenance-free. A maintenance-free battery has no means of measuring the specific gravity of the electrolyte solution other than a built-in hydrometer. The built-in hydrometer has all of the disadvantages of any other type of hydrometer in terms of temperature accurate readings, and it does not produce an actual reading. Instead, it produces one of three color readings, as shown in **Figure 5–46**. At best, this produces a ballpark estimate of the electrolyte specific gravity: However, it is the only means of reading electrolyte specific gravity in a maintenance-free battery.

84 Factor

The actual specific gravity of the electrolyte in any cell in a battery relates directly to its voltage. The 84 factor enables cell voltage to be calculated by adding 0.840 to the specific gravity measured. If the specific gravity in a cell was measured to be 1.260, cell voltage would be calculated as follows:

$$1.260 + 0.840 = 2.100 \text{ volts}$$

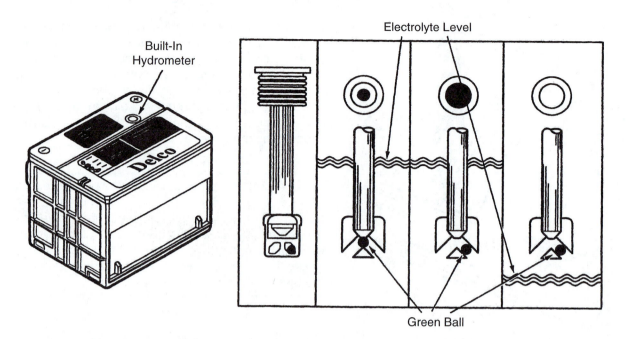

FIGURE 5–46   Built-in hydrometer operating principle. (Courtesy of AC Delco)

If all 6 cells in the battery were measured with the same specific gravity, battery voltage can be calculated as follows:

$$2.100V \times 6 = 12.6V$$

The 84 factor is a ballpark method of relating cell specific gravity to voltage. It does not apply when the battery electrolyte is in a completely discharged state.

Shop Talk

A 12V lead-acid battery has six cells connected in series. Each cell produces a voltage of a little more than 2V when fully charged.

BATTERY CHARGE/ DISCHARGE CYCLES

Next, we will take a look at the chemical interactions that occur within a lead-acid battery during the charge and discharge cycles. A charge cycle occurs when the battery is being regenerated by the alternator. A discharge cycle occurs when the battery power is being consumed by the vehicle electrical system, such as when cranking the engine.

Discharge Cycle

When a circuit is constructed by connecting the positive and negative battery terminals with a load, current will flow through the circuit. The lead peroxide

(PbO_2) on the positive plate reacts with the sulfuric acid solution electrolyte (H_2SO_4), with the result that its oxygen molecule (O_2) is released to the electrolyte, forming H_2O. In the reaction, the negative plate (Pb) reacts with the electrolyte to form lead sulfate ($PbSO_4$). **Figure 5–47** and **Figure 5–48** show the chemical action that takes place during the discharge cycle. This chemical action will continue until both the positive and negative plates are coated with lead sulfate ($PbSO_2$) and the electrolyte has been chemically reduced to water (H_2O). It is almost impossible to reduce battery electrolyte to a state of pure water that would produce a specific gravity reading of 1.000. In other words, batteries seldom become discharged to the point of being electrically neutral. Typically, what is described as a "dead" battery contains electrolyte with a specific gravity in the region of 1.140 to 1.160. It is also important to note that a fully charged battery, one with the correct ratio of sulfuric acid to water, provides a good measure of antifreeze protection. Water freezes at 32°F (0°C). The higher the concentration of sulfuric acid, the lower the freeze temperature. As the electrolyte solution degrades to water during discharge, the electrolyte freeze point rises. Batteries that freeze up in winter conditions are almost always in a discharged state.

Charge Cycle

When a generating device such as an alternator is connected in a circuit between the battery terminals, the battery can be recharged. This essentially

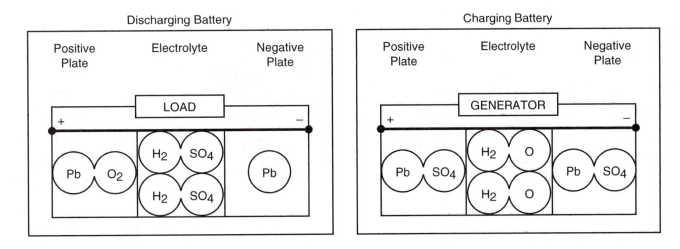

FIGURE 5–47   Chemical composition in batteries. (Courtesy of AC Delco)

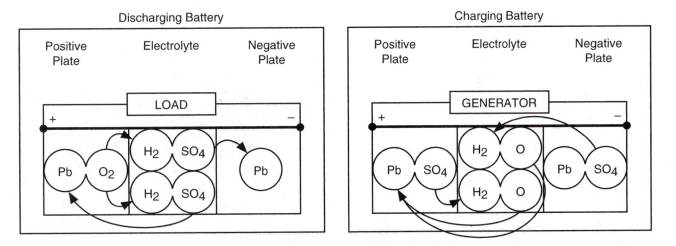

FIGURE 5–48   Changes to the chemical composition during the charge and discharge cycles. (Courtesy of AC Delco)

reverses the chemical reaction that takes place in the charging cycle. During the charge cycle, the sulfate coatings that have formed on both the positive and negative plates are reacted with to return them to the liquid electrolyte. This process corrects the ratio of sulfuric acid (H_2SO_4) and water (H_2O). During the charge process water molecules in the electrolyte reduce to hydrogen and oxygen. The hydrogen combines with the sulfate in the electrolyte to form sulfuric acid, whereas the oxygen is drawn to the positive plate to reconstruct the lead peroxide coating. **Figure 5–48** shows the chemical action that takes place during the charge cycle.

When a battery is in the charge cycle, gassing is caused by electrolysis. Gassing is the conversion of the water in the electrolyte to hydrogen and oxygen gas. In older batteries, these gases were vented from the battery housing, forming an explosive combina-

tion that, if ignited with a small spark, could destroy a battery or battery bank (two or more interconnected batteries) and could be highly dangerous. Current batteries have expansion or condensation chambers that prevent the escape of charging gases.

CAUTION: Hydrogen gas may be discharged during charging. Care should always be taken when connecting or disconnecting battery terminals not to create a spark that could result in a battery explosion. Always disconnect the ground cable first and reconnect it last.

Most vehicle batteries are not designed for deep cycling. **Deep cycling** occurs when a battery is brought close to a completely discharged state and then recharged. Recharging a fully discharged battery tends to be hard on the battery due to the heat created by internal resistance. This can cause plate buckling. Applications that use deep cycle (cycling by complete discharge, followed by full recharge) batteries, such as electric forklift trucks, use batteries constructed with much thicker plates when they do use lead-acid batteries. Vehicle batteries are not designed for repeated deep cycling and will fail prematurely if, for whatever reasons, they are repeatedly deep cycled.

Sulfation

When a battery becomes discharged, both plates are coated with lead sulfate ($PbSO_4$). During the charge cycle, this sulfate coating is converted as explained previously. However, when the sulfate coating hardens on the plates, it can no longer be converted. Its output becomes limited and the condition progresses to complete failure. A battery described as sulfated has failed due to **sulfation.** This condition is usually irreversible and requires the replacement of the battery.

MAINTENANCE-FREE BATTERIES

Maintenance-free and low-maintenance batteries have become widely used in the trucking industry in recent years. When a battery is described as maintenance-free, no provision is made for adding water to depleted electrolyte. A low-maintenance battery is one that uses essentially the same construction as the maintenance-free battery, but does provide for periodic inspection and replenishing of the electrolyte. In terms of their fundamental chemical operating principles, they are identical to the lead-acid batteries already described. The antimony used to strengthen lead plates in standard batteries also acts as a catalyst (substance that enables a chemical reaction without itself undergoing any change) for the electrolysis of electrolyte to gaseous oxygen and hydrogen. Maintenance-free batteries use substances such as calcium, cadmium, and strontium instead of antimony on the plates. These substances reduce gassing during the charge cycle. **Figure 5–49** and **Figure 5–50** show the differences in the plates of standard and maintenance-free batteries. Additionally, an expansion or condensation chamber is used to contain the gassing, condensing the hydrogen and oxygen to water and allowing it to drain back into the electrolyte. A vent provides a measure of safety to the battery housing, but only in the event of significant pressure rise will it trip to bleed to the atmosphere.

FIGURE 5–49   Lead antimony cast grid used on most standard lead-acid batteries. (Courtesy of AC Delco)

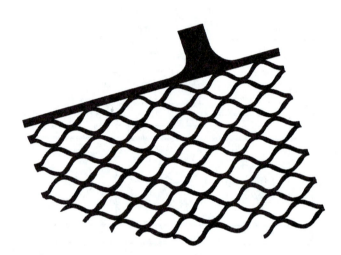

FIGURE 5–50   Lead calcium wrought grid used on many maintenance-free batteries. (Courtesy of AC Delco)

The main objective of the maintenance-free battery is to eliminate the periodic addition of water to the battery cells as it is boiled off by charging electrolyte, but there are advantages and disadvantages. The advantages are

- Elimination or reduction of the periodic topping up of the electrolyte required in standard batteries
- Ability to sustain overcharging without losing electrolyte
- Longer inactivated shelf life
- Generally high cold cranking amperage (CCA) ratings compared with similarly sized standard batteries

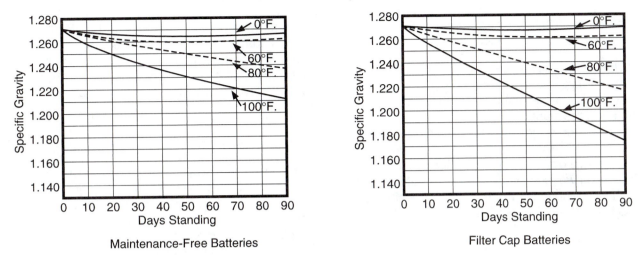

FIGURE 5–51   *Self-discharge graph. (Courtesy of AC Delco)*

The disadvantages are

- Less ability to sustain deep cycling
- Generally lower reserve capacity
- Shorter total service life expectancy
- Faster discharge rates

Some maintenance-free batteries have a built-in hydrometer. This is usually called a *visual state of charge indicator*. This was described previously in this chapter in the section on electrolyte and is shown in **Figure 5–46. Figure 5–51** shows a comparison of the rate of self-discharge of standard lead-acid batteries versus maintenance-free batteries. The electrolyte degrades at a faster rate in standard batteries, especially at higher ambient temperatures.

GELLED ELECTROLYTE BATTERIES

A gelled electrolyte battery is usually known as a **gel cell** battery. The gel cell is a type of lead-acid battery and uses all the fundamental operating principles of any other type of lead-acid battery. The main difference of the gel cell is the electrolyte. The gel cell battery uses a special thixotropic (a solid or gel that liquefies under certain conditions) electrolyte that, when stirred or shaken, liquefies, but returns to the gelled state when left at rest. The battery is pressurized and sealed using special vents. For this reason, it never requires filling. Gel cells therefore can be operated in almost any position although upside-down installation is not recommended. Freightliner Corporation has specified gelled electrolyte batteries in isolated battery, deep cycle applications. In vehi-

cles with multiple battery bank, isolated battery systems, the gel cell is used as the isolated battery.

Gel Cell Operating Principle

A gelled electrolyte battery can be classified as a **recombinant battery.** In the charge cycle of the lead-acid battery, the oxygen released at the positive plate recombines in the cell with the hydrogen that is released at the negative plate. The recombination of hydrogen and oxygen produces water, which is reabsorbed to the electrolyte. The gel cell is entirely maintenance-free and will never require the addition of water. Additionally, it does not cause corrosion around the terminals.

Gel cells are designed to be acid-limited. This means that the electrolyte is diluted to a discharge condition before the plates sulfate. This protects the plates from ultra-deep discharging, which can cause plate shedding and accelerated positive grid corrosion that can ultimately destroy the battery.

If a gel cell battery is subjected to electrical charging voltages that exceed 14.1V, it will release more hydrogen and oxygen than can be recombined. This excess of gas raises the pressure in the battery housing until it can trip the pressure-sensitive vents. After the excess of oxygen and hydrogen has been released, it is lost and cannot be recombined to water. When a gel cell is subjected to overcharging, the electrolyte dries out and the battery fails prematurely. Because gel cell batteries operate under pressure, they may appear to bulge slightly even during appropriate charging. However, extreme bulging usually indicates overcharging or blocked bleed vents and the battery should be immediately taken out of service. Gel cells, when used as an isolated battery in multibattery truck electrical systems, must be connected directly to the battery charger.

Some advantages of gel cells:

- Spillproof and leakproof
- Vent no oxygen or hydrogen during charging
- No current charge limitation at 13.8 volts
- Vibration-resistant
- Double or more the service life of an equivalent maintenance-free battery
- Can sustain deep cycling operation
- Operate in wet conditions including underwater
- If charged and used properly, are cost-effective due to much greater service life

Disadvantages:

- Weigh more
- Will fail if subjected to overcharging
- Require special chargers (automatic, temperature-sensing, voltage-regulated) that regulate charge pressure within a narrow range: 13.8V–14.1V
- Vulnerable to abuse

5.12 BATTERY RATINGS

All vehicle batteries are rated to standards established by the **Battery Council International (BCI)** in conjunction with the Society of Automotive Engineers (SAE). In terms of design, the current capacity of any battery depends primarily on the size of the plates and the quantity of chemically active material coated to the plates. The larger the surface area of the plate in contact with the electrolyte, the more chemical action can take place to produce current flow. Battery rating classification can be defined as follows:

Cold Cranking Rating

Cold cranking amperes (CCA) is the primary method of rating a vehicle battery. It specifies the current load a battery is capable of delivering for 30 seconds at a temperature of 0°F (−18°C) without dropping the voltage of an individual cell below 1.2V, or 7.2V across the terminals. Truck batteries may have CCA ratings exceeding 1,000A. The power required to crank an engine tends to go up as temperature drops, whereas the power available (battery power) goes down. **Figure 5–52** shows the relationship of power available and cranking power required. Engines that must be started in subzero conditions require batteries with high CCA ratings.

Ampere-Hour Rating

The **ampere-hour rating** of a battery is the amount of current that a fully charged battery can feed through a circuit before the cell voltage drops to 1.75V. In a typical 6-cell, 12V battery, this would equal a battery voltage of 10.5V.

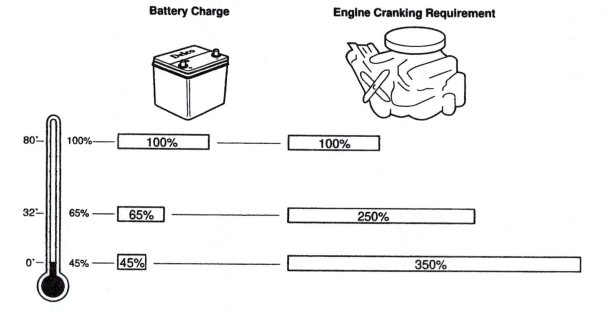

FIGURE 5–52   Power available versus power required. (Courtesy of AC Delco)

Reserve Capacity Rating

The **reserve capacity rating** of a battery system will determine the amount of time a vehicle can be driven with its headlights on in the event of a total charging system failure. The current output will vary with the type of vehicle electrical system, but the critical low voltage value is 10.5V.

BATTERY BANKS

Most current trucks use multiple batteries to fulfill the engine cranking and electrical chassis requirements of the vehicle. Most highway trucks use 12V electrical systems supplied by batteries arranged in banks. Although arrangements of 6V and 12V batteries have been used in the past, today it is most common for arrangements of 12V batteries to be used. A small number of highway trucks use a 12/24 system. A 12/24 system is a 12V chassis electrical system with a 24V cranking capability. The switching may be performed electromechanically or electronically. **Figure 5–53** shows how 12V batteries can be arranged in series or in parallel in a battery bank.

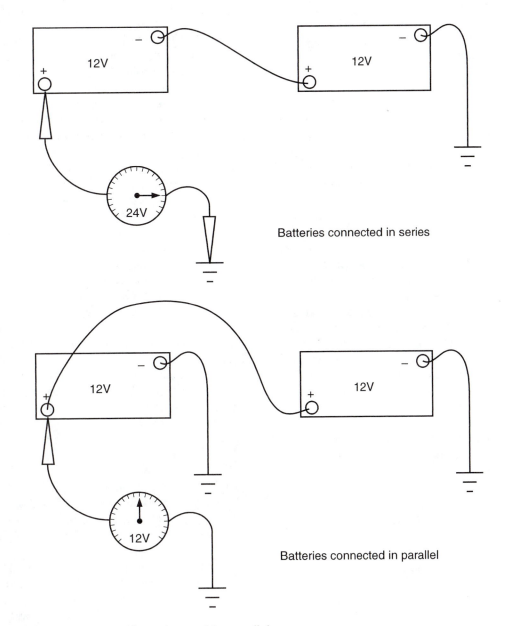

FIGURE 5–53 *Batteries arranged in series and in parallel.*

SUMMARY

- All matter is composed of atoms.
- All atoms have an electrical charge. When an atom is balanced (the number of protons match the number of electrons), the atom can be described as being in an electrically neutral state.
- All matter is electrical in essence. *Electricity* concerns the behavior of atoms that have become, for whatever reason, unbalanced or electrified.
- Electricity may be defined as the movement of free electrons from one atom to another.
- Current flow is measured by the number of free electrons passing a given point in an electrical circuit per second.
- Electrical pressure or charge differential is measured in *volts,* resistance in *ohms,* and current in *amperes.*
- The magnetic properties of some metals such as iron are due to electron motion within the atomic structure.
- A direct relationship exists between electricity and magnetism. Electromagnetic devices are used extensively on vehicles.
- Magnetomotive force (mmf) is a measure of electro-magnetic field strength: its unit is ampere-turns.
- Ohm's Law is used to perform circuit calculations on series, parallel, and series-parallel circuits.
- In a series circuit, there is a single path for current flow and all of the current flows through each resistor in the circuit.
- A parallel circuit has multiple paths for current flow: The higher the resistance in each path, the less current flows through it.
- Kirchhoff's Law of Voltage Drops states that the sum of voltage drops through resistors in a circuit must equal the source voltage.
- When current is flowed through a conductor, a magnetic field is created.
- Reluctance is resistance to the movement of magnetic lines of force: Iron cores have permeability and are used to reduce reluctance in electromagnetic fields.
- Capacitors are used to store electrons: They consist of conductor plates separated by a dielectric.
- Capacitance is measured in farads: Capacitors are rated by voltage and by capacitance.
- When current is flowed through a wire conductor, an electromagnetic field is created: When the wire is wound into a coil, the electromagnetic field strength is intensified.

- The principle of a transformer can be summarized by describing it as flowing current through a primary coil and inducing a current flow in a secondary or output coil.
- Transformers can be grouped into three categories: isolation, step-up, and step-down.
- The storage device for electrical energy on vehicles is the lead-acid battery.
- The battery stores energy as chemical energy, not in the form of electrons: It acts as a sort of electron pump in an electrical circuit.
- A standard battery consists of anode and cathode plates arranged in cells, grouped in series within the battery housing: The electrolyte used in lead-acid batteries is a solution of distilled water and sulfuric acid.
- The electrolyte in a lead-acid battery is both conductive and reactive.
- During the discharge cycle of the battery, lead peroxide on the positive plate combines with the electrolyte, releasing its O_2 into the electrolyte to form H_2O. The lead on the negative plate reacts with electrolyte and forms lead sulfate ($PbSO_4$). The result of discharging a battery is lead sulfate formation on both the positive and negative plates of the battery.
- During the charge cycle, the sulfate coating leaves both the positive and negative plates to be recombined into the electrolyte: This process re-establishes the correct proportions of sulfuric acid and water in the electrolyte.
- During the charge cycle, both oxygen and hydrogen are released in a process known as gassing. In maintenance-free and low-maintenance batteries, the results of charge gassing are contained in a condensation chamber.
- A fully charged battery will have an electrolyte mixture concentration of 36 percent sulfuric acid and 64 percent water that will produce a specific gravity reading of 1.260 at 80°F.
- Gel cell batteries are used in truck electrical systems requiring an isolated battery.
- When gel cell batteries are recharged, they must be connected directly to the battery charger.
- Only approved chargers may be used to charge gel cell batteries: These regulate the charging voltage to no less than 13.8V and no more than 14.1V. Regular battery chargers use a charging voltage of about 16V, which will destroy a gel cell battery.
- Batteries are performance rated by cold cranking amps, reserve capacity and ampere-hour rating.
- Most truck batteries are specified to an electrical system by their cold cranking amp (CCA) rating.

REVIEW QUESTIONS

1. A material described as an insulator has how many electrons in its outer shell?
   a. less than 4
   b. 4
   c. more than 4

2. Which of the following is a measure of electrical pressure?
   a. amperes
   b. ohms
   c. voltage
   d. watts

3. Which of following units of measurement expresses electron flow in a circuit?
   a. volts
   b. watts
   c. farads
   d. amps

4. How many electrons does the element silicon have in its outer shell?
   a. 2
   b. 4
   c. 6
   d. 8

5. Who originated the branch of electricity generally described as *electromagnetism?*
   a. Franklin
   b. Gilbert
   c. Thomson
   d. Faraday

6. Which of the following elements could be described as being electrically inert?
   a. oxygen
   b. neon
   c. carbon
   d. iron

7. Which of the following is a measure of charge differential?
   a. voltage
   b. wattage
   c. amperage
   d. ohms

8. An element classified as a semiconductor would have how many electrons in its outer shell?
   a. less than 4
   b. 4
   c. more than 4
   d. 8

9. Which of the following describes resistance to movement to magnetic lines of force?

   a. reluctance
   b. inductance
   c. counter electromotive force
   d. capacitance

10. Use Ohm's Law to calculate the current flow in a series circuit with a 12V power source and a total circuit resistance of 6Ω.

11. Calculate the power consumed in a circuit through which 3A are flowed at a potential difference of 24V.

12. A farad is a measure of
    a. inductance
    b. reluctance
    c. charge differential
    d. capacitance

13. Which of the following would be associated with the positive plate in a fully charged battery?
    a. H_2SO_4
    b. $PbSO_4$
    c. PbO_2
    d. Pb

14. Which of the following would be associated with the negative plate in a fully charged battery?
    a. H_2SO_4
    b. $PbSO_4$
    c. PbO_2
    d. Pb

15. Which of the following would be associated with the electrolyte in a lead-acid battery?
    a. H_2SO_4
    b. $PbSO_4$
    c. PbO_2
    d. Pb

16. When the term *gassing* is applied to battery operation, which gas (es) is being referred to?
    a. oxygen
    b. hydrogen
    c. both oxygen and hydrogen
    d. neither oxygen nor hydrogen

17. Which of the following terms describes the amount of time a vehicle can be operated with its headlights on before battery voltage drops to 10.5V?
    a. ampere-hours rating
    b. cold cranking amps
    c. reserve capacity
    d. wattage

18. The term *deep cycling* in the context of a battery means
    a. complete discharge, followed by full recharge
    b. prolonged overcharging

c. sulfation
d. recombination cycling

19. Technician A says that magnetism is a source of electrical energy in an automobile. Technician B says that a chemical reaction is a source of electrical energy in an automobile. Who is correct?
    a. Technician A
    b. Technician B
    c. both A and B
    d. neither A nor B

20. Which of the following is a measure of electrical circuit current flow?
    a. watts
    b. volts
    c. ohms
    d. amps

21. Resistance to current flow is measured in ____.
    a. amps
    b. volts
    c. watts
    d. ohms

22. Which of the following is the best insulator of electricity?
    a. copper
    b. aluminum
    c. distilled water
    d. tap water

23. Which type of circuit offers only one path for current flow?
    a. series
    b. parallel
    c. series-parallel
    d. open

24. A break in a circuit that causes a loss of continuity is called a _____.
    a. short
    b. ground
    c. open
    d. overload

25. Which of the following terms describes a *recombinant* battery?
    a. standard
    b. low-maintenance
    c. maintenance-free
    d. gel cell

26. Which of the following would cause high resistance in a closed electrical circuit?
    a. short to ground
    b. corroded terminals
    c. broken wire
    d. heavy gauge electrical wiring

27. If one branch of a parallel circuit has high resistance, how will the other branch(s) be affected?
    a. Current will increase.
    b. Current will decrease.
    c. Voltage will decrease.
    d. Voltage will increase.

28. When describing the negative post of a battery terminal in an energized electrical circuit, which of the following applies?
    a. It is the source of electrons flowing through the circuit.
    b. It has a deficit of electrons.
    c. It is more positive than the first load in the circuit.
    d. It is electrically neutral.

29. If an electrical connection in a closed electrical circuit heats up, which of the following would be the more likely cause?
    a. a blown fuse
    b. wire gauge thickness too large
    c. corroded connectors
    d. overcharged battery

30. When a dead short occurs, which of the following would be true?
    a. circuit resistance usually increases
    b. current flow usually increases
    c. voltage usually increases
    d. all of the above

Chapter

6

Prerequisite
Chapter 5

Fundamentals of Electronics and Computers

Objectives

After studying this chapter, you should be able to

- Outline some of the developmental history of electronics.
- Describe how an electrical signal can be used to transmit information.
- Define the term *pulse width modulation.*
- Define the principle of operation of N- and P-type semiconductors.
- Outline the operating principles and applications of diodes.
- Describe the construction and operation of a typical transistor.
- Describe what is meant by the *optical spectrum.*
- Identify some commonly used optical components used in electronic circuitry.
- Explain what is meant by an *integrated circuit* and outline its application in on-board vehicle electronics.
- Define the role of gates in electronic circuits.
- Describe the operating modes of some common gates used in electrical circuits including AND, OR, and NOT gates.
- Interpret a *truth table* that defines the outcomes of gates in an electronic circuit.
- Explain why the binary numeric system is used in computer electronics.
- Define the role of an electronic control module in an electronic management system.
- Outline the distinct stages of a computer processing cycle.
- Describe the data retention media used in vehicle ECMs.
- Demonstrate an understanding of input circuits on a vehicle electronic system.
- Describe the operating principles of the VORAD collision warning system.

Key Terms List

AND gate
anode
binary system
bipolar
bits
bytes
cathode
central processing unit (CPU)
chopper wheel
Darlington pair
data processing
doping

duty cycle
electronic control module (ECM)
electronically erasable programmable read-only
    memory (EEPROM)
electrons
fiber optics
Hall effect
integrated circuits (I/Cs)
input circuit
light-emitting diodes (LEDs)
NOT gate
OR gate
output circuit

photonic semiconductors
potentiometer
programmable read-only memory (PROM)
pulse wheel
pulse width (PW)
pulse width modulation (PWM)
random access memory (RAM)
read-only memory (ROM)

reluctor
semiconductors
thermistors
tone wheel
transistor
truth table
zener diode

In the previous chapter, basic electrical principles were introduced. We discovered in that chapter that electricity is the form of energy that results from charged particles such as **electrons** or protons being either static (not flowing) or dynamic (flowing) as current flow in a circuit. Electronics is the branch of electricity that addresses the behavior of flows of electrons through solids, liquids, gasses, and across vacuums. It is probably essential to have a sound understanding of basic electricity before attempting to understand basic electronics theory.

The science of electronics began with the discovery of the *electron* by J. J. Thomson (1856–1940) in 1897. This discovery quickly resulted in the invention of the diode (1904), the triode (1907) and, perhaps most importantly, the **transistor** in 1947 by three Bell Laboratories scientists, Bardeen, Brattain, and Shockley. A diagram of the world's first transistor is shown in **Figure 6–1**. Diodes and transistors are fundamental components in any electronic circuit. They are defined as solid-state components.

Solid-state devices are the building blocks of electronic circuits. They are called solid-state devices because they are manufactured from solid materials called **semiconductors**. A number of dif-

ferent types of semiconductor devices exist, but the previously mentioned diode and transistor are probably the most important. Though they are never repaired, the technician will benefit from having a fundamental knowledge of these devices, especially when troubleshooting.

The objective of this chapter is to deliver a foundation level understanding of electronics. It should be recognized that most of the concepts introduced can be studied in much greater depth, but this would probably exceed what the average truck technician is required to know at this time. The terms *electronics* and *electronic engine management* are generally used in automotive technology to describe any systems managed by a computer. In the final portion of this chapter, basic computers will be introduced in terms of their operating principles. The application of electronics theory will be explored in both Chapters 7 and 8.

6.1 USING ELECTRONIC SIGNALS

Electricity can be used to transmit signals or information as well as power. An electric doorbell is designed to deliver a simple electrical signal. If it is functioning properly, the circuit remains un-energized, that is, inactive, for most of its service life. A visitor signals his or her presence by mechanically depressing a switch that closes the electrical circuit and produces an audible chime. Electronic signals often operate in a somewhat similar manner using on-off signals at high speeds and of different durations. When an electronic circuit is constructed to manage information, low current and low voltage circuits are generally used. Electrical signals can be classified as analog or digital. An analog signal operates on variable voltage values. A graphic representation of an analog signal is shown in **Figure 6–2**.

Digital signals operate on specific voltage values, usually the presence or lack of voltage. Simple electronic circuits can be designed to transmit relatively complex data by using digital signals. A digital signal produces a square wave, which is shown in diagram form in **Figure 6–3**.

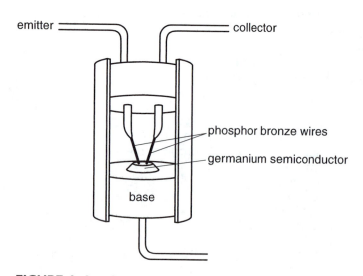

FIGURE 6–1   *Cutaway view of the world's first transistor, produced by Bell Laboratories.*

emitter
collector
phosphor bronze wires
germanium semiconductor
base

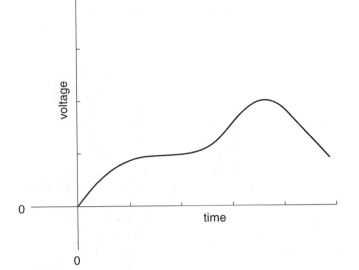

FIGURE 6–2   Analog signal.

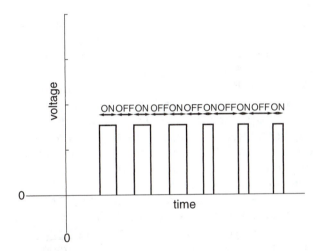

FIGURE 6–4   Varying frequency: The number of pulses per second is measured in hertz.

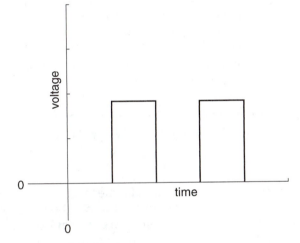

FIGURE 6–3   Square wave digital signal.

FIGURE 6–5   Pulse width modulation: Note how the on time of the pulses is varied. On time is referred to as duty cycle.

Shop Talk

A square wave is a train of high and low/no voltage pulses. Some types of digital sensors use varying square wave frequencies to transmit information.

A square wave is a train of voltage pulses with specific high and low values. When the high and low values do not change from the predetermined specific values, the elements that may be changed are *frequency* and **duty cycle**. Frequency is simply the number of pulses per second. It is expressed in hertz (from H. R. Hertz, a German physicist). Information can be transmitted by varying the frequency of a signal, as in **Figure 6–4**.

Varying duty cycle can also be used to transmit information. A square wave of fixed frequency but variable duty cycle is achieved by changing the percentage of on time. This is known as **pulse width modulation** or **PWM**. This term is used extensively in digital electronics. If a circuit consisting of a power source, light bulb, and switch is constructed, the switch can be used to "pulse" the on/off time of the bulb. This pulsing can be coded into many types of data such as alpha or numeric values. Pulses are controlled, immediate variations in current flow in a circuit. In fact, the increase or decrease in current would ideally be instantaneous. If this were so, the pulse could be represented graphically as in **Figure 6–5**.

👓 **Shop Talk**

Some computer-controlled devices operate on a fixed frequency but variable **pulse width (PW)** *duration— that is, on time versus off time. Percentage of on time is referred to as* duty cycle. *The commonly used diesel electronic unit injectors (EUIs) are switched in this way and it is referred to as* pulse width modulation *or PWM.*

True pulse shaping, however, results in graduated rise when the circuit is switched to the on state and in graduated fall when the circuit is switched to the off state. This is explains why a square wave is often not truly square when displayed on a labscope. We call the deviation from the "square" appearance *ringing*. Most true square waves usually display some ringing when viewed on a scope. **Figure 6–6** compares a theoretical square sine wave with an actual wave.

Waves are rhythmic fluctuations in circuit current or voltage. They are often represented graphically and are described by their graphic shapes. **Figure 6–7** shows some typical waveforms.

The term *signals* is therefore used to describe electrical pulses and wave forms that are spaced or shaped to transmit data. The mechanisms and processes used to shape data signals are called modulation. In vehicle electronics, the term *modulation* is more commonly used in reference to digital signaling. An example would be the signal that a diesel engine **electronic control module (ECM)** uses to control the solenoid in an electronic unit injector (EUI). A PWM signal can be divided into primary modulation, which controls the amount of on time, and secondary or sub-modulation, which controls the cur-

rent flow. Frequency is another element of an electronic signal. A signal modulated at a frequency of 50 Hz completes 50 cycles per second. If voltage is the electrical signal that is being modulated at a frequency of 50 Hz, then each second is divided into 50 segments within which the voltage will be in the on state for a portion of time. The percentage of on time is expressed as duty cycle or pulse width (PW). A 100 percent duty cycle would indicate the maximum on time signal. A 0 percent duty cycle would indicate no signal or maximum off time.

Electronic *noise* is unwanted pulse or wave form interference that can scramble signals. All electrical and electronic components produce electromagnetic fields that may generate noise. Note that all electronic circuits are vulnerable to magnetic and electromagnetic field effect. The ringing shown in the actual square sine wave in **Figure 6–6** is what electronic noise can look like in diagram form. When electronic noise becomes excessive, signals can be corrupted. A corrupted signal means that the information intended for relay is inaccurate. Most electronic circuits must be protected against electronic noise. This means they must be shielded from interference from low-level radiation such as other vehicle electrical systems, high tension electrical wiring, and radar.

6.2 SEMICONDUCTORS

Semiconductors are a group of materials with exactly 4 electrons in their outer shell. As such, they cannot be classified either as insulators or conductors. Silicon is the most commonly used semiconductor in the manufacture of electronic components, but other substances, such as germanium, are also used. The first transistor shown in **Figure 6–1** earlier in this chapter used a germanium semiconductor. Silicon and germanium both have 4 electrons in their valence shells, but they would "prefer" to have 8. This means that semiconductor atoms readily unite in clusters called crystals, sharing electrons in their outer shells. Silicon and germanium can be grown into large crystals by heating the elements to their melting temperature, followed by a period of cooling. **Figure 6–8** shows how electrons in the valence shell are shared in a semiconductor crystal, and **Figure 6–9** shows the crystal structure of a crystallized cluster of germanium atoms.

Pure silicon and pure germanium are of little use in electronic components. For a semiconductor to be useful it must be doped; that is, have small quantities of impurities added to it. The **doping** agents are usually phosphorus and boron. The doping intensity will

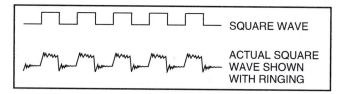

FIGURE 6–6 *Diagram of theoretical square sine wave compared with actual square sine wave shown with ringing.*

| SQUARE WAVE-LOW FREQUENCY ⎍⎍⎍ |
| SQUARE WAVE-HIGH FREQUENCY ⊓⊓⊓⊓⊓⊓ |
| SINE WAVE-LOW FREQUENCY ∿∿∿ |
| SINE WAVE-HIGH FREQUENCY ∿∿∿∿∿∿ |

FIGURE 6–7 *Waveforms.*

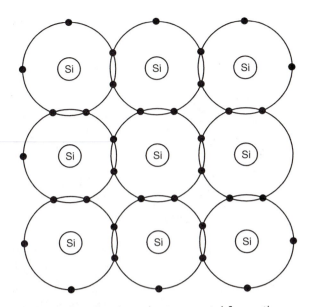

FIGURE 6–8   *Semiconductor crystal formation.*

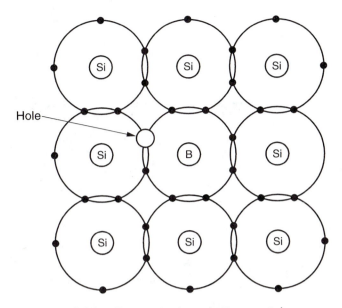

FIGURE 6–10   *P-type semiconductor crystal.*

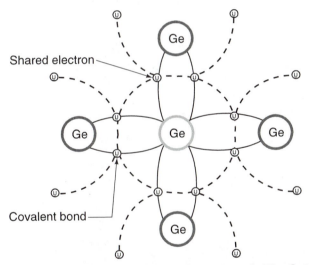

Germanium Semiconductor: 4 Electrons In The Outer Shell–Shown Crystallized With Shared Electrons

FIGURE 6–9   *A crystallized cluster of germanium atoms.*

define the electrical behavior of the crystal. After doping, silicon crystals may be sliced into thin sections known as *wafers*.

The type of doping agent used to produce silicon crystals will define the electrical properties of the crystals produced. Whereas semiconductors have 4 electrons in their valence, the doping agents have either 3 (trivalent) or 5 (pentavalent) electrons in their valence shells. When a semiconductor crystal is constructed, the doping agent produces a "bias" in the electrical character of the semiconductor crystal produced. For instance, a boron atom (common doping element) has

3 electrons in its outer shell. A boron atom in a crystallized cluster of silicon atoms will produce a valence shell with 7 electrons instead of 8. This "vacant" electron opening is known as a *hole*. The hole makes it possible for an electron from a nearby atom to fall into the hole. In other words, the holes can move, permitting a flow of electrons. Silicon crystals doped with boron (or other trivalent elements) form P-type (positively doped) silicon. **Figure 6–10** shows the hole created by doping a semiconductor crystal with a trivalent element such as boron.

A semiconductor crystal can also be produced with an N or negative electrical characteristic. A phosphorus atom has 5 electrons in its outer shell. It is *pentavalent*. In the bonding between the semiconductors and the doping material there is room for only 8 electrons in the center shell. Even when the material is in an electrically neutral state, the extra electron can move through the crystal. When a silicon crystal is manufactured using a doping material with 5 electrons in the valence shell it forms N-type (negatively doped) silicon. The first transistor manufactured used a germanium semiconductor. Because the element germanium is a semiconductor, just like silicon, it also has 4 electrons in its valence shell. **Figure 6–11** shows a silicon crystallized cluster doped with arsenic to form an N-type semiconductor.

Doping a semiconductor crystal always defines its electrical characteristics. We describe the electrical characteristics of a semiconductor as P type or N type. In **Figure 6–12**, germanium is used once again to show how the element forms a P-type semiconductor when doped with arsenic and an N-type semiconductor when doped with boron. P-type and N-type semiconductor crystals permit an electrical

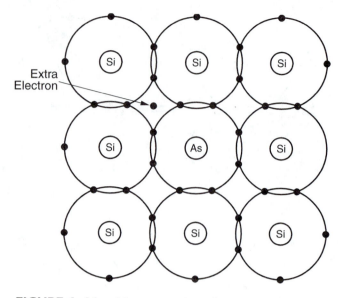

FIGURE 6-11 *N-type semiconductor crystal.*

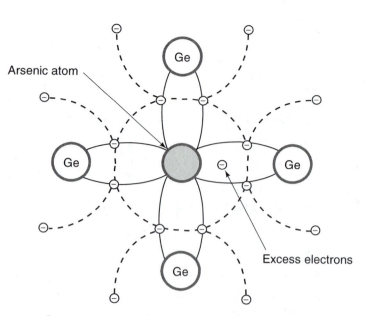

Germanium Doped With Arsenic To Form A N-Type Semiconductor

current to flow in different ways. In the P-type semiconductor, current flow is made possible by a deficit of electrons whereas in the N-type semiconductor, current flow is made possible by an excess of electrons. Whenever a voltage is applied to a semiconductor, electrons will flow towards the positive terminal and the holes will move towards the negative terminal.

6.3 DIODES

The suffix *-ode* literally means *terminal*. For instance, it is used as the suffix for the words **cathode** and **anode**. The word *diode* means literally "having two terminals." Diodes are constructed of semiconductor materials. A little earlier in this chapter, we discovered how both P-type and N-type semiconductor crystals can conduct electricity. The actual resistance of each type is determined by either the proportion of holes or surplus of electrons. When a chip is manufactured using both P and N-type semiconductors, electrons will flow in only one direction. The diode is used in electronic circuitry as a sort of one-way check valve that will conduct electricity in one direction (forward) and block it in the other (reverse). When a diode is connected in its proper polarity, that is, to permit current flow, it is said to be in forward bias. When it is connected in opposite polarity, that is, to block current flow, it is said to be in reverse bias. In a diagram of a diode, the arrow always points in the direction of conventional current flow when forward biased (**Figure 6–13**).

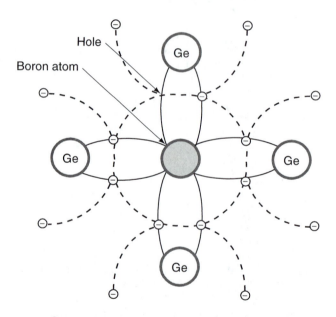

Germanium Doped With Boron To Form A P-Type Semiconductor

FIGURE 6–12 *Doped germanium semiconductor crystals of the P and N types.*

The positive terminal (+) is called the anode and the negative terminal (−) the cathode. As an electrical one way check valve, diodes will permit current flow only when correctly polarized. For instance, diodes are used in AC generators (alternators) to produce a DC characteristic from AC in a process known as rectification. They are also used extensively in electronic circuits. Diodes are static circuit elements because they neither gain nor store energy.

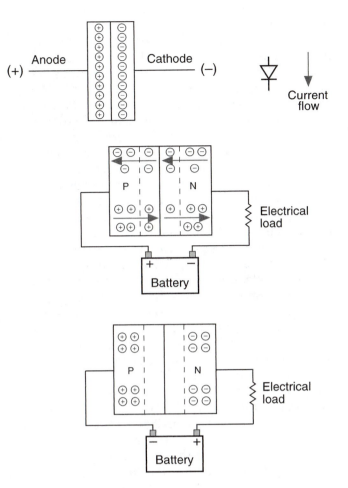

 Shop Talk _____

A diode is simply a one-way electrical check valve. When it is connected in the correct polarity to allow current flow, it is described as **forward biased**. *When connected in opposite polarity, that is, to block current flow, it is described as* **reverse biased**.

FIGURE 6–13 Diode operation and diodes in forward and reverse bias.

Diodes may be destroyed when subjected to voltage or current values that exceed their rated capacity. Excessive reverse current may cause a diode to conduct in the wrong direction, and excessive heat can melt the semiconductor material. The following abbreviations are used to describe diodes in a circuit:

V_F = forward voltage I_F = forward current
V_R = reverse voltage I_R = reverse current

TYPES OF DIODES

Numerous types of diodes exist, and they play a variety of roles in electronic circuits. The following is a sample of some of the more common types.

Small Signal Diodes

Small signal diodes are used to transform low current AC to DC (using a rectifier), perform logic data flows, and absorb voltage spikes. Small signal diodes are used to rectify the AC current produced by an AC generator's (alternator's) stator to the DC current required by the vehicle electrical system.

Power Rectifier Diodes

These function in the same manner as small signal diodes but are designed to permit much greater current flow. They are used in multiples and often mounted on a heat sink to dissipate excess heat caused by high current flow.

Zener Diodes

A **zener diode** (**Figure 6–14**) functions as a voltage-sensitive switch. They are named after their inventor (Clarence Zener in 1934). The zener diode is designed to block reverse bias current, but only up to a specific voltage value. When this reverse breakdown voltage is attained, it will conduct the reverse bias current flow without damage to the semiconductor material. Zener diodes are manufactured from heavily doped semiconductor crystals. They are used in electronic voltage regulators and in other automotive electronic circuitry. Zener diodes are rated by their breakdown or threshold voltage (V_Z), and this can range from 2V to 200V.

Light-Emitting Diodes (LEDs)

All diodes emit electromagnetic radiation when forward biased, but diodes manufactured from some semiconductors (notably gallium arsenide phosphide) emit it at much higher levels. **Light-emitting**

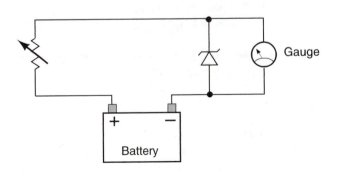

FIGURE 6–14 Zener diode.

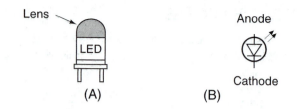

FIGURE 6–15   *Light-emitting diode (LED).*

diodes (LEDs) may be constructed to produce a variety of colors and are commonly used for digital data displays. For instance, a digital display with 7 linear LED bars arranged in a bridged rectangle could display any single numeric digit by energizing or not energizing each of the 7 LEDs. LED arrangements constructed to display alpha characters are only slightly more complex. LEDs convert electrical current directly into light (photons) and therefore are highly efficient because no heat losses occur. LEDs are already being used as clearance lights in trucks and trailers, and OEMs anticipate that their use will increase significantly over the next few years. They consume a fraction of the power of conventional bulbs and outlast them. **Figure 6–15** is a diagram of an LED in schematic form.

Photo Diodes

All diodes produce some electrical response when subjected to light. A photo diode is designed to detect light and therefore has a clear window through which light can enter. Silicon is the usual semiconductor crystal medium used in photo diodes.

Summary of Diodes

The following list summarizes this discussion of diodes:

- Diodes are 2-terminal devices.
- A diode always has an anode (positive terminal) and a cathode (negative terminal).
- Diodes act like one-way check valves, allowing current flow in one direction (forward bias) and blocking it in the other (reverse bias).
- Diodes can be regarded as static circuit elements in that they neither gain nor store energy.

6.4 TRANSISTORS

Transistors are 3-terminal semiconductor chips that are used extensively in electronic circuits. Unlike diodes, which are static elements in electronic circuits, transistors are active circuit elements capable of amplifying or transforming a signal level. A transistor consists of two P-N junctions. As with the diodes, P and N describe the semiconductors that define the transistor's electrical characteristics. That means two basic types of transistor exist, NPN and PNP. A terminal connects each of the semiconductor segments. A transistor's three terminals are known as *base, collector,* and *emitter.* To some extent, the names used to describe the terminals explain their functions. A small base current is used to control a much larger current flow through the collector and emitter. The base can be regarded as a "switch." The collector can be regarded as the input and the emitter as the output of the transistor.

In electricity, a relay is used to enable a small electrical current to control a much larger electrical current. In electronics, a transistor can be used in much the same way, that is, as a pilot relay, to switch a circuit on or off. However, the transistor is more versatile because it can be used to switch and amplify. Amplifying means that by varying the amount of base current, the collector-emitter current flow through the transistor can be controlled. An everyday use of an amplifying transistor is replacing the variable resistor that controls the brightness of display lighting. Audio amplifiers in stereo systems have used transistors for many years now. Transistors can be generally grouped into **bipolar** and *field effect* categories. (These will be explained later in this chapter.) Transistors offer many advantages over the electrical devices they replace, such as the now obsolete glass tubes. They are faster, much more compact, and have a significantly greater lifespan. **Figure 6–16** shows how transistors are represented in schematics and how they operate.

👓 Shop Talk

When a transistor is used as an amplifier, a small amount of base current can be used to control a larger amount of current flow through the collector and emitter. Today, we use transistors instead of series variable resistors to control the variable brightness on the dash instrument cluster.

BIPOLAR TRANSISTORS

A bipolar transistor functions very much like a sort of switched diode. It has 3 terminals known as collector, emitter, and base. Each terminal is connected to semiconductor media in what is known as a silicon sandwich. The sandwich may either be NPN or PNP. An NPN sandwich is constructed of a layer of P-type semiconductor material sandwiched between two

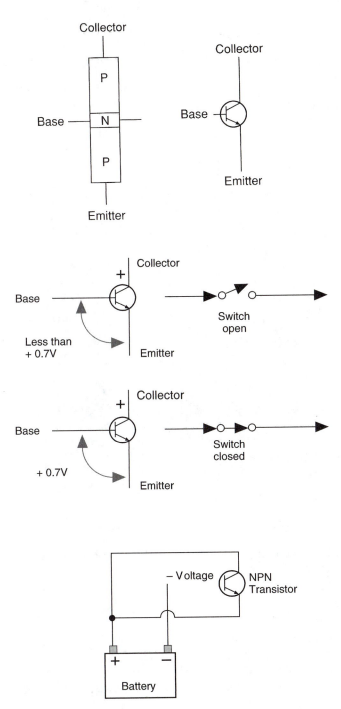

FIGURE 6–16 Transistor operation.

Looking at the operation of a typical NPN transistor, the base-emitter junction acts as a diode. Current flows forward into the base and out of the emitter but cannot flow in the reverse direction. The collector-base also acts as a sort of diode, but supply voltage is applied to it in the reverse direction. However, when the transistor is used as a switch in the on state, the collector-base junction may become forward biased. Under normal operation, current flow through the base-emitter terminals can be controlled by the current flowing through the base terminal. In this way, small base currents are used to control large collector currents. In fact, the amplification capability of transistors is typically 100 times (although it may be much less or much more). This allows a base current to control a collector current 100 times greater.

Facts About Bipolar Transistor Operation

- The base emitter junctions will not conduct until the forward bias voltage exceeds ±0.6V.
- Excessive current flow through a transistor will cause it to overheat or fail.
- Excessive voltage can destroy the semiconductor crystal media.
- A small base current can be used to control a much larger collector current.

Examples of Bipolar Transistors

Many different types of bipolar transistor exist, but a couple of note are small signal switching and power transistors. These two types are used extensively in vehicle electronic systems. Small signal transistors are used to amplify signals. Some may be designed to fully gate current flow in their off position and others may both amplify and switch. Power transistors may conduct high current loads and are often mounted on heat sinks to enable them to dissipate heat. They are sometimes known as *drivers* because they serve as the final or output switch in an electronic circuit used to control a component such as a solenoid or pilot switch. For instance, the solenoids in electronic unit injectors (EUIs) and hydraulically actuated electronic unit injectors (HEUIs) are controlled by power transistors located in injector driver units built into the electronic control modules (ECMs).

FIELD EFFECT TRANSISTORS (FETS)

FETs are more commonly used than bipolar transistors largely because they are cheaper to manufacture. They may be divided into *junction*-type and

layers of N-type semiconductor. The middle of the sandwich always acts as a *gate* capable of controlling the current flow through all three layers. The base is fairly thin and has comparatively fewer doping atoms than that of the semiconductor material on either side of it. They are designed so a small emitter-base current will ungate the transistor and permit a larger emitter-collector current flow.

metal-oxide semiconductors. Both types are controlled by a very small input (base/gate) voltage. **Figure 6–16** shows a typical FET transistor.

Junction FETs (JFETs)

There are two types of JFETs: N-channel and P-channel. The channel behaves as a resistor that conducts current from the source side to the drain side. Voltage at the gate will act to increase the channel resistance, thereby reducing drain to source current flow. This allows the FET to be used either as an amplifier or a switch. When gate voltage is high, high resistance fields are created, narrowing the channel available for conductivity current to the drain. In fact, if the gate voltage is high enough, the fields created in the channel can join and completely block current flow.

Facts About JFETs

You should be aware of a couple of facts regarding JFETs:

- JFET gate-channel resistance is very high, so the device has almost no effect on external components connected to the gate.
- Almost no current flows in the gate circuit because the gate-channel resistance is high. The gate and channel form a "diode," and as long as the input signal "reverse biases" this diode, the gate will show high resistance.

Metal-Oxide Semiconductor Field Effect Transistors (MOSFETs)

The MOSFET has become the most important type of transistor in microcomputer applications in which thousands can be photo-infused onto minute silicon wafers. They are easy to manufacture and consume fractional amounts of power.

MOSFETs are classified as P-type or N-type. However, in a MOSFET, there is no direct electrical contact between the gate with the source and the drain. The gate's aluminum contact is separated by a silicon oxide insulator from the remainder of the transistor material.

When the gate voltage is positive, electrons are attracted to the region around the insulation in the P-type semiconductor medium. This produces a path between the source and the drain in the N-type semiconductor material, permitting current flow. The gate voltage will define the resistance of the path or channel created through the transistor, permitting them to both switch and act as variable resistors. MOSFETs can be switched at very high speeds and,

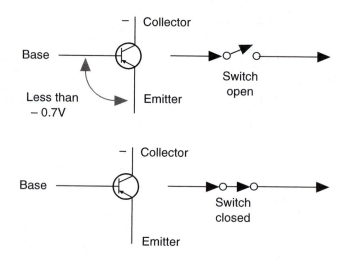

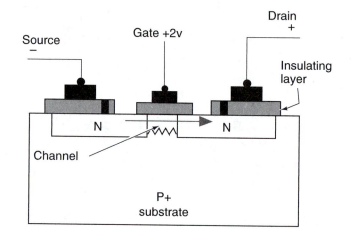

FIGURE 6–17 *Transistor operating principle and schematics.*

because the gate-channel resistance is high, almost no current is drawn from external circuits.

Figure 6–17 shows the operating principle of a transistor: Note how the device is switched by a voltage applied to the gate.

THYRISTOR

Thyristors are 3-terminal, solid-state switches. They are different from transistors in that they are only capable of switching. A small current flow through one of the terminals will switch the thyristor on and permit a larger current flow between the other two terminals. Thyristors are switches, so they are either in an on or off condition. They fall into two classifications depending on whether they switch AC or DC current. Some thyristors are designed with only two terminals and they will conduct current when a specific trigger or breakdown voltage is achieved.

Silicon-Controlled Rectifiers (SCRs)

SCRs are similar to a bipolar transistor with a fourth semiconductor layer added. They are used to switch DC. When the anode of an SCR is made more positive than the cathode, the outer two P-N junctions become forward biased: The middle P-N junction is reverse biased and will block current flow. A small gate current, however, will forward-bias the middle P-N junction, enabling a large current to flow through the thyristor. SCRs will remain on even when the gate current is removed. The on condition will remain until the anode-cathode circuit is opened or reversed biased. SCRs are used for switching circuits in vehicle electronic and ignition systems. **Figure 6–18** shows a forward direction SCR.

DARLINGTON PAIRS

A **Darlington pair** (named after the inventor Sidney Darlington) consists of a pair of transistors wired so the emitter of one supplies the base signal to a second, through which a large current flows. The objective once again is to use a very small current to switch a much larger current. This type of application is known as amplification. Darlington pairs are used extensively in vehicle computer control systems and in ignition modules. **Figure 6–18** shows a Darlington pair relationship.

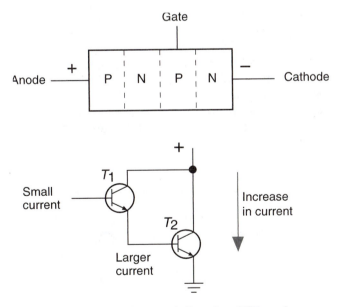

FIGURE 6–18 A forward direction SCR and a Darlington pair relationship.

6.5 PHOTONIC SEMICONDUCTORS

Photonic semiconductors emit and detect light or *photons.* A photon is a unit of light energy. Photons are produced electrically when certain electrons excited to a higher than normal energy level return to a more normal level. Photons behave like waves. The distance between the wave nodes and anti-nodes (wave crests and valleys) is known as *wavelength.* Electrons excited to higher energy levels emit photons with shorter wavelengths than electrons excited to lower levels. Photons are not necessarily visible and it is perhaps important to note that they may truly be described as *light* only when they are visible.

All visible light is classified as electromagnetic radiation. The specific wavelength of light rays will define its characteristics. Light wavelengths are specified in nanometers; that is, billionths of a meter.

THE OPTICAL LIGHT SPECTRUM

The optical light spectrum includes ultraviolet, visible, and infrared radiation. **Figure 6–19** shows a graphic representation of the optical light spectrum. Photonic semiconductors either emit or can detect near-infrared radiation frequencies. Near-infrared means that the frequency is slightly greater than the visible red end of the visible light spectrum and is therefore usually referred to as light. **Figure 6–20** shows the full optical light, or electromagnetic, spectrum. Note the portion of the spectrum that is classified as visible light.

OPTICAL COMPONENTS

Optical components may conduct, refract, or modify light. In the same way electrical signals can relay information by modulating duty cycle and frequency, light waves can be similarly used. The use of optics in vehicle technology is rapidly increasing. Some of the more common optical components are discussed in the following sections.

Filters

Filters transmit only a narrow band of the spectrum and block the remainder. They operate much like the filter lens used in a welding helmet.

Reflectors

Reflectors reflect a light beam, or at least most of it, in much the same way a mirror functions.

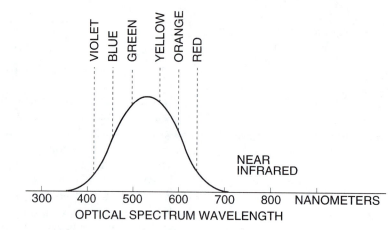

FIGURE 6–19   *The optical spectrum.*

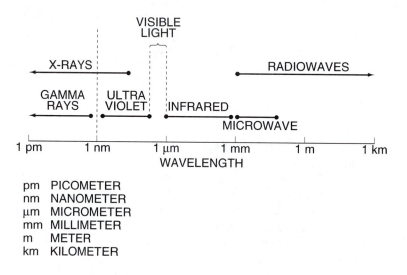

FIGURE 6–20   *The electromagnetic spectrum.*

Beam Splitters

Beam splitters transmit some of the optical wavelength and reflect back the rest of it.

Lenses

Lenses bend light waves. They are often used in conjunction with semiconductor light sources and detectors. They are often used to collect and focus light onto a detector.

Optical Fibers

Optical fibers are thin, flexible strands of glass or plastic that conduct light. The light travels through a core surrounded by conduit or cladding. Optical fibers are increasingly used to transmit digital data by pulsing light, and the use of **fiber optics** is expected to grow tenfold in the next decade.

Solar Cells

A solar cell consists of a P-N or N-P silicon semiconductor junction built onto contact plates. A single sili-con solar cell may generate up to 0.5V in ideal light conditions (bright sunlight), but output values are usually lower. Like battery cells, solar cells are normally arranged in series groups, in which case the output voltage would be the sum of cell voltages. They can also be arranged in parallel, in which case the output current would be the sum of the cell currents. They are sometimes used as battery chargers on vehicles.

6.6   TESTING SEMICONDUCTORS

The technician is seldom required to test an individual diode or transistor in a vehicle electronic circuit, but it is an activity that is often taught in technical training programs because it increases awareness of how electronic systems work and fail. Diodes and transistors are normally tested using an ohmmeter. The semiconductor to be tested should be isolated from the circuit. The digital multimeter (DMM) should be set to the diode test mode.

Diodes

Diodes should produce a high-resistance reading with the DMM test leads in the forward bias direction and low resistance when the leads are reversed. Low-resistance readings both ways indicate a shorted diode. High resistance both ways indicates an open diode. **Figure 6–21** and **Figure 6–22** introduce the basic testing of a diode: This is explored in a little more detail in Chapter 8.

Transistors

Again, the instrument used to bench test a transistor is the DMM in ohmmeter mode. The testing of transistors is covered in a little more detail in Chapter 8, but this should give you some idea of how to go about the procedure. Use **Figure 6–23** and **Figure 6–24**, a DMM, and some transistors to practice on. A functional transistor should test as follows:

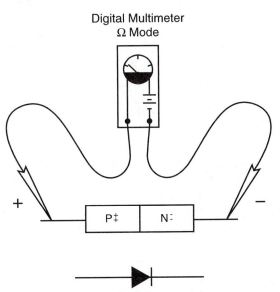

FIGURE 6–21   Testing a diode in forward bias.

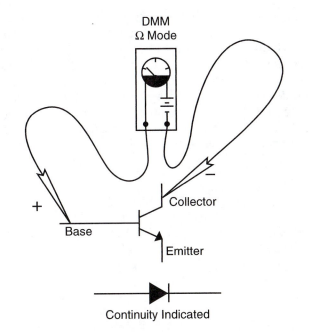

FIGURE 6–23   Testing transistor base to collector continuity.

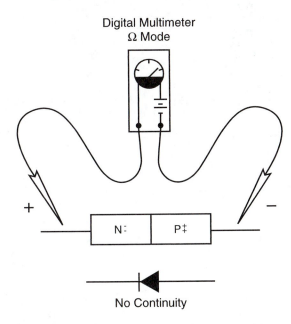

FIGURE 6–22   Testing a diode in reverse bias.

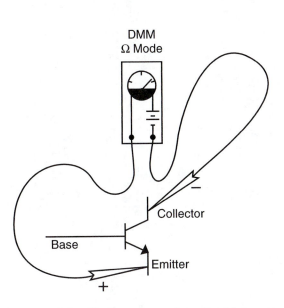

FIGURE 6–24   Testing a transistor for high resistance across the emitter-collector junction.

- Continuity between the emitter and base
- Continuity between the base and the collector when tested one way, and high resistance when DMM test leads are reversed
- High resistance in either direction when tested across the emitter and collector terminals

6.7 INTEGRATED CIRCUITS (I/Cs)

Integrated circuits or **I/Cs** consist of resistors, diodes, and transistors arranged in a circuit on a *chip* of silicon (**Figure 6–25**). The number of electronic components that comprise the I/C vary from a handful to hundreds of thousands, depending on the function of the chip. Integrated circuits have innumerable household, industrial, and automotive applications and are the basis of digital watches, electronic pulse wipers, and all computer systems.

Shop Talk

A chip is an integrated circuit. One chip can be smaller than a fingernail and contain many thousands of resistors, diodes, and transistors.

Integrated circuits fall into two general categories. Analog integrated circuits operate on variable voltage values. Electronic voltage regulators are a good vehicle example of an analog I/C. Digital integrated circuits operate on two voltage values only, usually *presence of voltage* and *no voltage.* Digital I/Cs are the basis of most computer hardware including processing units, main memory, and data retention chips. Integrated circuit chips can be fused into a motherboard (main circuit) or socketed. The latter has the advantage of easy removal and replacement. A common chip package used in computer and vehicle engine/electronic control modules (ECMs) is the dual in-line package (DIP). This package consists of a rectangular plastic-enclosed I/C with usually 14 to 16 pins arranged evenly on either side. DIPs may be fused (not removable) or socketed to the motherboard.

6.8 GATES AND TRUTH TABLES

In the process of outlining the operation of transistors in one of the previous sections, the importance of gates was emphasized. These are the electronically controlled switching mechanisms that manage the operating mode

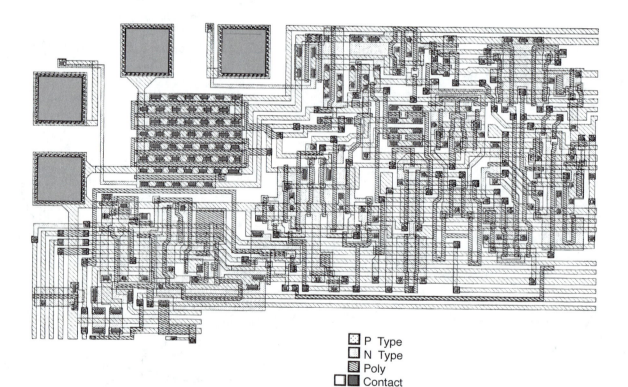

P Type
N Type
Poly
Contact
Metal

FIGURE 6–25 Integrated circuit.

of a transistor. Digital integrated circuits are constructed by using thousands of gates. In most areas of electronics, gates can be either open or closed: In other words, in-between states do not exist. Just like the toggle switch that controls the lights in a room, the switch is either on or off. The terms used to describe the state of a gate are *on* and *off*. In a circuit, these states are identified as *presence of voltage* or *no voltage*.

AND Gates

The best way to learn about the operation of gates in a digital circuit is to observe the operation of some electromechanical switches in some simple electrical circuits. In **Figure 6–26**, a power source is used to supply a lightbulb in a series circuit. In the circuit, there are two pushbutton switches that are in the normally open state. In such a circuit, the lightbulb will only illuminate when both switches are closed. We call this type of gate circuit an **AND gate**.

The operation of the AND gate can be summarized by looking at the circuit and coming to some logical conclusions. A table that assesses a gated circuit's operation is often called a **truth table**. A truth table applied to the AND gate circuit would read as follows:

Switch A	Switch B	Outcome
Off	Off	Off
Off	On	Off
On	Off	Off
On	On	On

A truth table is usually constructed using the digits *zero* (0) and *one* (1) because the **binary system** (outlined in detail following this section on gates) is commonly used to code data in digital electronics. Therefore, a truth table that charts the outcomes of the same AND switch would appear as follows:

Switch A	Switch B	Outcome
0	0	0
0	1	0
1	0	0
1	1	1

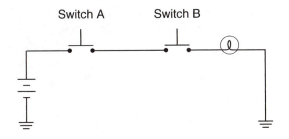

FIGURE 6–26 AND gate: Both switches are of the normally open type.

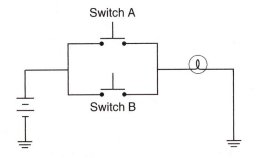

FIGURE 6–27 OR gate: Both switches are of the normally open type.

OR Gates

If another circuit is constructed using two normally open electromechanical switches, this time arranged in parallel, the circuit would appear as in **Figure 6–27**. This kind of gate is called an **OR gate**. Both switches are of the normally open type: If either one is closed, the gate outcome is on.

A truth table constructed to represent the possible outcomes of the OR gate circuit would appear as follows:

Switch A	Switch B	Outcome
0	0	0
1	0	1
0	1	1
1	1	1

NOT or Inverter Gates

A **NOT gate** circuit switch can be constructed by using a push button switch that is in the normally closed state. In other words, circuit current flow is interrupted when the button is pushed. In the series circuit shown in **Figure 6–28**, current will flow until the switch opens the circuit.

If a truth table is constructed to graphically represent the outcomes of a NOT gated circuit, this is what it would look like:

In	Out
0	1
1	0

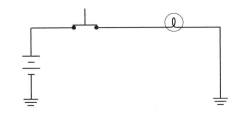

FIGURE 6–28 NOT gate: The single switch is of the normally closed type.

Although the examples presented here used electromechanical switches for ease of understanding, in digital electronics, circuit switching is performed electronically using diode and transistor gates. AND, OR, and NOT gates are 3 commonly used means of producing an outcome based on the switching status of components in the gate circuit.

GATES, TRUTH TABLES, AND BASIC DATA PROCESSING

Figure 6–29 shows some simple input logic gates and the outcomes that can be produced from them. When the number of inputs to a logic gate is increased, its outcome processing becomes more complex.

Gates are normally used in complex networks in which they are connected by buses, or connection points, to form a logic circuit. **Bits** of data can be moved through the logic circuit or highway to produce outcomes that evolve from the logical processing of inputs. By massing hundreds of thousands of logic circuits, the processing of information becomes possible. A computer works simply by processing input data and generating logical outcomes based on the input data. Although it may not be required that a vehicle technician understand the precise operation of a digital computer, these have become so much part of the workplace that it is certainly desirable to have some idea of the basics. Computer operation is explored in a little more detail later in this chapter.

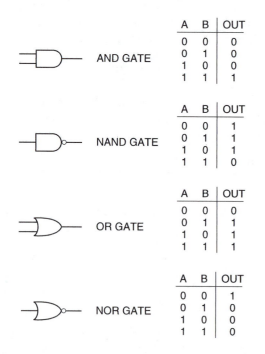

FIGURE 6–29 Logic gates and truth tables showing the switching outcomes.

The binary system is an arithmetic numeric system using two digits and it therefore has a base number of 2. The base number of a numeric system is the number of digits used in it. The decimal system is therefore a 10-base number system. The binary numeric system is often used in computer electronics because it directly corresponds to the *on* or *off* states of switches and circuits. In computer electronics, the binary system is the primary means of coding data using the digits 0 and 1 to represent alpha, numeric, and any other data. Numeric data is normally represented using the decimal system: Again, this is simple coding. The digit 3 is a representation of a quantitive value that most of us have been accustomed to decoding since early childhood. If decimal system values were coded into binary values, they would look like this:

Decimal digit	Binary digit
0	0
1	1
2	10
3	11
4	100
5	101
6	110
7	111
8	1000
9	1001
10	1010

In digital electronics, a *bit* is the smallest piece of data that a computer can manipulate. It is simply the ability to represent one of two values, either *on* (1) or *off* (0). When groups of bits are arranged in patterns, they are collectively described as follows:

4 bits = nibble
8 bits = byte

Most digital electronic data is referred to quantitively as **bytes**. Computer systems are capable of processing and retaining vast quantities of data: In most cases, millions (megabytes) and billions (gigabytes) of bytes are described. A byte is 8 bits of data. It has the capability of representing up to 256 data possibilities. For instance, if a byte were to be used to code numeric data, it would appear as follows:

Decimal digit	Binary coded digit
0	0000 0000
1	0000 0001
2	0000 0010
3	0000 0011

Decimal digit	Binary coded digit
4	0000 0100
5	0000 0101
6	0000 0110
7	0000 0111
8	0000 1000

and so on up to 256. Try it. Note that each time a one/on progresses one column to the left, it doubles the value of the column to its right.

```
0000 0001 = 1
0000 0010 = 2
0000 0100 = 4
0000 1000 = 8
0001 0000 = 16
0010 0000 = 32
0100 0000 = 64
1000 0000 = 128
```

A number of methods are used to code data. Those familiar with some computer basics may be acquainted with some American Standard Code for Information Interchange (ASCII) codes. This coding system has its own distinct method of coding values and would not be compatible with other coding systems without some kind of translation. Because on-off states can so easily be used to represent data, most digital computers and communications use this technology. It is also used in optical data processing, retention, and communications. Digital signals may be transmitted in series one bit at a time, which tends to be slower, or in parallel, which is much faster. If the numbers 0 through 3 had to be transmitted through a serial link, they would be signaled sequentially, as shown in **Figure 6–30**.

If the same numbers were to be transmitted through a parallel link, they could be outputted simultaneously as shown in **Figure 6–31**, which would be much faster.

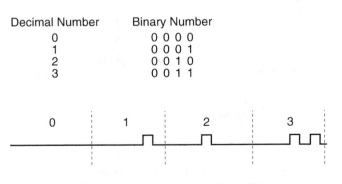

Decimal Number	Binary Number
0	0 0 0 0
1	0 0 0 1
2	0 0 1 0
3	0 0 1 1

FIGURE 6–30 *Sequential switching of binary coded numbers through a serial link.*

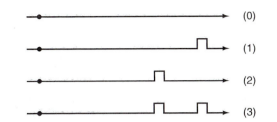

FIGURE 6–31 *Parallel link: Data transmission time is greatly reduced.*

6.10 MICROPROCESSORS

A microprocessor is a solid-state chip that contains many hundreds of thousands of gates per square inch. The microcomputer is the operational core of any personal or vehicle computer system. Most highway trucks and, increasingly, bus and off-highway equipment, use computers to manage engine and other onboard system functions. The majority of highway trucks manufactured in North America since 1994 are "drive by wire" with full authority chassis management. Onboard vehicle computers are referred to as engine/electronic control modules (ECMs) or electronic/engine control units (ECUs). ECM will be the acronym used in this text except when discussing product whose OEM (original equipment manufacturer) chooses to use the acronym ECU. ECMs normally contain a microprocessor, data retention media, and often the output or switching apparatus. The ECM can be mounted on the component to be managed (a transmission management ECM is often located on the transmission) or, alternatively, inside the vehicle cab. Increasingly as this technology develops, several system ECMs can be mastered by a single vehicle management ECM. An example would be a vehicle equipped with computerized engine management, transmission, and ABS/traction control. The three systems can be mastered by one *vehicle* management ECM for synchronized performance. Alternatively, each separate system ECM can be bussed (electronically connected) to the vehicle management computer. Some OEMs use the term *multiplexing* to refer to a vehicle management system using multiple, interconnected ECMs.

CAUTION: Static discharge can destroy sensitive computer equipment. Wear a ground strap or use common sense to avoid damaging computer equipment by careless static discharge.

A truck technician must have a basic understanding of both vehicle and personal computers to interact effectively with today's technology. This section will introduce the essentials of electronic management of vehicle systems. Computerized system management is simply summarized as a set of electronically connected components that enable an information processing cycle comprising 3 distinct stages:

1. Data input
2. Data processing
3. Outputs

The information processing cycle is shown in diagram form in **Figure 6–32**.

DATA INPUT

Data is simply raw information. Most of the data that has to be signalled in a truck electronic system ECM comes from monitoring sensors and command sensors. An example of a monitoring sensor would be a tailshaft speed sensor that signals road speed information to the

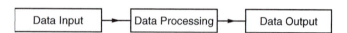

FIGURE 6–32 Information processing cycle.

various ECMs that need to know the vehicle road speed. An example of a command sensor would be the throttle position sensor, whose signal is used by the engine and transmission electronics. In fact, anything that signals input data to an electronic system can be described as a sensor. This means that sensors may be simple switches that an operator toggles open or closed, devices that ground or modulate a reference voltage (V-Ref), or devices that are powered up either by V-Ref or battery voltage (V-Bat). V-Ref is voltage conditioned by the system ECM. It is almost always 5V. Some components used in the **input circuit** of a vehicle computer system are discussed in the following sections.

Thermistors

Thermistors precisely measure temperature. There are two types. Resistance through a thermistor may decrease as temperature increases, in which case it is known as a negative temperature coefficient (NTC) thermistor. A thermistor in which the resistance increases as temperature rises is known as a positive temperature coefficient (PTC) thermistor. The ECM receives temperature data from thermistors in the form of analog voltage values. The coolant temperature sensor, ambient temperature sensor, and oil temperature sensor are usually thermistors. **Figure 6–33** shows a schematic of a thermistor.

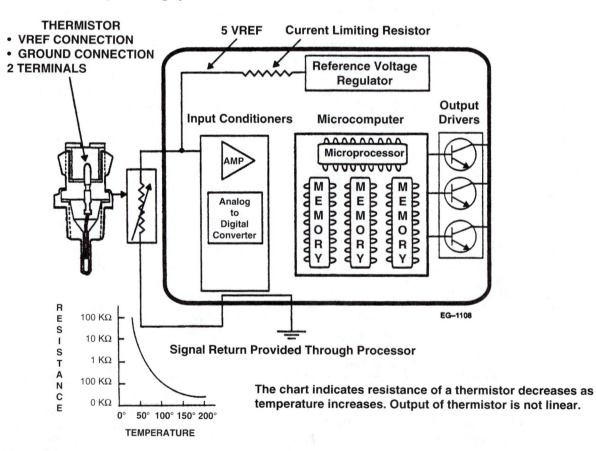

FIGURE 6–33 NTC-type thermistor. (Courtesy of International Truck and Engine Corp.)

Variable Capacitance (Pressure) Sensor

These are supplied with reference voltage and usually designed to measure pressure values. The medium whose pressure is to be measured acts on a ceramic disc and moves it either closer to or farther away from a steel disc. This varies the capacitance of the device and therefore the voltage value returned to the ECM.

Variable capacitance-type sensors are used for oil pressure sensing, MAP sensing (turbo boost pressure), barometric pressure sensing (BARO), and fuel pressure sensing. **Figure 6–34** shows an engine oil pressure sensor that uses a variable capacitance electrical operating principle.

Potentiometers

The **potentiometer** is a three-wire sensor (the wires used are V-Ref, ground, and return signal). The potentiometer is designed to vary its resistance in proportion to mechanical travel. The potentiometer receives a V-Ref or reference voltage from the system ECM. It outputs a voltage signal exactly proportional to the motion of a mechanical device. The signal returned to the system ECM is always less than V-Ref, so sometimes they can be referred to as voltage dividers. As the mechanical device moves, the resistance is altered within the potentiometer. Throttle position sensors (TPSs) are commonly potentiometers. **Figure 6–35** shows a TPS that uses a potentiometer principle of operation.

Piezo-Resistive Pressure Sensor

This type of sensor is often used to measure manifold pressure and can be known as a Wheatstone bridge sensor. A doped silicon chip is formed in a diaphragm shape so it measures 250 microns around its periphery and only 25 microns at its center: this means it is about 10 times thinner in the center. This permits the silicon diaphragm to flex at the center when subjected to pressure. A set of sensing resistors is formed around the edge of a vacuum chamber in which the diaphragm is located so when pressure causes the diaphragm to deflect, the resistance of the sensing resistors changes in proportion to the increase in pressure.

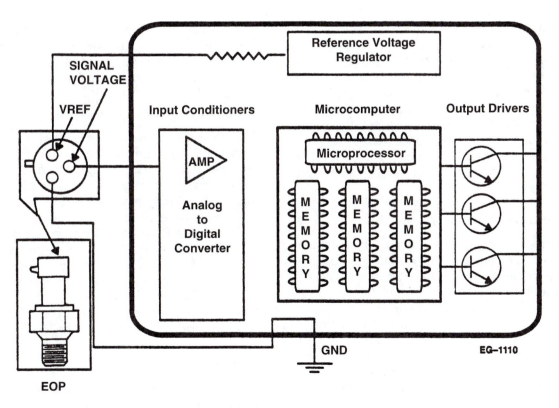

Variable Capacitance Sensor

FIGURE 6–34 _Variable capacitance-type sensor. (Courtesy of International Truck and Engine Corp.)_

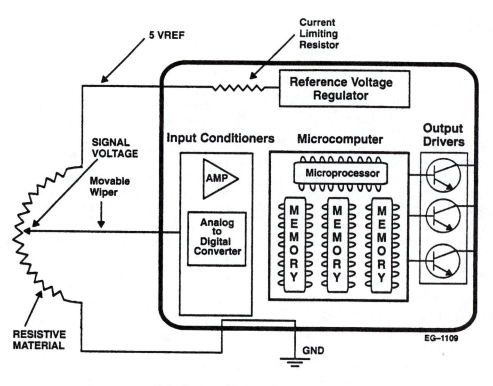

Potentiometer (Variable Resistance Voltage Divider)

FIGURE 6–35 *Potentiometer as used in a throttle position sensor. (Courtesy of International Truck and Engine Corp.)*

An electrical signal proportional to pressure is obtained by connecting the sensing resistors into a Wheatstone bridge circuit in which a voltage regulator holds a constant DC voltage across the bridge. When there is no pressure acting on the silicon diaphragm, all the sensing resistance will be equal and the bridge will be balanced. When pressure causes the silicon diaphragm to deflect, the resistance across the sensing resistors increases, unbalancing the bridge and creating a net voltage differential that can be relayed to the ECM as a signal.

Hall Effect Sensors

Hall effect sensors generate a digital signal as timing windows or vanes on a rotating disc passed through a magnetic field. The disc is known as a **pulse wheel** or **tone wheel,** terms that are both used to describe the rotating member of the induction pulse generator, so care should be taken to avoid confusion. The frequency and width of the signal provides the ECM with speed and position data. The disc incorporates a narrow window or vane for relaying position data. Hall effect sensors are used to input engine position data for purposes of event timing computations such as the beginning and duration

of the pulse width. The camshaft position sensor (CPS), timing reference sensor (TRS), and engine position sensors (EPS) are examples. **Figure 6–36** shows the signal output of a Hall effect sensor.

Induction Pulse Generator

A toothed disc known as a **reluctor,** pulse wheel, or tone wheel (also commonly known by the slang term **chopper wheel**) with evenly spaced teeth or serrations is rotated through the magnetic field of a permanent stationary magnet. As the field builds and collapses, AC voltage pulses are generated and relayed to the ECM. Reluctor-type sensors are used variously on modern truck chassis in applications such as ABS wheel speed sensors, vehicle speed sensors (VSS) located in the transmission tailshaft or a front wheel, and engine speed sensors. **Figure 6–37** shows the operating principle of an induction pulse generator.

Switches

Switches can either open or close a circuit by toggling (driver command) or ground-out (such as a radiator coolant level sensor).

HALL EFFECT SENSOR

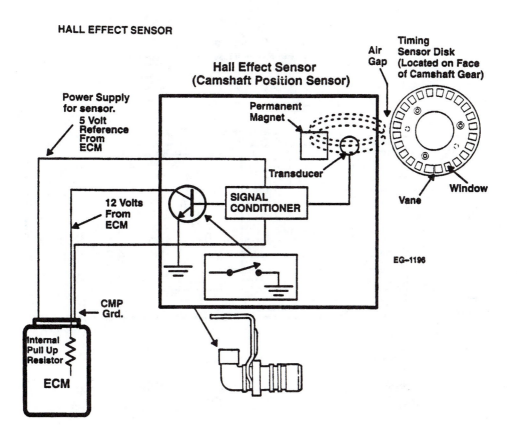

FIGURE 6–36 Hall effect sensor. (Courtesy of International Truck and Engine Corp.)

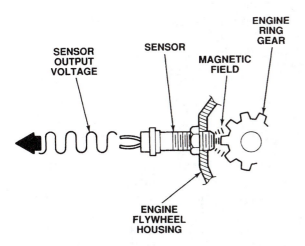

FIGURE 6–37 Induction pulse generator. (Courtesy of International Truck and Engine Corp.)

Data Processing

Data processing is the "thinking" function of a computer or microprocessor. This thinking function involves receiving inputs, consulting program instructions and memory, and then generating outputs. The term *data processing* is used to describe both the simple and complex processes that take place within the computer.

> **CAUTION:** Chassis electrical voltage pressure can destroy vehicle computers and sensors. When testing vehicle computer circuits, exercise caution when connecting break-out boxes and T's. Never use jumper wires and diagnostic forks across terminals unless the test procedure specifically requires this.

Probably the most important component in the microprocessor is the **central processing unit (CPU)**. The CPU contains a control unit that executes program instructions and an arithmetic logic unit (ALU) to perform numeric calculations and logic processing such as comparing data. It also clocks the processing frequency: The higher the frequency, the faster the processing speed of the computer. Most vehicle microprocessors run on voltage values lower than the 12V chassis voltage in much the same way that a home computer runs on much lower voltage than household supply voltage. Typically, this is

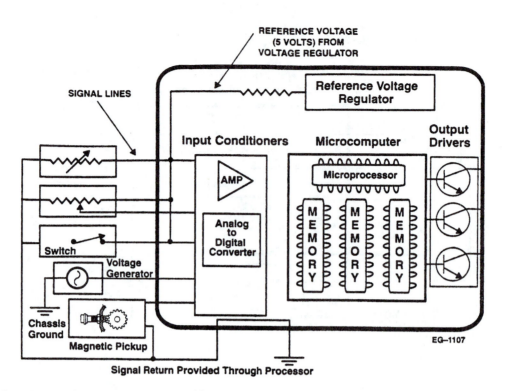

FIGURE 6–38 *Types of ECM input signals.* (Courtesy of International Truck and Engine Corp.)

5V. It is a function of the ECM to transform the chassis voltage to that required to run the microprocessor circuit. Additionally, a lower voltage value is required as reference voltage for the input circuit: Again, this is commonly 5V.

The role of the CPU is to manage the processing cycle. This requires receiving input data and locating it in the processing cycle. The various input devices discussed previously are shown in **Figure 6–38**. The CPU also fetches and carries information from the ECM memory compartments and loads this into the processing cycle. **Random access memory (RAM)** is data that is electronically retained in the ECM. Only this data can be manipulated by the CPU. Input data and magnetically retained data in **read-only memory (ROM), programmable read-only memory (PROM)** and **electronically erasable programmable read-only memory (EEPROM)** are transferred to RAM for processing. RAM can be called primary storage or main memory. Because RAM data is electronically retained, it is lost when its electrical power circuit is shut off. Because many of the signals into the ECM processing cycle are in analog format, these signals have to be converted to digital signals. The component required to perform this is known as an analog-to-digital converter (ADC). The same is true in reverse when the ECM processing generates an outcome to a component that requires an analog supply. A digital-to-analog converter (DAC) changes

a digital signal to an analog voltage signal. **Figure 6–39** is a simplified schematic of an ECM showing a basic processing cycle. The types of memory used in an ECM are explained in a little more detail later on in this chapter.

The following summarizes the functions of an ECM:

- Uses a CPU to clock and manage the processing cycle
- Contains in memory banks the data required to manage the system
- Conditions the processor circuit voltage
- Manages reference voltage
- Converts analog input data to a digital format using an ADC
- Converts digital outputs to analog voltages required to actuate electrical components

Outputs

The results of processing operations must be converted to action by switching units and actuators. In most (but not all) truck system management ECMs, the switching units are integral with the ECM. In an ECM managing a typical automatic transmission, clutch actuators or solenoid valves are among the primary output devices to be switched. ECM commands are converted to an electrical signal that is

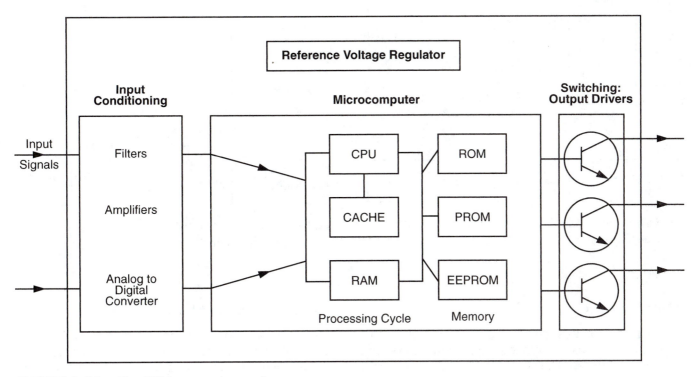

FIGURE 6–39 The ECM processing cycle.

used to energize clutch solenoids so the outcomes of the processing cycle are effected by shifting range ratios. The **output circuit** of any computing system simply effects the result of the processing cycle. Output devices on a home computing system would be the display monitor and the printer.

SAE Hardware and Software Protocols

Among the OEMs of truck and bus engines in the United States, there has been a generally higher degree of cooperation in establishing shared electronics hardware and software protocols than in the automobile manufacturing segment of the industry. To some extent this cooperation has been orchestrated by the Society of Automotive Engineers (SAE) and American Trucking Association (ATA). It has been necessary because most North American trucks are engineered by a truck manufacturer using major component subsystems supplied by an assortment of component OEMs. For instance, a Caterpillar engine could be specified to any truck chassis and be required to electronically interact with a Fuller AutoSelect transmission and Wabco ABS/ATC (antilock brake system/automatic traction control). Until recently, separate SAE J standards (surface vehicle recommended practices) controlled the hardware and software protocols that enabled a

TPS on a Cummins engine to provide the throttle position data required to manage an Allison WT transmission. The more recent SAE J 1939 standard attempts to incorporate the software and hardware standards covered in J 1587 and J 1708 into one standard. Some current trucks have J 1939 standards covered.

SAE J 1587	Electronic data exchange protocols used in data exchange between heavy duty, electronically managed systems.
SAE J 1708	Serial communications and hardware compatibility protocols between microcomputer systems in heavy vehicles.
SAE J 1939	The set of standards that incorporates both J 1587 and J 1708. Both software and hardware protocols and compatibilities are covered by J 1939. J 1939 will be updated by simply adding a suffix: Separate electronic systems that are J 1939-compatible can share rather than require duplication of common hardware and communicate using a common "language."

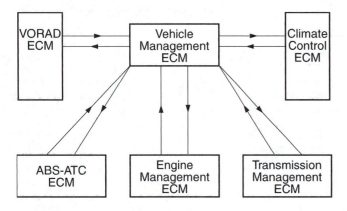

FIGURE 6–40   *Multiplexing—communication between multiple electronic management systems.*

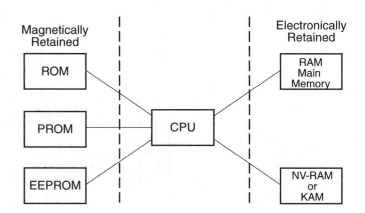

FIGURE 6–41   *ECM data retention: types of memory and the fetch-and-carry role of the CPU.*

Multiplexing

When a truck chassis uses multiple electronically managed circuits, bussing the system ECMs makes sense. Bussing the system ECMs avoids duplication of hardware and serves to synergize (a synergized system's components work in harmony) the operation of each system. For instance, both the engine management and transmission ECMs require accelerator position input to function. Enabling the two ECMs to communicate with each other would permit them to share input data and reduce the amount of input circuit hardware. The terms used to describe two or more ECMs connected to operate multiple systems are *multiprocessing* and *multiplexing.* Multiplexing will be used in this text. To connect two ECMs, they must use common operating protocols; that is, they must be speaking the same language. Many of the systems manufactured since 1995 are designed for multiplexing. However, when two electronic systems are required to operate with each other and their communication and operating protocols do not match, an electronic translator called an interface module is required. Interface modules are often not manufactured by either of the system OEMs. **Figure 6–40** shows how multiplexing is used in a modern truck integrated electronic system.

6.11   DATA RETENTION IN VEHICLE ECMS

Data is retained both electronically and magnetically in current generation truck and bus ECMs; however, optical data retention laser read systems will appear in the near future. The data categories discussed in the following sections are used by contemporary ECMs (**Figure 6–41**).

Random Access Memory (RAM)

The amount of data that may be retained in RAM is a primary factor in measuring the computing power of a computer system. It is also known as *main memory* because the CPU can only manipulate data when it is retained electronically. At start-up, RAM is electronically loaded with the management operating system instructions and all necessary running data retained in other data categories (ROM/PROM/ EEPROM).

RAM data is electronically retained, which means that it is always to some extent volatile (lost when the circuit is not electrically energized). In other words, data storage could be described as temporary—if the circuit is opened, RAM data is lost. It should be noted that most truck ECMs use only fully volatile RAM; therefore, when the ignition key circuit is turned off, all RAM data is lost.

However, a second type of RAM is used in some truck and many automobile systems, often those that have no EEPROM capability. This is nonvolatile RAM (NV-RAM) or keep alive memory (KAM) in which data is retained until either the battery is disconnected or the ECM is reset. The ECM is usually reset by depressing a computer reset button that temporarily opens the circuit. Codes and failure strategy (action sequence) would be written to NV-RAM and retained until reset.

Read-Only Memory (ROM)

ROM data is magnetically or optically retained and is designed ot to be overwritten. It is permanent although it can be corrupted (rarely) and, when magnetically retained, it is susceptible to damage when exposed to powerful magnetic fields. Low-level radiation such as that encountered routinely (from police radar and high tension electrical wiring while driving on a highway) will not affect any current ECMs. A majority of the data retained in the ECM is logged in ROM. The *master program* for the system management is loaded into ROM. Production standardization is permitted by constructing ROM architecture so it contains common necessary data for a number of different systems. For example, identical ROM chips can be manufactured to run a group of different transmissions in a series. However, to actually make a transmission operate in a specific drivetrain application, the ROM data would require further qualification from data loaded into PROM and EEPROM.

Programmable ROM (PROM)

PROM is magnetically retained data. It is usually a chip, a set of chips, or a card socketed into the ECM motherboard that can often be removed and replaced. PROM's function is to qualify ROM to a specific chassis application. In the earliest truck engine management systems, programming options such as idle shutdown time could only be altered by replacing the PROM chip. Customer programmable options are written to EEPROM in current systems where they can be easily altered without changing any hardware. Some OEMs describe the PROM chip as a personality module, an appropriate description of its actual function of trimming or fine-tuning the ROM data to a specific application. Newer personality modules also contain the EEPROM data, explained in the next section.

Electronically Erasable PROM (EEPROM)

The EEPROM data category contains customer data programmable options and proprietary data (programming data owned and controlled by the OEM) that can be altered and modified using a variety of tools ranging from a generic reader/programmer to a mainframe computer. A generic reader/programmer such as ProLink 9000 equipped with the correct software cartridge can be used to rewrite customer options such as tire rolling radius, governor type (LS or VS), progressive shifting, road speed limit and so on.

Only the owner password is required to access the ECM and make any required changes to customer data options.

Proprietary data is more complex both in character and methods of alteration. It contains data such as the fuel map in an engine system. The procedure here normally requires accessing a centrally located mainframe computer, usually at the OEM central location, via a modem. The appropriate files are then downloaded, via the phone lines, to a personal computer (PC) and subsequently to a diskette. The ECM is then reprogrammed (altering or rewriting the original data) from the diskette. Some proprietary reader/programmers—manufactured by one OEM for use only on their equipment—can act as the interface link between the vehicle ECM and the mainframe, but the procedure is essentially the same.

6.12 COLLISION WARNING SYSTEMS (CWSs)

Only one collision warning system is prominent in the trucking industry at this time, the Eaton VORAD system. VORAD is a loose acronym for Vehicle Onboard RADar. This is a collision warning system (CWS) that uses radar technology to sense some specific vehicle danger conditions. VORAD is used in trucks to sense straight-ahead closing velocities and proximity as well as blind side proximity. It has been proven to reduce accident rates and is appealing to fleet operations with higher than average accident rates because it can also reduce insurance payments.

OPERATING PRINCIPLE

Radar is an electronic system that is used to detect objects at greater distances or under conditions of poorer visibility than the human eye can handle. The word radar comes from the initial letters of RADio Detecting And Ranging. Modern radar is capable of precisely measuring both distance and relative speed.

Doppler Effect

Christian Doppler (1803–1853) was an Austrian physicist who theorized that because the pitch of sound is highest at a sound source and gets progressively lower with distance, the same should apply to other waves such as light waves. Change in frequency can be used to determine relative speed. For instance, if the light waves produced by a star were observed to shift toward the red end (lower frequency/longer wavelength) of the spectrum (see **Figure 6–20**), the star is becoming more distant from the observer; also, the rate of change in the frequency can be used to calculate the velocity of the star. In the case of an observed star accelerating toward an observer on Earth, the reverse would be true: The light wave produced would shift toward the violet (higher frequency/shorter wavelength) end of the spectrum. Again, the velocity of the star can be calculated by determining the rate of shift. Doppler effect is used extensively by astronomers and is the basis of radar systems.

Pulse Radar

Radar operates by transmitting electromagnetic waves toward an object and then monitoring the return wave bounced back by the object. The properties of the returned wave are then analyzed by a signal processor and converted to a form usable by the operator or system. For instance, an air traffic controller would have a CRT display that would have a map-like image of the area scanned indicating the location and velocity of aircraft in the area. The objectives of VORAD are much more simple, so the return signals are simply processed as warning signals.

Pulse radar requires that a transmitter produce short pulses of microwave energy and that a receiver monitor the return signals as they are bounced off objects. Close objects would produce an echo from the pulse sooner than distant objects. Microwaves travel at the speed of light, about 1,000 feet (300 meters) per microsecond, so the delay between the transmission of a pulse and the echo can be used to calculate distance. The Doppler effect, the change in observed frequency produced by motion described in the previous section, is used by the processor to determine speed.

VORAD COLLISION DETECTION SYSTEM

VORAD provides object detection on the front and right (blind) side of a truck and may optionally provide left (driver) side object detection. The system is electronically managed and consists of relatively few components. It must be connected into the vehicle electrical and electronic system so it can share sensor signals such as the input from the vehicle road speed sensor (tailshaft sensor). The components used in the VORAD system are shown in **Figure 6–42**.

Antenna Assembly (AA)

The forward antenna is mounted directly to the center of the bumper and is used to transmit and receive microwave pulses. Transmitted radar signals are

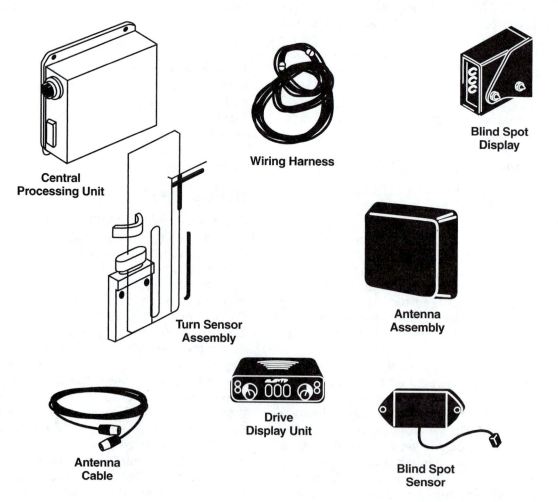

Central Processing Unit

Wiring Harness

Blind Spot Display

Turn Sensor Assembly

Antenna Assembly

Antenna Cable

Drive Display Unit

Blind Spot Sensor

FIGURE 6–42   VORAD components. (Courtesy of Roadranger Marketing. One great drive train from two great companies—Eaton and Dana Corporations)

Blind Spot Sensor
Optimal: 30"–36" Above Ground:
39"–45" Behind Mirror

Antenna Assembly
Optimal: 22"–26" Above Ground:
On Vehicle Centerline (required)

FIGURE 6–43   *Antenna and blind spot sensor location.* (Courtesy of Roadranger Marketing. One great drive train from two great companies—Eaton and Dana Corporations)

bounced off objects in a straight line path from the transmitting antenna. Return signals are received by the antenna and the Doppler effect (frequency shift) of the transmitted and received signals is calculated. This information is then converted to a digital format and sent to a central processing unit (CPU) that performs additional processing.

The mounting location ensures that the radar beam is aimed directly in front of the truck. The antenna assembly has a range of about 350 feet and is capable of monitoring up to 20 vehicles (the 20 closest to the front of the truck) in its scanned range. **Figure 6–43** shows the location of the antenna on a typical tractor application.

Blind Spot Sensors (BSS)

The blind spot sensor is a simple motion detector that senses objects in the driver's blind spot on the right side of the vehicle. It is mounted not more than 3 ft. above road level and about 4 ft. behind the mirror assemblies. Blind spot sensor information is sent to the ECU, which will output the appropriate driver alerts. **Figure 6–43** shows the required location of a blind spot sensor.

Turn Sensor Assembly (TSA)

The turn sensor monitors the movement of the steering wheel shaft. Because the antenna outputs its microwave signal in a straight line, during a turn it will sense objects located on the side of the road such as buildings, billboards, stationary vehicles, and so on. The turn sensor unit signals to the ECU that a turn is being made, and the ECU will disable the audible driver warnings for the duration of the turn. The turn sensor consists of a shaft mounted magnet that rotates past sensors when the shaft rotates more than 2 to 4 degrees. When the steering wheel is turned and the magnet passes across either sensor, the ECU receives a signal indicating that a turn is being made. **Figure 6–44** shows the turn sensor assembly components.

Central Processing Unit (CPU)

It is the CPU that receives VORAD input signals from the antenna and the turn and blind spot sensors, computes the input data and programmed instructions, and switches the system outputs, meaning the lighting of appropriate indicator lamps and the sounding of audible driver alerts. The CPU can be

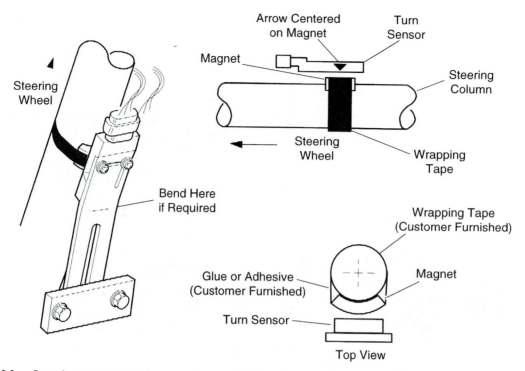

FIGURE 6–44 *Steering sensor and magnet mounted on the steering shaft.* (Courtesy of Roadranger Marketing. One great drive train from two great companies—Eaton and Dana Corporations)

located in a number of places on the truck, but is most commonly found inside the firewall (under the dash) or behind the driver's seat. The CPU contains a slot for an optional driver's card. The driver card has two functions:

- It can log up to 10 minutes of system data. This feature is typically used to reconstruct the events leading up to an accident.
- It can function as an ID card. It may be programmed as personal electronic ID, identifying the name and employee number of the driver.

The CPU has write-to-self memory and a clock. This permits VORAD to store trip recorder information, crash times, engine idle time, miles traveled, average speed, average following time, and distances. Like other electronic driver monitoring devices, VORAD can play a role in driver education. For instance, truck drivers who are inclined to tailgate will be less likely to do so if their driving habits are being electronically monitored. Note the location of the VORAD ECM in a typical highway tractor.

Driver Display Unit (DDU)

The DDU is mounted to the top of the dash in a location that is both easily visible and reachable by the driver. The unit houses the controls and indicators used by VORAD. The DDU controls system power-

up, speaker volume, the range thresholds of the vehicle alerts, and writing of data to the driver card. The indicator lights are used to signal system power, system failure, three stages of distance alerts, and the detection of stationary or slow-moving objects.

A light sensor on the DDU automatically adjusts the intensity of the display lights. Display light intensity is greatest in bright sunlight and is dimmed for night driving. The DDU also contains a small speaker used to warn the driver when second or third stage intervals to impact are detected, or if a vehicle is detected in the blind spot when a left turn is indicated. **Figure 6–45** shows the location of the DDU clearly in the driver's sight line.

Blind Spot Display (BSD)

The blind spot display is mounted to the right-side, inside-front cab pillar in a close-to-direct sight line from the driver and the right-side mirror. The unit contains red and yellow lights that are switched by the ECU, indicating whether or not a vehicle is detected in the blind spot. As with the driver display unit, a light sensor in the blind spot display unit automatically adjusts the light intensity of the display lights as light conditions vary.

When the blind spot motion sensor detects a vehicle in the blind spot, the red light is illuminated. When the blind spot is detected to be clear, the yellow lamp is illuminated. When left-side sensors are

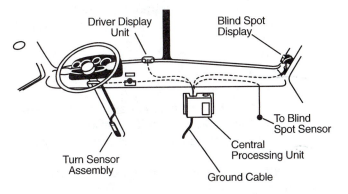

FIGURE 6–45 *Location of the VORAD compo-nents in a typical truck cab. (Courtesy of Roadranger Marketing. One great drive train from two great com-panies—Eaton and Dana Corporations)*

used on a VORAD system, the set-up is identical to the blind spot display except that blind spot sensors are used on both sides of the vehicle and display units are located on both sides of the cab.

The VORAD Cab

Figure 6–45 shows the location of the VORAD com-ponents in a typical truck cab. Note that there are alternatives to this arrangement, but the location of the driver alerts shown here is optimum.

SUMMARY

- Data can be transmitted electronically by means of electrical waveforms.
- Semiconductors are by definition elemental materials with 4 electrons in their outer shells.
- Silicon is the most commonly used semicon-ductor material.
- Semiconductors must be doped to provide them with the electrical properties that can make them useful as electronic components.
- After doping, semiconductor crystals may be classified as having N or P electrical properties.
- Diodes are 2-terminal semiconductors that often function as a sort of electrical one-way check valve.
- Zener diodes are commonly used in vehicle electronic systems: They act as a voltage-sensitive switch in a circuit.
- Transistors are 3-terminal semiconductor chips.
- Transistors can be generally grouped into bipo-lar and field effect types.
- Essentially, a transistor is a semiconductor sandwich with the middle layer acting as a con-trol gate. A small current flow through the base-

emitter will ungate the transistor and permit a much larger emitter-collector current flow.
- Many different types of transistors are used in vehicle electronic circuits, but their roles are primarily concerned with switching and amplifi-cation.
- The optical spectrum includes ultraviolet, visible, and infrared radiation.
- Optical components conduct, reflect, refract, or modify light. Fiber optics are being used increasingly in vehicle electronics, as are optical components.
- Integrated circuits consist of resistors, diodes, and transistors arranged in a circuit on a chip of silicon.
- A common integrated circuit chip package used in computer and vehicle electronic sys-tems is a DIP with either 14 or 16 terminals.
- Many different chips with different functions are often arranged on a primary circuit board, also known as a motherboard.
- Gates are switched controls that channel flows of data through electronic circuitry.
- AND, OR, and NOT gates are 3 commonly used means of producing an outcome based on the switching status of components in the gate circuit.
- The binary numeric system is a 2-digit arith-metic system that is often used in computer electronics because it directly corresponds to the on or off states of switches and circuits.
- A *bit* is the smallest piece of data that a com-puter can manipulate. It has the ability to show one of two states, either on or off.
- A *byte* consists of 8 bits.
- A byte of data can represent up to 256 pieces of coded data.
- Almost all current on-highway trucks use com-puters to manage the engine and usually other chassis systems as well.
- A truck with multiple ECM-managed systems may electronically connect them to a vehicle management ECM capable of mastering all the systems.
- A vehicle ECM information processing cycle comprises 3 stages: data input, data process-ing, and outputs.
- RAM or main memory is electronically retained and therefore volatile.
- The master program for system management is usually written to ROM.
- PROM data is used to qualify the ROM data to a specific chassis application.
- Some OEMs describe their PROM component as a personality module.
- EEPROM gives an ECM a read/write/erase memory component.

- Multiplexing is the term used to describe a system where two or more ECMs are connected to reduce input hardware and optimize vehicle operation.
- Input data may be categorized as command data and system monitoring data.

REVIEW QUESTIONS

1. An element classified as a semiconductor would have how many electrons in its outer shell?
   a. less than 4
   b. 4
   c. more than 4
   d. 8

2. How many electrons does the element silicon have in its outer shell?
   a. 2
   b. 4
   c. 6
   d. 8

3. Which of the following best describes the function of a simple diode?
   a. an electronic switch
   b. an electrical one-way check valve
   c. an electronic amplifying device
   d. an electrical storage device

4. The term *pulse width modulation* refers to
   a. waveforms shaped to transmit data
   b. unwanted voltage spikes
   c. electronic noise
   d. rectification of AC to DC

5. To form a P-type semiconductor crystal, the doping agent would be required to have how many electrons in its valence shell?
   a. 3
   b. 4
   c. 5
   d. 8

6. To form an N-type semiconductor crystal, the doping agent would be required to have how many electrons in its valence shell?
   a. 3
   b. 4
   c. 5
   d. 8

7. The positive terminal of a diode is correctly called a(n)
   a. electrode
   b. cathode
   c. anode
   d. emitter

8. Which of the following terms best describes the role of a typical transistor in an electronic circuit?
   a. check valve
   b. relay
   c. rectifier
   d. filter

9. When testing the operation of a typical transistor, which of the following should be true?
   a. high resistance across the emitter and base terminals
   b. continuity across the emitter and collector terminals
   c. continuity across the base and emitter terminals

10. What would be the outcome in an OR gate if one of two switches in the circuit was closed?
    a. off
    b. on

11. How many different data codes could be represented by a byte?
    a. 2
    b. 8
    c. 64
    d. 256

12. Which of the following transistor terminals controls output?
    a. anode
    b. emitter
    c. collector
    d. base

13. Which of the following components is typically used in a half wave rectifier?
    a. rheostat
    b. transformer
    c. diode
    d. zener diode

14. An N-type semiconductor crystal is doped with
    a. trivalent element atoms
    b. pentavalent element atoms
    c. carbon atoms
    d. germanium atoms

15. A PWM (pulse width modulated) signal is
    a. analog
    b. digital
    c. either analog or digital
    d. neither analog or digital

16. Which of the following is a 3-terminal device?
    a. zener diode
    b. transistor
    c. capacitor
    d. diode

17. The acronym that describes the component that executes program instructions is the
    a. CRT
    b. RAM
    c. CPU
    d. ROM

18. Which of the following data retention media is electronically retained?
    a. RAM
    b. ROM
    c. PROM
    d. EEPROM

19. Which of the following memory categories would the master program for system management be written to in a typical truck chassis ECM?
    a. RAM
    b. ROM
    c. PROM
    d. EEPROM

20. Which of the following memory categories would customer data programming be written to from a generic reader/programmer?
    a. RAM
    b. ROM
    c. PROM
    d. EEPROM

7

Charging, Cranking, and Chassis Electrical Circuits

Prerequisite
Chapters 2, 3, 5, and 6

Objectives

After reading this chapter, you should be able to

- Perform basic battery maintenance.
- Explain the procedures for jump starting, testing, charging, and storage.
- Identify the components of a charging system.
- Describe the procedures used to test a heavy-duty truck charging system.
- Identify the components of a truck cranking circuit.
- Outline the procedures used to test a truck cranking circuit.
- Repair lighting system components.
- Troubleshoot truck electrical circuit components.
- Explain the purpose of a truck warning and shutdown system.
- Explain electrical troubleshooting procedure.

Key Terms

armature
brushes
charging system
deep cycling
Department of Environmental Resources (DER)
halogen light
head-up display (HUD)
one-piece design
rotor
sensing voltage

slip rings
starter circuit
starter motor
starting safety switch
stator
sulfated
thermostat
voltage limiter
voltage regulator
windings
wraparound

This chapter will build on the electrical and electronic theory introduced in Chapters 5 and 6. Most of the content of this chapter is the practical application of electricity in a modern truck electrical system. To properly understand electricity, the technician must first know something about electrical theory. In deal-ing with truck charging, cranking, and electrical circuits, this chapter uses a practical approach. Remember that some procedures rely on theory introduced in Chapters 5 and 6, so refer back to these chapters when necessary. **Figure 7–1** shows a typical heavy-duty truck electrical system.

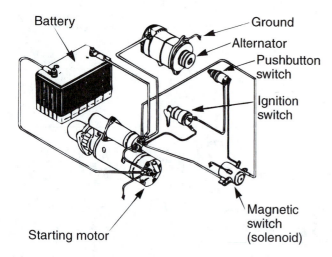

Figure 7-1 *Typical heavy-duty electrical system, consisting of a battery (usually three or four connected), a starting motor, an alternator, a magnetic switch, an ignition switch, a push button switch, and the required wiring.*

7.1 ELECTRICAL SYSTEM SAFETY

Always disconnect the battery's ground cable when working on the electrical system or engine. This prevents sparks from short circuits and accidental starting of the engine.

Never "short" across cable connections or battery terminals. Always disconnect the battery ground cable before fast-charging batteries in the vehicle, especially if the system is equipped with an alternator. Improper connection of charger cables to the battery can reverse the current flow and damage the alternator.

When removing a battery from a vehicle, always disconnect the battery ground cable first. When installing a battery, connect the ground cable last. Never attempt to polarize an alternator after reconnecting a battery. Any attempt to polarize might damage the alternator, regulator, or circuits.

Never reverse the polarity of the battery connections. All current vehicles use a negative chassis ground. Reversing this polarity will damage the alternator and circuit wiring.

Never attempt to use a fast charger as a boost to start the engine.

7.2 BATTERY MAINTENANCE

A lead-acid battery has a finite service life and requires maintenance to achieve it. With a reason-

able amount of care, the life of a battery can be appreciably extended. Neglect and abuse will invariably cause shorter battery service life. The battery should be inspected at each chassis lubrication or other periodic services. Battery maintenance includes the following:

- Inspect battery and mounting for corrosion, loose mounting hardware, and so on. (**Figure 7-2**). Loose holddown straps or covers allow the battery to vibrate or bounce during vehicle operation. This vibration can shake the active materials off the grid plates, considerably shortening battery life. It can also loosen the plate connections to the plate strap, loosen cable connections, and even crack the battery case. In heavy-duty trucks, the physical mounting and location of the battery is important in ensuring its long life. **Figure 7-3** illustrates the mounting recommendations of the Maintenance Council (TMC) of the American Trucking Associations, Inc.
- Battery corrosion is commonly caused by spilled electrolyte or electrolyte condensation from gassing. Corrosion occurs when sulfuric acid from the electrolyte corrodes, attacks, and damages battery components such as connectors and terminals. Corroded connections increase resistance at the battery terminals,

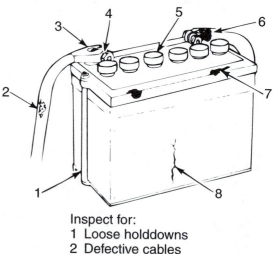

Inspect for:
1. Loose holddowns
2. Defective cables
3. Damaged terminal posts
4. Loose connections
5. Clogged vents
6. Corrosion
7. Dirt or moisture
8. Cracked case

FIGURE 7-2 *Battery maintenance inspection points.*

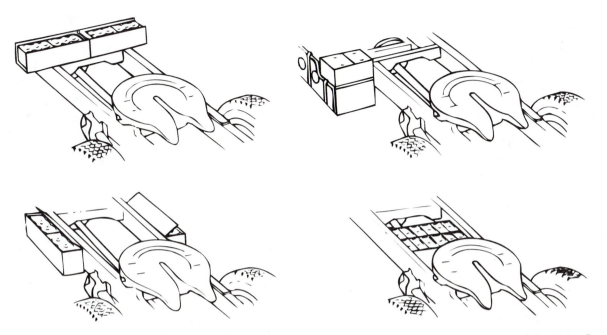

FIGURE 7–3 *Proper battery mounting recommendations. (Illustrations courtesy of the Maintenance Council of the American Trucking Associations, Inc.)*

which reduces voltage to the vehicle's electrical system. Corrosion in the battery cover can also create a path for current, which allows the battery to slowly discharge. Finally, corrosion can spread beyond connectors and terminals to the holddown straps and carrier box, resulting in physical damage to the battery.

■ If corrosion is found on the terminal posts, remove the cable terminals from the battery and clean them. Remove the ground cable first, using the proper sized wrench and a terminal puller. A terminal cleaning brush can be used to clean tapered posts and the mating surfaces of the cable clamps. Other types of terminals can be cleaned with a wire brush. The cable terminals should then be cleaned with a solution of baking soda and water to neutralize any remaining battery acid. Do not allow the baking soda and water solution to enter the cells. Clean dirt from the battery top with a cloth moistened with baking soda and water. Then wipe with a cloth moistened with clean water. A wire brush can be used to remove dirt, corrosion, or rust from the battery tray or holddown components. After rust is removed, rinse the tray and hardware with clear water, dry with air, and repaint.

After cleaning, reinstall the battery and holddown components. Coat the battery and cable terminals with terminal grease and connect the cables to the battery terminals. Connect the ground cable last.

■ Maintaining a proper electrolyte level in nonsealed batteries is essential to keeping them in good operating condition. Remember that in this type of battery, water is lost due to evaporation from heating and gassing during the charging process. If the electrolyte level is down, fill each cell with distilled water to bring the liquid level to the full level indicator. If the battery does not have a level indicator, bring the level to $1/2$ inch above the tops of the separators. Do not overfill any cell. This weakens the concentration of the acid (reducing the electrolyte's specific gravity), which decreases battery efficiency. Also, when a cell is overfilled, the excess electrolyte can be forced from the cell by the gas formed in the battery. This will cause corrosion of adjacent metals, reduced performance, and shorter battery life. Underfilling causes a greater concentration of acid, which deteriorates the plates' grids more rapidly.

■ Batteries can be overcharged by either the vehicle's **charging system** or battery charger. In either case, the result is an accelerated chemical reaction within the battery that causes gassing and a loss of electrolyte in the cells. This can remove the active materials from the plates, permanently reducing the capacity of the battery. Overcharging can also cause excessive heat, which can oxidize the positive plate grid material and even buckle the plates, resulting in a loss of cell capacity and early battery failure.

👓 Shop Talk

Frequent need for refilling battery cells might indicate that the battery is being overcharged. Check the charging system and readjust the voltage regulator as needed.

- An undercharged battery is operating in a partially discharged condition. This condition can be caused by excessive battery output, stop-and-go driving, or a fault in the charging system. An undercharged battery will become **sulfated** when the sulfate normally formed on the plates becomes dense, hard, and chemically irreversible. Sulfation occurs when the sulfate is allowed to remain in the plates for a long time. Sulfation causes two problems: First, it lowers the specific gravity levels and increases the danger of freezing at low temperatures. Second, in cold weather, a sulfated battery may not have the reserve power needed to crank the engine. The rapid discharging and recharging of the battery can cause a condition known as **deep cycling**. Repeated deep cycling can cause the positive plate material to break away from its grids and fall into the sediment chambers at the base of the battery case. Deep cycling can reduce battery capacity and lead to premature short circuiting between the plates. A recent envelope design innovation found in many batteries has reduced this problem.

BATTERY SERVICING PRECAUTIONS

The electrolyte in a battery contains sulfuric acid and is very corrosive. If splashed into eyes or onto the skin, it can cause serious burns and injury. The electrolyte is also poisonous; if swallowed, prompt medical attention must be given.

A battery produces hydrogen and oxygen gases during normal operation. These gases may escape through the battery vents. When present in sufficient amounts, the gases are explosive.

To avoid the potential for serious injury and damage, the following precautions should be observed whenever handling, testing, or charging a battery:

1. If electrolyte is splashed on the skin or in the eyes, rinse with cool, clean water for about 15 minutes and call a doctor immediately. Do not add eye drops or other medication unless advised by a doctor.

2. If electrolyte is swallowed, drink several large glasses of milk or water followed by milk of magnesia, a beaten raw egg, or vegetable oil. Call a doctor immediately.

3. Electrolyte can damage painted or unpainted metal parts. If electrolyte is spilled or splashed on painted or unpainted metal surfaces, neutralize it with a baking soda solution and rinse the area with clean water.

4. Use extreme care to avoid spilling or splashing electrolyte. Wear safety glasses or a face shield when working with batteries. Wear rubber gloves and an apron when handling or carrying batteries.

5. To prevent possible skin burns and electrical arcing, do not wear watches, rings, or other jewelry while performing maintenance work on the batteries or any other part of the electrical system.

6. When removing a battery from a vehicle, always disconnect the battery ground cable first. When installing a battery, connect the ground cable last.

7. Never reverse the polarity of the battery connections. Generally, all highway vehicles use a negative ground. Reversing this polarity damages the alternator and circuit wiring. A cause of battery explosion is reversed polarity jumper cables. Remember to connect negative to negative and positive to positive.

8. A battery that has either lead-calcium cells or one that is operated in a particularly warm climate may have a shorted cell. This problem occurs when the grid plates start to grow to the point where they touch each other, producing the shorted cell condition. This problem can be detected by visually examining the cells. If five cells have normal fluid levels and the sixth one is almost dry, a short condition probably exists.

WINTERIZING BATTERIES

As mentioned earlier, a battery decreases in cranking power as it gets colder (that is, the battery's available energy decreases). While a battery at 80°F has 100 percent of its power, the same battery at 32°F has only 83 percent power available. When the temperature drops to zero, the battery still has 61 percent of its power but it decreases to 45 percent at 20°F. At 30°F, it has only about 10 percent of the original power still available. Also, as the temperature of the battery and the engine decreases, it becomes more difficult for the starter to crank the engine. An engine at 20°F is $3\frac{1}{2}$ times more difficult to crank than an engine at 80°F (**Figure 7–4**).

When checking a battery for winterization (see Chapter 4), disconnect battery cables and load test

Power Available   Power Required

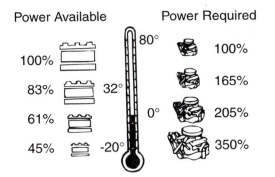

100% 83% 32° 61% 0° 45% -20°

100% 165% 205% 350%

FIGURE 7–4 _Relationship of battery power available and power required._

A

B

FIGURE 7–5 _(A) Blanket-type battery warmer and (B) battery box (plate-type) warmer. (Courtesy of Phillips Temro, Inc.)_

each battery individually at $\frac{1}{2}$ CCA rating for 15 seconds. (For example, a 625 CCA battery rating uses 300 amps.) After 1 minute, measure the voltage. A reading of 9.6 or higher is good at 70°F; 8.5 or higher is good at 0°F. Ambient temperature must be taken into account.

In cold weather, it is important to keep the battery warm. Various styles of battery heaters are available. The most common are the blanket type (**Figure 7–5A**) and the plate type (**Figure 7–5B**). The plate-

type battery heater usually is placed under the battery, but it can also be placed between batteries. The blanket type of heater comes in a variety of lengths and wattages. It also incorporates its own insulation.

It is best to heat a battery slowly. Also, a well insulated battery box helps, not only during the heating of the battery, but in maintaining warmth.

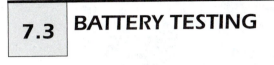

7.3 BATTERY TESTING

A battery test should be performed whenever battery trouble is suspected. Battery testing will determine

- If the battery is satisfactory and can remain in service.
- If the battery should be recharged before placing it back in service.
- If the battery must be replaced.

A complete battery test, as outlined here, includes these steps:

1. Visual inspection
2. State of charge test
3. Battery capacity (load) test

CAUTION: Be sure to follow all instructions and safety procedures suggested by the test equipment manufacturer.

Visual Inspection

Begin by inspecting the outside of the battery for obvious damage such as a cracked or broken case or cover that would allow electrolyte to leak out. Check for damage to the battery terminals. If obvious physical damage is found, replace the battery. If possible, determine the cause of damage and correct it. Next check the electrolyte level in the battery. On low-maintenance batteries, simply unscrew the vent caps or pull off the vent plugs and check the level of electrolyte in each cell. On maintenance-free batteries, use a knife blade to cut through the decal on top of the battery. Peel the decal off to expose the vent caps or vent manifold (**Figure 7–6**). Remove the caps or use a screwdriver to pry up the vent manifold.

As mentioned, if the electrolyte level is below the tops of the plates in one or more cells, add distilled water until the electrolyte level is just above the tops of the separators. Charge the battery 15 minutes at 15 to 25 amperes to mix the water with the electrolyte.

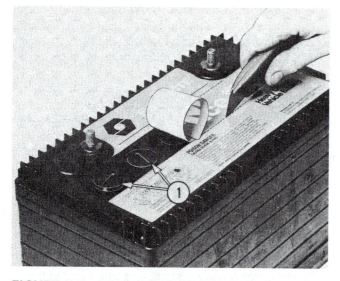

FIGURE 7–6 The electrolyte level can be checked in a maintenance-free battery by removing the decal to expose the vent manifold or caps. (Courtesy of International Truck and Engine Corp.)

FIGURE 7–7 Using a hydrometer to test specific gravity.

State of Charge Test

The battery's state of charge can be determined by either the specific gravity test or by the stabilized open circuit voltage test. Checking specific gravity on low-maintenance or conventional batteries requires the use of a hydrometer or refractometer. Maintenance-free batteries normally have a built-in hydrometer.

Specific Gravity (Hydrometer) Test. Perform the specific gravity (hydrometer) test on all cells (**Figure 7–7**). Measure and record specific gravity, corrected to 80°F, of each cell. Determine the battery's state of charge. If specific gravity readings are 1.225 or higher and are within 50 points (0.050 specific gravity) between highest and lowest cells, proceed to the load test. If specific gravity readings are low (below 1.225) or vary more than 50 points between highest and lowest cells, recharge battery as instructed under "Charging the Battery" and inspect the vehicle electrical system to determine the cause for low state of charge.

If, after charging, specific gravity readings are within 50 points between highest and lowest cells, proceed to step 3. If readings still vary more than 50 points after charging, replace the battery.

How to Use a Hydrometer. Figure 7–8 illustrates the correct method of reading a hydrometer. The barrel must be held vertically so the float is not rubbing against the side. The electrolyte should be drawn in and out of the hydrometer barrel a few

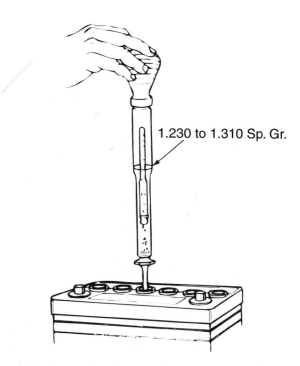

1.230 to 1.310 Sp. Gr.

FIGURE 7–8 Reading a hydrometer.

times to bring the temperature of the hydrometer float and barrel to that of the acid in the cell. Draw an amount of acid into the barrel so that with the bulb fully expanded, the float will be lifted free, touching neither the side, top, nor bottom stopper of the barrel. When reading the hydrometer, your eye should be on a level with the surface of the liquid in the hydrometer barrel. Disregard the curvature of the liquid where the surface rises against the float stem and the barrel due to surface tension. Keep the float clean. Make certain it is not cracked. Never take a hydrometer reading immediately after water is added to the cell. The water must be thoroughly mixed with the underlying electrolyte before hydrometer readings are reliable. This mixing is accomplished by charging the battery for a short period of time. If a reading is being taken immediately after the battery has been subjected to prolonged cranking, it will be higher than the true value. The water formed in the plates during the rapid discharge has not had time to mix with the higher specific gravity acid above the plates. Because there are many different types of battery hydrometers available, always follow the manufacturer's instructions.

Temperature Correction. Hydrometer floats are calibrated to give a true reading at a fixed temperature. A correction factor must be applied for any specific gravity reading made when the electrolyte temperature is not 80°F.

A temperature correction must be used because the electrolyte will expand and become less dense when heated. The float will sink lower in the less dense solutions and give a lower specific gravity reading. The opposite occurs if the electrolyte is cooled. It will shrink in volume, becoming more dense. The float will rise higher and read high.

A correction factor of 0.004 specific gravity (sometimes referred to as 4 points of specific gravity) is used for each 10°F change in temperature. Four points of specific gravity (0.004) are added to the indicated reading for each 10°F increment above 80°F and 4 points are subtracted for each 10°F increment below 80°F. This correction is important at extremes of temperature because it can become a substantial value.

Figure 7–9 illustrates the correction for hydrometer readings when the electrolyte temperature is above or below 80°F. Example 1, in cold weather, a partially discharged battery in a vehicle at 20°F might read 1.250, indicating it is almost fully charged. However, when the correction factor is applied, the true value is only 1.226. Example 2 in **Figure 7–9** could be encountered in a battery exposed to the sun in hot weather. Also, electrolyte frequently reaches 100°F in service in warm weather. The uncorrected specific gravity reading of 1.235 might indicate too low a state of charge to install in a vehicle or that there is a problem in the electrical system if the battery is in service.

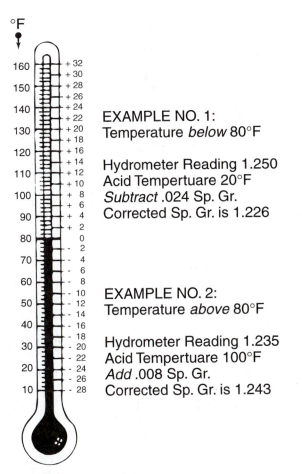

EXAMPLE NO. 1:
Temperature *below* 80°F

Hydrometer Reading 1.250
Acid Tempertuare 20°F
Subtract .024 Sp. Gr.
Corrected Sp. Gr. is 1.226

EXAMPLE NO. 2:
Temperature *above* 80°F

Hydrometer Reading 1.235
Acid Tempertuare 100°F
Add .008 Sp. Gr.
Corrected Sp. Gr. is 1.243

FIGURE 7–9 *Temperatures above 80°F require subtracting from the recorded specific gravity; temperatures below 80°F require adding to the specific gravity reading.*

However, the true reading of 1.246 might not be unreasonably low, depending on the length of storage of the battery or the type of service that it has been experiencing in the vehicle.

Built-in Hydrometers. On many sealed maintenance-free batteries a special temperature-compensated hydrometer is built into the battery cover. A quick visual check will indicate the battery's state of charge. The built-in hydrometer has a colored ball within a cage that is attached to a clear plastic rod. The colored ball will float at a predetermined specific gravity of the electrolyte that represents about a 65 percent state of charge. When the colored ball floats, it rises within the cage and positions itself under the rod. Visually, a colored dot then shows in the center of the hydrometer (**Figure 7–10A**). The built-in hydrometer provides a guide for battery testing and charging. In testing, the colored dot means the battery is charged enough for testing. If the colored dot

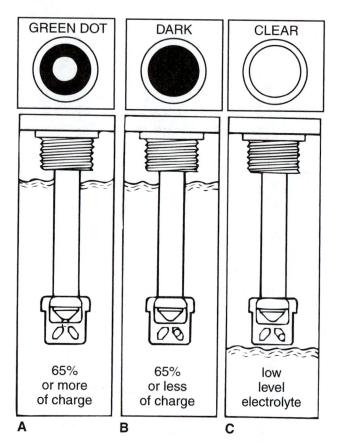

GREEN DOT	DARK	CLEAR
65% or more of charge	65% or less of charge	low level electrolyte
A	B	C

FIGURE 7-10 *Reading a built-in hydrometer: while this is the most common color code used with truck batteries, keep in mind that a few battery manufacturers use a slightly different one. Check the top of the battery case to find out the color codes for a given battery. (Illustrations courtesy of The Maintenance Council of the American Trucking Associations, Inc.*

is not visible and the hydrometer has a dark appearance (**Figure 7-10B**), it means the battery must be charged before the test procedure is performed.

The hydrometer on some batteries may be clear or light color (**Figure 7-10C**). This means the fluid level might be below the bottom of the rod and attached cage. This might have been caused by excessive or prolonged charging, a broken case, excessive tipping, or normal battery wearout. Whenever this clear or light color appearance is present while looking straight down on the hydrometer, always tap the hydrometer lightly with a small screwdriver to dislodge any gas bubbles that might be giving a false indication of low electrolyte level. If the clear or light color appearance remains, and if a cranking complaint exists that is caused by the battery, replace it. It is important when observing the hydrometer that the battery have a clean top to see the correct indication. A flashlight might be required in some poorly-lit areas. Always

look straight down when viewing the hydrometer. Complete hydrometer information for most batteries is printed on the label located on the top of the battery. By referring to this label, an accurate interpretation of the hydrometer appearance can be made.

WARNING: Never jump start or attempt to recharge a fully discharged maintenance-free battery. Jumping or recharging can create an explosion hazard. These batteries have no way to vent gas build-up and therefore should be replaced if fully discharged.

OPEN CIRCUIT VOLTAGE TEST

The open circuit voltage of the battery can also be used to determine its state of charge. Follow these steps to test voltage:

1. If the battery has just been recharged or has recently been in vehicle service, the surface charge must be removed before an accurate voltage measurement can be made. To remove surface charge, crank the engine for 15 seconds. Do not allow the engine to start. To prevent engine starting, apply the engine stop control or disconnect the fuel solenoid valve lead wire as required.
2. After cranking the engine, allow the battery to rest for 15 minutes.
3. Connect the voltmeter across the battery terminals (**Figure 7-11**) and observe the reading. Compare the reading obtained with **Table 7-1** to determine the battery's state of charge.
4. If the stabilized voltage is below 12.4 volts, the battery should be recharged. Also, inspect the vehicle's electrical system to determine the cause of the low state of charge. After charging the battery, proceed to the load test.

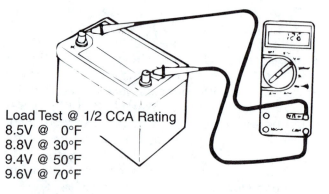

Load Test @ 1/2 CCA Rating
8.5V @   0°F
8.8V @ 30°F
9.4V @ 50°F
9.6V @ 70°F

FIGURE 7-11   *Performing an open circuit voltage test with a digital multimeter.*

TABLE 7-1: STATE OF CHARGE DETERMINED BY OPEN CIRCUIT VOLTAGE TEST

Stabilized Open Circuit Voltage	State of Charge
12.6 volts or more	Fully charged
12.4 volts	75% charged
12.2 volts	50% charged
12.0 volts	25% charged
11.7 volts or less	Discharged

Load Test

The load, or capacity, test determines how well any type of battery, sealed or unsealed, functions under a load.

The load test requires the use of a battery tester equipped with an adjustable carbon pile. The test can be performed with the battery either in or out of the vehicle, but the battery must be at or very near a full state of charge. Use the specific gravity test or open circuit voltage test to determine charge, and recharge the battery if needed before proceeding. To obtain accurate results, the electrolyte should be as close to 80°F as possible. Cold batteries will show considerably lower capacity. Never load test a sealed battery if its temperature is below 60°F.

After attaching the carbon pile (**Figure 7-12**), use the following steps to load test a battery:

1. If the battery was recently charged or in service, remove the surface charge of the battery by applying a 300-amp load for 15 seconds, or by disabling the engine and cranking the engine for that length of time.
2. Adjust all mechanical settings on the load tester to zero. Rotate the load control knob fully counterclockwise to the off position.
3. Connect the load tester to the battery. An inductive pickup must surround the wires from the ground terminal if used in the test. Observe correct polarity and be sure the test leads contact the battery posts. Batteries with sealed terminals require adapters to provide a place for attaching the tester's leads (**Figure 7-13**).
4. If the tester is equipped with an adjustment for battery temperature, turn it to the proper setting. For best results, use a thermometer to check electrolyte temperature in one of the cells. On sealed units, estimate the temperature as closely as possible.
5. Refer to the battery specifications to determine its cold cranking amps (CCA) rating.
6. Turn the load control knob on the tester to draw battery current at a rate equal to one-half ($1/2$) the battery's cold cranking amps (CCA) rating. For example, if a battery's CCA rating is 440, the load should be set at 220 amperes.

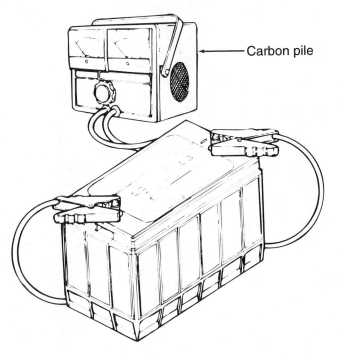

FIGURE 7-12 *Performing a capacity test on a battery.*

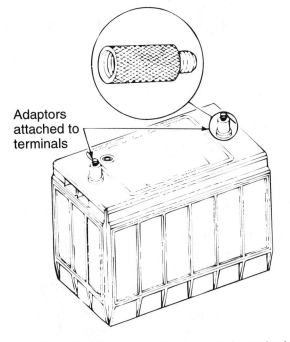

FIGURE 7-13 *Batteries with threaded terminals require the use of adapters for testing or charging.*

TABLE 7–2: **MINIMUM LOAD TEST VOLTAGES AS AFFECTED BY TEMPERATURE**

Battery Temperature (F)	Minimum Test Voltage
70°	9.6 volts
60°	9.5 volts
50°	9.4 volts
40°	9.3 volts
30°	9.1 volts
20°	8.9 volts
10°	8.7 volts
0°	8.5 volts

7. Maintain specified load for 15 seconds while observing the voltmeter of the tester. Turn the control knob off immediately after 15 seconds of current draw.

8. At 70°F or above (or on testers that are temperature corrected), the voltage at the end of the 15 seconds should not fall below 9.6 volts. If the tester is not temperature corrected, use **Table 7–2** to determine the adjusted minimum voltage reading for the battery temperature. If the voltage reading exceeds the specification by a volt or more, the battery is supplying sufficient current with a good margin of safety. If the voltage drops below the minimum specification, the battery should be replaced. If the reading is right on the spec, the battery might not have the reserve necessary to handle cranking under tough conditions, such as low temperatures. Keep in mind that this varies with the state of charge determined before the load test was made. If the battery was at 75 percent charge and fell right on the load spec, it is probably in good shape.

7.4 CHARGING THE BATTERY

There are two methods of recharging a battery: the slow charge method and the fast charge method. Either method can be used to recharge most batteries. However, some batteries must be charged slowly. Both methods are explained below.

The charge a battery receives is equal to the charge rate in amperes multiplied by the time in hours. Thus a 5-ampere rate applied to a battery for 10 hours would be a 50 ampere-hour charge to the battery. A 60 ampere-hour applied for 1 hour would be a 60-ampere charge.

The battery charging guides given in **Tables 7–3** and **7–4** show approximately how much recharge a fully discharged battery requires. For partially discharged batteries, the charging current or charging time should be reduced accordingly. For example, if the battery is 25 percent charged (75 percent discharged), reduce charging current or time by one-fourth ($1/4$). If the battery is 50 percent charged, reduce charging current or time by one-half ($1/2$). If time is available, lowering the charging rates in amperes is recommended. While the battery is being charged, periodically measure the temperature of the electrolyte. If the temperature exceeds 125°F or if violent gassing or spewing of electrolyte occurs, the charging rate must be reduced or temporarily halted. This must be done to avoid damage to the battery.

CAUTION: Do not overcharge batteries, particularly maintenance-free type batteries. Overcharging causes excessive and needless loss of water from the electrolyte. Overcharging also causes the battery to produce explosive combinations of hydrogen and oxygen.

TABLE 7–3: **BATTERY CHARGING GUIDE— 6- AND 12-VOLT LOW MAINTENANCE BATTERIES**

(Recommended rate* and time for fully discharged condition)

Rated Battery Capacity (Reserve Minutes)	Slow Charge	Fast Charge
80 minutes or less	14 hours @ 5 amperes 7 hours @ 10 amperes	1$3/4$ hours @ 40 amperes 1 hour @ 60 amperes
Above 80 to 125 minutes	20 hours @ 5 amperes 10 hours @ 10 amperes	2$1/2$ hours @ 40 amperes 1$3/4$ hours @ 60 amperes
Above 125 to 170 minutes	28 hours @ 5 amperes 14 hours @ 10 amperes	3$1/2$ hours @ 40 amperes 2$1/2$ hours @ 60 amperes
Above 170 to 250 minutes	42 hours @ 5 amperes 21 hours @ 10 amperes	5 hours @ 40 amperes 3$1/2$ hours @ 60 amperes
Above 250 minutes	33 hours @ 10 amperes	8 hours @ 40 amperes 5$1/2$ hours @ 60 amperes

*Initial rate for standard taper charger.

TABLE 7–4: BATTERY CHARGING GUIDE—12-VOLT MAINTENANCE-FREE BATTERIES

(Recommended rate* and time for fully discharged condition)

Rated Battery Capacity (Reserve Minutes)	Slow Charge	Fast Charge
80 minutes or less	10 hours @ 5 amperes 5 hours @ 10 amperes	2½ hours @ 20 amperes 1½ hours @ 30 amperes 1 hour @ 45 amperes
Above 80 to 125 minutes	15 hours @ 5 amperes 7½ hours @ 10 amperes	3¾ hours @ 20 amperes 2½ hours @ 30 amperes 1¾ hours @ 45 amperes
Above 125 to 170 minutes	20 hours @ 5 amperes 10 hours @ 10 amperes	5 hours @ 20 amperes 3 hours @ 30 amperes 2¼ hours @ 45 amperes
Above 170 to 250 minutes	30 hours @ 5 amperes 15 hours @ 10 amperes	7½ hours @ 20 amperes 5 hours @ 30 amperes 2½ hours @ 45 amperes
Above 250 minutes	20 hours @ 10 amperes	10 hours @ 20 amperes 6½ hours @ 30 amperes 4½ hours @ 45 amperes

*Initial rate for standard taper charger.

Slow Charging

The slow charging method uses a low charging rate for a relatively long period of time. The recommended rate for slow charging is 1 ampere per positive plate per cell. If the battery has nine plates per cell, normally four of the nine will be positive plates. Therefore, the slow charge rate would be 4 amperes. Charging periods as long as 24 hours might be needed to bring a battery to full charge.

The best method of making certain a battery is fully charged, but not overcharged, is to measure the specific gravity of a cell once per hour. The battery is fully charged when no change in specific gravity occurs over a 3-hour period or when charging current stabilizes (constant voltage type charger).

Batteries with charge indicators are sufficiently charged when the colored dot in the hydrometer is

visible. Gently shake or tilt the battery at hourly intervals during charging to mix the electrolyte and to see if the colored dot appears. Do not tilt the battery beyond a 45-degree angle.

If the colored dot does not appear after a 75 ampere-hour charge, continue charging for another 50 to 75 ampere-hours. If the colored dot still does not appear, replace the battery.

Shop Talk

Batteries with charge indicators cannot be charged if the indicator is clear or light in color; replace these batteries.

If a low-maintenance (conventional) battery is to be charged overnight (10–16 hours) use the specified Slow Charge rate in **Table 7–3**. Maintenance-free batteries must not be charged at rates greater than specified in the maintenance-free battery charging guide (**Table 7–4**). If a maintenance-free battery is to be recharged overnight (16 hours), a timer or voltage-controlled charger is recommended to avoid overcharging. If the charger does not have such controls, a 3-ampere rate should be used for batteries of 80 minutes or less capacity and 5 amperes for 80–125 minute reserve capacity batteries. Batteries charged for over 125 minutes should be charged at the specified slow charge rate (**Table 7–4**).

Batteries that have remained in a discharged condition for long periods of time without a recharge have sulfated; sulfur in the electrolyte combines with the lead in the plates, hardening plates and making them inactive. A sulfated battery must be recharged at a low rate to avoid overheating and excessive gassing. It might require two or three days of slow charging to bring a sulfated battery to a fully charged condition. Again, care should be taken not to overcharge maintenance-free type batteries.

Batteries can become so badly sulfated they cannot be restored to a normal operating condition, regardless of the rate of charge or the length of time the charge is applied. If a battery cannot be restored to a fully charged condition by slow charging, it should be rejected.

Fast Charging

The fast charge method provides a high charging rate for a short period of time. The charging rate should be limited to 60 amperes for 12-volt batteries. Maximum charging rate for 6-volt batteries (above 180 reserve capacity minutes) can be approximately doubled.

Ideally, fast charges should be limited to the charging times shown under fast charge in the battery charging guides (**Tables 7–3** and **7–4**). The battery generally cannot be fully charged within these time periods, but will receive sufficient charge (70 to 90 percent) for practical service. To completely recharge a battery, follow the fast charge with a slow charge until no change in specific gravity occurs over a 3-hour period (or until the colored dot is visible in the built-in hydrometer). When fast-charging a battery in the vehicle, always disconnect the battery ground cable before applying the charge.

A battery with a specific gravity of 1.225 or above should never be charged at a high rate. If the charger has not tapered to a low rate, adjust to a slow charge, preferably at a rate of 1 ampere per positive plate per cell.

Charging Instructions

Follow these steps to charge a battery:

1. Before placing a battery on charge, clean the battery terminals if necessary.
2. Add distilled water sufficient to cover the plates. Fill to the proper level near the end of charge. If the battery is extremely cold, allow it to warm before adding distilled water because the level will rise as it warms. In fact, an extremely cold battery will not accept a normal charge until it becomes warm.
3. Connect the charger to the battery, following the instructions of the charger manufacturer (**Figure 7–14**). Connect the positive (+) charger lead to the positive battery terminal and negative (−) lead to negative terminal. If the battery is in the vehicle, connect the negative lead to the engine block if the vehicle has a negative ground (negative battery terminals are connected to ground). Connect the positive lead to ground if vehicle has a positive ground. Rock the charger lead clamps back and forth to make certain a good connection has been made.
4. Turn the charger on and slowly increase the charging rate until the recommended ampere value is reached.

CAUTION: If smoke or dense vapor comes from the battery, shut off the charger and reject the battery. If violent gassing or spewing of electrolyte occurs, reduce or temporarily halt the charging.

5. When the battery is charged, turn the charger off and disconnect it.

6. Install the battery in the truck. If the engine does not crank satisfactorily when a recharged battery is installed, load test the battery. If the battery passes the load test, the vehicle's fuel, ignition, cranking, and charging systems should be checked to locate and correct the no-start problem. If it does not pass the load test, the battery should be replaced.

Charging Safety

When charging the batteries, gas forms in each cell and escapes through the vent holes. In poorly ventilated areas, the gas lingers around the battery several hours after it has been charged. The gas is explosive and could be ignited by sparks, flame, or heat; if ignited, the battery will explode. Follow these precautions when charging the batteries:

1. Make sure that the area is well ventilated.
2. Do not break live circuits at the terminals because a spark can occur at the point where a live circuit is broken. Use care when connecting or disconnecting booster leads or cable clamps on chargers. Poor connections are common causes of electrical arcs, which cause explosions.
3. Do not smoke near batteries that are being charged or have recently been charged. Keep the batteries away from open flames or sparks.
4. If the battery is frozen, let it reach room temperature before attempting to charge it. A dead battery may freeze at temperatures near 0°F. Never try to charge a battery that has ice in the cells. Passing current through a frozen battery can cause it to rupture or explode. If ice or slush is visible or the electrolyte level cannot be seen, allow the battery to thaw at room temperature before servicing. Do not take chances with sealed batteries—if there is any doubt, allow them to warm to room temperature.
5. Do not place or drop tools or metal objects on top of the batteries. This could short out the batteries, causing sparks and possible damage.
6. Never lean over a battery during charging, testing, or jump starting operations. When a battery explodes, the case cover and electrolyte is thrown upward.
7. When handling a plastic-cased battery, pressure placed on the end walls could cause electrolyte to spew through the vents. Therefore, always use a battery carrier to lift these batteries or lift with your hands placed at opposite corners.

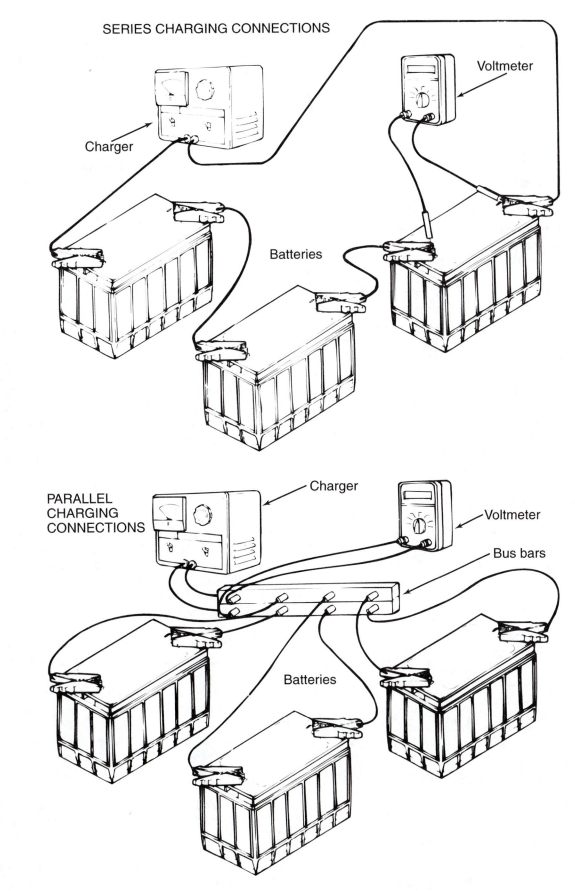

SERIES CHARGING CONNECTIONS

Voltmeter

Charger

Batteries

A

PARALLEL
CHARGING
CONNECTIONS

Charger

Voltmeter

Bus bars

Batteries

B

FIGURE 7–14   Battery charger connections.

JUMP STARTING

The procedure outlined here should be followed when it becomes necessary to use a booster battery to start a vehicle with a discharged battery. Make sure booster and discharged batteries are the same voltage (6-volt or 12-volt).

Shop Talk

Make certain the stalled vehicle and the one containing the booster battery do not touch. If the two vehicles are in contact, a ground connection could be established, which could cause sparking when jumper cables are attached.

1. Set the parking brake on both vehicles. Place both transmissions in neutral or park. Turn lights, heater, and other electrical loads off. Make certain the ignition key is turned off in both vehicles.
2. Attach an end of one jumper cable to the insulated terminal of the booster battery and the opposite end of the cable to the insulated terminal of the discharged battery (**Figure 7–15**).
3. Attach one end of the other jumper cable to the ground terminal of the booster battery, and the opposite end to a ground at least 12 inches from the battery of the vehicle being started. The vehicle frame is usually a good ground.

Shop Talk

The final ground connection must provide good electrical conductivity and current-carrying capacity. Do not connect directly to the ground post of the discharged battery.

4. Make sure that the clamps from one cable do not touch the clamps on the other cable. Do not lean over the batteries when making connections.
5. Make sure that everyone is standing away from the vehicles. Start the engine of the vehicle with the booster batteries. Wait a few minutes, and then attempt to start the engine of the vehicle with the discharged batteries. Do not operate the starter longer than 30 seconds. Wait at least 2 minutes between starting attempts to allow the starter to cool. If the engine does not start after several attempts, check for the cause.
6. After starting, allow the engine to idle. Disconnect the ground connection from the vehicle with the discharged battery. Then disconnect the opposite end of the cable.
7. Disconnect the other cable from the discharged battery first, and then disconnect the opposite end.

Jump Starting a 24-Volt System

The procedure for jump starting a truck with a 24-volt system is almost the same as the procedure for jump starting a vehicle with a 12-volt system. The only difference is in connecting the cables to the batteries. The proper procedure is given here and illustrated in **Figure 7–16**.

1. Attach an end of one jumper cable to the insulated terminal of the booster battery that is connected directly to the starter. Attach the other end of the jumper cable to the insulated terminal of the discharged battery that is connected directly to the starter.
2. Attach one end of a second jumper cable to the ground terminal of the booster battery that is connected directly to the series-parallel switch. Attach the other end of this jumper cable to the ground terminal of the discharged battery that is connected directly to the series-parallel switch.
3. Attach one end of a third jumper cable to the insulated terminal of the booster battery that is connected directly to the series-parallel switch. Attach the other end of this jumper

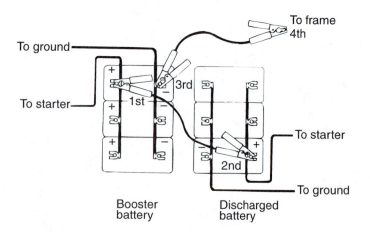

FIGURE 7–15   *Correct method of attaching jumper cables to a 12-volt system.*

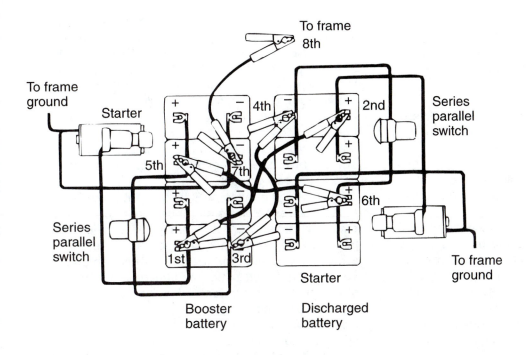

FIGURE 7–16 Correct method of attaching jumper cables to a 24-volt system.

cable to the insulated terminal of the discharged battery that is connected directly to the series-parallel switch.

4. Attach one end of a fourth jumper cable to the ground terminal of the booster battery that is connected directly to the starter. Attach the other end of this cable to a ground at least 12 inches from the battery of the vehicle being started. The vehicle frame is usually a good ground.

After starting, allow the engine to idle. Disconnect the ground connection from the vehicle with the discharged battery. Then disconnect the opposite end of the cable. Disconnect the other cables from the discharged battery first; then, disconnect the opposite ends.

Battery Storage and Recycling Procedure

The EPA recommends that lead-acid batteries be stored properly to prevent contamination or injury resulting from spills or leaks. Batteries should be stored indoors because this will reduce the possibility of cracks and leaks due to extreme temperature variations. An asphalt or concrete floor area can be protected from any spills by applying an acid-resistant coating such as epoxy. Acid-resistant curbing should be constructed around the storage area to contain any spills. Curbs for small areas are normally constructed of either asphalt or wood that is covered with a 20 to 40 mm sheet of acid-resistant polyethylene, polypropylene, or polyvinyl chloride.

It is best to store batteries in a cool, well ventilated area, placed upright on pallets, and stacked no more than three layers high (**Figure 7–17**). Inspect the batteries regularly, and make sure that the top posts of the storage area are protected with waffle board. Because used batteries could leak, place a sturdy, acid-resistant, leak-proof material such as plastic sheeting below them. Spills are considered hazardous waste because they are corrosive and may contain toxic levels of lead. Any spills that do not remain in the storage area should be reported to the local office of the **Department of Environmental Resources (DER)**.

FIGURE 7–17 Proper way to store batteries when not in use.

Most states have specific laws in effect for the safe disposal of lead-acid batteries. The Battery Council International (BCI) has the following recycling regulations that serve as a guideline for many of these states:

1. Used lead-acid batteries must be delivered to a battery retailer or wholesaler, or to a recycling center or lead smelter.
2. All battery retailers and wholesalers must accept, at the point of transfer, a quantity at least equal to the number of new batteries it sells.
3. All retailers of batteries should post a 5 × 7-inch sign that contains the universal recycling symbol and gives instructions on proper battery disposal along with retailer responsibilities. (Specific sign requirements vary according to state law.)
4. It is important to remember that violation of these regulations could result in fines and/or imprisonment.

7.5 CHARGING SYSTEMS

A truck's charging system consists of the batteries, alternator, **voltage regulator**, associated wiring, and the electrical loads of the truck. The purpose of the system is to recharge the batteries whenever necessary and to provide the current required to power the electrical components of the truck. It does this by converting the mechanical energy of the engine into electricity. A charging system can be compared to a compressed air system on a truck (**Figure 7–18A**). The compressor forces air into a reservoir or tank. When a pneumatic component needs air, air is drawn from the reservoir. Eventually, the air pressure in the tank will drop below a certain level and a pressure-sensitive switch will activate the compressor. Air will be pumped into the reservoir again until cut-out pressure is reached. Then, the switch will unload the compressor.

In a charging system (**Figure 7–18B**), the battery acts as a reservoir of electrical energy. Electrical energy is drawn from it to crank the engine and to power electrical systems. The electrical "pressure" in a battery is called voltage. When fully charged, a heavy-duty 12-volt battery will have 12.6 volts available across its terminals. As the battery is discharged, the voltage level will drop. When it reaches a set voltage level, an electrical switch, called a voltage regulator, activates the alternator.

The alternator acts as an electron pump. It sends electrical current into the battery to restore the voltage level. On 12-volt systems, charging is typically at 14.2 volts to get the battery up to 12.6 volts. Just as the air compressor in our example has to push against air already in the tank, the alternator has to push against the voltage in the batteries. When the batteries are discharged, it does not take much voltage to start current flowing into the battery.

A regulator might turn the alternator on and off as many as 600 times per minute. When demand is high, the alternator will stay on longer; when low, it will freewheel. This helps fuel economy since a 100-amp alternator can consume 6 to 7 horsepower.

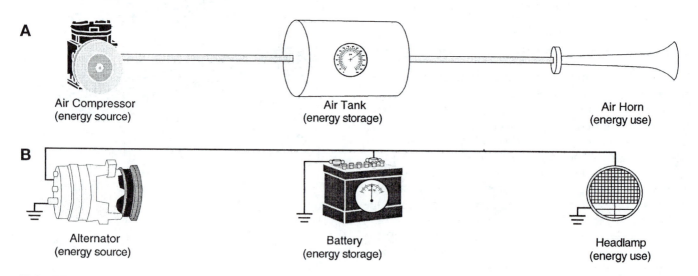

A

Air Compressor
(energy source)

Air Tank
(energy storage)

Air Horn
(energy use)

B

Alternator
(energy source)

Battery
(energy storage)

Headlamp
(energy use)

FIGURE 7–18   *A charging system can be compared to a compressed air system. Just as a compressor keeps air tanks full of air, an alternator keeps a battery "full" of voltage.*

ALTERNATOR CONSTRUCTION

To generate electricity, the alternator uses this basic law of physics: When magnetic lines of force move across a conductor (such as a wire or bundle of wires), an electrical current is produced in the conductor. This principle is illustrated in **Figure 7–19**. Electricity can be induced in an electrical circuit by either moving the circuit wiring through the magnetic field of a magnet or by moving the magnetic field past the wiring. In both situations, the direction of electrical current flow is determined by the polarity of the magnetic field. The magnetic forces of the north pole of a magnet will force electrons to flow in one direction and the south pole of a magnet will force electrons to flow in the opposite direction. If the wiring is exposed alternately to both north and south poles, an alternating current will be produced.

The strength of the current produced depends on several factors: the strength of the magnetic field, the speed of the wire passing through the field, and the size and number of wires.

In an alternator, the **rotor** provides the magnetic fields necessary to induce a current flow. The rotor spins inside a stationary coil of wires called a **stator**. The moving magnetic fields create a flow of electrons through the stator wiring that is pumped to the battery when it needs recharging. The following paragraphs explain how the rotor and stator and other alternator parts work together to provide charging current to the battery. The components of an alternator are shown in **Figure 7–20**.

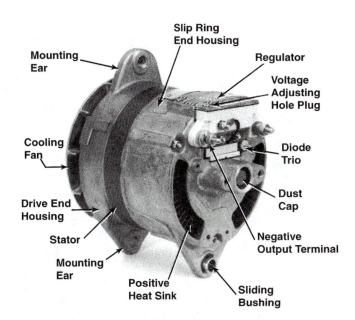

FIGURE 7–20 *Components of a heavy-duty truck alternator. (Courtesy of Prestolite Electric Corp.)*

Rotor

The rotor is the only moving component within the alternator. It is responsible for producing the rotating magnetic field. The rotor (**Figure 7–21**) consists of a coil, two pole pieces, and a shaft. The magnetic field is produced when current flows through the coil; this coil is simply a series of **windings** wrapped around an iron core. Increasing or decreasing the current flow through the coil varies the strength of the magnetic field, which in turn defines alternator output. The current passing through the coil is called the *field current.* It is usually 3 amperes or less.

The coil is located between the interlocking pole pieces. As current flows through the coil, the core essentially becomes a magnet. The pole pieces assume the magnetic polarity of the end of the core that they touch. Thus, one pole piece has a north polarity and the other has a south polarity. The extensions on the pole pieces, known as the fingers, form interlacing magnetic poles. A typical rotor has fourteen poles, seven north and seven south, with the magnetic field between the pole pieces moving from the north poles to the adjacent south poles (**Figure 7–22**). The more poles a rotor has, the higher the alternator output will be.

Slip Rings and Brushes

The wiring of the rotor coil is connected to **slip rings**. The slip rings and **brushes** conduct current to the rotor. Most alternators have two slip rings mounted directly on the rotor shaft; they are insulated from the shaft and from each other. A spring-loaded carbon

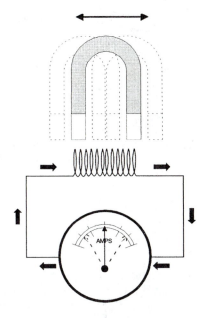

FIGURE 7–19 A magnetic field moving past a wire will produce current flow in the wire.

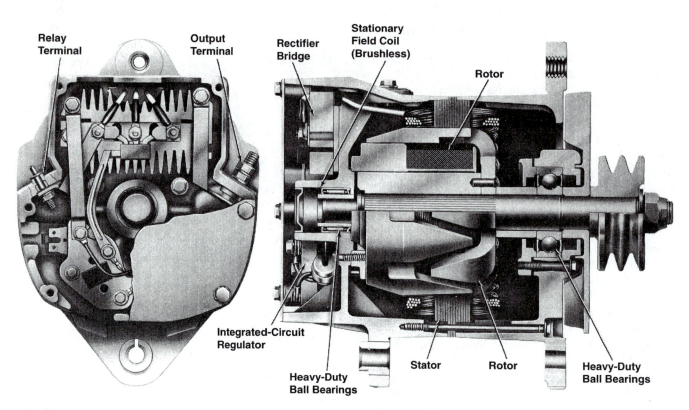

FIGURE 7-21 Cutaway view of an alternator. (Courtesy of Delco-Remy Co.)

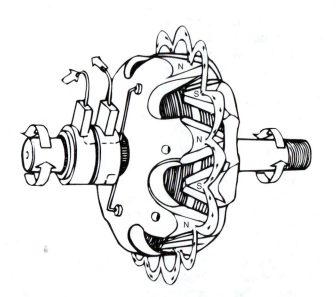

FIGURE 7-22 The interlacing fingers of the pole pieces create alternating north and south poles. (Courtesy of Delco-Remy Co.)

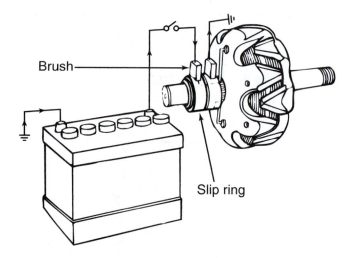

FIGURE 7-23 Slip rings and brushes conduct current to the spinning field coil. (Courtesy of Delco-Remy Co.)

brush is located on each slip ring to carry the current to and from the rotor windings (**Figure 7-23**). Because the brushes carry only 1.5 to 3.0 amperes of current, they do not require frequent maintenance. This is in direct contrast to generator brushes, which conduct all of the generator's output current and consequently wear rapidly.

Stator

The stator is made up of many conductors, or wires, into which the spinning rotor induces voltage. The wires are wound into slots in the alternator frame, with each wire forming several coils spaced evenly around the frame (**Figure 7-24**). There are as many

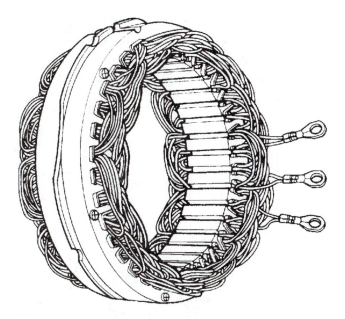

FIGURE 7–24   *A stator consists of three windings looped around the inside perimeter of the alternator frame or case. (Courtesy of Delco-Remy Co.)*

coils in each wire as there are pairs of north and south rotor poles.

The wires are grouped into three separate bundles, or windings. The coils of the three windings are staggered in the alternator frame so that the electrical pulses created in each coil will also be staggered. This produces an even flow of current out of the alternator. The ends of each winding are attached to separate pairs of diodes, one positive and the other negative. They are also wired together in one of two configurations: a wye shape or a delta shape (**Figure 7–25**).

End Frame Assembly

The end frame assembly, or housing, is made of two pieces of cast aluminum and contains the bearings for the end of the rotor shaft. The drive pulley and fan are mounted to the rotor shaft outside the drive end frames. Each end frame also has built-in air ducts so the air from the rotor shaft fan can pass through the alternator. A heat sink—called a rectifier, bridge, or diode holder—containing three positive diodes is attached to the rear end frame; heat can pass easily from these diodes to the moving air (**Figure 7–26**). Three negative diodes are often mounted in the end frame itself. Because the end frames are bolted together and then bolted directly to the engine, the end frame assembly provides the electrical ground path for many alternators. This means that anything connected to the housing that is not insulated from the housing is grounded.

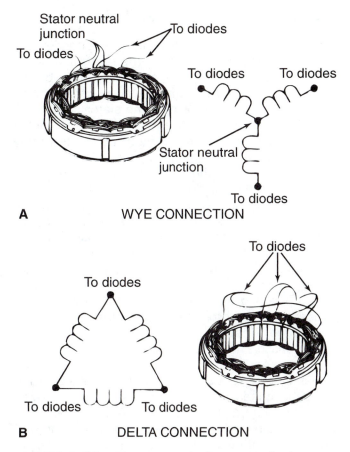

A   WYE CONNECTION

B   DELTA CONNECTION

FIGURE 7–25   *The stator windings are wired together in one of two arrangements: (A) a wye configuration; or (B) a delta configuration. (Courtesy of Delco-Remy Co.)*

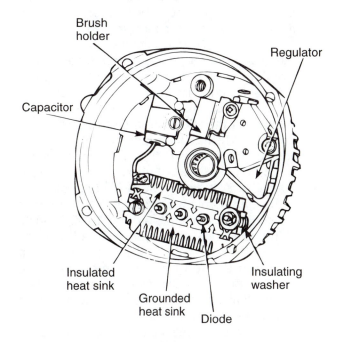

FIGURE 7–26   *Diodes are mounted in a heat sink to keep them cool. (Prestolite Electric Corp.)*

ALTERNATOR OPERATION

As the rotor, driven by a belt and pulley arrangement, rotates inside the alternator, the north and south poles (fingers) of the rotor alternately pass by the coiled windings in the stator. Only a very small air gap separates the rotor from the stator so that the windings are subjected to a maximum amount of the magnetic force fields. As each pole alternately passes by the coils, magnetic lines of force cause electrons to flow in the wires. As a north pole passes by a coil in the winding, electrons first flow in one direction. As the next pole, having a south polarity, passes by the coil, the flow of electrons changes direction. Thus, an alternating current is produced. When viewed on an oscilloscope, the alternating pulses of current are seen as a sine wave (**Figure 7–27**). The positive side of the wave is produced by a north pole and the negative side of the wave is produced by a south pole. Because a truck's alternator has three windings staggered around the rotor, a three-phase sine wave is produced.

However, a battery can only be charged using direct current. A truck's accessories also are powered by direct current. In order for the alternator to provide the electricity required, the alternating current (AC) must be converted, or rectified, to direct current (DC). While generators accomplish this using a mechanical commutator and brushes, alternators do it electronically with diodes. A diode can be thought of as an electrical check valve. It will permit current to flow in only one direction—positive or negative. When a diode is placed in a simple AC circuit, one-half of the current is blocked. In other words, the current can flow from X to Y as shown in **Figure 7–28**, but it cannot flow from Y to X. When the voltage reverses at the start of the next rotor revolution, the current again can pass from X to Y, but not back. This type of output is not efficient since current would be available only half of the time. Because only 50 percent of the AC voltage produced by the alternator is being converted to DC, this is known as half-wave rectification.

By adding more diodes to the circuit, the full wave can be rectified to DC. In **Figure 7–29A**, current flows from X to Y. It flows from X, through diode 2, through the load, through diode 3, and then to Y. In **Figure 7–29B**, current flows from Y to X. It flows from Y, through diode 4, through the load, through diode 1, and back to X. In both cases, the current flows through the load in the same direction. Because all of the AC is now rectified, this is known as full-wave rectification. However, there remain brief moments when the current flow is zero; therefore, most alternators use three windings and six diodes to produce overlapping current pulses, thereby ensuring that the output is never zero.

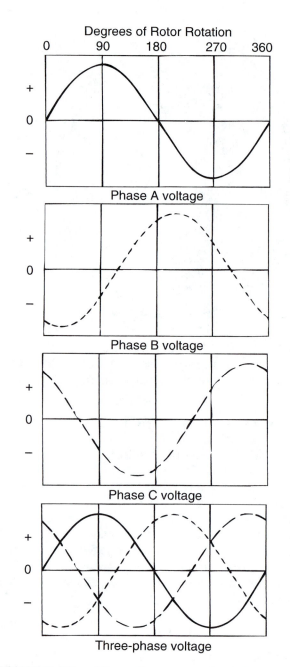

FIGURE 7–27 Alternating current is seen as a sine wave on an oscilloscope. (Courtesy of Prestolite Electric Corp.)

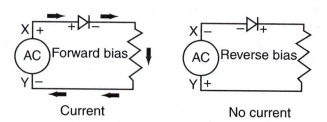
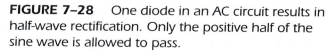

FIGURE 7–28 One diode in an AC circuit results in half-wave rectification. Only the positive half of the sine wave is allowed to pass.

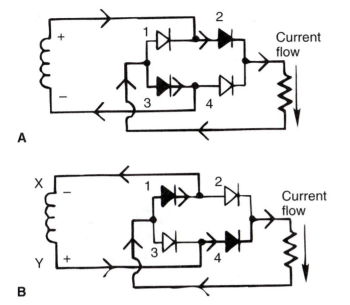

A

B

FIGURE 7–29   *By adding more diodes to the circuit, current can now flow (A) from X to Y, and (B) from Y to X, resulting in rectifying both halves of the sine wave to DC current.*

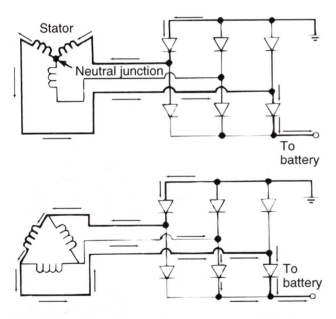

FIGURE 7–30   *A three-phase alternator has six diodes to provide full rectification.*

In a heavy-duty three-phase alternator, there are three coils and six diodes (**Figure 7–30**). At any time, two of the windings will be in series and the third winding will be neutral, doing nothing. Depending on the combination of coils and the direction of current flow, one positive diode will rectify current flowing in one direction and a negative diode will rectify the

current flowing in the opposite direction. This is true for both Y and delta configurations. Many heavy-duty truck alternators use a delta arrangement. The windings in a delta stator are arranged in parallel rather than series circuits. The parallel paths allow more current to flow through the diodes, thereby increasing the output of the alternator.

BRUSHLESS ALTERNATORS

Many alternators do not use brushes to deliver voltage to the field windings of the rotor. In this type of alternator, the field windings do not rotate with the rotor. It is held stationary while the rotor turns around it. The rotor itself retains sufficient residual magnetism to energize the stator when the truck engine is first started. Part of the voltage induced in the stator windings is then diverted to the field windings, which energizes the electromagnet. As the strength of the stationary field windings increases, the magnetic field of the rotor increases and the stator output reaches its specified potential.

VOLTAGE REGULATORS

For over 50 years prior to the introduction of solid state voltage regulators in the early 1970s, mechanical regulators were used to control alternator (or generator) output. Like so many other mechanical components that have been replaced with electronic parts, the mechanical voltage regulator is slow and imprecise when compared to transistorized regulators. A wiring schematic of a solid-state voltage regulator is shown in **Figure 7–31**.

A voltage regulator defines the amount of current produced by the alternator and the voltage level in the charging circuit. It does this by turning the field circuit off and on. Without a voltage regulator, the battery would be overcharged and the voltage level in the electrical systems would rise to a point where lights would burn out and fuses and fusible links would blow. Controlling the voltage level is particularly important as electronic components are added to heavy-duty trucks. Microprocessors and electronic sensors and switches are easily damaged by voltage spikes and high voltage levels.

The voltage regulator receives battery voltage as an input. This is called the **sensing voltage;** it allows the regulator to sense and monitor the battery voltage level. When the battery voltage rises to a particular level (approximately 13.5 volts), the regulator will turn the field current off. When the battery voltage drops below a set level, the regulator will again turn the field circuit back on. This on/off cycling of the field circuit takes place hundreds of times each second.

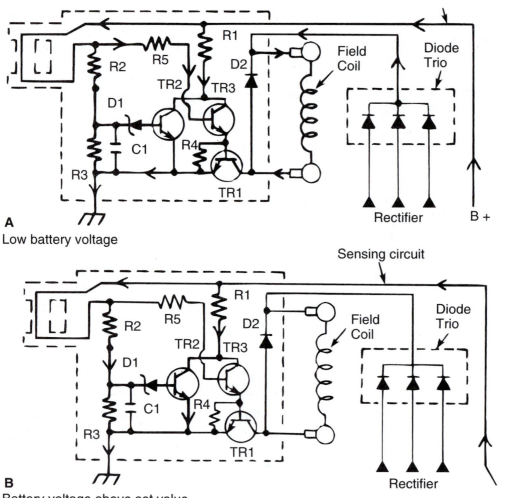

A
Low battery voltage

Sensing circuit

B
Battery voltage above set value

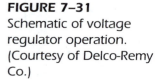

FIGURE 7–31
Schematic of voltage regulator operation. (Courtesy of Delco-Remy Co.)

The ability of a battery to accept a charge varies with temperature. A cold battery needs a higher charging current to bring it up to capacity. In warm hot weather, less current is needed. So, the voltage regulator has a thermistor to vary the voltage level at which the regulator will switch the field circuit on and off.

Types of Field Circuits

The field circuit, which is controlled by the voltage regulator, might be one of two types: if it is an "A" circuit, the regulator is on the ground side of the rotor. Battery voltage is picked up by the field circuit inside the alternator. The regulator turns the field circuit off and on by controlling a ground. A "B" type voltage regulator is positioned on the feed side of the alternator. Battery voltage is fed through the regulator to the field circuit, which is then grounded in the alternator. The regulator turns the field circuit off and on by controlling the current flow from the battery to the field circuit. Most voltage regulators are mounted

either on or in the alternator. Some charging systems might have a regulator mounted separately from the alternator. A typical electronic regulator is an arrangement of transistors, diodes, zener diodes, capacitors, resistors, and thermistors. Most alternators with internal regulators use current generated by the alternator stator to power the field circuit. The regulator controls a ground to turn the field circuit off or on. Since the stator generates AC current, a set of three diodes are used to rectify the current to DC. These three diodes are often referred to as a *diode trio*. A separate sensing circuit delivers battery voltage to the regulator to control the on/off cycle of the alternator.

Figures 7–31A and **B** demonstrate how current flows through a typical voltage regulator. Battery current is delivered to the regulator through the sensing circuit. While the battery voltage is low, it turns on transistor 3 (TR3), which then turns on transistor 1 (TR1) (**Figure 7–31A**). When TR1 is on, current flows from the stator, through the diode trio, through the field coil, through TR1 and to ground.

When battery current builds to approximately 13.5 to 14 volts, the zener diode (D1) trips to switch transistor 2 (TR2) on. This diverts current away from TR3, which then turns off. When TR3 turns off, TR1 must turn off also. This blocks the flow of current through the field circuit (**Figure 7–31B**). The magnetic field of the rotor collapses and the alternator ceases to generate current. The battery then begins to discharge until the voltage level drops to a point when the zener diode turns off and stops current flow to TR2. TR3 then turns TR1 on to allow current flow through the field circuit, energizing the rotor coil. This process repeats itself hundreds of times every second.

CAUSES OF CHARGING SYSTEM FAILURE

A malfunction in the charging system results in either an overcharged battery or an undercharged battery. An overcharged battery will experience water loss, eventually resulting in hardened plates and the inability to accept a charge. Overcharging can be caused by one or a combination of the following:

- Defective battery
- Defective or improperly adjusted regulator
- Poor sensing lead contact to the regulator or rectifier assembly

An undercharged battery will result in slow cranking speeds and a low specific gravity of the electrolyte. Undercharging is not always caused by a defect in the alternator. Undercharging can be caused by: a loose drive belt; loose, broken, corroded, or dirty terminals on either the battery or alternator; undersize wiring between the alternator and the battery; or by a defective battery that will not accept a charge.

Undercharging can also be caused by one or a combination of the following defects in the alternator field circuit:

- Poor contacts between the regulator and the carbon brushes
- Defective diode trio
- No residual magnetism in the rotor
- Defective or improperly adjusted regulator
- Damaged or worn brushes
- Damaged or worn slip rings
- Poor connection between the slip ring assembly and the field coil leads
- Shorted, open, or grounded rotor coil

Undercharging can also be the result of a malfunction in the generating circuits. One or more of

the stator windings (phases) can be shorted, open, or grounded. The rectifier assembly might be grounded, or one or more of the diodes might be shorted or open.

A visual inspection of the charging system is all that is needed to identify some of the conditions just described. Loose mounting bolts will cause a loose belt, which can result in an undercharged battery. The belt should be properly tensioned and the mounting hardware tightened. Loose or broken wires or corroded connections are visible and should be corrected before any further testing is done to the system.

Determining the battery's state of charge is the first step in testing the alternator. The battery must be at least 75 percent charged before the alternator will perform to specifications. (Some manufacturers specify that the battery must be 95 to 100 percent charged before testing the alternator.) Recharge the battery if the charge is low; if it will not accept a charge, a fully charged test battery should be installed in its place before testing the charging system.

CHARGING SYSTEM TESTING

After verifying that the battery is fully charged and is functioning correctly, the charging system can be performance tested. There are basically three tests, although some manufacturers might specify more. First, the output of the alternator is tested. If the output is below specifications, the voltage regulator is bypassed and battery current is wired directly to the field circuit of the rotor. This is called *full fielding* the rotor. If this corrects the problem, the fault is in the regulator; if the output remains low with the regulator bypassed, the alternator might be defective. However, before condemning the alternator or regulator, voltage drop tests should be performed on the system's wiring to determine if high resistance might be the cause of the charging troubles.

 Shop Talk

Alternators that power the field circuit with current produced by the stator windings rely on residual magnetism in the rotor to initially energize the stator when the engine is being started. During handling or repair, this residual magnetism can be lost. It must be restored before testing the system. This is done by connecting a jumper wire between the diode trio terminal and the alternator output terminal as shown in **Figure 7–32**.

The following procedures are general in nature and will not apply to every alternator. The major differences

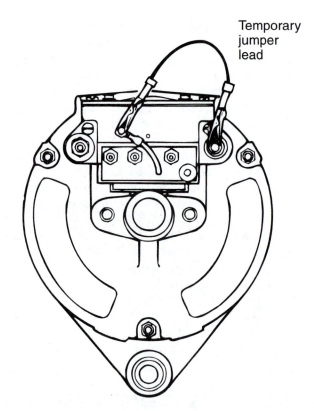

FIGURE 7–32   Restoring residual magnetism to the rotor. (Courtesy of Freightliner Trucks)

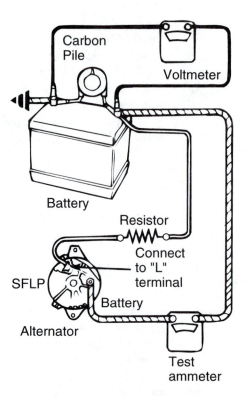

FIGURE 7–33   Test connections for an ammeter, voltmeter, and carbon pile. (Courtesy of Delco-Remy Co.)

from model to model are the meter test points and the specifications. Keep in mind that not all vehicle manufacturers require all of these tests to be performed, while others suggest even more. During any test, it is very important to refer to the vehicle manufacturer's specifications; even the most accurate test results are worthless if they are not matched against the correct specs. Before beginning, obtain a copy of the vehicle service manual and keep it handy for reference.

Alternator Output Testing

To test the maximum output of an alternator, follow these steps:

1. Disconnect the ground battery cable from the alternator.
2. Install an ammeter in series with the alternator output terminal and the ground battery cable or use an inductive pick-up. Also install a voltmeter and a carbon pile across the terminals of the battery (**Figure 7–33**).

Shop Talk

An inductive clamp ammeter is often used instead of a series ammeter. It is clamped onto the ground battery cable instead of connected in series with it. Therefore, the cable is not disconnected.

3. Start the engine and run fast enough to obtain maximum output from the alternator (typically about 2,000 rpm).
4. Turn on all accessories and increase the load on the carbon pile until the voltage in the systems drops to 12.7 volts. This will cause the regulator to send full fielded voltage to the rotor, maximizing the alternator output.
5. Compare the amperage reading with the manufacturer's specifications. If the output is within 10 percent of rated amperage, the alternator is good. If the output is more than 10 percent below specs, full field test the alternator.

Full-Field Testing the Alternator

By applying full battery voltage directly to the field windings in the rotor, it can be determined whether or not the regulator is the cause of the undercharging condition. There are two variations of this procedure that apply to alternators with external regulators. If the field circuit is grounded through the regulator (an A circuit), the regulator is disconnected from the field terminal on the alternator and a jumper is connected between the terminal and a ground. If the alternator receives battery voltage through the regulator (a B circuit), the regulator is disconnected from the field

(F) terminal and a jumper is connected to the terminal and to the insulated battery terminal. In either case, the regulator circuit is bypassed completely and full battery voltage is available to the rotor.

Now, repeat the procedure given for testing the alternator output. With the load applied, observe whether or not the alternator output rises to rated amperage. If it does, the alternator is functioning correctly and the regulator must be replaced. If it does not, the alternator must be tested further to determine the cause of undercharging.

CAUTION: When testing the output of a full-fielded alternator, watch the rise in system voltage carefully. Because the current output is not regulated, the battery voltage level can quickly reach an excessive level, high enough to overheat the battery, cause electrolyte to spew from the vent holes, and damage sensitive electronic components. Do not let the system voltage rise above the typical maximum regulator setting of 15 volts.

Shop Talk

Some alternators with remotely mounted electronic regulators are connected to the regulators by a wiring harness and a multipin connector. Full-fielding the field circuit is accomplished by removing the connector from the regulator and connecting a jumper wire between two pins (terminals) in the harness connector (consult the manufacturer's manual to correctly identify the proper test terminals). Doing so bypasses the regulator, sending battery current directly to the field circuit.

Full-Fielding Internally Regulated Alternators

Full-field testing is not possible on some alternators with internal voltage regulators. To isolate and test the regulator on these types of alternators, the alternator must be removed from the truck and disassembled. Other internally regulated alternators have a hole through which the field circuit can be tested. The test will require the use of a short jumper with insulated clips and a stiff paper clip wire or a $1/32$-inch drill bit (**Figure 7–34**).

1. With the engine off and all electrical accessories turned off, measure the voltage across the battery terminals. Make a note of the reading.
2. Start the engine and run it at the speed necessary to generate full alternator output.

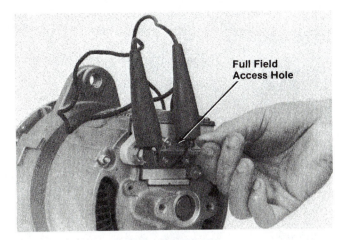

FIGURE 7–34 *Full-fielding an alternator with an integral voltage regulator. (Courtesy of Prestolite Electric Corp.)*

3. Connect a short jumper to the alternator negative output terminal and to the shank of the straightened clip wire or drill bit.
4. Insert the wire or bit into the full-field access hole as far as it will go and make a note of the voltage reading. If a fault in the regulator or diode trio is causing the undercharged condition, the alternator output should climb to within 10 percent of its rated output.

Stator Winding Testing

If full-fielding the alternator does not solve the undercharging condition, the regulator is probably okay and the problem is within the alternator. On some alternators, such as the one shown in **Figure 7–35**, the individual stator windings can be tested without removing the alternator from the vehicle. The windings are tested by connecting an AC voltmeter across the alternator AC terminals 1 and 2, 1 and 3, and 2 and 3. These readings should be approximately the same if the stator is okay. If the readings

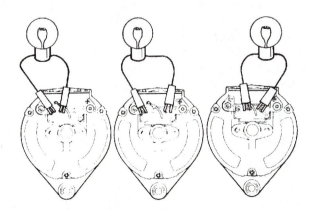

FIGURE 7–35 *AC terminals of the stator windings. (Courtesy of Prestolite Electric Corp.)*

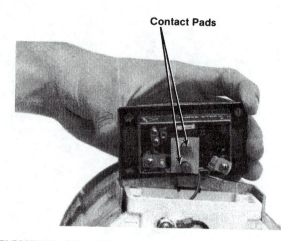

Contact Pads

FIGURE 7–36 *Contact pads should be cleaned with 600-grit grade sandpaper. (Courtesy of Prestolite Electric Corp.)*

vary, the stator has an open or short. If the readings are approximately the same, the stator is okay and some other fault exists in the alternator, causing the undercharging condition. The alternator will have to be removed from the vehicle for further testing if the problem is not found in the circuit wiring.

If an alternator with an internal regulator is undercharging, the problem could be a defective diode trio. If one or more of the diodes are defective, full-field voltage will not be delivered to the rotor; thus, the output of the alternator will be low. Some alternators have a diode trio that can be removed without removing the alternator from the vehicle. Then, a known good diode trio can be installed. If this corrects the undercharging problem, the fault has been found and corrected.

Undercharging might also be caused by worn or corroded brushes or contact pads in the regulator. On some alternators, the regulator can be removed to provide access to the brushes and the contact pads (**Figure 7–36**). If the brushes appear burned, cracked, or broken, or if they are worn to a length of $3/16$ inch or less, they should be replaced. The brushes should also be replaced if the shunt lead inside the brush spring is broken. If the brushes are okay, the contact caps on the brushes and the contact pads on the regulator should be cleaned, using 600-grit (or finer) grade sandpaper or crocus cloth.

Regulator Circuit Testing

In addition to providing current to the field windings of the rotor, the voltage regulator must also keep the system voltage within a predetermined range, typically 13.5 to 14.5 on a 12-volt system. The regulator must reduce the output of the alternator when necessary to keep the system voltage below the set

level. If it does not, the battery will be overcharged. To test the system voltage regulation, follow this procedure:

1. Connect the volt-amp tester to the battery as described earlier under "Alternator Output Testing."
2. Start the engine and run it at the speed necessary to achieve full output from the alternator.
3. Observe the voltage levels and current output. As the voltage level approaches 13.8 to 14.9, the alternator output should slowly decrease. When the battery is fully charged, the output of the alternator should be very small. If the output of the alternator remains high after the system voltage reaches its specified peak, the voltage in the system will rise above 15 volts. The overcharging battery will overheat, begin to lose water, and might even start spewing electrolyte through vent holes. The problem might be a defective regulator, a short in the field circuit, or high resistance in the wiring. For example, any unwanted resistance in the sensing wire will cause the voltage regulator to "think" the voltage level is lower than it really is. Thus, the alternator will continue to generate current even when the battery is fully charged.

To test the ground of the regulator, install a voltmeter between the case of the regulator (if externally mounted) and the battery ground. Start the engine and measure the voltage difference between the two points. With the engine running, there should not be any difference between the two ground points. The voltmeter should read 0.

If the ground side tests good, measure the voltage supplied to the regulator by the sensing circuit. It should be equal to battery voltage. If resistance is high on the ground side or if voltage is low on the sensing (supply) side of the regulator, the problem might be loose or corroded connections or a partial open in the circuit. If resistance and supply voltage meet specifications, the overcharging condition is the fault of either the alternator or the regulator.

Charging Circuit Testing

If there is excesive resistance in the charging circuit, the alternator might not fully charge the battery during peak load periods. The design of the alternator limits the current output to a maximum level. Resistance in the circuit will prevent full current from reaching the battery. Under peak load conditions, the battery will not be fully charged, and over time, can be damaged.

To test voltage drop in the charging circuit, connect the volt-amp tester as explained earlier and load the battery to full alternator output. Then, connect a voltmeter from the output terminal of the alternator and the insulated terminal of the battery. The voltage drop between the two points should be slight (0.2 volts on alternators rated at 14 volts and 0.5 volts on alternators rated at 28 volts is a typical maximum). If the voltage drop is higher than the manufacturer's specifications, there might be loose or corroded connections in the circuit or a fusible link might be degenerating.

The ground side of the circuit should also be tested. Move the voltmeter leads to the alternator casing and the battery ground pole. The voltage drop in this case should be 0. If not, the alternator mounting might be loose or corrosion might be built up between the casing and mounting bracket.

7.6 STARTING SYSTEMS

The vehicle's starting system is designed to turn or "crank" the engine over until it can operate under its own power. To do this, the **starter motor** receives electrical power from the vehicle's batteries. The starter motor then converts this energy into mechanical energy, which it transmits through the drive mechanism to the engine's flywheel.

A typical starting system, as described earlier, has five basic components and two distinct electrical circuits (**Figure 7–37**). The components are

- Battery
- Key switch (or starter button)
- Battery cables
- Magnetic switch
- Starter motor

The starter motor draws a great deal of electrical current from the battery. A large starter motor might require 300 to 400 amperes of current. This current flows through the heavy-gauge cables that connect the battery to the starter.

The driver switches the flow of current using the starting switch. However, if the cables were to be routed from the battery to the starting switch and then on to the starter motor, the voltage drop caused by resistance in the cables would be too great. To avoid this problem, the system is designed with two connected circuits: the **starter circuit** and the control circuit.

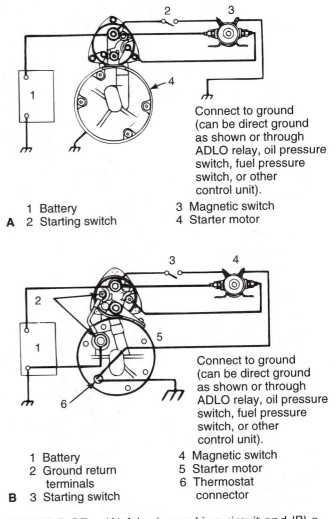

1 Battery   3 Magnetic switch
A 2 Starting switch   4 Starter motor

Connect to ground (can be direct ground as shown or through ADLO relay, oil pressure switch, fuel pressure switch, or other control unit).

1 Battery   4 Magnetic switch
2 Ground return terminals   5 Starter motor
B 3 Starting switch   6 Thermostat connector

Connect to ground (can be direct ground as shown or through ADLO relay, oil pressure switch, fuel pressure switch, or other control unit).

FIGURE 7–37 (A) A basic cranking circuit and (B) a cranking circuit with a thermostat. (Courtesy of International Truck and Engine Corp.)

CRANKING CIRCUIT

The starter circuit carries the high current flow within the system and supplies power for the actual engine cranking. Components of the starting circuit are the battery, battery cables, magnetic switch or solenoid, and the starter motor.

Battery and Cables

Earlier sections of this chapter detail the important role the battery plays in a vehicle's electrical system. Many problems associated with the starting system can be solved by first troubleshooting the batteries and related components. The batteries, of course, are the heart of the starting system; they supply all the power needed to crank the engine. They must be of sufficient size and capacity to meet the cranking demands in all weather conditions.

The cranking circuit requires two or more heavy-gauge cables. Two of these cables attach directly to the batteries. One of these cables connects between the battery's negative terminal and a good ground (for negatively grounded battery systems) or a ground stud on the starter case. The other cable connects the battery positive terminal with the starter relay. On vehicles where the switch does not mount directly on the starter motor, two cables are needed. One runs from the positive battery terminal to the solenoid and the second from the solenoid to the starter motor terminal. In any case, these cables carry the necessary heavy current load from the battery to the starter and from the starter back to the battery. All cables must be in good condition. Cables can be corroded by battery acid. Corrosion on the cables will act as a resistor and cause a voltage drop and decrease circuit amperage, reducing power available to the starter. Contact with the engine parts and other metal surfaces can fray the cable insulation. Frayed insulation can cause a dead short that can seriously damage electrical components. A short to ground is the cause of many dead batteries and can result in fire. Cables must also be heavy enough to comfortably carry the required current load.

When checking cables and wiring, always check any fusible links in the wiring. Almost all vehicles are equipped with fusible links to protect the wiring from overloads. These links are different in construction than a fuse, but operate in much the same way. The most common type is made of a wire with a special nonflammable insulation. Wire used to make fusible links is ordinarily two sizes smaller than the wire in the circuit they are designed to protect. Often, when a fusible link is subjected to a current overload, the insulation will become charred and give the appearance of a failed link. This is not always a true indication. The best test is to connect an ohmmeter across the link to check for continuity.

The largest fusible link is usually located at the starter solenoid battery terminal. From this terminal, current is distributed to all parts of the vehicle. Another large fusible link joins this battery terminal to the main body harness and protects the complete wiring of the vehicle. This link may take several forms, from a wire to a small piece of metal with terminal connections on each end. It may be located in its own special holder or be attached directly to the starter relay terminal. When a link has failed, always troubleshoot the system and locate the cause before replacing the link.

Starter Relay

A heavy-duty truck's starting system contains a magnetic switch that enables the control circuit to open and close the cranking circuit. This magnetic switch, usually called a starter relay, is either mounted at a remote location from the starter or built into the front end of the starter's solenoid assembly. Remote-mounted starter relays are located close to the battery or starter to keep the cables as short as possible. **Figure 7–38** shows a starter relay mounted on the firewall of a conventional truck near the windshield wiper motor on the driver's side. On cab-over-engine models, the starter relay is often mounted on the front frame cross member.

The starter relay consists of an electromagnet plunger, contact disc, and two springs (**Figure 7–39**). The electromagnet has a hollow core in which the plunger moves.

When the operator turns the starting switch to the crank position, battery current flows through the starting switch to a control circuit terminal on the starter relay. Control circuit current flows through the windings in the electromagnet, creating a magnetic force that pulls the plunger into the hollow core. Spring pressure can then force the contact disc against the starter circuit terminals, closing the circuit. High amperage current flows through contacts to the starter motor.

When the engine starts, the starting switch is released, and the control circuit is opened at the starting switch. This deactivates the electromagnet. The return spring forces the plunger out of the hollow core, which also forces the contact disc to move away from the starter circuit terminals, interrupting current flow to the starter motor.

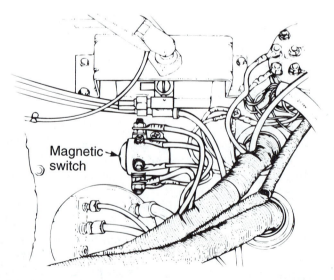

FIGURE 7–38 A magnetic switch, or starter relay, mounted on a firewall near a windshield wiper motor. (Courtesy of Freightliner Trucks)

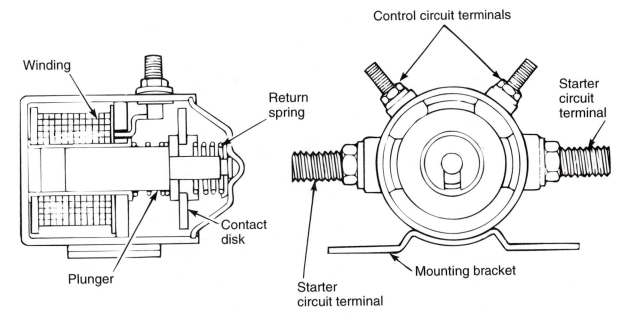

FIGURE 7–39   Component parts of a magnetic switch. (Courtesy of Freightliner Trucks)

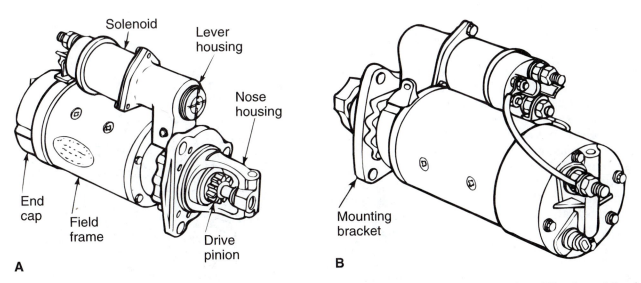

FIGURE 7–40   (A) Front view and (B) rear view of a starter motor. (Courtesy of International Truck and Engine Corp.)

Starter Motors

The starter motor (**Figure 7–40**) converts the electrical energy from the battery into mechanical energy for cranking the engine. The starter is an electric motor designed to operate under great electrical loads and to produce very high horsepower.

 Shop Talk _____

The starter can only operate for short periods of time without rest. The high current needed to oper-

ate the starter creates considerable heat, and continuous operation will cause serious heat damage to the unit. The starter must never operate for more than 30 seconds at a time and should rest for 2 minutes between these extended crankings. This permits the heat to dissipate without damage to the unit.

All starter motors are generally the same in design and operation. Basically the starter motor consists of a housing, field coils, an **armature**, a commutator

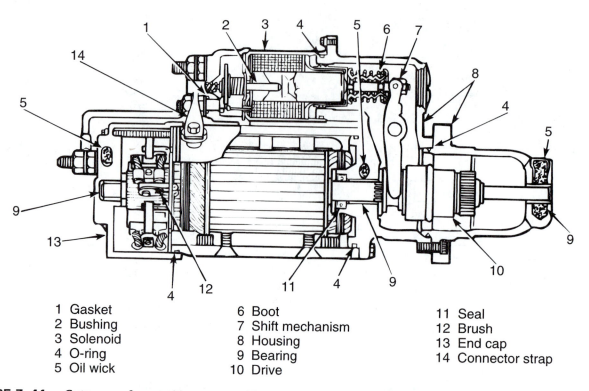

1 Gasket	6 Boot	11 Seal
2 Bushing	7 Shift mechanism	12 Brush
3 Solenoid	8 Housing	13 End cap
4 O-ring	9 Bearing	14 Connector strap
5 Oil wick	10 Drive	

FIGURE 7–41   *Cutaway of a cranking motor. (Courtesy of Delco-Remy Co.)*

and brushes, end frames, and a solenoid-operated shift mechanism. **Figure 7–41** shows cutaway of a typical starter used on heavy-duty trucks. The starter housing or frame encloses the internal starter components and protects them from damage, moisture, and foreign materials. The housing supports the field coils (**Figure 7–42**) and forms a conducting path for the magnetism produced by the current passing through the coils. The field coils and their pole shoes are securely attached to the inside of the iron housing. The field coils are insulated from the housing but are connected to a terminal that protrudes through the outer surface of the housing. The field coils and pole shoes are designed to produce strong stationary electromagnetic fields within the starter body as current is passed through the starter. These magnetic fields are concentrated at the pole shoes. Fields will have a N or S magnetic polarity depending on the direction the current flows. The coils are wound around respective pole shoes in opposite directions to generate opposing magnetic fields.

The field coils connect in series with the armature winding through the starter brushes. This design permits all current passing through the field coil circuit to also pass through the armature windings.

The armature is the rotating component of the starter. It is located between the drive and commutator end frames and the field windings (**Figure 7–43**).

When the starter motor operates, the current passing through the armature produces a magnetic

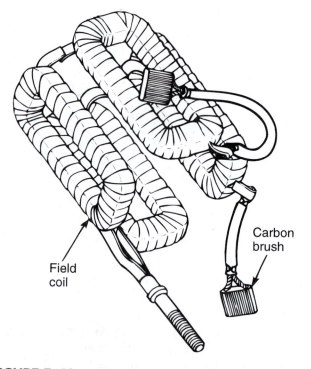

FIGURE 7–42   *Four field coils used in a starter motor. (Courtesy of Delco-Remy Co.)*

field in each of its conductors. The reaction between the armature's magnetic field and that of the field coils causes the armature to rotate. This creates the torque to crank the engine.

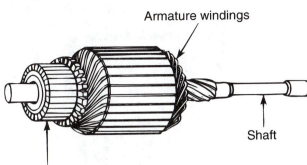

Armature windings

Shaft

Commutator

FIGURE 7–43 Starter motor armature and commutator. (Courtesy of Delco-Remy Co.)

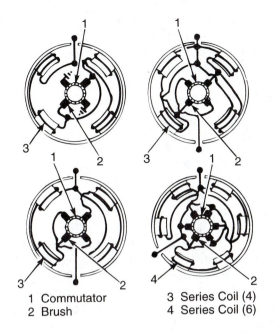

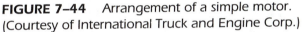

1 Commutator
2 Brush

3 Series Coil (4)
4 Series Coil (6)

FIGURE 7–44 Arrangement of a simple motor. (Courtesy of International Truck and Engine Corp.)

The armature has two main components: the armature windings and the commutator. Both mount to the armature shaft. The armature windings are not made of wire. Instead, heavy flat copper strips that can handle heavy current flow are used. The windings are constructed of several coils of a single loop each. The sides of these loops fit into slots in the armature core or shaft, but are insulated from it. Each slot contains the side of two of the coils. The coils connect to each other and to the commutator so that current from the field coils flows through all of the armature windings at the same time. This action generates a magnetic field around each armature winding, resulting in a repulsion force all around the conductor. This repulsion force causes the armature to turn.

The commutator assembly (**Figure 7–44**) presses onto the armature shaft. It is made up of heavy copper segments separated from each other and the armature shaft by insulation. The commutator segments connect to the ends of the armature windings.

Starter motors have four to twelve brushes that ride on the commutator segments and carry the heavy current flow from the stationary field coils to the rotating armature windings via the commutator segments. The brushes are held in position by a brush holder.

Operation. The starter motor converts electric current into torque or twisting force through the interaction of magnetic fields. It has a stationary magnetic field (created by passing current through the field coils) and a current-carrying conductor (the armature windings). When the armature windings are placed in this stationary magnetic field and current is passed through the windings, a second magnetic field is generated with its lines of force wrapping around the wire (**Figure 7–45**). Since the lines of force in the stationary magnetic field flow in one direction across the winding, they combine on one

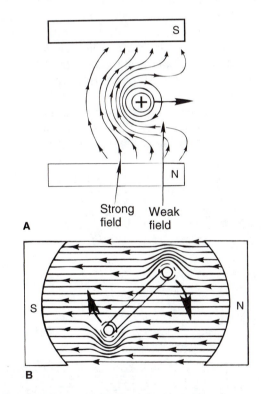

Strong field   Weak field

A

B

FIGURE 7–45 (A) When a current carrying conductor passes through a magnetic force field, the field is deflected to one side of the conductor. This creates pressure on one side of the wire and forces the conductor to move away from the strong force field. (B) When a loop of wire is placed between two magnets, the intersection between the force fields causes the loop to rotate around its axis.

side of the wire increasing the field strength but are opposed on the other side, weakening the field strength. This creates an unbalanced magnetic force, pushing the wire in the direction of the weaker field.

Since the armature windings are formed in loops or coils, current flows outward in one direction and returns in the opposite direction. Because of this, the magnetic lines of force are oriented in opposite directions in each of the two segments of the loop. When placed in the stationary magnetic field of the field coils, one part of the armature coil is pushed in one direction while the other part is pushed in the opposite direction. This causes the coil, and the shaft to which it is mounted, to rotate.

Each end of the armature windings are connected to one segment of the commutator. Two carbon brushes are connected to one terminal of the power supply. The brushes contact the commutator segments conducting current to and from the armature coils (**Figure 7–46**).

As the armature coil turns through a half revolution, the contact of the brushes on the commutator causes the current flow to reverse in the coil. The commutator segment attached to each coil end will have traveled past one brush and is now in contact with the other. In this way, current flow is maintained constantly in one direction, while allowing the segment of the rotating armature coils to reverse polarity as they rotate. This ensures that the armature will spin in one direction.

In a starter motor, many armature segments must be used. As one segment rotates past the stationary magnetic field pole, another segment immediately

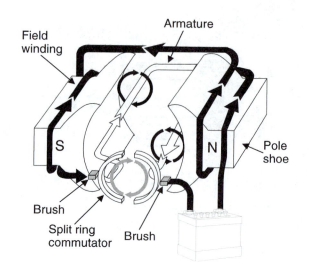

FIGURE 7–46 Current flow in the conductor must be reversed every 180 degrees of rotation so that the armature will continue to rotate in the same direction.

takes its place. The turning motion is made uniform and the torque needed to turn the flywheel is constant rather than fluctuating as it would be, if only a few armature coils were used. Starter motors are series motors. Current is first flowed through the field coils, then routed to the armature. This produces maximum torque at initial start.

Shift Mechanism

The solenoid-operated shift mechanism (**Figure 7–47**) is mounted in a solenoid housing that is sealed

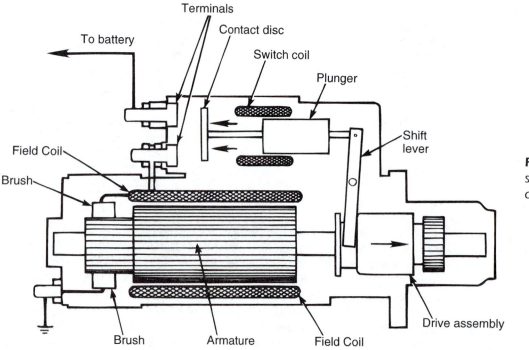

FIGURE 7–47 The solenoid engages the drive mechanism.

to keep out oil and road splash. The case is flange mounted to the starter motor housing. It contains an electromagnet with a hollow core. A plunger is installed in the hollow core, much like the starter relay described earlier. The solenoid performs two functions. When energized, plunger movement acts on a shift lever that shifts the starter motor drive pinion into mesh with the engine flywheel teeth so that the engine can be cranked. The solenoid also closes a set of contacts that allows battery current to flow to the starter motor; this ensures that the pinion and flywheel are engaged before the starter armature begins to revolve.

The solenoid assembly has two separate windings: a pull-in winding and a hold-in winding (**Figure 7–48**). The two windings have approximately the same number of turns but are wound from different gauge wire. Together these windings produce the electromagnetic force needed to pull the plunger into the solenoid coil. The heavier gauge pull-in windings draw the plunger into the solenoid, while the lighter gauge hold-in windings produce enough magnetic force to hold the plunger in this position.

Both windings are energized when the starting switch is turned to the start position. If the system has a remote starter relay, battery current passes through the relay to the "f" terminal on the solenoid. When the plunger contact disc touches the solenoid terminals, the pull-in winding is deactivated. At the same time, the plunger contact disc makes the motor feed connection between the battery and the starting motor, directing full battery current to the field coils and starter motor armature for cranking power.

Shop Talk _____

In some starters, the solenoid also performs the duty of a relay. The control circuit is wired to the windings of the solenoid. Battery current is routed directly to the starter motor through the drive solenoid.

_____

As this electrical connection is being made at the terminal end of the solenoid, the mechanical motion of the solenoid plunger is being transferred to the drive pinion through the shift lever, bringing the pinion gear into mesh with the flywheel ring gear (**Figure 7–49**). When the starter motor receives current, its armature starts to turn. This motion is transferred through an overriding clutch and pinion gear to the engine flywheel and the engine is cranked.

With this type of solenoid-actuated direct drive starting system, teeth on the pinion gear might not immediately mesh with the flywheel ring gear. If this occurs, a spring located behind the pinion compresses so that the solenoid plunger can complete its stroke. When the starter motor armature begins to turn, the pinion teeth will mesh with the flywheel teeth and the spring pressure.

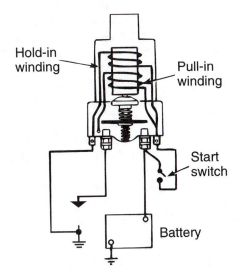

FIGURE 7–48 *The starter solenoid has two windings: a pull-in winding, and a hold-in winding. (Courtesy of International Truck and Engine Corp.)*

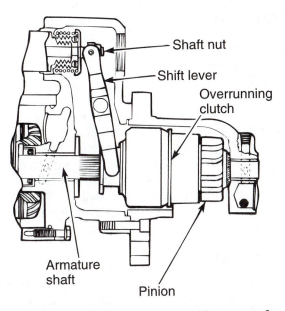

FIGURE 7–49 *Drive mechanism. (Courtesy of Volvo Trucks North America Inc.)*

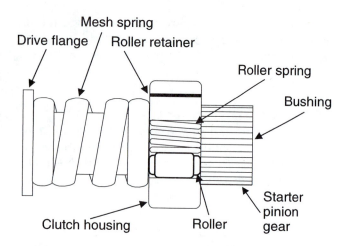

Mesh spring

Drive flange / Roller retainer

Roller spring

Bushing

Starter pinion gear

Clutch housing   Roller

FIGURE 7–50 *Override clutch operation.*

Override Clutches

The override clutch performs an important job in protecting the starter motor. When the engine starts and runs, its speed increases. If the starter motor remained connected to the engine through the flywheel, the starter motor would spin at very high speeds, destroying the armature windings. To prevent this, the starter must be disengaged from the engine as soon as the engine turns more rapidly than the starter. But with a solenoid-actuated drive mechanism, the pinion remains engaged with the flywheel until current stops flowing to the starter. In these cases, an overriding clutch is used to disengage the starter. A typical overriding clutch is shown in **Figure 7–50**.

The clutch housing is internally splined to the starting motor armature shaft. The drive pinion turns freely on the armature shaft within the clutch housing. When the clutch housing is driven by the armature, the spring-loaded rollers are forced into the small ends of their tapered slots and wedged tightly against the pinion barrel. This locks the pinion and clutch housing solidly together, permitting the pinion to turn the flywheel and, thus, crank the engine.

When the engine starts, the flywheel spins the pinion faster than the armature. This action releases the rollers, unlocking the pinion gear from the armature shaft. The pinion then "overruns" the armature shaft freely until being pulled out of the mesh without stressing the starter motor. Note that the overrunning clutch is moved in and out of mesh with the flywheel by linkage operated by the solenoid.

Thermostats

Some starter motors are equipped with a **thermostat.** The thermostat monitors the temperature of the motor. If prolonged cranking causes the motor temperature to exceed a safe threshold, the thermostat

will open and the starter current will be interrupted. The starter motor, then, will not operate until the motor cools and the thermostat closes.

CONTROL CIRCUIT

The control circuit allows the driver to use a small amount of battery current to control the flow of a large amount of current in the starting circuit. It consists of the following components:

Starting Switch

The circuit usually consists of a starting switch connected by light gauge wire to the battery and the starter relay. When the starting switch is turned to the start position, a small amount of current flows through the coil of the starter relay, closing it and allowing high current to flow directly to the starter motor. The starting switch performs other jobs besides controlling the starting circuit. It normally has at least four separate positions: accessory, off, on (run), and start.

Some older trucks have a push-button starter switch. Battery voltage is available to the switch when the starting switch is in the on position. When the pushbutton is depressed, current flows through the control circuit to the starter relay coil.

Starting Safety Switch

The **starting safety switch** prevents vehicles with automatic transmissions from being started in gear. Safety switches, often called neutral safety switches, can be located in either of two places in the control circuit. One position is between the starting switch and the starter relay. Placing the transmission in park or neutral will close the switch so current can flow to the starter relay. The safety switch can also be connected between the starter relay and ground so that the switch must be closed before current can flow from the starter relay to ground.

Starter Relays

The starter relay covered earlier in the chapter is actually the point in the cranking circuit where the control circuit and starter circuit come together. Low current in the control circuit passes through the starting switch and starting safety switch to energize the starter relay and activate the starter circuit.

CRANKING CIRCUIT TESTING

The cranking circuit requires testing when the engine will not crank, when the engine cranks slowly, or when the starter motor will not turn.

Preliminary Checks

Cranking output obtained from the motor is also affected by the condition and charge of the battery, the wiring circuit, and the engine's cranking requirements.

The battery should be checked and charged as needed before testing. Be sure the batteries are rated to meet or exceed the vehicle's manufacturer's recommendations. The voltage rating of the batteries must also match the voltage rating of the starter motor.

 Shop Talk _____

The starter should not be operated if the voltage at the battery is below 9.6 volts. Some leasing companies now use a voltage sensing module to prevent starter operation if voltage is below 9.6 volts.

Check the wiring for clean, tight connections. The starter motor may draw several hundred amperes during cranking. Loose or dirty connections will cause excessive voltage drop. Clean and tighten all connections as necessary. The cranking system cannot operate properly with excessive resistance in the circuit.

Also make certain the engine crankcase is filled with the proper weight oil as recommended by the engine manufacturer. Heavier than specified oil when coupled with low operating temperatures will lower cranking speed to the point where the engine will not start.

Check the starting switch for loose mounting, damaged wiring, sticking contacts, and loose connections. Check the wiring and mounting of the safety switch, if so equipped, and make certain the switch is properly adjusted. Check the mounting, wiring, and connections of the starter relay and starter motor. Also be sure that the starter pinion is properly adjusted.

Check the thermostat if the starter will not work. Use an ohmmeter to measure the resistance between the two thermostat terminals on the motor. The ohmmeter should read zero. If it does not, the thermostat is open-circuited. However, do not check the thermostat when the starter motor is hot: the thermostat is designed to open above a specific temperature.

Troubleshooting Procedures

A systematic troubleshooting procedure is essential when servicing the starting system. Nearly 80 percent of defective starters returned on warranty claims function to specification when tested. This is the result of poor or incomplete diagnosis of the starting and related charging systems. **Table 7–5** itemizes a systematic approach to starting system diagnosis. Testing the starting system can be divided into area tests, which check voltage and current in the entire system, and more detailed pinpoint tests, which target one particular component or segment of the wiring circuit.

Starter Relay Testing

The starter relay bypass test is a simple method of determining if the relay is operational. This test should be performed when the starter motor does not activate when the ignition key is in the start position (or when the starter button is depressed).

Connect a jumper cable around the starter relay (starter relay) as shown in **Figure 7–51**. This bypasses the relay. Then, crank the engine. If the engine cranks with the jumper installed, the starter

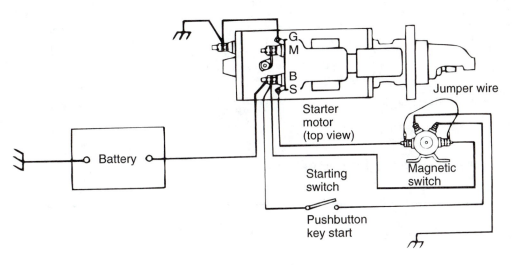

FIGURE 7–51 Check the starter relay operation by bypassing the switch with a jumper wire. (Courtesy of Freightliner Trucks)

TABLE 7–5:   TROUBLESHOOTING A CRANKING CIRCUIT

Problem	Possible Cause	Tests and Checks	Remedy
Engine cranks slowly or unevenly	1. Weak battery	1. Perform battery open circuit voltage and load voltage tests. Perform battery load tests (capacity). Check capacity and voltage ratings against engine requirements.	1. Service, recharge, or replace defective battery.
	2. Undersized or damaged cables	2. Perform visual inspection.	2. Replace as needed.
	3. Poor starter circuit connections	3. Perform visual inspection for corrosion and damage.	3. Clean and tighten. Replace worn parts.
	4. Defective starter motor caused by high internal resistance	4. Perform cranking current test and no-load test.	4. If cranking current is under specs, proceed with no-load bench testing.
	5. Engine oil too heavy for application	5. Check oil grade.	5. Change oil to proper specs.
	6. Seized pistons or bearings	6. Check compression and cranking torque.	6. Repair as needed.
	7. Overheated solenoid or starter motor	7. Check for missing or damaged heat shields.	7. Replace shield. Service as needed.
	8. High resistance in starter circuit	8. Use cranking current test, insulated circuit test, and ground circuit tests to pinpoint area of high resistance.	8. Replace defective components.
	9. Poor starter drive/ flywheel engagement	9. Perform visual inspection of drive and flywheel components.	9. Replace damaged items.
	10. Loose starter mounting	10. Perform visual inspection.	10. Tighten as needed.
Engine does not crank	1. Discharged battery	1. As listed above	1. As listed above
	2. Poor or broken cable connections	2. As listed above	2. As listed above
	3. Seized engine components	3. As listed above	3. As listed above
	4. Loose starter mounting	4. As listed above	4. As listed above
	5. Open in control circuit	5. Perform control circuit test to determine "open" condition or areas of high resistance.	5. Repair or replace components as needed.
	6. Defective starter relay	6. Perform starter relay bypass test.	6. Replace starter relay if engine cranks when these components are bypassed.
	7. Defective starter motor caused by internal motor malfunction	7. Perform starter relay bypass test.	7. Replace starter motor if engine will not crank when starter relay is bypassed.
Starter motor spins but does not crank engine	1. Defective starter drive	1. Perform starter drive test.	1. Replace starter drive.
	2. Worn or damaged pinion gear	2. Perform visual inspection of components.	2. Replace starter drive.
	3. Worn or damaged flywheel gears	3. Perform visual inspection of flywheel.	3. Replace as needed.
Starter does not operate or Movable pole shoe starter chatters or disengages before engine has started	1. Battery discharged	1. Perform battery load test.	1. Recharge or replace.
	2. High resistance in starting circuit	2. Perform cranking current test, insulated circuit test, and ground circuit tests to pinpoint area of high resistance.	2. Replace defective components.
	3. Open in solenoid or movable pole shoe hold-in winding		3. Replace solenoid or movable pole shoe starter.
	4. Worn solenoid unable to overcome return spring pressure		4. Replace solenoid. Install lighter return spring.
Noisy starter cranking	5. Defective starter motor Loose mounting	5. Perform visual inspection.	5. Replace starter motor. Tighten mounts. Correct alignment.

relay is defective and should be replaced. If the motor still will not crank, check the control circuit. The starter relay also can be checked by connecting a voltmeter across the winding, from the pushbutton key start connection to one of the mounting bolts that attaches the switch to the vehicle. Have someone hold the starting switch closed. If the voltmeter reading is 0 volts, check for an open circuit. If the voltmeter reading is less than 11 volts, check for corroded or loose connections (refer to "Control Circuit Testing"). Repair or replace any damaged wires. Then, check the voltage to the starter relay again. If the voltmeter reading is now 11 volts or more, the relay and control circuit is okay. If the reading is still less than 11 volts, replace the starter relay.

Cranking Current Testing

The cranking current test measures the amount of current, in amperes, that the starter circuit draws to crank the engine. This amperage reading will be useful in isolating the source of certain types of starter problems.

1. Connect the leads of a volt-amp tester (**Figure 7–52**).
2. Set the carbon pile to its maximum resistance (open).
3. Crank the engine and observe the voltmeter reading.
4. Stop cranking and adjust the carbon pile until the voltmeter reading matches the reading taken in step 3.
5. Note the ammeter reading.

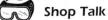 **Shop Talk**

If the analyzer uses an inductive pickup, be sure that the arrow on the inductive pickup is pointing in the right direction as specified on the ammeter. Then, crank the engine for 15 seconds and observe the ammeter reading.

Compare the reading obtained during testing to the manufacturer's specifications. **Table 7–6** summarizes the most probable causes of current draw that is either too high or too low. If the problem appears to be caused by too much resistance in the circuit, test the starting system resistance as covered in the circuit resistance tests in the next sections.

TABLE 7–6:	RESULT OF CRANKING CURRENT TESTING
Problem	**Possible Cause**
Low current	Undercharged or defective battery Excessive resistance in circuit due to faulty components or connections
High current draw	Short in starter motor Mechanical resistance due to binding engine or starter system component failure or misalignment

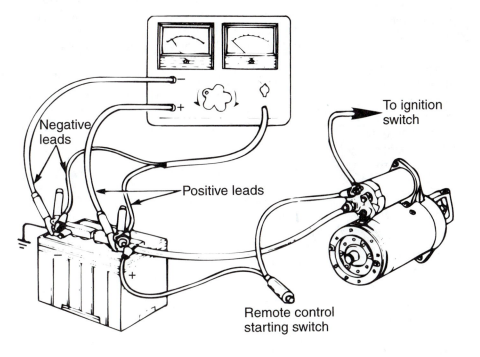

FIGURE 7–52 Test connections for a current draw test. (Courtesy of International Truck and Engine Corp.)

Negative leads

Positive leads

To ignition switch

Remote control starting switch

Control Circuit Testing

The control circuit test examines all the wiring and components used to control the starter relay.

The control circuit can also be checked by connecting a voltmeter across the coil terminals of the solenoid or starter relay. Remove the battery cable from the solenoid when performing the test to allow both windings of the solenoid to operate and to prevent the cranking motor from turning. If there are other leads connected to the battery terminal of the solenoid, remove these also and temporarily reconnect them to the battery cable. Operate the starting circuit and read the voltage across the solenoid coil terminals. This should be at least 10 volts for a 12-volt system and 20 volts for a 24-volt system.

> **WARNING:** When performing this test, do not operate the solenoid for extended periods of time as severe overheating will occur.

If the voltage available at the relay terminals is lower than specifications, check the control circuit wiring and components for high resistance.

High resistance in the solenoid switch circuit will reduce the current flow through the solenoid windings, which can cause improper functioning of the solenoid. In some cases of high resistance, it might not function at all. Improper functioning of the solenoid switch will generally result in the burning of the solenoid switch contacts, causing high resistance in the starter motor circuit.

Check the vehicle wiring diagram, if possible, to identify all control circuit components. These normally include the starting switch, safety switch, starter drive solenoid winding, or a separate relay drive.

While someone holds the starting switch in the start position, connect the voltmeter leads across each wire or component. A reading of more than 0.1 volt across any one wire or switch is usually an indication of trouble. If a high reading is obtained across the safety switch used on automatic transmissions, check the adjustment of the switch according to the manufacturer's service manual.

Starter Circuit Testing

If the control or energizing circuit is operating properly, voltage drop tests can be made on the starter circuit. Properly performed, this series of tests will pinpoint any source of excessive resistance in the starter circuit.

To perform voltage drop tests, an accurate low range voltmeter is required. The meter range should be 2 or 3 volts full scale and be equipped with leads that are long enough to reach the various points being checked. One of the leads should be equipped with a clean sharp probe so that when battery cables are being checked, it can be jabbed into the battery post. This will allow the drop across the clamp and post to be measured along with the cable drop. When connecting the voltmeter to the switch or motor, connect it to the terminal stud rather than to the terminal so that the drop across the connection will also be measured. Also, connect the positive voltmeter lead to the part of the circuit that is more positive and the negative lead to the more negative point.

The first test is the ground circuit resistance check. One lead of the voltmeter is connected to the ground terminal of the battery and the other lead is connected to the base of the starter (**Figure 7–53**). Then the voltmeter is read while the engine is cranked. If the voltage reading exceeds 0.2 volt, excessive resistance exists in the ground circuit. Further voltage checks must be made between the ground terminal and the base of the starter to pinpoint the specific source of the unwanted resistance. The problem can be isolated by measuring the voltage, in the same manner explained above, across the following connections:

- Ground cable connection at the battery
- Ground cable connection at the engine
- Starter bolt connection to the engine
- Contact between the starter main frame and the end frames
- Any connection between the ground terminal and the starter base. Dirt, acid corrosion, loose connections, or any other contaminant that prevents good metal-to-metal contact can cause excessive resistance at any of these points.

If the ground circuit resistance check does not uncover any problems with resistance, the insulated circuit resistance check can be performed. The positive lead of the voltmeter is connected to the positive

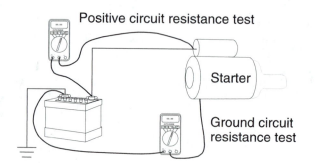

FIGURE 7–53 *Starter circuit testing.*

terminal of the battery. Then the engine is cranked. While the engine is being cranked, the other voltmeter lead is brought into contact with the starter input terminal. A voltmeter reading higher than 0.4 to 0.6 volt indicates there is a high resistance in one of the following components and connections of the insulated side of the circuit:

- Cable connection at the battery
- Cable connection at the solenoid
- Cable
- Starter solenoid

Isolate the cause of excessive resistance by performing additional voltage checks across these possible sources. Repair or replace any damaged wiring or faulty connections.

 Shop Talk

When performing starter circuit tests, prevent the engine from starting by putting the fuel shutoff in the shutoff position. Also, make the second connection of the insulated circuit resistance check only while the starter is running. When the starter is not operating, the voltage at the starter input terminal is 12 volts. This high voltage could damage a low range voltmeter.

NO-LOAD TESTS

When testing indicates that a starter malfunction is the cause of the no-start or hard-start condition, the starter must be removed from the vehicle for additional testing, first of which should be a no-load test. The no-load test is used to identify specific defects in the starter that can be verified with tests when disassembled. Also, the no-load test can identify open or shorted fields, which are difficult to check when the starter is disassembled. The no-load test also can be used to indicate normal operation of a repaired motor before installation.

To perform a no-load test, first clamp the starter motor in a bench vise. Then, connect the test equipment as shown in **Figure 7–54**. Connect a voltmeter from the motor terminal to the motor frame and use an rpm indicator to measure armature speed. Connect the motor and an ammeter in series with a fully charged battery of the specified voltage, and a switch in the open position from the solenoid battery terminal to the solenoid switch terminal. Close the switch and compare the rpm, current, and voltage reading with specifications. It is not necessary to obtain the exact voltage specified, because an accu-

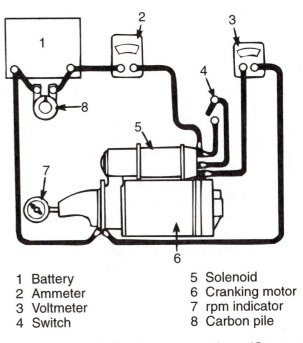

1	Battery	5	Solenoid
2	Ammeter	6	Cranking motor
3	Voltmeter	7	rpm indicator
4	Switch	8	Carbon pile

FIGURE 7–54 No-load test connections. (Courtesy of Delco-Remy Co.)

rate interpretation can be made by recognizing that if the voltage is slightly higher, the rpm will be proportionately higher, with the current remaining essentially unchanged. However, if the exact voltage is desired, a carbon pile connected across the battery can be used to reduce the voltage to the specified value. If more than one 12-volt battery is used, connect the carbon pile to only one of the 12-volt batteries. If the specified current draw does not include the solenoid, deduct from the ammeter reading the specified current draw of the solenoid hold-in winding. Make disconnections only with the switch open. Interpret the test results as follows:

1. Rated current draw and no-load speed indicates normal condition of the cranking motor.
2. Low free speed and high current draw indicate
    - Too much friction. Tight, dirty, worn bearings, bent armature shaft or loose pole shoes allowing armature to drag.
    - Shorted armature. This can be checked further after disassembly.
    - Grounded armature or fields. Check further after disassembly.
3. Failure to operate with high current draw indicates
    - A direct ground in the terminal or fields.
    - Seized bearings. This can be determined by turning the armature by hand.

4. Failure to operate with no current draw indicates
   - Open field circuit. This can be checked after disassembly by inspecting terminal connections and tracing the circuit with a test lamp.
   - Open armature coils. Inspect the commutator for badly burned bars after disassembly.
   - Broken brush springs, worn brushes, high insulation between the commutator bars or other causes that would prevent good contact between the brushes and commutator.
5. Low no-load speed and low current draw indicate high internal resistance due to poor connections, defective leads, dirty commutator and causes listed under step 4.
6. High free speed and high current draw indicate a shorted field. If shorted fields are suspected, replace the field coil assembly and check for improved performance.

Starting aids for cold weather are fully discussed in Chapters 4 and 31.

7.7 LIGHTING SYSTEMS

The lighting systems of the heavy-duty truck can be divided into two categories: exterior and interior. All of the vehicle's lighting systems have a common setup. The system's wiring proceeds from the power source to a circuit breaker and then to a switch and through a connector to the lamp assemblies.

Figures 7–55A and **7–55B** show the exterior lighting systems of a conventional and a COE truck, respectively. In the front of the conventional vehicle, there are headlights, turn signal lights, and side marker lights. Additional turn signal lights can be mounted on each side extender or on each of the outside door mirrors. COEs have skirt-mounted turn signal lights. Fog lights can also be mounted in the front bumper of both truck styles. On the roof of the cab there are clearance and identification lights. At the rear of the vehicle there are tail lights, stoplights, turn signals, and backup lights. Sometimes these rear lights can be consolidated in one assembly on each side of the vehicle. Utility lights are optional but must function if fitted.

Inside the cab are dome lights with high-intensity reading lights, and an instrument panel that features fully lighted gauges and labels for switches and controls. In the bunk or sleeper are accessory lights and high-intensity reading lights. Baggage compartments might be equipped with accessory lights.

MAINTENANCE

In addition to replacing all defective lamps and bulbs, periodically check to see that all wiring connections are clean and tight, that lamp units are

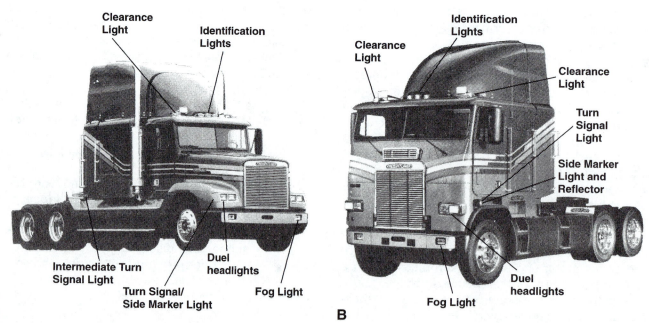

FIGURE 7–55 Exterior lighting (A) on conventional truck and (B) on COE truck. (Courtesy of Freightliner Trucks)

PHOTO SEQUENCE 3
SOLDERING TWO COPPER WIRES

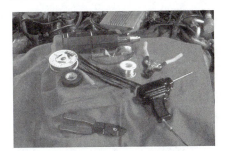

P3–1   Tools required to solder copper wire: 100-watt soldering iron, 60/40 rosin core solder, crimping tool, splice clip, heat shrink tube, heating gun, and safety glasses.

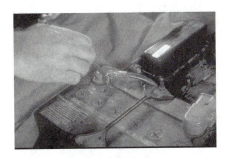

P3–2   Disconnect the fuse that powers the circuit being repaired. Note: If the circuit is not protected by a fuse, disconnect the ground lead of the battery.

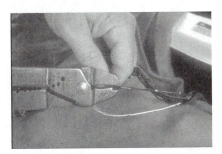

P3–3   Cut out the damaged wire.

P3–4   Using the correct-size stripper, remove about $1/2$ inch of the insulation from both wires.

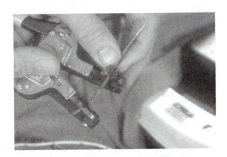

P3–5   Now remove about $1/2$ inch of the insulation from both ends of the replacement wire. The length of the replacement wire should be slightly longer than the length of the wire removed.

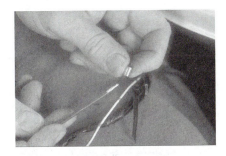

P3–6   Select the proper-size splice clip to hold the splice.

P3–7   Place the correct size and length of heat shrink tube over the two ends of the wire.

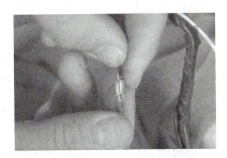

P3–8   Overlap the two splice ends and center the splice clip around the wires, making sure that the wires extend beyond the splice clip in both directions.

P3–9   Crimp the splice clip firmly in place.

PHOTO SEQUENCE 3 (Continued)
SOLDERING TWO COPPER WIRES

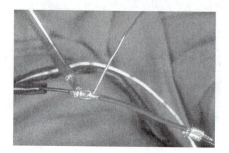

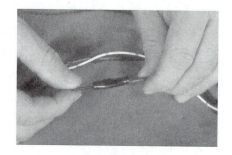

P3–10 Heat the splice clip with the soldering iron while applying solder to the opening of the clip. Do not apply solder to the iron. The iron should be 180 degrees away from the opening of the clip.

P3–11 After the solder cools, slide the heat shrink tube over the splice.

P3–12 Heat the tube with the hot air gun until it shrinks around the splice. Do not overheat the heat shrink tube.

tightly mounted to provide a good ground and that headlights are properly adjusted. Loose or corroded connections can cause a discharged battery, difficult starting, dim lights, and damage to the alternator and regulator.

Wires and/or harnesses must be replaced if insulation becomes burned, cracked, or deteriorated. Whenever it is necessary to splice a wire or repair one that is broken, use rosin flux solder to bond the splice and insulating tape to cover all splices or bare wires. A legitimate repair alternative is to use properly crimped and sealed butt splices. This can be done with heat-shrinkable tubing with sealant inside to cover the splice, or with connectors that come with tubing and sealant built in. Do not attempt to make a splice by twisting the wires together and taping or by using wire nuts. These methods make poor connections that deteriorate quickly.

Shop Talk

Lights should be turned off when cranking the engine to avoid transient voltage spikes. As the engine is cranked, 650 to 1,200 amps are drawn through the system. At the precise moment when the starter is disengaged, electricity may surge into any closed electrical circuits. This random surge can shorten the life of the lights, causing them to burn out prematurely.

7.8 HEADLIGHTS

Most headlight systems consist of either two or four sealed-beam tungsten headlight bulbs. In a two-headlight system, each light has two filaments, a high beam and a low beam. In a typical four-headlight arrangement, the two outer lights have a double filament, while the inner pair of lights has only the high-beam filament. Where the four headlights are vertically arranged, the upper lights have the double filament. In the past all headlights were round; however, many newer headlights are rectangular in shape for increased efficiency and to allow for aerodynamic vehicle designs.

CAUTION: Sealed beam lights are made of a single piece of molded glass containing tungsten filaments that are designed to operate in an atmosphere of inert gas. The slightest exposure to air will cause the filaments to burn out.

Halogen lights are often used as headlights. A halogen light has a small quartz-glass bulb that contains a fuel filament surrounded by halogen gas. This small, gas-filled bulb is contained within a larger metal reflector and lens element. A glass balloon

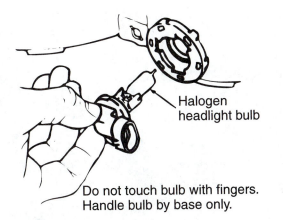

Halogen
headlight bulb

Do not touch bulb with fingers.
Handle bulb by base only.

FIGURE 7–56 Typical replacement bulb halogen headlights.

sealed to the metal reflector allows the halogen bulb to be removed without permitting water or dirt to damage the optics within the light (**Figure 7–56**). These bulbs may be replaced by unplugging the electrical connector, twisting the retaining ring about $1/8$ turn counterclockwise, and removing the bulb from its socket. To install a new bulb, simply reverse this procedure.

Sealed beam halogen lights are designed to give substantially more light on high beam, extending the driver's range of visibility for safer night driving. A halogen headlight emits approximately twice the candlepower (150,000 vs. 75,000) on high beam as the conventional incandescent-sealed lights. Both are limited by federal regulations (FMVSS 108) to a maximum of 200,000 candlepower output. Halogen bulbs have the advantage of producing a whiter light which helps improve visibility. Halogen bulbs also last longer, stay brighter, and use less wattage for the same amount of light produced.

HEADLIGHT ADJUSTMENT

Headlights must be kept in adjustment to obtain maximum illumination. Sealed beams that are properly adjusted cover the correct range and afford the operator the proper nighttime view. Out-of-adjustment headlights can cause other drivers discomfort and sometimes create hazardous conditions. In order to properly adjust the headlights, headlight aim must be checked first. Various types of headlight aiming devices are available commercially. When using aiming equipment, follow the instructions provided by the equipment manufacturer. Where headlight aiming equipment is not available, headlight aiming can be checked by projecting the upper beam of each light upon a screen or chart at a dis-

tance of about 25 feet ahead of the headlights. The truck should be exactly perpendicular to the chart.

The chart should be marked in the following manner (**Figure 7–57**). First, measure the distance between the centers of the matching headlights. Use this measurement to draw two vertical lines on the screen with each line corresponding to the center of a headlight. Then draw a vertical centerline halfway between the two vertical lines. Next, measure the distance from the floor to the centers of the headlights. Subtract 2 inches (50 mm) from this height and then draw a horizontal line on the screen at this new height. With headlights on high beam, the focal point of each projected beam pattern should be centered on the point of intersection of the vertical and horizontal lines on the chart. If necessary, adjust headlights vertically and/or laterally to obtain proper aim.

👓 **Shop Talk** _____

Headlight aim should always be checked on a level floor with the vehicle unloaded. In some states, the above instructions might conflict with existing laws and regulations. Where this is the case, legal requirements must be met. Modify the instructions accordingly.

Adjusting screws are provided to move the headlight assembly in relation to the hood (fender) to obtain correct headlight aim. Lateral or side-to-side adjustment is accomplished by turning the adjusting

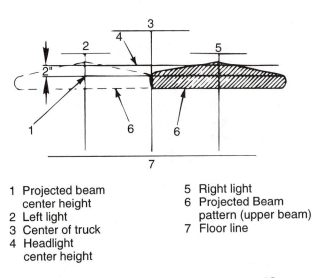

1 Projected beam center height	5 Right light
2 Left light	6 Projected Beam pattern (upper beam)
3 Center of truck	7 Floor line
4 Headlight center height	

FIGURE 7–57 Headlight aiming pattern. (Courtesy of International Truck and Engine Corp.)

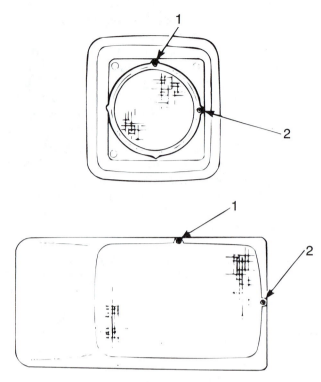

1 Vertical adjustment
2 Lateral adjustment

FIGURE 7–58 *Typical location of headlight adjusting screws. (Courtesy of International Truck and Engine Corp.)*

screw at the side of the headlight (**Figure 7–58**). Vertical or up-and-down adjustment is accomplished by turning the adjusting screw at the top of the headlight. Adjustments can be made without removing headlight bezels.

HEADLIGHT REPLACEMENT

There can be slight variations in procedures from one model to another when replacing the headlights. For instance, on some models the turn signal light assembly must be removed before the headlight can be replaced. Overall, however, the procedure will not differ much from the following typical instructions:

1. Remove the headlight bezel retaining screws (**Figure 7–59**). Remove the bezel. If necessary, disconnect the turn signal wires.
2. Remove the retaining ring screws from one or both lamps.
3. Remove the retaining rings.
4. Remove the light from the housing; disconnect the wiring connector from the back of the light.

A

B

FIGURE 7–59   Bulb removed from (A) front and (B) back. (Courtesy of Freightliner Trucks)

 Shop Talk

Some manufacturers recommend coating the prongs and base of the new sealed beam with terminal grease for corrosion protection. Use an electrical terminal protective approved by the manufacturer.

5. Push the wiring connector onto the prongs at the rear of the new light.
6. Place the new light in the headlight housing. Position it so that the embossed number in the lamp lens is on the top.

7. Place the retaining ring over the light and install the retaining ring screws; tighten them securely.

8. Check the aim of the headlight and adjust it in the manner described earlier.

9. Install the headlight bezel; secure it with the retaining screws. Connect the turn signal wiring (if it was disconnected). Do not over-tighten bezel retaining screws. Overtightening could cause damage (stripping) of threads in the hood (fender).

BULB REPLACEMENT

When replacing a bulb, inspect the bulb socket. If the socket is rusty or corroded, the socket or light assembly base should be replaced. Also, inspect the lens and gasket for damage when the lens is removed and replace any damaged part.

There are several types of construction designs for exterior lights: those in which the lens is removed and the bulb is removed from the front (**Figure 7–60**), those in which the light assembly must be removed as a unit (**Figure 7–61**), and LED lighting units. Removing the lens from the assembly of some types could cause damage, and wiping the reflector surface to clean it can reduce the light's brightness. Never remove the lens from the type in which the socket and bulb is removed from the back of the assembly.

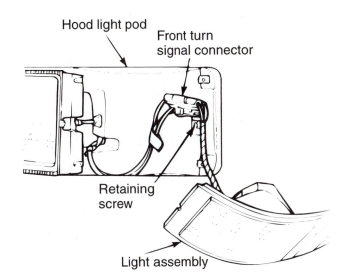

FIGURE 7–61 *Turn signal/marker light assembly. (Courtesy of International Truck and Engine Corp.)*

Bulbs are held in their sockets in a number of ways. Some bulbs are simply pushed into and pulled out of their sockets, and some are screwed in and out. To release a bayonet style bulb from its socket, the bulb is pressed in and turned counterclockwise. The blade mount style is removed by pulling the bulb off the mounting tab, then turning the bulb and removing it from the retaining pin. LED lighting units are replaced as units when tested to be defective.

🥽 Shop Talk

When replacing tungsten light bulbs, avoid touching the glass part of the new bulb assembly. Skin oil, present on even recently washed hands, will remain on the glass. This oil will inhibit heat from dissipating. Increased heat inside the bulb can make the filament burn out prematurely.

In order to replace the gauge, indicator, and warning lights, the instrument cluster or gauge panel must be removed. Sometimes the switch and control-knob labels in the dash are illuminated by means of fiber optics. The fiber optics light source is a single bulb attached by a nut to the back of the speedometer or tachometer mounting stud. LEDs (light-emitting diodes) are also used as warning and indicator lights. If one of the LEDs no longer works, replace the light bar assembly.

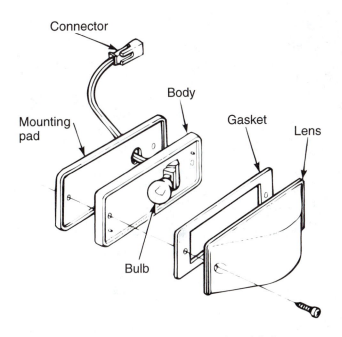

FIGURE 7–60 *Side marker/turn signal light. (Courtesy of International Truck and Engine Corp.)*

HEADLIGHT DIMMER SWITCH

The headlight dimmer switch, or courtesy switch, can be mounted on the floor or it can be a part of the turn signal assembly. To replace a floor-mounted unit, use the following procedure:

1. Remove the two screws that hold the dimmer switch to the mounting bracket (**Figure 7–62**).
2. Disconnect the dimmer switch wiring connector from the main wiring harness.
3. Remove the dimmer switch and pigtail as an assembly.
4. Position the new switch on the mounting bracket. Install and tighten the screws until snug.

5. Connect the dimmer switch wiring connector to the main wiring harness.
6. Check the dimmer switch operation.

Typical instructions for replacing a switch located in the turn signal switch are as follows:

1. Make sure the key switch is off and disconnect the battery negative cable.
2. Disconnect the switch wiring connector from the cab wiring harness connector (**Figure 7–63**).
3. Remove the switch mounting screws and the switch assembly from the steering column.
4. Remove the three screws holding the plastic grips together (**Figure 7–64**). Lift off the grip with the function symbols embossed on it.

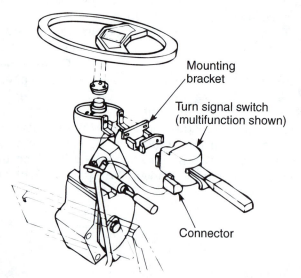

FIGURE 7–63   Typical parts of a turn signal/hazard switch. (Courtesy of International Truck and Engine Corp.)

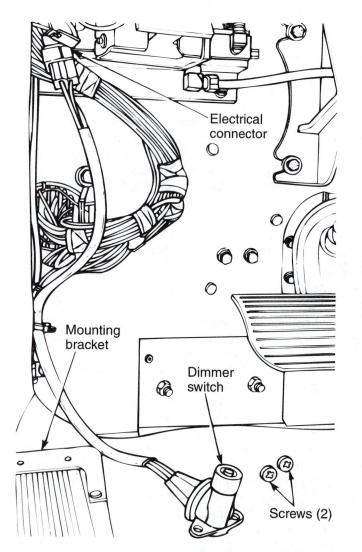

FIGURE 7–62   Typical components of the dimmer switch. (Courtesy of International Truck and Engine Corp.)

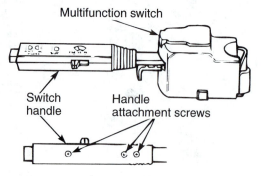

FIGURE 7–64   Three screws holding a turn signal grip together. (Courtesy of International Truck and Engine Corp.)

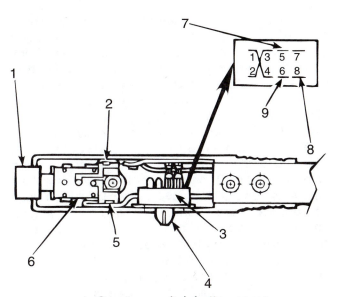

1. Courtesy switch button cover
2. Yellow wire terminal
3. Windshield wiper switch
4. Wiper switch button cover
5. Red wire terminal
6. Switch with circuit board facing up
7. Yellow wire from courtesy switch to no. 5 position
8. Green wire (high speed) to no. 8 position
9. Brown wire (low speed) to no. 6 position

FIGURE 7–65 Courtesy switch detail. (Courtesy of International Truck and Engine Corp.)

5. Lift out the courtesy switch, disconnecting the edge terminals of the yellow and red wires (**Figure 7–65**).
6. Carefully reassemble the edge terminals to the new switch.
7. Position the new courtesy switch inside the handle.
8. Attach the grip with the embossed symbols to the other grip and fasten it with the three screws. Tighten until snug.
9. Replace the battery negative cable and check the switch operation.

TURN SIGNAL SWITCH REPLACEMENT

Before replacing the turn signal switch, ensure that the trouble is in the switch and not elsewhere in the circuit. Check that the circuit breaker and fuse are functional, and inspect the signal light bulbs for bro-

ken filaments. Also, check the flasher relay, and replace it if necessary. If the turn signal switch must be replaced, make sure the key is off and the battery negative cable is disconnected.

TRAILER CIRCUIT CONNECTOR

The wiring terminal block for the trailer is usually located inside the cab, directly behind the driver's seat. Access is gained by removing the plastic cover held in place by screws. In sleeper models, the connector is in the luggage compartment.

A seven-wire core leads back from the terminal block to the trailer electrical connector cord and plug assembly. The standard wire color codes are given in **Table 7–7**. **Figure 7–66** shows an ATA (American Trucking Association) seven wire trailer receptacle and plug.

TABLE 7–7:	WIRE CONNECTION STANDARDS
Wire Color	**Light and Signal Circuits**
White	Ground return to towing vehicle
Black	Clearance, side marker, and identification lights
Yellow	Left-hand turn signal and hazard signal
Red	Stoplights and anti-wheel lock devices
Green	Right-hand turn signal and hazard signal
Brown	Tail and license plate lights
Blue	Auxiliary circuit

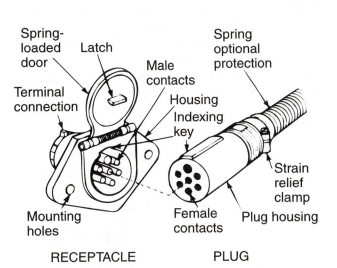

RECEPTACLE   PLUG

FIGURE 7–66 Light cord plug connector. (Courtesy of International Truck and Engine Corp.)

INSTRUMENT PANELS AND GAUGES

In horse-and-wagon days, the "dashboard" really was a vertical board placed in front of the driver to protect him/her from mud slung up from horse hooves just ahead. Today's dashboard is more properly called the instrument panel since it mounts an array of electrical gauges, switches, and controls, all connected to mazes of wiring, printed circuitry, and air hoses beneath stylishly finished sheets of plastic or metal.

BASIC DESIGN

As shown in **Figure 7–67**, there are some examples of instrument panel designs and layouts. How the "dash" is shaped and laid out affects ease of operation of the truck's systems. If the driver has to constantly grope for the light switch, for example, his/her eyes may be off the road at a dangerous time. Some truck models have a one-piece instrument panel, either flat or curved, and others employ a wraparound type. Each has advantages. The **one-piece design,** where the entire panel parallels the windshield, has a simplicity of appearance that many drivers prefer. It allows use of a bench seat for three-person operations (for example, garbage collection) and does not interfere with movement inside the cab (like from the driver's seat to the sleeper area in highway tractors).

The **wraparound** is actually a two-piece combination of a flat panel directly ahead of the steering wheel and a second panel angled outward and facing the driver that puts right-side switches and controls closer to the driver's hands. The wraparound is especially useful in a high cabover, where the engine cover inhibits interior movement. It is important to remember that the driver should be able to simply glance at the gauges. This requirement is essential in instrument cluster design.

With traditional "analog" gauges (where gauge needles swing through an arc formed by a circle of numerals), field practice has indicated that white numerals on black backgrounds are the easiest to read (**Figure 7–68**). That is why most gauge faces are white-on-black, with orange or red highlights. Because red lighting aids the operator's night vision, many OEMs now illuminate their instruments in red or orange rather than white. Most gauges indicate normal conditions when the indicator needles are somewhere between 10 and 2 o'clock, but this can vary.

Truck manufacturers have, mostly, converted from mechanically and electrically actuated gauges to electronic instrument displays. These are more reliable and easier to repair.

An electronic speedometer can be quickly reset to compensate for changes in tire and wheel size, and of course it will never suffer a broken cable. Quick and easy removal of gauges from the panel is another recently adopted feature made possible by electronic instrument clusters. Many gauges now are of the plug-in variety, and can be popped out of the panel to be fixed on a workbench, or replaced.

Digital instrument panels are now options from many builders (**Figure 7–69**). These designs, using colorful LED, LCD, or vacuum-fluorescent displays offer some operational advantages. The digital dash can present more information in a smaller space. Engine conditions, for example, can be monitored and displayed only when needed, one at a time and in the same location on the panel. Or the driver can push a button to see what is going on.

Most drivers find digital data displays easy to read. Many digital displays now feature both an analog-like climbing scale and a numerical readout.

COMPONENT PLACEMENT

The Maintenance Council (TMC) of the American Trucking Associations, Inc. has a Recommended Practice (RP 401) outlining the placement of gauges, switches, and controls (**Figure 7–70**). Truck OEMs have generally followed this guide since it was devised many years ago. This explains the similarities among the many instrument panels used in heavy trucks today.

Though RP 401 has undergone some updating over the years, the recommended location for key gauges and displays does not change much. Operational gauges, including the speedometer, air pressure, and fuel level, are ahead and slightly to the right of the driver; engine condition gauges, like tachometer, oil pressure, water temperature, and volt or amp meter, are ahead and to the left; switches for lights and windshield wipers are to the immediate right; parking brake valves are further to the right; and controls for differential lock and fifth wheel slider are to the upper right. This basic layout makes sense, since it places the most used items (for example, speedometer and light switches) where they can be easily seen and operated by the driver. Other gauges, switches, and controls are not as critical to safety, RP 401 states, so can be located where the OEM sees fit. RP 401 also suggests that noncritical conditions relating to the engine, like exhaust temperature, go to the left, and those relating to operation, like axle temperature, be placed to the right. Most OEMs observe RP 401.

Standard instrumentation in heavy trucks includes the following:

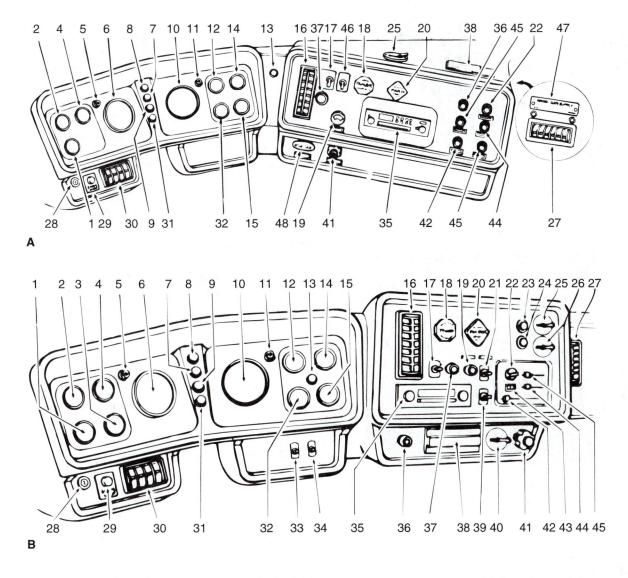

FIGURE 7–67   Two typical dash layouts. (Courtesy of Volvo Trucks North America Inc.)

1 Battery voltmeter gauge	17 Headlight switch	33 Cargo lamp switch
2 Engine oil pressure gauge	18 Trailer air supply valve knob	34 Heated mirror switch
3 Optional component gauge	19 Wiper/washer control knob	35 Radio
4 Engine Coolant temperature gauge	20 System park brake valve knob	36 Cigar lighter
5 Left-hand turn signal indicator	21 Spot light switch	37 Instrument panel rheostat switch
6 Engine tachometer gauge	22 Heater/AC blower motor knob	38 Ash tray
7 High temperature/low water indicator	23 Interdifferential differential indicator	39 Driving lamp switch
8 Headlight hi-beam indicator	24 Interwheel differential indicator	40 Air slide 5th wheel control lever
9 Low air pressure indicator	25 Interdifferential differential control lever	41 Hand throttle control knob
10 Speedometer gauge	26 Interwheel differential control lever	42 Panel, floor, defroster, control lever
11 Right-hand turn signal indicator	27 Heater/Air Conditioner Diffuser	43 Air conditioner ON/OFF switch
12 Fuel gauge	28 Ignition key switch	44 Fresh/recirculating air control lever
13 Preheater indicator	29 Starter pushbutton	45 Heat/cold control lever
14 Dual air pressure gauge	30 Heater/Air conditioner diffuser	46 Marker interrupter switch
15 Optional component gauge	31 Low oil pressure indicator	47 Radio power supply
16 Heater/air conditioner diffuser	32 Optional component gauge	48 Engine stop control knob

Oil Pressure Gauge. This gauge indicates engine oil pressure. The oil pressure typically should be between 30 and 90 psi when the engine is running at rated engine speed at normal operating temperature. A lower pressure is normal at low idle speed. Panels may

also contain an oil temperature gauge. A panel may have a separate readout for outside air temperature.

Coolant Temperature Gauge. This gauge indicates engine coolant temperature. It should nor-

FIGURE 7–68 *Typical analog gauge instrument panel.*

FIGURE 7–69 *Typical digital instrument panel. (Courtesy of Peterbilt Motors Co.)*

1 Engine gauges
2 Safety gauges
3 Light switches
4 Windshield wash and wipe
5 Brake controls
6 Differential lock and fifth wheel slide

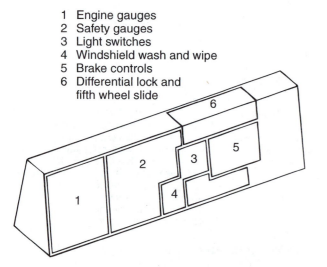

FIGURE 7–70 *The Maintenance Council's Recommended Practice 401 maps out suggested areas for engine and safety gauges, plus locations for many switches and controls. Most truck OEMs follow this layout. (Illustration courtesy of the Maintenance Council of the American Trucking Associations, Inc.)*

mally indicate between 170°–195°F. Higher temperatures may occur under certain conditions. Maximum temperatures range between 200°–220°Fahrenheit with the cooling system pressurized.

Tachometer. The tachometer indicates engine rpm (engine speed). The engine can be operated at high idle for short periods without damage, but should not be allowed to overspeed.

Fuel Level Gauge. This gauge indicates the fuel level from multiple tanks or an overall fuel level reading. Usually electrically operated, it registers only when the key is switched on.

Service Hour Meter. This gauge indicates the total number of clock hours the engine has operated. Some panels have a separate time-of-day clock and an elapsed timer.

Pyrometer. The pyrometer monitors the exhaust gas temperature.

Voltmeter. This gauge measures the battery voltage.

Odometer. An odometer records the distance traveled.

Speedometer. This gauge indicates road speed. In addition, some instrument panels are equipped with an audible speed alarm, cruise control set speed and cruise status.

Air Pressure Gauges. These gauges monitor the air pressure in the primary and secondary air reservoirs. Typically air pressure should not fall below 64 psi.

Ammeter. This gauge indicates the amount of charge or discharge in the battery charging circuit. Normal operation of the indicator should be slightly to the positive side of zero. With the engine running, during normal operation, if the indicator is constantly to the negative side of zero or shows excessive charge, the electrical system should be checked for malfunction.

Fuel Pressure Gauge. This gauge indicates fuel pressure on the pressure side of the transfer pump or the fuel rail pressure. Check fuel pressure data with values specified by the engine OEM.

Diagnostic Data. Diagnostic information about the instrument panel itself or information about an electronic engine is often displayed on the instrument panel. Some electronic panels are bussed to the SAE/ATA data communication network.

It is important for gauges to be in good working order since they monitor the engine and the truck's other components. Changes in gauge readings are indicative of potential gauge or vehicle problems. This also applies to gauge readings that have changed significantly but are still within specifications. The cause of any sudden or significant change in the readings should be determined and corrected.

To make instrument gauge reading easier, some commercial vehicles use a **head-up display (HUD)**. This is a technology (used in some passenger vehicles) that superimposes data on the driver's normal field of vision. The operator can view the information, which appears to "float" just above the hood at a range near the front of a conventional tractor or truck. This allows the driver to monitor conditions such as limited road speed without interrupting his normal view of traffic. The HUD offers two major advantages over conventional displays: first, it allows the driver to keep his head up and view the road while reading the HUD information. Second, the image appears to be floating at a distance equivalent to the front of a conventional vehicle. This means the driver will not have to refocus in order to read the HUD, as he would with a typical instrument panel display. These advantages provide for safer, more comfortable driving.

Voltage Limiter

Some instrument panel gauges require protection against heavy voltage fluctuations that could damage the gauges or cause them to give incorrect readings. A **voltage limiter** provides this protection by limiting voltage to the gauges to a preset value. The limiter contains a heating coil, a bimetal arm, and a set of contacts. When the ignition is in the on or accessory position, the heating coil heats the bimetal arm, causing it to bend and open the contacts. This action cuts the voltage from both the heating coil and the circuit. When the arm cools down to the point where the contacts close, the cycle repeats itself. Rapid opening and closing of the contacts produces a pulsating voltage at the output terminal that averages 5 volts. Voltage limiting may also be performed electronically.

AIR AND OIL PRESSURE GAUGE FAILURE

If pressure gauges do not show a reading, or the reading is suspected of being in error, a cluster gauge reading and master shop gauge reading must be taken by placing both a Y or T connector and shop gauge in the gauge line.

If air pressure is evident at either of the two air lines, check the air filter inside the air pressure gauge to make sure it is not plugged. If the filter is not plugged, and the cluster air gauge pressure reading is outside a ±5 psi of a master gauge reading, replace the cluster air gauge. If the cluster oil gauge pressure reading is outside a ±4 psi of a shop gauge reading, replace the cluster oil gauge.

ELECTRICAL TEMPERATURE AND FUEL DISPLAY FAILURE

If any of the displays on the electrical temperature gauges or the fuel gauge are indicating a problem, the vehicle wiring that connects the gauge in the circuit must be checked for sound connections. If the wiring and connector are okay, a thermistor-type temperature sensor's electrical resistance must be checked with a quality ohmmeter at the cluster harness connector. Always check for engine ECM fault codes before testing circuit components.

Thermistor Resistance Check

To check thermistor-type temperature sensors, the approximate temperature of the component being checked must be known to determine the amount of resistance that will be read at the cluster harness connector. Consult a chart such as the one shown in **Table 7–8**.

TABLE 7–8: THERMISTOR RESISTANCE CHECK	
Component Temperature (°F)	Thermistor Resistance (Ohms)
32	8,335
50	4,945
68	3,010
86	1,871
104	1,185
140	505
158	341
194	169.8
212	125.6
230	95.2
248	73.5
266	57.5

Restricted Needle Movement

If the sensor electrical resistance checks okay, the gauge should be checked for restricted pointer movement. Very gently move the pointer between the minimum and maximum dial positions. If the pointer moves freely, the gauge must be checked for an open or shorted winding.

Open or Shorted Winding Check

Remove the gauge and check for an open or shorted winding with an ohmmeter. All resistance values below 20 ohms should be considered a short. All resistance values above 1,000 ohms should be considered an open. All readings between 20 and 1,000 ohms should be considered within an acceptable range. If the resistance reading is not within the acceptable range, replace the gauge. This test cannot be performed accurately on a voltmeter because there is a semiconductor inside the gauge that will make resistance readings inaccurate.

CAUTION: Static electricity can cause permanent damage to the cluster. Before working on the cluster, be sure to remove all static electricity from your body by touching metal that is grounded. Do not wear clothing that causes static buildup (like nylon). Do not touch pin connectors during removal and installation of gauges. Work on the cluster in a clean environment to avoid dust.

Internal Connection Check

If the gauges' winding resistance is within an acceptable range, check the internal connections between the cluster's harness pins and the gauge pins for a good connection. Refer to the service manual for the proper gauge pins to use for this test procedure. Measure the resistance on the circuits that connect the cluster's harness ground and sensor line to the gauge pin ground and sensor line for the gauge in question. They should read under 1 ohm.

Voltage Check

Check the voltage with a voltmeter across the power pin and ground pin in the cluster for the gauge being checked. Accessory voltage is applied and all other gauges should be installed in the cluster. The voltage reading should be just below accessory voltage. If either the voltage measurement or the two electrical resistances are incorrect, replace the cluster with a new cluster without gauges.

The cluster's voltmeter should have about the same reading as the voltmeter used to make the gauge voltage check. If the cluster's voltmeter has a different reading, replace the cluster voltmeter.

TABLE 7–9: FUEL LEVEL OUTPUTS	
Tank Fuel Levels	Sensor Output (Ohms)
Full	88 (• 3)
3/4	66 (• 3)
1/2	44 (• 2)
1/4	22 (• 2)
Empty	1 (• 1)

Fuel Gauge

If the electric fuel gauge shows a reading and it is suspected of being in error, the fuel level sensor must be checked. The fuel level sensor used is a float/resistive potentiometer type. **Table 7–9** shows typical output characteristics at different fuel levels.

Shop Talk

Before checking the fuel gauge, be sure the cab interior is warmed up (during cold weather), and that the vehicle has been sitting still long enough to allow the fuel to settle.

When the sensor is shorted to ground, the gauge should read empty. Locate the fuel sensor gauge input in the cluster harness. If the resistance reading for the fuel tank level matches the **Table 7–9** or OEM spec and the gauge reading does not match, replace the fuel gauge.

VOLTMETER AND PYROMETER DISPLAY FAILURE

If either the voltmeter or pyrometer displays are indicating a problem, the wiring that affects that function must be checked for clean, tight, unbroken, and non-shorted connections. If the wiring is okay, measure the appropriate voltage on the vehicle's harness connector that mates to the cluster for the gauge in question. The vehicle's power must be on during this test. The voltmeter reads across accessory voltage and ground. Replace the voltmeter if it does not read near accessory voltage.

The pyrometer is supplied with two driving voltages. The first is the pyrometer's power and this is regulated at a specified value. The second is the voltage that varies according to the exhaust gas temperature. Its voltage range varies but is generally less than 3V. All voltages are measured with respect to the electronic ground pin.

TABLE 7–10: VOLTAGE LEVELS/TEMPERATURES	
Temperature (°F)	Voltage (DC)
600	0.965
900	1.590
1200	2.260
1500	2.900

Pyrometer Sensor Voltage

Table 7–10 shows typical voltage levels at different temperatures. If the voltages are within limits, the pyrometer should be checked for a sticky pointer. If the pointer sticks, replace the gauge. If the pointer moves freely, the meter's internal voltage must be checked.

Internal Voltage Check

To check the pyrometer's internal voltage, turn off the vehicle's electrical power. Remove the pyrometer and install the cluster without bezel and the meter. With the electrical power on, measure the voltage across the meter's pins. If the voltages are not present, replace the cluster with a gaugeless cluster. If the voltages are correct, turn accessory power off and reinstall the gauge. If the meter does not show a reading when accessory power is turned on, replace the pyrometer.

SPEEDOMETER/ODOMETER AND TACHOMETER/HOURMETER DISPLAY FAILURE

If the speedometer, odometer, tachometer, or hourmeter displays are indicating a problem, the vehicle's wiring must be checked for clean, tight, unbroken, and nonshorted connections. Also check engine ECM faults.

Make sure the accessory voltage to the cluster is correct as specified.

If all four meters are not working and the voltmeter and fuel gauge are working, replace the cluster with a new gaugeless cluster. If this is not the case, check for proper dip switch settings inside the cluster. The speedometer and odometer use the road speed sensor; the tachometer and hourmeter use the engine speed sensor. If either pair of meters does not operate, but the other pair does, the sensor that operates the failed pair must be checked. Both the correct signal amplitude and frequency must be as specified at the harness pins of the vehicle connector.

An hourmeter may not operate if the cluster does not receive the correct tachometer signal that tells it

the engine is running. An odometer may not operate without the correct speedometer signal.

If the appropriate sensor input into the cluster checks okay and both of the paired functions still do not work, the cluster must be replaced with a new cluster without gauges.

Odometer and Hourmeter Failure

If the vehicle speed sensor checks okay and only the odometer or hourmeter does not operate (speedometer and tachometer do operate), check the two electrical connections between the counter mechanism and the printed circuit board.

If the connections are not tight, remove the counter in question and gently pinch them together. Reinstall the counter mechanism and check for correct operation.

If the connections are tight, the counter must be removed and checked electrically. The counter can be checked by quickly tapping its two electrical leads to vehicle battery voltage. If the counter does not ratchet properly, replace it. If the counter mechanism does ratchet properly, replace the cluster with a new cluster without gauges.

Mechanical Engine Coolant Temperature Display Failure

If a mechanical water temperature gauge indicates a reading, but the reading is suspected to be in error, a test to check the gauge's accuracy must be completed.

Put the sensing bulb in boiling water and check the gauge reading against a master temperature gauge inserted in the water. If the cluster gauge reading and the master gauge readings differ by more than 5°F, the cluster gauge must be replaced. No cluster voltage is required for this test.

Other Instrumentation

The trucking industry is using new technology to eliminate blind spots and poor rear vision. Current additions to side-view mirrors are the following blind spot safety features:

Closed-Circuit Television (CCTV). CCTV operates on the same principle as security cameras in buildings. A camera is generally mounted on the rear of the vehicle and sends pictures along a cable to an in-cab monitor. Early systems had high maintenance costs, but current charge-coupled-device (CCD) systems are durable. Well-accepted as a backup device among garbage haulers and bus companies, they have little or no application as tractor/trailer blind spot viewing devices.

Fiber Optics. Fiber optics function like a flexible periscope. Thousands of optic fibers are arranged in a specific order along a tube and transmit a picture to an in-cab viewing screen. No energy source is required and it is virtually maintenance-free. Although still in prototype stage, it has potential for tractor/trailer blind side viewing.

Infrared. Infrared systems use rays of infrared light that are reflected off surrounding objects. Signals are sent to a microprocessor, which calculates distance and relays information to an in-cab monitor. Information comes in the form of an audible alarm and/or visual warning light. Excellent potential for blind side detection at highway speeds but few systems are currently commercially available.

Ultrasonic. Ultrasonic systems use sound waves much like a bat would to "see" objects. A transmitter bounces high-frequency waves off objects to a receiver. A microprocessor constantly calculates the distance between the vehicle and the object and a digital readout or LEDs tell the driver how close he/she is. Systems often have audible alarms that activate at preset distances. Some also automatically engage the vehicle's brakes at a preset distance when reversing. This method has been relatively successful in backup warning devices among garbage haulers but has limited application in tractor/trailer blind spot detection. This method is susceptible to wind at high speeds.

Vehicle On-Board Radar (VORAD). VORAD is a driver alert system. Can be used for blind side proximity of closing velocity alerts to the driver.

7.9 WINDSHIELD WIPERS

Windshield wipers may be electric or air operated. One motor may be used to drive both wipers, or in a COE each wiper may be operated separately by its own motor.

The components of a typical electric windshield wiper system are shown in **Figure 7–71**. The two-speed electric motor is activated by turning a control knob on the instrument panel to the right to the first detent for low speed and to the second detent for high speed. Sometimes an intermittent control is available to operate the wipers in the delay mode. A knob on the intermittent control box adjusts the seconds per stroke. Once the wiper control knob is turned to low or high speed, the intermittent control is no longer functioning. When air brake equipment is

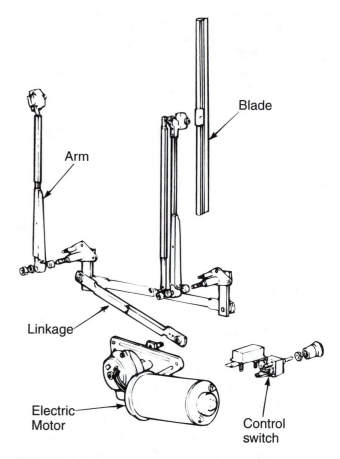

FIGURE 7–71 *Typical electric windshield wiper components. (Courtesy of International Truck and Freight Corp.)*

specified for trucks, an air-operated windshield wiper is available optionally and tends to be used in preference to electric wipers.

7.10 ALARM AND SHUTDOWN SYSTEMS

Alarm and shutdown systems can monitor parameters like coolant temperature and engine-mounted control devices, and can warn the driver of out-of-limit conditions such as low oil pressure. They can shut down an engine to protect it from possible damage. How much protection these systems offer depends on their cost and design. These systems are used mostly on engines that are not managed by ECMs.

A basic warning device can consist of a probe inserted into the engine radiator to monitor coolant level. When the coolant level falls below the probe sensor in the radiator, an electrical signal is sent to a light mounted on the cab dash. This probe can be connected to a second device that shuts off engine fuel after a preset time delay. Probes or sensors may also monitor pressure, temperature, and any on/off switch that can be connected to a computer. Sensors can measure the following:

- Air pressure in brake systems
- Hydraulic fluid level in fluid brake systems
- rpm
- Lubricant level in the transmission or rear
- Transmission or rear temperature
- Coolant level in the radiator
- Coolant temperature in the radiator
- Coolant level or flow in the engine
- Turbo temperature
- Oil temperature
- Oil level in the engine

Depending on the type and programming of the alarm/shutdown system, the out-of-normal range signal from a sensor can result in either a driver alert or an engine shutdown.

7.11 RAPID CHECKING OF A TRUCK ELECTRICAL CIRCUIT

The following sequence of tests is designed to produce a quick assessment of the truck electrical circuit. This type of sequential check is often performed as part of routine preventive maintenance.

GENERAL VOLTMETER TEST

These tests are used to determine the general condition of a vehicle electrical system. The idea is to produce a report card on the battery, the cranking circuit, and the charging circuit. In fact, the test is so fast and easy to perform, it should become part of routine service procedure.

Battery Voltage

Battery voltage can only be accurately measured when there is no surface charge: This may be removed by applying a load on the battery for a one-minute period. (Turning on the headlamps will usually suffice.) Now electrically isolate each battery by disconnecting the terminals. If the voltmeter is not auto-ranging, set the scale to read up to 18V and connect the positive lead to each positive battery post and the negative lead to each negative battery post. The

voltmeter readings produced at 70°F may be interpreted as follows:

12.6V or higher	100 percent charged
12.4V	75 percent charged
12.2V	50 percent charged
12.0V	25 percent charged
11.9V or lower	fully discharged

A fully charged lead-acid battery should produce 2.1 volts per cell: 6 cells in series will produce a reading of 12.6V.

Cranking Voltage

Ensure the engine is not capable of starting by no-fueling the engine. Connect the voltmeter leads across the cranking motor terminals. Crank the engine for 15 continuous seconds. The voltmeter reading should read above 9.6V for the full 15 seconds. If the reading falls below 9.6V, a problem with one of the following is indicated:

- Defective or corroded battery cables or terminals
- Defective or discharged batteries
- Defective cranking motor, solenoid, or relay

Charging Voltage

Start the engine and run at 75 percent of rated speed with no load. Now turn on all the electrical accessories on the vehicle. Use the voltmeter to test battery voltage. It should read between 13.5V and 15.0V. If the reading is lower than 13.5V, one of the following possible problems is indicated:

- Loose alternator drive belt(s)
- Defective or corroded electrical connections
- Defective voltage regulator
- Defective alternator

If the reading exceeds 15.0V, the problem is likely

- Defective voltage regulator
- Poor electrical connections

Voltage Drop Test

Voltage drop testing is a means of testing the condition of wires, cables, connections, and loads in the electrical circuit. It is a much better method of testing electrical circuit resistances because the testing is performed in a fully energized electrical circuit as opposed to checks with an ohmmeter, which must be performed on inactive circuits. The ohmmeter uses very low current flow and voltage and therefore does

PHOTO SEQUENCE 4

PERFORMING A RAPID ASSESSMENT OF A TRUCK ELECTRICAL CIRCUIT

The following procedure is designed to produce a quick report card on a truck electrical system and is performed as a PM procedure, rather than as a troubleshooting procedure.

PS–1 Check the battery voltage. Turn the headlights on for a minute to remove any surface charge. Now isolate each battery by disconnecting it from the chassis electrical circuit (remove the terminals). Use a voltmeter and select V-DC. If it is not auto-ranging, set it in the 18V scale. Connect the positive lead to the positive battery post and the negative lead to the negative battery post.

PS–2 Record the reading for each battery tested. Make sure that the voltage values meet the specifications after the temperature adjustment. Any battery out of specification requires further testing. Reconnect the batteries.

PS–3 Check cranking voltage. Ensure that the engine will not start during this test. Set the DMM to V-DC and connect the test leads across the cranking motor terminals observing the correct polarity. Now crank the engine for 15 continuous seconds. The reading should be above 9.6V for the test period. The DMM should then read at least 10.5V.

PS–4 Charging voltage test. Start the engine and run at 75 percent of its rated speed (1500 rpm if rated speed is 2000 rpm). Turn on all the electrical accessories and use a DMM on the V-DC scale to check the battery voltage. This should read between 13.5V and 15V. Note the DMM reading.

PS–5 If the voltage reading is lower than 13.5V, a problem with the alternator, alternator drive mechanism, voltage regulator, or electrical connections is indicated.

PS–6 If the voltage is higher than 15V, a problem with the voltage regulator or the electrical connections is indicated.

PHOTO SEQUENCE 4 (Continued)

PERFORMING A RAPID ASSESSMENT OF A TRUCK ELECTRICAL CIRCUIT

PS–7 If problems with the wiring or connections are indicated, they can be located by performing a voltage drop test. To perform a voltage drop test, set the DMM to V-DC on a scale so that readings of less than 1V can be read. To perform a voltage drop test on a starter cable, crank the engine with the positive test lead placed at the end of the cable closest to the positive terminal of the battery and the negative test lead placed at the other end. The DMM should read 0.1V. Voltage drop testing can also help identify defective terminals. Just remember that the circuit must be energized and the positive test lead placed at the more positive location in the circuit.

not test a circuit under active circuit conditions. In any electrical circuit, the cables should be of a gauge and length that represents minimal circuit resistance. A voltage drop test is a quick and easy way to determine the resistance of electrical wires and cables: Cables and connections in good condition should test at zero or very low voltage drop values. When the voltage drop value exceeds 0.2V across a cable, the cable is indicating high resistance. Electrical connections should not produce voltage drops greater than 0.1V and, perhaps most important, total voltage drop of all the cables and terminals in a cranking circuit should never exceed 0.5V.

Performing a Voltage Drop Test

The best instrument for performing a voltage drop test is a digital multimeter (DMM). If the DMM is not auto-ranging, select a low scale such as 0V to 3V for the test. The most common circuit tested for voltage drops is the cranking circuit. For instance, if an engine cranks slowly, two possible causes are the circuit cables and terminals: In fact, many cranking motors and batteries are needlessly changed every day in the trucking industry due to defective cables and terminals. Voltage drop tests are a quick method of verifying the integrity of any connections in the cranking circuit.

1. Set the DMM on the V-DC scale so readings of 1V or less can be easily read. Ensure that the DMM test leads and probes are inserted into the correct sockets. If they are color-code polarized, ensure that the red lead is identified as the positive and the black lead is identified as the negative.
2. The positive probe should always be connected to the most positive section of the circuit being tested: This is simply the section closest to the positive post of the battery. The black probe should always be connected to the most negative section of the circuit: This is always that closest to the negative post of the battery.
3. No-fuel the engine and crank for up to 30 continuous seconds. Test each terminal and wire sequentially through the circuit from the positive battery post, through the positive cables and connections, from the cranking motor to chassis ground, and from chassis ground back to the negative post of the battery.
4. While performing the voltage drop tests, using a remote starter switch makes the job easier. Also, crank for 30 seconds maximum, ensuring at least a 2 minute interval between each cranking cycle. Remember, a voltage drop value exceeding 0.2V through any individual terminal or cable indicates a possible problem. A voltage drop exceeding 0.5V in the cranking circuit is out of specification and requires repair.

Visual Inspection

Never overlook the importance of simply observing the condition of cranking circuit components. When voltage drop tests are performed on a cranking circuit, observe the condition of cables and terminals, taking note of corrosion, insulation damage, and signs of overheating. After the voltage drop test has been performed, touch sections of the cranking circuit to check for heat: High resistances in electrical circuits produce heat, which is easily detected. Any terminal that overheats should either be cleaned or replaced. A cable that overheats should always be replaced.

Testing a Solenoid or Relay

1. Initially set the voltmeter range to a value that exceeds the battery voltage value: 12V for most truck electrical systems and 24V on 12/24 systems and some buses.
2. Connect the voltmeter probes in sequence through the circuit so the positive probe is in the most positive location and the negative at the most negative. See **Figure 7–37** for the required test sequence.
3. While the engine is being cranked, switch the voltmeter to a low scale in order to read the voltage drop value.
4. A voltage drop that exceeds 0.2V through a relay or solenoid indicates that the component is defective and must be replaced.

BATTERY DRAIN TEST

The objective is to determine if a component or circuit in the vehicle is causing a drain on the battery when the ignition circuit is off. It is good practice to perform this test each time a battery or set of batteries is replaced.

1. Ensure that the ignition circuit is open, all accessories are switched off, and cab door and courtesy lights are off.
2. Disconnect the main negative terminal from the battery pack. Connect at test light (this is one of the few legitimate uses left for the test light) in series with the negative cable terminal and the negative battery terminal: The test light probe should contact the battery negative terminal and the alligator clamp should be attached to the cable terminal. The test light should not illuminate.
3. Test number 2 may be performed with an ohmmeter, but it is not quite as easy. Ensure that the positive battery cable is removed from the battery pack. Place one DMM test probe on the now disconnected positive battery terminal and the other on a good chassis ground location. A resistance of 100 ohms or less will result in draining the battery. To calculate the extent of current drain, use Ohm's law. Assuming a 12V chassis electrical system:

$I = E/R$
A = 12V divided by 100 ohms
A = 0.12 amperes or 120 milliamps

Shop Talk

When performing this test, the lower the resistance reading, the greater the amperage drain on the battery.

Note: When performing battery drain tests, it should be remembered that a number of electronic components draw some battery power when the ignition circuit is open, including some system ECMs, electronically tuned radios, and digital clocks. Also, diode leakage at the alternator rectifier can also account for battery drain-off.

SOURCING A BATTERY DRAIN

First, obtain a vehicle wiring schematic and become familiar with the circuits. Check for the presence of any circuits controlled by mercury switches (these can remain closed) and any electronic components that require a closed circuit to function. To identify the circuit responsible for the current draw, disconnect the circuit breakers (or fuses) from each subcircuit until the test light no longer illuminates. If, after performing this test, the subcircuit has not been identified, the drain is probably in the alternator or cranking motor circuits.

First, isolate the alternator. Electrically disconnect the alternator and observe the test light. If it extinguishes, the problem is in the alternator rectifier: Remove the alternator and recondition. If the test light remains lit, reconnect the alternator. Next, remove the positive terminal connection at the starter solenoid: If the test light extinguishes, the cranking motor/solenoid is defective and must be reconditioned.

SHORT CIRCUITS

A vehicle electrical short circuit usually results in tripping the circuit breaker or blowing the fuse protecting the subcircuit. A cycling circuit breaker will break the circuit at the point current flow through the circuit exceeds the amperage rating: When it cools, it will close the circuit. This will recur until the cause of the problem is located. A noncycling breaker will not automatically reset: The circuit must be opened before the breaker can close the circuit again. A fuse, when subjected to current overload, will blow. When a fuse blows immediately following replacement, the cause is often a short circuit.

The best method of locating a short circuit is to use a DMM in ohmmeter mode. Once again, locate the vehicle wiring schematic and use it as a guide through each subcircuit. Disconnect the positive wire to the breaker power strip. Connect one of the ohmmeter test probes to a good ground and the other to the circuit side of the circuit breaker. If the circuit is shorted to ground, the ohmmeter will read zero or close to zero (indicating low resistance). Next, using the vehicle wiring schematic as a guide, follow through the circuit, disconnecting one component at a time while observing the ohmmeter. If a component is unplugged and the resistance reading jumps to a high or infinity value, it is the likely source of the short.

FEEDBACK PROBLEMS

When current supplied to an electrical component lacks an adequate path to ground, it can seek an alternative route to complete the circuit. The dual filament bulbs used in truck stop/turn/taillight circuits are a common cause of feedback. These dual filament bulbs share a common ground in the socket. When the all of current fed to the bright filament (brake/turn light) exceeds the ground's ability to handle it (due to high resistance), some of the current backfeeds through the other filament (taillight). This can energize both the taillight and the clearance light circuit, resulting in all clearance lights flashing when a turn signal is actuated or clearance lights illuminated at braking. When troubleshooting this condition, remember it is caused by a high resistance ground connection. Restore the ground integrity of the component and the condition will be repaired.

SUMMARY

- High-powered components of the electrical system—the battery pack, alternator, starter (cranking) motor, and associated wiring—are normally designed as an integrated system that varies according to the vehicle's application.
- There are four types of lead-acid batteries used in heavy-duty trucks: conventional, low-maintenance, maintenance-free, and sealed maintenance-free.
- Measuring the specific gravity of the electrolyte is required to determine the sulfuric acid content of the electrolyte. The state-of-charge of a battery can be determined by knowing the specific gravity of the electrolyte.
- A fully charged battery has all of the sulfate in the electrolyte.
- As the battery discharges, some of the sulfate combines with the lead in the plates to form lead sulfate. The acid becomes more diluted and its specific gravity drops as water replaces some of the sulfuric acid.
- Temperature affects the output capacity of a battery with respect to cranking an engine. The cranking power of the battery is reduced as battery temperature is lowered.

- Disconnect the battery's ground cable when working on the electrical system or engine. Never "short" across cable connections or battery terminals.
- Disconnect the battery ground cable before fast-charging the batteries in a vehicle.
- When removing a battery from a vehicle, always disconnect the battery ground cable first. Never reverse the polarity of the battery connections.
- Never attempt to use a fast charger as a boost to start the engine.
- The battery should be inspected at the time of chassis lubrication or other periodic services.
- In cold weather it is important to keep the battery warm. There are various types of battery heaters available.
- There are two methods of recharging a battery: the slow charge method and the fast charge method. Either method can be used to recharge most batteries. However, some batteries must be charged slowly.
- Batteries should be stored indoors because this will reduce the possibility of cracks and leaks due to extreme temperature variations.
- A truck's charging system consists of the batteries, alternator, voltage regulator, associated wiring, and the electrical loads of the truck.
- The purpose of the charging system is to recharge the battery whenever necessary and to provide the current required to power the electrical components of the truck.
- A malfunction in the charging system results in either an overcharged battery or an undercharged battery.
- The vehicle's starting system is designed to turn or "crank" the engine over until it can operate under its own power.
- The starter circuit carries the high current flow within the system and supplies power for the actual engine cranking.
- The starter motor converts the electrical energy from the battery into mechanical energy for cranking the engine.
- The lighting systems of the heavy-duty truck can be divided into two categories: exterior and interior. Each system's wiring proceeds from the power source to a circuit breaker and then to a switch and through a connector to the lamp assemblies.
- Most headlight systems consist of either two or four sealed-beam tungsten headlight bulbs.
- The wiring junction block for the trailer is usually is located inside the cab, directly behind the driver's seat.
- Diagnostic information about the instrument panel itself or information about an electronic

engine is often displayed on the instrument panel. Some electronic panels can be connected to the SAE/ATA data communication network.
- Windshield wipers may be electric or air operated. One motor may be used to drive both wipers, or in a COE each wiper may be operated separately by its own motor.

REVIEW QUESTIONS

1. Which of the following are functions of the batteries of a truck?
   a. Produce voltage and a source of current for starting.
   b. Act as a voltage stabilizer for the entire electrical system of the vehicle.
   c. Provide current whenever the vehicle's electrical demands exceed the output of the charging system.
   d. all of the above

2. When performing a load test on a battery with 400 CCAs, Technician A sets the test load at 400 amps. Technician B removes the surface charge by applying a 300-amp load for 5 seconds. Who is correct?
   a. Technician A
   b. Technician B
   c. both Technician A and B
   d. neither A nor B

3. What is the state of charge of a battery that has a specific gravity of 1.2 at 80°F?
   a. completely discharged
   b. about 1/2 charged
   c. about 3/4 charged
   d. fully charged

4. What is the first step when removing an old battery from a vehicle?
   a. disconnect the ground cable
   b. remove the battery holddown straps and cover
   c. inspect and clean the area
   d. remove the heat shield

5. Battery acid has just been spilled on a person's clothing. Technician A says to wash the acid off with baking soda and water. Technician B says to wash the acid off with vinegar. Who is correct?
   a. Technician A
   b. Technician B
   c. both A and B
   d. neither A nor B

6. What part of the alternator produces the rotating magnetic field?
   a. stator
   b. rotor
   c. brushes
   d. poles

7. A typical rotor contains _____
   a. 14 north poles and 14 south poles
   b. 14 north poles or 14 south poles
   c. 7 north poles and 7 south poles
   d. a pair of poles, one north and one south

8. Alternating current is rectified to DC by the _____
   a. stator
   b. rotor
   c. diodes
   d. regulator

9. What is used to control the amount of current produced by the alternator?
   a. voltage regulator
   b. storage battery
   c. resistor
   d. two-pole switch

10. What tool is used to measure the current output of the alternator?
    a. ammeter
    b. ohmmeter
    c. diode tester
    d. voltmeter

11. Batteries, battery cables, a relay, solenoid, and the starter motor are found in what part of a truck's starting system?
    a. control circuit
    b. cranking circuit
    c. charging circuit
    d. regulator circuit

12. The control circuit of the truck's starting system is used to _____
    a. open and close the system's starter circuit
    b. regulate the amount of current passing through the alternator
    c. prevent voltage "spikes" from occurring
    d. prevent overcharging of the system's batteries

13. What component prevents the starter motor from spinning at speeds that could damage the motor?
    a. stator
    b. rotor
    c. override clutch
    d. flywheel

14. If a cranking current test indicates the starter circuit is drawing low amounts of current, which of the following is the most likely cause of the problem?
    a. short in starter motor
    b. excessive resistance in the circuit
    c. mechanical resistance due to binding engine or starter system parts
    d. none of the above

15. Which of the following can be used to drive a gauge?
    a. sensor
    b. mechanical connection
    c. microcomputer
    d. all of the above

16. In a heavy-duty truck, a pyrometer is used to monitor _____
    a. road speed
    b. exhaust gas temperature
    c. fuel pressure
    d. the battery charging circuit

17. To avoid the build-up of static electricity that can damage the instrument panel gauges _____
    a. touch grounded metal before beginning work
    b. work in a clean, dust-free area
    c. avoid wearing nylon and other materials that cause static build-up
    d. all of the above

18. A thermistor operates by changing _____
    a. resistance according to its temperature
    b. temperature according to resistance in its wiring
    c. resistance according to changes in diaphragm pressure
    d. none of the above

19. Which of the following correctly describes a typical truck engine starter motor?
    a. AC motor
    b. permanent magnet motor
    c. series motor
    d. output torque is greatest at high rpm

20. Which of the following opens at the moment a starter motor starts to rotate?
    a. starter relay
    b. hold-in winding
    c. starter solenoid
    d. pull-in winding

8

Diagnosis and Repair of Electronic Circuits

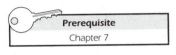
Prerequisite
Chapter 7

Objectives

After studying this chapter, you should be able to

- Explain what is meant by sequential electronic troubleshooting.
- Perform tests on some key electronic components including diodes and transistors.
- Define the acronym EST.
- Identify some types of EST in current use.
- Identify the levels of access and programming capabilities of each EST.
- Explain why electronic damage may be caused by electrostatic discharge and by using inappropriate circuit analysis tools.
- Describe the type of data that can be accessed by each EST.
- Identify what type of data may be read using the on-board flash codes.
- Perform some basic electrical circuit diagnosis using a DMM.
- Identify the function codes on a typical DMM.
- Test some common input circuit components such as thermistors and potentiometers.
- Test semiconductor components such as diodes and transistors.
- Describe the full range of uses of a ProLink 9000 reader/programmer.
- Connect a ProLink to vehicle ECM via the SAE/ATA data connector and scroll through the display windows.
- Update a ProLink software cartridge by replacing the PROM chip(s) or data cards.
- Define the objectives of snapshot test.
- Outline the procedure required to use a PC and OEM software to read, diagnose, and reprogram vehicle electronic systems.
- Understand the importance of precisely completing each step when performing sequential troubleshooting testing of electronic circuits.
- Interpret the SAE J 1587/1939 codes for MIDs, PIDs, SIDs, and FMIs using SAE interpretation charts.
- Repair the sealed electrical connectors used in most electronic wiring harnesses.

Key Terms

active codes
ATA data connectors
baud rate
blink codes
breakout box
breakout T

butt splice
continuity
current transformers
digital multimeter (DMM)
Deutsch connector
electronic service tools (ESTs)
flash codes
failure mode indicators (FMIs)

ground strap
Hall effect probe
handshake
historic codes
inactive codes
locking tang
meter resolution
Metri-Pack connectors
message identifier (MID)
multiprotocol cartridge (MPC)
open circuit
parity
personal computer (PC)

parameter identifiers (PIDs)
ProLink 9000
root mean square (rms)
reader/programmer
SAE J 1587
SAE J 1708
SAE J 1939
subsystem identifiers (SIDs)
snapshot test
stop bits
tang release tool
thermistors
three-way splice

The objective of this chapter is to provide the truck technician with a general introduction to some of the repair techniques required of truck electronic management systems other than the engine and transmission. The subject of electronic engine management systems is handled in some detail in a companion to this textbook, and electronically managed transmissions and brake systems are dealt with in later chapters in this book. Any system that is managed by a computer will operate on the principles outlined in Chapter 6. Because an engine is complex in terms of the subsystems that enable it to operate, a complex electronic management package is required to run it. By comparison, electric pulse wipers are also electronically managed but the circuit required to manage them is simple. The only input device is a multi-position switch. The output that results from its processing cycle depends on the selected switch position. The electronics required to manage chassis subsystems such as the transmission and ABS (anti-lock braking system), although more complex than electric pulse wipers, tend to be more simple than engine management electronics. These systems are not as comprehensively monitored, and the range of outputs is limited.

The first requirement of working on electronically managed chassis systems is to understand the fundamentals of the operating principles as outlined in Chapters 5 and 6. The next is to acquire the elementary skills required to operate the **electronic service tools (ESTs)** to perform circuit analysis. Since the trucking industry has entered the electronic age, the quality of OEM technical service manuals has necessarily greatly improved. However, the technician is required to accurately interpret the text in these service publications. Years ago, a diesel engine technician experienced in one OEM's product could probably successfully repair another OEM's equipment with little reference to service literature. This is simply not possible with today's technology. The technology is not necessarily complex, but each OEM is distinct. Each OEM publishes sequential troubleshooting procedures that must be accurately adhered to. Failure to exactly comply with each step or to scramble the troubleshooting sequence can make the outcome of the procedure meaningless.

The good news is that the manufacturers of trucks in North America have cooperated to a much larger extent than automobile manufacturers. Since 1995, all highway electronic systems have been manufactured in compliance with three SAE standards known as J 1584, J 1708 and J 1939. The first two SAE recommended practices, SAE J 1584 and J 1708, governed the electronic hardware compatibility and software protocols ("language" rules and regulations). The latest SAE J 1939 governs both electronic hardware and software protocols, the matter formerly covered by SAE J 1584 and J 1708. Complying with the hardware standards enables multiple on-board electronic systems to "share" hardware such as throttle position and vehicle speed sensors, avoiding unnecessary duplication of components. Compliance with software protocols means that multiple on-board electronic systems can "speak" to each other, allowing them to be connected to one another to optimize vehicle performance. It also means that one manufacturer's software and EST can at least read another's using the SAE MIDs, SIDs, PIDs, and FMIs. These terms will be fully explained later in this chapter. When two or more electronic systems are connected electronically to each other, it is known as a *handshake* connection. The practice of handshaking multiple electronic circuits in a chassis system is known as multiplexing.

Troubleshooting electronic circuits is not only a lot easier than troubleshooting hydromechanical circuits, it produces faster, more accurate results. The guess work that was always required with more complex hydromechanical systems has been almost eliminated. This is not to say that troubleshooting electronically managed circuits will not produce the odd problem that the sequential troubleshooting steps have not addressed. Every technician working on electronic circuits will have a nightmare tale about an intermittent failure that produced no codes or tattletales and defied every attempt in the book to solve it, but such instances are rare.

Two sets of skills are required to troubleshoot electronic circuits. The first is a knowledge of how to operate the ESTs recommended by the OEM. At minimum, a **digital multimeter (DMM)** and a microprocessor-based EST such as a ProLink or a **personal computer (PC)** with the OEM software is required. The next skill is simply the ability to read and accurately interpret wiring schematics and the OEM service literature. Armed with these skills, technicians are often able to repair a circuit malfunction without having truly understood the cause of the problem. Good technicians, however, should make it their business to fully understand the reasons for a circuit malfunction. This will enable the technician to build a set of diagnostic skills that, after some years, will be known as *experience.*

Most OEMs offer training courses in the diagnosis and repair of their electronic systems. This specialized training becomes more essential as the complexity of the system increases. The technician who is fully familiar with one of the truck ABS systems could probably diagnose and repair other truck ABS systems given the appropriate EST and troubleshooting literature. This would not be smart when working on electronically managed engines and transmissions, when OEM specialized training is usually essential to avoid system damage and costly down-time. Electronically managed vehicle systems are not more complicated, but it is important to avoid guesswork and trial and error approaches to troubleshooting.

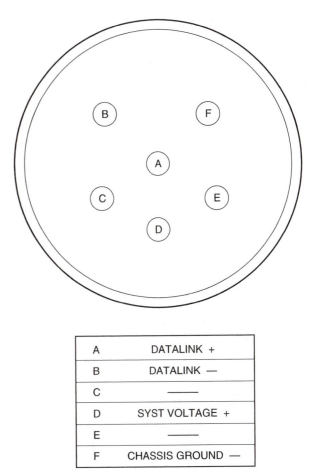

A	DATALINK +
B	DATALINK —
C	————
D	SYST VOLTAGE +
E	————
F	CHASSIS GROUND —

FIGURE 8–1 ATA 6-pin data connector.

8.1 TYPES OF EST

The term electronic service tool (EST) is generally used in the trucking industry to cover a range of electronic service instruments from on-board diagnostic/malfunction lights to sophisticated computer-based communications equipment. The use of generic ESTs and procedures will be reviewed in this section. Proprietary ESTs are designed to work with an OEM's specific electronics and will not be discussed in any great detail in this text because they are system-specific. All the electronic system OEMs provide courses that address their technology. The use of each proprietary EST is covered in these courses.

ESTs capable of reading ECM data are connected to the on-board electronics by means of a Society of

Automotive Engineers (SAE)/American Trucking Association (ATA) J 1584/J 1708/1939 connector in current systems. These data connectors are used by all the truck engine electronics OEMs and are most often referred to as **ATA data connectors.** The common connector and protocols by OEMs allows the proprietary software of one manufacturer to at least read the parameters and conditions of their competitors. In short, this means that if a Navistar powered truck has an electronic failure in a location where the only service dealer is Freightliner, some basic problem diagnosis can be undertaken. An ATA connector as used in J 1584/1708 system is a 6-pin **Deutsch connector** and is wired as indicated in **Figure 8–1.**

The more recent J 1939 connector uses a 9-pin Deutsch connector.

8.2 FLASH OR BLINK CODES

The simplest EST used for electronic circuit diagnosis is the system's own electronic warning light or lights.

The lights used vary depending on the system: Usually they will be mounted on the vehicle dashboard so the operator can be alerted, but in the case of some ABS systems, they may be located on the chassis controller module. They may also share their function with other circuit warning lights. Electronic malfunction lights may be used initially as a driver alert but, in most cases, they have the ability to "blink" out numeric codes. These codes relate to a specific circuit malfunction that enable anyone to read and report the problem without any specialized tools.

As on-board electronic systems become better integrated and their software interacts using SAE protocols, the need for electronic malfunction light(s) dedicated to a single circuit has diminished. Most current electronically managed truck chassis have a digital dash display that can alert the operator to system malfunctions using not numeric codes (which must be interpreted), but an alpha display that indicates what the problem is and often the corrective action required.

Flash or **blink codes** are in most cases designed to read, at minimum, active system fault codes. Depending on the system and its manufacturer, sometimes historic or **inactive codes** can also be read. In cases where multiple codes are displayed, it is essential that the OEM troubleshooting literature be consulted because certain types of circuit failure can trigger codes in functional circuits and components.

8.3 USING DIGITAL MULTIMETERS

A digital multimeter (DMM) is simply a tool for making electrical measurements. DMMs may have any number of special features, but essentially they mea-

sure electrical pressure or *volts,* electrical current or *amps,* and electrical resistance or *ohms.* A good quality DMM with minimal features may be purchased for as little as 100 dollars. As the features, resolution, and display quality increase in sophistication, the price increases proportionally. A typical DMM and its features are shown in **Figure 8–2**, **Figure 8–3**, **Figure 8–4**, **Figure 8–5**, and **Figure 8–6**.

As most electronic circuit testing *requires* the use of a DMM, this instrument should displace the analog multimeter and circuit test light in the truck/bus

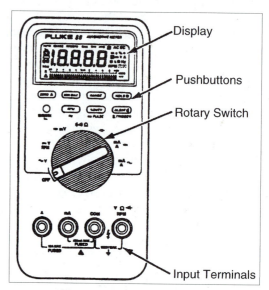

Fluke 88 Digital Multimeter

FIGURE 8–2 *Typical DMM. (Courtesy of International Truck and Engine Corp.)*

Input Terminals The digital multimeter has four input terminals as shown in the following illustration. The terminals are protected against overload to the limits shown in the User's Manual.

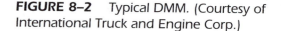

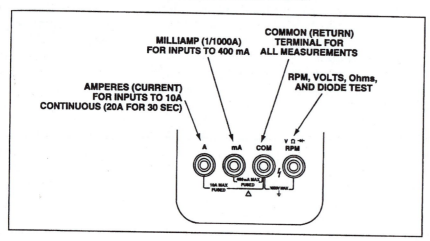

FIGURE 8–3 DMM input terminals. (Courtesy of International Truck and Engine Corp.)

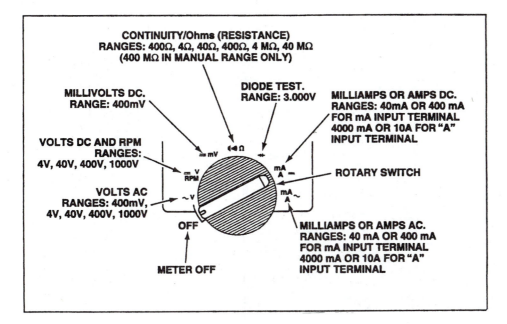

FIGURE 8–4   DMM rotary switch. (Courtesy of International Truck and Engine Corp.)

The pushbuttons are used to select meter operations. When a button is pushed, a display symbol will appear and the beeper will sound. Changing the rotary switch setting will reset all pushbuttons to their default settings.

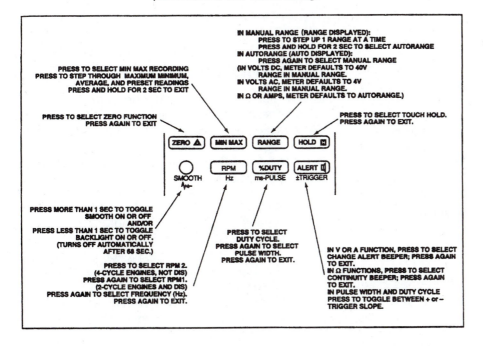

FIGURE 8–5   DMM push buttons. (Courtesy of International Truck and Engine Corp.)

technician's toolbox. Reliability, accuracy, and ease of use are all factors to consider when selecting a DMM for purchase. Some options the technician may wish to consider are a protective rubber holster (which will greatly extend the life of the instrument), analog bar graphs, and enhanced resolution. This section will deal with the use of DMMs. A knowledge of basic electricity will be assumed.

Resolution

Resolution shows how fine a measurement can be made with the instrument. Digits and counts are used to describe the resolution capability of a DMM. A 3½-digit meter can display three full digits ranging from 0–9, and one half digit that displays either 1 or is left blank. A 3½-digit meter will therefore display 1,999 counts of resolution. A 4½-digit DMM can display as many as 19,999 counts of resolution.

Display   The digital multimeter has a digital and analog display capability. The digital display should be used for stable inputs, while the analog display should be us ed for frequently changing inputs. If a measurement is to large to be displayed, "OL" is shown on the digital display.

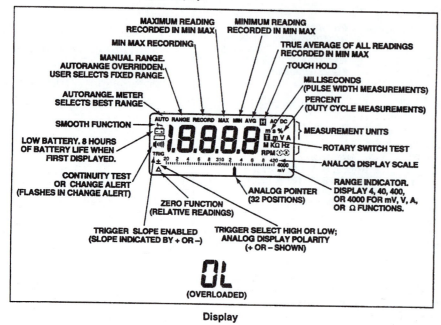

Display

FIGURE 8–6   *DMM data display. (Courtesy of International Truck and Engine Corp.)*

OPERATING PRINCIPLES

In any electrical circuit, voltage, current flow, and resistance can be calculated using Ohm's Law. A DMM makes use of Ohm's Law to measure and display values in an electrical circuit. A typical DMM will have the following selection options, chosen by rotating the selector:

off	Shuts down DMM
V	Enables AC voltage readings
V–	Enables DC voltage readings
m V–	Enables low-pressure DC voltage readings
Ω	Enables component or circuit resistance readings
α	Enables **continuity** (a circuit capable of being closed) testing. Identifies an open/closed circuit.
A	Checks current flow (amperage) in an AC circuit
A–	Checks current flow (amperage) in a DC circuit

Many DMM's have *enhanced resolution,* so the meter's reading power is usually expressed in counts rather than digits. For instance, a 3½-digit meter may have enhanced resolution of 4,000 counts. Basically, **meter resolution** is expressed in counts rather than digits. For example, a 3½-digit or 1,999 count meter will not measure down to 0.1V when measuring 200V or higher. However, a 3,200 count meter will display 0.1V up to 320V, giving it the same resolution as a 4½-digit, 19,999 count meter until the voltage exceeds 320V.

Accuracy

Accuracy in a meter specification describes how close the displayed reading on the DMM is to the actual value of the measured signal. This is usually expressed as a percentage of the reading. An accuracy rating of ±1% in a DMM reading a voltage value of 10V means the actual value could range between 9.9V and 10.1V. DMM accuracy can be extended by indicating how many counts the right display digit may vary. An accuracy rating of ± (1% + 2) would mean that a displayed voltage of 10V could have an actual value range between 9.88V and 10.12V.

Analog multimeters have accuracy ratings that vary between 2–3% of full scale. DMM accuracy ratings range between ± (.07% + 1) to ± (0.1% + 1) of the *reading.*

CAUTION: Whenever a truck circuit has an electronic control module, ensure that a digital multimeter (DMM) is used to make voltage measurements. Analog voltmeters can damage ECM circuits.

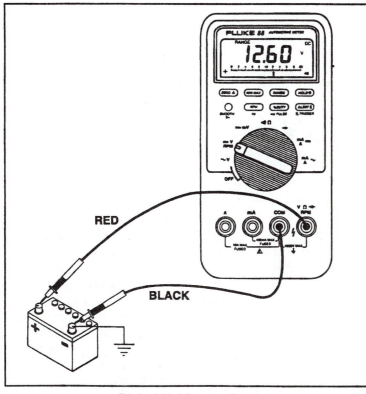

FIGURE 8–7 *DMM set up for making voltage measurements. (Courtesy of International Truck and Engine Corp.)*

Digital Multimeter Setup

MEASURING VOLTAGE

Checking circuit supply voltage is usually one of the first steps in troubleshooting. This would be performed in a vehicle DC circuit by selecting the V-setting and checking for voltage present or high/low voltage values. Most electronic equipment is powered by DC. For example, home electronic apparatus such as computers, televisions, and stereos use rectifiers to convert household AC voltage to DC voltage.

The waveforms produced by AC voltages can either be sinusoidal (sine waves) or non-sinusoidal (sawtooth, square, ripple). A DMM will display the **root mean square (rms)** value of these voltage waveforms. The rms value is the effective or equivalent DC value of the AC voltage. Meters described as *average responding* give accurate rms readings only if the AC voltage signal is a pure sine wave. They will not accurately measure non-sinusoidal signals. DMMs described as *true-rms* will measure the correct rms value regardless of waveform and should be used for non-sinusoidal signals.

A DMM's capability to measure voltage can be limited by the signal frequency. The DMM specifications for AC voltage and current will identify the frequency range the instrument can accurately measure (see **Figure 8–7**).

Voltage measurements determine:

- Source voltage
- Voltage drop
- Voltage imbalance
- Ripple voltage
- Sensor voltages

MEASURING RESISTANCE

Most DMMs will measure resistance values as low as 0.1Ω. Some will measure high resistance values up to $300M\Omega$ (megohms). Infinite resistance or resistance greater than the instrument can measure is indicated as OL on the display. For instance, an **open circuit** (one in which there is no path for current flow) would read OL on the display.

Resistance and continuity measurements should be made on open circuits only. Using the resistance or continuity settings to check a circuit or component that is energized will damage the test instrument. Some DMMs are protected against such "accidental" abuse—the extent of damage will depend on the model. For accurate low-resistance measurement, test lead resistance, typically between 0.2Ω–0.5Ω depending on quality and length, must be subtracted from the display reading (see **Figure 8–8**). Test lead resistance should *never* exceed 1 Ω.

If a DMM supplies less than 0.3V DC test voltage for measuring resistance, it is capable of testing resistors isolated in a circuit by diodes or semiconductor junctions, meaning that they do not have to be removed from the circuit board.

Resistance measurements determine:

- Resistance of a load
- Resistance of conductors
- Value of a resistor
- Operation of variable resistors

Continuity is a quick resistance check that distinguishes between an open and a closed circuit. Most DMMs have audible continuity beepers that beep when they detect a closed circuit, permitting the test to be performed without looking at the meter display. The actual level of resistance required to trigger the beeper varies from model to model.

Continuity tests determine:

- Fuse integrity
- Open or shorted conductors
- Switch operation
- Circuit paths

Diode Testing

A diode is an electronic switch that can conduct electricity in one direction while blocking current flow in the opposite direction. They are commonly enclosed in glass cylinders; a dark band identifies the cathode or blocking terminal. Current flows only when the anode is more positive than the cathode. Additionally, a diode will not conduct until the forward voltage pressure reaches a certain value, 0.3V in a silicon diode. Some meters have a diode test mode. When testing a diode with the DMM in this mode, 0.6V is delivered through the device to indicate continuity; reversing the test leads should indicate an open circuit in a properly functioning diode. If both readings indicate an open circuit condition, the diode is open. If both readings indicate continuity, the diode is shorted.

MEASURING CURRENT

Current measurements are made in series, unlike voltage and resistance readings, which are made in parallel. The test leads are plugged into a separate set of input jacks, and the current to be measured flows through the meter.

Current measurements determine

- Circuit overloads
- Circuit operating current
- Current in different branches of a circuit

When the test leads are plugged into the current input jacks and are used to measure voltage, a direct short across the source voltage through a low value resistor inside the DMM called a current shunt occurs. A high current flows through the meter. If not adequately protected, both the meter and the circuit can be damaged (see **Figure 8–9**).

A DMM should have current input fuse protection of high enough capacity for the circuit being tested. This is mainly important when working with high pressure (220V+) circuits.

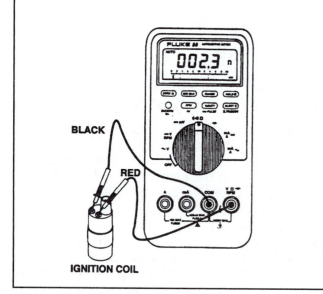

Measuring Resistance

To measure resistance, set your digital multimeter rotary switch to Ω. The meter will power up. Plug the black (negative) lead into the **COM** input jack and the red (positive) lead into the **VΩ** input jack. Since the digital multimeter measures resistance by passing a small current through the component, source voltage must not be present in the circuit. Place the meter in parallel with the component and note the reading.

BLACK

RED

IGNITION COIL

Digital Multimeter Setup

FIGURE 8–8   *DMM set up for measuring resistance. (Courtesy of International Truck and Engine Corp.)*

Measuring Current

Amperage is the measurement of current flow. To measure amperage, set the digital multimeter to the **mA/A** function. Plug the black (negative) lead into the **COM** input jack and the red (positive) lead into the **A** or **mA** input jack. Place the meter in <u>series</u> with the circuit so that current passes through the meter. Use the correct type of current probes for this purpose. Power the circuit and note the reading.

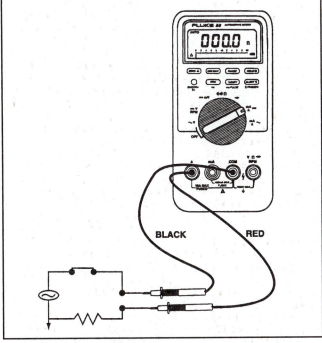

Digital Multimeter Setup

FIGURE 8–9   *DMM set up for measuring current flow. (Courtesy of International Truck and Engine Corp.)*

CURRENT PROBE ACCESSORIES

When making current measurements that exceed the DMMs rated capacity, a current probe can be used. There are two types discussed in the following sections.

Current Transformers

A **current transformer** measures AC current only. The output of a current transformer is 1mA per 1A. This means that a current flow of 100A is reduced to 100mA, which can be handled by most DMMs.

The test leads would be connected to the mA and common input jacks and the meter function switch set to mA AC.

Current transformers provide a rather inaccurate circuit test that is used for ballpark reckoning only. There are no applications for this tool in truck electrical systems.

Hall Effect Probe

The output of a **Hall effect probe** is 1 mV per ampere. It will measure AC or DC. The test leads are connected to the V and common input jacks. The DMM function switch should be set to the V or mV scale, selecting V-AC for AC current or V-DC for DC current measurements. The pick-up is clamped around the wire through which current flow is to be measured. Like the current transformer, the Hall effect probe is not especially accurate. It may be used to measure cranking motor current draw.

DMM Features

When considering a DMM for purchase, the following features should be considered:

- Fused current inputs
- Use of high energy fuses (600V+)
- High voltage protection in resistance mode
- Protection against high voltage transients
- Insulated test lead handles
- CSA and UL approval

SOME TYPICAL DMM TESTS

Always perform tests in accordance with truck and bus OEM specifications. Never jump sequence or skip steps in sequential troubleshooting charts. Most DMM tests on truck and bus electronic systems will be used in conjunction with a generic **reader/programmer (ProLink 9000)** or PC and proprietary software. The following tests assume the use of a Fluke 88 DMM, but do not differ much when other DMMs are used.

1. Engine Position: Fuel Injection Pump Camshaft, Engine Camshaft and Crank Position Sensors
   Hall effect sensors
   (a) Cycle ignition key and then switch off.
   (b) Switch meter to measure V DC/rpm.
   (c) Identify the ground and signal terminals at the Hall sensor. Connect the positive (+) test lead to the signal terminal and the negative (−) test lead to the ground terminal. Crank engine.
   At cranking speeds, the analog bar graph should pulse. At idle speeds or faster, the pulses are too fast for bar graph readout.
   (d) Press the duty cycle button once. Duty cycle can indicate square wave quality (poor quality signals have a low duty cycle). Functioning Hall sensors should have a duty cycle of about 50 percent, depending on the sensor. Check the specifications.
2. Potentiometer-Type TPS
   Resistance test:
   (a) Key off.
   (b) Disconnect TPS.

(c) Select Ω on DMM. Connect the test probes to the signal and ground terminals. Next, move accelerator through its stroke, observing the DMM display.

(d) The analog bar should move smoothly without jumps or steps. If it steps, there may be a bad spot in the sensor.

Voltage test:

(a) Key on, engine off.

(b) Set meter to read DC V. Connect the negative lead to ground.

(c) With the positive lead, check the reference voltage value and compare to specs. Next, check the signal voltage (to the ECM) value through the accelerator pedal stroke. Check the values to specification. Also observe the analog pointer: As with the resistance test, this should move smoothly through the accelerator stroke.

3. Magnetic Sensors

Magnetic or variable reluctance sensors function similarly to a magneto. The output is an AC voltage pulse, the voltage and frequency values of which rise proportionally with rotational speed increase. Voltage values range from 0.1V up to 5.0V, depending on the rotational speed and the type of sensor. Frequency varies with the number of teeth on the reluctor wheel and the rpm. Vehicle speed sensors (VSS), engine speed sensors (ESS), and ABS wheel speed sensors all use this method of determining rotational speed. Test using V-AC switch setting and locating test leads across the appropriate terminals.

4. Min/max Average Test for Lambda (λ), Exhaust Gas or O_2 Sensors

(a) With the key on, engine running, and DMM set at DC V, select the correct voltage range.

(b) Connect the negative test lead to a chassis ground and the positive test lead to the signal wire from the Lambda sensor. Press DMM min/max button.

(c) Ensure the engine is warm enough to be in closed loop mode (100 mV–900 mV O_2 sensor output). Run for several minutes to give the meter time to sample a scatter of readings.

(d) Press min/max button slowly three times while watching the DMM display. A maximum of 800 mV and a minimum of fewer than 200 mV should be observed.

The average should typically be around 450 mV.

(e) Next, disconnect a large vacuum hose to create a lean burn condition. Repeat steps c and d to read average voltage. Average voltage should be lower, indicating a lean condition.

(f) The same test can be performed using propane enrichment to produce a rich AFR condition and therefore higher voltage values.

(g) Lambda sensor tests can be performed while road testing the vehicle. 450 mV normally indicates stoichiometric fueling ($\lambda = 1$), but check to specifications.

5. Thermistors

Most **thermistors** used in computerized engine systems have a negative temperature coefficient (NTC), meaning that as sensor temperature increases, their resistance decreases. They should be checked to specifications using the DMM ohmmeter function and an accurate temperature measurement instrument.

OEM's seldom suggest random testing of suspect components. The above tests are typical procedures. Circuit testing in today's computerized engine management systems is highly structured and part of a sequential troubleshooting procedure. It is important to perform each step in the sequence precisely as instructed. Never forget that skipping a step can invalidate every step that follows.

BREAKOUT BOXES AND BREAKOUT Ts

The DMM is often used in conjunction with a **breakout box** or **breakout T.** Breakout devices are designed to be T'ed into an electrical circuit to enable circuit measurements to be made on both closed (active) and de-energized circuits. The idea is to access a circuit with a test instrument without interrupting the circuit. A breakout T normally describes a diagnostic device that is inserted into a simple two- or three-wire circuit such as that used to connect an individual sensor, whereas a breakout box accesses multiple wire circuits such as main harness connectors for diagnostic analyses of circuit conditions. Most of the electronic management system OEMs use a breakout box that is inserted into the interface connection between the engine electronics and chassis electronics harnesses. The face of the breakout box displays a number of coded sockets into which the probes of a DMM can be safely inserted to read circuit conditions. Electronic troubleshooting sequencing is often structured based on the data read by a DMM accessing a circuit. A primary advantage of breakout diagnostic devices is that they permit the reading of an active electronic circuit, for instance, while an engine is running.

> **CAUTION:** When a troubleshooting sequence calls for the use of breakout devices, always use the recommended tool. Never puncture wiring or electrical harnesses to enable readings in active or open electronic circuits. The corrosion damage that results from damaging wiring conduit will create problems later on and the electrical damage potential can be extensive.

Diagnostic Connector Dummies

Diagnostic connector dummies are used to read a set of circuit conditions in a circuit that has been opened by separating a pair of connectors. The dummies are manufactured by the electrical/ electronic connector manufacturer as a means of accessing the circuitry with a DMM without damaging the connector sockets and pins. Ensure that correct dummies are used.

> **CAUTION:** The terminals in many connectors are especially vulnerable to the kind of damage that can be caused by attempting to insert DMM probes, paper clips, and other inappropriate devices. Even more important, remember that it is always possible to cause costly electrical damage by shorting and grounding circuits in a separated electrical connector.

Shop Talk _____

When performing a multiple-step, electronic troubleshooting sequence on a large multi-terminal connector, photocopy the coded face of the connector(s) from the service manual and use it as a template. The alphanumeric codes used on many connectors can be difficult to read and using a template is a good method of orienting the test procedure.

TESTING SEMICONDUCTORS

Testing a diode is shown in **Figure 8–10**.

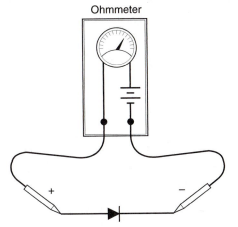

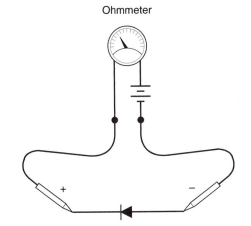

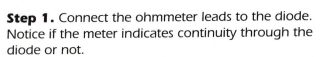

Step 1. Connect the ohmmeter leads to the diode. Notice if the meter indicates continuity through the diode or not.

Step 2. Reverse the diode connection to the ohmmeter. Notice if the meter indicates continuity through the diode or not. The ohmmeter should indicate continuity through the diode in only one direction. (Note: If continuity is not indicated in either direction, the diode is open. If continuity is indicated in both directions, the diode is shorted.)

FIGURE 8–10   *Testing semiconductors. (Courtesy of Utah Technical College)*

Testing a transistor

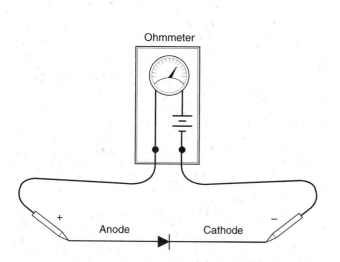

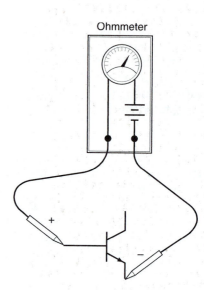

Step 1. Using a diode, determine which ohmmeter lead is positive and which is negative. The ohmmeter will indicate continuity through the diode only when the positive lead is connected to the anode of the diode and the negative lead is connected to the cathode.

Step 3. With the positive ohmmeter lead still connected to the base of the transistor, connect the negative lead to the emitter. The ohmmeter should again indicate a forward diode junction. (Note: If the ohmmeter does not indicate continuity between the base-collector or the base-emitter, the transistor is open.)

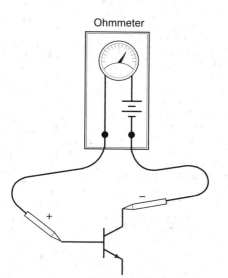

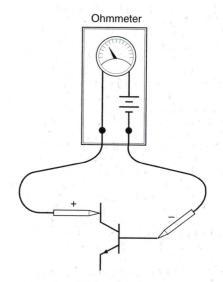

Step 2. If the transistor is an NPN, connect the positive ohmmeter lead to the base and the negative lead to the collector. The ohmmeter should indicate continuity. The reading should be about the same as the reading obtained when the diode was tested.

Step 4. Connect the negative ohmmeter lead to the base and the positive lead to the collector. The ohmmeter should indicate infinity or no continuity.

FIGURE 8–10   Testing semiconductors (Continued). (Courtesy of Utah Technical College)

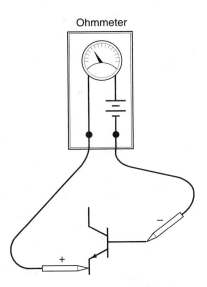

Step 5. With the negative ohmmeter lead connected to the base, reconnect the positive lead to the emitter. There should again be no indication of continuity. (Note: If a very high resistance is indicated by the ohmmeter, the transistor is "leaky" but may still operate in the circuit. If a very low resistance is seen, the transistor is shorted.)

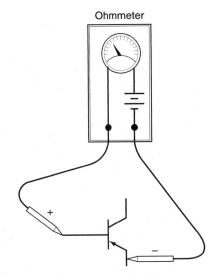

Step 7. If the positive ohmmeter lead is connected to the base of a PNP transistor, no continuity should be indicated when the negative lead is connected to the collector or the emitter.

Testing an SCR

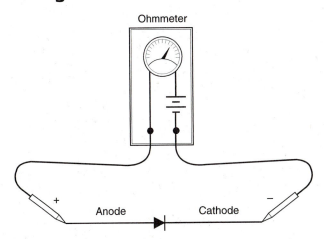

Step 1. Using a junction diode, determine which ohmmeter lead is positive and which is negative. The ohmmeter will indicate continuity only when the positive lead is connected to the anode of the diode and the negative lead is connected to the cathode.

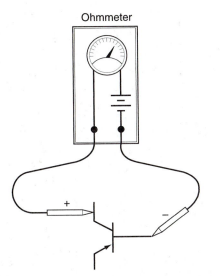

Step 6. To test the PNP transistor, reverse the polarity of the ohmmeter leads and repeat the test. When the negative ohmmeter lead is connected to the base, a forward diode junction should be indicated when the positive lead is connected to the collector or emitter.

FIGURE 8–10 Testing semiconductors (Continued). (Courtesy of Utah Technical College)

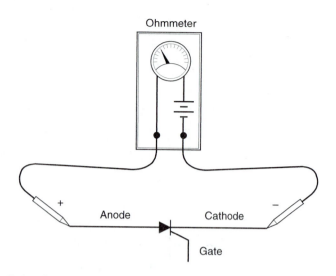

Step 2. Connect the positive ohmmeter lead to the anode of the SCR and the negative lead to the cathode. The ohmmeter should indicate no continuity.

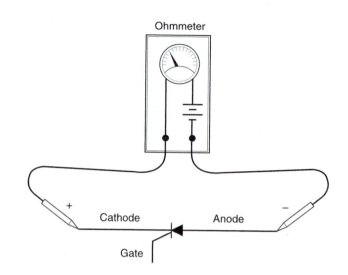

Step 4. Reconnect the SCR so that the cathode is connected to the positive ohmmeter lead and the anode is connected to the negative lead. The ohmmeter should indicate no continuity.

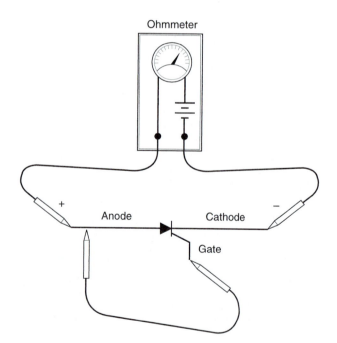

Step 3. Using a jumper lead, connect the gate of the SCR to the anode. The ohmmeter should indicate a forward diode junction when the connection is made. (Note: If the jumper is removed, the SCR may continue to connect or it may turn off. This will be determined by whether the ohmmeter can supply enough current to keep the SCR above its holding current or not.)

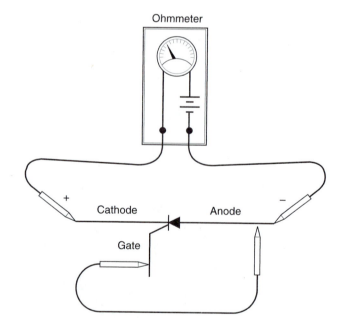

Step 5. If a jumper is used to connect the gate to the anode, the ohmmeter should indicate no continuity. (Note: SCRs designed to switch large currents [50 amperes or more] may indicate some leakage current with this test. This is normal for some devices.)

FIGURE 8–10 Testing semiconductors (Continued). (Courtesy of Utah Technical College)

8.4 PROLINK 9000

The MPSI ProLink 9000 is classified as a generic reader/programmer EST. This simple electronic data link has become the industry standard electronic service tool capable of reading most current systems with the correct software cartridge or card. Unlike a DMM, the ProLink 9000 is completely user friendly in that the novice could, in little time, acquire some expertise with the instrument simply by exploring menu through menu while connected to an ECM. These are microprocessor-based test instruments that will do the following (depending on the system):

(a) Access active and **historic codes**: erase historic (inactive) codes
(b) View all system identification data
(c) View data on system operation with the engine running or the truck moving
(d) Perform diagnostic tests on system subcomponents, such as EUI testing
(e) Reprogram customer data parameters on engine and chassis systems
(f) Act as a serial link to connect the vehicle ECM via modem to a centrally located mainframe for proprietary data programming (some systems only)
(g) Snapshot system data parameters to assist intermittent fault-finding solutions

This section will introduce some of the features of ProLink, first in a general sense, and then by detailing some of the most commonly performed procedures. The advantages of the ProLink 9000 reader/programmer are numerous, most notably its

(a) User friendliness. Technicians need very little time to develop a working knowledge of this EST. The best method of becoming familiar with it is simply to use it. Some training is recommended with some of the more complex electronic systems, but the curious technician can achieve some mastery of the instrument just by using it. Menus can be selected; data can be scrolled through. If you get stuck, simply unplug and reconnect. This can be done at any time without sustaining physical or electronic damage.
(b) Universality. This EST has become the industry standard and is used extensively even in OEM dealerships whose systems require the use of PCs for more detailed

diagnostic and programming functions. It will read most of the truck/bus engine and other onboard systems (and also automobile electronic systems) simply by having the correct software cartridge or card inserted.
(c) Easy updatability. Cartridges are updated simply by opening them up, removing, and discarding out-of-date PROM chips or data cards and inserting new one(s). At PROM chip replacement, the RAM chip may have to be momentarily disconnected from the motherboard to dump its nonvolatile data content. The process for upgrading the newer **multi-protocol cartridge (MPC)** is even more simple: An updated system card is inserted into the MPC unit. ProLink can also be upgraded by installing an MPC. No disassembly is required.
(d) Toughness. The tough plastic housing really does withstand the rough treatment some of these instruments are likely to receive in the typical truck/bus service facility.

The disadvantages of the ProLink 9000 are

(a) Viewer window is limited to four lines of data display.
(b) Inability to accept an alpha password (some systems), or a complex procedure for inputting an alpha password. It should be noted that this feature might also be considered an advantage by some truck fleets because it locks out the ability to alter customer programming data using ProLink, preventing drivers from requesting alterations to data such as maximum road speed when away from fleet terminals.
(c) As electronic management systems become more complex, the limitations of ProLink 9000 become more evident. PC-based systems are more often required to properly troubleshoot electronic problems.

Generic reader/programmers are connected to the on-board ECM by means of the SAE/ATA J 1584/1708/1939 Data Link; that is, the ATA data connector.

RS-232 SERIAL PORT

This port is located on the right side of the ProLink head. There are three choices when this menu item is selected:

1. Printer output
2. Terminal output
3. Port setup

There are four parameters that need to be set up for the device to communicate with a printer or PC terminal (see **Figure 8–11** and **Figure 8–12**):

1. **Handshake.** A **handshake** refers to how data will be transmitted between two electronic components. There are two commonly used methods: busy and xon/xoff. Select the method specified on the printer specifications.
2. **Baud rate.** The **baud rate** is the speed at which data is transmitted. Both the output device and the receiving device must agree on transmission speed.
3. **Parity.** The **parity** is the even or odd quality of the number of 0s and 1s. The menu options are *none, odd* or *even.*
4. **Stop bits.** The **stop bits** are the last elements in the transmission of a character. The output and receiving devices must agree on how many stop bits to expect to determine the end of a transmission character.

Shop Talk

ProLink will transmit to a printer or PC terminal any data that it can read itself from a system ECM. Printouts of data are especially helpful when analyzing the causes of a condition.

UPDATING A PROLINK CARTRIDGE

Depending on when the software cartridge was made, the update method varies. Older cartridges are updated by replacing the PROM chip or chips using the following procedure:

1. Ensure that during this procedure, there is no danger of a static charge being unloaded into the cartridge. On the cement floor of a truck shop, there is little danger of this happening, but if on a carpeted floor, wear a **ground strap.**

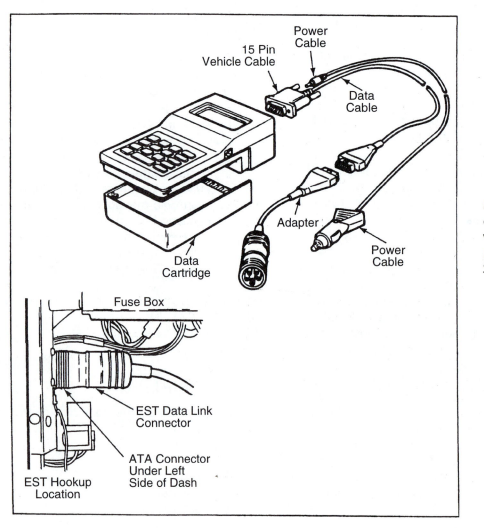

FIGURE 8–11 ProLink data connection hardware. (Courtesy of MPSI [Micro Processor Systems, Inc.])

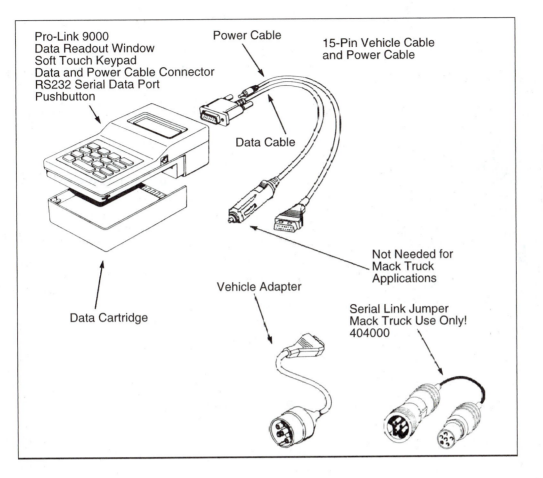

Pro-Link 9000
Data Readout Window
Soft Touch Keypad
Data and Power Cable Connector
RS232 Serial Data Port
Pushbutton

Power Cable

15-Pin Vehicle Cable
and Power Cable

Data Cable

Not Needed for
Mack Truck
Applications

Vehicle Adapter

Data Cartridge

Serial Link Jumper
Mack Truck Use Only!
404000

FIGURE 8–12
ProLink head, software
cartridge, and cables.
(Courtesy of MPSI
[Micro Processor
Systems, Inc.])

2. Remove the cartridge from the ProLink 9000 reader programmer. Remove the four screws fastening the cartridge cover plate, and then separate the motherboard from the cartridge casing. Place on a clean wooden bench.

3. Using a chip lever, carefully remove the PROM chip(s) (one or two depending on the system) from the motherboard. Using the schematic in the update chips packaging, install the updated chips. Special care is required to avoid bending/damaging the connecting prongs of the new chips when installing them. Next, it may be required to temporarily remove the RAM chip in the same manner as the PROM chips. This should be momentarily separated from the motherboard (there is a nonvolatile memory segment in some systems) and then carefully reinstalled. Ensure that the two or three chips are fully installed in their sockets in the motherboard.

4. Reassemble the cartridge. Note the date of the PROM transfer and the software version somewhere on the cartridge exterior. Then install the software cartridge into the ProLink 9000 head.

5. Power up the ProLink by connecting to an appropriate vehicle ECM. Scroll through the various menus to ensure everything is working. Note: When system cartridges are updated, they will in most cases read and function with all system software/hardware versions that preceded it.

The ProLink circuit board is fuse protected. **Figure 8–13** shows the location of the fuse.

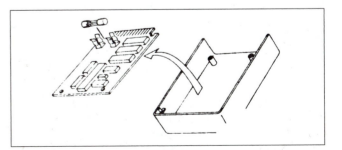

FIGURE 8–13 *Location of fuse on the ProLink circuit board. (Courtesy of MPSI [Micro Processor Systems, Inc.])*

FIGURE 8–14   *ProLink MPC cartridge. (Courtesy of MPSI [Micro Processor Systems, Inc.])*

MULTIPROTOCOL CARTRIDGE

The recently introduced ProLink MPC (multiprotocol cartridge) is updated simply by replacing the MPC card, a credit card-sized data retention card (see **Figure 8–14**). The MPC card is specific to one OEM system. Some of the MPC cards are equipped with a general heavy-duty reader that permits ProLink to read all the SAE-coded data. Replacement of an outdated MPC card by a current version means that no cartridge disassembly is required to update software. This simplifies the upgrade procedure for ProLink data cartridges.

Hot Keys

Hot keys or special function keys are a feature introduced with the MPC cartridge. They are accessed using the numeric keypad. They are function commands similar to the F commands on a PC keyboard. Some hot key commands are outlined in the following sections.

Hot Key 0. This is a Custom Data list. It allows data items from different lists to be combined into a single list. For instance, data from the transmission module could be displayed along with data from the chassis module. Use the up and down arrows to place the ProLink cursor on each data field to be

included in the list and push the 0 key. This will add the item to the list and display the list. To exit, push the Func key. To add more items, repeat the previous steps. To remove items, reverse the previous steps.

Hot Key 1. This feature allows the screen to display more lines of data by not displaying the title and arrows.

Hot Key 5. This is a Print Screen command. The user may print the data displayed on the screen at any time when the 5 key on the keypad is depressed.

Hot Key 7. This key displays MPC information such as its serial number, the length of time of the session, available memory, and when the "trigger" can be fired. To use the trigger option in hot key 7, press the Enter key. This will cause the unit to beep, record the time of the trigger, and freeze the current data. Data will not update after the trigger. To resume normal operation, push the 7 key again. Recorded data may be viewed by simultaneously holding the red side button and using arrow up/arrow down keys.

Hot Key 8. This is a bar graph feature. It can be used to view numeric data in a form other than numbers. To use this feature, use the up and down arrow keys to place the cursor on the data field, and then push the 8 key on the pad. The graph will display the lowest value received on the left side of the screen, the current value in the middle, and the highest value on the right side. The bar graph will fluctuate up and down as data is received by ProLink. To exit the bar graph mode, press the Func key and the data list will again be displayed.

8.5   PROLINK TASKS

This section outlines the procedure required to perform some common ProLink operations. The system used as a basis for this description is Mack Trucks' V-MAC III system. V-MAC III is a chassis management system designed to electronically coordinate the operation of all the on-board electronic systems, including the engine management system, so it is ideal for the purpose of demonstrating the many functions of the ProLink EST. The operations described here are the most frequently performed ProLink tasks. There may be minor differences in the performance of each procedure when using cartridges or PC cards (MPC systems) dedicated to other OEM electronic systems. For purposes of this description, a ProLink 9000 equipped with an MPC and dedicated Mack Trucks PC card is used.

INITIAL START-UP

This is an initializing process that permits the chassis electronic system to communicate with ProLink. It requires connection by means of the ATA data link (6-pin or 9-pin).

- Turn the vehicle ignition on. Connect the ProLink 9000 to the serial communication port and ensure that the unit powers up.
- The first information to be displayed is the system software identification, which should appear something like the following:

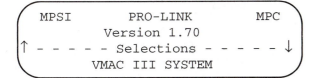

```
   MPSI        PRO-LINK        MPC
            Version 1.70
↑ - - - - - Selections - - - - - ↓
          VMAC III SYSTEM
```

Select the Enter button on the keypad. This will initiate the loading sequence, during which screens such as the following may appear:

```
            LOADING
        VMAC III SYSTEM
           V 1.0"
        (FUNC) TO CANCEL
```

- When the application has finished loading, a screen confirms this.

```
        CONNECTED TO A
        VMAC III SYSTEM

        (ENTER) TO CONTINUE
```

- Pressing the Enter key will produce a screen identifying the hot keys:

```
       SPECIAL FUNCTION
            KEYS
         0, 1, 5, 7, 8
        (ENTER) TO CONTINUE
```

- Pressing Enter on the preceding screen will produce a screen that prompts the user to consult the OEM manual for a description of the hot keys. Pressing Enter again will produce a menu offering five choices:

```
    ▪ Monitor parameters
    ▪ Fault codes
    ▪ Diagnostic tools
    ▪ Customer programming
    ▪ English/metric toggle
```

```
            VMAC III
↑ - - - - - Selections - - - - - ↓
        MONITOR PARAMETERS
```

Now the up and down arrow keys may be used to scroll through the five menu options listed above. The current selection is displayed on the lower line of the display window. When the Enter key is pressed, the displayed option is entered. For instance, in the preceding screen, pressing the Enter button would display the monitor parameters fields. To exit any application, the Func key would be depressed.

ENGLISH/METRIC TOGGLE

Data may be displayed in either English or metric values by ProLink. This is a toggle option so one or other must be selected. To change English readings to metric, use the following screen:

```
    HIT ENTER TO CHANGE
        FROM ENGLISH
    TO METRIC, FUNCT TO
    (CONTINUE) CANCEL
```

MONITOR PARAMETERS

Selecting this option results in data being downloaded from the system control module (ECM/ECU) to be displayed in the ProLink display window. Return to the 5th screen of the initializing process shown previously. Pressing Enter will produce a data list that will look something like this in the V-MAC system:

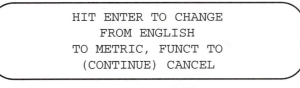

```
    ROAD SPEED MPH      0
    THROTTLE %          0
    ENGINE SPD RPM      0
    ENGINE LOAD %       0
```

Pressing the up and down keys permits the user to scroll through the data list line by line. Pressing the right or left arrow keys permit the data to be scrolled through four lines at a time. The list is not circular, so there is a beginning and end to scrolling through the data fields.

FAULT CODES

Use the initializing process to produce this screen and select Enter:

```
            VMAC III
↑ - - - - - Selections - - - - - ↓
          FAULT CODES
```

This will produce a display screen with three options below the Selections data line:

```
1. Active  →
2. Inactive (sometimes called "historic")
3. Clear Faults  →
```

If the user wishes to read **active codes**, the Enter key would be depressed on this screen:

```
         FAULT CODES

↑ - - - - - Selections - - - - - ↓
         ACTIVE CODES
```

To exit fault code selection, the Func key is pressed.

Active Codes

An active code is any system code that is currently active. If there are no active codes logged into the system, the display window will read Empty. The ProLink display window will indicate an alpha explanation of the active code and also identify the SAE PID and FMI and system codes logged. SAE and ATA MIDs, SIDs, PIDs, and FMIs are explained in a little more detail later in this chapter. The system code is the OEM code, usually the same code that would be flashed on the diagnostic lights. For an example of how a code is logged, if a TPS signal (used both by the engine and transmission management electronics) produced a voltage above normal (shorted high), this code would be displayed as follows:

```
      VEH MANAGEMENT SYS
          THROTTLE
      VOLTS ABOVE NORMAL
      PID: 91 FMI: 3 5-1
```

If more than one active code is logged into the system, the up and down arrow keys will allow the user to scroll the list. If only one active code is present, the up and down arrows will have no function. Pressing the Func key will exit the active codes function.

Inactive Codes

Inactive codes are codes that have been logged into the system in the past which are both no longer currently active and have not been cleared. Some OEMs use the term historic codes to describe inactive codes. Line 1 of the readout displays the name of the sensor, switch, or circuit that has produced the fault. Line 2 identifies the nature of the fault. Line 3 dis-

plays the PID, FMI, and the OEM fault code. Line 4 displays the number of separate occurrences of the fault code. So if the throttle position problem indicated in the active codes screen shown a little earlier had occurred on twelve occasions, the ProLink display window would read as follows:

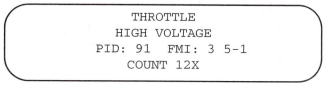

```
          THROTTLE
        HIGH VOLTAGE
      PID: 91  FMI: 3 5-1
          COUNT 12X
```

Clear Codes

Only inactive codes can be cleared from a chassis electronic system. To clear codes, be sure the engine is not running and have the ignition key on. It is also usually required that inactive codes be read before they are cleared.

Locate fault codes in the menu and press Enter. Next, select the clear fault option and, if there are multiple chassis electronic systems, the correct system (MID) must be selected. Scroll to the following screen display:

```
     CLEAR ALL INACTIVE
          CODES?

     (CONTINUE) CANCEL
```

Use the left and right arrow keys to select Continue or Cancel and press Enter. A prompt to confirm the clearing of codes will appear, followed by a confirmation that all inactive codes have been cleared.

Changing Passwords

It is usually necessary to key in a password to perform certain functions, especially those that affect performance programming. Different passwords may be used for customer data, vehicle data, and fleet data programming access to the system or chassis management control module—or the same password may be used for all functions. To select the change password function, locate program parameters in the menu and press Enter. The screen should appear as below:

```
     Choose Password Type
          To Change

        CUSTOMER DATA
```

Use the arrow keys to scroll through the types of passwords, customer, vehicle, fleet and so on. Many systems will only have one password, in which case the arrow keys will have no function. Next, it is necessary to enter the current password. If a numeric password is used, enter it directly through the keypad. If an alpha password is used, use the up and down arrows. If the wrong password is input, you will be instructed to press the Enter key and try again.

```
Enter Current
Password or PRESS
ENTER IF NONE
- - - - - - - - - -
```

Next, input the new password. It can be from one to 10 digits. In systems that accept alpha passwords, use the up and down arrows. The screen should appear as follows:

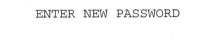

```
ENTER NEW PASSWORD


- - - - - - - - - -
```

 Shop Talk _____

Always double-check that the password has been correctly input. After a password has been input to a system, no future access can be achieved without it. Read the number back to yourself to ensure that it is correctly input.

The sequence is concluded by confirming the new password entry and the appearance of this screen:

```
ACCESS APPROVED

ENTER TO CONTINUE
```

Followed by

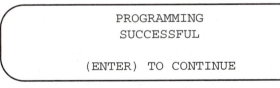

```
PROGRAMMING
SUCCESSFUL

(ENTER) TO CONTINUE
```

CUSTOMER PROGRAMMING

Customer data programming is all programming that the owner of the vehicle has ownership of (as opposed to proprietary data programming, which the OEM has ownership of). Customer data programming is explained in some detail in Chapter 6. It includes data that may have to be changed frequently, for example daily or weekly, or because some critical hardware such as a differential carrier

has been changed. The customer data programming on a vehicle management system may be subdivided into categories such as vehicle data, fleet data, change of ECM/ECU, and so on. However, the means of changing this programming is similar. You should exercise caution. Changing customer data can produce significant performance problems. Scroll to this screen from the program parameter function.

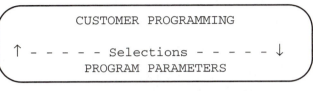

```
CUSTOMER PROGRAMMING

↑ - - - - - Selections - - - - - ↓
PROGRAM PARAMETERS
```

Press Enter. At this point, there will be three choices:

- View parameters →
- Past programming log/history →
- Program parameters →

This will produce the following screen:

```
CUSTOMER PROGRAMMING

↑ - - - - - Selections - - - - - ↓
VIEW PARAMETERS
```

When Enter is pressed, all the customer data fields currently programmed to the ECM/ECU can be read. This option is not password protected. The data can be scrolled through using the up and down arrow keys.

Program Parameters

This function permits customer data to be altered. Remember that this can create significant vehicle performance problems. For instance, an incorrectly programmed differential carrier ratio will result in each ECM/ECU that uses vehicle road speed data receiving incorrect road speed signals.

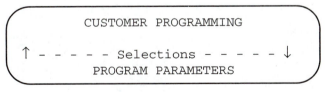

```
CUSTOMER PROGRAMMING

↑ - - - - - Selections - - - - - ↓
PROGRAM PARAMETERS
```

Press Enter. At this point, the following options are available by scrolling:

1. Vehicle ECU
2. Engine ECU
3. Date/time setting
4. Governor setting
5. Change password

More or fewer options may be available depending on the system. Use the up and down arrow keys to scroll the options and Enter to select. Always consult the OEM service literature before attempting to change any customer data. The OEM procedure should always be used. In all cases, a password is required to alter customer data programming. This may be the factory default password (if none has been programmed since the vehicle was manufactured) or any numeric or alpha password programmed.

> **CAUTION:** Exercise extreme care when altering customer data. Significant performance problems and component damage may result from incorrectly programming data to an ECM.

8.6 | PCs AND OEM SOFTWARE

Until recently, knowledge of PCs has been required of most technicians working with engine electronics, but to a much lesser extent when working on other chassis electronics systems. This is changing. The PC used with the electronic system OEM software is being used more and more in the trucking industry as the primary EST. A PC is a computer and, as such, most of what we said about vehicle computers in Chapter 6 also applies, so it would be a good idea to review that first. A brief description of a typical PC, its operating system, and how they are used in the trucking industry is provided here. **Figure 8–15** shows a typical PC station with the system housing, speakers, monitor, keyboard, and mouse.

Computer hardware is the simplest aspect of computer technology to understand. Computer hardware consists of the things that you can hold: Things

FIGURE 8–15 PC station. (Courtesy of International Business Machines)

such as system housings, monitors, a keyboard, a mouse, and a printer. Hardware also describes the guts of the system such as the motherboard, microprocessor, memory cards and chips, disk drives, and modems. Most people with a little technical inclination, the ability to read instructions, and a couple of hours to spare could assemble a current PC system without breaking a sweat. Serious problems with hardware in today's computers are not common.

The computer problems that can make life working with computer systems difficult are software problems. You cannot touch software. It consists of program instructions and commands written in various different types of programming language. The computer system needs software from the moment we switch it on. When we wish to use a computer as a word processor to type a letter, we need software to tell the hardware in the computer system how to behave. Software makes the hardware in the computer system produce results such as the production of a letter or a connection to the Internet.

THE HARDWARE

Computer hardware is simple and the source of few real problems in a current PC system. It actually gets blamed for malfunctions that are usually the fault of the software in the system. When diagnosed as defective, computer hardware is seldom repaired—it is simply replaced. Does this sound familiar? That is exactly what we tend to do with many vehicle systems today. The following is a description of some PC hardware from the user's point of view (see **Figure 8–16**).

System Housing

This is the term we use to describe the flat (desktop) or vertical (tower) box into which most of the guts of the computer are located. From the outside, we can observe an on/off switch, diskette slots and CD drives, buttons that control CD drives, and perhaps a speaker volume control. Inside are components that make the system operate as a computer. In older systems, some of the hardware components listed below would have been located outside the system housing. In a typical PC today, we will find the following components:

- **Motherboard.** This is the main circuit board. Most of the other components are attached to this either directly or indirectly.
- **Processor.** This is usually called the CPU or central processing unit. The processor is the brain of the processing cycle, and the speed at which it operates plays a big role in describing

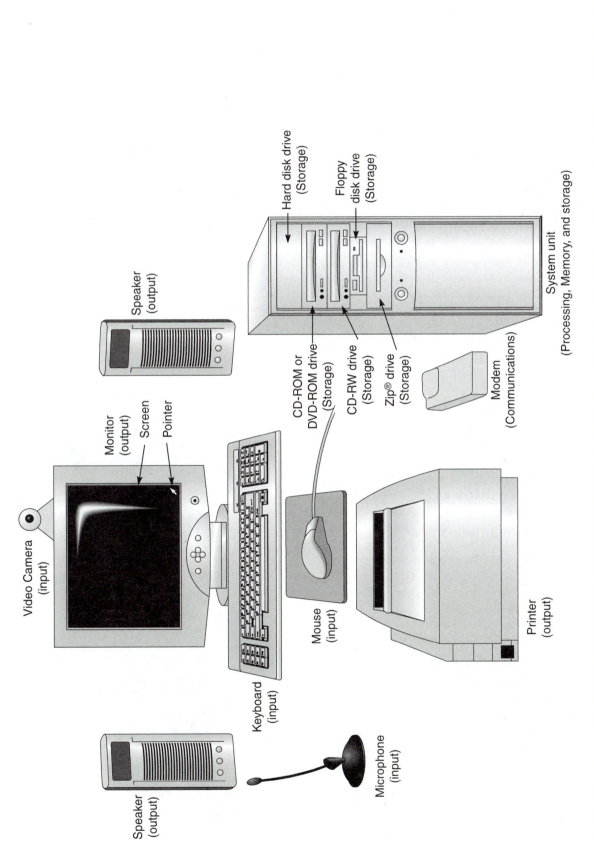

FIGURE 8–16   Common computer hardware components include a keyboard, mouse, microphone, system unit, disk drives, printer, monitor, digital camera, speakers, and a modem.

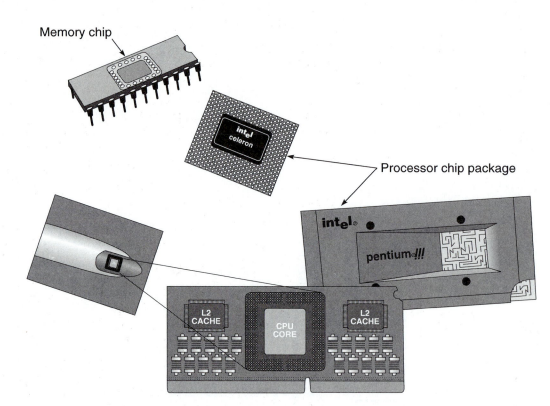

FIGURE 8–17   Processors and memory chips.

computer performance. Processors are manufactured by companies such as Intel and AMD and their speed is measured in hertz. **Figure 8–17** shows some examples of processors.

- **RAM (random access memory) chips.** RAM is main memory capability, so total RAM plays a big role in defining the computing power of the system. RAM is electronically retained and the CPU can only manipulate data in electronic format.

- **Memory chips.** Data can also be retained magnetically on chips that are attached to the motherboard. **Figure 8–17** shows an example of a memory chip.

- **Hard disk drive.** The hard disk is a means of magnetically retaining information within the computer housing. Consists of platters (many layers) of data retention disks that can be loaded with vast quantities of information. **Figure 8–18** shows self-contained and removable hard disks.

- **Floppy diskette.** Today, rigid plastic containers that measure $3^1/_2$ inches are used (see **Figure 8–19**). Data can be magnetically recorded to a diskette in the same way that information is written to an audio tape. They are a portable means of transferring/retaining data from one PC to another.

Some hard disks are self-contained devices housed inside the system unit. Removable hard disks, in contrast, are inserted and removed from a drive.

FIGURE 8–18   *Self-contained and removable hard disks.*

FIGURE 8–19 *A floppy disk is inserted and removed from a floppy disk drive.*

FIGURE 8–20 *CD tray from a PC.*

- **CD reader/writer.** This is a means of optically reading data from a CD (compact disc) using a laser. Most OEM software is provided to their dealers on CD today because they can hold very large quantities compared to diskettes. A CD is a essentially a metal disc coated with plastic. Many

computers today have CD "burners"; that is, they will write to CD format. At this moment in time, there are three categories of CD: CD-ROM (can only be "read," not recorded to), CD-RW (can be recorded to), and DVD-ROMs (digital video ROM). **Figure 8–20** shows a typical CD drive unit.
- **Modem/network card.** The means of linking a PC electronically to other computer systems by means of the telecommunications systems.
- **Ports.** A means of linking the PC to a power supply, peripherals, network lines, and the data connectors of a typical truck electronic system.

Peripherals

Peripherals describe the hardware components we attach to the system housing. Some of them are used to input information into the PC processing cycle, whereas others perform specific tasks such as printing out documents. Some peripherals are essential (such as a keyboard and mouse) whereas others are optional (such as printers and scanners).

- **Monitor.** A CRT (cathode ray tube) that displays information on a TV-like screen.
- **Mouse.** Moving a mouse on a flat pad positions a cursor (pointer) on the monitor. It contains a couple of command and function keys that make using computer systems much faster.
- **Keyboard.** The primary input device on a PC system. Consists of what appears to be a typewriter keyboard, but also has a number of command and function keys. These command and function keys may have different functions according to what software is being run.
- **Trackball or joystick.** Some PCs, especially portable ones described as notebook or laptop, use a trackball or joystick instead of a mouse. Used in conjunction with a couple of command and function keys, they do exactly what a mouse does.
- **Printer.** When on-paper documentation is required of onscreen or retained information, the printer is used to produce it. For instance, when using an "electronic service manual" from an OEM data hub, if the technician requires a printed copy of the procedure described electronically, the printer can generate it.
- **Scanner.** A device used to copy paper images or text into electronic format so it can be displayed, read, or electronically transferred using a PC system.
- **Modem.** This converts the digital signals of the computer processing cycle into the analog signals required to transmit them through the telecommunications system.

PC SOFTWARE

When any computer system is switched on, it goes through boot-up. During boot-up, the OS or operating system is transferred into the processing cycle of the computer. The OS is an essential set of commands that instruct the processor on how to handle the computer hardware and how to interpret program instructions as they come in from other software. All computers have an OS, and the system that is used by most of industry today is based on a specific Microsoft OS known as DOS. However, today DOS is almost always enhanced with a graphical user interface (GUI) that makes the computer much more user-friendly. The GUI used by most of industry today is also a Microsoft product: Windows. **Figure 8–21** shows the Windows GUI screen with its user-friendly icons.

Just about every PC used in a truck dealership, fleet, or service facility today will use one of several generations of Windows. They all do essentially the same thing: make the computer much easier to use and reduce the role of the keyboard in inputting information. Instead, icons represent commonly required commands and drop-down menus are used to expand on them. The mouse is used to view and select these options.

- **DOS.** Today, it is not necessary to know much about DOS because it loads automatically into the PC processing cycle when we boot the system up. On boot-up, we go through DOS to the GUI or Windows display.
- **Windows.** Several generations exist; Windows is the GUI used by most of industry today. This Microsoft product has competitors, but they are not used much by industry. After the boot-up procedure, we will see the Windows display on our screen. This uses icons to represent the software options already loaded onto the system hard drive. To make a selection, the mouse is used to move the cursor over the icon representing the desired software package, and the mouse is left-clicked. The left button of a mouse is a go-to command in Windows applications. The right button drops down menu options that vary according to the location of the cursor on the screen and the program.
- **Program software.** Program software are the instructions that get the computer processing cycle to process specific functions. For instance, if you wanted to type a letter and print it out, first you would need word processing software to turn the keypad into a typewriter. As keys are pressed on the keyboard, these are input into the processing cycle of the computer and displayed on the monitor as letters. The spelling of each word in the document can be checked by a function called spell-check. And when the user is happy with the appearance of the letter displayed onscreen, the system can be instructed to print it. The contents of the letter are digitally transferred to printer driver software, which then instructs the printer to print out a copy of the letter onto paper. Computers use hundreds of different types of program software to produce many different outcomes.
- **OEM software.** When a PC is used in conjunction with truck OEM software, it is generally loaded onto the hard drive of the computer, either from a CD or floppy diskettes. This means that after the PC has booted-up, an icon will be displayed on the Windows screen. Left-clicking on this icon will open the OEM software by loading it into the processing cycle of the computer. In most cases, a large number of choices will be available. Some of these choices may require that the PC be linked to the electronics on the vehicle by means of the ATA connector (J 1584/1708 or J 1939), the OEM data hub (by means of a modem and phone line handshake), or both. For instance, when performing a handshake connection between the vehicle electronics and the OEM data hub, in just about every case the chassis VIN and some kind of password are required to complete the connection. Most OEMs have taken a lot of trouble to make their systems easy to use, so do not get put off the first time you attempt to use this technology. Being familiar with Internet protocols helps: A hand-

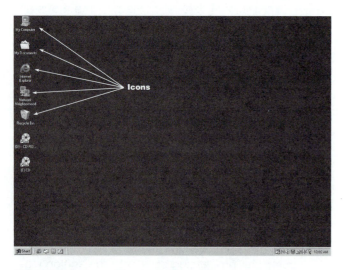

FIGURE 8–21 The Windows GUI (graphical user interface).

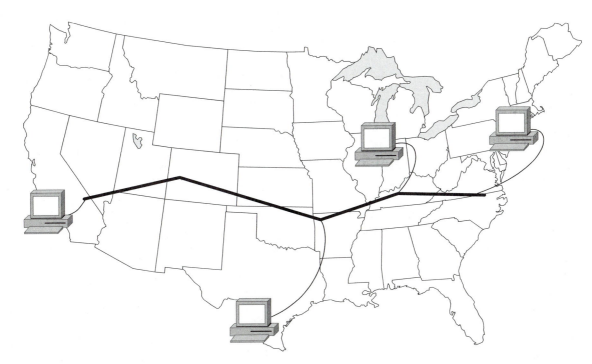

FIGURE 8–22   A network can be quite large and complex, connecting users all over the country.

shake connection with an OEM data hub that allows you to download data from it to a vehicle ECM is not much different from downloading a game file from an Internet site.

- **Networking and OEM LANs.** Almost all OEMs have their own limited access LANs or local area networks. These are usually driven by a data hub and are used to track such things as vehicle history, warranty claims, vehicle location, sales data, programming options, and so on. **Figure 8–22** shows the idea behind a network system linked by the telecommunications system.

8.7 | ELECTRONIC TROUBLESHOOTING

Electronically managed systems are capable of malfunctioning hydromechanically, so the technician must not assume that every problem is electronically based. The major OEMs produce excellent sequential problem-solving technical manuals. To use such manuals effectively, the technician should be capable of using a digital multimeter (DMM), have a basic understanding of electrical and electronic circuitry, and be capable of interpreting schematics. Sequential troubleshooting takes place in stages. It is critical that the instructions in each stage are precisely undertaken before proceeding to the next;

subsequent stages will generally be rendered meaningless if a stage is skipped. Some OEMs have PC-based sequential troubleshooting programs for their systems, which simplifies the sequencing of the troubleshooting stages and saves a lot of page-turning.

 **Shop Talk**

Diagnostic trees are often used by truck OEMs to troubleshoot malfunctions. The "root" of the tree is the problem. The "branches" are the various different paths that circuit testing will route the diagnostic technician. These tests are sometimes referred to as "leaves." Never skip tests or sections within a diagnostic tree. Some diagnostic trees are driven by OEM software and may have hundreds of steps.

It cannot be emphasized enough how important it is to use the correct tools. Generally, test light circuit testers should never be used, nor should analog test meters.

When the use of a breakout box is mandated, make sure that it is used. Also, it should be remembered that electrostatic discharge can damage microprocessor components, so it is good practice to wear a ground strap when opening up any housing containing a microprocessor. When testing separated Deutsch and Weatherpac type connectors with a DMM, ensure that the sockets have been identified before probing with the test leads. Use socket adaptors where necessary to avoid damaging terminal cavities. Never spike or strip back wires when testing circuits.

SNAPSHOT TESTS

Most systems will accommodate a **snapshot test** readout from the ECM to facilitate troubleshooting intermittent problems. Such problems often either do not generate codes, or codes are logged with no clear reason. Snapshot mode troubleshooting can be triggered by a variety of methods (codes, conditions, manually) and will record data frames before and after the trigger. These can be recorded to a reader/programmer instrument (ProLink) or a PC (laptop) while the vehicle is actually running. Because frames can be snapshot both before and after the trigger, it is possible to analyze a wide spectrum of data that may have contributed to a problem. Each data frame can be examined individually after the event. The portability of the ProLink unit lends itself to this type of testing.

8.8 SAE/ATA J 1587/ J 1708/J 1939 CODES AND PROTOCOLS

The following is a partial listing of SAE/ATA codes that have been adopted by all the North American truck engine/electronics OEMs. Generally, **SAE J 1584** covers common software protocols in electronic systems, **SAE J 1708** covers common hardware protocols, and the more recent **SAE J 1939** covers both hardware and software protocols. The acceptance and widespread usage of these protocols enables the interfacing of electronic systems manufactured by different OEMs on truck and bus chassis, plus provides any manufacturer's software to at least read other OEMs' electronic systems. The truck technician who works with multiple OEM systems may find it easier to work using these codes rather than use the proprietary codes.

The term **message identifier (MID)** is used to describe a major vehicle electronic system, usually with independent processing capability. For instance, the engine, transmission, ABS, and collision warning system are all major chassis systems and each is allocated its own MID. **Parameter identifiers (PIDs)** are usually primary subsystems common to all different OEM types covered by the MID. For instance, all electronically managed transmissions (MID: 130) are equipped with a tailshaft speed

sensor that is used to signal vehicle speed to all of the MIDs requiring vehicle speed data (engine, transmission, and so on). The tailshaft speed sensor is identified by PID 191 regardless of who manufactures the transmission. **Subsystem identifiers (SIDs)** are used to describe subsystems that fall within the major system (MID) and often include systems or components used specifically by one manufacturer. For instance, only one OEM transmission uses an autoshift low gear actuator, so this function is allocated a SID. Either a PID or a SID is displayed, not both.

Message Identifier (MID)	Description
128	Engine controller
130	Transmission
136	Brakes—antilock/traction control
137–139	Brakes—antilock, trailer #1, #2, #3
140	Instrument cluster
141	Trip recorder
142	Vehicle management system
143	Fuel system
162	Vehicle navigation
163	Vehicle security
165	Communication unit—ground
171	Driver information system
172	Service link
178	Vehicle sensors to data converter
179	Data logging unit
181	Communication unit—satellite
219	Collision warning unit (VORAD)

Parameter Identifier (PID)	Description
65	Service brake switch
68	Torque limiting factor
70	Parking brake switch
71	Idle shutdown timer status
74	Maximum road speed limit
81	Particulate trap inlet pressure
83	Road speed limit status
84	Road speed
85	Cruise control status
89	PTO status
91	Percent accelerator pedal position
92	Percent Engine Load
93	Output Torque
94	Fuel delivery pressure
98	Engine oil level
100	Engine oil pressure
101	Crankcase pressure
102	Manifold boost pressure
105	Intake manifold temperature
108	Barometric pressure

Parameter Identifier (PID)	Description
110	Engine coolant temperature
111	Coolant level
113	Governor droop
121	Engine retarder status
156	Injector timing rail pressure
157	Injector metering rail pressure
164	Injection control pressure
167	Charging voltage
168	Battery voltage
171	Ambient air temperature
173	Exhaust gas temperature
174	Fuel temperature
175	Engine oil temperature
182	Trip fuel
183	Fuel rate
184	Instantaneous MPG
185	Average MPG
190	Engine speed

The acronym SID is used to identify subsystems in an electronic circuit, especially those specific to one manufacturer's product or one type of system.

Subsystem Identifiers (SIDs) Common to all MIDs	Description
242	Cruise control resume switch
243	Cruise control set switch
244	Cruise control enable switch
245	Clutch pedal switch
248	Proprietary data link
250	SAE J 1708 (J 1587) 1939 data link

Subsystem Identifiers (SIDs) for MID 128, 143	Description
01–16	Injector Cyl #1 through #16
17	Fuel shutoff valve
18	Fuel control valve
19	Throttle bypass valve
20	Timing actuator
21	Engine position sensor
22	Timing sensor
23	Rack actuator
24	Rack position sensor
29	External fuel command input

Subsystem Identifiers (SIDs) for MID 130	Description
1–6	C1–C6 solenoid valves
7	Lockup solenoid valve
8	Forward solenoid valve
9	Low signal solenoid valve

Subsystem Identifiers (SIDs) for MID 130	Description
10	Retarder enable solenoid valve
11	Retarder modulation solenoid valve
12	Retarder response solenoid valve
13	Differential lockout solenoid valve
14	Engine/transmission match
15	Retarder modulation request sensor
16	Neutral start output
17	Turbine speed sensor
18	Primary shift selector
19	Secondary shift selector
20	Special function inputs
21–26	C1–C6 clutch pressure indicators
27	Lockup clutch pressure indicator
28	Forward range pressure indicator
29	Neutral range pressure indicator
30	Reverse range pressure indicator
31	Retarder response system pressure indicator
32	Differential lock clutch pressure indicator
33	Multiple pressure indicators

Subsystem Identifiers (SIDs) for MID 136–139	Description
1	Wheel sensor ABS axle 1 left
2	Wheel sensor ABS axle 1 right
3	Wheel sensor ABS axle 2 left
4	Wheel sensor ABS axle 2 right
5	Wheel sensor ABS axle 3 left
6	Wheel sensor ABS axle 3 right
7	Pressure modulation valve ABS axle 1 left
8	Pressure modulation valve ABS axle 1 right
9	Pressure modulation valve ABS axle 2 left
10	Pressure modulation valve ABS axle 2 right
11	Pressure modulation valve ABS axle 3 left
12	Pressure modulation valve ABS axle 3 right
13	Retarder control relay
14	Relay diagonal 1
15	Relay diagonal 2
16	Mode switch—ABS
17	Mode switch—traction control
18	DIF 1—traction control valve
19	DIF 2—traction control valve
22	Speed signal input
23	Warning light bulb
24	Traction control light bulb
25	Wheel sensor, ABS axle 1 average
26	Wheel sensor, ABS axle 2 average
27	Wheel sensor, ABS axle 3 average

Subsystem Identifiers (SIDs) for MID 136–139	Description
28	Pressure modulator, drive axle relay valve
29	Pressure transducer, drive axle relay valve
30	Master control relay

Subsystem Identifiers (SIDs) for MID 162	Description
1	Dead reckoning unit
2	Loran receiver
3	Global positioning system (GPS)
4	Integrated navigation unit

Failure mode indicators (FMIs) are indicated whenever an active or historic code is read using ProLink or a PC. The code actually displayed in the electronic service tool may be the OEM's code, but *all* North American truck electronics use FMIs so system failures can at least be read by their competitor's diagnostic software. FMIs help make life easier for the truck technician because any electronically detected failure must be categorized into one of the following:

Failure Mode Indicators (FMI)	Description
0	Data valid but above normal operating range
1	Data valid but below normal operating range
2	Data erratic, intermittent or incorrect
3	Voltage above normal or shorted high
4	Voltage below normal or shorted low
5	Current below normal or open circuit
6	Current above normal or grounded circuit
7	Mechanical system not responding properly
8	Abnormal frequency, pulse width, or period
9	Abnormal update rate
10	Abnormal rate of change
11	Failure mode not identifiable
12	Bad intelligent device or component
13	Out of calibration
14	Special instructions
15	Reserved for future assignment

Shop Talk

Common MID, SID, PID, and FMI codes are used by all the OEMs making trucks in North America.

Unlike OBD in automobiles, which covers only emissions-related components, the truck codes are fully comprehensive of all on-board electronic systems and allow one OEM's diagnostic instrument to at least read their competitor's system and diagnostic codes.

8.9 TROUBLESHOOTING VORAD

The Eaton VORAD collision warning system was studied in theory in Chapter 6. Now we can take a look at some simple troubleshooting of the system using both Eaton proprietary codes and the MIDs/SIDs/FMIs listed previously. As a collision warning system, the MID for VORAD is 219. This means that the fault codes can be displayed either directly from VORAD electronics and codes or by connecting to the ATA connector and using the SAE codes.

For any particular fault code, the count and duration can either be reset or cleared. To clear a code,

1. Select the fault code.
2. Position the display selection arrow on Clr Ct/Min display line.
3. Press Enter.

Displayed Fault	Code	SID	FMI	Description
RAM	11	254	12	CPU Internal scratch memory error
Flash	12	254	12	CPU Program memory error
FPGA	13	254	12	CPU gate array—will not program
DSP	14	254	12	CPU Digital signal processor—will not program
Battery	15	254	12	CPU internal battery low
NVRAM	16	254	12	CPU internal data memory corrupted
No DSP data	17	254	9	CPU processor not receiving antenna data
Frequency INJ	21	2	8	Antenna test signal bad
F/E data or CLK	22	2	2	Antenna serial link bad
Microwave	23	1	2	Antenna radar not transmitting
Brake	31	3	5	Brake input not corrected
Speaker	41	4	5	Speaker not connected

Displayed Fault	Code	SID	FMI	Description
	42			NO FAULTS FOUND
R. Steering	51	5	5	Right steering sensor not connected
L. Steering	52	5	5	Left steering sensor not connected
STR misaligned	53	3	13	Steering column sensor misaligned
Speedometer	61	6	5	Speedometer monitor
Right signal	71	7	5	Right turn signal input not connected
Right BSS	72	10	5	Right side sensor not connected
Left signal	73	8	5	Left turn signal input not connected
Left BSS	74	11	5	Left side sensor not connected
Volume	91	9	5	DDU Volume control bad
Range	92	9	5	DDU Range control bad
DRVR Disp	92	9	5	DDU not connected

VORAD OPERATING PARAMETERS

VORAD operates at a frequency of 24.725 GHz, and can "see" through fog, snow, smoke, and heavy rain. The system uses low power microwaves and is designed to not interfere with other CWS systems, police radar, or other computer management systems. It can detect objects as small as a motorcycle but not animals or people.

The cab display has 3 lights: red, orange, and yellow. The first warning light is the yellow light, which illuminates when a vehicle is within 3 seconds of the current speed of the host vehicle. The second warning light is orange, which is illuminated along with the yellow light when a vehicle is within 2 seconds of the host vehicle at the current speed. The red light is illuminated along with the orange and yellow when a vehicle is detected to be in a 1 second range. When the warning lights are illuminated, a distinct tone is delivered. There are five distinct tones, one for each of the following:

- 2 second alert
- 1 second alert
- Slow-moving vehicle alert
- Stationary vehicle alert
- Blind spot alert

The tones sound only once per incident, whereas the warning lights should stay illuminated throughout the incident. The idea is to alert the driver, not to annoy.

8.10 ELECTRICAL WIRING, CONNECTOR, AND TERMINAL REPAIR

Most of the wiring and connectors used on electronically managed chassis systems in North America are produced by a couple of manufacturers, so although the procedures outlined in this section are specific to one OEM, they are representative of those required for all the major OEMs.

METRI-PACK CONNECTORS

The following sections sequence the disassembly, repair, and reassembly procedure required for the **Metri-Pack** 150 **connectors** used in DDEC electronic circuits (Courtesy of DDC). This procedure is similar to that required for the assembly and repair of most electronic wiring and connectors.

Installation of Metri-Pack 150 Connectors

Metri-Pack 150 connectors are the "pull-to-seat" design. The cable is pushed through the seal and correct cavity of the connector before crimping the terminal to the cable. It should be stripped of insulation *after* it is placed through the seal and connector body. Use the following instructions for terminal installation:

1. Position the cable through the seal and correct cavity of the connector. See **Figure 8–23**.

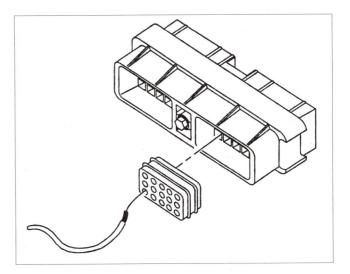

FIGURE 8–23 Inserting wire in connector.

2. Strip the end of the cable using wire strippers to leave 5.0 ± 0.5 mm (0.2 ± 0.02 in.) of bare conductor.

3. Squeeze the handles of the crimping tool together firmly to cause the jaws to automatically open.

4. Hold the "wire side" facing you.

5. Push the terminal holder to the open position and insert the terminal until the wire attaching portion of the terminal rests on the 20-22 anvil. Be sure the wire core wings and the insulation wings of the terminal are pointing toward the upper jaw of the crimping tool. See **Figure 8–24**.

6. Insert the cable into the terminal until the stripped portion is positioned in the wire core wings, and the insulation portion ends just forward of the insulation wings. See **Figure 8–25**.

7. Compress the handles of the crimping tool until the ratchet automatically releases and the crimp is complete.
Note: For faster, more efficient crimping operation, a bracket or bench rest may be used to cradle one handle of the tool. The operator can apply the terminals by grasping and actuating only one handle of the tool. See **Figure 8–26**.

8. Release the crimping tool with the lock lever located between the handles, in case of jamming.

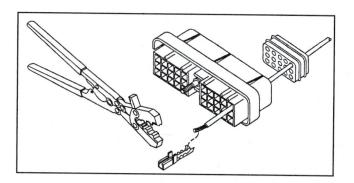

FIGURE 8–24 Terminal and crimping tool position.

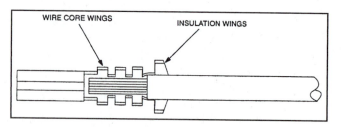

FIGURE 8–25 Cable to terminal alignment.

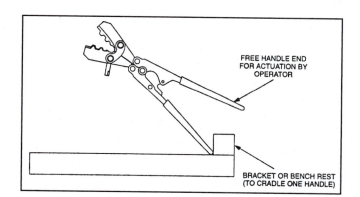

FIGURE 8–26 Crimping tool operation.

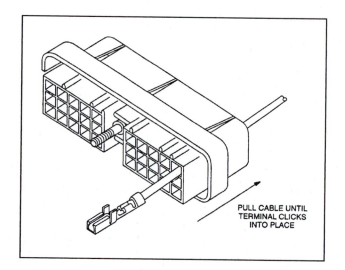

FIGURE 8–27 Pulling the terminal to seat.

9. Align the **locking tang** of the terminal with the lettered side of the connector.

10. Pull the cable back through the connector until a click is heard. See **Figure 8–27**. Position the seal into the connector.
Note: For ECM 30-pin connectors, put locking tang opposite lettered side.

Removal and Repair

A tang on the terminal locks into a tab molded into the plastic connector to retain the cable assembly. Remove Metri-Pack 150 terminals using the following instructions.

1. Insert the **tang release tool** into the cavity of the connector, placing the tip of the tool between the locking tang of the terminal and the wall of the cavity. See **Figure 8–28**.

2. Depress the tang of the terminal to release it from the connector.

3. Push the cable forward through the terminal until the complete crimp is exposed.

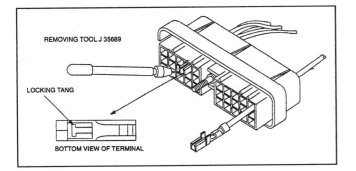

FIGURE 8–28   Terminal removal.

4. Cut the cable immediately behind the damaged terminal to repair it.
5. Follow the installation instructions for crimping the terminal and inserting it into the connector.

Installation of Metri-Pack 280 Connectors

Use the following instructions for terminal installation:

1. Insert the terminal into the locating hole of the crimping tool using the proper hole according to the gauge of the cable to be used. See **Figure 8–29**.
2. Insert the cable into the terminal until the stripped portion is positioned in the cable core wings, and the seal and insulated portion of the cable are in the insulation wings. See **Figure 8–30**.
3. Compress the handles of the crimping tool until the ratchet automatically releases and the crimp is complete. A properly crimped terminal is shown in **Figure 8–30**.

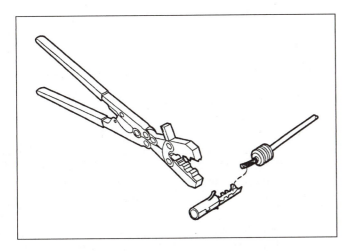

FIGURE 8–29   Terminal position.

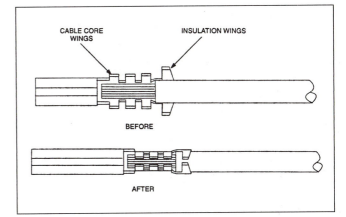

FIGURE 8–30   Cable and terminal position before and after crimping.

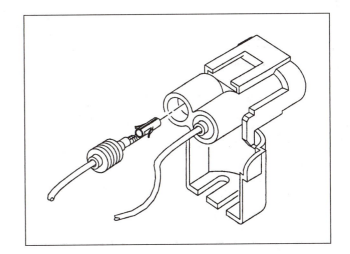

FIGURE 8–31   Inserting terminal in connector.

4. Release the crimping tool with the lock lever located between the handles, in case of jamming.
5. Push the crimped terminal into the connector until it clicks into place. Gently tug on the cable to make sure it is secure. See **Figure 8–31**.

Removal and Repair

Two locking tangs are used on the terminals to secure them to the connector body. Use the following instructions for removing terminals from the connector body.

1. Disengage the locking tang, securing the connector bodies to each other. Grasp one half of the connector in each hand and gently pull apart.
2. Unlatch and open the secondary lock on the connector. See **Figure 8–32**.

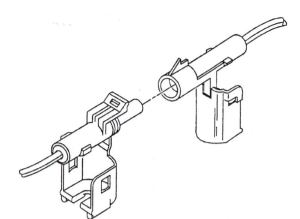

FIGURE 8–32   Unlatched secondary lock.

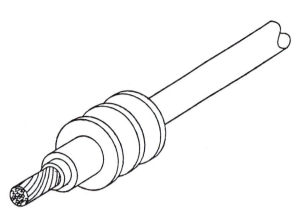

FIGURE 8–34   Proper cable seal position.

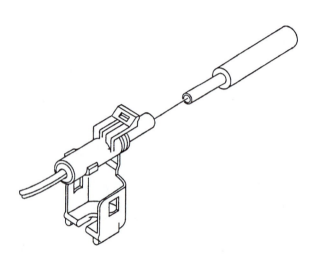

FIGURE 8–33   Removal tool procedure.

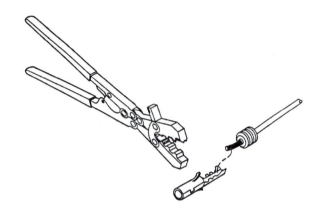

FIGURE 8–35   Crimping procedure.

DEUTSCH CONNECTORS

At least one Deutsch connector is used by all the OEMs because the industry standard data link, the ATA connector, is always a Deutsch connector. Deutsch connectors have cable seals that are integrally molded into the connector. They are push-to-seat connectors with cylindrical terminals. The diagnostic terminal connectors are gold plated.

Installation of Deutsch Connectors

Use the following instructions for installation:

3. Grasp the cable to be removed and push the terminal to the forward position.
4. Insert the tang release tool straight into the front of the connector cavity until it rests on the cavity shoulder.
5. Grasp the cable and push it forward through the connector cavity into the tool while holding the tool securely in place. See **Figure 8–33**.
6. The tool will press the locking tangs on the terminal. Pull the cable rearward (back through the connector). Remove the tool from the connector cavity.
7. Cut the wire immediately behind the cable seat and slip the new cable seal onto the wire.
8. Strip the end of the cable strippers to leave 5.0 ± 0.5 mm (0.2 ± 0.02 in.) of bare conductor. Position the cable seal as shown. See **Figure 8–34**.
9. Crimp the new terminal onto the wire using the crimp tool. See **Figure 8–35**.

1. Strip approximately 1/4 inch (6 mm) of insulation from the cable.
2. Remove the lock clip, raise the wire gauge selector, and rotate the knob to the number matching the gauge wire that is being used.
3. Lower the selector and insert the lock clip.
4. Position the contact so that the crimp barrel is 1/32 inch above the four indenters. See **Figure 8–36**. Crimp the cable.
5. Grasp the contact approximately one inch behind the contact crimp barrel.

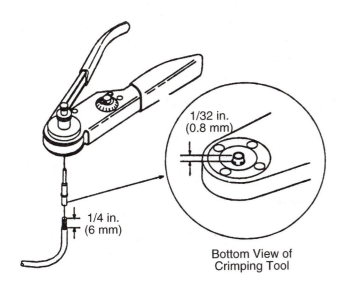

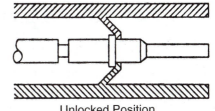

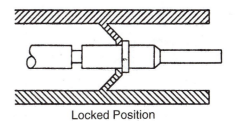

FIGURE 8–36 *Setting wire gauge selector and positioning the contact.*

FIGURE 8–38 *Locking terminal into connector.*

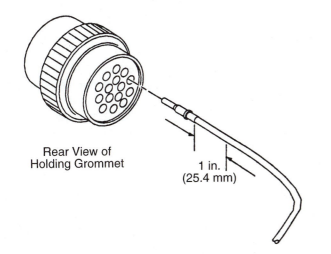

FIGURE 8–37 *Pushing contact into grommet.*

FIGURE 8–39 *Removal tool position.*

6. Hold the connector with the rear grommet facing you. See **Figure 8–37**.
7. Push the contact into the grommet until a positive stop is felt. See **Figure 8–38**. A slight tug will confirm that it is properly locked into place.

Removal

The appropriate size removal tool should be used when removing cables from connectors.

1. With the rear insert toward you, snap the appropriate size remover tool over the cable of contact to be removed. See **Figure 8–39**.
2. Slide the tool along the cable into the insert cavity until it engages and resistance is felt. Do not twist or insert tool at an angle. See **Figure 8–40**.

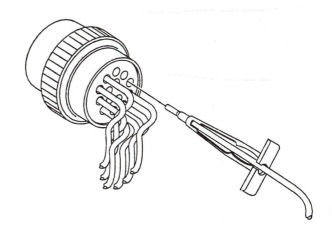

FIGURE 8–40 *Removal tool insertion.*

3. Pull contact cable assembly out of the connector. Keep reverse tension on the cable and forward tension on the tool.

8.11 SPLICING GUIDELINES

Not all wiring in electronic circuits can be spliced, so in all cases, the OEM service literature should be consulted before attempting such a procedure. The following may be used as a general guideline. The objective is to produce a high-quality, tight splice with durable insulation that should outlast the life of the vehicle (**Figure 8–41** and **Figure 8–42**). The selection of crimping tools and splice connectors will depend on the exact repair being performed.

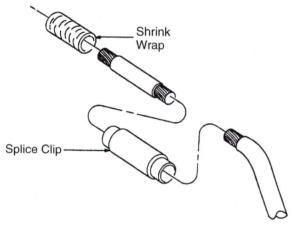

FIGURE 8–41   Spliced wire.

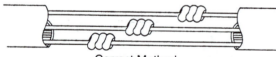

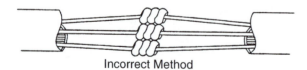

FIGURE 8–42   Multiple splices.

THREE-WIRE SPLICE

Three-way splice connectors are commercially available to accommodate three-wire splices. The technique is the same as a single **butt splice** connector. See **Figure 8–43**.

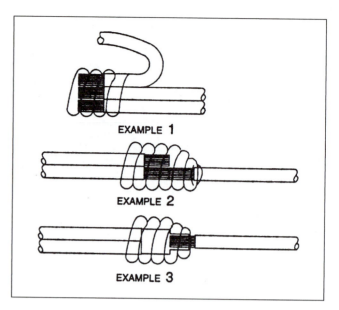

FIGURE 8–43   Three-way splice.

CIRCUIT SYMBOLS AND DIAGRAMS

The following circuit symbols are typical in electronic and chassis electrical wiring schematics (**Figure 8–44**). The technician should be able to easily identify most of the symbols displayed here because they are so commonly used. They do vary slightly depending on the OEM.

SUMMARY

- The ESTs used to service, diagnose, and reprogram truck engine management systems are onboard diagnostic lights, DMMs, scanners, generic reader/programmers, proprietary reader/programmers, and PCs.
- Flash codes are an onboard method of accessing diagnostic codes. Most systems will display active codes only. Some will display active and historic (inactive) codes.
- ProLink 9000 with the appropriate OEM software cartridge has become the industry-standard portable shop floor diagnostic and customer data programming EST.
- Most OEMs are either currently using or plan to begin using the PC and proprietary software as their primary diagnostic and programming EST.
- ESTs designed to connect with the vehicle ECM(s) do so via the SAE/ATA J 1708/1939 connector.
- Most electronic circuit testing requires the use of a DMM.
- A continuity test is a quick resistance test that distinguishes between an open and a closed circuit.

Circuit Diagram Symbols

Many circuit diagram components are represented by a symbol or icon. The most commonly used symbols are identified below. Note that switches are always shown in the "Key Off" position. Also, the twisted pair symbol represents two wires or cables that have been twisted together to prevent electrical interference.

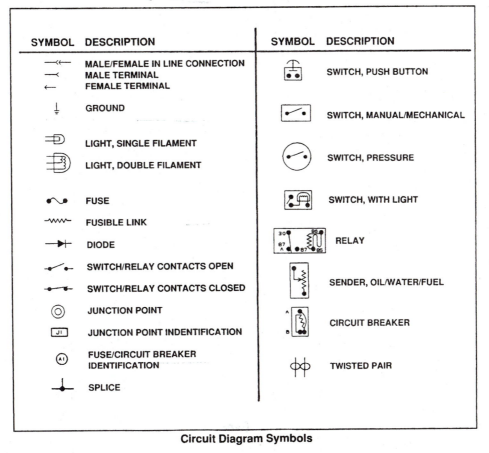

SYMBOL	DESCRIPTION
	MALE/FEMALE IN LINE CONNECTION
	MALE TERMINAL
	FEMALE TERMINAL
	GROUND
	LIGHT, SINGLE FILAMENT
	LIGHT, DOUBLE FILAMENT
	FUSE
	FUSIBLE LINK
	DIODE
	SWITCH/RELAY CONTACTS OPEN
	SWITCH/RELAY CONTACTS CLOSED
	JUNCTION POINT
	JUNCTION POINT INDENTIFICATION
	FUSE/CIRCUIT BREAKER IDENTIFICATION
	SPLICE

SYMBOL	DESCRIPTION
	SWITCH, PUSH BUTTON
	SWITCH, MANUAL/MECHANICAL
	SWITCH, PRESSURE
	SWITCH, WITH LIGHT
	RELAY
	SENDER, OIL/WATER/FUEL
	CIRCUIT BREAKER
	TWISTED PAIR

Circuit Diagram Symbols

FIGURE 8–44   Electrical circuit symbols. (Courtesy of International Truck and Engine Corp.)

- A dark band identifies the cathode on a diode.
- Circuit resistance and voltage are measured with the test leads positioned parallel to the circuit.
- Direct measurement of current flow is performed with the test leads located in series with the circuit.
- A Hall effect probe can be used to approximate high current flow through a DC circuit.
- The ProLink 9000 EST is used to access active and historic codes, read system identification data, perform diagnostic testing of electronic sub-components, reprogram customer data, act as a serial link for mainframe linkage, and perform snapshot data analysis.
- A ProLink OEM software cartridge is updated by replacing the PROM chips.
- A snapshot test is performed to analyze multiple data frames before and after a trigger, usually a fault code or manually keyed.
- SAE J 1584 and J 1939 codes and protocols numerically code all onboard electronic sys-

tems, parameters, and failure modes.
- System parameter (PID) failures are identified by one of 14 FMIs, making circuit diagnosis easy.
- Sealed electronic circuit connectors must be assembled using the correct OEM tooling and components.

REVIEW QUESTIONS

1. Which EST is used to read flash codes?
   a. dash diagnostic lights
   b. generic reader programmer
   c. diagnostic fork
   d. PC

2. The appropriate EST for performing a resistance test on a potentiometer-type TPS isolated from its circuit is a

a. DMM
b. scanner
c. ProLink 9000
d. PC

3. Which of the following is the correct means of electronically coupling a ProLink 9000 with a truck chassis ECM?
   a. modem
   b. ATA connector
   c. any Deutsch connector
   d. parallel link connector

4. When resistance and continuity tests are made on electronic circuit components, the circuit being tested should be
   a. open
   b. closed
   c. energized

5. DMM test lead resistance should never exceed
   a. 0.2 ohms
   b. 1.0 ohms
   c. 10 ohms per inch
   d. 100 ohms per inch

6. The output (signal) of a pulse generator rotational speed sensor is measured in
   a. V-DC
   b. ohms
   c. V-AC
   d. amperes

7. Which of the following cannot usually be performed by a ProLink EST?
   a. erase historic fault codes
   b. customer data programming
   c. snapshot testing
   d. proprietary data programming

8. Which of the following procedures cannot be performed using a PC and the OEM software?
   a. erase historic fault codes
   b. customer data programming
   c. erase active fault codes
   d. EUI cutout tests

9. Which of the following methods is the economic method of upgrading OEM ProLink software?
   a. replace the OEM software cartridge
   b. upgrade the cartridge PROM chips or cards
   c. reprogram the software cartridge from a mainframe
   d. replace the RAM chip

10. When reading an open TPS circuit, which FMI should be indicated?
    a. 0
    b. 3
    c. 5
    d. 9

11. How many counts of resolution can be displayed by a $4\frac{1}{2}$-digit DMM?
    a. 1,999
    b. 19,999
    c. 9,995
    d. 99,995

12. The specification that indicates how fine a measurement can be made by a DMM is known as
    a. root mean square (rms)
    b. percentage deviation
    c. resolution
    d. hysteresis

13. When a DMM is set to read AC, which is the term used to describe the averaging of potential difference to produce a reading?
    a. root mean square (rms)
    b. percentage averaging
    c. resolution
    d. saw toothing

14. Which port on a ProLink 9000 is used to connect the printer?
    a. ATA data connector
    b. SAE J 1939 connector
    c. Parallel output port
    d. RS-232 port

15. Which of the following components are inserted into terminal sockets to enable testing by DMM probes and minimize physical damage?
    a. cycling breakers
    b. noncycling breakers
    c. terminal dummies
    d. diagnostic forks

16. Which electrical voltage value does a DMM output when in diode test mode?
    a. 0.3V
    b. 0.6V
    c. 0.9V
    d. 1.1V

17. Technician A states that the accuracy of an analog meter is usually no better than 3%. Technician B states that most DMMs have an accuracy factor that is within 0.1% of the reading. Who is correct?
    a. Technician A
    b. Technician B
    c. both Technicians A and B
    d. neither Technician A nor B

18. The component that can be safely inserted into electronic circuits so they can be tested without interrupting them is called a
    a. test light
    b. diagnostic fork
    c. breakout box
    d. DMM test probe

19. Technician A states that using the snapshot test mode when using ProLink to troubleshoot an intermittently occurring fault code will help identify the conditions that produced the code. Technician B states that snapshot test mode can only analyze a historical code. Who is correct?
    a. Technician A
    b. Technician B
    c. Both Technicians A and B
    d. Neither Technician A nor B

20. When verifying the *signal* produced by a potentiometer type TPS, the DMM test mode selected should be
    a. V-AC
    b. V-DC
    c. diode test
    d. resistance

21. Describe the difference between an active and a historic code.

22. What is meant by the term GUI? Give some examples.

23. What is meant by the term networking?

24. Which of the following PC components would be usually described as a *peripheral*?
    a. system housing
    b. hard disk drive
    c. CPU
    d. printer

25. If a PC is to be networked to the Internet or an OEM data-hub, which of the following hardware devices is required?
    a. a digital camera
    b. speakers
    c. a modem
    d. a CD drive

Prerequisite
Chapters 2, 3

Clutches

Objectives

After reading this chapter, you should be able to

- **Understand the operation of a clutch.**
- **List the components of a clutch.**
- **Explain the differences in construction and operation between a pull type clutch and a push type clutch.**
- **Describe how both manual and self-adjusting clutches are adjusted.**
- **Explain how to adjust clutch linkage.**
- **Describe the function of a clutch brake.**
- **Troubleshoot a clutch for wear and damage.**
- **Give possible causes for specific clutch defects and explain how to correct the condition causing a defect.**
- **Outline the procedure for removing and replacing a clutch.**

Key Terms

adapter ring
adjusting ring
antirattle springs
clutch
clutch brake
dampened discs
lockstrap
National Institute for Occupational Safety and Health (NIOSH)

pull type clutch
push type clutch
quick release valve
release bearing
rigid discs
self-adjusting clutch
torque limiting clutch brake
total pedal travel
two-piece clutch brake
wear compensator

| 9.1 | ## CLUTCH FUNCTION |

The function of a **clutch** is to transfer the power of the truck engine from the flywheel to the transmission. At the point at which clutch engagement begins, the transmission input shaft may either be stationary, as when the truck is at a standstill, or rotating at a different speed than the flywheel, as in the case of upshifting or downshifting. Once the clutch is fully engaged, however, the flywheel and the transmission input shaft will rotate at the same speed.

The transmission of torque through the clutch is accomplished by gradually bringing one or more rotating drive members connected to the engine flywheel into contact with one or more driven members connected to the transmission input shaft. Contact between the driving and driven members is established and maintained by both spring pressure and friction surfaces. The pressure exerted by the springs on the driven members is relieved by the driver through the clutch pedal and linkage when the pedal is depressed.

To perform its job, a clutch is equipped with one or two discs that have friction surfaces known as facings (**Figure 9–1**). As the clutch pedal in the cab is depressed, the clutch's pressure plate is drawn away from the flywheel, compressing the springs and freeing the disc(s) from contact with the flywheel friction surface. At this point, the clutch is disengaged and the transfer of torque from the engine to the transmission is interrupted.

As the clutch pedal is released, the pressure plate moves toward the flywheel, allowing the springs to clamp the disc(s) between the flywheel and the pressure plate. The discs are designed to slip for a short period of time as they gradually accept the torque from the flywheel. This minimizes torsional or twisting shock to the drivetrain components. As pressure increases, the discs accept the full torque from the flywheel. At this point, engagement is complete, and engine torque is transferred to the transmission. Once engaged, the clutch must continue to transmit all the engine torque to the transmission without slippage.

9.2 BASIC COMPONENTS

The basic parts of a clutch are illustrated in **Figure 9–2**. Clutch components can be grouped into two basic categories: driving and driven members.

DRIVING MEMBERS

The driving members of a clutch are the cover assembly, pressure plate, and, if the clutch has two driven discs, an intermediate plate. Although it is not part of the clutch, the engine flywheel is considered a driving member of the clutch, and should be inspected and serviced whenever work is performed on the clutch.

Flywheel

The flywheel may either be flat or countersunk in what is known as a "pot type" design (**Figure 9–3**). For the clutch to work properly, the flywheel must be perpendicular to the crankshaft with little allowable runout (as little as 0.005 inch). The surface of the flywheel that contacts the driven discs is machined smooth and dry of lube oil or grease.

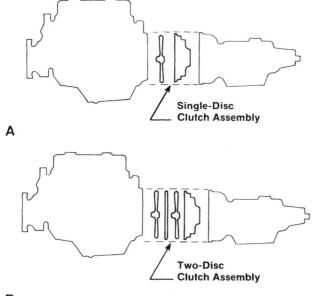

A

B

FIGURE 9–1   The clutch uses (A) one; or (B) two friction discs to couple the engine to the transmission. (Courtesy of Eaton Corp.—Eaton Clutch Div.)

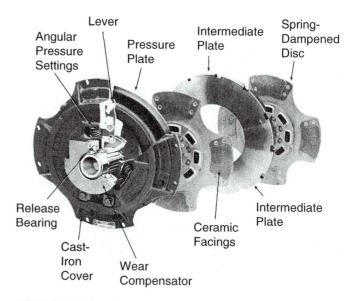

FIGURE 9–2   An angle spring clutch with two spring dampened discs. (Courtesy of Eaton Corp.— Eaton Clutch Div.)

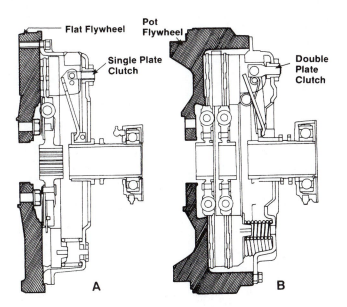

FIGURE 9–3 (A) Some clutches are installed on flat-faced flywheels and (B) others are installed in pot flywheels. (Courtesy of Haldex Brake Systems)

Clutch Cover

The clutch cover assembly is constructed of either cast iron or stamped steel (**Figure 9–4**). Cast iron covers offer maximum ventilation and heat dissipation and are normally used on clutches specified in heavy-duty Class 8 trucks. Stamped steel covers are more commonly used on Class 6 and 7 trucks.

The clutch cover is bolted to the flywheel and rotates at the same speed. The clutch cover contains the pressure plate, which is fitted to the cover with pressure springs. Most Class 8 truck clutch covers also contain the levers that move the pressure plate back and forth, thereby making and breaking contact with the disc assemblies. The **release bearings** can be either sealed or greaseable bearings.

Pressure Plate

The pressure plate is machined smooth on the side facing the driven disc (or discs). The pressure plate mounts on pins or lugs on the clutch cover and is free to slide back and forth on these pins or lugs.

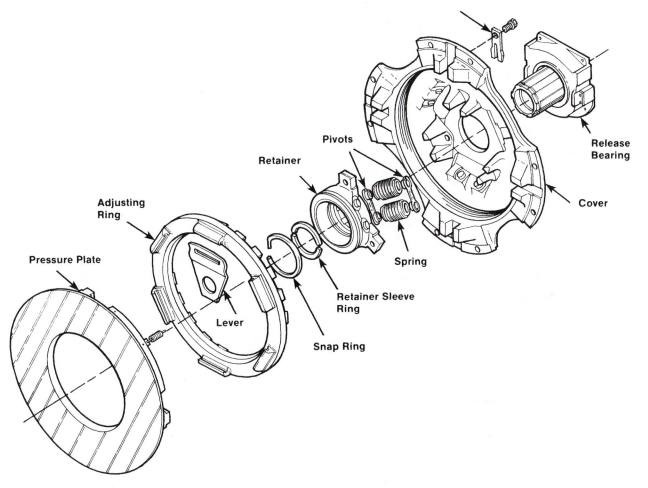

FIGURE 9–4 Exploded view of an angle spring clutch assembly. (Courtesy of Eaton Corp.—Eaton Clutch Div.)

When spring pressure is applied to the pressure plate, it clamps the driven friction discs between it and the flywheel.

Pressure Springs and Levers

Pressure springs are located between the clutch cover and the pressure plate. Both coil spring and diaphragm spring designs are used.

Clutches with Coil Springs. Most clutch designs use multiple coil springs to force the plate against the driven discs. In some clutches, coil springs are positioned perpendicular to the pressure plate and are equally spaced around the perimeter of the cover (**Figure 9–5**).

Other clutches use fewer coil springs and angle them between the cover and a retainer (**Figure 9–6**). Angle spring designs require 50 percent less clutch pedal effort. They also provide a constant plate load regardless of the clutch's age.

The angle spring clutch illustrated in **Figure 9–6** operates under indirect pressure. Three pairs of coil springs are located away from the pressure plate rather than directly on it. The spring load is applied through a series of six levers. In this particular clutch design, the plate load and release bearing load are not directly proportional to the spring load.

The engaged position of a new (nonworn) angle spring clutch is shown in **Figure 9–7A**. Pressure load is the result of the axial load of the springs multiplied by the lever ratio. The pressure springs are positioned at an angle to the center line of the clutch with their ends attached to the flywheel ring (cover) and the release sleeve retainer. As the release sleeve retainer, including the release sleeve and the release

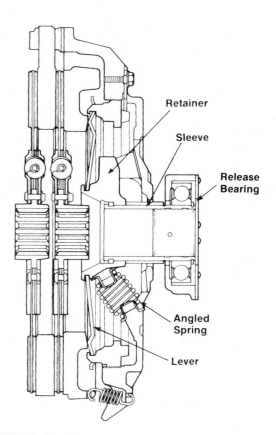

FIGURE 9–6 The coil springs in this clutch are angled between the cover and a retainer. Levers multiply the spring force against the pressure plate. (Courtesy of Eaton Corp.—Eaton Clutch Div.)

bearing, is moved toward the flywheel, the springs pivot freely without bending or buckling. Connected to the retainer are levers that, when forced forward by the retainer, multiply the force of the springs. Pivot points on the levers press the pressure plate against the driven discs.

When the clutch pedal is released (**Figure 9–7B**), spring load increases but axial load decreases, resulting in reduced pedal effort. The plate load is obtained from the axial component of the spring force multiplied by the lever ratio. The axial component of the spring force changes with release bearing movement, but not in direct proportion to the spring load.

When facing wear (**Figure 9–7C**), spring load reduces but axial load remains constant. As the clutch wears, the release sleeve assembly moves toward the flywheel, and the pressure springs elongate and lose load. The axial component of the reducing spring force, however, remains essentially constant. This means the pressure plate force remains constant throughout the life of the clutch. When the clutch is adjusted to compensate for friction facing wear, the pressure plate position does not move, but the rotating **adjusting ring** moves the

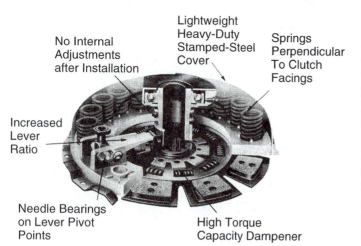

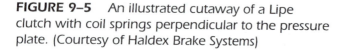

FIGURE 9–5 An illustrated cutaway of a Lipe clutch with coil springs perpendicular to the pressure plate. (Courtesy of Haldex Brake Systems)

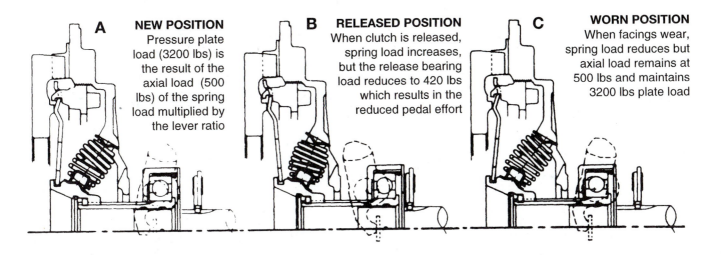

FIGURE 9–7 Clutch operation: (A) in the new position, pressure plate load (3200 lbs.) is the result of the axial load (500 lbs.) of the spring load multiplied by the lever ratio; (B) when the clutch is released, spring load increases, but the release bearing load reduces to 420 lbs., which results in the reduced pedal effort; (C) when facings wear, spring load reduces, but axial load remains at 500 lbs. and maintains 3200 lbs. plate load. (Courtesy of Eaton Corp.—Eaton Clutch Div.)

levers toward the transmission, pushing the release sleeve and bearing assembly in that direction also. This reestablishes the internal spring position to the original setting for continued clutch use.

Clutches with Diaphragm Springs. Clutches equipped with a diaphragm or Belleville spring assembly are often used in class 6 or 7, medium-duty trucks. Like coil spring-equipped clutches, diaphragm and Belleville spring clutches operate under indirect pressure using a retainer and lever arrangement to exert pressure on the pressure plate. The levers may also be referred to as *fingers* or *tapered fingers.* As with angle spring clutches, the result is low release bearing load and constant pressure plate load. Pressure plate load on the friction material's surface varies by the thickness of the diaphragm spring. This type of clutch employs either a single disc or multiple discs with the addition of an intermediate plate and the appropriate type of flywheel. It can also be designed either as a push type or pull type clutch. The actual design used is determined by the clutch linkage/vehicle design, space requirements inside the vehicle, and the torque load required from the clutch assembly.

In the diaphragm clutch assembly shown in **Figure 9–8**, the clutch cover assembly is bolted directly to the engine flywheel. The clutch cover assembly drives the pressure plate through three drive straps.

In a push type clutch, pressing the clutch pedal moves the release bearing towards the engine flywheel. As the clutch's diaphragm spring fingers are

FIGURE 9–8 Components of a diaphragm spring clutch. (Courtesy of Eaton Corp.—Eaton Clutch Div.)

depressed, the pressure plate retracts from contact with the driven disc(s) and the clutch is disengaged

When the clutch pedal is released, the release bearing and clutch diaphragm spring fingers move away from the engine flywheel. The diaphragm Belleville spring exerts pressure through its levers to the pressure plate. This results in the driven disc(s) being *locked up* between the friction surfaces of the pressure plate and the engine flywheel in the single disc type.

Torque flow through a multiple disc assembly is the same as with a coil spring or angle spring type.

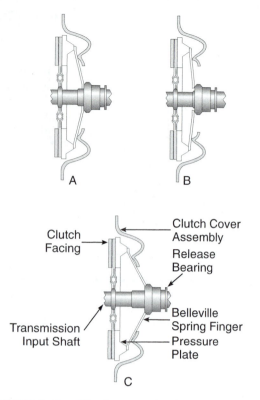

FIGURE 9–9 Diaphragm clutch operation: (A) new, engaged position; (B) released position; (C) worn, engaged position.

The pressure plate diaphragm Belleville spring exerts pressure through the rear disc to the intermediate plate and to the forward disc onto the flywheel friction face. This locks the clutch assembly together.

On a conventional direct pressure (D.P.) clutch, as the friction facings wear, pressure plate load loss occurs as a result of spring elongation. The diaphragm clutch shown in **Figure 9–8** uses a Belleville spring design that maintains the same plate load at the new and fully worn positions (**Figure 9–9A & C**).

Intermediate Plate

If the clutch has two driven discs, an intermediate, or center, plate will separate the discs. The plate is machined smooth on both sides since it is pressed between two friction surfaces. An intermediate plate increases the torque capacity of the clutch by increasing the friction area, giving more area for the transfer of torque.

Some intermediate plates have drive slots machined in their outer edge. These slots fit over and are driven by hardened steel drive pins press fit into holes in the flywheel rim. (See **Figure 9–2.**) Other intermediate plates have four or more tabs that fit into and are driven by slots in the clutch cover or **adapter ring** (**Figure 9–10**).

Some clutches with heavy-duty intermediate plates require **antirattle springs** (**Figure 9–11**). These reduce wear between the intermediate plate

FIGURE 9–10 This two-plate clutch uses an adapter ring mounted between the clutch cover and the flywheel to provide the needed depth for the intermediate plate and second clutch disc. (Courtesy of Haldex Brake Systems)

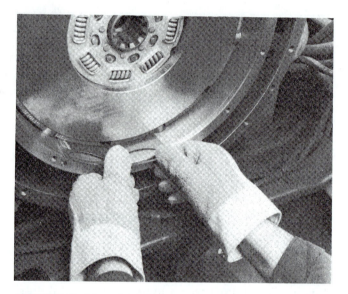

FIGURE 9–11 Antirattle springs placed between the intermediate plate and the flywheel reduce wear on the drive pins. (Courtesy of Eaton Corp.—Eaton Clutch Div.)

and the drive pins, and improve clutch release. Without the springs, the drive slots in the plate would wear excessively, resulting in poor clutch release. Three or four antirattle springs are positioned between the edge of the plate and the inside wall of the "pot type" flywheel. These are spaced equal distances apart (three springs—120 degrees apart; four springs—90 degrees apart).

Adapter Ring

Some two plate clutches use an adapter ring when the clutch is installed on a flat flywheel. The adapter ring is bolted between the clutch cover and the flywheel. It is sized to provide the needed depth to accommodate the second clutch disc and the intermediate plate. An adapter ring can be seen in the illustration in **Figure 9–10**.

Clutch Adjustment Mechanisms

As the friction linings wear, some method must be provided to adjust the clutch and compensate for friction material wear. Both manually adjusted and self-adjusting clutches are used.

Manually Adjusted Clutches. These clutches have a manual adjusting ring that permits the clutch to be adjusted to compensate for friction facing wear. The ring is positioned behind the pressure plate and is threaded into the clutch cover. A **lockstrap** or lock plate secures the ring so that it cannot move. The levers are seated in the ring. When the lockstrap is removed, the adjusting ring is rotated in the cover so that it moves toward the engine. This forces the pivot points of the levers to advance, pushing the pressure plate forward and compensating for wear in the linings. A manual adjusting ring is shown in **Figure 9–4.**

Self-Adjusting Clutches. Self-adjusting clutches automatically take up the slack between the pressure plate and clutch disc as wear occurs. The adjusting ring has teeth that mesh with a worm gear in a **wear compensator (Figure 9–12).** The wear compensator is mounted in the clutch cover and has an actuator arm that fits into a hole in the release sleeve retainer **(Figure 9–13).** As the retainer moves forward each time the clutch is engaged, the actuator arm rotates the worm gear in the wear compensator. The rotation of the worm gear is transferred to the adjusting ring in the clutch cover, taking up slack between the pressure plate and the driven disc.

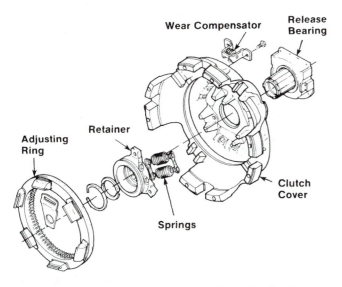

FIGURE 9–12 Exploded view of a self-adjusting, angle spring clutch cover assembly. (Courtesy of Eaton Corp.—Eaton Clutch Div.)

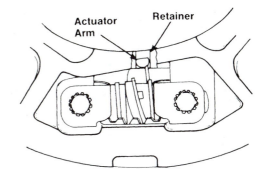

FIGURE 9–13 The actuator arm of the wear compensator is installed in a slot in the sleeve retainer. (Courtesy of Eaton Corp.—Eaton Clutch Div.)

DRIVEN MEMBERS

The driven components of a clutch consists of the clutch discs. Clutch discs are lined on both sides with friction material and have an internally splined hub. The hub fits over the splined end of the transmission input shaft. When the discs are clamped between the pressure plate and the flywheel, torque is transmitted through the discs to the splined input shaft, rotating the shaft at engine speed.

Disc Design

There are two basic disc designs: rigid and dampened **(Figure 9–14).** **Rigid discs** are steel plates to which friction linings, or facings, are bonded or riveted. Rigid discs are most often used in two-plate clutches where the torque loads can be distributed over a greater surface area. A rigid disc cannot absorb torsional shock loads and its misapplication can result in damage to the transmission, drive shaft, and axle. Torsional vibration can also cause a rigid disc to crack.

Most manufacturers specify the use of dampened discs in trucks equipped with high torque-rise engines. **Dampened discs** have coaxial dampening springs incorporated into the disc hub. When engine or driveline torque is first transmitted to the disc, the plate rotates on the hub, compressing the springs. This action absorbs the shock and torsional vibration caused by low rpm, high torque engines. The cushioning effect extends clutch life and protects other driveline components from torsional overloads.

Friction Facings

Friction facings are critical to clutch life and performance because they directly receive the torque of the engine each time the clutch is engaged. There are two types of friction facings: organic and ceramic **(Figure 9–14).**

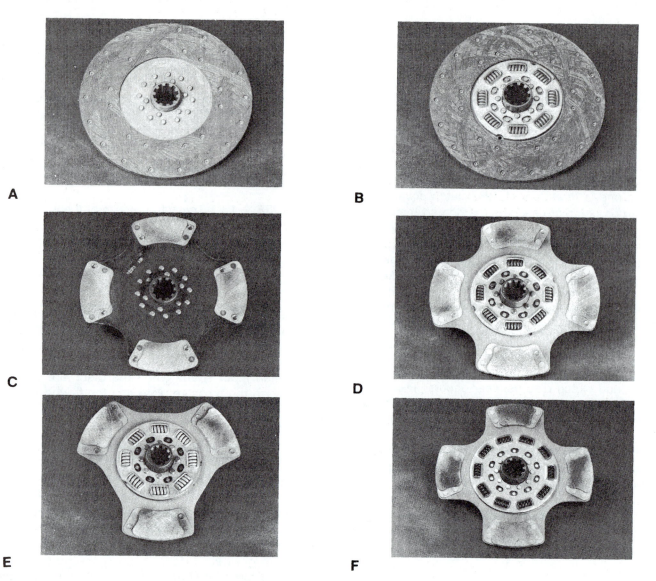

FIGURE 9–14   *Disc designs: (A) rigid organic; (B) dampened organic; (C) rigid 4-button ceramic; (D) dampened 4-button ceramic; (E) dampened 3-button ceramic; and (F) super dampened 4-button ceramic. (Courtesy of Eaton Corp.—Eaton Clutch Div.)*

Most organic friction facings are made from nonasbestos materials such as glass, mineral wool, and carbon. Organic friction facings are usually molded to the full surface of the disc and are called "full faced." Grooves in the facing allow worn particles to be thrown off rather than accumulating on the face of the disc. Organic friction facings are usually specified on linehaul trucks that are not subject to the wear caused by operating in stop-and-go traffic.

Ceramic friction facings are made from a mixture of ceramics and copper or iron. They have a higher coefficient of friction, heat tolerance, and torque capacity than organic friction facings. Ceramic friction facings grab quicker with less slip. They also offer a longer service life, making them popular on pick-up and delivery (P&D) vehicles, and more rugged applications either on- or off-highway.

Ceramic facings are small pads or buttons that are riveted to a disc, or isolator. The disc can be round with slots machined in it between each button or it might be a scalloped, paddle wheel configuration. Three, four, or more buttons can be installed on each face of the disc.

RELEASE MECHANISMS

There are major differences in the way clutches are released or disengaged. All clutches are disengaged through the movement of a release bearing. The release bearing is a unit within the clutch assembly that mounts on the transmission input shaft but does not rotate with it. The movement of the bearing is controlled by a fork attached to the clutch pedal linkage. As the release bearing moves, it forces the pressure plate away from the clutch disc. Depending on the design of the clutch, the release bearing will move in one of two directions when the clutch is disengaged. It will either be pushed toward the engine and flywheel, or it will be pulled toward the transmission input shaft.

Push Type Clutches

In a **push type clutch** (**Figures 9–15** and **9–16**), the release bearing is not attached to the clutch cover. Todisengage the clutch, the release bearing is pushed toward the engine. When the pedal of a push type clutch is initially depressed, there is some free pedal movement between the fork and the release bearing (normally about $1/8$ inch). After the initial movement, the clutch release fork contacts the bearing and forces it toward the engine.

As the release bearing moves toward the engine, it acts on release levers bolted to the clutch cover assembly. As the release levers pivot on a pivot point, they force the pressure plate (to which the opposite ends of the levers are attached) to move away from the clutch discs. This compresses the springs and disengages the discs from the flywheel, allowing the disc (or discs) to float freely between the plate and flywheel, breaking the torque between the engine and transmission.

When the clutch pedal is released, the spring pressure on the pressure plate forces the plate forward once again, clamping the plate, disc, and flywheel together and allowing the release bearing to return to its original position.

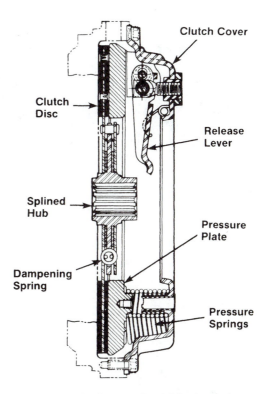

FIGURE 9–16 Cutaway of a 14-inch push type clutch assembly on a pot (recessed) flywheel. (Courtesy of Haldex Brake Systems)

Push type clutches are used predominantly in light- and medium-duty truck applications in which a **clutch brake** is not required. This type of clutch has no provisions for internal adjustment. All adjustments normally are made externally via the linkage system.

Pull Type Clutches

As its name implies, a **pull type clutch** does not push the release bearing toward the engine; instead, it pulls the release bearing toward the transmission. In clutches with angled coil springs or a diaphragm spring, the release bearing is attached to the clutch cover via a sleeve and retainer assembly (**Figure 9–17** and **Figure 9–18**). When the clutch pedal is depressed, the bearing, sleeve, and retainer are pulled away from the flywheel. This compresses the springs and causes the pivot points on the levers to move away from the pressure plate, relieving pressure on the pressure plate. This action allows the driven disc or discs to float freely between the plate(s) and the flywheel. On pull type clutches with coil springs positioned perpendicular to the pressure plate (**Figure 9–19**), the release levers are connected on one end to the sleeve and retainer; on the other end they are connected to pivot points (**Figure 9–20**). The pressure plate is connected to the levers near the pivot points. So, when the levers are pulled away

FIGURE 9–15 A Lipe single plate, push type clutch. (Courtesy of Haldex Brake Systems)

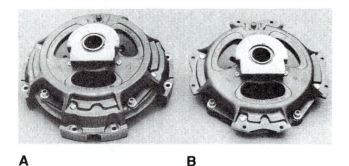

A   **B**

FIGURE 9–17   Pull type diaphragm spring clutches: (A) 15¹/₂ inch; and (B) 14 inch. (Courtesy of ArvinMeritor Inc.)

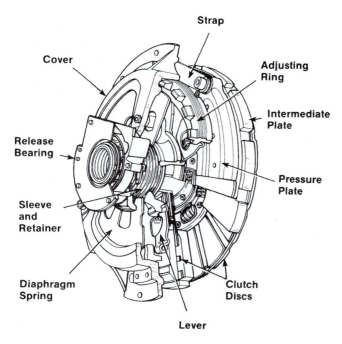

FIGURE 9–18   Components of a pull type clutch. (Courtesy of ArvinMeritor Inc.)

FIGURE 9–19   A single plate, pull type clutch. (Courtesy of Haldex Brake Systems)

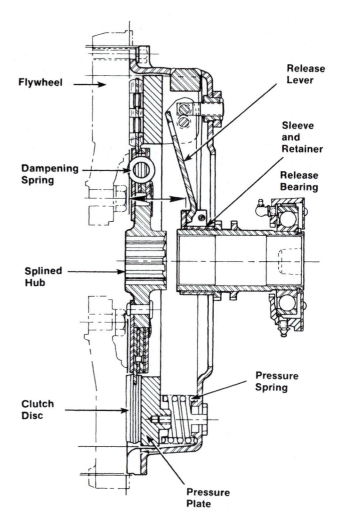

FIGURE 9–20   Cutaway of a Lipe single plate, pull type clutch. (Courtesy of Haldex Brake Systems)

from the flywheel, the pressure plate is also pulled away from the clutch discs and the clutch is disengaged. When the clutch pedal is released, spring pressure forces the pressure plate forward against the clutch disc and the release bearing, sleeve, and retainer return to their original position.

Pull type clutches are used in both medium- and heavy-duty applications and are adjusted internally.

CLUTCH BRAKES

Most pull type clutches have a component not found on push type clutches: a clutch brake. The clutch brake is a circular disc with a friction surface that is mounted on the transmission input spline between the release bearing and the transmission (**Figure 9–21A**). Its purpose is to slow or stop the transmission input shaft from rotating to allow gears to be engaged without clashing (grinding). Clutch brakes are used only on vehicles with nonsynchronized transmissions.

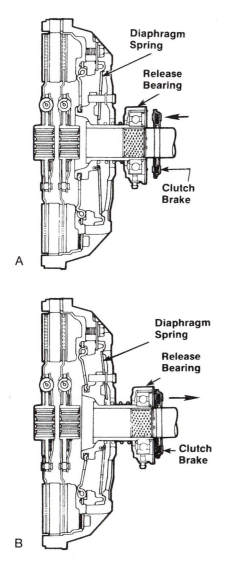

A

B

FIGURE 9–21 Clutch brake: (A) clutch engaged, brake neutral; and (B) clutch disengaged, clutch brake engaged. (Courtesy of ArvinMeritor Inc.)

Only 70 to 80 percent of clutch pedal travel is needed to fully disengage the clutch. The last inch or two of pedal travel is used to engage the clutch brake. When the pedal is fully depressed, the fork squeezes the release bearing against the clutch brake, which forces the brake disc against the transmission input shaft bearing cup (**Figure 9–21B**). The friction created by the clutch brake facing slows the rotation of the input shaft and countershaft. This allows the transmission gears to mesh without clashing.

Conventional Clutch Brake

Conventional clutch brakes are designed to be used when shifting from neutral to first or reverse. Unlike some clutch brakes, a conventional clutch brake is not used to aid upshifting. A conventional clutch

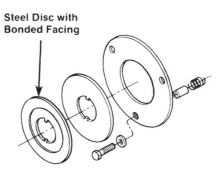

FIGURE 9–22 Limited torque clutch brake. (Courtesy of Eaton Corp.—Eaton Clutch Div.)

brake consists of a steel washer faced on both sides with friction material, or discs. The steel washer has two tangs that engage machined slots in the transmission input shaft. This mounting arrangement allows the brake to move back and forth along the shaft, while also turning at the same speed as the shaft.

Limited Torque Clutch Brake

This clutch brake (**Figure 9–22**) enables faster upshifting into other gear ranges as well as shifting into first and reverse from a stop. When the truck is moving and the clutch brake is engaged, it slows down the transmission input shaft, which allows the speed of the transmission input shaft to synchronize more quickly with that of the transmission countershafts. This allows quicker engagement, which means faster shifts.

Torque Limiting Clutch Brakes

A **torque limiting clutch brake** (**Figure 9–23**) is designed to slip when torque loads of 20 to 25

FIGURE 9–23 Torque limiting clutch brake. (Courtesy of Eaton Corp.—Eaton Clutch Div.)

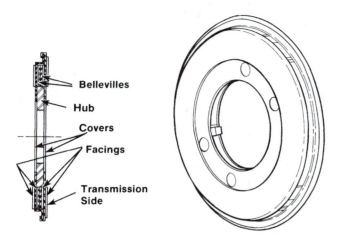

FIGURE 9–24 Torque limiting clutch brake. (Courtesy of Eaton Corp.—Eaton Clutch Div.)

pound-feet are reached. This protects the brake from overloading and the resulting heat damage. As shown in **Figure 9–24**, a torque limiting clutch brake has a hub and Belleville washers inside a cover. The hub and washers are designed to slip when a certain torque level is reached. This type of clutch brake is used only when shifting into first or reverse while the vehicle is stationary.

Two-Piece Clutch Brakes

The **two-piece clutch brake** design is sold by aftermarket suppliers (**Figure 9–25**). It can be quickly installed without removing the transmission. Great care should be taken not to damage the transmission input shaft when cutting out the defective clutch brake.

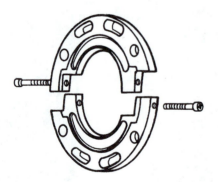

FIGURE 9–25 Two-piece clutch brake designed for aftermarket installation. (Courtesy of Eaton Corp.— Eaton Clutch Div.)

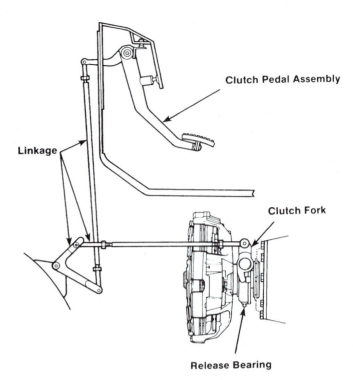

FIGURE 9–26 The clutch linkage connects the clutch pedal to the clutch release lever and fork. (Courtesy of ArvinMeritor Inc.)

CLUTCH LINKAGE

Most clutches on heavy-duty trucks are controlled by a mechanical linkage between the clutch pedal and the release bearing. Some trucks have hydraulic clutch controls. The linkage connects the clutch pedal to the fork, or yoke (**Figure 9–26**).

With the clutch pedal fully raised, there should always be $1/8$ inch free play between the fork, or yoke, and the release bearing. This free play is taken up by the first 1 to $1 1/2$ inches of pedal travel. Then, as the pedal is depressed farther, the fork acts against the release bearing and forces it along the input shaft of the transmission. On pull type clutches with a clutch brake, the last inch of pedal travel will force the bearing against the clutch brake.

Mechanical Clutch Linkage

There are two types of mechanical linkages used in heavy-duty trucks: one that uses a combination of levers and fulcrums to multiply the pedal pressure applied by the driver, and another that links the clutch pedal and release fork through a specially designed clutch control cable. Examples of both types are shown in **Figures 9–27** and **9–28**. Components in each type will vary, depending on the truck and clutch manufacturer.

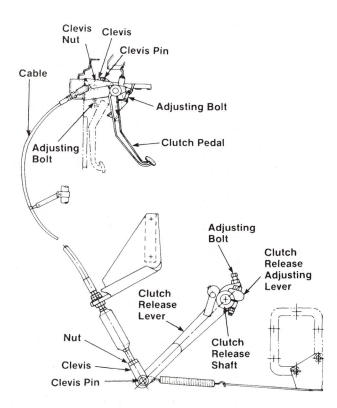

FIGURE 9–27 *Clutch control adjustment arm. (Courtesy of Eaton Corp.—Eaton Clutch Div.)*

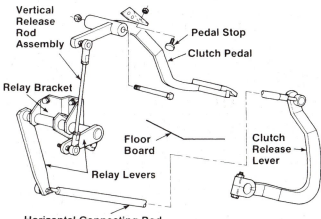

FIGURE 9–28 *Typical clutch mechanical linkage found on a conventional truck. (Courtesy of Volvo Trucks North America, Inc.)*

Hydraulic Clutch Linkage

Figure 9–29 shows the components of a typical hydraulic clutch control system. The clutch is controlled and operated by hydraulic fluid pressure and is assisted by an air servo cylinder. This clutch consists of a master cylinder, hydraulic fluid reservoir, and an air assisted servo cylinder. These are all connected by metal and flexible hydraulic lines.

When the truck driver presses down on the clutch pedal, the plunger forces the piston in the master cylinder to move forward. This closes off the reservoir and forces hydraulic fluid to move down the line to a reaction plunger and pilot valve in the servo cylinder.

As the hydraulic pressure increases, it forces the reaction plunger to move forward to close off an exhaust port and to seat the pilot valve. When the plunger is moved farther, it unseats the pilot valve, which allows air to enter the servo cylinder, exerting pressure onto the rear side of the air piston. The movement of the air piston assists in clutch pedal application. As the clutch pedal pressure increases, the air piston is moved farther forward and the air pressure overcomes the hydraulic pressure in the reaction plunger. This causes the pilot valve to reseat, preventing any more air from reaching the air piston. The pilot valve and reaction plunger will remain in this position until there is a change in the pressure.

When the hydraulic pressure decreases, the return spring returns the reaction plunger and the pilot valve seats itself, which in turn uncovers the exhaust port and allows the air to escape from the servo cylinder.

Air Clutch Control

This system consists of the components shown in **Figure 9–30**. Air is used to disengage the clutch and to apply the clutch brake. The clutch control valve design is similar to that of an air treadle valve. The clutch brake control valve is fastened to the bracket of the clutch control valve in such a way that the plunger is located underneath the clutch control pedal.

When the engine is operating, the same air pressure is present at the clutch reservoir and the inlet valve of the clutch control valve. As the clutch control valve is depressed, the exhaust valve closes and the inlet valve opens. This allows air to flow to the clutch brake valve. Simultaneously, air flows over a **quick release valve** to the clutch release air chamber, and from the quick release valve over a pressure-reducing valve to the oil level cylinder. The pressure in the air chamber moves the pushrod outward to disengage the clutch. The pressure at the oil level cylinder diaphragm causes it to move and to force the oil from the reservoir into the bell housing sump. This increases both the lubrication to the clutch components and the cooling capacity. If the clutch control valve pedal is further depressed, the brake control valve directs air to the clutch brake.

When the clutch control valve pedal is released, the inlet valve of the clutch brake control valve

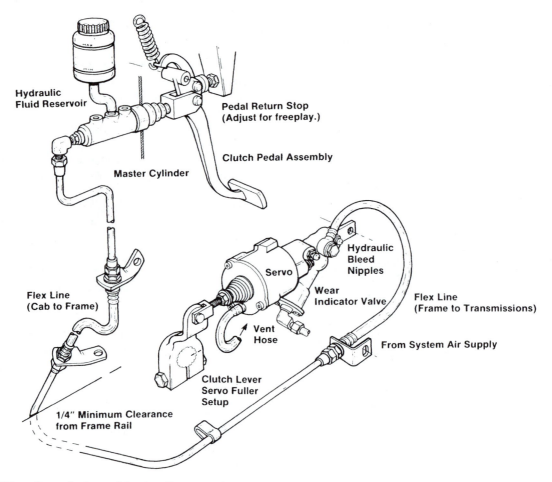

FIGURE 9–29   Overall view of hydraulic controls.

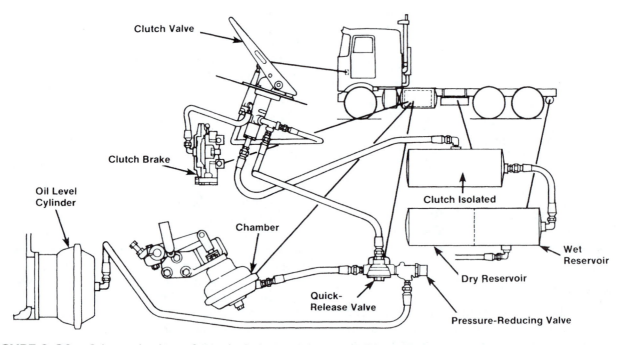

FIGURE 9–30   Schematic view of a typical air control system. (Photo/image(s) courtesy of Mack Trucks, Inc.)

closes and then the exhaust valve opens. This releases the air pressure from the clutch brake piston. Then the exhaust valve of the clutch control valve opens and the air pressure escapes from the clutch release air chamber and the oil level cylinder. As a result, the clutch engages and the diaphragm moves, drawing oil from the sump back into the reservoir.

9.3 TROUBLESHOOTING

The major cause of clutch failure is excess heat. The heat generated between the flywheel, driven discs, intermediate plate and pressure plate may be intense enough to cause surface melting and friction material failure.

Wear potential is practically nonexistent when the clutch is fully engaged. However, considerable heat can be generated at the moment of engagement when the clutch is picking up the load. An improperly adjusted or slipping clutch will rapidly generate sufficient heat to self-destruct.

Proper training of both drivers and service technicians is important in extending clutch life. The most critical points to cover in driver training and service technician awareness training are learning to start in the correct gear, proper clutch engagement, recognizing clutch malfunctions, and identifying the need for clutch readjustment. Service personnel must be aware of what driver abuse can do to clutch life. This type of training will help service technicians spot and analyze failures due to driver abuse. See **Table 9–1** for a quick guide to troubleshooting clutch operation.

Starting in the Correct Gear

An unloaded truck can be started in a higher transmission gear ratio than when partially or fully loaded. If auxiliary transmissions or multispeed axles are used, they must be in the lower ratios for satisfactory starts. Drivers should be shown what ratios can be used for safe starts when the truck is empty or loaded. Allowing a driver to experiment to find the proper starting gear can lead to burned-out clutches.

If the truck is diesel powered, a good general rule for both empty and loaded trucks is to select the gear combination that allows you to start moving with an idling engine, or, if needed, just enough throttle to prevent stalling. After the clutch is fully engaged, the engine should be accelerated to near governed speed for the upshift into the next higher gear.

Gear Shifting Techniques

Many drivers upshift into the next gear, or skip-shift into a higher gear, before the vehicle has reached the highway speed. This shifting technique can be almost as damaging as starting off in a gear that is too high. This can space the engine speed and driveline speeds too far apart, requiring the clutch to absorb the speed difference as friction.

Vehicle or Clutch Overloading

All clutches are designed and recommended for specific vehicle applications and loads. These limitations should not be exceeded. Overloading can damage both the vehicle's clutch and drivetrain. If the total gear reduction in the power train is not sufficient to handle overloads, the clutch will experience excessive wear, since it is forced to pick up the load at a higher speed differential.

Riding the Clutch Pedal

Riding the clutch pedal means operating the vehicle with the clutch partially engaged. This is very destructive to the clutch, as it permits slippage and generates excessive heat. Riding the clutch will also put constant thrust load on the release bearing, which can thin out the lubricant and also cause excessive wear on the release pads. Release bearing failures are often the result of this type of driving practice.

A similar problem occurs if a driver attempts to hold the vehicle on an incline with a slipping clutch. This practice forces the clutch to act like a fluid coupling. (See Chapter 10.) But the slipping clutch quickly generates more heat than can be dissipated and damage occurs.

Improper Coasting

Coasting with the clutch released and the transmission in gear can cause the driven discs to rotate at a very high rpm, through the multiplication of ratios from the final drive and transmission. This can result in the clutch facings being "thrown" off the clutch discs. Driven disc speeds as high as 10,000 rpm can occur during such simple procedures as coasting tractors down an unloading ramp. Although safety factors are built into the clutch facing for normal operation, abusing the clutch may exceed its burst strength (the maximum rotational speed the clutch can operate at without breaking apart).

Engaging the clutch while coasting can also result in a damaging shock load being placed on the clutch, as well as the remainder of the drivetrain.

Driver/Service Technician Communication

Drivers should be reminded to report erratic clutch operation as soon as possible. This will allow service technicians the opportunity to make the necessary inspection, internal clutch adjustment, linkage adjustments, and lubrication checks before total clutch failure and breakdown on the road.

The importance of free play, or pedal lash as it is sometimes called, cannot be overstressed. This is the amount of free play in the clutch pedal in the cab. A free pedal of $1\frac{1}{8}$ inch to 2 inches is ideal. Free play is directly related to another important clutch measurement—free travel. Free travel is the clearance distance

TABLE 9–1:   TROUBLESHOOTING CLUTCH OPERATION

Symptom	Probable Cause	Remedy
Clutch does not release or does not release completely	Clutch linkage and release bearing need adjustment	Adjust clutch linkage and release bearing.
	Worn or damaged linkage	Lubricate linkage. Make sure the linkage is not loose. If condition still exists, replace linkage.
	Worn or damaged release bearing	Replace release bearing.
	Worn or damaged splines on input shaft	Replace input shaft.
	Clutch housing loose	Tighten fasteners to specified torque. If necessary, replace fasteners.
	Worn or damaged pressure plate	Replace pressure plate and cover assembly.
	Worn or damaged center plate	Replace center plate.
	Center plate binding	14-inch clutch: Inspect drive pins in flywheel housing and slots in center plate. 15$\frac{1}{2}$-inch clutch: Inspect tabs on center plate and slots in cover. Service as necessary.
	Damaged hub in clutch discs	Replace discs.
	Linings worn below specified dimension	Replace discs.
	Linings damaged	Replace discs.
	Oil or grease on linings	Clean linings. If oil or grease cannot be removed, replace discs.
	Linings not as specified for vehicle operation	Install correct type of linings.
	Damaged pilot bearing	Replace pilot bearing.
Clutch pedal hard to operate	Damaged bosses on release bearing assembly	Replace release bearing assembly. Make sure clutch is correctly adjusted.
	Worn or damaged clutch linkage	Lubricate linkage. If condition still exists, replace linkage.
	Worn or damaged components	Replace pressure plate and cover assembly.
Clutch slips on engagement	Driver keeps foot on clutch pedal	Use correct vehicle operating procedure.
	Clutch linkage and release bearing need adjustment	Adjust release bearing and clutch linkage.
	Worn or damaged components	Replace pressure plate and cover assembly.
	Worn or damaged linings	Replace discs.
	Oil or grease on linings	Clean linings. If oil or grease cannot be removed, replace discs.
	Linings not as specified for vehicle operation	Use correct linings.
	Worn or damaged flywheel	Service flywheel as necessary. See procedure of engine or vehicle manufacturer.
Noisy clutch	Clutch linkage and release bearing need adjustment	Adjust clutch linkage and release bearing.
	Worn or damaged linkage	Lubricate linkage. If condition still exists, replace linkage.
	Worn or damaged release bearing	Lubricate release bearing. If condition still exists, replace release bearing.
	Worn or damaged clutch housing	Replace clutch housing and pressure plate assembly.
	Clutch housing loose	Tighten fasteners to specified torque. If necessary, replace fasteners.
	Damaged hub or broken spring(s) in clutch discs	Replace clutch discs.
	Linings worn below specified dimension	Replace clutch discs.
	Linings damaged	Replace clutch discs.
	Oil or grease on linings	Clean linings. If oil or grease cannot be removed, replace linings.
	Damaged pilot bearing	Replace pilot bearing.

Symptom	Probable Cause	Remedy
Vibrating clutch	Worn or damaged splines on input shaft	Replace input shaft.
	Pressure plate and cover assembly out of balance	Remove, check balance, and install pressure plate and cover assembly. If condition still exists, replace pressure plate and cover assembly.
	Worn or damaged splines in hub of clutch discs	Replace discs.
	Loose flywheel	Tighten fasteners to specified torque. If necessary, replace fasteners. Check flywheel mounting surface for damage. Replace if necessary.

between the release yoke fingers and the clutch release bearing pads. Because of this important relationship, the amount of free pedal should be included and commented on in the daily driver's report. Changes in free pedal measurement are an indicator to the service technician that problems may exist with internal clutch components.

9.4 PERIODIC MAINTENANCE

Clutches should be checked periodically for proper adjustment and lubrication. Actual maintenance varies with the design of the clutch. Some clutches are self-adjusting and once the clutch has been installed and the initial adjustment made, no further adjustment to bearing free play should be necessary over the life of the clutch. Other clutches are manually adjusted and must be adjusted periodically as the friction linings wear. All clutches require regular inspection and adjustment of the clutch linkage.

LUBRICATION

Some clutches have sealed release bearings that never need lubrication. Other clutches have release bearings fitted with grease fittings (**Figure 9–31**), and these must be periodically lubricated. Some clutches can be equipped with special lubrication tube assemblies that allow the release bearing to be greased without removing the inspection cover of the bell housing. These grease fitting extensions can help reduce maintenance and downtime.

A clutch release bearing should be lubricated according to the manufacturer's lubrication schedule. Typically, this means whenever the clutch is adjusted, once a month, every 6,000 to 10,000 miles,

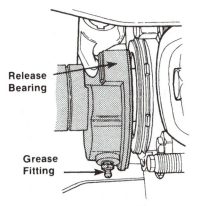

Release Bearing

Grease Fitting

FIGURE 9–31 *Some release bearings have a grease fitting and must be periodically lubricated. (Courtesy of ArvinMeritor Inc.)*

or whenever the chassis is lubricated, whichever comes first. Off-road operation or other severe applications require shorter service intervals. Only an extreme pressure (EP) grease such as a lithium soap-based grease having a temperature performance range of −10°F to +325°F should be used.

WARNING: Replacement release bearing housings are not prepacked with grease. They must be lubricated when the clutch is installed in the vehicle or premature failure will result.

A small amount of grease is also applied between the release bearing pads and the clutch release fork at normal service intervals.

Some clutches have grease fittings in the clutch housing (**Figure 9–32**) where the clutch release shaft passes through it. These fittings should be greased whenever the release bearing is lubricated.

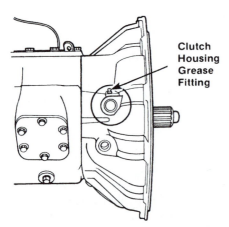

FIGURE 9–32 Some transmission housings have grease fittings where the clutch release shaft passes through the housing. (Courtesy of ArvinMeritor Inc.)

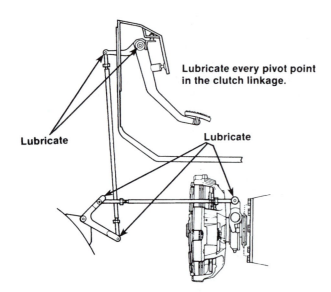

FIGURE 9–33 Clutch linkage lubrication points. (Courtesy of ArvinMeritor Inc.)

Whenever the release bearing and other lubrication points on the clutch are serviced, all pivot points on the clutch linkage should also be lubricated (**Figure 9–33**).

CLUTCH ADJUSTMENTS

As mentioned earlier, proper clutch free play, or the first easy movement of the clutch pedal, should be $1\frac{1}{2}$ to 2 inches for both push type and pull type clutches. The amount of free pedal present is easily determined by placing your hand or foot on the clutch pedal and gently pushing it down until an increase in pushing effort is felt. Any more pedal

movement after this point will cause the release bearing to begin disengaging the clutch.

In a push type clutch, free pedal is set to $1\frac{1}{2}$ to 2 inches to obtain the desired $\frac{1}{8}$-inch free travel clearance between the clutch release bearing and clutch release levers (**Figure 9–34**), or diaphragm spring, whichever is used.

In a pull type clutch, the $\frac{1}{8}$-inch free travel clearance occurs between the release yoke fingers and the clutch release bearing pads (**Figure 9–35**). This $\frac{1}{8}$" free travel should produce $1\frac{1}{2}$" of free pedal.

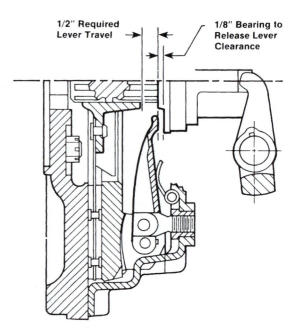

FIGURE 9–34 In a push type clutch, the desired free travel clearance is $\frac{1}{8}$ inch. Total release bearing travel on a push type clutch must be approximately $\frac{5}{8}$ inch. (Courtesy of Haldex Brake Systems)

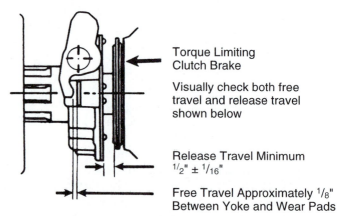

FIGURE 9–35 On a pull type clutch, there should be $\frac{1}{8}$-inch clearance between the release fork and the boss on the release bearing. (Courtesy of Eaton Corp.—Eaton Clutch Div.)

Free pedal specifications are greater than free travel specifications because as movement is transferred through the linkage, it moves through a large arc of travel. Too much free pedal can prevent complete disengagement of the clutch. Too little free pedal can cause clutch slippage, heat damage, and shortened clutch life.

As the disc friction facings wear through normal operation, the amount of free pedal will gradually decrease. If inspection indicates clutch pedal free-pedal travel is less than 1/2 inch, immediate adjustment of the clutch should be performed. DO NOT WAIT UNTIL NO FREE PEDAL EXISTS BEFORE MAKING THIS ADJUSTMENT.

It is important to remember that the method of setting free pedal and free travel is different between push type clutches and pull type clutches. Always use the correct method for the type of clutch being serviced and refer to the proper service manual for exact specifications and setting procedures.

Push Type Clutch Adjustment

In a push type clutch, adjusting the external clutch linkage to obtain 11/2 to 2 inches of free pedal will normally result in the specified 1/8-inch free travel clearance between the release bearing and the clutch release lever or diaphragm spring. Before making the linkage adjustment, inspect the clutch linkage for wear and damaged components. If excessive free play is present in the clutch pedal linkage due to worn components, repair as necessary. Excessive wear of the release linkage can give a false impression of the actual amount of release bearing clearance.

Once the free pedal is set to between 11/2 and 2 inches, it is recommended that the free travel clearance be double-checked. To do this:

1. Set the parking brakes and chock the wheels.
2. Remove the clutch inspection cover from the bottom of the transmission bell housing.
3. Measure the clearance between the release bearing and the clutch release levers or diaphragm spring. This clearance must be within specifications for the release bearing to release properly.
4. If this clearance is not present, adjust the linkage until the specified 1/8-inch clearance is obtained. Remember, the linkage must be in good condition to obtain accurate results.

Pull Type Clutch Adjustment

Pull type clutches may require a two-step adjustment to obtain the specified free travel and free pedal set-tings. The first step is a release bearing free travel adjustment that may not be required. The second step is a pedal or linkage adjustment. Remember that the free travel adjustment must be performed first. This free travel adjustment is usually an internal adjustment; however, some clutch models are equipped with an external quick-adjust mechanism.

Pull Type Clutch Preadjustment Considerations

Before making any adjustments to a pull type clutch, review the following conditions to ensure optimum clutch performance:

1. Clutch brake squeeze (increased resistance) begins at the point the clutch brake is initially engaged. Optimum clutch brake squeeze begins 1 inch from the end of the pedal stroke or above the floor board (**Figure 9–36**). This adjustment is made by shortening or lengthening the external linkage rod.
2. Optimum free pedal is 11/2 to 2 inches (**Figure 9–37**). This adjustment is made internally in the clutch, never with the linkage.
3. Release travel is the total distance the release bearing moves during a full clutch pedal stroke. A typical release travel distance of 1/2 to 9/16 inches is required to ensure that the release bearing releases sufficiently to allow the driven discs to spin freely, avoiding clutch drag. Optimum free travel is 1/8 inch (see **Figure 9–35**). These adjustments are accomplished by using the adjusting ring.

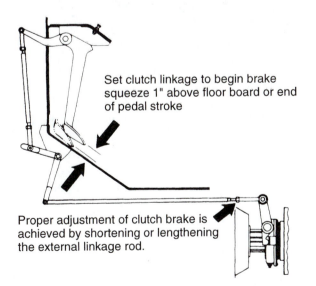

Set clutch linkage to begin brake squeeze 1" above floor board or end of pedal stroke

Proper adjustment of clutch brake is achieved by shortening or lengthening the external linkage rod.

FIGURE 9–36 The last 1 inch of clutch pedal travel should squeeze the clutch brake. (Courtesy of Eaton Corp.—Eaton Clutch Div.)

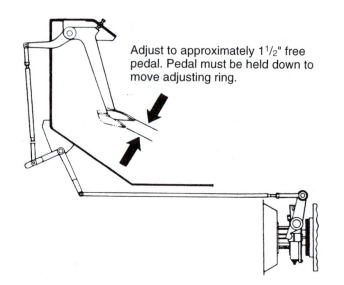

Adjust to approximately 1¹/₂" free pedal. Pedal must be held down to move adjusting ring.

FIGURE 9–37 The first 1¹/₂ inches of pedal travel should take up the clearance between the fork and the release bearing. (Courtesy of Eaton Corp.—Eaton Clutch Div.)

4. Internal adjustment of the adjusting ring must be made before any linkage adjustment is made.

5. All internal clutch adjustments should be made with the clutch pedal down (clutch released position).

6. Turning the adjusting ring clockwise moves the release bearing toward the transmission. Turning the adjusting ring counterclockwise moves the release bearing toward the engine.

7. Linkage adjustment on a pull type clutch should only be made:
   - At initial dealer preparation to set total pedal stroke and yoke throw.
   - To compensate for linkage wear, clutch brake wear, or transmission cap wear.
   - When worn or damaged linkage components are replaced

Shop Talk

Linkage adjustments are not normally required.

PHOTO SEQUENCE 5
CLUTCH ADJUSTMENT

The following is a general procedure used to adjust heavy-duty, double-disc 14-inch and 15¹/₂-inch pull type clutches.

P5–1 Assess the need for a clutch adjustment by first checking clutch brake squeeze. Clutch brake squeeze should begin one inch from the floor at the end of the pedal stroke. Clutch brake squeeze should be verified before attempting an internal adjustment of a clutch. In most cases, clutch brake squeeze will not require adjustment, but it should always be checked.

P5–2 To adjust the clutch brake, either lengthen or shorten the external linkage by loosening the locknut and turning the adjusting rod either clockwise or counterclockwise.

P5–3 Next, check the clutch pedal free play. This is the pedal travel that results before the clutch disengagement occurs. Free travel is always adjusted internally: Attempting to correct this externally will result in incorrect clutch brake squeeze. **Clutch pedal free play should be between 1¹/₂ and 2 inches.**

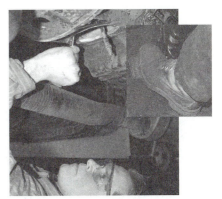

P5–4 Remove the clutch inspection plate to check the condition of the clutch brake, release travel, and cross-shaft yoke free play. Use a trouble light to inspect the clutch brake. Release travel (the total travel of the release bearing through a full stroke of the clutch pedal) should be between $^{1}/_{2}$ inch and $^{9}/_{16}$ inch. Internal free play is the distance between the cross shaft yoke and the release bearing when the clutch is fully engaged. It should measure $^{1}/_{8}$ inch.

P5–5 To perform a clutch adjustment, use a wrench to turn the Kwik-Adjust nut as shown.

P5–6 To make an adjustment, fit the adjusting tool to the clutch and have the clutch pedal fully applied by a second person. This releases the clutch, permitting the adjusting ring to be rotated by the adjusting tool. Three notches of CW travel will move the release bearing approximately $^{1}/_{16}$ inch.

Lockstrap Mechanism. The lockstrap is a manual adjustment mechanism that allows for adjustment of free travel. To make the adjustment:

1. Remove the inspection plate at the bottom of the clutch housing (**Figure 9–38**).
2. To adjust the threaded adjusting ring, the clutch assembly which is bolted to the flywheel must be manually rotated around until the internal adjusting ring lockstrap and its bolt are visible through the inspection plate opening (**Figure 9–39**).

P5–7 Install the clutch inspection plate. Drive the truck a short distance to verify the clutch performance after the adjustment.

Internal Adjustment Mechanisms— Angle Spring Clutch

There are three basic types of adjustment mechanisms currently in use on angle coil spring clutches used in heavy-duty truck applications. Two are manual adjusting mechanisms and the third is a self-adjusting mechanism.

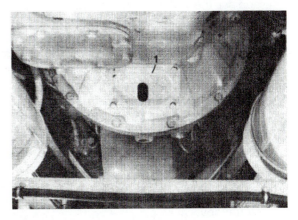

FIGURE 9–38 The clutch inspection cover must be removed to make adjustments to the clutch. (Courtesy of Freightliner Trucks)

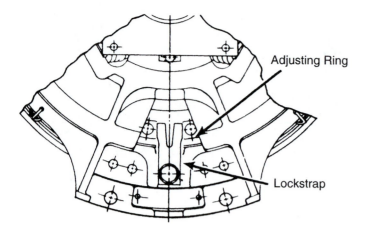

FIGURE 9–39   For adjustment, the lockstrap and bolt must be visible through the clutch inspection cover. (Courtesy of Eaton Corp.—Eaton Clutch Div.)

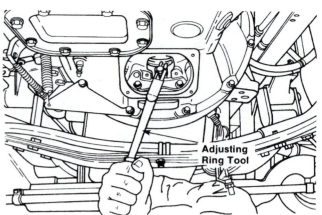

FIGURE 9–40   Special adjusting ring tools are available for adjusting the clutch. (Courtesy of ArvinMeritor Inc.)

 **Shop Talk**

The lockstrap must be located at BDC (bottom dead center) to adjust the clutch. Bar the engine to this position using an appropriate tool.

3. Remove the cap screw and lock washer that fastens the lockstrap to the clutch cover. Remove the lockstrap.

4. Push the clutch pedal to the bottom of pedal travel. Use another person or a block of wood to hold the pedal at the bottom of travel. Hold the pedal in this position when the adjusting ring is moved. The pedal should be down when turning adjusting ring and should be up when taking the measurement.

5. Rotate the adjusting ring to obtain the specified clearance of the release bearing. Use a screwdriver or an adjusting tool as a lever against the notches on the ring to move the adjusting ring (**Figure 9–40**). When the adjusting ring is moved one notch, the release bearing will move 0.023 inch. Moving the ring three notches will move the release bearing approximately $1/16$ inch. Turning the adjusting ring clockwise moves the release toward the transmission and increases the pedal free travel. Turning the adjusting ring counterclockwise moves the release bearing toward the engine and decreases the pedal free travel.

6. Install the lockstrap. Install and tighten the cap screw that fastens the lockstrap to the clutch cover.

7. Release the clutch pedal.

8. Check the clearance between the yoke and the wear pads. The distance must be $1/8$ inch

(**Figure 9–35**). If the distance is not at this dimension, adjust the clutch linkage according to the vehicle manufacturer's procedures.

9. Install the cover for the inspection hole. Install and tighten the cap screws to the manufacturer's specifications.

Kwik-Adjust Component. This manual adjust mechanism (**Figure 9–41**) allows for adjustment of free travel without the use of special tools or the need to remove any bolts. The adjustment is made using a properly sized socket wrench to turn the adjusting bolt.

1. Using a $3/4$-inch socket (12 point) or box-end wrench, depress the square headed bolt and rotate the bolt as needed to make the adjustment (**Figure 9–42**). The Kwik-Adjust will reengage at each quarter turn.

2. Assure that the bolt head is locked in position with the bolt flats aligned with the bracket.

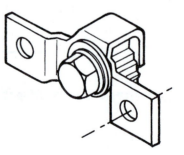

FIGURE 9–41   External manual adjustment mechanism for quick adjustment of free travel. (Courtesy of Eaton Corp.—Eaton Clutch Div.)

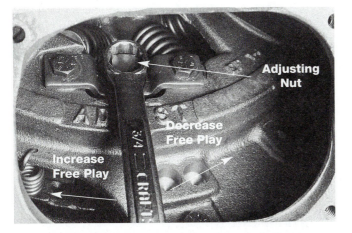

FIGURE 9–42 Performing the external manual adjustment. (Courtesy of Eaton Corp.—Eaton Clutch Div.)

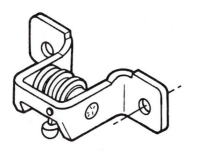

FIGURE 9–43 Replaceable wear compensator component. (Courtesy of Eaton Corp.—Eaton Clutch Div.)

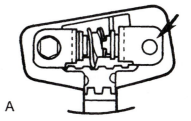

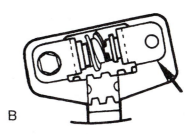

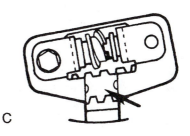

FIGURE 9–44 Steps in making a cast angle spring clutch adjustment on a clutch equipped with a wear compensator. (Courtesy of Eaton Corp.—Eaton Clutch Div.)

Wear Compensator. This is a replaceable component that automatically adjusts for facing wear each time the clutch is actuated (**Figure 9–43**). Once facing wear exceeds a predetermined amount, the adjusting ring is advanced and free pedal dimensions are returned to normal operating conditions. To make the wear compensator adjustment,

1. Manually turn the engine flywheel until the adjuster assembly is in line with the clutch inspection cover opening.
2. Remove the right bolt and loosen the left bolt one turn (**Figure 9–44A**).
3. Rotate the wear compensator upward to disengage the worm gear from the adjusting ring (**Figure 9–44B**).
4. Advance the adjusting ring as necessary (**Figure 9–44C**).

CAUTION: Do not pry on the innermost gear teeth of the adjusting ring. Doing so could damage the teeth and prevent the clutch from self-adjusting.

 Shop Talk

Each lower gear tooth of the adjusting ring represents about 0.010 inch of release bearing movement. Therefore, six notches moved means 0.060 inch or about $1/16$ inch release bearing movement.

5. Rotate the assembly downward to engage the worm gear with the adjusting ring. The adjusting ring may have to be rotated slightly to re-engage the worm gear.
6. Install the right bolt and tighten both bolts to specifications.
7. Visually check to see that the actuator arm is inserted into the release sleeve retainer. If the assembly is properly installed, the spring will move back and forth as the pedal is full stroked.

Note: The clutch will not compensate for wear if the actuator arm is not inserted into the release sleeve retainer, or if the release bearing travel is less than $1/2$ inch.

Whenever a self-adjusting clutch is found to be out of adjustment, check for the following:

- Actuator arm incorrectly inserted into the release bearing sleeve retainer
- Bent adjuster arm
- Clutch parts, such as the adjusting ring, are frozen or damaged

After identifying and repairing or replacing the defective condition or component, readjust the bearing setting.

Internal Adjustment: Nonclutch Brake Clutches

The following steps outline the typical internal adjustment procedure for manual and self-adjusting angle spring clutches not equipped with a clutch brake.

1. Remove the inspection cover at the bottom of the clutch housing.
2. With the clutch engaged (pedal up), measure the clearance between the release bearing housing and the clutch cover.
3. If clearance is not within service manual specifications—typically $1^7/_8$ inches for a single-plate clutch and $3/_4$ inch for a two-plate clutch (**Figure 9–45**)—continue with steps 4 and 5 below; otherwise skip to step 6 below.
4. Release the clutch by fully depressing the clutch pedal.
5. Using the internal adjustment procedure outlined previously for lockstrap, Kwik-Adjust, and wear compensator adjustment mechanisms, advance the adjusting ring until the proper clearance specification is obtained. If the clear-

ance between the release bearing housing and the clutch cover is less than specifications, rotate the adjusting ring counterclockwise to move the release bearing toward the engine. If the clearance is greater than specifications, rotate the adjusting ring clockwise to move the release bearing toward the transmission.
6. Apply a small amount of grease between the release bearing pads and the clutch release fork.
7. Proceed with linkage adjustment as needed.

Internal Adjustment: Clutch Brake Clutches

The following steps outline the typical internal adjustment procedure for manual and self-adjusting angle spring clutches equipped with a clutch brake.

1. Remove the inspection cover at the bottom of the clutch housing.
2. With the clutch engaged (pedal up), measure the clearance between the release bearing housing and the clutch brake. This is the release travel. If clearance (**Figure 9–46**) is less than $1/_2$ inch or greater than $9/_{16}$ inch (typical), continue with steps 3 and 4 below; otherwise continue with step 5 below.
3. Release the clutch by fully depressing the clutch pedal.
4. Using the internal adjustment procedures listed above for lockstrap, Kwik-Adjust, and wear compensator adjustment mechanisms, advance the adjusting ring until a distance of $1/_2$ to $9/_{16}$ inch is attained between the release bearing housing and the clutch brake with the clutch pedal released.

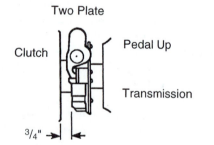

FIGURE 9–45 Two plate clutch: clearance between release bearing housing and clutch cover for the internal adjustment on manual or self-adjusting angle spring clutches without a clutch brake. (Courtesy of Eaton Corp.—Eaton Clutch Div.)

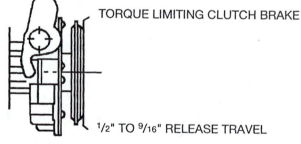

FIGURE 9–46 Clearance between release bearing housing and clutch cover for the internal adjustment on manual or self-adjusting angle spring clutches with a clutch brake. (Courtesy of Eaton Corp.—Eaton Clutch Div.)

5. If the clearance between the release bearing housing and the clutch brake is less than specifications, rotate the adjusting ring counterclockwise to move the release bearing toward the engine. If the clearance is greater than specifications, rotate the adjusting ring clockwise to move the release bearing toward the transmission.

6. Apply a small amount of grease between the release bearing pads and the clutch release fork.

7. Proceed with linkage adjustment as needed.

Clutches with Perpendicular Springs

Pull type clutches with perpendicular springs use a threaded sleeve and retainer assembly that can be adjusted to compensate for disc lining wear. The following adjustment procedure is also illustrated in **Figure 9–47**.

1. Use a drift and hammer or a special spanner wrench to unlock the sleeve locknut.

2. Turn the slotted adjusting nut to obtain the release travel clearance of $1/2$ inch if the clutch is equipped with a clutch brake and $3/4$ inch if it does not have a clutch brake.

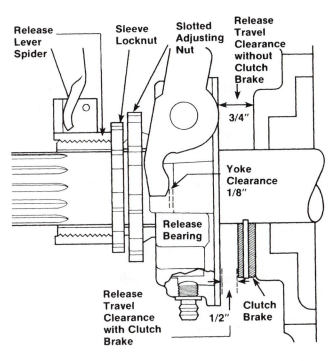

FIGURE 9–47 Clutch adjusting mechanism for a clutch with perpendicular springs. (Courtesy of Haldex Brake Systems)

3. Securely lock the sleeve lock nut against the release lever retainer, or spider, using the drift and hammer or special spanner wrench.

4. Adjust the clutch linkage to obtain a yoke-to-bearing free travel clearance of $1/8$ inch if the vehicle is equipped with a nonsynchronized transmission with a clutch brake. On vehicles equipped with synchronized transmissions (without a clutch brake) the yoke clearance should be $1/4$ inch and will result in approximately 3 inches of free pedal.

CLUTCH LINKAGE INSPECTION

The clutch linkage must be inspected carefully. The clutch will not operate correctly if the linkage is worn or damaged. Inspect the linkage according to the following procedures.

1. Push down on the clutch pedal and have another person check the release fork for movement. The smallest movement of the clutch pedal must cause movement at the release fork. If the release fork does not move when the clutch pedal moves, find and service the cause of the free play condition.

2. The linkage must move when the pedal is actuated. Make sure the linkage is not obstructed. Make sure every pivot point in the linkage operates freely. Make sure the linkage is not loose at any point. If the linkage does not operate freely, find and service the cause of the condition.

3. Inspect every part of the clutch linkage. Make sure the pedal, springs, brackets, bushings, shafts, clevis pins, levers, cables, and rods are not worn or damaged. If a hydraulic system is used, make sure that the system is not leaking and that the reservoir is filled to the specified level. Replace any parts that are missing or damaged. Do not straighten any damaged parts. Replace the parts.

4. Make sure every pivot point in the linkage is lubricated. Use the lubricant specified by the manufacturer of the vehicle. A NLGI #2 multipurpose lithium grease is typically specified for bushings and other pivot points.

CLUTCH LINKAGE ADJUSTMENT

As mentioned earlier, there are three types of linkage mechanisms currently used in heavy-duty truck applications: mechanical, hydraulic, and air controlled linkages. Adjustment procedures vary between truck manufacturers, so always follow the procedure listed in the truck service manual.

The following is general information on adjusting the clutch linkage. For specific adjustment procedures of the clutch linkage, see the procedure of the manufacturer of the vehicle.

Pedal Height. On some vehicles, the height of the travel of the clutch pedal is adjusted. The height is adjusted by stop bolts. If the pedal height is not correct, the amount of free pedal will not be measured and adjusted correctly (**Figure 9–48**). Consult the manufacturer's service manual for pedal height specifications.

Total Pedal Travel. The **total pedal travel** is the complete distance the clutch pedal must move. It can be adjusted with bumpers and stop bolts in the cab or with stop bolts and pads on the linkage. Total travel makes sure that there is enough movement of the pedal to correctly engage and disengage the clutch (**Figure 9–49**). Consult the manufacturer's service manual for specifications.

Free Pedal Travel. As explained earlier, the free pedal is the distance the pedal is depressed before the release bearing starts to move. It is typically $1\frac{1}{2}$ to 2 inches (See **Figure 9–36**). If the free travel is more than 2 inches, the clutch might not fully release. The clutch discs could be loaded to contact the flywheel continually, causing excessive wear. If the free travel is less than $\frac{1}{2}$ inch, the clutch might slip and load the throw-out bearing.

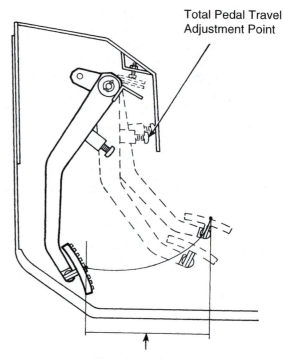

Total Pedal Travel Adjustment Point

Total Pedal Travel

FIGURE 9–49 *Total pedal travel adjustment point. (Courtesy of ArvinMeritor Inc.)*

Checking for Necessary Linkage Adjustment

The following procedure is used to determine if linkage adjustment is required on manual and self-adjusting angle spring clutches.

1. With the clutch disengaged, (pedal down), measure the free travel clearance between the release yoke fingers and the wear pads on the release bearing housing on both sides of the housing simultaneously. If the clearance is greater or less than $\frac{1}{8}$ inch, either shorten or lengthen the external linkage to obtain the $\frac{1}{8}$-inch clearance. (See **Figure 9–35**.)
2. This dimension should correlate to a free pedal of $1\frac{1}{2}$ to 2 inches. If it does not, trim the linkage adjustment to obtain these dimensions.
3. Apply a small amount of grease between the release bearing pads and the clutch release fork.
4. Tighten all lock nuts.

Adjustment

The procedure for adjusting the release yoke finger to release bearing wear pad clearance can differ,

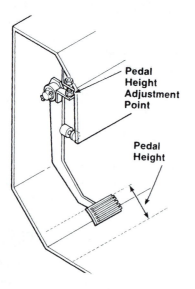

Pedal Height Adjustment Point

Pedal Height

FIGURE 9–48 *Pedal height adjustment point. (Courtesy of ArvinMeritor Inc.)*

depending on the type of truck and linkage design. A basic procedure is given here:

1. Disconnect the lower clutch control rod from the pedal shaft (**Figure 9–50**) or the bellcrank (**Figure 9–51**).
2. Place a spacer block between the pedal stop and the stop bracket or pedal shank. The actual size of the spacer varies with the vehicle manufacturer and the linkage design.
3. Pull the lower control rod forward until the release bearing contacts the yoke fingers.
4. Loosen the jam nuts on the threaded adjusting rod of the threaded end of the control rod until the holes in the pedal shaft or bellcrank align with the holes in the rod end or clevis.
5. Reconnect the linkage members and remove the spacer block from the pedal stop.
6. Operate the clutch pedal and recheck the yoke-to-bearing clearance. If the clearance is still insufficient, it might be necessary to adjust the pedal height or travel. By turning the pedal stop in, the pedal will return farther and the yoke-to-bearing clearance will be increased.

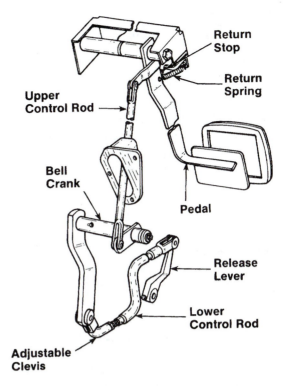

FIGURE 9–51 *Adjustments to this linkage are made by adjusting the clevis on the lower control arm. (Courtesy of International Truck and Engine Corp.)*

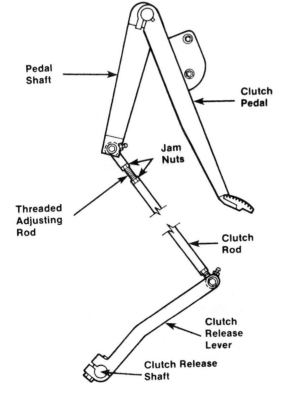

FIGURE 9–50 *This linkage is adjusted by lengthening or shortening the control rod. (Courtesy of Volvo Trucks North America Inc.)*

Clutch Brake Setting

To check the clutch brake setting, depress the clutch pedal in the cab and note the point during the pedal stroke when the clutch brake engages by viewing clutch brake movement through the inspection cover. With the release travel and free travel settings, the clutch brake squeeze should occur approximately 1 inch from the end of the pedal stroke (**Figure 9–37**). To check this:

1. Insert a 0.010-inch thickness gauge between the release bearing and the clutch brake.
2. Depress the clutch pedal to squeeze the thickness gauge.
3. Let the pedal up slowly until the gauge can be pulled out and note the position of the pedal in the cab. It should be $1/2$ to 1 inch from the end of the pedal stroke.
4. To adjust the clutch brake setting, shorten or lengthen the external linkage according to service manual procedures. If proper adjustment cannot be obtained, check the linkage for excessive wear.
5. Reinstall the inspection cover.

9.5 CLUTCH SERVICING

When it is determined that the clutch is not operating properly and is in need of servicing, the transmission and clutch cover assembly must be removed to access the clutch pack assembly. Parts that are worn or damaged must be replaced. Generally, components of the clutch pack assembly are not serviceable separately. As a rule, if the pressure plate, springs, release levers, etc., are damaged, the complete clutch assembly should be replaced. Today, clutch rebuilding is usually performed only by clutch specialty rebuilding shops.

The following sections explain how to remove a clutch, inspect clutch components for damage, and install a clutch.

SERVICING PRECAUTIONS

For many years, clutch facings were made with asbestos fibers. Asbestos has been proven to cause numerous health hazards when inhaled in sufficient amounts or over an extended period of time. The Occupation Safety and Health Administration has placed stringent regulations on working environments where asbestos brake linings or clutch facings must be handled. The very least a technician should do to protect himself or herself from exposure is wear a respirator when working on a clutch.

Many manufacturers have developed nonasbestos facings using natural and man-made materials in place of asbestos. Current OSHA regulations do not cover nonasbestos fibers. Medical experts do not agree about the possible long-term risks of working with and breathing nonasbestos fibers. However, some experts think that long-term exposure to some nonasbestos fibers could cause disease. Therefore, it is wise for the technician to always follow the precautions given here whenever working on any clutch.

1. Whenever possible, work on clutches in a separate area away from all other operations.
2. Consider wearing a respirator approved by the **National Institute for Occupational Safety and Health (NIOSH)** or OSHA during all clutch service procedures. Wear the respirator from removal through installation.
3. Never use compressed air or dry brushing to clean clutch parts or assemblies. OSHA recommends the use of cylinders that enclose the clutch. These cylinders have vacuums with high-efficiency filters and workman's arm sleeves. But if such equipment is not available, carefully clean parts and assemblies in the open air.
4. During disassembly, carefully place all parts on the floor to avoid getting dust into the air. Use an industrial vacuum cleaner with a high-efficiency filter system to clean dust from the clutches. After using the vacuum, remove any remaining dust with a rag soaked in water and wrung until nearly dry.
5. When cleaning the work area, never use compressed air or dry sweeping. Use an industrial vacuum with a high-efficiency filter and rags soaked in water and wrung until nearly dry. Used rags should be disposed of with care to avoid getting dust into the air. Use an approved respirator when emptying vacuum cleaners and handling used rags.
6. Workers should wash their hands before eating or drinking. Work clothes should not be worn home; they should be vacuumed after use and then should be laundered separately, without shaking, to prevent fiber dust from getting into the air.

TRANSMISSION/CLUTCH REMOVAL

The following is a general procedure for removing the transmission and clutch. Always consult a manufacturer's service manual for specific instructions and specifications.

Transmission Removal

1. Remove the shift lever from the transmission. If necessary, remove the shift assembly from the transmission.
2. Mark the yoke or flange of the drive shaft and the output shaft of the transmission. The marks on the drive shaft and the output shaft ensure the drive shaft is correctly reinstalled.
3. Remove the drive shaft.
4. Disconnect all the electrical connections from the transmission.
5. Disconnect all the air lines from the transmission.
6. If used, remove the return spring from the clutch lever on the transmission. Mark and disconnect the clutch linkage from the clutch housing on the transmission.
7. Rotate the release yoke so it will clear the release bearing when it is removed.
8. If a hydraulic system is used on the clutch, disconnect the pushrod and the spring from

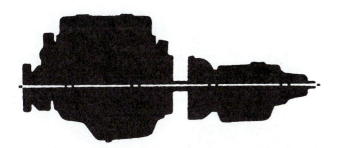

FIGURE 9–52 *Maintain proper angle when pulling clutch and transmission away from engine. (Courtesy of Eaton Corp.—Eaton Clutch Div.)*

the release fork. Remove the hydraulic cylinder from the bracket on the transmission. Use wires to support the cylinder on the frame.

9. Use a sling or jack to maintain transmission alignment (**Figure 9–52**). Do not allow the rear of the transmission to drop or allow the transmission to hang unsupported in the splined hubs of the driven discs as this could distort them and cause poor clutch operation or clutch release problems. Taking these precautions will prevent bending and distortion of the clutch discs and ensures trouble-free performance.

10. Remove the fasteners that attach the transmission to the brackets on the frame.

CAUTION: Make sure the transmission does not hang by the input shaft in the pilot bearing bore in the flywheel. The clutch assembly and the pilot bearing will be damaged if the transmission is supported by the input shaft.

11. Remove the bolts and washers that attach the bell housing to the engine. Pull the transmission and bell housing straight out from the engine. Remove the transmission from the vehicle (**Figure 9–53**).

12. If used, remove the clutch brake assembly from the input shaft of the transmission.

 Shop Talk _____

If the clutch is not being replaced, mark the cover of the clutch and the flywheel. The marks ensure that the clutch is correctly reinstalled on the flywheel.

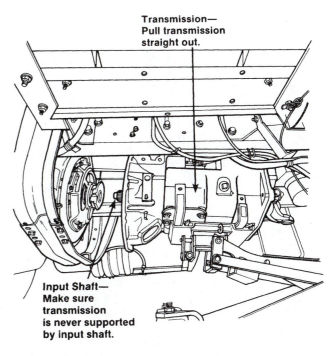

Transmission— Pull transmission straight out.

Input Shaft— Make sure transmission is never supported by input shaft.

FIGURE 9–53 *When removing the transmission, make sure that it is fully supported and pull it straight out. (Courtesy of ArvinMeritor Inc.)*

Removing Clutch From Flywheel

1. Install a clutch alignment tool through the release bearing and the driven discs and into the flywheel pilot bearing (**Figure 9–54**). The alignment tool supports the clutch assembly during removal. If an alignment tool is not available, use an input shaft from a manual transmission.

2. Pull the bearing back using the release tool shown and insert two $3/8$-inch spacers between the clutch cover and the release bearing (**Figure 9–55**). The spacers relieve the spring load of the clutch cover and allow for reinstallation.

3. Loosen the mounting bolts around the flywheel in a progressive crisscross pattern, but do not remove them.

4. Completely remove the top two bolts that fasten the pressure plate and cover assembly to the flywheel. Install two $3/8$-inch diameter, $2^1/2$-inch long guide studs in the holes (**Figure 9–56**).

5. Connect a lifting device to the pressure plate and cover assembly (**Figure 9–57**).

6. Remove the remaining bolts that fasten the pressure plate and cover assembly to the flywheel.

FIGURE 9–54 Install an alignment tool in the clutch before loosening the clutch cover assembly. (Courtesy of Eaton Corp.—Eaton Clutch Div.)

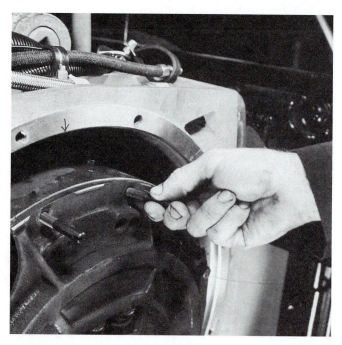

FIGURE 9–56 Install guide studs in the top of the cover assembly to support the clutch and guide it straight away from the flywheel during removal. (Courtesy of Eaton Corp.—Eaton Clutch Div.)

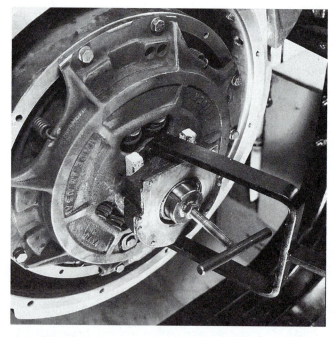

FIGURE 9–55 Place a spacer between the release bearing and the clutch cover to unload the clutch. (Courtesy of Eaton Corp.—Eaton Clutch Div.)

 Shop Talk

When removing a 15¹⁄₂-inch clutch, the discs and the pressure plate can stay in the cover.

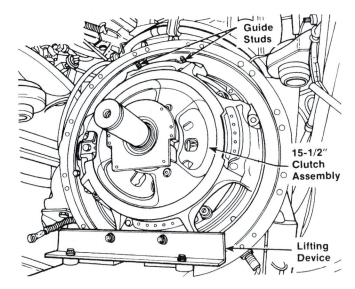

FIGURE 9–57 Use a lifting device to support the heavier 15¹⁄₂-inch clutches. (Courtesy of ArvinMeritor Inc.)

7. Lift the pressure plate and cover assembly over the alignment tool and off the flywheel.
8. Remove the rear disc.
9. Remove the center plate.
10. Remove the front disc.
11. Remove the alignment tool from the flywheel.

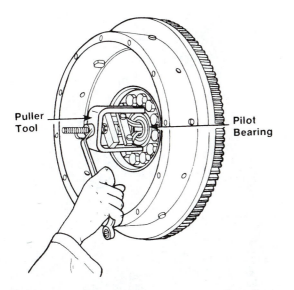

FIGURE 9–58 Use a puller to remove the pilot bearing. (Courtesy of ArvinMeritor Inc.)

Pilot Bearing Replacement

Every time the clutch assembly is serviced or the engine is removed, the pilot bearing in the flywheel should be removed and replaced. Use an internal puller or a slide hammer to remove the pilot bearing (**Figure 9–58**). Discard the pilot bearing.

CLUTCH INSPECTION

After removing the clutch from the flywheel, inspect the clutch components for damage or signs of wear. It is important that the cause of wear be determined and the problem corrected.

Transmission Clutch Housing and Flywheel Surfaces

The engine and transmission must be in alignment. To check this, perform the following inspection procedure. Surfaces being gauged or measured must be clean for accurate measurements.

1. Inspect the surface of the flywheel for wear or damage. Make sure that the flywheel is not cracked. Heat discoloration is a normal wear condition that can be removed with an emery cloth. Some wear or damage can be removed by grinding a new surface on the flywheel. If wear or damage on the surface of the flywheel cannot be removed, the flywheel must be replaced.
2. Inspect the teeth of the ring gear on the outer surface of the flywheel. If the teeth are worn or damaged, replace the ring gear or

the flywheel and inspect the starter drive teeth.
3. Inspect the mating surfaces of the transmission clutch housing and the engine flywheel housing (**Figure 9–59A**). Any appreciable wear on either housing flange will cause misalignment. Most wear will be found on the lower half of these surfaces, with the most common wear occurring between 3 o'clock and 8 o'clock positions (**Figure 9–59B**). Replace the clutch housing or flywheel housing, if worn.

👓 Shop Talk

When measuring runout dimensions on a flywheel, force the flywheel toward the engine so that the end play of the crankshaft is not measured during the flywheel runout check.

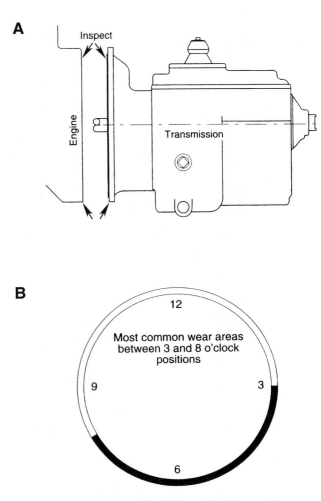

FIGURE 9–59 (A) Clutch housing inspection points; and (B) areas of maximum wear. (Courtesy of Eaton Corp.—Eaton Clutch Div.)

FIGURE 9–60   *Checking flywheel runout. (Courtesy of Eaton Corp.—Eaton Clutch Div.)*

FIGURE 9–61   *Checking runout of the flywheel housing bore. (Courtesy of Eaton Corp.—Eaton Clutch Div.)*

Flywheel Outer Surface Runout. To make this check:

1. Secure a dial indicator to the flywheel housing with the gauge finger on the flywheel near the outer edge (**Figure 9–60**).
2. Zero the dial indicator.
3. Rotate the flywheel by hand one revolution in the direction of engine rotation. On some engines the crankshaft can be rotated by putting a socket on the nut that holds the pulley on the front of the crankshaft. If access to the front of the crankshaft is difficult, use a spanner wrench on the teeth of the flywheel to rotate the crankshaft.
4. Record the reading on the dial indicator, marking the high and low points.
5. The acceptable runout on the outer surface of the flywheel is a specified amount multiplied by the diameter of the flywheel in inches. For example, maximum permissible runout may be listed as 0.0005 inch per inch of flywheel diameter with the total indicated difference between the high and low points being 0.007 inch or less for a 14-inch clutch, and 0.008 inch or less for a 15½-inch clutch. Check service manual specifications for exact tolerances.

Checking Flywheel Housing Pilot Runout. To make this check:

1. Secure the dial indicator to the crankshaft.

2. With the dial gauge plunger against the housing pilot, rotate the crankshaft one revolution in the direction of engine rotation (**Figure 9–61**).
3. Use a marker or soapstone to mark the high and low points and record dial readings. Total difference between high and low points should not exceed manufacturer's specifications, which normally range between 0.006 and 0.015 inch.
4. If runout exceeds specifications, service the flywheel as required.

Checking Flywheel Housing Face Runout. Proceed as follows:

1. Position the dial gauge plunger to contact the face of the engine flywheel housing (**Figure 9–62**).
2. As described above, rotate the crankshaft marking high and low points. Typically, the total difference between the high and low points should not exceed 0.008 inch.

Checking the Runout on the Bore of the Pilot Bearing. Proceed as follows:

1. Install a dial indicator so that the base of the indicator is on the mounting surface of the flywheel housing. Put the plunger of the dial indicator against the outer surface of the bore for the pilot bearing (**Figure 9–63**).

FIGURE 9–62   Checking runout on the outer surface of the flywheel housing. (Courtesy of Eaton Corp.—Eaton Clutch Div.)

FIGURE 9–63   Checking runout on pilot bearing bore. (Courtesy of Eaton Corp.—Eaton Clutch Div.)

2. Zero the dial indicator.
3. Manually turn the crankshaft one revolution in the direction of the engine rotation.
4. Record the reading on the dial indicator. The maximum allowable runout for the surface of the bore of the pilot bearing is typically 0.005

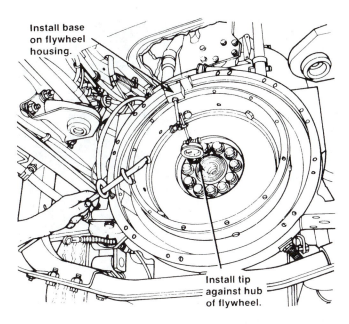

Install base on flywheel housing.

Install tip against hub of flywheel.

FIGURE 9–64   Checking end play of crankshaft. (Courtesy of ArvinMeritor Inc.)

inch. If the runout is more than manufacturer's specifications, service the crankshaft as required. If these limits are exceeded, the problem must be corrected or misalignment will cause premature wear to the drivetrain components.

Checking the End Play of the Crankshaft. Proceed as follows:

1. Install a dial indicator so that the base of the indicator is on the flywheel housing. Set the plunger of the dial indicator against the hub of the flywheel (**Figure 9–64**).
2. Put the dial indicator on the zero (0) mark.
3. Pull the flywheel away from the engine.
4. Record the reading on the dial indicator. Check the reading against the specification of the engine manufacturer. Service the crankshaft as required.

Flywheel Drive Pin Inspection/Removal

On 14-inch clutches, inspect the drive pins in the flywheel housing. Replace any worn or damaged drive pins using the following procedure:

1. Remove the flywheel. See the procedure of the manufacturer of the vehicle. Consult the manufacturer's specifications.
2. Remove the setscrew(s) that fasten each drive pin in the flywheel housing (**Figure 9–65**).

FIGURE 9–65   Remove the drive pin setscrew and drive the pins out with a drift and hammer. (Courtesy of Eaton Corp.—Eaton Clutch Div.)

3.   Use a hammer and punch to remove the drive pins from the flywheel housing.

WARNING: Wear eye protection when driving pins with a hammer and punch.

Release Fork and Shaft Inspection

Inspect the shaft and the release fork (**Figure 9–66**). Ensure the release fork is straight and the tips of the fork are not worn or damaged. Replace forks that are worn or damaged.

Make sure the shaft rotates freely in the clutch housing. The shaft should not have any side-to-side movement in the housing. If used, inspect the bushings for the shaft in the housing. Replace any components that are worn or damaged.

Input Shaft Inspection

Inspect the splines on the input shaft. Ensure the splines are not worn or damaged. Inspect the area of travel or the release bearing for damage. Use an emery cloth to remove small scratches from the input shaft. Replace input shafts that are worn or damaged. Wear or damage on the input shaft can cause the clutch to operate incorrectly.

Pressure Plate and Cover Assembly

Remove dirt and contamination from the pressure plate and cover assembly with nonpetroleum-based cleaning solvents.

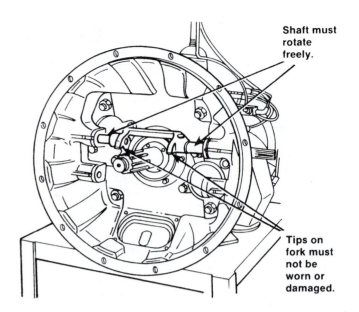

FIGURE 9–66   Check the release shaft and fork for wear and smooth operation. (Courtesy of ArvinMeritor Inc.)

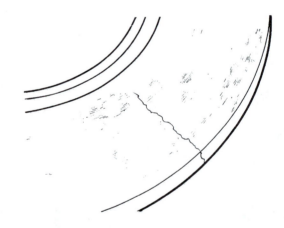

FIGURE 9–67   This pressure plate has cracked due to extreme heat. (Courtesy of Eaton Corp.—Eaton Clutch Div.)

Inspect the cover for wear and damage. Make sure the spring (or springs) inside the cover is not broken. If a spring is broken, the clutch cover must be disassembled to replace the spring.

Inspect the pressure plate for wear or damage (**Figure 9–67**). Replace plates that are cracked. Minor heat discoloration is normal. Heat marks can be removed with an emery cloth. If heat discoloration cannot be removed, replace the pressure plate. Continue to check the pressure plate according to the following procedure.

1.   Place the pressure plate and cover assembly on a bench so that the plate faces toward you.

2. Use a thickness gauge and a straight edge to measure any scratches or scoring on the pressure plate. A caliper is used to measure thickness (**Figure 9–68**). If damage to the surface of the plate is more than the manufacturer's specifications (typically 0.015 inch), replace the pressure plate.

3. Make sure the surface of the pressure plate is flat. Put a straightedge ruler across the complete surface of the pressure plate. Put a thickness gauge under each gap that appears between the ruler and the pressure plate (**Figure 9–69**). Measure the pressure plate at four positions. If the gap is more than the manufacturer's specifications, replace the pressure plate.

4. Measure the runout of the pressure plate to make sure the surface is parallel. Put the base of a dial indicator inside the center of the plate. Put the tip of the dial indicator on the surface of the plate (**Figure 9–70**). Set the dial indicator at zero. Rotate the dial indicator one complete turn around the surface of the pressure plate. If the reading on the dial indicator is more than 0.002 inch, replace the pressure plate.

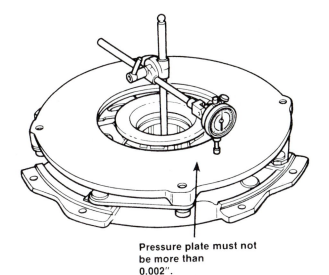

Pressure plate must not
be more than
0.002".

FIGURE 9–70 Check the pressure plate runout. (Courtesy of ArvinMeritor Inc.)

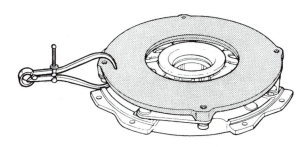

FIGURE 9–68 Measure the thickness of the pressure plate as well as the depth of any scratches. (Courtesy of ArvinMeritor Inc.)

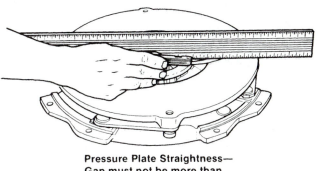

Pressure Plate Straightness—
Gap must not be more than
0.002".

FIGURE 9–69 Use a straightedge and a thickness gauge to check the pressure plate for straightness. (Courtesy of ArvinMeritor Inc.)

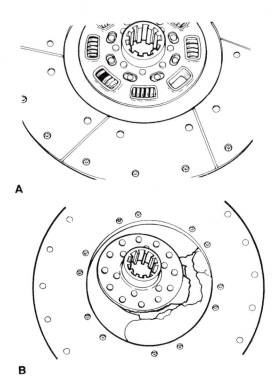

A

B

FIGURE 9–71 Check the driven discs for: (A) broken springs; and (B) cracked hub. (Courtesy of Eaton Corp.—Eaton Clutch Div.)

Clutch Discs

Inspect the clutch discs for wear or damage (**Figure 9–71**). Make sure the dampening springs are not loose in the hub (springs that rattle are not loose). If the springs have any movement, the springs are loose. Make sure the splines in the hub are not damaged. Make sure the hub is fastened to the disc. Replace discs that are worn or damaged. Use a cleaning solvent with a nonpetroleum base to remove grease and oil from the discs. If grease and oil cannot be removed, replace the disc.

Ceramic Lining

The ceramic lining is fastened to the disc with rivets. Replace the disc if the friction face is loose or damaged. Replace the disc if the friction face is worn to the top of the rivets or below the top of the rivets (**Figure 9–72**).

Molded Organic Lining

The organic friction face is integrally molded onto the disc. Replace the disc if the friction face is loose or damaged. Use a micrometer to measure the thickness of the friction face on the disc (**Figure 9–73**). If the thickness is less than the manufacturer's specifications, replace the disc.

Coasting with the clutch disengaged can cause the clutch linings to burst as a result of high centrifugal forces. **Figure 9–74** shows a disc with a ruptured lining.

Center Plate

On 14-inch clutches, inspect the slots for the drive pins in the center plate. If the slots are worn, replace the center plate. On 15$\frac{1}{2}$-inch clutches, inspect the tabs on the outer edge of the center plate. If the tabs are worn or damaged, replace the center plate.

Inspect the center plate for wear or damage. Make sure the plate is not cracked. Heat discoloration is a normal condition. The heat discoloration can be removed with an emery cloth. If heat discoloration cannot be removed, replace the center plate. Continue to check the center plate on each side of the plate according to the following procedure.

1. Use a micrometer or caliper to measure the thickness of the center plate (**Figure 9–75**). If

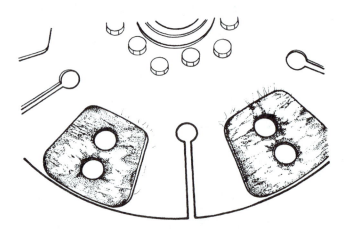

FIGURE 9–72   Check for excessive wear and heat damage. (Courtesy of Eaton Corp.—Eaton Clutch Div.)

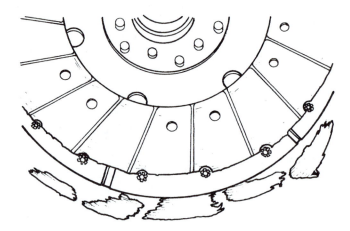

FIGURE 9–74   A burst friction facing is a sign that the truck has been coasting with the clutch disengaged. (Courtesy of Eaton Corp.—Eaton Clutch Div.)

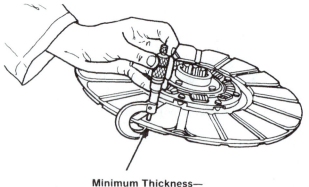

Minimum Thickness—
0.283"

FIGURE 9–73   Measure plate thickness and replace if worn thin. (Courtesy of ArvinMeritor Inc.)

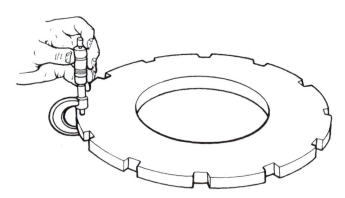

FIGURE 9–75   Measure the thickness of the center plate and replace it if it is worn smaller than the manufacturer's specifications. (Courtesy of ArvinMeritor Inc.)

the thickness is less than the manufacturer's specification, replace the center plate.

2. Make sure the center plate is flat. Put a straightedge ruler across the complete surface of the plate. Put a thickness gauge under each gap that appears between the ruler and the center plate (**Figure 9–76**). If the gap is more than the manufacturer's specifications, go to step 6.

3. Measure the runout of the center plate to make sure the surface of the plate is parallel. Put the base of a dial indicator inside the center of the plate. Put the tip of the dial indicator on the surface of the plate (**Figure 9–77**). Zero the dial indicator. Rotate the dial indicator one complete turn around the sur-

face of the center plate. If the reading on the dial indicator is more than the manufacturer's specifications, go to step 6.

4. If either the runout or the flatness of the center plate exceeds the manufacturer's specifications, machine the pressure plate. Do not grind more material from the plate than necessary. Do not machine the plate smaller than the manufacturer's specification for minimum plate thickness. If the plate cannot be ground flat with the minimum thickness specification, replace the plate.

Pilot Bearing

Although the pilot bearings are replaced when the clutch is removed, inspect the old pilot bearing for wear and damage. Determine and correct the cause of the wear and damage.

CLUTCH INSTALLATION

Installation procedures will vary according to clutch size and model. The following procedures for 14-inch diameter and 15$\frac{1}{2}$-inch clutches are general in nature. Follow service manual instructions.

Installing a 14-Inch Diameter Clutch

Begin the installation by installing new drive pins in the *previously unused* holes in the flywheel. The pins must be spaced out equally with their shanks press fit into the flywheel. It is recommended that a drive pin aligning tool is used during this installation. Install new drive pins according to the following procedure:

1. Put the drive pin in the flywheel housing. Use a machinist's square to make sure the flat sides of the pin are at a 90 degree angle to the top of the housing (**Figure 9–78**).

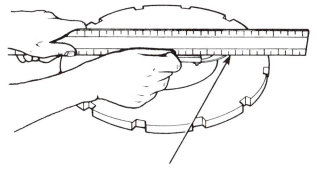

Gap must be 0.002″ or less.

FIGURE 9–76 Use a straightedge and thickness gauge to check the center plate for flatness. (Courtesy of ArvinMeritor Inc.)

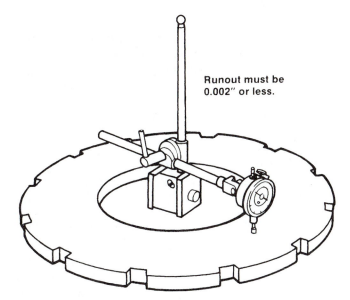

Runout must be 0.002″ or less.

FIGURE 9–77 Use a dial indicator to check the center plate for runout. (Courtesy of ArvinMeritor Inc.)

FIGURE 9–78 Use a machinist's square to make sure that the flat of the pin drive is square with the friction face of the flywheel. (Courtesy of Eaton Corp.—Eaton Clutch Div.)

2. Use a C-clamp to install the drive pin in the housing. Press the pin into the housing until the head of the pin touches the inner bore of the housing.

   Make sure that the heads are square with the friction surface as misaligned drive pins may cause a clutch release problem.

3. Put two small spacers on the friction surface of the flywheel and carefully set the intermediate plate over the drive pins. The spacers will prevent getting your fingers pinched by the intermediate plate and give you a finger hold when removing it.

4. Turn the intermediate plate in one direction as far as it will go. Use a 0.006-inch thickness gauge to check the clearance between the drive pin and the drive slot (**Figure 9–79**). Check the same side of each pin. The minimum clearance between the drive pins and the drive slots is 0.006 inch. If the proper clearance is not obtained, realign the drive pins and recheck the squareness.

5. Next, recheck the clearances. This check is necessary to ensure that the clutch will release properly when installed. *Do not file the drive pin slots* on the intermediate plate to obtain correct clearances. Doing so will cause an unequal load on the pins. This is a frequent cause of poor release or the clutch not releasing at all. It can also result in broken drive pins.

6. If the alignment and clearance are correct, lock each of the six pins in place with two new, $^3/_8$ inch by $^3/_8$ inch set screws (**Figure 9–80**).

FIGURE 9–80 *Installing setscrews to hold pins in place. (Courtesy of Eaton Corp.—Eaton Clutch Div.)*

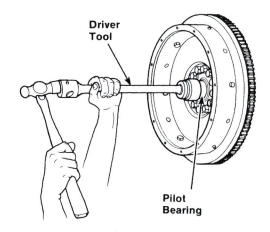

FIGURE 9–81 *Use a driver to install the pilot bearing. (Courtesy of ArvinMeritor Inc.)*

7. Reinstall the flywheel to the engine crankshaft, making sure the chalk marks are lined up. Refer to the engine manual for torque specs on the flywheel mounting bolts.

8. Remove the pilot bearing and replace with a new bearing (**Figure 9–81**).

CAUTION: Tap on the outer race of the pilot bearing only. Make sure it is seated properly in the bearing bore. This bearing must have a press fit within the pilot bearing bore.

9. Place the front driven disc against the flywheel with the side stamped "flywheel" facing the engine (**Figure 9–82**). Remember: It is essential that the side stamped "flywheel" faces the engine and the side stamped "pressure plate" faces the transmission.

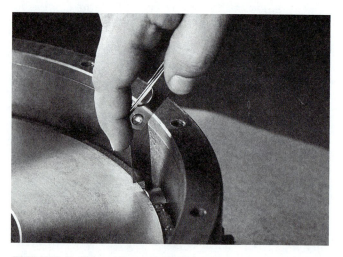

FIGURE 9–79 *Use a thickness gauge to check the pin-to-plate clearance. (Courtesy of Eaton Corp.—Eaton Clutch Div.)*

FIGURE 9–82   *Installing front driven disc.*
(Courtesy of Eaton Corp.—Eaton Clutch Div.)

FIGURE 9–83   *Installing intermediate plate.*
(Courtesy of Eaton Corp.—Eaton Clutch Div.)

10. The relative position of the buttons on the front and rear driven discs is not important. If so equipped, make sure that the Kwik-Adjust mechanism will be aligned with the opening in the clutch housing of the transmission.

11. Install the intermediate plate by positioning the drive slots on the drive pins and remove the aligning tool (**Figure 9–83**).

CAUTION: Some heavy-duty clutches have thicker intermediate plates and thinner super buttons than standard clutches. Do not intermix these components.

12. If you are installing a Super-Duty Clutch, be sure to install three antirattle springs. (See **Figure 9–11**.) Space them equally between the drive pins with the rounded sections toward the flywheel face. For safety reasons, you should wear heavy gloves when installing antirattle springs.

13. Insert the aligning tool through the hub of the rear disc with the side stamped "pressure plate" facing the transmission and install it after the intermediated plate (**Figure 9–84**).
    Note: It is essential that the side stamped "pressure plate" faces the transmission and the side stamped "flywheel" faces the engine.

14. Reinsert the aligning tool through the hub of the front driven disc and into the pilot bearing. The relative position of the buttons on the front and rear discs is not important.

15. Position the clutch cover over the guide studs installed at the top of the flywheel (**Figure 9–85**). Make sure that the Kwik-Adjust mechanism will be aligned with the opening in the bell housing of the transmission. Start six $3/8$-inch x $1\,1/4$-inch grade 5 or better mounting bolts with lockwashers and tighten them finger tight.

16. *Lightly* tap the aligning tool to make sure it is centered and seated into the pilot bearing. Remove the guide studs and replace them with bolts and lockwashers.

FIGURE 9–84   *Reinstalling aligning tool. (Courtesy of Eaton Corp.—Eaton Clutch Div.)*

FIGURE 9–85   *Installing clutch cover. (Courtesy of Eaton Corp.—Eaton Clutch Div.)*

FIGURE 9–86   *Torque sequence for tightening bolts. (Courtesy of Eaton Corp.—Eaton Clutch Div.)*

17. Tighten the bolts in a crisscross sequence to pull the clutch into its proper position in flywheel pilot (**Figure 9–86**). Start with the lower left-hand bolt. Failure to tighten the bolts in this manner can cause permanent damage to the clutch cover or create an out-of-balance condition.

18. To achieve final torque, progressively tighten all bolts 35 to 40 lbs. ft. As the bolts are tightened, the wooden spacers should fall out. If they do not fall free, be sure to remove them. You may have to lightly tap the aligning tool with a mallet to remove them.

Installation of a Typical 15¹/₂-Inch Clutch

1. Insert two ⁷/₁₆-inch (5 inch long) guide studs into the two upper mounting holes of the flywheel (**Figure 9–87A**).
2. Rotate the flywheel to level the guide studs (**Figure 9–87B**).
3. Insert the aligning tool through the release bearing sleeve in the new clutch (**Figure 9–87C**).
4. Put the rear driven disc on the aligning tool with the side stamped "pressure plate" facing the pressure plate (**Figure 9–87D**).
5. Place the intermediate plate in the clutch cover and align the driving lugs of the plate with the slots provided (**Figure 9–87E**). The positive separator pins in the intermediate plate must be flush with the cast lug on the pressure plate side.
6. Install the front disc on the aligning tool with the side stamped "flywheel" facing the engine (**Figure 9–87F**).
7. Remember: It is essential that the side stamped "flywheel" faces the engine and the side stamped "pressure plate" faces the transmission. The relative positions of the buttons on the front and rear driven discs is not important.
8. If equipped, make sure the Kwik-Adjust mechanism will be aligned with the opening in the bell housing of the transmission.
9. Position the clutch over the guide studs and slide it forward until contact is made with the flywheel surface (**Figure 9–88A**). Because a 15¹/₂-inch clutch assembly weighs about 150 pounds, a hoist may be required to lift it into place.
10. Start six ⁷/₁₆-inch x 2¹/₄-inch grade 5 or better mounting bolts with lockwashers and tighten them finger tight (**Figure 9–88B**).
11. *Lightly* tap the aligning tool to make sure it is centered and seated in the pilot bearing (**Figure 9–88C**). Remove the guide studs and replace them with bolts and lock washers.
12. Tighten the bolts in the crisscross pattern sequence shown in **Figure 9–86** to pull the clutch into its proper position in the flywheel pilot. You must start with the lower left-hand bolt. Failure to tighten the bolts in this manner

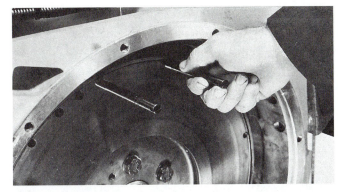

A

B

C

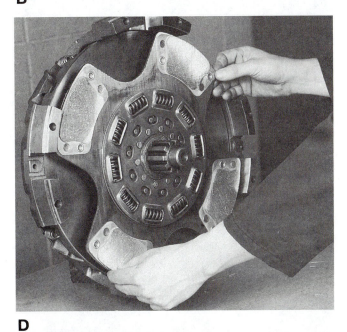

D

E

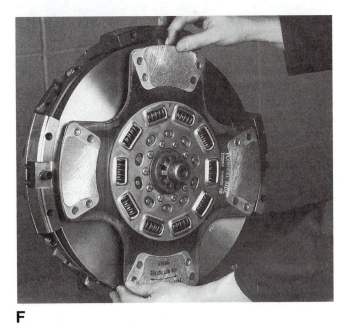

F

FIGURE 9–87 Installation of 15$\frac{1}{2}$-inch clutch: (A) install two guide studs; (B) align flywheel guide slots; (C) insert aligning tools; (D) install rear driven disc; (E) install intermediate plate; and (F) install front disc. (Courtesy of Eaton Corp.—Eaton Clutch Div.)

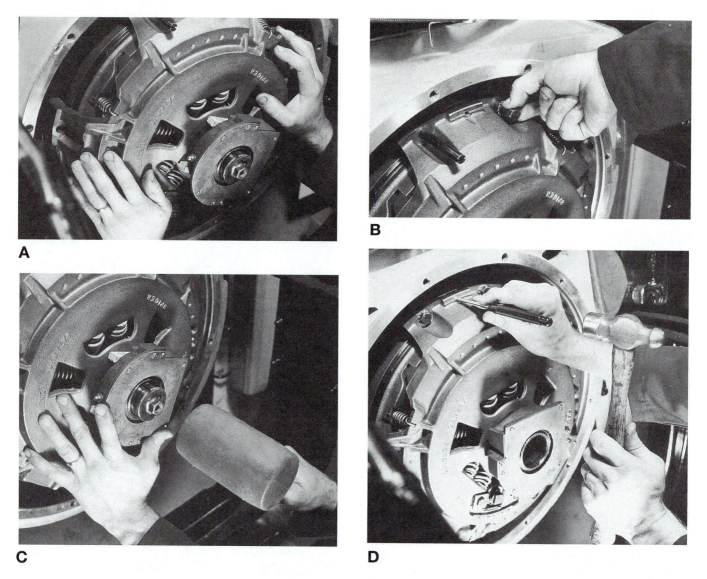

FIGURE 9–88   Installation of 15¹/₂-inch clutch (continued): (A) install clutch cover; (B) install mounting bolts; (C) install aligning tool; and (D) install positive separator pins after torquing mounting bolts. (Courtesy of Eaton Corp.—Eaton Clutch Div.)

can cause permanent damage to the clutch cover or create an out-of-balance condition. To achieve the final torque, progressively tighten all bolts to 45 to 50 lbs./ft.

Note: As the bolts are tightened, the wooden spacers should fall out. If they do not fall free, remove them. You may have to *lightly* tap on the aligning tool with a mallet to remove it.

13. Using a ¹/₄-inch diameter flat nose drift, lightly tap each of the four positive separator pins toward the flywheel (**Figure 9–88D**). After tapping, the pins should be flush against the flywheel.

REINSTALLING THE TRANSMISSION

To reinstall the transmission:

1. Shift into gear.
2. Inspect the transmission input shaft for wear. If worn, replace.
3. Using a clean, dry cloth wipe the shaft clean.
4. Check for wear on transmission bearing caps. See O.E. service manual if replacement is needed.
5. If a clutch brake is used, be sure to install it on the input shaft of the transmission at this time. (See **Figure 9–23**.)

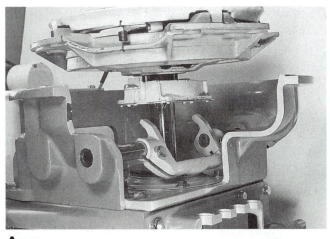

A

B

C

FIGURE 9–89 *Install the transmission being careful to rotate the release yoke so it clears the wear pads of the release bearings. (Courtesy of Eaton Corp.—Eaton Clutch Div.)*

6. Check for wear on the fingers of the clutch release yoke. Also, check the cross shaft bushings (**Figure 9–89A**). Replace them if necessary.

7. Check to be sure that neither cross shaft protrudes through the release fork since this could cause side loading of the release bearing (**Figure 9–89B**).

8. Rotate the release yoke so that it clears and is rotated over the wear pads of the release bearings as the transmission is moved forward (**Figure 9–89C**). *Be careful:* The release yoke fingers may not be elevated to the straight out position, and they could damage the clutch cover when moving the transmission forward. Make sure that the transmission is aligned with the engine when it is raised into position.

9. Do not allow the rear of the transmission to drop and hang unsupported in the splined hubs of the driven discs as this could distort them and cause poor clutch operation or clutch release problems. Taking these precautions will prevent bending and distortion of the finely tuned clutch discs and ensures trouble-free performance.

10. Move the transmission forward but never force the transmission into the flywheel housing. If it doesn't enter freely, investigate the cause and make the adjustments until it does.

11. Mate the transmission with the engine housing and install the mounting bolts.

12. Torque the bolts to the proper manufacturer's specifications.

13. Attach the clutch release linkage. Installation of the transmission is complete and adjustments can be made to the clutch and, if necessary, the linkage.

SUMMARY

- The function of a clutch is to transfer torque from the truck engine flywheel to the transmission.
- The components of a clutch can be grouped into two basic categories: driving members and driven members. Driving members include the flywheel, clutch cover, pressure plate, pressure springs and levers, intermediate plate, adapter ring, and adjustment mechanisms. Driven members include discs and disc facings.
- A clutch can be released or disengaged by one of two methods: push type clutch or pull type clutch.
- Clutch brakes are found in some pull type clutches and can be grouped into four types: conventional, limited torque, torque limiting, and two-piece clutch brakes.

- The clutch linkage, which connects the clutch pedal to the release fork or yoke, can be one of three types: mechanical, hydraulic, and air control.
- The major cause of clutch failure is excess heat. Damage caused by heat may be the result of starting in the incorrect gear, shifting or skip-shifting, riding the clutch pedal, and improper coasting.
- Free pedal, or pedal lash, is the amount of free play in the clutch pedal in the cab. This measurement is directly related to free travel, which is the clearance distance between the release yoke fingers and the clutch release bearing pads.
- Clutches should be checked periodically for proper adjustment and lubrication.
- There are three basic types of adjustment methods used on angle coil spring clutches. Two are manual adjusting mechanisms (lock-strap mechanism and Kwik-Adjust component) and the third is self-adjusting (wear compensator).
- Pull type clutches with perpendicular springs use a threaded sleeve and retainer assembly that can be adjusted to compensate for disc lining wear.
- The procedure for adjusting the release yoke finger to release bearing wear pad clearance can differ, depending on the type of truck and linkage design.
- Asbestos and nonasbestos fibers could pose a health risk. Technicians should take the appropriate safety precautions when servicing clutches.
- The following should be checked in a clutch inspection: transmission bell housing, flywheel housing, flywheel drive pin, release fork and shaft, input shaft, pressure plate and cover assembly, clutch discs, friction facings, center plate, and pilot bearing.

REVIEW QUESTIONS

1. Which of the following clutch components is a driven member?
   a. clutch cover
   b. clutch disc
   c. intermediate plate
   d. pressure plate

2. Which of the following is not a part of the clutch cover assembly?
   a. pressure spring
   b. release lever
   c. pressure plate
   d. adapter ring

3. Which of the following clutch components is splined to the input shaft of the transmission?
   a. clutch disc
   b. pressure plate
   c. release bearing
   d. flywheel

4. Which of the following is used to stop the rotation of the input shaft when upshifting into higher gears?
   a. conventional clutch brake
   b. torque limiting clutch brake
   c. clutch pedal
   d. none of the above

5. Which of the following driving procedures will result in clutch damage?
   a. driving with foot on the clutch pedal
   b. using the clutch as a brake to hold the truck on a hill and incline
   c. coasting downhill with the transmission in gear and the clutch disengaged
   d. all of the above

6. Which of the following could be a cause of poor clutch release?
   a. worn friction facings on the clutch discs
   b. worn clutch linkage
   c. worn release bearing
   d. all of the above

7. When a truck is brought in for servicing because of a vibration problem, Technician A suspects damaged splines on the transmission input shaft. Technician B says that the problem might be a driveline out of phase. Who is correct?
   a. Technician A
   b. Technician B
   c. both A and B
   d. neither A nor B

8. How often should a clutch be inspected and lubricated?
   a. every month
   b. every 6,000 to 10,000 miles
   c. any time the chassis is lubricated
   d. all of the above

9. When adjusting clutch linkage, Technician A says that pedal free travel should be about $1\frac{1}{2}$ to 2 inches. Technician B says that free travel should be less than $\frac{1}{2}$ inch. Who is correct?
   a. Technician A
   b. Technician B
   c. both A and B
   d. neither A nor B

10. When servicing clutches, Technician A always wears an OSHA-approved respirator. Technician B uses compressed air to clean off components. Who is following correct safety procedures?
    a. Technician A
    b. Technician B
    c. both A and B
    d. neither A nor B

11. Which of the following steps must be performed first when removing a clutch?
    a. unbolting the clutch from the flywheel
    b. removing the clutch brake
    c. installing a clutch alignment tool through the clutch
    d. supporting the transmission with a transmission jack

12. On which of the following types of flywheels would the center intermediate plate be driven by drive pins?
    a. flat flywheel
    b. pot flywheel
    c. torque converter
    d. all of the above

13. Upon inspection, it is discovered that the facing of a clutch disc has burst. Technician A says that the cause is the driver using the clutch as a brake. Technician B says the problem might be the driver coasting downhill with the transmission in gear and the clutch disengaged. Who is correct?
    a. Technician A
    b. Technician B
    c. both A and B
    d. neither A nor B

14. When installing a 14-inch clutch, Technician A installs the new drive pins in previously unused holes in the flywheel. Technician B files the drive pin slots to ensure proper clearance are met. Which technician is following the proper procedure for the type of clutch being installed?
    a. Technician A
    b. Technician B
    c. both A and B
    d. neither A nor B

15. When installing a 15$\frac{1}{2}$-inch clutch, which of the following components should be installed in the clutch cover before installing the clutch on the vehicle?
    a. release bearing
    b. front clutch disc
    c. intermediate plate
    d. clutch brake

16. Which of the following is true?
    a. excessive clutch pedal free travel can cause slippage and short clutch life
    b. insufficient clutch pedal free travel can prevent complete clutch disengagement
    c. both a and b
    d. neither a nor b

17. Which of the following adjustments is made on nonsynchronized transmissions only?
    a. pedal height
    b. total pedal travel
    c. clutch brake squeeze
    d. free travel

18. A driver complains that the clutch pedal is hard to operate. Technician A says that the problem may be damaged bosses on the release bearing assembly. Technician B says that the clutch linkage may be worn or damaged. Who is correct?
    a. Technician A
    b. Technician B
    c. both A and B
    d. neither A nor B

19. Technician A says that when the driver releases the clutch while coasting downhill, the clutch overheats. Technician B says that overheating results when the driver keeps a foot on the clutch pedal. Who is correct?
    a. Technician A
    b. Technician B
    c. both A and B
    d. neither A nor B

20. There are grooves worn in the pressure plate. Technician A says that the clutch discs are worn or damaged. Technician B says that the springs are worn or damaged. Who is correct?
    a. Technician A
    b. Technician B
    c. both A and B
    d. neither A nor B

21. Which of the following is used to remove minor heat discoloration from the pressure plate?
    a. cleaning solvent with a nonpetroleum base
    b. ammonia
    c. emery cloth
    d. none of the above

22. Which of the following is true?
    a. the pilot bearing should be replaced whenever the clutch is removed
    b. the pilot bearing should be inspected when the clutch is removed
    c. flywheel cracks can be removed by grinding a new surface
    d. all of the above

10

Standard Transmissions

Prerequisite
Chapter 9

Objectives

After reading this chapter, you should be able to

- Identify and describe the various gear designs used in heavy-duty truck transmissions and components.
- Define and calculate both gear pitch and gear ratios.
- Explain the relationship between speed and torque in various combinations of meshed gears.
- Identify the major components of a typical transmission including input and output shafts, main shaft and countershaft gears, and shift mechanisms.
- Identify and describe the major types of shift mechanisms used in heavy-duty truck transmissions.
- Explain the role of main and auxiliary gear sections in a transmission and trace power flows through these sections.
- Describe the operating principles of range shift and splitter shift air systems.
- Define the roles of transfer cases and PTOs in heavy-duty truck operation.

Key Terms

air filter/regulator assembly
axis of rotation
bottoming
cab-over-engine (COE) trucks
climbing
direct drive
drive gear
driven gear

gears
gear pitch
keyed
overdrive
revolutions per minute (rpm)
rotation
shift bar housing
shift fork or yoke
slave valve
transfer case

During normal operating conditions, torque from the engine is transferred through the clutch to the input shaft of the transmission. **Gears** in the transmission housing alter the torque and speed of this input before transferring it to the transmission's output shaft. The rotating output shaft then drives the propeller shaft, drive axle shafts, and drive axle wheels. The gears in a standard transmission function to give the truck's engine a mechanical advantage over the vehicle's dri-

ving wheels. Without this mechanical advantage, an engine would not be able to move a heavily loaded vehicle from a standing start to highway speeds.

In any engine, the crankshaft always rotates in the same direction. If the engine transmitted its power directly to the drive axles, the wheels could be driven only in one direction. The transmission also provides gearing to reverse torque drive direction so the vehicle can be reversed.

The standard or manual transmission installed in heavy-duty trucks commonly has from five to twenty separate speed ratios. This enables the driver to operate the vehicle at top engine efficiency under a variety of loads and road conditions. The low-speed/high torque gear ratios needed for start-up and climbing steep grades can also be used to provide engine braking on severe downgrades.

10.1 MAIN AND AUXILIARY GEARING

Transmissions may consist of a single housing or multiple housings. A single housing transmission may use single or multiple countershafts. As the requirement for gear ratios increases, the single housing can be compounded by adding an auxiliary section or sections to it.

Many heavy-duty truck transmissions consist of two sections with distinct sets of gearing. One is the main or front gearing. The second is the auxiliary gearing coupled directly to the rear of the main gearing (**Figure 10–1**). The main and auxiliary gearing may be contained in a separate housing, with the auxiliary housing bolted directly onto the main or front housing. However, in some transmissions, both the main and auxiliary gearing are contained in a single housing.

The major advantage of this gearing arrangement is the large number of gear ratios. The main section normally contains gearing for five forward speeds plus reverse. The auxiliary housing contains gearing for two or three speeds. Two-speed auxiliary gearing creates high and low-speed ranges. Three-speed auxiliary gearing adds an **overdrive** gearing. When multiplied by the ratios produced in the main section, the auxiliary gearing can be used to create 8-, 9-, 10-, 13-, 15-, 18-, and 20-speed transmissions.

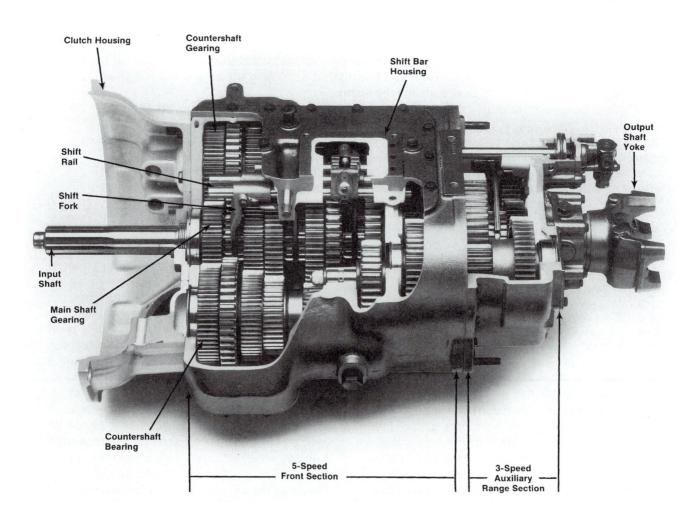

FIGURE 10–1 A thirteen-speed twin countershaft transmission with main and auxiliary gearing contained in separate housing. (Courtesy of ArvinMeritor Inc.)

10.2 GEARS

The purpose of a gear in a manual transmission transfer case is to transmit rotating motion. Gears are normally mounted on a shaft, and they transmit rotating motion from one parallel shaft to another. There are three different modes of operation. The shaft can drive the gear, the gear can drive the shaft, or the gear can be free to turn on the shaft. In this last case, the gear acts as an idler gear.

Sets of gears can be used to multiply torque and decrease speed, increase speed and decrease torque, or transfer torque and speed unchanged. The following are a number of basic gear definitions.

- **Drive Gear (Input).** A **drive gear** or driving gear is a gear that drives another gear or causes another gear to turn.
- **Driven Gear (Output).** A **driven gear** is a gear that is driven or forced to turn by a drive gear, by a shaft, or by some other device.
- **Rotation and Direction of Rotation. Rotation** is simply a term used to describe the fact that a gear, shaft, or other device is turning. The direction of rotation is described by comparing it to the rotation of the hands of a clock. All clocks that have hands rotate in the same direction. In order to tell what time it is, one must face the front of the clock. As time progresses the hands of the clock move in a clockwise (CW) direction. This can also be described as turning in a right-hand direction (as right-hand threads on a bolt or nut). Rotation in the opposite direction is described as counterclockwise (CCW) rotation or turning to the left (as in left-hand threads on a bolt or nut).

To determine the direction of rotation of gears in a transmission, face the front of the transmission. The front of the transmission is the end that is attached to the engine, or in a longitudinal manner. The direction of rotation of all transmission gears and shaft is, therefore, determined by viewing the end of the transmission that is attached to the engine.

The abbreviation of the term clockwise is CW, and for counterclockwise, it is CCW. Another term for clockwise rotation is forward rotation. Other terms used for counterclockwise rotation are backward rotation, turning backward, or turning in a reverse direction.

- **Speed of Rotation.** The term speed is used to describe the rotating frequency of gear shafts. When used to describe engine speed, it refers to the number of turns or revolutions the engine crankshaft makes in 1 minute (stated in **revolutions per minute,** or **rpm**).

When used to describe the rotation of transmission components, the word speed is used in a more general way to compare the rotating speed of one component to that of another. Terms used to describe speed comparisons include faster, slower, increased speed, decreased speed, **direct drive** (same speed) or one to one, underdrive and overdrive (output faster than input).

- **Axis of Rotation.** The **axis of rotation** is the center line around which a gear or part revolves.
- **Drive and Coast.** The drive side of gear teeth is the side that is in contact with teeth of another gear while torque is being applied. This is the side of gear teeth that is subject to the most wear. The coast side of gear teeth is the opposite side from the drive side. This side of the teeth is in contact when the drive wheels are driving the engine (for example, during deceleration).
- **Constant Mesh.** Gears that remain in mesh with each other and are not engaged or disengaged from each other by transmission or driver action when shifting gears.
- **Backlash.** This is the rotational movement of a gear when the gear with which it is in mesh is held stationary. All gears should have some backlash to allow for expansion of metal due to heat and for proper lubrication. Excessive backlash is an indication of gear tooth wear.
- **Bottoming. Bottoming** occurs when the teeth of one gear touch the lowest point between teeth of a mating gear. Bottoming does not occur in a two-gear drive combination but can occur in multiple-gear drive combinations. A simple two-gear drive combination tends to force the two gears apart; therefore, bottoming cannot occur in this arrangement.
- **Climbing. Climbing** is a gear problem caused by excessive wear in gears, bearings, and shafts in which the gears move sufficiently apart to cause the apex (or point) of the teeth on one gear to climb over the apex of the teeth on another gear with which it is meshed. This results in a loss of drive until other teeth are engaged; it also causes rapid destruction of the gears.

GEAR DESIGN

Figure 10–2 illustrates basic gear tooth geometry. **Gear pitch** is an important factor in gear design and

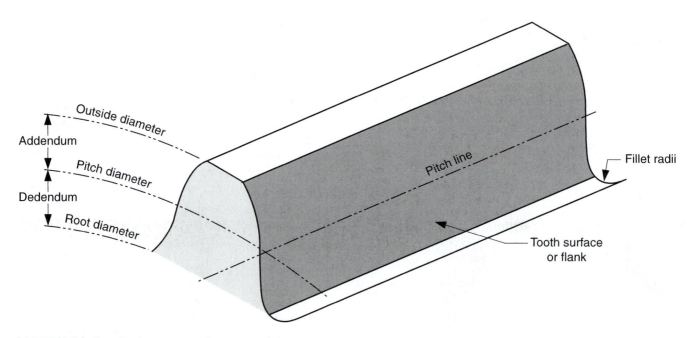

FIGURE 10–2 Basic gear tooth nomenclature.

operation. Gear pitch refers to the number of teeth per given unit of pitch diameter. A simple way of determining gear pitch is to divide the number of teeth by the pitch diameter of the gear. For example, if a gear has thirty-six teeth and a 6-inch pitch diameter, it has a gear pitch of six (**Figure 10–3**). The important fact to remember is that gears must have the same pitch in order to operate together. A five pitch gear will mesh only with another five pitch gear, a six pitch only with a six pitch, and so on.

The teeth of two gears, while in mesh, pass through the stages of contact (**Figure 10–4**): coming into mesh, at full mesh, and coming out of mesh. Coming into mesh is when the initial contact occurs in the dedendum (lower) portion of one tooth (on the

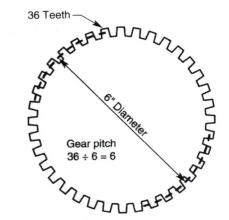

FIGURE 10–3 Determining gear pitch.

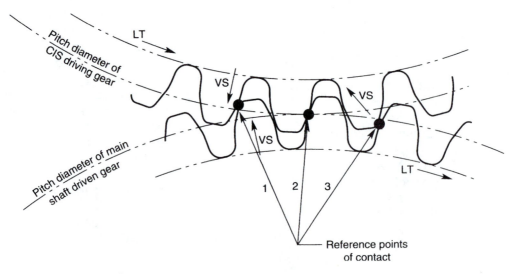

FIGURE 10–4 Three stages of tooth gear contact.

driving gear) and in the addendum (upper) portion of the mating tooth (on the driven gear). At this point of torque transfer, tooth loading (LT) is relatively light since most of it is carried by the teeth in full mesh and a portion by the teeth going out of mesh. Contact between the two teeth moves in a sliding action as they proceed through mesh. The sliding velocity (VS) decreases until it is zero when the contact points reach the intersection of their common pitch lines. At full mesh when the two teeth meet at their common or operating pitch line, there is only a rolling motion, no sliding. However, at this stage, tooth loading is at its greatest. Coming out of mesh, the two mating teeth again move in a sliding action, basically opposite of the initial contact stage.

An ideal tooth shape is designed for each gear. Since it is impossible to make a perfect component, manufacturers have established tolerances that will not shorten a gear's life. **Figure 10–5** is an example of these tolerances. The actual tooth profile can fall anywhere within the tolerance band.

A gear tooth is also designed to be parallel with the centerline of the shaft. As with the tooth profile, certain acceptable tolerances have been determined as shown in **Figure 10–6**. These close tolerances make it possible to successfully mate gears with lead variation in opposite directions without causing major failures.

Some manufactured transmission main shaft gearing is crowned. **Figure 10–7** is an exaggerated representation of the slight curvature machined from end to end on a main shaft gear tooth. Crowning prevents highly concentrated end loads, which might cause surface damage.

FIGURE 10–5   Gear tooth profile tolerance zone.

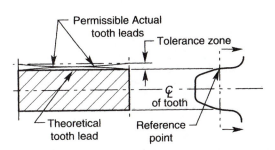

FIGURE 10–6   Gear tooth leads tolerance.

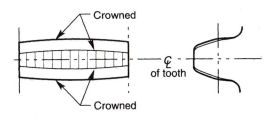

FIGURE 10–7   Gear tooth crowning.

GEAR RATIOS

As mentioned under "Gear Design" earlier in this chapter, only gears with matching pitches can be meshed together. When meshed, the relative speed of the meshed gears will depend on the ratio of the number of teeth on the driving (input) gear to the number of teeth on the driven (output) gear. In **Figure 10–8**, the two gears shown in **A** are the same size and have the same number of teeth. Since both gears are the same size, the speed of both gears is the same (direct drive). The two gears in **Figure 10–8B** show a small drive (input) gear driving a larger

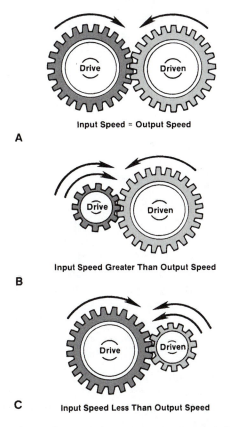

FIGURE 10–8   (A) When gears of equal size are meshed, output speed equals input speed; (B) a small gear driving a large gear reduces output speed; and (C) a large gear driving a small gear increases output speed (overdrive). In all cases, the output gear rotates in the opposite direction of the input gear.

driven (output) gear. Since the driving gear is smaller (fewer teeth) than the larger driven gear, the speed of the driven (output) gear will be slower than that of the input. Because the gear speed of the output shaft is reduced, this type of ratio (drive gear smaller than driven gear) is referred to as gear reduction. If, however, the driving gear is larger (more teeth) than the driven gear (as shown in **Figure 10–8C**), the speed of the driven (output) gear will increase. When output speed is greater than input speed, it is commonly referred to as "overdrive."

The relative speed of two meshed gears can easily be calculated by comparing the size and speed of one of the meshed gears to the size and speed of the other. Multiply the number of teeth of one gear by the speed of that gear, and divide that number by the number of teeth on the other gear.

For example, calculate the speed of the driven gear when the driving gear has 45 teeth and is rotating at 200 rpm while the driven gear has 75 teeth. The formula is as follows:

$$\frac{\text{teeth on driven gear}}{\text{teeth on driving gear}} = \frac{\text{rpm of driving gear}}{\text{rpm of driven gear}}$$

Putting in the known data leads to

$$\frac{75}{45} = \frac{200}{x}$$

therefore,

$$x = \frac{45 \times 200}{75} = \frac{9000}{75}$$

x = 120 rpm of driven gear

The input speed of 200 rpm is reduced to an output speed of 120 rpm. The relationship of input and output speeds is stated as gear ratio. Gear ratio can be calculated in a number of ways depending on the known data. One method is to divide the speed of the driving (input) gear or shaft by the speed of the driven (output) gear or shaft. In the above example this results in:

$$\frac{200 \text{ rpm}}{120 \text{ rpm}} = 1.66{:}1 \text{ gear ratio}$$

The gear ratio can also be found by dividing the number of teeth on the driven gear by the number of teeth on the driving gear. Again, in the above sample problem, this leads to

$$\frac{75 \text{ driven gear teeth}}{45 \text{ driving gear teeth}} = 1.66{:}1 \text{ gear ratio}$$

This means the driven (output) gear is rotating 1.66 times slower than the driving (input) gear.

Speed Versus Torque

A rotating gear also produces rotational force known as torque. Torque is calculated by multiplying the applied force by its distance from the centerline of rotation. A simple example of this law of physics is demonstrated whenever a torque wrench is used. As shown in **Figure 10–9**, when a force of 10 pounds is applied perpendicular to the centerline of the bolt's rotation at a distance 1 foot from this centerline, 10 pound-feet of torque are generated at the centerline of rotation. The torque wrench acts as a lever to apply this force.

Meshed gears use this same leverage principle to transfer torque. If two gears in mesh have the same number of teeth, they will rotate at the same speed, and the input gear will transfer an equal amount of torque to the output gear (**Figure 10–10A**).

It has already been explained that when a driven gear is larger than its driving gear, driven (output) speed decreases. But how is torque affected? **Figure 10–10B** illustrates a small gear with 12 teeth driving a large gear with 24 teeth. This results in a gear ratio of 2:1 with the output speed being half of the input speed.

In this example, torque at the input shaft of the driving gear is 10 pound-feet. The distance from the centerline of this input shaft to the gear teeth is 1 foot. This means the driving gear transfers 10 pounds of force to the teeth of the larger driven gear. The distance between the teeth of the driven gear and the centerline of its output shaft is 2 feet. This means the torque at the output shaft is 10 pounds × 2 feet, or 20 pound-feet. Torque has doubled. The amount of torque increase from a driving gear to a driven gear is directly proportional to the speed decrease. When speed is halved, torque doubles.

Review the example just discussed. With a gear ratio of 1.66:1, the driven gear is rotating 1.66 times

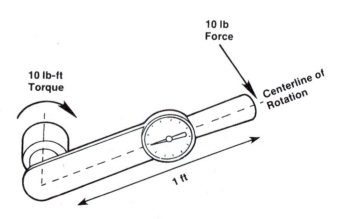

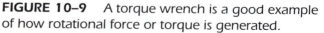

FIGURE 10–9 A torque wrench is a good example of how rotational force or torque is generated.

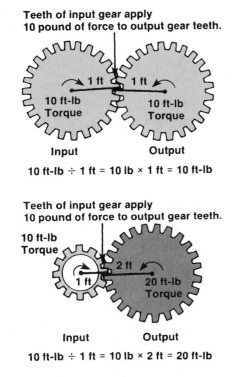

Teeth of input gear apply
10 pound of force to output gear teeth.

1 ft   1 ft
10 ft-lb Torque   10 ft-lb Torque

Input   **Output**

A   10 ft-lb ÷ 1 ft = 10 lb × 1 ft = 10 ft-lb

Teeth of input gear apply
10 pound of force to output gear teeth.

10 ft-lb Torque

2 ft
1 ft   20 ft-lb Torque

Input   **Output**

B   10 ft-lb ÷ 1 ft = 10 lb × 2 ft = 20 ft-lb

FIGURE 10–10   *Gear teeth are actually a series of levers used to transfer torque. (A) If gears are the same size, output torque equals input torque; and (B) when the input gear is smaller than the output gear, output torque increases.*

slower than the driving gear, but it is producing 1.66 times the torque of the driven gear. If torque at the driving gear input is 100 pound-feet, torque at the driven gear output is 1.66 × 100 pound-feet, or 166 pound-feet.

Most manual transmission gearing is speed reducing/torque increasing. In many transmissions, the top gear is a 1:1 gear ratio where speed and torque are transferred directly from the input to the output shaft. In some cases, the top gear is an overdrive gear combination. This means it is a speed-increasing/torque-reducing gearing.

Overdrive gear ratios are stated using a decimal point, such as 0.85:1. This means that for every 0.85 times the input shaft rotates, the output shaft rotates one complete revolution. Output speed is 1.176 times greater (1.00 divided by 0.85) than input speed. However, output torque is only 0.85 of input torque. So if input speed and torque are 100 rpm and 100 pound-feet, the output values are 117.6 rpm and 85 pound-feet.

GEAR MOUNTING

Gears can be mounted or fixed to shafts in a number of ways. Gears can be internally splined to a shaft, or

FIGURE 10–11   *An idler gear is used to transfer motion without changing rotational direction.*

they can be **keyed** to a shaft. Gears can also be manufactured as an integral part of the shaft.

Gears that must have the ability to freewheel around the shaft during certain speed ranges are mounted to the shaft using bushings or bearings.

IDLER GEARS

An idler gear is a gear that is placed between a drive gear and a driven gear to transfer motion without changing rotational direction or gear ratio (**Figure 10–11**). This allows the driven gear to rotate in the same direction as the drive gear. Idler gears are used in reverse gearing. If two idler gears are used, the driven gear will rotate in the direction opposite to that of the drive gear.

Idler gears can also serve as a transfer medium between the drive and driven gears in place of a chain drive or belt drive. Always remember that idler gears do not affect the relative speeds of either the drive or driven gears.

TYPES OF GEARS

The following are the major types of gear tooth designs used in modern transmissions and differentials.

Spur Gears

The spur gear is the simplest gear design used in manual transmissions. As shown in **Figure 10–12**, spur gear teeth are cut straight across the edge parallel to

FIGURE 10–12   *Spur gear and pinion arrangement.*

the gear's shaft. During operation, enmeshed spur gears have only one tooth in contact at a time.

Its straight tooth design is the spur gear's main advantage. It minimizes the chances of popping out of gear, an important consideration during acceleration/deceleration and reverse operation. For this reason, spur gears are often used in the reverse gear train.

The spur gear's major drawback is the clicking noise that occurs as teeth contact one another. At higher speeds, this clicking becomes a constant whine. Quieter gears, such as the helical design, are often used to eliminate this gear whine problem.

 Shop Talk

*When a small gear is meshed with a larger gear, as shown in **Figure 10–12**, the small gear is often called a pinion or pinion gear, regardless of its tooth design.*

Helical Gears

A helical gear has teeth that are cut at an angle or are spiral to the gear's axis of rotation (**Figure 10–13**). This allows two or more teeth to mesh at the same time. This distributes tooth load and produces a very strong gear. Helical gears also run more quietly than spur gears because they create a wiping action as they engage and disengage the teeth on another gear. One disadvantage is that helical teeth on a gear cause the gear to move fore or aft on a shaft, depending on the direction of the angle of the gear teeth. This axial thrust must be absorbed by thrust washers and other transmission gears, shafts, or the transmission case.

FIGURE 10–13   Helical gearing arrangement.

Helical gears can be either right- or left-handed depending on the direction the spiral appears to go when the gear is viewed face-on. When mounted on parallel shafts, one helical gear must be right-handed and the other left-handed. Two gears with the same direction spiral will not mesh in a parallel mounted arrangement. Helical and spur gears are used in heavy-duty truck transmission gearing.

10.3 GEAR TRAIN CONFIGURATIONS

A heavy-duty truck manual transmission consists of a main shaft and one, two, or three countershafts. All shafts are mounted parallel to one another. Torque is transmitted from a transmission input shaft at the front of the transmission through gearing to the countershaft(s). It travels along the countershaft(s) until it reaches the selected gearing. It then passes through this gearing back to the main shaft and out to the driveline components.

Single countershaft transmissions are used in some heavy-duty applications but are more common in medium-duty trucks. Most heavy-duty trucks use a double, or twin, countershaft design (**Figure 10–14**). In a twin countershaft transmission, the torque is split equally between two shafts. Each countershaft gear set carries only half the load of a similar gear set used on a single countershaft design.

Mack Trucks transmissions are equipped with a triple countershaft. As shown in **Figure 10–15**, the three countershafts are equally spaced around the mainshafts. This design distributes the torque load equally among the three countershafts, helping to keep deflection and gear tooth loading to a minimum.

10.4 MECHANICAL SHIFT MECHANISM

Manual transmissions can be further classified according to how shifts between gears are made. The three general classifications are sliding gear, collar shift, and synchronized shift mechanisms.

SLIDING GEARSHIFT MECHANISMS

A sliding gearshift design must have a main shaft and a countershaft mounted parallel to each other. To change from one speed/torque range to another, the gears on the main shaft are moved horizontally

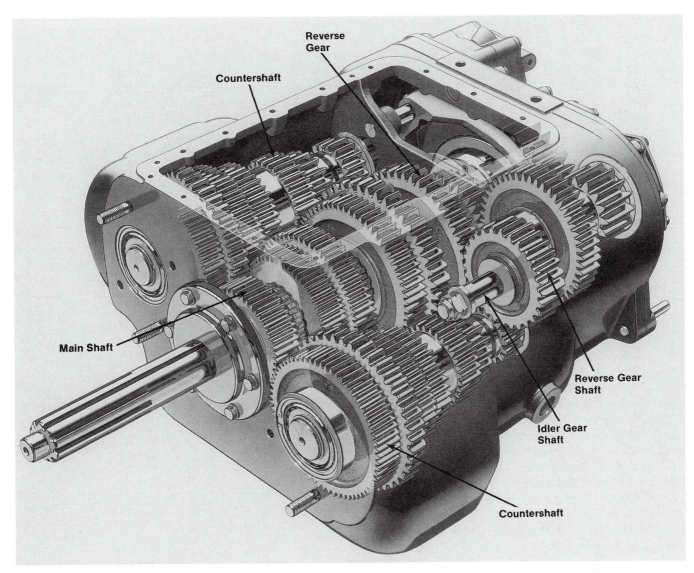

FIGURE 10–14   A nine-speed, twin countershaft transmission; note the use of an idler shaft and gear needed to generate proper reverse rotation. (Courtesy of ArvinMeritor Inc.)

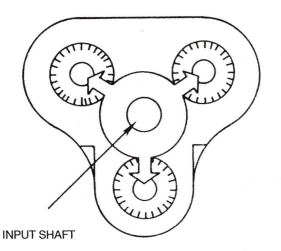

FIGURE 10–15   Layout of the triple countershaft transmission.

along the main shaft until they engage the proper gear on the countershaft.

To provide this movement, spur cut sliding gears are made with a shift collar integral with the gear. When the operator shifts gears, a **shift fork or yoke** engages with the collar and slides the gear along the main shaft until it engages with its mating gear on the countershaft.

Older heavy-duty truck transmissions used sliding gears for all speed ranges. However, these transmissions were unsynchronized, and it was difficult to precisely match the speed of meshing gears. As a result, grinding and gear clash were a constant problem, and all sliding gear units were prone to rapid wear, gear teeth chipping, and fracture.

In modern heavy-duty truck transmissions, the only gear ratios that can still use spur type sliding

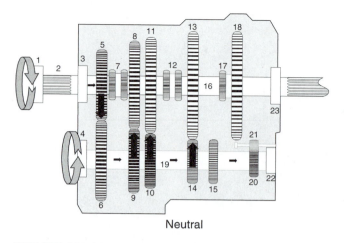

Neutral

FIGURE 10–16 *Typical components of a five-speed single countershaft transmission; power flow shown is neutral with the engine at idle.*

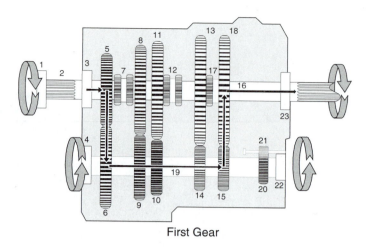

First Gear

FIGURE 10–17 *First-gear power flow in a typical five-speed single countershaft transmission.*

gears are first and reverse. All other gears in the transmission use some form of sliding clutch collar or synchronizer to engage each gear range. In addition, all nonsliding gears are constantly meshed with their mating countershaft gears.

Figure 10–16 shows a diagram of a five-speed single countershaft truck transmission that uses both sliding gears and collar shifting mechanisms to change ratios. The following passage explains the sliding gear operation for first and reverse gears.

Item 18 is the first and reverse sliding gear with item 17 as its integral shift collar. When the shift fork or yoke (not shown) that engages in the shift collar is moved by the gearshift lever in the cab, it pushes the collar and gear to the front or rear of the transmission. Sliding it to the front engages the first and reverse sliding gear with item 15, the first gear on the countershaft. This results in first gear power flow (**Figure 10–17**). Torque flows from the clutch splines to the transmission input shaft (2) through the clutch shaft gear (5) to the countershaft driven gear (6). Power is then transmitted along the countershaft to the first gear (15) and up to the first and reverse sliding gear (18) on the main shaft (16). Since the first and reverse sliding gear is splined to the main shaft, power is transferred through the main shaft to the vehicle's driveline.

When reverse is selected, the shift lever is moved from neutral to reverse. The shift fork forces the first and reverse sliding gear backward where it engages with the reverse idler gear (21). The reverse idler gear is required to allow the first and reverse sliding gear to rotate in the same direction as the reverse gear (20) on the countershaft. Power flow (**Figure 10–18**) is from items 2 to 5 and 6, then along the countershaft to gears 20 and 21. From 21, power flows to

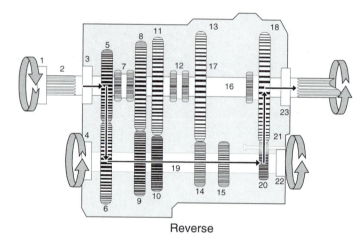

Reverse

FIGURE 10–18 *Reverse gear power flow in a typical five-speed single countershaft transmission.*

18, then to the main shaft (16) and out to the driveline.

COLLAR SHIFT MECHANISMS

This section takes a closer look at collar shift operation in the basic transmission shown in **Figure 10–16**. Collar shifting controls the engagement of second, third, fourth, and fifth gears.

In a collar shifting arrangement, all gears on the countershaft are fixed to the countershaft. Main shaft gears are not fixed to the main shaft. Instead, they are free to rotate or freewheel around either a bearing or bushing. These bearings or bushings are normally positioned on a nonsplined section of the main shaft. The main shaft gears are in constant mesh with their mating countershaft gears. This means they are turning at all times.

Consider the neutral power flow of the five-speed transmission as shown in **Figure 10–16**. The clutch splined shaft (input shaft) (2) rotates at engine speed whenever the clutch is engaged. The clutch shaft gear (5) is solidly attached to the clutch splined shaft and rotates with it. The clutch shaft gear meshes with the countershaft driven gear (6), so it, the countershaft, and all the gears fixed to the countershaft also rotate.

These countershaft gears transfer torque to their mating gears on the main shaft. But main shaft gears 8, 11, and 13 are not fixed to the main shaft; they are freewheeling on bearings or bushings. Since they cannot transmit torque to the main shaft, it does not turn, and there is no power output to the driveline.

To transfer torque to the main shaft, a freewheeling main gear must be locked to it. Truck transmissions accomplish this in a number of ways. One method uses a shift collar and shift gear. The shift gear is internally splined to the main shaft at all times. The shift collar is splined to the shift gear.

As shown in **Figure 10–19**, main gears that use shift collar engagement are designed with a short, toothed hub on their side. The teeth on this main gear hub align with the teeth on the shift gear. The internal teeth of the shift collar mesh with the external teeth of the shift gear and hub.

When a given speed range is not engaged, the shift collar rides on the shift gear. When the operator engages that speed range, the shift fork engages the shift collar and slides it over so it meshes with the teeth of the main gear hub. The shift collar is now riding on both the shift gear and main gear hub, locking them together. Power can flow from the main gear to the shift gear, to the main shaft, and out to the propeller shaft.

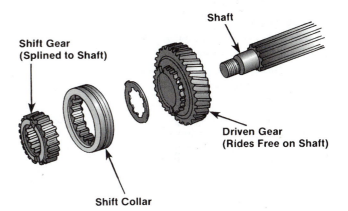

FIGURE 10–19 *Collar shift mechanism using a shifter gear, shift collar, and splined, hub driven-gear.*

A second, more common method of locking main gears to the main shaft does not use a shift gear. Instead, the shift collar is splined directly to the main shaft. This shift collar, which is also called a clutch collar or a sliding clutch, is designed with external teeth. These external teeth mesh with internal teeth in the main gear hub or body when that speed range is engaged. Most shift collars or sliding clutches are positioned between two gears so they can control two-speed ranges depending on the direction in which they are moved by the shift fork (**Figure 10–20**).

Second Gear Power Flow. To shift from first to second gear, the operator pushes the clutch pedal down to disengage the clutch. The gear select lever is then moved to neutral, disengaging the first and

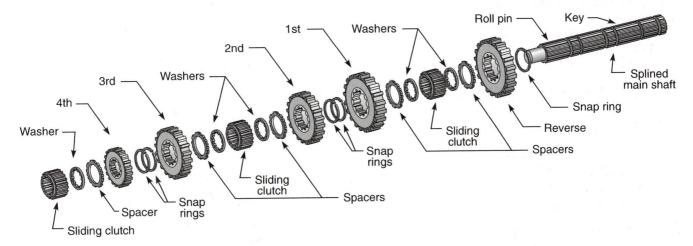

FIGURE 10–20 *Exploded view of a five-speed main shaft showing the relative position of the sliding clutches and gears: one sliding clutch controls first and reverse shifts, another controls second and third shifts, and a third controls fourth and main gear engagement. Note: The main gear is not shown but is located forward of the sliding clutch at the extreme front of the main shaft.*

reverse gear (18). Moving the shift lever from neutral to the second gear position causes the second and third shift collar (or sliding clutch) to move (12) back toward the second gear (13) on the main shaft. The shift collar engages second gear and locks it to the main shaft.

Power flows from 2 to 5 and 6, along the countershaft to 14. Gear 14 drives gear 13, which in turn drives main shaft output to the driveline (**Figure 10–21**).

Third Gear Power Flow. Once again the driver returns to neutral before shifting to third gear. Moving from neutral to third gear moves the second and third shift collar (or sliding clutch) (12) forward toward the third gear (11), locking it to the main shaft.

Power flows from 2 to 5 and 6, along the countershaft to 10, up to 11, through the shift collar (12) to the main shaft and out to the driveline (**Figure 10–22**).

Fourth Gear Power Flow. After shifting from third to neutral, the neutral to fourth gearshift causes the shifter fork to move the fourth and fifth shift collar (or sliding clutch 7) into mesh with the fourth gear (8). Power flows from 2 to 5 and 6, along the counter shaft to 9, through 8 and 7 to the main shaft and out to the driveline (**Figure 10–23**).

Fifth Gear Power Flow. After shifting from fourth to neutral, the neutral to fifth gearshift causes the shifter fork to move the fourth and fifth shift collar (or sliding clutch 7) into mesh with the clutch shaft gear (5). This locks the clutch shaft (2) directly to the main shaft (16). Input and output speeds are the same. So a 1:1 gear ratio is established. The power flow (**Figure 10–24**) is from 2 to 5 through 7 to the main shaft and out. In fifth gear the clutch shaft gear is still meshed with the countershaft driven gear (6). This

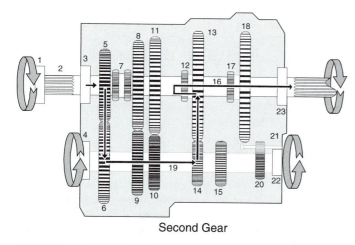

Second Gear

FIGURE 10–21   Second-gear power flow in a typical five-speed single countershaft transmission.

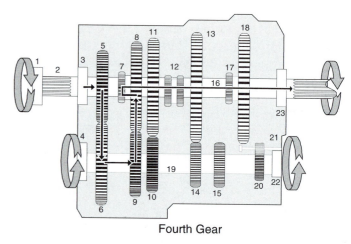

Fourth Gear

FIGURE 10–23   Fourth-gear power flow in a typical five-speed single countershaft transmission.

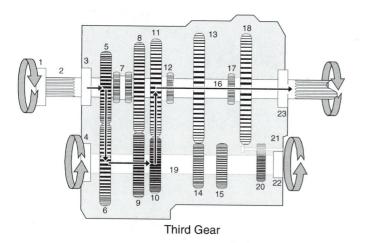

Third Gear

FIGURE 10–22   Third-gear power flow in a typical five-speed single countershaft transmission.

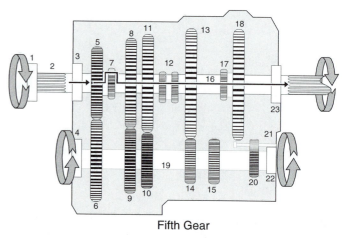

Fifth Gear

FIGURE 10–24   Fifth-gear power flow in a typical five-speed single countershaft transmission.

means that the countershaft and its gears are all turning. The main shaft gears 8, 11, and 13 are also freewheeling on the main shaft, but have no effect on the power flow.

DRIVING NONSYNCHRONIZED TRANSMISSIONS

The sliding gear and collar shift mechanisms outlined in the previous sections are used on nonsynchronized transmissions. Shifting nonsynchronized transmissions requires double-clutching between gears. This was mentioned in the previous discussion on power flows, but the proper procedure is outlined in more detail in the following paragraphs. The example used is a seven-forward-speed transmission with an engine governed speed of 2100 rpm.

To pull out from a standing start with the engine at idle, depress the clutch, and move the shift lever to the first gear position. Gradually release the clutch and accelerate the engine to governed speed.

 Shop Talk _____

As explained in Chapter 9, a clutch brake is used to stop gear rotation to complete a shift into first or reverse when the vehicle is stationary. The clutch brake is actuated by depressing the clutch pedal completely to the floor. For normal upshifts and downshifts, only partial disengagement of the clutch is needed to break engine torque.

Once governed speed has been reached in first gear, depress the clutch and release the accelerator. Move the shift lever to neutral. Next, engage the clutch and allow engine speed to drop approximately 750 rpm. Depress the clutch and shift to second gear. Re-engage the clutch and accelerate to governed speed. Continue shifting through to the top gear ratio in this manner.

 Shop Talk _____

The 750 rpm drop used in this example will vary according to a particular engine's governed speed and torque rise profile.

When downshifting from seventh gear, allow engine speed to drop approximately 750 rpm before depressing the clutch and moving the shift lever to neutral. Now engage the clutch and accelerate engine rpm to governed speed. Depress the clutch and move the shift lever into the sixth gear position.

Now re-engage the clutch. Continue downshifting in this manner. Once again, the 750 rpm drop is only an example.

SYNCHRONIZED SHIFTING

A synchronizer performs two important functions. Its main function is to bring components that are rotating at different speeds to one synchronized speed. A synchronizer ensures that the main shaft and the main shaft gear to be locked to it are rotating at the same speed. The second function of the synchronizer is to actually lock these components together.

The result of these two functions is a clash-free shift. Many manual transmissions having 5 to 7 forward speeds are completely synchronized except for first and reverse gears. These transmissions are not synchronized in first and reverse gears because these gears are normally selected when the vehicle is stationary. No synchronizing of engine speed and road speed is required to get in gear from a stop. In fact, a synchronizer could cause hard shifting in these gear positions because a synchronizer needs gear rotation to do its job.

Many transmissions use both main and auxiliary gearing to provide a large number of forward speeds. In these transmissions synchronizers are normally used only in the two- or three-speed auxiliary gearing.

Block or cone and pin synchronizers are the most common in heavy-duty truck applications. Both block or cone and pin synchronizers are often referred to as sliding clutches.

Block or Cone Synchronizers

Figure 10-25 illustrates a block or cone synchronizer. The synchronizer sleeve surrounds the synchronizer assembly and meshes with the external splines of the clutch hub. The clutch hub is splined to the transmission output (main) shaft and is held in position by two snap rings.

The synchronizer sleeve has a small internal groove and a large external groove in which the shift fork rests. Three slots are equally spaced around the outside of the clutch hub. Inserts fit into these slots and are able to slide freely back and forth. These inserts are designed with a ridge in their outer surface. Insert springs hold the ridge in contact with the synchronizer sleeve internal groove.

The synchronizer sleeve is precisely machined to slide on the clutch hub smoothly. The sleeve and hub sometimes have alignment marks to assure proper indexing of their splines when assembling to maintain smooth operation.

Brass or bronze synchronizing blocker rings are positioned at the front and rear of each synchronizer

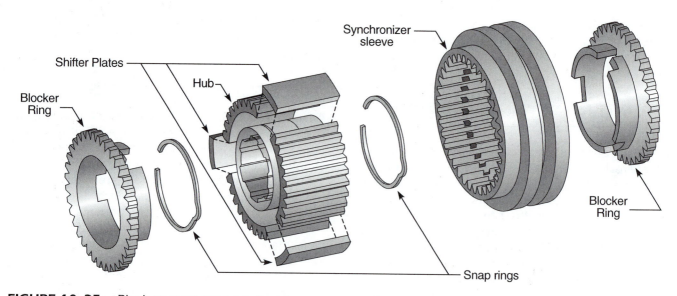

FIGURE 10–25 Block or cone type synchronizer components.

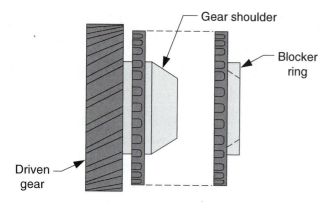

FIGURE 10–26 Gear shoulder and blocker ring mating surfaces.

assembly. Each blocker ring has three notches equally spaced to correspond with the three inset notches of the hub. Around the outside of each blocker ring is a set of beveled dog teeth, which are used for alignment during the shift sequence. The inside of the blocker ring is shaped like a cone. This coned surface is lined with many sharp grooves.

The cone of the blocker ring makes up only one-half of the total cone clutch. The second or mating half of the cone clutch is part of the gear to be synchronized. As shown in **Figure 10–26**, the shoulder of the main gear is cone shaped to match the blocker ring. The shoulder also contains a ring of beveled dog teeth designed to align with the dog teeth on the blocker ring.

Operation. In a typical five-speed transmission, two synchronizer assemblies are used. One controls second and third gear engagements, the other synchronizes fourth and fifth gear shifting.

When the transmission is in first gear, the second/third and fourth/fifth synchronizers are in their neutral position (**Figure 10–27A**). All synchronizer components are now rotating with the main shaft at main shaft speed. Gears on the main shaft are meshed with their countershaft partners and are free-wheeling around the main shaft at various speeds.

To shift the transmission into second gear, the clutch is disengaged and the gearshift lever is moved to the second gear position. This forces the shift fork on the synchronizer sleeve backward toward the second gear on the main shaft. As the sleeve moves backward, the three inserts are also forced backward because the insert ridges lock the inserts to the internal groove of the sleeve.

The movement of the inserts forces the back blocker ring coned friction surface against the coned surface of the second gear shoulder. When the blocker ring and gear shoulder come into contact, the grooves on the blocker ring cone clutch cut through the lubricant film on the second-speed gear shoulder and a metal-to-metal contact is made (**Figure 10–27B**). The contact generates substantial friction and heat. This is one reason bronze or brass blocker rings are used. A nonferrous metal such as bronze or brass minimizes wear on the hardened steel gear shoulder. If similar metals were used, wear would be accelerated. This frictional coupling is not strong enough to transmit loads for long periods. But as the components reach the same speed, the synchronizer sleeve can now slide over the external dog teeth on the blocker ring and then over the dog teeth on the second-speed gear shoulder. This completes the engagement (**Figure 10–27C**). Power flow is now from the second-speed gear, to the synchronizer sleeve, to the synchronizer clutch hub, to the main output shaft, and out to the propeller shaft.

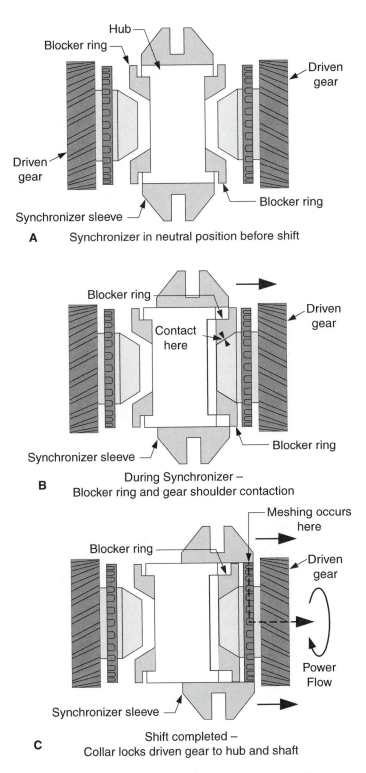

A Synchronizer in neutral position before shift

B During Synchronizer –
Blocker ring and gear shoulder contaction

C Shift completed –
Collar locks driven gear to hub and shaft

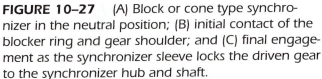

FIGURE 10–27 (A) Block or cone type synchronizer in the neutral position; (B) initial contact of the blocker ring and gear shoulder; and (C) final engagement as the synchronizer sleeve locks the driven gear to the synchronizer hub and shaft.

To disengage the second-speed gear from the main shaft and shift into third-speed gear, the clutch must be disengaged as the shift fork is moved to the front of the transmission. This pulls the synchronizer sleeve forward, disengaging it from second gear. As the transmission is shifted into third gear, the inserts again lock into the internal groove of the sleeve. As the sleeve moves forward, the forward blocker ring is forced by the inserts against the coned friction surface on the third-speed gear shoulder. Once again, the grooves on the blocker ring cut through the lubricant on the gear shoulder to generate a frictional coupling that synchronizes the gear and shaft speeds. The shift fork can then continue to move the sleeve forward until it slides over the blocker ring and gear shoulder dog teeth, locking them together. Power flow is now from the third-speed gear, to the synchronizer sleeve, to the clutch hub, and out through the main shaft.

Pin Synchronizers

Figure 10–28 shows the components of a typical pin synchronizing unit. The clutch drive gear has both internal and external splines. It is splined to the main (output) shaft and held to it with two snap rings. The sliding clutch gear is loosely splined to the external splines of the clutch drive gear. The sliding clutch gear is made with a flange that has six holes with slightly tapered ends. The shift fork fits into an external groove cut into this flange.

Aluminum alloy blocker rings have coned friction surfaces designed with several radial grooves. Three pins are fastened to each blocker ring, and each pin is machined with two distinctly different diameters. The outer stop synchronizer rings are machined with an internal cone friction surface. The bores of the outer stops have internal teeth that spline to the external teeth on the shoulders of the gears to be synchronized. Finally, the smaller diameter ends of the blocker ring pins rest in the holes of the sliding clutch gear.

Operation. When the pin synchronizer is in a neutral (disengaged) position, the clutch drive gear, sliding clutch gear, and the blocker ring all rotate with the main (output) shaft. The gears either side of the synchronizer, as well as the outer rings splined to the gear shoulders, are driven at varying speeds by the countershaft gears.

To shift into a pin synchronized gear, the clutch is disengaged, and the shift lever is moved to the desired speed ratio. This moves the shift fork and the sliding clutch gear toward the gear to be synchronized. The three holes in the sliding clutch gear butt against the side of the larger diameter ends of the three pins on the blocker ring. This forces the blocker ring into the adjoining outer stop ring. The two coned friction surfaces of the blocker and outer rings come into contact. The friction coupling equalizes the turning speed of these components. And

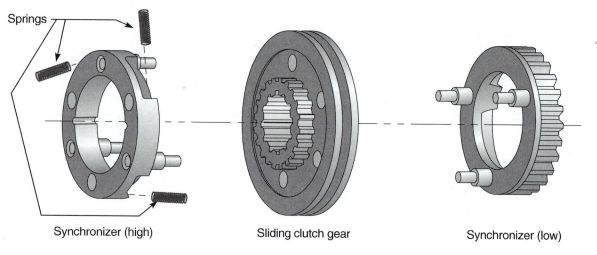

Springs

Synchronizer (high)   Sliding clutch gear   Synchronizer (low)

FIGURE 10–28   *Pin type synchronizer.*

since the driven gear is splined to the outer stop ring, it is now turning at main shaft speed.

As with the block or cone synchronizer, this frictional coupling cannot be held for long. When all components are turning at the same speed, the holes in the sliding clutch gear are able to slide over the larger diameter portion of the pins. The internal teeth of the sliding clutch gear slide over the external teeth of the gear, completing the clutch engagement.

To disengage the gear from the shaft, the clutch is disengaged and the shift lever is moved to neutral. This moves the shift fork and sliding clutch gear away from the gear, disengaging the sliding clutch gear teeth from the main gear teeth.

DISC AND PLATE SYNCHRONIZERS

The disc and plate synchronizer uses a series of friction discs alternated with steel plates, not unlike the "clutch pack" of alternating discs and plates used in many automatic transmissions. The primary difference is that, where automatic transmissions generally use hydraulic pressure to compress the clutch and

disc assembly, heavy-duty truck transmissions using disc and plate synchronizers depend upon the physical pressure of the shifter fork during shifting to compress the plates together. The discs have external tangs or splines around their outer perimeter. These external splines allow the discs (but not the plates between each disc) to engage with the slots machined inside the synchronizer drum that houses the discs and plates. Consequently, when the synchronizer drum turns, the clutch discs (with external tangs) turn with it while the plates (which do not have external splines) do not. The plates do, however, have internal tangs or splines that mate with the blocker ring and input shaft (**Figure 10–29**). Since the plates have no external splines and the discs have no internal splines, the only movement that can be transmitted between them comes from friction between the two. If minimal pressure is applied to the disc and plate assembly, the amount of friction between the two will be small, and the amount of movement transferred will be correspondingly small. As more pressure is applied to the disc and plate assembly, friction increases and more movement is transferred.

Input shaft with
synchronizer gear   Detent ball   Synchronizer drum   Output gear
and shaft

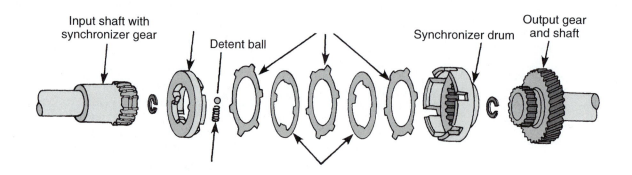

FIGURE 10–29   Disc and plate type synchronizer. (Reproduced by permission of Deere & Company ©2000)

The purpose of the disc and plate synchronizer is to match the speed of input gears with output gears during shifting. This occurs when the truck driver disengages the clutch and moves the gear shift lever, causing the shifter fork to move the synchronizer drum forward. This forward motion causes the discs and plates to rotate the blocker ring around the input shaft until the drive lugs on the blocker ring engage with the outer edge of the detent ball depression cut into the synchronizer gear.

Once the blocker ring has locked into the synchronizer gear, any additional forward motion of the shifter forks will compress the discs and plates together, causing them to rotate at the same speed. When the speeds of the input gear and blocker gear are synchronized to the speed of the output gear and both are turning at the same speed, the

- Thrust force is released.
- Blocker disengages from its locked position.
- Drum can be moved forward to engage both gears.

PLAIN SYNCHRONIZERS

A plain synchronizer uses fewer working parts and is very similar in design to a block synchronizer (**Figure 10–30**). As is typical with most synchronizers, the hub is internally splined to the main shaft. Mounted on the hub, and also connected to it, is the sliding sleeve that is controlled by the shift fork movement. In operation, the friction generated between the hub and the gear tends to synchronize their speeds. Pressure on the sliding sleeve prevents the sleeve

from moving off the hub and also prevents the teeth on the sleeve from engaging the gear teeth until sufficient pressure has been put on both friction areas to synchronize the components. When in sync, any further movement of the sleeve will allow the sleeve teeth to engage the gear teeth.

DRIVING SYNCHRONIZED TRANSMISSIONS

The purpose of the synchronizer is to simplify shifting and help the driver make clash-free shifts. But there is a right way and a wrong way to shift synchronized transmissions. When a shift is required, the driver should disengage the clutch and move the shift lever toward the desired gear position. When the synchronizer ring makes initial contact with the desired gear, the blockers automatically prevent the shift collar from completing the shift until the gear and main shaft speeds are matched. At that time, the blocker neutralizes automatically and a clash-free shift is the result. Steady pressure on the shift lever helps the synchronizer do its job quickly. When synchronized, the lever moves into gear smoothly and easily.

If the driver jabs at the synchronizer or "teases" it by applying pressure and releasing pressure, the synchronizer cannot do its job. It takes one or two seconds to match speeds, and constant shift lever pressure assures the faster synchronizer action. It is possible to override a blocker if the lever is forced into gear. This, of course, defeats the purpose of the synchronizer and can cause gear clash, which is damaging to the life of the transmission.

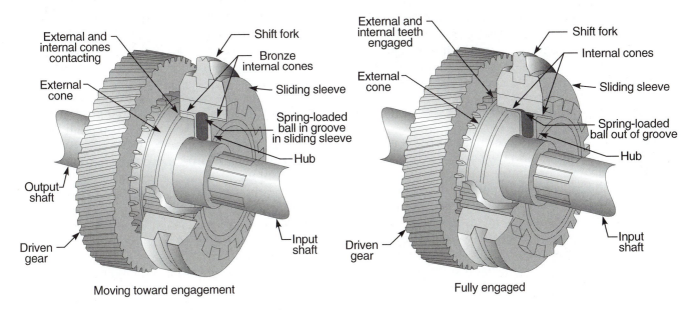

Moving toward engagement   Fully engaged

FIGURE 10–30   Plain type synchronizer. (Reproduced by permission of Deere & Company ©2000)

When pulling off from a standing stop, depress the clutch, wait for complete disengagement, move the shift lever to first gear position, and accelerate to an rpm that will allow enough vehicle and driveline momentum to select the next higher gear and still have vehicle acceleration after completing the shift into second gear. This is using a progressive shift technique, which saves fuel. It is not necessary to run the engine rpm to the rated (governed) rpm when upshifting.

When downshifting, a similar procedure is used. Consider shifting from top (fifth) gear to fourth gear. As the shift point is approached (normally about 100 rpm over the shift point), disengage the clutch and move the shift lever with a steady, even pressure toward fourth gear. The synchronizer will pick up fourth gear, get it up to vehicle speed, and allow a clash-free shift from fifth to fourth. After the shift, re-engage the clutch and at the same time accelerate the engine to keep the vehicle moving at the desired speed. If further downshifts are required, continue in a similar manner.

Remember that when downshifting into first gear, first is not synchronized and will require a double-clutch operation to complete a clash-free shift. It is possible to double-clutch on all shifts if desired. This aids the synchronizer in its function by getting the engine speed and driveline matched.

10.5 GEARSHIFT LEVER COMPONENTS

As shown in **Figure 10–31**, the gearshift tower can be installed directly on top of the transmission housing to provide direct shifting of the gears.

In **cab-over-engine (COE) trucks**, the shift tower is not usually located directly above the transmission. In this case, a remote shift lever with an additional linkage is needed to connect the shift tower with the shift bar housing on top of the transmission (**Figure 10–32**).

SHIFT BAR HOUSING

The top cover of the transmission, also known as the **shift bar housing,** contains the components used to actuate main shaft gearshifting for both forward and reverse operation. Design variations exist between manufacturers. **Figure 10–33** illustrates the shifter housing control components for a seven-speed heavy-duty transmission. **Figure 10–34** illustrates the shift bar housing assembly used to control shifting in the five-speed main transmission of an eighteen-speed transmission.

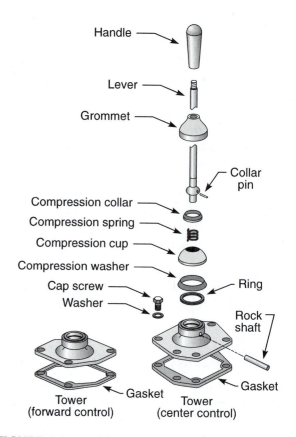

FIGURE 10–31   Typical direct shift lever and tower assembly.

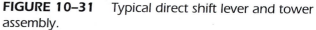

👓 **Shop Talk** _____

Auxiliary transmission gearing requires its own control system. The controls can be mechanical or air-operated. Air-operated controls will be discussed in detail later in this chapter.

_____

Operation. The gearshift lever (direct shift tower arrangement) or the inner shaft finger (remote linkage arrangement) fits into the slot of the shift bracket block. As shown in **Figures 10–33** and **10–34**, this bracket or block can be a separate item or an integral part of the shift fork or yoke. The shift forks, or yokes, as they might be called, are supported on rods or shift bars. Shift bars are sometimes referred to as rails. When the gearshift lever is moved, motion is transferred to the shift bracket or block and onto the shift bar and shift fork. The shift fork is fitted to the shift collar or synchronizer of the transmission so that the actual shift can be made.

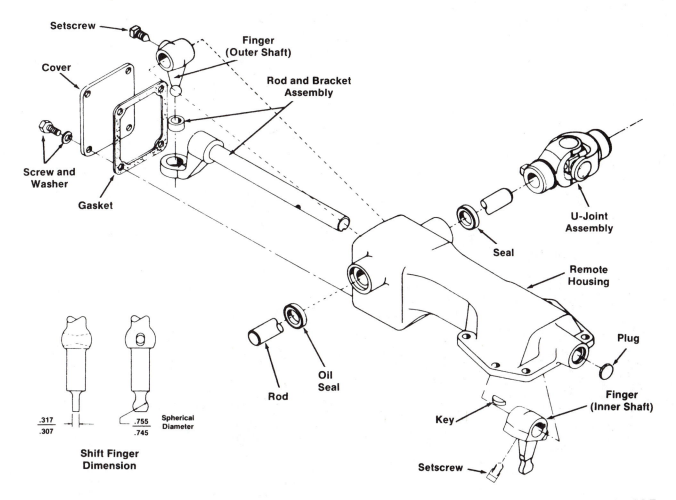

Setscrew

Finger
(Outer Shaft)

Cover

Rod and Bracket
Assembly

Screw and
Washer

Gasket

U-Joint
Assembly

Seal

Remote
Housing

Plug

Oil
Seal

Rod

.317
.307

.755
.745

Spherical
Diameter

**Shift Finger
Dimension**

Key

Finger
(Inner Shaft)

Setscrew

FIGURE 10–32 Remote control assembly used to connect the shift tower with the shift bar housing in COE application. (Courtesy of Eaton Corporation)

The shift bar must be held in position to prevent the transmission from slipping out of gear and to prevent the other bars from moving. This is accomplished by machining a recess onto the shift bar that will line up with a spring-loaded detent steel ball or poppet when the bar is in position. The detent ball holds the shift bar in position and prevents movement of the other bars. When the shift bar is moved, the detent balls are forced out of the recess and into the recess on the other bars, which holds them firmly in position.

10.6 COUNTERSHAFT TRANSMISSIONS

Now that the basics of gears, shift mechanisms, and power flow have been discussed, a closer examina-

tion of countershaft transmissions having both a main and auxiliary gear section should follow.

Twin countershaft transmissions having nine to eighteen forward speed ranges are among the more common heavy-duty truck transmissions. Fuller, Rockwell, and Eaton all produce heavy-duty truck transmissions based on these design principles. Mack Trucks builds its own transmissions having a triple countershaft design, but the operating principles discussed below can be applied.

Figure 10–35 shows a simplified twin countershaft transmission. The gear sets on each side share the load in such a way that the torque is split equally between the two shafts. Each countershaft has to carry only half the torque load that a single countershaft transmission would be subjected to. This allows the face width of the gears to be narrower than those used in a single countershaft transmission.

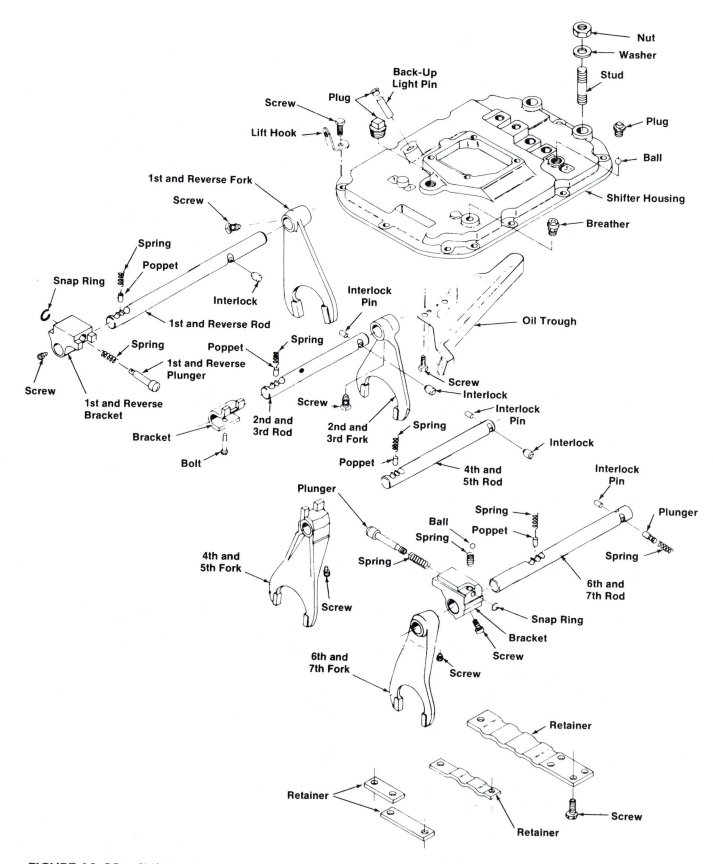

FIGURE 10–33 *Shift housing control components for a seven-speed transmission; shift brackets, rods, and forks are used. (Courtesy of Eaton Corporation)*

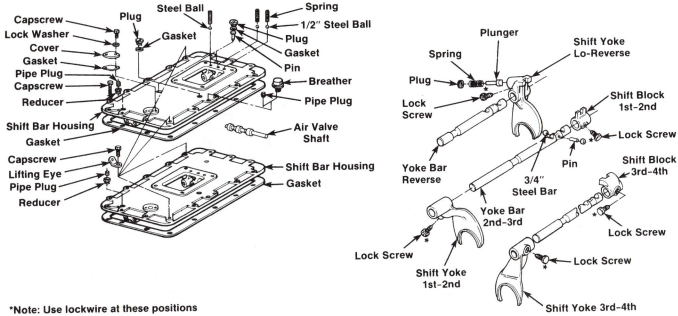

FIGURE 10–34 Shift housing control components for a five-speed main section of an eighteen-speed transmission; the shift blocks, yoke bar, and shift yokes correspond to the brackets, rods, and forks shown in **Figure 10–33**. (Courtesy of Roadranger Marketing. One great drive train from two great companies—Eaton and Dana Corporations)

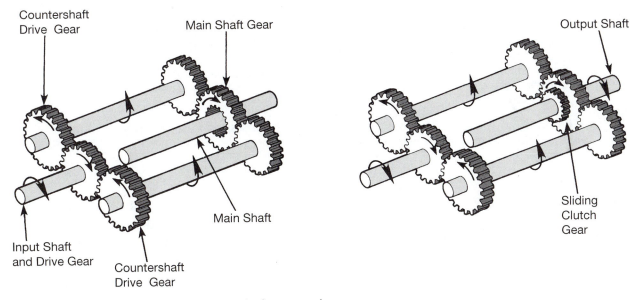

FIGURE 10–35 Simplified twin countershaft gear train.

Additionally, the main shaft gears float between the countershaft gears when disengaged, eliminating the need for gear bushings or sleeves. Transmissions that use the twin countershaft design typically use spur gears in constant mesh. When disengaged, the main shaft gears freewheel around the main shaft because they are in constant mesh with the countershaft drive gears. The motion is not transferred to the actual shaft itself, however, until the sliding clutch gear is moved into engagement. The output shaft will then turn at the same speed as the mainshaft gear. The sliding clutch gear that engages with the mainshaft gear is typically splined to the main shaft.

A simplified diagram (see **Figure 10–36**) of the power flow through a twin countershaft transmission will help show how torque and speed are changed, and how torque is divided between the two countershafts. The input shaft and drive gear (1) are in constant mesh with

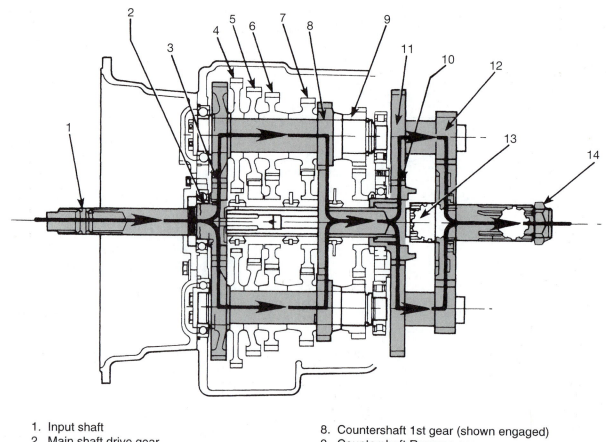

1. Input shaft
2. Main shaft drive gear
3. Countershaft drive gear
4. Countershaft PTO gear
5. Countershaft 4th gear
6. Countershaft 3rd gear
7. Countershaft 2nd gear

8. Countershaft 1st gear (shown engaged)
9. Countershaft Reverse gear
10. Auxiliary drive gear
11. Auxiliary countershaft drive (high range) gear
12. Auxiliary countershaft low range gear
13. Auxiliary low range gear
14. Output shaft

FIGURE 10–36   *Low range power flow in a typical ten-speed transmission having a five-speed main section and a two-speed auxiliary section; main section first gear engagement is shown.*

both countershaft drive gears (2); when the input shaft turns, the countershafts turn. The countershaft gears are in constant mesh with the floating main shaft gears (3). The main shaft gears are simply freewheeling on the main shaft (4). A sliding clutch gear (5), which is splined to the main shaft, is engaged into the internal clutching teeth of the main shaft gear, coupling it to the main shaft. The main shaft will now be turning at the selected gear ratio.

AUXILIARY GEAR SECTIONS

Countershaft transmissions having eight or more forward speeds consist of a five-speed main section and a two- or three-speed auxiliary section. The main and auxiliary gearing can be contained in a single casing or each section can have separate casings that are then bolted together. The auxiliary

gearing is always located behind or toward the rear of the main or front gearing section.

Figure 10–36 illustrates a ten-speed transmission. The main section is a five-speed design; the auxiliary section is a two-speed or two-range design. There is a direct (or high) range and a low range in this transmission.

Shifts in the main section are effected mechanically and controlled by shift collars. Selection of the low or high range in the auxiliary section is made by an air-operated piston, which moves to lock the low range or high range gear to the output shaft when required. The driver controls this range selection mechanism through the use of a master control valve switch mounted on the gearshift tower in the operating cab.

Figure 10–36 also illustrates the power flow through this ten-speed transmission when it is operating in the low range.

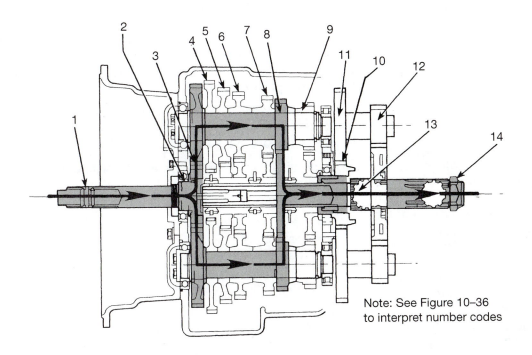

FIGURE 10–37 High range power flow in a typical ten-speed transmission having a five-speed main section and a two-speed auxiliary section; main section first gear is engaged.

Note: See Figure 10–36 to interpret number codes

1. Torque from the engine flywheel is transferred to the input shaft.
2. Splines of input shaft engage the internal splines in the hub or the drive gear.
3. Torque is split between the two countershaft drive gears.
4. Torque is delivered by the two countershaft gears to whichever main shaft gear is engaged. This particular power flow shows first gear engagement. Shifting and power flows in the main section are similar to those in the five-speed transmission discussed in the "Mechanical Shift Mechanisms" section in this chapter. The one difference in **Figure 10–36** is the presence of a PTO (power take-off) gear on the countershafts between the countershaft drive gears and fourth gears.
5. Internal splines in the hub of the main shaft gear transfer torque to the main shaft through a sliding clutch gear.
6. The main shaft transfers torque to the auxiliary drive gear through a self-aligning coupling gear located in the hub of the auxiliary drive gear.
7. Torque is split between the two auxiliary countershaft drive gears.
8. Torque is delivered by the two countershaft low range gears to the low range gear.
9. Torque is delivered to the output shaft through a self-aligning sliding clutch gear that locks the low range gear to the output shaft.
10. The output shaft is attached to the driveline.

Figure 10–37 shows the same transmission when power is directed through the high range or direct drive gearing of the auxiliary section. In this range, the sliding clutch gear locks the auxiliary drive gear to the output shaft. The low range gear on the output shaft is now the freewheeling gear.

The ten forward speeds are selected by using a 5-speed shifting pattern twice, the first time with the auxiliary section engaged in low gear or low range; the second time engaged in high gear or high range. By using the same shifting pattern twice, the shift lever position for sixth speed is the same as first, seventh the same as second, eighth the same as third, ninth the same as fourth, and tenth the same as fifth. **Figure 10–38** illustrates the gearshift lever pattern and range control button positions for this model transmission.

Shift 1-2-3-4-5 in low range.
Repeat pattern in high range and shift 6-7-8-9-10

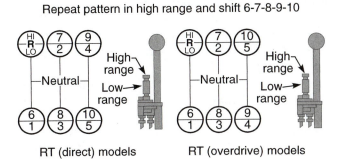

RT (direct) models   RT (overdrive) models

FIGURE 10–38 Gearshift lever pattern and range control button positions.

To operate the truck transmission in an upshifting mode, do the following:

1. Move the gearshift lever to the neutral position.
2. Start the engine.
3. Wait for the air system to reach normal line pressure.
4. Look at the range control button. If it is up, push it to the down position. (With the downward movement of the button, the auxiliary section will shift into low range.) If the button was down when the truck was last used, the transmission is already in low range.
5. Now start the vehicle and shift through first, second, third, and fourth to fifth.
6. When in fifth and ready for the next upward shift, pull the range control button up and move the lever to sixth speed. As the lever passes through the neutral position, the transmission will shift from low range to high range.
7. With the transmission in high range, shift progressively through seventh, eighth, and ninth to tenth.

Downshifting the truck transmission involves the following procedure:

1. When shifting down, move the lever from tenth through each successive lower speed to sixth.
2. When in sixth and ready for the next downward shift, push the range control button down and move the lever to fifth speed. As the lever passes through the neutral position, the transmission will shift from high range to low range.
3. With the transmission in low range, shift downward through each of the four remaining ratios.

During operation of the truck transmission the following precautions should be noted:

- Use range control button only as described.
- Do not shift from high range to low range at high speeds.
- Do not make range shifts with the truck moving in reverse gear.

Skip-shifting is also possible on most standard transmissions. Depending on the load, a 10-speed transmission can most likely be shifted as a 6-, 7-, 8-, or 9-speed transmission by selectively bypassing certain gears in the shifting sequence. When skip- shift-

ing the range control button should be placed in the high range before the shift that passes fifth gear. When skip-shifting down, it is necessary to push the range control button down to low range before the shift that passes sixth.

HIGH/LOW RANGE SHIFT AIR SYSTEMS

An air-operated gearshift system consists of the following:

- Air filter/regulator
- Slave valve
- Master control valve
- Range cylinder
- Fittings and connecting air lines

Figure 10–39 illustrates a typical air-operated gearshift control system used to engage high and low range gearing in the auxiliary section.

Air Filter/Regulator. The **air filter/regulator assembly** (**Figure 10–40**) minimizes the possibility of moisture-laden air or impurities from entering the system. The regulator is used to reduce the main air supply pressure to the range valve and the **slave valve**. As shown in **Figure 10–41**, air from the vehicle's air system is supplied to the air supply port on the air regulator. The air filter removes foreign matter from the air then allows it to pass through the air regulator where the pressure is adjusted to system specifications (typically between 57 and 62 psi). From there the air passes through the 1/4-inch ID supply air line and 1/8-inch OD range valve supply air line to the supply ports of the slave valve and range valve. Depending upon the position of the knob on the range valve, air will pass through either the low range air line or the high range air line to the range shift cylinder.

Range shifts can be made only when the gearshift lever is in, or passing through, neutral. Thus, the range desired can be preselected while the shift lever is in a gear position. As the lever is moved through neutral, the actuating plunger in the shift bar housing will release the slave valve, allowing it to move to the selected range position.

Slave Valve. The slave valve (**Figure 10–42**) can be of the piston or poppet type. The valve is used to distribute the inlet air pressure to both the low and high range air hoses that connect to the range cylinder. The piston (or plunger) is the controlling element as to when and where air pressure is distributed. Slave valve operation (low range and high range) is illustrated in **Figures 10–43** and **10–44**.

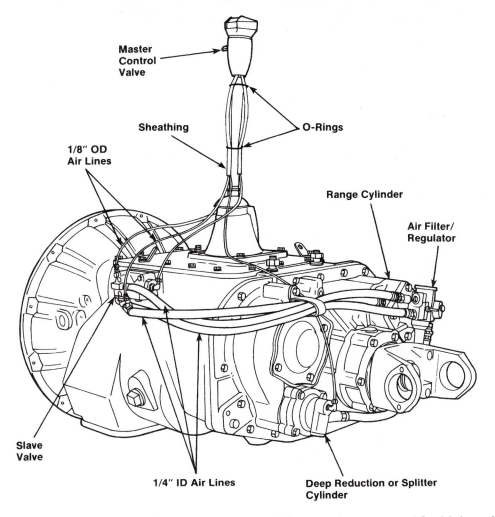

FIGURE 10–39 Components of a typical air operated gearshift control system used for high and low range shifting. (Courtesy of Eaton Corporation)

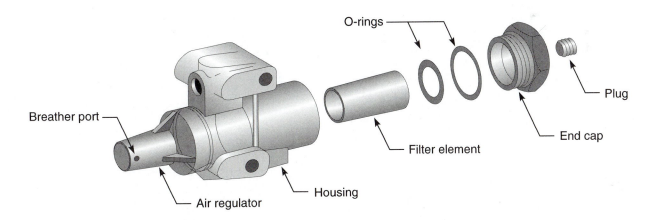

FIGURE 10–40 Air filter/regulator assembly.

An air valve shaft protruding from the shift bar-housing prevents the actuating piston in the slave valve from moving while the gear shift lever is in a gear position and releases the piston when the lever is moved to or through neutral (**Figure 10–45**).

Range Cylinder. The range cylinder is located at the auxiliary section at the rear of the transmission. In low range, air pressure is supplied from the slave valve to the low range port or the range cylinder (**Figure 10–46**). This air pressure forces the range cylinder piston to the rear, engaging the low range gearing. Air pressure on the back of the piston and in the high range air line is exhausted out the slave valve. In high range, air pressure is supplied from the slave valve to the high range port of the range cylinder

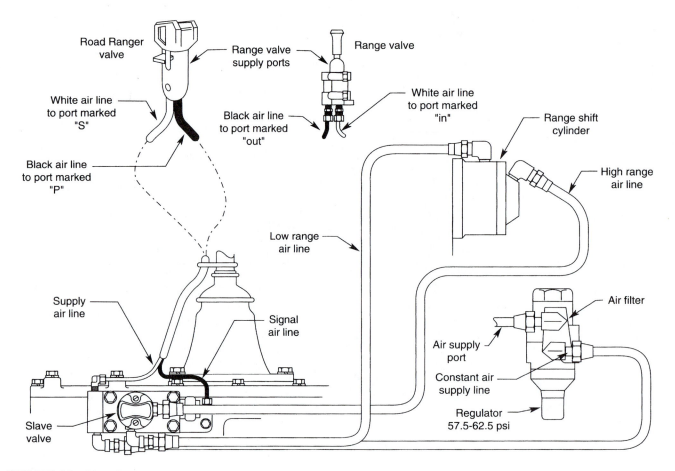

FIGURE 10–41   Typical range shift air system operation.

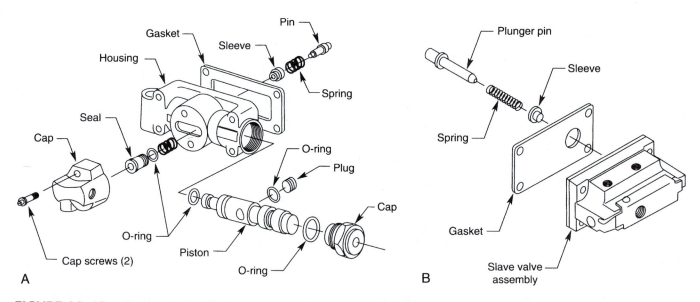

FIGURE 10–42   Slave valves: (A) piston type; and (B) poppet type.

cover. This air pressure forces the range cylinder piston forward, engaging the high range gearing. The air pressure that was on the front side of the piston and in the lower range air line during low range operation is exhausted out the slave valve.

Range Valve. The master control valve is sometimes referred to as the range valve. Constant air pressure is supplied to the inlet port of the range valve (**Figure 10–47**) from the slave valve. In low range (control knob down), air passes through the

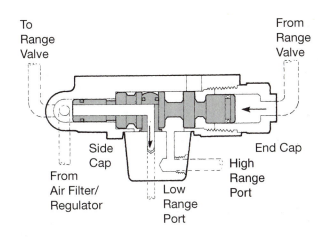

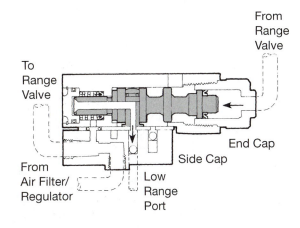

FIGURE 10–43 Two slave valves in operation low range.

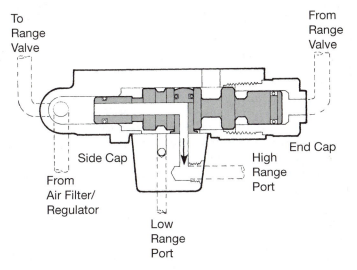

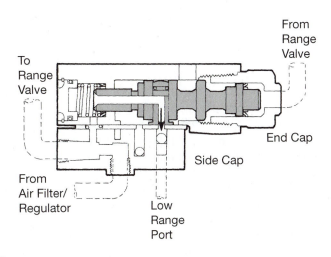

FIGURE 10–44 Two slave valves in operation high range.

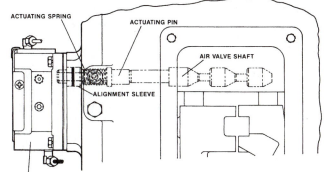

FIGURE 10–45 Components of the slave valve preselection system. (Courtesy of Roadranger Marketing. One great drive train from two great companies—Eaton and Dana Corporations)

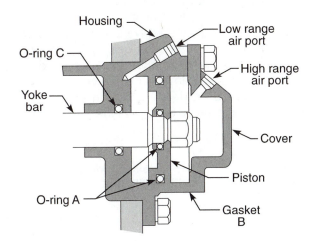

FIGURE 10–46 Typical range cylinder components that engage low gearing.

valve slide and out the outlet port. This air returns to the slave valve to the end port (or P-port), depending on the type of slave valve used. In high range (control knob up), the valve slide prevents the air from

passing through the range valve. Air pressure that was in the outlet line is now exhausted out the bottom of the range valve.

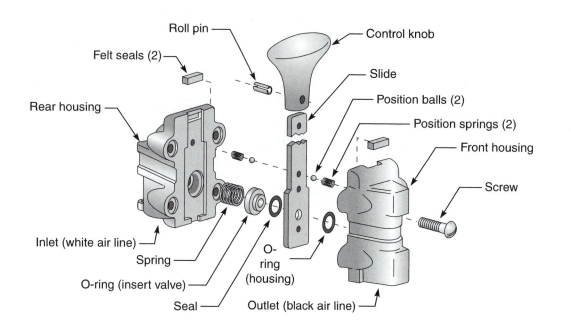

FIGURE 10–47
Master control valve (range valve) exploded view.

Labels: Roll pin, Felt seals (2), Rear housing, Inlet (white air line), Spring, O-ring (insert valve), Seal, Control knob, Slide, Position balls (2), Position springs (2), Front housing, Screw, O-ring (housing), Outlet (black air line)

ADDITIONAL AUXILIARY GEARING

The auxiliary gearing can be designed to include a third gear in addition to the high and low range gears. Depending on the design of the transmission this third gear can be used in a number of ways.

1. It can be an overdrive gear used to produce additional overdrive gear ratios in both the low and high gear ranges. This design is used in 18-speed overdrive transmissions.
2. It can be an overdrive gear used to produce additional overdrive gear ratios in the high gear range only. This design is used in 13-speed overdrive transmissions.
3. It can be an underdrive gear used to produce additional underdrive gear ratios in the high gear range only. This design is used in 13-speed underdrive transmissions.
4. It can be a deep reduction gear that produces a single additional low/low gear ratio.

Split Shifting

The third gear of the auxiliary section is engaged or disengaged from the power flow through the use of a splitter shift system that is air activated by a button on the shift tower. **Figure 10–48** illustrates details of a typical splitter air system that has both the high/low range selector and splitter selector mounted on the shift tower. The system shown is used on typical 13-speed overdrive or underdrive transmissions. The splitter gear system in a 13-gear transmission is used only while in high range and splits the high range gearing either into overdrive or underdrive ratios depending on the transmission model. Splitter systems used on eighteen gear transmissions are used to split both high and low range gearing.

Splitter Cylinder Assembly. In the basic system discussed here, constant air from the air filter/regulator assembly is supplied to the splitter cylinder through a

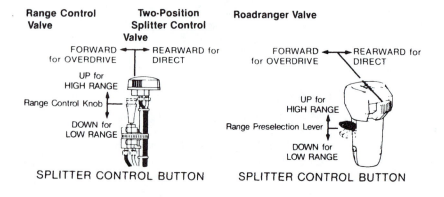

FIGURE 10–48 Shift tower using both air activated high/low range button and split shift button. (Courtesy of Roadranger Marketing. One great drive train from two great companies—Eaton and Dana Corporations)

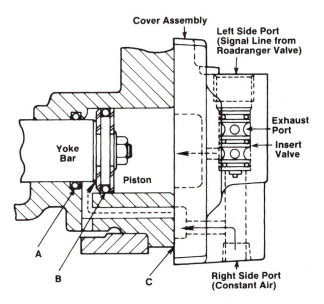

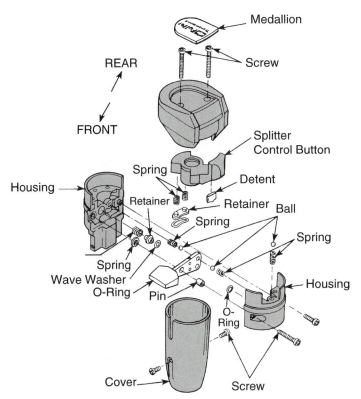

FIGURE 10–49 *Splitter piston components and operation. (Courtesy of Roadranger Marketing. One great drive train from two great companies—Eaton and Dana Corporations)*

FIGURE 10–50 *Exploded view of shift lever range selection and splitter valve assembly that splits torque between the two auxiliary drive gears. (Courtesy of Eaton Corporation)*

port on the cylinder cover (see **Figure 10–39**). Air flow is always supplied to the front side of the shift piston in the cylinder.

An insert valve installed in the cover provides the proper airflow needed to move the splitter piston in the cylinder either forward or rearward. This movement is transferred to the yoke or shift fork bar and onto the shift fork (**Figure 10–49**). The movement of the yoke or shift fork engages a sliding clutch, which in turn locks the third auxiliary gear to the main output shaft.

While operating in the high or low range, the air needed to make the splitter selection and complete the shift is supplied to the shift tower valve (**Figure 10–50**) from the fitting at the S-port of the slave valve mounted on the transmission housing. When an overdrive (button forward) selection is made (**Figure 10–51A**), the air passes through the shift tower valve and is supplied to the left port of the cylinder cover of the splitter cylinder. When the splitter control buttonon the shift tower is in the direct or rear position (**Figure 10–51B**), the S-port of the shift tower valve is closed and no air is supplied to the left port of the splitter cylinder cover.

EIGHTEEN-SPEED TRANSMISSION

Figure 10–52 illustrates a transmission with eighteen forward speeds. The main or front gearing is a five-speed transmission. In this particular transmission, main gears are referred to as low, first, second, third, and drive, rather than first, second, third, fourth, and drive. The auxiliary section contains an overdrive

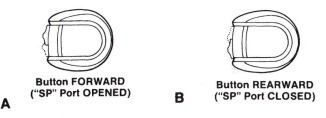

FIGURE 10–51 *Typical split shift operation. (Courtesy of Eaton Corporation)*

gear known as the front drive gear. The high range gear in this particular auxiliary section is called the rear drive gear. There is also a low range gear.

Power transferred through the front drive gear produces an overdrive condition; power transferred through the rear drive gear produces a direct drive. The direct and overdrive gearing is used to split the high and low range gear ratios to provide a greater number of speed and torque selections. **Table 10–1** lists the gear ratios for the eighteen forward and four reverse speeds possible with the transmission shown in **Figure 10–52**. It also includes the following data: main transmission gear engaged, auxiliary range (high/low) engaged, and auxiliary drive gear (direct/overdrive) engaged.

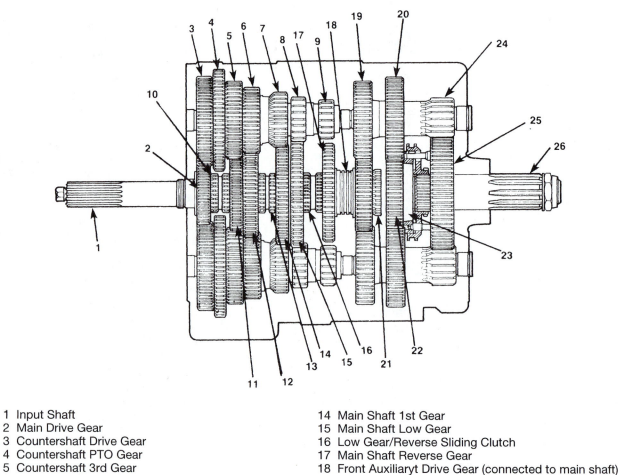

FIGURE 10–52 Components of an eighteen-speed transmission having a five-speed main section and a three-speed auxiliary section. (Courtesy of Eaton Corporation)

1 Input Shaft
2 Main Drive Gear
3 Countershaft Drive Gear
4 Countershaft PTO Gear
5 Countershaft 3rd Gear
6 Countershaft 2nd Gear
7 Countershaft 1st Gear
8 Countershaft Low Gear
9 Countershaft Reverse Gear (idler gear not shown)
10 Main Drive Gear/3rd Gear Sliding Clutch
11 Main Shaft 3rd Gear
12 Main Shaft 2nd Gear
13 2nd Gear/1st Gear Sliding Clutch
14 Main Shaft 1st Gear
15 Main Shaft Low Gear
16 Low Gear/Reverse Sliding Clutch
17 Main Shaft Reverse Gear
18 Front Auxiliaryt Drive Gear (connected to main shaft)
19 Auxiliary Countershaft Front Drive Gear
20 Auxiliary Countershaft Rear Drive (HI range) Gear
21 Auxiliary Drive Gear/Rear Drive Gear Sliding Clutch
22 Auxiliary Rear Drive Gear (HI range)
23 HI Range/Low Range Synchronizer
24 Auxiliary Countershaft Low Range Gear
25 Auxiliary Low Range Gear
26 Output Shaft

As discussed above, the splitter shift system for direct and overdrive engagement is air controlled by a button on the shift tower. While the transmission is in the high or low ranges, this splitter control button can be moved forward to shift the auxiliary transmission from a direct to overdrive condition.

Low Range Direct Power Flow

In these gears (first, third, fifth, seventh, ninth, and #1 Reverse in **Table 10–1**), power flow is through the rear auxiliary drive gear (direct) and auxiliary low range gear (**Figure 10–53**). A complete power flow description is as follows:

1. Power (torque) from the vehicle's engine is transferred to the transmission's input shaft.
2. Splines of the input shaft engage internal splines in the hub of the main drive gear.
3. Torque is split between the two countershaft drive gears.
4. Torque is delivered along both countershafts to the mating countershaft gears of the engaged main shaft gear.
5. Internal clutching teeth in the hub of the engaged main shaft gear transfer torque to the main shaft through the appropriate sliding clutch.

TABLE 10–1:	GEAR RATIOS FOR AN 18 SPEED FULLER TRANSMISSION			
Gear Selected	Main Section Gear Engaged	Range High/Low	Direct or Overdrive	Gear Ratio ___:1
1	Low	Low	Direct	14.71
2	Low	Low	Overdrive	12.45
3	First	Low	Direct	10.20
4	First	Low	Overdrive	8.62
5	Second	Low	Direct	7.43
6	Second	Low	Overdrive	6.21
7	Third	Low	Direct	5.26
8	Third	Low	Overdrive	4.45
9	Fourth (main)	Low	Direct	3.78
10	Fourth (main)	Low	Overdrive	3.20
11	First	High	Direct	2.70
12	First	High	Overdrive	2.28
13	Second	High	Direct	1.94
14	Second	High	Overdrive	1.64
15	Third	High	Direct	1.39
16	Third	High	Overdrive	1.18
17	Fourth (main)	High	Direct	1.00
18	Fourth (main)	High	Overdrive	0.85
#1 Reverse	Reverse	Low	Direct	14.71
#2 Reverse	Reverse	Low	Overdrive	12.45
#3 Reverse	Reverse	High	Direct	3.89
#4 Reverse	Reverse	High	Overdrive	3.29

Note: The relative speed of the PTO gear to input rpm is 0.696 for both the right and bottom PTO gears on the countershaft.

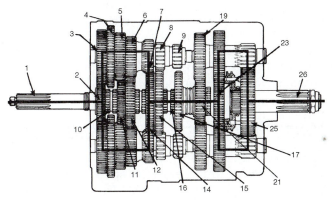

Note: See **Figure 10–52** to interpret number codes.

FIGURE 10–53 Low range direct power flow in an eighteen-speed twin countershaft transmission. (Courtesy of Eaton Corporation)

6. Main shaft transfers torque directly to the rear auxiliary drive gear, through its engaged sliding clutch.
7. The rear auxiliary drive gear splits torque between the two auxiliary countershaft drive gears.
8. Torque is delivered along both auxiliary countershafts to the engaged reduction gear on the output shaft.

9. Torque is transferred to the output shaft through the sliding clutch.
10. The output shaft delivers torque to the driveline.

Low Range Overdrive Power Flow

In these gear selections (second, fourth, sixth, eighth, tenth, and #2 Reverse in **Table 10–1**), power flow is through the front auxiliary gear (overdrive) and the auxiliary low range gear (**Figure 10–54**). Power flow through the main section is like that outlined in steps 1 through 5 under "Low Range Direct." However, the sliding clutch between the front and rear auxiliary drive gears moves forward as the driver engages the splitter control button on the shift tower. The sliding clutch locks the auxiliary front drive gear to the range main shaft. Then the power flow is as follows:

1. The front auxiliary drive gear splits torque between the two auxiliary countershaft drive gears.
2. Torque is delivered along both countershafts to the low range gear on the range main shaft or output shaft. Once again, the high/low synchronizer is used to lock this low range or reduction gear to the output shaft.

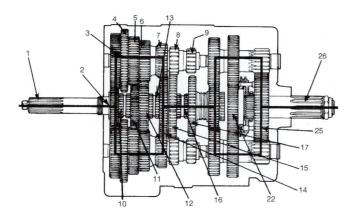

Note: See **Figure 10–52** to interpret number codes.

FIGURE 10–54 Low range overdrive power flow in an eighteen-speed twin countershaft transmission. (Courtesy of Eaton Corporation)

3. Torque is transferred to the output shaft through the sliding clutch of the synchronizer.
4. Torque is delivered to the driveline as low range overdrive.

High Range Direct Power Flow

In these gear selections (eleventh, thirteenth, fifteenth, seventeenth, and #3 Reverse in Table 10–1), power flow is through the rear auxiliary drive gear (**Figure 10–55**). This gear is locked to the auxiliary output shaft by the front/rear sliding clutch and the high side of the high/low range synchronizer. This locks the rear auxiliary drive gear directly to the output shaft. All other gears in the auxiliary section are freewheeling.

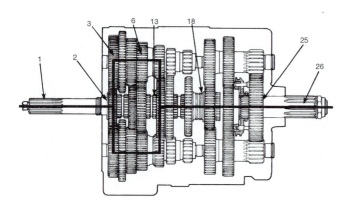

Note: See **Figure 10–52** to interpret number codes.

FIGURE 10–55 High range direct power flow in an eighteen-speed twin countershaft transmission. (Courtesy of Eaton Corporation)

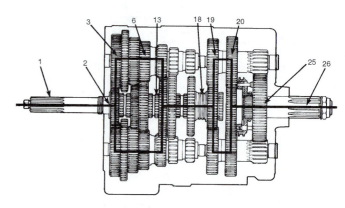

Note: See **Figure 10–52** to interpret number codes.

FIGURE 10–56 High range overdrive power flow in an eighteen-speed twin countershaft transmission. (Courtesy of Eaton Corporation))

High Range Overdrive Power Flow

In these gear selections (twelfth, fourteenth, sixteenth, eighteenth, and #4 Reverse in Table 10–1), power flow is through the front auxiliary drive gear, which is locked to the output shaft by the front/rear sliding clutch (**Figure 10–56**). This splits torque between the two auxiliary drive gears.

Torque is then delivered along both auxiliary countershafts to the engaged rear auxiliary drive gear. This rear auxiliary drive gear is locked to the output shaft by the high side of the high/low range synchronizer. Power flow is through the rear gear to the synchronizer sliding clutch and out through the output shaft to the driveline.

THIRTEEN-SPEED TRANSMISSIONS

The thirteen-speed transmission illustrated in **Figure 10–57** is similar to the eighteen-speed transmission described above in many ways. The main or front gearing is a five-speed transmission. The auxiliary section contains (moving from front to back) a high range gear, a low range gear, and an overdrive gear. In some models, this overdrive gear is replaced with an underdrive gear.

As shown in **Figure 10–58**, the first five gear ratios (low, first, second, third, and fourth) are found with the splitter control button in the rearward (direct drive) position and the range selector in its low range (down) position. Torque is delivered along both countershafts to the engaged low range gear on the range mainshaft or output shaft. This creates low range power flows through the auxiliary gearing for each of the five speeds of the main section (**Figure 10–57**).

The driver then shifts to the high range by pulling up on the range selector. This action moves a sliding

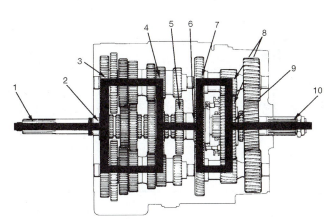

1. Input shaft
2. Main drive gear
3. Countershaft drive gear
4. Main shaft gear countershaft gear
5. Clutch on mainshaft
6. Auxiliary drive gear
7. Auxiliary countershaft drive gears
8. Low range main and countershaft gears
9. Range mainshaft and sliding clutch
10. Output shaft

FIGURE 10–57 *Components of a thirteen-speed transmission having a five-speed main section and two-speed auxiliary section. (Courtesy of Roadranger Marketing. One great drive train from two great companies—Eaton and Dana Corporations)*

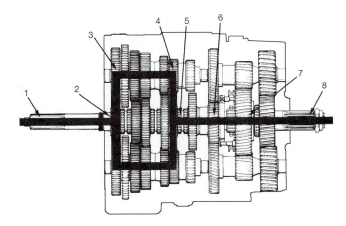

1. Input shaft
2. Main drive gear
3. Countershaft drive gear
4. Main shaft gear countershaft gear
5. Clutch on mainshaft
6. Auxiliary drive gear
7. Range mainshaft
8. Output shaft

FIGURE 10–59 *High range direct power flow in a thirteen-speed twin countershaft transmission. (Courtesy of Roadranger Marketing. One great drive train from two great companies—Eaton and Dana Corporations)*

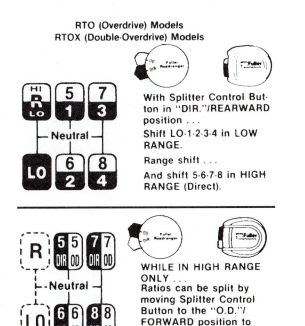

FIGURE 10–58 *Shift pattern for the thirteen-speed transmission shown in* **Figure 10–57.** *(Courtesy of Roadranger Marketing. One great drive train from two great companies—Eaton and Dana Corporations)*

clutch that locks the auxiliary drive gear directly to the range mainshaft or output shaft. Torque is delivered through the range mainshaft and/or output shaft as

high range direct power flows for the next four gear ratios—fifth, sixth, seventh, and eighth (**Figure 10–59**).

While in the high range only, the fifth, sixth, seventh, and eighth gear ratios can be "split" by moving the splitter control button on the cab shift lever to the O.D. (overdrive) or forward position. Activating the splitter shift system in this manner, moves a sliding clutch that locks the overdrive splitter gear in the auxiliary section to the output shaft. Torque is delivered along both auxiliary countershafts to the auxiliary overdrive gears to the output shaft overdrive gear and out through the output shaft (**Figure 10–60**). High range overdrive powerflows are possible for four gear ratios—overdrive fifth, overdrive sixth, overdrive seventh and overdrive eighth.

As you can see in **Table 10–2**, the fifth gear overdrive gear ratio falls between the fifth and sixth gear high range direct ratios. So if the driver wished to shift progressively through all gears, the correct sequence would be:

1. High range direct fifth gear with splitter button in rear position (disengaged).
2. Double clutch, engaging splitter button into forward position while in neutral and shifting into high range fifth overdrive.
3. Double clutch, disengaging splitter button into rear position while in neutral and shifting into high range direct sixth gear.

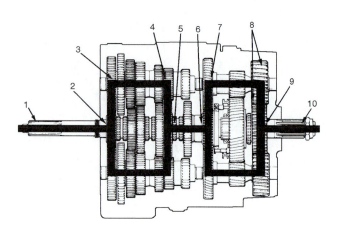

1. Input shaft
2. Main drive gear
3. Countershaft drive gear
4. Main shaft gear countershaft gear
5. Clutch on mainshaft
6. Auxiliary drive gear
7. Auxiliary countershaft drive gears
8. Engaged overdrive/underdrive gear
9. Sliding clutch
10. Output shaft

FIGURE 10–60 *High range overdrive power flow in a thirteen-speed twin countershaft transmission. (Courtesy of Roadranger Marketing. One great drive train from two great companies—Eaton and Dana Corporations)*

Thirteen-speed underdrive transmissions operate on the same principles except the underdrive gear ratio for a particular gear is selected before the direct drive ratio. Underdrive ratios for a typical thirteen-speed transmission are also listed in **Table 10–2**.

DEEP REDUCTION TRANSMISSIONS

Figure 10–61 illustrates the shift pattern for a ten forward speed transmission. The forward gear ratios are low-low, low, and first through eighth. Low-low is a special deep reduction gear for maximum torque and pull-out power generation. It is generated using a deep reduction gear in the auxiliary section. The low-low gear is engaged by activating a split shifter or dash mounted deep reduction valve while operating in the low range only.

Constant air from the air filter/regulator assembly is supplied to the reduction cylinder at the port on the right side of the cylinder cover. With the deep reduction lever in the "Out" position, the valve is opened and air is supplied to the center port of the cylinder cover, moving the reduction piston forward to disengage the deep reduction gearing. When the lever is moved to the "In" position, the valve is closed and no air is supplied to the center port. Constant air from the air filter/regulator assembly moves the reduction piston rearward to engage the reduction gearing.

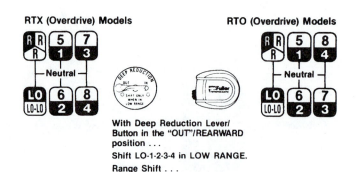

With Deep Reduction Lever/Button in the "OUT"/REARWARD position . . .
Shift LO-1-2-3-4 in LOW RANGE.
Range Shift . . .
Shift 5-6-7-8 in HIGH RANGE.

WHILE IN LOW RANGE ONLY
and shift lever in LO . . .
LO-LO can be obtained by moving Deep Reduction Lever/Button to the "IN"/FORWARD position.

FIGURE 10–61 *Shift pattern for ten-speed transmission having a deep reduction gear in the auxiliary housing. (Courtesy of Roadranger Marketing. One great drive train from two great companies—Eaton and Dana Corporations)*

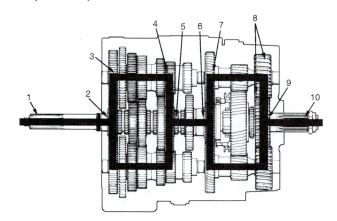

1. Input shaft
2. Main drive gear
3. Countershaft drive gear
4. Main shaft gear countershaft gear
5. Clutch on mainshaft
6. Auxiliary drive gear
7. Auxiliary countershaft drive gears
8. Deep reduction gear
9. Sliding clutch
10. Output shaft

FIGURE 10–62 *Deep reduction (low-low) power flow in a ten-speed transmission. (Courtesy of Roadranger Marketing. One great drive train from two great companies—Eaton and Dana Corporations)*

As shown in **Figure 10–62**, the powerflow is through both countershafts and countershaft deep

TABLE 10–2: GEAR RATIOS THIRTEEN SPEED AND DEEP REDUCTION TRANSMISSIONS (TYPICAL)				
Gear	Main Section	Range	Additional Auxiliary Gearing	Ratio
	Gear Engaged	Low or High	Direct or Special Gear	
Thirteen-Speed Overdrive Transmission				
LOW	Low	Low	Direct	12.56
1	First	Low	Direct	8.32
2	Second	Low	Direct	6.18
3	Third	Low	Direct	4.54
4	Fourth	Low	Direct	3.38
5	First	High	Direct	2.46
5 OD	First	High	Overdrive	2.15
6	Second	High	Direct	1.83
6 OD	Second	High	Overdrive	1.60
7	Third	High	Direct	1.34
7 OD	Third	High	Overdrive	1.17
8	Fourth	High	Direct	1.00
8 OD	Fourth	High	Overdrive	.87
R1	Reverse	Low	Direct	13.13
R2	Reverse	High	Direct	3.89
Thirteen-Speed Underdrive Transmission				
LOW	Low	Low	Direct	12.56
1	First	Low	Direct	8.32
2	Second	Low	Direct	6.18
3	Third	Low	Direct	4.54
4	Fourth	Low	Direct	3.38
5 UD	First	High	Underdrive	2.86
5	First	High	Direct	2.46
6 UD	Second	High	Underdrive	2.12
6	Second	High	Direct	1.83
7 UD	Third	High	Underdrive	1.56
7	Third	High	Direct	1.34
8 UD	Fourth	High	Underdrive	1.16
8	Fourth	High	Direct	1.00
R1	Reverse	Low	Direct	13.13
R2	Reverse	High	Direct	3.89
Deep Reduction 8LL (8 1 2 Speed) Transmission				
LOW-LOW	Low	Low	Deep Reduction	14.05
LOW	Low	Low	Direct	9.41
1	First	Low	Direct	6.23
2	Second	Low	Direct	4.62
3	Third	Low	Direct	3.40
4	Fourth	Low	Direct	2.53
5	First	High	Direct	1.83
6	Second	High	Direct	1.36
7	Third	High	Direct	1.00
8	Fourth	High	Direct	.74
R1	Reverse	Low	Deep Reduction	14.69
R2	Reverse	Low	Direct	9.83
R3	Reverse	High	Direct	2.89

reduction gears, to the output shaft deep reduction gear, which is locked to the output shaft by the sliding clutch.

In shifting from low-low to low, the driver double clutches, releasing the split shifter and moving to low range low. Low through fourth gears are low range gear ratios. The driver then range shifts into high range for gears five through eight.

10.7 TRANSFER CASES

A **transfer case** is simply an additional gearbox located between the main transmission and the rear axle. Its job is to transfer power from the truck's transmission to its front driving axles and its rear driving axles (**Figure 10–63**). The transfer case can transfer power directly using a 1:1 gear ratio, or it can provide gearing for an additional low gear reduction in the power train of the vehicle. The drop box design of the case allows the front driveline from the transfer case to clear the underside of the vehicle (**Figure 10–64**).

Most transfer cases are available with a power takeoff (PTO) and a front axle declutch. The front axle declutch is used to drive the front axle whenever the vehicle encounters deep grades or rough terrain. (This is slightly different from some off-highway vehicles that are designed for continuous all-wheel-drive operation.) Both the PTO and front axle drive declutch are activated by separate shift levers located in the cab of the vehicle.

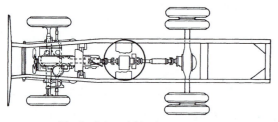

Typical 4 × 4 Hookup

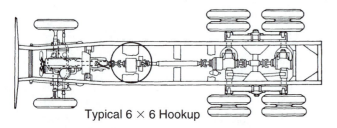

Typical 6 × 6 Hookup

FIGURE 10–63　Location of a transfer case on a typical (A) 4 × 4; and (B) 6 × 6 hookup. (Courtesy of ArvinMeritor Inc.)

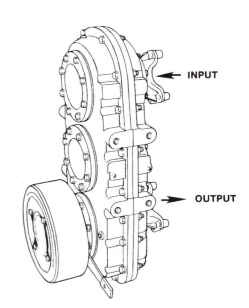

INPUT

OUTPUT

FIGURE 10–64　Single-speed, direct drive transfer case for front-wheel-drive 4 × 2 vehicles; drop box design allows for shaft clearance under the engine. (Courtesy of ArvinMeritor Inc.)

In addition, the transfer case might be equipped with an optional parking brake and a speedometer drive gear that can be installed on the idler assembly. Most transfer cases use the countershaft design and the gearing is of the constant mesh helical cut type. Most countershafts are mounted in ball or roller bearings, which are tapered. All rotating and contact components of the transfer case are lubricated by oil from gear throw-off during operation. However, some units are provided with an auxiliary oil pump, externally mounted to the transfer case.

Some manufacturers' models come equipped with a driver-controlled air-actuated differential lockout for extreme traction conditions. Most transfer cases have a large access panel to allow the technician access to some internal components for servicing. The torque range of a transfer case is typically from 3,300 to 15,000 pound-feet input torque capacity.

A conventional three shaft transfer case design is shown in **Figure 10–65**. This type of design is a two-speed, three-shaft transfer case. It can be adapted to incorporate a PTO and a front axle declutch to drive the front axle whenever the truck encounters deep grades or rough terrain. A cutaway view of the conventional three-shaft design is shown in **Figure 10–66**.

Another configuration of the transfer case is the cloverleaf four-shaft design as shown in **Figure 10–67**. This two-speed, four-shaft design can be adapted to incorporate a PTO and a mechanical type auxiliary brake. Sectional views of this cloverleaf type transfer case are shown in **Figure 10–68**.

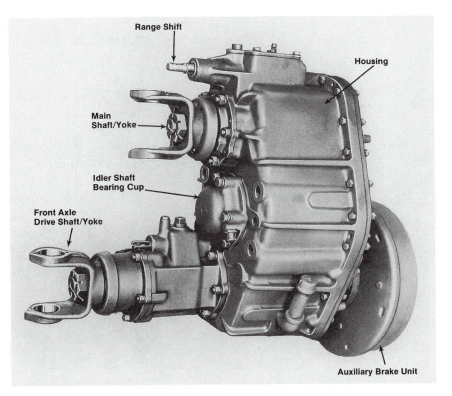

FIGURE 10–65 Three shaft design transfer case. (Courtesy of ArvinMeritor Inc.)

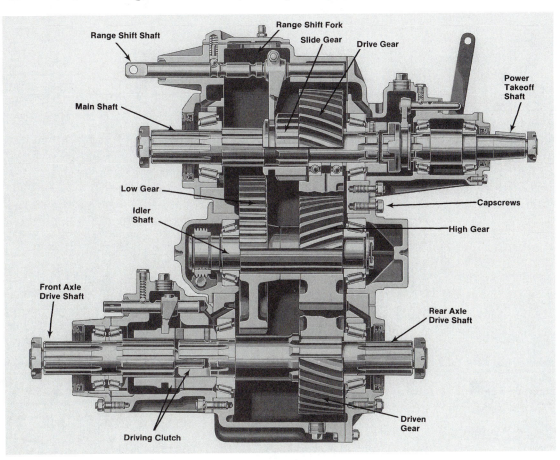

FIGURE 10–66 Cutaway view of a conventional type transfer case (three shift design). (Courtesy of ArvinMeritor Inc.)

FIGURE 10-67 Cloverleaf design transfer case. (Courtesy of ArvinMeritor Inc.)

10.8 POWER TAKEOFF UNIT

A variety of accessories on heavy-duty trucks have a need for an auxiliary drive. This auxiliary drive can be provided at the transmission by installing an optional PTO unit (**Figure 10–69**). A typical installation is shown in **Figure 10–70**.

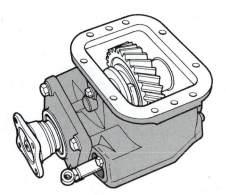

FIGURE 10-69 Power takeoff (PTO) unit essentially serves as an extension of the transmission.

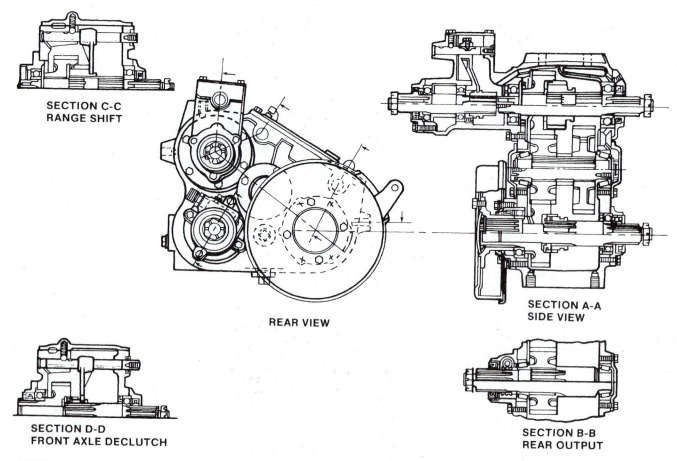

SECTION C-C
RANGE SHIFT

SECTION D-D
FRONT AXLE DECLUTCH

REAR VIEW

SECTION A-A
SIDE VIEW

SECTION B-B
REAR OUTPUT

FIGURE 10-68 Sectional views of a cloverleaf design transfer case. (Courtesy of ArvinMeritor Inc.)

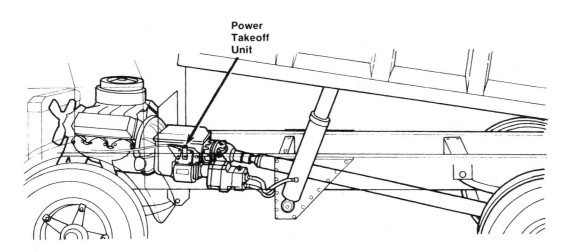

FIGURE 10–70
Typical installation of a PTO unit. (Courtesy of Eaton Corporation— Eaton Clutch Div.)

The PTO eliminates the need for a second, or auxiliary, engine to drive the truck's accessories. It requires no cooling system, has no ignition system, nor requires other routine engine maintenance. There are six basic types of PTOs, classified by their installation:

- Side mount
- Split shaft
- Top mount
- Countershaft
- Crankshaft driven
- Flywheel

The side mount PTO is attached to the side of the main transmission or auxiliary transmission and is the most common type. The split shaft PTO is designed to transmit engine power and torque from the drive shaft. It is attached within the truck's drivetrain, behind the transmission, and requires special mounting to the chassis frame. The top mount PTO is designed to operate from, and is attached to, the top of the auxiliary transmission. It is used in heavy-duty applications where full engine power is required.

The countershaft PTO is mounted behind certain transmissions and replaces the normal bearing cap at the rear of the countershaft. This design is used on twin countershaft transmissions. The crankshaft driven PTO is driven by the front of the engine crankshaft. They are used in heavy-duty and extra heavy-duty applications that require auxiliary power while the truck is in motion. A clutch type crankshaft driven PTO is preferable since it can be engaged without shutting off the engine. The flywheel PTO is sandwiched between the bell housing and the transmission. It allows full engine torque for continuous operation. No disconnect is normally provided since it runs directly off the transmission main shaft with no way to stop rotation, except to turn off the engine.

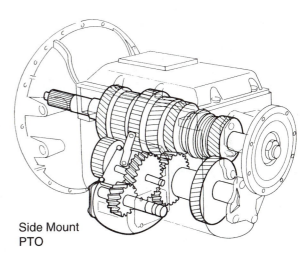

Side Mount PTO

FIGURE 10–71   PTO installation interface with the truck's transmission. (Courtesy of Eaton Corporation—Eaton Clutch Div.)

The typical PTO is designed to transfer engine torque and transfer it to another piece of equipment. This is done either by mounting the equipment directly to the PTO or by connecting the equipment with a small drive shaft. The PTO input gear meshes with one of the gears in the truck's transmission (**Figure 10–71**). Refer to Figures 10–36 and 10–52. The PTO ties into these transmissions through the PTO gear on the transmission countershaft. The rotation of the PTO countershaft gear drives the PTO input gear. The PTO input gear drives the PTO output shaft to drive the vehicle's optional equipment.

Gears in the PTO unit can be of either spur or helical design. Proper meshing between the PTO drive gear in the truck's transmission and the gears of the PTO unit is important for sound operation.

Gear ratio is another important part of PTO operation. The correct PTO installation should have the necessary torque capacity and operating speed that will meet the requirements of the application. An

**Exploded View
of Typical PTO**

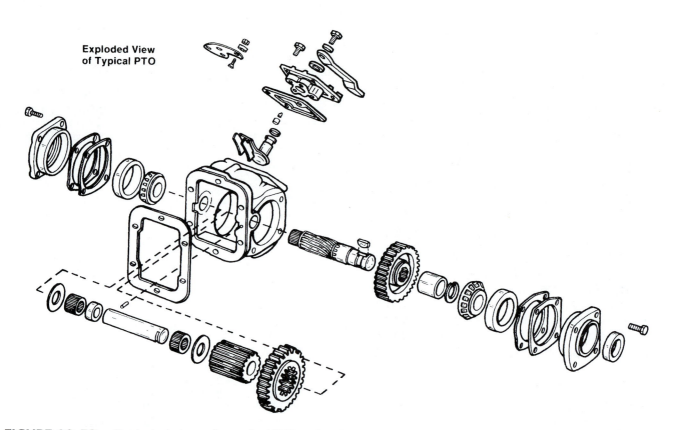

FIGURE 10–72   Exploded view of a typical PTO unit. (Courtesy of Eaton Corporation—Eaton Clutch Div.)

exploded view of a typical PTO unit is shown in **Figure 10–72.**

As already mentioned, there are some power takeoffs that function whenever a vehicle is operating. Some of these operate at a fixed ratio of engine speed; others vary depending on which gear the transmission is in. Other PTOs are designed to work only when the vehicle's transmission has been shifted to the neutral position. All transmission-mounted PTOs are clutch-dependent in that they only transmit torque when the vehicle's clutch is engaged. Flywheel and crankshaft drive (also known as front-of-engine) power takeoffs operate independently of vehicle clutch engagement. The correct techniques for using the various types of PTOs are explained in the owners' manuals.

There are several types of shift mechanisms that are used to connect the power takeoff with operator controls. There are several systems used: cable, lever, air, and electric-over-air. The cable type is the most common, while the electric-over-air is the most expensive.

Most PTOs feature simple designs and rugged construction. Dirty transmission fluid is just about all that will harm a PTO. The same fluid used to lubri-

cate the transmission is also lubricating the PTO. Regularly changing a vehicle's transmission fluid also will keep harmful contaminants from entering a PTO.

SUMMARY

- Torque from the engine is transferred through the engaged clutch to the input shaft of the transmission where gears in the transmission housing alter the torque and speed of the input. Various size gears in a standard transmission function to give the truck's engine the ability to move a heavily loaded vehicle from a standing start up to highway speeds.
- Transmissions having a limited number of gear ranges often have a single set of gears—called main gearing—contained in a single housing.
- Many heavy-duty truck transmissions consist of two distinct sets of gearing: the main or front gearing, and the auxiliary gearing located directly on the rear of the main gearing.
- Gear pitch refers to the number of teeth per unit of pitch diameter.

- The three stages of contact through which the teeth of two gears pass while in operation are coming into mesh, full mesh, and coming out of mesh.
- The relationship of input and output speeds is expressed as gear ratio.
- The amount of torque increase from a driving gear to a driven gear is directly proportional to the speed decrease.
- The major types of gear tooth design used in modern transmissions and differentials are spur gears and helical gears.
- A heavy-duty manual transmission consists of a main shaft and one, two, or three counter-shafts.
- The three general classifications of manual transmissions according to how shifts between gears are made are sliding gear, collar shift, and synchronized shift mechanisms.
- The two functions of a synchronizer are to bring components that are rotating at different speeds to one synchronized speed and to lock these components together. Block or cone and pin synchronizers are the most popular in heavy-duty trucks.
- An air-operated gearshift system consists of the following: air filter/regulator, slave valve, master control valve, range cylinder, and fittings and connecting air lines.
- Auxiliary section gearing can be designed to include a third gear in addition to the high and low range gears. This third gear is engaged or disengaged from the power flow through the use of a splitter shift system that is air activated by a button on the shift tower.
- A transfer case is an additional gearbox located between the main transmission and the rear axle. Its job is to transfer power from the truck's transmission to the front driving axles and the rear driving axles.
- A variety of accessories on heavy-duty trucks have a need for an auxiliary drive, which can be provided at the transmission by installing an optional PTO (power takeoff) unit.
- The six types of PTOs, classified by their installation, are side mount, split shaft, top mount, countershaft, crankshaft driven, and flywheel.

REVIEW QUESTIONS

1. Gears can
   a. transmit speed and torque unchanged
   b. decrease speed and increase torque
   c. increase speed and decrease torque
   d. all of the above

2. Which of the following gear ratios indicates the most amount of overdrive?
   a. 2.15:1
   b. 1:1
   c. 0.85:1
   d. 0.64:1

3. Which type of gear is more likely to develop gear whine at higher speeds?
   a. spur gear
   b. helical gear
   c. worm gear
   d. all of the above

4. When an idler gear is placed between the driving and driven gear, the driven gear
   a. rotates in the same direction as the driving gear
   b. rotates in the opposite direction of the driving gear
   c. remains stationary

5. When sliding gearshift mechanisms are used, which gears do they control?
   a. high and low gears in the auxiliary section
   b. first and reverse gears in the main section
   c. all forward gears except first in the main section
   d. direct and overdrive gears in the auxiliary section

6. Which type of transmission requires double-clutching between shifts?
   a. synchronized
   b. nonsynchronized

7. The component used to ensure that the main shaft and the main shaft gear to be locked to it are rotating at the same speed is known as a
   a. synchronizer
   b. sliding clutch
   c. transfer case
   d. PTO unit

8. Which of the following is the most widely used manual transmission gear train arrangement in heavy-duty trucks?
   a. single countershaft
   b. twin countershaft
   c. triple countershaft
   d. none of the above

9. The twin countershaft design
   a. splits torque between two separate shafts
   b. allows the face width of each gear to be reduced
   c. allows the main shaft to float between the countershafts
   d. all of the above

10. In a range shift air control system, which component distributes inlet air pressure to both the low and high range air hoses running to the range cylinder?
    a. slave valve
    b. splitter valve
    c. air regulator
    d. air compressor

11. To provide for overdrive gearing in the high and low auxiliary transmission ranges, which of the following components is added to the basic range shift air control system?
    a. splitter cylinder
    b. button actuated shift tower valve
    c. additional slave valve
    d. both a and b

12. Name the component used to retain the shift bar in position
    a. detent ball
    b. slave valve
    c. range cylinder
    d. clutch gear

13. An auxiliary drive for truck accessories that is used with the transmission is called
    a. a transfer case
    b. a power takeoff unit
    c. an auxiliary transmission
    d. a synchronizer

14. A transfer case can be used to
    a. provide an extra gear reduction (low) in the drive train
    b. transfer torque from the transmission to the front driving axle
    c. transfer torque from the transmission to the rear driving axle(s)
    d. all of the above

15. Which of the following components in a typical heavy-duty, twin countershaft transmission could be described as floating?
    a. countershafts
    b. mainshaft
    c. input shaft
    d. output shaft

11

Standard Transmission Servicing

Prerequisite
Chapter 10

Objectives

After reading this chapter, you should be able to

- Explain the importance of using the correct lubricant and maintaining the correct oil level in a transmission.
- List the preventive maintenance inspections that should be made periodically on a standard transmission.
- Explain how to replace a rear seal.
- Describe a procedure for troubleshooting a transmission.
- Identify potential causes of various transmission performance problems, such as unusual noises, leaks, vibrations, jumping out of gear, and hard shifting.
- Outline a basic procedure for overhauling a transmission.
- Analyze transmission components for wear and damage and pinpoint the cause.
- Troubleshoot an air shift system.

Key Terms

brinelling
fretting
jumpout

slipout
timing
yoke sleeve kit

Properly operated and maintained, a standard transmission will give hundreds of thousands of miles of trouble-free operation. At least one transmission manufacturer offers a limited warranty of 750,000 miles. But like all mechanical truck components, a transmission is subject to wear, fatigue, abuse, and manufacturing defects. Even when operated properly under ideal conditions, adjustments will eventually be necessary and parts will have to be replaced. This chapter will explain how to maintain, troubleshoot, and service a heavy-duty truck transmission.

11.1 LUBRICATION

Proper lubrication is the key to a good all-around maintenance program. If the oil is not doing its job, or if the oil level is ignored, all the maintenance procedures in the world will not ensure a long transmission life.

Standard transmissions are designed so that the internal parts operate in a bath of oil that splash

TABLE 11–1: RECOMMENDED LUBRICANTS		
Temperature	**Grade**	**Type**
Above 0°F Below 0°F	SAE 30, 40, or 50 SAE 30	Heavy-Duty Engine Oil Meeting MIL-L-2104C or MIL-L-46152 NOTE: Oils meeting MIL-L-2104B or MIL-L-45199 are also acceptable
Above 0°F Below 0°F	SAE 90 SAE 80 SAE 75W-90	Straight Mineral Gear Oil SHP Synthetic Gear Lubricant

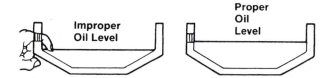

FIGURE 11–1 The oil level must be level with the bottom of the plug hole. (Courtesy of Roadranger Marketing. One great drive train from two great companies—Eaton and Dana Corporations)

lubricates by the motion of gears and shafts. Some units also use a pump to circulate oil to critical wear surfaces on the main shaft. All transmission components will be properly lubricated if the following procedures are followed:

1. Use the correct grade and type of oil.
2. Change oil regularly.
3. Maintain the proper oil level and inspect it regularly.

RECOMMENDED LUBRICANTS

It is vital that only the lubricants recommended by the transmission manufacturer be used in the transmission. Most manufacturers suggest a specific grade and type of transmission oil, heavy-duty engine oil, synthetic or straight mineral oil, depending on the ambient air temperature during operation. **Table 11–1** gives typical grade and type for temperature ranges above and below 0°F.

Do not use EP gear oil or multi-purpose gear oil when operating temperatures are above 230°F. Many of these gear oils break down above 230°F and coat seals, bearings, and gears with deposits that might cause premature failures. If these deposits are observed (especially coating seal areas and causing oil leakage), change to heavy-duty engine oil or mineral gear oil to ensure maximum component life.

OIL CHANGES

Most manufacturers recommend an initial oil change and flush after the transmission is placed in actual service. This change should be made any time following 3,000 miles but before 5,000 miles of over-the-road service. In off-highway use, the change should be made after 24 and before 100 hours of service have elapsed.

There are many factors that influence the regular schedule of oil change periods. In general, it is suggested that a drain period be scheduled every 50,000 miles for normal over-the-highway operations. Off-highway usually requires oil change every 1,000 hours. The oil level in the transmission should be checked at manufacturer's recommended intervals of 5,000 or 10,000 highway miles. When it is necessary to add oil, types and brands of oil should not be mixed. Top off the oil level so that it is even with the filler plug opening (**Figure 11–1**).

 Shop Talk

Make sure the oil is level with the filler opening. Because you can reach oil with your finger does not mean oil is at the proper level. One inch of oil level is equivalent to about one gallon of oil in most transmissions.

Draining Oil

Drain the transmission while the oil is warm. To drain oil, remove the drain plug at the bottom of the case. Clean the drain plug before reinstalling.

Refill

Remove dirt around the filler plug. Then remove the plug from the side of the case and refill with new oil of the grade recommended for the season service application. Fill to the bottom of the level plug threads. If the transmission housing has two filler plugs, leave both out until the oil is level with each plug hole.

CAUTION: Do not overfill the transmission. Overfilling usually results in oil breakdown due to aeration caused by the churning action of the gears. Premature breakdown of the oil will result in varnish and sludge deposits that plug up oil ports and build up on splines and bearings.

Shop Talk

When draining transmission oil, check for metal particles in the oil. Such particles indicate excessive metal wear and the probability of a problem occurring in the near future.

11.2 PREVENTIVE MAINTENANCE INSPECTIONS

A good preventive maintenance (PM) program can avoid breakdowns, minimize vehicle downtime, and reduce the cost of repairs. Often, transmission problems can be traced directly to poor maintenance. **Figure 11-2** identifies components that should be checked periodically. The following is a typical inspection schedule that can be followed in most cases. However, this schedule is not applicable to all trucks because inspection intervals will vary depending upon operating conditions.

1. Daily
    - **Air Tanks**. Bleed air tanks to remove water or oil.
    - **Oil Leaks.** Check for oil leaks around bearing covers, PTO covers and other machined surfaces. Also check for oil leakage on the ground before starting the truck each morning.
2. Every 10,000 miles
    - **Fluid Level.** Remove the filler plugs and check the level of lubricant. Fill if necessary and tighten plugs securely.
    - **Capscrews and Gaskets**
    a. Check all capscrews and plugs, especially those on PTO covers and rear bearing covers for looseness, which would cause oil leakage.

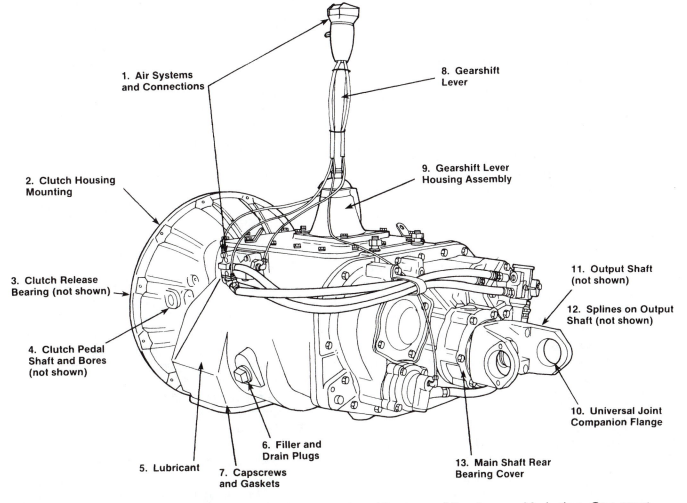

1. Air Systems and Connections
2. Clutch Housing Mounting
3. Clutch Release Bearing (not shown)
4. Clutch Pedal Shaft and Bores (not shown)
5. Lubricant
6. Filler and Drain Plugs
7. Capscrews and Gaskets
8. Gearshift Lever
9. Gearshift Lever Housing Assembly
10. Universal Joint Companion Flange
11. Output Shaft (not shown)
12. Splines on Output Shaft (not shown)
13. Main Shaft Rear Bearing Cover

FIGURE 11-2 Inspection points on standard transmission. (Courtesy of Roadranger Marketing. One great drive train from two great companies—Eaton and Dana Corporations)

b. Check PTO manifold and rear bearing covers for oil leakage due to defective gasket.

- **Main Shaft Rear Bearing.** Check around the rear bearing cover for oil leaks, especially if the transmission has recently been serviced or rebuilt.

3. Every 20,000 miles
   - **Air System and Connections (1).** Check for leaks, worn hoses and air lines, loose connections, and loose capscrews.
   - **Clutch Housing Mounting (2).** Check all capscrews in the bolt circle of the clutch housing for looseness.
   - **Clutch Pedal Shafts (4).** If the clutch pedal shafts are equipped with a zerk fitting, lubricate the shafts. Pry upward on the shaft to check for wear. If excessive movement is found, remove the clutch release mechanism and check for worn bushings.
   - **Remote Control Linkage.** Check the linkage U-joints for wear and binding. Lubricate the U-joints and check bolted connections for tightness. Check bushings for wear.
   - **Air Regulator and Air Filter.** Check and clean or replace the air filter element.
   - **Universal Joint Companion Flange (10).** Check the flange nut for proper torque. Tighten if necessary.
   - **Output Shaft (11).** Pry upward against the output shaft to check for radial clearance in the main shaft rear bearing. Check the splines on the output shaft (12) for wear from movement and chucking action of the universal joint companion flange.

4. Every 40,000 miles
   - **Clutch.** Check the clutch disc faces for wear if they can be inspected without removing the clutch. Check the dampening action of the clutch driven plate(s).
   - **Release Bearing (3).** Remove the inspection cover and check the axial and radial clearances in the release bearing. Check the position of the thrust surface of the release bearing with the thrust sleeve on push type clutches. Chapter 9 explains this procedure.

5. Every 50,000 miles
   - **Lubricant Change.** Drain and refill the transmission with the specified fluid or oil. Transmission oil analysis can be used to establish more precise oil change intervals suited to the truck's exact application and operating conditions.

- **Gearshift Lever (8).** Check for looseness and free play in the housing. If the lever is loose in the housing, check the housing assembly.
- **Gearshift Lever Housing Assembly (9).** Remove air lines at the slave valve and remove the gearshift lever housing assembly from the transmission. Check the tension spring and washer for set and wear. Check the gearshift lever spade pin and slot for wear. Check the bottom end of the gearshift lever for wear and the yokes and blocks in the shift bar housing for wear at contact points with shift lever.

Table 11–2 outlines this preventive maintenance schedule.

CHECKING FOR OIL LEAKS

Leakage in transmission rear seals (**Figure 11–3**) is perhaps the most common problem in truck transmissions. The problem is more than a nuisance because if not repaired, a leaking seal can lead to complete transmission failure. It can be very time consuming and expensive to replace a rear seal only to find the oil seal was not causing oil leakage. Using the following checklist, inspect the transmission to ensure proper identification of the leak path.

Before disassembling the rear seal housing, clean the rear bearing cover, rear seal, and output yoke. Do not use a high pressure spray washer to clean the rear seal housing. Use a clean dry cloth. Run the vehicle, then inspect these areas for oil leaks.

1. Inspect the speedometer connections. The speedometer sleeve should not be loose. There should be an O-ring or gasket between the mating speedometer sleeve and the rear bearing cover. If the speedometer connections are loose or the O-ring/gasket is damaged or missing, it will cause an oil leak. The speedometer sleeve should be torqued to specification and a hydraulic thread sealant applied to the threads.

2. Inspect the rear bearing cover retaining capscrews for tightness; they should be torqued to specification and thread sealant applied to the capscrew threads.

3. Anytime the capscrew, gasket, and collar assembly on the rear bearing cover is removed, the gasket and nylon collar must be replaced. Verify the collar and gasket are installed at the chamfered hole that is aligned near the mechanical speedometer opening. The capscrew must be torqued to

TABLE 11–2:   PREVENTIVE MAINTENANCE RECOMMENDATIONS**

PM Operation	Daily	5,000	10,000	20,000	30,000	40,000	50,000	60,000	70,000	80,000	90,000	100,000**
Bleed air tanks and listen for leaks	X											
Inspect for oil leaks	X											
Check oil level		X	X	X	X	X	X	X	X	X	X	X
Inspect air system connections				X		X		X		X		X
Check clutch housing capscrews for looseness				X		X		X		X		X
Lube clutch pedal shafts				X		X		X		X		X
Check remote control linkage				X		X		X		X		X
Check and clean or replace air filter element				X		X		X		X		X
Check output shaft for looseness				X		X		X		X		X
Check clutch operation and adjustment						X				X		
Change transmission oil		X*					X					X

*Initial fill on new units.
**REPEAT SCHEDULE AFTER 100,000 MILES.
Courtesy of Eaton Corporation

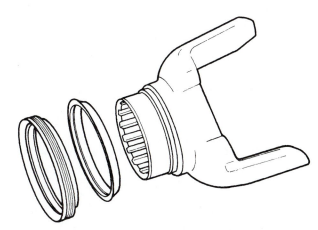

FIGURE 11–3 Yoke and rear seal that prevents leaks of transmission rear seals. (Courtesy of Roadranger Marketing. One great drive train from two great companies—Eaton and Dana Corporations)

specification with thread sealant applied to the capscrew threads. Use the same procedure to check an electronic tailshaft speed sensor.

4. Inspect the rear bearing cover. Make sure that a rear bearing cover gasket is in place. If the rear bearing cover was removed, a new gasket must be installed to prevent oil leakage.

5. Check the output yoke retaining nut for tightness. The retaining nut should be torqued to specification (typically 450 to 600 ft.-lb.).

CAUTION: Overtorquing the yoke retaining nut can also cause an output shaft bearing to fail.

6. Check the PTO covers and openings, making sure the capscrews are torqued to specification. Capscrews must have thread sealant applied to the threads. If oil leakage is present at the PTO covers, replace the PTO cover gaskets under the PTO covers.

7. Check all cast iron components such as the front bearing cover, front case, shift bar housing, rear bearing cover, and the clutch housing for cracks or breaks. These can cause oil leakage. Replace or repair parts found to be damaged.

8. Check the front bearing cover. If the oil return grooves have been damaged or destroyed by contact with the input shaft, it will cause oil leakage. If the grooves are damaged or destroyed, replace the front bearing cover. If the front bearing cover capscrews are not tightened to specifications, the front bearing cover will leak. If torquing the capscrews does not correct oil leakage from the front bearing cover, the front bearing cover gasket might be damaged or the front bearing cover and gasket might not be aligned with the case oil return hole and the hole in the cover. If the front bearing cover has an oil seal installed, inspect the oil seal. Damaged or worn seals must be replaced to prevent oil leakage.

9. An oil leak in the front bearing cover, the auxiliary housing, or in some cases the air breather in the shift bar housing, can be caused by a damaged O-ring in the range air shift system's cylinder. The damaged O-ring allows air to pressurize the transmission case causing oil to leak from the front bearing cover, auxiliary housing, or air breather.

10. If the transmission is equipped with an oil cooler and oil filter, inspect all the connectors, fittings, hoses, and filter element to make sure they are tight enough to prevent oil leakage.

11. Inspect the oil drain plug and oil fill plug for leakage. The oil drain plug and oil fill plug must be torqued to specification.

SEAL REPLACEMENT

If it is determined that the rear oil seal is leaking, the end yoke must be removed and the seal and slinger replaced. Check the yoke for signs of wear and damage (**Figure 11–4**). In particular, pay attention to the yoke seal surface. If there is any wear beyond what could be considered normal polishing, the yoke should be replaced or resleeved. There should be no rust, burrs, nicks, or grooves on the seal surface.

👓 Shop Talk _____

You cannot successfully repair a worn or damaged yoke by hand polishing. It is unlikely that the seal surface can be made smooth enough with emery paper or crocus cloth. Fine scratches, even those not visible to the naked eye, that are not perpendicular to the centerline of the yoke spline will either track oil under the seal or draw in contaminants.

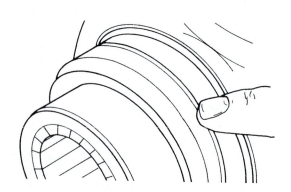

FIGURE 11–4 *Check the yoke for signs of wear and damage. (Courtesy of Roadranger Marketing. One great drive train from two great companies—Eaton and Dana Corporations)*

Yoke Sleeve Kits. A **yoke sleeve kit** can be installed instead of completely replacing the yoke. The sleeve is a heavy walled construction with a hardened steel surface; it is often called a wear ring. The outside diameter of the sleeve is the same as the original yoke diameter. Follow the kit manufacturer's exact instructions when turning down the original yoke and pressing the sleeve onto the yoke. Always use the oil seal recommended in the kit instructions.

👓 Shop Talk _____

Wear rings that increase the original yoke diameter can cause the new seal to wear more rapidly due to increased seal lip pressure.

To replace a rear seal, do the following:

1. Block the truck wheels and place the transmission in neutral.
2. Disconnect the drive shaft from the end yoke.
3. Loosen the end yoke nut and remove the nut, washer, and yoke.
4. Remove the oil seal from the rear bearing cover.
5. Wipe the bore of the bearing cover with a clean cloth to remove any contaminants or grit.
6. Install the new seal and slinger with the proper driving tool (**Figure 11–5**). Be sure that the seal (or slinger) is not cocked at installation. Even if straightened later, the

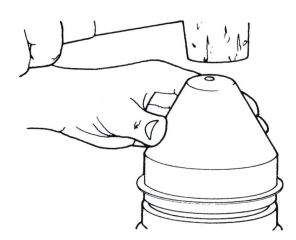

FIGURE 11–5 *Install the seal squarely in the bore with the proper driving tool. (Courtesy of Roadranger Marketing. One great drive train from two great companies—Eaton and Dana Corporations)*

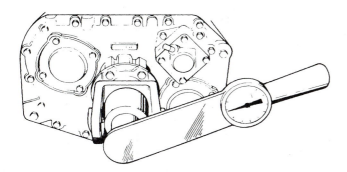

FIGURE 11–6 _Torque the yoke nut to manual specifications. (Courtesy of Roadranger Marketing. One great drive train from two great companies—Eaton and Dana Corporations)_

thin steel cage can be permanently distorted during the installation operation and leakage might occur.

7. Reinstall the end yoke over the output shaft and torque the nut to specification (**Figure 11–6**).

CAUTION: Do not overtorque the yoke nut. Overtorquing is one of the more common causes of bearing failure and the easiest to avoid with a little care.

11.3 TROUBLESHOOTING

To determine the location and nature of a transmission related problem, it is essential to follow a troubleshooting procedure. If possible, discuss the problem and symptoms with the driver. Gather information on the operating condition and the vehicle use, on the development and history of the problem, and on the shifting characteristics that are affected by the problem. It is also helpful to gather information on the history of the tractor itself, such as regular maintenance schedules and lubrication procedures, past transmission failures that might have occurred, and mileage or hours of use currently on the unit.

Inspect the transmission carefully. Look for signs of misuse, such as broken transmission mounts, fittings, or brackets. Look for signs of leakage. Check the air lines for leaks.

Road test the truck whenever possible. Technicians usually get second- or third-hand reports of trouble experienced with the unit and these reports do not always accurately describe the actual conditions. Sometimes symptoms seem to indicate trouble in the transmission, whereas actually the trouble might be caused by the axle, propeller shaft, universal joint, engine, or clutch. This is especially true of complaints of noise and vibration. Before removing the transmission or related components to locate the problem, always road test to check the possibility that trouble might exist in other systems or components. If the technician is licensed to drive a heavy-duty truck, road testing will be more effective; however, riding with the driver can also be informative.

The troubleshooting procedures just described will reveal certain symptoms of a problem, if not the problem itself. The following are possible causes for the performance problems that might be noticed during the troubleshooting process.

NOISY OPERATION

Technicians should road test to determine if a driver's complaint of noise is actually in the transmission. In some instances, drivers have insisted that a noise was in the transmission, but further investigation revealed the noise to be caused by one of the following conditions:

- Fan out of balance or blades bent
- Defective vibration dampers
- Crankshafts out of balance
- Flywheels out of balance
- Flywheel mounting bolts loose
- Engine rough at idle, producing rattle in gear train
- Clutch assembly out of balance
- Engine mounts loose or broken
- Power take-off engaged
- Universal joints worn out
- Propeller shafts out of balance
- Universal joint angles out of phase or at excessive angle
- Center bearings in driveline dry, not mounted properly, etc.
- Wheels out of balance
- Tire treads humming or vibrating at certain speeds
- Air leaks on suction side of induction system, especially with turbochargers

Technicians should try to locate and eliminate noise by means other than transmission removal or overhaul. However, if the noise appears to be in the transmission, try to break it down into the following classifications. If possible, determine what position the gearshift lever is in when the noise occurs. If the

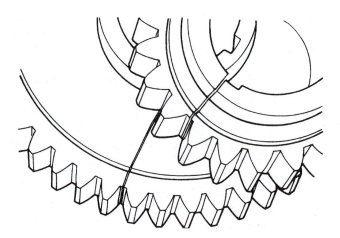

FIGURE 11–7 Cracked gears can cause such problems as whining, knocking, grinding, or thudding noises. (Courtesy of Roadranger Marketing. One great drive train from two great companies—Eaton and Dana Corporations)

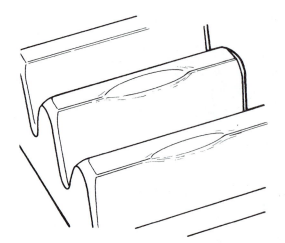

FIGURE 11–8 Bumps or swells on the gear teeth are caused by improper handling of gear before and during assembly. (Courtesy of Roadranger Marketing. One great drive train from two great companies—Eaton and Dana Corporations)

noise is evident in only one gear position, the cause of the noise is generally traceable to the gears in operation.

Growling

Growling and humming or a more serious grinding noise are caused by worn, chipped, rough, or cracked gears (**Figure 11–7**). As gears continue to wear, the grinding noise will be noticeable, particularly in the gear position that throws the greatest load on the worn gear. Growling is also caused by improper timing of the transmission during reassembly or improper timing due to a gear turning on the countershaft. Both conditions produce error in the tooth spacing.

Thumping or Knocking

As bearings wear and retainers start to break up, the noise could change to a thumping or knocking. This noise is heard at low shaft speeds in any gear position.

Irregularities on gear teeth can also cause thumping or knocking. Bumps or swells can be identified as highly polished spots on the face of the gear tooth (**Figure 11–8**). Generally, the noise is more prominent when the gear is loaded, making it easier to locate the problem gear. Bumps or swells are caused by improper handling of gears before or during assembly.

A gear cracked or broken by shock loading or by pressing on the shaft during installation will also produce this sound at low speeds. At high speeds a howl will be present.

Rattles

Metallic rattles within the transmission result from a variety of conditions. Engine torsional vibrations are transmitted to the transmission through the clutch. In heavy-duty equipment using clutch discs without vibration dampers often produces a rattle particularly in neutral. In general, engine speeds should be 600 rpm or above to eliminate rattles and vibrations noticeable during idle. A defective injector would cause a rough or lower idle speed and a rattle in the transmission. Rattle could also be caused by excessive backlash in the PTO unit mounting.

Vibration

Increased highway speed limits in most states permit sustained high speeds. The fact that engines and entire power trains can now cruise at a higher rpm can introduce vibration frequencies that were not critical in the past. At slower speeds these items might get by or only pass through a critical period while accelerating or decelerating through the gears.

At slower speeds, driveline vibrations such as bent tubes, joints out of phase or alignment, bad angles to short couples, clutches out of balance, and gears and shafts in the transmission out of balance, were fairly obvious. These items are more critical in vehicles running at sustained high speeds.

Critical vibrations associated with higher speeds are not the thumping or bumping type but are high frequency vibrations that sting or tingle the soles of your feet, tickle the end of your fingers, etc. This type of vibration will cause gear seizures, broken synchronizer pins, bearing failure due to retainer rivet failures, and so forth.

Whining

Gear whine is usually caused by insufficient backlash between mating gears such as improper shimming of a PTO. A high pitched whine or squeal can also be caused by mismatched gear sets. Such gear sets are identified by an uneven wear pattern on the face of the gear teeth. Pinched bearings, having insufficient axial or radial clearance, will also produce a whine.

NOISE IN NEUTRAL

Possible causes of noise while the transmission is in neutral include

- Misalignment of transmission
- Worn flywheel pilot bearing
- Worn or scored countershaft bearings
- Worn or rough reverse idler gear
- Sprung or worn countershaft
- Excessive backlash in gears
- Worn main shaft pilot bearing
- Scuffed gear tooth contact surface
- Insufficient lubrication
- Use of incorrect grade of lubricant

NOISE IN GEAR

Possible causes of noise while the transmission is in gear include

- Worn or rough main shaft rear bearing
- Rough, chipped, or tapered sliding gear teeth
- Noisy speedometer gears
- Excessive end play of main shaft gears
- Refer to conditions listed under "Noise in Neutral"

OIL LEAKS

Possible causes of oil leakage are

- Oil level too high
- Wrong lubricant in unit
- Nonshielded bearing used as front or rear bearing cap (where applicable), seals (if used), defective, wrong type seal used
- Transmission breather plugged
- Capscrews loose or missing from remote control, shifter housing, bearing caps, PTO or covers
- Oil drain-back openings in bearing caps or case plugged with varnish, dirt, silicone, covered with gasket material

- Broken gaskets, gaskets shifted or squeezed out of position, pieces still under bearing caps, clutch housing, PTO and covers
- Cracks or porosity in castings
- Drain plug loose
- Oil leakage from engine
- Speedometer adapter or connections

GEAR SLIPOUT AND JUMPOUT

When a sliding clutch is moved to engage with a main shaft gear, the mating teeth must be parallel. Tapered or worn clutching teeth will try to "walk" apart as the gears rotate, causing the sliding clutch and gear to slip out of engagement. **Slipout** will generally occur when pulling with full power or decelerating with the load pushing the drivetrain.

A number of other conditions can prevent full engagement of the sliding clutch and gear and lead to slipout. Inspect the shift lever, making certain it moves freely and completely into position. Make a visual check of the floorboard opening, driver's seat, or anything that could prevent full engagement.

- Improperly positioned forward remote control that limits full travel forward and backward from the remote neutral position
- Improper length shift rods or linkage that limit travel
- Loose bell cranks and sloppy ball and socket joints
- Shift rods, cables, etc., too spongy, flexible, or not secured properly at both ends
- Worn or loose engine mounts, loose on frame
- Setscrews loose at remote control joints or on shift forks inside remote or even inside transmission unit
- Shift fork pads or groove in sliding gear or collar worn excessively, taper wear on gear clutch teeth
- Transmission and engine out of alignment, insufficient pressure on the detent ball from a weak or broken detent spring (**Figure 11–9**)
- Excessive wear on the detent notch of the yoke bar (**Figure 11–9**)
- Excessive gear clashing that shortens the clutching teeth

Slipout in the auxiliary section can be caused by the clutching teeth being worn, tapered, or not fully engaged.

These conditions cause the clutch gear to "walk" out of engagement as the gears turn. Causes of these types of clutching defects are clashing or just normal wear. Vibrations set up by an improperly aligned driveline and low air pressure add to the slipout problem.

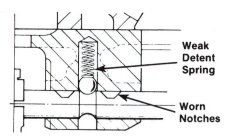

FIGURE 11–9 *Gear slipout or jumpout can be caused by weak or broken detent springs or worn yoke bar notches. (Courtesy of Roadranger Marketing. One great drive train from two great companies—Eaton and Dana Corporations)*

Jumpout occurs when a fully engaged gear and sliding clutch are forced out of engagement. It generally occurs when a force sufficient to overcome the detent spring pressure is applied to the yoke bar, moving the sliding clutch to a neutral position.

Conditions that result in jumpout include

- Extra heavy and long shift levers that swing, from operating over uneven terrain. Whipping action of the lever overcomes detent spring tension (**Figure 11–10**).

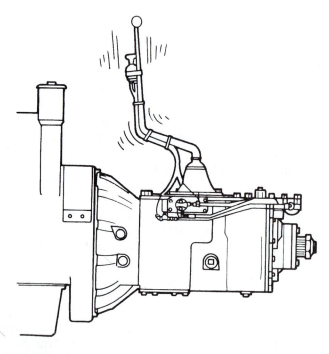

FIGURE 11–10 *When passing over rough road surfaces or uneven terrain, the whipping action of a long heavy-shift lever can overcome the detent spring tension, allowing the transmission to slip or jump out of gear. (Courtesy of Roadranger Marketing. One great drive train from two great companies—Eaton and Dana Corporations)*

- Mechanical remote controls with the master mounted to the frame. Relative movement between the engine-transmission package and frame can force the transmission out of gear. Worn or broken engine mounts increase the effects of this condition.
- Shift rod detent springs broken
- Shift rod detent notches worn
- Shift rod bent or sprung out of line
- Shift fork pads not square with shift rod bore
- Excessive end-play in drive gear, main shaft or countershaft, caused by worn bearings, retainers, etc.
- Thrust washers worn excessively or missing

Jumpout in the auxiliary section usually occurs with the splitter gear set. If torque is not sufficiently broken during splitter shifts, the sliding clutch gear might not have enough time to complete the shift before torque is reapplied to the gears. As torque is reapplied, the partially engaged clutch gear "jumps" out of the splitter gear. Since the gears have torque applied to them, damage will be done to the clutching teeth of the mating gears.

HARD SHIFTING

The effort required to move a gearshift lever from one gear position to another varies. If too great an effort is required, it will be a constant cause of complaint from the driver.

Most complaints are with remote type linkages used in cab-over-engine vehicles. Before checking the transmission for hard shifting, the remote linkage should be inspected. Linkage problems stem from worn connections or bushings, binding, improper adjustment, lack of lubrication on the joints, or an obstruction that restricts free movement.

To determine if the transmission itself is the cause of hard shifting, remove the shift lever or linkage from the top of the transmission. Then, move the shift blocks into each gear position using a prybar or screwdriver. If the yoke bars slide easily, the trouble is with the linkage assembly.

Listed below are some common causes of hard shifting.

- No lubricant in remote control units. Forward remote is isolated and is often overlooked. However, many remote controls used on transmissions and auxiliaries require separate lubrication.
- No lubricant in (or grease fittings on) U-joints or swivels of remote controls
- Lack of lubricant or wrong lubricant used, causing buildup of sticky varnish and sludge deposits on splines of shaft and gears

- Badly worn or bent shift rods
- Improper adjustment on shifter linkage
- Sliding clutch gears tight on splines of shaft
- Clutch teeth burred, chipped, or badly mutilated due to improper shifting, binding or interference of shift lever with other objects or rods inside the cab or near the remote control island
- Driver not familiar with proper shifting procedures
- Clutch or drive gear pilot bearing seized, rough, or dragging, clutch brake engaging too soon when clutch pedal is depressed, wrong lubricant
- Free running gears seized or galled on either the thrust face or diameters

TOWING OR COASTING PRECAUTIONS

It is extremely important for a truck to be towed properly to prevent damage to the transmission and frame. When towing a truck with a disabled transmission or other problem, manual transmissions require rotation of the main section countershaft and main shaft gears to provide adequate lubrication. These gears *do not* rotate when the vehicle is towed with the rear wheels on the ground and the drivetrain connected. The main shaft can be driven at high speed by the rear wheels. The friction between the main shaft splined washers, due to the lack of lubrication and the extreme difference in rotational speeds, will severely damage the transmission. Coasting with the transmission in neutral will produce the same damage.

To prevent this kind of damage during towing:

1. Pull the axle shafts or disconnect the driveline.

CAUTION: When the driveline or axle shafts are reinstalled, the nuts are tightened to the correct torques and the axle shafts are properly installed (RH and LH) and the driveshafts properly phased. Refer to the specific service manual for the correct torque values.

2. The recommended method of towing is to use a rigid tow bar and connect it to the front tow hook or pin (**Figure 11-11A**). This will prevent vehicle damage and can be used with a vehicle that is loaded or unloaded. If a wrecker service is obtained and a tow bar is not available, the following alternate method (**Figure 11-11B**) should be used. Use a lift

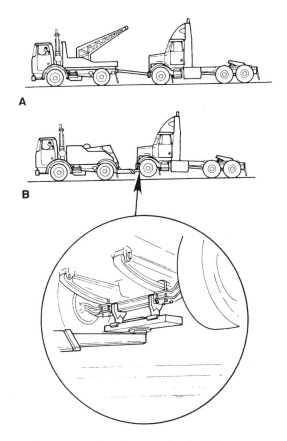

FIGURE 11-11 Towing methods: (A) recommended; and (B) alternate. (Courtesy of Volvo Trucks North America, Inc.)

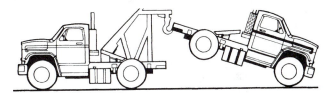

FIGURE 11-12 If the axle shafts or driveline cannot be pulled or disconnected, tow the truck with the drive wheels off the ground to avoid transmission damage. (Courtesy of Volvo Trucks North America, Inc.)

bar (a wrecker hitch cradle that fits under the front axle). The truck can then be raised and towed. This method is recommended to minimize possible damage to vehicles using air spoilers. Also, lifting the front axle will help reduce the possibility of frame and/or suspension damage.

3. Tow the truck with its drive wheels off the ground (**Figure 11-12**) if it is impractical to disconnect the axle shafts or driveline.

4. Also, never coast with the transmission in neutral, and never coast with the clutch depressed.

TEARDOWN INSPECTIONS

Identifying the transmission problem and determining its cause often is not possible without removing the unit from the truck and disassembling it. It is important that the transmission be carefully inspected before and during teardown to correctly diagnose the cause of the performance problem.

It is poor practice to disassemble a transmission without examining each part as it is removed. It happens many times that a technician has completely disassembled a unit and failed to find the cause of trouble by not bothering to examine the parts as they came apart. After the transmission is disassembled, check the lubricant for foreign particles, which often reveal sources of trouble that are overlooked during the disassembly.

Shop Talk

Prior to draining the transmission, make certain the drain pan is clean. Then you will be sure that any foreign particles found in the drained lubricant are from the transmission.

The troubleshooting guidelines given in **Table 11–3** list some of the more common transmission problems and their possible causes. To use the guidelines, perform the following:

1. Locate the transmission problem in the left-hand column.
2. Trace a line horizontally across the page until a rectangle with a number in it is reached.
3. Trace up the vertical column to find a possible cause. The number in the intersection of the vertical and horizontal lines indicates which corrections to use.

11.4 REMOVING THE TRANSMISSION

If troubleshooting procedures indicate that the problem is inside the transmission, it will have to be removed from the truck and disassembled. Refer to the vehicle manufacturer's recommended procedures for removal of the transmission. The following is a general procedure that will apply to most trucks:

1. After removing the battery ground cable, drain lubricant from the transmission.
2. Remove the shift lever from the transmission. If necessary, remove the shift assembly from the transmission.
3. Mark the yoke or the flange of the drive shaft and the output shaft of the transmission. The marks on the drive shaft and the output shaft will ensure the drive shaft is correctly installed on reassembly.
4. Remove the drive shaft (see Chapter 18 for additional information on removing drive shafts).
5. Disconnect all the electrical connections from the transmission.
6. Disconnect all the air lines from the transmission.
7. If used, remove the spring from the clutch lever on the transmission. Mark and disconnect the clutch linkage from the clutch housing on the transmission.
8. If a hydraulic clutch is used, disconnect the pushrod and the spring from the release fork. Remove the hydraulic cylinder from the bracket on the transmission. Use wires to support the cylinder on the frame.
9. Support the transmission with an overhead hoist or with a transmission jack. Make sure the transmission is securely supported.

CAUTION: Make sure the transmission does not hang by the input shaft in the pilot bearing bore in the flywheel. The clutch assembly, pilot bearing and input shaft can be damaged if the transmission is supported by the input shaft.

10. Remove the fasteners that attach the transmission to the brackets on the frame.
11. Remove the bolts and washers that attach the clutch housing to the engine. Pull the transmission straight out from the engine. Remove the transmission from the vehicle.
12. Set the transmission on a sturdy workbench or better yet attach it to a transmission stand so that all parts will be readily accessible.

11.5 OVERHAULING THE TRANSMISSION

It is important that the transmission manufacturer's disassembly, inspection, and reassembly instructions are followed when overhauling a transmission. The following sections present general transmission overhaul precautions, inspection tips, and reassembly guidelines. These general guidelines are followed by a more detailed breakdown and reassembly of a twin countershaft transmission that has both main

TABLE 11–3: TRANSMISSION TROUBLESHOOTING GUIDELINES

Possible Corrections

1. Instruct driver on proper driving techniques.
2. Replace parts (after trying other listed possible corrections).
3. Loosen lock screw and retighten to proper torque.
4. Look for resultant damage.
5. Smooth with emery paper.
6. Reset to proper specifications.
7. Install missing parts.
8. Check air lines or hoses.
9. Tighten part.
10. Correct the restriction.
11. Recheck timing.
12. Clean part.
13. Apply thin film silicone.
14. Apply sealant.

Possible Cause

Problem	WORN YOKE PADS	BENT YOKE BAR	WEAK OR MISSING DETENT SPRING	BURR ON YOKE BAR	INTERLOCK BALL OR PIN MISSING	TOO STRONG DETENT SPRING	CRACKED SHIFT BAR HOUSING	BREATHER HOLE PLUGGED	DAMAGED INSERT VALVE	DEFECTIVE REGULATOR	LOOSE HOSE OR FITTING	PINCHED AIR HOSE	STICKING SLAVE VALVE PISTON	DAMAGED "O" RING	IMPROPERLY MOUNTED GASKET	AIR HOSE HOOKED TO WRONG PLACE	PINCHED AIR LINE OR CONNECTOR	AIR CYLINDER PISTON NUT LOOSE	AIR CYLINDER PISTON CRACKED	GEAR TWISTED OUT OF TIME ON SHAFT	CRACKED GEAR OR BURR ON TOOTH	EXCESSIVE MAINSHAFT GEAR TOLERANCE	TWISTED MAINSHAFT	TAPERED CLUTCHING TEETH	WORN YOKE SLOT IN CLUTCH GEAR	BROKEN KEY (FRONT SECTION)	BROKEN KEY (AUXILIARY)	YOKE INSTALLED BACKWARDS	SYNCHRONIZER SPRING BROKEN	FAILED SYNCHRONIZER	INNER RACE LEFT OFF FRONT OF AUX. C/S	BEARING FAILURE
SLIP OUT (SPLITTER)	1,2								2	2	9						10							2,4	2	2,1					7,2	2
SLIP OUT (RANGE)										2	9						10							2,4	2							
SLIP OUT OR JUMP OUT (FRONT SECTION)	2		2,7																					2,4	2	2	2					
SLOW SHIFT (SPLITTER)									2	2	9	10	2	6			10															
SLOW SHIFT OR WON'T SHIFT (RANGE)										2	9	10	12,13	12,13,2			10	9,12	2,13								2,4		2,7	1,2		
HARD SHIFT OR WON'T SHIFT (FRONT SECTION)		2,3		5		2	2																	2,4		2						
ABLE TO SHIFT FRONT SECTION INTO 2 GEARS AT ONCE					7																					2						
GRINDING ON INITIAL LEVER ENGAGEMENT	2																							2,4								
LEVER LOCKS UP OR STICKS IN GEAR																							2									
NOISE																								2,4	5,2	6					7,4	2
GEAR RATTLE AT IDLE																						6										
VIBRATION																																
BURNED MAINSHAFT WASHER																																
INPUT SHAFT SPLINES WORN OR INPUT SHAFT BROKEN																																
CRACKED CLUTCH HOUSING																																
BROKEN AUXILIARY HOUSING																																
BURNED SYNCHRONIZER									2																			2,6	2,7			
BROKEN SYNCHRONIZER																																
HEAT																								2,4							2	2
TWISTED MAINSHAFT																																
DRIVE SET DAMAGE																																2,4
BURNED BEARING																															2	
OIL LEAKAGE								10																								
OVERLAPPING GEAR RATIOS																8,6																

TABLE 11-3:   TRANSMISSIION TROUBLESHOOTING GUIDELINES (CONTINUED)

Possible Corrections

1. Instruct driver on proper driving techniques.
2. Replace parts (after trying other listed possible corrections).
3. Loosen lock screw and retighten to proper torque.
4. Look for resultant damage.
5. Smooth with emery paper.
6. Reset to proper specifications.
7. Install missing parts.
8. Check air lines or hoses.
9. Tighten part.
10. Correct the restriction.
11. Recheck timing.
12. Clean part.
13. Apply thin film silicone.
14. Apply sealant.

Possible Cause

Problem	ROUGH RUNNING ENGINE	PRE-SELECTING SPLITTER	NOT USING CLUTCH	STARTING IN TOO HIGH OF GEAR	SHOCK LOAD	C/S BRAKE NOT WORKING (PUSH)	CLUTCH BRAKE NOT ADJUSTED (PULL)	CLUTCH BRAKE TANGS BROKEN (PULL)	NOT EQUIPPED WITH C/S OR CLUTCH BRAKE	LINKAGE OBSTRUCTIONS	IMPROPER LINKAGE ADJUSTMENT	BUSHINGS WORN IN CONTROL HOUSINGS	IMPROPER CLUTCH ADJUSTMENT	BROKEN ENGINE MOUNT	CLUTCH FAILURE	LOW OIL LEVEL	HIGH OIL LEVEL	POOR QUALITY OIL	TOO GREAT OPERATING ANGLE	INFREQUENT OIL CHANGES	NO SILICONE ON "O" RINGS	EXCESSIVE SILICONE ON "O" RINGS	MIXING OILS OR USING ADDITIVES	PIN HOLE IN CASE	DAMAGED REAR SEAL	LOOSE OR MISSING CAPSCREWS	IMPROPER TOWING OF TRUCK OR COASTING	MISALIGNMENT ENGINE TO TRANSMISSION	OUTPUT SHAFT NUT IMPROPERLY TORQUED	IMPROPER DRIVELINE SET UP	WORN SUSPENSION	TIRES OUT OF BALANCE, LOOSE LUG NUTS
SLIP OUT (SPLITTER)	1	1,4																														
SLIP OUT (RANGE)																																
SLIP OUT OR JUMP OUT (FRONT SECTION)		1,4								10	6			2													2,6					
SLOW SHIFT (SPLITTER)	1	1,4																			13	12,13										
SLOW SHIFT OR WON'T SHIFT (RANGE)																					13	12,13										
HARD SHIFT OR WON'T SHIFT (FRONT SECTION)		1,4								10			6	2,4																		
ABLE TO SHIFT FRONT SECTION INTO 2 GEARS AT ONCE																																
GRINDING ON INITIAL LEVER ENGAGEMENT						2,1,8	6,1	2,6	7,1	10	6	2	6																			
LEVER LOCKS UP OR STICKS IN GEAR										10	6																					
NOISE																2,4		2,4		2,4				2,4								
GEAR RATTLE AT IDLE	6																															
VIBRATION														2														6	6		2,6	2,6
BURNED MAINSHAFT WASHER															2,4,6												2,4,1					
INPUT SHAFT SPLINES WORN OR INPUT SHAFT BROKEN				1,2	1,2								2,6		2,6													6,2		6		
CRACKED CLUTCH HOUSING															2													2				
BROKEN AUXILIARY HOUSING																													2,6			
BURNED SYNCHRONIZER																2,6				2,6												
BROKEN SYNCHRONIZER														2														6	6		2,6	2,6
HEAT																6,4	6,4	2,6	2,6	2,6				2,6								
TWISTED MAINSHAFT				1,2	1,2																											
DRIVE SET DAMAGE															2,4,6				2,6													
BURNED BEARING															2,4,6		2,4,6		2,4,6					2,4,6								
OIL LEAKAGE																							6		2,4	2,4	6,7,14					
OVERLAPPING GEAR RATIOS																																

and auxiliary gear sections. The purpose of this detailed teardown is to present a clearer view of the parts of a typical transmission and how they fit and work together. Although the basic teardown and reassembly procedures might be similar on many transmissions, work should not be undertaken without the proper service manuals on hand.

PRECAUTIONS DURING DISASSEMBLY

Whenever disassembling a transmission, follow these general precautions:

- Carefully wash and relubricate all reusable bearings as removed and protectively wrap until ready for use. Remove bearings planned to be reused with pullers designed for this purpose.
- When disassembling the various assemblies, such as the main shaft, countershafts, and shift bar housing, lay all parts on a workbench in the same sequence as removed. This procedure will simplify reassembly and reduce the possibility of losing parts. Always place parts on clean paper to avoid contamination from dirt, paint chips, etc.
- Remove snap rings with pliers designed for this purpose. Snap rings removed in this manner can be reused providing they have not lost tension.
- Provide a clean place to work. It is important that no dirt or foreign material enters the unit during repairs. Dirt is an abrasive and can damage bearings. It is always good practice to clean the outside of the unit before starting the planned disassembly.
- Always apply force to shafts, housings, etc., with restraint. Never apply force to the part being driven after it stops solidly. Always use soft hammers, bars, and mauls for all disassembly work.

INSPECTION

Before reassembling the transmission, check each component carefully for abnormal or excessive wear and damage to determine reuse or replacement. Since the cost of a new component is generally a fraction of the cost of downtime and labor, avoid reusing a questionable component that could lead to additional repairs and expense soon after initial reassembly. To aid in determining the reuse or replacement of any transmission component, consideration should also be given to the unit's history, mileage, application, etc. Perform the following inspections during or after the teardown procedure:

Bearings

The service life of most transmissions is often governed by the life of the bearings. The majority of bearing failures can be attributed to vibration and dirt. Some of the more prominent reasons for unit removal with bearing failures are

- Worn out due to dirt
- Fatigue of raceways or balls
- Wrong type or grade of lubricant
- Lack of lubricant
- Vibration-breakup of retainer, brinelling of races, fretting corrosion, and seized bearings
- Bearings adjusted tight or loose
- Improper assembly brinelling bearing
- Improper fit of shafts or bore
- Acid etch of bearings due to water in lube
- Overloading of vehicle. Overload from engine or engine too large for transmissions used.
- Engine torsionals

Perform the following inspections:

1. Wash all bearings in clean solvent. Check balls, rollers, and raceways for pitting, discoloration, and spalled areas. Replace bearings that are pitted, discolored, or spalled.
2. Lubricate bearings that are not pitted, discolored, or spalled and check axial and radial clearances.
3. Replace bearings with excessive clearances.
4. Check bearing fits. Bearing inner races should be tight against the shaft. Outer races should be slightly tight to slightly loose in the case bore. If the bearing spins freely in the bore, however, the case should be replaced or repaired.

Dirt. More than 90 percent of all ball bearing failures are caused by dirt, which is abrasive. Dirt can enter the bearings during assembly of the units or be carried into the bearing by the lubricant while in service. Dirt can enter through the seals, breather, or even dirty containers used for addition or change of lubricant.

Softer material such as dirt, dust, etc., can form abrasive paste within the bearings themselves. The rolling motion tends to entrap and hold the abrasives. As the balls and raceways wear, the bearings become noisy. The abrasive action tends to increase rapidly as ground steel from the balls and race adds to the abrasives.

Hard, coarse materials, such as chips, can enter the bearings during assembly from hammers, drifts, or power chisels or be manufactured within the unit during operation from raking teeth. These chips produce small indentations in balls and races. Jamming of these hard particles between balls and races can cause the inner face to turn on the shaft, or the outer race to turn in the housing.

Fatigue. Bearing fatigue is characterized by flaking or spalling of the bearing race. Spalling is granular weakening of the hardened surface steel that causes it to flake away from the race. Because of their rough surfaces, spalled bearings will run noisy and produce vibration.

Normal fatigue failure occurs when a bearing lives out its life expectancy under normal loads and operating conditions. This type of failure is expected and is a result of metal breakdown due to the continual application of speed and load.

All bearings are subject to fatigue and must be replaced eventually. Operating experiences will dictate mileage replacement of bearings showing only normal wear.

Improper Lubrication. Bearing failure due to poor lubrication is characterized by discoloration of the bearing parts, spalling of the race, and possible breakage of the retainer. Failure can result not only from a low oil level, but also from contaminated oil, improper grade oil, or mixing of oil types (including the use of additives).

To prevent this type of failure, the transmission should always be filled to the proper level, using a recommended type and grade of oil, and changed at regular intervals.

Brinelling. Brinelling (Figure 11–13) can be identified as tiny indentations high on the shoulder or in the valley of the bearing raceway. They can be caused by improper bearing installation or removal. Driving or pressing on one race while supporting the other is the primary cause. To prevent brinelling, always support the race that has pressure applied to it. In addition to brinelling, damage can also occur to

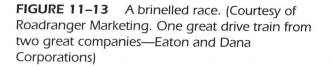

FIGURE 11–13   A brinelled race. (Courtesy of Roadranger Marketing. One great drive train from two great companies—Eaton and Dana Corporations)

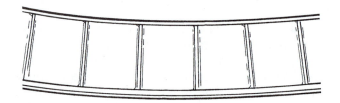

FIGURE 11–14   A fretted outer race. (Courtesy of Roadranger Marketing. One great drive train from two great companies—Eaton and Dana Corporations)

the bearing shields, retainers, and snap rings by using a hammer and chisel to drive bearings. This damage can be avoided by using correct drivers or pullers.

Fretting. The bearing outer race can pick up the machining pattern of the bearing bore as a result of vibration. This action is called **fretting (Figure 11–14)**. Many times a fretted bearing is mistakenly diagnosed as one that has spun in the bore. Only under extreme conditions will a bearing outer race spin in the bore.

Contamination. When bearings fail as a result of contamination, it is due either to contaminants entering the transmission case or the bearings having been improperly handled during service or storage. Bearings affected by contamination are identified by scoring, scratching, or pitting of the raceways and balls or rollers, or a buildup of rust or corrosion on the bearing parts. In addition, the presence of very fine particles in the oil, such as abrasive dust, or the use of older EP (extreme pressure) oils, will act as an abrasive and produce a highly polished surface on the raceways and balls or rollers. This abrasive action will significantly shorten the life of the bearing.

Impurities will enter the transmission during its normal breathing process. This will not seriously affect the bearings if the transmission oil is changed as recommended.

New bearings should be stored in their wrappers until ready for use. Used bearings should be thoroughly cleaned in solvent, light oil or kerosene, covered with a coat of oil, and wrapped until ready for use. Always use new wrappings after reapplying oil to cleaned bearings.

Gears

Check gear teeth for frosting and pitting. Frosting of gear tooth faces presents no threat of transmission failure. Often in continued operation of the unit, frosted gears will "heal" and not progress to the pitting stage. In most cases, gears with light to moderate pitted teeth have considerable gear life remaining and can be reused. Gears with advanced stage pitting should be replaced.

Check for gears with clutching teeth that have been abnormally worn, tapered, or reduced in length from clashing in shifting. Replace gears found in any of these conditions.

Check the axial clearance of gears. Where excessive clearance is found, check the gear snap ring, washer, spacer, and gear hub for excessive wear. Maintain the manufacturer's recommended axial clearance between the main shaft gears.

Splines

Check splines on all shafts for abnormal wear or damage. If the sliding clutch gears, companion flange, or clutch hub have worn into the sides of the splines, replace the specific shaft affected.

Tolerance/Limit Washers

Check surfaces of all limit washers. Washers scored or reduced in thickness should be replaced.

Reverse Idler Gear Assemblies

Check the reverse idler gear for excessive wear from the action of the roller bearings.

Gray Iron Parts

Check all gray iron parts for cracks and porosity. Replace or repair parts found to be damaged. Heavy castings may be welded or brazed provided the cracks do not extend into the bearing bores or bolting surfaces. When welding, however, never place the ground so as to allow current to pass through the transmission. Electrical arcing at bearings will pit bearing surfaces or, in extreme cases, weld the balls or rollers to the bearing races.

Clutch Release Parts

Check the clutch release components. Replace yokes worn at cam surfaces. Replace the bearing carrier if worn at the contact pads.

Check pedal shafts. Replace those worn at bearing surfaces.

Shift Bar Housing Assembly

Check for wear on the shift yokes and blocks at the pads and the lever slot. Replace excessively worn parts.

Check yokes for correct alignment. Replace sprung yokes.

Check lock screws in the yokes and blocks. Tighten and rewire those found loose.

If the housing has been disassembled, check the neutral notches of the shift bars for wear from the interlock (detent) balls.

Gearshift Lever Housing Assembly

Check the spring tension on the shift lever. Replace the tension spring and washer if the lever moves too freely.

If the housing is disassembled, check the spade pin and corresponding slot in the lever for wear. Replace both parts if excessively worn.

Bearing Covers

Check covers for wear from bearing thrust of the adjacent bearing. Replace covers damaged from the thrust of the bearing outer race. Check the bores of the bearing covers for wear. Replace those worn oversize.

Oil Return Threads and Seals

Check oil return threads in the front bearing cover. If the sealing action of the threads has been destroyed by contact with the input shaft, replace bearing cover.

Replace the oil seal in the rear bearing cover.

Sliding Clutches

Check all shift yokes and yoke slots in the sliding clutches for wear or discoloration from heat.

Check the engaging teeth of the sliding clutches for partial engagement pattern.

Synchronizer Assembly

Check the synchronizer for burrs, for uneven and excessive wear at contact surfaces, and for metal particles.

Check the blocker pins for excessive wear or looseness.

Check the synchronizer contact surfaces on the auxiliary drive and low range gears for excessive wear.

O-Rings

Replace all O-rings.

REASSEMBLY

The procedure for reassembling the transmission is basically the reverse of the disassembly procedure. During the reassembly process, make sure that the interiors of the case and housing are clean. It is important that dirt is kept out of the transmission during reassembly. Dirt is abrasive and can damage polished surfaces of bearings and washers.

Use certain precautions, as listed below, during reassembly.

- **Gaskets.** Use new gaskets throughout the transmission as it is being rebuilt. Make sure all gaskets are installed, because omission of gaskets can result in oil leakage or misalignment of bearing covers.
- **Capscrews.** To prevent oil leakage, use thread sealant on all capscrews. Be sure to correctly torque all capscrews to the manufacturer's specifications.
- **O-Rings.** Use new O-rings and lubricate them with a silicone lubricant before installation.
- **Initial Lubrication.** Coat thrust washers, shaft splines, and gear faces with a light coat of oil (or the manufacturer's recommended lubricant) during installation. This will provide initial lubrication, preventing scoring and galling.
- **Axial Clearances.** Maintain original axial clearances of the main shaft forward speed gears (typically 0.005 to 0.012 inch) and the main shaft reverse gear. Worn thrust washers should be replaced.
- **Bearings.** Use of flanged bearing drivers is recommended for the installation of bearings. These drivers prevent damage to the balls and races, maintaining correct bearing alignment with the shaft and bore. If a tubular or sleeve type driver is used, apply force only to the inner race.
- **Universal Joint Companion Flange.** Draw the companion flange into position with the main shaft nut, using the specified torque. Make sure the speedometer gear has been installed on the yoke. If a speedometer gear is not used, a replacement spacer of the same width must be used. Failure to properly install the yoke or flange will permit the shaft to move axially with resultant damage to the rear bearing.
- **Torque.** As you reassemble the transmission, refer to the manufacturer's specifications for proper installation and torque of all capscrews, setscrews, lock nuts, and plugs.

TIMING

All twin countershaft transmissions are "timed" at assembly. It is important that the manufacturer's timing procedures are followed when reassembling the transmission. Timing ensures the countershaft gears contact the mating main shaft gears at the same time, allowing main shaft gears to center on the main shaft and equally divide the load.

Timing is a simple procedure of marking the appropriate teeth of a gear set prior to installation and placing them in proper mesh during reassembly.

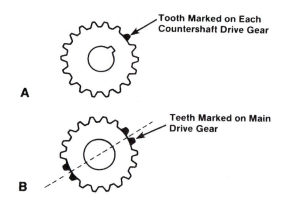

FIGURE 11–15   (A) On the countershaft, mark the tooth opposite the keyway; (B) on the main drive gear, mark two teeth on opposite sides of the gear. (Courtesy of Roadranger Marketing. One great drive train from two great companies—Eaton and Dana Corporations)

In the front section, it is necessary to time only the drive gear set. And depending on the model, the low range, deep reduction, or splitter gear set is timed in the auxiliary section. The following are examples of typical timing procedures.

Front Section

Before reassembling the front section of the transmission the drive gears of the main shaft and the countershafts must be marked in this way:

1. Prior to placing each countershaft assembly into the case, clearly mark the tooth located directly over the keyway of the countershaft drive gear as shown in **Figure 11–15A**. In some transmissions, this tooth is stamped with an "O" to aid in identification.
2. Mark any two adjacent teeth on the main drive gear.
3. Mark the two adjacent teeth located directly opposite the first set marked on the main drive gear. As shown in **Figure 11–15B**, there should be an equal number of unmarked gear teeth on each side between the marked sets.

When the main shaft and countershaft are installed in the case, the marked teeth should be meshed as shown in **Figure 11–16**.

Auxiliary Section

The gears in the auxiliary section must also be timed. The specific gears on the main shaft and countershafts that must be marked differ, depending on the auxiliary gearbox design. A general procedure is as follows:

1. Mark any two adjacent teeth on the main shaft gear of the set to be timed. Then mark the two adjacent teeth located directly opposite the first set marked as shown in **Figure 11–15B**.
2. Prior to placing each auxiliary countershaft assembly into the housing, mark the tooth stamped with an "O" on gear to mate with timed main shaft gear as shown in **Figure 11–15A**.
3. Install the main shaft gear in position on the range main shaft or output shaft.
4. Place the auxiliary countershaft assemblies into position and mesh the marked teeth of mating countershaft gears with the marked teeth of main shaft gear as shown in **Figure 11–16**.
5. Fully seat the rear bearings on each countershaft to complete the installation.

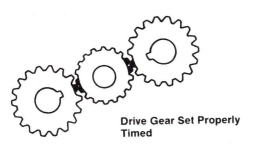

Drive Gear Set Properly Timed

FIGURE 11–16 When properly timed, the marked teeth on the main shaft and countershaft drive gears will mesh. (Courtesy of Roadranger Marketing. One great drive train from two great companies—Eaton and Dana Corporations)

Changing the Input Shaft

In some cases, it might become necessary to replace the input shaft of the transmission due to excessive clutch wear on the splines. The input shaft can usually be removed without further disassembly of the transmission (**Figure 11–17**).

Disassembly. To remove the input shaft, proceed as follows. These instructions apply to the input shaft only. Changing the main drive gear requires complete disassembly of the main section.

1. Remove the gearshift lever housing assembly (or remote control assembly) from the shift bar housing, and the shift bar housing assembly from the transmission case.
2. Remove the front bearing cover and gasket. If necessary, remove the O-ring from the cover of models that are so equipped.
3. Remove the bearing retaining snap ring from the groove in the shaft.
4. Push down on the input shaft to cock the bearing in the bore. Drive the input shaft toward the rear of the transmission, through the bearing as far as possible. Pull the input shaft forward to expose the snap ring of the bearing.
5. Use pry bars to leverage the snap ring and complete the removal of the bearing.
6. Remove the drive gear spacer and snap ring.
7. Pull the input shaft forward and out of the drive gear and case.

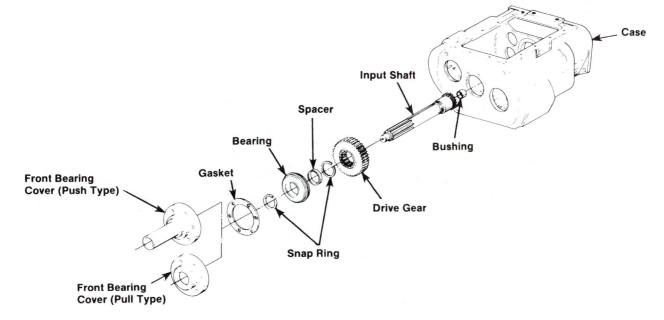

FIGURE 11–17 Components of typical input shaft assembly. (Courtesy of Roadranger Marketing. One great drive train from two great companies—Eaton and Dana Corporations)

Reassembly. To reinstall the new input shaft, proceed as follows:

1. If necessary, install the bushing in the pocket of the input shaft.
2. Install the new input shaft into the splines of the main drive gear, just far enough to expose the snap ring groove in the I D of the drive gear.
3. Install the snap ring in the snap ring groove inside the drive gear.
4. Install the drive gear spacer on the input shaft.
5. Install the drive gear bearing on the input shaft and into the case bore.
6. Install the bearing retainer snap ring.
7. Install the front bearing cover and gasket. Align the oil return hole in the case with the hole in the cover.
8. To facilitate proper reinstallation of the shift bar housing assembly on the case, make sure the main shaft sliding clutches are placed in the neutral position.
9. Reinstall the shift bar housing assembly, the front bearing cover, and all other components and assemblies previously removed, making sure to replace the gaskets used.

AIR CONTROL DISASSEMBLY

See **Figures 10–39** and **10–41** for the location of the air line and components in a typical air control system. Disassembly of the air control is as follows:

1. Disconnect the two air lines running from the shift tower valve to the slave valve on the transmission at the "S" port and "P" port on the slave valve.
2. Remove the air line at the splitter cylinder cover (if equipped).
3. Remove the valve cover from the shift tower and disconnect the three air lines.
4. Loosen the jam nut and turn the shift tower valve and nut from the gearshift lever. Remove the valve cover, air lines, sheathing, and O-rings from the lever.
5. Disconnect and remove the air line between the splitter cylinder and air filter/regulator.
6. Disconnect and remove the air line between the slave valve and the air filter/regulator.
7. Turn out the screws securing the air filter/regulator to the transmission and remove the air filter/regulator.
8. Disconnect the air hose between the slave valve and the range cylinder cover.
9. Remove the slave valve from the transmission case.

10. Remove the spring and plunger pin from the bore in the transmission case. Remove the slave valve gasket (see **Figure 10–42B**).

GEARSHIFT LEVER AND SHIFT HOUSING REMOVAL

Figures 10–32, 10–33, and **10–34** show the components of a typical gearshift lever and shift housing assembly. Basic disassembly is as follows:

1. Remove the screws securing the gearshift lever housing to the top of the transmission, jar the lever slightly to break the gasket seal, and then remove the gearshift lever housing and gasket from the shift bar housing. Remote control housings are removed from the shift bar housing in the same manner.
2. Remove the screws securing the shift bar housing assembly to the transmission, jar the top to break the gasket seal, and lift the shift bar housing from the transmission case. Remove the gasket.
3. Tilt the assembly and remove the three sets of tension springs and balls from the bores in the housing.
4. Secure the assembly in a vise (**Figure 11–18**). Move the third-fourth speed shift bar to the housing rear, and remove the yoke (fork) and block from the bar. It might be necessary on some models to cut yoke lock-wires and also remove an oil trough from the housing.

FIGURE 11–18 *Components of shift bar housing secured in vise and ready for disassembly. (Courtesy of Roadranger Marketing. One great drive train from two great companies—Eaton and Dana Corporations)*

5. Move the first-second speed shift bar to the housing rear; remove the yoke (fork) and block from the bar. As the neutral notch in the bar clears the rear boss, remove the small interlock pin from the bore.
6. Remove the actuating plunger from the center boss bore.
7. Move the low-reverse shift bar to the rear; remove the yoke from the bar. As the bar is pulled from the housing, two interlock balls will drop from the bottom off the rear boss bottom bore.
8. If needed, remove the plug, spring, and plunger from the low-reverse speed shift yoke bore.

Typical reassembly is as follows:

1. Install the reverse stop plunger in the low-reverse shift yoke. Install the spring in the bore on the plunger shank.
2. Install the plug and tighten it to compress the spring. Back the plug out one and a half turns and stake the plug through the small hole in the yoke.
3. With the housing in a vise, hold the notched end of the low-reverse shaft and install the bar in the lower bore of the shift bar housing bosses. Slide the bar through the low-reverse yoke and install the yoke lock screw. Tighten the lock screw and install the lockwire.
4. Install the actuating plunger in the center boss bore.
5. Install one interlock ball in the rear boss top bore so the ball rides between the low-reverse and the first-second shift bars.
6. Install the first-second shift bar in the middle bore of the housing bosses. Position the shift block on the bar between the center and rear bosses, and the first-second yoke on the bar between the front and center bosses. Just prior to inserting the notched end of the rear boss bar, install the small interlock pin vertically in the neutral notch bore. Install the block and yoke lock screws. Install the lockwire securely.
7. Install the second interlock ball in the rear boss top bore so it rides between the first-second and third-fourth shift bars.
8. Install the third-fourth speed shift bar in the housing boss upper bore with the third-fourth yoke positioned between the front and center bosses and the shift block between the center and rear bosses. Install the block and yoke lock screws and lockwire securely. Reinstall the oil trough, if so equipped.
9. Remove the assembly from the vise and install the three tension balls, one in each

bore, on the housing top. Install the three tension springs, one over each tension ball, in the housing bores.
10. Reinstall the shift bar housing and replace the gasket.
11. Reinstall the gearshift lever housing and replace the gasket.

OUTPUT YOKE REMOVAL

1. Prior to removing the output yoke, lock the transmission by engaging the two main shaft gears with their main shaft sliding clutches.
2. Use a large breaker bar to turn the output shaft nut from the output shaft.
3. Pull the yoke straight to the rear and off the output shaft (**Figure 11–19**).

AUXILIARY SECTION REMOVAL

To remove the auxiliary section from the transmission:

1. Remove the capscrews that attach the two sections.
2. Insert the puller screws in the puller threads in the auxiliary housing flange.

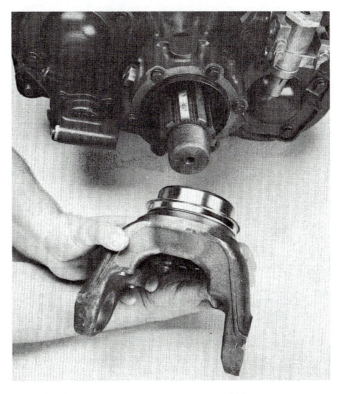

FIGURE 11–19 Removing the output yoke from the output shaft. (Courtesy of Roadranger Marketing. One great drive train from two great companies—Eaton and Dana Corporations)

FIGURE 11–20 Removing the auxiliary section using a chain hoist. (Courtesy of Roadranger Marketing. One great drive train from two great companies—Eaton and Dana Corporations)

3.   Tighten the puller screws evenly to move the auxiliary section straight to the rear and from the front section.
4.   Remove the capscrews and attach a chain hoist to the auxiliary section. Move the section to the rear until it swings free (**Figure 11–20**).

CLUTCH HOUSING REMOVAL

To remove the clutch housing and front bearing cover:

1.   Remove the clutch release mechanism, if so equipped.
2.   Remove the nuts from the studs and turn out the bolts that attach the clutch housing to the case.
3.   Tap on the clutch housing with a rubber mallet to break the gasket seal and remove the clutch housing from the case (**Figure 11–21**).

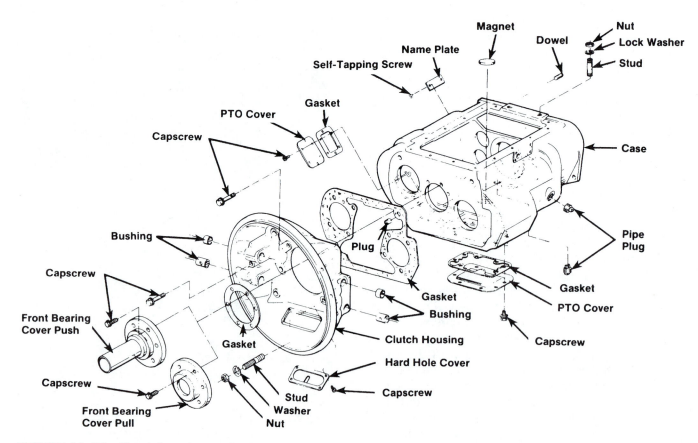

FIGURE 11–21 Clutch housing and other components of the main case assembly. (Courtesy of Roadranger Marketing. One great drive train from two great companies—Eaton and Dana Corporations)

AUXILIARY SECTION DISASSEMBLY

For ease of disassembly, mount the auxiliary section upright in a vise (**Figure 11–22**). The transmission shown in this example is an eighteen-speed model (see **Figure 10–52**) that has both range and splitter shift mechanisms in the auxiliary section.

1. Remove the lockwire and lock screw securing the splitter shift yoke to the yoke bar and piston (**Figure 11–23**).
2. Remove the splitter shift yoke and the sliding clutch on the auxiliary main shaft assembly (**Figure 11–24**).

FIGURE 11–22 _Auxiliary section in vise prior to disassembly. (Courtesy of Roadranger Marketing. One great drive train from two great companies— Eaton and Dana Corporations)_

3. Temporarily reinstall the output yoke on the output shaft and insert a breaker bar through the yoke to keep it from turning. Now loosen, but do not remove, the capscrew on the rear auxiliary drive gear.
4. Remove the capscrews holding the countershaft bearing covers in place (**Figure 11–25**).
5. Remove the snap rings from the rear of both auxiliary countershafts (**Figure 11–26**).
6. Use a soft bar and maul to drive the countershafts forward far enough to partially unseat the rear bearings.
7. Drive the countershafts to the rear to expose the bearing snap rings. Do not damage the bearing inner race.
8. Use a bearing puller to remove bearings from the countershafts.

CAUTION: Countershafts will fall as bearings are removed.

---

9. Remove the auxiliary countershafts from the auxiliary section.
10. Remove the bearings from the countershaft rear bearing bores.
11. Remove the capscrew and retainer that holds the rear auxiliary drive gear on the output shaft.
12. Remove the rear auxiliary drive gear from the output shaft.

To disassemble the range cylinder assembly (**Figure 11–27**), proceed as follows:

1. Remove the capscrews, range cylinder cover, and gasket.

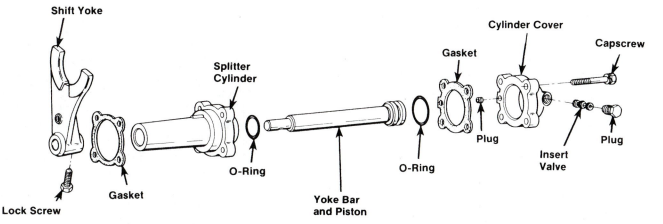

Shift Yoke

Lock Screw

Gasket

Splitter Cylinder

O-Ring

Yoke Bar and Piston

O-Ring

Gasket

Plug

Insert Valve

Cylinder Cover

Capscrew

Plug

FIGURE 11–23 _Splitter cylinder, yoke bar, and shift yoke assembly. (Courtesy of Roadranger Marketing. One great drive train from two great companies—Eaton and Dana Corporations)_

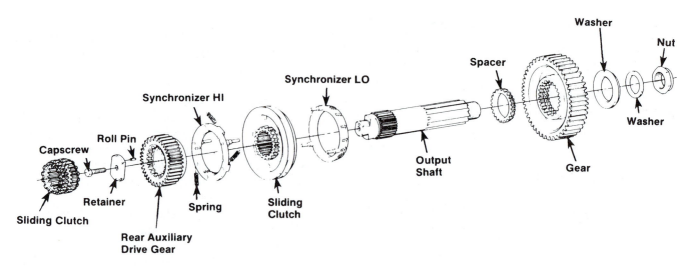

FIGURE 11–24   Auxiliary main shaft components. (Courtesy of Roadranger Marketing. One great drive train from two great companies—Eaton and Dana Corporations)

FIGURE 11–25   Removing the countershaft bearing cover. (Courtesy of Roadranger Marketing. One great drive train from two great companies—Eaton and Dana Corporations)

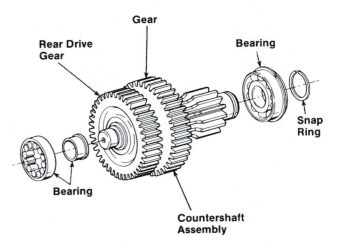

FIGURE 11–26   Auxiliary countershaft components. (Courtesy of Roadranger Marketing. One great drive train from two great companies—Eaton and Dana Corporations)

2.   Remove the nut from the yoke bar.
3.   Cut the lockwire and remove the yoke lock screws.
4.   Pull the yoke bar from the cylinder housing bore.
5.   Remove the shift yoke and synchronizer assembly from the output shaft.
6.   Remove the range piston from the cylinder bore.
7.   Remove the capscrews and range cylinder housing. If needed, remove the small O-ring from the range cylinder housing bore.

8.   To disassemble the synchronizer, place the low range synchronizer ring on the bench and cover the assembly with a rag to prevent losing the three springs released from the high range synchronizer at the pin locations.
9.   Pull the high range synchronizer from the blocker pins (see **Figure 10–25** for component details of a synchronizer).
10.   Remove the sliding clutch from the synchronizer ring low range pins.

To remove and disassemble the output shaft and rear bearing assemblies, proceed as follows:

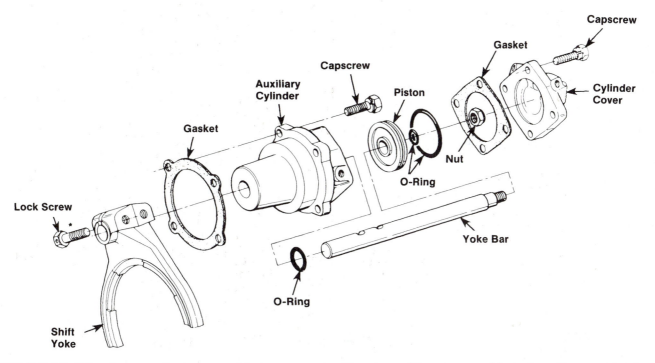

FIGURE 11–27 *Range cylinder assembly components. (Courtesy of Roadranger Marketing. One great drive train from two great companies—Eaton and Dana Corporations)*

1. Use a soft bar and maul to drive the output shaft forward and through the rear bearing assembly.
2. Remove the bearing inner spacer from the output shaft.
3. Using the front face of the low range reduction gear as a base, press the output shaft through the bearing and gear, freeing the bearing, low range gear, and splined washer.
4. Remove the stepped washer, low range gear, and splined washer from the shaft.
5. Remove the rear bearing retaining capscrews, rear bearing cover, and gasket from the auxiliary housing (**Figure 11–28**). The rear bearing cone will drop from the housing bore when the cover is removed. Remove the two bearing cups and spacer from the bearing bore.

To remove the splitter cylinder (see **Figure 11–23**), proceed as follows:

1. Remove the capscrews in the splitter cylinder cover and the splitter cylinder cover and gasket. If necessary, turn out the insert valve retaining nut and remove the insert valve from the bore.
2. Remove the splitter cylinder housing and yoke bar from the auxiliary housing.

3. Pull the yoke bar from the cylinder housing. If necessary, remove the O-ring from the piston O D groove.
4. Remove the small O-ring from the cylinder housing bore.

AUXILIARY SECTION REASSEMBLY

To reassemble and install the low range gear and output shaft, proceed as follows:

1. Mark the timing teeth on the low range gear. A highly visible color of toolmaker's dye is recommended. Mark any two adjacent gear teeth on the low range gear, front side. Then mark the two adjacent teeth that are directly opposite the first set marked. There should be the same number of teeth between the markings on each side of the gear (see **Figure 11–15B**).
2. Place the splined spacer on the output shaft shoulder (see **Figure 11–24**).
3. Install the low range gear on the output shaft, with the clutching teeth engaged with the splines down to engage the washer splines.
4. Install the low range gear rear washer on the output shaft and against the gear with the chamfer side facing up.

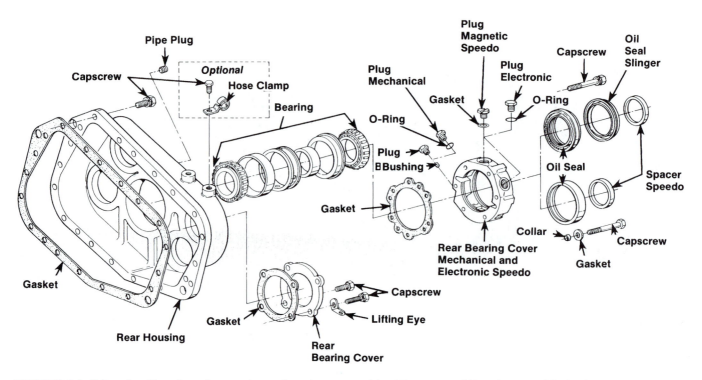

FIGURE 11–28 *Auxiliary housing and rear bearing assembly. (Courtesy of Roadranger Marketing. One great drive train from two great companies—Eaton and Dana Corporations)*

5. Using a heat lamp, oven, or hot plate, heat the output shaft rear bearing and install it on the output shaft. Seat the bearing securely on the shaft. The bearing can also be installed using the appropriate driver.

CAUTION: Do not heat the bearing above 275°F. Use Tempilstick (heat crayon) or thermostat controlled oven.

6. Install the bearing inner spacer on the output shaft.
7. Install the front bearing race in the auxiliary housing bearing bore.
8. Install the rear bearing race and rear bearing outer spacer in the auxiliary housing bearing bore. The bearing race will be at the proper depth when the face shoulder is seated on top of the bearing bore.
9. Place two main shaft spacers or flat steel stock of equivalent thickness (0.190 inch) on the rear face of the low range gear, 180 degrees from each other. Install the auxiliary housing over the output shaft assembly, allowing the housing to rest on the blocking.
10. Heat the rear bearing cone and install it on

the shaft, taper side down and inside the cup. Make sure the lip of the rear bearing cup is fully seated on the housing with the bearing installed.

CAUTION: Do not heat the bearing above 275°F. Use Tempilstick (heat crayon) or thermostat controlled oven.

11. If the oil seal was previously removed, install it in the rear bearing cover. The seal should be installed with an oil seal installation tool.
12. Position the corresponding new gasket on the cover mounting surface.
13. Install the rear bearing cover on the auxiliary housing. Use the nylon collar and brass washer with the capscrew at the chamfered hole that intersects the speedometer bore. Tighten the capscrews to the recommended torque ratings.

CAUTION: Because the collar becomes distorted when compressed, do not reuse the old nylon collar.

To reassemble and install the splitter cylinder assembly (see **Figure 11–23**), proceed as follows:

1. If previously removed, install the small O-ring in the cylinder housing bore.
2. If removed, install the O-ring on the piston O D. Insert the yoke bar in the cylinder housing bore.
3. Position the corresponding new gasket on the housing mounting surface and install the splitter cylinder housing with the yoke bar in the auxiliary housing.
4. If previously removed, install the insert valve, flat end to the outside, and the valve retaining nut in the bottom bore of the cylinder cover. Tighten to the recommended torque ratings. Prior to installation of the insert valve, make sure the three O-rings on the valve O D are not defective. Replace, if necessary.
5. Position the corresponding gasket on the splitter cylinder housing. Position the splitter cylinder cover and install four capscrews; tighten to the recommended torque.

To reassemble and reinstall the synchronizer (**Figure 11–29**), proceed as follows:

1. Place the larger low range synchronizer ring face down on the bench with the pins up. Place the sliding clutch, recessed side up, on the pins of the low range synchronizer. Pins on the low range synchronizer must line

FIGURE 11–29 _During reassembly, the high range synchronizer ring is placed over the pins on the low range ring. Springs seat against the pins. (Courtesy of Roadranger Marketing. One great drive train from two great companies—Eaton and Dana Corporations)_

up with the chamfered holes on the bottom of the sliding clutch.

2. Install the three springs in the bores in the high range synchronizer ring.
3. Place the high range synchronizer ring over the pins of the low speed synchronizer ring, seating the springs against the pins.
4. Apply downward pressure to the high range synchronizer ring while twisting counterclockwise to compress the springs and fully seat the ring on the blocker pins of the lower range synchronizer.

To reassemble the range cylinder (see **Figure 11–27**), proceed as follows:

1. Install the O-ring in the slot of the small bore in the cylinder. Apply silicone to all O-rings.
2. Position the corresponding new gasket on the housing mounting surface. Install the cylinder housing in the rear bore of the auxiliary housing, air fitting to the upper left. Tighten the capscrews to the recommended torque rating.
3. If previously removed, install the O-rings in the I D and O D of the range position.
4. Place the range shift yoke in the slot of the synchronizer sliding clutch, with the threaded hub of the yoke up and facing the range synchronizer.
5. Place the range shift yoke and synchronizer assembly into the auxiliary housing, engaging the splines of the sliding clutch with the output shaft, low range synchronizer to rear.
6. Insert the threaded-end of the yoke bar through the yoke and into the range cylinder housing bore, aligning the notches in the bar with the lock screw holes. Use caution not to damage the O-ring in the range cylinder.
7. Install two capscrews through the bottom of the range shift yoke and tighten the lockwire lock screws to the recommended torque.
8. In the cylinder housing bore, install the range piston on the yoke bar, with the flat side to the rear. Secure with a nut tightened to the recommended torque ratings.
9. Position the corresponding new gasket on the cover mounting surface and install the range cylinder cover on the housing, with the open port to the upper left. Then, tighten the capscrews to the recommended torque rating.

To install the rear auxiliary drive gear (see **Figure 11–24**), proceed as follows:

1. Install the rear auxiliary drive gear on the output shaft, with the clutching in teeth facing forward.

2. Install the retainer with the retainer pin facing the synchronizer; match the pin with the hole. Finger tighten the capscrew.

3. Temporarily install the output yoke. Insert and hold a breaker bar to keep the output shaft from turning.

4. Tighten the capscrew on the rear auxiliary drive gear to the recommended torque rating. Remove the output yoke and breaker bar.

TIMING AND INSTALLATION OF AUXILIARY COUNTERSHAFTS

During the reassembly of the auxiliary low range gear and output shaft, two sets of adjacent teeth were marked with toolmaker's dye. It is now time to match the appropriate marks on the countershaft low range gears with these teeth. Proceed as follows:

1. Use a proper bearing driver to seat the two auxiliary countershaft bearing races (see **Figure 11–26**).

2. Install the bearings in the countershaft rear bearing bores.

3. Temporarily install the rear bearing covers, using only two bolts and no gasket to hold the bearings in place as the countershafts are driven into the auxiliary housing.

4. On the low range gear of each auxiliary countershaft assembly, locate the tooth with "O" stamped on it and mark this tooth with toolmaker's dye (see **Figure 11–15A**).

5. Place the upper countershaft assembly into position in the auxiliary housing. Mesh the marked tooth on the auxiliary countershaft low range gear with one set of the two marked teeth on the output shaft low range gear.

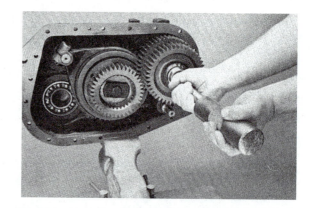

FIGURE 11–30 Driving the auxiliary section countershaft into the auxiliary bearing using a soft bar and maul. (Courtesy of Roadranger Marketing. One great drive train from two great companies—Eaton and Dana Corporations)

6. Use a soft maul to drive the countershaft into the auxiliary bearing (**Figure 11–30**).

7. Repeat steps 5 and 6 for the lower countershaft.

8. Remove the countershaft rear bearing covers and install the snap rings.

9. Install both rear bearing covers using new gaskets.

MAIN (FRONT) SECTION DISASSEMBLY

The following sections describe a general disassembly procedure for a five-speed front section. For any teardown, follow the exact manufacturer's procedures outlined in the transmission's service manual.

To remove the front auxiliary drive gear (**Figure 11–31**), proceed as follows:

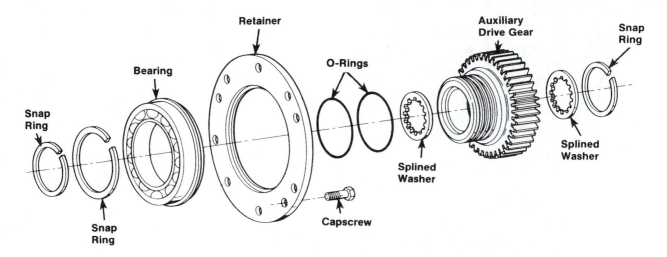

FIGURE 11–31 Auxiliary drive gear assembly. (Courtesy of Roadranger Marketing. One great drive train from two great companies—Eaton and Dana Corporations)

1. Cut the lockwire from the bearing retainer ring capscrews.
2. Remove the snap ring from the main shaft rear groove.
3. Remove the splined washer from inside the front auxiliary drive gear assembly.
4. Remove the capscrews from the auxiliary bearing retainer ring and turn in the puller screws to pull the front auxiliary drive gear assembly from the case bore (**Figure 11–32**).
5. Remove the front auxiliary drive gear, splined washer, and retaining snap ring from the main shaft.

To remove the reverse idler gear assembly (**Figure 11–33**), proceed as follows:

1. Move the main shaft reverse gear as far to the rear as possible and remove the snap ring from the I D of the gear.
2. Move the reverse gear on the main shaft forward and against the low speed gear, engaging the splines of the main shaft sliding clutch.
3. Using inside jaw pullers or an impact puller, remove the auxiliary countershaft front bearing from the left reverse idler gear bore (**Figure 11–34**). If needed, remove the front bearing from the right reverse idler gear bore as well.

FIGURE 11–32 Pulling the auxiliary drive gear from the bore in the rear of the main casing. (Courtesy of Roadranger Marketing. One great drive train from two great companies—Eaton and Dana Corporations)

FIGURE 11–34 Removing the auxiliary countershaft front bearing from the reverse idler gear bore. (Courtesy of Roadranger Marketing. One great drive train from two great companies—Eaton and Dana Corporations)

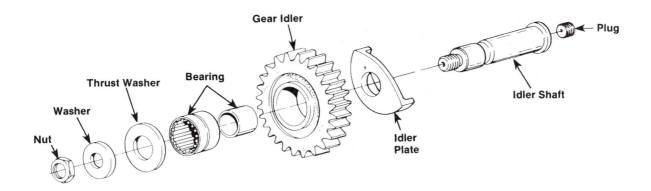

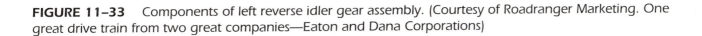

FIGURE 11–33 Components of left reverse idler gear assembly. (Courtesy of Roadranger Marketing. One great drive train from two great companies—Eaton and Dana Corporations)

4. Loosen the nut on the idler shaft. It may be necessary to insert a punch through the oil hole to retain the idler shaft. Remove the nut and washer from the shaft.

5. Remove the plug from the rear of the shaft and use an impact puller to remove the idler shaft from the main case bore.

6. As the shaft and idler plate are moved to the rear, remove the thrust washer and idler gear from the case. If needed, remove the inner race from the bearing and press the needle bearing from the idler gear.

To remove the countershaft bearings, do the following:

1. Temporarily reinstall the front auxiliary drive gear on the main shaft and remove the snap ring from each countershaft rear groove.

2. Remove the capscrews from the front bearing cover and remove the cover.

3. Special bearing puller tools are available that will remove bearings without damage. Use these tools to remove the countershaft rear bearings from the main transmission case. If replacement of the bearings is planned, a soft bar and maul can be used to drive the countershaft rear bearings to the rear and from the case (**Figure 11-35**). Drive the bearing from inside the case. Remember, this procedure will damage the bearings.

4. Remove the capscrew and remove the front bearing retainer plate from each countershaft.

5. Use a soft bar and maul to drive each countershaft to the rear as far as possible to partially unseat the front bearings.

6. From the case rear, use a soft bar and maul to drive each countershaft forward to unseat the front bearings from the case and expose the bearing snap rings.

7. Use a bearing puller or pry bar to remove the countershaft front bearings.

To remove and disassemble the main shaft assembly (**Figure 11-36**), proceed as follows:

1. Block the right countershaft assembly against the case wall and pull the main shaft assembly to the rear to free the pilot from the

FIGURE 11-35 Driving the countershaft rear bearings to the rear and from the case bores. This will damage the bearings, so only use this method if bearings are to be replaced. (Courtesy of Roadranger Marketing. One great drive train from two great companies—Eaton and Dana Corporations)

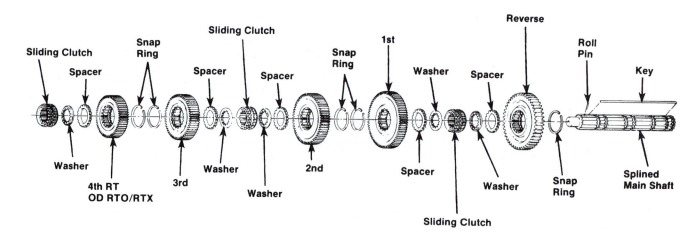

FIGURE 11-36 Main case main shaft assembly components. (Courtesy of Roadranger Marketing. One great drive train from two great companies—Eaton and Dana Corporations)

input shaft pocket. Tilt the front of the main shaft up and lift the assembly from the case. Use caution because reverse gear is free to slide off the main shaft.

2. Remove the sliding clutch from the front of the main shaft.
3. Remove the reverse gear and spacer from the rear of the main shaft.
4. Pull the key from the main shaft keyway.
5. Turn the reverse gear limit washer to align its splines with those of the main shaft and remove the washer.
6. Remove the low-reverse speed sliding clutch.
7. Use a small screwdriver to turn the limit washer in the hub of the low speed gear to align its splines with those of the main shaft. Pull the low speed gear from the rear of the main shaft to remove the limit washer, spacer, and gear. If needed, remove the snap ring from the I D of the low gear.
8. Remove each remaining gear, limit washer, spacer, and sliding clutch from the main shaft in a similar manner.

👓 Shop Talk

When removing limit washers, spacers, and gears, note their location on the main shaft to facilitate reassembly. Keep the internal splined washers and external splined spacers with the gear from which they were removed. Only one limit washer and spacer belong to each gear.

To remove the drive gear assembly (**Figure 11–37**), do the following:

1. Remove the bearing retaining snap ring from the groove in the shaft.

2. Use a suitable puller to remove the input shaft bearing.
3. Remove the drive gear spacer and snap ring.
4. Pull the input shaft forward and out of the drive gear and case.
5. Check the bushing in the pocket of the input shaft and replace it if worn.
6. Remove the main drive gear from the case.

To remove and disassemble the countershaft assemblies (**Figure 11–38**), proceed as follows:

1. Move the right countershaft assembly to the rear of the case as far as possible so the front of the shaft can be removed from the main case bore and moved to the center of the case. Now lift the entire countershaft assembly from the case.
2. If needed, remove and disassemble the right reverse idler assembly at this time. The procedure is identical to that used for the left reverse idler gear assembly outlined earlier in this chapter.
3. Remove the drive gear retaining snap ring from the front of each countershaft. Note that except for the PTO gears, the left and right countershaft assemblies are identical and disassembled in the same manner.
4. Using the rear face of the third gear as a base, press the drive gear, PTO gear, and third gear from each countershaft. This removes the front bearing inner race from the countershaft.

CAUTION: Never use the narrower PTO gear as a pressing base; it is prone to breaking if used in this way.

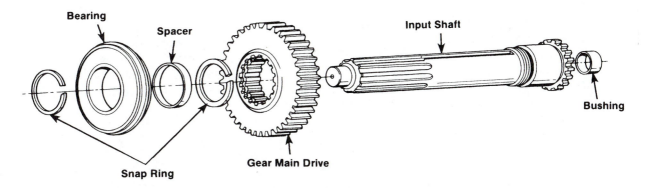

FIGURE 11–37 *Main drive gear assembly. (Courtesy of Roadranger Marketing. One great drive train from two great companies—Eaton and Dana Corporations)*

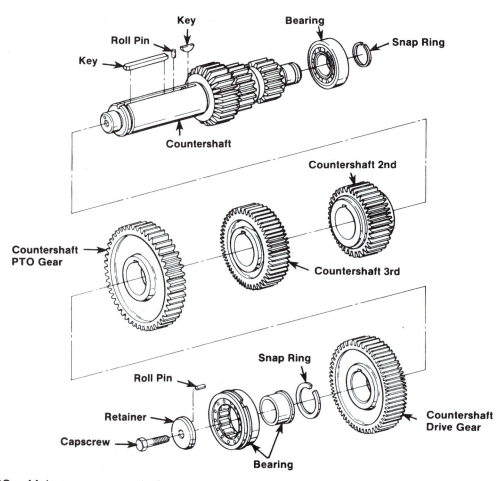

FIGURE 11–38   *Main case countershaft assembly components. (Courtesy of Roadranger Marketing. One great drive train from two great companies—Eaton and Dana Corporations)*

5. Use the rear face of the second gear as a base to press it off the countershaft.
6. If needed, remove the keys and roll pin from the countershaft.

To remove and disassemble the integral oil pump (**Figure 11–39**), proceed as follows:

1. Straighten the lock tang on the suction tube and remove the tube from the oil pump. Remove the O-ring if needed.
2. Remove the three Allen head capscrews and washer retaining the oil pump to the front of the case and remove the pump from the case.
3. Remove the retainer plate from the case.
4. Remove the outer oil pump element from the pump and remove the drive gear retaining snap ring from the drive shaft.
5. Remove the drive gear from the pump drive shaft, the key from the drive shaft keyway, the inner element retaining snap ring, and the internal oil pump element from the drive shaft.
6. Remove the two keys from the drive shaft keyways and remove the drive shaft from the integral oil pump housing.

7. Remove the relief valve roll pin from the pump housing and remove the relief valve spring from its bore along with the relief valve.
8. If needed, the front case plug can be driven through the front of the case for removal and replacement of the O-ring.

To disassemble the auxiliary drive gear (see **Figure 11–31**), do the following:

1. Remove the snap ring from the hub of the front auxiliary drive gear.
2. Using the rear face of the retainer ring as a base, press the drive gear through the bearing.
3. If needed, remove the O-ring from the hub O D.

MAIN (FRONT) SECTION REASSEMBLY

Refer to the appropriate illustrations under "Main Section Disassembly" for visual reference of the parts discussed below. To reassemble the auxiliary drive gear assembly, proceed as follows:

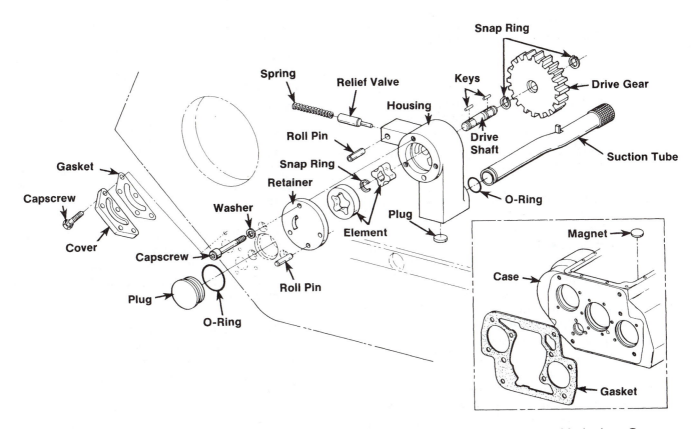

FIGURE 11–39 Components of the integral oil pump assembly. (Courtesy of Roadranger Marketing. One great drive train from two great companies—Eaton and Dana Corporations)

1. Install the O-rings on the extended front hub of the auxiliary drive gear.
2. Install the retainer ring on the front auxiliary drive gear with the snap ring groove facing the front hub, away from the gear teeth.
3. Start the drive gear bearing on the front hub with the bearing snap ring facing the groove in the retainer ring. Using both hands, press the bearing on the gear with the snap ring in the groove of the retainer ring.
4. Install the snap ring in the groove of the front gear hub to retain the bearing.

Reassembly of the integral oil pump (**Figure 11–39**) is basically the reverse of the disassembly procedure outlined earlier in this chapter.

To reassemble and install the right reverse idler gear, proceed as follows (see **Figure 11–33**).

1. If previously removed, thread the pipe plug in the rear of the reverse idle shaft and tighten. Install the idler plate on the shaft with the flat side to the front.
2. If previously removed, press the needle bearing into the bore of the reverse idler gear.

3. Install the bearing inner race on the idler shaft and insert the shaft into the case bore with the threaded end of the shaft to the front. As the idler shaft is moved forward, install the reverse idler gear on the shaft with the long hub to the front and seating on the bearing inner race. Position the thrust washer on the shaft between the gear and support boss in the case and continue with movement of idler shaft forward into the bore of the support boss.
4. Making sure that the reverse idler shaft is seated in the bore of the support boss and forward as far as possible, install the washer and stop nut on the front of the shaft. Tighten the nut to the recommended torque rating.
5. Install the outer race of the auxiliary countershaft front bearing into the case bore and against the idler plate. Note that the bearing inner race is installed on the front of the auxiliary countershaft and never with the outer race.

Except for the PTO gears, which differ in that one countershaft PTO gear has two more teeth than the other, reassembly of the left and right countershaft assemblies is identical (see **Figure 11–38**). The general procedure is as follows:

1. If previously removed, install the roll pin and key in the keyway of the countershaft.
2. Align the gear keyway with the countershaft key and press the second speed gear on the countershaft and the long hub of the gear to the countershaft rear.
3. Press the third speed gear on the countershaft and the long hub of the gear to the countershaft front.
4. Start the PTO gear onto the countershaft, bullet-nose side of the teeth facing up and toward the rear of the shaft. Align the keyway of the drive gear with the key in the countershaft and press both gears onto the shaft with the long hub of the drive gear against the PTO gear. Note that the left countershaft assembly has a 47-tooth PTO gear; the right countershaft assembly has a 45-tooth PTO gear.
5. To avoid confusion during installation, mark the end of the left countershaft with an "L" and the end of the right countershaft with an "R."
6. Install the drive gear retaining snap ring in the groove on the front of each countershaft.
7. Use a flanged driver to install the bearing inner race on the countershaft with the shoulder of the race against the shoulder of the countershaft. It is important to mark the countershaft drive gear for timing purposes. On the drive gear of each countershaft assembly, mark the tooth aligned with the keyway of the gear and stamped with an "O" for easy identification.

To install the countershaft assemblies in the main case, proceed as follows:

1. Place the left countershaft assembly into position in the case, making sure that the L-marked assembly has the larger 47-tooth PTO gear.
2. Place the right countershaft assembly into position in the case, making sure that the R-marked assembly has the smaller 45-tooth PTO gear.
3. Move the left countershaft assembly to the rear and insert a countershaft support tool or blocking to center the shaft in the rear case bore.
4. Use a flanged-end bearing driver to start the bearing in the case bore. Note that the inner

race of the roller-type front bearing is pressed on the front of the countershaft.

5. Center the front of the left countershaft in the bearing and move the assembly forward.
6. Use a flanged-end bearing driver to completely seat the front bearing or bearing outer race in the case bore.
7. Position the retainer plate on the front of the left countershaft, roll the pin in the hole at the end of the shaft, and secure with a capscrew tightened to the recommended torque ratings.
8. Remove the countershaft support tool or blocking from the rear case bore and install the left countershaft rear bearing with a larger I D lead chamfer to the front of the shaft and install the snap ring in the groove at the rear of the left countershaft.

Like the gears in the auxiliary section, the main drive gear assembly must be properly timed with the countershaft assemblies in the main casing. To time and reassemble the main drive gear assembly, do the following:

1. Install the snap ring in the I D of the main drive gear.
2. Mark the main drive gear for timing purposes. Mark any two adjacent teeth on the drive gear and repeat the procedure for the two adjacent teeth directly opposite the first set marked. A highly visible color of toolmaker's dye is recommended for making timing marks (see **Figure 11–15B**).
3. Mesh the marked tooth of the left countershaft drive gear with either set of two marked teeth on the main drive gear (**Figure 11–40**). Slide the input shaft through the main drive gear.
4. Install the drive gear spacer. Install the bearing on the input shaft with the external snap ring to the outside.
5. Install the drive gear bearing on the input shaft. Seat the bearing into the case bore with the proper bearing driver.
6. Temporarily install the front bearing cover.
7. Use a soft bar and maul to drive the input shaft through the bearing. Remove the front bearing cover.
8. Install the bearing retainer snap ring.
9. Install the front bearing cover and gasket, making sure to align the oil return hole in the case with the hole in the cover. Secure the front bearing cover on the case with six capscrews. Tighten to the recommended torque.

FIGURE 11-40 Timing the main drive gear with the countershaft drive gears. (Courtesy of Roadranger Marketing. One great drive train from two great companies—Eaton and Dana Corporations)

Reassemble the main shaft assembly (**Figure 11-38**).

1. If previously removed, install the corresponding snap rings in the I D of the main shaft gears.

2. Secure the main shaft in a vise equipped with brass jaws or wood blocks with the pilot end of the shaft down. If previously removed, install the roll pin in the keyway.

3. Install the third speed gear limit washer on the main shaft with the flat side of the washer up. Rotate the washer in the first or bottom groove of the main shaft to align the splines of the washer with those of the main shaft. Install the key in the main shaft keyway to lock the washer in place.

4. Install the spacer on the shaft against the washer.

5. Install the third speed gear on the main shaft with the clutching teeth down and engaged with the external splines of the spacer. Gear limit washers are internally splined and locked to the main shaft by the key. Gear spacers are externally splined to engage with the clutching teeth in the gear hubs. There is one limit washer and one spacer for each gear in the main shaft assembly.

6. Install the second speed gear on the shaft against the third speed gear with the clutching teeth up.

7. Install the spacer in the second speed gear, engaging the external splines of the spacer with the clutching teeth of the gear.

8. Remove the key from the keyway and install the second speed gear limit washer on the main shaft with the flat side of the washer down and against the spacer. Rotate the washer in the second groove of the main shaft to align the splines of the washer with those of the main shaft and reinsert the key in the keyway to lock the washer in place.

9. Set the axial clearance between the second and third gear according to the manufacturer's procedure and specifications. Clearances are adjusted by installing limit washers of different thicknesses.

10. Install the first-second speed sliding clutch, aligning the missing internal spline of the sliding clutch with the key in the main shaft.

11. Remove the key from the keyway and install the first speed gear limit washer on the main shaft with the flat side of the washer up. Rotate the washer in the second groove of the main shaft to align the splines of the washer with those of the main shaft and reinsert the key in the keyway to lock the washer in place.

12. Install the spacer and first speed gear on the main shaft with the clutching teeth down and engaged with the internal splines of the spacer.

13. Install the low speed gear on the shaft against the first speed gear with the clutching teeth up.

14. Install the spacer in the low speed gear, engaging the external splines of the spacer with the clutching teeth of the gear.

15. Remove the key from the keyway and install the low speed gear limit washer on the main shaft with the flat side of the washer down and against the spacer. Rotate the washer in the fourth groove of the main shaft to align the splines of the washer with those of the main shaft. Reinsert the key in the keyway to lock the washer in place.

16. Check the axial clearances and make the necessary adjustments between the low and first speed gears.

17. Install the low-reverse speed sliding clutch and align the missing internal spline of the sliding clutch with the key in the main shaft.

18. Remove the key from the keyway and install the reverse gear limit washer on the main shaft with the flat side of the washer up. Rotate the washer in the fifth groove of the main shaft to align the splines of the washer with those of the main shaft. Reinsert the key in the keyway to lock the washer in place.

19. Install the spacer on the shaft against the washer.

20. Install the reverse gear on the main shaft. Engage the clutching teeth of the gear with the splines of the spacer and sliding clutch and move the reverse gear against the low speed gear. Do not replace the reverse gear internal snap ring at this step.

21. Remove the main shaft assembly from the vise. Align the missing internal spline of third-fourth speed sliding clutch with the key in the main shaft and install on the front of the shaft engaging the external splines of the sliding clutch with the clutching teeth of third speed gear.

22. Block the right countershaft assembly against the case wall and lower the main shaft assembly into position with the reverse gear held against the low speed gear and the rear of the shaft moved into the case bore.

23. Move the pilot end of the main shaft into the pocket bushing of the input shaft.

24. With the reverse gear remaining against the low speed gear, mesh the corresponding forward speed gears of the left countershaft assembly. Check to make sure that the marked tooth on the left countershaft drive gear has remained in mesh with the marked set of teeth on the main drive gear.

25. Center the rear of the main shaft in the case bore and install the auxiliary drive gear assembly on the shaft, partially seating the bearing in the bore. Do not complete installation at this time.

It is now possible to time the right countershaft assembly.

CAUTION: The left countershaft assembly MUST remain in time with the main drive gear when timing the right countershaft.

Timing the Right Countershaft

1. Remove the blocking from the right countershaft assembly and place it parallel to the main shaft. Mesh the marked tooth of the right countershaft drive gear with the remaining set of two marked teeth on the main drive gear.

2. Insert the countershaft support tool or blocking into the rear bearing bore of the main housing.

3. Position the front countershaft bearing in the front bearing bore. Use the proper bearing driver to set the bearing.

4. Install the retainer washer and capscrew. Match the roll pin in the retainer with the inner bearing hole.

5. Tighten the capscrew to the recommended torque rating.

6. Position the countershaft rear bearing in the rear bearing bore. Use the proper bearing driver to seat the bearing in the bore.

7. With the bearing installation complete, install the snap ring in the right countershaft rear groove.

Shop Talk

Do not engage the sliding clutches with more than one gear at the same time. This will lock the gearing and prevent the main shaft and countershaft assemblies from rotating.

8. Move the reverse gear to the rear on the main shaft and use a screwdriver to engage the sliding clutches with all forward speed gears. A sliding clutch that will not engage with a gear indicates that the gear set is not in proper mesh. The bearings of the right countershaft must then be removed and the drive gear set retimed.

The left reverse idler gear assembly is installed at this time. The procedure is exactly like that described earlier for right reverse idler gear reinstallation.

To complete the reinstallation of the main shaft and auxiliary drive gear assemblies:

1. Move the reverse gear to the rear as far as possible, meshing the teeth of the gear with those of the reverse idler gears.

2. Align the external splines of the spacer with the clutching teeth of reverse gear and move the spacer forward on the main shaft and into reverse gear.

3. Install the snap in the hub of the reverse gear and move the reverse gear forward on the main shaft and into the proper position in the case.

4. Install the snap ring on the main shaft snap ring groove.

5. Install the splined washer on the main shaft splines behind reverse gear.

6. Reinstall the front auxiliary drive gear assembly on the rear of the main shaft. Use a flanged end driver and maul to seat the bearing in the case bore.

7. Align the six capscrew holes in the retainer with the tapped holes in the case. Install the capscrews. Tighten to recommended torque ratings and lockwire the capscrews in groups of three.

8. Install the splined washer on the main shaft splines behind the auxiliary drive gear.

9. Install the snap ring in the groove at the end of the main shaft.

The installation of the clutch housing, auxiliary section, and output yoke is basically a reversal of their disassembly. Use new gaskets and torque all capscrews to specification. Use a chain hoist to help position the auxiliary housing during its reinstallation. Always install the yoke slinger on the output yoke using a proper slinger driver.

When reinstalling the shift bar housing assembly, be sure all three main shaft sliding clutches are in their neutral position. With all three shift yokes on the shift bar housing also in the neutral position, fit the yokes into the slots of their corresponding sliding clutches. Use a new gasket and torque the shift bar housing capscrews to specifications.

When reinstalling the gearshift lever housing, make sure the shift bar housing assembly shift blocks and yoke notches are aligned in the neutral position. As the gearshift lever housing and gasket are positioned on the shift bar housing, fit the lever into the shift block and yoke notches. Install the capscrews and torque to specification.

Finally, reinstall all air lines.

11.6 TROUBLESHOOTING THE AIR SHIFT SYSTEM

If the transmission fails to make a range shift or shifts slowly, the fault might be in the range shift air system or in the actuating components of the shift bar housing assembly.

Troubleshooting procedures consist primarily of checking for leaks, for incorrect air line hookups, for correct airflow through the system, and for correct regulated air pressure. Air system troubleshooting guidelines are summarized in **Table 11–4**.

To locate trouble in the system, checks should be made with normal vehicle air pressure applied to the system, but with the engine off. The following are typical troubleshooting procedures.

TABLE 11–4: TROUBLESHOOTING AIR SYSTEM PROBLEMS

Problem: Slipout (Splitter)

Possible Cause	Possible Correction
Worn yoke pads	Instruct driver on proper driving technique. Replaced damaged parts.
Damaged insert valve	Replace.
Defective regulator	Replace.
Loose hose or fitting	Tighten.
Pinched air line or connector	Correct restriction.
Gear twisted out of time on countershaft	Check for damage and replace components.
Tapered clutching teeth	Replace damaged parts.
Worn yoke slot in clutch gear	Instruct driver on proper driving technique. Replace damaged parts.
Inner race left off front of auxiliary C/S	Install missing part.
Bearing failure	Replace damaged parts.
Improper preselection of splitter	Instruct driver on proper driving technique.
Not using clutch	Instruct driver on proper driving technique. Inspect for resulting damage and repair as needed.

Problem: Slipout (Range)

Possible Cause	Possible Correction
Defective regulator	Replace.
Loose hose or fitting	Tighten.

TABLE 11–4:   TROUBLESHOOTING AIR SYSTEM PROBLEMS (CONTINUED)

Possible Cause	Possible Correction
Pinched air line or fitting	Correct restriction.
Gear twisted out of time on countershaft	Check for damage and replace components.
Tapered clutching teeth	Replace damaged parts.
Improper range shifting technique	Instruct driver on proper driving technique. Inspect for resulting damage and repair as needed.

Problem: Slow Shift (Splitter)

Possible Cause	Possible Correction
Worn yoke pads	Instruct driver on proper driving technique. Replace damaged parts.
Defective regulator	Replace.
Loose hose or fitting	Tighten.
Pinched air hose	Correct restriction
Pinched air line or connector	Correct restriction.
Damaged O-ring in slave valve or in splitter cylinder assembly	Replace damaged part using thin film of silicon during reassembly.
Improperly mounted gasket	Replace part and set to proper specifications.
Improper preselection of splitter	Instruct driver on proper driving technique.
Not using clutch	Instruct driver on proper driving technique. Inspect for resulting damage and repair as needed.
Excessive silicone on O-rings	Clean part and apply proper amount.
Mixing oils or using additives	Check for damage and install proper grade oil.
Damaged O-ring	Replace.

Problem: Slow Shift or No Shift (Range)

Possible Cause	Possible Correction
Defective regulator	Replace.
Loose hose or fitting	Tighten part.
Pinched air hose	Correct restriction.
Sticking slave valve piston	Clean part and apply thin film of silicone.
Damaged O-ring in range cylinder or slave valve	Replace, applying thin film of silicone during reassembly.
Improperly mounted gasket	Replace part and set to proper specifications.
Improper air line hookup	Check air lines and hoses.
Pinched air line or connector	Correct restriction
Range cylinder piston nut loose	Tighten.
Range cylinder piston cracked	Replace part applying a thin film of silicone during reassembly.
Synchronizer spring broken	Replace broken or missing parts.
Failed synchronizer	Replace.
No silicone on O-rings	Apply thin film of silicone.
Excess silicone on O-rings	Clean part and apply thin film of silicone.
Damaged O-ring	Replace.
Restricted slave valve breather	Clean and/or replace part.
Improper range shifting techniques	Instruct driver on proper driving technique. Inspect for resulting damage and repair as needed.

Problem: Burned Synchronizer

Possible Cause	Possible Correction
Defective regulator	Replace part.
Yoke installed backwards	Replace part and set to proper specifications.
Synchronizer spring broken	Replace broken or missing parts.

Problem: Burned Synchronizer

Possible Cause	Possible Correction
Mixing oils or using additives	Check for damage and install proper grade oil.
Improper range shifting techniques	Instruct driver on proper driving technique. Inspect for resulting damage and repair as needed.

Problem: Overlapping Gear Ratios

Possible Cause	Possible Correction
Improper air line hookup	Check air lines and hoses and set to proper specifications.

Problem: Air Leak Out of Range Valve in High Range Only

Possible Cause	Possible Correction
Improper air line hookup	Check air lines and hoses and set to proper specifications.

Problem: Constant Air Leak Out of Air Range Valve

Possible Cause	Possible Correction
Damaged O-ring in range valve	Replace part applying a thin film of silicone during installation.

Problem: Air Leak Out of Slave Valve Breather

Possible Cause	Possible Correction
Damaged O-ring in slave valve or range cylinder	Replace part applying a thin film of silicone during installation.
Pinched air line or connector	Correct restriction.
Range cylinder piston nut loose	Tighten nut.
Range cylinder piston cracked	Replace part applying a thin film of silicone during reassembly.

Problem: Transmission Care Pressurized—(Low Range Only)

Possible Cause	Possible Correction
Damaged O-ring in range cylinder	Replace part applying a thin film of silicone during installation.

Problem: Transmission Care Pressurized—(All Range Positions)

Possible Cause	Possible Correction
Damaged O-rings in dash-mounted deep reduction valve or splitter cylinder assembly.	Replace part, applying a thin film of silicone during installation.

Problem: Constaint Air Leak Out of Insert Valve Exhaust Port

Possible Cause	Possible Correction
Damaged insert valve	Replace part, applying a thin film of silicone during installation.
Damaged O-ring in splitter cylinder assembly	Replace part, applying a thin film of silicone during installation.

Problem: Constaint Air Leak Out of Selector Valve Exhaust Port in "Direct" Position

Possible Cause	Possible Correction
Damaged insert valve	Replace part, applying a thin film of silicone during installation.
Damaged O-ring in splitter cylinder assembly	Replace part, applying a thin film of silicone during installation.

CAUTION: Take care when working under a vehicle while the engine is running because personal injury might result from the sudden and unintended movement of the vehicle. Place the transmission in neutral, apply the parking brake, and block the wheels.

INCORRECT AIR LINE HOOKUPS

Refer to the manufacturer's air system schematics to correctly locate and route air lines (see **Figures 10–39** and **10–41** for typical system components and routing). With the gearshift lever in neutral, move the control that provides range selection up and down.

- If the air lines are crossed between the control valve and the slave valve, there will be constant air flowing from the exhaust port of the control valve while in high range.
- If the air lines are crossed between the slave valve and the range cylinder, the transmission gearing will not correspond with the range selection. A low range selection will result in a high range engagement and vice versa.

AIR LEAKS

With the gearshift lever in neutral, coat all air lines and fittings with soapy water and check for leaks, moving the control that provides range selection up and down.

- If there is a steady leak from the exhaust port of the control valve, O-rings or components of the control valve are defective.
- If there is a steady leak from the breather of the slave valve, an O-ring in the valve is defective, or there is a leak past the O-rings of the range cylinder piston.
- If the transmission fails to shift into low range or is slow to make the range shift and the case is pressurized, see the "Range Cylinder" test procedure.

Tighten all loose connections and replace defective O-rings and parts.

CAUTION: The use of improper sealants and lubricants during reassembly of the air system can cause improper operation of the components and even severe transmission damage. Use manufacturer's *recommended silicone base* lubricant and sealants. RTV and other silicone sealants should not be used to seal the splitter cylinder cover or range cylinder cover. The use of this type of sealant can cause blockage of the air passages resulting in erratic operation of the air system or transmission damage.

AIR FILTER/REGULATOR

With the gearshift lever in neutral, check the breather of the air filter/regulator assembly (see **Figure 10–40**). There should be no air leaking from this port. The complete assembly should be replaced if a steady leak is found.

Cut off the vehicle air supply to the air filter/regulator assembly, disconnect the air line at the fitting in the supply outlet and install an air gauge in the opened port. Bring the vehicle air pressure to normal. Regulated air pressure should be set according to the manufacturer's specifications.

Shop Talk

Do not adjust the screw at the bottom of the regulator to obtain correct readings. The air regulator has been preadjusted within the correct operating limits. Any deviation from these limits, especially with regulators that have been in operation for some time, is likely to be caused by dirt or worn parts. If replacement or cleaning of the filter element does nothing to correct the air pressure readings, replace the complete assembly, because the air regulator is not serviceable.

SLAVE VALVES

The major problem encountered with slave valves is worn O-rings. Refer to **Figure 10–42** for the location of O-rings in a typical slave valve. If any of these O-rings are defective, there will be a constant air leak out of the exhaust on the slave valve. In normal operation, exhaust will occur only for an instant as the range shift is made.

If a constant leak is found, disconnect the air lines to the slave valve and remove the slave valve from the transmission. Disassemble the slave valve, noting the position of all components. It is recommended procedure to replace O-rings whenever the valve is disassembled. Lubricate O-rings with the recommended silicone lubricant before reassembly. Do not exceed recommended torque specifications during reassembly or the piston may bind in the valve bore.

Poppet type slave valves are usually not serviced. They are normally replaced when found defective.

Before replacing a poppet type slave valve, check for and correct any of the following conditions:

1. Restrictions in the air lines between the slave valve and the range valve.
2. Proper torque of the slave valve mounting cap screws (typically 8 ft.-lb.). If retorquing the cap screws eliminates the leak, remove the valve from the transmission and retorque the base plate screws on the back of the valve to the manufacturer's recommended torque (typically 45 ft.-lb.). Now, reinstall the slave valve, substituting flat mounting washers for the original star washers. Flat washers will not dig into and damage the valve housing. Torque mounting capscrews to recommended torque.

RANGE CYLINDER

If any of the seals in the range cylinder assembly (see **Figure 10–45**) are defective, the range shift will be affected. The amount of air being lost will affect the severity of the problem from slow shifting to a complete failure to shift.

-   An air leak at either O-ring "A" will result in complete failure to make a range shift and will result in a steady flow of air from the breather of the slave valve in both ranges.
-   A leak at gasket "B" will result in a steady flow of air to atmosphere while in _high range._
-   A leak at O-ring "C" will result in a slow shift to low range and will also pressurize the transmission case.
-   A gasket leak will result in a slow shift to the direct range.
-   Also, a cracked piston caused by improper installation will result in a failure to shift in either low or high range. It will also cause a steady leak out of the breather in the slave valve.

If any of the seals are defective, the range cylinder should be disassembled and new O-rings and gaskets installed. The following is the disassembly procedure for the range cylinder illustrated in **Figure 10–45**.

1. Remove the four capscrews and separate the range shift cylinder cover.
2. Remove the stop nut from the yoke bar.
3. Apply air to the low range air line to pop the piston from the cylinder housing.

WARNING: Do not direct the piston toward yourself or others during this step.

4. Remove the O-rings from the piston.
5. Install new O-rings, lubricate with recommended silicone lubricant, and reassemble the cylinder. Install a new gasket, but do not use Permatex™ or gasket sealant.

RANGE CONTROL VALVE

Troubleshooting the range valve (see **Figure 10–46**) will vary slightly between valve models. With the gearshift lever in neutral and the high range selected, the 1/8-inch outside diameter air line is disconnected from either:

1. the outlet or "P" port of the range control valve (some models); or
2. the slave valve (some models).

When low range is selected, a steady blast of air will flow from the opened P port or the disconnected line. Select high range to shut off the airflow. This indicates the control valve is operating properly. Reconnect the air line.

If the control valve does not operate properly, check for restrictions and air leaks. Leaks indicate defective or worn O-rings. If the O-rings or parts in the range valve are defective, there will be a constant air leak out of the exhaust located on the bottom of the valve. A defective insert valve O-ring will result in a constant leak through the exhaust in both ranges and the valve will not make range shifts. A defective housing O-ring will result in a constant, low volume leak through the exhaust in the low range only.

If the slide is assembled backwards, or the air lines between the slave valve and the range valve are hooked up backwards, there will be a constant leak through the exhaust in the high range. When installing the slide in the range valve, make sure that the slot in the slide faces the outlet port. When assembling a range valve, use a very small amount of silicone lubricant to avoid clogging the ports. A small amount of grease on the position springs and balls will help hold them in place during reassembly.

HIGH RANGE OPERATION

With the gearshift lever in neutral, select _low range_ and disconnect the 1/4-inch inside diameter air line at the port of the range cylinder cover. Make sure this line leads from the _high range_ or "H" port of the slave valve.

When _high range_ is selected, a steady blast of air should flow from the disconnected line. Select _low range_ to shut off the airflow.

Move the shift lever to a gear position and select the _high range._ There should be _no air_ flowing from the disconnected line. Return the gearshift lever to the neutral position. There should now be a steady flow of air from the disconnected line. Select the *low range* to shut off airflow and reconnect the air line.

If the air system does not operate accordingly, the slave valve or actuating components of the shift bar housing assembly are defective.

LOW RANGE OPERATION

With the gearshift lever in neutral, select *high range* and disconnect the 1/4-inch inside diameter air line at the fitting on the range cylinder housing. Make sure this line leads from the *low range* or "L" port of the slave valve.

When *low range* is selected, a steady blast of air should flow from the disconnected line. Select *high range* to shut off airflow.

Move the gearshift lever to a gear position and select *low range*. There should be *no air* flowing from the disconnected line. Return the gearshift lever to the neutral position. There should now be a steady flow of air from the disconnected line. Select *high range* to shut off the airflow and reconnect the air line.

If the air system does not operate accordingly, the slave valve or actuating components of the shift bar housing assembly are defective.

SPLITTER SHIFT SYSTEM

If the transmission fails to shift or shifts too slowly to or from the "split" position, the fault might be in the splitter shift air system or related components of the range shift system. Perform the following checks with normal vehicle air pressure supplied to the system. The engine should be off, the transmission in neutral, parking brake set, and wheels securely chocked. These checks should be made after proper air pressure has been verified at the air filter/regulator and all lines have checked out leak free.

Air Supply

With the gearshift lever in neutral, select the high or low range and loosen the connection at the "S" port of the valve mounted on the gearshift lever until it is determined that air is being supplied to this valve. If there is air to the valve, reconnect the fitting.

If there is no air supplied to this valve, check for air line restrictions between this valve and the slave valve. Make sure the line is properly connected to the slave valve.

Shift Tower Valve

With the gearshift lever in neutral, disconnect the air line running from the shift tower valve to the splitter cylinder cover at the splitter cylinder port.

While in high or low range, move the splitter control button on the shift tower forward. There should be air flowing from the disconnected line at the splitter cylinder cover. When the splitter control button is moved rearward, airflow should stop.

If the preceding conditions do not exist, the shift tower valve is defective or restrictions exist in the air line.

Splitter Cylinder

If any of the seals in the splitter cylinder assembly are defective, the splitter shift will be affected. There might be partial or total loss of shifting action depending on the severity of the air leakage. See **Figure 10–49** for the position of the various splitter cylinder O-rings. A leak at O-ring "A" results in a slow shift to engage the rear auxiliary drive gear, pressurizing of transmission case, or disengagement of the auxiliary gearing.

A leak at O-ring "B" results in slow shifting or complete failure to engage and disengage the front or rear auxiliary drive gearing, and a steady flow of air from the exhaust port of the shift tower valve and/or cylinder cover when the splitter control button is in the rearward position.

A leak at gasket "C" results in a slow shift to disengage the rear auxiliary drive gear, and a steady flow of air to the atmosphere.

Any constant flow of air from the exhaust port of the cylinder cover usually indicates a defective insert valve. Exhaust should occur only briefly when the splitter control button is moved rearward while in low or high range. A defective insert valve, leaking at the O-rings of valve O D or from inner seals, results in constant air leak and shift failure. Two indications of defective O-rings or seals are

- Constant air flowing from the exhaust port of the cylinder cover.
- Constant air flowing from the exhaust port of the control valve while splitter control button is rearward or forward (providing the control valve is operating properly).

The three O-rings in position on valve O D can be replaced. However, if an inner seal is damaged, the complete assembly must be replaced.

SUMMARY

- Proper lubrication procedures are a key to a good all-around maintenance program.
- Standard transmissions are designed so that the internal components operate in a bath of oil circulated by the motion of gears and shafts.

- It is vital that only the lubricants recommended by the transmission manufacturer be used in the transmission.
- An initial oil change should be performed after the transmission is placed in service. This change should be made any time after 3,000 miles, but before 5,000 miles of on-highway service.
- In general, it is suggested that a drain and flush period be scheduled every 50,000 miles for normal linehaul operation.
- A good preventive maintenance (PM) program can lower the incidence of breakdowns, minimize downtime, and reduce the cost of repairs.
- Leakage in transmission rear seals is a common problem in truck transmissions.
- It is important that the transmission manufacturer's disassembly, inspection, and reassembly instructions are followed when overhauling a transmission.
- Before reassembling the transmission, check each component carefully for abnormal or excessive wear and damage to determine reuse or replacement.
- More than 90 percent of all ball bearing failures are caused by dirt.
- Bearing failure due to poor lubrication is characterized by discoloration of the bearing parts.

REVIEW QUESTIONS

1. Which of the following lubricants is not recommended for use in a heavy-duty truck transmission when operating temperatures exceed 230°F?
   a. straight mineral gear oil
   b. SAE 50 grade heavy-duty engine oil
   c. multipurpose or EP gear oil
   d. SAE 50 grade transmission fluid

2. How often should the transmission gear oil be changed?
   a. after the first 3,000 to 5,000 miles of service
   b. after every 50,000 miles of over-the-highway service
   c. after every 1,000 hours of off-road operation
   d. all of the above

3. The transmission oil is at the proper level when it is
   a. visible through the filler opening
   b. reachable with a finger through the filler opening
   c. level with the filler hole
   d. at the proper level on the dipstick

4. Oil is leaking from the rear seal area of a transmission. Technician A says the problem is a damaged or missing speedometer sleeve O-ring or gasket. Technician B says the problem is caused by a cracked bearing cover. Who is correct?
   a. Technician A
   b. Technician B
   c. both A and B
   d. neither A nor B

5. Which of the following is an acceptable method of repairing a worn or damaged end yoke?
   a. hand finishing with crocus cloth
   b. turning down the yoke and installing a yoke sleeve kit
   c. replacing the yoke with a new one
   d. either b or c

6. Which of the following statements is true?
   a. the transmission must be disassembled to replace an input shaft
   b. an auxiliary section can be overhauled with the transmission installed in the chassis

7. Which of the following can generate a growling, humming, or grinding noise during transmission operation?
   a. worn, chipped, or cracked gears
   b. improper timing of the transmission during reassembly
   c. improper timing due to a gear turning on the countershaft
   d. all of the above

8. A driver complains that the vehicle jumps out of gear when driven over a bumpy road surface. Technician A says the problem could be worn detent notches in the shift rod. Technician B says the problem is probably worn gear clutch teeth. Who is correct?
   a. Technician A
   b. Technician B
   c. both A and B
   d. neither A nor B

9. Which of the following would most likely result in hard shifting?
   a. sliding clutch gears worn to a taper
   b. lack of lubricant in remote shift linkage
   c. broken detent springs
   d. worn shift fork pads

10. To prevent damage to the transmission due to lack of lubrication
    a. when towing the vehicle, pull the axle shafts or disconnect the driveline
    b. never coast with the transmission in neutral
    c. never coast with the clutch depressed
    d. all of the above

11. Most transmission bearing failures can be traced to
    a. lack of lubrication
    b. dirt
    c. improper shifting
    d. operation on rough highways

12. When removing a transmission from a truck, Technician A insists on supporting the weight of the transmission with a transmission jack. Technician B says that the transmission must be backed straight out of the engine to avoid damaging the flywheel pilot bearing or clutch components. Who is correct?
    a. Technician A
    b. Technician B
    c. both A and B
    d. neither A nor B

13. Which of the following precautions should be taken when disassembling a transmission?
    a. lubricate and wrap all reusable bearings as they are removed
    b. lay parts on a clean workbench in the order they are removed
    c. clean the unit before beginning disassembly
    d. all of the above

14. Which of the following are splined to the main shaft?
    a. first speed gear
    b. sliding clutch
    c. countershaft gears
    d. idler gear

15. Which two of the following must be marked for timing purposes before reassembling the front section?
    a. input shaft
    b. main drive gear
    c. second speed gear
    d. countershaft drive gear

Torque Converters

Objectives

After reading this chapter, you should be able to

- Explain the function of the torque converter in a vehicle equipped with an automatic transmission.
- Explain how the torque converter is coupled between the crankshaft and the transmission.
- Identify the three main elements of a torque converter and describe their roles.
- Define torque multiplication and explain how it is generated in the torque converter.
- Define both rotary and vortex fluid flow and explain how each affects torque converter operation.
- Describe the overrunning clutch, lockup clutch, and variable pitch stators.
- Outline torque converter service and maintenance checks.
- Remove, disassemble, inspect, and reassemble torque converter components.

Key Terms

coupling point
flex plate
lockup torque converter
overrunning clutch
roller clutch

rotary flow
stator assembly
turbine
variable pitch stator
vortex flow

As discussed in Chapter 9, a standard (manual) truck transmission uses a friction clutch that must be mechanically engaged and disengaged by the vehicle's driver to control the flow of power from the engine crankshaft to the gearbox and transmission. This type of clutch is controlled by a mechanical or a mechanical/hydraulic linkage between the clutch pedal and the release bearing.

Automatic truck (and passenger car) transmissions use a type of fluid coupling called a torque converter to transmit torque from the engine to the transmission.

The torque converter transmits twisting force hydraulically using automatic transmission fluid, often simply called transmission oil. The torque con-

verter transmits or multiplies the twisting force provided by the engine crankshaft. It then directs it through the transmission to provide a number of ratios suitable for the particular load and speed of the vehicle.

The amount of power transferred from the engine to the transmission by the torque converter is directly related to engine rpm. When the engine is turning at low idle speed, less power is transferred through the torque converter because there is insufficient oil flow for power transfer to occur. However, as engine speed increases, there is a corresponding increase in fluid flow. This increased flow creates enough force to transfer a greater amount of engine power through the torque converter and into the transmission.

12.1 DESIGN

Not all heavy-duty truck torque converters are serviceable. Many torque converters are welded together in such a way that a normal service shop cannot disassemble them for servicing and repair. The only types of torque converters (or T/Cs) that are readily serviceable are the types that are bolted together (**Figure 12–1**).

👓 Shop Talk

Torque converters can be confused with fluid couplings or flywheels since both use similar operating principles. The most fundamental difference is that torque converters use curved blades, while fluid couplings and fluid flywheels use straight blades. Torque converters also use stators and have the ability to multiply torque, neither of which is characteristic of fluid couplings.

FLEX PLATE (DISC)

A special **flex plate,** sometimes called a flex disc, is used to mount the torque converter to the crankshaft (**Figure 12–2**). The flex plate is positioned between

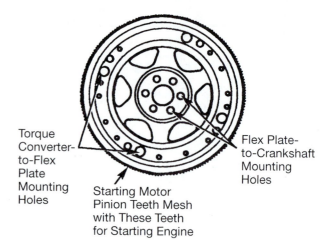

FIGURE 12–2   A special flex plate located between the crankshaft and torque converter transfers crankshaft rotation to the torque converter.

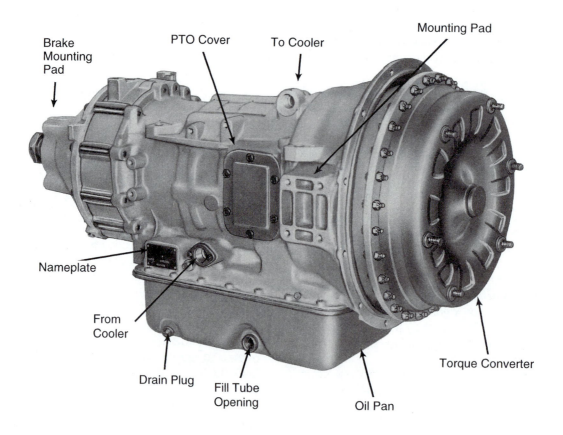

FIGURE 12–1   *The torque converter mounts to the front end of an automatic transmission. It transfers engine power from the crankshaft to the transmission gears. (Courtesy of Allison Transmission Division of GM)*

the engine crankshaft and the T/C. The purpose of the flex plate is to transfer crankshaft rotation to the shell of the torque converter assembly. The flex plate bolts to a flange machined on the rear of the crankshaft and to mounting pads located on the front of the torque converter shell.

The flex plate also carries the starter motor ring gear. Often when a torque converter is used, no flywheel is required because the mass of the torque converter and flex disc acts like a flywheel to smooth out the power pulses produced by the engine. Eliminating the flywheel is more common in gasoline engine vehicles; some larger diesel automatic transmissions use both a flex plate and flywheel. The flex plate also allows for a slight alignment tolerance between the engine and T/C assembly so exact alignment during installation is not critical to torque converter operation. Some applications use multiple flex plates.

TORQUE CONVERTER COMPONENTS

The torque converter provides a fluid coupling between the engine and the transmission gearing that can multiply torque. The torque converter has three main components (shown in **Figure 12–3**) and an optional lockup clutch.

Impeller or Pump

Directly coupled to the engine flywheel, the impeller rotates at engine rpm. The converter is full of fluid at all times. The impeller forms the rear half of the torque converter assembly and is designed with a series of internal, curved blades or vanes. One half of the split guide ring is attached to the blades. As the impeller rotates, fluid enters from around the pump hub, and centrifugal force throws this outward,

through the blades, around the split guided ring, and toward the turbine. This acts to propel the **turbine**, which can be compared to the similar action of a boat propeller driven in water. The impeller is mechanically driven by the engine.

Turbine

The turbine is coupled (splined) to the transmission turbine shaft and forms the front half of the torque converter assembly. The other half of the split guide ring is attached to the turbine vanes. Fluid from the impeller strikes the turbine vanes and flows around the split guide ring. In this way, the energy of the fluid from the impeller (an increase in impeller rpm increases the fluid velocity delivered to the turbine) is imparted to the turbine. As fluid exits the inner edge of the turbine vanes, it is directed back toward the impeller. When the turbine is rotated by this propelling action of the torque converter impeller, the turbine shaft is rotated and provides rotational input to the transmission gearing. No mechanical connection exists between the turbine and the impeller. Fluid acting on the turbine vanes exits near its hub, from which it is directed to the stator. The turbine is hydraulically driven by the impeller

Stator

The stator is located between the impeller and the turbine and is splined to the stator support. Fluid leaving the turbine does so in an opposite rotational direction to that of the turbine and impeller. This fluid acts on the stator vanes, locking it on the turbine shaft on a one-way clutch, and then is directed back to the impeller at an accelerated rate, increasing torque. This is known as **vortex flow,** and its effect is to multiply torque. However, as the turbine increases speed, the fluid it unloads into the stator is directed on the reverse side of the stator vanes causing it to *freewheel*. At this point, the stator ceases to multiply torque. As turbine rotational speed increases, flow through the stator smooths out and eventually stops. This is known as **rotary flow.** Torque multiplication, therefore, occurs only when pump and turbine rotational speeds are significantly different. It is, therefore, at a maximum at full stall and a minimum at the coupling phase.

Lockup Clutch

In any type of fluid coupling, a percentage of *slippage* will always exist. This means that turbine speed will never equal the impeller rotational speed. When rotary flow has been achieved and other speed and range criteria are met, the converter is *locked up,* meaning that the impeller and turbine speeds will be

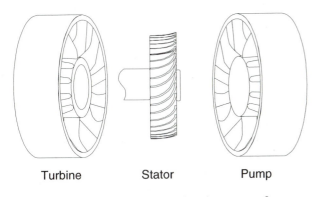

Turbine   Stator   Pump

FIGURE 12–3   *The three main elements of a torque converter: the turbine, the stator, and the pump (or impeller).*

equal. Lockup clutch components include a backing plate located in front of the turbine and bolted to the converter front cover; a lockup clutch plate splined to the turbine shaft, and a lockup, clutch piston splined to the converter front cover and, therefore, always rotated at engine speed. Whenever fluid is charged between the front cover and the lockup clutch piston, the clutch disc, which is splined to the turbine shaft, is *sandwiched* and forced to rotate at cover rpm (engine rpm).

T/C Exterior. The exterior of the torque converter shell is shaped like two bowls facing each other and welded or bolted together (**Figure 12–4**). To support the weight of the torque converter, a short stubby shaft projects forward from the front of the torque converter shell located in a socket at the rear of the crankshaft. This forms the frontal support for the torque converter assembly. At the rear of the torque converter shell is a hollow shaft with notches or flats at one end, ground 180 degrees apart. This shaft is called the pump drive hub; the notches or flats drive the transmission pump assembly. At the front of the transmission within the pump housing is a pump bushing supporting the pump drive hub and providing rear support for the torque converter assembly.

T/C Operation. The pump or impeller forms one section of the torque converter shell (**Figure 12–5**). The impeller has numerous curved blades that rotate as a unit with the shell. The impeller turns at engine speed, acting like a pump to start the transmission oil circulating within the torque converter shell.

While the impeller is positioned with its back facing the transmission housing, the turbine is positioned with its back to the engine. The curved blades of the turbine face the impeller assembly. The hub of the turbine is splined so the turbine can connect to and drive the turbine shaft. The turbine shaft transfers power flow to the main shaft of the transmission.

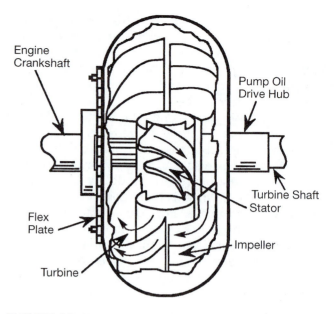

FIGURE 12–5   An interior view of a typical torque converter showing the pump/impeller, turbine, and stator. (Courtesy of Allison Transmission Division of GM)

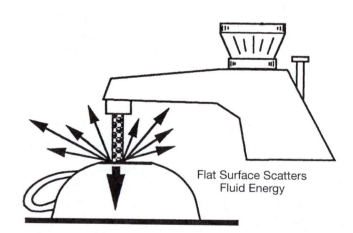

FIGURE 12–6   Contacting the flat back of the cup forces the water to splash away, scattering its energy.

The turbine blades have a greater curve than the impeller blades. This helps reduce oil turbulence between the turbine and impeller blades that would slow impeller speed and reduce the converter's efficiency.

Figure 12–6 illustrates how turbulence leads to a loss of energy and operating efficiency. When water strikes the flat surface of a teacup, it changes direction and moves back toward the oncoming flow. It splashes away from the flat surface in many directions; its energy is scattered. When the teacup is turned right side up as in **Figure 12–7**, the water flow strikes the curved inner surface of the cup, which is

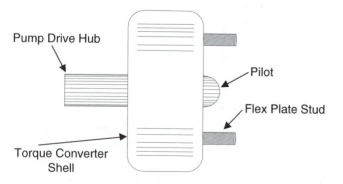

FIGURE 12–4   Torque converter input and output principle.

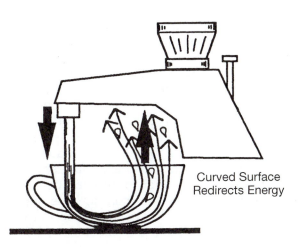

FIGURE 12–7 *Contacting the curved surface of the cup concentrates and redirects the water's energy.*

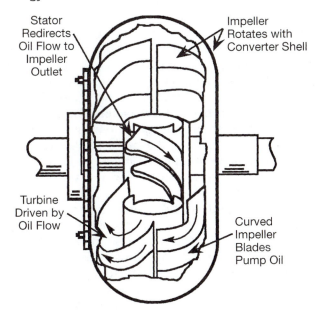

Stator Redirects Oil Flow to Impeller Outlet

Impeller Rotates with Converter Shell

Turbine Driven by Oil Flow

Curved Impeller Blades Pump Oil

FIGURE 12–8 *Converter operation.*

similar to the curved surface of the turbine blades. This curved surface helps direct and concentrate the water's energy. As the water flows around the curve of the cup body, it pushes against the surface and moves away from the cup. The push of the water makes the cup a reaction member, just as the turbine is the reaction member in the torque converter.

One of the fundamental laws of hydraulics states "the more the moving stream of fluid is diverted (changed), the more force it places on the curved reaction surface." So as oil in the torque converter moves around the turbine blades, it pushes against the blades as it moves away, placing additional force on the turbine blade driving the turbine shaft.

The stator is located between the pump/impeller and turbine. It redirects the oil flow from the turbine

back into the impeller in the direction of impeller rotation with minimal loss of speed or force. The blades of the stator are curved, so the stator side with the outward bulge is known as the *convex side.* The side of the stator blade with the inward curve is called the *concave side.* During torque converter operation, the stator must lock when forced to turn in one direction and rotate when moved in the opposite direction. To make this possible, the stator is mounted on an **overrunning clutch** (**Figure 12–8**).

12.2 BASIC OPERATION

As mentioned at the beginning of this chapter, transmission oil is used as the medium to transfer energy in the T/C and transmission. **Figure 12–9A** illustrates the T/C impeller or pump at rest, while **Figure 12–9B** shows it being driven. So, as the pump impeller rotates, centrifugal force throws the oil outward and upward. This is due to the curved shaped of the impeller housing. (Remember the previous example of the teacup and water.)

The faster the impeller rotates the greater the centrifugal force becomes. In **Figure 12–9B**, the oil is simply exiting the housing without producing any work. To harness some of this lost energy, the turbine assembly is mounted on top of the impeller (**Figure 12–10**). Now the oil thrown outward and upward from the impeller strikes the curved vanes of the turbine, causing the turbine to rotate.

The impeller and the turbine are not mechanically connected to each other. This is one of the major characteristics that differentiates a fluid coupling from a mechanical coupling. A hollow shaft at the

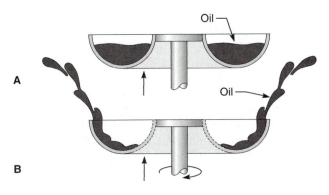

FIGURE 12–9 *(A) Transmission fluid (oil) at rest in the T/C impeller/pump. The fluid is level and the pump is not spinning; (B) the pump is spinning, turning oil up and outward.*

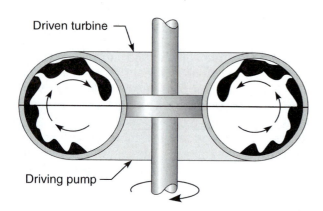

FIGURE 12–10 When the T/C rotates, oil is thrown up out of the pump into the turbine, and back into the pump. This action is called vortex flow.

center of the torque converter allows fluid under pressure to be continuously delivered from an oil pump. It is important to note that the oil pump delivering the fluid is driven by the engine. A seal or combination of seals prevents fluid from being lost from the system.

 Shop Talk

There can be a mechanical connection between the impeller and turbine by the use of a lockup clutch. A lockup clutch eliminates slippage between the impeller and turbine at certain speeds to help reduce the heat generated in the fluid and improve fuel mileage.

The turbine shaft is located within the hollow shaft at the center of the torque converter. It is splined to the turbine and transfers torque from the torque converter to the transmission's main drive shaft. Oil leaving the turbine is directed out of the torque converter to an external oil cooler and then to the transmission's oil sump or pan.

When the transmission is in neutral, torque cannot be transferred from the engine to the transmission output because the flow of power between these two points has been mechanically disconnected. This does not mean, however, that the driver must physically place the transmission in neutral each time the vehicle is stopped. With the transmission in gear and the engine at idle, the truck can be held stationary by applying the service brakes. At idle, engine speed is slow. Since the impeller is driven by engine speed, it too turns slowly, creating less centrifugal force within

the torque converter. The small amount of driveline torque generated can easily be held using the service brakes.

When the accelerator is depressed, engine speed increases. Since the impeller is driven directly by the engine, its speed increases with engine speed. This increase in speed increases the centrifugal force of the transmission oil, and the oil is directed against the turbine blades. The force of the oil against the blades transfers torque to the turbine shaft and transmission proportional to engine speed. The faster the engine turns, the more force is applied. Torque converter efficiency is always highest when input speed (engine speed) is greatest.

TYPES OF OIL FLOW

Understanding fluid flow within the torque converter during its operation is essential. The two types of oil flow that occur within the T/C are *rotary* and *vortex* (**Figure 12–11**). Rotary flow describes the centrifugal force applied to the fluid as the converter rotates around its axis. Vortex flow is the circular flow that occurs as the oil is forced from the impeller to the turbine and then back to the impeller. If a toy pinwheel was held at arm's length and swung in a large circle, air movement at the outer circle produces rotary flow, while the small circles cut by the pinwheel's propellers would produce vortex flow (**Figure 12–12**).

As the speed of the turbine increases, it approaches the speed of the impeller that is driving it. The point at which the turbine is turning at close to the same speed as the impeller is referred to as the **coupling point.** However, due to some fluid slippage that occurs between the turbine and the impeller, this is unobtainable unless some type of mechanical means is used to couple the two components. A lockup clutch is used to achieve this.

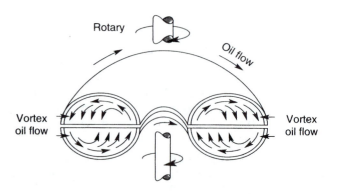
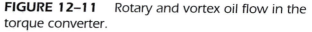

FIGURE 12–11 Rotary and vortex oil flow in the torque converter.

FIGURE 12–12 Rotary flow and vortex flow produces in the air by a pinwheel.

SPLIT GUIDE RINGS

Power only can flow through the converter when the impeller is turning faster than the turbine. When the impeller is turning at speeds much greater than the turbine, a great deal of oil turbulence occurs. Fast moving oil exits the impeller blades, striking the turbine blades with considerable force. It then has a tendency to be thrust back toward the center of both impeller and turbine.

To control this fluid thrust and the turbulence that results, a split guide ring is located in both the impeller and turbine sections of the T/C (**Figure 12–13**). The guide ring suppresses turbulence, allowing more efficient operation.

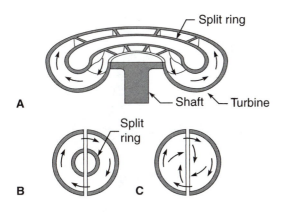

FIGURE 12–13 (A) The vanes are cut away to accommodate the split ring channel; (B) the vortex flow is guided and smoothed by the split ring; (C) without a split ring, fluid turbulence occurs at the center.

STATOR

The stator is a small wheel positioned between the pump/impeller and turbine (see **Figure 12–3**). The stator has no mechanical connection to either the impeller or turbine, but fits between the outlet of the turbine and the inlet of the impeller so that all the oil returning from the turbine to the impeller must pass through the stator.

The stator mounts through its splined center hub to a mating stator shaft, often called a ground sleeve. The ground sleeve is solidly connected to the transmission housing and therefore does not move. The stator can, however, freewheel when the impeller and turbine reach the coupling stage.

A stator can be either a rotating or fixed type. Heavy-duty trucks use the rotating type. Rotating stators are more efficient at higher speeds because less slippage occurs when the impeller and turbine reach the coupling stage.

The stator is the key to torque multiplication. It redirects the oil leaving the turbine back to the impeller, which helps the impeller rotate more efficiently (**Figure 12–14**). Torque multiplication can only

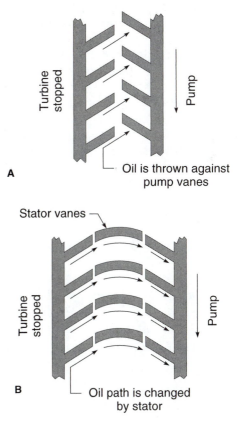

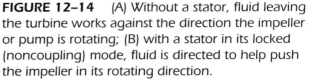

FIGURE 12–14 (A) Without a stator, fluid leaving the turbine works against the direction the impeller or pump is rotating; (B) with a stator in its locked (noncoupling) mode, fluid is directed to help push the impeller in its rotating direction.

occur when the impeller is rotating faster than the turbine. This results from the velocity or kinetic energy transferred to the transmission oil by the impeller's rotation, plus the velocity of the oil that is directed back to the impeller by stator action. During operation, the oil gives up part of its kinetic energy as it strikes the turbine vanes. The stator then redirects the fluid flow so the oil reenters the impeller moving in the same direction the impeller is turning. This allows the kinetic energy remaining in the oil to help rotate the impeller even faster, multiplying the torque produced by the converter.

Overrunning Clutch

An overrunning clutch keeps the stator assembly from rotating when driven in one direction and permits overrunning (rotation) when turned in the opposite direction. Rotating stators generally use a roller type overrunning clutch that allows the stator to freewheel (rotate) when the speed of the turbine and impeller reach the coupling point.

The **roller clutch (Figure 12–15)** is designed with an inner race, rollers, accordion (apply) springs, and outer race. Around the inside diameter of the outer race are several cam-shaped pockets. The clutch assembly rollers and accordion springs are located in these pockets.

As the vehicle begins to move, the stator stays in its stationary or locked position because of the wide difference between the impeller and turbine speeds. This locking mode takes place when the inner race rotates counterclockwise. The accordion springs force the rollers down the ramps of the cam pockets into a wedging contact with the inner and outer races. With the outer race held to the automatic

transmission housing, the stationary part connected to the inner race will be held stationary.

As vehicle road speed increases, turbine speed also increases until it approaches impeller speed. Oil exiting the turbine vanes strikes the back face of the stator, causing the stator to rotate in the same direction as the turbine and impeller. At this higher speed, clearance exists between the inner stator race and hub. The rollers in each slot of the stator are pulled around the stator hub. The stator freewheels or turns as a unit.

If the vehicle slows, engine speed also slows along with the turbine speed. This decrease in turbine speed allows the oil flow to change direction. It now strikes the front face of the stator vanes, halting the turning stator, and attempts to rotate it in the opposite direction.

As this happens, the rollers jam between the inner race and hub, locking the stator in position. In a stationary position, the stator now redirects the oil exiting the turbine so that torque is again multiplied.

Variable Pitch Stator

A **variable pitch stator** design is often used in torque converters in off-highway applications such as aggregate dump or haul trucks, or other specialized equipment used to transport unusually heavy loads in rough terrain. Each of a series of movable stator vanes has a crank rod that is fitted into a circular groove in the hydraulic piston. The movement of a hydraulic piston varies the angle of the stator vanes (**Figure 12–16**). A constant low pressure oil force is applied to one side of the piston while a valve directs a variable flow of high pressure main oil on the other side of the piston. As the flow of high

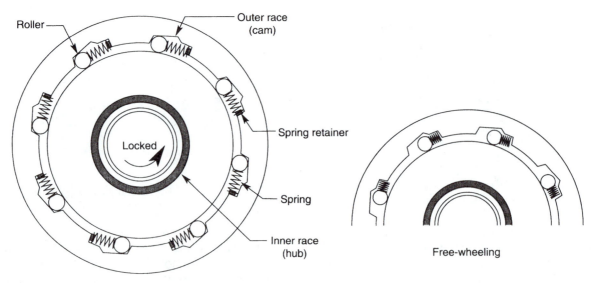

FIGURE 12–15 Roller-type overrunning clutch.

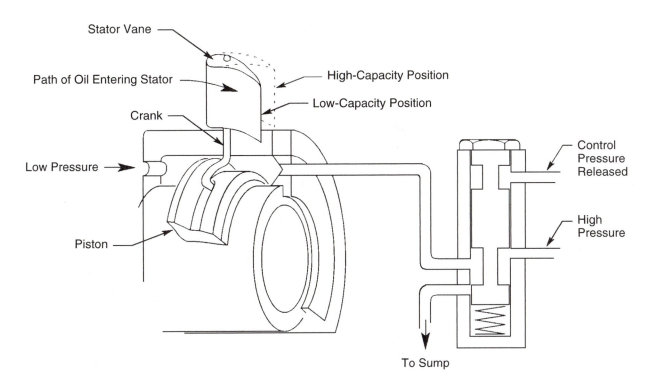

FIGURE 12-16 *Variable pitch stator assembly.*

pressure oil varies, the piston moves back and forth accordingly, turning the crank rods of each stator, which varies the stator angle.

When high pressure oil is applied to the piston assembly, the vanes rotate to a partially closed position (low capacity), which causes the torque converter to absorb less force. Since less of the available power is being absorbed by the torque converter, more of it is available to the power take-off gearing for operation of auxiliary equipment. When the oil pressure decreases to the point that the pressure-regulating valve reduces the amount of high pressure oil being delivered to the piston, the piston moves back to its normal position, allowing the stators to rotate to their usual position (high capacity, fully open). This directs the available fluid power so its energy can be most efficiently used.

LOCKUP CLUTCHES

Most modern heavy-duty trucks equipped with automatic transmissions have **lockup torque converters** (**Figure 12–17**). A lockup torque converter eliminates the 10 percent slip that takes place between the impeller and turbine at the coupling phase of operation. The engagement of a clutch between the engine crankshaft and the turbine assembly has the advantage of improving vehicle fuel economy and reducing

torque converter operating heat and engine speed. The lockup torque converter is a four-element (impeller, turbine, stator, lockup clutch), three-stage (stall, coupling, and lockup phase) unit.

There are two types of lockup torque converters: centrifugal and piston. Piston lockups tend to be used in heavy-duty truck applications. A piston lockup clutch consists of three main elements: a piston, clutch plate, and back plate. These components are located between the torque converter impeller cover and the torque converter impeller. The piston and back plate rotate with the converter impeller. The clutch plate is located between the piston and the back plate and is splined to the converter turbine.

The engagement of the lockup clutch is controlled by the lockup relay valve that receives its lockup signal from the modulated lockup valve. The clutch apply pressure compresses the lockup clutch plate between the piston and the back plate, locking all three together. When the converter impeller and turbine are locked together, they provide a direct drive one-to-one from the engine to the transmission gearing. The result is top vehicle speed performance and improved fuel mileage.

As rotational speed of the output shaft decreases, the relay valve automatically releases the lockup clutch and standard torque converter operation resumes.

1 Flanged Nut (6)
2 Torque Converter Assembly
3 Spacer Retainer (6)
4 Spacer (6)
5 Self-locking Nut
6 Converter Cover Assembly
7 Bushing
8 Seal Ring Retainer
9 Piston Inner Seal Ring
10 Lockup Clutch Piston
11 Piston Outer Seal
12 Lockup Clutch Plate
13 Seal Ring
14 Lockup Clutch Back Plate
15 Snap Ring
16 Bearing Race
17 Thrust Bearing Assembly
18 Bearing Race
19 Thrust Bearing Spacer
20 Converter Turbine Assembly
21 Turbine Hub
22 Converter Turbine
23 Rivet (8)
24 Converter Stator Assembly
25 Stator Thrust Washer
26 Cam Washer (2)
27 Stator
28 Cam
29 Side Plate Washer
30 Rivet (6)
31 Needle Bearing Assembly
32 Freewheel Roller Race
33 Freewheel Roller Spring (10)
34 Freewheel Roller (10)
35 Needle Bearing Assembly
36 Bearing Race
37 Bolt (8)
38 Lock Strip (4)
39 Converter Pump Hub
40 Gasket
41 Seal Ring
42 Converter Pump Assembly
43 Bolt (24)
44 Converter Pump Hub
45 Lock Strip (4)
46 Bolt (8)
47 Roller Bearing
48 Bearing Race
49 Seal Ring

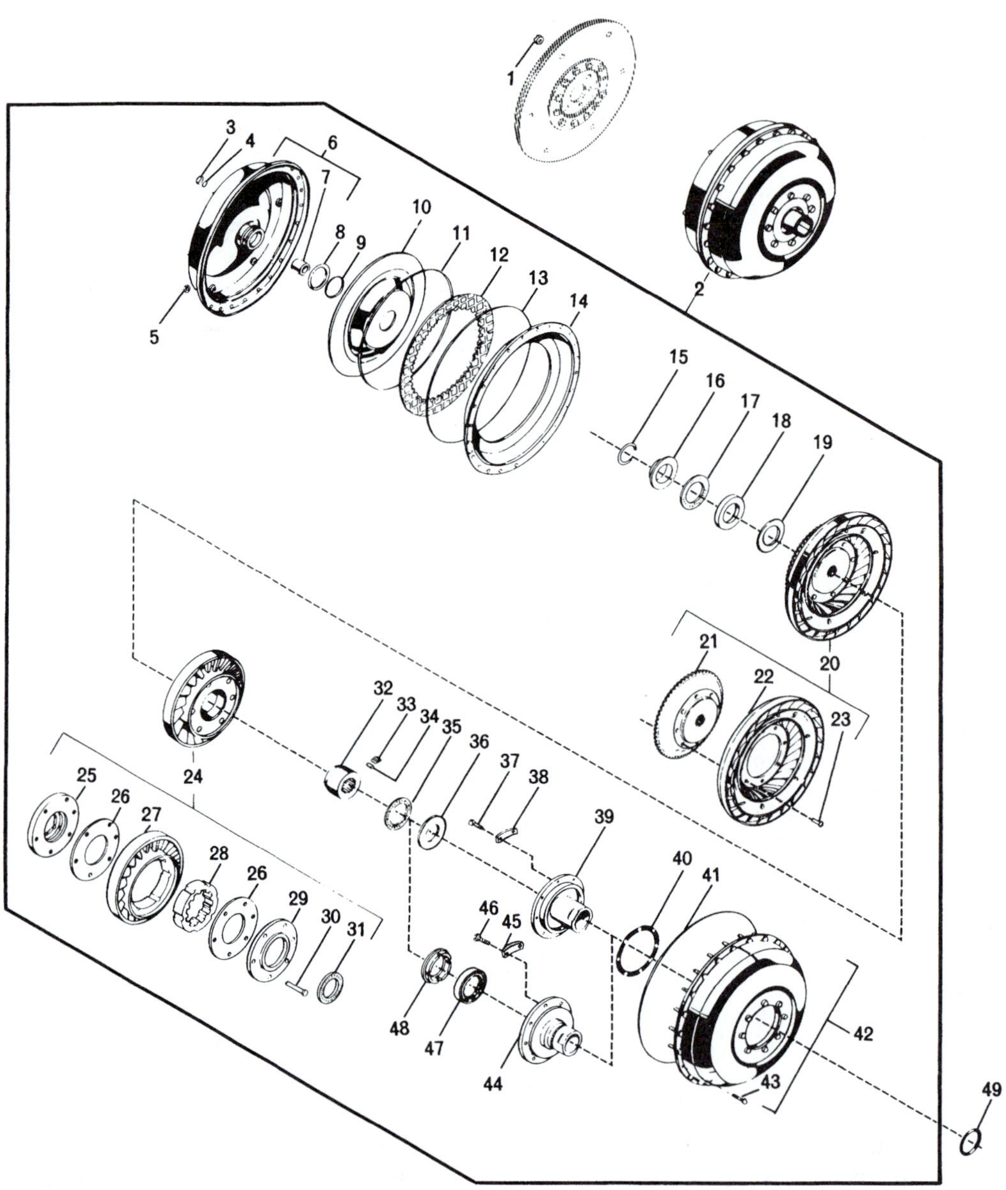

FIGURE 12–17   *Components of typical modern lockup torque converter. (Courtesy of Allison Transmission Division of GM)*

Summary of Torque Converter Operation

Torque multiplication phase

- This phase begins the moment the engine is started and the transmission is in gear.
- Vortex oil flow peaks at full stall; that is, the impeller (driven by the engine) is turning at maximum rpm, and the turbine is stalled.
- Vortex oil flow locks the stator one-way clutch, causing rotary oil flow to be at a minimum.
- The rate of vortex oil flow and torque multiplication decreases as the turbine speed increases and begins to approximate the impeller rpm.

- As turbine speed approaches 90 percent of the impeller speed, vortex oil flow drops to a minimum, and the converter enters the coupling phase.

Coupling phase

- The converter enters the coupling phase when vortex oil flow is at a minimum and rotary oil flow is at a maximum.
- In the coupling phase, turbine rpm is within 10 percent of impeller rpm.
- Rotary flow causes the stator one-way clutch to unlock, enabling it to freewheel.

- In the coupling phase, a slippage factor is always present, meaning that the turbine will always rotate at a slightly lower speed than the impeller.
- Torque converters equipped with a lockup clutch may apply it in the coupling phase.

Lockup phase

- A lockup clutch locks the turbine to the front cover.
- Lockup phase means that the impeller and turbine rotate at the same speed.
- Lockup increases the 90 percent turbine to impeller drive efficiency of the coupling phase to 100 percent drive efficiency.

Vortex flow

- Oil enters the impeller, passes around the split guide rings, and exits at the outer edge of the impeller vanes.
- The oil is next directed to act on the outer edge of the turbine vanes, to pass around the other half of the split guide ring, and to exit at the inside edges of the turbine vanes. This fluid is flowed in an opposite direction to the rotation of the impeller and turbine.
- Fluid exits the inboard edge of the turbine vanes and strikes the face of the stator vanes, locking it and redirecting it to the impeller.
- The fluid velocity is increased in the impeller in the vortex oil flow or torque multiplication phase.
- Vortex flow is at a maximum at full stall and at a minimum in the coupling phase.

Rotary flow

- Rotary flow occurs when the direction of oil flow in the stator is the same as that of the impeller and turbine.
- Rotary flow occurs when the turbine is being driven within 10 percent of the impeller rpm; that is, with 90 percent efficiency or better.
- Rotary flow is minimum at stall and at a maximum during the coupling phase.

12.3 MAINTENANCE AND SERVICE

Because any truck fitted with a torque converter will be matched to an automatic transmission, certain checks and tests relating to the maintenance of the torque converter are best covered in detail under general transmission service procedures. These checks and precautions include

- Using the manufacturer's specified transmission oil.
- Changing transmission oil at proper intervals and keeping the oil free of contamination.
- Using the proper oil changing procedure and regularly checking oil levels.
- Maintaining the torque converter or transmission breather.
- Maintaining the proper oil operating temperature.

See Chapter 14 for complete procedures on these maintenance items. The following sections cover general checks and tests for torque converters. The information here is not intended to replace manufacturer's repair manuals.

REMOVING TRANSMISSION/TORQUE CONVERTER

Drain the oil from the transmission before removal from the vehicle. For better drainage, the transmission should be warm. Since removal procedures can vary between vehicle models, consult the vehicle service manual for specific instructions for transmission/torque converter removal and installation.

Make sure all linkages, controls, cooler lines, modulator actuator cables, temperature connections, input and output couplings, oil filler tubes, parking brake linkages, and mounting bolts are disconnected before the transmission is removed. Oil lines should be carefully placed out of the way and openings covered to keep out dirt.

Position the jack or hoist sling relative to the transmission's center of gravity so the weight of the unit will be balanced as it is lifted.

CAUTION: The torque converter is free to move forward when the transmission is disconnected from the engine. Be sure the torque converter is not allowed to separate from the transmission while the transmission is being removed from the vehicle. Install a retaining strap to hold the T/C in place as soon as the transmission is clear of its mountings.

For transmission overhaul work, including removal and servicing of the torque converter, the transmission is normally mounted to an overhaul stand

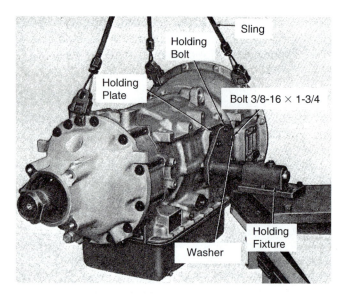

FIGURE 12–18 *Setup for mounting transmission and torque converter to overhaul stand. (Courtesy of Allison Transmission Division of GM)*

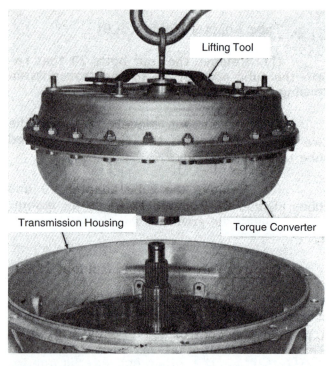

FIGURE 12–19 *Lifting the torque converter from the transmission body. (Courtesy of Allison Transmission Division of GM)*

(**Figure 12–18**). The transmission's power take-off (PTO) cover is removed and replaced with a holding plate that is attached to a fixture mounted to the stand. The fixture allows the transmission to be fully rotated so all parts are easily accessible.

Once the transmission is mounted to the overhaul stand, position the transmission front upward and

remove the retaining strap used to hold the T/C to the transmission. Attach the proper lifting tool and lift the torque converter assembly from the transmission (**Figure 12–19**). All components must be clean to permit inspection. Use mineral spirits or steam cleaning. Do not use caustic soda solutions for steam cleaning. Clean internally splined clutch friction plates with transmission fluid. Dry all parts with compressed air. Oil those parts that can rust. Do not allow water or cleaning solvent to enter sealed, welded units.

INSPECTION AND TESTS

Inspect the torque converter housing for damaged slots, scoring, cracks at welded seams, missing weights, dents, missing lugs or damaged threads. Replace the unit if these conditions exist.

Pressure Testing (Closed Welded T/Cs). Before performing this leak test on sealed units, make certain all oil is drained from the torque converter. Leaks are detected by pressurizing the T/C housing with compressed air and submerging it in water. **Figure 12–20** illustrates the basic tools and setup used for leak testing. Pressurize the housing to a maximum air pressure of 75 psi and submerge it in water. Bubbles are a sign of leakage; the T/C must be replaced.

CAUTION: Release the air pressure at the valve stem before loosening the nut holding the test fixture in place.

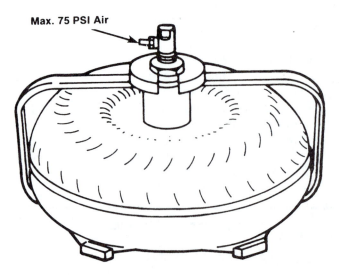

FIGURE 12–20 *Tools and setup for leak testing sealed weld torque converters. (Courtesy of Allison Transmission Division of GM)*

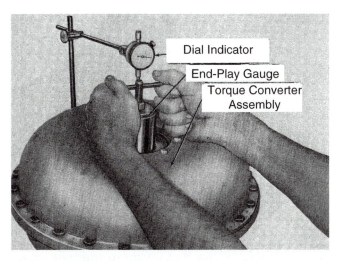

FIGURE 12–21 *Tools and setup for torque converter end-play inspection and testing. (Courtesy of Allison Transmission Division of GM)*

End-Play Check. To check if wear between internal parts is excessive, special tools are installed to test the amount of play in the thrust washer or flat needle roller bearing.

After the end-play gauge is installed and tightened, use a magnet base dial gauge and zero the gauge (**Figure 12–21**). Now lift up on the center screw as far as possible and record the gauge reading.

If the play is not within manufacturer's specifications on a welded closed unit, the unit must be replaced. On a bolted together unit the converter can be disassembled and rebuilt to correct this problem. In fact, standard servicing of bolted units involves disassembly and inspection of interior components even when end play is not out of spec.

DISASSEMBLY

The following disassembly and inspection procedures are typical for medium- and heavy-duty transmission torque converters. Parts are number keyed to **Figure 12–17**.

1. Remove the six rubber ID retainers (and spacers if used) from the converter cover assembly.
2. Remove the twenty-four nuts (5) from cover (6).
3. Remove, as a unit, the converter cover, lockup clutch piston, and related parts (**Figure 12–22**).
4. Place cover assembly on the work table with the lockup clutch piston up. Remove bearing race (16). Compress the center of piston (10) and remove snap ring (15).
5. Turn cover assembly over (piston down) and bump the cover sharply on a wood surface to remove the piston. Remove seal ring

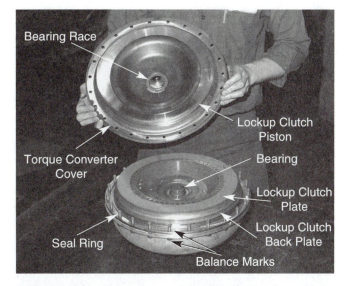

FIGURE 12–22 *Removing (or installing) converter cover assembly.*

retainer (8) and seal ring (9) from cover (6). Remove seal ring (11) from piston (10).
6. Remove bushing (7) only if replacement is necessary.
7. Remove lockup clutch plate (12).
8. Remove lockup clutch back plate (14) from torque converter pump (42). Remove seal ring (13) from plate (14).
9. Remove roller bearing assembly (17), bearing race (18), and spacer (19) from the hub of the turbine (20).
10. Remove the converter turbine assembly (**Figure 12–23**).

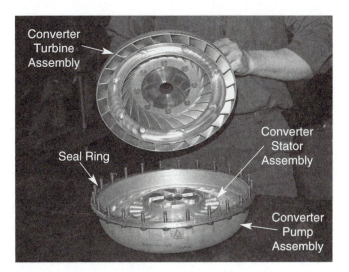

FIGURE 12–23 *Removing (or installing) converter turbine assembly.*

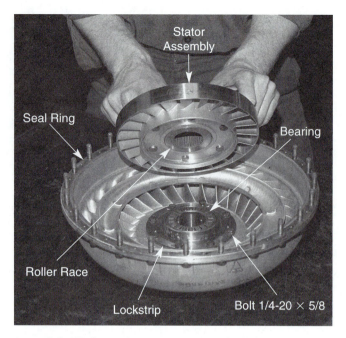

FIGURE 12–24   *Removing (or installing) converter stator and race assembly.*

11. Grasp the stator and the roller race as shown in **Figure 12–24** and remove as a unit.
12. Position the **stator assembly** (24) on the work table so that the freewheel roller race (32) is upward. Remove the roller race by rotating it clockwise while lifting it out of the converter stator.
13. Remove the ten rollers (34) and ten springs (33) from stator assembly (24).
14. Check needle bearing assembly (35).Wash and flush the needle bearing assembly thoroughly with dry-cleaning solvent or mineral spirits. Dry it, and lubricate with transmission oil. Replace the freewheel race only and rotate the bearing while pressing upon the freewheel race. If there is no roughness or binding, the needle bearing assembly can be left in the stator and cam assembly and reused. Do not mistake dirt or grit for a damaged needle bearing. Clean and oil the needle bearing if dirt is suspected. Check the needle bearing end of the freewheel race for a smooth finish. Replace the freewheel race if the bearing end is scratched or contains chatter marks.

 Shop Talk

If it is necessary to rebuild the stator assembly (24), remove the needle bearing assembly (35) before

starting the stator rebuild. Always follow the T/C manufacturer's rebuild procedures for all T/C subassemblies. Do not disassemble stator assembly unless parts replacement is needed.

15. If only the needle bearing assembly requires replacement, remove it carefully to avoid nicking the aluminum bore in which it is held.
16. Place the new needle bearing assembly, thrust race first, into the aluminum bore or the stator. Use a bearing installer and handle to install the thrust bearing (**Figure 12–25**).

CAUTION: Only apply load to the outer shell of the bearing during installation.

17. Drive the bearing assembly into the stator until the top of the outer shell is 0.025 to 0.035 inch above the shoulder in the side plate (**Figure 12–26**). The installing tool will seat on the stator area surrounding the bearing when the bearing is properly installed.
18. On some earlier units remove bearing (35) and bearing race (36) from the converter pump hub (39). On later model units, remove needle bearing (35) and bearing race (48), and remove roller bearing (47) from converter pump hub (44). Remove seal ring (41).

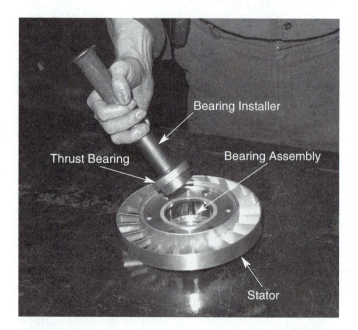

FIGURE 12–25   Installing the stator thrust bearing.

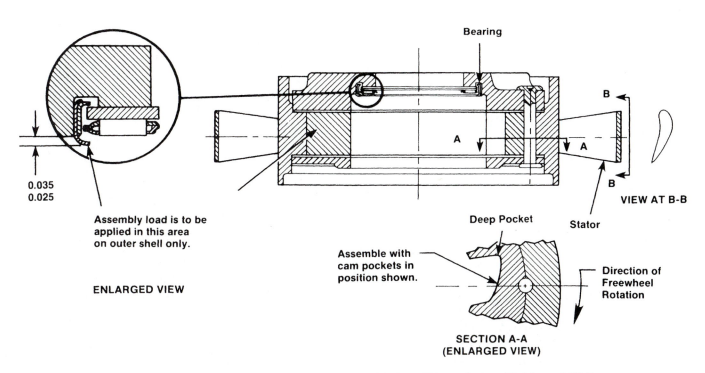

0.035
0.025

Assembly load is to be applied in this area on outer shell only.

ENLARGED VIEW

Bearing

VIEW AT B-B

Deep Pocket

Stator

Assemble with cam pockets in position shown.

Direction of Freewheel Rotation

SECTION A-A (ENLARGED VIEW)

FIGURE 12–26 *Typical stator thrust bearing. (Courtesy of Allison Transmission Division of GM)*

19. Flatten the corners of lock strips (38 or 45) and remove eight bolts (37 or 46) and four lock strips from converter pump hub (39 or 44).

20. Remove hub (39 or 44) and gasket (40) from pump (42). Remove seal ring (49).

INSPECTION AND ANALYSIS

When performing a torque converter overhaul, the following inspections and checks should be made on the converter and its related components.

Flywheel Assembly (if used). Inspect the flywheel (**Figure 12–27**) for cracks, impact damage, and signs of overheating. Inspect the flex plate mounting holes for pulled, stripped, or crossed threads. Damaged bolt threads can be repaired. Any signs of cracks or severe impact damage will require component replacement.

■ Inspect the flywheel crank adapter pilot for burrs, heat damage, or signs of misalignment. Light honing is permitted to remove slight irregularities.

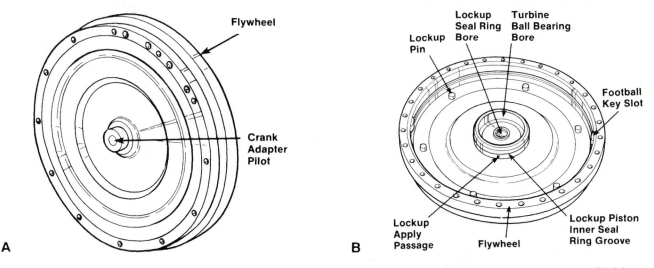

Flywheel

Crank Adapter Pilot

A

Lockup Seal Ring Bore

Turbine Ball Bearing Bore

Lockup Pin

Football Key Slot

Lockup Apply Passage

Flywheel

Lockup Piston Inner Seal Ring Groove

B

FIGURE 12–27 Flywheel inspection points: (A) front and (B) back. (Courtesy of Allison Transmission Division of GM)

- On flywheel assemblies with welded or pressed lockup pins, inspect for cracked welds, or evidence of fluid leakage around the pin. Cracked welds can be repaired if the lockup pin is not damaged excessively on the backside of the flywheel. Leakage of the pressed pins can be repaired by welding the pin diameter to the flywheel face. A heli-arc or TIG (tungsten inert gas) welding process should be used.
- The seal ring bore should be free of scratches, nicks, scoring, and signs of overheat. Slight irregularities can be corrected with crocus cloth. Inspect seal ring bore for any signs of excessive grooving from the rotating seal ring. Any grooving or excessive wear noted can be repaired by reworking with the available service bushing.
- Inspect the flywheel assembly for cracked, battered, or broken lockup pins. Any damage noted will require component replacement. Lockup apply passages should be open and free of debris.
- Inspect the ball bearing bore for scratches, grooves, nicks, scoring, and signs of overheat. Light honing is permitted to remove slight irregularities.
- Inspect the seal ring groove for cracked edges, nicks, burrs, and sharp edges. Light honing is permitted to remove slight irregularities. Any cracked or chipped edges noted will require component replacement.
- The football key slot should show no evidence of wear from movement of the lockup backing plate and football key. Excessive wear or elongation of the football key slot will require component replacement.
- Inspect the flywheel assembly for pulled, stripped, or crossed threads. Inspect the gasket surface for nicks, scratches, or burrs that could cause fluid leakage between the flywheel assembly and pump assembly. Light honing is permitted to remove slight irregularities. Damaged bolt threads can be repaired.

Lockup Clutch Back Plate. Inspect the lockup clutch back plate surface that contacts the friction plate for wear, scoring, scratches, and signs of overheat (**Figure 12–28**). Back plate should be flat to within 0.006 inch. Light honing is permitted to remove slight irregularities.

Inspect the football key slot for evidence of wear from movement of the lockup back plate and football key. Excessive wear or elongation of the football key slot will require component replacement. Inspect back plate for cracks from key slot to ID of plate. Any damage noted will require part replacement.

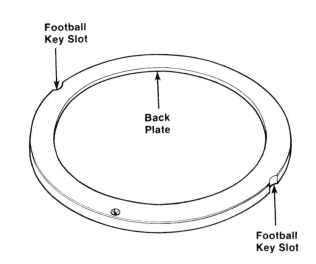

FIGURE 12–28 Lockup clutch back plate inspection. (Courtesy of Allison Transmission Division of GM)

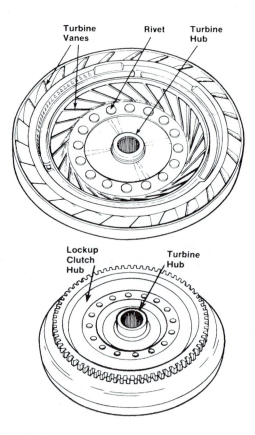
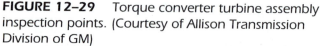

FIGURE 12–29 Torque converter turbine assembly inspection points. (Courtesy of Allison Transmission Division of GM)

Turbine Assembly (Figure 12–29). Inspect the turbine assembly for cracked or broken vanes and signs of overheat. If turbine assembly is taken apart, inspect the rivet holes for signs of wear or elongation. Any damage noted will require component replacement.

The turbine assembly hub must be checked for stripped, twisted, or broken splines. Any damage noted will require component replacement. Inspect the roller bearing journal for scoring, pitting, scratches, metal transfer, and signs of overheat. Slight irregularities can be removed with crocus cloth. Inspect the selective turbine thrust bearing spacer surface for scores, scratches, burrs, and signs of overheat. Light honing is permitted to remove slight irregularities.

Inspect the spline condition of the lockup clutch hub for notching due to lockup clutch plate movement. Replace lockup clutch hub if splines are notched deeper than 0.015 inch on any one side. If the turbine assembly is taken apart, inspect the rivet holes for signs of wear or elongation. Any damage noted will require component replacement.

Check rivets for cracks or loose fit. Any damage noted will require component replacement. Replacement of turbine, turbine hub, or lockup clutch hub will require replacement of rivets.

Lockup Clutch Piston (Figure 12–30). Inspect the seal ring groove of the lockup clutch piston for cracked edges, nicks, burrs, and sharp edges and remove any slight irregularities with light honing. Pistons with cracked or chipped groove edges must be replaced.

Check the lockup pinholes for signs of battering and elongation wear. Excessive elongation wear of the holes could indicate a wear condition of the lockup pins in the flywheel assembly. Excessive wear will require component replacement.

Inspect the piston for cracks, warpage, and signs of overheat. Inspect the surface in contact with the friction plate for wear, scoring, scratches, signs of overheat, and flatness. The surface in contact with the friction plate should be flat to within 0.003 inch TIR. Any damage noted will require component replacement. Inspect the seal surface for scratches, burrs, and nicks. Slight irregularities can be corrected with crocus cloth.

Stator (Figure 12–31). Inspect the stator assembly for cracked or broken vanes and signs of overheat. Any damage noted will require replacement of the stator assembly.

Inspect the rivets for cracks or loose fit. Any damage noted will require component replacement. Replacement of stator, stator thrust washer, cam, side plate washer, or stator cam washers will require replacement of rivets.

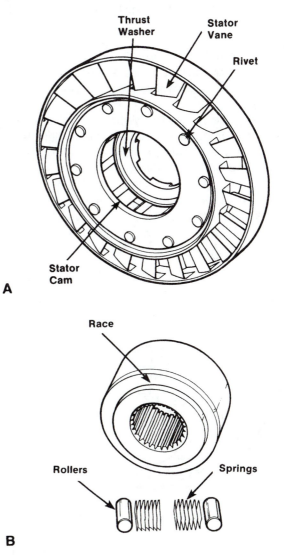

A

B

FIGURE 12–31 Stator assembly inspection points. (Courtesy of Allison Transmission Division of GM)

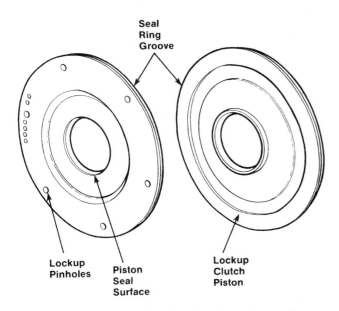

FIGURE 12–30 Lockup clutch piston inspection. (Courtesy of Allison Transmission Division of GM)

Check the stator cam roller pockets for pitting, scoring, signs of overheat, and wear. Emery cloth can be used to clean up slight irregularities.

Inspect the stator thrust bearing race surface for scoring, scratches, nicks, and grooves. Light honing is permitted to remove slight irregularities. Signs of overheat and metal transfer will require component replacement.

Check the stator freewheel roller surface, thrust bearing surface, and roller bearing surface for scoring, scratches, nicks, and grooves. Light honing is permitted to remove slight irregularities. Signs of overheat and metal transfer will require component replacement.

Inspect the stator freewheel rollers for scoring, scratches, grooves, and signs of overheat at bend. Any damage noted will require component replacement. The freewheel roller springs should show no signs of cracks, distortion, overheat, or breakage. Any damage noted will require component replacement.

Pump/Impeller (Figure 12–32). Inspect the pump for cracked or broken vanes and signs of overheat. Any damage noted will require component replacement.

Inspect the pump-to-flywheel surface for nicks, scratches, or burrs that could cause fluid leakage between the flywheel assembly and pump assembly. Inspect the pump-to-pump hub gasket surface for nicks, scratches, or burrs that could allow fluid leakage when assembled. Light honing is permitted to remove slight irregularities.

Pump (Impeller) Hub (Figure 12–33). Check the seal ring grooves on the pump hub for cracked edges, nicks, burrs, and sharp edges. Light honing is permitted to remove slight irregularities. Any cracked or chipped edges noted will require component replacement.

Inspect the front seal surface for scoring, scratches, nicks, and grooves. No rework of any irregularities noted is allowed on the seal surface. The use of crocus cloth or light honing on this surface could promote leakage past the front seal. Any irregularities noted will require component replacement.

There should be no signs of cracks, scoring, metal transfer, or heat damage on the pump drive flange. Remove any slight irregularities with light honing. Inspect the snap ring groove for burrs, cracks, and nicks (**Figure 12–34**). The snap ring must be able to snap tight in its groove for proper functioning. Light honing is permitted to remove slight irregularities.

Check the bearing race surface for scoring, scratches, nicks, and grooves. Light honing is permitted to remove slight irregularities. Signs of overheat and metal transfer will require component replacement.

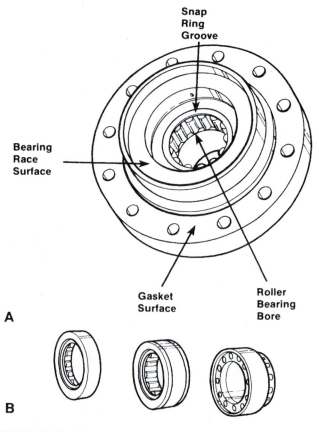

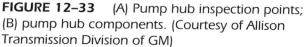

FIGURE 12–32   Pump/impeller assembly inspection points. (Courtesy of Allison Transmission Division of GM)

FIGURE 12–33   (A) Pump hub inspection points; (B) pump hub components. (Courtesy of Allison Transmission Division of GM)

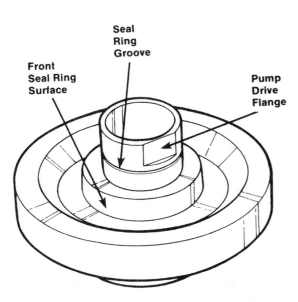

Seal
Ring
Groove

Front
Seal Ring
Surface

Pump
Drive
Flange

FIGURE 12–34 *Pump hub inspection points.*
(Courtesy of Allison Transmission Division of GM)

Inspect the roller bearing bore for scratches, grooves, nicks, scoring, and signs of overheat. Light honing is permitted to remove slight irregularities. Inspect the gasket surface for nicks, scratches, or burrs that could allow fluid leakage between the pump hub and pump assembly. Light honing is permitted to remove slight irregularities. Inspect the pump hub for pulled, stripped, or crossed threads. Damaged bolt threads can be repaired.

SUMMARY

- Automatic truck (and passenger car) transmissions use a type of fluid coupling known as a torque converter to transfer engine torque from the engine to the transmission.
- A flex plate, sometimes called a flex disc, is used to connect the torque converter to the crankshaft.
- Transmission oil is used as the medium to transfer energy from the engine driven impeller to the turbine, which in turn drives the transmission.
- Two types of oil flow take place inside the torque converter: rotary flow and vortex flow.
- A converter lockup clutch enables a mechanical coupling of the engine and transmission.

REVIEW QUESTIONS

1. The main difference between a torque converter and a standard fluid coupling is

   a. the torque converter uses a stator, the fluid coupling does not
   b. the torque converter has curved blades or vanes, while the fluid coupling uses straight vanes
   c. the torque converter can multiply engine torque, while the fluid coupling cannot
   d. all of the above

2. A _____ is used to mount the torque converter to the crankshaft.
   a. lockup clutch
   b. stator
   c. flex plate
   d. turbine

3. Which of the major torque converter components is the driving member?
   a. pump/impeller
   b. stator
   c. turbine
   d. lockup clutch

4. Which torque converter component is splined to a shaft that connects to the forward clutch of the transmission?
   a. pump/impeller
   b. stator
   c. turbine
   d. flex plate

5. Which of the major torque converter components is responsible for torque multiplication?
   a. pump/impeller
   b. stator
   c. turbine
   d. lockup clutch

6. What is used to transfer energy through the torque converter components?
   a. engine oil
   b. transmission fluid
   c. distilled water
   d. brake fluid

7. The stator is located
   a. between the flex plate and pump/impeller housing
   b. between the pump/impeller and turbine
   c. between the turbine and the first clutch of the transmission
   d. on the crankshaft

8. In a torque converter, _____ fluid flow is the flow of fluid around the circumference of the torque converter.
   a. rotary
   b. vortex
   c. primary
   d. secondary

9. In a torque converter, _____ fluid flow is the flow of fluid from the pump/impeller to the turbine and back to the pump by way of the stator.
   a. rotary
   b. vortex
   c. primary
   d. secondary

10. Which type of fluid flow must occur for torque multiplication to take place?
    a. rotary
    b. vortex
    c. primary
    d. secondary

11. The vanes of the torque converter are fitted with _____ to reduce fluid turbulence that can interfere with the power transfer through the unit.
    a. splash guards
    b. overrunning clutch
    c. split guide ring
    d. variable pitch stator

12. In order for torque multiplication to take place, the
    a. pump/impeller must turn faster than the turbine
    b. turbine must turn faster than the pump/impeller
    c. stator must freewheel
    d. lockup clutch must be engaged

13. At torque converter coupling phase,
    a. the speed of the turbine is almost as fast as the speed of the pump/impeller
    b. the speed of the pump/impeller is almost as fast as the turbine
    c. the overrunning clutch of the stator allows it to freewheel or turn
    d. both a and c
    e. both b and c

14. At torque converter coupling phase,
    a. vortex flow is at a minimum
    b. rotary flow is at a minimum
    c. the lockup clutch engages (if equipped)
    d. both a and c
    e. both b and c

15. Which is the most common lockup clutch design used in heavy-duty trucks?
    a. split ring
    b. centrifugal
    c. piston
    d. overrunning

13

Automatic Transmissions

Prerequisite
Chapter 12

Objectives

After reading this chapter, you should be able to
- Identify the components of a simple planetary gear set.
- Explain the operating principles of a planetary gear train.
- Define a compound planetary gear set and explain how it operates.
- Describe a multiple disc clutch and explain its role in the operation of an automatic transmission.
- Outline torque path power flow through a typical four- and five-speed automatic transmission.
- Describe the hydraulic circuits and flows used to control automatic transmission operation.
- List the two types of hydraulic retarders used on Allison automatic transmissions and explain their differences.

Key Terms

annulus
carrier
clutch pack
input retarder
multiple-disc clutches

output retarder
planetary gear sets
planetary pinion gears
priority valve
ring gear
sun gear

Automatic transmissions upshift and downshift with no direct assistance from the driver. Factors such as road speed, throttle position, and governed engine speed control and trigger shifting between gears.

Automatic transmissions rely on **planetary gear sets** to transfer power and generate torque from the engine to the driveline. Planetary gear sets are quite different from the cluster gear arrangements used in manual shift gearboxes. The shifting of planetary gears is actuated by the use of hydraulic force. An intricate system of valves is used to control and direct pressurized fluid in the closed system. The force generated by this fluid is used to apply and release the various clutches and brakes that control planetary gear operation. Most systems now use

electronic controls to obtain optimum operating performance and these will be studied in later chapters.

13.1 SIMPLE PLANETARY GEAR SETS

The planetary gear (**Figure 13–1**) is an old invention. Its first use was in early manual transmissions. Although it was replaced by the cluster gear arrangements discussed in Chapters 10 and 11, transmission engineers did not forget the many advantages planetary gear sets offered.

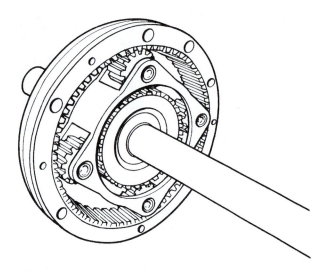

FIGURE 13–1   Typical simple planetary gear set.

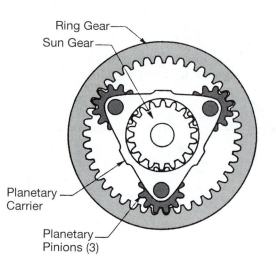

FIGURE 13–2   Planetary gear configuration is similar to the solar system, with the sun gear surrounded by the planetary pinion gears. The ring gear surrounds the complete gear set.

A Ford Motor Company engineer named Howard W. Simpson spent his retirement exploring the use of planetary gearing in automatic transmissions. Between 1948 and 1955, Simpson was granted close to two dozen patents pertaining to planetary gear design and configurations.

Simpson geartrains remain the standard in heavy-duty units and many passenger car automatics. The Simpson geartrain combines two simple planetary gear sets so that load can be spread over a greater number of teeth for strength and also to obtain the largest number of gear ratios possible in a compact area. But before looking at these compound systems, an introduction to simple planetary gear design and operation is required.

13.2 PLANETARY GEAR SET COMPONENTS

The simple planetary gear set acts as a mini-transmission. It can provide overdrive, reverse, forward reduction, neutral, and direct drive. The simple planetary gear set can also supply fast and slow speeds for each operating range, with the exception of neutral and direct drive.

A simple planetary gear set consists of three components: a **sun gear**, a **carrier** with planetary pinions mounted to it, and an internally toothed **ring gear** or **annulus**. The sun gear is located in the center of the assembly (**Figure 13–2**). It is the smallest gear in the assembly and is located at the center of the axis. The sun gear can be either a spur or helical gear design. It meshes with the teeth of the **planetary pinion gears**. Planetary pinion gears are small gears

fitted into a framework called the planetary carrier. The planetary carrier can be made of cast iron, aluminum, or steel plate and is designed with a shaft for each of the planetary pinion gears. (For simplicity, planetary pinion gears will be simply called planetary pinions.) Planetary pinions rotate on needle bearings positioned between the planetary carrier shaft and the planetary pinions. The number of planetary pinions in a carrier depends on the load the gear set is required to carry. Passenger vehicle automatic transmissions might have three planetary pinions in the planetary carrier. Heavy-duty highway trucks can have as many as five planetary pinions in a planetary carrier. The carrier and its pinions are considered one unit—the mid-size gear member.

The planetary pinions surround the sun gear's center axis and they themselves are surrounded by the annulus or ring gear, which is the largest component of the simple gear set. The ring gear acts like a band to hold the entire gear set together and provide great strength to the unit. The ring gear is located the greatest distance from the center of the axis and therefore exerts the most leverage on the center of the axis. To help remember the design of a simple planetary gear set, use the solar system as an example. The sun is the center of the solar system with the planets rotating around it. Hence the name planetary gear set.

DESIGN ADVANTAGES

Some advantages of the simple planetary gear set include

- Constantly meshed gears. With the gears constantly in mesh there is little chance of damage to the teeth. There is no grinding or missed shifts. The gear forces are divided equally.
- Planetary gear sets are very compact.
- Extreme versatility. Seven combinations of speed and direction can be obtained from a single set of planetary gears.
- Additional variations of both speed and direction can be added through the use of compound planetary gears.

HOW PLANETARY GEARS WORK

Each member of a planetary gear set, the sun gear, pinion gear carrier, and ring gear can revolve or be held at rest. Power transmission through a planetary gear set is only possible when one of the members is held at rest, or if two of the members are locked together.

Any one of the three members—sun gear, pinion gear carrier, or ring gear—can be used as the driving or input member. At the same time, another member might be kept from rotating and thus becomes the held or stationary member. The third member then becomes the driven or output member. Depending on which member is the driver, which is held, and which is driven, either a torque increase or a speed increase is produced by the planetary gear set. Output direction can also be reversed through various combinations.

Table 13–1 summarizes the basic laws of simple planetary gears. It indicates the output speed, torque, and direction of the various combinations available. It is also helpful to remember the following two points with regard to direction of rotation:

1. When an external-to-external gear tooth set is in mesh, there will be a change in the direction of rotation at the output (**Figure 13–3A**).
2. When an external gear tooth is in mesh with an internal gear, the output rotation for both gears will be the same (**Figure 13–3B**).

Figures 13–4 through **13–11** illustrate the eight possible gear rotation combinations as listed in **Table 13–1**.

Combination 1: Maximum Forward Reduction. With the ring gear stationary and the sun gear turning clockwise, the external sun gear will rotate the planetary

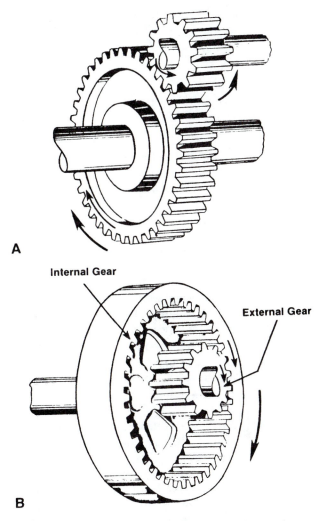

A

Internal Gear

External Gear

B

FIGURE 13–3   (A) With external teeth in mesh, there is a change in direction at the output; (B) when an external gear is meshed with an internal gear, both turn in the same direction.

		TABLE 13–1:	LAWS OF SIMPLE PLANETARY GEAR OPERATION		
Sun Gear	**Carrier**	**Ring Gear**	**Speed**	**Torque**	**Direction**
1. Input	Output	Held	Maximum reduction	Increase	Same as input
2. Held	Output	Input	Minimum reduction	Increase	Same as input
3. Output	Input	Held	Maximum increase	Reduction	Same as input
4. Held	Input	Output	Minimum increase	Reduction	Same as input
5. Input	Held	Output	Reduction	Increase	Reverse of input
6. Output	Held	Input	Increase	Reduction	Reverse of input
7. When any two members are held together, speed and direction are the same as input. Direct 1:1 drive occurs.					
8. When no member is held or locked together, output cannot occur. The result is a neutral condition.					

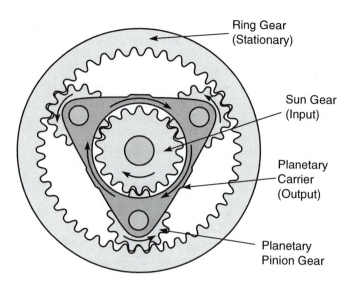

FIGURE 13–4   *Maximum forward reduction (greatest torque, lowest speed) is produced with the sun gear as input, ring gear stationary, and carrier as the output.*

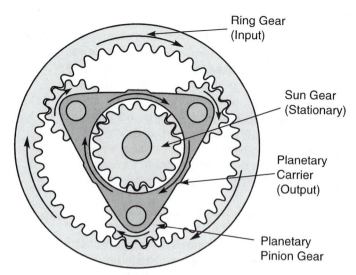

FIGURE 13–5   *Minimum forward reduction (good torque, low speed) is produced with the ring gear as input, the sun gear stationary, and carrier as the output.*

pinions counterclockwise on their shafts (**Figure 13–4**). The inside diameter of each planetary pinion pushes against its shaft, moving the planetary carrier clockwise. The small sun gear (driving) will rotate several times, driving the middle size planetary carrier one complete revolution, resulting in an underdrive. This combination represents the most gear reduction or the maximum torque multiplication that can be achieved in one planetary gear set. Input speed will be high, but output speed will be low.

Combination 2: Minimum Forward Reduction. In this combination, the sun gear is stationary and the ring gear rotates clockwise (**Figure 13–5**). The ring gear drives the planetary pinions clockwise and walks around the stationary sun gear. The planetary pinions drive the planetary carrier in the same direction as the ring gear forward. This results in more than one turn of the input as compared to one complete revolution of the output. The result is torque multiplication. But since a large gear is driving a small gear, the amount of reduction is not as great as in combination 1. The planetary gear set is operating in a forward reduction with the large ring gear driving the small planetary carrier. Therefore, the combination produces minimum forward reduction.

Combination 3: Maximum Overdrive. With the ring gear stationary and the planetary carrier rotating clockwise (**Figure 13–6**), the three planetary pinion shafts push against the inside diameter of the planetary pinions. The pinions are forced to walk around the inside of the ring gear, driving the sun gear

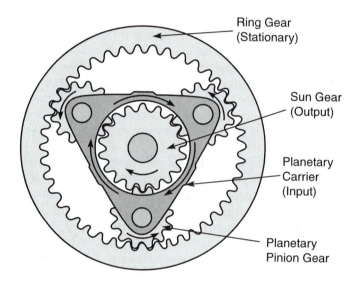

FIGURE 13–6   *Maximum overdrive (lowest torque, greatest speed) is produced with the carrier as input, ring gear stationary, and sun gear as output.*

clockwise. The carrier is rotating less than one turn input compared to one turn output, resulting in an overdrive condition. In this combination, the middle size planetary carrier is rotating less than one turn and driving the smaller sun gear at a speed greater than the input speed. The result is a fast overdrive with maximum speed increase.

Combination 4: Slow Overdrive. In this combination, the sun gear is stationary and the carrier rotates clockwise (**Figure 13–7**). As the carrier rotates, the pinion shafts push against the inside

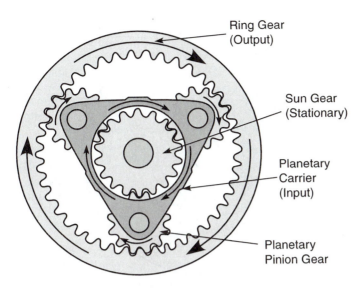

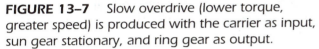

FIGURE 13–7   Slow overdrive (lower torque, greater speed) is produced with the carrier as input, sun gear stationary, and ring gear as output.

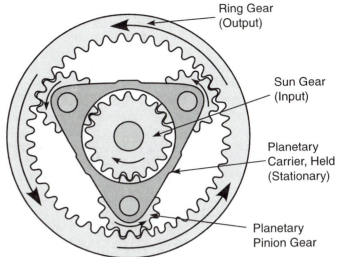

FIGURE 13–8   Slow reverse (opposite output direction, greatest torque, lowest speed) is produced with the sun gear as input, planetary carrier stationary, and the ring gear as output.

diameter of the pinions and they are forced to walk around the held sun gear. This drives the ring gear faster and the speed increases. The carrier turning less than one turn causes the pinions to drive the ring gear one complete revolution in the same direction as the planetary carrier. As in combination 3, an overdrive condition exists, but the middle size carrier is now driving the larger size ring gear so only slow overdrive occurs.

Combination 5: Slow Reverse. Here, the sun gear is driving the ring gear with the planetary carrier held stationary (**Figure 13–8**). The planetary pinions, driven by the external sun gear, rotate counterclockwise on their shafts. The external planetary pinions drive the internal ring gear in the same direction. While the sun gear is driving, the planetary pinions are used as idler gears to drive the ring gear counterclockwise. This means the input and output shafts are operating in the opposite or reverse direction to provide a reverse power flow. Since the driving sun gear is small and the driven ring gear is large, the result is slow reverse.

Combination 6: Fast Reverse. For fast reverse, the carrier is still held as in slow reverse, but the sun gear and ring gear reverse roles, with the ring gear now being the driving member and the sun gear driven (**Figure 13–9**). As the ring gear rotates counterclockwise, the pinions rotate counterclockwise as well, while the sun gear turns clockwise. In this combination, the input ring gear uses the planetary pinions to drive the output sun gear. The sun gear

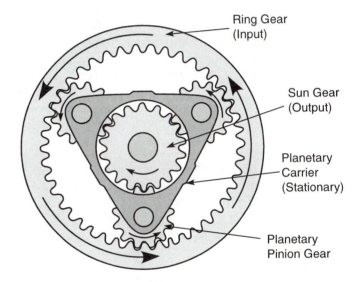

FIGURE 13–9   Fast reverse (opposite output direction, greater torque, low speed) is produced with the ring gear as input, carrier stationary, and sun gear as output.

rotates in reverse to the input ring gear. In this operational combination, the large ring gear rotating counterclockwise drives the small sun gear clockwise, providing fast reverse.

Combination 7: Direct Drive. In the direct drive combination, both the ring gear and the sun gear are input members (**Figure 13–10**). They turn clockwise at the same speed. The internal teeth of the clockwise turning ring gear will try to rotate the planetary

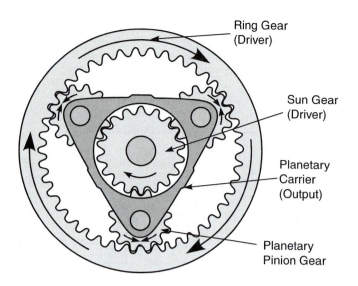

FIGURE 13–10   *Direct drive is produced if any two gear set members are locked together.*

Ring Gear (Driver)
Sun Gear (Driver)
Planetary Carrier (Output)
Planetary Pinion Gear

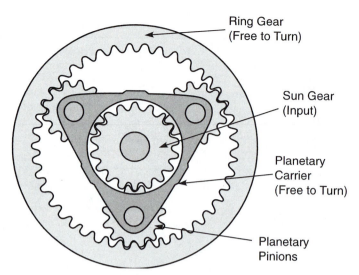

FIGURE 13–11   *With no held or stationary member, there can be no output. The result is a neutral position.*

Ring Gear (Free to Turn)
Sun Gear (Input)
Planetary Carrier (Free to Turn)
Planetary Pinions

pinions clockwise as well. But the sun gear, an external gear rotating clockwise, will try to drive the planetary pinions counterclockwise. These opposing forces lock the planetary pinions against rotation so that the entire planetary gear set rotates as one complete unit. This ties together the input and output members and provides a direct drive. For direct drive, both input members must rotate at the same speed.

Combination 8: Neutral Operation. Combinations 1 through 7 all produce an output drive of various speeds, torques, and direction. In each case, one member of the planetary gear set is held or two members are locked for the output to take place.

When no member is held stationary or locked, there will be input into the gear set, but no output. The result is a neutral condition (**Figure 13–11**).

Summary of Simple Planetary Gear Set Operation

- When the planetary carrier is the drive (input) member, the gear set produces an overdrive condition. Speed increases, torque decreases.
- When the planetary carrier is the driven (output) member, the gear set produces a forward reduction. Speed decreases, torque increases.
- When the planetary carrier is held, the gear set will produce a reverse.

To determine if the speed produced is fast or slow, remember the rules regarding large and small gears.

- A large gear driving a small gear increases speed and reduces torque of the driven gear.
- A small gear driving a large gear decreases speed and increases torque of the driven gear.

PLANETARY GEAR CONTROL

As we have seen, for a planetary gear set to produce an output, there must be some method of holding one gear member planetary gear member stationary and applying drive torque to it. In heavy-duty truck transmissions, this job is performed by multiple-disc clutches, which serve as both braking and power transfer devices.

The clutch uses a series of hollow friction discs to transmit torque or apply braking force. The discs have internal teeth that are sized and shaped to mesh with splines on the clutch assembly hub. In turn, this hub is connected to a planetary geartrain component so gear set members receive the desired braking or transfer force when the clutch is applied or released.

Design. Multiple-disc clutches have a large drum-shaped housing that can be either a separate casting or part of the transmission housing (**Figure 13–12**). This drum housing holds all other clutch components: the cylinder, hub, piston, piston return springs, seals, pressure plate, clutch pack (including friction plates), and snap rings.

The cylinder in a multiple-disc clutch is relatively shallow. The cylinder bore acts as a guide for piston travel. The piston is made of cast aluminum or steel

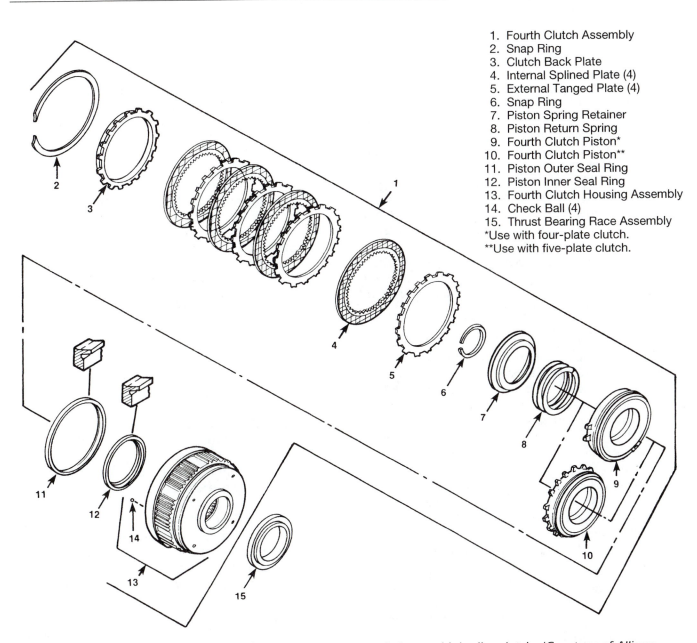

1. Fourth Clutch Assembly
2. Snap Ring
3. Clutch Back Plate
4. Internal Splined Plate (4)
5. External Tanged Plate (4)
6. Snap Ring
7. Piston Spring Retainer
8. Piston Return Spring
9. Fourth Clutch Piston*
10. Fourth Clutch Piston**
11. Piston Outer Seal Ring
12. Piston Inner Seal Ring
13. Fourth Clutch Housing Assembly
14. Check Ball (4)
15. Thrust Bearing Race Assembly
*Use with four-plate clutch.
**Use with five-plate clutch.

FIGURE 13–12 Components of a typical heavy-duty transmission multiple disc clutch. (Courtesy of Allison Transmission Division of GM)

with a seal ring groove around the outside diameter. A seal ring seats in the groove. This rubber seal retains fluid pressure required to stroke the piston and engage the clutch pack. The piston return springs overcome the residual fluid pressure in the cylinder and move the piston to the disengaged position when clutch action (holding or transfer) is no longer needed.

The **clutch pack** consists of clutch plates, friction discs, and one thick plate known as the pressure plate. The pressure plate has tabs around the outside diameter to mate with grooves in the clutch drum. It is held in place by a large snap ring. An actuated piston forces the engaging clutch pack against the fixed pressure plate. Because the pres-

sure plate cannot move or deflect, it provides reaction to the engaging clutch pack.

The number of clutch plates and friction discs contained in the pack depends on the transmission size and application. The clutch plates are made from carbon steel and have several tabs in their outside diameter. These tabs are located in grooves machined into the inside diameter of the clutch drum housing. Clutch plates must be perfectly flat, and although the surface of the plate might appear smooth, it is machined to produce a friction surface to transmit torque.

The friction discs are sandwiched between the clutch plates and pressure plate. Friction discs are designed with a steel core plate center with friction material bonded to either side. Cellular paper fibers, graphites, and ceramics are used as friction facings.

Cooling. During clutch engagement, friction takes place and develops heat between the clutch plates and friction discs. The transmission fluid absorbed by the paper-based friction material transfers heat from the discs to the plates. The plates then transfer the heat to the drum housing where it can be cooled by the surrounding transmission fluid. Operating temperatures can reach 1,100°F, so clutch disc cooling is an important design consideration in transmission manufacturing.

To assist cooling and provide other performance advantages, friction disc surfaces are normally grooved. Grooving is generally thought to provide the following advantages:

- Grooving allows the disc to store fluid to lubricate, cool, and quiet clutch engagement.
- When the clutch is engaging, fluid between the disc and plates should be squeezed out. Fluid remaining between these parts lowers the coefficient of friction between them. Grooves in the friction disc surface provide channels through which the fluid can escape and drain away quickly.
- When a clutch is disengaging, any fluid on the disc and plate surfaces can create a suction effect, making the two surfaces more difficult to separate. Grooved surfaces help eliminate this problem and allow for clean disengagement at high speeds.

13.3 | COMPOUND PLANETARY GEAR SETS

Automatic transmissions for both passenger car and heavy-duty truck applications use several simple planetary gear sets tied together to generate the required gear ratios and direction. The number of planetary gear sets that can be combined is limited by practical weight, space, and cost considerations. The four-speed heavy-duty transmissions discussed in this chapter have three simple gear sets front, center, and rear. Five-speed transmissions add a low planetary gear set to this basic configuration for a total of four gear sets.

Any one of the three members of the simple gear set can serve as the input or drive member. And any of the three members could be held stationary. In a working transmission, this is not practical. There is

only one input shaft from the engine to supply drive power. This input torque is directed to one of the planetary sun gears through the use of clutches.

Similarly, having a clutch (brake) to hold each member of each planetary gear set is not practical. A four-speed, three gear set transmission uses only five clutches, excluding the TC lockup clutch. A five-speed, four gear set transmission uses six clutches. They function as follows:

- Forward clutch locks converter turbine to main input drive shaft.
- First clutch anchors rear planetary ring gear.
- Second clutch anchors front planetary carrier.
- Third clutch anchors sun gear shaft.
- Fourth clutch locks main input drive shaft to center sun gear shaft.
- Low clutch (five speeds only) anchors low planetary carrier.

The three planetaries are connected by a sun gear shaft assembly, main shaft assembly, and a planetary connecting drum. This interconnection of planetary input, reaction, and output components produces the forward and reverse speeds and torque ranges desired.

Because the simple planetary gear sets are interconnected by common shafts and drums, the output of one planetary gear set can be used as the input of another planetary gear set. Therefore one gear set can take engine input torque, modify speed, torque, and direction, and pass this onto a second gear set. The second gear set then further modifies speed, torque, and direction before directing the power flow to the output shaft and vehicle drivetrain. With two simple planetaries working together, torque input is compounded. The gear sets used to produce forward and reverse ranges in typical four-speed and five-speed transmissions are summarized in **Table 13-2. Table 13-3** summarizes the clutches applied to produce the required power flow paths.

GEAR RANGE POWER FLOWS

In all automatic transmissions, input torque is directed from the torque converter through the transmission's planetary gearing and to the output shaft. Torque converter operation was covered in Chapter 12. As shown in the power flow illustrations in this chapter, Allison converts their four-speed units into five-speed units by simply adding an extra "low" planetary gear set.

In typical five-speed, one reverse models, power flows as follows:

- **Deep Ratio (DR) Models.** First gear power flow is through the low planetary gear set to the out-

TABLE 13–2: PLANETARY GEAR SET COMBINATIONS

Gear	Four-Speed Transmissions	Five-Speed Transmissions
First	Rear	Low and rear
Second	Front and center	Rear
Third	Center	Front and center
Fourth	Front, center, and rear	Center
Fifth	—	Front, center, and rear
Reverse	Center and rear	Center and rear

Note: In CR (close ratio) five-speed transmissions, all power flows must pass through the low planetary gear set before reaching the output shaft.

TABLE 13–3: CLUTCH APPLICATION IN VARIOUS GEAR RANGES

Gear	Four-Speed Transmissions	Five-Speed Transmissions
Neutral	First	First
First	First and forward	Low and forward
Second	Second and forward	First and forward
Third	Third and forward	Second and forward
Fourth	Fourth and forward	Third and forward
Fifth	—	Fourth and forward
Reverse	First and fourth	First and fourth

put shaft. In all other DR forward gears, power flow is through the rear planetary gear set to the output shaft.

- **Close Ratio (CR) Models.** In all forward gear ranges, power must flow through the low planetary gear set to get to the output shaft.

Oil Pump

An oil pump is used to circulate transmission fluid within the transmission for both lubrication and the hydraulic application of clutches. This pump is driven from the torque converter and is located between the torque converter and the transmission gearing and clutches (**Figure 13–13**). When the torque converter rotates, its rear hub drives the pump drive gear, which is in mesh with the driven gear. As the gears rotate, they draw oil into the pump and move the oil into the hydraulic system.

Because the pump is located ahead of the transmission gearing and clutches, it will not be driven when the vehicle is towed or pushed. Therefore, any time the vehicle must be towed or pushed more than a short distance (one-half mile maximum) the driveline must be disconnected or the driving wheels raised. Failure to do this will result in premature wear and damage to internal components.

FOUR-SPEED TRANSMISSION POWER FLOWS

The following power flow paths are found in a typical four-speed transmission. Four-speed transmissions use front, center, and rear gear sets to generate four forward gears and one reverse gear. Five clutches are used. A sun gear shaft assembly and connecting drum provide connecting links between the planetaries.

Neutral

Figure 13–14 shows a cross section of a four-speed transmission. The arrows indicate the transmission's neutral gear power flow. When the engine starts, the torque converter rotates, driving the oil pump in the transmission. With the range selector (gear shifter) in neutral, oil pressure is sent to the T/C, all points needing lubrication, and the first clutch at the back of the transmission. When the first clutch is applied, the rear planetary ring gear is locked against rotation. However, the forward clutch is not engaged. Because two clutches must be applied to produce output shaft rotation and the forward clutch is not engaged, torque is not transmitted beyond the fourth clutch hub.

 Shop Talk

Transmissions are fitted with a power take-off (PTO) gear that can be used to drive auxiliary equipment using engine power. The PTO gear is driven from the torque converter turbine. It can be engaged in neutral; its speed is regulated with the accelerator pedal.

First Gear

Figure 13–15 illustrates first gear torque path. When the transmission is in neutral, only the first clutch is applied. But when any forward gear is selected, oil pressure is directed to the forward clutch. When applied, the forward clutch locks the turbine shaft of the torque converter to the transmission main input shaft. The main or input shaft is now turning at turbine speed.

The first clutch is also applied and anchors the rear planetary ring gear against rotation. The rear sun gear is splined to the main shaft and rotates with it. In turn the rear sun gear rotates the rear planetary pinions. With the rear planetary ring gear locked, the planetary pinions and their carrier are forced to walk around internally within the held ring gear. Since the rear carrier is splined to the output shaft, the output shaft rotates at a maximum speed reduction.

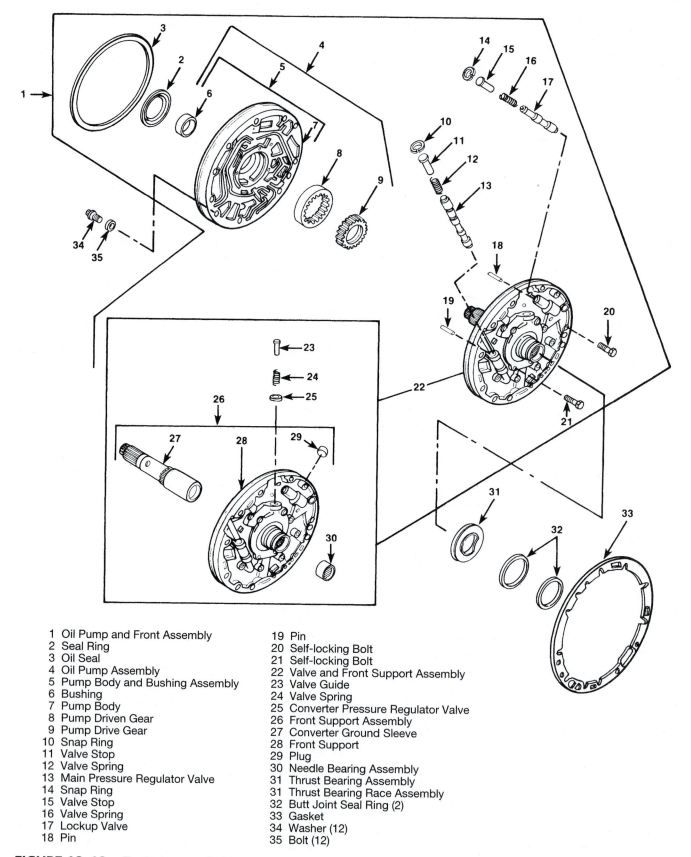

1 Oil Pump and Front Assembly
2 Seal Ring
3 Oil Seal
4 Oil Pump Assembly
5 Pump Body and Bushing Assembly
6 Bushing
7 Pump Body
8 Pump Driven Gear
9 Pump Drive Gear
10 Snap Ring
11 Valve Stop
12 Valve Spring
13 Main Pressure Regulator Valve
14 Snap Ring
15 Valve Stop
16 Valve Spring
17 Lockup Valve
18 Pin

19 Pin
20 Self-locking Bolt
21 Self-locking Bolt
22 Valve and Front Support Assembly
23 Valve Guide
24 Valve Spring
25 Converter Pressure Regulator Valve
26 Front Support Assembly
27 Converter Ground Sleeve
28 Front Support
29 Plug
30 Needle Bearing Assembly
31 Thrust Bearing Assembly
31 Thrust Bearing Race Assembly
32 Butt Joint Seal Ring (2)
33 Gasket
34 Washer (12)
35 Bolt (12)

FIGURE 13–13 Typical transmission oil pump components. (Courtesy of Allison Transmission Division of GM)

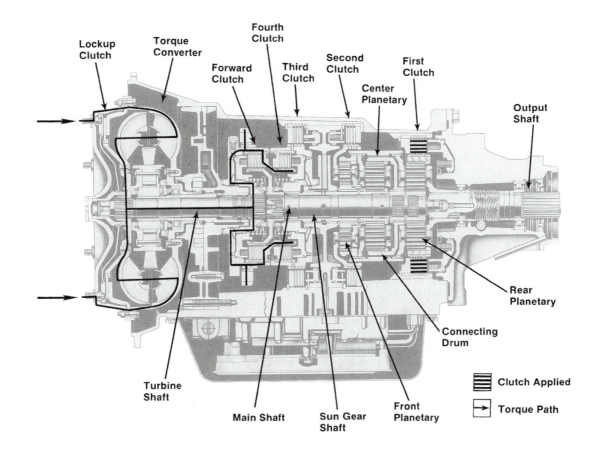

FIGURE 13–14   Typical four-speed neutral torque path power flow. (Courtesy of Allison Transmission Division of GM)

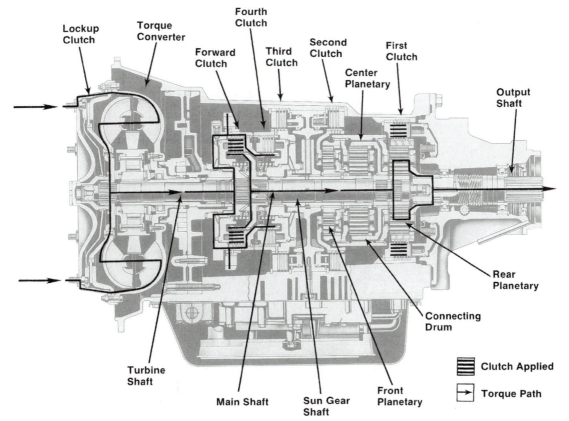

FIGURE 13–15   Four-speed first gear torque path power flow. (Courtesy of Allison Transmission Division of GM)

The clockwise rotating torque converter turns the turbine shaft and main shaft in a clockwise direction. The rear sun gear also turns clockwise. With the sun gear as the input, the ring gear held, and the carrier the output, the output rotational direction is the same as the input, and reduction is at a maximum. (See **Table 13–1**.)

Second Gear

Power flow through the second gear is a compound gear reduction since it runs through two planetary gear sets before reaching the output shaft (**Figure 13–16**).

The forward clutch and the second clutch are both applied. The forward clutch locks the turbine and main shafts together so they rotate clockwise. The first clutch is released in second gear, so the rear planetary assembly is now freewheeling and has no effect on second gear operation.

Torque from the main input shaft is transmitted to the rear sun gear and the center ring gear immediately in front of the sun gear. This center ring gear is splined to the main shaft and becomes the driving member, rotating clockwise. It drives the center planetary pinions in a clockwise direction. In turn the pinions are in mesh with the center sun gear, turning it in a clockwise direction.

The center sun gear is splined to the sun gear shaft. A forward sun gear is splined to the sun gear shaft immediately in front of the center sun gear. This forward or front sun gear drives the front planetary pinions whose carrier is being held by the applied second clutch. This causes the pinions to rotate, driving the front ring gear. Remember, the input force from the center sun gear and sun gear shaft is in a counterclockwise direction. In the front planetary, the sun gear is the input, the carrier is held, and the ring gear is the output. This corresponds to combination 5 in **Table 13–1**. Notice that this combination produces a reverse in the output direction. This means the front ring gear is now turning in a clockwise direction. The front ring gear is splined to the planetary connecting drum, which connects through the rear carrier to the output shaft. The end result is the output shaft rotating clockwise at a medium reduction.

Third Gear

Figure 13–17 shows third gear torque path. In third gear, the forward and third clutches are those applied. The third clutch anchors the sun gear shaft against rotation. This prevents the integral center sun gear from rotating. The forward clutch again locks the turbine and main shafts together so they can rotate clockwise as a unit. The rear sun gear is splined to

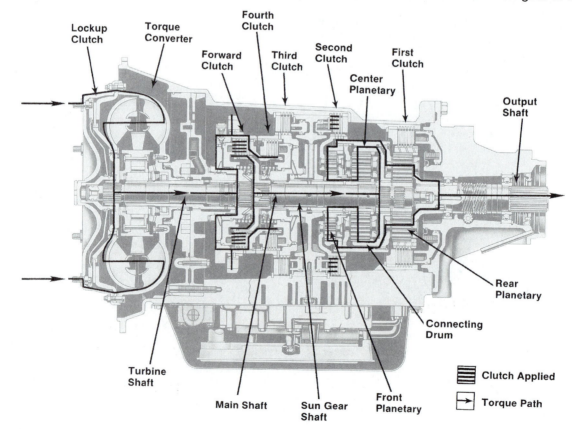

FIGURE 13–16 *Four-speed second gear torque power flow. (Courtesy of Allison Transmission Division of GM)*

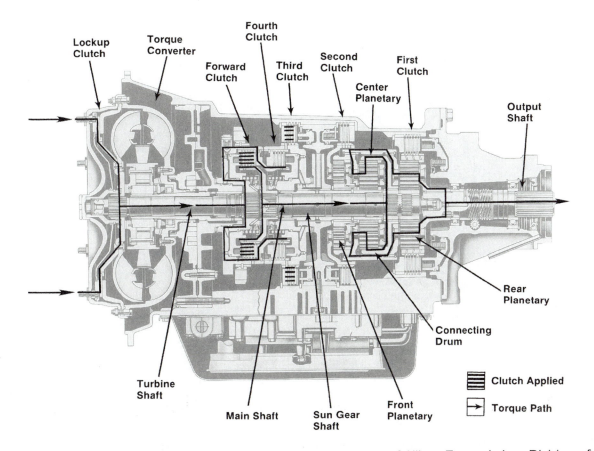

FIGURE 13-17 *Four-speed third gear torque power flow. (Courtesy of Allison Transmission Division of GM)*

both the main shaft and the center ring gear and rotates at input speed. With the center sun gear held stationary and the center ring gear rotating, the ring gear drives the center planetary pinions. This action rotates the center carrier at a minimum speed reduction. Both the center and rear carriers are splined to the planetary connecting drum and rotate as a unit. Since the output shaft is splined to the rear carrier, it rotates at the same speed as the center planetary carrier.

In this gear, the center ring gear is the input, the sun gear is held, and the carrier is the output. This is combination 2 in **Table 13-1**.

Fourth Gear

In this range (**Figure 13-18**), the forward and fourth clutches are applied. With these clutches applied, the transmission main shaft and center sun gear shaft are locked together and rotate as a unit at input speed. When this happens, both the center and rear sun gears will rotate at the same speed, also rotating their respective carriers, which are both splined to the planetary connecting drum and output shaft.

With any two members of a planetary gear set held together, the result is direct 1:1 drive. In this

case, although the front sun gear is being driven, no front planetary gear set member is being held. Therefore, the front planetary freewheels, with the power transfer being split to the center and rear planetary sets.

Reverse Gear

Reverse gear is the only gear in which the forward clutch is not applied (**Figure 13-19**). In reverse, the fourth clutch is applied and this rotates the center sun gear shaft with the front sun gear splined to it, at input speed.

The first clutch is applied also. It anchors the rear ring gear against rotation. The center sun gear rotates the center carrier pinions, which, in turn, rotate the center ring gear in the opposite direction. The center carrier is splined to the planetary connecting drum that connects to the rear carrier. The reverse rotation of the center ring gear rotates the rear sun gear. This causes the rear planetary pinions to drive the rear carrier in a reverse direction around the internal gears of the locked rear ring gear. Thus the output shaft that is splined to the rear carrier rotates in a reverse direction at a significant speed reduction.

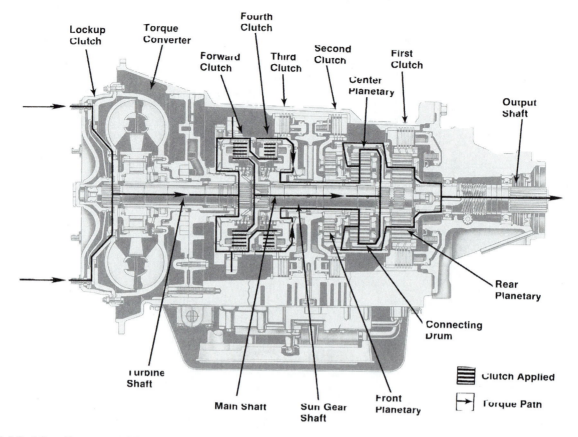

FIGURE 13–18   Four-speed fourth gear torque power flow. (Courtesy of Allison Transmission Division of GM)

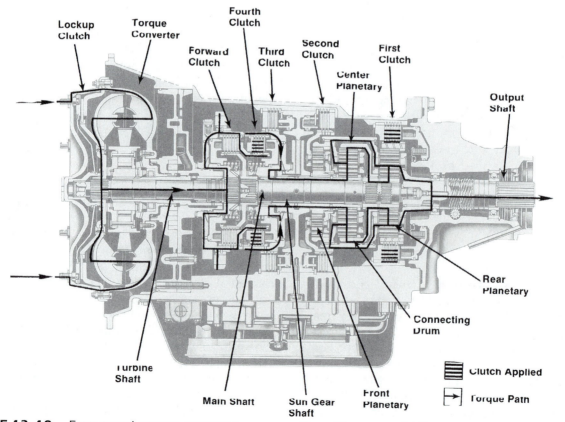

FIGURE 13–19   Four-speed reverse gear torque power flow. (Courtesy of Allison Transmission Division of GM)

FIVE-SPEED TRANSMISSION POWER FLOW

It is important to remember that the only difference between five-speed transmissions and four-speed units is that an additional low clutch and low planetary gear set are added to the back of the transmission. In deep ratio models, only first gear power flow is through the low planetary gear set. On close ratio models, the output shaft is splined to the low planetary carrier, so power must flow through the low gear set in all forward gears.

DR Neutral

Figure 13–20 shows a deep ratio transmission. Note the position of the low clutch and low planetary gear set at the rear of the transmission. All other components shown in **Figures 13–20** to **13–25** are identical to the four-speed transmissions. The neutral torque path is indicated in **Figure 13–20**. It is identical to the neutral torque path of four-speed transmissions discussed earlier. The forward clutch is not engaged. The first clutch is engaged, but power transfer is not possible. As before, the power take-off gear can be engaged and operated with accelerator action.

DR Low (First) Gear

In low or first gear, both the forward and low clutches are applied (**Figure 13–21**). The forward clutch locks the turbine shaft and main shaft together so they rotate as a unit. The low clutch holds the low planetary carrier stationary. The rear sun gear splined to the main shaft rotates the rear pinions. The rear carrier resists the pinion rotation because the rear carrier is splined to the output shaft. In other words, the weight of the vehicle load keeps the rear carrier from turning.

With the rear carrier unable to overcome the load placed on it, the rear planetary ring gear is forced to rotate. The low planetary sun gear is splined to the rear ring gear and also rotates with it. The low sun gear rotates the low pinions. Since the low carrier is anchored against rotation by the applied low clutch, the pinions rotate the low planetary ring gear. The low ring gear is splined to the output shaft and rotates it at an extreme reduction.

DR Second through Fifth Gears

Five-speed deep ratio transmission power flows for second, third, fourth, and fifth gears are identical to four-speed transmission first, second, third, and fourth gears, respectively. The low planetary is not used. (See **Figures 13–15** through **13–18**.)

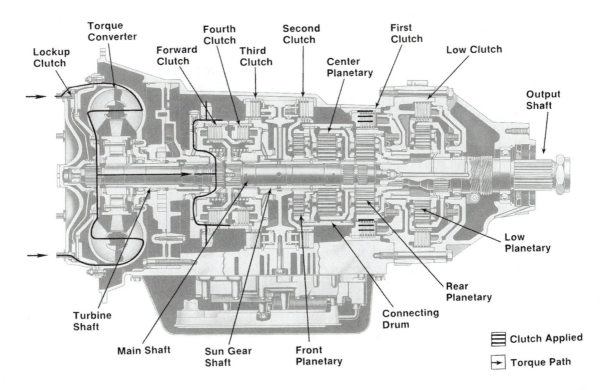

FIGURE 13–20 *Five-speed neutral torque power flow in a deep ratio, 5-speed transmission. (Courtesy of Allison Transmission Division of GM)*

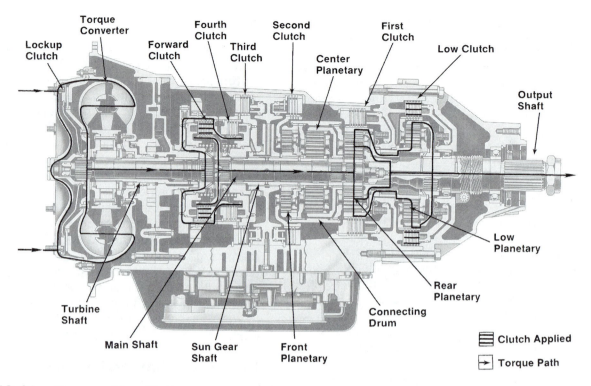

FIGURE 13-21   Five-speed low (first) gear torque path power flow in a deep ratio, 5-speed transmission. (Courtesy of Allison Transmission Division of GM)

DR Reverse Gear

Five-speed, deep ratio transmission reverse gear operation is identical to four-speed reverse. (See **Figure 13-19**.) The low planetary supply freewheels and does not affect output.

CR Neutral

Figure 13-22 illustrates the neutral position in a close ratio transmission. Instead of the first clutch being applied as in the four-speed and five-speed DR neutral flows, the low clutch is applied in the CR transmissions.

CR Low (First) Gear

The forward and low clutches are applied (**Figure 13-23**). Torque flows from the turbine shaft, forward clutch and housing, main shaft, low sun gear, low planetary pinions, and low carrier, that is forced to walk around the internal gears of the locked low ring gear held by the low clutch. The carrier therefore drives the output shaft in a clockwise direction.

Low clutch application also locks the rear carrier stationary since the rear carrier is splined to the low ring gear. The rear sun gear rotates since it is connected to the main shaft. However, no drive exists out to the output from this planetary set. Likewise, all

other rotating members in the front and center planetary gear sets add nothing to the drive.

CR Second through Fifth Gear

The power flow path for CR second, third, fourth, and fifth gears is the same as a four-speed transmission's first, second, third, and fourth gears, respectively, except that the flow must pass through the low planetary gear set carrier to reach the output shaft.

CR Reverse Gear

The CR transmission has a very low reverse gear, much more so than that used on DR models. In CR reverse (**Figure 13-24**), both the fourth and low clutches are applied. Torque travels from the turbine shaft, forward clutch housing (clutch released), to the fourth clutch, which locks onto the rear of the forward clutch housing, to the sun gear shaft, which becomes the driving member turning clockwise, to the center sun gear, center pinions, and center ring gear, which is now turning counterclockwise. The rear sun gear is splined to the center ring gear. The rear planetary gears simply rotate around their pinions because the rear carrier is held by the applied low clutch.

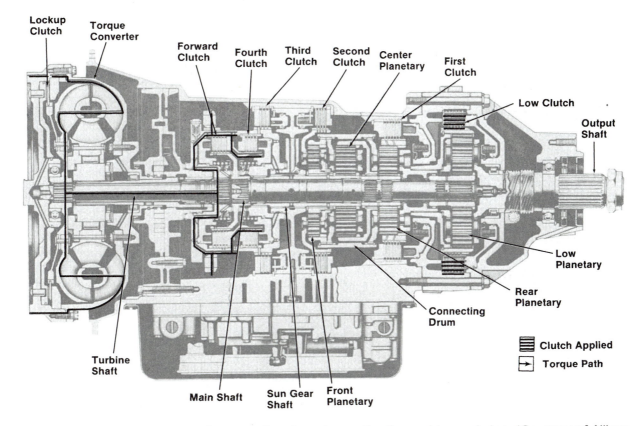

FIGURE 13–22   Neutral torque path power flow in a close ratio, 5-speed transmission. (Courtesy of Allison Transmission Division of GM)

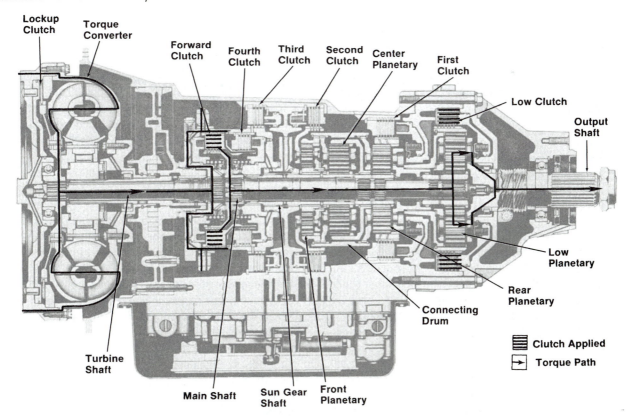

FIGURE 13–23   Five-speed low (first) gear torque path power flow in a close ratio, 5-speed transmission. (Courtesy of Allison Transmission Division of GM)

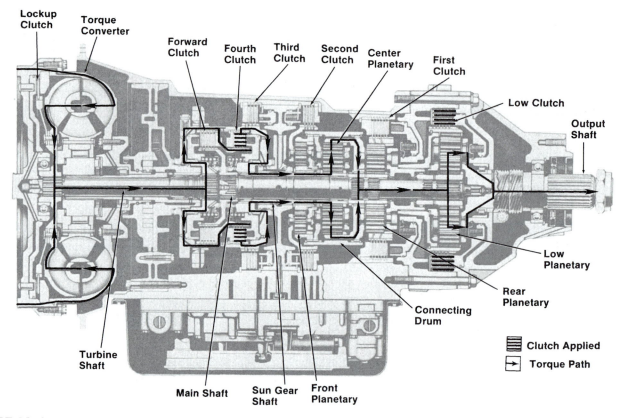

FIGURE 13–24 Five-speed reverse gear torque path power flow in a 5-speed, close ratio transmission. (Courtesy of Allison Transmission Division of GM)

The drive to the output shaft is through the center ring gear to the main shaft and low sun gear to the low planetary pinions. The low carrier is forced to walk around within the held low ring gear in a counterclockwise direction. The carrier is splined to the output shaft.

13.4 TRANSMISSION HYDRAULIC SYSTEMS

The hydraulic systems used to generate, direct, and control the pressure and flow of transmission oil within most transmissions are similar in design and operation. As with clutches and planetary gearing, modifications are made to basic systems to incorporate and control a fifth gear or special deep ratio or close ratio needs.

Transmission oil is the power transmitting medium in the torque converter. Its velocity drives the converter turbine and its flow cools and lubricates the transmission. Its pressure operates the various control valves and applies the clutches.

Figures 13–25 and **13–26** show flow-coded schematic diagrams of four-speed and five-speed automatic transmissions. The diagrams show the

systems as they would function in neutral with the engine idling.

OIL FILTER AND PUMP CIRCUIT

Oil is drawn from the sump (transmission oil pan) through a filter by the input driven oil pump. The oil discharge by the pump flows into the bore of the main pressure regulator valve.

MAIN PRESSURE CIRCUIT

Main pressure is regulated by the main-pressure regulator valve. Oil from the pump flows into the valve bore, through an internal passage in the valve, to the upper end of the valve. Pressure at the upper end of the valve forces the valve downward against the spring until oil flows into the converter-in circuit. When flow from the pump exceeds the circuit requirement, the converter regulator valve opens and allows the excess to exhaust to sump.

Although main pressure is controlled primarily by the spring at the bottom of the valve, it is also affected by forward regulator pressure and lockup pressure. These pressures are not present at the regulator valve during reverse operation and main pressure is regulated at a higher value.

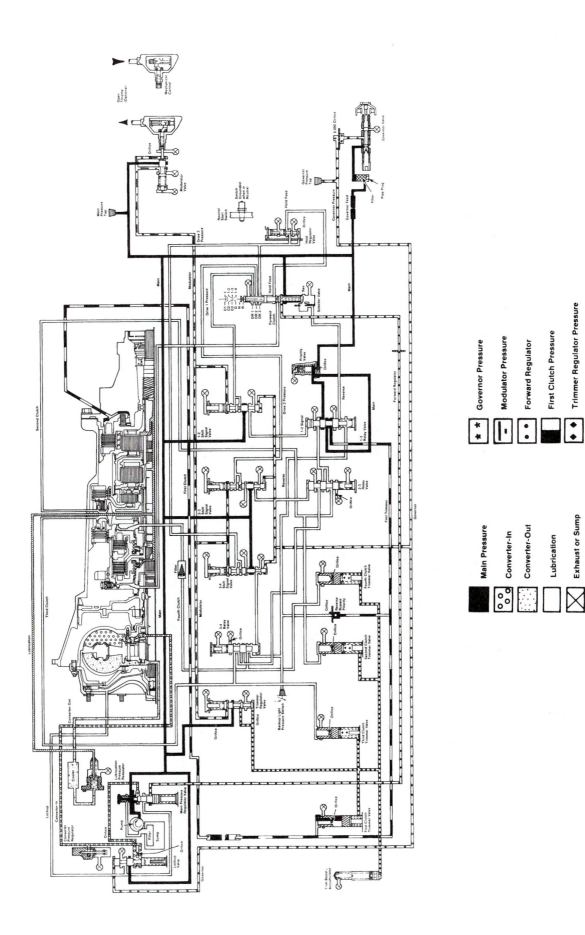

FIGURE 13-25   Hydraulic control circuit of four-speed transmission. (Courtesy of Allison Transmission Division of GM)

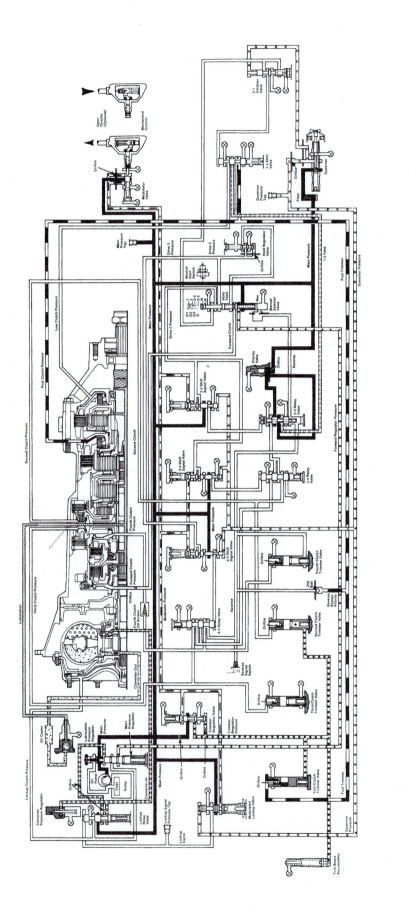

FIGURE 13-26 Hydraulic control circuit of five-speed transmission. (Courtesy of Allison Transmission Division of GM)

CONVERTER/COOLER/ LUBRICATION CIRCUIT

This circuit originates at the main pressure regulator valve. Converter-in oil flows to the torque converter. Oil must flow through the converter continuously to keep it filled and to carry off the heat generated by the converter. The converter pressure regulator valve regulates converter-in pressure by exhausting excessive oil to sump.

Converter-out oil, leaving the torque converter, flows to an external cooler (supplied by the vehicle or engine manufacturer). A flow of air or water over or through the cooler heat exchanger removes the heat from the transmission oil.

Lubrication oil is directed through the transmission to components requiring continuous lubrication and cooling. Oil in excess of that required by the lubrication circuit drains to the sump through the lubrication pressure regulator.

SELECTOR VALVE/ FORWARD REGULATOR CIRCUIT (FOUR SPEED)

The selector valve is manually shifted to select the operating range desired. It can be shifted to any one of six positions. These are: neutral (N), reverse (R), drive (D), drive 3 (D3), drive 2 (D2), and drive 1 (D1). At each of these positions the selector valve establishes the hydraulic circuit for operation in the condition indicated.

D, D3, D2, and D1 are forward ranges. The lowest gear attainable is first gear. In each of these positions, the transmission will start in first gear (except on second gear start). Shifting within any range is automatic, depending upon vehicle speed and throttle position.

The forward regulator circuit originates at the manual selector valve and terminates at the main-pressure regulator valve. In every forward range the circuit is pressurized. This pressure aids in reducing main pressure to a lower value. In reverse, forward regulator pressure is not present, allowing for higher main pressure to control the higher torque produced in reverse operation.

SELECTOR VALVE/FORWARD REGULATOR CIRCUIT (FIVE SPEED)

The selector valve is manually shifted to select the operating range desired. It can be shifted to any one of seven positions. These are: neutral (N), reverse (R), drive (D), drive 4 (D4), drive 3 (D3), drive 2 (D2), and drive 1 (D1). At each of these positions the selector valve establishes the hydraulic circuit for operation in the range selected.

D, D4, D3, D2, and D1 are forward ranges. The lowest gear is first gear, which is available in D1 range only. D2, D3, D4, or D position will start the transmission in second gear and automatically shift to range, depending on vehicle speed and throttle position. Some models do not upshift in D2 position under normal operating conditions.

Again, the forward regulator circuit originates at the manual selector valve and terminates at the main pressure regulator valve. The circuit is pressurized in every forward range. This pressure aids in reducing main pressure to a lower value.

In reverse, the forward regulator pressure is not present. This allows for higher main pressure required for the higher torque produced in reverse range.

GOVERNOR CIRCUIT

Main pressure is directed to the governor valve. The speed of the transmission output controls the position of the governor valve. The position of the governor valve determines the amount of pressure in the governor circuit. When the transmission output is not rotating, governor pressure is approximately 2 psi. Governor pressure increases with the speed of output rotation.

In a typical four-speed, governor pressure is directed to the 1–2, 2–3, and 3–4 shift signal valves, and the modulated lockup valve. In earlier models that do not include the modulated lockup valve, governor pressure is directed to the top of the lockup valve.

In the five-speed, governor pressure is directed to the 2–3, 3–4, and 4–5 shift signal valves, 2–1 inhibitor valve, and the modulated lockup valve. In models that do not include the modulated lockup valve, governor pressure is directed to the top of the lockup valve.

At the 2–1 inhibitor valve, governor pressure acts upon the top of the valve. When governor pressure is sufficient to push the valve downward, road speed is too high to permit a downshift to first gear. So a 2–1 shift cannot occur until governor pressure (and road speed) is reduced. DR1 pressure is blocked from reaching the top of the 1–2 shift valve until road speed (and governor pressure) is decreased to a value that will allow the inhibitor valve to move upward.

In transmissions designed to start in second gear, the presence of governor pressure at the bottom of the 1–2 shift valve and forward regulator pressure on the 1–2 modulator valve, plus the assistance of spring force under the 1–2 shift valve, causes the shift valve to move upward against its stop. This allows main pressure to pass through the 1–2 shift signal valve to the 1–2 relay valve. The 1–2 relay valve is forced downward in its bore, allowing apply pressure to be directed to the second clutch, permitting second gear start.

MODULATOR PRESSURE CIRCUIT

The modulator valve assists the governor pressure in moving any of the shift signal valves during an upshift. It can also delay a downshift. The modulator valve controls a regulated pressure derived from main pressure. The modulator valve receives main pressure and regulates it as a direct result of the accelerator position.

The modulator actuator varies modulator pressure as the accelerator pedal moves. It can be controlled in a number of ways. On gasoline engines, vacuum pressure is used. On diesel engines, air pressure or a mechanical linkage can be used to operate the modulator. The mechanical linkage is most common. Modulator pressure is highest when the engine is running at idle and decreases in proportion to accelerator pedal travel. In a mechanical linkage system, the modulator is controlled by a push-or-pull cable connected to the throttle.

When there is no throttle pedal pressure, the valve is moved to the right by a spring at the left end of the valve. As the driver depresses the throttle pedal, a wedge-shaped actuator forces the modulator valve to the left, decreasing modulator pressure. When spring force at the left of the valve is in balance with the spring force and modulator pressure at the right of the valve, modulator pressure is regulated.

When the accelerator is depressed, the movement of the actuator cam forces the modulator valve to the left. This leftward movement reduces modulator pressure. When the pedal travel is reduced, the downward movement of the actuator cam allows the spring at the left to push the valve to another regulating position. Because the throttle pedal angle varies with load and engine speed, modulator pressure also varies. This varying pressure is directed to the 1–2, 2–3, and 3–4 shift signal valves (four speed); 2–3, 3–4, and 4–5 shift signal valves (five speed); trimmer regulator valve; and modulated lockup valve. Earlier models do not include the modulated lockup valve.

At the shift signal valves, modulator pressure assists governor pressure to overcome shift valve spring force. Each one of the shift valves and springs is so calibrated that the valves will shift in the proper sequence at the proper time. A decrease in modulator pressure will cause a downshift if governor pressure is not enough to oppose the shift valve spring force. The trimmer boost and trimmer regulator valve are controlled by modulator pressure.

At the modulated lockup valve, modulator pressure assists governor pressure in moving (or holding) the valve downward. The variation in modulator pressure will vary the timing of lockup clutch engagement and release. The primary purpose of lockup modulation is to prolong the engagement of the lockup clutch (at closed throttle) in lower ranges to provide greater engine braking.

All modulated lockup valves are adjustable, but not all transmissions have modulated lockup. Models with adjustable lockup can be adjusted by rotating the adjusting ring at the bottom of the valve until the desired shift point is attained (see Chapter 14).

Trimmer Regulator Valves

Newer transmissions are equipped with trimmer regulator valves. The trimmer regulator valve reduces main pressure to a regulated pressure. The regulated pressure is raised or lowered by changes in modulator pressure.

Trimmer regulator pressure is directed to the lower sides of the first and second trimmer plugs on some models, or to the lower sides of the first, second, third, and fourth trimmer plugs on models equipped with smooth shift. This varies the clutch apply pressure pattern of the trimmer valves. A higher modulator pressure at zero accelerator pedal travel will reduce trimmer regulator pressure. This results in a lower initial clutch apply pressure. Conversely, a lower modulator pressure (open throttle) results in higher regulator pressure and a higher initial clutch apply pressure.

TRIMMER VALVES

The purpose of the trimmer valves is to avoid shift shock. These valves reduce pressure to the clutch apply circuit during initial application, then gradually return the pressure to operating maximum. This applies the clutch gently, and harsh shifts are prevented.

Although each trimmer valve is calibrated for the clutch it serves, all four trimmers function in the same manner. Each trimmer includes (top to bottom) an orificed trimmer valve, trimmer valve plug, one or two trimmer springs, and a stop pin.

When the clutch (except forward) is applied, apply pressure is sent to the top end of the valve. Initially, the plug and valve are forced downward against the spring until oil spills to exhaust. The escape of oil, as long as it continues, reduces clutch apply pressure. Oil flows through an orifice in the trimmer valve to the cavity between the trimmer valve and the trimmer valve plug. Pressure in this cavity forces the plug down to the stop. The pressure below the trimmer valve, because it is acting upon a greater sectional area than at the upper end, pushes the trimmer valve to the upper end of the valve bore. This stops the spilling of oil to the exhaust. When escape of oil stops, clutch pressure is at the maximum. The plug remains downward until the clutch is released.

Upon the release of the clutch, the spring pushes the trimmer components to the top of the valve bore. In this position, the trimmer is reset and ready to repeat the action at the next application of that clutch.

LOCKUP CIRCUIT

Lockup clutch engagement and release are controlled by the modulated lockup valve and the lockup relay valve. The modulated lockup shift valve is actuated by governor pressure or governor plug modulator pressure. At full throttle, governor pressure will move the valve downward. At part throttle, governor and modulator pressures will move the valve downward. The purpose of the modulated lockup valve is to prolong lockup clutch engagement while vehicle speed decreases (closed throttle). This feature provides engine braking action at speeds lower than the normal lockup disengagement point.

In its downward position, the modulated lockup valve directs main pressure to the top of the lockup valve. Main pressure pushes the lockup valve downward. In its downward position, the valve directs main pressure to the lockup clutch. This engages the clutch.

The position of the lockup valve affects the volume of oil flowing to the torque converter. When the valve is upward (lockup clutch released), there is maximum flow to the converter. When the valve is downward (clutch engaged), converter-in flow is restricted by an orifice, and flow is reduced.

When the lockup clutch is engaged, a signal pressure is sent to the main-pressure regulator valve. This signal pressure reduces main pressure to a lower pressure schedule.

On most models with adjustable lockup, modulator assistance at the modulated lockup valve is not present. By setting the adjusting ring to the shift point specifications and with the assistance of governor pressure, lockup engagement is initiated. Models that do not have modulated lockup, do not include the modulated lockup valve. Governor pressure is directed to the top of the lockup valve.

Priority Valve

The **priority valve** ensures that the control system upstream from the valve will have sufficient pressure during shifts to perform its automatic functions. In many models, first clutch pressure can bypass the priority valve when the transmission is in reverse.

Without the priority valve, the filling of the clutch might require a greater volume of oil (momentarily) than the pump could supply and still maintain pressure for the necessary control functions.

Inhibitor Valve

In five-speed transmissions, a 2–1 inhibitor valve prevents a downshift from second gear to first gear when road speed is too high. It will also protect the engine from overspeeding during downgrade operation in first gear by making a 1–2 upshift if road speed exceeds that which is safe for first gear operation.

When road speed is great enough to produce a governor pressure that forces the 2–1 inhibitor valve downward, D1 pressure to the 1–2 shift valve is blocked, thus, if a manual shift to drive 1 (first gear) is made, D1 pressure cannot reach the 1–2 shift valve. The transmission will remain in second gear until road speed (and governor pressure) decreases to a value that permits the 2–1 inhibitor valve to move upward. In its upward position, D1 pressure can reach the 1–2 shift valve and move it downward to the first gear position.

While operating in first gear, the 2–1 inhibitor valve will move downward if road speed (and governor pressure) exceeds that which is safe for first-gear operation. This will block D1 pressure and exhaust the cavity above the 1–2 shift valve. The shift valve will move upward to give second gear operation.

CLUTCH CIRCUITS

There are five separate clutches in four-speed transmissions and six clutches in five-speed models. As covered earlier, various clutch combinations are used in various drive ranges (see **Table 13–3**).

Each clutch has its own circuit. The first, second, third, and fourth clutches are each connected to a relay valve and to a trimmer valve. The forward clutch is connected directly to the selector valve and does not connect to a trimmer valve. Since the vehicle is not moving, the application of a trimmer valve is not required. The low clutch is not trimmed because there is no automatic upshifting from or downshifting to this clutch.

The low clutch circuit (five speeds) connects the clutch to the 1–2 shift valve. The 1–2 shift valve also receives 1–2 (low and first) feed from the 2–3 relay valve. In neutral, the 1–2 shift valve is held upward by spring pressure. It cannot move downward unless drive 1 line is charged. In the upward position, 1–2 (low and first) feed pressure is sent to the first clutch. The first clutch circuit connects the clutch to the first trimmer valve and to either the 1–2 relay valve (four speeds) or to the 1–2 shift valve (five speeds). In neutral the 1–2 shift valve is held upward by spring pressure. The 1–2 shift valve cannot move downward unless the 2–1 signal line is charged. This will not occur in neutral (vehicle standing) because there is no governor pressure to shift the 2–1 inhibitor valve. Only the first clutch is applied, so the transmission output cannot rotate. A bypass and check ball are provided between the reverse and first trimmer passages to ensure a rapid and more positive shift from first gear to reverse.

The first clutch is also applied in first gear operation (four-speeds), second gear operation (five-speeds), and reverse gear operation. Shifting the selector from neutral to D, D3, D2, or D1 charges the forward clutch circuit and applies the forward clutch. Shifting from neutral to reverse (R) charges the fourth clutch, while the first clutch remains charged. In reverse, fourth clutch (reverse signal) pressure is also directed to the bottom of 1–2 relay valve (four speeds) or 2–3 relay valve (five speeds). The pressure at this location prevents either relay valve from moving downward.

When the circuits are charged and the selector valve is at D, four-speed models will automatically shift from first to second, second to third, and third to fourth. Five-speed models will automatically shift from second to third, third to fourth, and fourth to fifth. These shifts occur as a result of governor pressure, and (if throttle is not fully open) modulator pressure. The position of the selector valve determines the highest gear that can be reached automatically while the vehicle is moving. Four-speed models that do not incorporate a second gear start will automatically shift 1–2, 2–3, and 3–4 when the selector valve is in drive 4 position. In drive 3, automatic 1–2, 2–3 shifts occur. In drive 2, an automatic 1–2 shift occurs. Drive 1 does not permit an upshift to occur unless an overspeed of the transmission output occurs.

The positions of the selector valve in a second gear start transmission is the same; only the transmission performance is altered. The transmission always starts in second gear when any forward range other than drive 1 is selected. When drive 4 is selected, the transmission starts in second gear and automatically shifts 2–3 and 3–4. In drive 3, an automatic 2–3 shift occurs. Drive 2 does not permit an upshift to occur unless an overspeed of the transmission output occurs. Drive 1 is the only position in which first gear is available and upshifting is not permitted unless an overspeed of the transmission output occurs.

In five-speed transmissions drive (D) automatic 2–3, 3–4, and 4–5 shifts can occur. In drive 3 (D3), automatic 2–3 and 3–4 shifts can occur. With some early transmissions an automatic 2–3 shift can occur in drive 2 (D2). In recent models, no shift from 2 can occur unless overspeed occurs. In drive 1 (D1), no shift from 1 can occur unless overspeed occurs.

The various drive ranges limit the highest gear attainable by introducing a pressure that prevents governor pressure from upshifting the signal valves (unless governor pressure is well above that normally attained). This pressure is a regulated, reduced pressure sourced from the main pressure circuit at the hold regulator valve. Main pressure is directed to the hold regulator valve through the hold feed line when the selector lever is in drive 1, drive 2, or drive 3 position. The pressure produced in the hold regulator valve is directed to the 3–4 shift valve (four-speeds) or the 4–5 shift valve (five-speeds) when the selector is in drive 3. The hold pressure is directed to the 2–3 and 3–4 (four-speeds) or the 3–4 and 4–5 (five-speeds) when the selector is in drive 2 position. In some newer five-speed models, hold pressure is directed to the 2–3, 3–4, and 4–5 shift valves. In other four- and five-speeds, pressure can be directed to all shift signal valves when the selector is in drive 1.

Hold regulator pressure at each shift signal valve will push the modulator valve upward and exert downward force on the shift valve. Thus, when hold pressure is present, upshifts are inhibited except in extreme overspeed conditions.

Automatic Upshifts

When the transmission is operating in first gear (four speed) or second gear (five speed), with the selector valve at drive (D), a combination of governor pressure and modulator pressure, or governor pressure alone, will upshift the transmission to second or third gear, respectively. At closed or part throttle, modulator pressure exists and will assist governor pressure. At full throttle, there is no modulator pressure. Thus, upshifts occur at a higher road speed. When the throttle is closed, shifts occur at lower wheel speeds.

Governor pressure is dependent upon the rotational speed of the transmission output. The greater the output speed (vehicle speed), the greater the governor pressure. When governor pressure is sufficient, the first upshift will occur. With a further increase in governor pressure (and vehicle speed), the second upshift will occur. A still further increase will cause a third upshift. Modulator pressure, when present, will assist governing pressure in overcoming shift valve spring force.

In any automatic upshift, the shift signal valve acts first. This directs a shift pressure to the relay valve. The relay valve shifts, exhausting the applied clutch and applying a clutch for a higher gear.

Automatic Downshifts

Automatic downshifts, like upshifts, are controlled by rear governor and modulator pressures. Downshifts occur in sequence as rear governor pressure and/or modulator pressure decrease. Low modulator pressure (full accelerator pedal travel) will hasten the downshift; high modulator pressure (zero accelerator pedal travel) will delay downshifts.

In any automatic downshift, the shift signal valve acts first. This exhausts the shift signal holding the relay valve downward. The relay valve then moves upward, exhausting the applied clutch and applying the clutch for the new lower gear.

Downshift and Reverse Inhibiting

The system is designed, as a result of valve sectional areas and spring forces, to prevent downshifts at too rapid a rate or to prevent a shift to reverse while moving forward. For example, if the vehicle is traveling at a high speed in fourth gear (4) and the selector valve is inadvertently moved to D1, the transmission will not immediately shift to first gear. Instead it will shift 4–3—2–1 as speed decreases (it will remain in fourth gear if speed is not decreased sufficiently to require an automatic downshift).

The progressive downshift occurs because the hold regulator pressure is calibrated along with the valve sectional areas to shift the signal valves downward against governor pressure only when governor pressure decreases to a value corresponding to a safe downshift speed. So if speed is too great, governor pressure is sufficient to hold the shift signal valve upward against drive 3, drive 2, or drive 1 pressure (all of which are the regulated holding pressures originating in the hold regulator valve). As governor pressure decreases, all shift signal valves move downward in sequence.

13.5 HYDRAULIC RETARDERS

Hydraulic retarders are one of several types of auxiliary braking systems used on heavy-duty trucks.

Some retarder systems are designed as an integral component of the automatic transmission. They provide some vehicle braking on severe downgrades and help maximize service brake life by providing braking action at the drive line. The retarder is applied by the driver as needed.

Integral retarders are usually available in two automatic transmission configurations: the **input retarder** and the **output retarder**.

INPUT RETARDERS

The input retarder is located between the torque converter housing and the main housing. It is designed primarily for over-the-road operations. The unit employs a "paddle wheel" type design with a vaned rotor mounted between stator vanes in the retarder housing.

Retardation or brake mode occurs when transmission oil is directed into the retarder housing. The oil causes resistance to the rotation of a vaned rotor. The retarding power is transmitted through the transmission to the vehicle driveline, thus slowing the vehicle's drive wheels. Retarding capacity is enhanced by downshifting.

Variable control is achieved by moving a hand lever or activating a separate foot pedal to regulate oil flow to the retarder. With the controls in the "off" position, the oil is evacuated from the retarder, leaving no drag on the rotor. Heat generated by the absorption of horsepower is dissipated by circulating the oil through a high-capacity transmission oil cooler.

OUTPUT RETARDERS

Most output retarders are mounted on the rear end of the transmission without adding additional length to it. It applies retarding force directly to the drive line. The output retarder uses a two-stage principle. The first stage consists of a rotor/stator hydraulic design. The second stage is a powerful, oil-cooled, friction clutch pack.

Upon activation of the output retarder, quick initial response is provided by a momentary application of the friction clutch pack while the hydraulic section is being charged with oil. The hydraulic section provides best retardation at higher speeds. As the vehicle slows, the second stage friction clutch pack phases in, providing the low-speed retardation power capable of slowing the vehicle to a virtual stop.

Since it is located at the rear of the transmission and transmits retarding force directly to the driveline, the output retarder functions independently of engine speed or transmission range for greater effectiveness. Heat generated by retardation is dissipated through the transmission cooling system.

Activation of the output retarder can be by a hand lever or separate foot pedal for highway downhill speed control. Retarder control can also be incorporated into the service brake pedal for stop-and-go traffic operations. This method provides a smooth transition from retarder to service brakes, depending on braking effort needed. For flexibility in varied driving conditions both systems can be used together.

SUMMARY

- Automatic transmissions upshift and downshift with no direct assistance from the driver.
- Factors such as road speed, throttle position, and governed engine speed control shifting between gears.
- The simple planetary gear set consists of three components: a sun gear, a carrier with planetary pinions mounted to it, and an internally toothed ring gear or annulus.
- Advantages of the planetary gear set are as follows:
  - constantly meshed gears
  - gear forces are divided equally
  - planetary gear sets are very compact
  - extreme versatility
  - additional variations of both speed and direction can be added through the use of compound planetary gears
  - any one of the three members—sun gear, pinion gear carrier, or ring gear—can be used as the driving or input member. Depending on which member is the driver, which is held, and which is driven, either a torque increase or a speed increase is produced.
- For gear set output, there must be some way of holding a planetary gear member stationary or of applying drive power to it. In heavy-duty truck transmissions this is done with multiple-disc clutches, which serve as both braking and power transfer devices.
- Compound planetary combinations are several planetary gear sets coupled to produce the required gear ratios and direction. Four-speed heavy-duty transmissions have three simple planetary gear sets—front, center, and rear. Five-speed transmissions add an additional low planetary gear set for a total of four gear sets.
- In all automatic transmissions, power flows from the torque converter through the transmission's planetary gearing and to the output shaft.
- Transmission oil is drawn from the sump, through a filter by the input driven oil pump.
- Oil discharge by the pump flows into the bore of the main pressure regulator valve.
- The converter/cooler/lubrication circuit originates at the main pressure regulator valve.
- The selector valve/forward regulator circuit is manually shifted to select the operating range desired.
- Main pressure is directed to the governor valve. The speed of the transmission output shaft controls the position of the governor valve, the position of which determines the amount of pressure in the governor circuit.
- The modulator actuator varies modulator pressure as the accelerator pedal moves.
- Lockup clutch engagement and release are controlled by the modulated lockup valve and the lockup relay valve.
- There are five separate clutches in four-speed transmissions and six clutches in five-speed models, each clutch having its own circuit.
- Hydraulic retarders are one of several types of auxiliary braking systems used on heavy-duty trucks and are applied by the driver as needed.

REVIEW QUESTIONS

1. Which is the largest member of a simple planetary gear set?
   a. sun gear
   b. carrier
   c. ring gear
   d. pinions

2. When any two members of a simple planetary gear set are locked together or held stationary,
   a. there is a reverse in rotation direction at the output member
   b. direct 1:1 drive occurs
   c. an overdrive situation results
   d. an underdrive situation results

3. When the carrier of a simple planetary gear set is the held member,
   a. there is a reverse in rotation direction at the output member
   b. direct 1:1 drive occurs
   c. an overdrive situation results
   d. an underdrive situation results

4. If a large external tooth gear is meshed with and driving a small external tooth gear,
   a. output speed increases; output direction is same as input
   b. output speed decreases; output direction is the reverse of input
   c. output speed increases; output direction is the reverse of input
   d. output speed decreases; output direction is the same as input

5. In heavy-duty automatics, multiple-disc clutches are used to
   a. hold components stationary
   b. lock components together to transfer power
   c. open and close hydraulic valves
   d. both a and b

6. In a close ratio transmission, all power flows must pass through the _____ before reaching the output shaft.
   a. front planetary gear set
   b. rear planetary gear set
   c. low planetary gear set
   d. center sun gear shaft

7. The _____ assists governor pressure in moving any shift signal shafts during an upshift.
   a. priority valve
   b. modulator valve
   c. trimmer valve
   d. inhibitor valve

8. The _____ helps avoid shift shocks by reducing pressure to the clutch apply circuits during initial application and gradually increasing this pressure.
   a. priority valve
   b. modulator valve
   c. trimmer valve
   d. inhibitor valve

9. Which type of hydraulic retarder is located between the torque converter and the transmission gearing?
   a. input retarder
   b. output retarder
   c. torque retarder
   d. none of the above

10. Which type of hydraulic retarder applies a friction clutch while the unit is charging with hydraulic fluid?
    a. input retarder
    b. output retarder
    c. torque retarder
    d. all of the above

11. Which of the following will do the most to determine the position of the governor valve?
    a. speed of the input shaft
    b. speed of the transmission output
    c. main pressure
    d. accelerator position

12. Which of the following will do the most to determine modulator pressure?
    a. engine speed
    b. torque converter turbine speed
    c. range shift position
    d. accelerator pedal position

13. Which of the following devices has an operating principle similar to a hydraulic retarder?
    a. torque converter
    b. planetary gearset
    c. hydraulic clutch
    d. s-cam brake

14. How many planetary gearsets are used in an Allison five-speed automatic transmission?
    a. 2
    b. 3
    c. 4
    d. 5

15. How is the main pressure regulated?
    a. transmission input speed
    b. main pressure regulator
    c. trimmer valve
    d. governor valve

Automatic Transmission Maintenance

Prerequisite
Chapter 13

Objectives

After reading this chapter, you should be able to
- Perform hot and cold level transmission oil checks.
- Name the types of hydraulic fluid suitable for use in heavy-duty truck transmissions.
- Change automatic transmission oil and filters.
- Inspect transmission oil for signs of contamination.
- Adjust the manual gear selector linkage, mechanical modulator control linkage, and air modulator control.
- Perform a transmission stall test.
- Perform engine speed and vehicle speed shift point tests.
- Describe basic test stand procedure.
- Test the transmission valve body.
- Summarize basic inspection and troubleshooting procedures for automatic transmissions.

Key Terms List

adjusting ring
auxiliary filter
breather

engine stall point
stall test
valve body and governor test stand
valve ring adjusting tool

This chapter covers the periodic maintenance procedures required to keep heavy-duty automatic transmissions operating smoothly. It includes guidelines for inspection, care of the oil system and **breather**, checking temperatures and pressures, procedures for storage, linkage adjustment, and the care of external coolers and piping. Procedures for adjusting shift points are included along with some general considerations for overhaul work.

Specific disassembly, overhaul, and assembly instructions are beyond the scope of this chapter. Manufacturers publish detailed service manuals and other support literature specifying overhaul procedures for each model in their transmission line. Transmission work, including shift point adjustment, requires specialized test equipment, tools, and training. For example, one service manual lists 88 special tools used in the disassembly, overhaul, and reassembly procedures for one type of transmission. For this reason, repair facilities specializing in transmission overhauls are often the best choice for this type of work. The procedures outlined in this chapter are those required for most heavy-duty Allison 4- and 5-speed transmissions.

14.1 INSPECTION AND CARE

The transmission should be kept clean to make inspection and servicing easier. Inspect the transmission for loose bolts, loose or leaking oil lines, oil leakage, and the condition of the control linkage and cables. Check the transmission oil level at the intervals specified in the vehicle operator's manual.

TRANSMISSION OIL CHECKS

Maintaining the proper oil level in an automatic transmission is very important. As explained in Chapter 13, transmission fluid or oil is used to apply clutches and to lubricate and cool the transmission components. If the oil level is too low or too high, a number of problems can occur.

A low oil level will not completely cover the oil filter. This causes oil and air to be drawn into the pump inlet, which is then directed to the clutches and converter. The air in the system causes converter aeration and irregular shifting. Aeration of the oil also changes the viscosity and color of the oil to a thin, milky liquid.

At normal oil levels (FULL mark on dipstick), the oil is slightly *below* both the top of the oil pan and the planetary gear sets (**Figure 14–1**). If additional oil is added bringing the oil level above the FULL mark, the planetary gears run in the oil, causing it to foam and aerate. Both overheating and irregular shifting can occur. For this reason, excess oil must be drained from the transmission if accidental overfilling occurs during servicing. It should be noted here that a defective oil filler tube seal ring will also allow the oil pump to draw oil and air from the sump, resulting in aeration of the oil.

When checking the oil level, check at least two times to ensure that an accurate reading is made. If the dipstick readings are inconsistent (some high, some low), check for proper venting of the transmission breather and/or oil filler tube. A clogged breather can force oil up into the filler tube and cause inaccurate readings. If the filler tube is unvented, the vacuum produced will cause the dipstick to draw oil up into the tube as it is pulled from the tube. Again the result is an inaccurate reading.

Transmission input speed and oil temperature significantly affect the oil level. An increase in input speed will lower the oil level, while an increase in oil temperature will raise it. For these reasons, always check the oil level with the engine at idle (600–650 rpm) and the transmission in neutral. Both cold and hot level checks must be taken. A cold level check

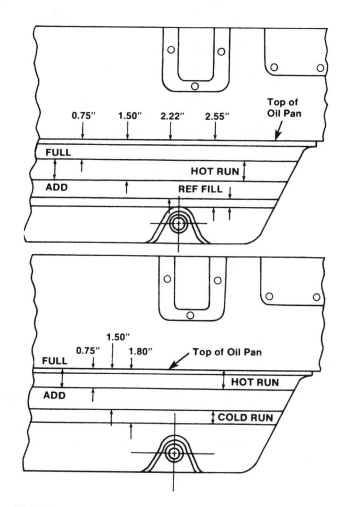

FIGURE 14–1   Oil levels on 4.3-inch transmission oil pan and 5.1-inch oil pan. (Courtesy of Allison Transmission Division of GM)

is taken to ensure that there is enough oil in the transmission to operate the vehicle until normal operating temperature is reached. The hot check is made when the transmission oil reaches normal operating temperature (160°–200°F). The hot check ensures that the oil level is at the proper level needed for operation.

When checking or adding oil to the transmission, dirt must not enter the filler tube. So, before removing the dipstick, clean around the filler tube opening. For all checks, park the vehicle on a level surface and apply the parking brake.

Cold Check

Operate the engine for approximately one minute to purge air from the system. To fill clutch cavities and circuits with oil, shift the transmission into drive and then to reverse. Allow the engine to idle, and then shift back to neutral.

Be sure the oil temperature is between 60°F and 120°F for an accurate cold check. Wipe the dipstick clean and take a level reading. If the oil level registers in the REF FILL (COLD RUN) band of the dipstick, the oil level is adequate to operate the transmission until a hot check can be taken (**Figure 14–2**).

Shop Talk _____

The REF FILL (COLD RUN) level is an approximate level and can vary among individual transmission applications. To ensure proper operating levels, a hot oil level check must be performed.

If the oil level registers at or below the lower line of the REF FILL (COLD RUN) band, add oil to bring the level within the REF FILL (COLD RUN) band. Do not fill above the upper line of this band. If the oil level is above the upper line of the REF FILL (COLD RUN) band, drain oil from the transmission to bring the level within this band. Now operate the vehicle until the hot check temperature is reached and proceed with the hot level check.

Hot Check

Be sure the oil temperature is between 160°F and 200°F. With the engine at idle and the transmission in neutral, wipe the dipstick clean and check the oil level. If the oil level registers in the HOT RUN band (between ADD and FULL), the oil level is adequate to continue operating the vehicle. If the oil level registers on or below the bottom line of the HOT RUN band or the ADD line, add oil to bring the level to the middle of the band. One quart of oil will raise the level from the bottom of the band to the top of the band (from the ADD line to the FULL line).

HYDRAULIC FLUID RECOMMENDATIONS

Always follow the manufacturer's exact hydraulic fluid specifications. For example, several transmission manufacturers recommend Dexron, Dexron II, and type C-3 (ATD approved SAE 10W or SAE 30) oils for their automatic transmissions. Type C-3 fluids are the only fluids usually approved for use in off-highway applications. Type C-3 SAE 30 is specified for all applications where the ambient temperature is con-

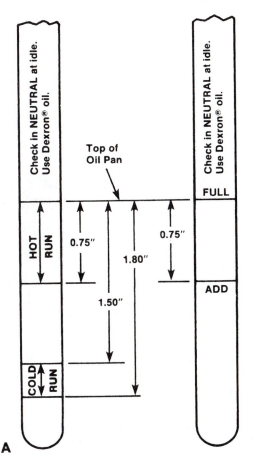

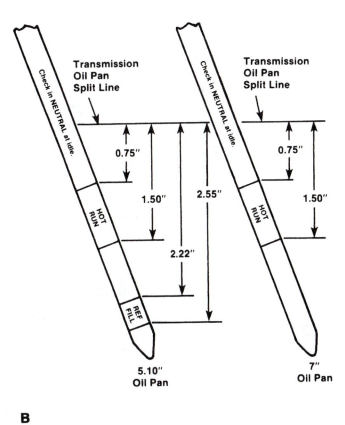

FIGURE 14–2 (A) Typical current and early 4.3-inch oil pan dipstick markings; (B) typical current 5.1-inch and 7.0-inch oil pan dipstick markings. (Courtesy of Allison Transmission Division of GM)

sistently above 86°F. Some but not all Dexron II fluids also qualify as type C-3 fluids. If type C-3 fluids must be used, be sure all materials used in tubes, hoses, external filters, seals, etc., are C-3–compatible.

COLD STARTUP

The transmission must not be operated in forward or reverse gears if the transmission oil falls below a certain temperature. Minimum operating temperatures for recommended fluids are as follows:

Dexron or Dexron II	−10°F
Type C-3 SAE 10W	10°F
Type C-3 SAE 30	32°F

When the ambient temperature is below the minimum fluid temperatures listed and the transmission is cold, preheat is required by either using auxiliary heating equipment or by running the engine in neutral at idle for at least 20 minutes to reach the minimum operating temperature. Failure to observe the minimum temperature limit can result in transmission malfunction or reduced transmission life.

OIL AND FILTER CHANGES

Change intervals are specified by the vehicle manufacturer. Typical filter and oil change intervals are 25,000 miles or twelve months for on-highway trucks and 1,000 hours or twelve months for off-highway applications.

Certain operating conditions may require shorter oil and filter change intervals. Oil must be changed whenever there is evidence of dirt or high-temperature conditions indicated by discoloration, strong odor, or oil analysis. Local conditions, severity of operation, or duty cycle might dictate more or less frequent service intervals. If an oil analysis program is being followed, the oil should be changed if the analysis shows the oil to be oxidized beyond the limits listed in **Table 14–1**.

TABLE 14–1:	FLUID OXIDATION MEASUREMENT LIMITS
Measurement	**Limit**
Viscosity	±25 percent change from new fluid
Carbonyl absorbance	+0.3 A*/0.1 mm change from new fluid
Total acid number	+3.0 change from new fluid
Solids	2 percent by volume maximum

*A = Absorbance units

Changing Procedure

To prevent contaminants from getting into the transmission, always handle transmission oil in clean containers and use clean fillers, funnels, etc.

CAUTION: Containers or fillers that have been used for any antifreeze or engine coolant solution must not be used for the transmission fluid. Antifreeze and coolant solutions contain ethylene glycol. If introduced into the transmission, ethylene glycol can cause the clutch plates to fail.

Always lay the dipstick on a clean surface when filling the transmission and keep replacement filters, seals, etc., in their cartons until ready for installation. The transmission should be at an operating temperature (160°–200°F) when the oil is drained. Shift the gear selector to neutral before draining the oil from the sump pan.

Transmissions can be equipped with either a standard or heavy-duty oil pan (**Figure 14–3**). When changing oil and filters on a standard pan, the entire oil pan is dismounted from the base of the transmission to access the filter. On the deeper, higher capacity heavy-duty oil pan, the pan is not dismounted. The filter is accessed through an opening in the side of the oil pan.

Standard Pan. The following is a step-by-step procedure for oil and filter changes on a typical standard transmission oil pan:

1. Remove the oil drain plug and gasket from the right side of the oil pan. Allow the oil to drain.
2. Remove the oil pan, gasket, and oil filler tube from the transmission. Discard the gasket. Clean the oil pan.
3. Remove the screw that retains the oil filter. Remove the filter and discard it. The oil filter intake pipe is not part of the filter and must not be discarded. Keep it for reinstallation.
4. Install the filter tube into the new filter assembly. Install a new seal ring onto the filter tube. Lubricate the seal ring with transmission oil. Install the new oil filter, inserting the filter tube into the hole in the bottom of the transmission. Secure the filter with the screw tightened to 10–15 pound-feet of torque.

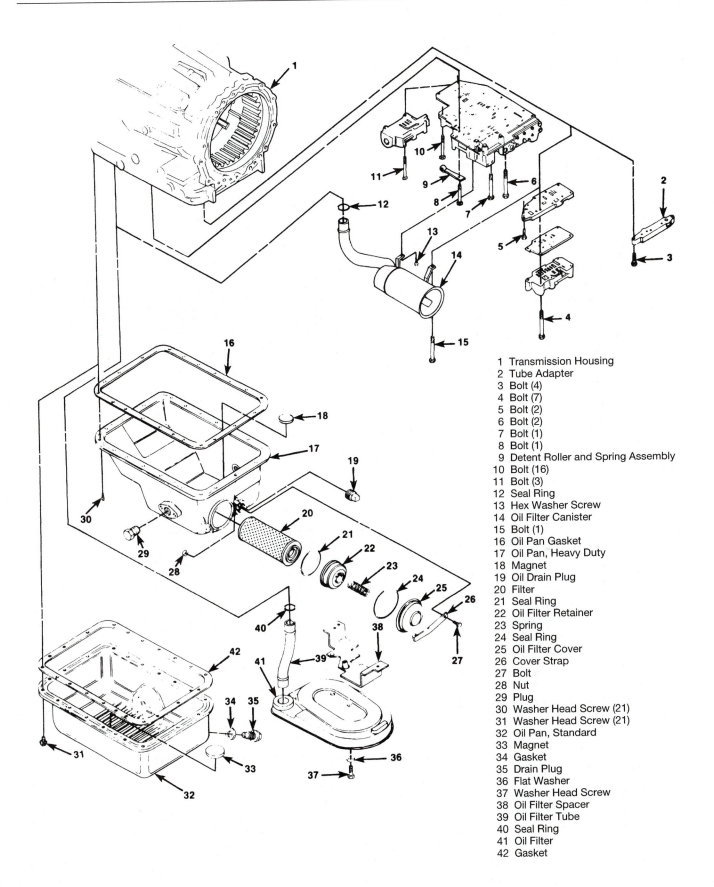

1 Transmission Housing
2 Tube Adapter
3 Bolt (4)
4 Bolt (7)
5 Bolt (2)
6 Bolt (2)
7 Bolt (1)
8 Bolt (1)
9 Detent Roller and Spring Assembly
10 Bolt (16)
11 Bolt (3)
12 Seal Ring
13 Hex Washer Screw
14 Oil Filter Canister
15 Bolt (1)
16 Oil Pan Gasket
17 Oil Pan, Heavy Duty
18 Magnet
19 Oil Drain Plug
20 Filter
21 Seal Ring
22 Oil Filter Retainer
23 Spring
24 Seal Ring
25 Oil Filter Cover
26 Cover Strap
27 Bolt
28 Nut
29 Plug
30 Washer Head Screw (21)
31 Washer Head Screw (21)
32 Oil Pan, Standard
33 Magnet
34 Gasket
35 Drain Plug
36 Flat Washer
37 Washer Head Screw
38 Oil Filter Spacer
39 Oil Filter Tube
40 Seal Ring
41 Oil Filter
42 Gasket

FIGURE 14–3 Typical transmission housing with standard and heavy-duty oil filter and oil pan configuration. (Courtesy of Allison Transmission Division of GM)

CAUTION: Do not use gasket-type sealing compounds or cement anywhere inside the transmission or anywhere they might be washed into the transmission. Never use insoluble vegetable-based cooking compounds or fibrous grease inside the transmissions. If grease must be used for internal assembly of transmission parts, use only low-temperature grease soluble in Dexron or C-3 transmission fluid, such as petrolatum.

5. Place the oil pan gasket onto the oil pan. A sealer or cement may be applied only to the areas of the oil pan flange outside the raised bead of the flange.
6. Install the oil pan and gasket, carefully guiding them into place. Be certain dirt or other material does not enter the pan. Turn in the oil pan retaining screws by hand and then torque all retaining screws to specification.
7. Install the filler tube at the side of the pan and tighten the tube fitting to specification.
8. Install the drain plug and gasket and tighten the plug.
9. Fill the transmission with oil to the proper level. Follow the manufacturer's specs for oil capacity. Recheck the oil level as described earlier in this chapter.

External Access, Heavy-Duty. The following are basic step-by-step procedures for oil and filter changes on a typical heavy-duty oil pan.

1. Remove the oil drain plug from the rear or side of the pan. Allow the oil to drain.
2. Remove the bolt and nut securing the strap holding the filter unit in place. Remove the filter cover, spring, and retainer.
3. Remove the filter. Discard the filter retainer and cover O-ring gaskets.
4. Install the seal ring on the filter retainer and install the filter and retainer into the external access canister.
5. Install the cover seal ring onto the lip of the oil pan.
6. Install the spring into the cover and place it over the filter retainer.
7. Install the strap, bolt, and nut to the pan and torque to specification.
8. Install the drain plug and tighten.
9. Fill the transmission with oil to the proper level. Follow the manufacturer's specs for oil capacity. Recheck the oil level.

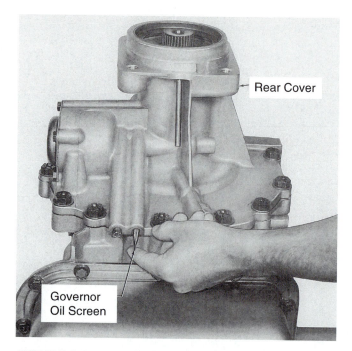

FIGURE 14-4 Location of governor filter on early model transmissions. (Courtesy of Allison Transmission Division of GM)

GOVERNOR FILTER CHANGE

The governor feed filter should be inspected and/or replaced at every oil/filter change. A pipe plug can be used to retain the governor oil screen in the transmission's rear cover (**Figure 14-4**). Remove the pipe plug and inspect the filter. If it is undamaged, clean it in mineral spirits and reinstall it. If it is damaged, replace it. Install the filter open end first into the transmission cover and reinstall the pipe plug.

Other models use a hexagon plug and O-ring seal to retain the governor oil filter in the transmission's rear cover (**Figure 14-5**). On some, the hexagon plug and filter location is on the output retarder. In both cases, the O-ring and filter are changed at every oil change. The plug is torqued to specs.

OIL CONTAMINATION

At each oil change, examine the oil that is drained for evidence of dirt or water. A normal amount of condensation will emulsify in the oil during operation of the transmission. However, if there is evidence of water, check the cooler (heat exchanger) for leakage between the water and oil areas. Oil in the water side of the cooler (heat exchanger) is another sign of leakage.

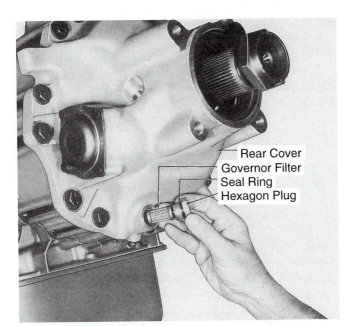

Rear Cover
Governor Filter
Seal Ring
Hexagon Plug

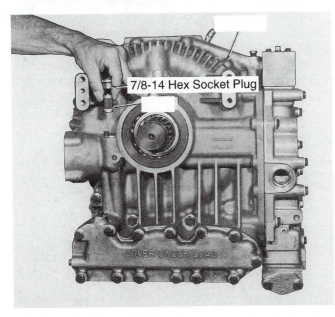

7/8-14 Hex Socket Plug

FIGURE 14–5 Location of governor filter on current model transmissions. (Courtesy of Allison Transmission Division of GM)

Metal Particles

Metal particles in the oil or on the magnetic drain plug (except for the minute particles normally trapped in the oil filter) indicate damage has occurred in the transmission. When these particles are found in the sump, the transmission must be disassembled and closely inspected to find the source. Metal contamination will require complete disassem-

bly of the transmission and cleaning of all internal and external circuits, cooler, and all other areas where the particles could lodge.

> **CAUTION:** If excessive metal contamination has occurred, replacement of all bearings within the transmission is recommended.

Coolant Leakage

If engine coolant leaks into the transmission oil system, immediate action must be taken to prevent malfunction and possible serious damage. The transmission must be completely disassembled, inspected, and cleaned. All traces of the coolant and varnish deposits resulting from coolant contamination must be removed.

Test kits are available that will detect traces of glycol in the transmission oil. However, certain additives in some transmission oil might cause a positive reading. If test results are questionable, test a clean (unused) sample of the transmission oil to confirm test results.

BREATHER CARE

The breather is located at the top of the transmission housing (**Figure 14–6**). It serves to prevent pressure buildup within the transmission and must be kept clean and the passage open. Exposure to dust and dirt will determine the frequency at which the breather requires cleaning. Use care when cleaning the transmission. Spraying steam, water, and/or cleaning solution directly on the breather can force water and cleaning solution into the transmission.

AUXILIARY FILTERS

Some vehicles are equipped with an **auxiliary filter** installed in the oil return line between the oil cooler and the transmission. This auxiliary filter is normally installed after there has been a failure in the system that introduces debris or dirt into the oil system. When such a failure or accident occurs, complete cleanup of the system cannot be assured with repeated system flushing.

The auxiliary filter prevents debris from being flushed into the transmission and causing a repeat failure. An auxiliary filter must be installed before the vehicle is placed back in service. This recommendation applies whether the transmission is overhauled or replaced with a new or rebuilt unit.

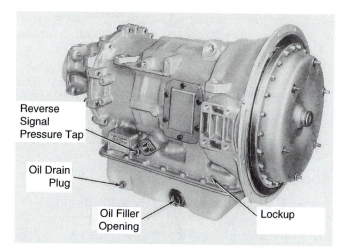

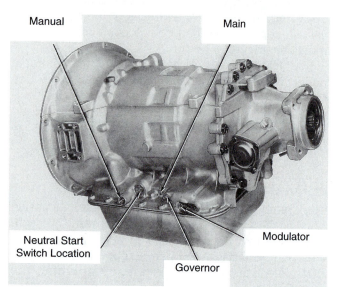

FIGURE 14–6 *Breather location on top of a transmission housing; also note the position of the manual selector shaft, neutral start switch, modulator control opening, pressure taps, plugs, and other key components. (Courtesy of Allison Transmission Division of GM)*

Most auxiliary oil filters are changed 5,000 miles after their initial installation and at regular oil change intervals thereafter. High-efficiency filters are available with no mileage limitations. They are changed when they become clogged or at three-year intervals, whichever occurs first. High-efficiency filters have a differential pressure switch that monitors pressure drop across the filter. It triggers a dash-mounted warning light if pressure exceeds a certain limit, indicating the filter is clogged.

EXTERNAL LINES AND OIL COOLER INSPECTION

Inspect all lines for loose or leaking connections, worn or damaged hoses or tubing, or loose fasten-

ers. Examine the radiator coolant for traces of transmission oil. This condition might indicate a defective heat exchanger.

Extended operation at high operating temperatures can cause clogging of the oil cooler and lead to transmission failure. The oil cooler system should be thoroughly cleaned after any rebuild work is performed on the transmission.

MANUAL SELECTOR LINKAGE INSPECTION

Proper adjustment of the manual gear range selector valve linkage is important. The shift tower detents must correspond exactly to those in the transmission. Failure to obtain proper detent in drive, neutral, or reverse gears can adversely affect the supply of transmission oil at the forward or fourth (reverse) clutch. The resulting low-apply pressure can cause clutch slippage and decreased transmission life.

Periodic inspections of the linkage should be made for bent or worn parts, loose threaded connections, loose bolts, and the accumulation of dirt and grease. Clean and lubricate all moving joints in the linkage.

The manual selector lever should move easily and give a crisp detent feel in each position. The linkage should be adjusted so the stops in the shift tower match and are positioned by the detents in the transmission. When the linkage is correctly adjusted, the pin that engages the shift lever linkage at the transmission can be moved freely in each range.

To check and adjust the linkage:

1. Place the operator's range selector control at the neutral position (**Figure 14–7**). Place the selector lever (on transmission) at the neutral position. Adjust the linkage so that it

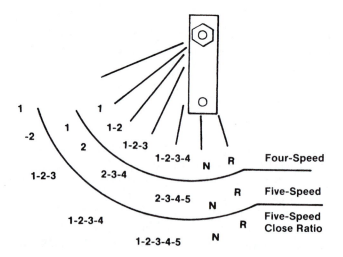

FIGURE 14–7 *Manual gear selector range positions. (Courtesy of Allison Transmission Division of GM)*

matches the selector lever on the transmission. Connect the linkage to the selector lever.

2. Shift through all selector positions, checking at each to ensure that the valve body detent positions correspond to the respective selector positions.

3. Check the neutral-start switch (see **Figure 14-6** for location). The starter should not operate, except when the range selector is at neutral.

Mechanical Modulator Control Linkage Installation/Adjustment

To install and adjust this type of linkage, perform the following:

1. Connect the engine (throttle pedal) end of the modulator cable housing to its mounting (**Figure 14-8**). Also see **Figure 14-6** for the modulator control location on the transmission.

2. Fully depress the accelerator pedal and check whether the throttle linkage will push or pull the cable core when the linkage is moving toward full travel position. If it pushes the cable core, push the cable core until it reaches the end of its travel. If movement of the throttle linkage toward full throttle position pulls the cable, pull the cable to the end of its travel.

3. Adjust the clevis or rod end on the cable core until it registers with the hole in the throttle linkage lever and the connecting pin can be freely inserted. With the pin removed, rotate the clevis or rod end one additional turn counterclockwise (viewing cable core from its end) for the pull type arrangement, or one additional turn clockwise for the push type arrangement. This ensures that the modulator does not prevent the throttle lever from reaching the full on position. Install the clevis pin or rod end to connect the throttle linkage and cable. Tighten the locknut against the clevis or rod end.

4. Check the travel of the cable core when the throttle is moved from the fully open to the fully closed position. A typical system travel distance is approximately $1\frac{1}{8}$ inches minimum and $1\frac{9}{16}$ inches maximum.

5. Some mechanical controls include a support bracket also secured by the retainer bolt.

6. Check the cable routing. Bends must not be less than an inch radius. The cable should not be nearer than 6 inches to the engine exhaust pipe or manifold. The cable must follow the movements of the throttle linkage; it might be necessary to add a spring to ensure that the movement occurs smoothly.

7. Adjust other types of mechanical controls as outlined in the vehicle manufacturer's instructions.

Air Modulator Control Connection/Inspection

To test this type of modulator control, perform the following:

1. Connect the air line (hose) that comes from the "tee" or junction of the accelerator-to-engine fuel control air line to the transmission modulator control.

2. Loosen the bolt that retains the modulator control on the transmission to readjust the angle of the modulator control if necessary. Retighten the bolt to 13–16 pound-feet.

3. Check the routing of the air line to ensure that the line is not twisted or kinked, will not chafe, and is not too near the exhaust manifold or exhaust pipe.

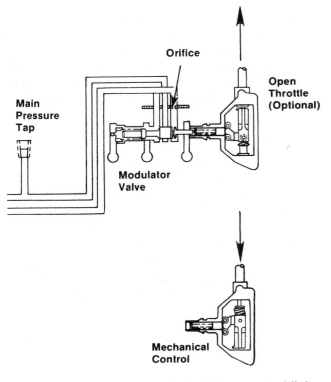

FIGURE 14-8 *Mechanical modulator control linkage details. (Courtesy of Allison Transmission Division of GM)*

STALL TESTING

When there is an obvious malfunction in the vehicle's power train (engine and transmission), a **stall test** is performed to determine which of the components is at fault. The condition created in a stall test is one where the converter pump is turning at its maximum speed with the turbine held or stalled. Excessive or maximum torque (vortex flow) is occurring within the torque converter during this time. To prevent the vehicle from breaking loose and causing personal injury or equipment damage, the vehicle must be prevented from moving.

WARNING: When conducting a transmission stall test, both the parking and service brakes must be applied and be in good working order. The wheels must be blocked against movement. Many experienced technicians also chain the vehicle to the floor pegs as an added precaution. As a further safeguard, no one must be allowed to stand in front of or behind the vehicle during the stall test.

The **engine stall point** (rpm) under load is compared to the engine manufacturer's specified rpm for the stall test. In order to do this, the engine manufacturer's test data must be available. This data can be obtained from the engine manufacturer, dealer, or distributor.

Temperature readings at the converter-out (to cooler) line are also taken during a 2-minute cool down period after the test to ensure the transmission cooling system is operating properly.

CAUTION: Never maintain full stall for more than 30 seconds at any one time because of the rapid rise in oil temperature. Do not allow the converter-out temperature to exceed 300°F. Do not rely on the converter-out oil temperature to limit stall duration. During the stall condition, internal temperatures rise much faster than converter-out oil temperatures. If the stall test must be repeated, allow for cool down. Do not let the engine overheat.

The following are general steps in performing a stall test on a heavy-duty automatic transmission. Always follow the specific instructions outlined in the service manual for the exact model you are working on. There can be slight, but important differences between models.

1. Connect a tachometer of known accuracy to the engine.

2. Install a temperature probe into the converter-out (to cooler) line.
3. Start the engine and bring the transmission oil up to its normal operating range temperature (160°–200°F).
4. Firmly apply the parking and service brakes. Block the drive wheels to prevent movement during the test. If floor pegs are available, chain the vehicle to the floor. Ensure all personnel are out of the way in the event of brake or restraint failure.
5. With the vehicle secured against movement, shift the selector control to any forward range with the exception of D1 or DR deep ratio models. The extremely high torque in this gear might damage the transmission and/or the vehicle drive line. The stall test can also be performed in reverse, if needed.
6. Accelerate the engine to rated speed. The converter-out temperature will begin to climb rapidly. When the temperature reaches 255°F minimum, record the engine rpm while at rated speed. Release the throttle and shift to neutral.

WARNING: Never allow the converter-out temperature to exceed 300°F.

7. Immediately after the engine rpm has been recorded, note the converter-out temperature.
8. With the transmission remaining in neutral, run the engine at 1200 to 1500 rpm for two minutes and record the converter-out temperature at the end of this time period.

Stall Test Results

Problems are pinpointed based on how much the stall test rpm reading varies from the engine manufacturer's specification for this test. When interpreting test results remember that environmental conditions such as ambient temperature, altitude, engine parasitic loads, etc., affect power input to the converter. Under such conditions, a stall speed deviation up to ±150 rpm from specification can be accepted as within normal range. Compare the test results with the following:

- If the engine stall speed is more than 150 rpm below the stall speed specified by the engine manufacturer, an engine problem is indicated, such as the need for a tune-up.
- If the engine stall speed is more than 150 rpm above specification, a transmission problem is

indicated, such as slipping clutches, cavitation, or torque converter failure.

- An extremely low stall speed, such as 33 percent of the specified engine stall rpm, during which the engine does not smoke, could indicate a freewheeling torque converter stator.
- If the engine stall speed conforms to specification, but the transmission oil overheats, refer to the cool-down check. If the oil does not cool during the two-minute cool-down check, a stuck torque converter stator could be indicated.

14.2 SHIFT POINT ADJUSTMENT

As covered in Chapter 13, automatic transmissions are designed to both upshift and downshift within predetermined engine rpm and vehicle road speed ranges. Two key adjustments determine when these automatic shifts occur: the shift signal valve spring pressure rate, and the modulator spring force adjustment.

Making adjustments to these springs involves dropping the oil pan and removing the valve body. The spring force of each individual gear range shift signal valve is increased or decreased using a special tool. The modulator valve spring is adjusted in a similar way.

Following are three methods of checking transmission shift points:

1. Calibration by engine rpm
2. Calibration by speedometer readings
3. Calibration by test stand (transmission output shaft speed)

All three methods are acceptable for troubleshooting shift point problems and making adjustments. The engine rpm and speedometer reading test methods involve taking a test drive and noting when shift occurs in relation to optimum engine or vehicle speeds. Valves are then adjusted to bring any out-of-range shift points into line with desired rpm or speed levels.

The test stand method is the most accurate. It involves using a **valve body and governor test stand,** a specialized piece of test equipment. The valve body of the transmission is removed from the vehicle and mounted into the test stand. The test stand duplicates all vehicle running conditions, so the valve body can be thoroughly tested and calibrated.

Transmission shift points cannot be properly adjusted if the transmission has the wrong governor installed. Check the part number on the governor with the listing in the current parts catalog for the exact model of transmission. It must be the correct governor or test results will be inaccurate.

Regardless of whether testing and calibration is to be performed using a road test or a test stand that simulates road operation, the following preparations must be made:

1. Warm the transmission or test stand setup to normal operating temperature (160°–200°F).
2. Check the engine no-load governor setting, and adjust if required, to conform to the transmission's engine speed requirements.
3. Check engine performance before making any shift point adjustments.
4. Check the accelerator pedal linkage that controls the modulator valve mechanical actuator on diesel engines.
5. Check the shift selector linkage for proper range selection.
6. Test the accuracy of the dash-mounted tachometer using a test tachometer. Adjust dash readings to account for any error. Check the accuracy of the speedometer against another vehicle.

ENGINE SPEED (RPM) METHOD

Ideally, the vehicle should be loaded to its normal operating weight. During the initial part of the test drive, note and record the full-load governed speed (rpm) of the engine. This is the reference speed from which checks and adjustments are made. Shifts should occur at specific engine rpm below full-load governed speed. For example, on a typical four-speed automatic, the shift from first to second gear may be designed to occur at 400 rpm less than the full-load engine governed speed. The second to third shift may ideally occur at 300 rpm less than full-load governed speeds, and the third to fourth shift at 200 rpm less. Always consult the service manual for the exact shift point for the model being tested (**Table 14–2**).

On close ratio models, the 1–2 upshift should occur at approximately 2 mph below the lockup engagement speed. To establish this lockup engagement point, install a pressure gauge at the transmission lockup pressure tap (see **Figure 14–6**). With the selector in D1 (hold), note the instant the gauge needle moves and record the engine rpm. This is the engagement point for the 1–2 shift.

The first step in adjusting shift points is to establish an accurate full-load governed speed for the vehicle. Once this is done, bring the vehicle to a stop and place the selector in drive. Begin the test drive and

TABLE 14–2:   SHIFT POINT ENGINE SPEEDS FOR SELECT TRANSMISSIONS

(All speeds listed are subtracted from rated or full-load engine governed speed.)

Shift	AT Series	MT Series (4-Speed)	MT Series (5-speed)	HT Series (all)
1–2	less 400 rpm	less 600 rpm	**	
2–3	less 300 rpm	less 200 rpm	less 600 rpm	less 100 rpm
3–4	less 200 rpm	less 200 rpm	less 200 rpm	less 100 rpm
4–5	less 200 rpm	less 100 rpm		

*Selected models and examples only. Check appropriate service manual to confirm speeds.

On close ratio models, the 1–2 upshift should occur at approximately 2 mph below the lockup engagement speed. To establish lockup engagement point, install pressure gauge at transmission lockup pressure tap (see **Figure 14–6). With the selector in D1 (hold), note the instant the gauge needle moves and record engine rpm. This is the engagement point.

quickly bring the engine to full throttle. Note the engine speed (at full throttle) at which the various shifts occur. Compare these shift points to the desired shift points listed in the transmission service manual.

Adjustments are made based on the rpm engine speed readings taken during the test drive. Shift speeds are altered by changing the positions of the adjusting rings that determine the retaining force of valve springs in the valve body (**Figure 14–9**).

Adjustment of the valves is performed using a **valve ring adjusting tool.** This tool has two small pins mounted parallel to each other. The pins are placed against the **adjusting ring** and pressure is applied inward to push in the spring and ring.

The adjusting ring is held in the shift signal valve bore by a pin that is press-fit through the valve body housing. When the ring is pushed in by the adjusting tool, the slots on the ring that engage the pin are released (**Figure 14–10**). The adjusting ring can then be turned to adjust spring pressure. Clockwise rotation increases spring pressure; counterclockwise rotation decreases spring pressure. The slots on the adjusting ring that engage the pin are on a sloping ramp around the circumference of the adjusting ring. This produces a change in spring pressure as the slots re-engage the pin in the new ring position.

Each notch in the adjustment ring will alter the shift point by a certain rpm value. For example, one notch might be equal to 25, 35, 40, or 50 rpm of engine speed based on the particular model transmission. Confirm the notch values by checking the correct service manual.

If the tested upshift engine speed fails to reach the specified rpm before the shift occurs, the shift point can be raised by increasing the spring force on the required shift signal valve (1–2, 2–3, 3–4, 4–5). If the upshift speed exceeds the specified rpm, or the upshift does not occur at all, the spring force must be reduced. Only adjust spring force on valves that do not upshift at the appropriate speeds.

If more than one shift signal valve spring requires adjustment in the same direction, it might be necessary to adjust the spring force on the modulator valve in the same direction. If the modulator valve is not adjusted, closed throttle downshifts can be abnormally high or low, depending on the direction the shift signal adjusting rings were rotated. If all full throttle upshift points are too low by approximately the same amount, check the adjustment on the modulator external linkage.

VEHICLE SPEED (SPEEDOMETER) METHOD

When a tachometer is not available for checking shift points, the vehicle's speedometer can be used. Begin the test by checking the top speed the vehicle can achieve in each gear when the selector lever is placed in those specific gears. On four-speed transmissions check top speed in first, second, and third gears. On five-speed transmissions, check top speed in second, third, and fourth gears. (On some five-speeds, there might not be a third gear selector position.)

Once top vehicle speeds are recorded for each gear, drive the vehicle at full throttle from a standing start and note the speed at which each upshift occurs. Compare the upshift rpms with the selected shift speeds recorded in the first part of the test drive.

On four-speed transmissions, the 2–3 upshift should occur at approximately 2 mph below the top speed for second gear. The 3–4 upshift should occur at about 2 mph below the third gear top speed. The 3–4 and 4–5 upshifts for five-speed transmission should occur about 2 mph below third and fourth gear top speeds, respectively.

The 1–2 shift in four-speeds and the 2–3 shift in five-speeds are not adjusted relative to the top speeds. Instead, the 2–1 downshift (four-speeds) and

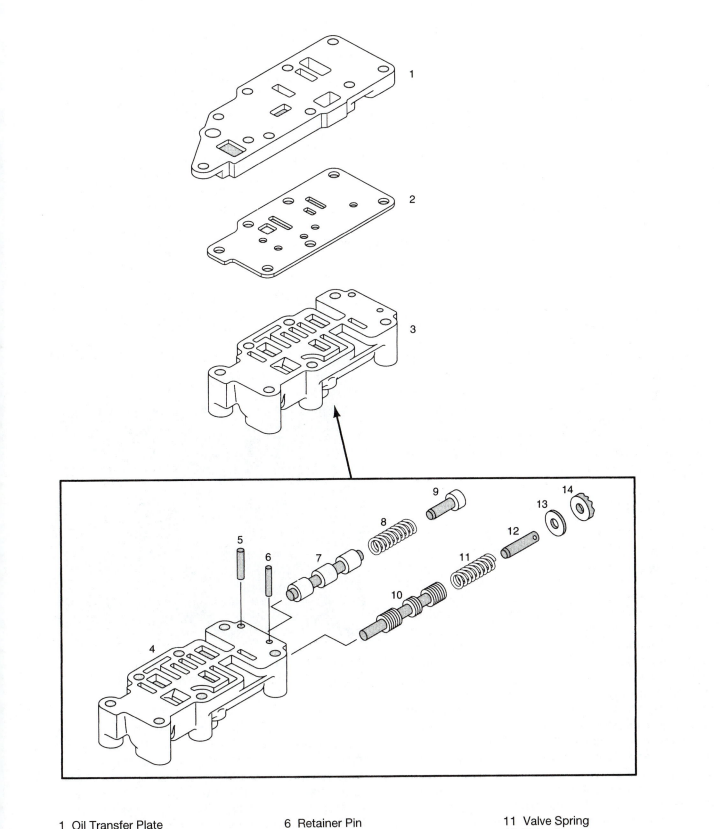

FIGURE 14–9 *Low shift valve body assembly.*

1 Oil Transfer Plate	6 Retainer Pin	11 Valve Spring
2 Separator Plate	7 1–2 Relay Valve	12 Valve Stop
3 Low Shift Valve Assembly	8 1–2 Relay Valve Spring	13 Washer
4 Valve Body	9 Valve Stop	14 Adjusting Ring
5 Retainer Pin	10 1–2 Shift Valve	

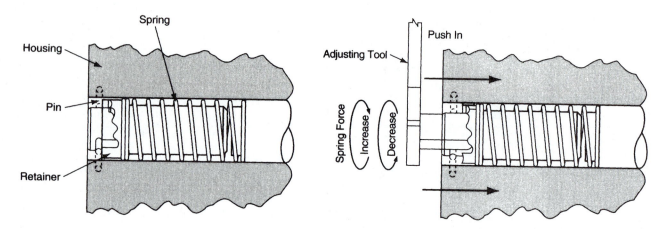

FIGURE 14-10   *Adjusting spring retainer position using special tool.*

3–2 downshift (five-speeds) should occur, with closed throttle, at 3 to 5 mph. Speedometer or vehicle speed is a less direct indicator of engine speed than a tachometer reading. Although no exact relationship between vehicle speed and notch increments exists, the same general adjustment principles apply as in rpm adjustment. If the upshift vehicle speed is below the levels described above, the shift point should be raised by increasing spring force on the necessary valve. If the vehicle fails to shift or shifts at the maximum rpm for that gear, the shift point should be lowered by decreasing spring pressure.

TEST STAND CALIBRATION

Figure 14–11 illustrates the valve body and governor test stand used to check the operation of some transmission valve bodies. The valve body is bolted onto a manifold that resembles the lower portion of the transmission housing. This manifold is drilled and tapped to accept all the fittings and hoses needed to properly route oil through the valve body for testing.

The manifold is secured in the test stand and all lines connected to it. The test stand can now be used to check five principal transmission valve body functions: governor pressure, modulator pressure, hold regulator pressure, shift points (up-down inhibit), and trimmer regulator operation.

To test the operation of the valve body components, the test stand duplicates the transmission's output shaft speed. The governor is gear driven from the test stand. A variable resistor control is used to increase or decrease output shaft rpm, which is clearly indicated on a digital tachometer. All clutch pressures are also indicated on easy-to-read gauges. The test stand top is constructed of heavy-duty clear plastic so all operating conditions can be observed during testing.

Service manuals give tables listing the desired output shaft rpm levels for each shift (**Table 14–3**). These

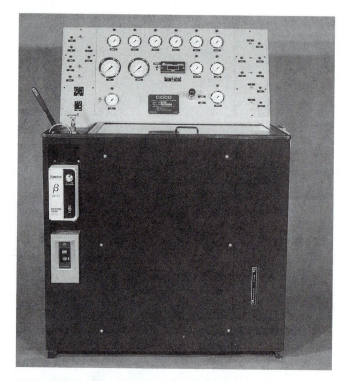

FIGURE 14–11   *Valve body and governor test stand. (Courtesy of Aidco International, Inc.)*

tables provide the information needed for adjusting shift points on transmissions matched to engines having governed speeds from 2200 to 4000 rpm.

The operator gradually increases output shaft speed using the variable resistor control and notes the rpm levels at which shifts occur. Adjustments are then made to individual valve springs as needed. The adjustment procedure is the same as described earlier under engine speed calibration.

Main hydraulic pressure to the clutch shift signal and relay valves in the valve body is controlled by the test stand main pressure regulator. An electric

TABLE 14–3:   SHIFT POINTS WITHOUT MODULATED OR ADJUSTABLE LOCKUP

Engine Full Load Governed Speed (rpm)			2200	2400	2600	2800	3000	3200	3400	3600	4000
Throttle Position	Range	Shift	Transmission Output Speed (rpm) at Start of Shift								
Full	DR4 or DR2	1–2	530–590	530–590	570–630	625–685	685–740	740–795	795–855	850–910	965–1020
	DR4 or DR3	2–3	875–1015	875–1015	970–1110	1110–1250	1240–1380	1255–1400	1420–1565	1445–1590	1640–1780
		3C–3LU	1050–1400	1190–1600	1190–1600	1360–1615	1310–1600	1580–1870	1580–1870	1745–2155	2160–2450
	DR4	3–4	1475–1625	1580–1725	1725–1870	1870–2010	2085–2230	2160–2300	2300–2445	2445–2590	2735–2880
	DR3	4–3	1560 min	1850 min	1910 min	2020 min	2370 min	2440 min	2550 min	2720 min	3090 min
	DR2	3–2	945–1300	1190–1470	1280–1580	1340–1580	1460–1900	1530–1930	1620–2000	1730–2120	2080–2460
	DR1	2–1	660–760	770–960	740–1080	775–1025	910–1100	930–1130	960–1200	1040–1280	1200–1590
Closed	DR4	4–3*	150–390	150–390	250–675	20–675	225–640	425–780	555–855	20–590	570–950
		3–2*	370–470	370–470	390–615	355–640	445–700	510–780	550–840	350–700	600–815
		2–1	20–300	20–300	20–470	20–535	205–530	285–550	365–550	20–535	345–550

*4–2 Downshift is acceptable.

TABLE 14–4:   PRESSURE SCHEDULE FOR SELECT GOVERNOR MODELS

	Governor Output Pressure (psi)			
Governor Assembly	440 rpm	800 rpm	1500 rpm	2200 rpm
#1	3.5–6.5	17.5–20.5	42.8–48.4	58.6–64.8
#2	3.5–7.5	16.0–20.0	39.1–44.2	59.0–65.0
#3	8.5–11.5	27.5–30.5	54.3–60.8	82.6–91.0
#4	9.0–13.0	38.0–42.0	63.8–70.2	103.0–112.0
#5	8.5–11.5	27.5–30.5	54.3–60.8	82.6–91.0
#6	3.5–7.5	16.0–20.0	39.1–44.2	59.0–65.0
#7	9.0–13.0	38.0–42.0	63.8–70.2	103.0–112.0
#8	8.5–11.5	27.5–30.5	54.3–60.8	82.6–91.0

Note: This table provides an overview of general pressure ranges. Consult the proper manufacturer's parts catalog and service literature to determine exact governor assembly being tested and proper pressure schedule.

heat circuit in the tester is used to raise the fluid to normal operating temperature.

With the main pressure at 150 psi and oil temperature at 100°F, the governor output pressure is checked at four separate shaft output speeds: 440, 800, 1500, and 2200 rpm (**Table 14–4**).

While no adjustment is possible for the governor assembly internal valve, this valve can be removed and cleaned or replaced to ensure specified performance.

14.3 TROUBLESHOOTING

Troubleshooting is the systematic search for and location of malfunctions in the engine or transmission. During troubleshooting, the engine and transmission must be regarded as a single package. When troubleshooting, do not operate the vehicle before performing the following checks:

- Visually inspect for oil leakage.
- Visually inspect all split lines, plugs, and all hose and tube connections at the transmission and cooler. Oil leakage at split lines can be caused by loose mounting bolts or broken gaskets.
- Tighten all bolts, plugs, and connections where leakage is found.
- Check to ensure the modulator control cable and linkage are free.
- Check the shift linkage for proper operation.

If the above inspection does not reveal the cause of trouble, and the vehicle is operational, test drive and observe transmission and general vehicle performance. Make certain the engine is properly tuned and the vehicle is in good running condition. Also ensure the oil level in the transmission is correct. **Table 14–5** outlines the possible causes and corrective action for the most common automatic transmission problems.

14.4 POWER TAKEOFFS

Automatic transmissions have special PTO application considerations, since they transmit power differently than manual transmissions. Engine torque enters the automatic transmission through the torque converter. The converter's ability to multiply engine torque

TABLE 14–5: AUTOMATIC TRANSMISSION TROUBLESHOOTING		
Symptom	**Probable Cause**	**Remedy**
Harsh downshifts	1. Improper oil level	1. Adjust oil level.
	2. Engine idle speed too high	2. Set engine idle speed to engine manufacturer's spec for an automatic transmission (500 rpm min).
	3. Manual selector linkage out of adjustment	3. Adjust linkage.
	4. Mechanical modulator out of adjustment	4. Adjust mechanical modulator.
	5. Valve sticking in governor	5. Clean governor and governor screen. (Replace only if necessary.)
	6. Downshift points too high	6. Adjust valve body. (Refer to service manual.)
	7. Valves sticking in control valve body	7. Clean valve body. (Refer to service manual.)
Shift cycling	1. Improper oil level	1. Adjust oil level.
	2. Engine governed speed too low for transmission shift calibration	2. Set engine no-load governed speed to the engine manufacturer's spec for an automatic transmission (2200 rpm min).
	3. Governor screen clogged	3. Remove governor screen and clean or replace.
	4. Valve sticking in governor	4. Clean governor and screen. (Replace only if necessary.)
	5. Improper shift valve adjustment	5. Adjust valve body. (Refer to service manual.)
	6. Valves sticking in control valve body	6. Clean valve body. (Refer to service manual.)
Harsh upshifts	1. Improper oil level	1. Adjust oil level.
	2. Manual selector linkage out of adjustment	2. Adjust linkage.
	3. Mechanical modulator out of adjustment	3. Adjust mechanical modulator.
	4. Valve sticking in governor	4. Clean glvernor and screen. (Replace only if necessary.)
	5. Improper shift valve adjustment	5. Adjust valve body. (Refer to service manual.)
	6. Valves sticking in control valve body	6. Clean valve body. (Refer to service manual.)
Downshifts occur at wide open throttle	1. Bent or loose modulator can	1. Replace modulator and seal ring.
	2. If used, check mechanical modulator for proper adjustment	2. Adjust mechanical modulator.
	3. Shift points set too high	3. Adjust valve body. (Refer to service manual.)

Symptom	Probable Cause		Remedy	
	4.	Flanges out of parallel	4.	Flanges should be parallel within 1 degree.
	5.	Output flange retaining bolt improperly torqued	5.	Install flange bolt and lock tab washer correctly.
	6.	Slip joint splines seized	6.	Lubricate slip joints properly. See vehicle manufacturer's specification.
	7.	Drive line yoke needle bearings worn	7.	Replace.
	8.	Drive line out of balance or bent	8.	Rebalance.
Excessive retarder response time	1.	Sticking low-speed valve	1.	Clean valve. (Refer to service manual.)
	2.	Sticking high-speed valve	2.	Clean valve. (Refer to service manual.)
	3.	Sticking priority valve	3.	Clean valve. (Refer to service
	4.	Leaking clutch piston seals	4.	Replace seals. (Refer to service manual.)
	5.	Leaking accumulator piston seals	5.	Replace seals. (Refer to service manual.)
No retarder operation at low speed	1.	Sticking charging valve	1.	Clean valve. (Refer to service manual.)
	2.	Sticking clutch valve	2.	Clean valve. (Refer to service manual.)
	3.	Sticking regulator charging valve	3.	Clean or replace valve. (Refer to service manual.)
	4.	Leaking clutch piston seals	4.	Replace seals. (Refer to service manual.)
	5.	Failed retarder clutch	5.	Replace clutch. (Refer to service manual.)
No retarder operation at high speed	1.	Low air supply	1.	Check air supply.
	2.	Ruptured air line to retarder control valve	2.	Repair air line.
	3.	Sticking air control	3.	Repair or replace control. (Refer to service manual.)
	4.	Sticking retarder control valve	4.	Clean or replace valve. (Refer to service manual.)
Low main and/or retarder pressure	1.	Sticking charging valve	1.	Clean or replace valve. (Refer to service manual.)
	2.	Sticking clutch valve	2.	Clean or replace valve. (Refer to service manual.)
	3.	Leaking clutch piston seals	3.	Replace seals. (Refer to service manual.)
	4.	Leaking accumulator piston seals	4.	Replace seals. (Refer to service manual.)
	5.	Leaking rotor hub seals	5.	Replace seals. (Refer to service manual.)

through fluid provides an infinite number of ratios to get a vehicle started quickly, free from lugging and stalling. Thus, as the power takeoff receives the engine torque through the torque converter, it is cushioned. Because of the infinitely variable ratio characteristics of the torque converter, the speed of the PTO output shaft changes with load and engine speed.

Speed (rpm) and load data are available from PTO manufacturers. To interpret the relationships, the torque converter model, drive equipment's rpm and

horsepower requirements and tentative choice of PTO must be known.

SUMMARY

- The transmission should be kept clean to make inspection and servicing easier.
- Inspect the transmission for loose bolts, loose or leaking oil lines, oil leakage, and the condition of the control linkage and cables.

- Maintaining the proper oil level in an automatic transmission is very important.
- At each oil change, examine the oil that is drained for evidence of dirt or water.
- Metal particles in the oil or on the magnetic drain plug (except for the minute particles normally trapped in the oil filter) indicate that damage has occurred in the transmission.
- If engine coolant leaks into the transmission oil system, immediate action must be taken to prevent malfunction and possible serious damage.
- Transmission disassembly, testing, and reassembly requires the use of OEM literature and service tools.
- During troubleshooting, the engine and transmission must be regarded as a single package.

REVIEW QUESTIONS

1. Which of the following can cause aerated transmission oil?
   a. low oil levels
   b. high oil levels
   c. clogged breather
   d. both a and b

2. At normal oil level (FULL mark on dipstick) the transmission oil level is
   a. slightly below the planetary gear sets
   b. slightly below the top of the oil pan
   c. above the ring gears
   d. both a and b

3. Which of the following can cause an inaccurate dipstick reading?
   a. failure to wipe the dipstick clean before taking the reading
   b. a clogged breather
   c. an unvented oil filler tube
   d. all of the above

4. When checking the transmission's cold check level, the oil should be
   a. between 0°–60°F
   b. between 60°–120°F
   c. between 120°–160°F
   d. between 160°–200°F

5. An increase in the transmission input speed will
   a. lower the oil level
   b. raise the oil level
   c. have no effect on the oil level
   d. aerate the oil

6. An increase in oil temperature will
   a. lower the oil level
   b. raise the oil level
   c. have no effect on the oil level
   d. aerate the oil

7. The _____ is located at the top of the transmission housing and serves to prevent pressure buildup within the transmission
   a. auxiliary filter
   b. governor
   c. breather
   d. release valve

8. Technician A installs an auxiliary filter after there has been a failure in the system that introduces debris or dirt into the oil system. Technician B changes the auxiliary filter every 5,000 miles. Who is correct?
   a. Technician A
   b. Technician B
   c. both A and B
   d. neither A nor B

9. When there is a performance malfunction in either the engine or transmission, a _____ is performed to determine which is at fault.
   a. stall test
   b. shift point adjustment
   c. oil analysis
   d. all of the above

10. An automatic transmission upshifts prematurely. Technician A says the mechanical modulator might have to be adjusted. Technician B says that the shift valve might have to be adjusted. Who is correct?
    a. Technician A
    b. Technician B
    c. both A and B
    d. neither A or B

11. If the engine stall speed is more than 150 rpm above the stall speed specified by the engine manufacturer
    a. an internal transmission problem is likely
    b. an engine problem is likely
    c. a torque converter problem is likely
    d. either a or b

12. A stall speed that conforms to specs, but with a transmission oil overheat problem, indicates
    a. an engine problem
    b. a stuck torque converter stator
    c. a freewheeling torque converter stator
    d. either a or c

13. A stall speed that is significantly (33 percent or more) below the manufacturer engine specs likely indicates
   a. an engine problem
   b. a stuck torque converter stator
   c. a freewheeling torque converter stator
   d. either b or c

14. Which of the following are acceptable methods of checking transmission shift points?
   a. calibration based on engine speed (rpm)
   b. calibration based on vehicle speed (speedometer reading)
   c. calibration based on test stand readings
   d. all of the above

15. Valve body and governor test stands use as a basis for analyzing valve body performance
   a. engine speed (rpm)
   b. transmission input shaft speed
   c. transmission output shaft speed
   d. either b or c

15

Electronically Automated Standard Transmissions

Prerequisites
Chapters 5, 6, 8, 9, 10, and 11

Objectives

After studying this chapter, you should be able to

- Explain how an Eaton standard transmission is adapted for AutoSelect or AutoShift operation.
- Interpret the Eaton serial number codes.
- Identify the hardware changes required in an automated version of the RoadRanger twin-countershaft transmission.
- Outline the electronic circuit components that are used to manage AutoSelect and AutoShift.
- Describe the main box and auxiliary box actuator components required for Eaton electronically automated, standard transmissions.
- Outline how the electronic circuit components work together to perform the system functions.
- Perform some basic diagnostic troubleshooting of Eaton AutoSelect and AutoShift electronics.
- Describe the Rockwell Meritor Engine SynchroShift transmission and the Meritor ZF 12 and 16-speed NoClutch systems.

Key Terms

AutoSelect
AutoShift
Engine SynchroShift (ESS)
gear select motor
Hall effect
inertia brake

logic power
rail select motor
power module
SAE J 1939
speed sensors
System Manager ECU
Transmission Controller ECU

For several decades, Eaton Fuller has dominated the heavy highway truck market with its range of standard, manually shifted transmissions. In the 1990s, Rockwell Meritor began to market heavy-duty truck standard transmissions to provide some competition for Eaton Fuller. A higher degree of driver skill and training is required to operate the manually shifted compounded gearing used in conventional truck transmissions, and surveys indicate that constant shifting of gears adds to driver fatigue. A prolonged industry-wide shortage of truck drivers has existed for some time, so truck fleet managers identified the need for a lower cost automatically shifted transmission a decade ago. Eaton responded to this industry need by developing an electronically automated standard transmission. **AutoShift** represents the second generation of this technology.

Rockwell Meritor introduced their **Engine SynchroShift (ESS)** standard transmission in 1997 to compete with Eaton's AutoShift. This system consists of input and output shaft speed sensors and uses the engine electronics to perform the processing required of these inputs to synchronize the engine speed for a shift. Because the ESS transmission has no processing or switching (output) circuit, it is cost-effective when compared with AutoShift. Some brief coverage of Meritor ESS will be provided at the end of this chapter.

For the technician with a basic understanding of Eaton standard transmissions, the operating principles of Eaton AutoShift transmissions will not present any problems. In engine technology, the term *partial authority* is used to mean a hydromechanical system that has been adapted for computer controls, and that is the idea behind this system. Both Eaton **AutoSelect** and AutoShift transmissions adapt a standard mechanical transmission platform for electronic management and partially automated operation. Each transmission uses a standard clutch and all the operating principles associated with Eaton mechanical transmissions. The earliest versions were designed for ten-speed transmissions, but the system is currently available on up to 18-speed transmissions. A real advantage of the AutoSelect and AutoShift transmissions is the extensive field knowledge of Eaton standard transmissions developed by truck technicians over many years. There are 3 primary functions associated with AutoSelect and AutoShift transmissions:

- **Selecting the starting gear.** All loads do not require a first gear start, and AutoShift permits the selection of a higher start gear. The driver is responsible for disengaging the clutch when the vehicle is stationary, and then selecting a shift lever position.
- **Engaging the clutch for starting and stopping.** When operating the AutoSelect/AutoShift transmission, the driver is required to use the clutch to get the vehicle rolling only after initial gear selection or to bring the vehicle to a stop.
- **Initiating a shift.** The AutoSelect prompts the driver when a shift is required. The driver breaks driveline torque to begin the shift and then the shift is completed while engine speed synchronizes with the transmission speed. The driver resumes accelerator input after the completion of the shift. AutoShift differs from AutoSelect in that it initiates all shifts other than at starting or stopping without input from the driver. It breaks driveline torque and completes the shift by itself. The latest version of AutoShift is designed to default to neutral if road speed drops too low. This prevents driveline and engine damage.

AutoSelect and AutoShift transmissions use a two-ECU management system. The system is mas-

tered by, and multiplexed to, other vehicle electronics by a **System Manager ECU**. Multiplexing is the term used to describe two separate electronic management systems that are connected so they "talk" to each other, allowing them to share input hardware and information. In an AutoShift transmission, a second ECU called the **Transmission Controller ECU** manages most of the switching functions of the transmission, effecting the processing outcomes of the System Manager. In simple terms, this means the System Manager does the "thinking" and the Transmission Controller converts the thinking into action. The transmission electronics are also designed to communicate with other chassis electronic systems.

This chapter will address only the electronic and electrical characteristics of AutoSelect and AutoShift. There will be much more emphasis on the AutoShift system than AutoSelect, simply because it has performed much better in the market place. AutoShift has received rapid acceptance from the truck driving community, which does not miss manual shifting through as many as 18 speeds as much as transmission manufacturers thought it might. Throughout this chapter, the assumption will be made that the operation of a truck standard mechanical transmission is fully understood.

15.1 TRANSMISSION IDENTIFICATION

The product coding used by AutoShift transmissions is more or less identical to that used in other Eaton RoadRanger twin-countershaft transmissions. An example of an Eaton serial number appears in **Figure 15–1**.

R	RoadRanger
T	Twin countershaft
A	Automated
O	Overdrive
14	× 100 + 50 = nominal torque capacity
7	Design level (5 = spur gear, 6 = multi-mesh gearing, 7 = multi-mesh front, helical auxiliary gearing)
10	Number of forward speeds
B/C	Ratio set designation
AC	AutoSelect
AS	AutoShift

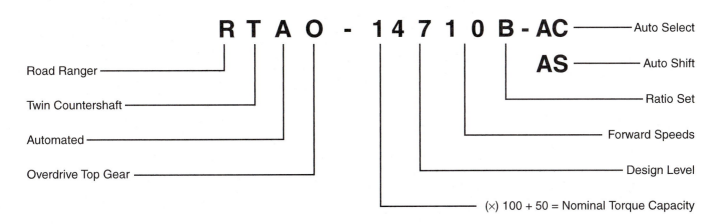

FIGURE 15–1   Eaton serial number codes. (Courtesy of Roadranger Marketing. One great drive train from two great companies—Eaton and Dana Corporations)

15.2 AUTOSHIFT COMPONENTS

The Eaton AutoSelect/AutoShift transmission components can be grouped as follows:

- Base standard gearbox
- Transmission automation components
- Vehicle automation components

For study purposes, the components in each subsection will be examined separately.

BASE TRANSMISSION

The AutoShift transmission is based on a Roadranger twin-countershaft platform. For purposes of this description, a transmission with 10 forward speeds will be used as an example. An Eaton 10-speed transmission has a 5-speed main or front case, with a 2-speed auxiliary section. Although the base transmission has a similar appearance to non-automated versions, the following changes are required:

- **Main box.** Two screw bosses have been added to the case so the transmission controller can be mounted to the side.
- **Range cylinder cover.** The range cylinder cover is altered to house the range shift control mechanism.
- **Shift bar housing.** The shift bar housing has been machined to accept the **speed sensors**.
- **Shift yokes.** The shift yokes have been hardened at the fork pads to improve durability during shifts.

TRANSMISSION AUTOMATION COMPONENTS

Several components have been added to the base transmission to automate main box and auxiliary section shifting. Because the AutoShift transmission performs electronically-over-electrically, exactly what the standard transmission accomplishes during a shift (fore and aft, right and left movement) is required of the shift finger in order to select a gear. Also, the engine, road, and gear speeds have to be computed to ensure that a non-binding shift is accomplished. Shifts are further optimized by giving the transmission electronics the capability to communicate with the other chassis electronic systems. The transmission automation components include

- X–Y shifter
- Range valve
- Speed sensors
- Reverse switch
- Power module
- Transmission Controller
- Transmission harness

ELECTRIC SHIFTER ASSEMBLY

The electric shifter assembly replaces the manual gear shift lever on the standard transmission. This can have the appearance of either a touch keypad or a paddle. Its function is to effect shifts in the transmission main box by using a shift finger. The shift finger is moved to select the correct rail and gear position using the normal standard shift bar housing. The electric shifter components are the shift finger, gear select yokes, ball screw assembly, electric motors, and position sensors. These components are shown in **Figure 15–2**.

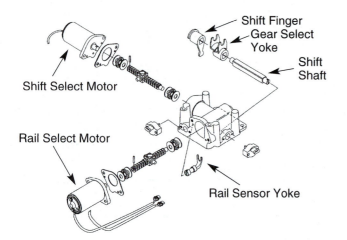

FIGURE 15–2 *Electric shifter assembly. (Courtesy of Roadranger Marketing. One great drive train from two great companies—Eaton and Dana Corporations)*

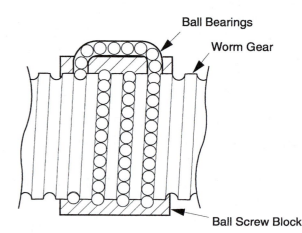

FIGURE 15–3 *Ball screw assembly. (Courtesy of Roadranger Marketing. One great drive train from two great companies—Eaton and Dana Corporations)*

The shift finger is the component that actually selects the commanded rail and gear. It appears similar to the end of a manual Roadranger shift lever. The shift finger is positioned first in the commanded rail, and next moved either forward or backward to select the commanded gear.

The shift shaft is a square section shaft within the electric shifter housing. The shift finger rides on the shift shaft. This permits the shift finger to be moved in the rail and gear select directions. The shift shaft has a machined area on the end to rotate the position sensor.

The gear select yoke also rides on the shift shaft. The yoke is driven by the gear select ball screw, which rotates the shift shaft for gear selection. The gear select sensor provides position data to the Transmission Controller.

Ball Screw Assemblies

Shift finger position is located by a pair of ball screws. One ball screw is used for the lateral (left to right) movement that is required to select the correct rail. The other ball screw is used for the forward-backward positioning required to select a gear.

Ball Screw Operation

The ball screw assembly is made up of a worm gear, ball screw nut, and ball bearings. Its operation resembles that of a recirculating ball steering gear. When the worm gear rotates, the ball bearings ride in the worm threads. When a ball reaches the end of a ball screw block, it is forced into a tube to be returned to the beginning of the worm threads in the ball screw block. See **Figure 15–3**.

Gear Select Ball Screw

The gear select ball screw nut has a pair of tabs, one on either side. These tabs drive the gear select yoke, which is located on the shift shaft with the shift finger. When gear selection is commanded, the ball screw block moves the gear select yoke, which in turn moves the shift shaft.

Rail Select Ball Screw

The rail select ball screw has a slot machined into it. The shift finger has a half moon protrusion designed to interface with the slot cut into the rail select ball screw. When a rail selection is made, the rail ball screw drives the shift finger that slides up and down the shift shaft. The rail select and gear select ball screws are not interchangeable despite their similarity in appearance.

The gear select and rail select ball screw assemblies are shown in **Figure 15–4**.

Electric Motors

Each of the ball screw assemblies is driven by its own electric motor. The motors use a permanent magnet operating principle and are capable of forward or reverse rotation depending on the direction of current flow. Changing the polarity of the current flowed through the motor is used to drive the ball screw assemblies to a specific position to effect gear shifts.

The **rail select motor** is pinned directly to the rail select ball screw and the **gear select motor** is pinned directly to the gear select ball screw. Once again, the electric motors used in each ball screw assembly are not interchangeable (see **Figure 15–5**).

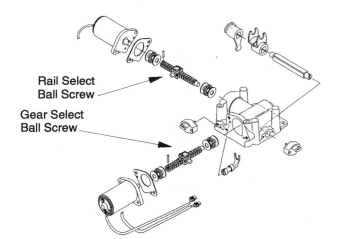

FIGURE 15–4   Gear select and rail select ball screw assemblies. (Courtesy of Roadranger Marketing. One great drive train from two great companies—Eaton and Dana Corporations)

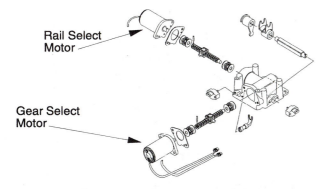

FIGURE 15–5   Gear select and rail select motors. (Courtesy of Roadranger Marketing. One great drive train from two great companies—Eaton and Dana Corporations)

Rail Select Motor

The rail select motor is a reversing DC motor managed by the Transmission Controller. Its function is to move the shift finger transversely (side to side) to select one of the 3 rails in the Roadranger shift bar housing.

Gear Select Motor

The gear select motor is a reversing DC motor managed by the Transmission Controller. Its function is to move the shift finger forwards and backwards: This movement is responsible for actually moving the shift rail into and out of a selected gear position.

Position Sensors

AutoSelect/AutoShift position sensors use a potentiometer operating principle. That is, they are 3-wire voltage dividers. A complete description of a poten-

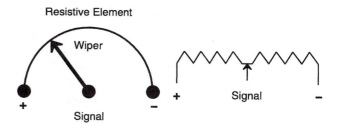

FIGURE 15–6   Position sensor potentiometer principle. (Courtesy of Roadranger Marketing. One great drive train from two great companies—Eaton and Dana Corporations)

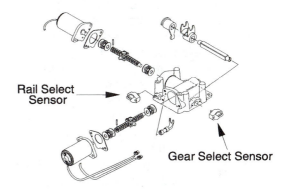

FIGURE 15–7   Location of position sensors. (Courtesy of Roadranger Marketing. One great drive train from two great companies—Eaton and Dana Corporations)

tiometer is provided in Chapter 6. The position sensor is supplied a reference voltage of 5V, modulated by the ECU. The function of the position sensor is to convert a mechanical position to a signaled voltage value which is returned to the ECU. The potentiometer consists of a resistor between the 5V input terminal and the ground connection. A wiper contacts the resistor and, depending on its location, determines the signal voltage value returned to the ECU. The potentiometer principle used in the position sensors is shown in **Figure 15–6**.

The shifter assembly incorporates 2 position sensors:

- The gear select sensor monitors shift finger position during gear selection. The signal returned to the Transmission Controller confirms whether a shift has been successfully effected.
- The rail select sensor is connected to the rail select ball screw. The position sensor yoke arm fits over a pin on the side of the rail select screw block (see **Figure 15–7**). Its signal allows the Transmission Controller to determine whether the shift finger is located over the correct rail before effecting a shift.

ELECTRIC SHIFTER OVERVIEW

The electric shifter is the means used by the Transmission Controller to effect shifts. A pair of motors are used to effect the shifts, while a pair of position sensors are used to help verify that the shifts have taken place. A shift sequence takes place as follows:

1. The rail select motor is energized.
2. The rail select motor turns the ball screw, which in turn moves the ball screw block.
3. The ball screw block moves the shift finger into the commanded rail select position; that is, one of the three rails is selected.
4. The rail select sensor signals the Transmission Controller that the shift finger is correctly located over the selected rail.
5. The gear motor is energized.
6. The gear motor turns its ball screw which in turn, moves the ball screw block.
7. The ball screw block moves the gear select yoke on the shift shaft, moving the shift shaft.
8. When the shift shaft moves, the shift finger is moved forward or backward for gear selection.
9. The gear select sensor confirms that the desired shift has been effected by signaling the Transmission Controller.
10. The shift sequence is completed.

Shop Talk

Current versions of AutoShift are designed to break driveline torque when a predetermined driveline (road) speed is sensed. This prevents driveline component stall damage to the engine.

RANGE VALVE

On AutoShift transmissions, the range valve takes the place of the HI-LO flipper switch mounted on the gear lever assembly. The HI-LO flipper switch was a master/slave air valve that controlled auxiliary section shifts on a typical Roadranger standard transmission. Similarly, the AutoShift range valve controls range cylinder air for auxiliary section shifts. The range valve contains a pair of solenoids, one for HI range and one for LO range. The range valve is shown in **Figure 15–8**.

Range Valve Components

The range valve consists of the following components:

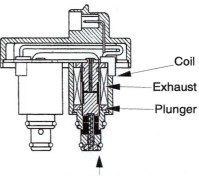

FIGURE 15–8 *Range valve components. (Courtesy of Roadranger Marketing. One great drive train from two great companies—Eaton and Dana Corporations)*

- **Plunger.** The plunger located at the center of the solenoid has a sealing surface at either end and 2 grooves cut in its side to allow for air travel.
- **Air Ports.** The solenoid contains 3 air ports:
  - The regulated air port is located at the bottom of the solenoid. It receives air from the air filter regulator. It has a seat to seal regulated air.
  - The exhaust port is located at the top of the solenoid. It is equipped with a seat used by the plunger to seal off the exhaust.
  - The range cylinder port is located at the side of the solenoid. It supplies air to the range cylinder.
- **Bias Spring.** The bias spring is located inside the solenoid cartridge. It forces the solenoid plunger downward, sealing off the regulated air port.
- **Coil.** The Transmission Controller uses the coil to control air at the range cylinder. The bias spring helps control the air. When the Transmission Controller energizes the coil, the resulting magnetic field pulls the plunger into the center of the coil. When the coil is no longer energized, the bias spring returns the plunger to its original position.

Range Valve Operation

Range valve operation is best understood by referring to diagrams. **Figure 15–9**, **Figure 15–10**, and **Figure 15–11** show range valve operation through each phase.

- **All Off:** When the vehicle is at rest, both solenoid coils are off, or electrical voltage is not applied to the coils. This causes the Bias Spring to push the solenoid Plunger down, closing off constant air supply and opening the Exhaust Port. This removes regulated air and exhausts to atmosphere, both sides of the Range Cylinder.

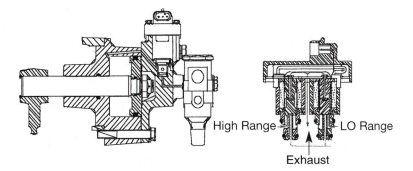

FIGURE 15–9　*Range valve: all-off position. (Courtesy of Roadranger Marketing. One great drive train from two great companies—Eaton and Dana Corporations)*

- **LO Range:** When a LO range shift is required, voltage is applied to the LO range coil. The magnetic field in the LO range coil pulls the Plunger up, opening the Constant Air Port and sealing off the Exhaust Port. This allows regulated air to flow into the LO range side of the Range Cylinder, shifting the transmission into LO range. Note that the HI range solenoid is still off, allowing the HI range side of the cylinder to exhaust.

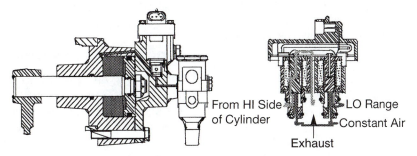

FIGURE 15–10　*Range valve: LO range operation. (Courtesy of Roadranger Marketing. One great drive train from two great companies—Eaton and Dana Corporations)*

- **HI Range:** When a HI range shift is required, voltage to the LO range coil is turned off and voltage is applied to the HI range solenoid. The Bias Spring pushes the LO range solenoid Plunger back down, opening the Exhaust Port and sealing off the Regulated Air Port. This evacuates all the air from the LO range side of the cylinder.

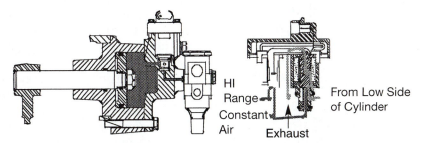

The coil's magnetic field pulls up the HI range Plunger, opening the Regulated Air Port and closing the Exhaust Port. This allows regulated air to push on the HI range side of the Range cylinder, shifting the transmission into HI range.

FIGURE 15–11　*Range valve: HI range operation. (Courtesy of Roadranger Marketing. One great drive train from two great companies—Eaton and Dana Corporations)*

SPEED SENSORS

Speed sensors are part of the input circuit. Their function is to signal shaft speed data to the Transmission Controller. Three speed sensors are used in the trans-

mission. They are responsible for signaling input shaft speed, mainshaft speed, and output shaft speed data. All three use an inductive pulse generator principle. A ferrous metal pulse wheel (chopper or tone wheel) with external teeth cuts through a magnetic field at a

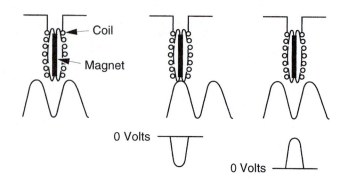

FIGURE 15–12 *AC pulse generator principle.*
(Courtesy of Roadranger Marketing. One great drive
train from two great companies—Eaton and Dana
Corporations)

sensor coil; when the teeth of the pulse wheel cut
through the magnetic field in the sensor, an AC voltage is induced in pulses. This principle is shown in
Figure 15–12 and explained in some detail in Chapter
6. The AC voltage value and frequency produced will
rise proportionally with rotational speed increase; in
this way, the Transmission Controller can determine
shaft speed by correlating the frequency and voltage
value with rpm. Testing of these sensors is by measuring voltage output.

Input Shaft Speed Sensor

The input shaft speed sensor is located at the right
front corner of the shift bar housing. The speed sensor
magnet and coil produce pulses from the upper countershaft PTO gear. The AC voltage and frequency rise
proportionally with shaft speed, and this data is delivered to the ECU. The upper countershaft is direct
driven by the input shaft through the head set gears,
so input shaft speed is calculated by multiplying countershaft speed by the head set ratio. **Figure 15–13**
shows the location of the input shaft speed sensor.

 Input Speed

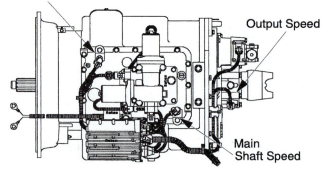

FIGURE 15–13 *Location of speed sensors.*
(Courtesy of Roadranger Marketing. One great drive
train from two great companies—Eaton and Dana
Corporations)

Mainshaft Speed Sensor

The mainshaft speed sensor is located at the left rear
corner of the shift bar housing at the auxiliary countershaft drive gear. Again, a pulse generator principle
is used, so the AC voltage value and frequency fed
to the ECU increase proportionally with shaft rotational speed. The auxiliary countershaft drive gear is
meshed to the mainshaft through the auxiliary drive
gear and driven at the same speed. **Figure 15–13**
shows the location of the mainshaft speed sensor.

Tailshaft Speed Sensor

The output shaft speed sensor picks up pulses produced by a tone ring located on the output shaft.
This tailshaft located sensor inputs road speed data
to the Eaton Transmission Controller and to other
chassis electronic systems. Road speed is determined by factoring the output shaft speed with the
drive axle carrier ratio and the rolling radius dimension of the tires. A standard 16 tooth pulse wheel is
used. **Figure 15–13** shows the location of the tailshaft speed sensor.

Function of the Speed Sensors

The input shaft speed and mainshaft speed data is
used by the Transmission Controller for synchronizing shifts in the transmission main box. The data
generated by the mainshaft and output shaft speed
sensors is used by the Transmission Controller to
verify range engagement. The output or tailshaft
speed sensor may also be shared with other chassis
electronic systems as vehicle road speed data.

REVERSE SWITCH

The reverse switch functions to signal the
Transmission Controller when the transmission is in
reverse. It consists of a pair of normally open terminal contacts, a bias spring, a plunger contact, and an
actuating ball. The reverse ball switch rides on a
plunger that follows the reverse rail in the shift bar
housing. When the transmission is not in reverse, the
switch pin rests down in a groove on the reverse rail.
The bias spring acts on the plunger contacts and the
switch remains open. When the transmission is
shifted into reverse, the switch pin acts on the switch
ball—forcing the plunger contacts onto the terminals.
Reverse switch operation is shown in **Figure 15–14**.

One of the reverse switch terminals is connected
to battery voltage. The other is connected to a
Transmission Controller input. When the switch is
closed (reverse gear selected), battery voltage is
input to both the Transmission Controller (as a signal) and the vehicle backup lights.

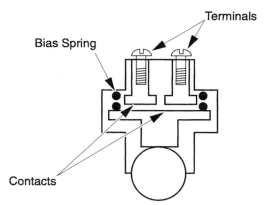

FIGURE 15–14 *Reverse switch operation. (Courtesy of Roadranger Marketing. One great drive train from two great companies—Eaton and Dana Corporations)*

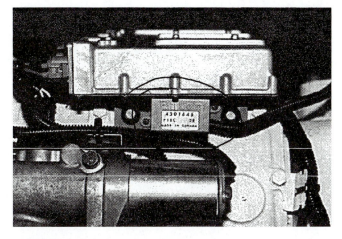

FIGURE 15–15 *Power module location. (Courtesy of Roadranger Marketing. One great drive train from two great companies—Eaton and Dana Corporations)*

POWER MODULE

The **power module** connects the vehicle electrical system with the transmission electrical and electronic systems and protects the transmission electrical system from current overloads with cycling-type circuit breakers. The unit will also protect the transmission circuits from voltage spikes and reverse voltage problems. The power module is located on the Transmission Controller next to the electric shifter unit (see **Figure 15–15**). It comprises the following components:

- Cycling circuit breakers
- Reverse voltage diode
- Reverse voltage Zener diode

Cycling Circuit Breakers

A circuit breaker protects circuits or portions of circuits from excessive current flows and high voltage spikes such as might be caused by boost generators or chargers. The AutoShift power module contains two SAE #1 cycling circuit breakers, one current-rated at 25 amps and the other at 10 amps. A cycling circuit breaker is an automatically resetting type. The 25-amp breaker protects the circuit supplying the electric shifter motors and the logic battery power from current overloads. The 10-amp breaker is located in series in the circuit supplying logic battery power after the 25-amp breaker. A cycling circuit breaker has a bimetal arm and two contacts as shown in **Figure 15–16**. The breaker is normally closed. When current flow is within its range specification, the bimetal arm has some spring tension and closes the contacts permitting current flow. When the current load through the breaker exceeds the specified rating, the bimetal arm heats up; the different metals subjected to heat expand at different rates, causing the arm to break contact. This opens the circuit and stops current flow. As the bimetal strip cools, it once again closes the contacts, permitting current flow.

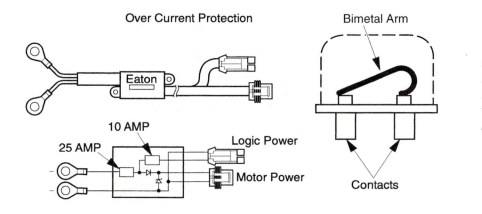

FIGURE 15–16 *Cycling circuit breakers. (Courtesy of Roadranger Marketing. One great drive train from two great companies—Eaton and Dana Corporations)*

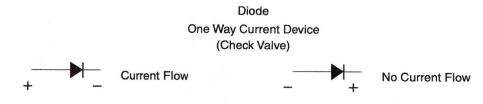

Diode
One Way Current Device
(Check Valve)

+ —   Current Flow   — +   No Current Flow

FIGURE 15–17   *Reverse voltage diode schematic. (Courtesy of Roadranger Marketing. One great drive train from two great companies—Eaton and Dana Corporations)*

Zener Diode
Almost A One Way Current Device
(Voltage Regulator)

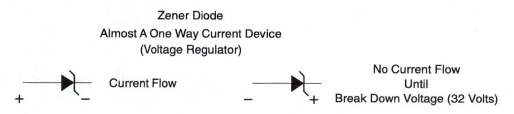

+ —   Current Flow

No Current Flow
Until
— +   Break Down Voltage (32 Volts)

FIGURE 15–18   *Reverse voltage Zener diode schematic. (Courtesy of Roadranger Marketing. One great drive train from two great companies—Eaton and Dana Corporations)*

Reverse Voltage Diode

The reverse voltage diode is located just past the 25-amp circuit breaker. The diode permits current flow in one direction only. In the event of a reverse polarity condition, the reverse voltage diode blocks current flow into the transmission electrical circuit.

A reverse voltage diode schematic is shown in **Figure 15–17**.

Reverse Voltage Zener Diode

The reverse voltage Zener diode is located between the negative terminal on the reverse voltage diode and battery ground. The Zener diode functions as a normal diode but acts to regulate voltage should it exceed a specified level. The reverse voltage Zener diode used in the Eaton power module will permit current flow while not permitting the voltage value to exceed 32V.

A reverse voltage Zener diode schematic is shown in **Figure 15–18**.

Power Module Functions

The power module has the following functions (see **Figure 15–19**):

- **Motor Power.** Connected to the vehicle cranking motor, the circuit passes through the 25-amp circuit breaker and the reverse voltage diode and connects to the motor terminals. This means that there is a constant power supply to the power module.
- **Motor ground.** The motor is direct grounded through the power module by means of a wire to the cranking motor ground terminal.
- **Logic power. Logic power** refers to the electric feed required to power up the system ECUs. It is routed from the vehicle cranking

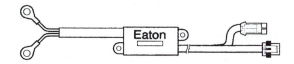

FIGURE 15–19   Power module connections. *(Courtesy of Roadranger Marketing. One great drive train from two great companies—Eaton and Dana Corporations)*

motor battery terminal, through both the 25-amp and 10-amp circuit breakers, and out to the logic power connector.
- **Logic ground.** The logic ground is routed direct from the cranking motor ground terminal, through the power module, and out to the logic power connector.

15.3 TRANSMISSION CONTROLLER

The Transmission Controller is one of two ECUs used by the Eaton AutoShift transmission. It has some limited processing capability but is mainly responsible for managing the switching requirements of the system. The Transmission Controller ECU is multiplexed; that is, it is electronically connected with the System Manager ECU. A full explanation of vehicle management computers is provided in Chapter 6; a brief overview of the Eaton Transmission Controller follows.

The Transmission Controller is located on the left side of the transmission, held in place by a pair of dowels in the shift bar housing and two bolts threaded into the case. It is connected to the system

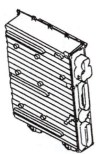

FIGURE 15–20 The Transmission Controller and its location. (Courtesy of Roadranger Marketing. One great drive train from two great companies—Eaton and Dana Corporations)

electrical/electronic circuit by means of a harness called the transmission harness. This connects logic power and ground to the transmission systems. All of the input and output circuits are connected to the Transmission Controller by means of this harness, including the data link to the System Manager. The Transmission Controller and its location are shown in **Figure 15–20**.

TRANSMISSION CONTROLLER FUNCTIONS

The Transmission Controller performs the following functions:

- **Receiving Input.** The Transmission Controller receives command inputs (such as shifter position) and monitoring sensor (such as speed and position sensor) signals and converts analog data to digital for processing.
- **Processing data.** On the basis of data input from the command and monitoring circuits, the Transmission Controller processes desired outcomes from the System Manager and determines whether they can be effected.

- **Driving outputs.** The Transmission Controller is responsible for switching the shifter motors, the range valve, and the optional inertia brake. It also handshakes data with the System Manager by outputting speed, gear range, and diagnostic information.

INERTIA BRAKE SYSTEM (AUTOSHIFT ONLY)

The **inertia brake** system on an AutoShift transmission helps stop the countershafts on initial engagements when the clutch is either out of adjustment or defective. The inertia brake consists of a solenoid and brake. The solenoid controls the inertia brake by switching regulated air pressure to act on the inertia brake piston and load the friction and reaction discs, which brake the countershaft. The solenoid is energized at 12V. The inertia brake is mounted at the 6- or 8-bolt PTO and, when applied by air pressure, connects the PTO gear with the grounded shaft, which in turn connects the countershaft to the transmission case, stopping it. When the inertia brake needs to be released, the Transmission Controller ceases to energize the solenoid and the air supply acting on the inertia brake piston is exhausted. This frees up the friction and reaction discs, permitting the countershafts to turn freely. A sectional view of the inertia brake assembly is shown in **Figure 15–21**.

CAUTION: When removing the inertia brake solenoid, ensure that system air pressure is completely relieved and the air/filter regualtor is removed.

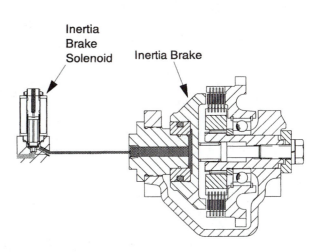

Inertia Brake Solenoid   Inertia Brake

FIGURE 15–21 Sectional view of the inertia brake. (Courtesy of Roadranger Marketing. One great drive train from two great companies—Eaton and Dana Corporations)

15.4 VEHICLE AUTOMATION COMPONENTS

The vehicle automation components connect the transmission electrical and electronic circuits with those of the vehicle. Those components include

- System Manager
- Gear display
- Shift lever
- Power connect relay
- Start enable relay
- Vehicle harnesses

SYSTEM MANAGER

The System Manager ECU is a second microprocessor module with the dual functions of interfacing with the rest of the vehicle and overall management of the transmission systems. It performs the following functions:

- **Controlling logic power.** The System Manager is responsible for switching system logic power on and off.
- **Managing shifting.** The primary processing functions of the AutoSelect/AutoShift transmission take place in the System Manager. The System Manager processes desired outcomes based on programmed shift logic maps in its data (memory) banks and communicates these as commands to the Transmission Controller. Shift commands are made on the basis of data input from the Transmission Controller, signal data from the command and monitoring circuits, and data received from other chassis systems.
- **Processing fault conditions.** System Manager receives and can act on diagnostic fault codes where required.
- **Processing shift lever commands.** When the driver selects a range position, the System Manager processes the command and effects it as long as it will not result in transmission or drivetrain damage. The System Manager communicates with the shift lever by means of the shift lever data link.
- **Controlling Service and Wait lights.** The System Manager can illuminate the Service and Wait lights. The signal is delivered through the shift lever data link.
- **Providing information to lever digital display.** Gear and engine speed information are provided by the System Manager to the digital gear display. This status information is also sent through the shift lever data link.

The System Manager is usually located in the shift tower assembly. It can be located elsewhere if the chassis application requires it.

> **CAUTION:** The shifter module must be calibrated before a vehicle is placed in operation.

Electric Shifter Calibration

The System Manager commands the Transmission Controller to calibrate the electric shifter each time the ignition key circuit is shut down. The calibration procedure accounts for the clicking noise heard each time the key is turned off. When the calibration is complete, the Transmission Controller stores the shifter calibrations in its nonvolatile RAM and signals the System Manager that the calibration is complete.

The calibration procedure requires that the neutral position be located by performing a rail sweep. The rail select and rail position motors move the shift finger, and the Transmission Controller records the point at which the sensors stop moving at the extremity of each sweep.

Data Retention

The System Manager stores fault code data in nonvolatile RAM before turning off the power connect relay during each shutdown.

SHIFT LEVER

The shift lever signals driver commands to the System Manager. It also displays data to the driver by means of Service and Wait lights. The chassis OEM determines the location of the shift lever assembly, but typically it is located within comfortable reach of the driver beside the driver seat. This can be a touch keypad, steering column paddle, or shift lever and console. A shift lever assembly and its console are shown in **Figure 15–22**.

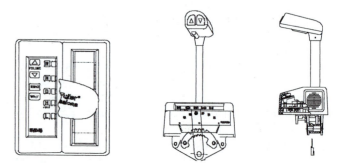

FIGURE 15–22 *Shift lever and console. (Courtesy of Roadranger Marketing. One great drive train from two great companies—Eaton and Dana Corporations)*

Shift Lever Operation

The shift lever is moved mechanically by the operator. **Hall effect** principle switches are used to signal the range request position of the shift lever to the System Manager. A magnet is located at the bottom of the shift lever below the display console. Its magnetic field will have the effect of closing the normally open Hall effect switch, sending a current signal to the System Manager. In this way, a range request command is made to the System Manager.

Service Light

The Service light provides fault information to the driver or technician. A flashing or continuous illumination of the Service light indicates that a fault code has been logged. Logged fault codes can then be pulse displayed by flashing numeric codes on the Service light.

Wait Light

The Wait light is only found on the earliest versions of AutoSelect/AutoShift. It is located on the shift console just below the Service light. It is illuminated on power-up while the System Manager and Transmission Controller perform a self-check sequence. If the transmission checks OK, the Wait light turns off.

Speaker

The speaker tone is used to alert the driver to break torque on the driveline so a shift can be effected. The tone signal is output by the System Manager.

Detents

The detent mechanism locates the transmission shift lever in the correct location in each range select position.

GEAR DISPLAY

The gear display is on the light emitting diode (LED) digital driver display located in the dashboard (see **Figure 15–23**). The display signals are output from the System Manager. The information displayed tells the driver the shift status in range and target gears during shifting. When a shift is initiated, the target gear is displayed continuously on the gear display LEDs. When the transmission shifts to neutral, the target gear begins to flash. When the shift sequence is completed, the display ceases to flash and the range status is displayed.

With an AutoShift transmission, up and down arrows in the display are used to inform the driver of

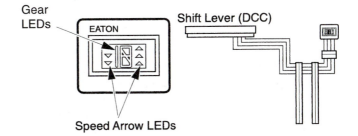

FIGURE 15–23 Digital gear display unit. (Courtesy of Roadranger Marketing. One great drive train from two great companies—Eaton and Dana Corporation)

what is required to synchronize the input shaft speed (that is, engine rpm, providing the transmission is in gear) with the transmission main shaft speed. Up arrows indicate the input shaft speed must be increased until it matches main shaft speed. Down arrows indicate the input shaft speed must be decreased to match the main shaft speed.

VEHICLE HARNESSES

The vehicle harnesses function to interface the transmission with the rest of the vehicle (see **Figure 15–24**). Three vehicle harnesses are required:

- *The transmission interface harness* (**Figures 15–25** and **15–26**) connects the transmission harness to the tower harness and System Manager.
- *The tower harness* (**Figures 15–27** and **15–28**) connects the System Manager to the shift lever, and power connect relay and provides a service port for voltage checks. It connects all the components in the shift tower.
- *The vehicle interface/dash harness* (**Figure 15–29**) connects the tower harness and System Manager to components inside the dash.

Tower Harness Service Port

The service port is located in the tower harness. It is used to access circuits for voltage checks during troubleshooting. The circuits can be checked without separating harness connectors.

Service port connections (see **Figure 15–28**):

- Pin A: ignition power.
- Pin B: dummy.
- Pin C: measures voltage at the power connect relay.
- Pin D: ground.
- Pin E: start enable relay coil.

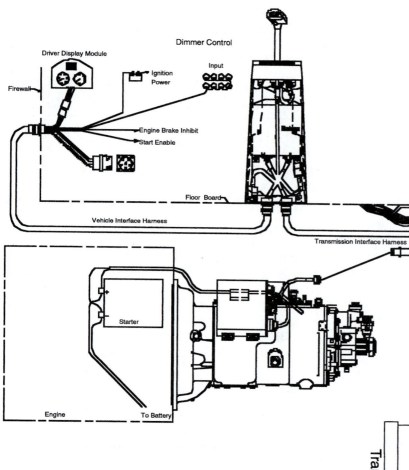

FIGURE 15–24   Vehicle harnesses. (Courtesy of Roadranger Marketing. One great drive train from two great companies—Eaton and Dana Corporations)

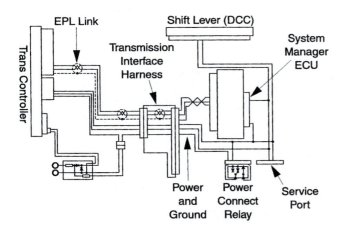

FIGURE 15–25   Transmission interface harness. (Courtesy of Roadranger Marketing. One great drive train from two great companies—Eaton and Dana Corporations)

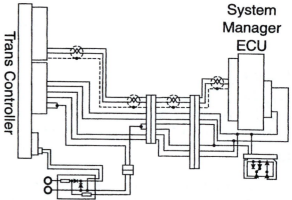

FIGURE 15–26   Revised transmission interface harness. (Courtesy of Roadranger Marketing. One great drive train from two great companies—Eaton and Dana Corporations)

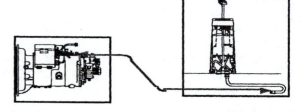

FIGURES 15–27   Tower harness. (Courtesy of Roadranger Marketing. One great drive train from two great companies—Eaton and Dana Corporations)

- Pin F: dummy.
- Pin G: power connect relay coil. Output from the System Manager.

Figures 15–29 shows the vehicle interface, dash harness, and gear display wiring.

- A Ignition voltage
- B
- C Battery Voltage
- D Ground
- E Start Enable Relay Coil
- F
- G Power Connect Relay Coil

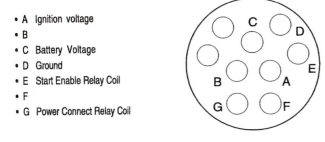

FIGURES 15–28 Tower harness service port identification. (Courtesy of Roadranger Marketing. One great drive train from two great companies—Eaton and Dana Corporations)

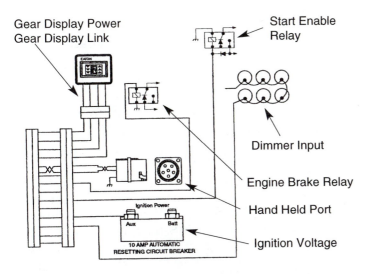

FIGURE 15–29 Vehicle interface, dash harness and gear display wiring. (Courtesy of Roadranger Marketing. One great drive train from two great companies—Eaton and Dana Corporations)

15.5 FAULT DETECTION AND DATA LINKS

The AutoSelect/AutoShift transmission electronics are designed to detect faults within the transmission electrical/electronic circuits as well as faults that occur in other vehicle systems on which it is dependent. **SAE J 1939**–compatible electronics permit the Eaton electronics to interact with other onboard vehicle electronic systems, as shown in **Figure 15–30**. For readout and diagnostic purposes, faults are coded in both Eaton and SAE J 1939 formats.

FALLBACK MODES

Whena fault code is logged, the transmission electronics are designed to assess all remaining functionality and then default to a limited operational mode.

Four fallback modes may be used. Each is classified by the number of mechanical shifts that can be performed.

Five-Speed Fallback

In a 5-speed fallback, the detected fault has occurred in the auxiliary section. The main section is fully operational. Shifting is therefore limited to the 5 speeds in the main box, in whichever range gear in the auxiliary box that can be achieved. For instance, when only LO range is available, the transmission will shift from first through fifth gear ratios. When only HI range is available, shifting is limited to sixth through tenth gear ratios.

One-Speed Fallback

One-speed fallback indicates a transmission fault where all shifting is inhibited. The transmission remains in one gear ratio, usually the one in which the fault was logged.

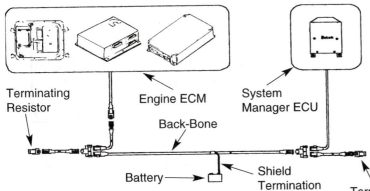

FIGURE 15–30 SAE J 1939 Data link connection to engine ECM. (Courtesy of Roadranger Marketing. One great drive train from two great companies—Eaton and Dana Corporations)

In-Place Fallback

An in-place fallback results in the transmission remaining in whatever state it is at the time of the occurrence. This could be in gear or in neutral.

Downshift Fallback

The transmission will downshift when appropriate until the vehicle is brought to a stop and the transmission is in neutral.

Fault Codes

11	System Manager ECU
12	Transmission Controller ECU
13/17/31	Relays
14	Shift lever
15	Shift lever communications
16	EPL communications
33	Battery voltage
35	Engine control (AutoShift only)
43	Range solenoid
44	Inertia brake solenoid (AutoShift only)
51/52	Position sensors
53	Reverse switch
61/63	Shifter motors
65	Shifter voltage
83	Missing lever
41	Range failed to engage
56/57/58	Speed sensors
71	Stuck engaged
72	Failed to select rail
73	Failed to engage gear
74	Failed to synchronize initial engagement

FAULT CODE STATUS

An active code indicates a problem active at the time of reading the system. An inactive code (historic code) indicates the occurrence of a fault that has either occurred in the past, or is not detected at the moment of reading the system. The presence of an inactive code may indicate that the vehicle has not been driven in the proper manner. All AutoSelect/AutoShift system troubleshooting should be performed using the Eaton troubleshooting guide.

Pretests

Transmission pretests should be performed before progressing to specific fault code isolation. Prechecks require that electrical inputs and pneumatic supply to the transmission be functioning properly. The pretests eliminate obvious high-level

problems immediately before progressing to a detailed, sequentially stepped diagnostic procedure.

Symptom-Driven Diagnostics

A transmission is, above all, a mechanical device, and problems will occur in which active or inactive electronic codes are not logged. Never assume that every fault must be electronically monitored.

Fault Isolation Procedure

The Eaton troubleshooting service literature is required. The objective of any diagnostic procedure is to isolate the failure from all its possible causes. In an Eaton AutoSelect/AutoShift transmission, there are the mechanical components of the main gearbox and the auxiliary box; the pneumatic control system; the electrical components, such as shift motors; and a comprehensive electronic management system. It is critically important to always follow the troubleshooting guide so the diagnostic approach is systematic.

15.6 ROCKWELL MERITOR ESS AND NOCLUTCH

Rockwell Meritor introduced Engine SynchroShift (ESS) in 1997 and has recently announced a second generation of the technology, a joint venture with the German company ZF. ESS functions overall very much as does the AutoShift system. Nevertheless, ESS will be briefly described here.

ESS is designed to automatically synchronize engine rpm to road speed during any shift sequence. The ESS transmission has no independent processing capability, and it essentially uses the engine electronics to operate. The ESS input circuit consists of input and output shaft speed sensors which talk to the engine ECM by means of a data bus and can make the shifting of a non-synchronized manual transmission as easy as shifting a fully synchronized transmission. Some of the features of ESS are

- Drivers use the clutch only for starting and stopping the vehicle. This reduces the number of clutch engagements, increasing clutch life and reducing driver fatigue.
- Auxiliary range shifts are automatic, reducing driver effort and synchronizer damage.
- ESS features a "break torque" button that enables the driver to power upshift and downshift. This reduces the driver skill required to make quick shifts while climbing a steep grade.

- ESS permits the driver to make brake pedal applications while downshifting. This feature makes downhill braking much safer and can result in more effective use of an engine compression brake.
- No transmission electronic controller is required. The transmission electronics simply input data to the engine ECM and get it to perform the processing required to make the system work.
- Rockwell Meritor also claims that ESS can improve fuel economy.

Operation

The ESS shift lever includes an on/off switch. This permits the system to perform in automated shift mode or with the system entirely deactivated. When the system is turned off, the transmission functions exactly as any manually shifted transmission. When ESS is switched on, the transmission operates as follows:

- **Upshift.** The driver applies force to the gear lever towards neutral while in gear and then depresses the "shift intent" switch up. Next, the gear lever actually moves into the neutral position, allowing the engine to automatically synchronize and the upshift to take place.
- **Downshift.** The driver applies force to the gear lever towards neutral while in gear and then depresses the "shift intent" switch down. Next, the gear lever is moved into the neutral position, allowing the engine to automatically synchronize and the downshift to take place.
- **Clutch action.** Upshifts and downshifts require no clutch action. In other words, the clutch remains fully engaged for all shifts.

Shop Talk

The ESS transmission is truly smart in that it uses multiplexing technology to get the engine electronics to perform the "thinking" (logic processing) required. There is no processing ability in the transmission. ESS provides inputs to the engine ECM and effects outputs by switching the required shifts.

System Electronics

The ESS system is intended as a Class 8 truck transmission and is compatible with Caterpillar, Cummins, and DDC engine electronics. ESS requires no transmission processing capability and simply has an input circuit that feeds data to the engine electronics.

The transmission input electronics communicate with the engine electronics by means of the J 1922/1939 data bus. Transmission sensors read the speed of the input and output shafts during all phases of operation, and this information is communicated to the engine ECM. When a shift is required, the shaft speed signals from the transmission are used by the engine electronics to either increase or decrease rpm and synchronize the engine speed to the transmission. Because of the simplicity of the system and the absence of processing capability, the cost of ESS makes it competitive.

SUMMARY

- The Eaton AutoSelect/AutoShift transmission is based on a standard Roadranger twin-countershaft mechanical transmission platform.
- The system electronics adapt the transmission for electronic management and semi-automated shifting.
- The transmission electronic system uses two ECUs to perform logic processing and output switching operations in the system.
- The System Manager performs most of the processing: It manages transmission functions and information, controls shifting through its handshake connection with the Transmission Controller, acts on driver command signals, performs diagnostic self-checks, and logs fault codes.
- The Transmission Controller manages shifting based on command signals received from the System Manager, monitors the input circuit components physically located within the transmission, exchanges speed and gear ratio status data with the System Manager, and provides the System Manager with diagnostic information.
- The System Manager ECU is normally located in the shift tower assembly.
- The Transmission Controller ECU is located on the left side of the transmission.
- Shifts are effected by an electric shifter assembly that replaces the gear shift lever in a standard Roadranger transmission: A pair of reversing electric motors actuated by the Transmission Controller precisely locates a shift finger in the recesses in the shift rails.
- The rail select motor is used to move the shift finger transversely (side to side) with the objective of locating it precisely in the shift recess of one of the three shift rails.
- The gear select motor moves forward and backward. After the shift finger has been positioned on the correct shift rail, the gear select motor is energized to move the rail into gear.

- The Eaton AutoSelect and AutoShift transmissions perform self-diagnostic tests and log fault codes when performance parameters of any electronically monitored component or system are out of specification.
- Readouts of fault codes can be made using the AutoSelect/AutoShift diagnostic (Service) lights or by using a handheld diagnostic EST.
- AutoSelect/AutoShift electronics are SAE J 1939–compatible, which permits multiplexing with other chassis electronics systems.

REVIEW QUESTIONS

1. Which letters in the suffix of the Eaton Roadranger transmission would indicate that the transmission was an AutoShift electronically shifted version?
   a. AS
   b. AC
   c. RT
   d. B

2. Which component on the AutoShift transmission performs the mechanical function of the shift lever on the standard Roadranger transmission?
   a. gear select yoke
   b. shift shaft
   c. worm gear
   d. shift finger

3. Which of the following correctly describes an AutoSelect/AutoShift gear select motor?
   a. reversing DC motor
   b. unidirectional DC motor
   c. reversing AC motor
   d. unidirectional AC motor

4. What type of sensor is used to signal shift finger position to the Transmission Controller?
   a. piezoelectric
   b. photoelectric
   c. potentiometer
   d. inductive pulse

5. Which of the following components is responsible for effecting shifts in the transmission auxiliary box?
   a. range valve solenoid
   b. rail select motor
   c. power module
   d. reverse switch

6. Which component in the AutoShift system protects the circuit electrical and electronic systems from voltage spikes by acting as an electronic voltage regulator?
   a. reverse voltage diode
   b. reverse voltage Zener diode
   c. 25-amp circuit breaker
   d. 10-amp circuit breaker

7. What is the usual location of the AutoShift System Manager ECU?
   a. shift tower
   b. left side of the transmission
   c. on top of the transmission
   d. under the cab dash

8. What is the function of the inertia brake used in the AutoShift transmission?
   a. slows the input shaft during driveline torque release
   b. slows the countershaft during initial engagement
   c. used to synchronize all shifts
   d. effects driveline retarding

9. To which AutoShift component is the driver shift lever command directed?
   a. Transmission Controller ECU
   b. gear select motor
   c. System Manager ECU
   d. power module

10. What should the driver do when the speaker tone in the shift lever assembly is sounded?
    a. upshift
    b. downshift
    c. accelerate
    d. break torque on the driveline

16

Allison Commercial Electronic Control (CEC) Transmission

Learning Objectives

After reading this chapter, you should be able to

- Outline the construction and critical components in an Allison transmission with CEC management electronics.
- Define the acronym CEC.
- Identify the components used in the CEC transmission to convert electronic processing into hydromechanical outcomes.
- Outline the electronic circuit components used in the CEC transmission.
- Describe the types of ECU that may be used to manage the CEC electronics.
- Identify the input signals required by the CEC ECU.
- Identify the outputs produced by the CEC ECU and describe how they effect the results of logic processing.
- Track the power flows produced in each gear range in the CEC transmission.
- Interpret CEC diagnostic codes.
- Perform a snapshot test to diagnose an intermittent transmission problem.

Key Terms

Commercial Electronic Control (CEC)
digital multimeter (DMM)
digital data reader (DDR)
electro-hydraulic valve body
hard codes

latching solenoids
nonlatching solenoids
ProLink
snapshot mode
soft codes

A number of electronically controlled automatic transmissions are available in North America. Many of these are found in buses. General Motor's Allison Transmission Division has dominated the heavy-duty highway automatic transmission market, and the only real threat to this dominance has come from off-shore, specifically from the German companies ZF and Voith. However, ZF and Voith transmissions are more likely to be found in bus applications than

trucks and currently hold only a small share of the market. This chapter will focus on the first of two Allison transmission series likely to be found in medium- and heavy-duty truck chassis.

Allison has used computerized management of their automatic transmissions since 1982. These controls have become more comprehensive over the years. In the current versions of Allison transmissions known as **Commercial Electronic Control (CEC)**,

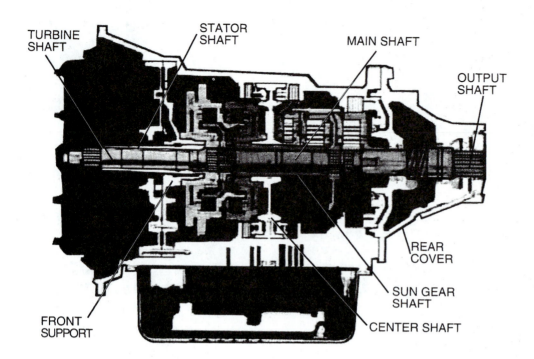

FIGURE 16–1
Transmission component overview. (Courtesy of Allison Transmission Division of General Motors)

the degree of control is probably best described as "partial authority. " CEC controlled transmissions are really just a means of adapting a hydromechanical Allison transmission for electronic management, whereas World transmissions (WTs), the subject of Chapter 17, are "full authority" transmissions. CEC was previously known as Allison Transmission Electronic Control (ATEC).

Automatic transmissions have a high unit cost compared to standard transmissions, a factor that leads them to be used in applications, such as buses and highway coaches, in which they may be fully reconditioned several times over during their service life. This high unit cost, rather than inappropriateness for the application, tends to keep them out of highway truck applications, although they are commonly found in city pick-up and delivery truck chassis and vocational applications such as fire trucks and garbage packers. A major advantage of the automatic transmission is the fact that a lower level of driver skill is required. Becoming accustomed to properly shifting a fifteen speed transmission requires some ability to understand the mechanics of the relationship between the engine and chassis drivetrain, a skill the highway truck driver develops over time. Handling automatic transmissions requires less skill. Because they are driven through a fluid coupling or torque converter, this provides some forgiveness when a shift ratio selection mistake is made. When electronic controls are added to a hydromechanical transmission, range shift selection becomes

more precise and the transmission is more protected from driver abuse.

The objective of this chapter will be to examine the Allison CEC transmission. A knowledge of the operating principles of Allison hydromechanical automatic transmissions will be assumed. This chapter will explore the differences between the hydromechanical Allison transmissions covered in Chapters 13 and 14 and the CEC electronically controlled version. In other words, we will take a look at the management electronics and hardware. **Figure 16–1** shows the major components of an Allison CEC transmission.

16.1 ELECTRONIC CONTROL OF ALLISON CEC TRANSMISSIONS

As we said in the introduction, Allison transmissions denoted as CEC use the same clutching and planetary gear sets as non-electronically managed Allison transmissions. In fact, in terms of operating principles, the CEC has much more in common with a non-computerized Allison transmission than with the newer WT electronic transmission described in the next chapter. The CEC system manages the transmission's hydraulics, so it is essential to fully understand the basic operation of an Allison hydromechanical automatic. The CEC transmission uses an **electro-hydraulic valve body** that controls the hydraulic circuits using solenoids. These solenoids replace the shift signal valves described in Chapter 13. Switching signals are output commands from the Allison electronic control unit (ECU).

The ECU receives inputs from a sensor circuit. Electronic tailshaft speed sensing replaces governor pressure, a TPS (throttle position sensor) signal replaces modulator pressure, and, depending on the application, the ECU may receive other sensor data regarding temperature. A push button or shift lever may be used to input command data to the ECU.

Several on-highway CEC transmissions are available, including the following:

> MT(B) 648
> HT(B) 741
> HT(B) 748
> HT(B) 755 CR
> HT(B) 755 DR
> V731 AND VR731

ELECTRONIC CONTROL UNIT (ECU)

The ECU is an onboard microcomputer that receives sensor and command data and manages shifting. Three ECU types are used with CEC transmissions; these are diagrammed in **Figure 16–2**.

- **Splashproof.** This is the earliest model, no longer in production.
- **Sealed Standard.** This model replaced splashproof.
- **Sealed Plus II.** This model has secondary mode capabilities and includes an additional connector to accommodate a secondary shift selector and other functions.

Allison suggests that the ECU be mounted in a location protected from direct exposure to weather and high concentrations of dust and sunlight with regard to the following:

- The ECU should be located away from road splash. Although it can stand up to the moisture a truck chassis is subjected to during a rainstorm, it should never be fully immersed in water.
- The ECU should be mounted in the coolest practical location, with good air flow and bolted to metal to enable good heat dissipation. All microcomputers and their switching apparatus (the means of generating output signals) create heat, and this heat can damage the components if it is allowed to build up.
- The ECU should be positioned with its electrical wiring connectors facing downwards if possible to avoid direct contact with moisture. Sufficient clearance should be afforded to permit ECU connectors to be removed without dismantling the unit.

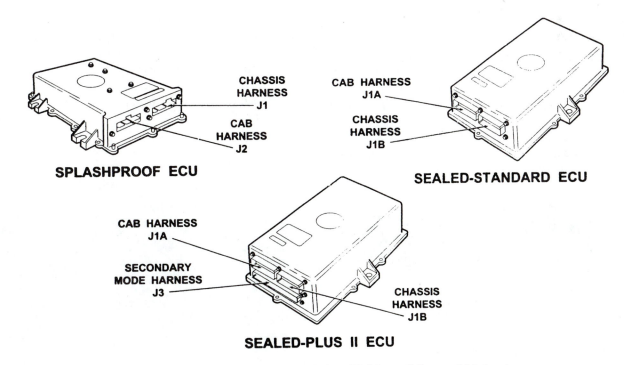

FIGURE 16–2   ECU types. (Courtesy of Allison Transmission Division of General Motors)

- The ECU should be located in an accessible position that nevertheless minimizes operator and service personnel incidental contact.
- The ECU should be located close to the source battery power with power leads no longer than necessary. The main power and ground inputs should be dedicated; that is, not shared with other electrical equipment and components. The technician should know that a minimum of 10V is required to operate the unit; 16V continuous and 19V intermittent are the maximum voltages to which the unit should ever be subjected. This means that great care should be exercised when blast charging; that is, using a generator to start a truck with dead batteries.
- The ECU requires continuous power to retain nonvolatile RAM in which diagnostic codes and TPS calibration data are retained. Nonvolatile RAM is lost if the battery is disconnected from the unit.
- The ECU is a sealed component and is not serviceable in the field. PROM removal/replacement is the only field service permitted on Allison ECUs.

PROGRAMMABLE READ-ONLY MEMORY (PROM)

Allison PROM chips log data that permit CEC systems to

- Be programmed for a variety of vehicle and equipment options
- Have flexibility of operating characteristics

The PROM is located inside the ECU and can be accessed through a cover in the ECU housing. To determine which options are programmed on a specific PROM chip, the Electronic Control Inquiry System (ECIS) should be accessed. This is a computer program that offers online access to ECU and PROM information. CEC PROM chips are shown in **Figure 16–3**. PROM data is magnetically written to the chip in the same way that audio data is recorded to a cassette tape.

Output Speed Sensor

The output speed sensor provides the ECU with tailshaft speed data and replaces the hydro-mechanical governor. The device has a magnetic sensor triggered by a pulse/reluctor wheel on the transmission output shaft. Output shaft rotation drives the pulse wheel through a magnetic field, generating an AC voltage signal that is relayed to the ECM. As the unit rotates at higher speeds, both the frequency and signal voltage increase. A full explanation of pulse generators is provided in Chapter 6. The pulse wheel uses a sixteen-tooth gear for on-highway applications. This sixteen-tooth wheel has become standard in all highway truck tailshaft speed sensors.

Throttle Position Sensor (TPS)

A TPS is a simple potentiometer device actuated by pull cable when a CEC transmission is fitted to a hydromechanical engine. The pull cable is attached to the engine fuel control/throttle lever. When the chassis is fitted with an electronic engine, usually only one TPS is required to signal both the Allison ECU and the engine ECM. When an Allison CEC is coupled to a DDEC II, III, or IV, an interface module may be required. This interface module may be minimum interface, that is, TPS only, or maximum interface (TPS, engine data, cruise control). The function of an interface module is to make the engine electronics "talk" to the transmission electronics so input signals like that of the TPS can be shared. It should be noted that CEC is not compatible with all electronic management systems.

The ECU recognizes accelerator travel as percent of throttle. Actual shift parameters will be determined by PROM calibration. The TPS and the shift logic programmed to the ECM replace the role played by the conventional modulator in managing the hydraulic circuit of the transmission. A typical TPS is shown in **Figure 16–4**.

FIGURE 16–3 CEC PROM chips. (Courtesy of Allison Transmission Division of General Motors)

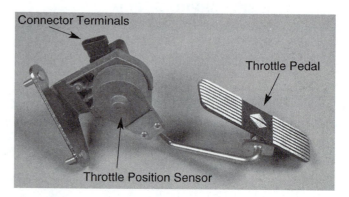

Connector Terminals

Throttle Pedal

Throttle Position Sensor

FIGURE 16–4 Throttle position sensor.

TPS Operation

The TPS converts throttle movement into a voltage signal to the ECM. It receives reference voltage and divides it in proportion to the linear travel of the throttle pedal. As the TPS wiper moves across a resistive strip, the resistance value changes, varying the voltage sent to the ECU as "counts" (**Figure 16–5**). In this way, throttle angle or accelerator pedal travel is converted to a voltage signal that is proportional to mechanical travel.

Throttle Position Linkage

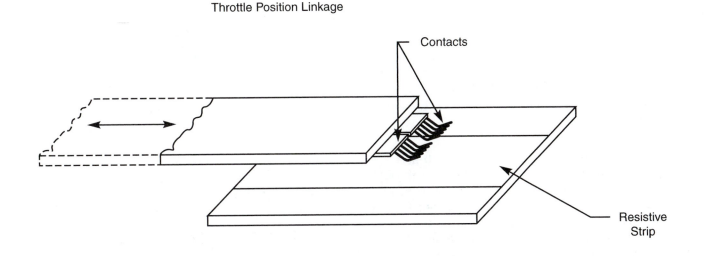

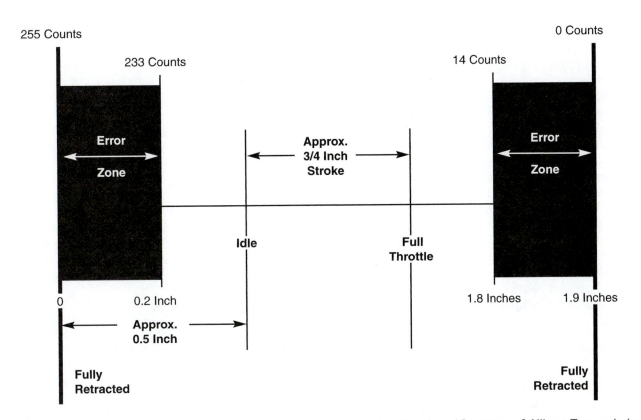

FIGURE 16–5   *Throttle position sensor operating principle and calibration. (Courtesy of Allison Transmission Division of General Motors)*

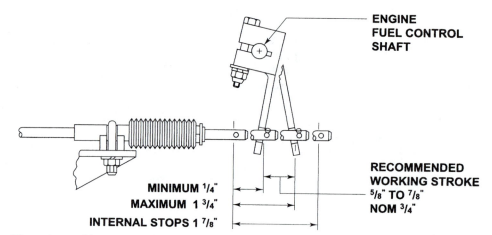

FIGURE 16–6 *Throttle position sensor initial adjustment. (Courtesy of Allison Transmission Division of General Motors)*

The TPS produces anywhere between 0 and 255 counts. Actual movement required to scan through these parameters is small, about 3/4 inch **Figure 16–6**). The TPS must therefore be adjusted to stay within the serviceable area and out of the "error zones." No optimum wide open or closed throttle count exists. As long as the counts stay in a safe zone, the TPS should operate properly. The target zone is 50 to 200 counts. If the reference voltage is 5 volts, the ECM will look for a return signal voltage value that must be more than 0 volts and must not exceed 4.9 volts. A return signal voltage of 0 volts would be produced only in the event that the TPS was open, whereas a signal return voltage that equalled the reference voltage could mean a short-circuited return.

The TPS self-adjusts to remain within the normal count area each time the vehicle is started. At start-up, the ECU is initialized and the sensor is recalibrated. When the ECU is powered up, idle counts are increased by 15 from the previous reading to compensate for linkage wear. This narrow count is widened when the operator steps on the accelerator. The ECM reads actual sensor travel and continually readjusts to the highest and lowest counts. The portion of the sensor's travel that is read by the ECU is small compared to the actual available counts.

Generally, the sensor stays within the serviceable area and does not require adjustment other than at initial installation. **Figure 16–7** shows the TPS mounting.

Shift Selectors

Either push-button or shift-lever types can be used with on-highway CEC transmissions. The push-button selector has the Do Not Shift light located on the shift pad. This light will cycle on/off as soon as the system is activated and during a "hard" failure.

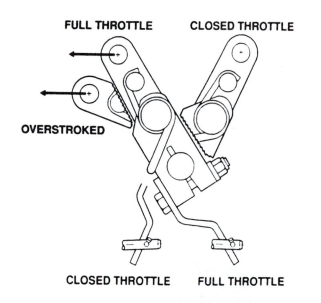

FIGURE 16–7 *Throttle position sensor mounting. (Courtesy of Allison Transmission Division of General Motors)*

Allison recommends that the push-button selector be mounted at no less than a 20 degree angle from horizontal to minimize the damage potential of an upset coffee cup. Lever selectors use Hall effect switches. The Hall effect switch units are not serviceable, but the mechanical components of the shift unit are.

Temperature Sensors

Temperature sensors monitor sump oil temperature for the ECU when fitted. A temperature sensor is mounted on the solenoid control circuit and is of the thermistor type. An NTC thermistor is used. A full explanation of thermistor operation is provided in Chapter 6. The data the temperature sensor inputs to the ECU processing cycle results in the following outputs:

- The blocking of all shifts when the transmission is –25°F or below. This protects the transmission from damage that could be caused by heating up too quickly.
- The limiting of transmission shifting to neutral, first range, and reverse (N, 1st, R) when temperature is between –25°F and +25°F.
- When temperature exceeds 270°F, the "hot" light is energized (if equipped), a trouble code is logged (temperature code # 24, sub-code hot # 23), and top gear is inhibited. Emergency vehicles such as fire trucks are usually programmed not to inhibit top gear in these circumstances.

Pressure Switches

Three types of pressure switch are used (see **Figure 16–8**):

- **Forward pressure switch.** This switch signals the ECU when the transmission is in forward ranges. One of two types is used, and is plumbed into the clutch apply circuit.
- **Reverse pressure switch.** This switch signals the ECU when the transmission is in reverse gear. One of two types is used, and is plumbed into the clutch apply circuit.
- **Oil pressure switch.** This switch signals the ECU when low fluid pressure or level exists. Three different types exist, one of which is used:
  - Lube pressure switch—normally open
  - Low oil level/pressure sensor (**Figure 16–9**)
  - Fluid oil level sensor—normally closed

The PROM must be calibrated for the type of switch used.

HARNESSES

CEC transmissions use either 2 or 3 harnesses.

- **Chassis wiring harness.** This connects the TPS, OSS (oil sensing switch) and the electro-hydraulic valve body to the ECU. Wires are coded with 100-series numbers.
- **Cab wiring harness.** This connects the shift selector assembly and interface units to the ECU. DDEC II and III engine management systems require the use of an interface module. The Diagnostic Data Link connector is part of the cab harness. The 12-pin connector in recent applications has been replaced by the standard ATA 6-pin (Deutsch) connector. Wires are coded with 300-series numbers.
- **Secondary wiring harness.** The Sealed Plus II ECU is equipped to handle a number of additional options that connect to the unit by means of a secondary wiring harness. These options may include
  - Fire truck special logic
  - Manual/automatic controls
  - Range hold features
  - Lock-up on and off
  - Dual shift selectors
  - Dual shift calibration
  - Bed hoist interlock

Secondary harness wire numbers are 300-series numbers.

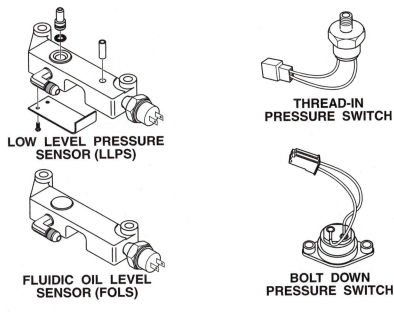

LOW LEVEL PRESSURE SENSOR (LLPS)

FLUIDIC OIL LEVEL SENSOR (FOLS)

THREAD-IN PRESSURE SWITCH

BOLT DOWN PRESSURE SWITCH

V06736

FIGURE 16–8 CEC pressure switches. (Courtesy of Allison Transmission Division of General Motors)

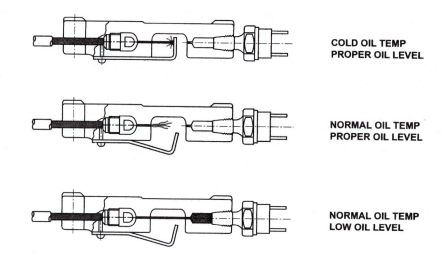

COLD OIL TEMP
PROPER OIL LEVEL

NORMAL OIL TEMP
PROPER OIL LEVEL

NORMAL OIL TEMP
LOW OIL LEVEL

FIGURE 16–9   Low oil level/pressure sensor. (Courtesy of Allison Transmission Division of General Motors)

ELECTRO-HYDRAULIC VALVE BODIES

ECU commands are converted to action by the electro-hydraulic valve body. These actions could be described as electronic-over-electric-over-hydraulic in CEC transmissions. The transmission's hydraulic circuits and valves are controlled by a series of solenoids that are switched by the ECU. The unit also contains pressure-sensing switches whose function is to report a given operating condition to the ECU.

Solenoids

Solenoids in Allison electro-hydraulic valve bodies serve to direct hydraulic pressure into specific circuits and passages or exhaust the circuit. There are two types:

- **Nonlatching. Nonlatching solenoids** have a plunger that is spring loaded to its "off" position. This plunger loads a check ball to close flow through the solenoid passage; in its "off" position, the solenoid exhaust port is open. When the plunger is energized, it retracts, closing the exhaust port and unseating the check ball, permitting flow through the solenoid passage (**Figure 16–10**). The moment the solenoid is de-energized, the spring closes the plunger. Nonlatching solenoids, therefore, permit hydraulic flow only when electrically energized. They can be secured to the valve body by a clamp plate (old-style) or Torx-head screws (current).
- **Latching. Latching solenoids** must be momentarily energized to be switched, after which the plunger will remain in position until energized again with reverse polarity. Clamp-

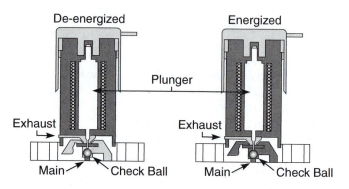

FIGURE 16–10   Solenoid operation.

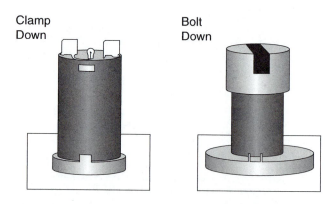

FIGURE 16–11   Types of latching solenoids.

down latching solenoids are identified by a single tab at the solenoid base (**Figure 16–11**). Bolt-down latching solenoids also have a tab located at the solenoid flange. When a latching circuit solenoid is in the open position, it will permit hydraulic flow through its circuit until it receives an ECU switch command to close it.

OVERVIEW OF ELECTRO-HYDRAULIC VALVE BODY OPERATION

Latching solenoids replace the shift signal valves found on nonelectronic transmissions; these control the position of the shift valves, which in turn control the clutch apply circuits. Latching solenoids can therefore both apply and exhaust clutch apply circuits.

The neutral range valve is controlled by one latching and one nonlatching solenoid; this valve controls the transmission's shifts from neutral to range and from range to neutral.

The forward–reverse valve is controlled by one latching solenoid. This determines whether the shift is to a forward range or reverse. Trimmer valves regulate the oncoming clutch application or the harshness of the shift. The trimmer regulator valve is controlled by a nonlatching solenoid. This essentially controls the pressure under the trimmer valve plug, which regulates trimmer valve operation. The lockup relay is controlled by a nonlatching solenoid, which means that lockup will only occur when this solenoid is energized. The solenoid priority valve and direction priority valve ensure a continuous flow of main pressure regardless of transmission range.

Shift signal valves	Latching solenoids
Neutral–range valve	1 latching, 1 nonlatching solenoid
Forward–reverse valve	Latching solenoid
Trimmer regulator valve	Nonlatching solenoid
Lockup relay	Nonlatching solenoid

Solenoid Designations

Each solenoid receives a constant flow of main pressure—it is also given a letter designation.

Solenoids A, B, C, and D are latching solenoids that replace the conventional shift signal valves. Solenoids are all directly switched by the ECU.

- **5-Speed Transmission**
  - Solenoid A controls low–1 shift valve.
  - Solenoid B controls 1–2 shift valve.
  - Solenoid C controls 2–3 shift valve.
  - Solenoid D controls 3–4 shift valve.
- **4-Speed Transmission**—no solenoid A, as there is no low shift range. Solenoids B, C, and D function as in the 5-speed.
- **3-Speed Transmission**—no solenoids A and D
  - Solenoid B controls 1–2 shift valve.
  - Solenoid C controls 2–3 shift valve.
  - Solenoid F (latching) switches main pressure to the bottom of the forward–reverse valve.

- Solenoid H (nonlatching) switches main pressure to the bottom of the neutral–range valve when energized.
- Solenoid J (latching) directs main pressure to the top of the neutral–range valve.
- Solenoid G (nonlatching) directs main pressure to the lockup relay valve when energized.
- Solenoid E (nonlatching) supplies the trimmer regulator valve to control shift quality.

👓 Shop Talk

An Allison CEC transmission, although electronically controlled, is still a hydromechanical device. As such, the importance of correct transmission fluid level cannot be emphasized enough. If fluid level is low, the converter and clutches will not receive enough fluid. If the converter level is high, the fluid will aerate and the transmission will overheat.

16.2 ELECTRO-HYDRAULIC CIRCUIT MANAGEMENT

This is the switching function of the ECU, also known as the output circuit. Shift logic computations made by the ECU based on the input circuit signals are converted to output signals that determine the outcomes. These switched outcomes will dictate exactly what the transmission does with the input torque that is delivered to it. It is important that the technician knows what output switching is required to produce each gear range, so the system can be effectively diagnosed when it malfunctions. Take some time with the following schematics and ensure that you understand exactly what is happening in each.

NEUTRAL FLOW SCHEMATIC

See **Figure 16–12**. When the engine is started, the torque converter drives the transmission pump, which feeds main pressure to the solenoid circuits and the solenoid priority valve. It then cascades through the 2–3 shift valve and the 1–2 shift valve and to the 1st clutch apply circuit. Only solenoid J permits flow-through. This directs main pressure to the top of the neutral-range valve, keeping the spool positioned downwards and preventing main pressure from entering the forward clutch apply circuit. Because only the 1st clutch is applied, the transmission is in neutral.

Neutral summary:
Solenoid J "on"   First clutch applied

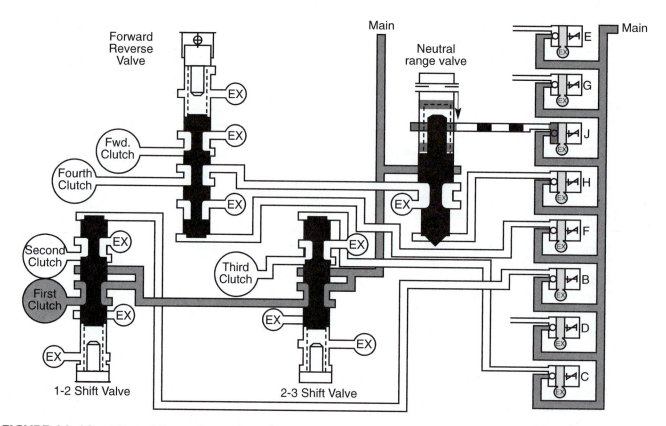

FIGURE 16–12   *Neutral flow schematic. (Courtesy of Allison Transmission Division of General Motors)*

FIRST RANGE

See **Figure 16–13**. Shifting to 1st range switches the ECU latching solenoid J to the "off" position and solenoids F (latching) and H (nonlatching) to the open position. Main pressure passing through solenoid H shuttles the neutral–range valve up, permitting main pressure to flow to the forward–reverse valve. Solenoid F directs main pressure to the forward-reverse valve, moving it up. This permits the main pressure originating from solenoid H to cascade through the forward–reverse valve and actuate the forward clutch apply circuit. The forward clutch apply circuit also directs main pressure to the main pressure regulator, helping to lower main pressure after the forward clutch has been applied. The 1st clutch remains applied, receiving its feed from the 2–3 shift valve, which remains in the up position.

1st Range summary:
Solenoid F on	Forward clutch applied
Solenoid H on	First clutch applied

SECOND RANGE

See **Figure 16–14**. Shifting to the 2nd range causes solenoid F to remain open because it is latching. Nonlatching Solenoid H continues to be energized and remains open, keeping the forward clutch applied. Latching Solenoid B is energized to its open position, which shuttles the spool in the 2–3 shift valve down, exhausting the 1st clutch apply circuit and directing main pressure to the second clutch apply circuit.

2nd range summary:
Solenoid F on	Forward clutch applied
Solenoid H on	
Solenoid B on	2nd Clutch applied

THIRD RANGE

See **Figure 16–15**. Shifting the 3rd range causes the latching solenoid F to remain open; nonlatching solenoid H continues to be energized, keeping the forward clutch applied. Solenoid C is energized to open and directs main pressure to the top of the 2–3 shift valve, causing it to shuttle and exhaust the 2nd clutch apply circuit and permitting main pressure to be directed to the 3rd clutch.

3rd range summary:
Solenoid F on	Forward clutch applied
Solenoid H on	
Solenoid B on	3rd Clutch applied
Solenoid C on	

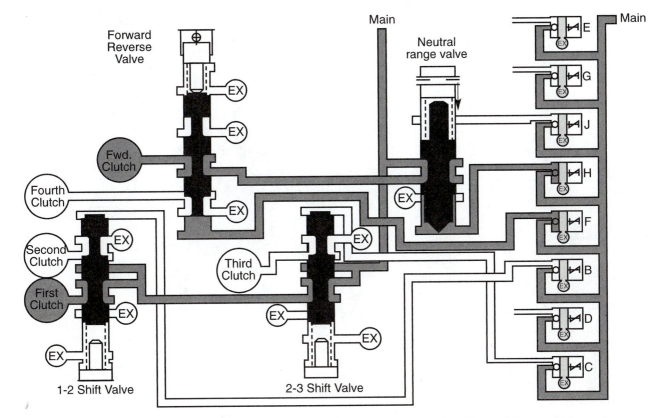

FIGURE 16–13   *First range schematic. (Courtesy of Allison Transmission Division of General Motors)*

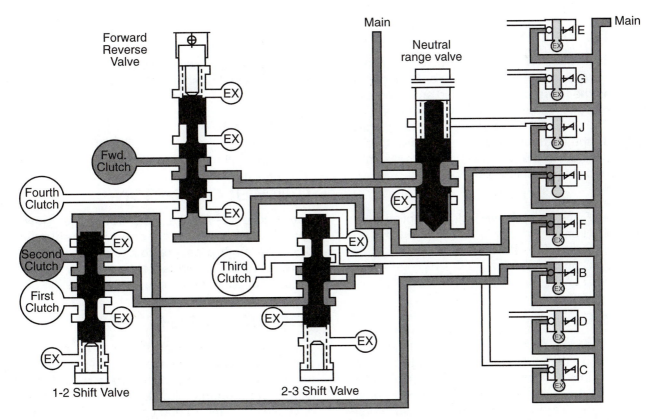

FIGURE 16–14   *Second range schematic. (Courtesy of Allison Transmission Division of General Motors)*

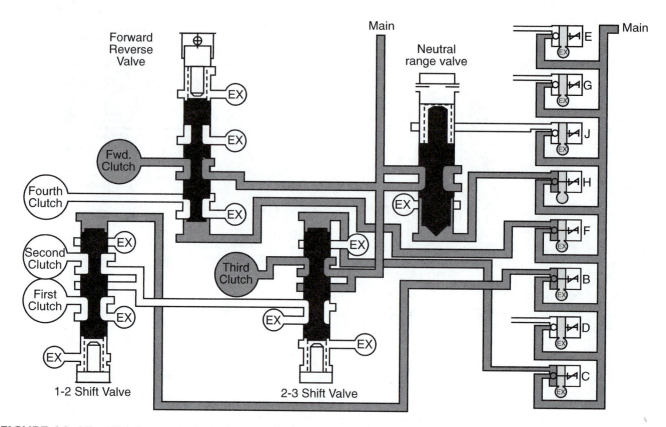

FIGURE 16-15 *Third range schematic. (Courtesy of Allison Transmission Division of General Motors)*

FOURTH RANGE

Reference **Figure 16-16**. Shifting to 4th range causes the following events: as before, Solenoids H and F keep the forward clutch applied and Solenoids B & C keep their respective shift valves down (1–2 and 2–3). Solenoid D is energized directing main pressure to the bottom of the 3–4 shift valve forcing it up; this exhausts the 3rd clutch and directs main pressure into the 4th clutch apply circuit.

4th range summary:
Solenoid F on   Forward clutch applied
Solenoid H on
Solenoid B on   4th clutch applied
Solenoid C on
Solenoid D on

REVERSE

See **Figure 16-17**. Shifting to reverse will cause all "on" latching solenoids to be energized to close. Only nonlatching solenoid H will be on, positioning the neutral–range valve up. This permits main pressure to flow through the forward–reverse valve into the 4th clutch apply circuit. Main pressure from the solenoid priority valve flows through the 2–3 shift valve and the 1–2 shift valve and into the 1st clutch apply circuit.

Reverse summary:
Solenoid H on   4th clutch applied
1st clutch applied

TRIMMER OPERATION

See **Figure 16-18**. The ECU determines how fast some clutches apply by energizing solenoid E, directing main pressure to the trimmer regulator valve. Trimmer action is determined by controlling the pressure under the trimmer plug.

LOCKUP CLUTCH

See **Figure 16-19**. When the ECU determines the appropriate conditions have been met, solenoid G is energized and main pressure is directed into the lockup clutch apply circuit. This converts the fluid drive action of the torque converter to a mechanical coupling.

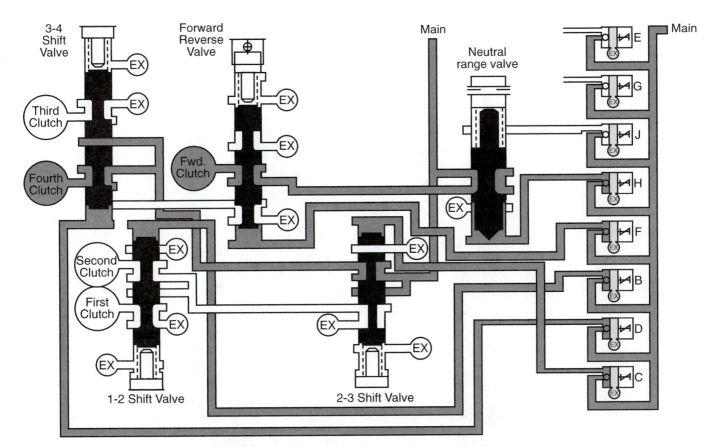

FIGURE 16–16 Fourth range schematic. (Courtesy of Allison Transmission Division of General Motors)

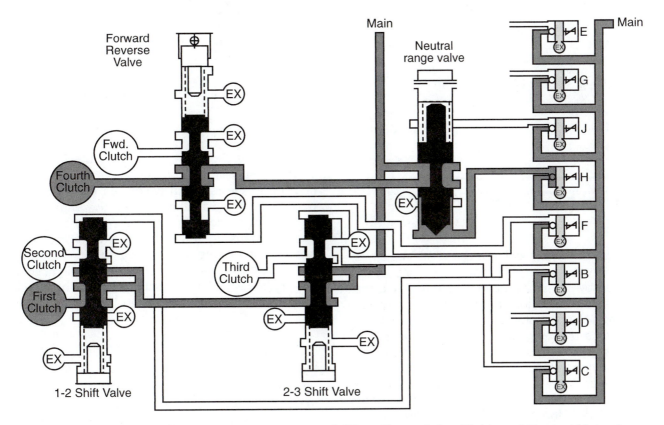

FIGURE 16–17 Reverse range schematic. (Courtesy of Allison Transmission Division of General Motors)

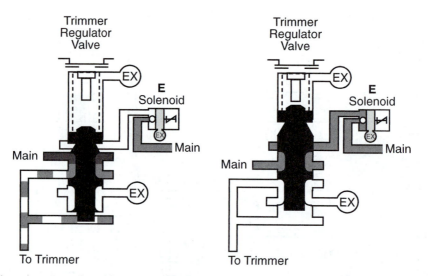

FIGURE 16–18 Trimmer operation (Courtesy of Allison Transmission Division of General Motors)

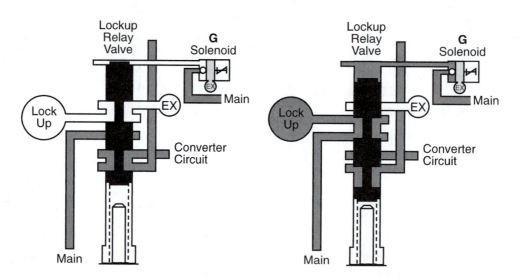

FIGURE 16–19 Lock-up clutch operation. (Courtesy of Allison Transmission Division of General Motors)

16.3 ELECTRICAL FAILURE

In the event of an electrical failure, the CEC transmission will operate in a sort of "limp home" mode; this is achieved by the latching and nonlatching solenoid configuration. At the moment of the electrical failure, all latching solenoids stay in position while nonlatching solenoids de-energize and revert to the "off" or "closed" position. This will lock the transmission in whatever range it is in, inhibiting all shifts. If solenoid G is energized, it will close, exhausting the lockup clutch apply circuit. Solenoid E no longer directs main pressure to the trimmer regulator valve. As long as the engine remains running, the neutral–range valve will stay up and permit operation of the transmission in the range in which the electrical failure occurred. However, the moment the engine stops, main pressure flow through the neutral–range valve ceases, causing the spool to shuttle down and exhaust the forward clutch apply circuit. On restart, if the electrical failure still exists, the latching solenoids will continue to direct pressure through the circuitry, but the forward clutch will no longer be applied, resulting in a neutral condition.

👓 **Shop Talk** _____

Severely low fluid will cause two things to happen: The transmission Check Trans light will illuminate, and upshifts to the highest range will be prevented.

16.4 VEHICLE INTERFACE

This term is normally used to describe the electrical connections between the electronic system, in this case the CEC, and the vehicle. The 6-pin ATA data link would be part of the vehicle interface. The purpose of the vehicle interface is to ensure that the various on-board electronic components perform synergistically. Interface modules are not supplied by Allison, so they are indicated as an OEM component. DMR Electronics manufactures one such device.

The PROM chip will define the specific mode of operation of an Allison transmission, including such data as upshift and downshift points for primary mode, secondary mode operation if applicable, and details such as oil level/pressure sensor types. Allison's Electronic Control Inquiry System (ECIS) is an on-line data system that indicates what options are programmed into a particular PROM. CEC can interface with other electronic engines to maximize performance, but this requires the use of an interface component such as the DMR interface module.

16.5 TROUBLESHOOTING

CEC Electronics will occasionally produce running problems, as does any other system; in defining and repairing these problems, Allison workshop manuals should be used. The following is a brief description of some basics of troubleshooting practice.

It should be always remembered that a CEC Allison transmission is a hydromechanical device capable of developing problems other than electronic. The system can therefore develop problems that do not set trouble codes or cause intermittent trouble codes. The ECU's self-diagnostic capabilities are capable of generating the following types of codes:

- **Soft codes** will cause the Check Transmission light to illuminate but will not cause the Do Not Shift light to illuminate.
- **Hard codes** cause both the Check Transmission and Do Not Shift lights to illuminate.

Any ECU-logged codes can be accessed in the following manner:

1. Retrieve codes
   a. Start engine—run with idle-shift selector in neutral.

b. Hold Electronic Control test switch in the on position.
   c. The Check Transmission light will flash the most important code logged. All codes are two-digit; the first digit is separated from the second by a pause. Only one code can be flashed.
2. Clear trouble codes
   a. Open ignition circuit.
   b. Hold Electronic Control test switch in the on position.
   c. Ensure brakes are applied and restart vehicle.
   d. Select reverse, hold in range for a couple of seconds, and then return to neutral.

👓 Shop Talk

The electronic inhibitor function is designed to prevent transmission damage under severe low temperature conditions. At temperatures of −25°F (−32°C) or below, the CEC transmission will not shift out of neutral. Between −24°F (−31°C) and 19°F (−7°C), shifts are limited to lowest drive and reverse only. Be patient when starting up and allow the transmission to warm up before attempting to drive away.

PROLINK

The **ProLink** (also known as the **digital data reader (DDR)**) and the appropriate software cartridge are the instruments of choice when diagnosing CEC management transmission electronic problems. They would generally be used in conjunction with a good-quality **digital multimeter (DMM)**.

1. To retrieve codes using ProLink,
   a. Connect to the digital diagnostic link (DDL).
   b. Select transmission from readout. Press "Function" to enter the Function Selection Menu.
   c. Scroll until "Diagnostic Codes" appears in the window and then press "Enter".
2. Clear trouble codes using ProLink.
   a. Leave ProLink connected to DDL.
   b. Cycle ignition key off and then on.
   c. Move shift selector to reverse, hold two seconds, and then move it back to neutral.

Using ProLink's **snapshot mode** can be especially useful in diagnosing intermittent or no-code problems. This mode can record selected frames of data before and after the snapshot sequence is triggered. The trigger can be a code or input manually. The advantage of snapshot mode is that all critical system parameters before and after the trigger event can be read afterwards.

DIAGNOSTIC CODES

12	Low lube pressure/oil level
13	Low battery voltage
14	Forward pressure switch
15	Reverse pressure switch
21	Throttle sensor
22	Speed sensor
23	Shift selector
24	Oil temperature
31	Secondary shift selector
32	Direction signals
33	Oil temperature sensor
34	PROM check
41	J solenoid/circuitry
42	F solenoid/circuitry
43	D solenoid/circuitry
44	C solenoid/circuitry
45	B solenoid/circuitry
46	A solenoid/circuitry
51	G (lockup) solenoid/circuitry
52	E (trim boost) solenoid/circuitry
53	H (neutral) solenoid/circuitry
54	Latching solenoid chain
66	Transmission-Engine Link (TECL)
69	ECU Test

👓 **Shop Talk**

When performing an Allison CEC stall test, be sure to consult the OEM's engine electronics directives for performing the test. It is usually not necessary to connect an out-of-chassis tachometer when performing stall tests when an electronic engine is driving the CEC transmission.

SUMMARY

- Allison CEC is best described as a partial-authority electronic management system that adapts a hydromechanical Allison transmission to computerized monitoring and management.
- The acronym CEC stands for Commercial Electronic Control.
- CEC consists of an input circuit, a microcomputer, and an output circuit.

- An electro-hydraulic valve body is used to convert the results of logic processing into hydro-mechanical outcomes.
- Primary input signals to CEC are the shift signal, TPS signal, and road speed data.
- Three types of ECU are used to manage the CEC electronics: Splashproof, Sealed Standard and Sealed Plus II .
- Understanding the powerflows obtained in each ratio range can help the technician troubleshoot Allison CEC electronics.
- CEC diagnostic codes can be read with a diagnostic instrument such as a DDR/ProLink, and the most important code can be flashed by the dash lights.
- A snapshot test is a means of diagnosing intermittent transmission problems.

REVIEW QUESTIONS

Use the information in Chapters 12, 13, and 14 as well as this chapter to answer these questions.

1. Which clutch, when applied, connects the turbine shaft to the mainshaft?
   a. forward clutch
   b. 1st clutch
   c. 2nd clutch
   d. 4th clutch

2. Which clutch, when applied, connects the turbine shaft to the sun gear shaft?
   a. forward clutch
   b. 1st clutch
   c. 2nd clutch
   d. 4th clutch

3. For the CEC transmission to produce output torque from input rotation, which of the following must occur?
   a. 2 clutches must be applied
   b. 3 clutches must be applied
   c. The converter must be in lockup phase.
   d. 2 members of one planetary must be driven together

4. How many reluctor teeth are on the tailshaft vehicle speed sensor used on highway applications of CEC transmissions?
   a. 8
   b. 16
   c. 32
   d. 64

5. What is the function of the reverse pressure switch?
   a. to signal the ECU when in reverse range
   b. to limit rpm when in reverse range
   c. to alert the operator when reverse is selected
   d. to modulate reverse clutch apply pressure

6. When the CEC sump oil temperature is between –25°F and 25°F, how is shifting likely to be affected?
   a. No shifting will occur.
   b. Shifts in all ranges will be modulated to higher tailshaft speeds.
   c. Shifts will be limited to neutral, first and reverse ranges.
   d. A trouble code will be logged, with unaffected shifting.

7. After a latching solenoid has been energized, what is required to return the solenoid to the original position?
   a. energize again
   b. energize with reverse polarity
   c. retraction spring pressure
   d. main hydraulic pressure

8. In the event of a complete chassis electrical failure, which of the following conditions is true?
   a. As long as the vehicle keeps running, main pressure locks shift valves in range for "limp home."
   b. The vehicle will limp home provided it is first shut down and then restarted.
   c. All latching solenoids will deactivate and CEC will default to 1st range.
   d. The transmission will revert to normal hydromechanical operation.

9. Which of the following best describes the role of the CEC interface module?
   a. enables the transmission and engine electronics to communicate
   b. replaces the TPS unit
   c. replaces the throttle unit
   d. switches the electro-hydraulic valve body

10. If battery power to the CEC ECU is interrupted and then reconnected, which of the following is the likely result?
    a. All ECU memory is lost.
    b. Shifts will be confined to 1st and reverse.
    c. Nonvolatile RAM is lost.
    d. The ECU should be replaced.

11. When the "hot" light on the dash illuminates, what else is likely to happen?
    a. Top range shifts can be inhibited.
    b. 1st range shifts are inhibited.
    c. The engine can be shut down.
    d. The transmission reverts to hydromechanical operation.

12. Which of the following input devices uses a Hall effect sensor?
    a. tailshaft sensor
    b. shift lever
    c. TPS
    d. interface module

Allison World Transmission

Prerequisite
Chapter 16

Objectives

After reading this chapter, you should be able to

- Describe the modular design used in the WT transmission.
- Outline how the WT transmission uses full authority management electronics to effect shifting and communicate with other vehicle electronic systems.
- List some of the service and repair advantages of the modular construction of the WT transmission.
- Group the WT modules into input, gearbox, and output categories.
- Describe how the WT electronic control module masters the operation of the transmission.
- Define the terms pulse width modulation, primary modulation, and secondary modulation.
- Outline how the WT transmission uses three interconnected planetary gearsets to stage gearing to provide 6 forward ranges, reverse, and neutral.
- Identify the overdrive ranges.
- Describe the WT integral driveline retarder components and operating principle.
- Outline the function of the dropbox option in one WT model.
- Describe the role of, and the components within, the electro-hydraulic control module.
- Outline the essential components in the WT electronic circuit and classify them as input circuit, processing, and output circuit components.
- Describe how SAE J 1939 compatible hardware and software allows WT to share componentry and data with other on-board electronic systems to optimize vehicle performance.
- Describe how the WT clutches are controlled.
- Outline the torque routes through the WT transmission in each range selected.
- Describe how diagnostic codes are logged in the Allison WT ECU and the manner in which they are displayed.
- Interpret some of the WT diagnostic codes.
- Perform some basic diagnostic troubleshooting using the Allison recommended tooling.

Key Terms List

accumulator
adaptive logic
digital display (DD)
digital diagnostic reader (DDR)
dropbox
Drop-7
duty cycle

electro-hydraulic control module
fast adaptive mode
frequency
main pressure regulator
modules
planetary gearset
planetary pinions
primary modulation

ProLink
pulse width (PW)
pulse-width modulation (PWM)
retarder
ring gear
scavenging pump
secondary modulation
serial communications interface (SCI)

slow adaptive mode
sun gear
throttle position sensor (TPS)
transfer case
vehicle interface module (VIM)
World Transmissions (WTs)
World Transmission Electronic Control (WTEC)

17.1 INTRODUCTION

World Transmissions (WTs) use a modular construction. This means that the transmission has been divided into distinct subcomponents that can be separated and repaired in isolation from the remainder of the transmission subcomponents. The modular construction of the WT transmission has simplified service and repair procedures compared to a CEC transmission and has also neatly divided up the subcomponents into packages that make it easy to study. Recently, Allison has broadened the range of their WT transmission series to address medium- and light-duty truck and bus applications with the 1000 series. This could result in the use of WT transmissions in more than just the vocational vehicle applications Allison has specialized in. Allison has classified their WT range by torque range into the 3000 series, 2000 series, and 1000 series.

The WT shares some common operating principles with the CEC transmission, so this section will focus more on the fundamental differences between the two transmissions. Straight-through configured WTs, that is, models with no **transfer case,** have a capacity for 6 forward speeds, neutral, and reverse. These transmissions can, however, be programmed for 4- or 5-speed ratios to meet specific vocational requirements. In this case, no hardware changes are required: it is simply a matter of software programming. A WT may also have an additional output module such as a transfer case, which can increase the forward speed ratios from 6 to 7 speeds.

Shop Talk

The term full authority *is borrowed from the way we categorize engines. In the early days of engine management electronics the term partial authority was used to describe a hydromechanical system that was adapted for electronic management. This description would fit an Allison CEC transmission perfectly. A full authority engine was one that was designed to be managed by electronics. Again, if we use these terms to categorize transmissions, full authority perfectly describes the WT transmission. It is capable of more comprehensively monitoring and controlling all transmission functions.*

The WT series contains hydraulically actuated spring-released clutches. Compensation for clutch wear occurs automatically. The system electronic controls can be programmed to provide optimum driving efficiency for every engine and driveline package. **Figures 17–1** through **17–5** show different WT models.

MODULAR DESIGN

The WT series transmissions are made up of major components and subsystems called **modules.** Each module is an integral subassembly that may be removed for service or repair as separate unit, providing a significant servicing advantage. The modules plus the external management electronics combine to form the complete transmission package. These modules may be grouped into the following four sections:

Input Modules

There are 3 input modules:

- Torque converter module
- Converter housing module
- Front support/charging pump module

Gearbox Modules

There are 5 gearbox modules:

- Rotating clutch module
- Converter housing module
- P1 planetary module
- P2 planetary module
- Main shaft module

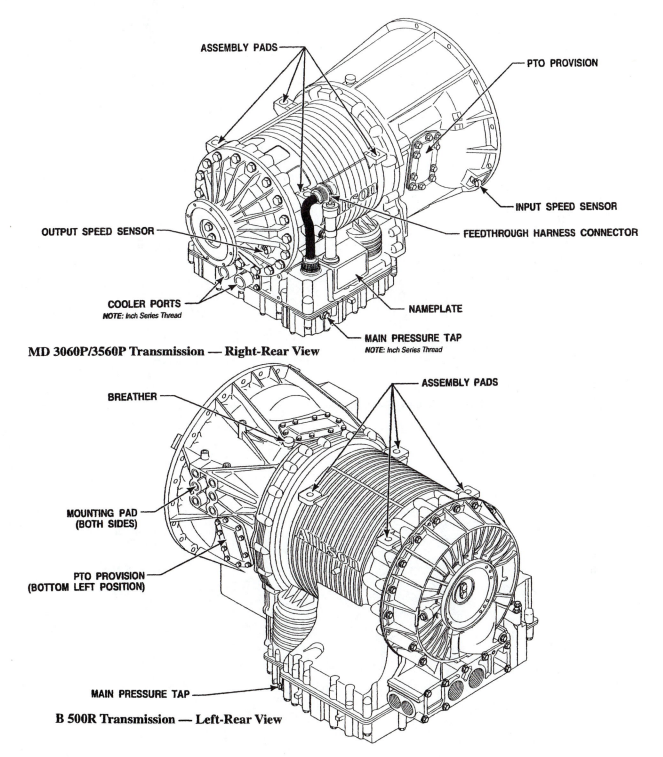

ASSEMBLY PADS

PTO PROVISION

OUTPUT SPEED SENSOR

INPUT SPEED SENSOR

FEEDTHROUGH HARNESS CONNECTOR

COOLER PORTS
NOTE: Inch Series Thread

NAMEPLATE

MAIN PRESSURE TAP
NOTE: Inch Series Thread

MD 3060P/3560P Transmission — Right-Rear View

BREATHER

ASSEMBLY PADS

MOUNTING PAD
(BOTH SIDES)

PTO PROVISION
(BOTTOM LEFT POSITION)

MAIN PRESSURE TAP

B 500R Transmission — Left-Rear View

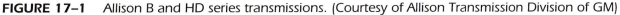

FIGURE 17–1 Allison B and HD series transmissions. (Courtesy of Allison Transmission Division of GM)

Output Module

There is one output module:

- Rear cover module, output **retarder** module, or transfer gear module (depends on model)

Electronic Control Module

There is one electronic control module (ECM):

- The ECM is located outside of the main transmission housing. This unit is responsible for

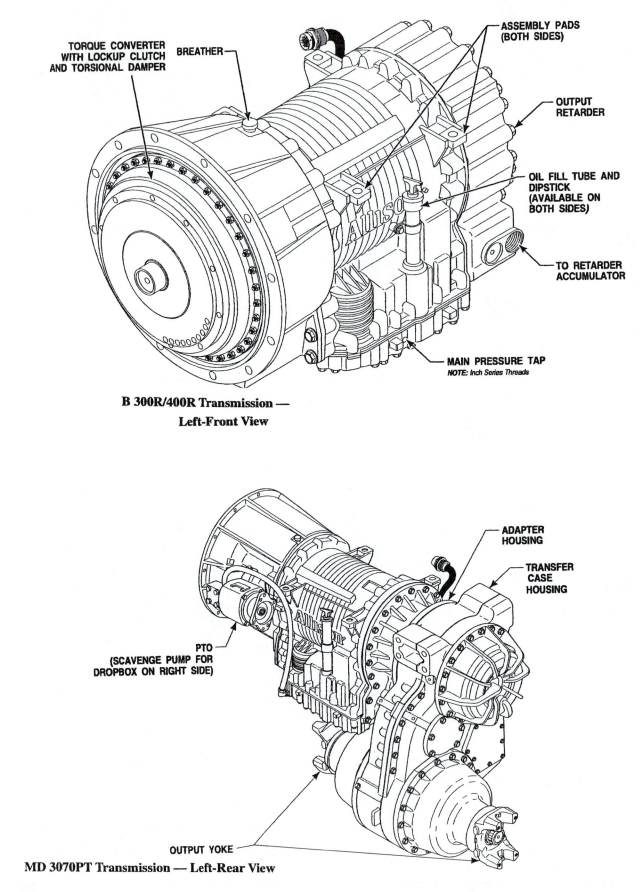

**B 300R/400R Transmission —
Left-Front View**

TORQUE CONVERTER
WITH LOCKUP CLUTCH
AND TORSIONAL DAMPER

BREATHER

ASSEMBLY PADS
(BOTH SIDES)

OUTPUT
RETARDER

OIL FILL TUBE AND
DIPSTICK
(AVAILABLE ON
BOTH SIDES)

TO RETARDER
ACCUMULATOR

MAIN PRESSURE TAP
NOTE: Inch Series Threads

MD 3070PT Transmission — Left-Rear View

ADAPTER
HOUSING

TRANSFER
CASE
HOUSING

PTO
(SCAVENGE PUMP FOR
DROPBOX ON RIGHT SIDE)

OUTPUT YOKE

FIGURE 17–2 Allison B and MD with dropbox transmissions. (Courtesy of Allison Transmission Division of GM)

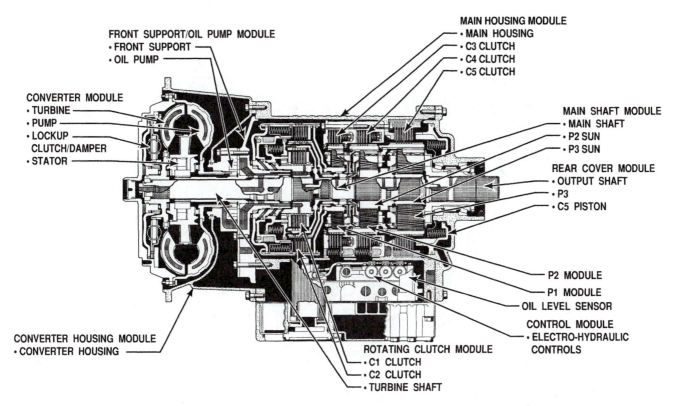

FIGURE 17–3 Cross-sectional view of MD 3060, B 300, and B 400 transmissions. (Courtesy of Allison Transmission Division of GM)

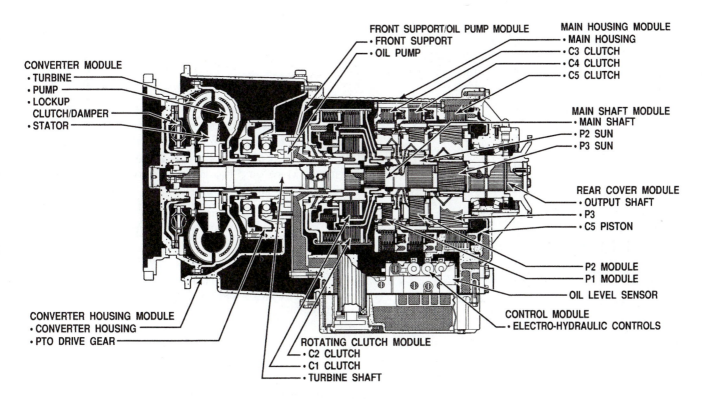

FIGURE 17–4 Cross-sectional view of an HD 4560P transmission. (Courtesy of Allison Transmission Division of GM)

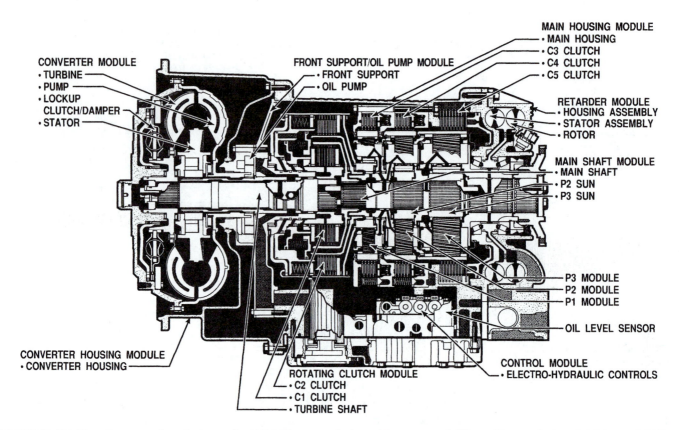

FIGURE 17–5 Cross-sectional view of a B 500R transmission. (Courtesy of Allison Transmission Division of GM)

receiving input signals, performing logic processing, and producing the switching signals to effect the outcomes of processing.

WT TRANSMISSION IDENTIFICATION

A transmission identification plate is located on the right rear side of the transmission (see **Figure 17–6**).

FIGURE 17–6 World transmission identification plate. (Courtesy of Allison Transmission Division of GM)

This plate shows the transmission serial number, part number (assembly number), and model number. All three sets of numbers are required when ordering parts or identifying the transmission. The model data can be decoded as follows:

MD, HD or B	Automatic Truck Transmission Automatic Bus/Coach Transmission
3, 4, or 5	Indicates Transmission Series
0	Close Ratio
5	Wide Ratio
6 or 7 (not present in B Series)	Number of forward speed ratios available
0 or 6	Reserved for major model changes
P	Power Takeoff (PTO)
R	Output Retarder
T (not present in B or HD Series)	Transfer case (dropbox)

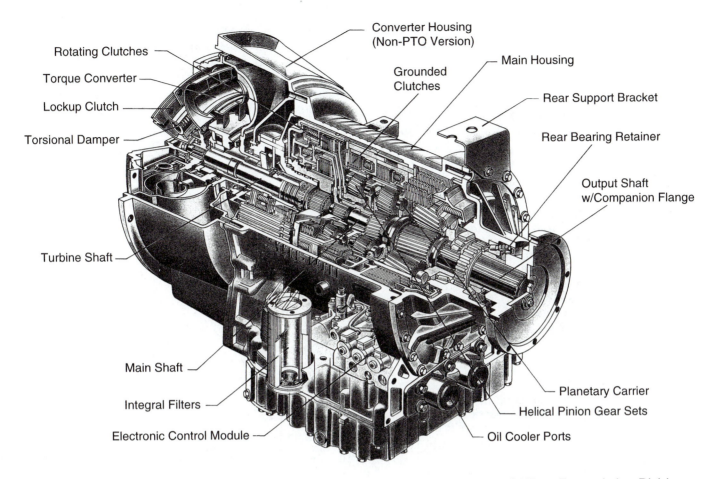

FIGURE 17–7 Offset cross-section of a typical Allison transmission. (Courtesy of Allison Transmission Division of GM)

Figure 17–7 shows a cross-sectional view of a WT transmission that should be referenced to locate the modules.

<h2>17.2 INPUT MODULES</h2>

WT input modules are located between the engine and the transmission gearing. WT transmissions are designed to couple directly to the engine without the use of adaptation hardware.

Torque Converter Module

The function of the torque converter is to couple the engine flywheel to the transmission and transmit engine torque. The operating principles are described in detail in Chapter 12. The WT torque converter can be locked up. The torque converter lockup clutch disc has an integral torsional damper that is only active during lockup.

Converter Housing Module

The converter housing module physically connects the engine to the gearbox modules. The module may be specified for a PTO mount: In this case, the converter housing is approximately four inches longer than a non-PTO optioned housing. The turbine shaft passes through the converter housing to the transmission gearing. The helical PTO drive gear is contained in the converter housing and is driven at engine rpm by the torque converter pump drive hub: The PTO gear hub drives the charging pump.

Front Support/Charging Pump Module

The front support is bolted directly to the transmission main housing and incorporates the converter ground sleeve, through which the turbine shaft passes to the rotating clutch assembly. The charging pump functions to pressurize the transmission fluid. It is direct-driven by the engine and therefore charges the transmission hydraulic circuits at any time the engine is running. This supply of pressurized hydraulic fluid is required for transmission operation.

17.3 GEARBOX MODULES

The gearbox modules combined contain five clutch assemblies and three **planetary gearsets.** All WT transmissions have the hardware for 6 forward ratios, neutral, and reverse. The MD and HD Series transmissions are available with either close-ratio or wide-

ratio gearing: Chassis application will determine which is used. All B Series transmissions are built with close-ratio gearing. Close- or wide-ratio gearing is determined by the physical characteristics of the planetary gearsets used in the assembly of the transmission. The upper two ranges of WT model transmissions have overdrive gear ratios. **Figure 17–8** identifies the internal components of a WT HD transmission.

Legend for interpreting Figure 17–8

Torque converter module
1. Torque converter turbine
2. Torque converter pump
3. Lockup clutch damper
4. Torque converter stator

Front support/oil pump module
5. Front support
6. Oil pump

Main housing module
7. Main housing
8. C3 clutch
9. C4 clutch
10. C5 clutch

Retarder module
11. Retarder housing assembly
12. Retarder stator assembly
13. Rotor

Mainshaft module
14. Mainshaft
15. P2 sun gear
16. P3 sun gear

P3 module
17. P3 planetary assembly

P2 module
18. P2 planetary assembly

P1 module
19. P1 planetary assembly

Control module
20. Hydroelectric controls

Rotating clutch module
21. C2 clutch
22. C1 clutch
23. Turbine shaft

Converter housing module
24. Converter housing

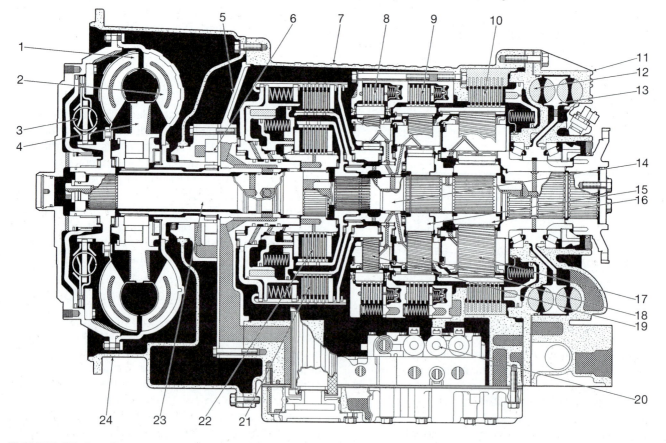

FIGURE 17–8 Components in a WT HD transmission. (Courtesy of Allison Transmission Division of GM)

ROTATING CLUTCH MODULE

The rotating clutch module is splined to the turbine shaft and therefore rotates with the turbine. This module contains the turbine shaft assembly, the rotating clutch hub assembly, the C1 and C2 clutch assemblies, the rotating clutch drum, and the **sun gear** drive hub assembly for the P1 planetary gearset. The P1 sun gear supplies constant rotational input to the P1 planetary set. This means that whenever the turbine shaft rotates, so does the P1 sun gear.

The C1 and C2 clutches consist of pistons, return spring assemblies, a drive hub, and clutch reaction and friction plates. The reaction plates in both C1 and C2 clutches are splined to rotate with the rotating clutch hub assembly.

The C1 clutch is applied by hydraulic fluid acting on the C1 piston: this action forces the piston and the clutch reaction plates against the clutch friction plates. Because the clutch friction plates are splined to the C1 drive hub, this will rotate providing turning force or torque at turbine speed to the main shaft assembly.

The means used to apply the C2 clutch is identical to that used to apply the C1 clutch. Because the C2 reaction plates are splined to the C2 drive hub, when the C2 piston is hydraulically actuated, the C2 clutch transmits torque to the P2 planetary assembly.

A balance piston is used to enhance control of off-going and on-coming rotating clutches. The balance piston is the third piston in the rotating clutch assembly and is located between the C1 spring assembly and the C1 pressure plate. Its function is to entrap fluid (lubrication pressure) between itself and the C1 piston, thereby balancing piston movement against exhaust backfill pressure behind the C1 piston.

Figure 17–9 identifies the internal components of a WT MD series transmission.

MAIN HOUSING MODULE

The main housing module consists of the C3, C4, and C5 clutches, the **ring gear** component of the P1 planetary gearset, and the transmission main housing. All three clutches in the main housing module are stationary and therefore do not rotate when hydraulically applied as in the case of the C1 and C2 clutches. The three clutches are applied and released by ECU output signals.

The purpose of the C3 clutch when actuated is to hold the P1 ring gear. The C3 clutch friction plates mesh with the P1 ring gear: These friction plates rotate with the P1 ring gear when it is not engaged. The C3 reaction plates are splined to the C3 clutch housing so when the C3 clutch is applied, the P1 ring gear is held stationary.

The function of the C4 clutch when applied is to hold the P2 ring gear. The P2 planetary ring gear is spline coupled to the P1 carrier assembly. The C4 clutch friction plates mesh with the P2 ring gear and will rotate with the P1 planetary carrier when the clutch is not applied. The C4 reaction plates are splined to the C4 clutch housing and therefore are held stationary: When the C4 clutch is applied, the P2 ring gear will therefore be held stationary. Holding the P2 ring gear will prevent P1 planetary carrier rotation.

The function of the C5 clutch is to hold the P3 ring gear. The P3 planetary ring gear is spline coupled to the P2 planetary carrier assembly. The C5 friction plates mesh with the P3 ring gear and therefore rotate with it whenever the C5 clutch is not applied. The C5 reaction plates are splined with the main housing and are therefore stationary. This means that when the C5 clutch is applied, the P3 ring gear is held: Preventing the P3 ring gear from rotating results in the P2 planetary carrier being held.

P1 PLANETARY GEAR MODULE

The P1 is the first of the three planetary gearsets in the transmission. The three subcomponents of the gearset are arranged as follows:

- **Sun gear.** The sun gear is driven by the rotating clutch drum.
- **Planetary carrier.** Its pinions mesh internally with the P1 sun gear and externally with the P1 ring gear. The ring gear of the P2 carrier is splined to the P1 planetary carrier.
- **Ring gear.** The ring gear is housed in the C3 clutch assembly.

Rotation of the P1 planetary gears is controlled by the application of the C3 or C4 clutches.

P2 PLANETARY GEAR MODULE

The P2 planetary gearset is the second in the WT transmission. Its three subcomponents are arranged as follows:

- **Sun gear.** The sun gear is splined to the transmission mainshaft.
- **Planetary carrier.** Its pinions are meshed with the sun gear internally and with the ring gear externally.
- **Ring gear.** The ring gear is spline coupled to the P1 carrier.

Legend for interpreting Figure 17–9

Torque converter module
1. Torque converter turbine
2. Torque converter pump
3. Lockup clutch damper
4. Torque converter stator

Front support/oil pump module
5. Front support
6. Oil pump

Main housing module
7. Main housing
8. C3 clutch
9. C4 clutch
10. C5 clutch

Retarder module
11. Retarder housing assembly
12. Retarder stator assembly
13. Rotor

Mainshaft module
14. Mainshaft
15. P2 sun gear
16. P3 sun gear

P3 module
17. P3 planetary assembly

P2 module
18. P2 planetary assembly

P1 module
19. P1 planetary assembly

Control module
20. Hydroelectric controls
21. Transmission oil level sensor

Rotating clutch module
22. C2 clutch
23. C1 clutch
24. Turbine shaft

Converter housing module
25. Converter housing

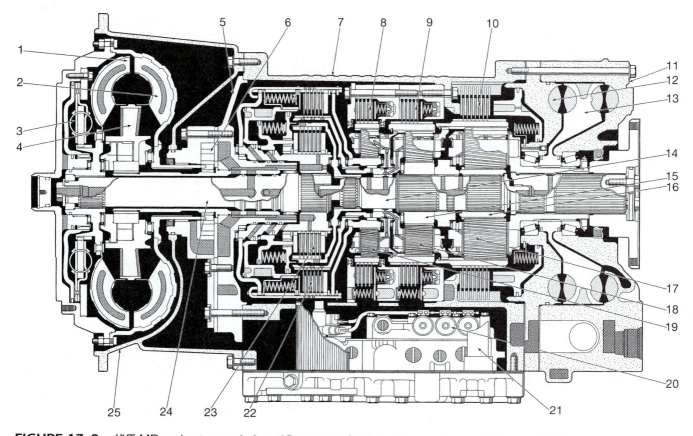

FIGURE 17–9 *WT MD series transmission. (Courtesy of Allison Transmission Division of GM)*

P2 planetary gearset rotation is controlled by the action of the P1 planetary gearset and by the application of either the C4 or C5 clutches. Input rotation to the P2 planetary carrier can either be provided by the P2 ring gear meshing with the P2 planetary carrier or by the P2 sun gear which is splined to the transmission mainshaft.

MAINSHAFT MODULE

The mainshaft passes through the center of the transmission and provides the primary path for transmitting torque through the transmission. The P2 and P3 sun gears are integral with the mainshaft, enabling it to transmit torque to both the P2 and P3 planetary gearsets. The mainshaft is also splined to the rotating clutch module. Input torque can be provided either by the rotating clutch module or the P2 planetary carrier.

OUTPUT MODULES

The output modules contain the components responsible for transmitting output torque to the vehicle driveline. There are some differences in the output module arrangements in WT transmissions, dependent on the specific model.

Rear Cover Module

The rear cover module houses the planetary carrier component of the P3 planetary gearset, the output shaft assembly, the C5 clutch piston, the speed sensor, and the output flange. When the parking brake option is installed, this is also located in the rear cover module. The P3 planetary gearset is arranged as follows:

- **Sun gear.** The sun gear is splined to the transmission mainshaft.
- **Planetary carrier.** The planetary carrier is splined to the output shaft of the transmission.
- **Ring gear.** The ring gear is splined to the P2 carrier assembly. The C5 clutch can hold the ring gear stationary, which will also prevent P2 planetary carrier rotation.

Engine torque is transferred through the transmission and output at the rear cover module.

Output Retarder Module

The retarder functions as an enhancement to the overall vehicle braking system and may help extend the life of the vehicle service brakes. A large percentage of vehicle brake applications involve relatively light service application pressures, so skillful driver management of the retarder can greatly extend foundation brake life. The retarder consists of a stator, rotor, housing, and control valve body. The rotor is splined to the transmission output shaft and is therefore driven at the transmission output shaft speed: It rotates between the stator and retarder housing.

Retardation will occur when the retarder housing is charged with pressurized transmission fluid: This will slow the rotation of the rotor and output shaft and therefore the vehicle driveline. The action of the retarder can be compared to that of a torque converter operating in reverse.

When the retarder is not in use, transmission fluid is drained from the retarder housing to an external **accumulator.** When the retarder is activated, it is charged with transmission fluid stored in the external accumulator. The control valve body is mounted to the right side of the transmission. Three types of retarder capacities are available in Allison WT transmission: low, medium, and high capacity configurations. The specific retarder fitted to a transmission is dependent on vehicle application.

For the retarder to be applied, the following is required:

- The vehicle dash mounted Retarder Enable switch must be on.
- The vehicle must be moving.
- The TPS signal must indicate that throttle travel is close to zero (below 10 percent).
- The ABS system must not be activated.

Retarder Controls

WT retarders are fitted with a variety of controls that vary with vehicle application. Some of the controls used on current transmissions follow:

- **Hand lever.** The driver selects either the off position or one of 6 levels of retardation. Each successive level will increase the voltage return signal to the ECU, with maximum retardation (braking) occurring at level 6.
- **Foot pedal.** A dedicated foot pedal that uses a potentiometer supplied with reference voltage returns a voltage signal to the ECU proportional with pedal travel.
- **Pressure switch.** This option integrates the operation of the WT retarder with the vehicle service brake system. The pressure switch is a normally open switch plumbed into the service brake application circuit that closes at a set trigger value. One pressure switch can be used to provide 100 percent retardation at the preselected service brake application pressure (trigger value). Pressure switches are manufactured with 2, 4, 7, and 10 psi trigger values. Pressure switches are often used in multiples, usually 2 or 3, to provide increased retardation proportional with service brake application pressure.
- **Auto apply—auto full-on.** When this option is specified, the ECU can switch to full retardation when certain conditions are met (usually, the Retarder Enable switch is on, throttle position is close to zero travel, and the transmission output speed is above a preset value).

- **Auto apply—auto percent-on.** This option permits the ECU to switch the retarder to a percent on (33 percent and 50 percent are typical) when certain preprogrammed conditions are met.
- **Combination.** Allison states that almost any of the above control options can be used together in a vehicle.

Retarder Electronic Controls

The retarder electronic controls consist of

- **The ECU.** The ECU receives and interprets speed, resistance module and temperature signals and then computes the appropriate retarder application response.
- **One or more resistance modules.** A resistance module is required for all retarder apply control units, so when multiple control units are used, each must have its resistance module.
- **Two pulse-width-modulated solenoids.** These solenoids control the regulation of retarder cavity pressure and the application of vehicle air pressure to the retarder accumulator. Both solenoids are normally closed, which means they must be electrically energized to perform their function.
- **A retarder temperature sensor.** This sensor inputs retarder temperature data to the ECU, which, when processed, could result in reducing retarder capacity in an over-temperature condition.

Retarder Operation

The retarder control valve solenoid is switched by the transmission ECU at a calibrated rate; when it is actuated, it permits main pressure to act on top of the retarder control valve. This action results in the following:

1. Main pressure enters a passage to charge the retarder charging circuit.
2. Vehicle system air pressure contacts the accumulator piston, discharging the stored transmission fluid into the retarder housing.

The fluid charged to the retarder housing by the accumulator, continuously supplemented by retarder charging pressure, effects the retarding action of the rotor. The transmission fluid circulates through the retarder housing during the retarding cycle and exits to be routed to the oil cooler before being rerouted to the retarder. During the retarding cycle, the torque converter-out fluid is routed directly into the transmission lubrication circuit. The ECU manages the retarder effective cycle. When output shaft speed is within the calibrated range, retarder capacity is at its maximum allowable value. When output shaft speed is either above or below the calibrated value, the ECU will modulate the retarder charging pressure at a lower value. The ECU manages the deactivation of the retarder at a low output shaft speed (typically 350 rpm) to smooth the transition of retarding effort to the vehicle service brakes.

When the retarder is deactivated, the apply process is reversed. The accumulator solenoid is de-energized, which exhausts the vehicle air pressure acting on the accumulator piston, and the retarder control valve solenoid is deactivated at a fixed rate. This controlled deactivation allows pressurized converter-out fluid to recharge the accumulator before it is rerouted to the oil cooler.

Transfer Gear or Dropbox Module

The MD 3070 PT uses a transfer gear module. This transmission is also known as a **Drop-7** transmission. It consists of an output adapter housing, transmission output shaft adapter, transfer case, transfer case charging oil pump, C6 clutch assembly, and a transfer case **scavenging pump** mounted at the right side PTO provision. The addition of the **dropbox** (see **Figure 17–10**) adds a forward gear to the WT transmission and provides for full-time, all-wheel drive.

The transfer case is mounted to the adapter housing and is underslung below the main transmission to provide for forward and rear drive yokes in line with the transmission. The transmission output shaft adapter splines to the P3 carrier hub. The transfer case charging pump and transfer case drive gear are mounted on, and driven by, the output shaft adapter. The function of the drive gear is to direct torque downward into the transfer case. When the stationary C6 clutch is actuated, the mainshaft is held to provide the dropbox with its extra low-range capability.

The transfer case idler gear is installed between the transfer case drive gear and driven gear and acts to transfer torque from one to the other. The driven gear transmits torque to the P4 planetary gearset, which functions to drive the forward and rear driveshafts. When the C7 clutch is engaged, differential action is locked out, meaning that the front and rear outputs are locked. When the C7 clutch is not engaged, dropbox output torque is split 30/70, with 30 percent to the front and 70 percent to the rear drivelines.

The transmission hydraulic circuit provides the pressure to apply the C6 and C7 clutches and to lubricate the dropbox components. The lubricating oil is returned from the dropbox sump to the main transmission housing by a scavenging pump. The dropbox oil pump functions to lubricate the dropbox components in the event that the vehicle is towed with the wheels on the ground.

Legend for interpreting Figure 17–10

Torque converter module
1. Torque converter turbine
2. Torque converter pump
3. Lockup clutch damper
4. Torque converter stator

Front support/oil pump module
5. Front support
6. Oil pump

Main housing module
7. Main housing
8. C3 clutch
9. C4 clutch
10. C5 clutch

Dropbox module
11. C6 clutch
12. C7 clutch
13. Torque proportioning differential

P3 module
14. P3 planetary assembly

Mainshaft module
15. P3 sun gear
16. Mainshaft
17. P2 sun gear

P2 module
18. P2 planetary assembly

P1 module
19. P1 planetary assembly

Control module
20. Hydroelectric controls

Rotating clutch module
21. C2 clutch
22. C1 clutch
23. Turbine shaft

Converter housing module
24. Converter housing
25. PTO drive gear

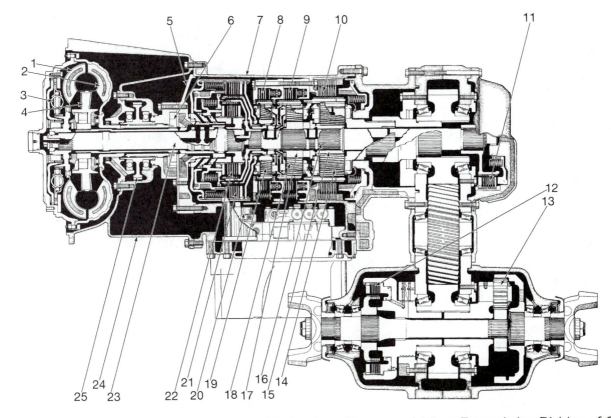

FIGURE 17–10   WT MD series transmission with dropbox. (Courtesy of Allison Transmission Division of GM)

Figure 17–10 identifies the internal components of a WT MD series transmission with a dropbox.

Electro-Hydraulic Control Module

The **electro-hydraulic control module** is mounted under the main transmission housing and, in fact, forms its bottom enclosure. This module houses the solenoids, sensors, valves, and regulators that manage the pressure and flow of transmission fluid to the clutches, the torque converter, and the lubrication/cooling circuits. The transmission filters are also located in this module.

17.4 ## WTEC II AND III ELECTRONICS

Electronic management of the WT transmission is similar to the CEC, with some of the refinements that might be expected in a more advanced version of the system management electronics. Most notable is an increased ability to interact with other on-board electronic systems and better programmability to suit specific vehicle applications. **World Transmission Electronic Control (WTEC)** components may be divided into input circuit components, a microcomputer and output circuit components: A wiring harness connects the various components in the system.

INPUT CIRCUIT COMPONENTS

- Shift selector(s)
- Throttle position sensor
- Engine speed sensor
- Turbine speed sensor
- Output speed sensor
- C3 pressure switch
- Sump temperature sensor
- Coolant temperature
- Vehicle interface module (VIM)

Input Signals

The following is a partial list of the input signals the WT ECU may be programmed to process:

- Secondary shift schedule
- PTO enable
- Shift selector transition
- Engine brake and preselect request
- Fire truck pump mode
- Automatic neutral
- Automatic neutral for PTO
- Automatic neutral for refuse packer
- Two-speed axle enable
- Manual lockup
- Antilock brake response
- Retarder enable
- Service brake status
- Differential clutch request
- Kickdown

MICROCOMPUTER

The ECU receives input signals, performs logic processing, generates outcomes, and performs output switching.

OUTPUT CIRCUIT COMPONENTS

- A–E clutch solenoids
- Lockup clutch solenoid F
- Forward latch solenoid G
- Vehicle interface module (VIM)
- PTO functions

Output Signals

The following is a partial list of the output signals WT electronics may be programmed to supply other chassis electronic management systems with:

- Engine brake enable
- Sump/retarder temperature indicator
- Range indicator
- Output speed indicator
- PTO enable
- Engine overspeed indicator
- Two-speed axle enable
- Lockup indicator
- Service indicator
- Shift-in-progress indicator
- Retarder indicator
- Neutral indicator for PTO

INPUT CIRCUIT

The WTEC III input circuit is responsible for supplying the ECU with monitoring and command data it needs to compute the shift logic. Input from the operator is sent to the ECU by means of the shift selector. Other input signals come from sensors located in the transmission itself and from the **vehicle interface module (VIM)**. The VIM handshakes data exchange (both ways) between WT electronics and other vehicle electronic management systems using SAE J 1939 communication protocols and hardware. All input signals feed data into the processing cycle of the ECU. Much of this data can be classified as monitoring data. Temperature sensors, shaft speed sensors, and pressure switches all monitor conditions within the transmission that are required to compute an outcome. Command input signals are usually operator-generated and often require the ECU to make a change in outcome. The **throttle position sensor (TPS)** range shift selector (see **Figure 17–11**), retarder controller, and PTO switches are all examples of command input signals.

Shift Selectors

A lever and two types of pushbutton shift selector are available on WT transmissions. **Figure 17–11** shows the typical appearance of the selectors. The strip-type pushbutton selector is not full function,

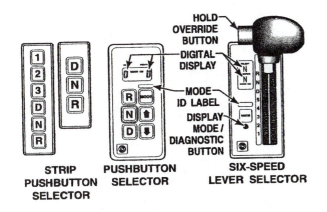

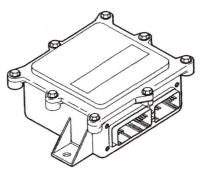

FIGURE 17–12 *Vehicle interface module (VIM). (Courtesy of Allison Transmission Division of GM)*

FIGURE 17–11 *WT shift selectors. (Courtesy of Allison Transmission Division of GM)*

allows the operator the least amount of input as to range selection, and has no illuminated **digital display (DD).** It is not often used.

The full function non-strip type pushbutton selector allows the operator to select any range programmed to WTEC by selecting the drive button and using arrowed buttons to downshift or upshift. Using these buttons does not override the automatic shifting function of the transmission: If a higher or lower range is selected by the operator, the transmission will shift through the ranges to the selected range. This shifter has a digital display which is explained later in this section.

A number of different types of lever selectors are used with WT transmissions, the most common being the six-position lever selector shown in **Figure 17–11.** The lever selector is an electromechanical device that simply inputs command data to the ECU: It will not permit the operator to select a shift schedule that could result in damage to either the transmission or drivetrain. When the D range position is selected, WTEC will manage shift schedules using all the ranges programmed to WTEC. Selection of any of the numeric ranges will confine shifting within the ranges below the selected range. For instance, if the numeric position 3 were selected, only ranges 1–3 would be used by WTEC.

Vehicle Interface Module (VIM)

The VIM is a splashproof (as opposed to watertight) housing that provides the necessary relays, fuses, and connection points to connect the WT ECU with the vehicle electrical system. The VIM contains two 10-amp fuses and somewhere between 2 and 6 relays. WT electronics will operate on either 12- or 24-volt chassis electrical systems. The relays used must match the ignition voltage.

One of the 10-amp fuses protects the main power connection to the ECU. The other protects the feed from the ECU to ignition circuit.

The VIM will always have at least two relays. One is used to output a signal to the reverse warning circuit and the other is used for the neutral start circuit. As many as four other relays may be used for special function outputs. **Figure 17–12** shows a VIM unit.

Throttle Position Sensor (TPS)

WT electronics are designed to handshake with other chassis electronic systems, so providing an electronically managed engine is used, a dedicated TPS is only required when a WT transmission is used with a hydromechanical engine. In cases where an electronically managed engine is used, WT electronics simply shares the TPS signal with the engine by means of a J 1939 data bus (connection), described by Allison as a **serial communications interface (SCI).**

The TPS, as with most truck TPSs, should be of the potentiometer type. The idea is to convert a reference voltage value output from the ECU into a signal voltage value modulated proportionally with mechanical travel of the accelerator pedal. The potentiometer is a three-wire variable resistor that converts position to a voltage value by sliding contacts across a resistive strip. **Figure 17–13** shows a TPS, its operating principle, and its calibration parameters. The SCI is really a means of both inputting and outputting data to the WT ECU, but it is covered under outputs later in this chapter.

Speed Sensors

Shaft speed sensors are required to input rotational speed data to the WT ECU. They use an induction pulse generator principle. **Figure 17–14** shows some examples of WT speed sensors.

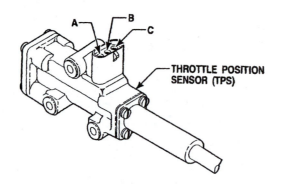

Throttle Position Sensor (TPS)

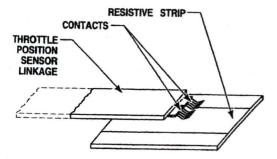

Throttle Position to Voltage Conversion

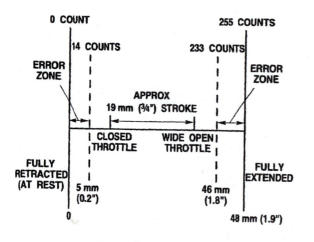

Throttle Position Sensor Determination

FIGURE 17–13 *Throttle position sensor, principle of operation, and calibration parameters. (Courtesy of Allison Transmission Division of GM)*

THE WT ECU

As with any vehicle electronic management system, the ECU can be regarded as the brain of the system.

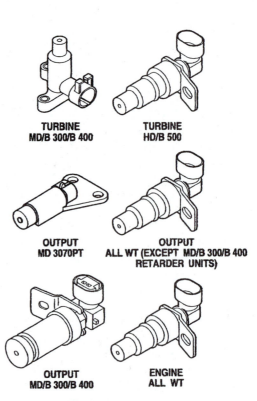

FIGURE 17–14 *WT speed sensors. (Courtesy of Allison Transmission Division of GM)*

A full account of microprocessing cycles is provided in Chapter 6 of this book for reference. The WTEC III ECU is responsible for processing the command, monitoring input data, plotting shift sequences and timing, and effecting the results of its computations into outcomes by switching clutches. The WT ECU has an EEPROM chip, which means that both Allison and customer preference or application data can be reprogrammed.

The significant improvement in WT shift logic is accounted for by replacing hard parameter processing outcomes with soft logic processing, better called **adaptive logic.** WT electronics continually compare actual transmission shift characteristics with the optimum shift calibration stored in ECU memory. When these actual shift characteristics fall outside the optimum parameters, the ECU alters solenoid shift modulation to correct the shift character. This means that WT electronics can produce close to optimum shift quality by constantly adapting to changes in operating conditions.

In simple terms, WT electronics constantly "learn" from actual performance conditions, factoring in variables such as grade, load, and engine power. After every shift is completed, the ECU will compare the shift to what ideally should have occurred during the shift and the next time a shift is made under similar conditions, the adjustments are effected.

Adaptive logic is therefore the establishing of the initial conditions for shifts. It can be categorized into two types, fast adaptive logic and slow adaptive logic. When the ECU is programmed to operate in **fast adaptive mode,** it makes large changes in initial shift conditions to adjust for major system tolerances. Major system tolerances are solenoid-to-solenoid, main pressure, and clutch-to-clutch variations. Any new or newly rebuilt transmission should be programmed to operate in fast adaptive mode.

Slow adaptive mode programming makes small changes in initial shift conditions in response to minor system changes. Minor system changes are conditions that occur over a period of time and are often associated with wear. As internal transmission components wear and solenoids electrically degrade or drift, WT electronics will change from fast adaptive to slow adaptive mode to compensate.

The Allison WTEC III ECU also has an autodetect feature. Autodetect is active in the system for the first 30 seconds of the first 24 to 49 engine starts (dependent on the component or sensor being detected). This means that WTEC searches for the following components or data inputs:

Retarder	Present/not present
Oil level sensor	Present/not present
TPS	Analog (potentiometer type) /J 1939 compatible
Engine coolant temperature	Analog/J 1939 compatible

Note: TPSs that output a **pulse-width modulation (PWM)** signal (variable capacitance type or current Caterpillar) are not auto-detected and must be programmed to WTEC using the **Digital Diagnostic Reader (DDR).**

After autodetect has been completed, it can be reset at any time using the DDR. During the first 24 starts, autodetect does not retain what was detected on any previous start. For example, if a retarder was detected on every engine start up to number 23, H solenoid diagnostics would be performed by the ECU until the system was shut down each time. If the H solenoid failed at start-up number 24, autodetect would not recognize that a retarder was present and no diagnostics would be logged for the H solenoid.

Figures 17–15 and **17–16** show WT electronic circuits without and with a retarder module.

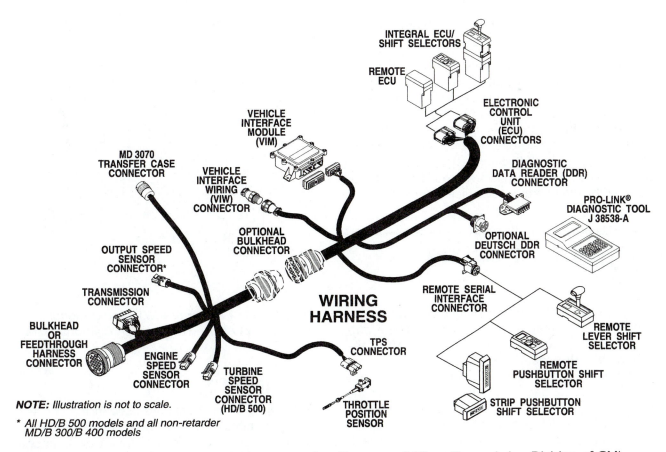

FIGURE 17–15   WT electronic circuit without retarder. (Courtesy of Allison Transmission Division of GM)

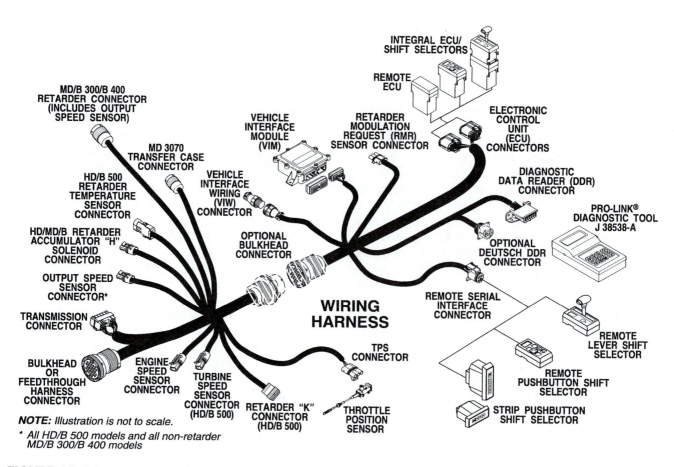

FIGURE 17–16 WT electronic circuit with retarder. (Courtesy of Allison Transmission Division of GM)

The WTEC ECU is also programmed with Allison data designed to protect the transmission and drive train components from abuses. This type of programming would include inhibiting action when selected to a full-throttle, neutral to range shift command, or a high speed direction change. The WTEC ECU uses SAE J 1939 diagnostic codes and can retain them as active or inactive codes that can be accessed by the operator or service technician.

17.5 OUTPUT CIRCUIT

As in any vehicle electronic management system, the output circuit is responsible for effecting the outcomes of the processing cycle of the ECM. In the Allison WT, the results of ECM processing are effected by an electro-hydraulic control module. This is mounted at the base of the transmission, and the channel plate forms its bottom enclosure. **Figure 17–17** shows a WT control module. Solenoids are used to convert electrical signals into hydromechanical outcomes.

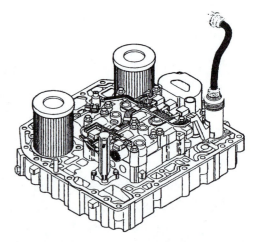

FIGURE 17–17 Typical WT control module. (Courtesy of Allison Transmission Division of GM)

SOLENOIDS

Two types of solenoids are used in the WT electro-hydraulic control module. Both are somewhat similar in construction, but differ in their switching characteristics. Both types receive main pressure from a

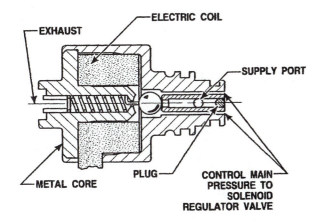

FIGURE 17–18 _WT normally closed solenoid._
(Courtesy of Allison Transmission Division of GM)

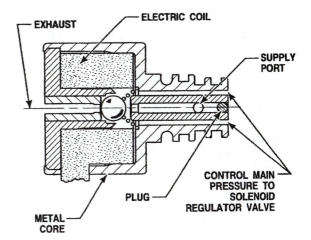

FIGURE 17–19 _WT normally open solenoid._
(Courtesy of Allison Transmission Division of GM)

supply port and may either route that pressure to exit through the solenoid regulator valve or to an exhaust port, depending on the switch status.

Normally Closed (NC) Solenoids

An NC solenoid remains closed until energized by an ECM electrical signal. When the solenoid is not energized, main pressure fluid is blocked by the seated check ball, routing it directly to the exhaust passage. When energized, the check ball unseats, blocking the exhaust port and permitting main pressure fluid to exit to the solenoid regulator valve. Refer to **Figure 17–18**.

Normally Open (NO) Solenoids

An NO solenoid remains open until energized by an ECM electrical signal. When the solenoid is not energized, the check ball is unseated, allowing main pressure fluid to flow into the solenoid through the supply passage and be routed to the solenoid regulator valve. Refer to **Figure 17–19**. When the normally open solenoid is energized, the check ball is forced onto its seat: This blocks the passage to the regulator valve and routes main pressure to the exhaust.

Pulse-Width Modulation (PWM)

The solenoids are controlled by PWM signals from the ECM. A PWM signal can be divided into **primary modulation,** which controls the amount of on time, and secondary or submodulation, which controls the current flow. The WT ECU controls the transmission clutch apply-and-release characteristics by varying the primary and **secondary modulation** signals to the control solenoids.

The primary modulation signal to WT solenoids is delivered at a **frequency** (see Chapter 6 for a full explanation of frequency) of 63 Hz. This means that each second is divided into 63 equal segments or cycles, during which the voltage will be in the on state for a period of time. This period of time is expressed as a percentage. The percentage of time a voltage is present within each $1/63$ second is referred to as the solenoid **duty cycle.** A 100 percent duty cycle would represent the maximum signal that could be delivered to the solenoid, whereas a zero percent duty cycle would indicate no signal. As the **pulse width (PW)** or duty cycle is increased, the percentage of solenoid on time is increased. This permits the solenoid regulator valve to apply or release a clutch with optimum shift quality. **Figure 17–20** shows a PWM waveform in graphic form, expressing it as percentage of cycle.

Secondary or submodulation supplies a constant optimum average current to the solenoid at high frequency (7,812 Hz): this permits the solenoid ball to be first pulled in, and then held. This means that the heat in the solenoid coil can be minimized. Secondary modulation, by managing current to a variable optimum requirement, permits consistent

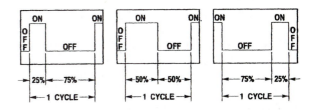
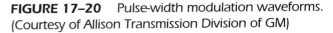

FIGURE 17–20 _Pulse-width modulation waveforms._
(Courtesy of Allison Transmission Division of GM)

solenoid performance when variables such as system battery voltage and sump temperature are factors.

WT Digital Display (DD)

The data that appears on DD is controlled by the ECU. In other words, it does not display data directly as a result of mechanically pushing buttons or moving the shift lever. The following information can be displayed:

- **Selector position.** The display is the acknowledgement by the ECU of the selected shift position.
- **DD Flashing.** Flashing indicates that the requested shift is inhibited because of an inappropriate condition such as inappropriate actual engine or road speed, but that the shift will occur when those preconditions have been met. The flashing display shows the numeric value of the current range.
- **DD blank.** A blank indicates that there is a problem in the electrical circuit to the shift selector or that the data link has failed.
- **DD displays "cat's-eye".** Communication with the ECU has been lost.

Serial Communications Interface

The SCI permits the transmission management electronics to interact with all other vehicle electronics systems using SAE J 1939 communication protocols and compatible hardware. This optimizes transmission operation with that of the rest of the powertrain and chassis systems. The SCI provides two-way data transfer between the WT ECU and other chassis electronic system, so it is really both an input and output mechanism in the WT electronic circuit.

Data that can be input through the SCI include:

- Throttle position
- Percent load
- Cruise control status
- Road speed limit status
- Engine coolant temperature
- ABS active
- Kickdown signal
- Anti-slip logic

Data that can be output through the SCI include:

- Transmission sump fluid temperature
- Retarder temperature
- Transmission output speed
- Engine rpm (input rpm)
- Range position selected

- Actual range
- Lockup clutch status
- Retarder active status

ECU Managed PTO Mode

WT provides for three main types of PTO management modes. Constant drive PTOs provide continuous PTO power whenever the transmission input is rotated. Clutched-drive PTOs provide PTO power when activated by the operator; activating the PTO causes a solenoid in the PTO to apply main pressure to the rotating PTO clutch, which transmits torque to the PTO output. Clutch drive PTOs may also be controlled by the ECU. The ECU can therefore be programmed to permit or inhibit PTO operation when specific transmission or chassis operating conditions exist. An example would be an interlock/shift status or parking brake status during operation. This feature allows PTO operation to be safely integrated with other chassis functions.

The WTEC III Digital Diagnostic Reader (DDR)

The Allison EST required to read and reprogram WTEC III electronics is the MPSI **ProLink**. Two levels of update are required to upgrade an Allison DDR cartridge to WT electronics. The first-level update requires the replacement of the PROM chip socket attached to the ProLink motherboard. This first level upgrade will only permit WTEC diagnostic data to be viewed. The second level of upgrade is accomplished by installing a multiprotocol cartridge (MPC), which includes an insertable MPC application card. The MPC upgrade will permit full function usage of the DDR. Full function usage of the DDR includes data listing, reprogramming, and rewriting of customer data options.

Reprogramming a WTEC III ECU currently involves removing all the data from the EEPROM memory component and then rewriting both the unchanged and revised data to EEPROM. When a DDR is upgraded either by replacing the DDR PROM chips or by a MPC cartridge and card, the DDR may be used with all previous versions of Allison management electronics.

🥽 **Shop Talk**

In a general sense, the WT electronic management circuit is no different from that used to manage most of today's engines. It consists of input, processing, and output circuits. It must also interact with other chassis electronic systems to optimize vehicle performance.

17.6 THE WT HYDRAULIC CIRCUIT

The Allison WT hydraulic system is responsible for

- Storing hydraulic fluid
- Generating hydraulic pressure
- Acting as the drive medium in the torque converter
- Controlling system pressures
- Directing flow through the various subcircuits
- Cooling and lubricating the transmission

The primary components of the hydraulic system are

- Transmission fluid
- Charging pump
- Three integral filters
- Electro-hydraulic control module
- Breather
- C3 pressure switch

There are 7 subcircuits that make up the hydraulic circuit:

- Main pressure circuit
- Control main circuit
- Torque converter circuit
- Cooler/lubrication circuit
- Clutch apply circuit
- Exhaust circuit
- Exhaust backfill circuit

Torque converter operation is covered in Chapter 12. Transmission fluid pressure and flow through the hydraulic circuit is provided by the engine-driven charging pump. The charging pump pulls fuel through a primary filter from the transmission sump. This fluid is then pumped through the main filter to supply the hydraulic circuit. Solenoids and valves located in the electro-hydraulic control module and managed by the ECU control the flow and pressure of the fluid, directing it to specific clutches to achieve range shifts. Fluid for the cooler and lubrication circuit flow through a third filter known as a lubrication filter.

The C3 pressure switch is located in the C3 clutch circuit. Its function is to monitor circuit pressure. An accumulator is located in the C3 pressure switch cavity. Its function is to dampen hydraulic pulsing in the clutch-apply circuit.

THE MAIN PRESSURE CIRCUIT

The main pressure circuit is responsible for supplying the hydraulic pressure required to manage the hydraulic circuit of the transmission. Main pressure is managed by the **main pressure regulator** valve. The primary function of the main pressure regulator valve is to modulate charging pressure values to main pressure values. Actual main pressure value is modulated based on input from the overdrive knockdown valve, the converter flow valve, and the control main pressure.

The main pressure regulator valve is a spool-type valve held in an upward position by spring force. Main pressure acts on the top of the main pressure regulator valve spool and forces it downward against spring pressure. When this pressure is high enough to drive the spool downward enough, main pressure is spilled to exhaust, thereby reducing main pressure.

Regulated main pressure is routed to 7 areas in the hydraulic circuit. Each of the 6 solenoid regulator valves plus the control main regulator valve receive regulated main pressure. Pressure at the outlet of each solenoid regulator valve is clutch feed pressure. Pressure at the outlet of the control main regulator valve is control main pressure.

Solenoids and Solenoid Regulator Valves

The WT ECU switches the solenoids in the electro-hydraulic control module. The solenoids direct control main pressure to the top of solenoid regulator valves, causing each to move inboard against spring pressure. This blocks the exhaust backfill circuit and allows main pressure to move through the valve passage into the clutch feed circuit, applying the clutch. When control main pressure ceases to act on the top of the solenoid regulator valve, the spring forces the valve back to the top of its travel, exhausting the clutch-apply circuit. This releases the clutch.

Clutch Application

In order to obtain any forward or reverse range, two clutches must be applied. **Table 17–1** shows which clutches are applied in each transmission range, the corresponding energized solenoids, the C1 and C2 latch valve positions, and the converter flow valve position.

Hydraulic Circuit Operation During an Electrical Failure

The WT transmission is managed electrically. In the event of a chassis electrical failure, the ECU and its switching actuators and solenoids will cease to function. Such an interruption of electrical power will result in the solenoid regulator valves locking in their NO or NC states. To minimize the impact of an electrical failure, the WT transmission defaults to totally

TABLE 17–1:   CLUTCH APPLICATION CHART

Range	Clutches applied	Solenoids energized	C1 latch position	C2 latch position	Conv. flow valve position
Neutral	C5	A, B, E	Up	Up	Down
Reverse	C3, C5	A, B, C, E	Up	Up	Down
First	C1, C5,	B, E, G	Down	Down	Down
Second	C1, C4, lockup	B, D, F, G	Down	Down	Up
Third	C1, C3, lockup	B, C, F, G	Down	Down	Up
Fourth	C1, C2, lockup	F, G	Down	Down	Up
Fifth	C2, C3, lockup	A, C, G, F	Down	Down	Up
Sixth	C2, C4, lockup	A, D, F	Up	Down	Up

Notes:

1. A and B solenoids are normally open when not energized.

2. C, D, E, F, and G solenoids are normally closed when not energized.

3. G solenoid is on in 1st and 2nd ranges but is switched off for the duration of the 1–2 upshift and the 2–1 downshift.

hydraulic operation. The C1 and C2 latch valves are used to accomplish this default mode of operation.

When a clutch is applied, clutch feed pressure from the solenoid regulator valve is routed through the latch valves to the clutch piston to actuate the clutch. Clutch-apply pressure acting against the lands of the latch valves holds them in position or, in the case of a normally closed valve, permits the fluid to pass through the valve. When an electrical failure occurs, the latch valves cause the transmission to engage specific clutches based on the range the transmission was in when the failure occurred.

The latch valves are activated by the normally closed solenoid G. When G solenoid is energized, control main pressure flows to the top of the C1 and C2 latch valves. This pressure forces the valves downward to supply the flow passages required for clutch engagement. In case of an electrical failure, the latch valves and the two NO solenoids, A and B, will enable the transmission to operate in this default or limp-home mode by reverting to total hydraulic operation.

17.7 ALLISON WT TORQUE PATHS AND HYDRAULIC RANGE MANAGEMENT

The general operating principles and power flow characteristics of torque converters and planetary gearsets are covered extensively in previous chapters and are not covered again in this section. The torque path routing in each gear range will be described, referencing a figure for consultation. Toward the end of this chapter, the range power

flows are shown in diagram form. In each range, the gearing is staged through the transmission. The stages are sequenced numerically, but the stages of gearing do not necessarily correlate with the numbering of the planetary gearsets. For instance, first range gearing is single stage and takes place in the P3 planetary gearset. As many as three stages of gearing may be used, depending on the range selected.

Shop Talk

Learning hydraulic circuit and torque power flows helps visualize transmission operation. However, they are difficult to recall and learning them is probably not necessary for the general truck technician. For the Allison specialist tech, knowing the hydraulic circuit factors and torque power flows can help in diagnosing performance problems.

HYDRAULIC CIRCUITS

The following sequence of schematics shows the hydraulic circuits in a WT transmission.

Neutral Operation

See **Figure 17–21**.

The transmission is in neutral when torque from the torque converter is not transmitted beyond the rotating clutch module and the P1 planetary sun gear. Both the CI and C2 clutches are released. The C5 clutch is applied, but no output movement occurs because a driving clutch is not applied to move the

WT HYDRAULIC SCHEMATIC – NEUTRAL

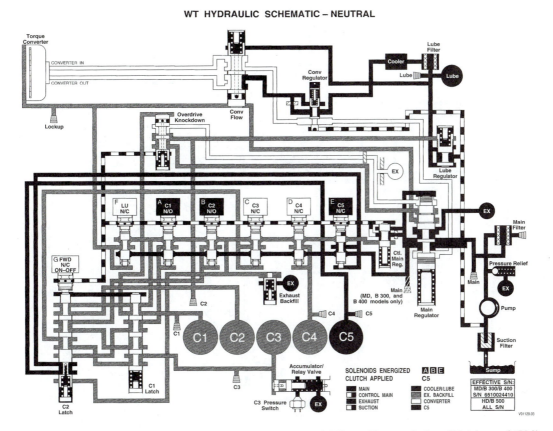

FIGURE 17–21   *WT hydraulic schematic—neutral. (Courtesy of Allison Transmission Division of GM)*

planetary gears. The result is that no output torque occurs. If the output shaft is rotating while in neutral, the stationary clutch applied will depend on the shaft rotational speed. This varying application corresponds to the neutral ranges identified as N1, N2, N3, and N4. These ranges are dependent on the application of the clutches C5, C4, C3, and C2 respectively. The application of a specific clutch during neutral operation means that one of the two clutches required to ensure a range selection when the transmission is shifted to D is already engaged.

First Range Operation

See **Figure 17–22**.

Clutches C1 and C5 are applied. Torque is applied by the torque converter to the turbine shaft. Because the C1 is applied, the rotating clutch assembly acts to lock the turbine shaft and mainshaft, causing them to rotate as a unit.

Torque is transmitted through the mainshaft to the P3 sun gear, which is splined to the mainshaft. The P3 sun gear acts to rotate the P3 pinion carrier, which is splined to the output shaft. The C5 clutch being engaged prevents the P3 ring gear from being rotated.

Torque route: converter turbine shaft, mainshaft, P3 sun gear, P3 pinion carrier, output shaft.

First Range Operation in WT with Dropbox/Transfer Case.
Clutches C3 and C6 are applied. The C6 clutch holds the mainshaft stationary. A WT transmission with a transfer case has a mainshaft that extends from the main housing through the P3 planetary gear assembly and output adapter to the C6 clutch. When the C3 clutch is applied, the P1 ring gear is held. Because the mainshaft cannot rotate, torque input is through the P1 sun gear. Because the P1 ring gear is held, output torque is transferred through the pinion carrier.

The P1 planetary carrier is splined to the P2 ring gear, which acts on the P2 pinion carrier. The P2 sun gear is integral with the mainshaft, which is being held by the C6 clutch. Torque is routed from the P2 pinion carrier to the P3 ring gear, which is splined to it. As the P3 sun gear is integral with the mainshaft, which is being held by the C6 clutch, torque is output from the P3 pinion carrier, which is splined to the output shaft adapter. The output shaft adapter applies torque to the transfer drive gear.

Torque is transmitted to the transfer case by means of the transfer drive gear. This drives an idler gear, reversing the direction of rotation. The driven gear is splined to the P4 (differential) pinion carrier. The P4 sun gear is integral with the transfer case

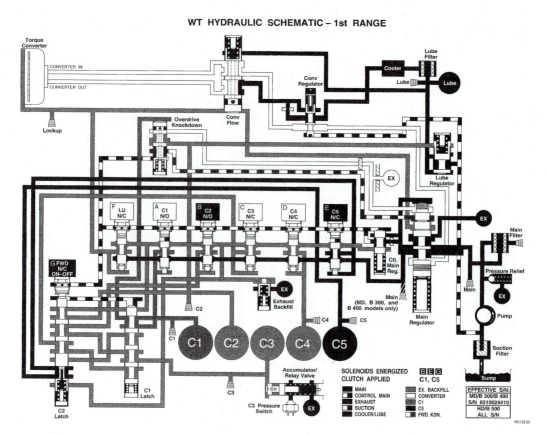

WT HYDRAULIC SCHEMATIC – 1st RANGE

FIGURE 17–22 WT hydraulic schematic—1st range. (Courtesy of Allison Transmission Division of GM)

output shaft. Torque is therefore transferred from the P4 pinion carrier to the P4 sun gear, rotating the transfer case output shaft.

When the C7 (differential lock) clutch is disengaged, transfer case output torque is split, with 70 percent directed at the rear output shaft and 30 percent at the front output shaft. When the C7 clutch is engaged, output torque is split evenly between the front and rear output shafts.

Torque route: converter turbine shaft, P1 sun gear, P1 planetary carrier, P2 ring gear, P2 pinion carrier, P3 ring gear, P3 pinion carrier, output shaft adapter, transfer drive gear, P4 pinion carrier, P4 sun gear, transfer case output shafts.

Second Range Operation

See **Figure 17–23**.

Clutches C1 and C4 are applied. The applied C1 clutch locks the turbine shaft and mainshaft so they rotate as a unit. Torque is applied to the P2 sun gear (integral with mainshaft), rotating the P2 pinion carrier. The applied C4 clutch holds the C2 ring gear. Torque applied to the P3 planetary is now split, being transferred from the P2 pinion carrier to the C3 ring gear and by the C3 sun gear that is splined to and rotates with the mainshaft. The P3 ring gear

rotates at a slower speed than the P3 sun gear. This results in the P3 ring gear action resembling that of a held member. The P3 sun gear becomes the main input torque path for the P3 planetary. This torque is applied to the P3 pinion carrier, which provides output torque to the transmission output shaft.

Torque route: converter turbine shaft, mainshaft, P2 sun gear, P2 pinion carrier, P3 ring gear/mainshaft, P3 sun gear, P3 pinion carrier, output shaft.

Third Range Operation

See **Figure 17–24**.

Clutches C1 and C3 are applied. The C1 clutch locks the turbine shaft and mainshaft so they rotate as a unit. The P1 planetary provides first stage gearing with torque passing from the P1 sun gear (part of the rotating clutch) to the P1 pinion carrier. The P3 clutch holds the P1 ring gear. The P1 carrier is splined to the P2 ring gear. The P2 planetary is therefore subject to input from the torque applied to both its ring gear and its sun gear, which is splined to and rotates with the mainshaft. The P2 ring gear will rotate at a lower rate than its sun gear, so the action of the P2 ring gear will resemble that of a held member in the planetary gearset. The P2 sun gear is therefore the main path for torque transmitted to the

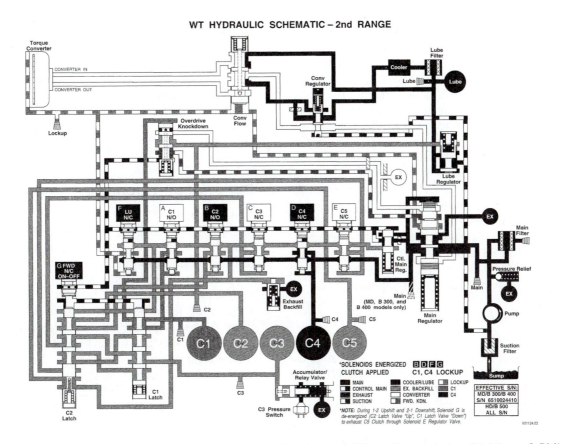

FIGURE 17–23 WT hydraulic schematic—2nd range. (Courtesy of Allison Transmission Division of GM)

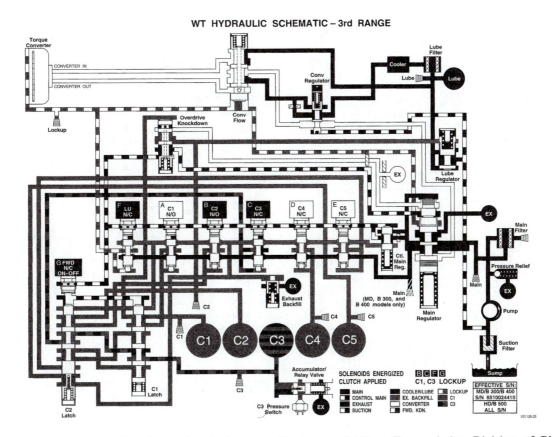

FIGURE 17–24 WT hydraulic schematic—3rd range. (Courtesy of Allison Transmission Division of GM)

P2 pinion carrier. The P2 pinion carrier is the output from the P2 planetary in the third gear range.

The P2 pinion carrier is splined to the P3 ring gear and provides one of two inputs to the P3 planetary gearset. The other input torque path is the P3 sun gear, which is splined to and therefore rotates with the mainshaft. The P3 ring gear rotates at a lower speed than the P3 sun gear, so P3 ring gear action will resemble that of a held member. The P3 sun gear becomes the main torque path for torque transmitted to the P3 planetary gearset. This torque is transmitted to the P3 pinion carrier, which is splined to the output shaft of the transmission.

Torque route: converter turbine shaft, mainshaft, P1 ring gear and P1 sun gear, P1 pinion carrier, P2 sun gear and P2 pinion carrier, P3 ring gear and P3 sun gear, P3 pinion carrier, output shaft.

Fourth Range Operation

See **Figure 17–25**.

Clutches C1 and C2 are applied. The C1 and C2 clutches rotate with the turbine shaft. The C1 drive hub is splined to the mainshaft so that when it is actuated, the mainshaft will rotate at turbine shaft (input) speed. The C2 drive hub is splined to the P3 ring gear (through the P2 pinion carrier), which results in the P3 ring gear rotating at turbine speed.

Because the C3, C4, and C5 stationary clutches are all disengaged, the remaining pinion carrier and ring gears rotate at the same speed and direction as the turbine shaft. The P3 ring gear drives the P3 pinion carrier which is splined to the transmission output shaft. The result is direct drive.

Torque route: converter turbine shaft, mainshaft, P3 sun gear and P3 ring gear, P3 pinion carrier, output shaft.

Fifth Range Operation

See **Figure 17–26**.

Clutches C2 and C5 are applied. The P1 planetary provides first stage gearing in fifth range with input torque routed from the P1 sun gear (part of the rotating clutch assembly/turbine shaft). The C3 stationary clutch locks the P1 ring gear, holding it stationary. Torque input to the P1 planetary gearset is output through the P1 pinion carrier. The application of the C2 clutch locks the turbine shaft and the P2 pinion carrier so they rotate together.

The P2 planetary gearset provides second stage gearing in fifth range. The P2 pinion carrier is subject to torque applied from the C1 drive hub and the P2 ring gear that is splined to and rotates with the P1 pinion carrier. Because the P2 ring gear rotates around the P2 pinion carrier at a slower rate than the

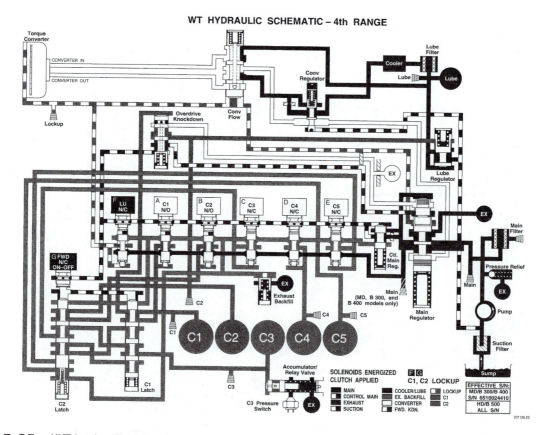

FIGURE 17–25 *WT hydraulic schematic—4th range. (Courtesy of Allison Transmission Division of GM)*

WT HYDRAULIC SCHEMATIC – 5th RANGE

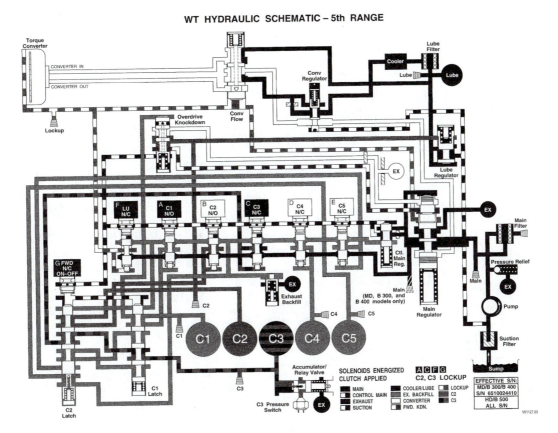

FIGURE 17–26 WT hydraulic schematic—5th range. (Courtesy of Allison Transmission Division of GM)

pinion carrier speed, P2 ring gear action resembles that of a held member. Torque is transmitted from the P2 carrier to the P2 sun gear, which provides second stage output to the mainshaft.

Third stage gearing uses the P3 planetary gearset, which receives torque input from both the mainshaft (both P2 and P3 sun gears are splined to the mainshaft) and the P3 ring gear, which is splined to the P2 pinion carrier. The P3 ring gear rotates around the P3 carrier at a slower speed than the driven speed of the P3 pinion carrier, so it will produce gear action that resembles that of a held member. This means that the P3 sun gear provides the main path for torque transmitted to the P3 pinion carrier. The P3 pinion carrier is splined to the output shaft.

Fifth range is an overdrive range. This means that transmission output shaft speed exceeds the input speed. When overdrive gearing is selected, an increase in fuel economy may be achieved, but a reduction in torque advantage occurs.

Torque route: converter turbine shaft, P1 sun gear, P1 pinion carrier, P2 ring gear and P2 planetary carrier, P2 sun gear, mainshaft, P3 ring gear, P3 sun gear, P3 pinion carrier, output shaft.

Sixth Range Operation

See **Figure 17–27**.

Clutches C1 and C4 are applied. The application of the C2 clutch locks the turbine shaft and the P2 pinion carrier, causing them to rotate as a unit. The P2 planetary gearset provides the first stage gearing in sixth range. The C4 clutch holds the P2 ring gear. With the P2 ring gear held and torque input to the P2 carrier, torque is transmitted to the P2 sun gear, which is splined to the mainshaft.

Second-stage gearing in sixth range takes place in the P3 planetary gearset. Torque is input to the P3 planetary gearset by the P3 sun gear (splined to the mainshaft) and the P3 ring gear that is splined to the P2 pinion carrier. The P3 ring gear rotates around the P3 pinion carrier at a slower rate than the rotational speed of the P3 sun gear, so its action resembles that of a held member. The result is that the sun gear becomes the main path for delivering torque to the P3 pinion carrier. The P3 pinion carrier transmits torque to the transmission output shaft.

Sixth range is also an overdrive range. The input shaft rotational speed is exceeded by the output shaft to a greater extent than when in the fifth range.

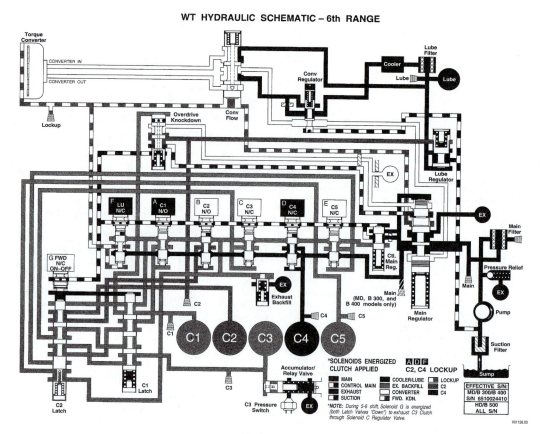

FIGURE 17–27 WT hydraulic schematic—6th range. (Courtesy of Allison Transmission Division of GM)

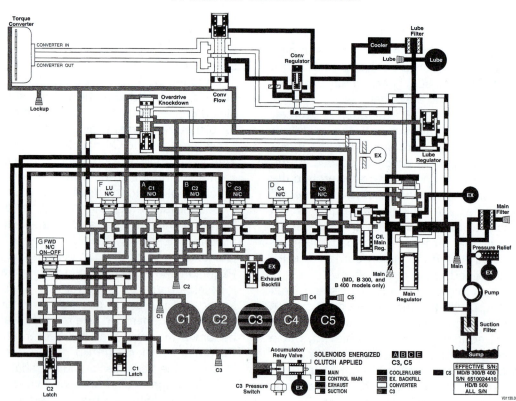

FIGURE 17–28 WT hydraulic schematic—reverse. (Courtesy of Allison Transmission Division of GM)

Torque route: converter turbine shaft, P2 pinion carrier, P2 sun gear, mainshaft, C3 ring gear/mainshaft, C3 sun gear, C3 pinion carrier, output shaft.

Reverse Operation

See **Figure 17–28**.

Clutches C3 and C5 are applied. The C3 clutch holds the P1 ring gear. The C5 clutch holds the P3 ring gear. First stage reverse gearing takes place in the P1 planetary gearset. The P1 sun gear rotates with the rotating clutch assembly. Torque is transmitted from the P1 sun gear to the P1 pinion carrier. Because the P1 ring gear is being held, torque is output from first stage gearing by the P1 pinion carrier.

Second stage gearing in reverse takes place in the P2 planetary gearset. Torque is input at the P2 ring gear, which is splined to the P1 pinion carrier. The P2 pinion carrier is held stationary because it is splined to the P3 ring gear, which is held by the C5 clutch. This means that torque is transmitted from the P2 ring gear through the **planetary pinions** in the held P2 planetary carrier, to the P2 sun gear. The P2 sun gear is splined to the mainshaft. This action through the P2 planetary gearset causes the mainshaft to rotate in an opposite direction (counterclockwise) to the rotation of the turbine shaft input.

Third-stage gearing takes place in the P3 planetary gearset. The mainshaft turns in an opposite direction to input and rotates the P3 sun gear that is splined to it. The P3 ring gear is held by the C5

clutch. Torque is transmitted from the P3 sun gear to the P3 planetary pinions and output through the pinion carrier to the output shaft.

Torque route: converter turbine shaft, P1 sun gear, P1 pinion carrier, P2 ring gear, P2 pinions (carrier held, reversing direction), P2 sun gear/mainshaft, P3 sun gear, P3 pinion carrier, output shaft.

POWER FLOWS

The following sequence of schematics shows the power flow routing through a WT transmission.

Neutral Power Flow

Only the C5 clutch is applied (see **Figure 17–29**).

- Torque from the turbine shaft is not transmitted beyond the rotating clutch module and the P1 sun gear.
- If the vehicle is rolling while in neutral, different clutches may be applied. This controls the speed of rotating components.

First Range Power Flow

In 1st range, the C1 and C5 clutches are applied (see **Figure 17–30**).

- C1 locks the turbine shaft and main shaft together. The P3 sun gear is part of the mainshaft module, so it becomes input for the P3 planetary gearset.

NEUTRAL

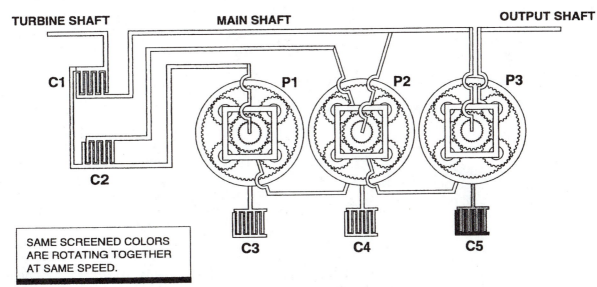

FIGURE 17–29   Neutral power flow. (Courtesy of Allison Transmission Division of GM)

1ST RANGE

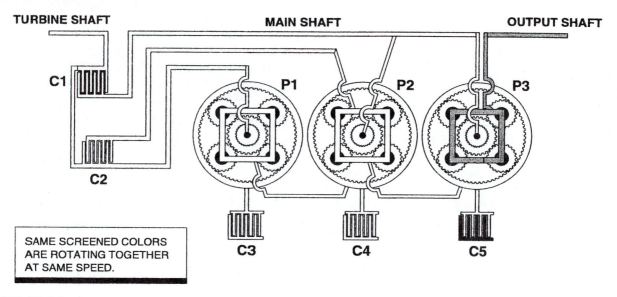

FIGURE 17-30 First range power flow. (Courtesy of Allison Transmission Division of GM)

2ND RANGE

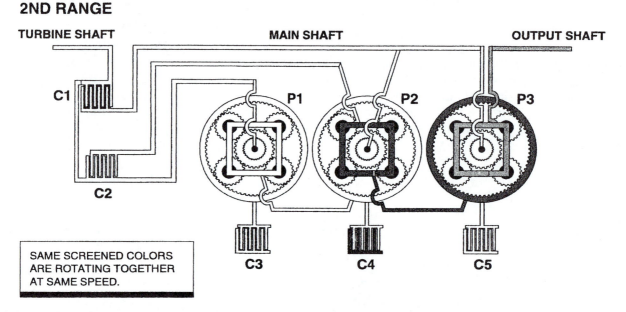

FIGURE 17-31 Second range power flow. (Courtesy of Allison Transmission Division of GM)

- C5 holds the P3 ring gear. The P3 carrier is splined directly to the output shaft. The P3 sun gear is input and the P3 ring gear is held, so the P3 carrier becomes the output.

Second Range Power Flow

In 2nd range, the C1 and C4 clutches are applied, making the P2 and P3 planetary sets work together to provide the output (see **Figure 17-31**).

- The C1 clutch locks the turbine shaft and main-shaft together and this drives the P2 sun gear.
- The C4 clutch holds the P2 ring gear. The P2 sun gear is input and the P2 ring gear is held, making the P2 carrier the output.
- The P3 sun gear is splined to the mainshaft so it rotates.
- The P3 ring gear is splined to the P2 carrier, so it rotates, but at a slower speed than the P3 sun gear.

- The P3 ring gear acts like a held member and the P3 sun gear becomes input. This makes the P3 carrier the output, and it is splined to the output shaft.

Third Range Power Flow

The C1 and C3 clutches are applied, and all three planetary gearsets work together to provide the appropriate output (see **Figure 17–32**).

- C1 clutch locks the mainshaft to the turbine shaft and the rotating clutch module rotates the P1 sun gear.
- The C3 clutch holds the P1 ring gear. Because the P1 sun gear is input and the P1 ring gear is held, the P1 carrier becomes the output.
- The P2 sun gear rotates with the mainshaft.
- The P2 ring gear is splined to the P1 carrier, so it rotates. Because the P2 ring gear rotates more slowly than the P2 sun gear, it acts like a held member. This makes the P2 sun gear the input, the P2 ring gear "held," and the P2 carrier the output.
- The P3 sun gear rotates with the mainshaft.
- The P3 ring gear is splined to the P2 carrier, so it rotates with the P2 carrier at a speed slower than the P3 sun gear. This makes the P3 sun gear input, the P3 ring gear held, and the P3 carrier the output. The P3 carrier is splined to the output shaft.

Fourth Range Power Flow

In 4th range, both rotating clutches C1 and C2 are applied (see **Figure 17–33**).

- The C1 clutch locks the turbine shaft to the mainshaft.
- The C2 clutch locks the turbine shaft to the P2 carrier.
- Because no stationary clutches are applied, all three planetary sun gears, carriers, and ring gear rotate at the same speed and in the same direction as the input from the turbine shaft.
- This results in a direct drive or 1:1 ratio.

Fifth Range Power Flow

In 5th range, the C2 and C3 clutches are applied and all three planetary sets work together to provide the appropriate output (see **Figure 17–34**).

- The P1 sun gear is rotating with the rotating clutch module.
- The C3 clutch is holding the P1 ring gear stationary, making the P1 carrier the output.
- The P2 carrier is rotating at turbine speed because the C2 clutch locks the turbine to the P2 carrier.
- The P2 ring gear is splined to and rotating with the P1 carrier. The P2 carrier is rotating faster than the P2 ring gear, so it is the input. The ring gear acts like a held member. This makes the P2 sun gear the output.

3RD RANGE

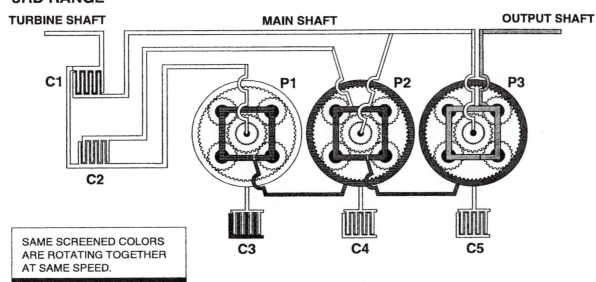

FIGURE 17–32   Third range power flow. (Courtesy of Allison Transmission Division of GM)

4TH RANGE

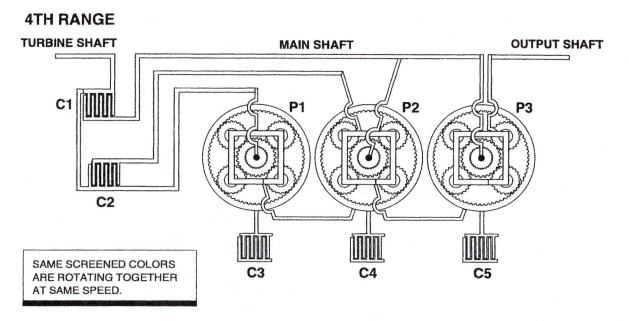

FIGURE 17-33 Fourth range power flow. (Courtesy of Allison Transmission Division of GM)

5TH RANGE

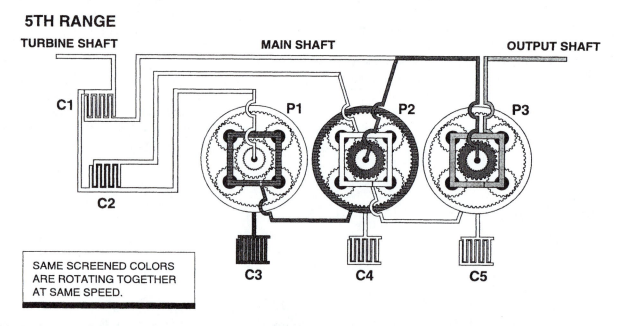

FIGURE 17-34 Fifth range power flow. (Courtesy of Allison Transmission Division of GM)

- The P2 sun gear rotates the mainshaft and the P3 sun gear.
- The P3 ring gear is splined to and rotating with the P2 carrier. However, the P3 sun gear is rotating faster than the P3 ring gear, so the sun gear is input and the ring gear acts as a held member. This makes the P3 carrier the output, and it is splined to the output shaft.
- This gear range produces an overdrive.

Sixth Range Power Flow

The C2 and C4 clutches are applied, and the P2 and P3 planetary sets work together to produce the appropriate output (see **Figure 17–35**).

- The C2 clutch locks the P2 carrier to the turbine shaft.
- The C4 clutch holds the P2 ring gear. The P2 carrier is the input (from the turbine shaft) and

6TH RANGE

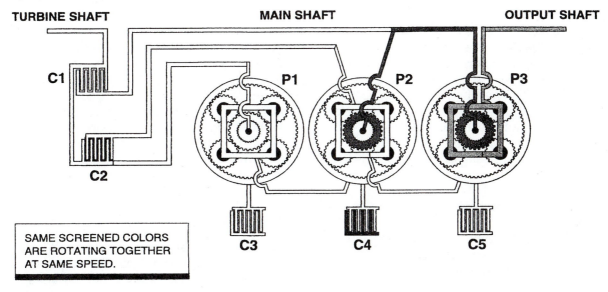

FIGURE 17–35   Sixth range power flow. (Courtesy of Allison Transmission Division of GM)

REVERSE RANGE

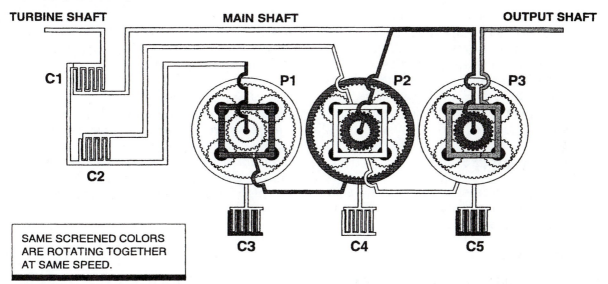

FIGURE 17–36   Reverse range power flow. (Courtesy of Allison Transmission Division of GM)

the P2 ring gear is held so the P2 sun gear becomes the output.

- The P2 sun gear rotates the mainshaft and the P3 sun gear. The P3 sun gear is input and the P3 ring gear acts like a held member. This makes the P3 carrier the output, and it is splined to the output shaft.
- This gearing produces an overdrive.

Reverse Range Power Flow

The C3 and C5 clutches are applied, and all three planetary gearsets work together to produce the appropriate output (see **Figure 17–36**).

- The P1 sun gear rotates with the rotating clutch module.
- The P1 ring gear is held by the C3 clutch, making the P1 carrier the output.

- The P1 carrier is splined to the P2 ring gear. The P2 ring gear becomes the input for the P2 planetary set.
- The C5 clutch holds the P3 ring gear, which is splined to the P2 carrier. Because the P2 ring gear is the input and the P2 carrier is held, the P2 sun gear becomes the output. Because the carrier is the held member, the P2 sun gear rotates in an opposite direction (counterclockwise) to the input direction of rotation.
- The P2 sun gear rotates the mainshaft in an opposite direction. Because the P3 sun gear rotates with the mainshaft, it also turns in an opposite direction.
- The P3 sun gear becomes reverse input for the P3 planetary set.
- The P3 ring gear is held by the C5 clutch. The P3 carrier becomes reverse output, and it is splined to the output shaft.

17.8 TROUBLESHOOTING WT TRANSMISSIONS

WT electronics are equipped with an extensive self-diagnostic capability. Fault codes are used to isolate the nature, source, and severity of a problem in the electronic circuit. To effectively troubleshoot WT transmissions, the Allison WT troubleshooting manuals should always be used. The following section outlines some of the diagnostic techniques recommended by Allison.

RANGE/SHIFT TESTS

Every time WT electronics effect a shift, the shift logic is verified and the transmission range is checked. The following verification tests are used:

Range Verification

WT verifies the transmission range continually whether or not a shift is in progress. Essentially, this involves verifying that the current range is the range commanded by the ECU. The test involves checking the current gear ratio by comparing the turbine and output shaft speeds. This ratio is then compared with the ratio logged in the ECU memory for the commanded range. If the two ratios do not match, a diagnostic code is logged.

Off-Going Ratio Test

The off-going ratio test is performed while a shift is in progress. Within a set time after the ECU has com-

manded a range shift, the ECU calculates the ratio between input and output speed, comparing this ratio with that of the previous range. For instance, if the speed ratio of the previous range is still present after the range shift has been commanded, the ECU will deduce that the off-going clutch failed to release. The shift commanded will be repeated twice to verify the fail to shift condition. If this does not result in the shift being effected, a diagnostic code is logged and the ECU commands the transmission in the previous range. The off-going ratio test is applied during the interval between the turbine speed shift initiation point and the pull-down detected point.

Oncoming Ratio Test

The oncoming ratio test is performed by WT electronics near the end of a shift in progress sequence. The oncoming ratio test checks turbine speed and output shaft speed to determine whether the transmission is in the new range commanded by the ECU. When the ratios fail to match, the ECU assumes the oncoming clutch did not actuate and will log a code.

Shop Talk

The off-going and oncoming ratio test can detect clutch slippage almost immediately.

This provides a considerable advantage to the technician troubleshooting WT transmissions. WT electronics can command a hold in the previous range to protect the transmission when severe clutch slippage warrants this failure strategy. A real problem in Allison CEC transmissions was the fact that by the time the driver complained of a slippage problem, significant transmission damage had already occurred.

C3 Pressure Switch Tests

The C3 pressure switch is an important diagnostic indicator. The C3 pressure switch signals the ECU as to whether pressure is present or not present in the C3 clutch. For instance, when C3 pressure is not present and should be, a diagnostic code is logged and a Do Not Shift (DNS) condition is immediately put into effect, with the result that the transmission remains in the current range.

If the opposite occurs, that is, C3 pressure is signalled in a range when it should not be, the result is the same. A diagnostic code is logged and a Do Not Shift condition results.

DNS LIGHT

Whenever WT electronics sense a condition that could damage the transmission, transmission electronics, or the vehicle drivetrain, shifting is restricted and the DNS light is illuminated.

 Shop Talk

The DNS light is not necessarily illuminated every time a diagnostic code is logged.

The ECU should be checked for fault codes whenever there is a transmission-related problem. The DNS light should illuminate every time the engine is started. It will remain lit for a few seconds only. If the DNS light fails to illuminate at start-up, the system should be checked immediately.

Continuous illumination of the DNS light indicates the ECU has logged a diagnostic code. When the DNS light is first illuminated, eight seconds of short beeps are emitted from the shift selector. The beeps are an audible alert to the operator that shifts are being restricted. Additionally, the SHIFT digit on the shift selector display will remain blank and the ECU may not respond to shift selector requests.

Limp Home Mode

When the DNS light is illuminated, the intent is to inform the driver that the transmission is not operating properly and requires repair. However, the transmission can usually be operated for a short time in the selected range in "limp home" mode, provided the ignition key is not switched off.

When the DNS light illuminates when the ignition key is cycled from off to on, the transmission will remain in neutral until the fault code responsible for the DNS condition is cleared. The converter lock-up clutch is always disengaged whenever transmission shifting is restricted or during any critical transmission malfunction.

ACCESSING DDR DATA CODES

The diagnostic data reader is the means the service technician uses to access the diagnostic codes logged in the WT ECU. WT diagnostic codes are numeric and consist of a two-digit main code and a two-digit sub code (see **Figure 17–37**). The WT ECU logs these codes and produces them for readout by sequencing either the most severe code or the most recent code first, followed by the order in which they were logged, beginning with the most recent and working backward. A maximum of 5 codes may be logged. As codes are added, the oldest non-active code is dropped from the list first. Should all the logged codes be active, the codes are listed in order of severity. When the number exceeds 5, the least severe is dropped. Codes may be accessed either by the DDR or by the shift selector mechanism. The process required to display and clear stored codes is almost identical to that described for the Allison CEC transmission in the previous chapter and will not be outlined here. All Allison troubleshooting should be undertaken with the appropriate Allison service literature.

CAUTION: Always use the appropriate Allison troubleshooting procedure and never make any assumptions based on a previous WT problem.

SUMMARY

- The WT transmission uses a modular design.
- The WT transmission uses full authority management electronics to effect shifting and communicate with other vehicle electronic systems.
- The WT modules may be removed from the transmission and serviced as separate units.
- WT modules may grouped in input, gearbox, and output categories.
- The WT input modules are the torque converter module, the converter housing module, and the front support/charging pump module.
- The WT gearbox modules are the rotating clutch module, the main housing module, the P1 planetary module, the P2 planetary module, and the mainshaft module.
- The WT output modules are the rear cover module, output retarder module, or the transfer gear/drop box module (depending on the model).
- The electronic control module is a computer that masters the operation of the transmission.
- The WT transmission uses three interconnected planetary gearsets to provide up to 3 stages of gearing, depending on the range selected.
- A WT transmission has the mechanical gearing to provide 6 forward ranges, 1 reverse, and neutral.
- Gear ranges 5 and 6, if programmed into WT electronics, are overdrive ranges.
- The WT transmission may have an integral driveline retarder that can enhance the vehicle service brake applications, especially those that require 10 percent or less of the peak service application pressure.

TROUBLESHOOTING WITH DIAGNOSTIC CODES
(Additional troubleshooting information can be found in the appropriate Allison Troubleshooting Manual.)

CODE LISTINGS AND PROCEDURES

Main Code	Sub Code	IF CODES READ	RECOMMENDED PROCEDURES
13	12	ECU Input Voltage Low	1. Check: a. Battery direct ground and power connections are tight and clean. b. Vehicle batteries are charged. c. Vehicle charging system is not over- or under-charging. d. VIM fuse is good. e. VIM connections are tight, clean, and undamaged. f. Vehicle manufacturer supplied wiring is correct. g. ECU connectors are tight, clean, and damaged. 2. If all points check, call distributor.
13	13	ECU Input Voltage Low (Medium)	
13	23	ECU Input Voltage High	
14	12, 23	Oil Level Sensor	1. Check: a. Is transmission equipped with oil level sensor? b. Engine speed sensor, output speed sensor, temperature sensor, and oil level sensor are working correctly. c. Wiring harness has no opens, shorts to ground, or shorts to battery. 2. If all points check, call distributor.
21	12, 23	Throttle Position Sensor	1. Check: a. TPS connector is properly connected. b. End of TPS cable is pulled out properly. c. Engine fuel lever is in idle position. d. Engine fuel lever provides proper amount of stroke on TPS cable. e. Wiring harness to TPS has no opens, shorts between wires, or shorts to ground. 2. If able, replace TPS. 3. If all points check, call distributor.
22	14, 15, 16	Speed Sensors	1. Check: a. Speed sensor connectors are tight, clean, and undamaged. b. Speed sensor mounting bolts are properly torqued (24–29 N·m (18–21 lb ft)). c. Wiring harness to sensors has no opens, shorts between wires, or shorts to ground. 2. If all points check, call distributor.
23	12, 13, 14, 15	Shift Selectors	1. Check: a. ECU connectors are tight, clean, and undamaged. b. Remote shift selector connector is tight and jumper wire is cut. 2. If able, replace shift selector(s). 3. If all points check, call distributor.
24	12	Sump Oil Temperature Cold	1. Check: a. Air temperature is below −32°C (−25°F). 1) If yes, this is a correct response for temperature. 2) If no, check that main transmission connector is tight, clean, and undamaged. b. ECU connectors are tight, clean, and undamaged. 2. If all points check, call distributor.
24	23	Sump Oil Temperature Hot	1. Let vehicle idle with parking brake applied, wheels chocked, and vehicle level. Check: a. Correct dipstick is installed. b. Fluid level is correct. 1) If fluid level is incorrect — correct fluid level. 2) If fluid level is correct — check for engine system overheating, causing transmission to overheat. 2. Check if ECU and transmission connectors are tight, clean, and undamaged. 3. If all points check, call distributor.
25	00, 11, 22, 33, 44, 55, 66, 77	Output Speed Sensor and Reading	1. Check: a. Speed sensor connector is tight, clean, and undamaged. b. ECU connectors are tight, clean, and undamaged. c. Fluid level is correct. d. Sensor mounting bolt torque is correct (24–29 N·m (18–21 lb ft)). e. Wiring harness to sensor has no opens, shorts between wires, or shorts to ground. 2. If all points check, call distributor.
32	00, 33, 55, 77	C3 Pressure Switch Open	1. Let vehicle idle with parking brake applied, wheels chocked, and vehicle level. Check: a. Correct dipstick is installed. b. Fluid level is correct. 2. Check: a. Main transmission connector is tight, clean, and undamaged. b. ECU connectors are tight, clean, and undamaged. c. Wiring harness has no opens, shorts between wires, or shorts to ground. 3. If all points check, call distributor.
33	12, 23	Sump Oil Sensor Failure	1. Check: a. Main transmission connector is tight, clean, and undamaged. b. ECU connectors are tight, clean, and undamaged. c. Wiring harness has no opens, shorts between wires, or shorts to ground. 2. If all points check, call distributor.
34	12, 13, 14, 15, 16	EEPROM	1. If able, recalibrate ECU; if not, replace ECU. 2. If cannot replace ECU, call distributor.
35	00, 16	Power Interruption EEPROM Write Interruption	1. Check: a. ECU connectors are tight, clean, and undamaged. b. VIM connectors are tight, clean, and undamaged. c. Vehicle manufacturer supplied wiring has correct power and ground connections. d. Power connections are battery direct. e. Ground connections are battery direct. f. Ignition switch connections are correct. 2. If all points check, call distributor.
36	00	Hardware/Software Not Compatible	1. If able, reprogram ECU. 2. If able, replace ECU. 3. If ECU cannot be reprogrammed or replaced, call distributor.
41	12, 13, 14, 15, 16, 21, 22, 23, 24, 25, 26	Open or Short to Ground in Solenoid Circuit	1. Check: a. Main transmission connector is tight, clean, and undamaged. b. ECU connectors are tight, clean, and undamaged. c. Wiring harness is not pulled too tight, and there is no damage, chafing, or screws through harness. d. Wiring harness has no opens, shorts between wires, or shorts to ground. 2. Change wiring harness (optional). 3. If all points check, call distributor.
42	12, 13, 14, 15, 16, 21, 22, 23, 24, 25, 26	Short to Battery in Solenoid Circuit	1. Check: a. Main transmission connector is tight, clean, and undamaged. b. ECU connectors are tight, clean, and undamaged. c. Wiring harness is not pulled too tight, and there is no damage, chafing, or screws through harness. d. Wiring harness has no opens, shorts between wires, or shorts to ground. 2. Unauthorized repairs have not been made. Change harness (optional). 3. If all points check, call distributor.
43	21, 25, 26	Solenoid Low Side Circuit, Open Driver, or Wire Shorted to Ground	1. Check: a. Transmission connector is tight, clean, and undamaged. b. ECU connectors are tight, clean, and undamaged. c. Wiring harness has no opens, shorts between wires, or shorts to ground. 2. If able, replace ECU. 3. If all points check, call distributor.
44	12, 13, 14, 15, 16, 21, 22, 23, 24, 25, 26	Solenoid Circuit Short to Ground	1. Check: a. Transmission connector is tight, clean, and undamaged. b. ECU connectors are tight, clean, and undamaged. c. Wiring harness has no opens, shorts between wires, or shorts to ground. 2. If all points check, call distributor.

FIGURE 17-37   WT diagnostic codes. (Courtesy of Allison Transmission Division of GM)

Code	Sub-codes	Name	Check Procedure
56	00, 11, 22, 33, 44, 55, 66, 77	Range Verification Test (Between Shifts)	1. Check: a. Turbine and output speed sensor connectors are tight, clean, and undamaged. b. Speed sensor wiring harness has no opens, shorts between wires, or shorts to ground. c. Transmission connector is tight, clean, and undamaged. d. ECU connectors are tight, clean, and undamaged. 2. Let vehicle idle with parking brake applied, wheels chocked, and vehicle level. Check: a. Correct dipstick is installed. b. Fluid level is correct. 3. If all points check, call distributor.
57	11, 22, 44, 66, 88, 99	Range Verification Test (C3)	1. Let vehicle idle with parking brake applied, wheels chocked, and vehicle level. a. Correct dipstick is installed. b. Fluid level is correct. 2. Check: a. Output and turbine speed sensor connectors are tight, clean, and undamaged. b. Speed sensor wiring harness has no shorts between wires or shorts to ground. 3. Check: a. Transmission connector is tight, clean, and undamaged. b. ECU connectors are tight, clean, and undamaged. c. C3 pressure switch wiring has no opens, shorts between wires, or shorts to ground. 4. If all points check, call distributor.
61	00	Retarder Over Temperature	1. Check: a. Fluid level is correct. b. Retarder apply system is not allowing retarder and throttle to be applied at the same time. c. Fluid cooler is adequately sized for load. 2. If all points check, call distributor.
62	12, 23	Retarder Temperature Sensor	1. Check: a. Retarder temperature measured with DDR is consistent with code; or determine if code is active using shift selector. b. Ambient temperature is above 178°C (352°F) or below –45°C (–49°F). c. Sensor connector is tight, clean, and undamaged. d. ECU connectors are tight, clean, and undamaged. e. Temperature sensor circuit has no opens, shorts, or shorts to ground. 2. If all points check, call distributor.
63	00, 26, 40	Input Function Fault	1. Check input wiring, switches, and connectors to determine why input states are different. 2. If all points check, call distributor.
64	12, 23	Retarder Modulation Request Device Fault	1. Use DDR to read retarder counts and identify problem wires. Check wiring for short to battery, ground wire open, or short to ground. 2. If all points check, call distributor.
65	00	Engine Rating High (Computer-Controlled Engines Only)	1. Check if engine rating or governor speed is too high for transmission. 2. If all points check, call distributor.
66	00	Serial Communications Interface Fault	1. Check: a. Serial connection to engine computer is tight, clean, and undamaged. b. SCI wiring harness has no opens, shorts, or shorts to ground. c. If DDR is not available, also be sure that transmission ECU connections are tight, clean, and undamaged. 2. If all points check, call distributor.
69	12, 13, 14, 15, 16, 21, 22, 23, 24, 25, 26, 32, 33, 34, 35, 36, 41	ECU Malfunction	1. Clear diagnostic code and retry vehicle start. 2. If code recurs, reprogram ECU. 3. If able, replace ECU. 4. If code continues to recur, call distributor.
45	12, 13, 14, 15, 16, 21, 22, 23, 24, 25, 26	Solenoid Circuit Open	1. Check: a. Transmission connector is tight, clean, and undamaged. b. ECU connectors are tight, clean, and undamaged. c. Wiring harness has no opens. 2. If all points check, call distributor.
51	01, 10, 12, 21, 23, 24, 35, 42, 43, 45, 46, 53, 64, 65	Offgoing Ratio Test (During Shift)	1. Check: a. Output and turbine speed sensor connectors are tight, clean, and undamaged. b. Speed sensor wiring harness has no opens, shorts between wires, or shorts to ground. 2. Let vehicle idle with parking brake applied, wheels chocked, and vehicle level. Check: a. Correct dipstick is installed. b. Fluid level is correct. 3. If all points check, call distributor.
52	01, 08, 32, 34, 54, 56, 71, 72, 78, 79, 99	Offgoing C3 Pressure Switch (During Shift)	1. Check: a. Output and turbine speed sensor connectors are tight, clean, and undamaged. b. Speed sensor wiring harness has no opens, shorts between wires or shorts to ground. c. Main wiring harness to transmission has no shorts between wires or shorts to ground. 2. Let vehicle idle with parking brake applied, wheels chocked, and vehicle level. Check: a. Correct dipstick is installed. b. Fluid level is correct. 3. If all points check, call distributor
53	08, 18, 28, 29, 38, 39, 48, 49, 58, 59, 68, 69, 78, 99	Offgoing Speed Test (During Shift)	1. Check: a. Turbine and engine speed sensor connectors are tight, clean, and undamaged. b. Speed sensor wiring harness has no opens, shorts between wires, or shorts to ground. 2. Let vehicle idle with parking brake applied, wheels chocked, and vehicle level. Check: a. Correct dipstick is installed. b. Fluid level is correct. 3. If all points check, call distributor.
54	01, 07, 10, 12, 17, 21, 23, 24, 27, 32, 34, 35, 42, 43, 45, 46, 53, 54, 56, 64, 65, 70, 71, 72, 80, 81, 82, 83, 85, 86, 92, 93, 95, 96, 97	Oncoming Ratio Test (After Shift)	1. Check: a. Turbine and output speed sensor connectors are tight, clean, and undamaged. b. Speed sensor wiring harness has no opens, shorts between wires or shorts to ground. 2. Let vehicle idle with parking brake applied, wheels chocked, and vehicle level. Check: a. Correct dipstick is installed. b. Fluid level is correct. 3. EEPROM calibration is correct for the transmission. 4. If all points check, call distributor.
55	17, 27, 87, 97	Oncoming C3 Pressure Switch (After Shift)	1. Let vehicle idle with parking brake applied, wheels chocked, and vehicle level. Check: a. Correct dipstick is installed. b. Fluid level is correct. 2. Check: a. Output and turbine speed sensor connectors are tight, clean, and undamaged. b. Speed sensor wiring harness has no opens, shorts between wires, or shorts to ground. 3. Check: a. Transmission connector is tight, clean, and undamaged. b. ECU connectors are tight, clean, and undamaged. c. C3 pressure switch wiring has no opens, shorts between wires, or shorts to ground. 4. If all points check, call distributor.

FIGURE 17-37 WT diagnostic codes. (Courtesy of Allison Transmission Division of GM)

- The driveline retarder consists of a stator, rotor, housing, and control valve assembly and functions much as a torque converter driven in reverse.
- One WT model is available with a dropbox or transfer gear for 4 wheel drive vehicles.
- The electro-hydraulic control module houses the solenoids, sensors, valves, and regulators required to effect the results of ECU processing into outcomes.
- WT electronics require an input circuit, a microcomputer, and an output circuit.
- SAE J 1939–compatible hardware and software allows WT to share components such as the throttle position sensor with other onboard electronics systems and exchange information to optimize vehicle performance.
- WT has an EEPROM chip that permits programming to tailor transmission performance to operating requirements.
- All WT transmissions have the same integral hardware and the potential for 6 ratio ranges.
- Programming will determine the actual number of ranges available in a chassis.
- Clutches are controlled by solenoids switched by the ECM.
- Pulse width modulation is the controlling of an output signal by varying the percentage of on and off time.
- WT solenoids are switched by PWM signals.
- The term *primary modulation* is used to describe the amount of time a solenoid is energized. *Secondary modulation* describes the current flow through a solenoid.
- Understanding the torque routes through the WT transmission can be a useful diagnostic tool.
- Codes logged into the WT ECU can be classified as active or historic (inactive).
- Codes are usually displayed in the order in which they are logged with the most recent displayed first. In cases where one code can produce more severe consequences, this is displayed first.
- A maximum of 5 codes can be logged into the WT ECU at one time. As codes are added, the oldest non-active code is dropped from the list. If all codes are active, that with the lowest priority is dropped first.

REVIEW QUESTIONS

1. Which letter(s) in the transmission identification code would indicate that a WT transmission was equipped with a dropbox?

   a. P
   b. HD
   c. DB
   d. T

2. How many clutch assemblies are used in the basic WT gearbox?
   a. 1
   b. 3
   c. 5
   d. 7

3. How many planetary gearsets are used in the basic WT gearbox?
   a. 1
   b. 3
   c. 5
   d. 7

4. Which of the following is responsible for converting a command electrical signal into mechanical movement?
   a. actuator transistor
   b. clutch piston
   c. solenoid
   d. main pressure

5. What is the sun gear in the P1 planetary gearset driven by?
   a. the P1 pinion carrier
   b. the mainshaft
   c. the rotating clutch drum
   d. the P2 pinion carrier

6. What force is used to effect the mechanical movement of clutch pistons in a WT transmission?
   a. PWM signal
   b. main pressure
   c. electrical pressure
   d. clutch feed pressure

7. Which of the following ranges in a WT transmission could be described as overdrive?
   a. 6 only
   b. 5 and 6
   c. all ranges above 3
   d. no overdrive unless a dropbox is used

8. In which of the planetary gearsets in the WT transmission does the reversal of input rotation take place?
   a. P1
   b. P2
   c. P1 and P2
   d. P2 and P3

9. Which of the following components is splined to the WT output shaft?

a. P3 pinion carrier
b. P3 sun gear
c. P3 ring gear
d. mainshaft

10. Which of the following is required to connect the WT transmission to the chassis electrical system?
a. VIM
b. transmission harness
c. DDR connector
d. TPS

11. When first put into service, for how many starts is the WT AutoDetect enabled?
a. 1
b. 12
c. 24
d. always enabled

12. In which of the following applications would WT electronics require a dedicated TPS signal?
a. WT is the only chassis electronic system.
b. WT is used with J 1939 electronics.
c. WT is used with a Caterpillar engine.
d. WT is used with J 1584/1708 electronics.

13. Technician A says that WT electronics will always display the most recently logged trouble code first. Technician B says that WT electronics are capable of logging up to five active codes. Who is right?
a. Technician A
b. Technician B
c. both Technician A and B
d. neither Technician A nor B

14. Technician A says that WT trouble codes can only be accessed by using a DDR with the appropriate software card or cartridge. Technician B says that the shift selector mechanism may be used to read active codes. Who is right?
a. Technician A
b. Technician B
c. both Technician A and B
d. neither Technician A nor B

15. Technician A says that to put a WT transmission into reverse, clutches C3 and C5 must be applied. Technician B states that in reverse range, the P3 pinion carrier becomes the reverse output. Who is right?
a. Technician A
b. Technician B
c. both Technicians A and B
d. neither Technician A nor B

18

Drive Shafts

Objectives

After reading this chapter, you should be able to
- Identify the components in a truck driveline.
- Explain the procedures for inspecting, lubricating, and replacing a universal joint.
- Describe the various types of wear a universal joint might experience.
- Outline the procedure for sourcing chassis vibration.
- Define and explain the importance of phasing.
- Explain the importance of driveline working angles and how to calculate them.
- Describe the procedure for balancing a drive shaft.
- Describe the operation of a hydraulic retarder.
- Explain how an electric retarder operates.

Key Terms List

axle shims
brinelling
drive shaft
eddy current
electronic inclinometer
forged journal cross
galling
in-phase
magnetic base protractor
National Lubricating Grease Institute (NLGI)
nonparallel ("broken back") driveline
out-of-phase

ovality
parallel joint driveline
pitting
PTO drive shaft
runout
spalling
spider
stand pipe
torque rod shims
transverse vibrations
trunnions
universal joint

Drive shafts—or propeller shafts, as they are sometimes called—have one basic function: transferring torque from one driveline component to another in a smooth, continuous fashion (**Figure 18–1**). In a heavy-duty truck, that means transmitting engine torque from the output shaft of the transmission to a rear axle or to an auxiliary transmission. Drive shafts are also used to connect forward and rear axles on 6 × 4 tractors.

In most cases, a drive shaft must transfer torque at an angle to the centerlines of the driveline components it is fastened to. Because the rear axle is not

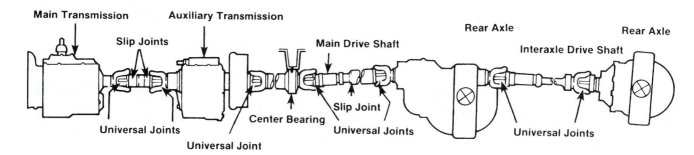

FIGURE 18–1   The drive shaft transmits torque between driveline components. (Courtesy of Spicer Universal Joint Division/Dana Corporation)

connected to the rigid rails of the truck frame but is mounted to the rear suspension system, the drive shaft must also be able to change angles as the rear suspension jounces and rebounds.

The drive shaft must also be able to change in length while transmitting torque. As the rear axle reacts to road surface changes, torque reactions, and braking forces, it tends to pivot forward or backward, requiring a corresponding change in the length of the drive shaft.

18.1   CONSTRUCTION

The drive shaft is able to contract, expand, and change operating angles while transmitting torque through the use of **universal joints** and slip joints. **Figure 18–2** shows the various components of a typical heavy-duty truck drive shaft.

DRIVE SHAFTS

In order to transmit engine torque to the rear axles, the drive shaft must be durable and strong. An engine producing 1,000 pound-feet of torque, when multiplied by a 12:1 gear ratio in the transmission, will deliver 12,000 pound-feet of torque to the drive shaft. The shaft must be strong enough to deliver this twisting force to a loaded axle without deforming or cracking under the strain.

Drive shafts are constructed of high-strength steel tubing to provide maximum strength with minimum weight. The diameter of the shaft and wall thickness of the tubing is determined by several factors: maximum torque and vehicle payload, type of operation, road conditions, and the brake torque that might be encountered. One-piece, two-piece, and three-piece drive shafts are used, depending on the length of the driveline.

The truck manufacturer's recommended drive shaft rating should always be adhered to when replacing a drive shaft. Oversized or undersized driveline parts will

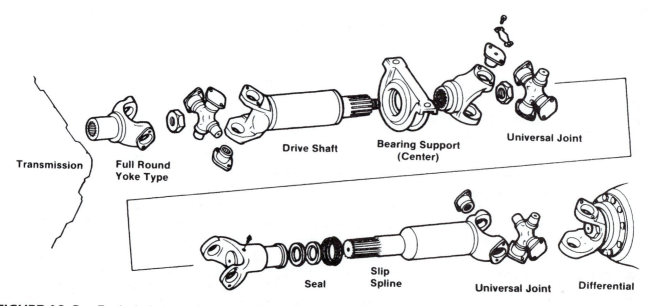

FIGURE 18–2   Exploded view of a heavy-duty truck drive shaft. (Courtesy of Volvo Trucks North America, Inc.)

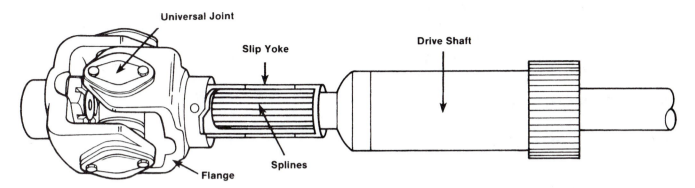

FIGURE 18–3 A drive shaft with a slip joint. (Courtesy of Chicago Rawhide)

lead to expensive or frequent driveline failures. Oversized drive shafts can place excessive strain on transmission and axle components, resulting in damage that is costly to repair. If the drive shaft is undersized, it will fail quickly and often.

Each end of the drive shaft has a yoke used to connect the shaft to other driveline components. The yoke might be rigidly welded to the shaft tube or it might be a spline, or slip, yoke (**Figure 18–3**). The tube yokes are connected through universal joints to end yokes on the output and input shafts of the transmission and drive axle.

A typical slip joint consists of a hardened, ground splined shaft welded to the drive shaft tube that is inserted into a slip yoke with matching internal splines. The splines between a slip joint and a permanent joint must support the drive shaft and be capable of sliding under full torque loads. To aid in this axial or slip movement, these splines are typically phosphate or polymer nylon-coated to increase slippage while reducing wear. A dust cap and washer made of synthetic rubber, cork, or felt screws over the slip yoke assembly to exclude contaminants and to keep lubricants in.

UNIVERSAL JOINTS

The purpose of the universal joint is to link the driveline components and to allow the drive shaft to operate at different and constantly changing angles. A universal joint consists of a **forged journal cross** (sometimes called a **spider**) with grease fittings, four sets of full compliment needle bearings, four seals, and four bearing cups (**Figure 18–4**). The ends of the universal cross are called **trunnions.** The trunnions are case-hardened ground surfaces on which the needle bearings ride.

Grease Fittings

Heavy-duty universal joints have cross-drilled oil passages and grooves on the end of the trunnions to

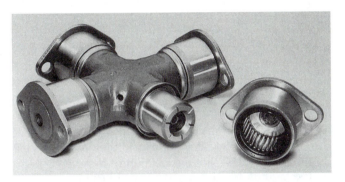

FIGURE 18–4 Parts of a universal joint. (Courtesy of Chicago Rawhide)

allow for easy needle bearing lubrication. One or two grease fittings allow the U-joint to be lubricated.

Designed into the center of each trunnion on some U-joints is a **stand pipe** (a type of check valve), which prevents reverse flow of the hot liquid lubricant generated during operation. When the universal joint is at rest, one or more of the cross ends will be up (**Figure 18–5**). Without the stand pipe, lubricant would flow out of the upper passageways

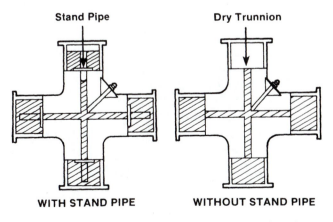

WITH STAND PIPE   WITHOUT STAND PIPE

FIGURE 18–5 A stand pipe on the end of each trunnion ensures constant lubrication and prevents dry startup. (Courtesy of Chicago Rawhide)

and trunnions, leading to partially dry startup. The stand pipe ensures lubrication of the trunnions and needle bearings each time the universal joint is put into service. Other U-joints have rubber check valves in each cross that perform the same function.

Bearing Assemblies

Each cross has four bearing assemblies. Each assembly consists of needle bearings in a bearing cup and a rubber seal around the open end of the cup.

Needle bearings are used because of their strength and durability. They can carry the high radial loads that occur in the oscillating action experienced during operation of the drive shaft.

They also require less radial space, which reduces their overall diameter and makes them ideal for universal joint applications. Universal joints are factory lubricated, with just enough lubricant to hold the rollers in place during assembly. However, additional lubrication should be applied before placing the U-joint in service.

The needle bearings are contained in bearing cups (**Figure 18–6**). The bearing cups fit over the trunnions on the U-joint. Most bearing cups are round and fit in the bore of a yoke. Other universal joints have wing bearing cups rather than the common round cups. The wing bearing does not fit in a bore in a yoke. Rather, it bolts to shoulders on the yoke (**Figure 18–7**).

Each bearing cup is lubricated and has a seal designed to keep the bearings free of dirt and other contaminants. **Figure 18–8** shows several seal designs.

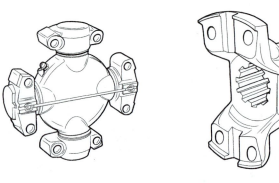

WING-TYPE U-JOINT   **WING-TYPE END YOKE**

FIGURE 18–7   Wing-type U-joint and yoke.

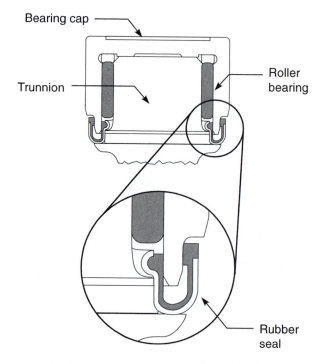

Bearing cap

Trunnion

Roller bearing

Rubber seal

FIGURE 18–8   Bearing seal. (Courtesy of Spicer Universal Joint Division/Dana Corporation)

FIGURE 18–6   Bearing cup and needle rollers. (Courtesy of Chicago Rawhide)

Mounting Hardware

Universal joints are connected to yokes in different ways. **Figure 18–9** shows four of the most common methods of securing U-joints with round bearing cups. Half round end yokes are fastened to the U-joints with bearing straps or U-bolts. Full round end yokes use snap rings or bearing plates to secure the joint in the yoke bore.

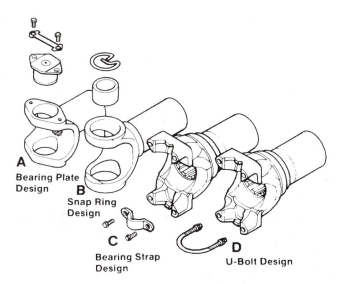

FIGURE 18–9 *Four methods of fastening a universal joint in a yoke: (A) bearing plate; (B) snap ring; (C) strap; and (D) U-bolts. (Courtesy of Spicer Universal Joint Division/Dana Corporation)*

CENTER SUPPORT BEARINGS

Center support bearings (**Figure 18–10**) are used when the distance between the transmission (or auxiliary transmission) and the rear axle is too great to span with a single drive shaft. The center support bearing is fastened to the frame and aligns the two connecting drive shafts. This bearing also absorbs any vibrational shock created in the frame or driveline.

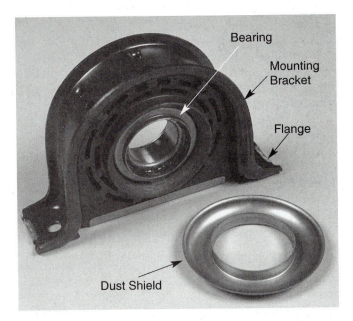

FIGURE 18–10 *Center bearing. (Courtesy of Chicago Rawhide)*

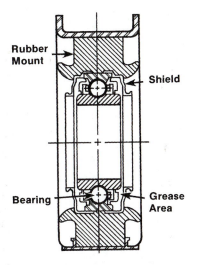

FIGURE 18–11 *Cross-section of a typical center bearing. (Courtesy of Volvo Trucks North America, Inc.)*

The center support bearing consists of a stamped steel bracket that is used to align and fasten the bearing to the frame. A rubber mount inside the bracket surrounds the bearing (**Figure 18–11**). The sealed bearing supports the two connecting drive shafts, while the synthetic rubber bearing mount absorbs the vibrational shock. Center supports are usually found on drive shafts over 60 inches. They are also known as hanger bearings.

18.2 DRIVE LINE ARRANGEMENTS

There are two types of driveline arrangements that will satisfactorily transmit power to the driving wheels: the **parallel joint driveline** and the **nonparallel ("broken back") driveline.**

In the parallel joint type, all companion flanges and/or yokes in the complete driveline are parallel to each other with the working angles of the joints ("A" and "B," **Figure 18–12**) of a given shaft being equal

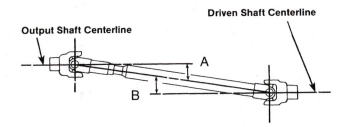

FIGURE 18–12 *Parallel joint driveline arrangement. (Courtesy of International Truck and Engine Corp.)*

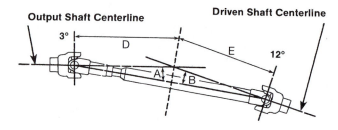

FIGURE 18–13 *"Broken back" type drive shaft; angles "A" and "B" are equal. (Courtesy of International Truck and Engine Corp.)*

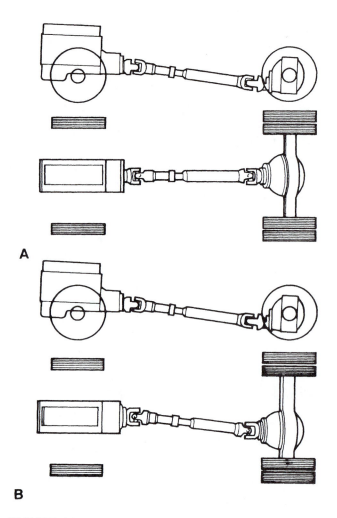

A

B

FIGURE 18–14 *(A) One plane angle drive shaft; and (B) two plane angle drive shaft. (Courtesy of Volvo Trucks North America, Inc.)*

and opposite. For instance, if the transmission main shaft centerline at the rear is down 5 degrees from a true horizontal plane, the centerline at the front of the auxiliary main shaft or rear axle pinion shaft must be 5 degrees up.

With the nonparallel or "broken back" type installation, the working angles of the joints ("A" and "B," **Figure 18–13**) of a given shaft are equal; however the companion flanges and/or yokes are not parallel. For example, the flange or yoke of the main transmission is 3 degrees down from true horizontal plane, while the rear axle pinion flange is up at 12 degrees. The universal joint working angles of this propeller shaft are equal; the shaft will run smoothly.

18.3 UNIVERSAL JOINT WORKING ANGLES

Proper universal joint working angles are necessary for trouble-free and long-lasting driveline operation. Most drivelines are angled in the vertical plane (**Figure 18–14A**), but on some trucks, the drivelines are also offset, or angled, in the horizontal plane (**Figure 18–14B**). When the drive shaft is angled in both the vertical and horizontal planes, a compound angle exists.

Any given universal joint has a maximum angle at which it will still transmit torque smoothly. This angle depends in part on the joint size and design. Exceeding the maximum recommended working angle will greatly shorten or immediately destroy the joint service life.

High angles combined with high rpm is the worst combination, resulting in a reduced U-joint life. Too large and unequal U-joint angles can cause vibrations and contribute to U-joint, transmission, and differential problems. Improper U-joint angles must be corrected.

Ideally, the operating angles on each end of the drive shaft should be equal to or within 1 degree of each other, have a 3 degree maximum operating angle, and have at least a $1/2$ degree of continuous operating angle.

Drive shaft rpm is the main factor, though, in determining maximum allowable operating angles. As a guide to maximum normal operating angles, refer to the chart given in **Table 18–1**. As noted, the faster the driveshaft spins, the less the allowable angle between each component. Angles are formed at the U-joints as the driveshaft angles down from the transmission to join the input yoke on the rear axle differential or, with a tandem drive, the inter-axle differential. For example, a 2,100-rpm engine running through a fast overdrive transmission and a fairly slow axle ratio might spin the driveshaft as fast as 3,000 to 3,100 rpm. This would limit the permissible angle to 5 degrees, measured as the truck sits on level pavement.

TABLE 18-1: MAXIMUM U-JOINT ANGLES	
Drive Shaft rpm	Maximum Normal Operating Angles
5000	3° 15'
4500	3° 40'
4000	4° 15'
3500	5° 0'
3000	5° 50'
2500	7° 0'
2000	8° 40'
1500	11° 30'

With equal working angles, the rear joint will slow down by the same amount that the forward joint speeds up, resulting in joint cancellation. The driving and driven shafts will turn at constant and identical speeds.

If the working angles of two opposed universal joints vary by more than 1 degree, the drive shaft will not rotate smoothly. Vibrations will be noticeable and joint failure will be the result.

The angles become greater as the vehicle's suspension flexes over uneven road surfaces. This happens at slow speeds, so it is not a major problem unless driveline angles are high to begin with. Driveline angle is less of a problem these days than when short-wheelbase tractors were needed to keep vehicles within tight overall length restrictions. Relaxation of those legal restrictions has led to use of longer wheelbases, which means more room for driveshafts and correspondingly lower angles.

Shop Talk

Remember that a U-joint, operating at an angle, causes a cyclical speed variation in the shaft where it speeds up and slows down twice every revolution. The non-uniform output speed produces pulsations in the shaft anytime the driveline is operating at an angle. Working the same way, but in reverse, is the U-joint at the rear of the driveline that connects to the axle input shaft. In order to avoid harmful torsional vibrations that result from the non-uniformity, the rear U-joint needs to cancel out the non-uniform rotation of the forward U-joint. This can be accomplished by having the transmission output shaft and axle input shaft at relatively similar angles and thus resulting in equal U-joint angles. This is known as drive shaft U-joint angle cancellation where both the forward and rear U-joints have equal angles. When U-joint angles at either end of the drive shaft are kept within at least 1 degree of each other, uncancelled, non-uniform rotation will be minimized. Gear clatter, gear seizures, synchronizer failures, and torsional vibration, caused by excessive rotation, will be eliminated.

The longer the drive shaft, the more its tube material weighs and the greater the rotational force, especially as drive shaft rpm increases. In extreme cases, balance becomes more critical. This is why manufacturers limit tube length. For example, at 3,000 rpm, the length of any single drive shaft section, measured between the center line of the U-joints at either end, should not usually exceed 70 inches.

18.4 DRIVE SHAFT PHASING

Heavy-duty universal joints have a unique characteristic. Because they are always operating on an angle, they do not transmit constant torque or turn at a uniform speed during their 360-degree rotation (**Figure 18–15**).

With the drive yoke running at a constant rpm, the drive shaft will increase and decrease speed once each revolution. To counteract this fluctuation in drive shaft speed, two universal joints are used and are placed **in-phase** (**Figure 18–16**).

In effect, **out-of-phase** conditions act somewhat like one man snapping a rope held by a man at the opposite end. The result is a reaction at the opposite

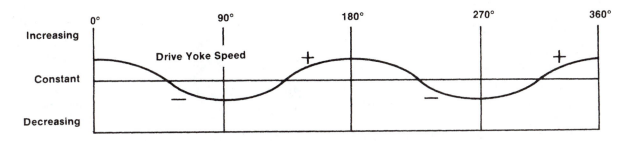

FIGURE 18–15 *The drive shaft will increase and decrease speed once each revolution. (Courtesy of Spicer Universal Joint Division/Dana Corporation)*

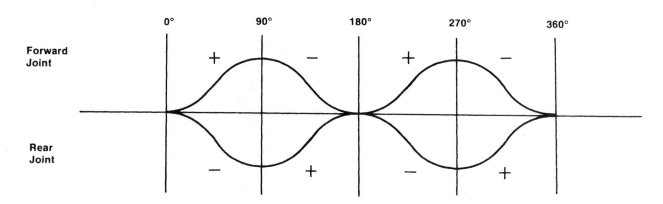

FIGURE 18-16   Two universal joints in phase will cancel out the speed fluctuations in the drive shaft. (Courtesy of Spicer Universal Joint Division/Dana Corporation)

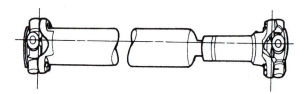

FIGURE 18-17   A drive shaft is in phase when the lugs on the tube yoke and slip yoke line up with each other. (Courtesy of International Truck and Engine Corp.)

end. If both men snap the rope at the same time, the resulting waves cancel each other and neither feels the reactions.

When in-phase, the slip yoke lugs (ears) and tube yoke lugs are in line (**Figure 18–17**). Normally, this is the ideal condition and gives the smoothest running shaft. There should be an alignment arrow stamped on the slip yoke and on the tube shaft to assure proper phasing when assembling these components. If there are no alignment marks, they can be added before disassembly of the shaft to ensure proper reassembly.

18.5 DRIVE SHAFT INSPECTION

To keep a vehicle operating smoothly and economically, the drive shaft must be carefully inspected and lubricated at regular intervals. Vibration and U-joint and shaft support (center) bearing problems are caused by such things as loose end yokes, excessive radial (side-to-side or up-and-down) looseness, slip spline radial looseness, bent shaft tubing, or missing plugs in the slip yoke.

To prevent performance problems from occurring, make the following inspections each time the drive shaft is lubricated.

1. Check the output and input end yokes on both the transmission and axle, or axles, for looseness (**Figure 18–18**). If loose, disconnect the drive shaft and retorque the end yoke retaining nut to the manufacturer's specification. If this does not correct the problem, yoke replacement might be necessary. If yoke replacement is required, check for the manufacturer's recommendations regarding replacement frequency of the end yoke retaining nut.

2. If the end yokes are tight, check for excessive radial looseness of the transmission output shaft and axle input and output shafts in their respective bearings (**Figure 18–19**). Consult transmission and axle manufacturer's specifications for acceptable radial

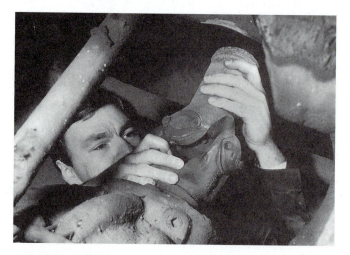

FIGURE 18-18   Check the drive shaft output and input yokes for excessive radial play. (Courtesy of Chicago Rawhide)

FIGURE 18–19 Check the transmission output shaft and the rear axle input shaft for excessive radial play. (Courtesy of Chicago Rawhide)

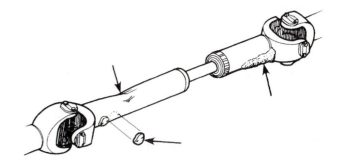

FIGURE 18–21 Points to inspect when checking the drive shaft tube for twisting, dents, missing weights, and buildup of foreign material. (Courtesy of Volvo Trucks North America, Inc.)

looseness limits and method of checking. If the radial play exceeds the specifications, the bearing should be replaced.

3. Check for looseness across the ends of the bearing assemblies and trunnions (**Figure 18–20**). This looseness should not exceed the manufacturer's specification (typically as small as 0.006 inch).

4. Check the slip spline for excessive radial movement. Radial looseness between the slip yoke and the tube shaft should not exceed the manufacturer's specification (typically as little as 0.007 inch).

5. Check the shaft (**Figure 18–21**) for damaged, bent tubing or missing balance weights. Make certain there is no buildup of foreign material on the shaft, such as undercoat or concrete. If found, they should be removed.

6. Check the center bearing to be sure it is mounted securely and correctly. Check for leaking lubricant, damaged seals, or other

damage. If damage is evident, replace the bearing. Do not attempt to repair it or to relubricate it.

18.6 LUBRICATION

Among the most common causes of universal joint and slip joint problems is lack of proper lubrication. Properly sized U-joints that are adequately relubricated at recommended intervals will normally meet or exceed their projected operating life span. Not only does regular lubing ensure that the bearings have adequate grease, but it also flushes the joints, thus removing abrasive contaminants from the bearings.

 **Shop Talk** _____

Always clean grease fittings thoroughly to remove accumulated grease and abrasives. Otherwise, these contaminants will be forced into the bearings during relubrication.

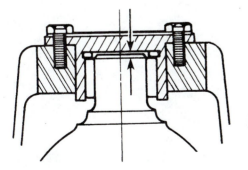

FIGURE 18–20 Check for excessive play in the U-joint bearings. (Courtesy of Volvo Trucks North America, Inc.)

UNIVERSAL JOINT LUBRICANTS

Heavy-duty drive shafts typically use a good quality lithium soap-based, extreme pressure (EP) grease meeting the **National Lubricating Grease Institute (NLGI)** grades 1 or 2 specifications. Grades 3 and 4 are not recommended because of their greater thickness.

When lubricating U-joints exposed to severe operating conditions, use a good quality lithium soap-based or equivalent EP grease having an operating temperature range of +325° to −10°F. In addition, the grease should meet the NLGI grades 1 or 2 specifications.

Initial Lubrication and Relube Cycles

Replacement universal joint kits contain only enough grease to provide needle roller bearing protection during storage. It is therefore necessary to completely lubricate each replacement kit prior to assembly into the yokes. Each cross lube reservoir should be fully packed with a recommended grease and each bearing assembly should also be wiped with the same grease. Fill all the cavities between the needle rollers and apply a liberal grease coating on the bottom of each bearing assembly.

CAUTION: Applying too much grease may cause hydraulic lock, making installation difficult.

After the kits are installed into the yokes and prior to placing into service, they should be relubed, through the lube fitting, using the same grease.

Relubrication cycles vary depending on the service requirements and operating conditions of the vehicle. A recommended relube cycle for various types of service is shown in **Table 18–2**.

TABLE 18–2: LUBRICATION SCHEDULE		
Type of Service **Time**	**Miles**	**or**
City	5,000/8,000	3 months
On highway	10,000/15,000	1 month
On/off highway	5,000/8,000	3 months
Extended (linehaul)	50,000	3 months
Severe usage off highway (4 × 4)	2,000/3,000	1 month

 Shop Talk

On-highway is defined as all applications requiring less than 10 percent of operating time on gravel, dirt, or unimproved roads. If longer than 10 percent operating time is classified as off-highway.

U-JOINT LUBRICATION

Follow these procedures for lubricating universal joints.

WARNING: Rotating drive shafts can snag clothes, skin, hair, hands, etc., causing serious injury or death. Do not work on a shaft (with or without a guard) when the engine is running. Do not go under the vehicle when the engine is running.

1. Using the proper lubricant, apply grease through one of the two grease fittings (**Figure 18–22**). Keep applying until the seals pop and the old grease is forced out of the bearings. This flushes abrasive contaminants from each bearing assembly and assures all four are filled. Bearings are purged when clean grease appears at all four seals.

2. If any seals fail to purge, move the drive shaft from side to side while applying gun pressure. This allows greater clearance on the thrust end of the bearing assembly that is not purging. If the U-joint has two grease fittings, try greasing from the opposite fitting.

3. Because of the superior sealing capability of some seals, there might occasionally be one or more bearing assembly seals that will not purge.

   When that occurs, release seal tension by loosening the bolts holding the bearing assembly that does not purge. It might be necessary to loosen the bearing assembly approximately $1/16$ inch minimum. If loosening it does not allow the grease to purge, remove the bearing assembly to determine the cause of the blockage.

4. Retorque the bolts to the manufacturer's specification.

FIGURE 18–22 *Apply lubricant until the old grease is forced out of the bearings. (Courtesy of Spicer Universal Joint Division/Dana Corporation)*

CAUTION: Half-round end yoke self-locking retaining bolts should not be reused more than five times. Follow instructions to prevent serious personal injury from loss of drive shaft function. If in doubt as to how many times bolts have been removed, replace with new bolts.

SLIP SPLINE LUBRICATION

The lubricant used for U-joints is satisfactory for use on slip splines: use a good EP grease that meets NLGI grade 1 or 2 specifications.

Relube splines at the same intervals recommended in **Table 18-2** for U-joints. Follow these steps.

1. Apply grease gun pressure to the lube fitting until lubricant appears at the pressure relief hole in the plug at the slip yoke end of the spline.
2. Now cover the pressure relief hole with your finger and continue to apply pressure until grease appears at the slip yoke seal (**Figure 18-23**). Sometimes it might be easier to purge the slip yoke by removing the dust cap and reinstalling it once grease appears.

CAUTION: In cold temperatures, be sure to drive the vehicle immediately after lubricating the drive shaft. This activates the slip spline and removes any excessive lubricant. Failure to do so could cause the excess lubricant to stiffen in the cold weather and force the plug out. The end of the spline would then be open to collect contaminants and cause the spline to wear and/or seize.

FIGURE 18-23 Apply grease to the slip joint until grease is forced out of the relief hole. Then, cover the hole and continue to apply lubricant until grease begins to ooze out around the seal. (Courtesy of Spicer Universal Joint Division/Dana Corporation)

CENTER SUPPORT BEARING LUBRICATION

Center bearings are lubricated by the manufacturer and are not serviceable. However, when replacing a support bearing assembly, be sure to fill the entire cavity around the bearing with waterproof grease to shield the bearing from water and contaminants. Enough grease must be put in to fill the cavity to the extreme edge of the slinger surrounding the bearing. Only use waterproof lubricants. Consult a grease supplier or the bearing manufacturer for recommendations.

👓 **Shop Talk** _____

When replacing a center bearing, be sure to look for and not lose track of shims that might be located under the center bearing for setting drive shaft angles.

18.7 UNIVERSAL JOINT REPLACEMENT

Universal joint replacement is easy, and in most cases requires no special tools. However, an arbor press or universal joint puller can make the procedure much easier and reduce the risk of damaging the yoke. In most truck applications, it will be unnecessary to raise the vehicle to perform the joint replacement. It is generally necessary to remove grease fittings before attempting to remove the U-joint. This will prevent destroying the grease fitting.

CAUTION: Always block wheels before working on any part of the driveline.

CAUTION: When removing a drive shaft with half round or flange type yokes, support the shaft with nylon straps before loosening the retainer strap or flange mounting bolts.

👓 **Shop Talk** _____

Before removal of the drive shaft, mark the slip yoke assembly and tube shaft with a marking stick or paint to assure proper alignment when reassembled. If the shaft assembly is to be cleaned before reassembly, use a center punch or scribe to prevent alignment marks from being removed.

BEARING PLATE U-JOINT

1. If the universal joint is connected to full round yokes with bearing cups, use a hammer and chisel to bend the tabs of the lock straps away from the bolt heads.
2. Then, remove the four bolts connecting each bearing assembly to the transmission and axle end yokes.
3. There are several ways to remove the bearing assemblies from the end yoke bores. A hydraulic jack can be used to apply a pressure under the drive shaft, raising the bearing cup up out of the yoke bore (**Figure 18–24A**). Lightly tap on the yoke if the bearing cup does not loosen when initial pressure is applied. When the bearing cup is exposed above the bore, it can be pulled out by hand (**Figure 18–24B**). Rotate the yoke and repeat the procedure to remove the other bearing from the yoke bore. Be sure to keep bearing cups matched with the trunnion from which they were removed. Do not reinstall caps on any other trunnion than their original.

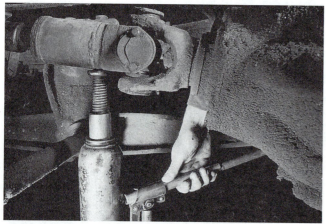

A

B

FIGURE 18–24   (A) Use a jack to force the bearing cup out of the yoke; (B) once the top of the cup is exposed, it can be removed by hand. (Courtesy of Chicago Rawhide)

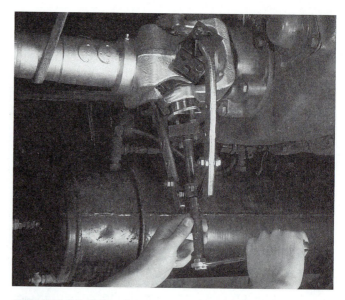

FIGURE 18–25   Removing a bearing plate type bearing with a puller. (Courtesy of Spicer Universal Joint Division/Dana Corporation)

4. A safer method of removing the bearing cup assembly is with the use of a two-jaw puller. Simply bolt the puller to the bearing assembly and pull each bearing out (**Figure 18–25**).
5. Free the trunnion from the end yoke by tilting the U-joint until the trunnions clear the yoke bore.
6. Collapse the drive shaft and lower the end to the ground.
7. Repeat on the opposite end of the shaft.

SNAP RING U-JOINT

If the universal joints are secured in the yoke with snap rings, remove the snap rings and raise the drive shaft with a jack to force the bearing cup out of the yoke bore, as described above.

HALF-ROUND YOKE

If the drive shaft has half round end yokes, remove the strap retaining bolts or the U-bolts. Then, collapse the shaft to move the bearing cups out of the yoke.

FLANGE TYPE YOKE

If the drive shaft has flange type yokes, loosen and remove the nuts and bolts securing the flange yoke to the transmission or axle companion flange. Hold the shaft firmly while tapping the flanges free. Compress the shaft from one end and lower the shaft to the floor.

FIGURE 18–26   Bearing plate type bearings can be removed with a puller. (Courtesy of Spicer Universal Joint Division/Dana Corporation)

FIGURE 18–27   Support the cross on vise jaws, then tap the yoke to drive the bearing up. (Courtesy of Chicago Rawhide)

 **Shop Talk** _____

If only one end of the drive shaft requires service, disconnect that end, unscrew the slip shaft seal (dust cap) from the slip yoke assembly, and then pull apart or disconnect the spline joint. When removing the entire drive shaft, disassemble one end at a time, laying the disconnected end on the floor carefully. When reassembling, be sure that the arrows or marks on the shaft and slip joint are in line to keep the drive shaft yokes in phase.

CAUTION: Do not distort the shaft tube by applying excessive clamping force.

Using an appropriate puller such as that shown in **Figure 18–26**, remove any bearing plate type bearings. If a puller is not available or if the universal joint is not equipped with bearing plates, use an arbor press or hammer and soft, round drift to remove the bearings. Another way to remove the universal joint from the yoke is shown in **Figure 18–27**. By supporting the cross on the jaws of a vise, the upper bearing can be removed by tapping on the yoke with hammer blows. When the bearing can be removed by hand, reverse the yoke and tap out the opposite bearing.

U-JOINT INSPECTION

Inspect the U-joints and bearing assemblies for signs of wear and damage. **Figure 18–28** shows several examples of U-joint deterioration.

- *Cracks* are stress lines due to metal fatigue. Severe and numerous cracks will weaken the metal until it breaks.
- **Galling** occurs when metal is cropped off or displaced due to friction between surfaces. Galling is commonly found on trunnion ends.
- **Spalling** (surface fatigue) occurs when chips, scales, or flakes of metal break off due to fatigue rather than wear. Spalling is usually found on splines and U-joint bearings.
- **Pitting**—small pits or craters in metal surfaces—is due to corrosion and can lead to surface wear and eventual failure.
- **Brinelling** is a type of surface failure evidenced by grooves worn in the surface. Brinelling is often caused by improper installation of the U-joints. Do not confuse the polishing of a surface (false brinelling), where no structural damage occurs, with actual brinelling.

YOKE INSPECTION

After removing the cross and bearings from both ends of the shaft, inspect the yoke bores for damage or raised metal. Raised metal can be removed with a rat tail or half-round file and emery cloth (**Figure 18–29**).

Check the yoke lug bores for excessive wear, using a go-no-go wear gauge. Use an alignment bar (a bar with approximately the same diameter as the yoke bore) to inspect for misalignment of the yoke lugs. Slide the bar through both lug holes simultaneously (**Figure 18–30**). If the bar will not slide through the bore of the yoke lugs simultaneously, the yoke has been distorted by excessive torque and should be replaced.

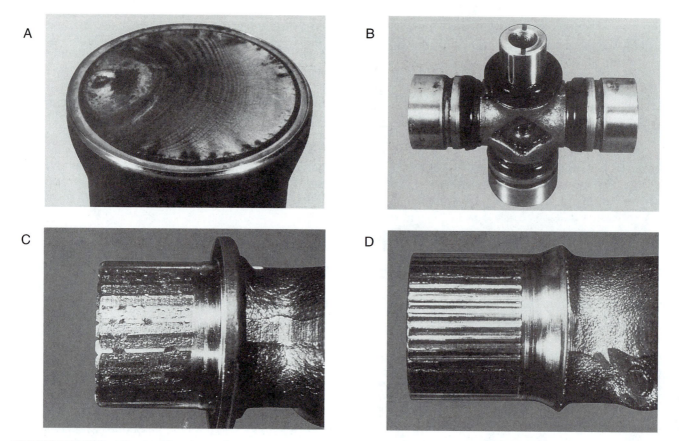

FIGURE 18–28   Types of universal joint damage: (A) cracks; (B) galling; (C) spalling; and (D) brinelling. (Courtesy of ArvinMeritor Inc.)

FIGURE 18–29   Clean the yoke bore with an emery cloth. (Courtesy of Chicago Rawhide)

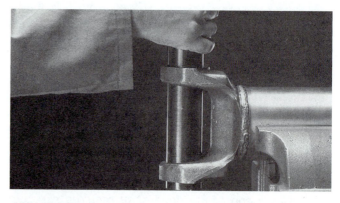

FIGURE 18–30   To inspect misalignment, slide an alignment bar through both holes simultaneously. (Courtesy of Spicer Universal Joint Division/Dana Corporation)

At this time, clean the bore of the end yokes on the transmission and axle and inspect them with an alignment bar gauge as just described. If after proper cleaning of the cross holes the alignment bar will not pass through both holes simultaneously, the end yoke is distorted and should be replaced before reinstalling the drive shaft.

U-JOINT REASSEMBLY

1. Place each end of the drive shaft, less cross and bearing kits, in a bench vise. Check the paint marking placed on the tube and slip yoke assembly prior to removing from the vehicle to be sure they are lined up or in phase.

2. Remove the cross and bearings from the box and remove all four bearing assemblies. Rotate the cross to inspect for presence of the one way check valve in each lube hole of all four trunnions. Then position the cross into the end yoke with its lube fitting in line as near as possible with the slip spline lube fitting. Keep the lube fitting on the inboard side.

3. Apply an antiseize compound to the outside diameter of four bearing assemblies. Move one end of the cross to cause a trunnion to project through the cross hole beyond the outer machined face of the yoke lug. Place a bearing assembly over the trunnion diameter and align it to the cross hole (**Figure 18–31A**).

4. Holding the trunnion in alignment with the cross hole, press the bearing assembly flush to the face of the end yoke by hand (**Figure 18–31B** and **C**).

5. If the bearing assembly binds in the cross hole, tap with a ball peen hammer directly in the center of the bearing assembly plate. Do not tap the outer edges of the bearing plate. Exact fit of all driveline components is extremely important.

6. When the bearing assembly is completely seated, put the lock plate tab in place and use the capscrews that are furnished with the kit and insert them through the capscrew holes in both the lock strap and bearing assembly (**Figure 18–31D**). Thread them by hand or with a wrench into tapped holes in the yoke. Do not torque down the bolts yet.

7. Move the cross laterally to the opposite side and through the cross hole beyond the machined surface of the yoke lug. Place a bearing assembly over the cross trunnion and slide it into the cross hole, seating the plate to the face of the lug. Put the lock plate tab in place and screw the bolts by hand or wrench into threaded holes in the yoke.

 Shop Talk

Projecting the trunnion through a cross hole beyond the machined surface of the lug will provide a surface to help align the bearing assembly with the cross hole. This method should also be used when assembling the drive shaft to the yokes of the vehicle at the transmission and axle or axles.

A

B

C

D

FIGURE 18–31 *Align the bearing with the trunnion when installing a universal joint. (Courtesy of Spicer Universal Joint Division/Dana Corporation)*

8. Repeat the process at the opposite end of the drive shaft. Make sure to position the cross in the yoke so that the lube fitting is in line with the lube fitting at the other end.

9. For flange yoke applications, install the flange yoke, bearing assemblies, and bolts at this time.

DRIVE SHAFT INSTALLATION

The installation of a drive shaft does not present any unusual mechanical difficulties. Before actual installation, the drive shaft should be checked for the following items:

- Damage or dents on the drive shaft tubing
- Splines should slide freely with slight drag from the slip shaft seal
- Cross should turn in bearing cup and be free from excessive bind. A slight drag is the most desirable condition on a new cross and bearing kit. Excessive looseness is not desirable and will result in an unbalanced drive shaft.
- Mounting flanges and pilots should be free from burrs, paint, and foreign substances, which would not allow proper seating at assembly.

CAUTION: Worn bearing assemblies used with new cross or new bearing assemblies used with a worn cross will wear rapidly, causing premature failure of the U-joint. Always replace the cross, four bearing assemblies, and mounting hardware as a unit.

Full Round End Yoke

1. Rotate the transmission end yoke by putting the transmission in neutral and the axle end yoke by jacking up one rear wheel, so that the cross holes are in a horizontal position. Be sure the wheels are chocked.

2. Swivel the cross of the drive shaft so that the trunnions are pointing toward each other from end to end. Position the shaft so that the slip joint is nearest the source of power. Use nylon support straps to aid in handling the drive shaft.

3. Holding the drive shaft firmly, place a trunnion between the lugs of either the axle or the transmission end yoke (**Figure 18–32A**). Repeat at the opposite end. The drive shaft is now being supported at each end by one

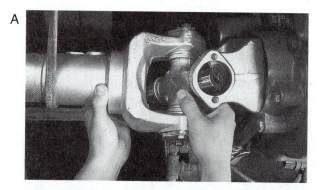

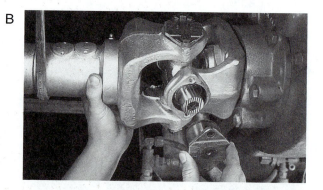

FIGURE 18–32 Installing a drive shaft: (A) rotate the end yoke so that the lugs are horizontal; (B) center a trunnion in the lug bore and install a bearing cup; (C) using a ballpeen hammer, carefully tap the center of the bearing cup to seat it in the yoke bore; (D) install capscrews, lubricate, and torque screws to specs. (Courtesy of Spicer Universal Joint Division/Dana Corporation)

trunnion in a yoke and the nylon support strap.

4. Shift the drive shaft to one side so that the opposite cross trunnion can be inserted through the remaining yoke hole. Repeat at the opposite end of the shaft. The drive shaft is now being supported at each end by two trunnion surfaces in the cross holes and the nylon support strap.

CAUTION: Be careful not to break off the grease fitting (if applicable) during installation.

5. Apply an antiseize compound to the outside of the remaining four bearing cups. Move one end of the shaft to cause a trunnion to project through the cross hole beyond the outer machined face of the yoke lug. Place a bearing cup over the trunnion and align it to the yoke hole (**Figure 18–32B**).

6. Holding the trunnion in alignment with the yoke hole, press the bearing assembly flush to the face of the yoke by hand.

7. If the bearing assembly binds in the cross hole, tap with a hammer directly in the center of the bearing cup plate (**Figure 18–32C**). Do not strike the outer edges of the bearing plate.

8. Slide the shaft sideways to project the opposite trunnion through the opposite yoke hole. Again, place a bearing cup over the trunnion and align it with the yoke bore. Place your hands on the opposite bearing assembly and press the sides of the bearing into place. If the assembly binds, tap it with a hammer.

9. Put the lock plate tab in place and insert the capscrews through the holes in the lock plates and bearing cups (**Figure 18–32D**). Thread the capscrews into the end yokes. Tighten them with a wrench until the plates are flush against the end yoke faces. Then, back the screws off slightly.

10. Lubricate the cross and bearing assembly until the lube appears to all four seals. (See "U-joint Lubrication.") Also, check the slip yoke for proper lubrication.

11. Torque all eight bolts to specification. Bend the manufacturer's lock plate tabs against the flat of the capscrew heads to lock the screws in place.

12. Repeat at the opposite end of the shaft. Remove the nylon support strap.

Half-Round End Yoke

For half-round end yokes, first hang the drive shaft in line with the end yokes using nylon straps. Then place the bearing assemblies on the cross trunnions. Install the bearing cups into the end yoke shoulders. Place straps over the bearing assemblies, thread the special self-locking capscrews into the threaded holes, and torque the bolts to specification. Lubricate the cross and bearing assemblies.

Flange Yoke

With the nylon support straps in place and holding the drive shaft firmly, align the (permanent end) flange pilots of the drive shaft flange yoke and axle companion flange with each other. Align bolt holes and install bolts, lockwashers, and nuts to temporarily secure the drive shaft to the axle. Compress the slip assembly to position the opposite end of the drive shaft to the transmission companion flange. Align bolt holes and install bolts, lockwashers, and nuts. Torque all mounting bolts to the manufacturer's specifications.

18.8 CHASSIS VIBRATION DIAGNOSIS

Vibration that is felt during vehicle operation can originate in an area other than the driveline. Therefore, before condemning the drive shaft as the cause of vibration, the vehicle should be road tested (or tested on floor stands) in an attempt to isolate the vibration cause.

First, determine the source of the problem. The cause of the vibration could be in the steering or suspension systems, in the engine or transmission mounting or operation, in the wheels or tires. To assist in finding the source, ask the operator to determine what, where, and when the vibration is encountered.

Next, road test the vehicle loaded and unloaded while recording engine rpm and road speed. Make note of any irregularities. If the problem is prevalent when pulling a trailer, it will be necessary to road test with a different trailer and load to assist in defining the cause of the disturbance.

Table 18–3 outlines a procedure for isolating a source of vibration.

DRIVE LINE BALANCING

An unbalanced driveline causes **transverse vibrations,** or bending movements, in the drive shaft.

TABLE 18–3: CHASSIS VIBRATION DIAGNOSIS FLOWCHART

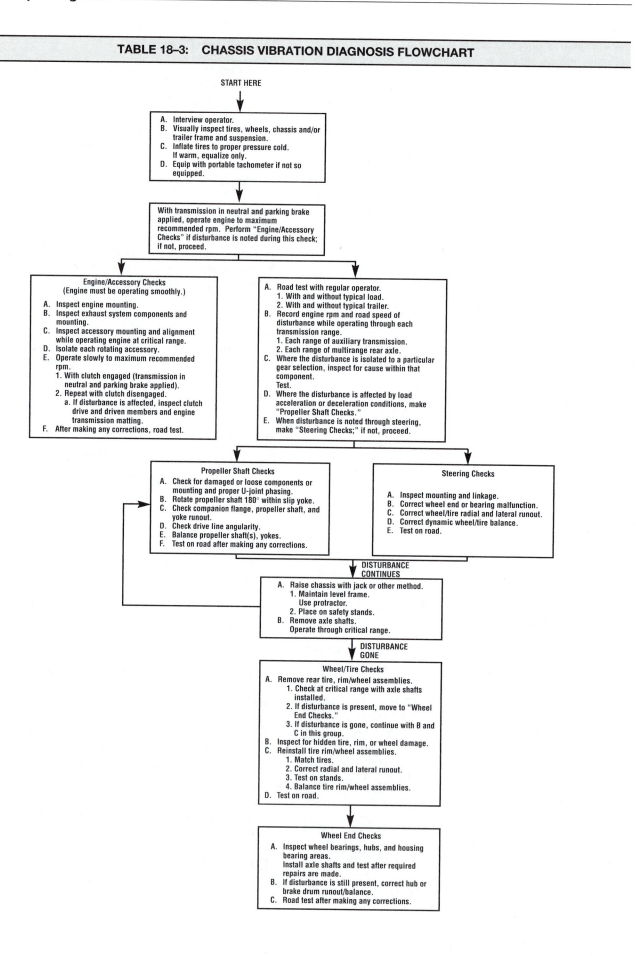

START HERE

A. Interview operator.
B. Visually inspect tires, wheels, chassis and/or trailer frame and suspension.
C. Inflate tires to proper pressure cold. If warm, equalize only.
D. Equip with portable tachometer if not so equipped.

With transmission in neutral and parking brake applied, operate engine to maximum recommended rpm. Perform "Engine/Accessory Checks" if disturbance is noted during this check; if not, proceed.

Engine/Accessory Checks
(Engine must be operating smoothly.)

A. Inspect engine mounting.
B. Inspect exhaust system components and mounting.
C. Inspect accessory mounting and alignment while operating engine at critical range.
D. Isolate each rotating accessory.
E. Operate slowly to maximum recommended rpm.
   1. With clutch engaged (transmission in neutral and parking brake applied).
   2. Repeat with clutch disengaged.
      a. If disturbance is affected, inspect clutch drive and driven members and engine transmission matting.
F. After making any corrections, road test.

A. Road test with regular operator.
   1. With and without typical load.
   2. With and without typical trailer.
B. Record engine rpm and road speed of disturbance while operating through each transmission range.
   1. Each range of auxiliary transmission.
   2. Each range of multirange rear axle.
C. Where the disturbance is isolated to a particular gear selection, inspect for cause within that component.
   Test.
D. Where the disturbance is affected by load acceleration or deceleration conditions, make "Propeller Shaft Checks."
E. When disturbance is noted through steering, make "Steering Checks;" if not, proceed.

Propeller Shaft Checks

A. Check for damaged or loose components or mounting and proper U-joint phasing.
B. Rotate propeller shaft 180° within slip yoke.
C. Check companion flange, propeller shaft, and yoke runout.
D. Check drive line angularity.
E. Balance propeller shaft(s), yokes.
F. Test on road after making any corrections.

Steering Checks

A. Inspect mounting and linkage.
B. Correct wheel end or bearing malfunction.
C. Correct wheel/tire radial and lateral runout.
D. Correct dynamic wheel/tire balance.
E. Test on road.

DISTURBANCE CONTINUES

A. Raise chassis with jack or other method.
   1. Maintain level frame. Use protractor.
   2. Place on safety stands.
B. Remove axle shafts. Operate through critical range.

DISTURBANCE GONE

Wheel/Tire Checks

A. Remove rear tire, rim/wheel assemblies.
   1. Check at critical range with axle shafts installed.
   2. If disturbance is present, move to "Wheel End Checks."
   3. If disturbance is gone, continue with B and C in this group.
B. Inspect for hidden tire, rim, or wheel damage.
C. Reinstall tire rim/wheel assemblies.
   1. Match tires.
   2. Correct radial and lateral runout.
   3. Test on stands.
   4. Balance tire rim/wheel assemblies.
D. Test on road.

Wheel End Checks

A. Inspect wheel bearings, hubs, and housing bearing areas.
   Install axle shafts and test after required repairs are made.
B. If disturbance is still present, correct hub or brake drum runout/balance.
C. Road test after making any corrections.

These vibrations are related to drive shaft rpm and are usually most noticeable within a particular engine rpm range.

There are two methods of balancing the driveline:

1. By rotating the propeller shaft 180 degrees in relation to the companion flange.
2. Dynamic balancing.

Rotating the Shaft 180 Degrees

Before any balance weights are added to the propeller shafts, disconnect the rear propeller shaft at the center bearing, and rotate the shaft 180 degrees in relation to the companion flange (for vehicles without center bearing, rotate the shaft at the transmission). Reconnect the shaft, then road test the vehicle or spin the driveline and check for vibration.

Dynamic Balancing

This method requires the use of a balance sensor and strobe light. These units may be designed specifically for the task of balancing driveshafts or may be multipurpose. Fit the unit to the drive shaft assembly to be tested: The test procedure is similar to that used when balancing tire and wheel assemblies. Using the strobe light and white machinist's crayon, mark the driveshaft and spin up. Use hose clamps to clamp weights into position with the clamp located over the weight washer (see **Figure 18–33**). When zero imbalance is achieved, tack weld the weight or weights into position. The weight of the weld tacks should be approximately equivalent to that of the hose clamp worm.

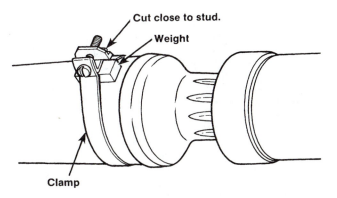

FIGURE 18–33 *Balancing a propeller shaft.* (Courtesy of International Truck and Engine Corp.)

Cut close to stud.

Weight

Clamp

CAUTION: Use small tack welds to attach weights. Larger welds can create an imbalance and distort driveshaft tubing.

CHECKING OPERATING ANGLES

To determine whether vibrations are caused by improper driveline angles, do the following:

1. Inflate all tires to the pressure at which they are normally operated. Park the vehicle on a surface which is level both from front to rear and from side to side. Do not attempt to level the vehicle by jacking up the front or rear axles. Shift the transmission to neutral and block the front tires. Jack up a rear wheel.
2. Rotate the wheel by hand until the output yoke on the transmission is vertical and lower the jack. This simplifies measurement later. Check drive shaft angles in the same loaded or unloaded condition as when the vibrations or noise occurred. Always try to check driveline angles in both loaded and unloaded conditions.
3. To determine drive shaft angles, use a **magnetic base protractor** or an **electronic inclinometer.**
   To use the magnetic base protractor or an electronic driveline inclinometer, place the sensor on the component to be measured. A display window will show the angle and in which direction it slopes.
   Always measure the slope of the drive train going from front to rear. A component slopes downward if it is lower than the front. A component slopes upward when it is higher at the rear than it is in front.
4. Check and record the angle on the main transmission. This reading can be taken on the end yoke lug, with the bearing assembly removed, or on a flat surface of the main transmission parallel to the output yoke lug plane. Record your readings on a sketch of the driveline.
5. Now check the drive shaft angle between the transmission and axle or forward axle. On short tube length drive shafts, check the angle of the drive shaft on either the tube or slip yoke lug with the bearing assembly removed (**Figure 18–34A**). On long tube length drive shafts, measure the angle on the tube at least 3 inches away from the yoke welds and at least 1 inch away from any balance weights

A

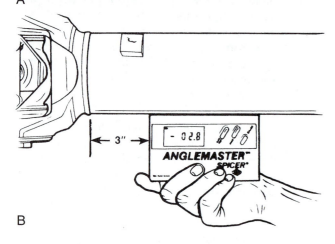

B

FIGURE 18–34   *Checking the drive shaft angle: (A) on a yoke with a bearing assembly removed; or (B) 3 inches from the tube yoke weld. (Courtesy of Spicer Universal Joint Division/Dana Corporation)*

(**Figure 18–34B**). Be sure to remove any rust, scale, or flaking from the drive shaft tube to obtain an accurate measurement.

6. Check the forward axle input yoke angle by removing a bearing assembly and measuring

the angle on the yoke lugs or on a flat surface of the axle housing parallel to the input yoke lug plane.

7. If applicable, measure the output yoke angle of the forward axle, the angle of the tandem drive shaft between the forward axle and rear axle, and the rear axle input yoke angle. With all of the angles recorded, complete a drawing as shown in **Figure 18–35**. There are no U-joint operating angles in the drawing at this time, just the inclination of the components. To determine U-joint operating angles, simply find the difference in inclination of the components.

When the inclination occurs in the same direction on two connected components, subtract the smaller number from the larger to find the U-joint operating angle. When the inclination occurs in the opposite direction on two connected components, add the measurements to find the U-joint operating angle. Now compare the U-joint operating angles on the drawing to the rules for ideal operating angles given earlier in **Table 18–1**.

CORRECTING U-JOINT OPERATING ANGLES

The recommended method for correcting severe U-joint operating angles depends on the vehicle suspension or drive shaft design. On vehicles with a leaf spring suspension, thin wedges called **axle shims** can be installed under the leaf springs of single axle vehicles to tilt the axle and correct the U-joint operating angles. Wedges are available in a range of sizes to change pinion angles.

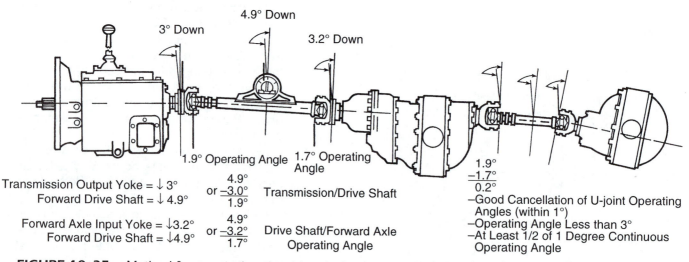

FIGURE 18–35   *Method for computing the drive shaft operating angles between the transmission and forward axle. (Courtesy of Spicer Universal Joint Division/Dana Corporation)*

On vehicles with tandem axles, the torque rods can be shimmed. **Torque rod shims** rotate the axle pinion to change the U-joint operating angle. A longer or shorter torque rod might be available from the manufacturer if shimming is not practical. Some torque rods are adjustable.

As a general rule, the addition or removal of a ¼-inch shim from the rear torque arm will change the axle angle approximately ¾ degree. A ¾-degree change in the pinion angle will change the U-joint operating angle about ¼ degree.

Other factors that can cause the U-joint operating angle to change are

- Suspension changes caused by worn bushings in the spring hangers, worn bushings in the torque rods, or incorrect airbag height
- Revisions in the components of the drive shaft
- Stretching or shortening the chassis
- Adding an auxiliary transmission or transfer case in the main drive shaft
- Worn engine mounts

These faults must be identified and corrected to eliminate driveline vibrations and to avoid premature U-joint failure.

CHECKING SHAFT RUNOUT

Vibrations will occur if the drive shaft is not straight. The amount of bend, or **runout**, in a shaft can be checked using a dial indicator. **Figure 18–36** shows where to check a drive shaft for runout and the allowable limits for total indicator runout (TIR). Before measuring, be sure to clean the surfaces where the dial indicator readings are to be taken.

Runout versus Ovality

When checking for runout, it is important to distinguish between runout and **ovality (Figure 18–37)**.

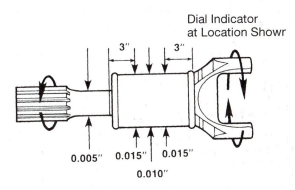

FIGURE 18–36 Allowable limits for total indicator shaft runout. (Courtesy of Spicer Universal Joint Division/Dana Corporation)

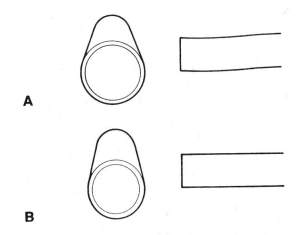

FIGURE 18–37 Determining ovality of shaft: (A) a circular bent shaft is not acceptable; (B) an oval, straight shaft may be acceptable. (Courtesy of Spicer Universal Joint Division/Dana Corporation)

Runout is when the tube is slightly bent but still maintains its circularity throughout the tube. During dynamic balancing, a dial indicator will show runout once per revolution.

Ovality occurs when the tube is not circular but oval in shape. During dynamic balancing, a dial indicator will display ovality twice per revolution. Even though a tube may be straight, ovality will make it seem bent. A tube with ovality can be used up to a 0.010 TIR reading. Beyond this limit the tube is not acceptable and must be discarded for drive shaft purposes.

YOKE RUNOUT AND ALIGNMENT

Not only must the drive shaft be straight but the yokes attached to the transmission, auxiliary transmission, and drive axles must be true and straight and in alignment with the shafts to which they are attached.

Checking Yoke Runout

Mount a dial indicator to the transmission or axle housing as shown in **Figures 18–38** and **18–39**. If the yoke is the half round type, place the tip of the indicator probe against the machined surface of the yoke shoulder. If the yoke is the full round type, insert the bearing assembly into the yoke cross holes, leaving just enough of the bearing retainer exposed to accept the indicator stylus. Then, rotate the yoke 180 degrees. If the runout is more than 0.005 inch, the yoke should be replaced.

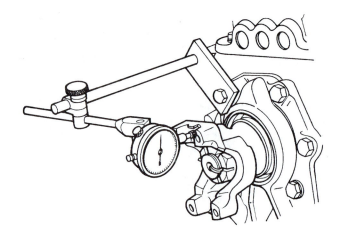

FIGURE 18–38 *When checking half-round end yoke for runout, place the indicator stylus on smooth surface of shoulder. (Courtesy of International Truck and Engine Corp.)*

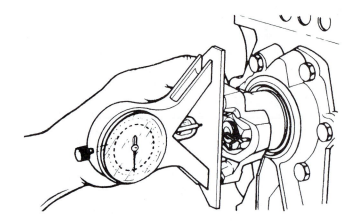

FIGURE 18–40 *When checking half-round end yoke for vertical alignment, place the protractor on the smooth surface in a vertical position. (Courtesy of International Truck and Engine Corp.)*

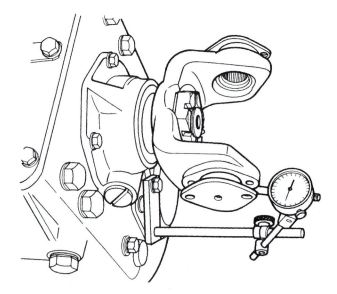

FIGURE 18–39 *When checking full-round end yoke for runout, be sure to leave enough surface for the indicator probe. (Courtesy of International Truck and Engine Corp.)*

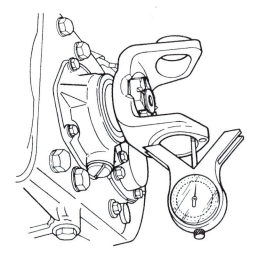

FIGURE 18–41 *When checking full-round end yoke for vertical alignment, place the protractor on the outside of the shoulder. (Courtesy of International Truck and Engine Corp.)*

Checking Yoke Vertical Alignment

The machined shoulder of a half-round yoke or the yoke bores in a full-round yoke should be 90 degrees to the centerline of the shaft to which the yoke is attached. If the shaft is horizontal (0 degrees), the yoke should be vertical (90 degrees); if the shaft is angled away from a true horizontal, the yoke should be at an angle equal to 90 degrees minus the declination of the shaft.

Before checking for vertical alignment, park the truck on a level surface. Make sure the engine and

transmission mountings are secure. Also check the yoke for looseness and tighten, if necessary.

When measuring the alignment of a yoke attached to the main transmission, first disconnect the drive shaft from the end yoke. Then, place the transmission in neutral and rotate the transmission main shaft until the yoke lugs are positioned vertically one above the other. Then, if the yoke is a half-round type, place a protractor with a magnetic base in a vertical position across the machined surfaces of the yoke (**Figure 18–40**). The dial should read 90 degrees minus the declination of the engine and transmission.

If the yoke is the full-round type, place the protractor base on the outside surface of the machined shoulder (**Figure 18–41**). The protractor should read the angle of the engine/transmission declination. If the end yoke is more than $1/2$ degree out of vertical alignment, the yoke is either loose or it is distorted and should be replaced.

When checking the end yoke attached to the pinion shaft of a drive axle, jack up one wheel and rotate the wheel until the lugs of the yoke are aligned vertically. Then, test the vertical alignment of the yoke, following the procedure described above for checking the vertical alignment of transmission end yokes.

Table 18–4 outlines a drive shaft troubleshooting guide.

	TABLE 18–4: DRIVE SHAFT TROUBLESHOOTING GUIDE	
Symptom	**Probable Cause**	**Remedy**
Vibration —low gear shudder —at certain speeds under full drive or full coast —under light loaded conditions	1. Secondary couple load reaction at shaft support bearing	1. Reduce U-joint continuous running angle. Check with transmission or axle manufacturer—replace shaft bearing.
	2. Improper phasing	2. Check drive shaft for correct yoke phasing.
	3. Incompatible drive shaft	3. Use larger diameter tube.
	4. Drive shaft weight not compatible with engine transmission mounting or drive shaft too long for speed.	4. Install two piece drive shaft with shaft support bearing.
	5. Loose outside diameter fit on slip spline	5. Replace spline or yoke.
	6. Excessively loose U-joint for speed	6. Replace U-joint.
	7. Drive shaft out of balance; not straight	7. Straighten and balance shaft.
	8. Unequal U-joint angles	8. Shim drive train components to equalize U-joint angles.
	9. U-joint angle too large for continuous running	9. Reduce U-joint continuous running angle.
	10. Worn U-joint	10. Replace U-joint.
	11. Inadequate torque on bearing plate capscrews	11. Inspect U-joint flex effort for looseness; torque to specification.
	12. Torsional and/or inertial excitation	12. Straighten and balance shaft. Check drive shaft for correct yoke phasing.
Premature Wear —low mileage U-joint wear —repeat U-joint wear —end galling of cross trunnion and bearing assembly —needle rollers brinelled into bearing cup and cross trunnion —broken cross and bearing assemblies	1. End yoke cross hole misalignment	1. Use alignment bar to check for end yoke cross hole misalignment; replace end yoke if misaligned.
	2. Excessive angularity	2. Check U-joint operating angles with an electronic driveline inclinometer; reduce excessive U-joint operating angles.
	3. Improper lubrication	3. Lubricate according to specifications.
	4. Excessive U-bolts torque on retaining nuts	4. Replace U-joint kit.
	5. Excessive continuous running load	5. Reduce U-joint continuous running angle.
	6. Continuous operation at high angle/high speed	6. Replace with higher capacity U-joint and drive shaft.
	7. Contamination and abrasion; worn or damaged seals	7. Check U-joint flex effort; replace joint or yoke if necessary. Clean and relubricate U-joint.
	8. Excessive torque load (shock loading) for U-joint and drive shaft size	8. Realign to proper running angle minimum 1 degree. Torque bearing retention method to specification.
Slip Spline Wear —seizure —galling	1. Improper lubrication	1. Lubricate slip spline according to specifications.

TABLE 18–4:	DRIVE SHAFT TROUBLESHOOTING GUIDE (CONTINUED)	
Symptom	**Probable Cause**	**Remedy**
—outside diameter wear at extremities and at 180 degrees —spline shaft or tube broken in torsion	2. Worn or damaged part 3. Tube size inadequate 4. Excessive torque load for U-joints and drive shaft size 5. Male spline head engagement length too short for application 6. Excessively loose outside diameter fit 7. Slip member working in extreme extended or fully collapsed position 8. Contamination	2. Replace spline; check design for application. Use nylon-coated slip spline. 3. Replace with higher capacity U-joint and drive shaft (use caution here). Use larger diameter tube. 4. Check U-joint flex effort; replace joint or yoke if necessary. 5. Check for male slip member with longer spline. 6. Replace spline; check design for application. Use nylon-coated slip spline. 7. Increase drive shaft assembly length to position slip spline head toward U-joint. 8. Clean and relubricate according to specification.
Shaft and/or Tube —shaft support bearing wear or fracture —shaft support rubber insulator wear or fracture —tube circle weld fracture	1. Balance weight located in apex of weld yoke lug area 2. Balance weight too close to yoke weld 3. Improper circle weld 4. Bending fatigue due to secondary couple loads 5. Drive shaft too long for operating speeds 6. Worn or damaged parts 7. Excessive torque load (shock loading) for U-joint and drive shaft size 8. Improper lubrication of bearings 9. Shaft support bearing misaligned (interferes with slinger)	1. Replace drive shaft. 2. Replace drive shaft. 3. Replace drive shaft. 4. Reduce U-joint continuous running angle. 5. Install two-piece drive shaft with shaft support bearing. Use larger diameter tube. 6. Normal bearing wear; replace. 7. Replace with higher capacity U-joint and drive shaft (use caution here). 8. Normal bearing wear; replace. 9. Realign mounting bracket to frame cross member to eliminate interference with slinger.
Yoke Fracture —yoke broken in hub —yoke broken at ear tip	1. Mating yoke lug interference at full jounce rebound 2. Excessive torque load for U-joint and drive shaft size 3. Improper shaft length and slip 4. Bending fatigue due to secondary couple loads	1. Reduce U-joint with continuous running angles. 2. Replace with higher capacity U-joint and drive shaft (use caution here). 3. Check installed lengths and adjust drive shaft lengths to provide slip conditions. 4. Replace yoke; check design for application.

18.9 PTO DRIVE SHAFTS

In some applications, an auxiliary unit, such as a pump, will be directly mounted to the power takeoff assembly. However, remotely mounted units are driven by the PTO through a drive shaft (**Figure 18–42**). A **PTO drive shaft** transmits torque from the PTO output shaft to the input shaft of the driven accessory. The drive shaft must be capable of transmitting the maximum torque and rpm required by the accessory, plus allowing for any shock loads that might develop.

PTO drive shafts are similar in design to the vehicle's power train drive shaft. Universal joints and slip joints allow the working angles between the PTO and the driven accessory to change due to movements in

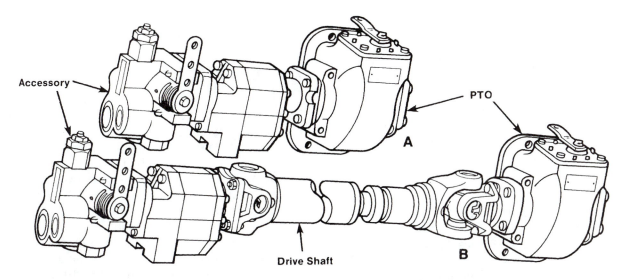

FIGURE 18–42 *(A) Some accessories are mounted directly to the PTO; (B) others are remotely mounted and driven by a drive shaft.*

the power train from torque reactions and chassis deflections. PTO drive shafts designed for high torque applications operating at more than 1200 rpm are made of tubular steel. Shafts designed for applications where rpm is less than 1200 are of solid shaft design using hex, square, or round bar-stock. Since most PTO drive shaft applications are for intermittent service, a precisely balanced shaft is rarely used.

PTO drive shafts are available with end fittings of the end yoke or flange construction. Drive shafts using a flange type construction are best suited for applications where the complete drive shaft is removed often. This is easily accomplished by removing the bolts, nuts, and washer used to connect the flange yoke and companion flange.

18.10 DRIVELINE RETARDERS

All devices known as retarders supplement the chassis braking system. That is, they are never intended to bring a vehicle to a standstill but instead relieve the chassis brakes of some of their work. This help can greatly extend the intervals between foundation brake service and overhaul. Most chassis braking requires brake applications using 20 percent or less of the service brake potential. Auxiliary braking can relieve the main brake system of much of this light-duty braking.

There are five general categories of retarders used on trucks today. Three of these are associated with the engine: internal engine compression brakes, external engine compression brakes, and hydraulic engine retarders. These engine based retarders are

explained in a companion textbook. The other two categories are transmission retarders and driveline retarders. Transmission retarders are addressed in the chapters on automatic transmissions earlier in this textbook. There follows an introduction to driveline retarders.

Driveline retarders can be either hydraulic or electric. A driveline hydraulic retarder operates on the same principles as those already discussed in the chapters on automatic transmissions. The difference is that it is not built into the transmission, but located elsewhere in the driveline. Electric driveline retarders brake or "retard" driveline rotation using an electromagnetic field. When the retarder is energized, a powerful magnetic field is created, causing a dragging effect on driveshaft rotation. Driveline retarders have an advantage over engine brakes in that the efficiencies are not related to engine rpm. **Figure 18–43** shows a sectional view of a hydraulic retarder.

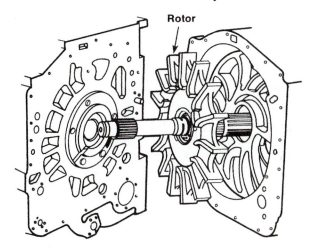

FIGURE 18–43 *Typical hydraulic retarder assembly. (Courtesy of Voith Transmissions Inc.)*

HYDRAULIC RETARDERS

All hydraulic retarders operate on the same principles, so the only thing that differentiates the driveline hydraulic retarder from the transmission hydraulic retarder is its location. Hydraulic retarders located in the driveline tend to be used more in Europe than America. They are uncommon in typical linehaul applications, so when they are seen it is in vocational applications, very often those off road.

Operation

To retard or brake the vehicle's speed, the driver/operator activates a control valve by using a hand valve or foot pedal (**Figure 18–44**). Depressing the control pedal will open the control valve and allow transmission oil to flow into the hydraulic retarder housing. Oil flow into the housing is directly proportional to pedal travel and, therefore, also to the degree of braking desired. The oil in the hydraulic retarder housing is under pressure and works against the vanes on the rotor. See **Figure 18–43**. As the rotor turns with the torque converter output shaft, the vanes on the rotor force the transmission oil into the openings between the fixed stator vanes. This action tends to work against the rotor, slowing it down and providing braking power to the vehicle's driving wheels. The pumping action of the rotor circulates transmission oil between the hydraulic retarder and the transmission heat exchanger to provide the necessary cooling. Engagement and use of this engine brake results in the transmission oil sump, charging pump, and heat exchanger effectively becoming part of the hydraulic retarder system as shown in **Figure 18–44**. When the hydraulic retarder is engaged and operated, the engine throttle will be in an idle position, reducing the load on the vehicle's cooling system. Once this driving load is reduced, the cooling system can now readily absorb the heat generated by the hydraulic retarder.

MID-MOUNT HYDRAULIC RETARDER

The mid-mount retarder is a type of hydraulic retarder in use on both on- and off-highway trucks. It is typically located between the vehicle's frame rails and activated from the cab. The mid-mount retarder is not built into the transmission but is installed external to it. This remote type of installation requires that oil lines be provided to carry transmission oil to and from the retarder. Operation of the mid-mount is identical to that presented for typical hydraulic retarders. An exploded view of a typical mid-mount hydraulic retarder is shown in **Figure 18–45**.

ELECTRONIC HYDRAULIC RETARDER

The retarder shown in **Figure 18–46** operates on the principle of hydrodynamics providing engine braking power. It is controlled electropneumatically either by a hand lever mounted in either the steering column or dash, or by the vehicle service brake pedal. It consists of

- Retarder unit with heat exchanger
- Retarder hand switch
- Electronic control unit (ECU)

To operate the system properly also requires the following inputs from the host truck:

- Air pressure source
- Electrical power source
- Cooling source

The retarder is designed for installation in the drivetrain of the vehicle and various designs for mounting the retarder on the gearbox are used. One method of installation involves the use of a flange that allows mounting directly onto the transmission.

Compressed air from the truck's supply reservoir is used to control the retarder. The ECU controls the pressure regulator in accordance with the position of the retarder hand lever at the steering column. The electrical power required by the ECU is readily supplied by the truck's existing electrical system. Heat generated by the retarder in operation is dissipated through the chassis cooling system.

The retarder control system uses an ECU and a hand lever or the truck's service brake pedal, or both. With the hand lever, the driver can select up to six different braking steps for the truck. More precise

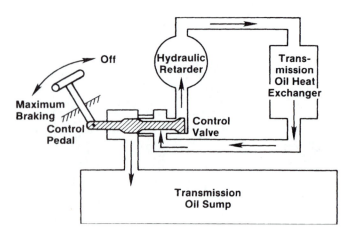

FIGURE 18–44 *Typical layout of the hydraulic retarder circuit. (Courtesy of Voith Transmissions Inc.)*

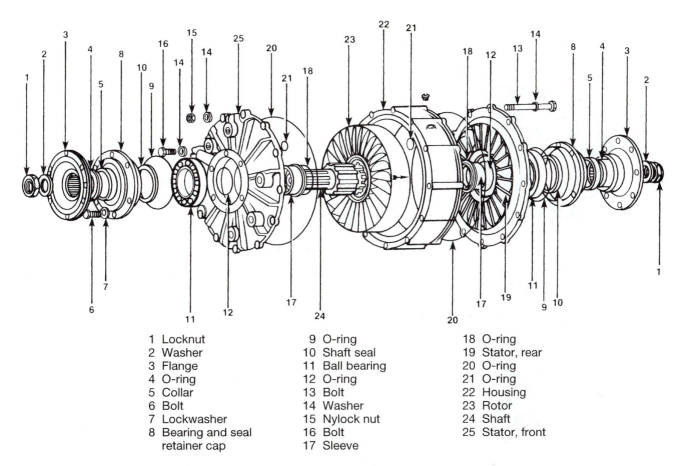

1 Locknut	9 O-ring	18 O-ring
2 Washer	10 Shaft seal	19 Stator, rear
3 Flange	11 Ball bearing	20 O-ring
4 O-ring	12 O-ring	21 O-ring
5 Collar	13 Bolt	22 Housing
6 Bolt	14 Washer	23 Rotor
7 Lockwasher	15 Nylock nut	24 Shaft
8 Bearing and seal	16 Bolt	25 Stator, front
retainer cap	17 Sleeve	

FIGURE 18–45 *Typical mid-mount hydraulic retarder. (Courtesy of Voith Transmissions Inc.)*

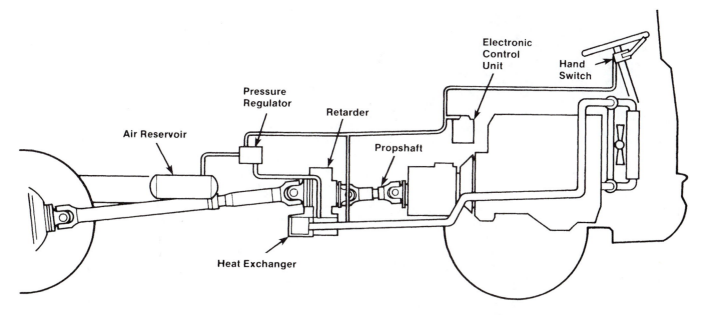

FIGURE 18–46 *Typical installation of a Voith retarder. (Courtesy of Voith Transmissions Inc.)*

control of the retarder braking action is provided by a pushbutton on the lever. When this button is pressed, the ECU processes the actual driving speed of the truck and can prevent the truck from exceed-ing a pre-programmed speed. The ECU also has the capability to monitor itself, automatically sensing trouble in the system by activating a warning light in the cab. **Figure 18–47** shows a Voith hydraulic retarder assembly.

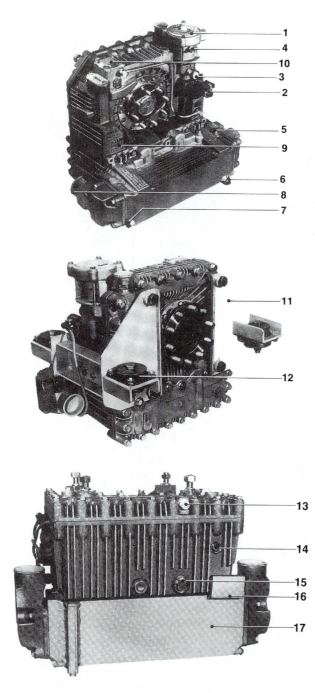

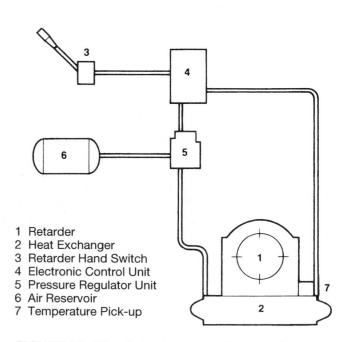

1 Retarder
2 Heat Exchanger
3 Retarder Hand Switch
4 Electronic Control Unit
5 Pressure Regulator Unit
6 Air Reservoir
7 Temperature Pick-up

FIGURE 18–48 *Retarder electronic control system. (Courtesy of Voith Transmissions Inc.)*

Some retarder control systems are electronically connected with the ABS controller to provide better stability and safety for the entire braking system. This connection allows the ABS controller to temporarily "disable" the retarder during certain modes of operation. While most ABS controllers are capable of this function, and connections are provided, they are not always used. **Figure 18–48** shows a typical retarder electronic control system.

ELECTRIC RETARDERS

Like the hydraulic driveline retarder, the electric retarder has seen more widespread use in Europe than in America. The electric retarder has been used on trucks, buses, and various on- and off-highway vehicles, including vehicles equipped with diesel or gasoline engines. The electric retarder offers noise-free braking with or without the use of the vehicle's service brakes.

This type of retarder or engine braking system works independently of the engine, transmission, or hydraulic system. The vehicle's 12- or 24-volt electrical system provides the power for the electric retarder. The retarder is available in various sizes and capacities.

DESCRIPTION

The electric retarder is mounted between the side frames of the vehicle (**Figure 18–49**). The major components of the retarder include a cast steel frame for

1 Control Valve
2 Electromagnetic Solenoid
3 Connection for Supply Air
4 Connection for Housing Pressure
5 Connection for Temperature Sensor
6 Oil Drain Plug—Heat Exchanger
7 Cooling Water Drain Plug
8 Cooling Water Inlet— Heat Exchanger
9 Nameplate
10 Ventilation
11 Brake Side Mounting
12 Connection Cone
13 Sump Cutoff Valve
14 Oil Check Plug—Sand Cast
15 Oil Drain Plug— Retarder
16 Cover—Ventilation Duct
17 Heat Exchanger

FIGURE 18–47 *Components of a Voith retarder. (Courtesy of Voith Transmissions Inc.)*

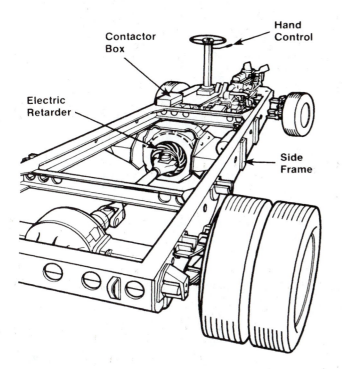

FIGURE 18–49 Electric retarder components. (Courtesy of Jacobs Vehicle Systems)

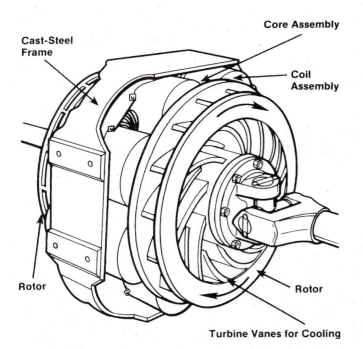

FIGURE 18–50 Typical electric retarder. (Courtesy of Jacobs Vehicle Systems)

the components, core assembly, coil assembly, and two rotors equipped with turbine vanes for cooling purposes (**Figure 18–50**). A master control switch and a multiposition control switch are used to activate and control the retarder. These switches are located in the vehicle's cab.

OPERATION

Approximately sixteen coils or electromagnets are mounted within the cast steel frame of the retarder. A typical arrangement includes eight electromagnets on each side of the frame. Energizing the retarder will cause the electromagnets to exert a strong dragging force on the two rotors in the frame. Because the electric retarder is mechanically in line with the drive shaft, the drag force is transmitted directly to the drive shaft, providing retarding action to slow down the vehicle. This operation is an application of the **eddy current** principle.

To understand the eddy current principle, the electric retarder can be considered as a large generator. The transmission drives the retarder rotor assembly. As the rotor rotates within the electromagnetic fields, small currents (or eddy currents) are generated within the rotor, somewhat similar to a power generator. However, the developed eddy currents have no place to flow from the rotor so they circulate via paths not influenced by the electromagnets. These eddy currents create a magnetic reactive force that tends to stop or retard rotor rotation, hence an engine braking system.

The electrical energy generated in an electric retarder cannot be outputted to a load and, therefore, must dissipate itself in the form of heat. Turbine vanes cast on the rotors create an airflow to dissipate this heat.

Activation of the retarder electromagnets is progressively controlled by a multiposition switch or lever mounted on the steering column or the dash (**Figure 18–51**). Operation of the control switch varies the degree of braking desired by the driver/operator. The multiposition control switch also has the advantage of an automatic time delay that will gradually increase the deceleration rate (retarding effort) without the driver engaging each switch position until the desired braking level is reached. A low-speed detector can be utilized with the electric retarder to deenergize the retarder whenever vehicle speed is below 5 mph.

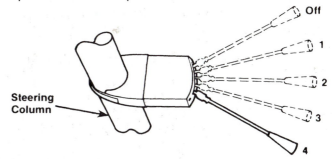
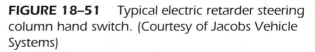

FIGURE 18–51 Typical electric retarder steering column hand switch. (Courtesy of Jacobs Vehicle Systems)

Maintenance

Maintenance of an electric retarder involves the removal and replacement of the retarder from the chassis or frame rails. The frame of the retarder is bolted to the rails and the rotors are connected in line with the vehicle's drive shaft. The electric retarder can be rebuilt in much the same manner as electric motors or alternators are rebuilt. This could involve rewinding the coil assembly, replacing the electromagnetics, or replacing the rotor assembly. Electrical wiring between the system switches and the retarder can also cause problems that can be repaired using standard electrical circuit troubleshooting.

SUMMARY

- The components of a drive shaft assembly include the following:
  - yoke
  - universal joint
  - slip spline
  - seal
  - bearing support
  - the propeller shaft
- The purpose of the universal joint is to link the driveline components and to allow the drive shaft to operate at different or changing angles.
- There are two types of driveline arrangements that will transmit power to the driving wheels: the parallel joint type and the nonparallel or "broken back" type.
- With the drive yoke running at a constant rpm, the drive shaft will increase and decrease speed once per revolution. To counter this fluctuation in drive shaft speed, two universal joints are used and placed in phase.
- Vibrations and U-joint and shaft support bearing problems are caused by such things as loose end yokes, excessive side-to-side or up-and-down looseness, slip spline radial looseness, bent shaft tubing, or missing plugs in the slip yoke.
- After removing the cross and bearings from both ends of the shaft, inspect the yoke bores for damage or raised metal. Raised metal can be removed with a rat tail or half round file and emery cloth.
- Check the yoke lug bores for wear, using a go-no-go wear gauge.
- Vibration that is felt during vehicle operation can originate in an area other than the driveline.
- A hydraulic driveline retarder uses hydraulic oil or transmission fluid friction as the means to retard driveline rotation.

- The principal components of a hydraulic driveline retarder are vaned rotors and stators.
- Electric driveline retarders operate by absorbing the energy of motion by converting it first to electrical energy using an "eddy current" principle, and then dissipating it as heat.
- Both hydraulic and electric driveline retarder operation can be ECM-controlled.

REVIEW QUESTIONS

1. U-joint failures, center bearing failures, and vibrations are caused by
   a. loose end yokes
   b. bent shafts
   c. missing balance weights
   d. all of the above

2. A truck is experiencing drive shaft vibrations. Technician A says that the problem might be a loose end yoke. Technician B says that the problem might be worn splines in the slip joint. Who is correct?
   a. Technician A
   b. Technician B
   c. both A and B
   d. neither A nor B

3. Which of the following will not affect the drive shaft balance?
   a. missing balance weights
   b. U-joint lubrication
   c. foreign material
   d. dents

4. If, while lubricating a U-joint, fresh lubricant appears at all four bearing seals
   a. the bearings are worn and should be replaced
   b. the seals are worn and should be replaced
   c. the trunnions are worn and the U-joint should be replaced
   d. the U-joint has been properly lubricated

5. The U-joints on a truck used for normal on-highway operation should be lubricated
   a. every 10,000 to 15,000 miles
   b. once a year
   c. 3,000 to 5,000 miles
   d. every 50,000 miles

6. Technician A says that some universal joints are factory lubricated and do not have to be lubricated after they are installed. Technician B says that if a seal "pops" when the U-joint is being lubricated, the joint must be replaced. Who is correct?
   a. Technician A
   b. Technician B

c. both A and B
d. neither A nor B

7. Which of the following might be the cause of U-joint failure after low-mileage use?
a. excessive torque on the U-bolt retaining nuts
b. excessive spline joint lubrication
c. low-speed, low-angle operation
d. none of the above

8. A truck experiences repeated U-joint failure caused by end galling of the cross trunnion and bearing cup. Technician A says the problem might be corrected by using high-temperature grease. Technician B says the problem might be caused by a distorted end yoke. Who is correct?
a. Technician A
b. Technician B
c. both A and B
d. neither A nor B

9. A drive shaft is in phase when
a. the shaft is in line with the output shaft of the transmission
b. the slip yoke and tube yoke lugs are aligned with each other
c. the shaft is balanced
d. all of the above

10. The working angle on most heavy-duty truck applications should not exceed
a. 3 degrees
b. 5 degrees
c. 8 degrees
d. 10 degrees

11. The drive shaft size rating is determined by
a. shaft length
b. working angle
c. rpm
d. all of the above

12. Technician A says that the working angle of the drive shaft can be corrected by adding wedge shims to the spring seats of a leaf spring suspension system. Technician B says that loose engine or transmission mountings or a distorted end yoke might be the cause of excessive angularity. Who is correct?
a. Technician A
b. Technician B
c. both A and B
d. neither A nor B

13. Grooves worn into the trunnions by the bearings are called
a. brinelling
b. galling

c. spalling
d. pitting

14. Technician A says that driveline vibration could be caused by worn clutch components. Technician B says that the truck should be road test loaded and unloaded to help determine the source of vibration. Who is correct?
a. Technician A
b. Technician B
c. both A and B
d. neither A nor B

15. A driveshaft spline joint
a. allows the drive shaft to slide axially
b. is often coated with a polymer nylon
c. is protected by a dust cap and washer
d. all of the above

16. The braking force of a hydraulic retarder is produced by
a. oil acting on rotor vanes
b. oil action in the heat exchanger
c. compressed air acting on the driveline
d. compressed air acting on a diaphragm

17. Which of the following is used to remove heat generated by an electric retarder?
a. heat exchanger
b. air flow and vanes
c. hydraulic fluid
d. water flow

18. Which of the following principles is used by an electric driveline retarder?
a. electromagnetism
b. electromagnetic induction
c. static electricity
d. hydrodynamics

19. Which of the following devices uses an "eddy current" principle?
a. hydraulic retarder
b. torque converter
c. electric retarder
d. universal joint

20. To allow retarding action to be limited during ABS operation, which of the following is required in a hydraulic or electric driveline retarder system?
a. electronic control
b. pneumatic control
c. treadle control
d. lever control

Heavy-Duty Truck Axles

Prerequisites
Chapters 2 and 3

Objectives

After reading this chapter, you should be able to
- Identify the axles in use on heavy-duty trucks.
- Name the components on drive axles and explain how they function.
- Describe the operation of the various drive axle configurations.
- Explain the purpose of a power divider or interaxle differential and describe the location and operation of its components.

Key Terms List

amboid gearing
axle range interlock
controlled traction differential assembly
differential
dead axles
differential carrier assembly
double reduction axles
driver-controlled main differential lock
dual range unit

friction plate assembly
fully floating type axle
hypoid gearing
power divider
semi-floating type axle
single reduction axle
spinout
spiral bevel gearing
tandem drive axle
three-speed differential
two-speed axle assembly

19.1 TYPES OF AXLES

Truck axles can be grouped into three types: drive axle, steering axle, and trailer axle. By far the most complex is the drive axle (**Figure 19–1**). Drive axles contain complex gearing not in the other two types. A typical drive axle contains axle shafts and a **differential carrier assembly**. The differential contains the gearing to accommodate the different wheel speeds that occur when turning corners. Drive axles are usually the rear axle assemblies on heavy-duty trucks and are often referred to as live axles.

Steering and trailer axles are used to carry and support the weight of the vehicle and provide a mounting surface for suspension systems. Steering axles (**Figure 19–2**) are typically used in the front of the heavy-duty truck and are used to steer the vehicle. Trailer axles (**Figure 19–3**) carry and support the trailer and its load.

A

B

FIGURE 19–1   *Typical heavy-duty axle assemblies: (A) rear driving; and (B) front driving. (Courtesy of ArvinMeritor Inc.)*

FIGURE 19–2   *Typical nondriving steering axle. (Courtesy of ArvinMeritor Inc.)*

FIGURE 19–3   *Rectangular axle beam trailer axle. (Courtesy of ArvinMeritor Inc.)*

19.2   DRIVE AXLES

In the early days of motoring, the driving wheels of a vehicle were rigidly geared to a continuous, one-piece axle shaft. As a result, when the axle was turned by engine torque, the wheels at each end of the axle turned together and at the same speed.

This was an effective way to move the vehicle. But wheels tended to scrub or scuff on the road surface whenever the vehicle traveled in any direction but straight or when the drive wheels were of mismatched diameters. That was tolerable as long as speeds were slow, loads were light, and distances traveled were short. But as technology improved, vehicle speeds, cargo loads, and journey lengths increased. Scrubbing of tires became more and more costly and inefficient.

Engineers searched for ways to allow each driving wheel to operate at its own speed. Many ideas were tried with mixed results before the design for the present-day, standard differential was developed. The idea was to split driveline torque dividing the axle in two—attaching each driving wheel separately to its own half-shaft and placing in between a pinion and gear arrangement. The arrangement was called the **differential** because it differentiates between the actual speed of each wheel and splits torque from the engine into equal force to each wheel.

Power delivery in a heavy-duty truck begins at the engine. It is transmitted through a transmission and a drive shaft, rotating at right angles to the direction of the wheels. Therefore, power flow must be turned 90 degrees (at a right angle) to have the axle shafts turn in the same direction as the wheels (**Figure 19–4**). The gear set in the drive axle carrier achieves this.

Differential Design

A differential carrier assembly consists of a number of major components (**Figure 19–5**). These include

- Input shaft and pinion gear
- Ring gear
- Differential with two differential case halves, a differential spider (or cross), four pinion gears, and two side gears with washers

Figure 19–6 shows an exploded view of a differential. This differential assembly fits between the axle shafts, with the shafts being splined to the differential side gears. **Figure 19–7** shows an exploded view of a drive axle carrier assembly. The components of a differential carrier are held in position by a number of bearings and thrust washers.

The drivetrain, or path of power in a motor vehicle is ...

1 Engine
2 Transmission
3 Drive Shaft
4 Differential
5 Axle Shafts
6 Wheels

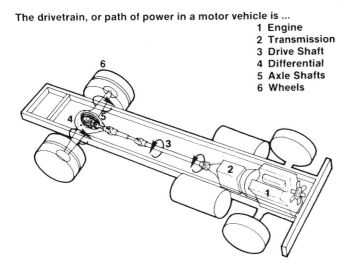

FIGURE 19-4 The drive train or path of power flow in a heavy-duty truck equipped with a single rear drive axle.

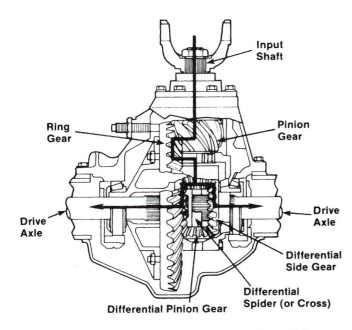

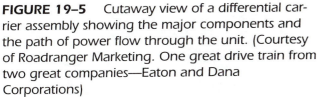

FIGURE 19-5 Cutaway view of a differential carrier assembly showing the major components and the path of power flow through the unit. (Courtesy of Roadranger Marketing. One great drive train from two great companies—Eaton and Dana Corporations)

The forward end of the input shaft is connected to the drive shaft by a yoke and universal joint. The pinion gear on the other end of the input shaft is in constant mesh with the ring or crown gear. The ring gear is bolted to a flange on the differential case. Inside the case, the legs of the spider are held in matching grooves in the case halves. The legs of the spider also support the four pinion gears. In addition, the

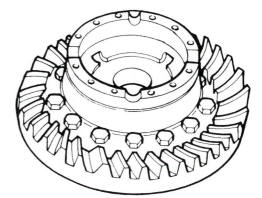

DIFFERENTIAL CASE HALVES

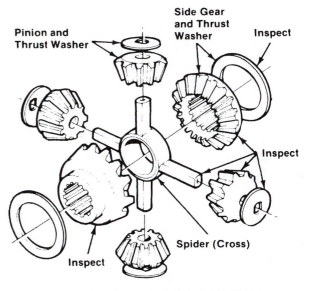

DIFFERENTIAL GEAR NEST ASSEMBLY

FIGURE 19-6 Components of a typical differential. (Courtesy of International Truck and Engine Corp.)

case houses the side gears, which are in mesh with the pinions and splined to the axle shafts.

Differential Operation

When drive shaft torque is applied to the input shaft and drive pinion, the input shaft and pinion rotate in a direction perpendicular to the truck drive axles. The drive pinion is beveled at 45 degrees and meshes to the ring gear, which is also beveled at 45 degrees, causing the ring gear to rotate at 90 degrees to the input shaft. This allows the torque flow to change direction and drive the axles and wheels.

As the ring gear rotates, the differential case rotates with it since the ring gear and case are bolted together. The spider or cross mounted in the casing halves must also rotate with the case and the ring gear.

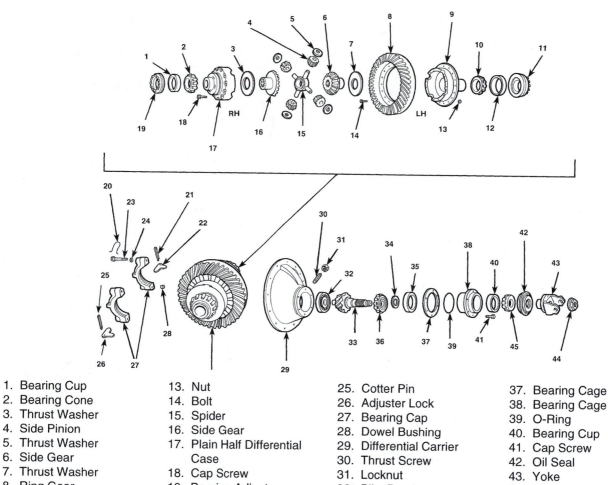

1. Bearing Cup
2. Bearing Cone
3. Thrust Washer
4. Side Pinion
5. Thrust Washer
6. Side Gear
7. Thrust Washer
8. Ring Gear
9. Flanged Half Differential Case
10. Bearing Cone
11. Bearing Adjuster
12. Bearing Cup
13. Nut
14. Bolt
15. Spider
16. Side Gear
17. Plain Half Differential Case
18. Cap Screw
19. Bearing Adjuster
20. Lock Wire
21. Cotter Pin
22. Adjuster Lock
23. Cap Screw
24. Flat Washer
25. Cotter Pin
26. Adjuster Lock
27. Bearing Cap
28. Dowel Bushing
29. Differential Carrier
30. Thrust Screw
31. Locknut
32. Pilot Bearing
33. Drive Pinion
34. Bearing Spacer (Variable)
35. Bearing Cup
36. Bearing Cone
37. Bearing Cage Shim
38. Bearing Cage
39. O-Ring
40. Bearing Cup
41. Cap Screw
42. Oil Seal
43. Yoke
44. Pinion Nut
45. Bearing Cone

FIGURE 19–7 *Components of a differential carrier assembly. (Courtesy of Roadranger Marketing. One great drive train from two great companies—Eaton and Dana Corporations)*

The differential pinions mounted on the legs of the spider revolve around the center axis of the spider as the differential case rotates (**Figure 19–8A**). These four pinions transfer torque to the side gears. As the side gears rotate, they transfer torque to the drive axles, which in turn transfer it to the wheels. As long as the truck is moving in a straight line, the ring gear, differential gears, and drive axles rotate at the same speed.

Operation During Turns. When a truck moves in a straight line, all wheels turn at the same speed, providing the wheels and tires are the same size. But when a truck turns, each wheel rotates at a different speed whether the wheels and tires are the same size or not. When two wheels run parallel to one another, when turning, the outside wheel revolves

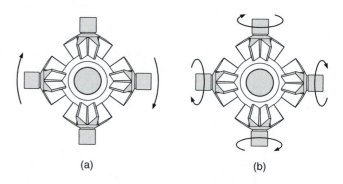

(a)   (b)

FIGURE 19–8 (A) When traveling in a straight line, the pinions rotate with the spider, acting as a bridge between the spider and side gears. (B) When turning a corner, the side gear on the inside slows down, causing the pinions to rotate on the spider legs, which adds speed to the opposite side gear.

faster than the inside wheel since it must travel through a longer arc. Non-driven wheels rotate independently of one another on their own wheel ends, but driven wheels are in mesh with the driveline. Ideally, they should receive equal driving force from the engine, while still being able to rotate at different speeds to compensate for the greater distance the outer wheel must travel. It is the differential that allows this to happen.

When the truck turns a corner, the side gear on the inside of the turn slows as the vehicle enters the turn. This causes the pinion gears to "walk" around or rotate on their spider shafts. The pinions rotate in a direction that adds speed to the opposite side gear. The reduction in speed of one side gear is equal to the increase in speed of the other (**Figure 19–8B**). At the same time, equal driving force is transmitted to each wheel. This prevents the rapid tire wear that would otherwise occur. The differential action makes the vehicle much easier to control while turning. The differential also allows the wheels on opposite ends of the axle to rotate at different speeds to compensate for uneven road conditions and slightly mismatched tires.

Whenever a truck turns, the wheels on the inside of the turn rotate slightly slower than vehicle speed, and its outer wheels rotate slightly faster than vehicle speed. Each wheel is connected by the axle shafts to side gears so that the side gears rotate at wheel speed. When a side gear on the inside of the turn travels at a slower speed than the spider, the pinions "walk" around the side gear. When a side gear on the outside of the turn travels at a higher speed than the spider, a "reverse-walking" effect is produced. The slowdown of one side gear is inversely proportional to the increase in speed of its opposite side gear. Differential action is explained in **Figure 19–8** and **Figure 19–9**.

DRIVE AXLE CONFIGURATIONS

Heavy-duty trucks may be equipped with a single rear drive axle. In most cases, dual rear drive axles are used to accommodate the gross vehicle weight (GVW) of larger trucks. Whether a single axle or a dual (tandem) axle configuration is used, both function in the same basic manner.

The next design factor in drive axles is the number of gear reductions and/or speeds built into the differential carrier gearing. The following paragraphs briefly describe the more common drive axle configurations and gearing used today.

- A **single reduction axle** is an axle assembly that employs only one gear reduction through its differential carrier assembly. The gear reduction is a function of the pinion gear and ring gear ratio (**Figure 19–10**).
- A planetary double reduction axle employs two separate gearing reductions. One gearing reduction is obtained by the driving pinion gear and ring gear arrangement, while the other gear reduction is achieved by spur pinion assembly.
- A **two-speed axle assembly** (often referred to as a **dual range unit**) has two different output ratios from the differential. The driver selects the ratios from the controls located in the cab of the truck.
- A **tandem drive axle** (**Figure 19–11**) combines two single axle assemblies through the use of an interaxle differential or **power divider** and a

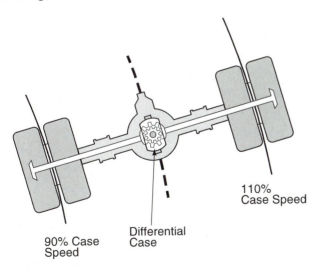

FIGURE 19–9   Turning a corner differential action.

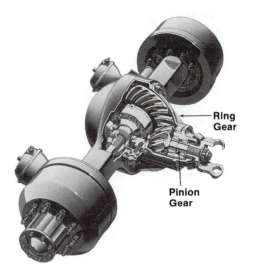

FIGURE 19–10   Gear reduction is accomplished with a pinion gear and ring gear arrangement. (Courtesy of ArvinMeritor Inc.)

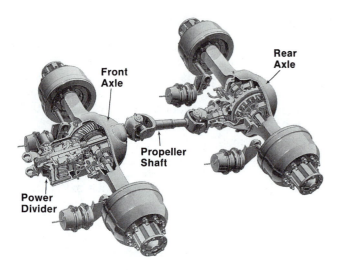

FIGURE 19-11 *Tandem drive axle assembly. (Courtesy of ArvinMeritor Inc.)*

short shaft that connects the two axles. The power divider distributes driveline torque to both the forward and rear drive axles of the tandem set. The power divider is mounted on the forward axle of the tandem drive unit. Like single axles, a tandem unit can employ single reduction gearing, double reduction gearing, or two-speed gearing.

Some axle arrangements use a **controlled traction differential assembly** or a traction equalizer. The controlled traction differential uses a **friction plate assembly** to transfer drive torque from a slipping wheel to the wheel that has good traction. The friction plate assembly consists of a multiple disc clutch designed to slip when a predetermined torque value is reached.

Lockout assemblies control the operation of an interaxle differential used on a tandem drive axle. Operation of the lockout assembly is controlled by the truck driver from the cab. Activation is either by air pressure or vacuum.

Selecting the axle ratio of a two-speed differential carrier assembly is controlled by an axle shift system. Two types of axle shift are used: air shift system and electric shift system.

The axle housing encases the differential gears and protects them from road contaminants and from possible damage. Most housings use a breather to vent air from the differential to relieve pressure buildup.

DIFFERENTIAL GEARING TYPES

A variety of gear types and arrangements are used in differential carrier assemblies for heavy-duty trucks. The common gear types include the following:

- Hypoid gearing
- Amboid gearing
- Spiral bevel gearing
- Spur bevel gearing
- Underslung worm drive gearing
- Overhead worm drive gearing

The most widely used gears are the **hypoid**, **amboid**, and **spiral bevel gearing**. Of these, the most common is the hypoid, which is used primarily on the forward axle of a tandem axle assembly. The amboid gear style is often used on the rear axle of a tandem axle assembly.

Bevel gears are used extensively in rear axle differentials, where they intersect at right angles to transfer power from the driveline to the axle half shafts. The teeth can be cut straight, like a spur gear, or with a spiral, like a helical gear (**Figure 19-12**).

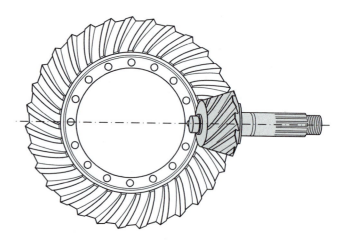

Spiral bevel gear set

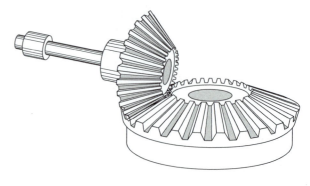

Bevel gear set

FIGURE 19-12 *Bevel gear sets. (Courtesy of ArvinMeritor Inc.)*

Hypoid Gearing

Like bevel gears, hypoid gears intersect at right angles when meshed (**Figure 19–13A**). Hypoid gearing uses a modified spiral bevel gear principle that allows several gear teeth to absorb the driving force

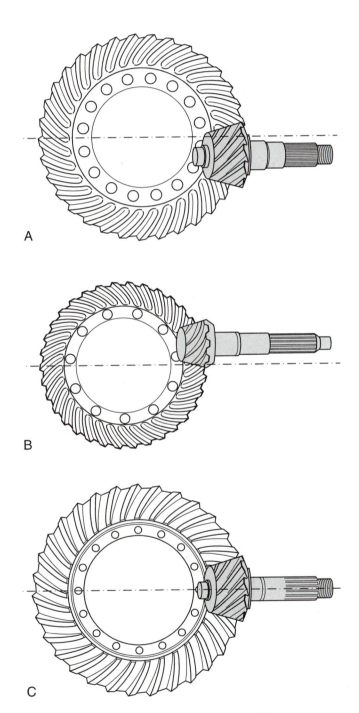

FIGURE 19–13 *(A) Hypoid gearing; (B) amboid gearing; and (C) spiral bevel gearing arrangements. (Courtesy of ArvinMeritor Inc.)*

and allows the gears to run quietly. A hypoid gear set is typically found at the drive pinion gear and ring gear interface. This gear arrangement is designed so that the axis of the pinion gear does not intersect the axis of the ring gear. In addition, the teeth on the drive pinion gear and the ring gear are cut spirally. Because of the curvature of the gear teeth, a sliding wiper action results between the two gear interfaces.

The pinion gear axis sets below the centerline of the ring gear axis, allowing the drive shaft to be mounted in a lower position. Because of this offset, hypoid differential carriers have a greater torque capacity than a spiral bevel gear arrangement. The hypoid pinion gear is typically larger in diameter and has larger gear teeth when compared with the construction of a spiral bevel gear. Hypoid pinion gears have a larger tooth contact area and have more teeth in longer contact with the ring gear teeth. Hypoid gear sets drive forward on the convex side of the ring gear teeth. These design characteristics provide strength and quieter operation.

Hypoid gears require the use of extreme pressure, or EP, gear oil because of the extensive sliding action between the gear teeth. Proper adjustment of backlash and tooth contact is critical to gear life.

Amboid Gearing

An amboid gear (**Figure 19–13B**) is similar to the hypoid except the axis of the drive pinion is located above the centerline of the ring gear. This requires the drive shaft to be mounted in a higher position than with the hypoid gearing arrangement. Amboid gear sets drive forward on the concave side of the ring gear teeth. This gearing arrangement is typically used on the rear axle of a tandem axle assembly.

Spiral Bevel Gearing

The spiral bevel gear arrangement (**Figure 19–13C**) has a drive pinion gear that meshes with the ring gear at the centerline axis of the ring gear. This gearing provides strength and allows for quiet operation. These characteristics are achieved because of the overlapping tooth contact; for example, prior to one tooth rolling out of contact with another, a new gear tooth comes in contact.

SINGLE REDUCTION AXLE

The single reduction axle is the most common and basic type of gearing arrangement used in most drive axles on heavy-duty trucks. The axle assembly has only one gear reduction through its differential ring gear and drive pinion. The single reduction axle can use either a hypoid, amboid, or spiral bevel pinion and ring gear arrangement.

The single reduction axle can be used as an individual assembly on a single axle truck, as the rear axle on a tandem axle assembly, or as a front drive steering axle.

Operation

Figure 19-14 shows a single reduction axle with a differential carrier mounted to the front of the axle housing. The carrier has a hypoid drive pinion and ring gear set and uses bevel gears in the differential. The pinion gear is typically positioned off-center from the ring gear centerline. The drive pinion gear is straddle-mounted on two tapered roller bearings and a straight roller type pilot bearing. The straight roller bearing (often called a spigot) is mounted on the head of the drive pinion. The differential is mounted on two tapered roller bearings. Power flow through this type of axle is through the drive pinion and ring gear, through the four side pinions, to the two side gears to the axle shafts and wheels. On this type of axle, when the carrier operates, there is normal differential action between the wheels all the time.

No-Spin Locking Differential

A variety of heavy-duty trucks equipped with single reduction axles use what is commonly called a **driver-controlled main differential lock (Figure 19-15)**. This type of axle assembly has greater flexibility over the standard type of single reduction axle because it provides equal amounts of driveline torque to each driving wheel, regardless of changing road conditions. This design also provides the necessary differential action to the road wheels when the truck is turning a corner. The no-spin feature is used when operating in mud, snow, ice, or wet road conditions because it prevents wheel spin-out.

The typical single reduction carrier with differential lock has the same type of gears and bearings as standard type carriers. However, the differential lock

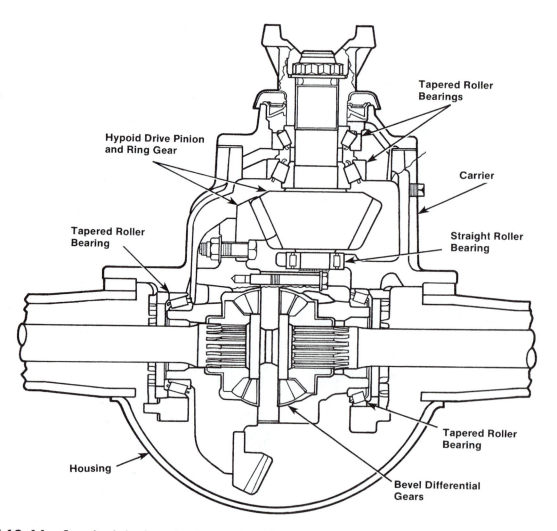

FIGURE 19-14 *Standard single reduction carrier. (Courtesy of ArvinMeritor Inc.)*

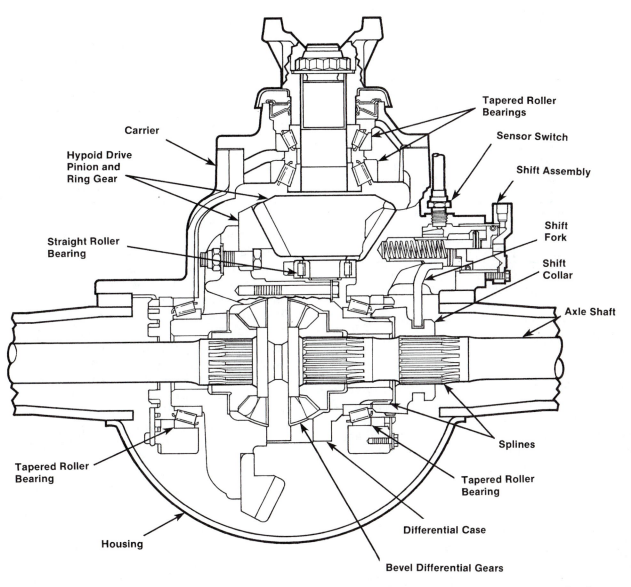

Carrier

Hypoid Drive Pinion and Ring Gear

Straight Roller Bearing

Tapered Roller Bearing

Housing

Tapered Roller Bearings

Sensor Switch

Shift Assembly

Shift Fork

Shift Collar

Axle Shaft

Splines

Tapered Roller Bearing

Differential Case

Bevel Differential Gears

FIGURE 19–15 *Standard single reduction carrier with differential lock. (Courtesy of ArvinMeritor Inc.)*

is operated by an air-actuated shift assembly mounted on the carrier and operated from the truck cab. By actuating an air or electric switch, mounted on the instrument panel, the driver can lock or unlock the differential to achieve positive traction when highway conditions require.

The drive axle equipped with a differential lock consists of the following (**Figure 19–15**):

- Sensor switch
- Splined differential case
- Shift fork
- Shift collar
- Air shift mechanism

Air pressure from the truck air supply system is directed through an input hose on the differential and flows into the air shift mechanism. When the differential lock is set, an electrically operated air solenoid opens (**Figure 19–16**) to activate the air shift mechanism on the carrier. The air shift mechanism will then transmit force to the shift fork, causing it to move forward. The shift fork in turn will move the shift collar into engagement with the splined differential case. At this time, the splines of the shift collar, the differential case half, and the axle half shaft will engage, locking the differential case, gearing and axle shafts together, maximizing traction to both wheels (**Figure 19–17A**).

Note: The sensor/switch for the lock indicator light.is actuated by mechanical interface between the sensor detent and the shift fork.

When the control switch in the cab is set to the unlocked position, the electric circuit to the air

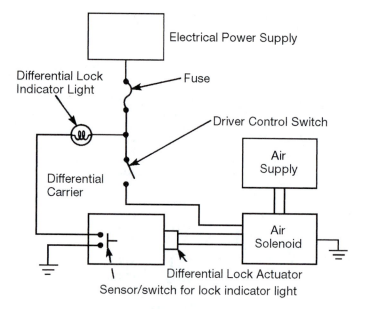

FIGURE 19–16 *Electrical and pneumatic schematic for the differential lock. (Courtesy of ArvinMeritor Inc.)*

operated solenoid is opened, exhausting the compressed air to the differential air shift mechanism. A spring inside the mechanism will then pull the shift fork and collar outward and unlock the differential (**Figure 19–17B**). Normal differential action between both wheels will then continue.

Controlled Traction Differential

Another type of limited slip differential is the controlled traction differential (**Figure 19–18**). A controlled traction differential is designed to inhibit normal differential action during uneven or slippery road conditions. It does this with a set of clutch discs and plates, which are called a clutch pack.

As shown in **Figure 19–19**, the clutch pack consists of twenty-one friction plates. (Other limited slip differential designs might use a different number of friction plates.) Ten of them are splined to a driver and eleven of them have two tangs each that fit into slots in the gear support case. Coil springs keep the clutch plates pressed tightly together. A sliding clutch is controlled by a shift fork that is controlled by a cab-mounted air valve.

As long as the air valve is in the off position, normal differential action occurs. However, when the driver actuates the traction control switch, the shift fork slides the clutch forward so that it moves inside the driver and engages the internal splines in the driver and the splines on the end of the side gear. This couples the side gear to the gear support case and ring gear through the clutch pack. While the clutch plates are engaged, the differential will remain locked up and both axle shafts will rotate at the same speed. This will prevent one set of wheels from spinning faster than the other.

The clutch plate is designed to slip when a prede-termined torque value is reached. This enables the

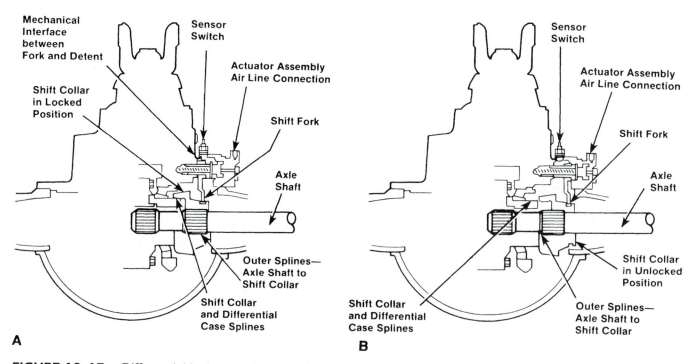

FIGURE 19–17 *Differential lock assembly: (A) locked; and (B) unlocked. (Courtesy of Roadranger Marketing. One great drive train from two great companies—Eaton and Dana Corporations)*

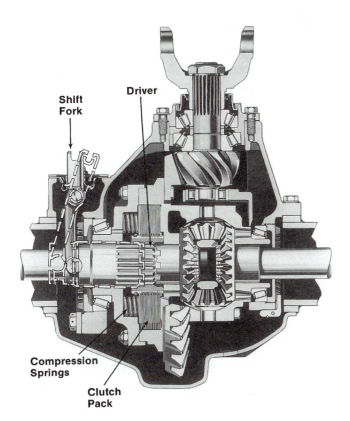

FIGURE 19–18 Heavy-duty controlled traction differential. (Courtesy of Roadranger Marketing. One great drive train from two great companies—Eaton and Dana Corporations)

truck to negotiate turns in a normal manner. The high torque created by wheels on the outside radius of a turn or curve rotating faster than the differential causes the plates in the clutch pack to slip, allowing the side gear to increase in speed and the pinions to walk around the side gear as in normal differential operation.

Controlled traction differentials are available with a driver-controlled shift unit (as just described), and with a permanent driver that permits full-time limited slip operation. If the axle is equipped with a permanent control traction differential, the regular driver is substituted with a driver designed to engage with the side gear at all times.

DOUBLE REDUCTION AXLE

On/off-road vehicles and other trucks required to haul heavy loads are sometimes equipped with **double reduction axles**. A double reduction axle uses two gear sets for greater overall gear reduction and higher torque. This design is favored for severe-service applications, such as dump trucks, cement mixers, and other heavy haulers.

The double reduction axle shown in **Figure 19–20** uses a heavy-duty spiral bevel or hypoid pinion and ring gear combination for the first reduction. The second reduction is accomplished with a wide-faced helical spur pinion and gear set. The drive pinion and ring gear function just as in a single reduction axle. However, the differential case is not bolted to the ring gear. Instead, the spur pinion is keyed to and driven by the ring gear. The spur pinion is in turn constantly meshed with the helical spur gear to which the differential case is bolted.

Another double reduction gear arrangement is shown in **Figure 19–21**. This design uses planetary gears (**Figure 19–22**) to achieve the second reduc-

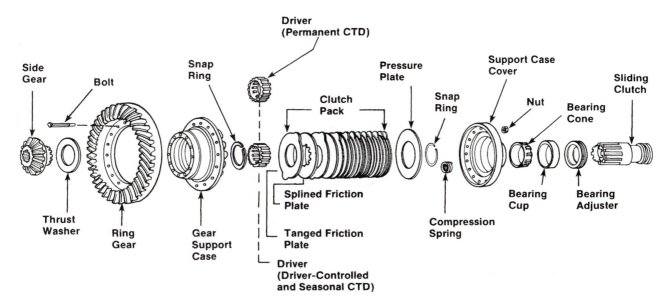

FIGURE 19–19 Exploded view of a heavy-duty controlled traction differential. (Courtesy of Roadranger Marketing. One great drive train from two great companies—Eaton and Dana Corporations)

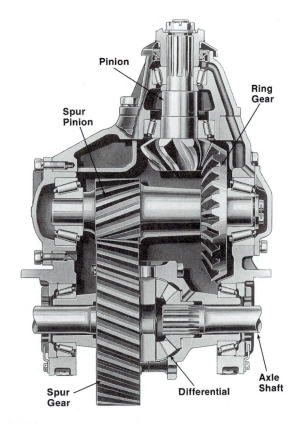

Pinion

Ring Gear

Spur Pinion

Spur Gear

Differential

Axle Shaft

FIGURE 19–20 *Cutaway view of a hypoid double reduction differential carrier. (Courtesy of Roadranger Marketing. One great drive train from two great companies—Eaton and Dana Corporations)*

tion. In an axle with planetary double reduction, the ring gear has internal gear teeth. These teeth mesh with planetary pinion gears (idler pinions), which are mounted on pins fastened to the differential case. A third gear, the sun gear, is splined to the left-hand bearing adjuster and is held in position by a retainer.

In this design, the differential casing is not bolted to the ring gear. The casing is driven by the ring gear through the idler pinions. The drive pinion drives the ring gear. As the ring gear rotates, it drives the planetary pinions, which rotate around the stationary sun gear. This results in a gear reduction and torque increase.

The actual gear ratio of the planetary reduction is calculated by adding the teeth on the sun gear and the internal ring gear, and dividing that sum by the number of teeth on the ring gear. The combined ratio of the double reduction axle ranges typically from approximately 5.04:1 to approximately 10.13:1. For additional information on planetary gearing, refer to Chapter 13, "Automatic Transmissions."

TANDEM DRIVE AXLES

Many heavy-duty trucks are equipped with two rear drive axles (**Figure 19–23**). These tandem axle trucks require a special gear arrangement to deliver power to both the forward and rearward, rear drive axles. The gearing must also be capable of allowing speed differences between the axles.

Interaxle Differentials

To accomplish the above requirements, tandem axle trucks use an interaxle differential, which is also called a power divider.

As shown in **Figures 19–24** and **19–25**, the interaxle differential is an integral part of the front rear axle differential carrier. Components of the interaxle differential are basically the same as in a regular differential: a spider (or cross), differential pinion gears, a case, washers, and the side gears (**Figure 19–26**).

During normal interaxle operation, torque from the vehicle driveline is transmitted to the input shaft splined to the interaxle differential spider. At this point, the interaxle differential distributes torque equally to both axles (**Figure 19–27**).

Forward Drive Axle. A helical gear is combined with the front interaxle differential side gear. The side gear is pressed onto a bushing that floats on the input shaft. As the input shaft rotates and drives the interaxle differential, the helical side gear is driven by the differential pinion gears. The helical side gear in turn drives the second helical gear fastened to the drive pinion. The drive pinion delivers torque to the front rear axle ring gear. This ring gear delivers torque to the front wheel differential. This conventional differential then delivers drive torque to the wheels on the forward drive axle, as discussed earlier in this chapter.

Rear Drive Axle. The rear interaxle differential side gear is splined to the output shaft. So for the rear drive axle, torque is transmitted from the output shaft side gear, through the output shaft, through the interaxle driveline (also called the interaxle propeller shaft), to the drive pinion at the rear axle.

This drive pinion drives the rear axle ring gear, which then delivers torque to the rear wheel differential.

Axle Speed Variations

The interaxle differential compensates for axle speed variations in the same way the carrier differential functions between the two wheels of a single drive axle. As previously shown in **Figure 19–26**, the interaxle differential has a spider and four pinions

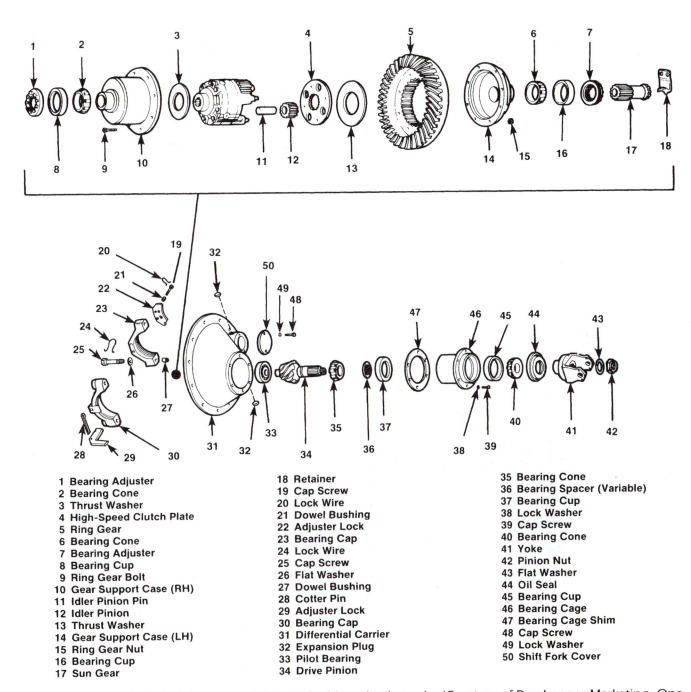

FIGURE 19–21 Exploded view of a planetary double reduction axle. (Courtesy of Roadranger Marketing. One great drive train from two great companies—Eaton and Dana Corporations)

1 Bearing Adjuster	18 Retainer
2 Bearing Cone	19 Cap Screw
3 Thrust Washer	20 Lock Wire
4 High-Speed Clutch Plate	21 Dowel Bushing
5 Ring Gear	22 Adjuster Lock
6 Bearing Cone	23 Bearing Cap
7 Bearing Adjuster	24 Lock Wire
8 Bearing Cup	25 Cap Screw
9 Ring Gear Bolt	26 Flat Washer
10 Gear Support Case (RH)	27 Dowel Bushing
11 Idler Pinion Pin	28 Cotter Pin
12 Idler Pinion	29 Adjuster Lock
13 Thrust Washer	30 Bearing Cap
14 Gear Support Case (LH)	31 Differential Carrier
15 Ring Gear Nut	32 Expansion Plug
16 Bearing Cup	33 Pilot Bearing
17 Sun Gear	34 Drive Pinion

35 Bearing Cone
36 Bearing Spacer (Variable)
37 Bearing Cup
38 Lock Washer
39 Cap Screw
40 Bearing Cone
41 Yoke
42 Pinion Nut
43 Flat Washer
44 Oil Seal
45 Bearing Cup
46 Bearing Cage
47 Bearing Cage Shim
48 Cap Screw
49 Lock Washer
50 Shift Fork Cover

held between two case halves. The side gears connected to the helical gear and to the end of the output shaft are meshed with these pinion gears.

One axle might want to turn faster than the other in two situations. When tires of the two axles are mismatched so that one or more are smaller in diameter than the others, and when there is reduction or loss of traction for one axle such as on slippery or loosely packed road surfaces, while the other axle retains good traction. Since the pinions drive both side gears when turning with the spider and can also rotate on the spider just as in a carrier differential, the two side gears can operate at different speeds when necessary. This enables the interaxle differential to compensate for mismatched tire problems. It also permits **spinout**, where one axle spins on a poor traction surface while the other is immobilized on good pavement. Spinout can cause serious damage to the differential in seconds.

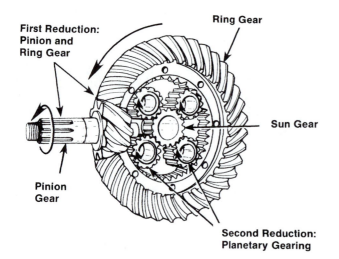

First Reduction:
Pinion and
Ring Gear

Ring Gear

Sun Gear

Pinion
Gear

Second Reduction:
Planetary Gearing

FIGURE 19–22 *Planetary double reduction gearing. (Courtesy of Roadranger Marketing. One great drive train from two great companies—Eaton and Dana Corporations)*

FIGURE 19–23 *Typical tandem drive axle. (Courtesy of Roadranger Marketing. One great drive train from two great companies—Eaton and Dana Corporations)*

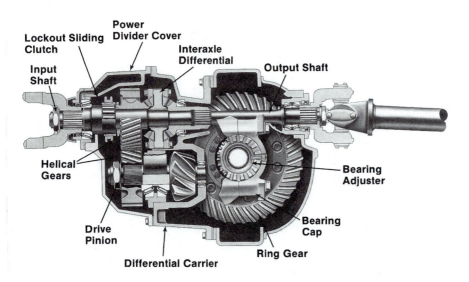

Lockout Sliding
Clutch

Power
Divider Cover

Interaxle
Differential

Output Shaft

Input
Shaft

Helical
Gears

Bearing
Adjuster

Drive
Pinion

Bearing
Cap

Differential Carrier

Ring Gear

FIGURE 19–24
Components of a tandem forward rear drive axle. (Courtesy of Roadranger Marketing. One great drive train from two great companies—Eaton and Dana Corporations)

Drivers should understand the three causes of spinout: backing under a trailer, starting on a slippery surface, and traveling on a slippery surface. In every case of spinout, it is important to remember that the interaxle differential is much more susceptible to spinout than a basic differential. That is because the interaxle differential rotates at much higher prop shaft rpm, which can throw lubrication away from the cross and pinion bearing surfaces and generate intense heat. Before spinout occurs, the driver can lock up the axles using the interaxle lockout mechanism.

Interaxle Differential Lockout

Most power dividers have a lockout mechanism that will prevent the interaxle differential from allowing the front and rear axles to rotate at different speeds. The lockout mechanism is incorporated into the power divider assembly. It enables the truck driver to lock out the interaxle differential to provide maximum traction under adverse road conditions. Lockout should only be engaged when the wheels are not spinning, that is, before traction is actually lost.

The lockout mechanism (**Figure 19–28**) consists of an air-operated lockout unit, a shift fork and pushrod assembly, and a sliding lockout clutch. The driver activates the lockout unit from a switch in the cab. When activated, the lockout unit extends the pushrod and shift fork. The shift fork is engaged with the sliding clutch, and when the fork is extended it forces the clutch against the helical side gear. Because the clutch is splined to the input shaft and

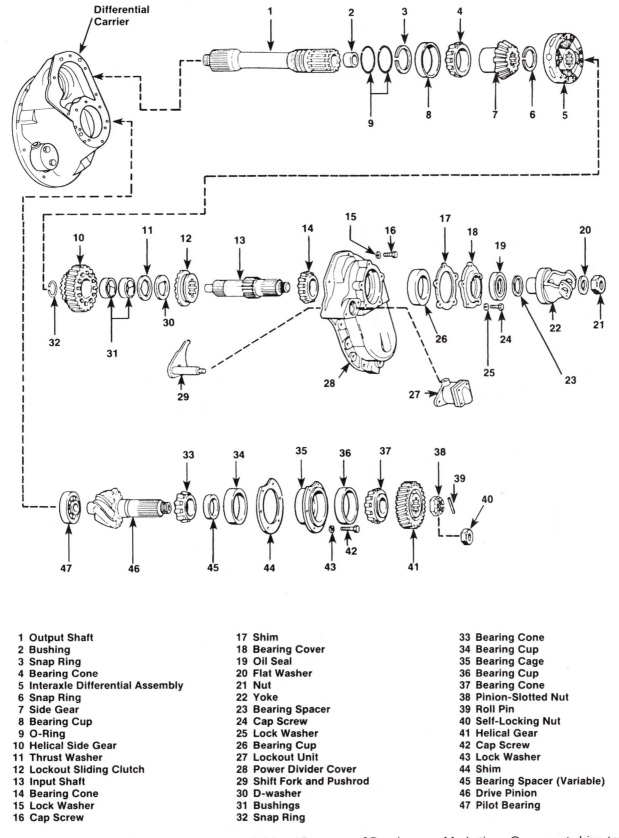

Differential Carrier

1 **Output Shaft**
2 **Bushing**
3 **Snap Ring**
4 **Bearing Cone**
5 **Interaxle Differential Assembly**
6 **Snap Ring**
7 **Side Gear**
8 **Bearing Cup**
9 **O-Ring**
10 **Helical Side Gear**
11 **Thrust Washer**
12 **Lockout Sliding Clutch**
13 **Input Shaft**
14 **Bearing Cone**
15 **Lock Washer**
16 **Cap Screw**

17 **Shim**
18 **Bearing Cover**
19 **Oil Seal**
20 **Flat Washer**
21 **Nut**
22 **Yoke**
23 **Bearing Spacer**
24 **Cap Screw**
25 **Lock Washer**
26 **Bearing Cup**
27 **Lockout Unit**
28 **Power Divider Cover**
29 **Shift Fork and Pushrod**
30 **D-washer**
31 **Bushings**
32 **Snap Ring**

33 **Bearing Cone**
34 **Bearing Cup**
35 **Bearing Cage**
36 **Bearing Cup**
37 **Bearing Cone**
38 **Pinion-Slotted Nut**
39 **Roll Pin**
40 **Self-Locking Nut**
41 **Helical Gear**
42 **Cap Screw**
43 **Lock Washer**
44 **Shim**
45 **Bearing Spacer (Variable)**
46 **Drive Pinion**
47 **Pilot Bearing**

FIGURE 19–25 Exploded view of a power divider. (Courtesy of Roadranger Marketing. One great drive train from two great companies—Eaton and Dana Corporations)

the input shaft is splined to the differential, moving the clutch into engagement with the helical side gear locks the side gear to the differential, preventing any differential action from occurring. Torque is delivered from the input shaft through the meshed clutch and helical side gear to the helical gear on the drive pinion. An equal amount of torque is delivered directly through the locked-up interaxle differential to the opposite side gear and output shaft (**Figure 19–29**).

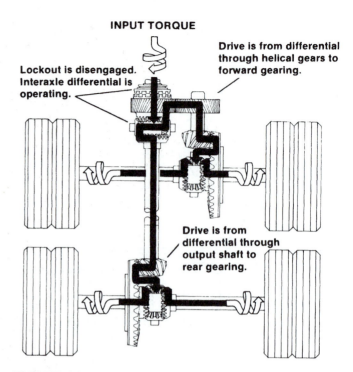

FIGURE 19–27 Torque distribution (power flow) in a tandem rear drive axle when the interaxle differential is operating (lockup disengaged). (Courtesy of Roadranger Marketing. One great drive train from two great companies—Eaton and Dana Corporations)

FIGURE 19–26 Components of an interaxle differential. (Courtesy of Roadranger Marketing. One great drive train from two great companies—Eaton and Dana Corporations)

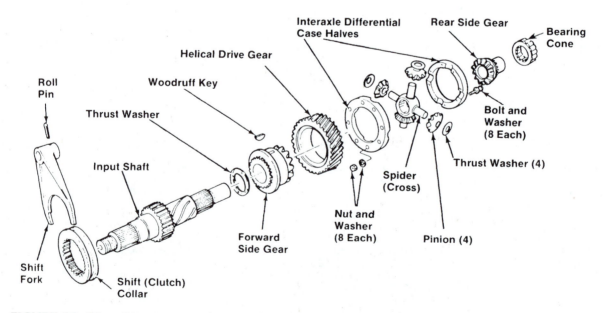

FIGURE 19–28 Components of a power divider showing lockout shift fork and shift clutch or collar. (Courtesy of Roadranger Marketing. One great drive train from two great companies—Eaton and Dana Corporations)

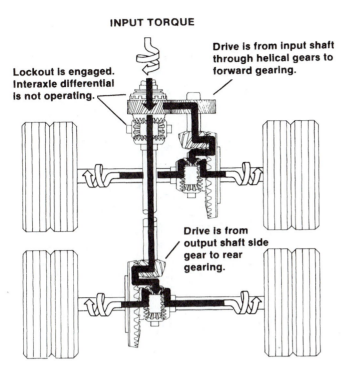

INPUT TORQUE

Lockout is engaged.
Interaxle differential
is not operating.

Drive is from input shaft
through helical gears to
forward gearing.

Drive is from
output shaft side
gear to rear
gearing.

FIGURE 19–29 Torque distribution (power flow) in a tandem rear drive axle when the interaxle differential is not operating (lockup engaged). (Courtesy of Roadranger Marketing. One great drive train from two great companies—Eaton and Dana Corporations)

19.3 LUBRICATION

Most single reduction axles are lubricated by the splashing of oil as the differential components rotate. Baffles and reservoirs are designed into the differential carrier to hold and direct lubricant to critical friction points (**Figure 19–30**).

Other axles—particularly those with a through-shaft type power divider and interaxle differential—use an oil pump to deliver lubricant to bearings, gears, and other lubrication points. The lube pump shown in **Figure 19–31** is operated by a drive gear engaged with the input shaft splines. When the truck is moving in a forward direction, a gear type pump delivers pressurized oil to the interaxle differential

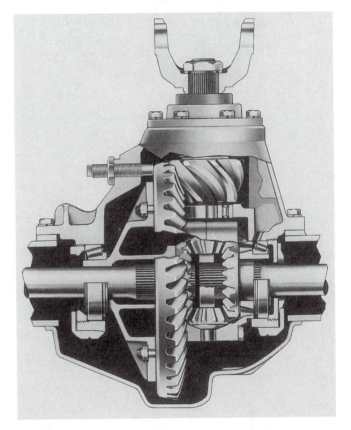

FIGURE 19–30 Power divider with lube pump. (Courtesy of Roadranger Marketing. One great drive train from two great companies—Eaton and Dana Corporations)

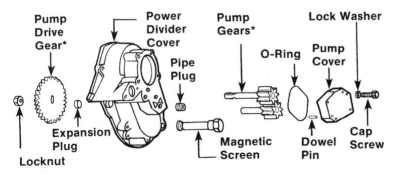

Pump Drive Gear* Power Divider Cover Pump Gears* Lock Washer O-Ring Pump Cover Pipe Plug Cap Screw Dowel Pin Magnetic Screen Expansion Plug Locknut

FIGURE 19–31 Differential lube system. (Courtesy of Roadranger Marketing. One great drive train from two great companies—Eaton and Dana Corporations)

*NOTE: The drive shaft on early pump models is equipped with a woodruff key. On late pump models, the key is eliminated. The drive shaft end has two machined flats and the drive gear mounting hole is shaped to accommodate these flats.

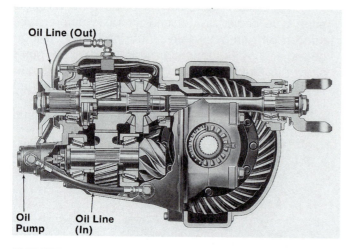

FIGURE 19–32 *Power divider with an externally mounted oil pump. (Courtesy of ArvinMeritor Inc.)*

and other vital power divider components. A magnetic strainer screen keeps the system clean by trapping minute particles with the magnet and by blocking larger particles with the screen.

The drive axle shown in **Figure 19–32** has an externally mounted oil pump, which is directly driven by the drive pinion shaft. Oil is drawn from the bottom of the carrier through an oil line to the pump assembly. The filtered oil is pumped through another external oil line to the top of the carrier where it is routed to the forward bearing and to the interaxle differential. A ball check valve in the outlet line traps oil in the system when the vehicle is stationary.

Some axles have a rotor type pump instead of the pump gears. Some axles have external, spin-on oil filters. Still others have oil passages drilled in the input gear and in the gear cover to deliver the oil to critical lubrication points.

TWO-SPEED AXLES

Drive axles are also available with a shift mechanism that allows the differential to be operated as either a single reduction axle or a double reduction axle. This type of axle is called a two-speed or dual range axle. It is similar to the double reduction axle discussed earlier, with one major difference. A double reduction axle always operates in the double reduction mode. The two-speed axle can be operated in either the single reduction or double reduction mode. Both are similar in design in that they use a planetary gear set as the second reduction gear set. The major difference between them is that the planetary sun gear on the two-speed carrier can be moved into two positions (high and low). The sun gear on a double reduction carrier is always held in one position (low).

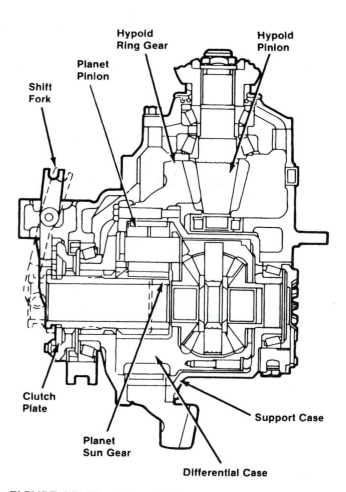

FIGURE 19–33 *Hypoid planetary two-speed drive unit. (Courtesy of ArvinMeritor Inc.)*

A typical two-speed drive unit has a first reduction hypoid pinion and ring gear set and a second reduction four pinion planetary gear set (**Figure 19–33**). Bevel gears are used in the main differential.

The hypoid pinion is supported by three bearings in the carrier. Two tapered roller bearings are mounted on the pinion shaft in front of the gear head. The third bearing is a straight roller or spigot bearing that is mounted on the nose of the pinion.

The main differential and the planetary gear set are mounted in differential case halves, which are contained in the support case. The support case is mounted in the carrier on two tapered roller bearings.

An air or electric shift unit is mounted on the carrier, and it is actuated by a switch in the cab of the vehicle. The switch can be actuated by the driver to shift the drive unit into high or low gear range.

In high gear range, the sun gear is engaged with the support case and the planetary pinions are locked in place. When the planetary pinions are locked, there is no second reduction, and the power is sent directly from the hypoid gear set to the main differential (**Figure 19–34A**). This hypoid single

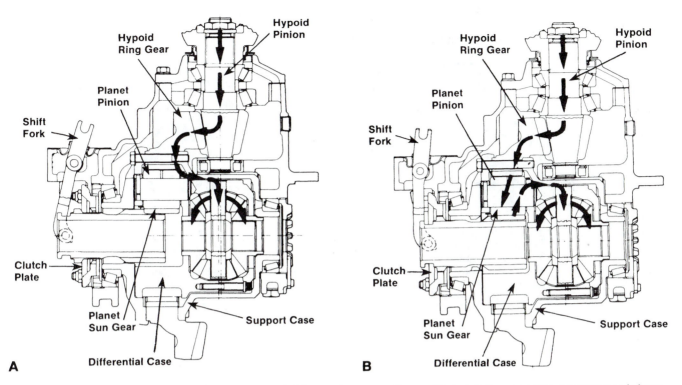

FIGURE 19–34   Hypoid planetary two-speed drive unit power flows: (A) in high range the planetary pinions are locked, there is no second reduction, and power flows directly from the hypoid gear set to the main differential. (B) In low range the sun gear engages the clutch plate, and power flow is through the planetary pinions (second reduction) to the main differential. (Courtesy of ArvinMeritor Inc.)

reduction drive provides a lower numerical ratio for higher road speeds but lower torque.

In low gear range, the sun gear is shifted to engage with the clutch plate. The four planetary pinions are then free to rotate and become the second reduction gear set (**Figure 19–34B**). The hypoid gear set ratio is now multiplied by the planetary gear set ratio. This gives the drive unit a higher numerical ratio for increased torque but decreases the top speed. In some two-speed drive axles, the sun gear is referred to as a sliding clutch gear.

Two-Speed Tandem Axles

Tandem drive axles can also be equipped with two-speed differential gearing on both the front and rear drive axles. In the high range mode, the power flow is through the pinion and ring gear only on both axles. The second reduction planetary gearing is bypassed (**Figure 19–35**).

However, when the low range mode is engaged, the planetary gearing is not bypassed. It generates the second reduction, resulting in increased torque and reduced output speed (**Figure 19–36**).

A two-speed axle doubles the number of drive gears the truck can be operated in. A typical progressive shift pattern might be first gear, low range;

first gear, high range; second gear, low range; second gear, high range; and so on. The two-speed axle gives the truck the high-reduction ratios needed for low speed hauling of heavy loads and low-reduction ratios needed for highway cruising speeds.

THREE-SPEED DIFFERENTIAL

The **three-speed differential** is basically a tandem two-speed axle arrangement with the capability of operating the two drive axles in different speed ranges at the same time. The third speed is actually an intermediate speed between the high and low range. The three speeds are obtained as follows:

1. High speed: both axles are shifted into high range (single reduction)
2. Low speed: both axles are shifted into low range (double reduction)
3. Intermediate speed: the forward axle is shifted into high range (single reduction) and the rear axle is shifted into low range (double reduction). The interaxle differential equalizes the differences in rpm between the two axles to give a speed midway between the high and low ranges.

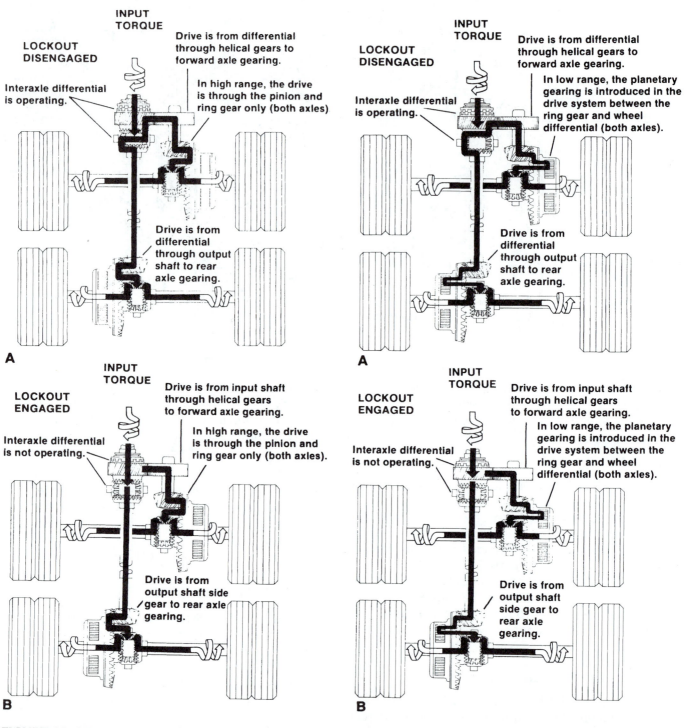

FIGURE 19–35 Torque distribution in the high range mode of a tandem rear axle: (A) lockup disengaged; and (B) lockup engaged. (Courtesy of Roadranger Marketing. One great drive train from two great companies—Eaton and Dana Corporations)

FIGURE 19–36 Torque distribution in the low range mode of a tandem rear axle: (A) lockup disengaged; and (B) lockup engaged. (Courtesy of Roadranger Marketing. One great drive train from two great companies—Eaton and Dana Corporations)

DUAL RANGE AXLE SHIFT SYSTEMS

Although some vehicles are equipped with electrical shift units most axles are equipped with pneumatic shift systems. There are two air-actuated shift system designs used to select the ratio of a dual range tandem axle or to engage a differential lockout:

1. **Standard System.** For range selection, a cab-mounted air shift valve operates two air shift units mounted on the axles. The interaxle differential lockout is also of the straight-air type and air operated to engage lockout and spring released to disengage lockout.

2. **Axle Range Interlock System.** This system has an added feature to prevent axle shifting when the interaxle differential is locked out. The axle range air shift valve for this system includes an interlock pin assembly to provide the interlock feature. The interaxle differential lockout is of the straight-air type. It is equipped with a control valve, which controls air pressure delivery to the shift valve interlock pin. Another lockout design is the reverse-air type interaxle differential lockout. It is spring operated to engage or lock the differential and air operated to disengage or unlock the differential.

Standard System

The standard dual range shift system (**Figure 19–37**) consists of

- A manually operated air shift valve to change axle ratio.
- A quick-release valve which provides for fast release of air pressure from the axle shift units.
- Two air shift units mounted on the axles. These shift units are mechanically connected to the axle shift forks and sliding clutches which, in turn, shift axles into low or high range.

For vehicles not equipped with automatic safety brakes, an ignition circuit-controlled solenoid valve exhausts the system and downshifts the axles when the ignition switch is turned off. The electrical circuit is protected by a circuit breaker.

For vehicles equipped with transmission-driven speedometers, the system must include a speedometer adapter that compensates speedometer readings when the axle is in low range. The adapter is operated by an electrical switch mounted on or near the quick-release valve. The switch is normally closed and is opened by air pressure.

With the axle ratio in low range, the switch is closed and the adapter is energized. The adapter operates with a ratio compatible with the axle low range for proper speed readings. With the axle in

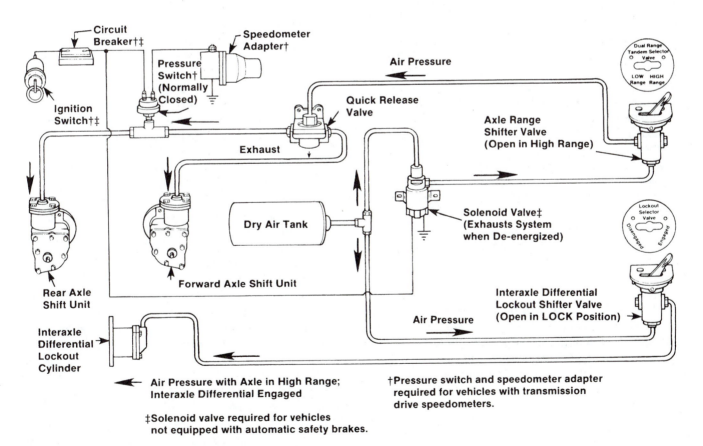

FIGURE 19–37 *Standard system for axle range selection and interaxle differential lockout. (Courtesy of Roadranger Marketing. One great drive train from two great companies—Eaton and Dana Corporations)*

high range, the air lines are pressurized and the pressure switch is open. The adapter now operates with 1:1 ratio for proper speed readings in high range.

Axle Shift System Operation. When the shift valve is moved to the high position, the valve is opened and air pressure is supplied through the quick release valve to both axle shift units. When drive line torque is interrupted, the shift units will shift both axles to the high range.

When the shift valve is moved to the low position, the valve is closed. Air pressure in the shift units is exhausted through the quick release valve. When driveline torque is interrupted, both axles are shifted to the low range and held in this position by the shift unit return springs.

Interaxle Differential Lockout System. The lockout air shift system (**Figure 19–37**) consists of

- A manually operated air shift valve that controls engagement or disengagement of the interaxle differential.
- A lockout cylinder that operates under air pressure. This cylinder is mechanically connected to a shift fork and sliding clutch. The clutch engages or disengages a differential helical side gear to lock or unlock the differential.

The standard lockout unit is air operated to engage lockout and spring released to disengage lockout. The piston is mechanically connected to the shift fork and sliding clutch. The clutch engages or disengages the helical side gear to lock or unlock the interaxle differential.

When the air shift valve lever is moved to the disengaged or unlock position, the valve closes and air pressure in the cylinder is exhausted. Air pressure at the piston is then discharged. Spring pressure moves the piston, shift fork, and sliding clutch. At this time the clutch is disengaged from the helical side gear and the interaxle differential is unlocked and functions normally.

When the air shift valve lever is moved to the engaged or lock position, the valve is opened and supplies air to the lockout cylinder (**Figure 19–38**). Air pressure enters the cover and moves the piston, shift fork, and sliding clutch. At this time the clutch engages the helical side gear and the interaxle differential is locked out, preventing differential actions.

Axle Range Interlock System

The **axle range interlock** feature in the system shown in **Figure 19–39** is designed to prevent axle shifting when the interaxle differential is locked out

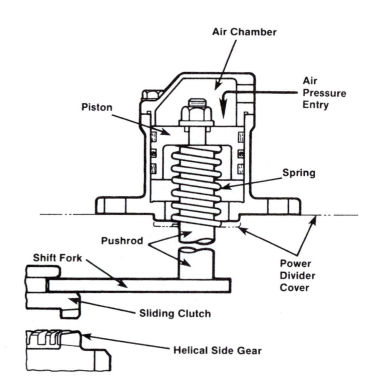

FIGURE 19–38 Air shift valve operation. (Courtesy of Roadranger Marketing. One great drive train from two great companies—Eaton and Dana Corporations)

(or when lockout is engaged). The basic shift system operates the same as the standard shift system to shift the axle and engage or disengage the lockout. However, it varies by adding an interlock pin assembly to the axle range shift valve and an interlock control valve to the lockout cylinder. These two components are interconnected with air lines.

Figure 19–40 illustrates the operation of the axle range lockout. When air is supplied to the lockout cylinder from the lockout shifter valve in the cab, the lockout piston forces the shaft and fork forward so that the sliding clutch engages with the helical side gear (**Figure 19–40A**). Movement of the piston allows the interlock control valve to open. Air pressure is then directed to the interlock pin assembly on the axle range shift valve. The interlock pin moves to block the shift valve lever, preventing the axle range from being changed.

When the differential lock is disengaged, air pressure is exhausted through the control valve port. The drop in air pressure allows the return spring in the lockout cylinder to force the lockout piston back into the cylinder (**Figure 19–40B**). The piston acts against the interlock control valve to exhaust air pressure to the interlock pin assembly. This allows the return spring on the interlock piston and pin to force the pin away from the shift valve lever. The axle range can now be shifted.

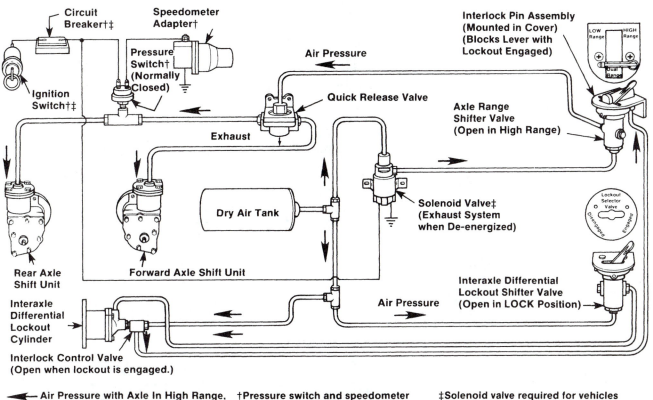

FIGURE 19–39 Axle range interlock system and interaxle differential lockout. (Courtesy of Roadranger Marketing. One great drive train from two great companies—Eaton and Dana Corporations)

AXLE HUB ARRANGEMENTS

Two drive axle hub arrangements are available to provide support between the axle hub and the truck's wheels: the **semi-floating type axle** and the **fully floating type axle**. Of the two, the semi-floating is the simplest design, but the fully floating axle is more popular in heavy-duty trucks.

Semi-Floating Axles

In the semi-floating type axle, drive torque from the differential carrier is delivered by each axle half-shaft directly to the wheels. A single bearing assembly, located at the outer end of the axle, is used to support the axle half-shaft. The part of the axle extending beyond the bearing assembly is either splined or tapered to a wheel hub and brake drum assembly. The main disadvantage of this type of axle is that the outer end of each axle shaft must support the weight of the truck over the wheels. If an axle half-shaft should break, the truck's wheel will fall off.

Fully Floating Axles

In the fully floating type axle, the axle half-shafts transmit only driving torque to the wheels. Typically, the wheels and hubs are mounted on a pair of opposed tapered bearing assemblies located in the axle housing. With this design, the truck weight is supported by the bearings, relieving the axle shafts from any bending stresses. This design also allows the axle half-shafts to be removed without jacking up the truck's wheels. If a fully floating axle shaft should break, there is no danger of the wheel falling off.

19.4 NONDRIVE AXLES

There are two types of non-driving or **dead axles:** steering axles and trailer axles. Steering axles are used on the front of most heavy-duty trucks.

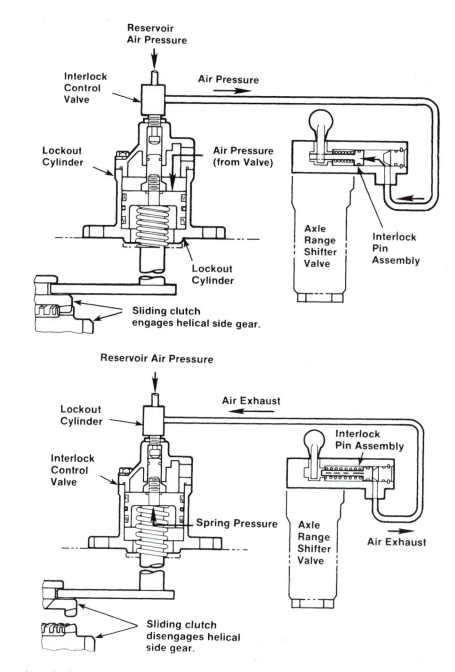

FIGURE 19–40 Axle range interlock operation: (A) lockout engaged; and (B) lockout disengaged. (Courtesy of Roadranger Marketing. One great drive train from two great companies—Eaton and Dana Corporations)

The most common type of steering axle is called an I-beam assembly because it is forged in the shape of an "I" (**Figure 19–41A**). The front suspension is directly mounted to flat machined pads on the ends of the I-beam. The steering mechanism is linked to steering levers, steering knuckles, and kingpins. I-beam steering axles are machined from steel forgings. Two other steering axle designs are illustrated in **Figure 19–41B** and **19–41C**.

The steering knuckle or kingpin can be either of the tapered or straight type. The tapered kingpin can

be drawn into the axle center by tightening a nut at the upper pin end. The straight kingpin type (**Figure 19–42**) can use either one or two flats and is held in the axle center by a tapered draw key. The steering knuckle that supports the kingpin contains bushings at the top and bottom and has grease grooves. Each steering knuckle is mounted to the other through a cross tube or tie-rod, threaded at each end and held in position by clamp bolts. The cross tube has right- and left-hand threads at opposite ends to allow for toe-in adjustments. More information about steering axle components is given in Chapter 21.

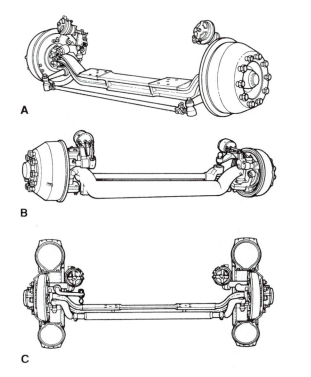

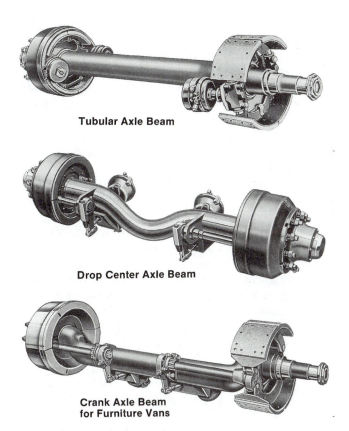

Tubular Axle Beam

Drop Center Axle Beam

Crank Axle Beam for Furniture Vans

FIGURE 19–41 Various steering axle designs: (A) standard I-beam axle; (B) tubular axle; and (C) Center-Point™ steering axle. (Courtesy of Roadranger Marketing. One great drive train from two great companies—Eaton and Dana Corporations)

FIGURE 19–43 Trailer axle configurations. (Courtesy of ArvinMeritor Inc.)

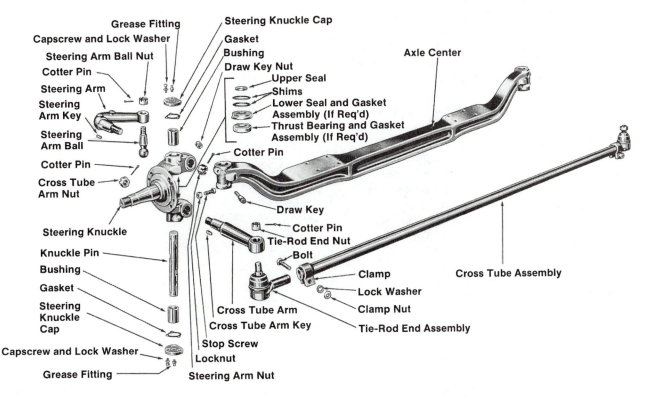

Grease Fitting
Capscrew and Lock Washer
Steering Arm Ball Nut
Cotter Pin
Steering Arm
Steering Arm Key
Steering Arm Ball
Cotter Pin
Cross Tube Arm Nut
Steering Knuckle
Knuckle Pin
Bushing
Gasket
Steering Knuckle Cap
Capscrew and Lock Washer
Grease Fitting

Steering Knuckle Cap
Gasket
Bushing
Draw Key Nut
Upper Seal
Shims
Lower Seal and Gasket Assembly (If Req'd)
Thrust Bearing and Gasket Assembly (If Req'd)
Cotter Pin

Axle Center

Draw Key
Cotter Pin
Tie-Rod End Nut
Bolt
Clamp
Cross Tube Arm
Cross Tube Arm Key
Stop Screw
Locknut
Steering Arm Nut
Lock Washer
Clamp Nut
Tie-Rod End Assembly

Cross Tube Assembly

FIGURE 19–42 Permanently sealed straight knuckle pin design. (Courtesy of ArvinMeritor Inc.)

Trailer axles are used to support the trailer and provide a means to mount the trailer suspension system and foundation braking components. Trailer axles are manufactured with capacities for single axles from 17,000 to 30,000 pounds. Axles are available with round or rectangular beams, straight and drop center designs, and with cam, wedge, or disc brakes in many sizes (**Figure 19–43**). Most brake spiders are welded to the axle beam for stronger brake installation.

Some tractors are also equipped with rear dead axles. These axles provide additional load-carrying capacity. If mounted ahead of a drive axle, they are called pusher axles; if mounted behind a live axle, they are referred to as tag axles.

Dead axles are often mounted in lift suspensions. These hold the axle off the road when the vehicle is traveling unloaded or turning, and lower it when a load is being carried. They are also used as air suspension third axles on heavy straight trucks where low over-axle weights are in effect.

SUMMARY

- There are three main types of axles: drive axles, steering axles, and trailer axles.
- Most single reduction axles are splash lubricated by gear oil as the differential components rotate. Baffles and reservoirs are designed into the differential carrier to hold and direct lubricant to critical lubrication points.
- Drive axles are also available with a shift mechanism that allows the differential to be operated as either a single reduction axle or a double reduction axle.
- Tandem drive axles can be equipped with two-speed differential gearing on both the front and rear drive axles.
- The three-speed differential is basically a tandem two-speed axle arrangement with the capability of operating the two drive axles in different speed ranges at the same time.
- There are basically two types of non-driving (dead) axles: steering axles and trailer axles.

REVIEW QUESTIONS

1. Which of the following is a function of a rear drive axle?
   a. provides gear reduction necessary to drive the truck
   b. redirects the torque from the drive shaft to the drive wheels
   c. carries and supports the weight of the truck and its load
   d. all of the above

2. Which of the following types of axle must allow for a difference in the speed of each of its wheels?
   a. steering axle
   b. drive axle
   c. trailer axle
   d. all of the above

3. Which of the following gear combinations perform the first reduction in a double reduction carrier?
   a. side pinion and side gear
   b. idler pinion and sun gear
   c. pinion gear and ring gear
   d. none of the above

4. Which of the following gear combinations allows for a speed differentiation between wheels on the opposite ends of an axle?
   a. drive pinion gear and ring gear
   b. differential gears and side gear
   c. idler pinions and sun gear
   d. all of the above

5. Which of the following actions permits the drive wheels on either end of an axle to turn at different speeds?
   a. rotation of the ring gear
   b. rotation of differential pinion gears
   c. rotation of the sun gears
   d. rotation of the differential case

6. Which of the pinion and ring gear arrangements place the pinion gear below the centerline of the ring gear?
   a. hypoid
   b. amboid
   c. spiral bevel
   d. spur bevel

7. Which of the following restricts normal differential action, except during cornering?
   a. no-spin locking differentials
   b. two-speed axles
   c. controlled traction differential
   d. double reduction differential

8. Which of the following axles could be equipped with a pneumatic shift mechanism?
   a. two-speed differential
   b. tandem axle power divider
   c. no-spin locking differential
   d. all of the above

9. In which of the following differentials would you find friction plates alternately splined to a driver and tanged to the gear support case?

a. controlled traction differentials
b. no-spin locking differentials
c. two-speed tandem axles
d. all of the above

10. In which of the following would you find a planetary gear arrangement?
a. single reduction differentials
b. trailer axles
c. double reduction differentials
d. none of the above

11. In a tandem axle power divider, which component drives the helical gear set?
a. input shaft
b. drive pinion
c. ring gear
d. forward side gear

12. When the drive pinion is loocated above the centerline of the ring gear, which term is used to describe the gearing?
a. hypoid
b. amboid
c. spiral bevel
d. spur bevel

20

Heavy Duty Truck Axle Service and Repair

Prerequisite
Chapter 19

Objectives

After reading this chapter, you should be able to

- Describe the lubrication requirements of truck and trailer dead axles.
- Outline the lubrication service procedures required for truck drive axle assemblies.
- Perform some basic level troubleshooting on differential carrier gearing.
- Outline the procedure required to disassemble a differential carrier.
- Disassemble a power divider unit.
- Perform failure analysis on power divider and differential carrier components.
- Reassemble power divider and differential carrier assemblies.

Key Terms List

banjo housing
fatigue failures
linehaul

shock failures
spinout failure
torsional failures
tooth contact pattern

Today, differential carrier assemblies tend not to be rebuilt very frequently in the field. The function of the technician is usually to diagnose a malfunction, and, if a major repair is required, remove the differential carrier assembly and replace it with a rebuilt/exchange unit. Servicing of heavy-duty truck axles consists of lubrication, inspection, diagnosis of malfunctions, and, when required, disassembly and reassembly.

It is essential for truck technicians to understand the lubrication requirements of differential carrier assemblies, steering, and trailer axles. A large percentage of all axle failures have their origin in the lubricant and the preventive maintenance practices used in an operation. It must always be remembered that axles support the vehicle and its load, so an axle failure can produce fatal consequences. For instance, the lubrication of the wheel bearings is directly dependent on the oil in the axle differential carrier, so low oil level can result in a vehicle that is dangerous to operate.

This chapter will focus primarily on the lubrication and servicing requirements of heavy-duty truck axles. It should be studied along with Chapters 22 and 23. Chapter 23 deals specifically with the critical adjustments of wheel-end procedure, so wheel bearing lubrication and adjustment is mentioned only in passing in this chapter.

20.1 TRUCK AXLE LUBRICATION

The efficiency and life of mechanical equipment is as dependent upon proper lubrication as it is upon proper engineering design. Mechanical components rely on lubrication to

- Provide a lubrication film between the moving parts to reduce friction
- Help cool components subject to friction
- Keep dirt and wear particles away from mating components

Proper lubrication depends upon using the right type of lubricant, at the proper intervals and maintaining the specified capacities. Experience suggests that many service problems can be traced to improper lubricant or lubrication practices.

The recommended lubrication practices and specifications covered in this section are general in nature and typical of manufacturers' procedures. However, it is advisable to refer to the manufacturers' service manuals for specific instructions.

NEW AND RECONDITIONED AXLE LUBE SERVICE

Drain and flush the factory-fill axle lubricant of a new or reconditioned axle after the first 1,000 miles but never later than 3,000 miles. This is necessary in order to remove fine wear particles generated during break-in that would cause accelerated wear on gears and bearings if not removed. Drain the lubricant while still warm from the carrier housing, and from the interaxle differential of the forward carrier of tandem axles. Avoid mixing lubricants.

 Shop Talk _____

Draining lubricants while warm ensures that contaminants are still suspended and also reduces drain time.

_____

Fill axles to the bottom of the level hole (in carrier or housing) with a specified lubricant while the truck is on level ground. If the axle employs an interaxle differential of the type that can be directly filled through a top filler plug hole (see **Figure 20–1**), pour an additional two pints of the same lubricant into the interaxle differential housing. The angle of the drive pinion, as mounted under the vehicle, determines which oil fill/level plug hole is to be used (**Figure 20–2**).

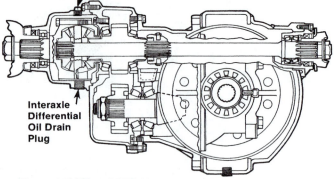

Interaxle Differential Oil Fill Plug

Interaxle Differential Oil Drain Plug

(Current SLHD or SQHD Model Type Shown)

FIGURE 20–1 *Location of oil fill and oil drain plugs on a typical forward rear tandem axle carrier. (Courtesy of ArvinMeritor Inc.)*

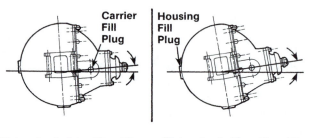

Carrier Fill Plug

Housing Fill Plug

Pinion Angle Less Than 7°— Fill to carrier fill plug hole.

Pinion Angle More Than 7°— Fill to housing fill plug hole.

FIGURE 20–2 *Proper lubricant levels. (Courtesy of ArvinMeritor Inc.)*

Measure the drive pinion angle. If the angle is less than 7 degrees (above horizontal), use the fill hole located in the side of the carrier. If the angle is more than 7 degrees (above horizontal), use the hole located in the **banjo housing**. Note: Some axle models might have only one lube fill hole, located in the banjo housing. With these models use this lube fill hole for all pinion angles. Some axle models have a small tapped and plugged hole located near, but below the housing lubricant level hole. This smaller hole is for the use of a lubricant temperature sensor only and must not be used as a fill or level hole.

After filling the axle housing to the correct level, drive the truck unloaded for 1 or 2 miles to circulate the lubricant throughout the axle and carrier assemblies. Follow the "New and Reconditioned Axle Lube Service" procedures and change lubricant at recommended intervals.

APPROVED LUBRICANTS

All lubricantsused in current differential carrier assemblies must meet the American Petroleum Institute

TABLE 20–1: AXLE GEAR LUBE VISCOSITY	
Ambient Temperature Range	**Proper Grade**
−40°F to −15°F	75W
−15°F to 100°F	80W-90
−15°F and above	80W-140
10°F and above	85W-140

(API)/ Society of Automotive Engineers (SAE) GL-5 standard. API-GL-5 ("GL" stands for gear lube) is the current gear lube classification that is accepted by the manufacturers of heavy duty differential carriers. In other words, gear lubricants classified as GL-1, GL-2, GL-3, GL-4, and GL-6 are either obsolete or no longer approved for use. GL-5 is an extreme pressure gear lubricant that is suitable for use in hypoid gear arrangements. It is consistent with the military specification known as MIL-L-2105-D.

API-GL-5 is available in several viscosities. The viscosity used in an application depends mostly on the expected operating temperatures. **Table 20–1** shows appropriate gear lube grades to use for the operating temperatures.

All the OEMs currently approve of the use of synthetic lubricants meeting the GL-5 performance classification. Synthetics adapt better to operating temperature extremes and can extend service intervals. They have superior centrifugal throw-off resistance, so they also have the potential to reduce the incidence of spinout failures.

OEMS do not generally approve the use of additives in differential carrier gear lube. Additives are most often applied to gear lube to suppress the symptoms of a mechanical condition such as gear whine. Although an additive can work to suppress noise, the severity of the mechanical damage is likely to be greater when the repairs are eventually undertaken.

LUBE CHANGE INTERVAL

After the initial lube change at 1,000 to 3,000 miles, subsequent lube changes should be made at 100,000-mile intervals for **linehaul** operation: that is, terminal to terminal, highway operation. For other types of operations, change the lube at 40,000-mile intervals. If the truck does not accumulate enough mileage to require a lube change, change the lubricant once yearly.

If the lube level falls below its proper level between changes, it should be replenished as needed. If loss is excessive, troubleshoot the problem. Use an API-GL-5 gear lube. To maintain proper viscosity levels, however, do not mix lube grades when adding to an existing supply.

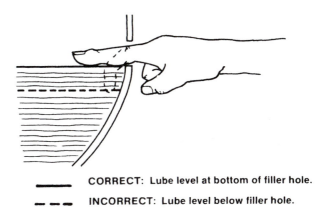

——— **CORRECT:** Lube level at bottom of filler hole.

– – – **INCORRECT:** Lube level below filler hole.

FIGURE 20–3 *Correct and incorrect ways of checking the lube level. (Courtesy of Roadranger Marketing. One great drive train from two great companies—Eaton and Dana Corporations)*

CHECKING LUBE LEVEL

Remove the fill hole plug located in the banjo housing. The lube should be level with the bottom of this hole (**Figure 20–3**). To be seen or touched is not sufficient; it must be level with the hole. When checking the lube level, also check and clean the housing breathers.

CHANGING LUBE

Draining

Drain when the lube is at normal operating temperature. It will run freely and minimize the time necessary to fully drain the axle. Unscrew the magnetic drain plug on the underside of the axle housing (**Figure 20–4**) and allow the lube to drain into a suitable

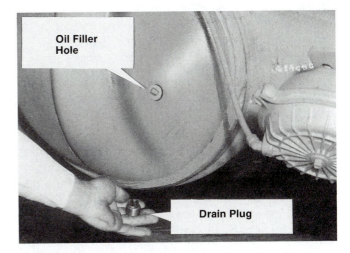

Oil Filler Hole

Drain Plug

FIGURE 20–4 *Removing the drain plug from the axle housing. (Courtesy of Roadranger Marketing. One great drive train from two great companies—Eaton and Dana Corporations)*

container. Inspect the drain plug for metal particles. After the initial oil change, these are signs of damage or extreme wear in the axle, and inspection of the entire unit may be required. Clean the drain plug and replace it after the lube has drained completely.

CAUTION: Be sure to direct compressed air into a safe area. Wear safety glasses.

To drain axles equipped with a lube pump, remove the magnetic strainer from the power divider cover and inspect for wear material in the same manner as the drain plug (**Figure 20–5**). Wash the magnetic strainer in solvent and blow dry with compressed air to remove oil and metal particles.

FIGURE 20–5 Location of magnetic strainer for axles with lube pump. (Courtesy of Roadranger Marketing. One great drive train from two great companies—Eaton and Dana Corporations)

Filling

Replace the drain plug. Remove the filler hole plug from the center of the axle housing cover and fill the axle with an approved lubricant until level with the bottom of the hole. On forward axles, add 2 pints of lubricant through the filler hole at the top of the differential carrier near the power divider cover (**Figure 20–6**).

Lube fill capacities in the various service manuals are good guidelines, but will vary somewhat on the basis of the angle the axle is installed in a particular chassis. Always use the filler hole as the final reference. If lube is level with the bottom of the hole, the axle is properly filled. Axles installed at angles

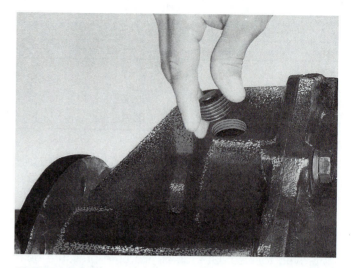

FIGURE 20–6 Location of oil filler hole at top of power divider. (Courtesy of Roadranger Marketing. One great drive train from two great companies— Eaton and Dana Corporations)

exceeding 10 degrees or operated regularly in areas of continuous and lengthy grades may require standpipes to allow proper fill levels.

WHEEL BEARING LUBRICATION

Under normal operating conditions, axle wheel bearings are protected by lubricant carried into the wheel ends by the motion of axle shafts and gearing. Lube becomes trapped in the cavities of the wheel end (**Figure 20–7**) and remains there.

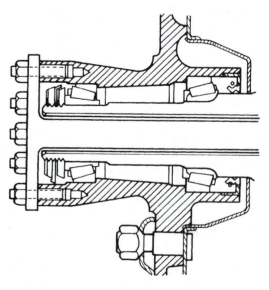

FIGURE 20–7 Location of wheel hub lube cavity. (Courtesy of Roadranger Marketing. One great drive train from two great companies—Eaton and Dana Corporations)

When wheels are being installed, the hub cavities are empty. Bearings must be prelubed with adequate lubricant or they will be severely damaged before the normal motion of gearing and axle shafts can force lube to the hub ends of the housing.

To avoid the risk of premature damage to wheel bearings, they must be prelubed any time wheels are being installed. The correct method will depend on the type of wheel assembly.

Hubs Equipped with Lube Fill Holes

Pour a pint of standard axle lubricant into the hub through the fill hole provided. This type of hub is not common.

Lubrication when Hubs Have No Fill Holes

Follow the procedure outlined here:

1. Fill the axle with lube through the axle housing cover fill hole.
2. Jack up the left side of the axle. Maintain this position for 1 minute to allow lube to flow into the wheel end at the right side.
3. Jack up the right side of the axle. Maintain this position for 1 minute to allow lube to flow into the wheel ends at the left side.
4. Lower the vehicle and add lube through the axle housing cover filler hole. The axle should require an additional 2 pints of lube to bring the level up to the bottom of the filler hole.

20.2 TRUCK AXLE CLEANING

The differential carrier assembly can be pressure washed or steam cleaned while mounted in the housing as long as all openings are tightly plugged. Once removed from its housing, the correct way to clean the assembly is to disassemble it completely. Wash steel parts with ground or polished surfaces in solvent. There are many suitable commercial solvents available that meet current environmental and fire safety standards.

WARNING: Gasoline is not an acceptable solvent because of its extreme combustibility. It is unsafe in the workshop environment.

Wash castings in solvent or clean in hot solution tanks using mild alkaline solution. Dry with compressed air and lightly coat with lubricant to prevent rusting.

20.3 TRUCK AXLE INSPECTION

Inspect steel parts for damage and wear. Look for pitting or cracking along gear contact surfaces. Scuffing, deformation, or discoloration are signs of excessive heat in the axle, usually related to low lubricant levels or improper lubrication practices.

Before reusing a gear set, inspect teeth for signs of excessive wear. Check **tooth contact pattern** for evidence of incorrect adjustment. Inspect all machined surfaces. They must be free of cracks, scoring, and wear. Look for elongation of drilled holes, wear on bearing bores, and nicks or burrs in mating surfaces. Inspect fasteners for rounded heads, bends, cracks, or damaged threads. The axle housing should be examined for cracks or leaks. Also look for loose studs or cross-threaded holes. Inspect machined surfaces for nicks and burrs.

20.4 CARRIER, AXLE, AND GEARING IDENTIFICATION

Identification of the differential carrier by the manufacturer is either stamped on the carrier itself or on a metal tag attached to the carrier (**Figure 20–8**).

The complete axle is usually identified by the specification number stamped on the side of the banjo housing (**Figure 20–9**). This number identifies all component parts of the axle, including special OEM requirements such as yoke or flange. In addition, some axles might include a metal identification tag.

The ring gear and drive pinion are matched components and must be replaced in sets. Check the appropriate manufacturer's axle parts book for part numbers. To identify gear sets, both parts are stamped with such information as the number of pinion and ring gear teeth, individual part number, and matched set number (**Figure 20–10**).

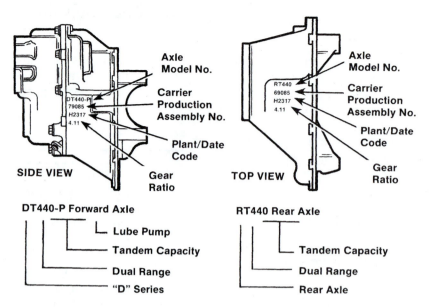

FIGURE 20-8 Differential carrier identification. (Courtesy of Roadranger Marketing. One great drive train from two great companies—Eaton and Dana Corporations)

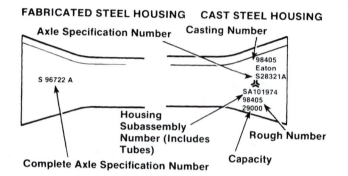

FIGURE 20-9 Axle specification number identification. (Courtesy of Roadranger Marketing. One great drive train from two great companies—Eaton and Dana Corporations)

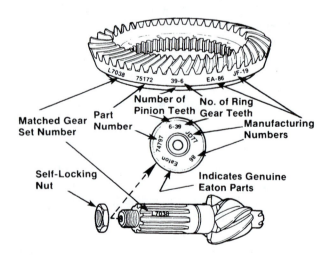

FIGURE 20-10 Ring gear and pinion identification. (Courtesy of Roadranger Marketing. One great drive train from two great companies—Eaton and Dana Corporations)

20.5 DRIVE AXLE FAILURE ANALYSIS/DIAGNOSTICS

When a drive axle fails, failure analysis should answer two key questions: What happened? How can a recurrence be avoided? Sometimes failure analysis is complicated and it takes lab equipment to determine the cause. But most failures can be identified in the shop. This section is designed to identify and analyze a failure, determine the cause and help prevent recurrences. This can help in warranty decisions and prevent other failures.

TYPES OF FAILURES

Drive axle carrier components usually fail because of one or more of the following reasons:

- Shock load
- Fatigue
- Spinout
- Lubrication
- Normal wear

Shock Failures

Shock failures occur when the gear teeth or shaft have been stressed beyond the strength of the material. The failure can be immediate from a sudden shock or it could be a progressive failure after an initial shock cracks the teeth or shaft surface. An immediate failure could be recognized by total failure of the gear teeth that were in mesh at the time of the shock.

Shock impacts subject shafts to torsional or twisting forces. An instantaneous break is usually indicated by a smooth, flat fracture pattern (**Figure 20–11A**) or a rough pattern broken at an angle (**Figure 20–11B**). When the initial shock impact is not great enough to cause an immediate fracture, the tooth or shaft will crack and a progressive failure can occur (**Figure 20–11C**). Common causes of shock impact failures are rough trailer hookups, spinning wheel grabbing on firm road surface, and misuse of interaxle differential lockouts.

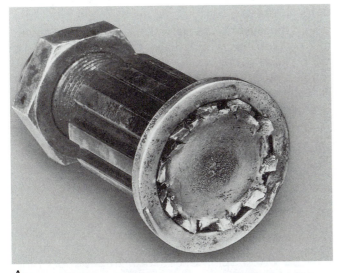

A

B

C

FIGURE 20–11 (A) Smooth and (B) rough fracture patterns; and (C) a fractured tooth as a result of instantaneous shock. (Courtesy of Roadranger Marketing. One great drive train from two great companies—Eaton and Dana Corporations)

Fatigue Failures

Fatigue failures are usually defined as a progressive destruction of shaft or gear teeth material. This type of failure is usually caused by overloading. For longer service life, the surface of gears and shafts is case hardened for wear resistance. A lower hardness ductile core is retained for toughness. Fatigue failures could occur in either or both of these areas. They can be classified in three types:

- **Surface Failures.** Surface failures are identified by cracked or pitted and spalled tooth surfaces and generally caused by repeated light overloading. As the failure progresses, surface material flakes away. In some instances an elliptical design (or "beach" marks) will be created in the fracture pattern. **Figure 20–12** illustrates typical surface failures.
- **Torsional (or Twisting) Failures. Torsional failures** in shafts are usually indicated by a star-shaped fracture pattern (**Figure 20–13**). An extremely high load could cause the shaft surface to crack initially. Then the crack (or cracks) progress to the center of the core. Repeated overloads finally fracture the shaft.

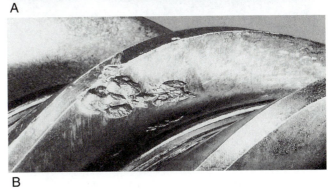

A

B

FIGURE 20–12 Surface failure patterns as a result of fatigue: (A) cracks and spalling; and (B) pitting and spalling. (Courtesy of Roadranger Marketing. One great drive train from two great companies—Eaton and Dana Corporations)

FIGURE 20–13 *Star-shaped torsional failure. (Courtesy of Roadranger Marketing. One great drive train from two great companies—Eaton and Dana Corporations)*

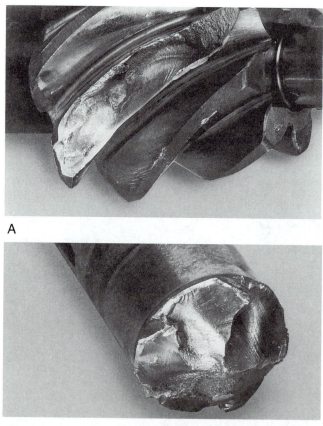

A

B

FIGURE 20–14 *Bending failure patterns as a result of fatigue: (A) beach mark; and (B) spiral pattern. (Courtesy of Roadranger Marketing. One great drive train from two great companies—Eaton and Dana Corporations)*

■ **Bending Failures.** In shafts, rotating and bending forces will usually cause a scalloped (or spiral) type fracture pattern (**Figure 20–14**). Overloads progressively crack the shaft and gradually weaken it until complete fracture occurs.

In gears, bending forces will generally crack the tooth surface. As the crack continues to grow, a "beach" mark design is formed in the fracture pattern (**Figure 20–14**). The tooth is finally weakened to a point where it will break. Fatigue failures are generally caused by overloading the truck beyond the rated capacity or abusive operation of the truck over rough terrain.

Spinout Failures

Spinout or excessive wheel spinning can cause severe damage to the wheel or interaxle differentials. On a single rear axle equipped vehicle main differential, **spinout failure** occurs when one wheel remains stationary (**Figure 20–15A**). On a tandem rear axle, spinout occurs in the interaxle differential when one of the axles remains stationary while the other axle turns (**Figure 20–15B**). Heat is the damaging factor. When a wheel spins because of traction loss, the speed of the differential gears increases greatly. The lube film is thrown off, and metal-to-metal contact occurs, creating friction and heat. If spinout is allowed to continue long enough, the axle can self-destruct.

Other failures could occur during spinout. If a spinning wheel is subjected to grabbing on a firm surface, a shock impact will occur. If shock is severe enough, this could result in broken gear teeth or shaft fracture.

Spinout failures can be prevented or minimized by proper handling of the truck when traction loss is encountered. When operating under a spinout condition, stop the wheels and engage the lockout. If adverse conditions cannot be overcome, use sand, salt, chains, or other procedures to improve traction.

Lubrication

The following types of incorrect lubrication in drive axles greatly affect the life of bearings, gears, and thrust washers.

■ *Contaminated lube* (usually moisture, normal wear, or break-in particles) can cause etching, scoring, or pitting of the contact surfaces. Foreign material in the lube acts as an abrasive and shortens service life.
■ *Inadequate lubrication* could create friction, cause overheating, break down the protective

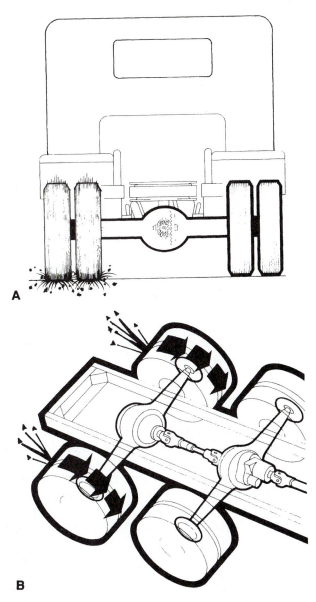

FIGURE 20–15 (A) Spinout on single rear axles; and (B) spinout on tandem rear axles. (Courtesy of ArvinMeritor Inc.)

film, and finally result in seizure or welding of mating components. This can be recognized by severe scoring or galling and actual melting of the material under severe conditions. Components are usually black and discolored.

■ *Improper lube* will not provide the lubricating film that is required to prevent metal-to-metal contact.

■ *Incorrect viscosity lubricant* for the operating temperatures could result in breakdown of the lube film. Mixing incompatible lubricants may damage internal components.

Normal Wear

Drive axles are engineered and manufactured to achieve a reasonable useful life. All components will wear under normal use. During break-in, wear will occur in any mechanical assembly. In many instances, this type of wear is beneficial. Mating components improve their mesh or surface contact. The important factor is to change lube (as recommended by the manufacturer), after break-in to remove any accumulated particles or metallic dust. Preventive maintenance is important to minimize wear.

Some marks and patterns on surface finishes, such as hobbing, forging, and so forth are the result of manufacturing processes. These should be recognized and not classified as a failure condition.

It is important to recognize normal wear to eliminate unnecessary component replacement. The technician should become acquainted with, and be able to identify, normal wear patterns of components. Scoring of surfaces and out-of-round shafts are conditions that will warrant replacement. Washers and seals should always be replaced when the axle is being repaired. The cost of these components is much less than the premature failure of a repair job.

DIAGNOSIS

Problems associated with different models of axles and types of gearing can be specific to one model only. However, one problem area can generally be caused by the same malfunction in most cases. The technician must always consider that universal joints, transmissions, tires, and drivelines can create noises that are often blamed on the drive axles. Typical troubleshooting symptoms and their probable causes are listed in **Table 20–2**.

20.6 DIFFERENTIAL CARRIER OVERHAUL

Overhaul of the differential carrier will vary between different makes and models of the same units. However, the general procedure will follow a similar pattern. The overhaul procedure presented here is based on a single reduction differential carrier. Refer to the manufacturer's service manual for the procedure to overhaul a differential carrier.

REMOVING DIFFERENTIAL CARRIER FROM AXLE HOUSING

To remove the differential carrier from the axle housing, do the following:

TABLE 20–2: DRIVE AXLE TROUBLESHOOTING

Symptom	Probable Cause	Remedy
Noisy on turns only	Differential pinion gears tight on spider	Overhaul drive axle and make necessary adjustments.
	Side gears tight in differential case Differential or side gears defective Excessive backlash between side gears and pinion	
Intermittent noise	Drive gear not running true	Overhaul axle and replace defective drive gear or differential bearings.
	Loose or broken differential bearings	
Constant noise	1. Lubricant incorrect 2. Lube level low 3. Drive gear teeth chipped or worn; loose or worn bearings 4. Drive gear and pinion not in adjustment for correct tooth contact 5. Too much or too little pinion-to-gear backlash or overlap of wear pattern	1. Verify type and class of lubricant used. 2. Check lube level and fill if needed. 3. Overhaul axle and replace defective drive gear, pinion, or bearings. 4. Adjust drive gear and pinion for correct tooth contact. 5. Adjust gear backlash.
Rear wheels do not drive (driveline rotating)	1. Broken axle shaft 2. Drive gear teeth stripped	1. Replace broken axle. 2. Overhaul axle and replace defective drive gear, pinion, or spider.
	Differential pinion or side gear broken Differential spider broken	

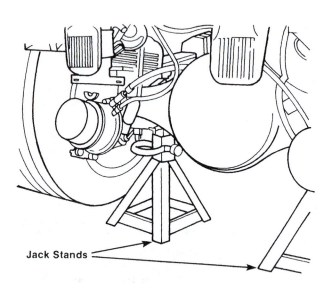

Jack Stands

FIGURE 20–16 *Supporting truck with jack stands under each spring seat. (Courtesy of ArvinMeritor Inc.)*

1. If the truck is equipped with a dual range axle, shift the axle to the low range.
2. Use a jack to raise the end of the truck where the axle is mounted.
3. Put jack stands under each spring seat of the axle to hold the truck in the raised position (**Figure 20–16**).

4. Remove the plug from the bottom of the axle housing and drain the lubricant from the assembly. Dispose of the lube properly.
5. Disconnect the driveline universal joint from the pinion yoke or flange on the carrier.
6. Remove the capscrews and washers or stud nuts and washers from the flanges of both axle shafts.
7. Loosen the tapered dowels in the flanges of both axle shafts. Use a brass drift and a large hammer to loosen the dowels (**Figure 20–17**).
8. Remove the tapered dowels and both axle shafts from the axle assembly.
9. On dual range units disconnect shift unit air lines. Remove shift unit, catching oil that will escape from reservoir.
10. Place a hydraulic roller jack under the differential carrier to support the assembly (**Figure 20–18**).
11. Remove all but the top two carrier-to-housing capscrews or stud nuts and washers.
12. Loosen the top two carrier-to-housing fasteners and leave them attached to the assembly. The fasteners will hold the carrier in the housing.
13. Loosen the differential carrier from the axle housing. Use a leather mallet to hit the mounting flange of the carrier at several points.

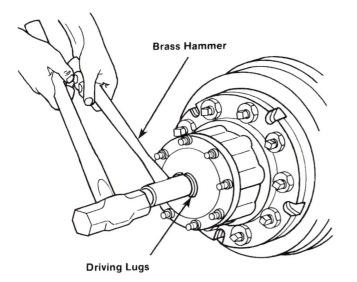

FIGURE 20–17 Loosen the tapered dowels with a brass drift and large hammer. (Courtesy of ArvinMeritor Inc.)

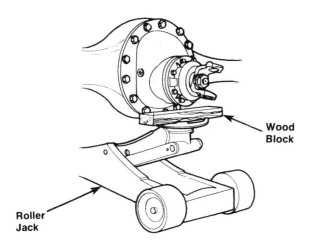

FIGURE 20–18 Support the differential carrier with a hydraulic jack. (Courtesy of ArvinMeritor Inc.)

14. Carefully remove the carrier from the axle housing using the hydraulic roller jack. Use a pry bar that has a round end to help separate the carrier from the housing.

CAUTION: The unit is heavy. Use jacks and hoists.

15. Remove and discard the carrier-to-housing gasket.
16. Lift the differential carrier by the input yoke or flange and put the assembly in a repair stand (**Figure 20–19**). Use a lifting tool or hoist for this procedure.

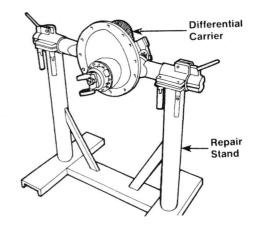

FIGURE 20–19 Place the removed differential carrier in a repair stand. (Courtesy of ArvinMeritor Inc.)

REMOVING DIFFERENTIAL AND RING GEAR FROM CARRIER

Before disassembling the carrier, inspect the hypoid gear for damage. If the inspection shows no damage, the same gear set can be reused. Measure the backlash of the gear set and make a note of the dimension (see adjustment section in this chapter). The backlash must be adjusted to the same dimension after the gear set is reinstalled into the carrier. Best overhaul results are obtained when used gearing is adjusted to run in established wear patterns. Omit this procedure if gear set is to be replaced.

To remove the differential and ring gear from the carrier, do the following:

1. Loosen the jam nut on the thrust screw (if applicable; some carriers do not have a thrust screw).
2. Remove the thrust screw and jam nut from the differential carrier (**Figure 20–20**).
3. Rotate the differential carrier in the repair stand until the ring gear is at the top of the assembly.
4. Mark one carrier leg and bearing cap for the purpose of correctly matching the parts when the carrier is reassembled. A center punch and hammer can be used to mark the parts (**Figure 20–21**). Do not use paint as it may contaminate the unit.
5. Remove the cotter keys, pins, or lock plates that hold the two bearing adjusting rings in position. Use a small drift and hammer to remove the pins. Each lock plate is held in position by two capscrews (**Figure 20–22**).
6. Remove the capscrews and washers that hold the two bearing caps on the carrier. Each cap is held in position by two capscrews and

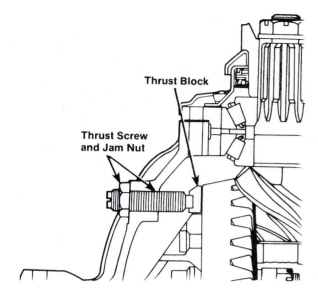

FIGURE 20–20 Thrust screw, jam nut, and thrust block. (Courtesy of ArvinMeritor Inc.)

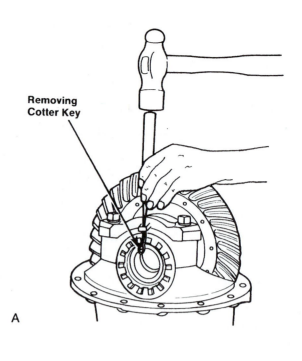

A

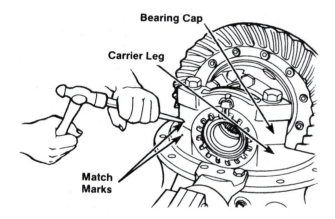

FIGURE 20–21 Marking the carrier parts for reassembly. (Courtesy of ArvinMeritor Inc.)

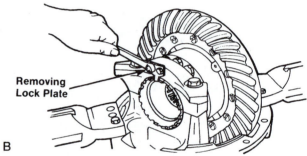

B

FIGURE 20–22 Removing the (A) cotter keys or (B) lock plate from adjusting rings (Courtesy of ArvinMeritor Inc.)

washers. When reusing the gear set, remove the left-hand bearing cap, adjuster, and lock as a unit. This will help return the gear set to its original adjustment during reassembly.

7. Remove the bearing caps and bearing adjusting rings from the carrier (**Figure 20–23**).

8. Safely lift the differential and ring gear assembly from the carrier and put it on a work bench.

9. Remove the thrust block (if provided) from inside the carrier. (The thrust block will fall into the carrier when the thrust screw is removed.)

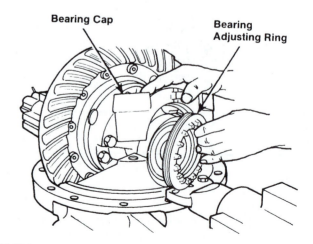

FIGURE 20–23 Removal of the bearing cap and adjusting ring. (Courtesy of ArvinMeritor Inc.)

DISASSEMBLING DIFFERENTIAL AND RING GEAR ASSEMBLY

To disassemble the differential and ring gear assembly, do the following:

1. If the matching marks on the case halves of the differential assembly are not visible, mark each case half with a center punch and hammer. The purpose of the marks is to match the plain half and flange half correctly when reassembling the carrier.
2. Remove all hardware that holds the case halves together.
3. Separate the case halves. Use a brass, plastic, or leather mallet to loosen the parts.
4. Remove the differential spider (cross), four pinion gears, two side gears, and six thrust washers from inside the case halves (**Figure 20–24**).
5. If the ring gear needs to be replaced, remove the hardware that holds the gear to the flange case half. If rivets hold the ring gear to the flange case half, remove them as follows:
   • Carefully center punch each rivet head in the center, on the ring gear side of the assembly.
   • Drill each rivet head on the ring gear side of the assembly to a depth equal to the thickness of one rivet head. Use a drill bit that is $1/32$ inch smaller than the body diameter of the rivets (**Figure 20–25**).

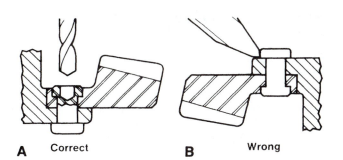

FIGURE 20–25   (A) Drill and punch out rivets. (B) Never use a hammer and chisel to remove rivets. (Courtesy of ArvinMeritor Inc.)

   • Press the rivets through the holes in the ring gear and flange case half. Press from the drilled rivet head.

> **CAUTION:**   Do not remove the rivet heads or rivets with a chisel and hammer because this can damage the flange case half or enlarge the rivet holes, resulting in loose rivets.

6. Separate the case half and ring gear using a press. Support the assembly under the ring gear with metal or wooden blocks and press the case half through the gear (**Figure 20–26**).
7. If the differential bearings need to be replaced, remove the bearing cones from the case halves. Use a bearing puller or press to remove them.

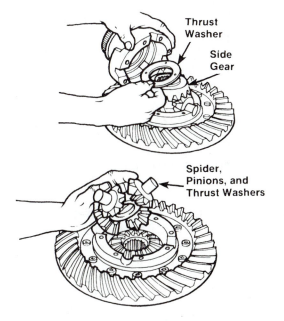

FIGURE 20–24   Disassembling the differential and ring gear. (Courtesy of ArvinMeritor Inc.)

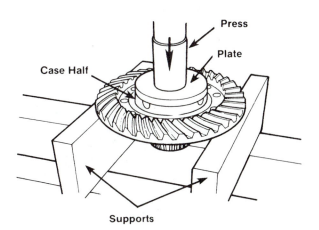

FIGURE 20–26   Pressing the flange case half out of the ring gear. (Courtesy of ArvinMeritor Inc.)

REMOVING DRIVE PINION AND BEARING CAGE

To remove the drive pinion and associated bearing cage from the carrier, do the following:

1. Fasten a yoke or flange bar to the input yoke or flange. The bar will hold the drive pinion in position when the nut is removed (**Figure 20–27**).
2. Remove the hardware from the drive pinion. Then, remove the yoke or flange bar.
3. Remove the yoke or flange from the drive pinion. If the yoke or flange is tight on the pinion, use a puller for removal.
4. Remove the hardware that holds the bearing cage in the carrier (**Figure 20–28**).

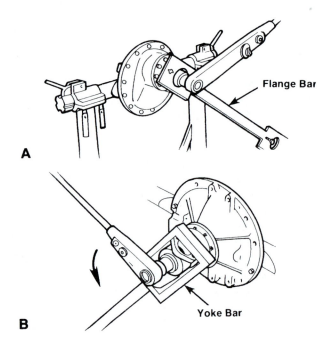

A

B

FIGURE 20–27   Removing the input (A) flange or (B) yoke. (Courtesy of ArvinMeritor Inc.)

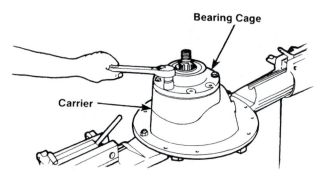

FIGURE 20–28   Bearing cage removal. (Courtesy of ArvinMeritor Inc.)

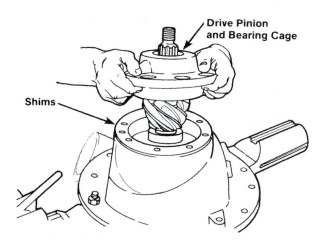

FIGURE 20–29   Removing the drive pinion and bearing cage. (Courtesy of ArvinMeritor Inc.)

5. Remove the cover and seal assembly and the gasket from the bearing cage. If the cover is tight on the bearing cage, use a brass drift and hammer for removal.
6. If the pinion seal is damaged, remove the seal from the cover. Use a press and sleeve or seal driver. If a press is not available, use a screwdriver or small pry bar for removal. Discard the pinion seal.
7. Remove the drive pinion cage and shims from the carrier (**Figure 20–29**).
8. If the shims are in good condition, keep the shims together for use later during reassembly of the carrier.
9. If the shims are to be discarded because of damage, first measure the total thickness of the pack. Make a note of the dimension. The dimension will be needed to calculate the depth of the drive pinion in the carrier when the gear set is installed.

DISASSEMBLING DRIVE PINION AND BEARING CAGE

To disassemble the drive pinion and bearing cage, do the following:

1. Put the drive pinion and bearing cage in a press. The drive shaft must be toward the top of the assembly.
2. Support the bearing cage under the flange area with metal or wooden blocks.
3. Press the drive pinion through the bearing cage (**Figure 20–30**). Do not allow the pinion to fall from the press when the bearing is free.

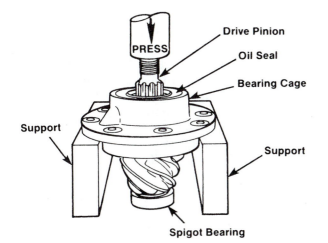

FIGURE 20-30   Pressing the drive pinion from the bearing cage. (Courtesy of ArvinMeritor Inc.)

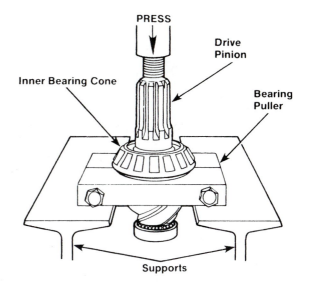

FIGURE 20-31   Pinion bearing removal. (Courtesy of ArvinMeritor Inc.)

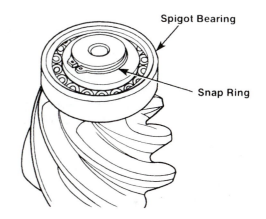

FIGURE 20-32   A snap ring secures the spigot bearing to the pinion shaft. (Courtesy of ArvinMeritor Inc.)

4. If the pinion oil seal is mounted directly in the outer bore of the bearing cage, remove the seal at this time. Be careful that the mounting surfaces of the bearing cage are not damaged.

5. If the seal is a one piece design (without mounting flange), discard the seal. If the seal is a triple-lid design (with flange), inspect the seal for damage. If the surfaces of the seal and yoke or flange *are smooth and not worn or damaged*, it is possible to reuse the seal again during reassembly of the unit. However, replacement is recommended to prevent the cost associated with premature seal failure in the rebuilt unit.

6. If the pinion bearings need to be replaced, remove the inner and outer bearing cups from the inside of the cage. Use a press and sleeve, bearing puller, or a small drift and hammer.

7. If the pinion bearings need to be replaced, remove the inner bearing cone from the drive pinion with a press or bearing puller. The puller must fit under the inner race of the cone to remove the cone correctly without damage (**Figure 20-31**).

8. If the spigot bearing needs to be replaced, put the drive pinion in a vise. Install a soft metal cover over each vise jaw to protect the drive pinion.

9. Remove the snap ring from the end of the drive pinion (**Figure 20-32**).

10. Remove the spigot bearing from the drive pinion with a bearing puller.

ASSEMBLING DRIVE PINION AND BEARING CAGE

To reassemble the drive pinion and bearing cage, do the following:

1. Put the bearing cage in a press. Support the bearing cage with metal or wooden blocks.

2. Press the bearing cup into the bore of the bearing cage until the cup is flat against the bottom of the bore. Use a sleeve of the correct size to install the bearing cup.

3. Put the drive pinion in a press, with the gear head (teeth) toward the bottom.

4. Press the inner bearing cone onto the shaft of the drive pinion until the cone is flat

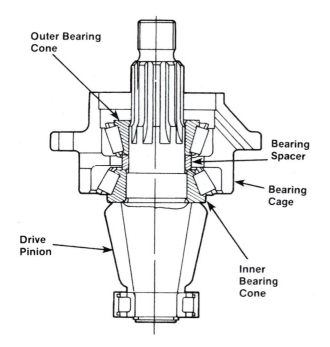

FIGURE 20–33   Drive pinion assembly. (Courtesy of ArvinMeritor Inc.)

against the gear head. Use a sleeve against the bearing inner race if necessary.

5. To install one-piece spigot bearing assemblies, put the drive pinion in a press, gear head teeth toward the top. Press the spigot bearing onto the end of the drive pinion until the bearing is flat against the gear head. Install snap ring to secure bearing. Be sure the snap ring is securely seated.
6. Apply axle lubricant on the bearing cups in the cage and bearing cones.
7. Install the drive pinion into the bearing cage.
8. Install the bearing spacer(s) onto the pinion shaft against the inner bearing cone (**Figure 20–33**).
9. Install the outer bearing cone onto the pinion shaft against the spacer(s).
10. At this time adjust the pinion bearing preload. (Refer to section 20.7, Axle Adjustments, Checks, and Tests.)
11. Once the preload has been set and adjusted, adjust the thickness of the shim pack in the pinion cage. (Refer to section 20.7 of this chapter for adjustment procedure.)

INSTALLING DRIVE PINION AND BEARING CAGE INTO CARRIER

After the drive pinion and bearing cage are assembled and properly adjusted, they can be installed into the carrier. Install the assembly into the carrier as follows:

1. If a new drive pinion and ring gear set is to be installed, or if the depth of the drive pinion has to be adjusted, calculate the thickness of the shim pack. (Refer to section 20.7 of this chapter for the procedures.)
2. Install the correct shim pack between the bearing cage and carrier.
3. Align the oil slots in the shims with the oil slots in the bearing cage and carrier. The use of guide studs will help align the shims (**Figure 20–34**).
4. Install the drive pinion and bearing cage into the carrier. If necessary, use a rubber, plastic, or leather mallet to hit the assembly into position.
5. If used, install the cover and seal assembly and gasket over the bearing cage (**Figure 20–35**).

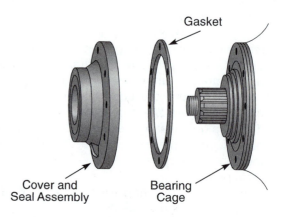

FIGURE 20–34   Shim pack installation. (Courtesy of ArvinMeritor Inc.)

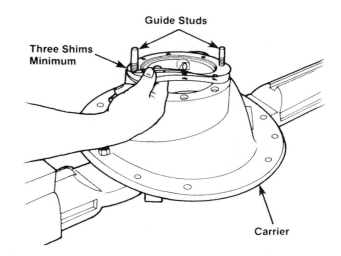

FIGURE 20–35   The cover and seal assembly and gasket should be mounted to the bearing cage.

6. Align the oil slots in the cover and gasket with the oil slot in the bearing cage.

7. Install the bearing cage to the carrier and tighten all hardware to the correct torque value.

8. Install the input yoke or flange onto the drive pinion. The yoke or flange must be against the outer bearing for proper installation.

9. Tighten the pinion nut to the correct torque value. Use the yoke or flange bar during the torquing operation.

ASSEMBLING DIFFERENTIAL AND RING GEAR

Shop Talk

During assembly, do not press a cold ring gear onto the flange case half. A cold ring gear will damage the case half because of the tight fit. Metal particles between the parts will cause gear runout that exceeds specifications.

To assemble the differential and ring gear, do the following:

1. Expand the ring gear by heating it in a tank of water to a temperature of 160–180°F for 10–15 minutes. Do not use a torch for this operation.

2. Safely lift the ring gear onto the flange case half immediately after the gear is heated. If the ring gear does not fit easily on the case half, reheat the gear.

3. Align the fastener holes of the ring gear and the flange case half. Rotate the ring gear as needed.

4. Install the proper hardware that holds the ring gear to the flange case half. Install the bolts from the gear side of the assembly. The bolt heads must be against the ring gear (**Figure 20–36**).

5. Tighten the bolts to the correct torque value. If rivets are used to hold the ring gear to the flange case half, install the correct size rivets in pairs opposite each other from the case half side of the assembly. The rivet heads must be against the flange case half (**Figure 20–37A**). Press the rivets into position from the ring gear side of the assembly using a riveter machine and the manufacturer's specified pressure. Hold riveting pressure for at least 1 minute and check for gaps between the back surface of the ring gear and the case flange using a 0.003 inch thickness gauge (**Figure 20–37B**).

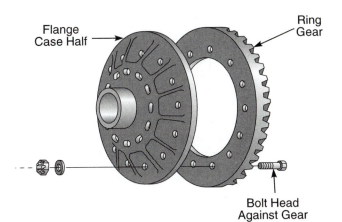

FIGURE 20–36   Ring gear installation to flange case half.

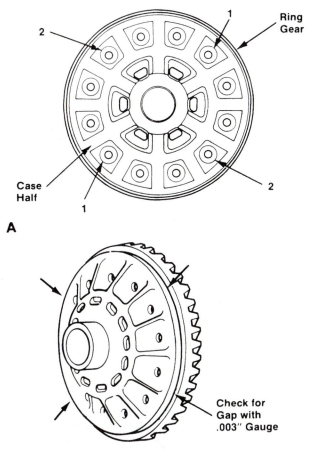

FIGURE 20–37   (A) Install rivets in pairs; (B) check for caps between the rear gear and case flange. (Courtesy of ArvinMeritor Inc.)

6. Install the bearing cones on both of the case halves (**Figure 20–38**). Use a press and sleeve if necessary to install the cones.

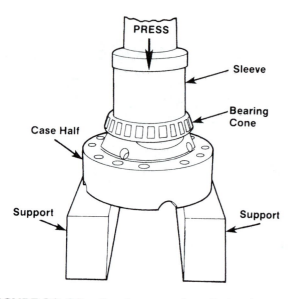

FIGURE 20–38 Bearing cone installation in case half. (Courtesy of ArvinMeritor Inc.)

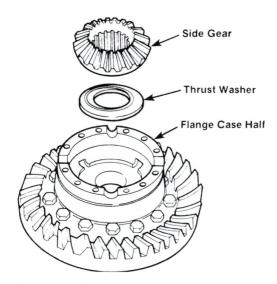

FIGURE 20–39 Side gear installation. (Courtesy of ArvinMeritor Inc.)

7. Apply axle lubricant on the inside surfaces of both case halves, spider, thrust washers, side gears, and differential pinions.
8. Put the flange case half on a bench, with the ring gear teeth toward the top.
9. Install one thrust washer and side gear into the flange case half (**Figure 20–39**).
10. Install the spider, differential pinions, and thrust washers into the flange case half (**Figure 20–40**).
11. Install the second side gear and thrust washer over the spider and differential pinions (**Figure 20–40**).

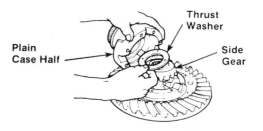

FIGURE 20–40 Spider, differential pinion, and side gear installation. (Courtesy of ArvinMeritor Inc.)

12. Put the plain half of the differential case over the flange half and gears. Rotate the plain half as needed to align the match marks.
13. Install hardware into the case halves. The distance between fasteners must be equal. Tighten the fasteners to the correct torque value in a pattern opposite each other.
14. Check the differential rolling resistance of the differential gears. (Refer to section 20.7 of this chapter for the procedure.)

INSTALLING DIFFERENTIAL AND RING GEAR INTO CARRIER

To install the assembled differential and ring gear set into the carrier, do the following:

1. Clean and dry the bearing cups and bores of the carrier legs and bearing caps.
2. Apply axle lubricant on the inner diameter of the bearing cups and onto both bearing cones that are assembled on the case halves.
3. Apply a suitable adhesive in the bearing bores of the carrier legs and bearing caps.
4. Install the bearing cups over the bearing cones that are assembled on the case halves.
5. Safely lift the differential and ring gear assembly and install into the carrier.
6. Install both of the bearing adjusting rings into position between the carrier legs (**Figure 20–41**). Turn each adjusting ring hand-tight against the bearing cup.
7. Install the bearing caps over the bearings and adjusting rings in the correct location as marked before removal (**Figure 20–42**).
8. Tap each bearing cap into position with a light leather, plastic, or rubber mallet. The caps must fit easily against the bearings, adjusting rings, and carrier.
9. Install the hardware that holds the bearing caps to the carrier. Tighten all hardware by

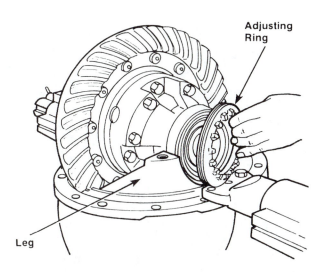

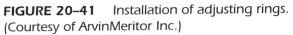

FIGURE 20–41 *Installation of adjusting rings.* (Courtesy of ArvinMeritor Inc.)

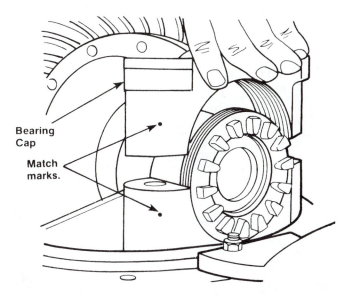

FIGURE 20–42 *Installation of bearing caps using match marks. (Courtesy of ArvinMeritor Inc.)*

hand first and then torque to the correct values.

10. Do not install the cotter keys or lock plates that hold the bearing adjusting rings in position. Continue the overhaul procedure by performing the following checks and/or adjustments:
   a. Adjust preload of differential bearing.
   b. Check runout of ring gear.
   c. Adjust backlash of ring gear.
   d. Check and adjust tooth contact pattern.
   e. Adjust the thrust screw.

These checks and adjustments are described in section 20.7 of this chapter.

INSTALLING DIFFERENTIAL CARRIER INTO AXLE HOUSING

To install the assembled differential carrier into the axle housing, do the following:

1. Clean the inside of the axle housing and the carrier mounting flange. Using a cleaning solvent and clean shop cloths to remove dirt and foreign matter. Blow dry the cleaned areas with air. Wear appropriate eye protection.
2. Inspect the axle housing for damage. Repair or replace if necessary.
3. Check for loose studs in the mounting surface of the housing where the carrier fastens. Remove and clean the studs that are loose.
4. Apply liquid adhesive to the threaded holes and install the studs into the axle housing. Tighten the studs to the correct torque value.
5. Apply silicone gasket material to the mounting of the housing where the carrier fastens (**Figure 20–43**).
6. Install hardware in the four corner locations around the carrier and axle housing. Tighten the fasteners hand-tight at this time.
7. Carefully push the carrier into position. Tighten the four fasteners two or three turns each in a pattern opposite each other.
8. Repeat step 7 until the four fasteners are tightened to the correct torque value.
9. Install the other fasteners that hold the carrier in the axle housing. Tighten the fasteners to the correct torque value.
10. Connect the driveline universal joint to the pinion input yoke or flange on the carrier.
11. Install the gaskets and axle shafts into the axle housing and carrier (**Figure 20–44**). The

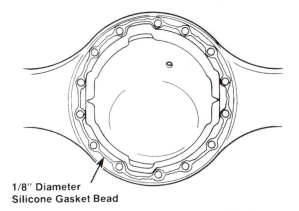

FIGURE 20–43 *Application of silicone gasket material to the mounting surface of the axle housing. (Courtesy of ArvinMeritor Inc.)*

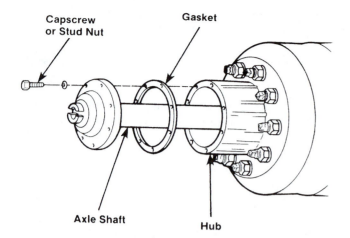

FIGURE 20-44   *Installing the gaskets and axle shafts into the axle housing and carrier. (Courtesy of ArvinMeritor Inc.)*

FIGURE 20-45   *Attaching power divider with chain hoist and sling. (Courtesy of Roadranger Marketing. One great drive train from two great companies—Eaton and Dana Corporations)*

gasket and flange of the axle shafts must fit flat against the wheel hub.

12. Install the hardware that fastens the axle shafts to the wheel hubs. Tighten to the correct torque value.

13. If the wheel hubs have studs, install the tapered dowels at each stud and into the flange of the axle shaft. Use a punch or drift and hammer if needed. Install the hardware on the studs and tighten to the correct torque value.

POWER DIVIDER REPLACEMENT

The power divider can be replaced with the axle assembly in or out of the truck chassis and with the differential carrier in place in the banjo housing. During removal and installation, the power divider must be supported as a safety precaution. Exercise caution to prevent injury or damage.

If the differential carrier is out of chassis, use a chain hoist to remove the power divider. Fasten the chain to the input yoke or flange. If the differential carrier is in chassis, use a transmission jack or a chain hoist and sling to remove the power divider. Wrap the sling strap around the power divider and attach it to the chain hoist hook (**Figure 20-45**).

To replace a power divider, do the following:

1. Disconnect the drive shaft from the differential carrier.
2. Loosen, but do not remove, the input yoke or flange nut.
3. Disconnect the lockout air line.
4. Position a drain pan under the power divider cover.

5. To remove the power divider assembly, remove cover capscrews and lock washers. Support the power divider with a suitable jack or chain hoist.
6. Tap the back face of the input yoke or flange to separate the cover from the differential carrier. If the cover does not separate easily, tap the sides of the cover near the dowel pin locations (**Figure 20-46**).
7. Drain lube from the differential.

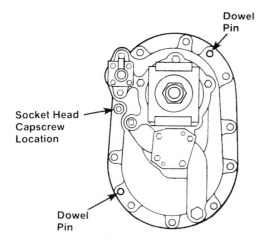

FIGURE 20–46 Power divider cover dowel pin locations. (Courtesy of Roadranger Marketing. One great drive train from two great companies—Eaton and Dana Corporations)

FIGURE 20–47 Installation of the shaft assembly into the differential carrier. (Courtesy of Roadranger Marketing. One great drive train from two great companies—Eaton and Dana Corporations)

FIGURE 20–48 Installation of the interaxle differential. (Courtesy of Roadranger Marketing. One great drive train from two great companies—Eaton and Dana Corporations)

8. Pull the power divider assembly forward until it is completely free of the carrier.

9. With the power divider removed, the interaxle differential can be lifted off the output shaft side gear.

10. If necessary, remove the output shaft by disconnecting the interaxle driveline. Remove the nut and output shaft yoke. Pull the output shaft assembly out of the carrier.

11. After repairing and/or replacing all defective components, lubricate them before replacing the power divider into the carrier.

12. If the output shaft was removed, lubricate the O-rings and then install the shaft assembly into the differential carrier and housing cover (**Figure 20–47**). Be sure to lubricate the seal lip.

13. Make sure the yoke or flange is clean and dry. Then, install the yoke or flange self-locking nut. Torque the nut to the correct torque value.

14. Install the interaxle differential on the output shaft side gear (with the nuts facing away from the side gear) (**Figure 20–48**).

15. Use a silicone rubber gasket compound on the differential carrier mating surfaces.

16. Make certain that the dowel pins are installed in the carrier. Then, install the power divider to the carrier. Use a transmission jack or chain hoist and sling to support the power divider.

17. During installation, rotate the input shaft to engage the input shaft splines with the interaxle differential. After installation, again rotate the input shaft to check for correct assembly. The output shaft should turn when the input shaft is rotated.

18. Install the power divider cover capscrews and lock washers, torquing capscrews to the correct torque value.

19. Check and adjust the input shaft end play. With the power divider assembled to the differential carrier, check end play with a dial indicator. End play typically should be 0.003–0.007 inch, but check the service manual. If necessary, adjust the end play.

20. After the input shaft end play is within specifications, complete the assembly procedures by connecting the driveline to the differential.
21. Connect the lockout air lines.
22. Fill the axle to the proper lube level.

End Play Adjustment

Input shaft end play requirements will vary with operating conditions, mileage, and rebuild procedures. To measure and adjust end play, do the following:

1. With the power divider assembled to the differential carrier, measure end play with a dial indicator positioned at the yoke end of the input shaft (**Figure 20–49**).
2. Move the input shaft axially and measure the end play. A new power divider or a used unit rebuilt with new parts should read an end play of 0.003–0.007 inch. A rebuild job with reused parts should read from 0.013–0.017 inch.
3. If the end play is not correct, remove the input shaft nut, flat washer, and yoke. Remove the bearing cover capscrews and lock washers. Then remove the cover and shim pack.
4. To increase end play, add shims to the shim pack. To decrease end play, remove shims from the shim pack.
5. To reassemble the input shaft, install the adjusted shim pack and bearing cover. Install capscrews and lock washers and torque to correct value.

FIGURE 20–49 *Measuring end play with a dial indicator. (Courtesy of Roadranger Marketing. One great drive train from two great companies—Eaton and Dana Corporations)*

6. Install the yoke, flat washer, and nut. Tighten nut snugly. Tap the end of the input shaft lightly to seat the bearings.
7. Measure the input shaft end play again with the dial indicator. If end play is still incorrect, repeat steps 3–7.
8. With the end play correct, seal the shim pack to prevent lube leakage. Then, torque the input shaft nut and cover capscrews to the correct value.

20.7 AXLE ADJUSTMENTS, CHECKS, AND TESTS

This section of the chapter introduces the axle adjustments, checks, and tests that the truck technician must be capable of performing. For the most part, the procedures described here are general in nature. The truck technician should refer to the manufacturer's service manuals for specific procedures.

PINION BEARING PRELOAD ADJUSTMENTS

Most late-model axles are provided with a press fit outer bearing on the drive pinion gear. Some of the early model axles use an outer bearing, which slips over the drive pinion. The procedures for adjusting both types are given in the following paragraphs.

Press Method Adjustment

To adjust the pinion bearing preload using the press-fit method, do the following:

1. Assemble the pinion bearing cage, bearings, spacer, and spacer washer (without drive pinion or oil seal). Center the bearing spacer and spacer washer between the two bearing cones (**Figure 20–50**).
2. When a new gear set or pinion bearings are used, select a nominal size spacer based on manufacturer's specifications. If original parts are used, use a spacer removed during disassembly of the drive.
3. Put the drive pinion and cage assembly in a press, with the gear teeth toward the bottom.
4. Apply and hold the press load to the pinion bearing. As pressure is applied, rotate the bearing cage several times so that the bearings make normal contact.
5. While pressure is held against the assembly, wind a cord around the bearing cage several times.

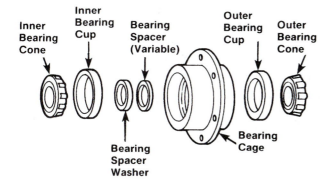

FIGURE 20–50 *Assembly of the pinion bearing cage. (Courtesy of Roadranger Marketing. One great drive train from two great companies—Eaton and Dana Corporations)*

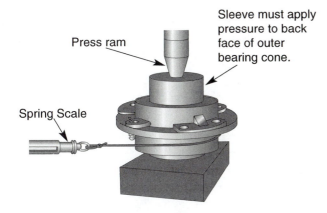

FIGURE 20–51 *Cage in press to check bearing preload.*

6. Attach a spring scale to the end of the cord (**Figure 20–51**). Pull the cord with the scale on a horizontal line.
7. As the bearing cage rotates, read the value indicated on the scale.
8. Preload is normally specified as torque required to rotate the pinion bearing cage, so only take a value reading when the cage is rotating. Starting torque will give a false reading.
9. To calculate the preload torque, measure the diameter of the bearing cage where the cord was wound. Divide this dimension in half to get the radius.
10. Use the following procedure to calculate the bearing preload torque:
    Pull (lb.) × radius (inches) = preload (in.-lb.) or
    Preload (in.-lb.) × 0.113 (a conversion constant) = preload (N•m.)
    Pull (kg) × radius (cm) = preload (kg-cm) or
    Preload (kg-cm) × 0.098 (a conversion constant) = preload (N•m.)

For example:
7.5 lb. × 3.31 in. = 24.8 in.-lb. (preload) or
24.8 in.-lb. × 0.113 = 2.8 N•m. (preload)
3.4 kg × 8.4 cm = 28.6 kg-cm (preload) or
28.6 kg-cm × 0.098 = 2.8 N•m. (preload)

11. If necessary, adjust the pinion bearing preload by changing the pinion bearing spacer. A thicker spacer will decrease preload while a thinner spacer will increase the preload.
12. Once the correct bearing preload has been established, note the spacer size used. Select a spacer 0.001 inch larger for use in the final pinion bearing cage assembly procedures. The larger spacer compensates for slight growth in the bearing, which occurs when pressed on the pinion shank. The trial spacer pack will result in proper pinion bearing preload in three of four cases.

Yoke or Flange Method of Adjustment

To adjust the pinion bearing preload using the yoke or flange method, proceed as follows:

1. Assemble the complete pinion bearing cage as recommended in the press method.
2. A forward axle pinion is equipped with a helical gear. For easier disassembly during bearing adjustment procedures, use a dummy yoke (if available) in lieu of the helical gear.
3. Install the input yoke or flange, nut, and washer on the drive pinion. The yoke or flange must be against the outer bearing. If the fit between the yoke or flange splines and drive pinion splines are tight, use a press to install the yoke or flange (**Figure 20–52**).

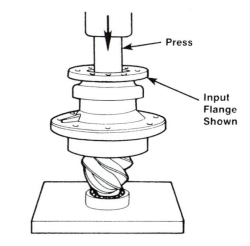

FIGURE 20–52 *Using a press to install the yoke or flange to the drive pinion. (Courtesy of ArvinMeritor Inc.)*

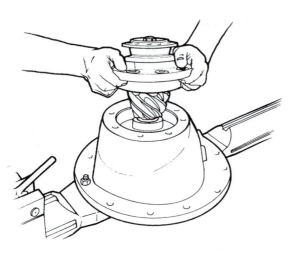

FIGURE 20–53 *Install the pinion and cage assembly in the carrier housing. (Courtesy of ArvinMeritor Inc.)*

4. Temporarily install the drive pinion and cage assembly in the carrier (**Figure 20–53**). Do not install shims under the bearing cage.
5. Install the bearing cage to the carrier capscrews. Washers are not required at this time. Tighten the capscrews hand-tight.
6. Fasten a yoke or flange bar to the yoke or flange (**Figure 20–54**). The bar will hold the drive pinion in position when the nut is tightened.
7. Tighten the nut on the drive pinion to 10–20 in.-lb.
8. Remove the yoke or flange bar.
9. Attach a torque wrench to the drive pinion nut. Rotate the drive pinion and read the value indicated on the torque wrench. Preload is correct when the torque required to rotate the pinion bearing cage is from 15–35 in.-lb.
10. To adjust the pinion bearing preload, disassemble the pinion bearing cage and change

the pinion bearing spacer size. A thicker spacer will decrease preload while a thinner spacer will increase preload.

DIFFERENTIAL ROLLING RESISTANCE CHECK

A check to measure and establish the differential's resistance to rotation is given in the following procedure. To perform this check, a special tool must be made. The tool can be made from an old axle shaft that matches the spline size of the differential side gear. **Figure 20–55** illustrates the fabrication details for this special tool.

To check the differential's resistance to rotation, do the following:

1. Install soft metal covers over the vise jaws to protect the ring gear (**Figure 20–56**).
2. Put the differential and ring gear assembly in the vise.

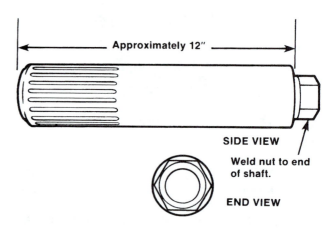

FIGURE 20–55 *Fabrication details for a special tool to check the rolling resistance. (Courtesy of ArvinMeritor Inc.)*

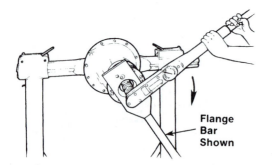

FIGURE 20–54 *Using a flange bar to hold the drive pinion in position. (Courtesy of ArvinMeritor Inc.)*

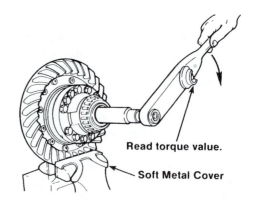

FIGURE 20–56 *Reading the torque value to check the rolling resistance. (Courtesy of ArvinMeritor Inc.)*

3. Install the special tool into the differential until the splines of the tool and one side gear are engaged.

4. Attach a torque wrench to the nut of the special tool and rotate the differential gears. As the differential gears rotate, read the value indicated on the torque wrench (**Figure 20–56**). Typical value is 50 pound-feet maximum applied to one side gear.

5. If the torque value exceeds the specification, disassemble the differential gears from the case halves.

6. Check the case halves, spider, gears, and thrust washers for the problem that caused the torque value to exceed specifications. Repair or replace defective parts as required. Remove any foreign debris.

CHECKING AND ADJUSTING SHIM PACK FOR THE PINION CAGE

This procedure checks and allows for the adjustment of the thickness of the shim pack used in the pinion bearing cage. Use this procedure if a new drive pinion and ring gear set is to be installed, or if the depth of the drive pinion has to be adjusted (**Figure 20–57**).

To check/adjust the shim pack thickness, do the following:

1. Measure the thickness of the old shim pack that was removed from under the pinion cage with a micrometer (**Figure 20–58**). Record the measurement for later use.

2. Look at the pinion cone (PC) variation number on the old drive pinion that is being replaced (**Figure 20–59**). Record this number also for later use.

3. If the old pinion cone number is a plus (+), subtract the number from the old shim pack thickness that was recorded in step 1.

4. If the old pinion cone number is a minus (−), add the number to the old shim thickness that was measured in step 1.

5. The value calculated in step 3 or 4 is the thickness of the standard shim pack, without variation.

6. Look at the pinion cone (PC) variation number on the new drive pinion that will be installed. Record the number for later use.

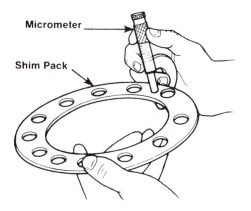

FIGURE 20–58 Measuring the thickness of the old shim pack. (Courtesy of ArvinMeritor Inc.)

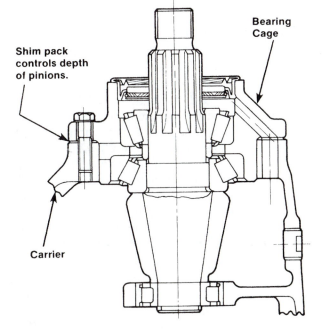

FIGURE 20–57 Drive pinion depth controlled by shim pack thickness. (Courtesy of ArvinMeritor Inc.)

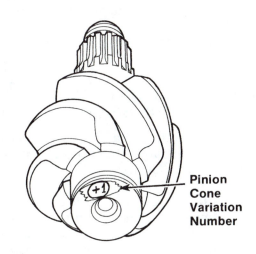

FIGURE 20–59 Location of the pinion cone (PC) variation number. (Courtesy of ArvinMeritor Inc.)

EXAMPLES:

		Inches	mm
1.	Old Shim Pack Thickness	.030	.76
	Old PC Number, PC + 2	− .002	− .05
	Standard Shim Pack Thickness	.028	.71
	New PC Number, PC + 5	+ .005	+ .13
	New Shim Pack Thickness	.033	.84
2.	Old Shim Pack Thickness	.030	.76
	Old PC Number, PC − 2	+ .002	+ .05
	Standard Shim Pack Thickness	.032	.81
	New PC Number, PC + 5	+ .005	+ .13
	New Shim Pack Thickness	.037	.94
3.	Old Shim Pack Thickness	.030	.76
	Old PC Number, PC + 2	− .002	− .05
	Standard Shim Pack Thickness	.028	.71
	New PC Number, PC − 5	− .005	− .13
	New Shim Pack Thickness	.023	.58
4.	Old Shim Pack Thickness	.030	.76
	Old PC Number, PC − 2	+ .002	+ .05
	Standard Shim Pack Thickness	.032	.81
	New PC Number, PC − 5	− .005	− .13
	New Shim Pack Thickness	.027	.68

FIGURE 20–60   Examples of determining shim pack thickness. (Courtesy of ArvinMeritor Inc.)

7. If the new pinion cone number is a plus (+), add the number to the standard shim pack thickness that was calculated in step 3 or 4.

8. If the new pinion cone number is a minus (−), subtract the number from the standard shim pack thickness that was calculated in step 3 or 4.

9. The value calculated in step 7 or 8 is the thickness of the new shim pack that will be installed. **Figure 20–60** illustrates several examples of determining shim pack thickness.

10. Install the drive pinion, bearing cage, and new shim pack into the differential carrier.

ADJUSTING PRELOAD OF DIFFERENTIAL BEARINGS

One of two methods can be used to check and adjust the preload of the differential bearings. Method one is as follows:

1. Attach a dial indicator on the mounting flange of the carrier and adjust the indicator so that the plunger or pointer is against the back surface of the ring gear (**Figure 20–61**).

2. Loosen the bearing adjusting ring that is opposite the ring gear so that a small

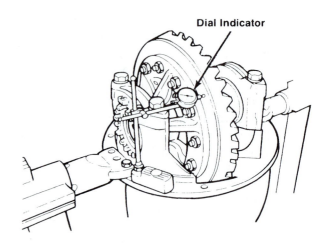

FIGURE 20–61   Dial meter attached to carrier-mounted flange. (Courtesy of ArvinMeritor Inc.)

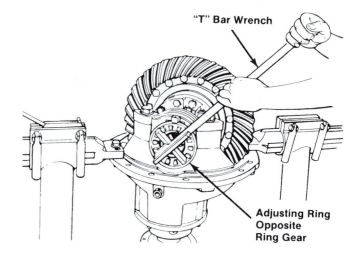

FIGURE 20–62   Turning the adjusting ring using a T-bar wrench. (Courtesy of ArvinMeritor Inc.)

amount of end play shows on the dial indicator. To turn the adjusting rings, use a T-bar wrench that engages two or more opposite notches in the ring (**Figure 20–62**).

3. Move the differential and ring gear to the left and right using pry bars as you read the dial indicator. Use two pry bars that fit between the bearing adjusting rings and the ends of the differential case (**Figure 20–63**). Or you can use two pry bars between the differential case or ring gear and the carrier at locations other than those just described. In either case, the pry bars must not touch the differential bearings.

4. Tighten the same bearing adjusting ring so that no end play shows on the dial indicator. Move the differential and ring gear to the left

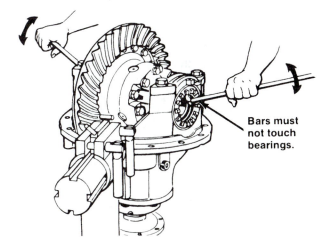

FIGURE 20–63   *Using pry bars to adjust play in the ring gear. (Courtesy of ArvinMeritor Inc.)*

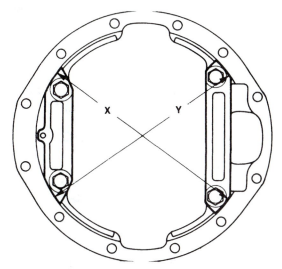

A

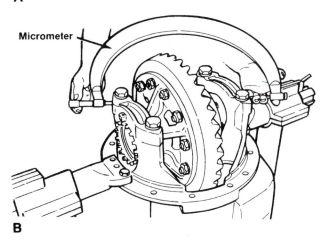

B

FIGURE 20–64   *(A) Location of distances measured to check expansion between bearing caps after tightening adjusting rings. (B) Measuring this distance. (Courtesy of ArvinMeritor Inc.)*

and right as needed. Repeat step 3 until zero end play is reached.

5. Tighten each bearing adjusting ring one notch from the zero end play measured in step 4.

A second method of checking preload is to measure the expansion between the bearing caps after you tighten the adjusting rings. Use the following procedure.

1. Turn both adjusting rings hand-tight against the differential bearings.
2. Measure the distance X or Y between opposite surfaces of the bearing caps (**Figure 20–64A**) using a large micrometer of the correct size (**Figure 20–64B**). Make a note of the measurement.
3. Tighten each bearing adjusting ring one notch.
4. Measure the distance X or Y again. Compare the dimension with the distance X or Y measured in step 2. The difference between the two dimensions is the amount the bearing caps have expanded.
   Example: Measurements of a carrier.
   Distance X or Y before tightening adjusting rings = 15.315 inches (389.00 mm).
   Distance X or Y after tightening adjusting rings = 15.324 inches (389.23 mm).
   15.324 inches − 15.315 inches = 0.009 inches (0.23 mm) difference.
   If the dimension is less than specification, repeat step 3 and 4 as needed.

RING GEAR RUNOUT CHECK

To check the runout of the ring gear, do the following:

1. Attach a dial indicator on the mounting flange of the differential carrier (**Figure 20–65**).
2. Adjust the dial indicator so that the plunger or pointer is against the back surface of the ring gear.
3. Adjust the dial of the indicator to zero.
4. Rotate the differential and ring gear while reading the dial indicator. The runout of the ring gear must not exceed 0.008 inch (a typical value; refer to the applicable manufacturer's service manual for the specific value).

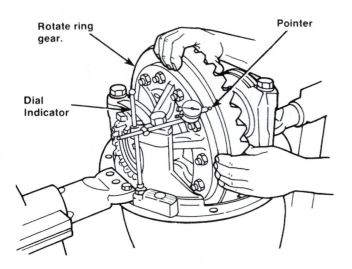

FIGURE 20–65 Checking ring gear runout. (Courtesy of ArvinMeritor Inc.)

5. If runout of the ring gear exceeds the specification, remove the differential and ring gear assembly from the carrier. Check the differential parts including the carrier for the problem that caused the runout of the gear to exceed specifications. Repair or replace defective components.

6. After the components are repaired or replaced, install the differential and ring gear into the carrier.

7. Repeat the preload adjustment of the differential bearings. Then, repeat this runout procedure.

CHECKING AND ADJUSTING RING GEAR BACKLASH

If the old gear set is installed, adjust the backlash to the setting that was measured before the carrier was disassembled. If a new gear set is to be installed, adjust the backlash to the correct specification for the new gear set.

To check and adjust ring gear backlash, do the following:

1. Attach a dial indicator onto the mounting flange of the carrier (**Figure 20–65**).

2. Adjust the dial indicator so that the plunger is against the tooth surface at a right angle.

3. Adjust the dial of the indicator to zero.

4. Hold the drive pinion in position.

5. While reading the dial indicator, rotate the differential and ring gear a small amount in both directions, against the teeth of the drive pinion (**Figure 20–66**). If the backlash reading is not within specification (typically rang-

ing from 0.010–0.020 inch), adjust the backlash as given in steps 6 and 7.

6. Loosen one bearing adjusting ring one notch, then tighten the opposite ring the same amount. Backlash is increased by moving the ring gear away from the drive pinion (**Figure 20–67**). Backlash is decreased by moving the ring gear toward the drive pinion (**Figure 20–68**).

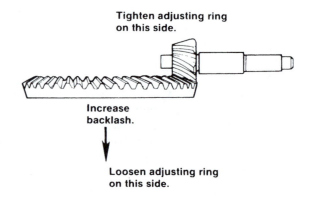

FIGURE 20–66 Check ring gear backlash. (Courtesy of ArvinMeritor Inc.)

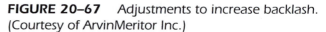

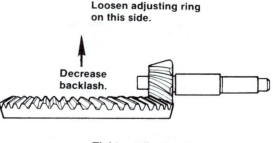

Tighten adjusting ring on this side.

Increase backlash.

Loosen adjusting ring on this side.

FIGURE 20–67 Adjustments to increase backlash. (Courtesy of ArvinMeritor Inc.)

Loosen adjusting ring on this side.

Decrease backlash.

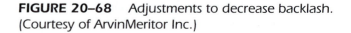

Tighten adjusting ring on this side.

FIGURE 20–68 Adjustments to decrease backlash. (Courtesy of ArvinMeritor Inc.)

7. Repeat steps 2–5 until the backlash is within specifications.

CHECKING AND ADJUSTING RING GEAR AND PINION TOOTH CONTACT

Correct tooth contact between the pinion and the ring gear cannot be overemphasized because improper tooth contact can lead to early failure of the axle and noisy operation. The tooth contact pattern consists of the lengthwise bearing (along the tooth of the ring gear) and the profile bearing (up and down the tooth). **Figure 20–69** shows ring gear tooth nomenclature.

Checking Tooth Contact Pattern on New Gear Set

Paint twelve ring gear teeth with a marking compound (**Figure 20–70**) and roll the gear to obtain a tooth contact pattern. The correct pattern is well centered on the ring gear tooth with lengthwise con-

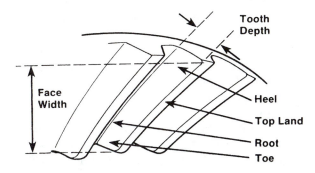

FIGURE 20–69 _Ring gear tooth nomenclature. (Courtesy of Roadranger Marketing. One great drive train from two great companies—Eaton and Dana Corporations)_

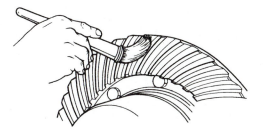

FIGURE 20–70 _Application of a marking compound to check tooth contact. (Courtesy of Roadranger Marketing. One great drive train from two great companies—Eaton and Dana Corporations)_

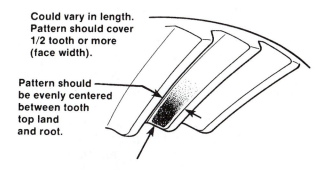

FIGURE 20–71 _Correct tooth contact pattern for new gearing. (Courtesy of Roadranger Marketing. One great drive train from two great companies— Eaton and Dana Corporations)_

tact clear of the toe (**Figure 20–71**). The length of the pattern in an unloaded condition is approximately one-half to two-thirds of the ring gear tooth in most models and ratios.

Checking Tooth Contact Pattern on Used Gear Set

Used gearing will not usually display the square, even contact pattern found in new gear sets. The gear will normally have a pocket at the toe-end of the gear tooth (**Figure 20–72**) that tails into a contact line along the root of the tooth. The more use a gear has had, the more the line becomes the dominant characteristic of the pattern.

Adjusting the Tooth Contact Pattern

When disassembling, make a drawing of the gear tooth contact pattern so that when assembling it is possible to display the same tooth contact pattern observed before disassembly. A correct pattern is clear of the toe and centers evenly along the face width between the top land and the root. Otherwise,

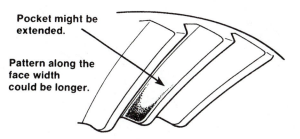

FIGURE 20–72 _Correct tooth contact pattern for used gearing. (Courtesy of Roadranger Marketing. One great drive train from two great companies— Eaton and Dana Corporations)_

the length and shape of the pattern are highly variable and are considered acceptable as long as it does not run off the tooth at any time. If necessary, adjust the contact pattern by moving the ring gear and drive pinion. Ring gear position controls the backlash. This adjustment also moves the contact pattern along the face width of the gear tooth (**Figure 20–73**). Pinion position is determined by the size of the pinion bearing cage shim pack. It controls contact on the tooth depth of the gear tooth (**Figure 20–74**).

These adjustments are interrelated. As a result, they must be considered together even though the pattern is altered by two distinct operations. When making adjustments, first adjust the pinion and then the backlash. Continue this sequence until the pattern is satisfactory.

THRUST SCREW ADJUSTMENT

For those differential carriers equipped with a thrust screw, perform the following procedure. (If the carrier does not have a thrust block, proceed to step 4 of this procedure.)

1. Rotate the carrier in the repair stand until the back surface of the ring gear is toward the top.
2. Put the thrust block on the back surface of the ring gear. The thrust block must be in the center between the outer diameter of the gear and the differential case.
3. Rotate the ring gear until the thrust block and hole for the thrust screw, in the carrier, are aligned.
4. Install the jam nut on the thrust screw, one half the distance between both ends (**Figure 20–75**).
5. Install the thrust screw into the carrier until the screw stops against the ring gear or thrust block.
6. Loosen the thrust one-half turn, 180 degrees.
7. Tighten the jam nut to the correct torque value against the carrier (typical values range from 150–295 pound-feet) (**Figure 20–76**).

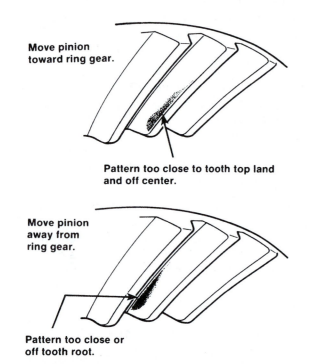

FIGURE 20–73 Two incorrect patterns when adjusting pinion position. (Courtesy of Roadranger Marketing. One great drive train from two great companies—Eaton and Dana Corporations)

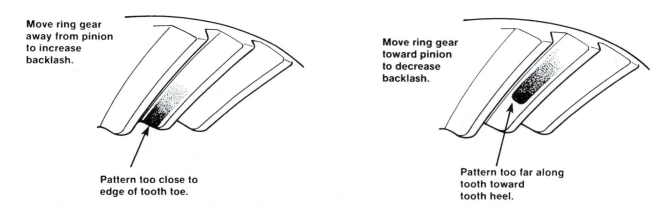

FIGURE 20–74 Two incorrect patterns when adjusting backlash. (Courtesy of Roadranger Marketing. One great drive train from two great companies—Eaton and Dana Corporations)

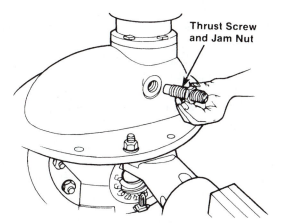

FIGURE 20-75 *Installing the jam nut on the thrust screw. (Courtesy of ArvinMeritor Inc.)*

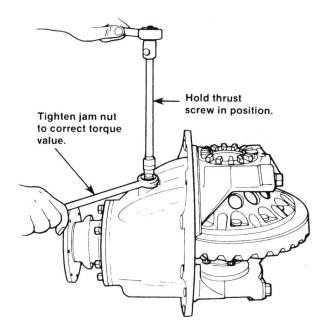

FIGURE 20-76 *Tighten the jam nut to the correct torque value. (Courtesy of ArvinMeritor Inc.)*

AXLE ALIGNMENT AND TRACKING

The procedures for proper axle alignment and tracking can be found within this textbook in Chapter 19.

SUMMARY

- Adhering to the manufacturer-recommended lubrication schedules is the key to ensuring the longest service life from both drive and dead axles.
- Servicing of axles on heavy duty trucks consists of routine inspection, lubrication, cleaning,

and, when required, troubleshooting and component overhaul.
- Failure analysis is required to prevent recurrent failures.
- Drive axle carrier components usually fail for one of the following reasons:
  - Shock load
  - Fatigue
  - Spinout
  - Lubrication
  - Normal wear
- Most differential carriers are replaced as rebuilt/exchange units, so the role of the technician is, more often than not, to diagnose the problem, and then, if necessary, replace the defective assembly.
- The technician that has disassembled and reassembled a differential carrier will find this troubleshooting procedure easier to follow.

REVIEW QUESTIONS

1. Technician A says that the gear oil in a differential carrier axle should be changed every 3,000 miles. Technician B says that the differential carrier should be flushed with kerosene at every oil change. Who is correct?
   a. Technician A
   b. Technician B
   c. both Technicians A and B
   d. neither Technician A nor B

2. A truck is producing rear axle noise only during turns. Technician A says that a likely cause of the problem is excessive backlash between the side gears and the differential pinions. Technician B says that the problem is probably excessive backlash between the drive pinion and the ring gear. Who is correct?
   a. Technician A
   b. Technician B
   c. both Technicians A and B
   d. neither Technician A nor B

3. Technician A says that ring gear backlash is correctly adjusted by alternately loosening and tightening the bearing adjusting rings. Technician B says that the tooth contact pattern can be moved closer to the root by increasing the ring gear backlash. Who is correct?
   a. Technician A
   b. Technician B
   c. both Technicians A and B
   d. neither Technician A nor B

4. Which of the following would indicate that the level was correct when checking lube oil level in a differential carrier housing with the full plug removed?
   a. oil spilling from fill hole
   b. oil level exactly even with bottom of fill hole
   c. oil level within one finger joint of fill hole
   d. oil level within one inch of fill hole

5. A highway truck is used in a linehaul application. Typically, when should the gear oil in the differential carrier be changed after the initial oil change?
   a. every 3,000 miles
   b. every 25,000 miles
   c. every 50,000 miles
   d. every 100,000 miles

6. Which gear lube viscosity rating is recommended for use in a differential carrier in a truck operating in temperatures of minus 30°F?
   a. 75W
   b. 80W
   c. 80W-90
   d. 80W-140

7. Which type of lubricant is recommended when hypoid gearing is used in a differential carrier?
   a. synthetic lube
   b. API GL-4
   c. any EP rated lubricant
   d. any API rated lubricant

8. When planning to drain and replace the lubricant in a differential carrier, which of the following is preferred?
   a. drain when the lubricant is at operating temperature
   b. flush with kerosene
   c. drain after allowing to cool overnight
   d. flush with solvent

9. After a differential carrier overhaul, which of the following methods would best ensure that the wheel bearings receive proper initial lubrication?
   a. fill the differential carrier to the inter-axle differential fill hole
   b. fill the differential carrier through the axle breather
   c. jack the vehicle at each wheel for one minute, then level and check fill hole level
   d. remove each axle shaft rubber fill plug and fill to the indicated oil level

10. Which of the following measuring tools should be used to check ring gear runout?
    a. fish scale
    b. torque wrench
    c. thickness gauges
    d. dial indicator

21

Steering Systems

Objectives

After reading this chapter, you should be able to:

- Identify the components of a heavy-duty truck's steering system.
- Describe the procedure for inspecting front axle components for wear.
- Explain how toe, camber, caster, axle inclination, turning radius, and axle alignment affect tire wear, directional stability, and handling.
- Describe the components and operation of a worm and roller and a recirculating ball type steering gear.
- Explain how to check and adjust a manual steering gear's preload and backlash.
- Identify the components of a power steering gear and pump and explain the operation of a power steering system.
- Describe the components and operation of a pneumatic steering system.
- Describe the components and operation of an electronically variable power steering system.
- Describe the components and operation of a load sensing power assist steering system.

Key Terms List

camber
caster
cavitation
dog-tracking
drag link

light beam or laser alignment
Pitman arm
pressure relief valve
tie-rod assembly
toe
toe-in
toe-out

The steering system in a Class 8 heavy-duty truck must deliver precise directional control of more than 50 tons of combined vehicle and cargo weight. And, it must do so requiring little driver effort at the steering wheel. Truck steering systems are either manual or power assisted, with power assist units using either hydraulic or air assist setups to make steering effort easier. Some hydraulic power assisted sys-

tems are available with electronic or load sensing controls.

In addition to its vital role in vehicle control, the steering system is closely related to front suspension, axle, and wheel/tire components (**Figure 21–1**). Improper steering adjustment can lead to alignment and tire wear problems. Suspension, axle, and wheel problems can affect steering and handling.

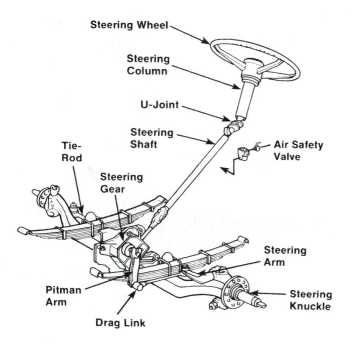

Steering Wheel

Steering Column

U-Joint

Steering Shaft

Tie-Rod

Steering Gear

Air Safety Valve

Steering Arm

Pitman Arm

Steering Knuckle

Drag Link

FIGURE 21–1 *Parts of a manual steering gear system. (Courtesy of Air-O-Matic Power Steering, Div. of Sycon Corp.)*

21.1   SYSTEM COMPONENTS

The key components that make up the steering system are the steering wheel, steering column, steering shaft, steering gear, **Pitman arm**, **drag link**, steering arm, ball joints, and **tie-rod assembly**.

- **Steering Wheel.** This is the driver's link to the entire system. The wheel is formed of a strong steel rod shaped into a wheel (**Figure 21–2**). Spokes extend from the wheel to the wheel hub, which is fastened securely at the top of the steering column. The wheel assembly is covered with rubber or plastic. The steering column transfers driver input to the steering gear. In other words, driver effort applied to the steering wheel at the rim becomes torque in the steering shaft. The larger the steering wheel diameter, the more torque is generated from the same amount of driver effort. Most steering wheels on heavy-duty trucks are 22 inches in diameter. A turn signal switch is usually clamped just below the steering wheel, onto the left side of the steering column. If the vehicle is equipped with a trailer brake control valve, it is clamped onto the right side of the steering column.
- **Steering Column.** The major components of the steering column assembly are a jacket

(tube), bearing assemblies (staked in place in the top and bottom of the jacket), a steering column shaft, and wiring and contact assemblies for the electric horn. At the upper end of the steering column shaft are threads to accept a wheel nut, and straight external splines to match the internal splines of the steering wheel hub. The lower end has straight external splines to match the internal splines of the steering driveline upper end-yoke. The steering column assembly is attached to the dash steering column bracket with two clamps that are hidden by the under-dash cover. A lower steering column cover extends from the under-dash cover to the floor. The steering column assembly is generally not repairable; if any steering column components are damaged or badly worn, the steering column assembly must be replaced.

- **Steering Gear.** This gearbox multiplies steering torque and changes its direction as received through the steering shaft from the steering wheel. There are two general types of heavy-duty steering gears: worm and roller and recirculating ball. These are explained later in this chapter.
- **Pitman Arm.** The Pitman arm is a steel arm clamped to the output shaft of the steering gear. The outer end of the Pitman arm moves through an arc in order to change the rotary motion of the steering gear output shaft into linear motion. The length of the Pitman arm affects steering response. A longer Pitman arm will generate more steering motion at the front wheels for a given amount of steering wheel movement.
- **Drag Link.** This forged rod connects the Pitman arm to the steering arm. The drag link can be a one- or two-piece component. The two-piece design is adjustable in length, a fact that makes it easy to center the steering gear with the wheels straight ahead. One-piece drag links are used in systems with very close tolerances. Other components are used to make adjustments to the system when a one-piece drag link is used. The drag link is connected at each end by ball joints. These ball joints isolate the steering gear and Pitman arm from axle motion. **Figure 21–3** shows the drag link in both a manual and power steering gear installation.
- **Steering Arm.** Sometimes called a steering lever, this forged steel component connects the drag link to the top portion of the driver's side steering knuckle and spindle. As the steering arm moves, it changes the angle of the steering knuckle and so alters the direction of the steering knuckle spindle.

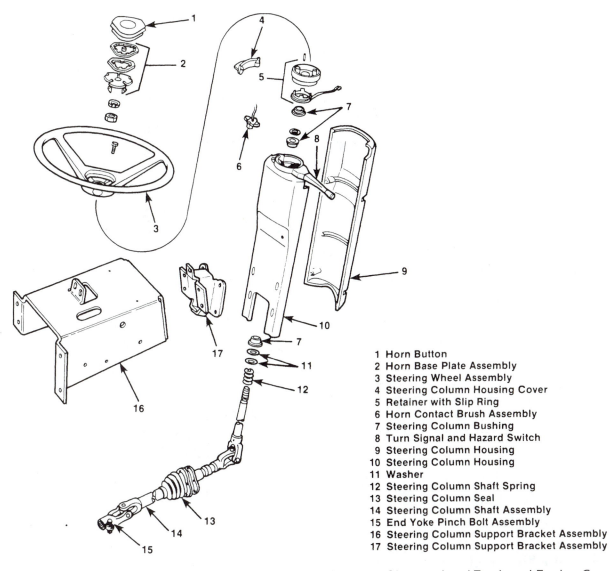

1 Horn Button
2 Horn Base Plate Assembly
3 Steering Wheel Assembly
4 Steering Column Housing Cover
5 Retainer with Slip Ring
6 Horn Contact Brush Assembly
7 Steering Column Bushing
8 Turn Signal and Hazard Switch
9 Steering Column Housing
10 Steering Column Housing
11 Washer
12 Steering Column Shaft Spring
13 Steering Column Seal
14 Steering Column Shaft Assembly
15 End Yoke Pinch Bolt Assembly
16 Steering Column Support Bracket Assembly
17 Steering Column Support Bracket Assembly

FIGURE 21–2   Steering column for a conventional cab. (Courtesy of International Truck and Engine Corp.)

- **Ball Joints.** This ball-and-socket assembly consists of a forged steel ball with a threaded stud attached to it. A socket shell grips the ball. The ball stud moves around to provide the freedom of movement needed for various steering links to accommodate relative motion between the axle and the frame rail when the front axle springs flex. A ball stud is mounted in the end of each steering arm and provides the link between the drag link and the steering arm.
- **Tie-Rod Assembly.** The steering arm or lever controls the movement of the driver's side steering knuckle. The method of transferring this steering motion to the opposite, passenger side steering knuckle is done through the use of a tie-rod assembly. The tie rod links the two

steering knuckles so that they act in unison. The tie-rod assembly is also called a cross tube.

The assembly consists of a tie-rod and two tie-rod arms. The tie-rod arms are rigidly bolted to the lower end of the steering knuckles. (Some manufacturers refer to tie-rod arms as Ackerman arms or cross tube arms.) The tie-rod is a long steel rod or tube that runs parallel to the front axle. Ball joints (called tie-rod ends) on either end of the tie-rod connect it to the socket joints in the tie-rod arms. So when steering action is applied to the driver's side steering knuckle, it is transferred by the tie-rod arm through the tie-rod to the opposite tie-rod arm and steering knuckle.

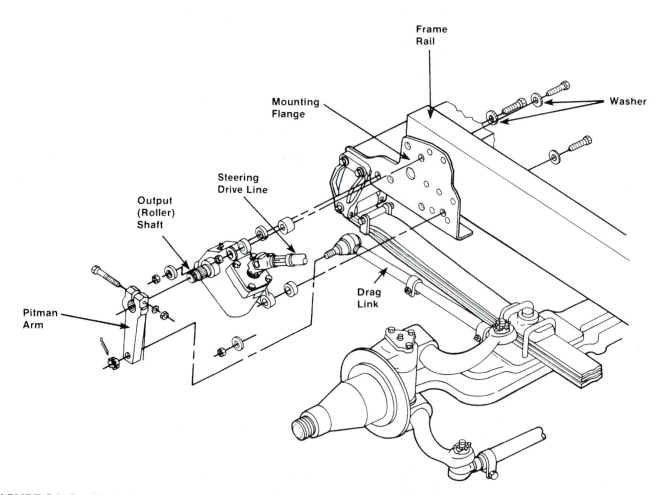

FIGURE 21-3 Typical manual steering gear installation. (Courtesy of Freightliner Trucks)

STEERING KNUCKLE

The steering knuckle (**Figure 21-4** and **Figure 21-5**) mounts to the front axle through the use of a heavy steel pin known as a kingpin or knuckle. The steering knuckle includes the spindle on which the wheel bearings and wheel hubs are mounted, a flange to which the brake's backing plate is bolted, and an upper and lower knuckle to which the steering arm and tie-rod arm are attached.

The kingpin can be either tapered (**Figure 21-5A**) or straight (**Figure 21-5B**). Tapered pins are drawn into the axle center and secured by tightening a nut at the upper pin end. Straight kingpins are secured to the axle with tapered draw keys that bear against flats on the pin. Tapered pins are usually sealed and do not require periodic lubricating. Straight pins have a cap on either end in which grease fittings are mounted. A heavy-duty, multipurpose lithium base (#2 grade) grease is typically specified for use in the steering knuckle.

The steering arm is fastened to the upper knuckle of the steering knuckle with a key and nut. The tie-

rod arm is fastened to the lower knuckle in the same way.

STEERING SYSTEM INSPECTION

WARNING: All steering mechanisms are critical safety items. As such, it is imperative that instructions in the service manual are followed to the letter. Failure to observe these procedures may result in loss of steering.

When steering problems exist, inspect the vehicle for road-worthiness; then, if necessary, verify the steering problem by test driving the vehicle or taking a ride with the driver. If the problem occurs only when the vehicle is loaded, make the test drive with the vehicle loaded.

After the problem is verified, check the simple causes before moving on to the more complicated possibilities. Perform the following checks in the order given whenever rapid tire wear, hard steering,

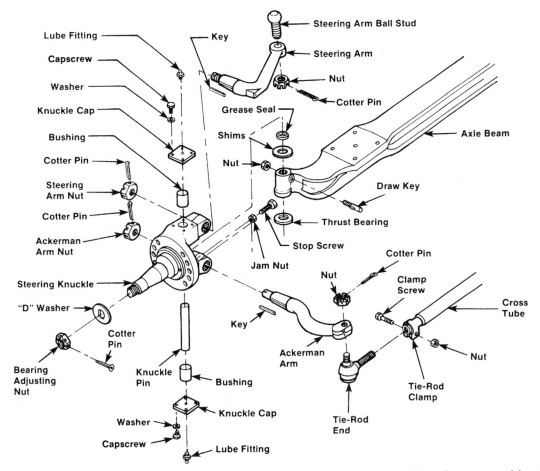

FIGURE 21–4 *Components of a steering axle. (Courtesy of Roadranger Marketing. One great drive train from two great companies—Eaton and Dana Corporations)*

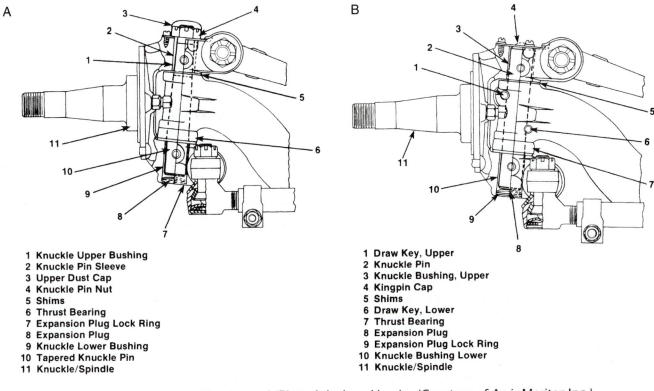

A

1 Knuckle Upper Bushing
2 Knuckle Pin Sleeve
3 Upper Dust Cap
4 Knuckle Pin Nut
5 Shims
6 Thrust Bearing
7 Expansion Plug Lock Ring
8 Expansion Plug
9 Knuckle Lower Bushing
10 Tapered Knuckle Pin
11 Knuckle/Spindle

B

1 Draw Key, Upper
2 Knuckle Pin
3 Knuckle Bushing, Upper
4 Kingpin Cap
5 Shims
6 Draw Key, Lower
7 Thrust Bearing
8 Expansion Plug
9 Expansion Plug Lock Ring
10 Knuckle Bushing Lower
11 Knuckle/Spindle

FIGURE 21–5 *(A) Tapered knuckle pin; and (B) straight knuckle pin. (Courtesy of ArvinMeritor Inc.)*

or erratic steering indicates a problem in the steering system.

1. Check that the front tires are the same size and model. Be sure they are equally and adequately inflated. Underinflated tires will cause hard steering. Refer to the wheel and tire information given in Chapter 23.

2. If the steering problem occurs only when the vehicle is loaded, make sure that the fifth wheel is adequately lubricated (see Chapter 30).

3. Check the steering linkage for loose, damaged, or worn parts. The steering linkage includes the tie rod, steering arms, bushings, and other components that carry movement of the Pitman arm to the steering knuckles. The wheels should turn smoothly from stop to stop.

4. Inspect the drag link, steering driveline, and upper steering column for worn or damaged parts.

5. Make sure that the steering driveline components, especially the universal joints, are adequately lubricated.

6. Check the front axle wheel alignment including wheel bearing adjustment, **caster** and **camber** angles, and **toe-in** as described later in this chapter.

7. Check the rear axle alignment. Rear axle misalignment can cause hard or erratic steering. If needed, align the rear axle(s).

8. Inspect the front axle suspension for worn or damaged parts.

👓 **Shop Talk** _____

Before doing the remaining checks, apply the parking brakes and chock the rear tires. Raise the vehicle until the front tires leave the pavement and then place safety stands under the frame rails. Be sure that the stands will support the weight of the vehicle.

9. With the front wheels straight ahead, turn the steering wheel until motion is seen at the wheels. Align a reference mark on the steering wheel with a mark on a ruler and then slowly turn the steering wheel in the opposite direction until motion is again seen at the wheels. Measure the lash (freeplay) at the steering wheel. Too much lash exists if the steering wheel movement exceeds:
   - 2¾ inches for a 22-inch steering wheel
   - 2¼ inches for a 20-inch steering wheel

10. Turn the steering wheel through a full right and full left turn. If the front wheels cannot be turned to the right and left axle steering stops without binding or noticeable interference, one or more of the following may be the cause.
   - The worm shaft bearings or the roller shaft mesh are adjusted too tightly.
   - The worm shaft or worm shaft bearings are worn or damaged.
   - The roller shaft is worn or damaged.
   Instructions on repair of these problems can be found later in this chapter.

11. Secure the steering wheel in the straight-ahead driving position. Move the front wheels from side to side. Any play in the steering gear bearings will be felt in the drag link ball stud at the Pitman arm. If any bearing play exists, make adjustments in the following order:
   - Adjust the worm shaft bearings.
   - Adjust the roller shaft total mesh preload.

WARNING: Do not drive the vehicle if too much lash or freeplay is in the steering gear. Excessive lash is a sign of an improperly adjusted steering gear or worn or otherwise damaged steering gear components. Driving the vehicle in this condition could result in a loss of steering control.

12. Remove the safety stands, lower the vehicle, and remove the chocks.

21.2 FRONT-END ALIGNMENT

The components of a truck's front axle, steering system, and suspension are carefully aligned by the vehicle manufacturer to balance all the forces created by friction, gravity, centrifugal force, and momentum while the vehicle is in motion. When properly aligned, the wheels of a loaded vehicle will contact the road correctly, allowing the wheels to roll without scuffing, dragging, or slipping on the road. A front end that is properly aligned will result in

- Easier steering
- Longer tire life
- Directional stability
- Less wear on front-end parts
- Better fuel economy
- Increased safety

The primary alignment angles are toe, caster, and camber. Kingpin inclination, turning angle, Ackerman geometry, and axle alignment will also affect tire wear and steering characteristics. Front-end alignment should be checked at regular intervals and particularly after the front suspension has been subjected to extremely heavy service or severe impact loads. Before checking and adjusting alignment, components such as wheel bearings, tie-rods, steering gear, shock absorbers, and tire inflation should be inspected and corrected where necessary.

TOE

Toe is the angle of the tires from the true straight-ahead position. That is, if a line were drawn through the centerline of the tire and compared to the centerline of the vehicle (the true straight-ahead position), the amount of deviation between the two lines indicates the toe angle of the tire. When the tire centerline is parallel with the vehicle centerline, the toe is equal to zero. When the front end of the tire points inward toward the vehicle, the tire has toe-in. When the front of the tire points outward from the vehicle, the tire has **toe-out (Figure 21–6)**.

Normally, incorrect toe angles accelerate tire wear but also have an adverse effect on directional stability of the vehicle. However, incorrect toe angles have the potential to cause more front tire wear than any other incorrect alignment angle. Too much toe-in produces a scuffing, or a featheredge, along the inner edges of the tires. Too much toe-out produces a similar wear pattern along the outer edge of the tires. In extreme cases of toe-in or toe-out, feathered edges develop on the tread across the entire width of the tread face of both radial and bias-ply tires.

Most trucks use a zero or a slight toe-in setting. They are also designed to use a positive camber angle setting; the tires lean away from the truck at the top. These settings are designed to compensate for a tendency of the tires to toe-out when the vehicle is moving. The tendency to toe-out is caused by at least two forces. First, a tire will always try to turn in the direction it is leaning. Second, the rotational forces acting on the tires will tend to make them toe-out as they spin. Looseness in the steering linkage and tie rods contributes to the toe-out tendency.

Toe is calculated in either degrees or inches. Because tires have a natural tendency to toe-out while a vehicle is moving, typical steering axles call for toe-in settings of $1/16$ inch $\pm 1/32$ inch. Actual toe settings also depend on whether radial or bias-ply tires are used. Adjustment for toe is done by adjusting the length of the tie-rod ends or replacing parts as needed.

When attempting to determine the causes of excessive tire wear, first check kingpin inclination, camber, and caster and correct, if necessary, in this order. No change should be made in toe-in until the other factors of front-wheel alignment are known to be within specifications.

When setting toe-in adjustment, the front suspension must be neutralized; that is, all component parts must be in the same relative position when making the adjustment as they will be in operation. To neutralize the suspension, the vehicle must be rolled forward 12 to 15 feet. By rolling the vehicle forward, all tolerances in the front suspension are taken up, and the suspension is then in normal operating position. Neutralizing the front suspension is extremely important, especially if the vehicle has been jacked up in order to scribe the tires; otherwise, the front wheels

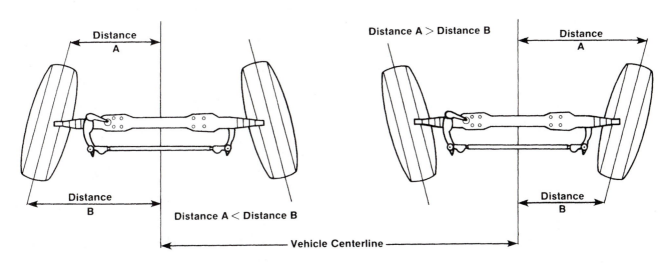

FIGURE 21–6 Toe-in (left) and toe-out (right). (Courtesy of International Truck and Engine Corp.)

will not return to normal operating position due to the tires gripping the floor surface when the vehicle is lowered.

Begin by turning the front wheels to the exact straight-ahead position. Actual toe-in measurements should be taken at hub height between the two points on the center of the tread at the rear of the tires (**Figure 21–7**). Mark the point and roll the truck ahead so that the points are in the front at hub height and measure the distance between the same two points on the tire treads.

The difference in the two measurements is the actual toe-in or toe-out.

1. To adjust the toe-in, turn the steering wheel so that the gear is in the mid-position.
2. Loosen the clamping bolts on the tie-rod.
3. Turn the tie-rod in the direction necessary to bring toe-in within the specified limits.
4. Torque the clamp bolts on the tie-rod to the manufacturer's specifications.

Always recheck toe-in after any change in caster or camber angle or after any alteration in tie-rod adjustment.

 Shop Talk

When tie-rod, drag link, or power steering linkage ends are replaced, they must be threaded into the tie-rod sufficiently so that when the clamp is applied, the clamping action will be directly over the threads on the ball joint end. Be sure that the end is in far enough (past the clamp) to provide adequate clamping and that the bolt in the clamp is installed next to (over) the slot in the tie-rod.

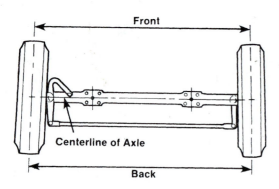

FIGURE 21–7   To determine toe, measure between the centerlines of the tires, both front and back position. The differences between the two measurements is the toe. (Courtesy of International Truck and Engine Corp.)

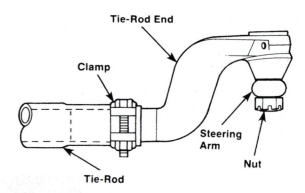

FIGURE 21–8   A typical deep drop tie-rod end. Note top-mounted curve. (Courtesy of International Truck and Engine Corp.)

DEEP DROP TIE-ROD ENDS

The toe-in will increase on certain axles as they are loaded. Therefore, it is important that vehicles with set back axles and deep drop tie-rod ends are given special attention. The deep drop tie-rod ends can be identified by their top-mounted curve (**Figure 21–8**).

Due to the turning geometry of these axles, it is possible that the tie-rod clamps, if facing forward, might interfere with the I-beam at full turn. This interference will not impede steering control but will reduce the effective turn angle of the vehicle. Refer to the appropriate service manual section for diagnosis and repair procedures.

CASTER

Caster is the forward or rearward tilt of the kingpin centerline when viewed from the side of the vehicle. Zero caster indicates that the centerline of the kingpin is straight up and down or vertical. Positive caster indicates the kingpin is tilted rearward as shown in **Figure 21–9**. Negative caster indicates that the kingpin is tilted forward.

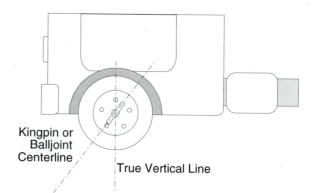

FIGURE 21–9   Caster is the forward or rearward (shown) tilt of the kingpin when viewed from the side of the vehicle.

Caster is a directional stability angle only. Incorrect caster by itself does not affect tire wear. Most heavy-duty trucks are designed with some degree of positive caster. Positive caster creates a "trailing" force in the front wheels, which tends to keep them tracking straight ahead. This makes it easier to return the wheels to a straight-ahead position after a turn has been made. The principle is exactly the same as that used in tilting the front fork of a bicycle, which makes it possible to ride the bicycle without holding the handle bars.

Also, when the front wheels of the truck are turned, caster lifts one side of the vehicle and lowers the other. When the steering wheel is released, the weight of the vehicle forces the lifted side down, causing the wheels to return to the straight-ahead position. In other words, positive caster produces a condition in which the right wheel resists a right turn and the left wheel resists a left turn. The farther into a turn, the more this resistance increases.

Caster settings generally affect steering in the following ways:

- Too little caster can cause wheel instability, wandering, and poor wheel recovery.
- Too much caster can result in hard steering, darting, oversteering, and low speed shimmy.

Recommended Caster Settings

Vehicle manufacturers will specify varying caster settings, depending on the truck model and whether it is loaded or unloaded, tandem or single axle, and manual or power steering. Generally speaking, just enough positive caster should be provided to ensure wheel stability and good wheel recovery. The following settings are satisfactory in most cases for unloaded vehicles with manual steering, when measured on a level floor:

- Tandem drive: 1/2–1 1/2 degrees positive
- Single drive: 1 1/2–2 1/2 degrees positive
- No more than 1/2 degree difference between the left and right wheels
- Positive caster on the left wheel should not be greater than on the right.

Caster specifications are based on vehicle design load (no payload), which will usually result in a level frame. If the frame is not level when alignment checks are made, this must be considered in determining whether the caster setting is correct.

With the vehicle on a smooth, level surface measure the frame angle with a machinist's protractor placed on the frame rail (**Figure 21–10**). The degree of tilt from the level frame position must be used in

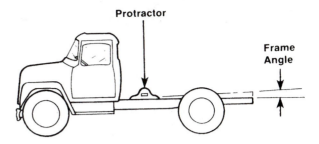

FIGURE 21–10 *Measuring frame angle. (Courtesy of International Truck and Engine Corp.)*

correcting the caster setting. Positive frame angle is defined as forward tilt (front end down) and negative angle as tilt to rear (front end high).

The measured frame angle should be added or subtracted, as required, from the specified level frame caster setting to obtain the caster that should actually be measured on the vehicle.

1. Positive frame angle should be subtracted from specified setting.
2. Negative frame angle should be added to specified setting.

As an example, if the specified caster setting is a positive 1 degree and it is found that the vehicle has a positive 1 degree frame angle, the measured caster should be 0 ±1/2. This would result in the desired 1 degree ±1/2 degree caster angle when the chassis settles to level the frame under load.

Caster should be measured with the vehicle on a level floor. Also, if the vehicle is equipped with a manually controlled air lift, adjust the lift so that the frame is at normal operating height. Any replacement of worn parts should be done before caster is measured.

Measuring Caster

Caster can be measured in a number of ways. One way is to use a machinist's protractor. Another is to use a radius gauge and caster gauge. Modern computerized alignment equipment is the fastest, most accurate method of measuring caster (and other alignment angles).

Machinist's Protractor. The machinist's protractor measures the angle between its base and true horizontal, as determined by a bubble cylinder on one side of the dial (similar to that used in a carpenter's level).

To measure caster angle,

1. Place the protractor on the machined surface of the front axle pad (**Figure 21–11**).

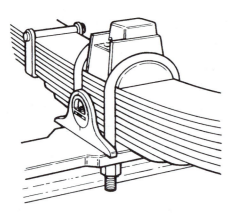

FIGURE 21-11   To measure caster, place the protractor on a clean, flat surface of the axle pad. Be sure that the axle pad is clean and the protractor touches the U-bolts. (Courtesy of Air-O-Matic Power Steering, Div. of Sycon Corp.)

The axle pad must be free of dirt and debris. Make sure that the protractor touches both U-bolts so that it is parallel with the truck frame.

2. Center the bubble in the cylinder by rotating the dial of the protractor. Lock the dial (if so equipped) and lift the protractor to read the angle.

3. It must then be determined whether this caster angle is positive or negative when the protractor is turned end for end. The original orientation of the protractor must be remembered when determining the direction of the caster angle. The following visualization can help prevent confusion: If the axle pad surface is parallel with the floor, there is zero caster. If it tilts toward the front of the truck, the caster is negative; toward the rear, it is positive.

4. Repeat the preceding procedure on the opposite side of the vehicle. If the caster difference from side to side exceeds 1/2 degree, the axle is probably twisted. Also, remember that the right side should not have less positive caster than the left. If the right side is less, it will tend to lead the vehicle to the right, particularly on crowned roads.

Camber/Caster Gauges

Figure 21-12 shows a caster/camber gauge. The following steps must be performed before making both camber (to be explained later) and caster measurements. The wheel runout must be checked and adjusted using a toe-in gauge to achieve a true camber/caster reading.

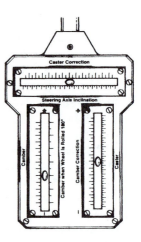

FIGURE 21-12   One type of camber/caster gauge.

1. Drive the truck onto a hard, level surface.
2. Raise the front end and position a radius gauge, with the lock pins installed, under each front wheel. The degree scale should be pointing outward.
3. Turn the steering wheel so that the front wheels face straight ahead.
4. Lower the tires onto the radius gauges. Be sure that the tires contact the center of the turntables (**Figure 21-13**).
5. Remove the lock pins, which hold the turntables in a fixed position.
6. Adjust both radius scales to zero.
7. Apply the brakes when using a full-floating turning gauge, but if a semi-floating design is used, do not apply them.

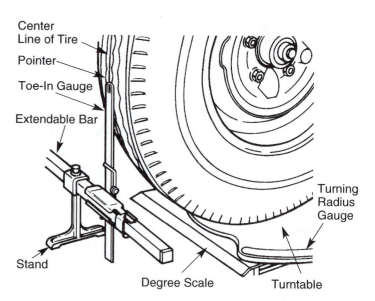

FIGURE 21-13   Simple wheel alignment equipment including a turning radius gauge and toe-in gauge.

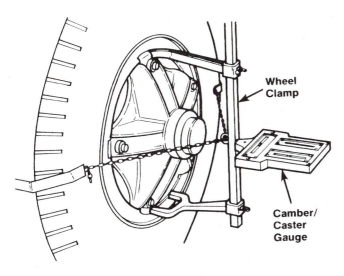

FIGURE 21–14   *The camber/caster gauge attached to the wheel clamp.*

8. Install the wheel clamp to the wheel and the camber/caster gauge to the wheel clamp (**Figure 21–14**). On smaller wheels the camber/caster gauge is fitted directly to the wheel hub. Follow the manufacturer's directions.

9. To check and adjust wheel runout, first check and make sure that the wheel clamp fingers rest firmly against the rim and that the gauge rests firmly against its seat.

10. Raise the front wheel off the turning radius gauge.

11. Note the reading on the camber scale.

12. While turning the wheel, note the point of the highest wheel runout and mark the tire at that point. Then position this point in the forward or rearward direction. In this way, the true wheel centerline is brought into the vertical position, splitting the runout.

13. Lower the front end so that the tires rest on the turning radius gauges.

 Shop Talk _____

Some caster/camber gauges have a runout adjustment that can be used to achieve the same results as the preceding procedure. To use these adjustments, follow the manufacturer's instructions.

To measure caster, follow this procedure:

1. Set both turning radius scales at zero.

2. Make a right-hand turn with the steering wheel so that the left front wheel is turned to 20 degrees.

3. Turn the adjusting screw to center the left wheel caster gauge bubble.

4. Turn the steering wheel to the left until the left front wheel is at the 20-degree left-hand turn position.

5. Rotate the gauge to level it.

6. Visually align the center of the bubble with the left- or right-hand caster graduation and read the degree of caster. The caster is positive if the bubble is toward the positive sign and negative if the bubble is toward the negative sign.

7. Repeat this procedure on the opposite side of the truck.

If the caster difference from side to side exceeds 1/2 degree, the axle is probably twisted. Also, remember that the right side should not have less positive caster than the left. If the right side is less, it will tend to lead the vehicle to the right, particularly on crowned roads.

Caster Shims

Caster angle is changed through the use of a caster shim, a metal plate inserted between the springs and the axle pad (**Figure 21–15**). Shims are manufactured in half degree increments, from 1/2 degree through 4 degrees, and in a choice of 3-, 3 1/2-, or 4-inch widths.

The following points should be observed in the use of caster shims:

- Use only one shim on each side. Do not stack them.
- The width of the shim must be the same as the spring.
- Do not use shims of more than 1 degree difference from side to side.

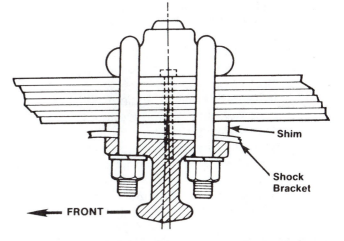

FIGURE 21–15   *Caster angle is changed by inserting metal shims between the spring and axle pad. (Courtesy of Air-O-Matic Power Steering, Div. of Sycon Corp.)*

Changing Caster

After determining the caster desired on each side of the axle, use the following procedure to install the caster shims.

CAUTION: When changing caster, be sure to block the wheels so that the truck will not roll. Also, use safety stands to hold up the springs while changing the shims.

---

1. Loosen all spring U-bolts on each side just enough to relieve tension (**Figure 21–15**).
2. On one side, fully loosen the U-bolts with the nuts remaining on the ends of the bolts.
3. With a frame jack, lift the springs from the axle. Do not lift the wheel from the floor.
4. If a caster shim has been previously installed, remove it and make sure that the angle of this shim is accounted for in the calculation of desired caster angle. (For safety's sake, use a tool to remove the shim. Never put fingers between the axle and the springs.)
5. Remove all dirt and debris from the exposed axle pad and clean thoroughly with a rag.
6. Check to be sure that the center pin is long enough to pass through the caster shim and enter the axle.
7. Insert the shim, aligning the hole with the center pin, and slowly lower the springs onto the axle, making sure the pin enters the axle through the shim. The shim must be installed with the thick part of the shim in the right position to obtain the correct reading.
8. Tighten the U-bolt nuts.
9. Repeat steps 2 through 8 on the other side of the axle.
10. Torque the U-bolt nuts to manufacturer's specifications in a cross pattern in at least three steps.
11. Measure the caster again. Changing shims does not always produce the exact caster change desired.
12. Normally, the caster angle will not change significantly when a load is added to the vehicle.
    But if steering difficulties persist, caster should be rechecked with the vehicle loaded.

Checking for a Twisted Axle

In a few cases, steering problems can be traced to a twist in the front axle. This can be an actual structural twist, or one induced by improper use of caster shims. A twisted axle should be suspected whenever

- The difference in caster exceeds 1/2 degree from side to side
- The caster shims in place differ by 1 degree or more
- A low speed shimmy exists, and there is no evidence of looseness in the steering system
- A leading and/or darting condition persists

To detect a twisted axle, use the following procedure:

1. Pull the vehicle onto a level floor, leaving the steering wheel free for at least the last 10 feet to permit the tires to seek a true straight-ahead position.
2. Remove the capscrews, which hold the kingpin cap in place.
3. Lift the cap from the knuckle and clean away any excess grease.
4. Partially insert two capscrews about 1/4 inch deep. Place the machinist's protractor on the knuckle and against the two capscrews to ensure that it is parallel with the frame (**Figure 21–16**).
5. Center the bubble, lift carefully, and record the angle. If this measurement is more than 1/2 degree different than the angle measured at the axle pad, it can indicate a twisted condition between the pad and the kingpin.
6. Repeat steps 2 through 5 on the other end of the axle.

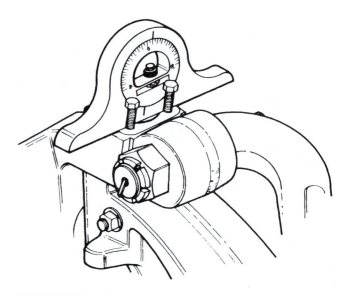

FIGURE 21–16 *Checking for a twisted axle using a machinist's protractor on the steering knuckle. (Courtesy of Air-O-Matic Power Steering, Div. of Sycon Corp.)*

It sometimes happens that the previous check will not indicate a twisted axle at either end, but there remains a side-to-side caster difference exceeding $1/2$ degree and/or the steering problems previously mentioned still exist. This can be due to a twist in the axle that can be detected only in its natural or unloaded state. This can be determined by the following check:

1. Loosen all U-bolts on the front axle enough to relieve tension while keeping center pins in place.
2. Using a bumper or frame jack, lift the vehicle weight from the axle.
3. Measure the angle at both kingpins and axle pads. Any differences in these readings will locate a twisted area in the axle.

CAMBER

Camber is the inward or outward tilt of the top of the tires when viewed from the front of the vehicle (**Figure 21–17**). If the centerline of the tire is straight up and down or perfectly vertical, the tire has a camber setting of zero degrees. If the top of the tire tilts outward, it has positive camber, and if the top of the tire tilts inward, it has negative camber (**Figure 21–18**). Camber settings of $+1/4$ degree for passenger side wheels and $+3/4$ degree for driver side wheels are typical.

The amount of camber used depends on the amount in degrees the kingpin is inclined. An incorrect camber angle causes the side of the tread to wear, resulting in abnormal tire wear. Excessive posi-

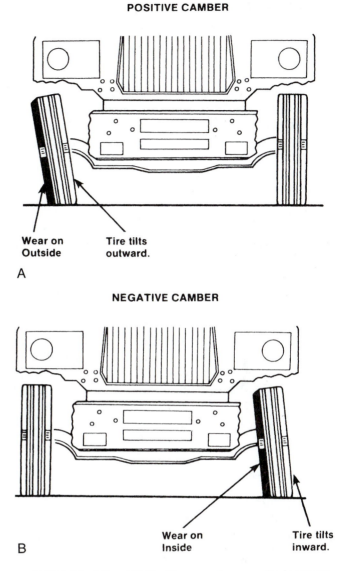

POSITIVE CAMBER

Wear on Outside

Tire tilts outward.

A

NEGATIVE CAMBER

Wear on Inside

Tire tilts inward.

B

FIGURE 21–18 (A) Positive camber results in wear on the outer edge of the tire; (B) negative camber results in tire wear on the inner edge. (Courtesy of Air-O-Matic Power Steering, Div. of Sycon Corp.)

tive camber causes the tire to wear on its outside shoulder. Likewise, excessive negative camber causes the tire to wear on its inside shoulder. Unequal camber in the front wheels might also cause the truck to lead to the right or left. The truck will lead to the side that has the most positive camber.

Camber adjustments are not often made on heavy-duty truck axles. Some service facilities adjust camber by cold bending axle beams, but some axle manufacturers do not approve of this practice. Bending an axle can damage the structural integrity of the beam, increase the risk of metal fatigue, and lead to the axle breaking. Heating and bending the axle should never be attempted as it removes the heat tempering.

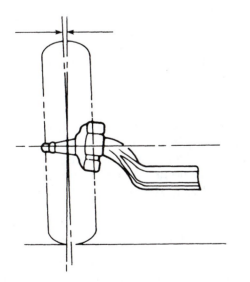

FIGURE 21–17 Camber is the inward or outward tilt of the tire when viewed from the front of the vehicle. (Courtesy of Air-O-Matic Power Steering, Div. of Sycon Corp.)

Measuring Camber

The preparatory procedure that must be performed before measuring camber has already been discussed in the section on caster. Perform those steps first. Then stabilize the front end. Next, visually align the center of the bubble with the left- or right-hand camber graduation in order to read the camber from the camber scale. The camber is positive if the bubble is toward the positive sign and negative if toward the negative.

Kingpin Inclination

Kingpin inclination (KPI) is the amount in degrees that the top of the kingpin inclines away from the vertical toward the front-to-back center of the truck as viewed from the front of the truck (**Figure 21–19**). Kingpin inclination working together with the camber angle places the approximate center of the tire tread in contact with the road. Kingpin inclination has the effect of reducing steering effort and improves the directional stability in the vehicle.

There is no means of adjusting this angle; therefore, it will not change unless the front axle has been bent. Corrections or changes to this angle are accomplished by replacement of broken, bent, or worn parts.

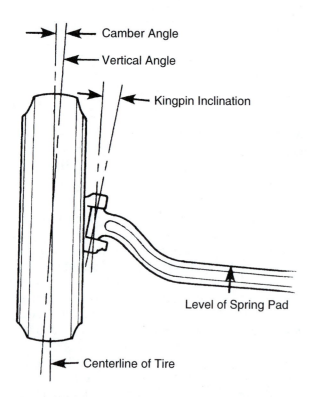

FIGURE 21–19   Kingpin inclination and camber angle as viewed from the front of the truck. (Courtesy of International Truck and Engine Corp.)

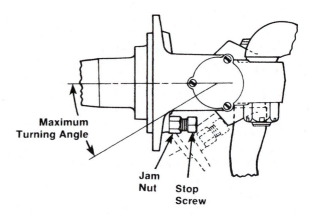

FIGURE 21–20   The maximum turning angle of a wheel is determined by a stop screw. (Courtesy of International Truck and Engine Corp.)

Turning Angle or Radius

Turning angle or radius is the degree of movement from a straight-ahead position of the front wheels to either an extreme right or left position. Two factors limit the turning angle: tire interference with the chassis and steering gear travel. To avoid tire interference or bottoming of the steering gear, adjustable stop screws are located on the steering knuckles (**Figure 21–20**). Turning radius or angle should be checked using the radius gauge described earlier. If turning angle does not meet specifications, it must be corrected.

To adjust the turning angle, loosen the jam nuts and turn in the steering knuckle stop screws. Position support stands under the front axle so that the wheels are off the floor and place wheel blocks behind the rear wheels. Turn the wheels to an extreme right turn until the steering gear bottoms or contact of the tire to chassis is made. Then back off the steering wheel one-quarter turn or back off the steering wheel until $1/2$–1 inch clearance is obtained between the tire and chassis. Be sure to check both front tires for clearance. When the proper clearance is determined, back the wheel stop screw out until the screw contacts the stop and tighten the jam nut. Repeat the same procedure on the left extreme turn and adjust the left steering knuckle stop screw.

 Shop Talk

*The power steering unit's **pressure relief valve** should open before the steering stop screw contacts the axle. If necessary, adjust the power steering unit so that the power assist stops approximately 1 degree before reaching the steering stop screws.*

Driving Front Axles

There is a stop screw located on each end of the axle housing for the purpose of limiting the amount of the turning angle of the wheels. These screws are not adjusted in accordance with the frame and tire interference as in conventional front axles. Instead, these screws are provided to limit the turning angle of the universal joints in the axles.

ACKERMAN GEOMETRY

Ackerman geometry involves the arrangement of the suspension components so that the tires roll freely during turns. During a turn, the inner wheel must follow a tighter circle than the outer wheel. Or, as it is sometimes expressed, the outer wheel must toe-out during a turn. This allows the inner and outer wheel to turn at different angles so that both wheels can roll without scrubbing. To achieve this, the steering linkage must be arranged so the axis of the steering wheels intersect on the axis of the rear axle (**Figure 21–21**). On single drive axles, the centerline of the drive axle is the rear axis. On tandem drive axles, the intersecting axis is centered between the two drive axles (**Figure 21–22**).

The Ackerman steering arrangement provides perfect rolling angles for both front wheels at only one turning angle. Other turning angles introduce an error, the size of which depends on the length and inclination of the tie-rod arms. As a result, the lengths of the cross tube and tie-rod arms must be selected so as to minimize the error throughout the steering range. In trucks, the steering error must be smallest at small turning angles; this is where most turns are made and where the highest operating speeds are encountered.

Toe-out on turns is accomplished by having the ends of lower steering arms (end at tie-rod) closer together than the kingpins (**Figure 21–23**). The amount of toe-out depends on the length and angle

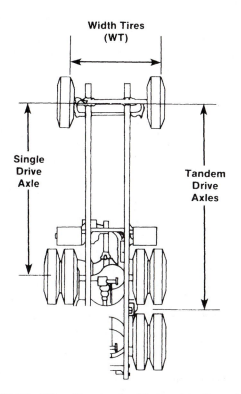

FIGURE 21–22   On single drive axles, the wheel base is measured from the centerline of the front axle to the centerline of the rear axle. On tandem drive axles, the wheelbase is measured from the centerline of the front axle to the center point between the two drive axles. (Courtesy of ArvinMeritor Inc.)

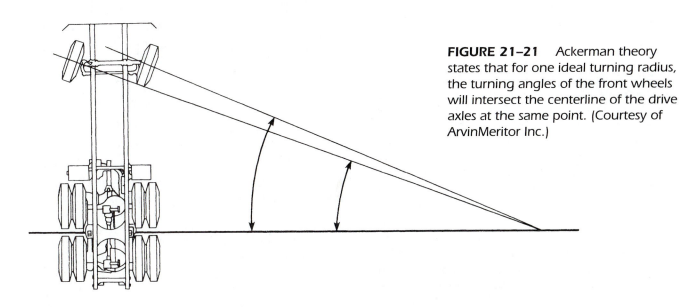

FIGURE 21–21   Ackerman theory states that for one ideal turning radius, the turning angles of the front wheels will intersect the centerline of the drive axles at the same point. (Courtesy of ArvinMeritor Inc.)

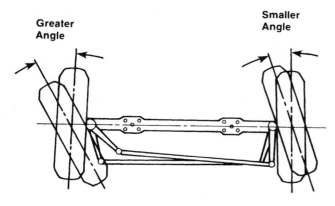

FIGURE 21–23 In a toe-out condition on turns, the inside wheel turns at a greater angle than the outside wheel. (Courtesy of ArvinMeritor Inc.)

of the steering arms and the length of the cross tube. Even though the toe-in with the wheels in the straight-ahead position can be adjusted correctly, a bent steering arm can cause the toe-out on a turn to be incorrect, causing the scuffing of tires.

The turning radius angle is checked using turning radius plates. To check the turning radius angle, position the front wheels on the plates and in the straight-ahead position. After removing the locking pins from each plate, adjust the scale on the edge of the plates so that the pointers read zero. Turn the wheels to the right until the gauge at the left wheel reads 20 degrees. Then read and record the angle of the right wheel. The right wheel should then be turned to an angle of 20 degrees. The left wheel should be at the same angle as was the right wheel when the wheels were turned to the right. If the angle is different, check the steering arms for damage and replace any one that is bent.

AXLE ALIGNMENT

Ideally, the vehicle's axles should be perpendicular to the vehicle centerline, and the rear wheels should track directly behind the front wheels when the vehicle is moving straight ahead. When this happens, the thrustline created by the rear wheels is parallel to the vehicle centerline (**Figure 21–24A**). However, if the axles are not perpendicular to the vehicle centerline, the rear wheels do not track directly behind the front wheels, and the thrustline of the rear wheels deviates from the centerline of the vehicle (**Figure 21–24B**). This can cause the steering wheel to rest in an uncentered position, the front tires to wear quickly, and the vehicle to oversteer when turning in one direction and understeer when turning in the other direction. Incorrect track can occur on single-axle vehicles, tandem-axle vehicles, and trailers. On a single-axle vehicle, the rear-axle thrustline can be off if the entire

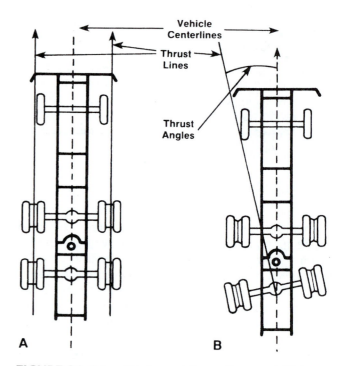

FIGURE 21–24 (A) Correct thrust line; and (B) incorrect thrust line. (Courtesy of ArvinMeritor Inc.)

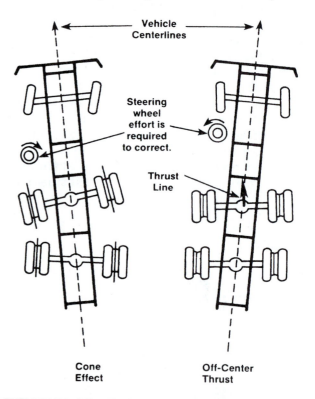

FIGURE 21–25 Typical tracking problems of tandem-axle vehicles. (Courtesy of ArvinMeritor Inc.)

axle is offset or if only one wheel has an improper toe angle. On a tandem axle, there are a number of different combinations that can cause incorrect tracking. **Figure 21–25** illustrates some of those combinations.

One method of checking a single axle for misalignment is to clamp a straightedge across the frame so that it is square with the frame rails on each side. Then simply measure from the straightedge to the center of the hub. The distances on each side must be within $1/8$ inch of each other. If not, the axle must be aligned.

Trailer Tracking

It is also possible for the trailer axles to be out of alignment and cause a tracking problem. Depending on the severity of the trailer misalignment, it might be possible to see the effects of the misalignment as the trailer travels down the road. Usually, the trailer will travel at an angle to the tractor. Misalignment also makes it very hard to back up the trailer. This is commonly called **dog-tracking (Figure 21–26)**.

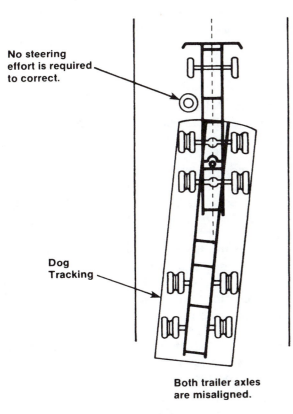

FIGURE 21–26 A dog-tracking trailer. (Courtesy of ArvinMeritor Inc.)

Measurement

For a tandem axle, the setup, measurement, and specifications for the forward-rear axle are the same as a single drive axle. The forward-rear axle must be square within $1/8$ inch. It is also necessary to check and make sure that the rear-rear axle is parallel with the forward-rear axle. A trammel bar or tape measure can be used to measure from the hub center to the

hub center on each side of the vehicle. These measurements must be within $1/4$ inch of each other. If either axle or both axles are misaligned, steering and tire wear problems will result. (Additional information on aligning tandem axles is given in Chapter 22.)

Axle Offset

Another problem is an axle that is not centered with the centerline of the vehicle (**Figure 21–27**). When the axle is offset and the vehicle is driven straight down the road, the steering wheel will be straight, and the vehicle will not dog-track, but when going around a corner, the vehicle will oversteer in one direction and understeer in the other direction.

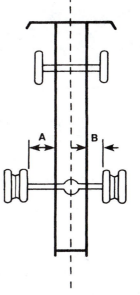

Dimension "A" should equal dimension "B."

FIGURE 21–27 Axle offset. (Courtesy of ArvinMeritor Inc.)

21.3 ELECTRONIC ALIGNMENT EQUIPMENT

As mentioned earlier in this chapter, wheel and axle alignment measurements can be taken in a number of ways. Two electronic methods are light or laser beam-type wheel alignment systems.

LIGHT BEAM ALIGNMENT SYSTEMS

Light beam or laser alignment systems use wheel-mounted instruments to project light beams onto charts and scales to measure toe, caster, and camber

FIGURE 21–28 Light beam alignment equipment. Light projectors are mounted to the wheels to project alignment angles onto special charts and scales. (Courtesy of Hunter Engineering Company)

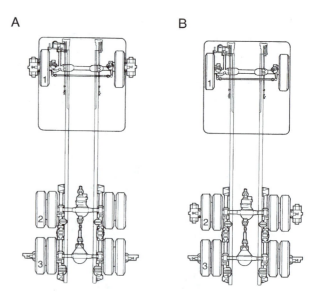

FIGURE 21–29 Typical light beam alignment setup for multiwheel alignment with all axles adjustable. (A) Setup to align front steering axle and rearmost tandem axle; and (B) setup to align forward tandem axle. (Courtesy of Hunter Engineering Company)

and to note the results of alignment adjustments (**Figure 21–28**). The wheel-mounted light projectors can be focused to handle any length wheelbase. The system checks all alignment angles. Two- and four-wheel models are available.

Four-wheel models reference all measurements to the rear wheel thrustline to ensure precise alignment. They can check rear camber and toe without moving the vehicle or switching the instruments. This is done by redirecting the light beam through the use of mirrors and precisely positioned scales built into the projectors and the rear wheel instruments.

Figure 21–29A shows the initial setup for multiwheel alignment on vehicles with all axles adjustable. The light projectors are mounted on the front steering axle, and the rear wheel instruments are mounted on the rear axle of the tandem drive axle.

The projectors project light in two directions: forward onto the wheel alignment chart positioned in front of the vehicle and rearward onto scales built into the rear wheel instruments. This setup enables you to measure and record the camber of the rear axle of the tandem drive axle. To set the thrust angle of the rearward tandem axle, the front steering wheels are positioned straight ahead so the light beam is equal on the rear scales of the rear wheel instruments. The steering wheel is then locked in place with a special tool, and the toe of the front wheel is adjusted to specification. Front wheel toe

readings are taken off the chart positioned in front of the vehicle. To adjust the rear axle of the tandem drive axle, the rear scales of the rear wheel instruments are lifted to expose mirrors. These mirrors redirect the light beam from the projector back onto scales mounted in the rear toe boxes of the projectors. The rearward tandem axle can then be aligned, based on the readings shown on the rear toe box scales.

To align the front steering axle, the front wheels are set straight ahead and camber and caster measurements are taken off the chart positioned in front of the vehicle. Adjustments are made as needed. With the wheels still in the straight-ahead position, toe is measured and adjusted as needed.

To align the forward axle of the tandem axle, the projectors are now mounted on this axle (**Figure 21–29B**). The rear scales of the rear wheel instruments are lifted to expose the mirrors. The beams from the projectors are now directed back into the rear toe boxes of the projectors. Adjustments are then made based on the reading on the rear toe box scales.

TRAILER AXLES

The light beam alignment system can also be used to align trailer axles. A gauge bar attached to the trailer kingpin works with the system to set trailer wheels straight ahead as referenced to the trailer centerline.

The light projectors are easily adjusted to compensate for wheel distortion, runout, and mounting discrepancies. During operation, light beams are projected onto the trailer gauge scales to show the position of the wheels. The wheels are adjusted to obtain equal measurement side to side on the scales. An air control assembly allows the operator to control the trailer brakes during the alignment operation.

The kingpin flag assembly is mounted on the kingpin approximately perpendicular (90 degrees) to the trailer's geometric centerline (**Figure 21–30A**). The projectors are mounted on the rearmost wheels of the trailer with the rear light beams directed toward the kingpin flag. The forward wheels are chocked, the air valve assembly is connected and the trailer brakes are released. The projectors are then adjusted by jacking up the rearmost axle and focusing the projector light beams on a special compensating panel positioned between the projector and the kingpin flag. This adjustment compensates for any wheel distortion, runout, or mounting discrepancies.

When this adjustment is made, the rearmost axle of the trailer can be adjusted until the light beam readings are equal on both kingpin flags. The rear wheel instruments are then installed on the front axle wheel with the mirrors facing the projectors (**Figure 21–30B**). The rearmost wheels are chocked. The front axle is then raised, and the projectors are adjusted to compensate for irregularities. The mirrors of the rear wheel instruments are exposed to redirect the light beams back into the rear toe boxes of the projectors. The front axle is then adjusted until the readings on the rear toe box scales are equal.

This procedure can equalize the thrustline of all axles to the geometric centerline. Individual toe and total toe cannot be determined.

COMPUTERIZED ALIGNMENT SYSTEMS

The most advanced alignment systems in use today are controlled by microprocessors. They use a sensor mounted at each wheel for fast precise alignment (**Figure 21–31**). Alignment readings, specifications, and step-by-step instructions are displayed on a cathode ray tube (CRT) screen. Keyboard-entered specifications are automatically compared against the actual angles of the vehicle with the results displayed on the CRT screen. Specifications can be retained in the computer's memory for future use.

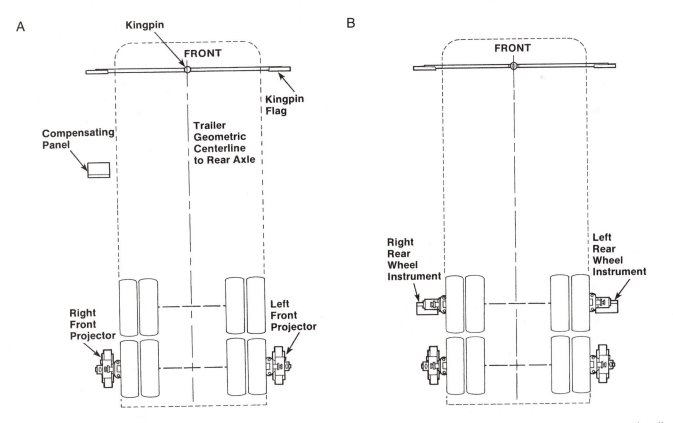

FIGURE 21–30 *Typical light beam alignment setup for trailer axle alignment: (A) setup for rearmost axle alignment; and (B) setup for forward axle alignment. (Courtesy of Hunter Engineering Company)*

FIGURE 21-31 *Computer-controlled alignment equipment. This system uses computer-controlled wheel-mounted sensors, toe lines, and a microprocessor. (Courtesy of Hunter Engineering Company)*

Special instructions and graphics on the screen guide the operator as adjustments are made on the truck (**Figure 21-32A**). As adjustments are made the block on the CRT screen moves toward its target. When the block aligns with the target, as on the right in **Figure 21-32A**, the adjustment is within one-half of the specified tolerance. Alignment angles can also be more precisely adjusted by selecting a screen display that shows numbers rather than bar graphs (**Figure 21-32B**). In this mode, the system offers complete flexibility for the expert alignment technician.

The system provides for four-wheel alignment with four sensors (two axles simultaneously) or for two-wheel alignment with two sensors (one axle at a time). All wheels are aligned to a common centerline for precise alignment.

By moving instruments, the system can also check both rear axles of a tandem drive axle, as well as the front steering axle. This means all axles can be aligned relative to a common centerline. **Figure 21-33A** shows the initial setup for multiwheel alignment on vehicles with all axles adjustable. Sensors are mounted on the front steering axles and the rear axle of the tandem drive axle. Toe lines are run between the sensors. The system is then adjusted or compensated according to manufacturer's instructions before the alignment checks are made.

After the truck specifications and tolerances are keyed into the system computer, alignment work can begin. By pressing the appropriate key, the thrust angle of the rear drive axle is computed and displayed on the CRT. Adjustments to this angle can then be made until it is within specifications. The caster and toe is then measured and adjusted on the front steering axle. Once again, a few simple keystrokes compute the data and track alignment progress.

After the front axle is aligned, the front sensors are moved to the forward axle of the tandem pair. The toe lines are attached as shown in **Figure 21-33B**. The thrust angle of this axle is then measured and adjusted as needed. Like the light beam system explained earlier, this computer-controlled system can be adapted to align trailer axles using a gauge bar attached to the trailer kingpin. Toe lines connect the gauge to the sensors (**Figure 21-34**).

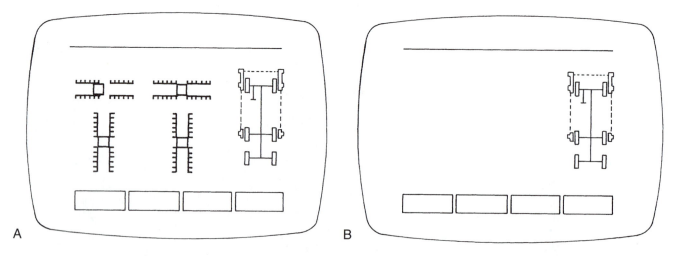

FIGURE 21-32 *In this computer-controlled system, two different visual modes can be used to track adjustments. (A) Sliding blocks move into position when adjustments reach specifications. (B) Precise numerical readouts allow fine tuning of adjustments. (Courtesy of Hunter Engineering Company)*

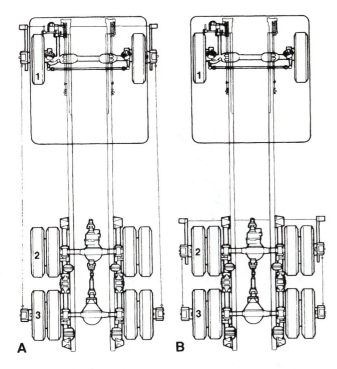

FIGURE 21–33 *Sensor mounting setup for (A) front steering axle and rearmost tandem axle alignment; and (B) forward tandem axle alignment using a computer-controlled alignment system. (Courtesy of Hunter Engineering Company)*

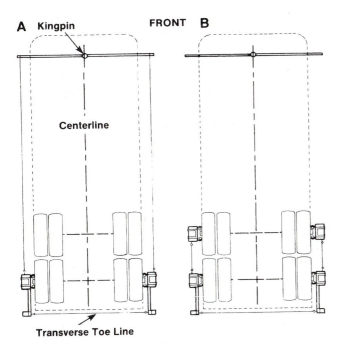

FIGURE 21–34 *Computer sensor mounting for trailer axle alignment: (A) rearmost axle alignment; and (B) forward axle alignment. (Courtesy of Hunter Engineering Company)*

STEERING AXLE INSPECTION

More specific servicing procedures and intervals can be found in the vehicle manufacturer's service manual. The following procedures are general guidelines for periodic steering axle inspection:

- **General Inspection.** A thorough visual inspection for proper assembly, broken parts, and looseness should be performed each time the vehicle is lubricated. In addition, make certain that the spring-to-axle mounting nuts and the steering connection fasteners are secure.
- **Wheel Alignment.** Front steering wheel alignment should be checked periodically according to the vehicle manufacturer's recommended service interval (typically 3 months or 25,000 miles). However, if excessive steering effort, vehicle wander, or uneven and excessive tire wear is evident, the wheel alignment should be checked immediately.
- **Steering Axle Stops.** While steering axle stops should be checked periodically, they seldom need adjustment. However, if the steering turning radius is insufficient or excessive (resulting in the tires or wheels rubbing against other truck parts), the stops should be adjusted.
- **Tie-Rod Ends.** Tie-rod ends should be inspected each time the axle is lubricated. Check for torn or cracked seals and boots, worn ball sockets, or loose fasteners.
- **Knuckle Thrust Bearings.** Theknuckle thrust bearings should be checked each time the hub/drum is removed. The knuckle vertical play should be adjusted each time the knuckle pin is removed for service, at each axle overhaul, or whenever excessive knuckle vertical movement is noted.
- **Knuckle Pins.** The knuckle pin that fits in the bushings should be inspected when the pin is removed for service, at axle overhaul, or whenever looseness is noted.
- **Wheel Bearings.** The wheel bearings should be inspected for damage or wear each time the hub or drum is removed. Check for signs of wear and distress.

Lubrication

The steering axle components should be lubricated at least every 25,000 miles. A heavy-duty multipurpose lithium-based grease should be used. Knuckle pins, thrust bearings, and tie-rod ends should be pressure lubricated at each PM service interval. Those without grease fittings are permanently lubricated.

The recommended steering gear lubricants are given in **Table 21–1**.

TABLE 21–1:	STEERING GEAR LUBRICANTS
Temperature	Hypoid Gear Oil
continuously above +30° F	SAE 90 API Service Classification GL-5
continuously below +30° F	Hypoid Gear Oil SAE 75 API Service Classification GL -5

WARNING: If the vehicle is operated at temperatures below +30°F with SAE 90 weight oil in the steering gear, the steering can become sluggish and resistant to the point where the driver may have difficulty performing an evasive maneuver. When operating in temperatures continuously below +30°F, install SAE 75 weight oil.

Wheel bearings are generally lubricated at each brake overhaul or whenever the hub or drum is removed for service. Procedures for lubricating wheel bearings and hubs are given in Chapter 23.

STEERING AXLE PROBLEM DIAGNOSIS

If general troubleshooting and inspecting of a steering system points to a problem in the steering axle, perform the following procedures to identify parts that are worn or in need of adjustment.

CAUTION: Before making any test on the steering axle or performing any servicing procedure, set the parking brake and block the drive wheels to prevent the vehicle from moving. After jacking the truck up until the steering axle wheels are raised off the ground, support the truck with safety stands. Never work under a vehicle supported only by a jack.

Knuckle Vertical Play Inspection

1. Mount a dial indicator on the axle beam. Position the indicator plunger on the knuckle cap (**Figure 21–35**).
2. Pry the steering knuckle downward.
3. Zero the dial indicator.
4. Lower the front axle to obtain the dial indicator reading. If the reading exceeds the manu-

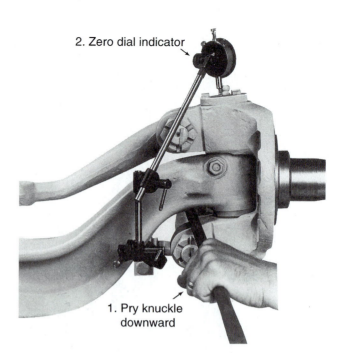

2. Zero dial indicator

1. Pry knuckle downward

FIGURE 21–35 *Inspecting knuckle vertical play. (Courtesy of Roadranger Marketing. One great drive train from two great companies—Eaton and Dana Corporations)*

facturer's specifications (typically 0.04 inch), inspect the thrust bearings. Replace them if necessary.

Knuckle Pin Fit Inspection— Upper Bushing

1. Mount a dial indicator on the axle as shown in **Figure 21–36**. Reference the plunger on the upper part of the knuckle.
2. Move the top of the wheel in and out with a push/pull motion. Have an assistant monitor the dial indicator.

Reference dial indicator on knuckle

FIGURE 21–36 *Inspecting freeplay in upper bushing. (Courtesy of Roadranger Marketing. One great drive train from two great companies—Eaton and Dana Corporations)*

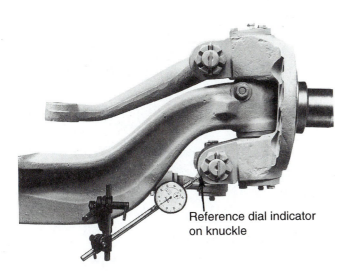

FIGURE 21–37 Measuring freeplay in lower bushing. (Courtesy of Roadranger Marketing. One great drive train from two great companies—Eaton and Dana Corporations)

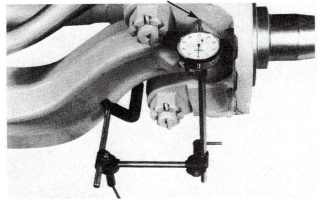

FIGURE 21–38 Measuring upper bushing torque deflection. (Courtesy of Roadranger Marketing. One great drive train from two great companies—Eaton and Dana Corporations)

3. Readings in excess of the manufacturer's specifications (typically 0.015 inch) indicate the need for bushing replacement.

Lower Bushing Inspection

1. Mount the dial indicator on the axle. Reference the plunger on the lower tie-rod end socket of the steering knuckle (**Figure 21–37**).
2. Move the bottom of the wheel in and out with a push/pull motion. Have an assistant read the dial indicator.
3. A dial indicator reading in excess of the manufacturer's specifications (typically 0.015 inch) indicates that the lower bushing should be replaced.

Upper Bushing Torque Deflection Test

1. Mount the dial indicator to the axle, referencing the upper knuckle steering arm socket area (**Figure 21–38**).
2. Have an assistant apply the foot brake. Try to roll the wheel forward and backward and note the deflection.
3. Readings in excess of the manufacturer's specifications (typically 0.015 inch) indicate that the top bushing should be replaced.

Lower Bushing Torque Deflection Test

1. Mount the dial indicator on the axle and the plunger on the lower bushing area.

2. Have an assistant apply the foot brake. Try to roll the wheel forward and backward and note the deflection.
3. Readings in excess of the manufacturer's specifications (typically 0.015 inch) indicate that the lower bushing should be replaced.

Tie-Rod Inspection

1. Shake the tie-rod or cross tube. Movement, or looseness, between the tapered shaft of the ball and the cross tube socket members indicates that the tie-rod end assembly should be replaced.
2. The threaded portion of both tie-rod ends must be inserted completely in the tie-rod split (**Figure 21–39**). This is essential for adequate clamping. Replace components if this fit cannot be obtained.
3. Ball and socket torque (exclusive of boot resistance) should be 5 inch-pounds or more on disconnected tie-rod end assemblies. Replace assemblies that test less than 5 inch-pounds. Loose assemblies will adversely affect the steering system performance and might prevent adjustment of the steering assembly to the vehicle manufacturer's recommended alignment specifications.
4. If the taper shank-to-tie-rod arm connection is loose or the cotter pin is missing, disconnect and inspect these parts for worn contact surfaces. If either one is worn, replace it.

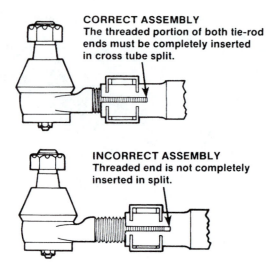

CORRECT ASSEMBLY
The threaded portion of both tie-rod ends must be completely inserted in cross tube split.

INCORRECT ASSEMBLY
Threaded end is not completely inserted in split.

FIGURE 21–39 Correct and incorrect tie-rod installation. (Courtesy of Roadranger Marketing. One great drive train from two great companies—Eaton and Dana Corporations)

Wheel Bearing Inspection

1. Remove the wheel bearings from the spindle bearings.
2. Thoroughly clean the bearings, spindle, hub cap, and hub cavity. Parts can be washed in a suitable commercial solvent. Be certain that the parts are free of moisture or other contaminants.
3. Clean hands and tools to be used before assembly.
4. Closely examine the bearing cup, rollers, and cage (**Figure 21–40** and **Figure 21–41**).

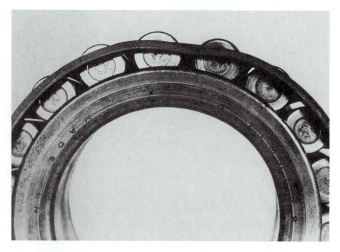

FIGURE 21–40 Inspect bearings for damage, nicks, and gouges. (Courtesy of Roadranger Marketing. One great drive train from two great companies—Eaton and Dana Corporations)

FIGURE 21–41 Inspect bearing races for grooving, etching, staining, and scoring. (Courtesy of Roadranger Marketing. One great drive train from two great companies—Eaton and Dana Corporations)

5. Replace any damaged or distressed bearings with mated bearing assemblies.
6. Lubricate bearings, after making certain they are free of moisture or other contaminants. Refer to Chapter 23 for additional information on installing bearings.

Adjusting Knuckle Vertical Play

If the vertical play in the knuckle exceeds the manufacturer's specifications, adjustments can be made by removing or installing shims between the axle beam and the upper part of the knuckle. To do so, follow these steps:

1. Remove the capscrews, lockwashers, and knuckle caps.
2. Remove the nut and draw key.
3. Partially drive out the knuckle pin, using a hammer and brass drift. Drive from the top down. Remove the pin far enough to facilitate removal and replacement of the shims.
4. Adjust the shim pack (**Figure 21–42**) to obtain the manufacturer's specified vertical play (typically 0.005–0.025 inch).
5. With the correct shims installed, drive the knuckle pin back into its original mounting position.

FIGURE 21–42 With the knuckle pin partially removed, remove or install shims to change camber. (Courtesy of Roadranger Marketing. One great drive train from two great companies—Eaton and Dana Corporations)

Remove or install shims.

Knuckle pin partially removed

6. Recheck the vertical play and readjust if necessary to stay within specifications.
7. With the adjustment complete, make sure that the knuckle pin flat is aligned with the draw key opening.
8. Install the draw key and nut. Torque the nut to the manufacturer's specification.
9. Install the knuckle cap gasket (if applicable) or install a silicone rubber gasket compound on the cap mounting surface. Install caps.
10. Install the capscrews and the lockwashers. Torque the capscrews to the manufacturer's specifications.

STEER AXLE OVERHAUL

When worn or bent components require the steer axle to be overhauled, follow the procedures given here.

Overhaul Preliminary Steps

Prior to performing overhaul procedures, the following preliminary steps should be accomplished.

1. Set the parking brake and block the drive wheels to prevent vehicle movement.
2. Jack the vehicle until steering axle wheels are raised from the ground. Support the raised vehicle with safety stands.

WARNING: Never work under a vehicle supported by only a jack.

Steering Knuckle Disassembly

1. Back off slack adjuster-to-return brake shoes to the released position and clear of the drum.
2. Remove the hub cap from the wheel and then remove the cotter pin, nut, washer, and outer bearing cone assembly.
3. Remove the wheel and hub assembly.
4. Disconnect the air (or hydraulic) line from the brake assembly.

CAUTION: Do not allow contamination of hydraulic fluid.

5. Remove brake assembly.
6. To disconnect the tie-rod end from the Ackerman arm, remove the cotter pin and slotted nut (**Figure 21–43**).
7. Disconnect the tie-rod end using a suitable tool such as a pickle fork (**Figure 21–44**).

CAUTION: Never use heat on any steering system components.

Remove cotter pin and nut.

FIGURE 21–43 To remove a tie-rod, first remove the cotter pin and nut. (Courtesy of Roadranger Marketing. One great drive train from two great companies—Eaton and Dana Corporations)

PHOTO SEQUENCE 6
MEASURING STEERING KINGPIN WEAR AND VERTICAL END-PLAY

P6–1 Apply the parking brakes, block the rear wheels, and use a floor jack to raise the front axle until the tires are off the shop floor.

P6–2 Lower the vehicle so that it is supported securely on safety stands with the front tires still off the shop floor.

P6–3 Mount a dial indicator on the front axle I-beam and position the dial indicator plunger on the inner side of the upper end of the steering knuckle. Zero the dial indicator.

P6–4 While a coworker moves the top of the wheel and tire inward and outward, observe the dial indicator reading. If the total movement on the dial indicator exceeds the specified knuckle pin bushing movement, the knuckle pin bushing must be replaced.

P6–5 Mount the dial indicator on the front axle I-beam with the dial indicator stem touching the inner side of the lower end of the knuckle. Zero the dial indicator. While a coworker moves the bottom of the tire inward and outward, observe the dial indicator reading. If the dial indicator reading exceeds the specified knuckle pin bushing movement, replace the lower knuckle pin bushing.

P6–6 Mount the dial indicator on the axle I-beam and position the dial indicator plunger on top of the upper knuckle pin cap.

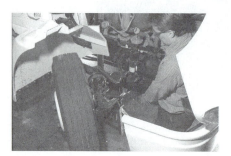

P6–7 Use a pry bar to force the steering knuckle downward. Check the reading on the dial indicator.

P6–8 Next observe the dial indicator reading while a coworker uses a large pry bar to lift upward on the tire and wheel. If the dial indicator reading exceeds the truck manufacturer's specifications, remove the steering knuckle and inspect the thrust bearing. Replace this bearing if necessary and install the required shim thickness.

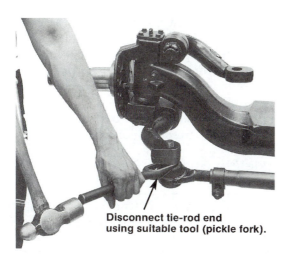

Disconnect tie-rod end
using suitable tool (pickle fork).

FIGURE 21–44 Use a pickle fork to separate the tie-rod end from the tie-rod arm. *(Courtesy of Roadranger Marketing. One great drive train from two great companies—Eaton and Dana Corporations)*

8. Disconnect the drag link from the steering arm by removing the cotter pin and slotted nut.

Knuckle Pin Removal

1. Remove the capscrews and lockwashers to remove the knuckle caps.
2. Remove the nut from the draw key; then, drive the key out using a hammer and brass drift (**Figure 21–45A**).
3. Drive the knuckle pin out with a hammer and brass drift (**Figure 21–45B**).
4. Remove the steering knuckle from the axle beams and discard the thrust bearing.

1. Remove nut
2. Drive out draw key.

A

Cleaning

After disassembly and before attempting inspection, clean the parts as follows:

1. Wash the steel parts with ground or polished surfaces in suitable cleaning solvent. Dry parts with clean lintless rags.

WARNING: Gasoline is not an acceptable cleaning solvent because of its extreme combustibility. It is unsafe in the workshop environment.

2. Wire brush or steam-clean castings, forgings, and other rough-surface parts that are susceptible to accumulation of mud, road dirt, or salt.

Knuckle Pin Grease Seal Replacement

1. Remove the grease seal from the knuckle upper arm.
2. Install a new seal, using a suitable pilot drift or a similar device that will not damage the seal as it is installed.

Shop Talk _____

*When installing the grease seal, be sure the lip is pointing toward the axle side of the knuckle (**Figure 21–46**). This is essential for correct seal operation.*

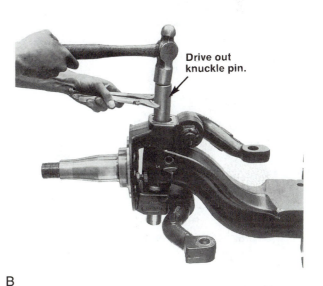

Drive out
knuckle pin.

B

FIGURE 21–45 To remove the kingpin, (A) drive out the draw key and (B) drive out the knuckle pin. *(Courtesy of Roadranger Marketing. One great drive train from two great companies—Eaton and Dana Corporations)*

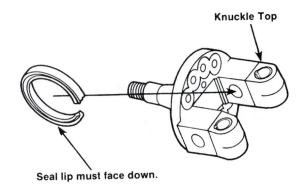

Knuckle Top

Seal lip must face down.

FIGURE 21–46   *The seal lips must face the axle when installed in the knuckle. (Courtesy of Roadranger Marketing. One great drive train from two great companies—Eaton and Dana Corporations)*

Ackerman or Steering Arm Replacement

1. Disconnect the tie-rod end from the Ackerman arm (or drag link from the steering arm).
2. Remove the cotter pin and nut from the arm at the knuckle end.
3. Drive the arm out of the knuckle using a suitable brass drift and hammer.

> **CAUTION:**   Never use heat on any steering components.

4. Install a new key in the slot provided in the Ackerman arm (or steering arm).
5. Install the nut and tighten it to the correct torque. Install the cotter pin.

 Shop Talk _____

Tighten to the minimum torque first; then continue tightening to align the cotter pin hole. Verify that the torque does not exceed the maximum specification. If it does, replace the nut.

6. Reconnect the tie-rod ends. Install the nut and tighten it to the correct torque. Install the cotter pin.

Tie-Rod End Replacement

1. Disconnect the tie-rod end.
2. Loosen the clamp nut and unscrew the tie-rod end.

3. Install the new tie-rod end. Thread the end into the cross tube past the tube split (see **Figure 21–39**).
4. Attach the rod end to the Ackerman arm. Tighten to the correct torque. Install the cotter pin.
5. Adjust toe-in. Tighten the cross tube clamp screw to correct the torque.

Knuckle Pin Bushing Replacement

> **WARNING:**   During removal and installation procedures, never use a steel hammer or tool to strike hardened metal parts.

1. Drive the upper bushing out of the knuckle using suitable piloted drift.
2. Drive the lower bushing out in the same manner.
3. Remove all foreign material from the knuckle pin areas of the steering knuckle and axle beam.
4. Wire-brush machined surfaces taking care not to damage them. Clean the knuckle pin bores.
5. Lightly lubricate the outside diameter of the bushing.
6. Align the bushing seam as shown in **Figure 21–47**.
7. To install the bushings, first hand start the bushing in the bore. Then, using a suitable piloted drift, drive bushings in until they are properly located (see **Figure 21–47**). The bushings are usually recessed below the surface of the knuckle. On some models, the lower bushing might be recessed until it is flush with the inside surface of the knuckle.

Steering Knuckle Assembly

 Shop Talk _____

Always replace the knuckle pin if the pin finish is damaged. Always replace the pin and bushings when either are faulty. Some bushings must be reamed to fit the kingpin after installation.

1. Lightly lubricate the thrust bearing areas of the steering knuckle, axle end, and knuckle pin bore of the axle beam (**Figure 21–48**).
2. Position the thrust bearing on the knuckle with the seal facing up (**Figure 21–49**).
3. Mount this assembly on the axle beam.

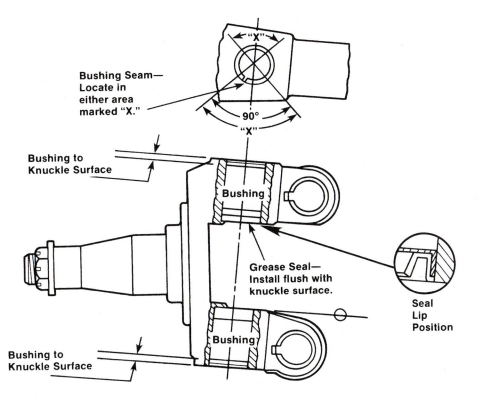

Bushing Seam—Locate in either area marked "X."

Bushing to Knuckle Surface

Bushing

Grease Seal—Install flush with knuckle surface.

Bushing

Seal Lip Position

Bushing to Knuckle Surface

FIGURE 21–47 *Bushing and seal installation. (Courtesy of Roadranger Marketing. One great drive train from two great companies—Eaton and Dana Corporations)*

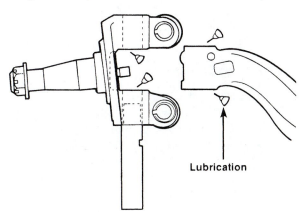

Lubrication

FIGURE 21–48 *Lubricate the steering knuckle and axle before assembling the two. (Courtesy of Roadranger Marketing. One great drive train from two great companies—Eaton and Dana Corporations)*

Remove knuckle assembly.

FIGURE 21–49 *Position the thrust bearing with the seal up. (Courtesy of Roadranger Marketing. One great drive train from two great companies—Eaton and Dana Corporations)*

4. Wedge up the steering knuckle and fill the gap (at the knuckle upper end) with shim(s). This is a preadjustment of knuckle vertical play.

5. Position the knuckle pin flat, aligned with the draw key hole, and start installing the pin by hand.

6. Install the pin in the knuckle and axle beam from the top. If necessary, tap the pin in place using a hammer and brass drift.

 Shop Talk _____

At this point in reassembly, check the knuckle vertical play and adjust if necessary (see Steps 7 through 10).

_____

7. To check knuckle vertical play, center the steering components.

8. Mount the dial indicator to the flange and reference top of the knuckle pin (**Figure 21–50A**). Zero the dial indicator.

9. Simulate axle loading with a jack (**Figure 21–50B**) and note the dial indicator reading.

10. If the vertical play is not according to specifications, add or remove shims as necessary to obtain the correct end play.

11. With the adjustment complete, check the draw key opening and pin flat alignment (**Figure 21–51**).

12. Install a new draw key, using a soft face hammer.

13. Install the draw key nut and tighten it to the correct torque.

14. Apply silicone rubber gasket compound (**Figure 21–52**) to the knuckle cap mounting surface or install a gasket if applicable.

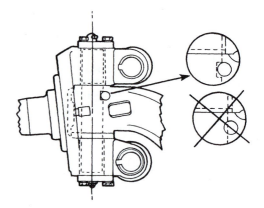

FIGURE 21–51 Align the draw key hole with the knuckle pin flat. (Courtesy of Roadranger Marketing. One great drive train from two great companies—Eaton and Dana Corporations)

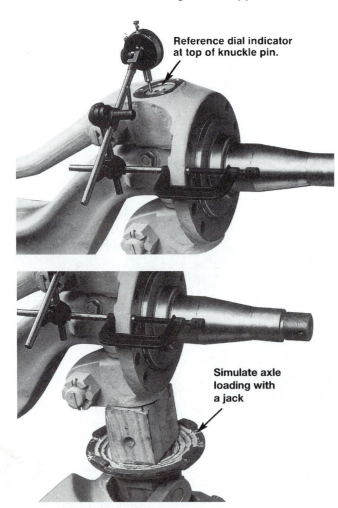

FIGURE 21–50 To check knuckle vertical play, (A) reference the dial indicator at the top of the knuckle pin, and (B) simulate axle loading with a jack. (Courtesy of Roadranger Marketing. One great drive train from two great companies—Eaton and Dana Corporations)

FIGURE 21–52 Apply silicone rubber gasket compound to the knuckle flat before installing the cap. (Courtesy of Roadranger Marketing. One great drive train from two great companies—Eaton and Dana Corporations)

15. Install capscrews and lockwashers to secure the cap.

16. Tighten the capscrews to the manufacturer's specification.

17. Attach the drag link to the steering arm. Install and torque the nut to specifications. Install the cotter pin.

18. Attach the tie-rod end to the Ackerman arm. Install and torque. Install the cotter pin.

19. The brake and wheel assemblies can now be reinstalled, following the manufacturer's instructions.

MANUAL STEERING GEARS

As was mentioned earlier in this chapter, basically two types of manual steering gears are used in heavy-duty trucks: the worm and roller type and the recirculating ball and worm type.

Worm and Roller Gears

A standard worm gear arrangement consists of a worm or driving gear and a driven gear called a roller or sector gear. The tooth pattern on a worm is called a thread, and the number of teeth it has determines whether it is a single-thread or double-thread worm.

The worm gearing used in steering systems is slightly modified. The driving worm gear has more of an hourglass shape (**Figure 21–53**) and is attached

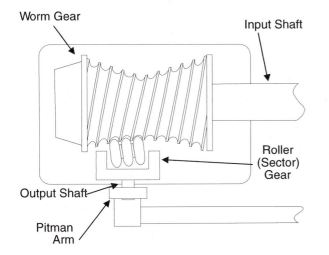

FIGURE 21–53 An hourglass-shaped worm gear and the sector roller manual steering gear.

to the input shaft linked to the steering shaft. The threads of the driving worm gear are meshed with the threads of a sector gear, commonly called a roller. The driven sector gear transfers the rotary motion of the worm gear input shaft to the output shaft. The output shaft is splined to the Pitman arm, which translates the rotary motion of the shaft to linear motion. Movement in the Pitman arm is transmitted through a drag link to the steering arm and knuckle and thus to the wheels.

Recirculating Ball Gears

The input shaft of this type of steering gear (**Figure 21–54**) is also connected to a worm gear, but the worm gear in a recirculating ball type unit is straight. Mounted on the worm gear is a ball nut. The ball nut has interior spiral grooves that mate with the threads of the worm gear. The ball nut also has exterior gear teeth on one side. These teeth mesh with teeth on a sector gear and shaft.

In the grooves between the ball nut and the worm gear are ball bearings. The ball bearings allow the worm gear and ball nut to mesh and move with little friction. When the steering wheel is turned, the input shaft will rotate the worm gear. The ball bearings will transmit the turning force from the worm gear to the ball nut, causing the ball nut to move up and down the worm gear. Ball return guides are connected to each end of the ball nut grooves. These allow the ball bearings to circulate in a continuous loop.

As the ball nut moves up or down on the worm gear, it causes the sector gear to rotate, which, in turn, causes the Pitman arm to swivel back and forth. This motion is transferred to the steering arm and knuckle to turn the wheel.

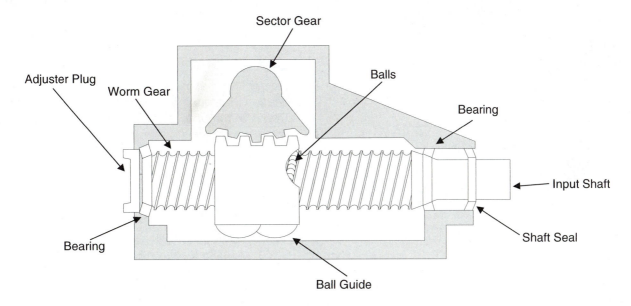

FIGURE 21–54 A recirculating ball type steering gear.

Manual Steering Gear Troubleshooting

A manual steering gear is a very reliable component. It is filled with lubricant but has few moving parts. However, due to wear and road shocks, it will occasionally need adjustments.

Mesh and Backlash

When driving straight ahead, the roller (or sector gear) should be positioned at the center of the worm gear. Maximum mesh between the gears is achieved in this position, which is considered the mesh, or nonbacklash area (**Figure 21–55A**). The minimum clearance in the center of the gear results in good response and road feel at the steering wheel during highway speeds.

As the steering wheel is turned and the roller moves away from center, some of the mesh begins to be lost. That is, clearances begin to develop between the roller and worm gear teeth. These clearances permit a reaction known as backlash, which begins to be felt approximately one-half turn each way from the center position (**Figure 21–55B**). Backlash can be felt by wiggling the Pitman arm.

Backlash is necessary because of the normal wear pattern of the steering gear. Because most driving is done at or near the straight-ahead position, it follows that most of the wear occurs in the center of the worm gear. In order to maximize the service life of the gear assembly, manufacturers have designed

them with increasing gaps between gear teeth in the off-center positions. This allows the gear to be tightened to adjust for wear in the center without causing a bind in the outer areas that receive less wear.

Preloads

All steering gears are assembled with a specified amount of resistance to the turning of the input shaft. This preload is necessary to control the various elements in the steering linkage and to prevent random movement within the system.

There are two types of preloads: worm bearing preload and total mesh preload. The *worm bearing preload* is an end-to-end pressure on the worm shaft and its bearings. The *total mesh preload* is the sum of pressure of the roller against the worm gear and the worm bearing preload. Both types of preloads are originally set to factory specifications (expressed in inch-pounds of torque) and can be reset by the user as required.

When preloads are out of adjustment, steering can be affected significantly. Insufficient preload permits lost motion within the gear, which means that the driver must turn the steering wheel farther than normal to get any response from the front wheels. Excessive preloads result in hard steering, and sometimes in a "breaking loose" effect that produces a darting and oversteering condition.

Measuring and Adjusting Preloads

Incorrect preloads are a common cause of the steering problems mentioned, but by no means are they the only cause. Before checking the preloads, it is wise to check the entire steering linkage for worn or binding areas. If all other parts of the linkage operate smoothly, proceed with the preload check.

👓 **Shop Talk** _____

Remember that there are two distinct preload specifications (frequently called high point adjustment). Do not attempt to shortcut the process by working with the total mesh preload only.

_____

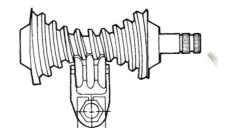

A

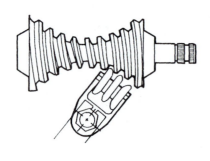

B

FIGURE 21–55   (A) Mesh area; and (B) backlash area. (Courtesy of Air-O-Matic Power Steering, Division of Sycon Corp.)

Worm Bearing Preload Check (Measured in the Backlash Area)

1.   Position the front wheels straight ahead. If possible, drive the vehicle in a straight line for a short distance, stopping at the spot where service operations will be performed. Apply the parking brakes and chock the tires.

2. Disconnect the drag link from the Pitman arm and the steering shaft from the input shaft.

3. Slowly turn the input shaft until the roller (sector gear) is almost at the end of the worm gear.

CAUTION: Do not force the roller all the way to the end of the worm, or damage can result.

---

4. Reverse the rotation of the input shaft a quarter turn. In this position, there should be no contact between the roller and worm gear teeth.

5. Place an inch-pound torque wrench on the end of the input shaft. Turn the torque wrench back and forth within about a quarter turn arc to record the torque in the area where the gear teeth are not touching. The output shaft should not turn.

6. If the preload reading of the moving shaft is within the manufacturer's recommended limits proceed to the total mesh preload check. If the preload is above or below these limits, it will need to be adjusted. This is done by adding or removing shims between the gear housing and the cover plate for the gears shown in **Figure 21–56** or by turning the worm bearing adjuster plug on the gears shown in **Figure 21–57**. Full instructions are contained in the manufacturer's service manual.

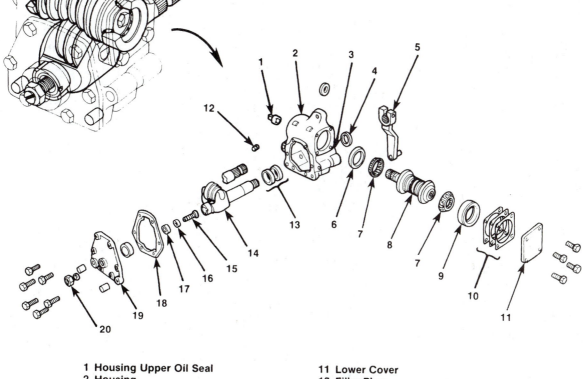

1 Housing Upper Oil Seal	11 Lower Cover
2 Housing	12 Filler Plug
3 Bushing	13 Mounting Spacer
4 Side Oil Seal	14 Lever Shaft
5 Steering Arm	15 Lever Shaft Adjusting Screw
6 Upper Bearing Cup	16 Thrust Washer
7 Bearing Cone	17 Snap Ring
8 Tube with Cam	18 Side Cover Gasket
9 Lower Bearing Cup	19 Side Cover
10 Lower Cover Shim Pack	20 Jam Nut

FIGURE 21–56 Exploded view of a typical manual steering gear. (Courtesy of International Truck and Engine Transportation Corporation)

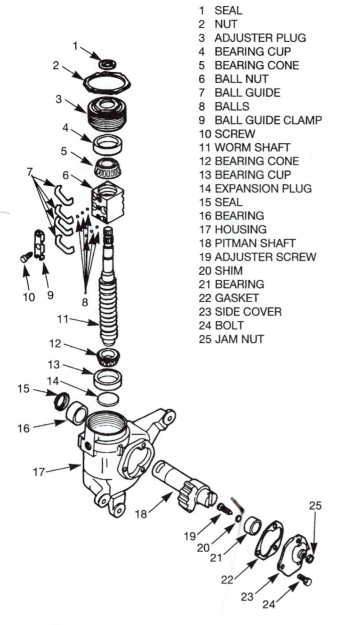

1 SEAL
2 NUT
3 ADJUSTER PLUG
4 BEARING CUP
5 BEARING CONE
6 BALL NUT
7 BALL GUIDE
8 BALLS
9 BALL GUIDE CLAMP
10 SCREW
11 WORM SHAFT
12 BEARING CONE
13 BEARING CUP
14 EXPANSION PLUG
15 SEAL
16 BEARING
17 HOUSING
18 PITMAN SHAFT
19 ADJUSTER SCREW
20 SHIM
21 BEARING
22 GASKET
23 SIDE COVER
24 BOLT
25 JAM NUT

FIGURE 21–57 Exploded view of a typical recirculating ball type manual steering gear.

Worm Shaft Bearing Preload Adjustment (Worm and Roller Gear)

1. Clean the steering gear.
2. Remove the fill plug and the drain plug from the steering gear. Drain the lubricant and dispose of it in an environmentally safe manner.
3. Remove the worm shaft cover screws and the worm shaft cover.
4. If worm shaft preload is loose or less than the torque specification required, remove the shims. Removal of shims will increase preload; adding shims will decrease preload. The shims can be in five sizes: 0.002, 0.003,

0.005, 0.0075, and 0.010 inch. Remove only one shim at a time, as necessary, using a knife blade. The size of the shim to be removed or added will depend on the amount of lash. Use a micrometer to identify the thickness of the shims.

5. Assemble the worm shaft cover and shims.
6. Install the worm shaft cover screws. Obtain a light preload before final tightening of the screws.

Shop Talk

Do not final torque the screws immediately. Rotate the worm shaft assembly with one hand, while slowly tightening the worm cover screws with the other hand. This will prevent damaging the bearing surfaces if the initial amount of shims is inadequate.

7. Using a crisscrossing pattern or sequence, tighten the worm shaft cover screws to the manufacturer's torque specification.
8. Recheck the worm bearing preload. If necessary, repeat the adjustment procedure using a different selection of shims.

Roller Shaft Total Mesh Preload Check

Do not check the roller shaft total mesh preload unless the worm shaft bearing preload is within the specifications. Checking the roller shaft total mesh preload when the worm shaft bearing preload is out of adjustment will result in inaccurate measurements.

1. After making the correct worm shaft bearing preload adjustment, centralize the steering gear by turning the worm shaft from stop to stop, counting the number of turns required. Starting at the end of travel, turn back half that number of turns. This will place the roller tooth of the roller shaft in the approximate center of the worm gear.
    Check the alignment mark on the output shaft of the steering gear. If the alignment mark is at a right angle to the worm shaft, the gear is centered (**Figure 21–58**). If not, turn the worm shaft backward or forward until the output shaft alignment mark is at a right angle to the worm shaft.
2. Rotate the worm shaft with a torque wrench approximately 180 degrees through the center position. Compare the torque reading with the manufacturer's specification. If the roller shaft total mesh preload measurement is not within specifications, adjust the preload.

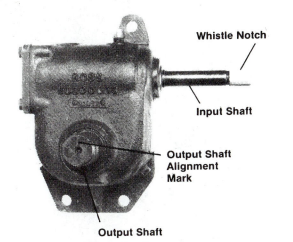

Whistle Notch

Input Shaft

Output Shaft Alignment Mark

Output Shaft

FIGURE 21–58 *When the gear is centered, the output shaft alignment mark is at a right angle to the input shaft. (Courtesy of Freightliner Trucks)*

Roller Shaft Total Mesh Preload Adjustment

1. If adjustment is necessary, loosen the adjustment screw locknut (**Figure 21–59**).
2. Center the steering gear.
3. Turn the slotted adjustment screw clockwise to increase the preload or counterclockwise to decrease the preload; then check the adjustment. Repeat until the preload is within specifications.
4. Hold the adjustment screw in place with a screwdriver while tightening the adjustment screw locknut to specifications.

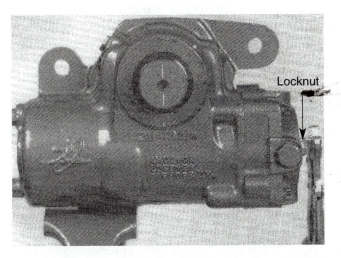

Locknut

FIGURE 21–59 *To adjust the total mesh preload, first loosen the adjustment screw locknut. Then, turn the adjustment screw as necessary. (Courtesy of TRW Chassis Systems)*

5. Recheck the preload.
6. Fill the steering gear with lubricant, as follows:
   a. If not already done, remove the filler plug (and pressure relief fitting) from the steering gear housing. Thoroughly clean the plug.
   b. Add the recommended lubricant until it is within $1/2$ inch of the filler hole.
   c. Install the filler plug (and pressure relief fitting) and tighten it to the manufacturer's specifications.
7. Center the steering wheel and reconnect the steering driveline and linkage.

Operation and Backdrive Check

The steering gear should turn smoothly, whether operated by the steering wheel through the input shaft or by the front wheels through the output shaft. Reversing the gear through the output shaft is termed backdrive and is an important factor in proper wheel recovery.

To check the backdrive, turn the gear to each end (again being careful not to bottom out the gear). Return the gear to center by rotating the Pitman arm by hand. This will take a fair amount of effort, but the average technician should be able to move the gear. If any roughness or binding is noted or if the gear cannot be moved at all, the gear should be repaired in accordance with the manufacturer's service manual.

Centering the Steering System

A surprising number of trucks are being driven with the steering gear "off center." That is, the front wheels point straight ahead, but the roller is not in the center of the worm gear. This causes both erratic steering and unnecessary wear on the steering gear.

It is useful to understand the concept of a centered steering system. When the front wheels are in a natural straight-ahead position, the steering arm can be located in one, and only one, point in space. Likewise, when the roller is positioned exactly midway between the outer edges of the mesh (nonbacklash) area on the worm gear, the lower end of the Pitman arm can be located in only one place. The distance between these two points is, therefore, a precise and measurable value (**Figure 21–60**). The mechanism that connects the two points is, of course, the drag link, and if it is not adjusted to the exact length required, the gear will be forced to operate off center.

Use the following procedure to center the steering system.

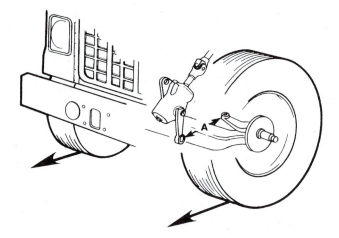

FIGURE 21–60 With the steering gear centered and the front wheels straight ahead, the drag link length must equal dimension A. If not, the gear could be forced off center. (Courtesy of Freightliner Trucks)

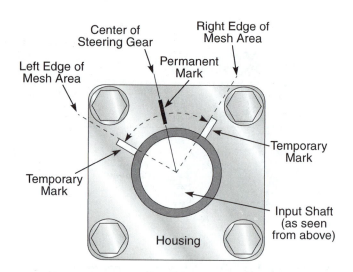

FIGURE 21–61 The midway point between the ends of the mesh area is the center of the steering gear. It should be noted with a permanent mark on the gear housing.

 Shop Talk

If caster adjustment is required, this should be done before centering the steering system.

1. Point the front wheels straight ahead. The preferred method is to pull the vehicle onto a level floor, leaving the steering wheel free for at least the last 10 feet so that the front wheels will seek their natural straight-ahead position. Other methods include sighting the front wheels with the rear wheels and equalizing the distances from the front wheel (not the tire) to the spring at identical points on each side of the vehicle.

2. Place chocks around the wheels.

3. Remove the drag link and disconnect the steering shaft from the gear. Starting with the steering gear in either backlash area, turn the input shaft toward the center until the Pitman arm will no longer wiggle. This is one edge of the mesh area and should be identified temporarily with a mark on the housing in line with a mark on the input shaft (**Figure 21–61**).

4. Locate the other edge of the mesh area and put another temporary mark on the housing in line with the mark previously made on the input shaft.

5. The midway point between the two marks on the housing is the center of the steering gear. (If the mesh area is more than one full turn of the input shaft, be sure to consider this when locating the midway point.) Mark

this point on the housing. This mark should line up with the mark on the input shaft when the vehicle is moving straight ahead on a level road.

6. Double check the center mark by measuring the amount of input shaft turn on either side of the mark within the mesh area. These arcs should be equal, and a minimum of one-third turn on each side of center.

7. Permanently mark the center position with paint on the housing and on the input shaft. Also make a permanent mark on the Pitman arm in line with a mark on the housing. These two sets of marks will serve as reference points to identify the center of the gear.

8. With the front wheels straight and the steering gear centered, adjust the drag link to the exact length needed to connect it with the Pitman arm and the steering arm without disturbing their straight-ahead position. Then install the drag link in accordance with the vehicle's service manual.

9. With the wheels and gear still straight and centered, reconnect the steering shaft with the input shaft. Align the steering wheel kingspoke to this straight-ahead position, removing and repositioning the wheel on the steering column if necessary. (Do not attempt to align the steering wheel by changing the length of the drag link.)

10. As a final check, drive the vehicle on a level surface and bring it to a stop with the steering wheel free. The wheels will be in their

natural straight-ahead position, the steering wheel should be centered, and the center marks on the steering gear should be in alignment. If not, readjust the length of the drag link. If the center marks are in alignment, but the kingspoke of the steering wheel is not, remove and reposition the wheel on the steering column.

21.4 POWER STEERING SYSTEMS

The heavy-duty truck's power steering system is in many ways identical to the manual steering system. The same steering axle and steering linkage is used, and both have a steering column and steering gear. The difference is in the addition of a hydraulic system. A power steering system contains a pump that delivers hydraulic fluid under pressure to a power steering gear (**Figure 21–62**). This gear is manufactured with hydraulic valving and passages that allow the hydraulic pressure to assist in turning the wheels.

POWER STEERING PUMP

The power steering pump is used to develop the hydraulic flow needed to provide the power assist to the steering gear. The pump on a heavy-duty truck can be either gear driven or belt driven, depending on the engine application. It is usually mounted near the front of the engine.

Power steering pumps are very similar in construction. Two basic types are most commonly used on heavy-duty trucks: roller and vane (**Figure**

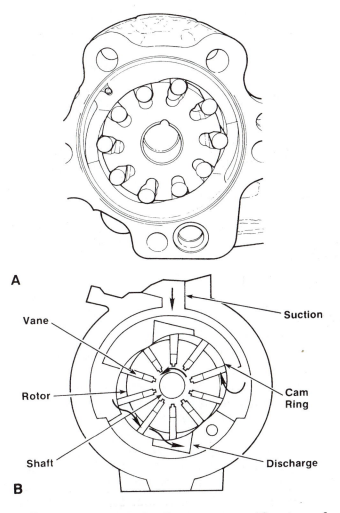

A

B

FIGURE 21–63 (A) A roller type pump (Courtesy of International Truck and Engine Corp.) and (B) a vane type pump. (Courtesy of Volvo Trucks North America, Inc.)

21–63A and **Figure 21–63B**). (Some vehicles are equipped with gear or rotary pumps.) When the pump shaft is driven by the engine, the rotor turns. The vanes turn with the rotor and are pressed outward against the cam ring by a combination of centrifugal force and hydraulic pressure. The inside of the cam ring is oval. When the rotor turns, the volume of the space changes between the rotor, the cam ring chamber, and two of the vanes. When two vanes move past the suction side, the area between them increases, creating suction. When the vanes approach the outlet side, the area between the vanes becomes smaller, the pressure rises, and oil is forced out into the delivery line. The pump has two inlet and two outlet channels to smooth pump operation and to increase pump displacement by providing two delivery pulses per revolution.

Hydraulic fluid used in the pump is stored in a reservoir that might be remotely mounted or might

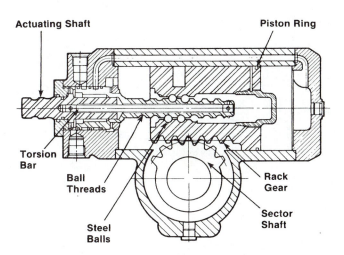

FIGURE 21–62 Cutaway view of a typical power steering gear. (Courtesy of R.H. Sheppard Co., Inc.)

be an integral part of the pump assembly. Fluid is routed between the pump and reservoir and between the pump and the steering gear via hoses and lines. Peak pressure is controlled by a relief valve.

FLOW CONTROL AND PRESSURE RELIEF VALVES

The flow control valve in the pump regulates pressure and flow output from the pump (**Figure 21–64**). This valve is necessary because of the variations in engine rpm and the need for consistent steering ability in all ranges from idle to highway speeds. It is positioned in a chamber that is exposed to pump outlet pressure at one end and supply hose pressure at the other. A spring is used at the supply pressure end to help maintain a balance.

As the fluid leaves the pump rotor, it passes the end of the flow control valve and is forced through an orifice that causes a slight drop in pressure. This reduced pressure, aided by the springs, holds the flow control valve in the closed position. All pump flow is sent to the steering gear.

When the engine speed increases, the pump can deliver more flow than is required to operate the system. Because the outlet orifice restricts the amount of fluid leaving the pump, the difference in pressure at the two ends of the valve becomes greater until pump outlet pressure overcomes the combined force of supply line pressure and spring force. The valve is pushed down against the spring, opening a passage that returns the excess flow back to the inlet side of the pump.

A spring and ball contained inside the flow control valve form a pressure relief valve and are used to relieve pump outlet pressure. This is done to protect the system from damage due to excessive pressure when the steering wheel is held against the stops. Because flow in the system is severely restricted in this position, the pump would continue to build pressure until a hose ruptured or the pump destroyed itself.

When outlet pressure reaches a preset level, the pressure relief ball is forced off its seat, creating a greater pressure differential at the two ends of the flow control valve. This allows the flow control valve to open wider, permitting more pump pressure to flow back to the pump inlet, and pressure is held at a safe level.

POWER STEERING GEARBOX

A power steering gearbox is basically the same as a manual recirculating ball gearbox with the addition of a hydraulic assist. A power steering gearbox is filled with hydraulic fluid and uses a rotary control valve to control the flow of fluid into, through, and out of the gear. Some gears are also equipped with a pressure relief valve.

In a power steering gear, the movement of the ball nut, also called a piston or rack gear, is assisted by hydraulic pressure. When the wheel is turned, the rotary valve changes hydraulic flow to create a pressure differential on either side of the piston. The unequal pressure causes the piston to move toward the lower pressure, reducing the effort required to turn the wheels.

The integral type power steering has the rotary control valve and a power piston integrated with the gearbox. The rotary valve directs the oil pressure to the left or right chamber to steer the vehicle. The spool valve is actuated by a lever or a small torsion bar located in the worm gear (**Figure 21–65**).

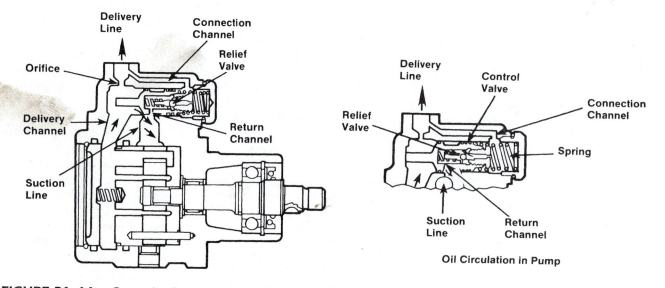

FIGURE 21–64 Control valve operation. (Courtesy of Volvo Trucks North America, Inc.)

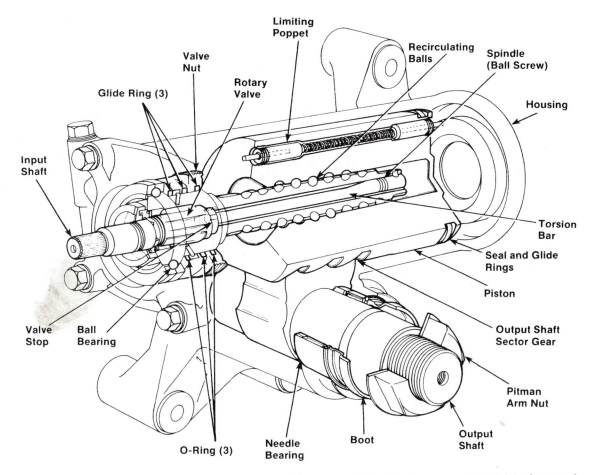

FIGURE 21-65 *Power steering gear components. (Courtesy of Bendix Commercial Vehicle Systems)*

Hydraulic Operation

The driven end of the worm gear (called a ball screw in power steering gears) rotates on a ball bearing contained in the valve body. Hydraulic pressure enters and exits the power steering gear through lines connected to threaded ports in the valve body. A pressure relief valve contained in the valve body prevents over-pressurization of the power steering gear. Hydraulic pressure in excess of the setting of the relief valve causes the valve to open a channel to the reservoir return side of the gear.

The ball screw assembly is held in the valve housing by a valve nut, which forms the outermost element of the rotary control valve. The valve nut contains circular channels and radial passages, which serve to direct hydraulic oil into and out of the rotary control valve. The ball screw assembly forms the rotary control and consists of three parts: the input shaft, torsion bar, and ball screw (**Figure 21-65** and **Figure 21-66**).

One end of the input shaft is finely splined for connection to the steering column, and the other end has a coarse spline that fits loosely with a similar spline inside the worm screw. The coarse splines form mechanical stops that limit the amount or relative rotation between the ball screw and input shaft. A torsion bar connects the input shaft to the ball screw. Six evenly distributed longitudinal grooves are machined into the outer surface of the input shaft and correspond to six grooves machined into the bore of the ball screw. Holes extend from the outside surface of the ball screw into the six grooves in the bore. These holes allow pressurized oil to enter and exit the two inner elements of the rotary control valve. The six grooves in the bore of the ball screw are connected alternately to each side of the piston through three pairs of the drilled holes. The other three holes admit pressurized oil directly to three of the six grooves in the input shaft. The other three grooves in the output shaft carry oil to the return line connection. The length of the six pairs of grooves cut into the ball screw and input shaft allows large pressure changes to be achieved with a small rotational displacement of the valve elements.

The rotary control valve is an open center type that allows a continuous flow of oil (through the longitudinal grooves in the input shaft and bore of the ball

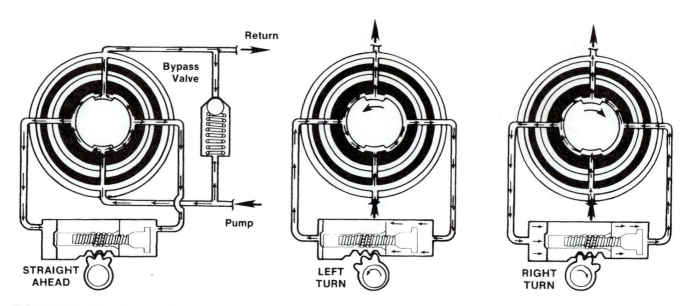

FIGURE 21–66   *Schematic of a rotary valve operation. (Courtesy of Bendix Commercial Vehicle Systems)*

screw) when held in the neutral position by the torsion bar (**Figure 21–66**). When steering effort is applied, the input shaft and ball screw tend to turn in unison; however, the spring action of the torsion bar results in the input shaft rotating slightly in advance of the ball screw. The six pairs of grooves that form the rotary control valve are displaced from their neutral flow position. As steering effort increases so does the amount of displacement. Depending on the direction steered, the groove displacement of the input shaft directs hydraulic oil through the appropriate drilled passages in the ball screw to one side or the other of the piston. Hydraulic pressure acting upon the piston surface eliminates much of the piston's resistance to movement. Spring force exerted by the torsion bar causes the ball screw to rotate as piston resistance is removed. As the ball screw rotates, the relative groove displacement is eliminated, and the rotary valve returns to a neutral position.

Moderate effort at the steering wheel produces smaller valve displacements and lower power assist, thus providing good steering feel. At increased displacements the pressure rises more rapidly giving increased power assistance and quicker response. Maximum pressure is developed after approximately 3 degrees displacement giving a direct feel to the steering. Groove displacement is limited by the freeplay of the stop spline mesh between the input shaft and ball screw. The splines take up the steering movement while allowing the torsion bar to hold the groove displacement. The torsion bar and stop splines form two parallel means of transmitting the steering torque. When no steering torque is applied, the torsion bar returns the valve grooves to a neutral position allowing the pressurized oil to flow to the return line.

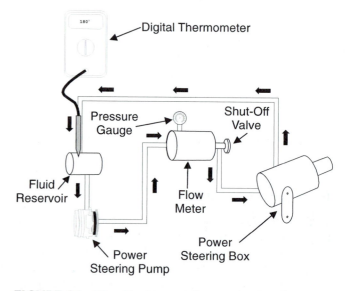

FIGURE 21–67   *Equipment for test hydraulic supply system.*

Power assisted movement of the valve nut within its bore is limited by poppet valves installed in both piston faces (see **Figure 21–67**). When the piston approaches its extreme travel in either direction, the stem of the limiting poppet valve makes contact at the end of the piston bore. As piston travel continues, the limiting poppet is unseated and some hydraulic power assist is removed as pressurized oil is diverted to the return line. As more and more power assist is removed by the action of the limiting poppet valves, steering effort increases. The piston can travel to the extreme ends of its bore; however, the maximum steering assistance available is reduced to protect the steering components in the axle.

The bypass valve is located in the valve body and permits oil to flow from one side of the piston to the other when it is necessary to steer the vehicle without the hydraulic pump in operation. Oil displaced from one side of the piston is essentially transferred to the other side, which prevents reservoir flooding and **cavitation** (bubble collapse) in the pressure line.

The pressure relief valve is located in the valve body and limits internal hydraulic pressure to a preset maximum. The pressure relief valve can be set to various pressures; however, its setting is typically 150 psi lower than the power steering pump relief valve setting.

TROUBLESHOOTING POWER STEERING SYSTEMS

If there is a complaint about a hydraulic steering system the first step is always to talk to the driver. Get a detailed explanation of what the problem is and when it happens. Ask about any repairs, modifications, or prior adjustments. Then, if the truck is roadworthy, test drive it or ride with the driver and try to duplicate the problem.

In most cases, the steering gear and pump are not the cause of the steering problem. These should be the last components you check. Visually inspect the tires, suspension components, steering linkage, and axle for damage or misalignment. Check the tires for mismatched sizes. Make sure they are properly inflated. Make sure that the suspension is not sagging or shifting. If the rear axles are out of line, the truck will be difficult to steer. Also check for proper location and lubrication of the fifth wheel.

Inspect all front axle components and suspension components for wear, looseness, or seizure. Check the steering gear mounting to be sure it is tight and not shifting on the chassis or axle. Also check for worn or binding steering column parts.

TROUBLESHOOTING SAFETY

When testing or servicing power steering systems, follow these precautions:

- Always block vehicle wheels. Stop the engine when working under a vehicle. Keep hands away from pinch points.
- Never connect or disconnect a hose or line containing pressure. Never remove a component or pipe plug unless all system pressure has been depleted.
- Never exceed recommended pressure and always wear safety glasses. Never attempt to disassemble a component until after reading and understanding recommended procedures.

- Use only the proper tools and observe all precautions pertaining to the use of those tools.
- Use only the manufacturer's suggested replacement parts and components. Only components, devices, and mounting and attaching hardware specifically designed for use in hydraulic systems should be used. Replacement hardware, tubing, hose, fittings, etc., should be of equivalent size, type, and strength as the original equipment.
- Devices with stripped threads or damaged parts should be replaced. Repairs requiring machining should not be attempted.

HYDRAULIC TESTS

If a visual inspection of the vehicle indicates a fault in the power steering system, test the hydraulic supply system before rebuilding or replacing the gear or pump. Proper power steering gear performance is dependent on an adequate supply of oil pressure and oil flow. When diagnosing power steering problems, oil pressure and oil flow must meet the manufacturer's specifications. Pressure and flow specifications vary considerably, and the vehicle manufacturer recommendations must be followed carefully at all times.

High system oil temperatures reduce the overall efficiency of the steering pump and the steering gear. High temperatures are caused by restriction to flow or inadequate system oil capacity to allow for heat dissipation during normal operation. A supply pump that constantly operates at maximum pressure relief will also generate more heat than can be dissipated.

Various types of pressure gauges and flow meters are available and can be used to diagnose power steering problems. A pressure gauge that reads at least 3,000 psi and a flow meter with a capacity to 10 gpm are used to check pressures and oil flow. A shutoff valve downstream from the pressure gauge makes it possible to isolate the steering pump from the steering gear and by closing the valve, maximum pump relief pressure can be read.

A simple thermometer in the reservoir will indicate system oil temperatures. **Figure 21–67** illustrates a typical installation of test equipment in the hydraulic system.

The following procedure is general in nature. Always consult the vehicle manufacturer's service manual for specifications and detailed instructions.

1. Make necessary gauge/meter connections.
2. Start engine and check system oil level, ensuring that oil flow is in the proper direction through the flow meter.

PHOTO SEQUENCE 7

FLOW AND PRESSURE TEST A HYDRAULIC-ASSIST STEERING GEAR SYSTEM

P7–1   Use a high pressure washer or steam cleaner to thoroughly clean the area around the steering gear.

P7–2   Park the vehicle in a stall and ensure that the front wheels are tracking in the straight-ahead position. Fit a shop exhaust pipe to the truck's exhaust stack.

P7–3   Place a drain pan under the steering gear to catch power steering fluid spilled during this procedure.

P7–4   Separate the high-pressure supply hose from the steering gear. Be aware that the fluid in this line may have some residual high pressure. Some fluid will drain from the system, so the drain pan should be placed directly under the steering gear.

P7–5   Connect the power steering analyzer in series with the high-pressure supply hose and the nipple on the inlet side of the steering gear. Now, the system will function as normal except that the hydraulic fluid will be routed through the power steering analyzer.

P7–6   Start the engine. Run it at idle speed and record the pressure and flow readings on the power steering analyzer. Check to specifications.

P7–7   Next close down the flow valve until the pressure increases to the test value specified in the OEM service literature. This may be between 500 and 1000 psi depending on the system. Record the flow in gpm at this pressure value. If the flow is lower than specified, the power steering pump may be defective and require replacement.

P7–8   Test the pump relief valve and flow control pressures by closing down the flow valve until the flow gauge reads zero. Record the peak pressure reading on the gauge. Perform this 2 or 3 times but for no longer than 5 seconds separated by a 30 second interval. Check the gauge pressure.

P7–9   Compare all test results to OEM specifications and perform repairs as required. Ensure that when the power steering analyzer is removed, the engine is run and the steering turned from lock to lock and back to center before attempting to remove the truck: this will ensure that no air remains in the high-pressure circuit.

3. Place thermometer in reservoir.
4. Run the engine at correct idle speed and steer from lock to lock several times to allow system to warm up (140–160°F).
5. With the engine running at a specified idle speed, slowly turn the shutoff valve until closed and read the pressure at which the pressure relief valve opens. (Open the shutoff valve as quickly as possible to avoid heat buildup or possible damage to the steering pump.) This pressure reading should equal the maximum pump pressure specified by the manufacturer of the chassis. Check your specifications.

CAUTION: A malfunctioning pressure relief valve might not relieve pump pressure, and closing the shutoff valve might cause severe pump damage or high-pressure hoses to rupture. Constantly observe the pressure gauge while closing the shutoff valve. If pressure rises rapidly or appears to be uncontrolled, do not completely close the valve before inspecting the pump and pressure relief valve.

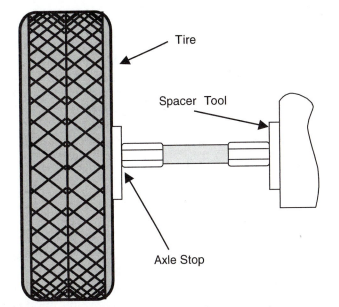

FIGURE 21–68 *Install a spacer between the axle stop and the axle when making a leakage test.*

6. With the engine running at a specified idle speed, vehicle stationary on the shop floor, and with a normal load on the front axle, steer the wheels from full right to full left turn and observe the flow meter. The flow must not fall below the minimum gpm flow specification.
7. Increase the engine speed to approximately 1,500 rpm and note the flow rate with the steering wheel stationary. Check this reading against the maximum flow rate specifications. Excessive oil flow can cause high operating temperature and sluggish heavy steering response.

 Shop Talk

It is important that the flow be checked at a normal operating temperature and with a load on the front axle. Inadequate flow will cause binding and uneven or intermittently hard steering.

8. If the supply pump is performing to specification, install a $1/2$-inch spacer between the axle stops on one side (**Figure 21–68**) and turn the steering wheel hard in the direction necessary to pinch the spacer block. Record

the maximum pressure reading. The maximum pressure reading should be within 100 psi as was recorded in step 5 for pump relief pressure when the shutoff valve was closed. Remove the spacer and repeat the test in the opposite direction. Record pressure. If the pressure does not meet the recorded maximum pressure reading, the steering gear is worn internally and must be repaired or replaced.

9. Normal system back pressure will be 50–75 psi with the engine idling and the steering wheel stationary. Back pressure is checked with the system at a normal operating temperature.
10. Steering system oil temperature is best checked after 2 hours of normal operation. The ideal operating temperature should range between 140–160°F. Normal operation in this range will allow for intermittently higher temperatures that will be encountered during periods of heavy steering usage.
11. Visually check for the presence of air mixed with the oil in the steering system. The oil should be clear. Any signs of frothing indicate air entry, and steering performance will be affected. Carefully check for leakage on the suction side of the steering pump. Drain and refill the system and bleed for air.

Shop Talk

Before any steering gear repairs are attempted, the preceding hydraulic supply tests must be completed and corrections made as required. Many times steering gears have been repaired or replaced needlessly because a hydraulic supply system had not been tested thoroughly.

21.5 AIR ASSISTED STEERING SYSTEMS

In <$pneumatic steering systems> addition to the integral power steering systems already discussed in this chapter, some heavy-duty trucks are equipped with an air assisted manual steering system (**Figure 21–69**).

An air powered steering assist system uses the vehicle's compressed air supply to ease the manual steering operation. The system consists of three major components: a torque valve, power cylinder, and air safety valve.

1. A sealed torque valve is combined with a replacement drag link (**Figure 21–70**). The torque valve senses the steering forces and automatically meters the appropriate amount of air to the power cylinder.
2. The power cylinder attaches to either a left-hand (as shown in **Figure 21–69**) or right-hand wheel arm bracket or to the tie-rod (**Figure 21–71**), depending on the type of

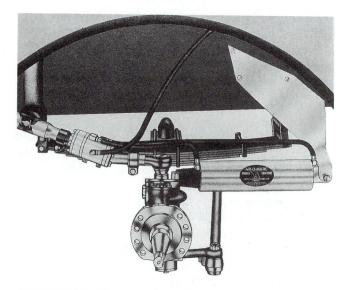

FIGURE 21–69   An air assist system. (Courtesy of Air-O-Matic Power Steering, Div. of Sycon Corp.)

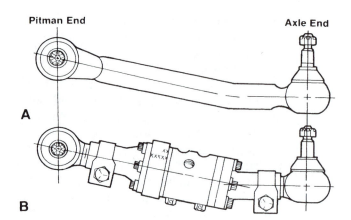

FIGURE 21–70   (A) Regular drag link; and (B) drag link with torque valve. (Courtesy of Air-O-Matic Power Steering, Div. of Sycon Corp.)

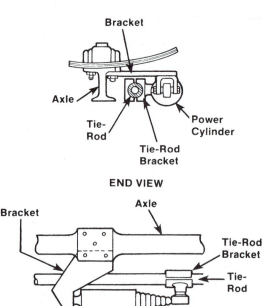

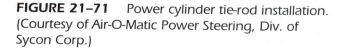

FIGURE 21–71   Power cylinder tie-rod installation. (Courtesy of Air-O-Matic Power Steering, Div. of Sycon Corp.)

vehicle. The power cylinder receives the metered air from the torque valve and, in turn, provides the power that assists in turning the wheels.

3. An air safety valve is installed in the air supply line to the torque valve. The function of the safety valve is to conserve air in the event the air pressure in the reservoir falls below approximately 65 psi. When this occurs, steering reverts to manual automati-

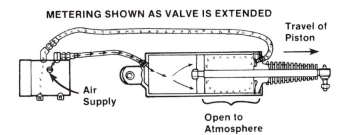

METERING SHOWN AS VALVE IS EXTENDED

Travel of Piston

Air Supply

Open to Atmosphere

FIGURE 21–72 *Operation of power cylinder and torque valve. (Courtesy of Air-O-Matic Power Steering, Div. of Sycon Corp.)*

cally. Then when air pressure builds up again (to approximately 12 psi above cutoff setting), the valve will reopen, restoring power steering assist. The safety valve is not a regulating valve that can be adjusted to provide more or less power. A manual shutoff valve is also installed in the air supply line and is used to turn off the air supply manually. The valve must be shut off when servicing the air assist system.

The sealed torque valve is installed into the drag link where it senses the steering forces. It has a spring that extends or compresses, as required, in response to the steering forces and meters the appropriate amount of air to the power cylinder.

The metering of air begins after the valve has extended or compressed approximately 0.040 inch. As air is being metered to one side of the power cylinder, the other side of the power cylinder is open to the atmosphere to permit the piston to move (**Figure 21–72**).

When the turn is completed and the direction of the steering effort is reversed or the steering wheel is released, the torque valve repositions itself to the neutral position, which opens the pressurized side of the power cylinder to atmosphere and releases the air.

MAINTENANCE

Daily

As a normal part of a daily or pretrip inspection, do the following:

1. Drain water from all air tanks.
2. Examine all steering components, including the pneumatic components for loose, damaged, cracked, or worn parts.
3. Check for loose or damaged air lines.

Weekly

Once a week, the following maintenance tasks should be performed:

1. Check all fasteners (screws, bolts, nuts, etc.) for proper torque requirements.
2. Lubricate power cylinder piston rod socket and both drag link sockets.

Quarterly (or Every 25,000 Miles)

Every three months or 25,000 miles, the system must be lubricated internally. To do so, follow these steps:

1. Be careful to keep dirt and other foreign matter out of the torque valve and air supply line.
2. With the vehicle standing still and the air supply to the torque valve shut off, disconnect the air supply line from the torque valve or the safety valve, whichever is more convenient.
3. Add approximately a $1/2$ ounce of clean engine oil (10W-40) or pneumatic air tool oil to the supply port or line and reconnect it. Open the air supply to the torque valve.
4. With the weight of the vehicle on the ground, oscillate the steering wheel to distribute the added oil throughout both sides of the torque valve and power cylinder.

TROUBLESHOOTING

Pneumatic steering system components develop few problems. Eventually, the system might begin to leak air as the power cylinder piston wears (**Figure 21–73**). In the event air blows by the cylinder piston,

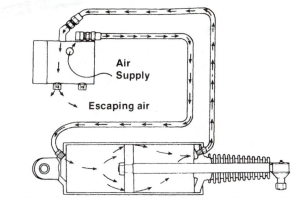

Air Supply

Escaping air

Note: This diagram shows a typical system only. The actual air line connections might vary.

FIGURE 21–73 *Blowby condition due to worn power cylinder piston. (Courtesy of Air-O-Matic Power Steering, Div. of Sycon Corp.)*

the air being metered will escape through the opposite exhaust port. This blowby gives the impression of a leaking torque valve, when the torque valve might be in normal working order.

Checking Torque Valve

Before checking, lubricate the system in accordance with the quarterly maintenance procedure given previously. To determine whether or not the torque valve leaks, build up the vehicle air pressure, shut down the engine, and check the following:

1. There should be minimal or no air coming out the exhaust ports when the vehicle is sitting.
2. At the torque valve, disconnect the air lines that go to the power cylinder and insert pipe plugs into the torque valve.

CAUTION: Care must be taken to keep dirt and other foreign matter out of the torque valve and air lines.

3. Observe the torque valve while an assistant extends and compresses the torque valve by turning the steering wheel. Minimal or no air should escape from the torque valve while being extended or compressed (approximately $1/16$ inch each direction.)

👓 **Shop Talk** _____

Some air might escape from the exhaust port when the torque valve first starts to feed. This is normal and should disappear as the torque valve continues to extend or compress. Also, a small burst of air will be noticed when the direction is reversed.

4. If the preceding proves the torque valve to be in normal working order, proceed to check the power cylinder. If the torque valve exhausts a substantial amount of air in any of the previous tests, replace with a new or rebuilt one.
5. In the event several torque valves have failed on one vehicle, consult the manufacturer's service manual to check for proper installation and proper air source.

Checking Power Cylinder

1. If the power cylinder is attached to the tie-rod, check for any binding condition in the

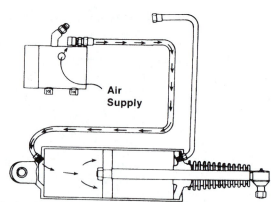

Note: This diagram shows a typical system only. The actual air line connections might vary.

FIGURE 21–74 Checking for blowby condition. (Courtesy of Air-O-Matic Power Steering, Div. of Sycon Corp.)

power cylinder as the wheels are turned in each direction. Adjust the tie-rod bracket assembly to eliminate binding, if necessary.

2. Reconnect the air line that goes to the torque valve port opposite the rotating end, leaving the other disconnected (**Figure 21–74**).
3. With full reservoir pressure, engine shut down, and wheels on the ground, have an assistant turn the steering wheel to cause the torque valve to extend.
4. While observing the disconnected hose, have the assistant continue to turn the wheels to a full turn. If a substantial quantity of air escapes from the disconnected hose, this would indicate a worn piston.
5. Reverse this procedure and turn the wheels in the opposite direction.
6. If air escapes from the disconnected hose in either direction, replace the power cylinder with a rebuilt one.

21.6 HIGH-PERFORMANCE TRUCK STEERING

Power steering in North America has become standard on Class 8 trucks with an application rate of greater than 95 percent. Manual steering has become yesterday's technology, and speed-proportional steering will be tomorrow's. The use of electronics can provide driver-selectable controls and automatic integrated adaptive/active vehicle steering, suspension, braking, and traction control systems.

Two systems that provide reduced parking effort combined with improved high-speed road feel are covered in the following sections. Both the electronic variable and load-sensing systems discussed operate with a conventional integral steering gear.

ELECTRONICALLY VARIABLE STEERING

To better understand the advantages offered by this new technology, the limitations of conventional power steering must be understood. In a conventional power steering system, the demand for power assistance is opposite to its availability. This requires designers to compromise the power steering pump optimization pressures. This results in a parasitic power loss that can amount to 40 to 50 percent of the power consumed by the steering system. It also results in light steering effort at highway speeds because the flow control setting is often high, as dictated by dry park conditions.

Optimized power steering control strives to

- Improve steering wheel road feel with directional stability enhanced by a steering effort proportional to driving speed.
- Equalize or reduce dry park effort.
- Reduce system power consumption.

Because conventional power steering systems do not sense or respond to these variables, the electronically controlled version can provide a correction towards the ideal. Ideally such a system changes the driver's road feel (effort) as a function of road speed and lateral acceleration.

The electronic system shown in **Figure 21–75** offers single-switch, operator-selectable steering modes. The system consists of a modified, integral power steering gear that includes a revised high-gain rotary control valve; a pulse width modulated (PWM), solenoid-controlled, variable-flow control pump; a reluctance-type speed sensor (similar to that used for electronic speedometers); and an electronic control module for closed-loop solenoid current control.

Variable assist is provided by changing the integral-pump-control flow supplied to the steering gear. The steering gear provides the driver a variable torque input based upon the variable flow control. Steering gear features include: freeplay control, a high-efficiency manual section, low mechanical ratio, and a direct rotary valve. The power assist is controlled by the steering gear's adaptive features such as torsion-bar rate and valve "feel" characteristics.

Because the system modulates the pump's control flow to the steering gear, the gear's rotary control valve design helps achieve appropriate effort or

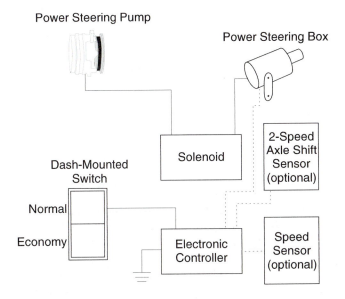

FIGURE 21–75 *Electronic variable power steering system that uses a dash mounted switch to allow drivers to operate in any one of three modes for maximum road feel and control.*

"feel" changes. The potential to bypass pump flow internally, combined with the low power steering gear total loop drop at reduced flow, is the source of power consumption improvements.

Pump Design

This electronic steering system uses a solenoid-controlled variable-flow control concept. The pump assembly includes a high-efficiency vane pump, flow control and proportional solenoid-type throttle valve.

The system is designed with fail-safe features. When the solenoid has no power, flow is low (at around 60 percent of its normal capacity) similar to conventional power steering systems.

Electronic Control Unit (ECU)

The electronic controller's primary purpose is to change the PWM drive signal to the pump solenoid with respect to the road-speed sensor signal. The controller incorporates switches that allow for three potential operating modes with an LED display to assist in troubleshooting.

In economy mode, flow control is at maximum below 24 km/h, at its minimum above 80 km/h, and decreases proportionally from 24 to 80 km/h. The normal mode has high flow at low speeds and decreases as speed increases (similar to economy mode) except the minimum flow is set at 13.6 L/min and is reached at about 61 km/h. In this mode, the steering system acts much like a conventional power steering system with 13.6 L/min flow control. If the

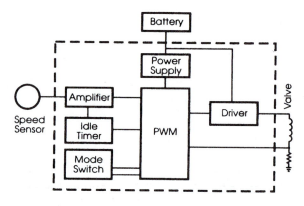

FIGURE 21–76 *Speed sensor input and battery power is fed to the PWM control unit. The control unit output signal drives the control valve.*

vehicle is stationary for longer than 10 minutes, the system switches into a standby mode. This mode reduces flow to minimum power assist as long as the speedometer indicates zero. When road speed is detected, the system automatically switches to its original mode.

The PWM drive circuit operates the solenoid valve by a fixed-frequency, pulse-width-modulated signal with closed-loop duty cycle control that uses the solenoid current as feedback (**Figure 21–76**). This ensures that the solenoid drive is relatively constant with changes in vehicle battery voltage. It also allows the controller to provide current to the solenoid directly from the battery without need for high-current voltage regulators.

LOAD-SENSING POWER STEERING

The nonelectronic, hydraulically controlled load-sensing system shown in **Figure 21–77** has a demand-type flow control that uses steering load pressure to control its variable-flow power assist. This

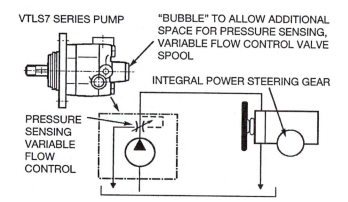

FIGURE 21–77 *This hydraulically controlled load-sensing steering pump incorporates a pressure-sensing variable flow control feature that varies power assist according to demand.*

system provides for speed proportioning of flow to the steering gear from a flow-controlled pump. However, this system does not provide driver-selectable operational modes, nor programmable flow settings. Instead, the system relies on a control strategy in which the pressure requirement for power steering is a function of vehicle speed.

The load-sensing system is similar to a conventional version, except that a modification has been made to the flow-control section of the pump as shown. Compared to the electronic version, the load-sensing system provides increased flow only to the steering gear "on demand" in response to a steering-pressure increase designated as "load sensing."

The load-sensing primary pump components include a high-efficiency, balanced vane pumping cartridge and a combination flow/relief/load-sense valve. The valve construction and operation within the circuit are shown in **Figure 21–78**. The valve spring has two functions. The first function is to

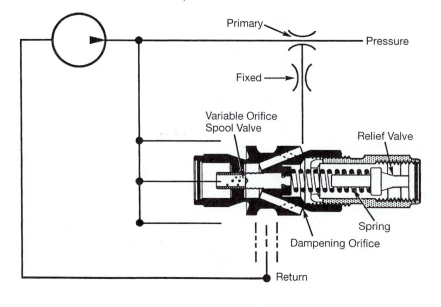
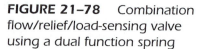

FIGURE 21–78 *Combination flow/relief/load-sensing valve using a dual function spring*

restrict the movement of the variable-orifice spool valve according to the steering load. The second function is to load the relief poppet valve at the relief pressure setting.

The small variable-orifice valve is exposed to return pressure on the opposite end of the spool. As pump outlet pressure increases, a progressively larger opening is produced by the moving spool. This larger opening allows a greater volume of fluid to flow into the spring-chamber end of the large flow-control valve. Because the fluid must pass through a fixed orifice to escape this spring chamber, more fluid flow produces a higher pressure. The higher pressure adds to the load imparted to the flow-control spool by the spring and increases the restriction of flow to the return circuit. As a result, more flow is forced through the primary control orifice and delivered to the steering gear.

A conventional pressure-compensated flow-control valve maintains a constant flow rate regardless of pressure; whereas the pump used on this system increases flow as pressure is increased.

SUMMARY

- The steering system in a Class 8 heavy-duty truck must deliver precise directional control of more than 50 tons of combined vehicle and cargo weight.
- Truck steering systems are either manual or power assisted, with power assist units using either hydraulic or air assist setups to reduce steering effort.
- Improper steering adjustment can lead to alignment and tire wear problems.
- A front end that is properly aligned will result in
  - Easier steering
  - Longer tire life
  - Directional stability
  - Less wear on front-end components
  - Better fuel economy
  - Increased safety
- Ackerman geometry involves the arrangement of suspension components so that the tires roll freely during turns.
- Axle alignment measurements can be taken in a number of ways. Two such advanced systems are light or laser beam-type wheel alignment systems and fully computerized sensor systems.
- The most advanced alignment systems in use today are controlled by microprocessors.
- Alignment readings, specifications, and step-by-step instructions are displayed on a cathode-ray tube (CRT) screen.

- Keyboard-entered specifications are automatically compared against the actual angles of the vehicle with the results displayed on the CRT screen.
- The steering axle components should be lubricated at least every 25,000 miles.
- Two general types of manual steering gear are used in heavy-duty trucks. They are the worm and roller type and the recirculating ball and worm type.
- Preload is necessary to control the various elements in the steering linkage and to prevent random movement within the system.
- There are two types of preloads: worm bearing preload and total mesh preload.
- Insufficient preload permits lost motion within the gear, which means that the driver must turn the steering wheel farther than normal to get any response from the front wheels. Excessive preloads result in hard steering and sometimes in a "breaking loose" effect that produces a darting and oversteering condition.
- A power steering system contains a pump that delivers hydraulic fluid under pressure to a power steering gear. This gear is manufactured with hydraulic valving and passages that allow the hydraulic pressure to assist in turning the wheels.
- A power steering gearbox is basically the same as a manual recirculating ball gearbox with the addition of a hydraulic assist.
- An air-powered steering assist system uses the vehicle's compressed air supply to ease the manual steering operation.
- Power steering in North America has become standard on Class 8 trucks with an application rate of more than 95 percent. Manual steering has become yesterday's technology, and speed-proportional steering will be tomorrow's.

REVIEW QUESTIONS

1. Which of the following components uses universal joints?
   a. steering wheel
   b. steering shaft
   c. Pitman arm
   d. tie-rod assembly

2. Which of the following components contains ball joints?
   a. steering gear
   b. Pitman arm
   c. steering knuckle
   d. tie-rod end

3. Which of the following components connects the Pitman arm to the steering arm?
   a. drag link
   b. tie-rod assembly
   c. cross tube
   d. steering column

4. Straight kingpins are secured to the axle with
   a. a nut
   b. cotter pins
   c. draw keys
   d. tapered bearings

5. An out-of-alignment condition can result in
   a. directional instability
   b. front tire wear
   c. hard steering
   d. all of the above

6. Which of the following alignment angles has the potential to cause the most tire wear?
   a. incorrect toe
   b. incorrect camber
   c. incorrect caster
   d. insufficient turning angle

7. When making corrections to front-end alignment, Technician A always makes corrections to toe angle first. Technician B measures toe angle between points centered in the tire tread at hub height. Who is correct?
   a. Technician A
   b. Technician B
   c. both A and B
   d. neither A nor B

8. Which of the following is defined as the forward or rearward tilt of the kingpin centerline when viewed from the side of the vehicle?
   a. camber
   b. caster
   c. toe
   d. kingpin inclination

9. Which of the following can be defined as the inward or outward tilt of the top of the tires when viewed from the front of the vehicle?
   a. turning radius
   b. Ackerman geometry
   c. camber
   d. toe

10. Which of the following alignment angles can be corrected using shims?
    a. toe
    b. caster
    c. camber
    d. none of the above

11. Which of the following alignment angles could be corrected only by bending the axle?
    a. caster
    b. camber
    c. kingpin inclination
    d. none of the above

12. When inspecting a steering axle, Technician A uses a dial indicator to check knuckle vertical play. Technician B uses a machinist's protractor to check caster setting. Who is correct?
    a. Technician A
    b. Technician B
    c. both A and B
    d. neither A nor B

13. Which of the following steering gear arrangements is normally found on a heavy-duty truck equipped with power steering?
    a. worm and roller
    b. recirculating ball
    c. rack and pinion
    d. all of the above

14. Which of the following adjustments can be made to a worm and roller type steering gear?
    a. worm bearing preload
    b. total mesh preload
    c. centering the steering system
    d. all of the above

15. Technician A states that in a conventional power steering system, demand for power assistance is opposite to its availability. Technician B says newer load-sensing hydraulically controlled power steering systems provide increased flow from the steering pump in response to an increase in steering pressure. Who is correct?
    a. Technician A
    b. Technician B
    c. both A and B
    d. neither A nor B

Prerequisites
Chapters 2 and 3

Suspension Systems

Objectives

After reading this chapter, you should be able to
- Identify and describe the four types of suspension systems used on today's heavy-duty trucks.
- Name the components of a spring suspension system and explain how they work.
- Describe a fiber composite spring.
- Name the components of an equalizing beam suspension system and explain how they work.
- Name the components of a torsion bar suspension system and explain how they work.
- Name the components of an air spring suspension system and explain how they work.
- Perform troubleshooting techniques to isolate and locate defective suspension system components.
- Execute replacement procedures to repair a defective suspension system.
- Explain the relationship between axle alignment and suspension system alignment.
- Perform suspension system alignments.
- Describe the operation of the cab air suspension system.

Key Terms List

air spring suspension
bogie
constant rate springs

leaf springs
rubber cushions
shock absorbers
torsion bar suspension
vari-rate springs

+ Alignment

The primary purpose of the suspension system on a heavy-duty truck is to support the weight of that truck. Additional requirements placed upon the suspension system include the ability to stabilize the truck when traveling over normal as well as over rough terrain; to cushion the chassis from road shock while simultaneously allowing the driver/operator to steer the truck; to maintain the proper axle spacing and alignment; and to function efficiently over a wide range of speed and load conditions. Any attempts to overload or exceed the maximum suspension load rating will damage not only the suspension system but also the truck frame, axles, and tires.

Four general types of suspension systems are employed on today's heavy-duty trucks:

- Spring
- Equalizing beam
- Torsion bar
- Air spring

The systems are used not only on heavy-duty trucks and tractors but also on the trailers they haul. Each type of suspension will be discussed in this chapter, including its servicing requirements.

22.1 LEAF SPRING-TYPE SUSPENSIONS

Three types of **leaf springs** are used on heavy-duty trucks and trailers: constant rate, progressive or vari-rate, and auxiliary.

CONSTANT RATE

Constant rate springs are leaf-type spring assemblies that have a constant rate of deflection. For example, if 500 pounds deflect the spring assembly 1 inch, 1,000 pounds would deflect the same spring assembly 2 inches.

Constant rate springs are mounted to the axle with U-bolts, nuts, and lockwashers. The front end of the spring is mounted to a stationary bracket. The rear end of the spring is mounted to a spring shackle. The shackle allows for variations in spring length during compression or jounce and rebound. Spring leaves are held together by a long bolt that passes through the spring center. The spring assembly is secured to the axle housing by U-bolts. Leaf alignment is maintained by spring clips that are typically located at the two ends of the spring leaves.

This type of spring assembly is used in both front and rear axle applications on heavy-duty trucks. **Figure 22–1** and **Figure 22–2** show typical constant rate spring assemblies.

The tapered leaf spring is a spring design with leaves that are thicker in the center than at the ends (**Figure 22–3**). This design results in fewer leafs, providing lighter weight. Tapered leaf springs are similar to conventional springs in usage and installation.

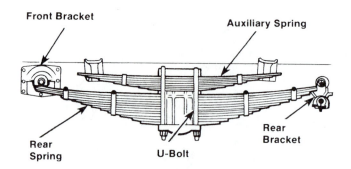

FIGURE 22–2   Rear axle application—constant rate spring with auxiliary pack. (Courtesy of International Truck and Engine Corp.)

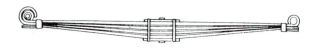

FIGURE 22–3   Taper leaf spring assembly. (Courtesy of International Truck and Engine Corp.)

PROGRESSIVE (VARI-RATE)

Progressive or **vari-rate springs** are leaf-type spring assemblies with a variable deflection rate obtained by varying the effective length of the spring assembly. This is accomplished by using a cam bracket. As the spring assembly deflects, the point of contact on the bracket moves toward the center of the spring assembly, shortening the effective length (**Figure 22–4**). Vari-

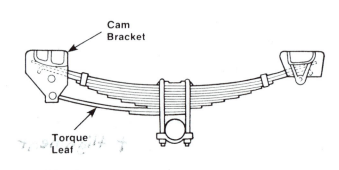

A UNLOADED

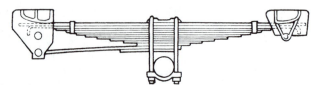

B LOADED

FIGURE 22–4   Progressive spring assembly. (Courtesy of International Truck and Engine Corp.)

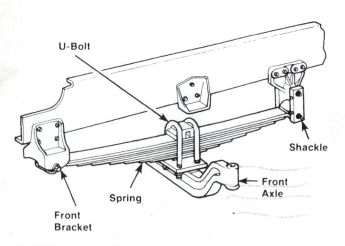

FIGURE 22–1   Front axle application—constant rate spring. (Courtesy of International Truck and Engine Corp.)

rate spring assemblies also incorporate a progressive feature in that the lower spring leaves are separated at the ends. As the spring assembly deflects, these leaves come in contact, providing increased capacity and stiffness.

AUXILIARY SPRINGS

Auxiliary springs, also called overload springs, are leaf-type spring assemblies usually mounted on top of the main rear spring assemblies. Auxiliary springs factor only when the truck is under heavy load. The auxiliary spring assumes part of the load by contacting cam brackets attached to frame rail slides when the load is sufficient to compress the rear spring assemblies to the point of contact. **Figure 22–2** illustrates an auxiliary spring assembly.

APPLICATIONS

Spring-type suspensions can be found on both front and rear suspension systems. They are specifically designed to handle the load requirements for which the vehicle is rated. If the specified load weights are exceeded, the life of the suspension components will be greatly reduced.

The primary function of the spring assemblies and the walking beams is to support the load of the chassis. In addition, springs cushion the chassis from road shock, providing acceptable ride characteristics from empty vehicle to full rated load. Walking beams and some types of spring also maintain axle spacing and alignment.

Most heavy-duty spring assemblies, front or rear, are manufactured from special heat-treated alloy steel. The walking beams are either fabricated steel beam sections or cast I-beam sections.

Front Suspension

Front spring assemblies are of various types. The more common types, including *taperleaf eye and slipper, multileaf eye and slipper, taperleaf shackle,* and *multileaf shackle* are shown in **Figure 22–5**. In all of these front suspension arrangements, the spring assemblies are mounted to the axle with U-bolts, hardened washers, and nuts. The forward end of each spring mounts to a stationary frame bracket. The rear of each spring mounts to a pair of spring shackles, which are suspended from a frame-mounted bracket. The spring shackles allow for variations in spring length during spring flexing.

The leaf spring assembly serves to absorb and store energy and then to release it. Individual leaves are held together by a center bolt, with the center bolt acting as a locating dowel during installation on

the vehicle. On the ends of the leaf spring assembly are alignment clips. The alignment clips are used to limit sideways spread and vertical separation of the individual leaves.

Rear Suspension—Single Axle

The single rear axle spring assemblies generally are one of two types: *eye and slipper spring* or *spring-type radius rod* (**Figure 22–6**). Semielliptical spring assemblies are attached to the axles with U-bolt assemblies. The spring ends ride in brackets that are mounted on the frame rails. Steel wear shoes are cast into each bracket. Radius rods attached to the axle seats and the forward spring brackets hold the axle in alignment.

Rear Suspension—Tandem Axle

The tandem axle rear spring suspension (**Figure 22–7**) uses a six-point equalizing leaf spring design, which compensates for axle articulation, side to side and front to rear. Four semielliptical spring assemblies are attached to the axles with U-bolts. On both sides of the vehicle, the forward end of the forward spring and the rear end of the rear spring ride in brackets mounted on the frame rails. Steel wear shoes are cast into each bracket. At the center, between the forward and rear springs, the springs ride on an equalizer, which pivots on a sleeve in the equalizer bracket. Each axle is held in alignment by a pair of torque rods that extend forward from the axle seats to the forward spring brackets for the forward-rear axle and to the equalizer brackets for the rearmost axle.

Bogie Suspensions

Bogie spring assemblies include the camelback, the taperleaf, and the walking beam. The ends of both camelback spring assemblies and walking beams are mounted in rubber shock insulators. The current taperleaf spring bogie has springs with eyes at both ends to accommodate pin type connections at the axle housings. Metal-backed rubber insulators are installed in the spring ends to minimize road shock.

Spring Suspension with Shock Absorbers

Most front spring suspension systems use **shock absorbers** (**Figure 22–8**). Shock absorbers dampen spring oscillation. They control body sway, eliminate excessive tire wear, front wheel shimmy, and spring breakage. They also improve the riding qualities of the truck and are especially useful when the truck is empty or only partially loaded.

Spring Suspension with Torque Rods

The spring suspension system with torque rods is one of the less complicated suspensions available.

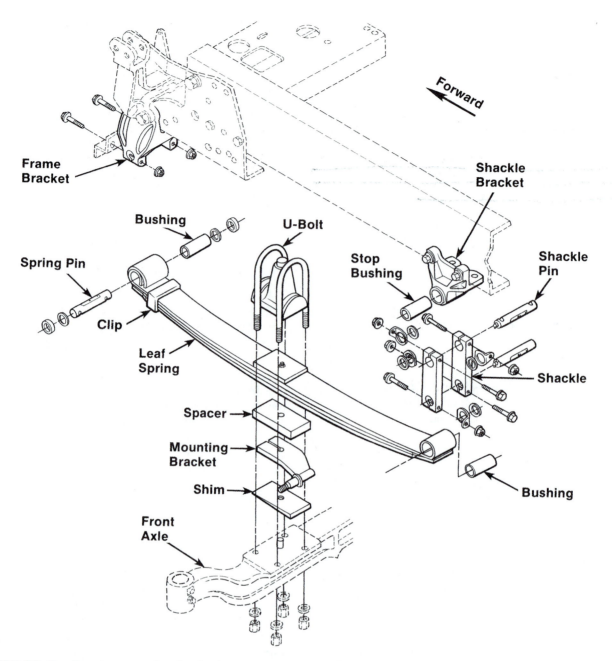

FIGURE 22–5 Front suspension (spring) arrangement on a heavy-duty truck. (Courtesy of Freightliner Trucks)

Consequently, it is less difficult to service. Construction is simple, consisting of only two major components: leaf spring assemblies and torque rods. The leaf springs are mounted to the axles. On a 6 × 4 chassis, the leaf spring assemblies mount at three different locations on each frame rail, distributing the load over a large area of the frame (**Figure 22–9**). On a 4 × 2 chassis, the leaf spring assemblies mount at two locations. The leaf spring assembly brackets and frame cross members are fastened to the frame rail with the same hardware, resulting in the suspension being an integral part of the chassis.

Torque rods provide stability by minimizing axle windup caused by driveline torque and load transfer during braking. Suspension alignment is accomplished by inserting shims at the torque rod end mounting.

Some four-spring suspension systems use six torque rods instead of four (**Figure 22–9**). This system design uses two additional torque rods that extend from the center of each axle housing to a frame cross member on the truck. The installation of the center torque rods provides added stability. Only four of the six torque rods are adjustable to allow for

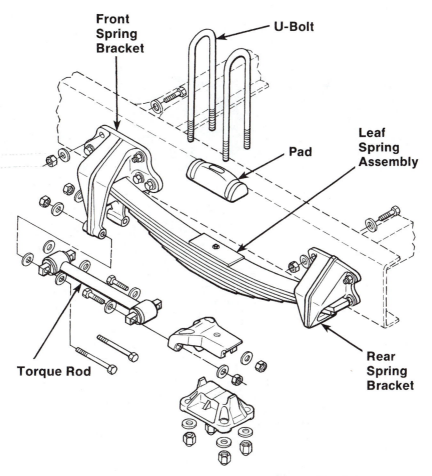

Front
Spring
Bracket

U-Bolt

Leaf
Spring
Assembly

Pad

Torque Rod

Rear
Spring
Bracket

FIGURE 22–6 Single axle spring suspension. (Courtesy of Freightliner Trucks)

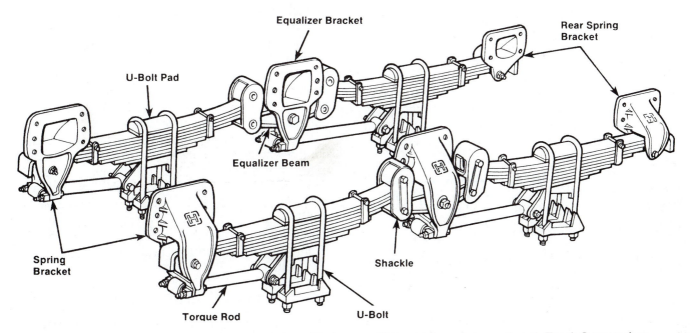

Equalizer Bracket

Rear Spring
Bracket

U-Bolt Pad

Equalizer Beam

Spring
Bracket

Shackle

Torque Rod

U-Bolt

FIGURE 22–7 Tandem axle spring suspension. (Courtesy of Hendrickson International, Truck Suspension Systems)

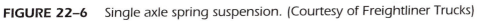

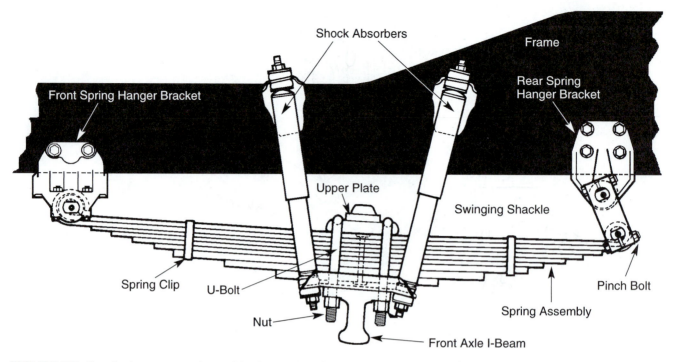

FIGURE 22–8   *Spring suspension with shock absorbers.*

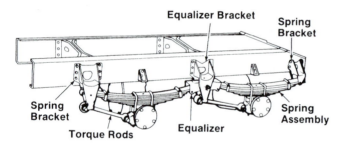

FIGURE 22–9   *Four-spring suspension system with torque rods. (Courtesy of Dayton-Walther Corporation)*

alignment of the suspension system. The extra torque rods also permit the truck axles to accept all of the driveline torque by minimizing axle windup and braking load transfer.

22.2 EQUALIZING BEAM SUSPENSIONS

Tandem drive axles require a special suspension that permits flexibility between the axles but still withstands rugged use with long service life. Equalizing beam type suspensions provide these characteristics. The types of equalizing beam suspensions used on heavy-duty trucks are the leaf spring-type and the rubber cushion-type.

The equalizing beam suspension system is designed to lower the center of gravity of the axle load. This is accomplished by placing the beam on or below the axle centerline. This design provides additional leverage to work with the torque rods in absorbing axle torque and road shock.

LEAF SPRING-TYPE SUSPENSION

The leaf spring-type suspension (**Figure 22–10**) uses semielliptic leaf springs to cushion load and road shock. The springs are mounted on saddle assemblies above the equalizer beams and pivot at the front on spring pins and brackets. The rear of the springs have no rigid attachment to the spring brackets but are free to move forward and backward to compensate for spring deflection.

Equalizing beams use a lever principle to distribute the load equally between axles and to reduce the effect of road irregularities. The torque rods permit absorption of wind-up torque, which is the tendency of the axles to turn backward or forward on their axis due to starting or stopping forces. The cross tube connecting the equalizing beams ensures correct alignment of the tandem and prevents damaging load transfer.

Fiber Composite Leaf Springs

Used for a few years in autos, fiber composite leaf springs have found their way into heavy-duty trucks. They are made of fiberglass which is laminated and

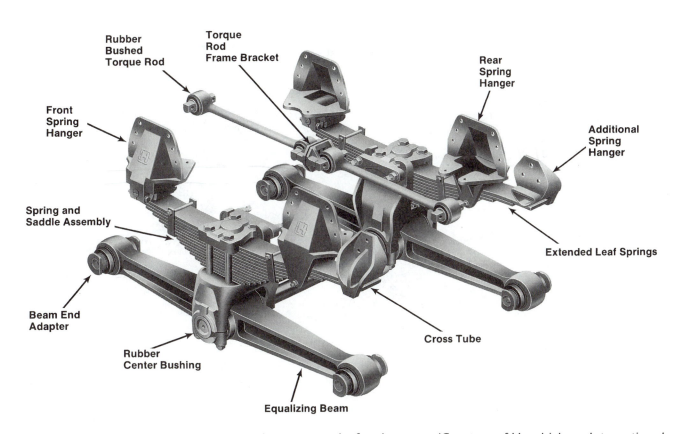

FIGURE 22–10 *Equalizing beam suspension system—leaf spring-type. (Courtesy of Hendrickson International, Truck Suspension Systems)*

bonded together by tough polyester resins. The long strands of fiberglass are saturated with resin and bundled together by wrapping (a process called filament winding) or squeezed together under pressure (compression molding).

Fiber composite leaf springs are lightweight and possess some unique ride control characteristics. Springs are dead weight. Reducing the weight of the suspension not only reduces the overall weight of the vehicle, but also reduces the unsprung mass of the suspension itself. This reduces the spring effort and amount of shock control required to keep the wheels in contact with the road. The result is a smoother-riding, better-dampening, and faster-responding suspension.

Other advantages of using fiber composite springs include the following:

- **Quieter ride.** A fiber composite spring does not resonate or transmit sound like a steel spring. In fact, it actually dampens noise.
- **No spring sag.** All steel springs sag with age, whether they are leaves or coils. Spring sag affects ride height, which in turn alters wheel alignment, handling, steering, and braking. A weak spring can load the suspension unevenly,

allowing the wheel under the weak spring to lose traction when accelerating or braking. However, according to the manufacturers of fiber composite springs, there is no sag with age.

- **Less body roll.** In applications where the leaf springs are mounted transversely, the spring also acts like a sway bar to limit body sway and roll when cornering. This load transfer characteristic also permits softer than normal spring rates to be used for a smoother ride.

With fiber composite leaf springs, there is little or no danger of the spring suddenly snapping. The laminated layers create a built-in fail-safe mechanism for keeping the spring intact should a problem arise.

RUBBER CUSHION-TYPE SUSPENSION

This type of suspension system (**Figure 22–11**) uses **rubber cushions** in place of leaf springs to absorb load and road shock. On these units, rubber cushions are mounted on a saddle assembly at each side. Mounted between frame brackets and the suspension, each rubber block unit is secured by four

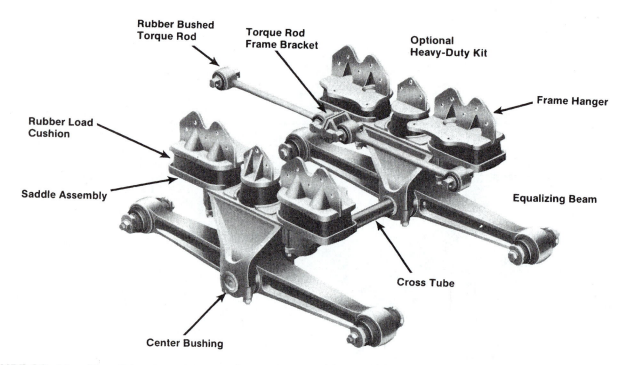

FIGURE 22–11 Equalizing beam suspension system—rubber cushion-type. (Courtesy of Hendrickson International, Truck Suspension Systems)

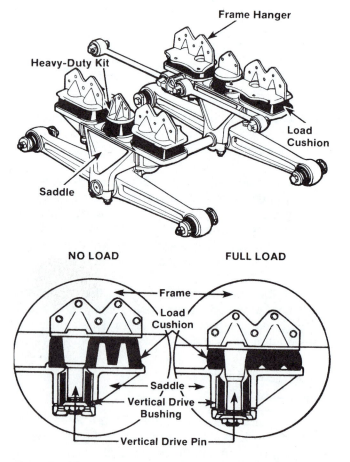

FIGURE 22–12 Sectional view of the rubber cushion-type equalizing beam suspension. (Courtesy of Hendrickson International, Truck Suspension Systems)

rubber-bushed drive pins, each of which passes through the rubber cushion. All driving, braking, and cornering forces are transmitted through these pins.

Without load, the spring rides on the outer edge of the spring load cushions. As the load increases, the crossbars of the cushions are progressively brought into contact to absorb the additional load. Cushioning and alignment are accomplished by the four drive pins encased in rubber bushings. The bushings permit the drive pins to move up and down in direct relation to movement of the load cushions (**Figure 22–12**).

22.3 TORSION BAR SUSPENSIONS

This type of suspension system uses torsion bars instead of steel leaf spring arrangements. The typical **torsion bar suspension** consists of a torsion bar, front crank, and rear crank with associated brackets, a shackle pin, and assorted bushings and seals. Axle torque and loads are supported by hardened steel bushings located at the end of each crank assembly. End thrusts are normally absorbed by a center-threaded bushing and a center-threaded area shackle pin. When the central area of the shackle pin is located within the mating threaded bushing, the axles can be aligned correctly. Seals are used at the bushing ends and at the shackle links to inhibit water seepage and highway debris.

Typical torsion bar suspensions use prestressed bars that acquire a permanent right-hand or left-hand twist. Normal manufacturing practice is to stamp the bars with an R or L to indicate on which side of the truck the bars are to be installed.

22.4 AIR SPRING SUSPENSIONS

Many heavy-duty trucks and trailers are equipped with **air spring suspension** systems (**Figure 22–13**). The air bag or air spring suspension system provides a smooth, shock- and vibration-free ride with a preset constant frame height. The air springs on the suspension system take the place of mechanical leaf-type springs. This design helps to minimize road shock transfer to the truck frame, the cargo, and the driver/operator. The air spring suspension system adjusts to load conditions automatically, providing a low rate suspension with light or no loads and a high rate suspension with heavier loads.

Advantages of the air spring suspension include the following:

- Front pivotal bushing controls truck roll and axle alignment, yet allows easy up and down travel.
- Structural frame brackets attach by welding or bolting.
- Trailing arms are steel box beams.

- The axle connection is a seat assembly, which is welded as part of the beam and U-bolted for double safe operation.
- The air spring usually incorporates a preset check valve to prevent any leaks in the suspension system from affecting the truck's air brake system.

Many current suspensions combine steel leaf springs and air springs as shown in **Figure 22–13** and **Figure 22–14**.

The major components of an air spring suspension system are as follows: a height control valve, air lines, tapered leaf springs, shock absorbers, torque rods, and wear pads. Air suspensions are currently more often used than any other type of suspension in heavy-duty trucks.

COMPONENTS

The brain of the air spring suspension system is the height control valve (**Figure 22–15**). The valve will automatically maintain truck ride height. The height control valve is usually mounted at the rear of the truck frame (**Figure 22–16**). The valve has a lever connected to the rear axle assembly by means of a linkage rod. This height control lever is sensitive to frame height and will automatically fill or exhaust the air springs to maintain the truck ride height.

Figure 22–17 illustrates the height control valve in the neutral position. At this point, the lever is parallel to the ground, and the air suspension has been properly inflated to level the vehicle. No air can pass through the valve at this time.

FIGURE 22–13 _Combination leaf and air spring suspension system. (Courtesy of Freightliner Trucks)_

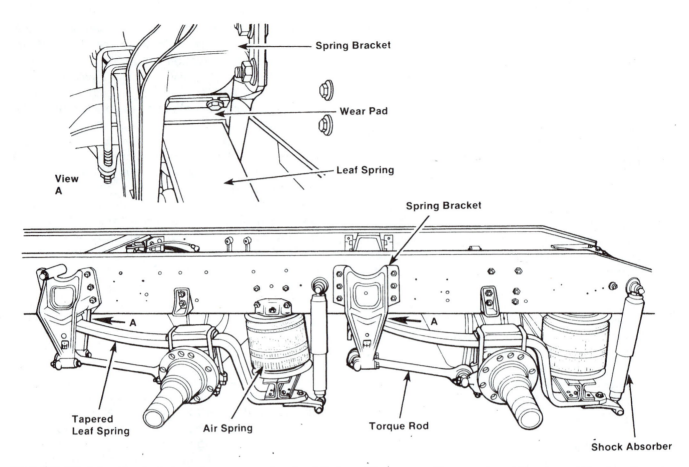

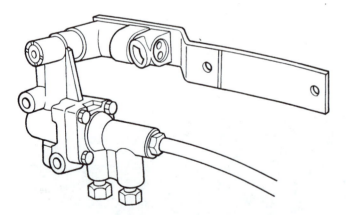

View
A

Spring Bracket

Wear Pad

Leaf Spring

Spring Bracket

Tapered
Leaf Spring

Air Spring

Torque Rod

Shock Absorber

FIGURE 22–14   Typical air spring suspension installation. (Courtesy of International Truck and Engine Corp.)

FIGURE 22–15   Height control valve. (Courtesy of International Truck and Engine Corp.)

During tractor operation, a hydraulic delay feature dampens random input from typical road shock (bumps and potholes) to keep the valve from rapid cycling between opening and closing.

When a load is applied, such as coupling a trailer to the tractor, the frame lowers. The lever is moved through an angle above horizontal, and the valve opens after the time delay is surpassed. As air

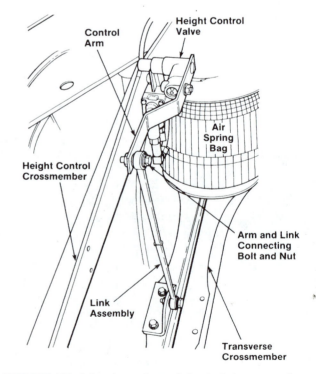

Control
Arm

Height Control
Valve

Air
Spring
Bag

Height Control
Crossmember

Arm and Link
Connecting
Bolt and Nut

Link
Assembly

Transverse
Crossmember

FIGURE 22–16   Location of the height control valve and components. (Courtesy of International Truck and Engine Corp.)

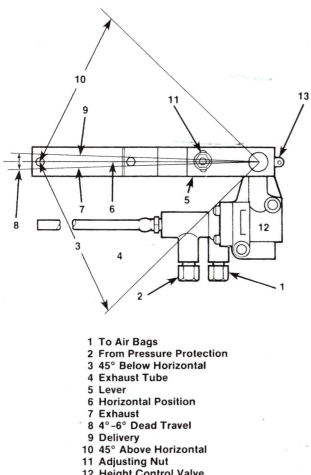

1 To Air Bags
2 From Pressure Protection
3 45° Below Horizontal
4 Exhaust Tube
5 Lever
6 Horizontal Position
7 Exhaust
8 4°–6° Dead Travel
9 Delivery
10 45° Above Horizontal
11 Adjusting Nut
12 Height Control Valve
13 Centering Pin

FIGURE 22–17 Height control valve operation. (Courtesy of International Truck and Engine Corp.)

passes through the valve and the air suspension is inflated, the frame resumes its set ride height.

The frame rises when the trailer is uncoupled. The lever moves through an angle below horizontal and after a time delay, the valve exhausts air until the frame once again resumes its set ride height.

Air lines, located between the front and rear axle air springs, transfer air pressure between the axles. The air pressure is used to maintain a constant frame rail height and an equal load distribution between the axles.

The tapered leaf spring (see **Figure 22–14**) acts as a lightweight one-piece beam to transfer loads directly to the axle. The spring is also used to absorb road shock, tractive, and braking forces.

Axle oscillation is dampened by hydraulic shock absorbers (see **Figure 22–14**), which are mounted where shock travel is greatest. The shock absorbers are designed so that in combination with the air springs, the best possible ride and stability will be obtained.

Lateral and longitudinal torque rods are used to minimize axle windup, caused by driveline torque and load transfer during braking.

Low-friction wear pads (see **Figure 22–14A**) are located in the spring brackets, allowing the tapered leaf spring to move forward and rearward with a minimum amount of friction, reducing ride harshness.

AIR SPRING MECHANICS

The air springs used in air suspension systems are of either the reversible sleeve type or the convoluted type (**Figure 22–18**). The most popular style of air spring in use today on heavy-duty trucks is the reversible sleeve version. The convoluted air spring is also available in three styles: single, double, and triple convolutions. The style shown in **Figure 22–18** is classified as a double convoluted-type air spring. The major components of an air spring are listed and described here:

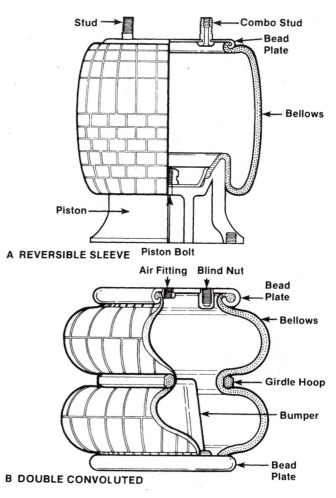

A REVERSIBLE SLEEVE

B DOUBLE CONVOLUTED

FIGURE 22–18 Components of air springs. (Courtesy of Firestone Industrial Products Co.)

- **Stud.** Used to attach the spring to the suspension; usually manufactured as a part of the bead assembly.
- **Combo Stud.** Serves the dual purpose of mounting the spring to the suspension and providing an air port.
- **Bead Plate.** Crimped onto the bellows at the factory, it allows complete testing before the spring leaves the factory.
- **Bellows.** The heart of the air spring, manufactured of rubber. Includes at least four plies, or layers, of materials: an inner layer, two or four plies of cord-reinforced fabric, and an outer cover. The rubber construction withstands temperatures down to minus 65°F.
- **Bumper.** Also known as a jounce block, this item is a solid rubber fail-safe device used on many truck suspension systems. It is used to prevent damage to the suspension in the event of sudden air pressure loss or jounce.
- **Piston.** Only found on the reversible sleeve-type spring; usually made from aluminum, steel, fiberglass, or hard rubber. The piston provides a lower mounting arrangement for the air spring in the form of tapped holes or studs. Also called a *pedestal*.
- **Piston Bolt.** Attaches the piston to the bellows assembly. In some cases, it may be extended to serve as a means of attaching the spring to the suspension.
- **Blind Nut.** A permanent part of the bead plate assembly; it provides an alternate mounting system to the stud.
- **Air Fitting.** Provides an air port for the spring.
- **Girdle Hoop.** A ring between the convolutions of the convoluted type spring; used to provide lateral stability.

22.5 SPRING SUSPENSION SYSTEM SERVICING

Servicing a spring suspension system consists of periodic checks and inspections, troubleshooting to identify defective components, and component replacement. System alignment is discussed separately at the end of this chapter.

SERVICE CHECKS

On spring suspension systems, U-bolts must be tightened at periodic intervals. This applies to both new trucks/trailers and to trucks or trailers on which spring service has been performed. Maintenance is limited to periodic inspections to ensure the torque values on U-bolts are maintained, spring pin bushings are lubricated, and proper spring end clearance exists.

The U-bolt nuts must be retightened after the truck or trailer has operated under load for 1,000 miles or six months, whichever occurs first. Thereafter, the spring U-bolt nuts should be retightened every 36,000 miles.

Also inspect spring ends to ensure that they are not in contact with the sides of the equalizer or hanger brackets (see **Figure 22–19**). Spring end contact indicates that the spring assemblies are not seated on the axle housing or there is a need for suspension alignment. Check U-bolt seats, U-bolts, and spring assemblies for proper installation.

CAUTION: When checking U-bolts always tighten to original specifications. Rusty joints must be disassembled, cleaned, and lubricated to ensure a like-new condition prior to tightening.

Use a wire brush and solvent or pressure wash equipment to clean the suspension system and remove dirt, grease, and scale. When using solvents, follow the safety precautions recommended by the solvent manufacturer.

Inspect the shackle bracket and the spring shackles for cracks, wear, and other damage. Replace any damaged parts. Inspect the spring for cracks and corrosion. If any cracked or broken leaves are dis-

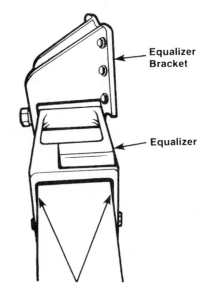

FIGURE 22–19 Spring end clearance. (Courtesy of International Truck and Engine Corp.)

covered, replace the entire spring as a unit or rebuild the spring assembly. Never paint leaf springs. This will lower inter-leaf friction and reduce its dampening capability. If rusting or corrosion is severe, replace the spring.

Using a micrometer, check the spring pins and the shackle pin for wear. Replace a pin if the diameter at any point on the pin is less than the specified minimum. Inspect the spring and bracket bushings. Replace the bushings if gouged, cracked, or otherwise damaged. If the bushings are not damaged, inspect them for wear. Using a micrometer, check the inside diameter of each bushing. The inside diameter of any bushing should not exceed the diameter of its pin by more than 0.020 inch. Replace any bushing that exceeds this limit.

SYSTEM DIAGNOSIS

The troubleshooting chart in **Table 22–1** covers the most frequent causes of problems that might occur in a leaf spring suspension system. However, different symptoms can produce similar complaints relating to axle and wheel alignment problems. Be sure that these related areas are in proper working order before attempting to troubleshoot the suspension system.

Rough Ride Diagnosis

There are two ways of describing rough ride conditions: harmonic and harsh. Harmonic ride problems are those in which the once-per-revolution energy input from such things as bent or imbalanced wheels match the natural frequency of the frame flexing. This produces rhythmic vibration, which continues as long as the critical road speed is maintained. Harmonic ride problems can occur on smooth roads. The main sources of harmonic ride problems are typically wheels, rims, and brake drums or hubs.

Harsh ride problems are those in which the suspension transfers, rather than absorbs, momentary energy inputs produced when the tires hit bumps or holes in the road. Wavy asphalt, or a series of bumps, might cause repetition of the harsh, jarring motion in the cab, but the motion stops after the tires pass over the bumps. Harsh ride problems occur on rough roads.

This section contains general information on correcting rough ride problems. It is not intended as a replacement for the component manufacturer's service manuals.

Visually check the vehicle for signs of damaged or missing suspension components. Repair or replace

TABLE 22–1:	LEAF SPRING SUSPENSION TROUBLESHOOTING GUIDE	
Symptom	**Probable Cause**	**Remedy**
Vehicle leans to one side	1. One or more broken spring leaves	1. Replace the spring assembly.
	2. Weak or fatigued spring assembly	2. Replace the spring assembly.
	3. Unmatched spring design/load capacity spring assemblies	3. Install the correct spring assemblies as originally specified for the vehicle.
	4. Bent or twisted frame rail	4. Check the frame rails for bends and twists. Correct as needed.
Vehicle wanders	1. One or more broken spring leaves	1. Replace the spring assembly.
	2. Wheels out of alignment	2. Adjust the wheel alignment.
	3. Caster incorrect	3. Install correct caster shims.
	4. Steering gear not centered	4. Adjust steering.
	5. Drive axles out of alignment.	5. Align the drive axles.
Vehicle bottoms out	1. Excessive weight on the vehicle causing an overload	1. Reduce the loaded vehicle weight to the maximum spring capacities.
	2. One or more broken spring leaves	2. Replace the spring assembly.
	3. Weak or fatigued spring assembly	3. Replace the spring assembly.
Frequent spring breakage	1. Vehicle overloaded or operated under severe conditions	1. Reduce the loaded vehicle weight to the maximum spring capacities. Caution the driver on improper vehicle handling.
	2. Insufficient torque on the U-bolt nuts	2. Torque the U-bolt nuts to the manufacturer's value.
	3. Loose center bolt allowing the spring leaves to slip	3. Check the spring leaves for damage. If damaged, replace the spring assembly. If not, tighten the center bolt nut to the manufacturer's value.
	4. Worn or damaged spring pin bushings allowing spring end-play	4. Replace the spring pin and bushing.

TABLE 22–1:   LEAF SPRING SUSPENSION TROUBLESHOOTING GUIDE (CONTINUED)

Symptom	Probable Cause	Remedy
Noisy spring	1. Loose U-bolt nut or center bolt allowing spring leaf slippage	1. Inspect the components for damage. Replace damaged components as necessary. Torque all fasteners to the manufacturer's values.
	2. A loose, bent, or broken spring shackle or front spring bracket impairing the spring flex	2. Inspect the shackle and bracket for damage. Replace damaged components as necessary. Torque the fasteners to the manufacturer's values.
	3. Worn or damaged spring pins allowing spring end-play	3. Replace any worn or damaged spring pins.
Rough ride	Refer to applicable subject in text.	

TABLE 22–2:   ROUGH RIDE DIAGNOSIS

Symptom	Probable Cause	Remedy
Harsh ride, tires off the ground	Seized front spring shackle pins not allowing the springs to flex	Replace seized shackle pins.
Harsh ride, tires on the ground	1. Tires improperly inflated	1. Adjust the tire pressure.
	2. The frame is bottoming out against the suspension.	2. Check the suspension for weak or damaged springs or components. Inspect the springs for "gull-winging" when the vehicle is loaded. Replace the spring assembly as necessary. Reduce the overall loaded weight on each axle to conform with the maximum spring load capacities on the vehicle specification sheet. Do not exceed the maximum spring load capacities.
	3. The vehicle normal loaded weight is markedly below the spring load capacity.	3. Contact the manufacturer for the correct application of a lower rated spring. Replace the spring assembly.
	4. When the vehicle is loaded, the front axle spring shackle angle is not within the rearward 3 to 18 degree angle.	4. Contact the manufacturer for shackle angle corrective measures.
	5. The weight on the tractor fifth wheel is causing overloading on the front axle springs.	5. If possible, move the fifth wheel toward the rear of the vehicle; otherwise, change the loading pattern on the trailer.
	6. There is a loaded weight differential between the rear axles greater than 800 pounds.	6. Contact the manufacturer for corrective measures.
	7. Forces from the trailer suspension are pushing on the tractor fifth wheel, causing a rough ride condition.	7. Review the ride problems in this subject that might apply to the trailer suspension. Perform the corrections, as necessary.

the components as required. Then test-drive the truck. When test-driving the truck, be sure to duplicate as closely as possible those conditions under which the problem occurs. Note the area(s) of the truck where the problem(s) seems to be coming from.

Raise the truck until the tires are off the ground and all of the weight is removed from the leaf springs. Be sure to block the axle and frame with safety stands. Then perform the corrections given in **Table 22–2**.

WARNING: Safety stands must securely support all of the wheel and frame weight during suspension repairs. Unsecured components might drop when the fasteners are loosened or removed, causing component damage and serious personal injury.

When troubleshooting and/or repair procedures have been completed, remove the safety stands from under the frame and axle and lower the truck.

LEAF SPRING REPLACEMENT

Although the exact replacement procedures are different for each type of leaf spring assembly, the basic procedures are similar. The following is a general outline leaf spring replacement.

To remove a spring assembly,

1. Chock the tires, place a floor jack under the truck frame, and raise the truck sufficiently to relieve the weight from the spring to be removed. Then place safety stands under the frame.
2. Remove shock absorbers where used.
3. Remove U-bolts, spring bumper, and retainer or U-bolt seat.
4. Remove the lubricators (not used on springs equipped with rubber bushings).
5. Remove the nuts from the spring shackle pins or bracket pins.
6. Slide the spring off the bracket pin and shackle pin.

Although the actual installation procedures will vary with each spring type, the pivot end of the spring is usually fastened to the frame bracket first. The shackle end then can be fastened by aligning the shackle to the other frame bracket. When installing the U-bolts or spring clips to secure the axles, do not final tighten until springs have been placed under normal load. Spring failures may occur at the center bolt hole if U-bolts become loosened. These bolts must be retightened at specified intervals or during periodic maintenance checks.

Tightening of the front and rear axle U-bolt nuts at proper intervals on new vehicles and on vehicles with replaced springs is important to prevent spring shifting, breaking of leaves, and center bolt shearing. Loose spring clips or U-bolts will also cause axle misalignment, which could adversely affect tire wear. When tightening U-bolt nuts, follow the recommended tightening sequence after lubricating the spring clip threads with the recommended lubricant.

The lubricant will ensure adequate clamping of the spring pack with the nuts tightened to the specified torque.

To correctly torque U-bolt nuts, first tighten all the nuts until they are snug. Next, tighten the nuts until approximately one-third of the recommended torque is achieved. Repeat tightening of the nuts, using the same sequence, gradually increasing the torque through a second, third, and fourth stage, until the recommended final torque is attained.

SHOCK ABSORBER INSPECTION

Shock absorbers should be checked every 12,000 miles. To check shock absorbers on the vehicle,

1. To check for noise, make certain that the noise is not coming from some other part of the vehicle. Then check shock absorber brackets and mounting stud nuts to make sure that they are tight and that the shock absorber is not striking or rubbing on the frame or some other part of the chassis.
2. Check the rubber mounting bushings and replace if worn.
3. Disconnect the lower end and work the shock absorber by hand to see that the outer tube is not striking or rubbing against the fluid reservoir tube.
4. While the shock absorber is disconnected, check the piston movement by pulling and pushing the absorber down and up slowly to make sure that the piston does not bind in the pressure tube. There should be considerable resistance when extending the absorber, but only slight resistance when collapsing the unit. Also note the rate of effort for distance of travel. This rate should not vary.
5. If it is determined that the shock absorber is defective, replace it.

CAUTION: Do not operate a vehicle with a shock absorber removed or defective because this places undue stress on the rear springs.

SHOCK ABSORBER REPLACEMENT

1. Remove the nuts from the top and bottom bolts from the eyes of the shock absorber.
2. Remove the shock absorber from the vehicle.
3. Place rubber bushings into the eyes of the new shock absorber.

4. Install the top bolt through the top mounting bracket on the frame rail.

5. Place a flat washer over the bolt and thread the bolt into the upper rubber bushings of the new shock absorber.

6. Place a second flat washer over the bolt and then screw on the nut finger tight.

7. Extend the lower part of the shock absorber until the mounting holes line up in the lower bracket and shock absorber eye.

8. Install the bottom bolt through the lower mounting bracket on the axle.

9. Place a flat washer over the bolt and thread the bolt into the lower rubber bushing of the new shock absorber.

10. Place a second flat washer over the bolt and screw on the nut finger tight.

11. Check to make sure that the rubber bushings are seated properly and tighten the nut to the specified torque.

👓 Shop Talk

Some shock absorber mount brackets have a stud welded to the bracket, rather than a nut and bolt.

RADIUS ROD REPLACEMENT

Before replacing a radius rod on a leaf spring suspension system, secure the truck's parking brakes and chock the tires. Be sure the replacement radius rod is the correct length because they come in many different sizes. Then replace the radius rod by performing the following:

1. Note the number of axle alignment washers (refer to **Figure 22–20**, item No. 4) at the forward end of the radius rod that is being removed.

2. Remove the fasteners that attach the radius rod to the forward spring bracket and to the axle seat.

3. Remove the radius rod and any axle alignment washers. Retain the axle alignment washers for use when installing the new radius rod.

4. At all points where steel components (including bolts, washers, and nuts) contact an aluminum forward spring bracket, apply Alumilastic™ compound, or an equivalent, on the mating surfaces. Failure to apply Alumilastic™ compound, or an equivalent, to areas where aluminum and steel parts contact each other could lead to corrosion of the metals, resulting in damage to the suspension.

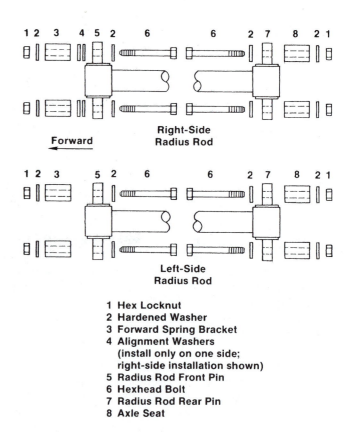

1 Hex Locknut
2 Hardened Washer
3 Forward Spring Bracket
4 Alignment Washers
  (install only on one side;
  right-side installation shown)
5 Radius Rod Front Pin
6 Hexhead Bolt
7 Radius Rod Rear Pin
8 Axle Seat

FIGURE 22–20 Radius rod attachment (top view). (Courtesy of Freightliner Trucks)

5. Position the replacement radius rod so that the radius rod pins are between the rear side of the forward spring bracket and the front side of the axle seat.

6. Install the hexhead bolts, hardened washers, and locknuts in the axle seat and the radius rod rear pin.

7. Install any previously removed axle alignment washers between the radius rod front pin and the forward spring bracket. Install the hexhead bolts, hardened washers, and locknuts in the radius rod front pin and the forward spring bracket.

8. Tighten the radius locknuts to the torque value specified by the manufacturer.

9. Check the axle alignment. If necessary, adjust the rear axle alignment.

SPRING BRACKET REPLACEMENT

Failure to replace a worn, cracked, or damaged spring bracket could result in breakage of the bracket itself, which in turn could cause a loss of vehicle control, resulting in personal injury or property damage.

Prior to replacing the spring bracket, chock the front tires. Then raise the rear of the truck and block the front or rear axle with safety stands. Raise the truck frame so that all the weight is removed from the leaf springs. Then, block the frame with safety stands. Make sure that the safety stands will securely support the weight of the axle and the frame. Then do the following:

1. If removing the forward spring, note the number of axle alignment washers, if any, between the bracket and the radius rod front pin. Remove the fasteners that attach the radius rod to the bracket and remove any axle alignment washers.
2. Remove the fasteners that attach the spring bracket to the frame rail and remove the spring bracket.
3. Place the new spring bracket on the frame rail. Align the mounting holes and install the spring bracket bolts, hardened washers, and locknuts.
4. If replacing the forward spring bracket, install the nuts for the top two bolts on the outboard side of the frame rail and install the nuts for the bottom four bolts on the inboard side of the frame rail (see **Figure 22–21A**).
5. If replacing the rear spring bracket, install the nuts for the top two bolts on the outboard side of the frame rail and install the nuts for the bottom two bolts on the inboard side of the frame rail (see **Figure 22–21B**).

6. Tighten the locknuts to the torque value specified by the manufacturer.

> **CAUTION:**   Failure to properly torque the suspension fasteners can result in abnormal tire wear and damage to the springs, spring brackets, and frame rail.

7. Check the axle alignment. If necessary, adjust the rear axle alignment.

EQUALIZER REPLACEMENT

Before attempting to replace the equalizer assembly, make sure that the truck's front tires are chocked. Then raise the rear of the truck and block the axles with safety stands. Raise the truck frame so that all the weight is removed from the leaf springs; then block the frame with safety stands. Make sure that the stands will securely support the weight of the axles and frame. To allow access to the equalizer, remove the wheel assemblies on that side of the vehicle. Then refer to **Figure 22–22** and replace the equalizer as follows:

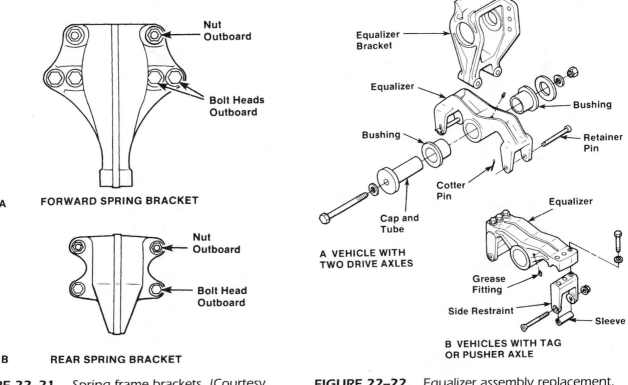

A   FORWARD SPRING BRACKET

B   REAR SPRING BRACKET

FIGURE 22–21   Spring frame brackets. (Courtesy of Freightliner Trucks)

A VEHICLE WITH TWO DRIVE AXLES

B VEHICLES WITH TAG OR PUSHER AXLE

FIGURE 22–22   Equalizer assembly replacement. (Courtesy of Freightliner Trucks)

1. If removing an equalizer from a truck with two drive axles, remove the cotter pin from the outboard end of each spring retainer pin; then remove the retainer pins.

2. Remove the equalizer cap and tube assembly locknut, inboard bearing washer, bolt, and outboard bearing washer.

3. Insert a bar between the bottom of the equalizer and the equalizer bracket. Gently lever the weight of the equalizer off the equalizer cap and tube assembly. Insert a piece of bar stock through the inboard equalizer cap and tube assembly bolt hole and lightly tap the cap and tube assembly out of the equalizer.

4. Thoroughly clean the equalizer bushings and inspect them for damage or defects. Replace the bushings if necessary.

5. Apply multipurpose chassis grease to the inside of the equalizer bushings; then install the bushings in the equalizer.

6. Install the replacement equalizer in the equalizer bracket.

7. Apply multipurpose chassis grease to the equalizer cap and tube assembly. Start the cap and tube assembly into the equalizer, through the equalizer bracket.

8. Push the equalizer cap and tube assembly partway through the equalize and then place the inboard wear washer(s) between the inboard equalizer bushing and the equalizer bracket. Push the cap and tube assembly the rest of the way into the equalizer bracket.

9. Place the outboard bearing washer on the equalizer cap and tube assembly bolt and install the bolt in the cap and tube assembly.

10. Install the inboard bearing washer and locknut on the cap and tube assembly bolt. Tighten the locknut to the torque value specified by the manufacturer.

CAUTION: Failure to apply Alumilastic™ compound, or an equivalent, to areas where aluminum and steel contact each other could lead to corrosion of the metals, resulting in suspension damage.

11. Install the wheel assemblies. Remove the safety stands from under the frame and axle and lower the truck.

12. If the radius rods have been loosened or the equalizer bracket has been removed, check the rear axle alignment. If necessary, adjust the axle alignment.

22.6 EQUALIZER BEAM SUSPENSION SYSTEM SERVICING

Clean all dirt from suspension parts and inspect all parts carefully for cracks or damage. Inspect rubber bushings for damage or deterioration. If the truck has been in service for a long time, replace all bushings.

Most repairs to the suspension system consist of replacing worn or damaged parts. The major item that will concern the technician is removal and replacement of the rubber bushings.

Although the bushings have long life and replacement will be limited, they can be replaced if damaged or deteriorated. Special service tools for performing this task are available but, though recommended, are not absolutely necessary. If press equipment is available, standard steel tubing having diameters to match the bushing sleeves (metal bands surrounding rubber bushings) can be used as adapters for removing and installing the bushings. Press pressures required to remove the bushing and sleeve assemblies will generally be between 35 and 50 tons.

Special tool equipment is available from some manufacturers that can be used to replace the equalizer beam bushings without completely removing the equalizer beam from the chassis. The special equipment can also be used to replace the bushings with the equalizer beams removed.

SPRING SHOCK INSULATORS

The purpose of the spring shock insulator (sometimes called *spring load cushions*) is to secure the spring to the chassis and to reduce the transmission of road shock to the vehicle. To do this, the insulator must be in good condition and firmly bonded to the metal insert of the cushion.

The spring shock insulators are generally made of butyl and rubber and are not resistant to petroleum products. If the cushions become saturated with these products, they will split and disintegrate. In an operation where these products are used, an oil-resistant compound load cushion is available.

Visual inspection of the shock insulators will determine the necessity for replacement. Check them for cracks, deterioration, and abnormal wear by removing the insulator cap (if so equipped). Also check the height of cushion against the recommended height given in the service manual. A permanent set of not more than 1/4 inch would still make the insulator acceptable for use. Remember that the purpose of this inspection is to determine whether

PHOTO SEQUENCE 8
TRAILER HEIGHT CONTROL VALVE ADJUSTMENT

P8–1 Ensure that the trailer is unloaded. The trailer must be connected to a tractor or supported on a trailer kingpin stand. Be sure the trailer parking brakes are applied and the trailer wheels are blocked. The trailer air system pressure should be at 70 psi (482.65 Kpa) or greater throughout the adjustment procedure.

P8–2 Disconnect each height control valve linkage at the lower bracket. Move each height control valve arm to the upward position to raise the trailer frame. Install steel spacer blocks of the specified thickness between the rigid beams and the trailer frame. An alternate method is to install safety stands under the trailer frame to obtain the specified block thickness at the proper location between the rigid beam and the trailer frame.

P8–3 Move each height control valve arm to the downward position until all the air is exhausted from the air springs and the trailer frame is supported on the spacer blocks. Be sure the spacer blocks or safety stands are supplying the specified ride height between the center of the trailer axle and the bottom of the trailer frame.

P8–4 Move each control valve arm 45 degrees downward for 10 to 15 seconds, and then return the control valve arms to the center position.

P8–5 Install the wood locating pins in each height control valve adjusting block and bracket.

P8–6 Loosen the 1/4 in. (6.35 mm) adjusting nut on the adjusting block on each height control valve.

PHOTO SEQUENCE 8 (CONTINUED)
TRAILER HEIGHT CONTROL VALVE ADJUSTMENT

P8–7 Reconnect the height control valve linkage to the lower bracket and tighten the retaining nut to 24 or 48 in.=lb (2.68–5.37 N•m).

P8–8 Tighten the ¹/₄ in. (6.35 mm) adjusting nut to 24 to 48 in. lb (2.68 to 5.37 N•m) on each height control valve and remove the wood locating pins.

P8–9 Use a floor jack to raise the trailer frame and remove the spacer blocks or safety stands. An alternate method of raising the trailer frame is to disconnect the height control valve linkage at the lower bracket and move the height control arm upward to raise the trailer frame. Be sure the linkage to the lower bracket retainer is tightened to the specified torque.

P8–10 With the air pressure in the trailer air brake reservoirs greater than 70 psi (482.65 Kpa), be sure the air springs are inflated to supply the specified ride height between the center of the axles and the lower side of the frame.

the insulator is suitable for further use, or whether it should be replaced. Arbitrary replacement of useable insulators is wasteful, and reuse of worn insulators can result in unnecessary downtime.

A vehicle equipped with spring shock insulators must not be operated with worn or defective shock insulators. Faulty insulators can cause axle misalignment or floating, which adversely affects tire life. Continued operation with damaged or worn insulators will result in damage to the "T" slot area of the shock insulator cap and eventual cap failure. Spring damage is also possible.

To remove and replace spring shock insulators or load cushions, follow the procedure given in the service or chassis manual.

SYSTEM OVERHAUL

The following procedure should be undertaken before performing repairs on an equalizing beam suspension system.

Preparation for Overhaul Repair

1. Block the wheels on both axles of the truck.
2. Remove the axle shafts from both rear axles.
3. Remove the four saddle caps that lock the center bushing of the equalizer beams to the saddle assemblies.
4. Disconnect the torque rods from the axle housing by loosening the locknuts and driving the shaft from the bushing and the axle housing bracket.
5. Using an overhead crane, raise the truck frame sufficiently so that the lower part of the saddle clears the top of the axle housings. Block the frame securely. Block the axles to prevent them from pivoting on the wheels.
6. Roll axles, with the beams attached, out from under the truck.

Overhaul

There are three main types of equalizer beam mounting configurations:

- Bolt type beam end mounting (**Figure 22–23**)
- Tube type beam end mounting (**Figure 22–24**)
- Ball and socket beam end mounting (**Figure 22–25**)

The replacement procedure given here will cover the bolt type beam end.

1. Disconnect the equalizer beam from the axle housing, referencing **Figure 22–23**.
   a. Remove the equalizer beam end bolt.
   b. Drive the bushing adapters and sleeve (if so equipped) out of the bushing and axle housing brackets. Adapters are notched to help keep the chisel in place between the axle bracket and adapter. Driving the chisel first on one side and then on the other side of the adapter will wedge out the adapter.
   c. After the first adapter is removed, the opposite adapter can be driven out with an impact hammer or heavy bar and hammer.
2. Separate the equalizer beams from the cross tube.
3. After both equalizer beams have been separated from the axles, separate the beams from the cross tube by pulling them apart by hand. The cross tube floats in the inner sleeve of the center bushings. The slide float is 3 inches plus, which will polish the cross tube where it enters the center bushing on each side. This visible polished area on the cross tube is normal.

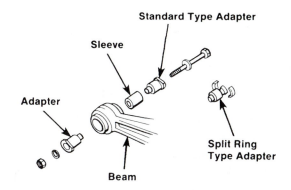
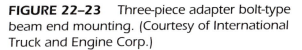

FIGURE 22–23 Three-piece adapter bolt-type beam end mounting. (Courtesy of International Truck and Engine Corp.)

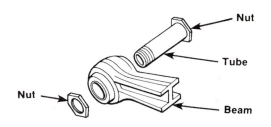

FIGURE 22–24 Tube-type beam end mounting. (Courtesy of International Truck and Engine Corp.)

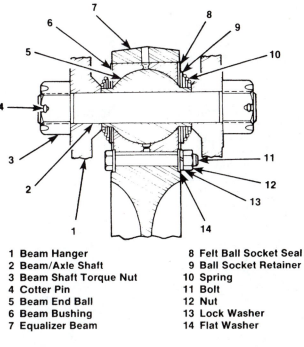

1 Beam Hanger	8 Felt Ball Socket Seal
2 Beam/Axle Shaft	9 Ball Socket Retainer
3 Beam Shaft Torque Nut	10 Spring
4 Cotter Pin	11 Bolt
5 Beam End Ball	12 Nut
6 Beam Bushing	13 Lock Washer
7 Equalizer Beam	14 Flat Washer

FIGURE 22–25 Ball and socket beam end mounting. (Courtesy of International Truck and Engine Corp.)

4. Remove the saddle assemblies, springs, or cushions. See the *leaf spring type* (**Figure 22–26**).

   a. Support the spring and saddle assembly on a floor jack and safety stand. Loosen the spring-aligning setscrew. Remove the spring top pad saddle bolts and nuts. Remove the top pad. Lower the floor jack and remove the saddle.

   b. Remove the spring assembly by repositioning the floor jack under the spring. Loosen the locknut on the spring pin draw key (item 3).

   c. Back off the nut sufficiently to protect the draw key threads and then strike the nut with a soft hammer to loosen the draw key. Remove the draw key and drive the spring pin out of the spring and spring bracket. Lower the spring assembly from the frame and remove it from the jack. See the *rubber cushion* or *shock insulator type* (**Figure 22–27**).

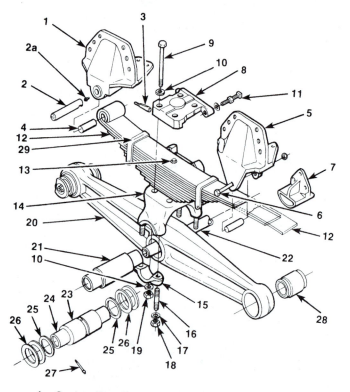

1	Spring Bracket	15	Saddle Cap
2	Spring Pin	16	Stud
2a	Zerk Fitting	17	Washer
3	Draw Key	18	Nut
4	Bushing	19	Nut
5	Spring Bracket	20	Equalizer Beam
6	Rebound Bolt with Spacer	21	Center Bushing
		22	Cross Tube
7	Spring Bracket (extended leaf only)	23	Center Bushing Bronze
8	Top Pad	24	Sleeve
9	Top Pad Bolt	25	Seals
10	Washer	26	Washer
11	Setscrew	27	Zerk Fitting
12	Leaf Spring Assembly	28	End Bushing
13	Center Bolt	29	Part Number
14	Saddle Assembly		

FIGURE 22–26 *Leaf spring-type suspension system.* (Courtesy of International Truck and Engine Corp.)

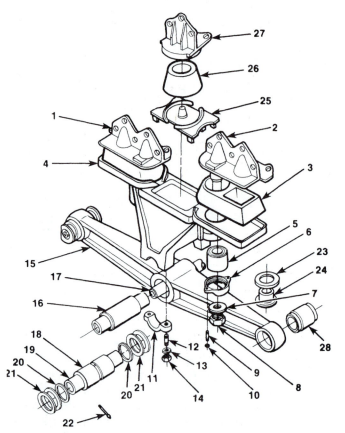

1	Frame Hanger	17	Cross Shaft
2	Frame Hanger	18	Center Bushing Bronze
3	Rubber Cushion	19	Sleeve
4	Saddle Assembly	20	Seal
5	Drive Pin Bushing	21	Thrust Washer
6	Bushing Retainer Cap	22	Zerk Fitting
7	Rebound Washer	23	Rebound Bumper*
8	Drive Pin Nut	24	Stop Nut*
9	Stud	25	Plate (optional with RS 380)
10	Nut	26	Rubber Cushion (optional with RS 380)
11	Saddle Cap		
12	Stud	27	Frame Hanger (optional with RS 380)
13	Washer		
14	Nut	28	End Bushing
15	Equalizer Beam		
16	Center Bushing		

FIGURE 22–27 *Rubber cushion-type suspension system.* (Courtesy of International Truck and Engine Corp.)

a. Remove the four vertical drive pin bushing retainer caps (item 6).

b. Remove the drive pin nuts (item 8). Separate the saddle (item 4) and load cushions from the vertical drive pins. In extreme cases, it might be necessary to cut the drive pin nuts to remove them.

c. Replace the drive pin bushings if required.

5. Remove the torque rods from the cross member frame bracket.

6. Assemble or install new spring, cushion, or insulator and saddle assemblies. See the *leaf spring type* (**Figure 22–26**).

a. Seat the spring in the spring saddle with the head of the spring center bolt positioned in the hole provided in the saddle.

b. Position the spring top pad over the "cup" on the main spring leaf.

c. Install the saddle-to-top pad bolts and nuts. Run the nuts up snug but do not tighten completely at this time.

d. To properly position the spring in the saddle, tighten the spring aligning setscrews to the manufacturer's specified torque. Tighten the aligning screw locknuts.

e. Tighten saddle-to-top pad bolt nuts to the manufacturer's specified torque.

f. Using a roller jack, position the spring (with saddle) in the front and rear spring mounting brackets.

g. Align the spring eye with the spring pin bore in the front bracket. Install the spring pin, aligning the draw key slot in the pin with the draw key bore in the bracket.

h. Install the draw key, lock washer, and nut. Tighten the nut to the manufacturer's specified torque.

i. Install the lubricator fitting in the spring pin and lubricate. See the *rubber cushion-type mounting* (**Figure 22–27**).

a. With the drive pin bushings assembled to the saddle and the bearing caps installed, assemble or install the rubber load cushion on the saddle.

b. Apply a light coat of multipurpose grease to the drive pins on the frame brackets.

c. Assemble the frame brackets to the saddle by inserting the drive pins through the load cushions and the drive pin bushings.

d. Install drive pin nuts and washers. Washers are installed with the flange down. Tighten the nuts snug. If the frame hangers were removed, tighten the nuts to the specified torque after the hangers have been assembled to the chassis frame.

e. Assemble the preassembled saddles, load cushions, and frame hangers to the chassis frame. Install the mounting bolts, lockwashers, and nuts.

f. Tighten mounting bolt nuts to the specified torque.

g. Tighten drive pin nuts to the specified torque.

7. Assemble or install new cross tube and equalizer beams to the axle assemblies (**Figure 22–26** and **Figure 22–27**). When assembling the suspension, the equalizer beam, mounting nuts, and the equalizer beam saddle cap, mounting nuts should have final torque applied with the wheels of the vehicle on the ground.

a. Align the beam end bushings with the hanger brackets and insert bushing adapters. Lubricate the adapters with a light coat of multipurpose grease. The flat surface on adapters must be in the vertical position.

b. Tap the adapters into the bushing until the adapter flanges contact the outer faces of the hanger brackets.

c. Install the bolt and locknut. Do not tighten the nut at this time.

CAUTION: When reinstalling the "rubber-bushed" equalizing beams to the axles and saddles, do not tighten the nuts until the torque rods are installed and the equalizing beams are level with the frame. (This is to ensure that the rubber is in a neutral position, eliminating the possibility of torsional windup, which might affect ride quality.) Then tighten all nuts to recommended torque.

8. Place one end of the cross tube into the center sleeve of the installed equalizer beam. Be sure that the tube seats into the sleeve.

9. Assemble the other equalizer beam to the cross tube and position the beam in the spring saddle.

10. Roll the axle assemblies with the equalizer beams under the center of the saddle. Make sure that the outer bushings are lined up with the center of the saddle legs.

11. Lower the vehicle frame centering saddles on the beam end bushings.

12. Install the saddle cap and nuts. Do not tighten until the torque rods are installed and the equalizer beams are level with the frame. Torque the nuts to the recommended torque.

13. The cross center tube is free to rotate or float in the bushing assemblies at each side. The function of the tube is to ensure alignment between the left and right equalizer beams.

14. The saddle caps serve to secure the sleeve with the bushing. The clamping effect of the caps against the sleeve and bushing has no effect upon the movement of the cross center tube.

15. Install the torque rods to the axle brackets and the frame brackets. When tightening, rap the bracket with the hammer to drive the taper of the torque rod stud into the bracket. If the ends of torque rods have straddle mount ends (2 holes) as shown in **Figure 22–28**, use spacers between the bracket and cross member for proper axle setting. Do not exceed $1/2$ inch to obtain major adjustment. Tighten the shaft nuts to specified torque.

16. Tighten the equalizer beam end fittings to the specified torque.

17. Install the axle shafts. Connect the propeller shafts, brake lines, interaxle lock control linkage, or piping.

18. On vehicles equipped with air brakes, check the brake piping for air leaks.

👓 Shop Talk

Three types of equalizer beams might be used in the manufacturing of trucks equipped with equalizer beam suspensions. Although the appearance of the equalizer beams is similar, a substantial difference exists in the material of which they are manufactured.

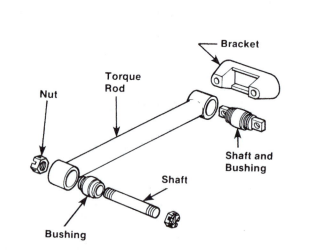

FIGURE 22–28   Torque rod disassembly. (Courtesy of International Truck and Engine Corp.)

The materials of the equalizer beams involved are aluminum, cast steel, and nodular iron. Aluminum and cast steel equalizer beams are manufactured in such a way that they can be stressed in either direction. Nodular iron equalizer beams are manufactured in such a way that they can be stressed in one direction only. Thus it is imperative that nodular iron equalizer beams are installed with the correct side up.

To aid in identification and to facilitate installation of nodular iron beams, an arrow and the word UP should be molded on the side of the nodular iron beams. Also, reinforcing gate pads are usually designed into each end and middle of the top side of nodular iron beams.

Nodular iron equalizer beams must be installed with the gate pads on the top side of the beams and the arrow molded on the beam side pointing upward.

22.7 AIR SUSPENSION SYSTEM SERVICING

Proper service and repair is important to the safe and reliable operation of an air suspension system. The servicing procedures recommended and described in this section are effective methods of performing air suspension maintenance. Basically, this maintenance consists of routine checks and services, component replacement, and height control valve adjustment.

SERVICE CHECKS AND TECHNIQUES

A daily quick look to verify that the truck or trailer is level and riding at the correct height will help the truck technician find any obvious problems. A closer inspection can detect broken or loose parts before any serious problems arise.

There should be monthly inspections of clearances around the air springs, tires, shock absorbers, and all other moving parts in the suspension system. Any part interferences should be tended to immediately. Bolts should be secure, axle connections tight, and signs of wear should be noted. All U-bolt locknuts should be retightened every 100,000 miles.

Every quarter, or 90 days, inspect all welded connections for signs of deterioration. Inspect frame attachment joints, cross member structures, and all pivoting and clamping connections.

Air springs will last almost indefinitely in most applications. However, they fail quickly if rubbed, scuffed, or punctured. If an air spring fails, the truck will settle down on the internal rubber bumpers, allowing the driver to proceed to the closest service facility. Always determine what caused the air spring to fail. Shock absorbers absorb energy to prevent suspension oscillations. Shock absorbers are also used as rebound stops in some air suspension systems. Air springs can be pulled apart if their stroke is not limited. Replace shock absorbers only if operational problems are detected. When replacing a shock absorber, do not lift the vehicle without the shock absorbers in place. Air spring damage can occur if the air springs are overextended.

Many types of air controls are available with air suspension systems. The most common system automatically regulates frame height by controlling air supplied to the air springs. When used in conjunction with other types of suspensions, such as leaf spring suspension, an operator-controlled pressure regulator is often used. It is important to periodically drain or purge the air suspension system to prevent the buildup of moisture and contaminants.

Often a height control valve is improperly diagnosed as defective and unnecessarily replaced. There are basic steps to determine whether the valve is defective or whether proper adjustment has not been performed. To determine the status of a height control valve, perform the following service checks:

1. Remove the bolt that attaches the height control link to the height control valve lever.
2. Ensure that air reservoir pressure is above 100 psi.
3. Raise the height control valve lever 45 degrees above horizontal. Air pressure at the air springs should begin to increase within 15 seconds, which will cause the truck frame to rise.
4. Lower the height control valve lever 45 degrees below horizontal. Air pressure at the air springs should begin to decrease within 15 seconds. Let air escape until air spring height is approximately 10 inches.
5. Raise the valve lever to 45 degrees above horizontal again, until air spring height is approximately 12.5 inches. Release valve lever.
6. Check the valve body, all tubing connections, and air springs for leaks with soapy water.
7. Check air spring height 15 minutes after performing Step 5.
8. If no leaks are found, the valve functions as just described, and air spring height has not changed in 15 minutes, the valve is functioning properly.
9. Install the bolt that attaches the height control lever to the height control link. Torque nut to specification.

Ride Height Adjustment

Some versions of height control valves have a centering pin and bosses. The pin is positioned in the bosses after setting the height. To adjust the height control valve, do the following:

1. Remove the plastic center pin from the bosses (**Figure 22–29**).
2. Ensure that approximately 100 psi air pressure is available in the reservoir (check cab gauges).
3. Disconnect the vertical link from the height control valve lever.
4. The height adjustment is checked at the rearmost axle.
   Place a straightedge on the centerline of the top metal plate on the air spring (**Figure 22–30**). Do not position the straightedge on

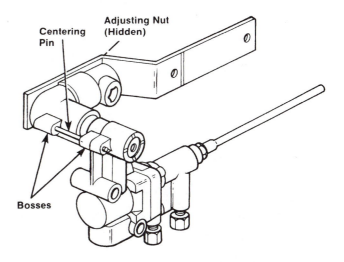

FIGURE 22–29 *Height control valve centering pin. (Courtesy of International Truck and Engine Corp.)*

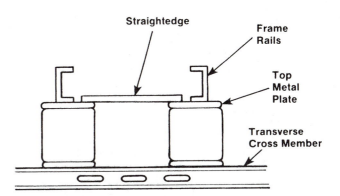

FIGURE 22–30 *Ride height adjustment. (Courtesy of International Truck and Engine Corp.)*

the frame rails. Measure the distance from the bottom edge of the straightedge to the top edge of the transverse cross member. Refer to the manufacturer's service manual for the correct specification distance.

5. If the distance is less than specified, raise the valve control lever and hold it up until the distance is correct. Release the lever. Loosen the lever adjusting nut, which is located 1 7/8 inch from the lever pivot point, so that the steel portion of the lever moves but the plastic portion does not. Position the lever so that the link-to-lever bolt can be inserted through the lever and link.

6. If the distance is more than specified, lower the valve lever and hold it down until the distance is correct. Release the lever. Loosen the lever adjusting nut. Position the lever and link so that the lever bolt can be inserted through the lever and link.

7. Install the lever link bolt nut. Tighten the nut 115 to 130 inch-pounds.

8. Center the plastic portion of the valve lever and insert the alignment pin through the boss on the valve body and plastic arm.

9. Tighten the lever adjusting nut to lock the steel and plastic portions of the lever together; then tighten to specification.

10. Remove the centering pin (if equipped) and discard.

AIR SPRING REPLACEMENT

The following procedure is recommended for the replacement of an air spring (**Figure 22–31**):

1. Chock the wheels and dump the air pressure from the air spring.

2. Raise the truck frame to remove the load from the suspension. Block the frame with safety stands.

3. Remove the locknuts and washers that connect the air spring to the air spring mounting bracket (**Figure 22–32**).

4. Remove the air lines connected to the air spring.

5. Remove the brass air fittings from the air spring.

6. Remove the locknuts and washers that connect the air spring to the frame hanger. Then remove the air spring.

7. Install the replacement air spring to the frame hanger by inserting the studs on the air spring into the appropriate holes on the hanger.

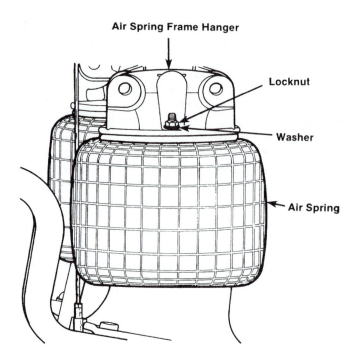

FIGURE 22–31   Air spring. (Courtesy of Hendrickson International, Truck Suspension Systems)

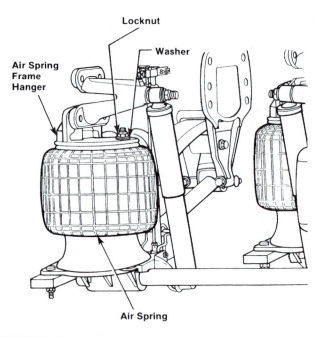

FIGURE 22–32   Air spring replacement. (Courtesy of Hendrickson International, Truck Suspension Systems)

8. Install the air spring to the spring mounting bracket by inserting the air spring studs into the holes in the bracket.

9. Install the locknuts and washers to secure the air spring to the bracket. Tighten and torque the locknuts.
10. Install washers and locknuts to secure the air spring to the frame hanger. Tighten and torque the locknuts.
11. Install the brass air fitting to the air spring using a thread sealant.
12. Install the air lines to the brass fitting.
13. Supply air pressure to the air spring. Adjust the height of the truck or trailer to specification.

SHOCK ABSORBER REPLACEMENT

To replace a defective shock absorber, refer to **Figure 22–33** and do the following:

1. Remove the locknut and washer that connect the shock absorber to the shock absorber frame hanger.
2. Remove the washer and locknut that connect the shock absorber to the air spring mounting bracket.
3. Remove the shock absorber from the mounting studs.
4. Install the replacement shock absorber to the studs on the hanger and the bracket.
5. Install washers and locknuts to the studs. Tighten and torque all locknuts.

LEVELING VALVE REPLACEMENT

Refer to **Figure 22–34** and do the following to replace a defective leveling valve:

1. Purge the air pressure from the suspension system and the truck's brake system. Then remove the air lines from the leveling valve.
2. Remove the leveling valve link from the valve by removing a nut and lock washer.
3. Remove locknuts, washers, and bolts that connect the leveling valve to the truck frame.
4. Remove the brass air fittings from the defective leveling valve.
5. Install the brass air fittings to the replacement leveling valve.
6. Install bolts, washers, and locknuts to secure the valve to the truck frame.
7. Reinstall the air lines to the leveling valve.
8. Install the threaded extension rod (see **Figure 22–34**) to the leveling valve using a lock washer and nut.
9. Apply air pressure to the truck's braking system and then to the suspension system.
10. Adjust the leveling for valve proper ride height.

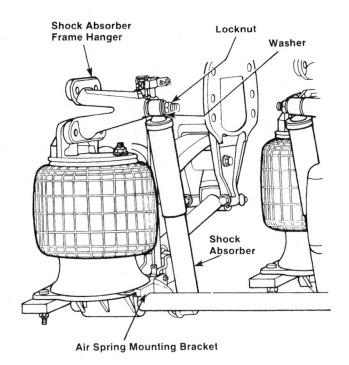

FIGURE 22–33 Shock absorber replacement. (Courtesy of Hendrickson International, Truck Suspension Systems)

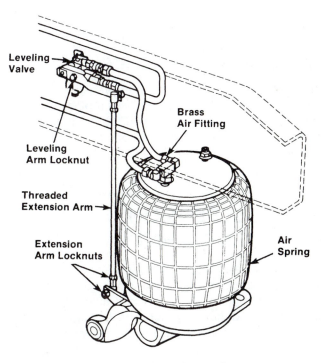

FIGURE 22–34 Leveling valve replacement. (Courtesy of Hendrickson International, Truck Suspension Systems)

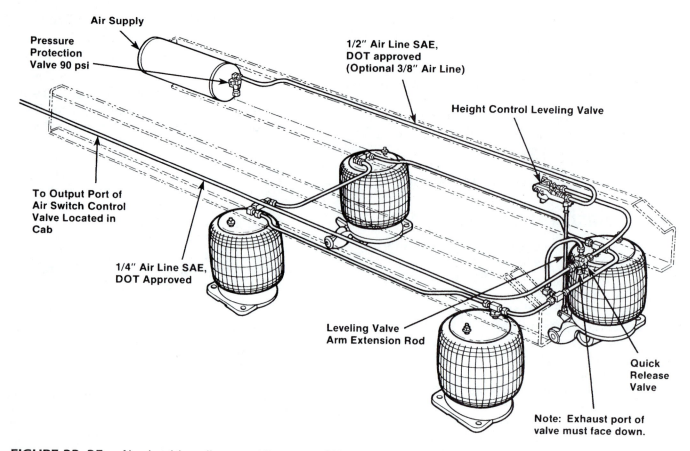

Air Supply

**Pressure
Protection
Valve 90 psi**

**1/2" Air Line SAE,
DOT approved
(Optional 3/8" Air Line)**

Height Control Leveling Valve

**To Output Port of
Air Switch Control
Valve Located in
Cab**

**1/4" Air Line SAE,
DOT Approved**

**Leveling Valve
Arm Extension Rod**

**Quick
Release
Valve**

**Note: Exhaust port of
valve must face down.**

FIGURE 22–35 *Air plumbing diagram. (Courtesy of Hendrickson International, Truck Suspension Systems)*

AIR LINE MAINTENANCE

Air lines provide the necessary air pressure to the air springs. Repairs consist basically of replacing defective lines and/or fittings. **Figure 22–35** illustrates a typical air plumbing installation for an air suspension system.

22.8 SUSPENSION ALIGNMENT

Alignment of a truck's suspension system is closely allied to axle alignment. Proper operation of the suspension system relies on correct axle alignments. This section describes methods of aligning truck axles to ensure correct suspension alignment.

AXLE ALIGNMENT

Several different types of equipment and methods can be used to check axle alignment. These meth-

ods include light beam alignment equipment and computer-controlled sensor systems. Basic procedures for checking axle alignment using these systems are included in Chapter 21.

Another accurate method of checking axle alignment is by using laser alignment equipment. A straightedge and trammel also can be used to check axle alignment. This last method is the most difficult and inaccurate method of checking axle alignment and should be used only when no other alignment equipment is available.

Before performing axle alignment checks or making torque rod adjustments, remove road debris from the suspension system and perform these preliminary procedures:

1. Park the vehicle on a level surface.
2. Release the parking brake.
3. Check the rear wheels for runout and correct if necessary.
4. Turn the wheels in a straight-ahead position.
5. Move the vehicle forward and rearward several times to relieve suspension stress.

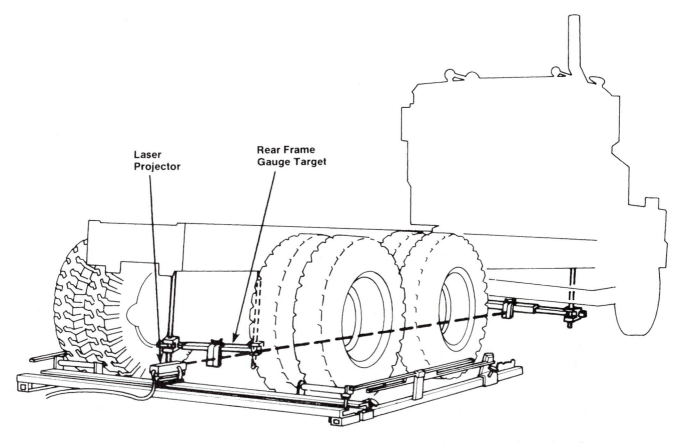

FIGURE 22–36 Typical laser aligner setup. (Courtesy of International Truck and Engine Corp.)

Laser Alignment

Alignment equipment using laser technology can be used to check rear axle alignment on a tractor or trailer. **Figure 22–36** shows a typical laser aligner connected to a tandem axle tractor. Always refer to the manufacturer's service manual for the recommended operating procedures.

Laser aligners use a "soft" laser light that is projected from the center of the rear frame gauge target to the front frame gauge target. The self-centering frame gauge targets are used to find the centerline of the chassis. The rear axle is in precise alignment when the beam of light is centered between the front and rear frame gauge targets. If the beam of light is not centered, the rear axle must be adjusted. When the rear axle is properly aligned, the position of the forward rear axle is found with a tram bar (**Figure 22–37**). The forward rear axle is then adjusted until it is parallel with the rear axle. The alignment procedures are the same for a trailer, except that an extender target is clamped to the kingpin. The extender target ensures that the rear axle is aligned with the center of the kingpin.

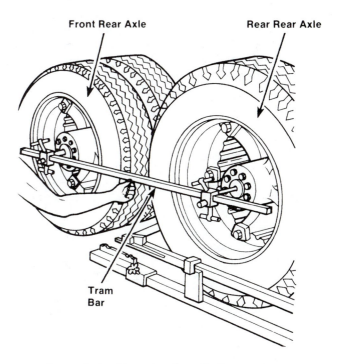

FIGURE 22–37 Tram bar installation and use. (Courtesy of International Truck and Engine Corp.)

Straightedge and Trammel Method

The following alignment check (measuring of the suspension) applies to all tandem suspensions covered in this chapter, all 4 × 2 vehicles having spring suspensions with torque rods, and air suspensions with tapered leaf springs. In the event axle adjustment is necessary, refer to the axle adjustment section in Chapter 21 and the appropriate procedure for the suspension being serviced.

1. Clamp a straightedge to the top of the frame rail ahead of the forward rear axle on 6 × 4 vehicles and ahead of the rear axle on 4 × 2 vehicles. Use a framing square against the straightedge and outside surface of the frame side member to ensure the straightedge is perpendicular to the frame (**Figure 22–38**).
2. Suspend a plumb bob from each end of the straightedge in front of the tire and on the outboard side of the forward rear axle on 6 × 4 vehicles or the rear axle on 4 × 2 vehicles. (**Figure 22–39**).
3. Position a slotted bar so that the pointers are engaged in the center hole of (both) rear axle(s).
4. Measure the distance between the cord of the plumb bob and the pointer on the forward rear axle of 6 × 4 vehicles or the rear axle on 4 × 2 vehicles. Record this dimension as dimension "A" (**Figure 22–40**).
5. Position the slotted bar with pointers on the opposite side of the vehicle in the same manner and measure corresponding distance as in step 4. Record this dimension as dimension "A" also.

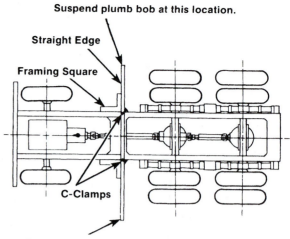

FIGURE 22–39 Installation and location of plumb bob. (Courtesy of International Truck and Engine Corp.)

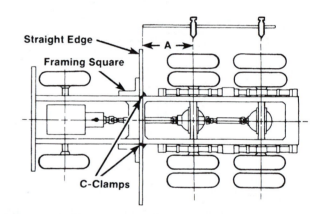

FIGURE 22–40 Measuring dimension "A" axle alignment. (Courtesy of International Truck and Engine Corp.)

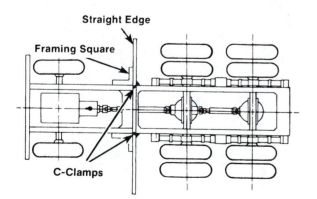

FIGURE 22–38 Straightedge location on the frame. (Courtesy of International Truck and Engine Corp.)

6. Any difference in dimensions from side to side in excess of 0.125 inch must be equalized.
7. Position the slotted bar on one side of the vehicle so that the pointers are engaged in the center hole of both rear axles.
8. Measure the distance between the plumb bob and the pointer on the rear axle and record this dimension as dimension "B" (**Figure 22–41**).
9. Position the slotted bar on the opposite side of the vehicle with the same pointer being used on the rear axle for both sides and measure the distance between the cord of the plumb bob and the pointer on the rear

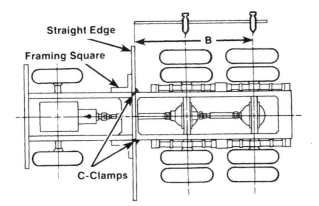

FIGURE 22–41 *Measuring dimension "B" axle alignment. (Courtesy of International Truck and Engine Corp.)*

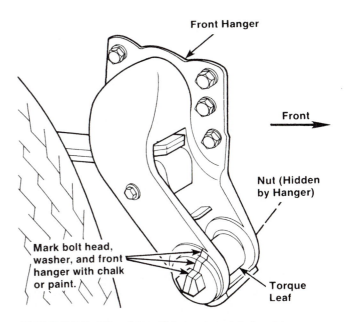

FIGURE 22–42 Identifying eccentric bushing movement for dimension "A." (Courtesy of International Truck and Engine Corp.)

axle. Record this dimension as dimension "B" also.

10. Any difference in dimensions from side to side must be equalized if in excess of 0.125 inch.

11. It is recommended that the vehicle be driven a short distance and the alignment rechecked.

Axle Adjustments. Adjustment of the axle location for alignment purposes on certain suspension systems is accomplished by means of eccentric bushings located at the torque leaf mounting eyes.

1. To equalize dimension "A" (**Figure 22–40**) obtained from the straightedge and trammel alignment procedure, loosen the torque leaf bolt nuts enough to free the rubber bushings in the castings. Mark the adjusting bolt, washer, and front hanger with chalk or paint (**Figure 22–42**). This provides a means of visually identifying the eccentric bushing movement.

2. To adjust the axle, place a wrench on the torque leaf bolt and turn the bolt in the opposite direction of the desired axle movement.

3. After adjustment is made, tighten the torque leaf bolt nuts to the manufacturer's specified torque and recheck alignment dimensions on the forward rear axle.

4. To equalize dimension "B" (**Figure 22–43**) obtained from the straightedge and trammel alignment procedure, loosen the torque leaf bolt nut at the equalizer bracket enough to free the rubber bushings in the castings. Mark the adjusting bolt, washer, and equalizer bracket with chalk or paint. This pro-

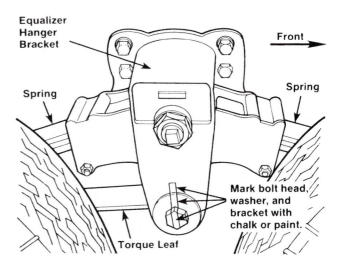

FIGURE 22–43 Identifying eccentric bushing movement for dimension "B." (Courtesy of International Truck and Engine Corp.)

vides a means of visually identifying the eccentric bushing movement.

5. To adjust the axle, place a wrench on the torque leaf bolt and turn the bolt in the opposite direction of the desired axle movement.

6. After the adjustment is made, tighten the torque leaf bolt nuts to the manufacturer's specified torque and recheck the alignment dimension on the rear axle.

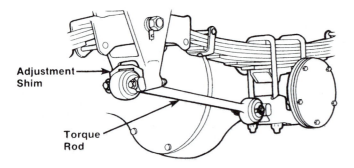

FIGURE 22–44 *Use and location of alignment shims. (Courtesy of International Truck and Engine Corp.)*

Axle Adjustment for Spring Suspensions with Torque Rods and Air Suspensions with Tapered Leaf Springs. Adjustment of the axle location for alignment purposes is accomplished by means of shims installed at the torque rod end mounting.

1. Equalize dimension "A" in the straightedge and trammel alignment procedure (**Figure 22–40**) by installing or removing shims at the forward rear axle torque rod end mounting (**Figure 22–44**). Installation of shims at the torque rod forward end mounting moves the axle forward. Installation of shims at the torque rod axle end mounting moves the axle rearward. Tighten the torque rod mounting nuts to the specified torque and check alignment dimensions on the forward rear axle.

2. Equalize dimension "B" (6 × 4 vehicles only) in the straightedge and trammel alignment procedure (**Figure 22–41**) by installing or removing shims at the rear axle torque rod end mounting (**Figure 22–44**). Installation of shims at the forward end mounting moves the axle forward. Installation of shims at the torque rod axle end mounting moves the axle rearward. Tighten torque rod mounting nuts to specified torque and check alignment dimensions on the rear axle.

3. Following the installation of the alignment shims, inspect the torque rod mounting nuts. A minimum of two threads of the mounting bolt must extend through the nut to allow the mounting nut locking feature to function properly.

4. When performing axle alignment, it is preferred to have alignment shims installed at the torque rod forward mounting. However, if axle alignment cannot be obtained by installing shims at the forward mounting, it is permissible to install additional shims at the axle end mounting.

22.9 | CAB AIR SUSPENSIONS

Cab air suspension systems have become the common method of mounting a cab. A typical system has two air springs, two shock absorbers, and a leveling valve to maintain ride height (**Figure 22–45**). In some cases a panhard (transverse) rod is fitted for lateral stability. Suspension travel is not great, usually within the range of 1 to 3 inches.

Because no two truck cabs are alike, there are many variations in size and mounting bracketry. A leading independent cab air suspension manufacturer has 40 models in its brochure, covering 10 different makes of truck. Some are single-point mounts, featuring a single air bag and shock absorber, and some are dual-point (**Figure 22–46**). Depending on their design and the weight of the mounts they replace, they add anywhere from 7 to 96 pounds to the vehicle.

RETROFITTING

When retrofitting a cab air suspension to a used truck, follow the manufacturer's installation instruc-

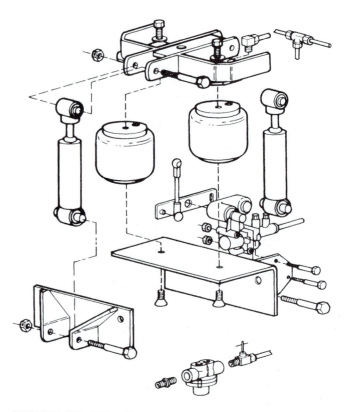

FIGURE 22–45 *Exploded view of typical two-point cab air suspension.*

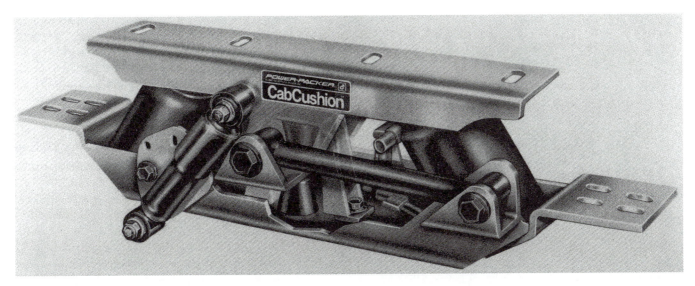

FIGURE 22–46 Cab air suspension system with a transverse rod, two air springs, one shock absorber and a leveling valve. (Courtesy of APA Engineered Solutions [Power-Packer])

tions. Here are some points to keep in mind when making the installation:

- Cabovers are a natural for air suspension, because they already have front pivots, and their height tends to produce more pitching motion than a conventional.
- Conventionals may need to have the hood adjusted forward away from the cowl of the cab, to provide clearance from cab motion. The sleeper boot could restrict motion somewhat but should have enough flexibility not to tear.
- If the exhaust mounts to the back of the cab, check whether flexible exhaust piping needs to be added underneath to prevent damage to the system. One solution is to switch to a frame-mounted tower bracket and eliminate cab contact.

MAINTENANCE

Keeping moving parts lubricated and the air system clean are about all the maintenance a cab air suspension system should require, except for periodic shock absorber replacement. Realistic mileage figures for shock life range from 90,000 to 350,000 miles.

Electronic cab suspensions that read vibrations and activate solenoids to cancel them are now being introduced by cab suspension manufacturers. It is also possible to use electronically controlled air valving on a more conventional system. The cab suspension ECU could even "talk" to the axle suspension over the truck data bus and coordinate cab ride improvement.

22.10 DRIVER AIR SUSPENDED SEATS

Driver comfort is to a large extent dependent on having a comfortable seat. There are three types: mechanically suspended, pneumatically (air) suspended, and solid mounted. Mechanical suspension seats use a mechanical free-moving support system to isolate the driver from jolts transferred from the chassis to the cab. Air suspension seats do the same thing with an air bag/cushion and shock absorber assembly, using compressed air from the vehicle's air system (**Figure 22–47**). Mechanical suspension types are the

FIGURE 22–47 Typical air-suspended driver's seat. (Courtesy of Bostrom Seating, Inc.)

FIGURE 22–48 *Solid mounting seat. (Courtesy of Bostrom Seating, Inc.)*

best choice if the truck does not have an air system, although at least one manufacturer offers a completely self-contained air type that requires no air supply. Solid-mounted seats are the cheapest, and, as the name implies, provide the least amount of isolation from the road (**Figure 22–48**).

22.11 RIDE ANALYSIS

The following suspension ride analysis should be undertaken to ensure the best ride possible. It is based on material given in this chapter.

Quick Vehicle Checks

1. Look for corroded or damaged threads on the cab mounting bolts, which may be contributing to fore-and-aft pitching.
2. Check the condition of the cab mounting insulators for evidence of collapsing or checking due to weather deterioration. Always replace insulators when necessary. Then replace corroded mounting bolts and bolts with damaged threads. Torque them to specification.
3. Where applicable, check the cab latch mechanism for sufficient tension. Any loose play in the latch mechanism can result or contribute to chassis resonance. Also check for elongated latches or a badly fretted latch bar.
4. Check the condition of the seat. If it is in poor condition, recommend an air suspen-

sion seat replacement with vertical as well as fore-and-aft isolation. If a new seat is required, it should be installed after ride analysis procedures are completed.

5. Check for proper tire inflation, including both tractor and trailer tires. Under- or overinflation could be contributing to a bad ride. When tires have been properly inflated, follow the recommended procedure for checking lateral runout.

Test Drive

1. Use a chase car with a driver and an observer to determine whether bogie hop is occurring. Ideally, the test drive should be made with the loaded trailer that is normally used. A second test drive should be done with a differently loaded trailer. If the fifth wheel is adjustable, put it in the optimum position with relation to the load during the road test and then continue the test drive. This will determine whether the fifth wheel position was causing or influencing the ride complaint.
2. If throttle sensitivity is a suspected cause, try the hand throttle during the test drive. Inspect for relative motion between cab, engine, and throttle linkage. Also check for loose rod pins and brackets and for corrosion at pivot points that would create a binding condition. Always make sure that the geometry of the entire linkage meets specifications.
3. The observer in the chase car should look for front axle shimmy, as well as bogie hop. Bogie hop can be determined even though tires do not leave the road surface. There may be a very noticeable tire deflection that indicates that bogie hop is present and contributing to chassis resonance and rough ride. If a front axle shimmy is detected, align and balance the wheels on the front axle. If bogie hop is noticed, check for tire balance and remember to inspect the brake drum. Because a rough ride may be the result of more than one circumstance, test drive the vehicle after each repair or alteration has been made to make certain all problems have been corrected.

Check List for Wheel Hop and Vibration Problems

1. Check first for any loose, worn, or damaged components.
2. Check wheel balance.

3. Check the shock absorbers in both the front and back.

4. Check wheel and tire runout—laterally and radially. This also includes hub and drum runout.

5. Be sure that the rear spring shackle or movable shackle has a rearward slant to it and is not vertical. If it is vertical, this can cause a hop problem due to the fact that the spring does not know whether it should move forward or backward.

6. If the springs are inadequate, this can cause the vehicle to bounce. This can cause axle roll, giving a steering problem. A tolerance of 1,000 pounds in a highway application and 1,500 pounds for off-highway should be allowed over maximum load. Axles are quite often spec'd out a little higher than expected load, and so should the suspension system.

7. Check for unmatched tires, especially on the duals. Also check for proper tire pressure.

8. Be sure that the front suspension is in alignment.

9. Check all crossmembers and spring shackles that support or hold the suspension system to the frame. Looseness in these can cause vibrations and hops.

10. Check the driveline angle, U-joint phasing, and any looseness in hanger bearings or U-joints. Check the drive shaft between housings. Also, check to see whether the short shaft is bottoming in the slip yoke. This can cause a hop when the housing goes over a bump. Also, check for worn or twisted splines on the shaft or slip yoke.

11. Check all torque arms on the upper part of the housing. Make sure that they are tight and not worn out. Also, check the angle of these torque arms. If they are at a fairly sharp angle, they can cause the housing to rotate forward and backward as the tandem oscillates, which, in turn, causes the U-joints to bind and results in a hop problem.

12. On the rear housings, check the mounting of the wheels and matching of the tire, brake drum, and hub balance.

13. Check all rear suspension bushings. They should be in good shape.

14. Be sure the housings are in alignment. Check camber and toe. Also, be sure that the housings are tracking properly.

15. On any walking beam tandem assemblies, check the center shaft that goes from one beam to the other to see whether it is bent. If it is, it can cause the rear suspension to bind. Another point to check on these hous-

ings is the mounting of the saddles to the frame, especially if the frame has been reinforced. The movement of the saddles out another $3/8$ or $1/2$ inch on each side can cause a binding in the center pivots, thereby causing oscillation of the rear suspension (**Figure 22–49**).

16. The frame can cause a problem by having inadequate crossmembers or strength (**Figure 22–50**). If a frame has too much flexibility, it

FIGURE 22–49 A great deal of pressure (load) is placed on the suspension system when the dump trailer is in a raised position. (Courtesy of American Trailer Industries Inc.)

FIGURE 22–50 C-channel cross-members and full-length gussets stabilize the web of the trailer frame. (Courtesy of American Trailer Industries Inc.)

can flex, causing a rhythmic vibration, which can resonate throughout the frame. This can cause a vibration that seems like it is coming from the front wheels. The reason it feels more severe in the front is because it is the most flexible part of the suspension, as the rear is being held down by a greater load. Other items to check are downward flexing from overload and rolling of the rails in and out due to the fuel tanks that are hung on the side of the rail (especially where there are not adequate crossmembers). Check these possibilities by installing a false crossmember and then road test. Also, clamp a reinforcement underneath the frame to check for a downward flex problem. If the frame has been extended and sleeved to accommodate a long dump body, (grain type or gravel type) the frame itself might have enough strength with the added channel to hold the load, but because of the added distance hop problems are caused. Another problem that can be encountered when stretching out the frame and moving rear axles to the end of the frame is overloading the front axle.

17. Fifth wheel position can be too far forward or too far back and can cause hops and vibrations.

18. Do not overlook the trailer, especially if the complaint is that the hop occurs only when the trailer is hooked up. A trailer with the rear axle badly out of line, mismatched tires, out-of-round tires, worn-out suspension systems, or broken springs can carry vibration on hop problems.

19. Frame misalignment also can cause a binding in the whole drivetrain and contribute to hop problems.

20. If cast spoke wheel front axles are heavily loaded, flex of the rim between the spokes can cause an out-of-round effect.

SUMMARY

- The main purpose of the suspension system is to support the weight of the vehicle.
- Constant rate, progressive or vari-rate, and auxiliary are three types of leaf spring suspensions.
- Shock absorbers are used in spring suspension systems to eliminate excessive tire wear, front wheel shimmy, and spring breakage.
- Leaf spring and rubber cushion are both equalizing beam types of suspension used on heavy-duty trucks.

- Many heavy-duty trucks are equipped with air spring suspension systems. The springs are either the reversible sleeve type or the double convoluted type.
- *Harmonic* and *harsh* are both terms used to describe rough ride conditions.
- Bolt-type, tube-type, and ball and socket-type are all types of equalizer beam mounting configurations.
- Proper operation of the suspension system relies heavily on correct axle alignment.
- The air suspension cab system is the most popular method of mounting a cab.
- Keeping moving parts lubricated and the air system clean are practically all the maintenance a cab air suspension system should need.
- Mechanically suspended, pneumatically (air) suspended, and solid mounted are three types of driver seat suspensions used to isolate the driver from chassis vibration.

REVIEW QUESTIONS

1. A truck is using a progressive spring suspension system. Technician A says that as the load increases, the inner rear hangers will come in contact with the springs and increase the spring rate. Technician B says that the inner rear hangers support the full load of the truck when the truck is empty. Who is correct?
   a. Technician A
   b. Technician B
   c. both A and B
   d. neither A nor B

2. The driver of a truck equipped with a rear air suspension system complains that the truck rides too high. Technician A says that this could be caused by overpressurized air springs. Technician B says that the leveling valve is out of adjustment. Who is correct?
   a. Technician A
   b. Technician B
   c. both A and B
   d. neither A nor B

3. A tractor with an air suspension system leans to the left. Technician A says that the left leveling valve could be the cause. Technician B says that the right leveling valve could be the cause. Who is correct?
   a. Technician A
   b. Technician B
   c. both A and B
   d. neither A nor B

4. Axle alignment on a tandem rear leaf spring suspension is adjusted by
   a. changing the length of the torque rods
   b. rotating the eccentric bolts on the forward drive axle
   c. both a and b
   d. neither a nor b

5. A truck recently had its front springs and hangers replaced. The driver of the truck complains that it is difficult to keep the truck going straight ahead. Technician A says that this could be caused by installing the caster shims backward. Technician B says that this could be caused by installing the wrong size hangers. Who is correct?
   a. Technician A
   b. Technician B
   c. both A and B
   d. neither A nor B

6. Loose U-bolts on a suspension system can cause
   a. hard steering
   b. leaf spring breakage
   c. abnormal tire wear
   d. all of the above

7. Torque arms are used on a suspension system to
   a. retain axle alignment
   b. control axle torque
   c. both a and b
   d. neither a nor b

8. A device that is used to reduce road shock and support loads on a vehicle is a
   a. equalizer bracket
   b. torque rod
   c. spring
   d. frame hanger

9. A leaf-type spring suspension that has a variable deflection rate obtained by varying the effective length of the spring assembly is known as a(n)
   a. progressive spring suspension
   b. constant spring suspension
   c. auxiliary spring suspension
   d. none of the above

10. What type of spring is usually mounted on top of the truck rear spring assemblies and is used only when a truck is under a heavy load?
    a. progressive spring
    b. constant spring
    c. auxiliary spring
    d. air spring

11. What device is used to absorb energy and dampen suspension oscillation?
    a. equalizer bracket
    b. torque rod
    c. springs
    d. shock absorber

12. The "brain" of the air spring suspension system is the
    a. leveling valve
    b. air spring
    c. hydraulic shock absorber
    d. quick release valve

13. A saddle and cap assembly is being installed on a truck's equalizing beam suspension system. Technician A says that if the assembly uses self-locking nuts they should never be reused. Technician B says that the weight of the truck must be on its wheels when torquing the saddle and cap assembly. Who is correct?
    a. Technician A
    b. Technician B
    c. both A and B
    d. neither A nor B

14. An equalizing beam suspension
    a. distributes the load equally between the axles
    b. reduces wheel hop
    c. uses lever principle
    d. all of the above

15. A truck is equipped with an air suspension system. After unloading the truck the driver notices that one side rises. Technician A says that this could be caused by a plugged quick release valve. Technician B says that a defective leveling valve could cause this problem. Who is correct?
    a. Technician A
    b. Technician B
    c. both A and B
    d. neither A nor B

Wheels and Tires

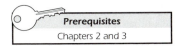

Prerequisites
Chapters 2 and 3

Objectives

After reading this chapter, you should be able to

- Identify the major wheel configurations used on heavy-duty trucks.
- Explain the difference between standard and wide-base wheel systems and stud- and hub-piloted mountings.
- Identify the common types of tire-to-rim hardware and describe their functions.
- Explain the importance of proper matching and assembly of tire and rim hardware.
- Outline the safety procedure for handling and servicing wheels and tires.
- Describe brake drum mounting configurations.
- Perform wheel runout checks and adjustments.
- Properly match tires in dual and tandem mountings.
- List the major components of both grease- and oil-lubricated wheel hubs.
- Perform bearing and seal service on grease-lubricated front and rear wheel hubs.
- Perform bearing and seal service on oil-lubricated front and rear wheel hubs.
- Perform front and rear bearing adjustment.
- Describe THC wheel-end procedure.

Key Terms List

aspect ratio
bias ply
footprint
radial
rims

section height
section width
The Maintenance Council (TMC)
wheels
wheel balance
wheel-end

Often considered part of the vehicle's suspension system, **wheels** and tires play a vital role in the safe operation of all heavy-duty trucks and trailers. They carry all the weight of the vehicle and operate on a variety of road conditions. After fuel costs, tires represent one of the biggest cost factors of operating a truck.

Wheels and tires and must be properly maintained and serviced. Improperly mounted, matched, aligned, or inflated tires can create a dangerous on-road situation. Poorly maintained tires will also wear unevenly and at a faster rate. Chances of a blowout or other major failure are also increased.

Wheel bearings, lubricant seals, and other components in the wheel hub that keep the wheel and tire turning smoothly on the axle spindle also require regular maintenance and service.

For many years, there was little difference in the wheel configurations used on heavy-duty trucks. Today, advances in technology and changing customer needs have increased the number of wheel configurations available. It is now possible to operate on cast spoke or disc wheel systems in steel or aluminum. **Rims** can be removable or part of the wheel body. They can be single or multipiece.

Wheels can be clamp mounted, stud piloted, or hub piloted. Dual-wheel or wide-base single wheels are available. Wheel bearings can be either grease or oil lubricated. Several different wheel seal designs can be used. The tires mounted to the wheels can be bias ply or radial, tube or tubeless, standard or low profile. Brake drums can be inboard or outboard mounted.

The wheel system chosen will affect payload, fuel efficiency, tire mileage, and hardware requirements. It will also have a major effect on service and maintenance procedures. Regardless of the system used, truck wheels and tires are extremely heavy. Inflation pressures can also exceed 100 psi. This combination of weight and pressure can create dangerous work situations if proper safety and working procedures are not followed.

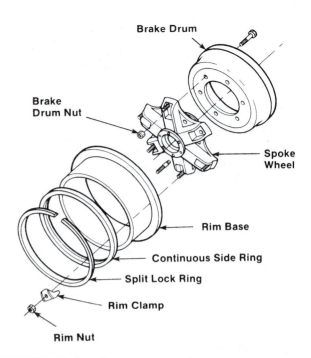

FIGURE 23–1 *Components of a cast spoke wheel. (Courtesy of Freightliner Trucks)*

WHEELS AND RIMS

There is often confusion over the terms *wheel* and *rim*. Many use the two terms interchangeably, but this is incorrect. The rim supports the tire. On a spoke wheel, the rim and wheel are separate components. On a disc wheel, the rim is a distinct section of a wheel assembly.

CAST SPOKE WHEELS

Although disc wheel configurations are increasing in popularity, cast spoke wheels are far from obsolete. Their record of high-strength performance make cast spoke designs a popular choice on dump, construction, refuse, and leased trucks and trailers. Spoke wheels are often chosen when heavy front axle loadings of 16,000 pounds and above are encountered.

A spoke wheel consists of a one-piece casting that includes the hub and spokes (**Figure 23–1**). Spokes are made of ductile iron, cast steel, or aluminum. Tires are mounted on a separate rim that is clamped onto the spokes.

In dual-wheel applications, a spacer band is positioned between the inner and outer rims (**Figure 23–2A**). This spacer band holds the two rims apart and provides for exact spacing of the tires (**Figure 23–2B**).

Spoke wheels are manufactured in three styles: three-, five-, and six-spoke configurations. Six-spoke

designs are often used on heavily loaded front axles. Five or six spokes are used on drive axle duals, but six-spoke designs are often preferred because added wheel clamping force is on the rim, which reduces the chance of rim slippage. Three-spoke wheels have wider spokes, using two wheel clamps per spoke. Trailers are their most popular application.

Spoke wheels use multipiece rims that clamp to the spokes with wheel clamps (See **Figure 23–2B**). If the clamps are not installed correctly, the wheel might go out-of-round and wobble. The proper installation and torquing sequence is critical to true running. Generally, spoke wheels experience greater alignment and balance problems than disc designs, but with proper installation and torquing, it is possible that spoke wheels can run virtually trouble free. They produce significantly fewer "wheel-off" incidents than disc wheel assemblies.

DISC WHEELS

In disc wheels the rim is permanently attached to a center disc (**Figure 23–3**). Holes in the center disc allow mounting to the hub studs with nuts.

These one-piece forged steel or aluminum wheels run extremely true. The result is reduced vibration with longer tire life, less overall maintenance, easier handling, and improved driver comfort. Aluminum disc wheels are lighter than steel, so payloads can be increased. They also dissipate heat faster so tires run cooler.

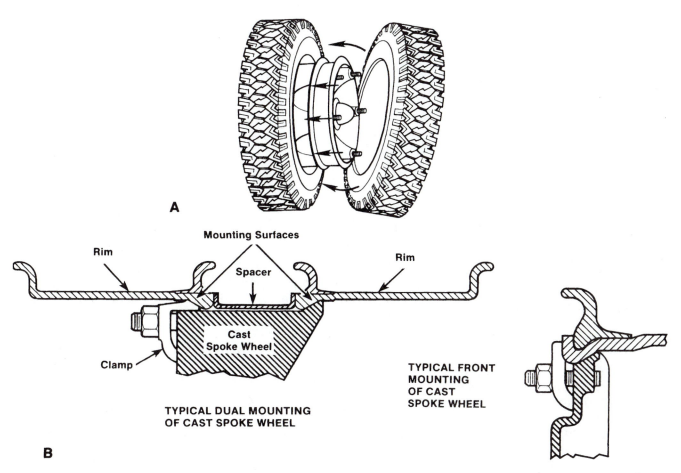

FIGURE 23–2 (A) Position of the spoke wheel dual mounting spacer band, and (B) cross-section view of mounted dual wheels. (Courtesy of International Truck and Engine Corp.)

FIGURE 23–3 Components of a typical disc wheel. (Courtesy of The Budd Company)

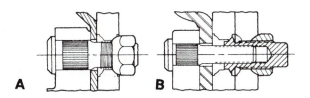

FIGURE 23–4 Aluminum disc wheel stud-piloted mounting configurations: (A) single wheel, and (B) dual wheel. (Courtesy of Alcoa Wheel Products International)

Like spoke wheels, disc wheels can be used in single and dual configurations. There are also two different mounting systems: stud piloted and hub piloted.

Stud-Piloted Wheel

Figure 23–4A illustrates a stud-piloted mounting for a single disc wheel. The wheel simply mounts onto studs on the hub and is secured using single cap nuts. **Figure 23–4B** shows a dual-disc wheel configuration.

Inner cap nuts screw onto the hub studs. The flange on the nut caps rests in the seat created between the inner and outer wheel. This helps center the two wheels and clamp them together. Finally, outer cap nuts screw onto the threaded ends of the inner cap nuts, thereby securing the entire assembly to the hub.

Hub-Piloted Wheel

The hub-piloted system simplifies centering and clamping wheels to hubs. In a hub-piloted system,

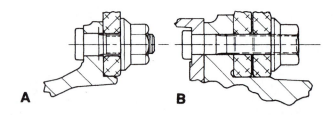

FIGURE 23–5 *Aluminum disc wheel hub-piloted mounting configurations: (A) single wheel and (B) dual wheel. (Courtesy of Alcoa Wheel Products International)*

the hub centers the wheel. The nuts and studs provide only clamping force (**Figure 23–5**).

A hub-piloted wheel uses one cone locknut per stud, eliminating the need for inner cap nuts. This significantly reduces the amount of wheel fastening hardware compared to stud-piloted wheels.

Over- or undertorquing of stud-piloted wheels can cause broken studs and cracked or loose wheels. The single flange nuts of hub-piloted wheels are less susceptible to this problem. In stud-piloted systems, a loose inner nut can easily go undetected, eventually pounding out the nut's ball seat. With hub-piloted systems, both the inner nut and its ball seat are eliminated. With the cone locknut design, a flat washer is seated directly against the wheel face. The nonrotating washer prevents galling of the wheel surface.

WIDE-BASE WHEELS

Wide-base wheels can also be referred to as *high flotation*, *super single*, *wide body*, *duplex*, or *jumbo wheels*. One wide-base wheel and tire replaces traditional dual wheels and tires (**Figure 23–6**). Instead of 18 tires, a tractor/trailer needs only 10. Two traditional wheels are used at the steering axle, and eight wide base wheels are used at the other axle positions.

Interest in wide-base wheels today is operating costs. Wide-base performance has steadily improved and the system is now widely available in the U.S. Wide-base tires have been specifically designed to replace duals. Depending on the dual tires replaced and the load capacity required, retrofitting with wide-base tires need not greatly affect vehicle gear ratios or top speed capability.

Compared to steel dual wheels and tires, aluminum wide-base wheels and tires are significantly lighter in weight. This can increase payload capacity. Wide-base wheels also increase fuel efficiency by decreasing weight and rolling resistance. One factor in rolling resistance is sidewall flexing. A wide-base tire has two flexing sidewalls, compared to four side-

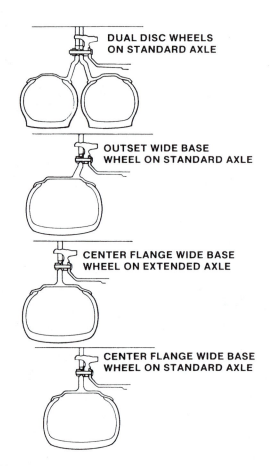

FIGURE 23–6 *Wide-base wheel mounting configurations. (Courtesy of International Truck and Engine Corp.)*

walls in a dual configuration. In addition, duals can consume more energy when they are mismatched by even slight diameter differences or by air pressure differences. These are not factors when a single wide-base wheel is used.

23.2 TIRE-TO-RIM HARDWARE

The tire can be held on the rim in a number of ways. The simplest is the single piece rim. A fixed flange built into the edge of the rim supports both sides of the tire (**Figure 23–7**). Single piece rims are used in combination with tubeless tires.

Tube-type tires are held on the rim using various side ring and/or lock ring combinations. Side ring and lock ring designs vary from manufacturer to manufacturer, so it is important to always use properly matched components.

Two types of side rings are used:

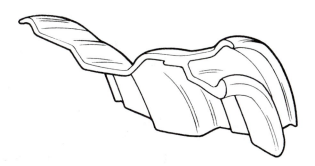

FIGURE 23–7   *Single piece demountable rim used with tubeless tires. (Courtesy of Accuride Corporation)*

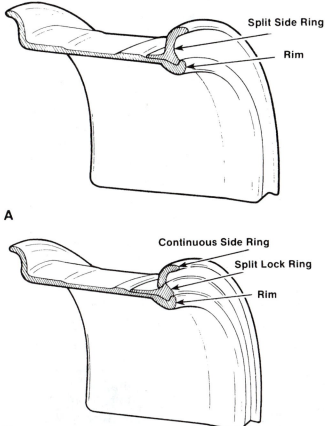

A

B

FIGURE 23–8   *Side ring configurations: (A) split side ring; and (B) continuous side ring with separate split lock ring to secure it to the rim. (Courtesy of Accuride Corporation)*

- **Split Side Rings.** In two-piece assemblies, the side ring retains the rim on one side of the rim. The fixed flange supports the other side (**Figure 23–8A**). The split side ring is designed so that it acts as a self-contained lock ring as well as a flange.
- **Flange or Continuous Side Rings.** In three-piece assemblies, the flange, or continuous

side ring, supports the tire on one side of the rim. The continuous side ring is, in turn, held in place by a separate split lock ring (**Figure 23–8B**).

23.3 TIRES

There are two basic types of tire construction used in heavy-duty applications: **bias ply** and **radial**. It is critical that radial and bias ply tires not be installed on the same axle. Radial and bias ply tires differ in their tread profile, surface contact, and handling characteristics. The best tire performance can be achieved by using one type of tire construction on all axles.

Dual configurations must never use mismatched tires. All tires on an axle must be of the same construction; mismatched tires on opposite sides of the same axle can cause drive axle failure. If the left duals are radial, the right duals must also be radials. If the vehicle has two or more drive axles, the tires on the drive axles must be either all bias ply or all radial. It is best if all eight drive axle tires on a tandem unit are matched.

One of the reasons for matching radials is that they deflect more than bias ply tires under load. Mixing tires can cause overloading of the bias ply tires on the vehicle. Typical contact profiles or **footprints** made by a nonradial tire and radial truck tire are shown in **Figure 23–9**.

The basic differences between radial and bias tire types are as follows (**Figure 23–10**):

- **Radial Tires.** The body ply cords are placed perpendicularly across the tread from bead to bead. In addition, radial tires have belt plies

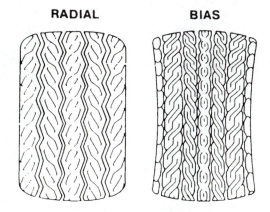

RADIAL   **BIAS**

FIGURE 23–9   Comparison of footprint between a radial and bias ply-type truck tire (Courtesy of Bridgestone/Firestone, Inc.)

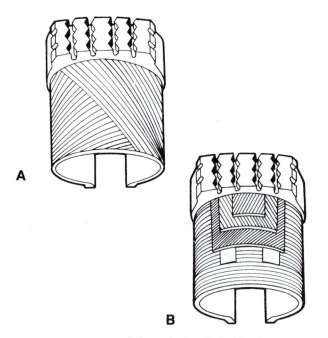

FIGURE 23–10   Construction of (A) bias ply, and (B) radial tires. (Courtesy of Volvo Trucks North America, Inc.)

that run circumferentially around the tire, under the tread. They constrict the radial ply cords and give rigidity to the tread.

- **Bias Ply Tires.** The body ply cords lie in a diagonal direction from bead to bead. The tires may also have narrow plies under the tread, called breakers, with cords strung in approximately the same direction as the body ply cords.
- **Body Ply, Breaker, and Belt Materials.** Tire body plies, breakers, and belts may be of poly-

ester, rayon, nylon, fiberglass, steel, or aramids. In radial ply tires, these materials can be used in various combinations such as steel body steel belt, polyester body fiberglass belt, or nylon body steel belt.

The basic types of highway truck tread designs are rib, lug, and special service mud and snow lug (**Figure 23–11A–C**).

- **Rib Type Tread.** Tires with rib type tread are all position tires. They can be used on all wheel positions at legal highway speeds. These tires are always recommended for front wheel use on tractors and large, straight trucks in long haul highway service. The open groove design provides maximum steering control and good skid resistance.
- **Lug Type Treads.** Cross lug or cross rib and rib lug type tires are designed for drive wheel service and are suitable for most highway operations. These tread designs provide maximum resistance to wear and better traction. The tires are suitable for some off-road operations but do not provide maximum off-road traction.
- **Special Service Mud and Snow Lug Treads.** Special service mud and snow lug tires are designed for traction on drive wheels for on- and off-road service. They should be selected only when maximum traction in mud or snow is required.

The type of tire tread design used is determined by vehicle application and in some cases by personal preference. The rib type tread is most common

FIGURE 23–11   Three typical tire tread designs. (Courtesy of Bridgestone/Firestone, Inc.)

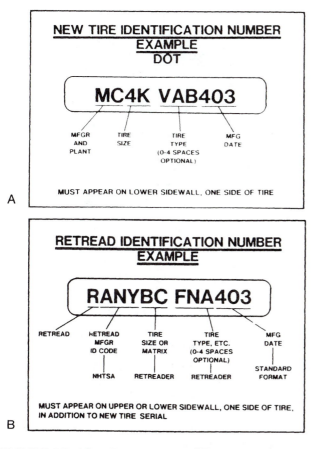

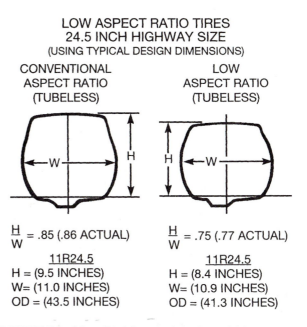

FIGURE 23-13 *Dividing height by width equals the profile.*

FIGURE 23-12 *Department of Transportation (DOT) tire coding identification for (A) a new tire, and (B) a retread.*

for on-highway operations and is used almost exclusively on steering (front) axles.

All tires (new and retread) sold in the United States must have a DOT number cured into the lower sidewall on one side of the tire. **Figure 23-12** gives the standard format designated by the federal government for both new and retread truck tires.

TIRE SIZE

The 22.5-inch wheel/tire size has been gaining popularity over 24.5-inch sizes and is predicted to be the more widely used wheel/tire size in coming years.

Size 19.5-inch tires are gaining a growing share of the high-cube truck/trailer market because they are suited for applications in which low trailer floor and fifth wheel heights are needed to get 110- to 120-inch door openings. Size 17.5-inch tires are used on some high-cube applications, but these smaller wheel/tire combinations can accommodate only a $12^1/_{16}$-inch brake drum, and the 19.5-inch wheel can take a 15-inch brake drum.

LOW-PROFILE TIRES

Low-profile radial truck tires enhance the radial design to produce even lower costs per mile. The low-profile name comes from the tire's **aspect ratio** (**Figure 23-13**), which, for any tire, is calculated by dividing the tire's **section height** (tread center to bead plane) by its **section width** (sidewall to sidewall). Simply put, low-profile tires are "squatter" than conventional radials. Advantages offered by low-profile radials include lower weight (up to 10 percent less than standard radials), lower rolling resistance (again about 10 percent less), greater vehicle stability due to a lower center of gravity, a better footprint as a result of improved pressure distribution, high retreadability, improved fuel economy, better traction, and increased tread life.

23.4 | TIRE CARE AND MAINTENANCE

Proper tire care and maintenance is second only to increasing fuel mileage in reducing overall cost per mile of truck operation. Although this subject appears to be a simple one, it has been established that most tires wear fast or fail early due to neglect. Tire care and maintenance is relatively simple, takes little time, and pays off.

Even with the best maintenance of truck tires and their related equipment, the service that the tire delivers is largely in the hands of the driver. Careless driving habits can result in tire damage and shorten the life of a tire.

Tire Inspection

Regular inspection of tires is the first step in increasing mileage. An inspection will help to spot troubles such as underinflation, overinflation, and misalignment early. Minor damage that can normally be detected and repaired during an inspection could save a tire that would otherwise blow out.

Proper tire inflation will increase tire mileage. Underinflation causes abnormal wear at the sides of the tread because the outer edges of the tire carry the load, while the center tends to flex up, away from the road (**Figure 23–14**). This also causes the tire to run hotter. Any tire that is determined to be underinflated should be inflated to the specified pressure.

Driving on an underinflated or a nearly flat tire, even for a short distance, can cause severe cord damage. This, in turn, could cause the plies of fabric in the casing to separate, the tread to separate from the fabric, or both. Underinflation reduces tire life and is also the number one cause of blowouts. Overinflation causes abnormal wear in the center of the tread, because it has to carry more than its share of the load. Overinflation also shortens the life of the tire (**Figure 23–15**).

Check for correct pressure when tires are cool. When a tire is in use and becomes heated, the air in the tire expands, and the air pressure is raised. Normal operating pressure buildup is 20 psi or less. Never bleed tires to relieve buildup of pressure. If excessive pressure buildup occurs, one or more factors such as load distribution, underinflation, or speed could be responsible.

Overinflation reduces the capability of the tire to absorb ordinary shock and causes fabric or tread

FIGURE 23–15 Overinflated tire. (Courtesy of Mack Trucks, Inc.)

separation, or both, resulting in tire failure. It will not compensate for overloading. An overinflated tire is more vulnerable to snags, cuts, and punctures.

Mechanical Irregularities

Tires in various positions on the truck will wear differently. Due to alignment factors, the front tires are more likely to experience abnormal wear than the rear ones. When toe-in is excessive, tire wear appears as feathered edges on the inside edge of the tread design (**Figure 23–16**). A toe-out condition will show feathered edges on the outer edge of the tire tread design (**Figure 23–17**).

Camber is the inward or outward tilt at the top of the rim wheel. Too much positive camber, (**Figure 23–18A**) in which the wheel tilts outward will cause the outside edge of the tire to wear prematurely. Too much negative camber, (**Figure 23–18B**) in which the wheel tilts inward will cause the inside edge of the tire to wear first.

FIGURE 23–14 Underinflated tire. (Courtesy of Mack Trucks, Inc.)

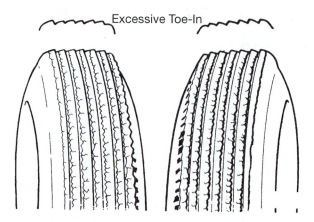

FIGURE 23–16 Excessive toe-in. (Courtesy of Mack Trucks, Inc.)

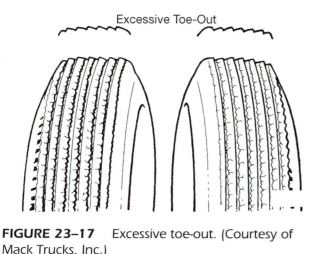

FIGURE 23–17 Excessive toe-out. (Courtesy of Mack Trucks, Inc.)

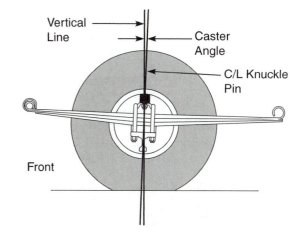

FIGURE 23–19 Caster angle. (Courtesy of Mack Trucks, Inc.)

As shown in **Figure 23–19**, positive caster is the backward tilt of the axle, or inclination of the kingpin. Too little positive caster will cause the wheel to wander or weave, resulting in spotty wear. Excessive caster may cause shimmy wear. Unequal caster from side to side causes the wheel to pull to one side, resulting in uneven wear.

Other mechanical problems that can cause excessive wear include a sprung or sagging axle, brakes that are out of adjustment, or out-of-round brake drums. A sprung or sagging axle will cause the inside dual tire to carry a greater load. Improper brake adjustments will lead to spotty tire wear, and out-of-round drums will usually wear tires in a single spot. Rapid or uneven tire wear may also be caused by a sprung or twisted frame, worn wheel bearings, loose spring clips, or loose torque rods.

Proper Load Distribution

Improper loading can cause the tires on one side of the truck or trailer to carry a greater load than those on the other side (**Figure 23–20**). This may affect starting from standstill causing the driving wheels to slip on the light side or the tires to wear faster. Although the gross load may not be excessive, one wheel, one axle, or one side of the truck may be overloaded due to improper distribution of the load. Equalize the load on each axle. On semitrailers, distribute the load so that each axle and the fifth wheel are carrying their shares of the load.

Other Causes of Tire Wear

All tires are speed rated. Exceeding rated speed creates heat. Excessive heat produced by running a vehicle at higher speeds will shorten tire life. At

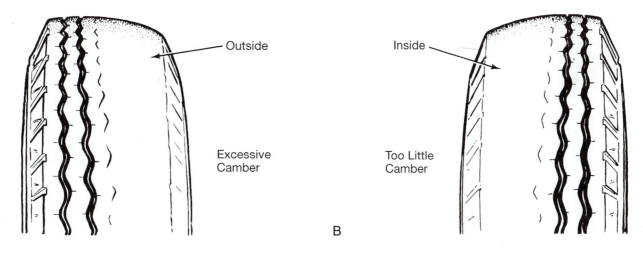

FIGURE 23–18 Tire wear from (A) excessive positive camber; and (B) excessive negative camber. (Courtesy of Mack Trucks, Inc.)

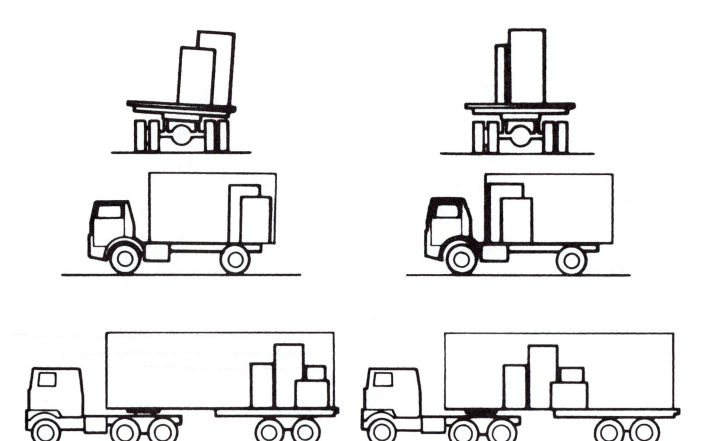

FIGURE 23–20 Load distribution. (Courtesy of Mack Trucks, Inc.)

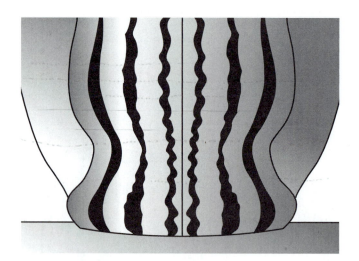

FIGURE 23–21 High-speed distortion.

higher speeds, the tire can become distorted, as shown in **Figure 23–21**. Higher speeds can cause a small cut to enlarge, causing a blowout. High-speed distortion exerts a strain on the tire fabric which may cause tire failure.

Shop Talk

*The manufacturers of speed restricted tires (**Figure 23–22**) on a vehicle caution the driver not to exceed the speed limits and guidelines in their manuals. To alert the operator, a decal (**Figure 23–23**) located in the cab within the operator's field of vision advises that the vehicle is equipped with speed restricted tires.*

Tire Rotation and Replacement

Tire rotation procedures are a matter of shop and PM practices. A regularly scheduled rotation plan can help reduce overall tire costs. The relocation of tires from the front to rear wheel positions depends on the type of truck being operated and the size and type of tires (**Figure 23–24**).

Federal Motor Carrier Safety Regulations require a vehicle to have at least $4/32$ inch of tread depth on the front tires. These tires can be rotated to driving or trailer axles and used until $2/32$ inch of tread remains. Any tire with less than $2/32$ inch of tread (as measured in a major groove and not over wear bars)

FIGURE 23–22 Speed restricted tire. (Courtesy of Mack Trucks, Inc.)

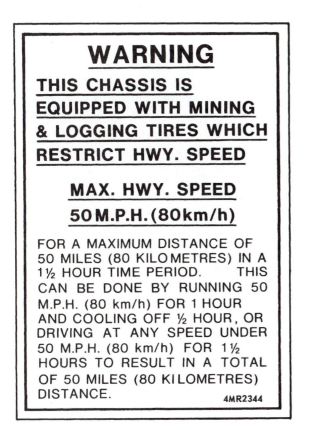

WARNING

THIS CHASSIS IS EQUIPPED WITH MINING & LOGGING TIRES WHICH RESTRICT HWY. SPEED

MAX. HWY. SPEED
50 M.P.H.(80km/h)

FOR A MAXIMUM DISTANCE OF 50 MILES (80 KILOMETRES) IN A 1½ HOUR TIME PERIOD. THIS CAN BE DONE BY RUNNING 50 M.P.H. (80 km/h) FOR 1 HOUR AND COOLING OFF ½ HOUR, OR DRIVING AT ANY SPEED UNDER 50 M.P.H. (80 km/h) FOR 1½ HOURS TO RESULT IN A TOTAL OF 50 MILES (80 KILOMETRES) DISTANCE.

4MR2344

FIGURE 23–23 Speed restricted tire decal. (Courtesy of Mack Trucks, Inc.)

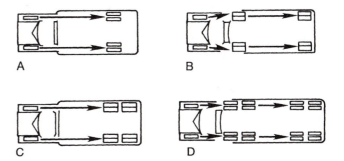

A   B

C   D

FIGURE 23–24 Suggested tire rotation patterns for (A) two-axle truck, (B) tractor and semitrailer, (C) three-axle truck, and (D) tractor with tandem axle drive and semitrailer with tandem axle.

must be removed from service. Tires with the word *regroovable* molded on the sidewall may be regrooved. These tires, along with recapped and retreaded tires, should not be used on front wheels. However, these tires can be used in other wheel positions until $2/32$ inch of tread remains. At this point, they should be replaced.

Any tire with an unrepaired fabric break, or one that has been repaired with a blowout patch or boot, should be replaced, as should any tire with exposed or damaged body cords, or a bump, bulge, or knot caused by cord separation.

See **Table 23–1** for the load range designations to replace ply ratings that tire manufacturers have adopted.

CAUTION: The load and cold tire inflation pressure imposed upon a particular wheel or rim must not exceed the wheel or rim manufacturer's recommendation, even though the tire may be approved for a higher load or inflation.

TABLE 23–1: LOAD RANGE EQUIVALENTS FOR PLY RATINGS

Load Range	Ply Rating
F	12
G	14
H	16
J	18
L	20

23.5 TIRE, RIM, AND WHEEL SERVICE

On trucks equipped with disc wheels, tire service involves removing the disc wheel and tire assembly from the wheel hub. On vehicles with cast spoke wheels, the rim and tire assembly is removed from the spoke wheel by dismounting the rim clamps. The spoke wheel/hub is not disturbed.

The variety of wheel systems and wheel/rim combinations makes a detailed illustration of tire-to-rim mounting and dismounting beyond the scope of a textbook. Detailed tire-to-wheel/rim changing procedures are available from all major tire and wheel manufacturers in service manuals, wall charts, and video programs (**Figure 23–25**).

The Occupational Safety and Health Administration (OSHA) has established rules and regulations pertaining to servicing multipiece rim wheels. The regulations state that it is the employer's responsibility to provide employees with training on service procedures and safety precautions for the tires, wheels, and rim combinations on which they will work.

A current rim manual containing instructions for types of rims being serviced must be available in the service area, and current safety and parts machining charts must be posted in the work area. The employer must also ensure that only tools recommended in the rim manual are used to service single and multipiece rim wheels. Never attempt to service any tire without hands-on training from a qualified tire service person. Be sure to follow the procedures demonstrated and printed in the OSHA regulations and wall charts.

OSHA has reprinted the NHTSA charts as part of a continuing campaign to alert rim wheel servicing personnel of the industry accepted procedures for servicing multipiece rim wheels. Reprints of these charts are available through OSHA's area offices. To find the address and telephone number of the nearest OSHA office, look in the telephone directory under U.S. Government, U.S. Department of Labor, or Occupational Safety and Health Administration. Single copies of the charts are available without charge.

Shop Talk

Federal and state regulations have made operators and garages liable for ensuring that tires are disposed of legally.

The safety and inspection information in this chapter summarizes the OSHA regulations. For complete details, refer to the complete OSHA standard 1910.177.

TIRE AND RIM SAFETY

WARNING: Air pressure in an inflated truck tire mounted on a rim/wheel creates explosive energy. This pressure can cause the tire/rim components to burst apart with great force, which can cause personal injury or death.

All personnel working with tires should be familiar with and follow some basic tire and rim safety rules prior to demounting and mounting tires. These rules are outlined as DOs and DON'Ts in **Table 23–2**.

WARNING: Stay out of the trajectory (danger) zone as indicated by the shaded area in **Figure 23–26**. Under some circumstances the trajectory may deviate from its standard path. Never attempt to seat rings while a tire is totally or partially inflated.

Removal from a Vehicle

Before removing any tire and wheel assembly or tire and rim assembly from the vehicle, set the spring brakes and chock the wheel assemblies not being removed. Jack up the vehicle using a heavy-duty jack.

FIGURE 23–25 Technicians must always refer to current rim mounting instruction charts supplied by manufacturers and federal safety agencies.

CAUTION: Never jack up a vehicle with a jack placed under a leaf spring.

TABLE 23–2:   TIRE AND RIM SAFETY

DO	DON'T
Instruct all tire and rim-handling personnel how to mount tires safely.	Overinflate tires.
Remove valve core to completely deflate tire before disassembling tire from rim.	Overload rims.
Use proper tools both to mount and demount tires.	Remove tire from rim before completely deflating.
Use approved rust-retarding compounds to keep rims clean and free from rust and corrosion.	Attempt to correct seating of side or lock rings by hammering while tire is inflated. *Always* remove air pressure first.
Use correct size rim for specified tire.	Use corroded, damaged, or distorted rims, rings, or rim parts.
Avoid rim damage when changing tires.	Fail to identify different types of wheel clamps;
Examine inside of tire before mounting and dry thoroughly if any moisture is found.	Use petroleum oil or grease on tire beads or rims because they can ruin the tires.
Use proper tubes and flaps with radial tires.	Use mismatched side or lock rings.
Install side or lock ring split directly opposite (180 degrees) from valve stem slot.	Inflate tire unless it is placed in a safety cage or a portable lock ring guard.
Use correct tire lubricant, but sparingly, to minimize the possibility of fluid entering the tire, especially in radial tire applications.	Inflate tire before all side and lock rings are in place.

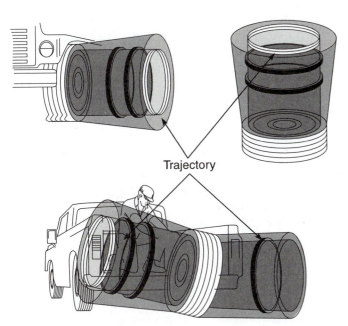

Trajectory

FIGURE 23–26   *Stay out of the trajectory (danger zone) as indicated by the shaded area. (Photo/Image(s) courtesy of Mack Trucks, Inc.)*

Whenever a vehicle is jacked up, use proper heavy-duty axle stands under the chassis to prevent personal injury or vehicle damage in the event of jack malfunction or vehicle tipping. Block front wheels when removing rear wheel and tire assemblies. Block rear wheels when removing front wheel and tire assemblies.

Raise the axle where the wheel and tire assembly is being removed and support it on adequate capacity jack stands. Cage the parking brake chamber

manually when working on rear brakes. Apply the parking brake when working on front brakes.

When removing the wheel nuts from spoke wheels, loosen all nuts 1/4 to 1/2 inch and then rap on the clamps with a hammer to free them from the wheel. It is important to free the clamps from the wheel before removing the wheel nuts. Failure to free the clamps may cause them to spring from the wheel under extreme pressure, which could cause serious personal injury.

👓 Shop Talk

Disc wheel nuts for right side wheels generally have right-hand threads, and wheel nuts for left side wheels usually have left-hand threads.

It is very important, when demounting aluminum wheels, to make sure that the end of the wheel wrench or socket is smooth. Burrs on the end of the wrench or socket will tear grooves in the disc around cap nuts which, in turn, may eventually cause the disc to crack.

CAUTION: When handling a tire assembly, remember to lift properly, using your legs as well as your body. Use caution when servicing the assemblies, wheels, or other components to prevent injury to your hands, feet, or body. Do not allow wheels, rings, rims, or other components to fall or drop to the floor.

Deflation and Disassembly

Always examine the tire/rim assembly for proper component seating prior to removing it from the vehicle. Also be sure to deflate the tire completely before removing it from the vehicle for any type of service. Deflate both tires of a dual assembly. When deflating a tire, first reduce tire pressure by pushing the tire valve core plunger; then remove the entire valve core (**Figure 23–27**). Be sure to keep your eyes away from the valve.

CAUTION: Always use protective glasses or face shields when working on wheels and tires.

Never position your body in front of the rim during deflation. Stand as far as possible away from the valve stem. Avoid positioning yourself so that your face or body is immediately over the work being done on any tire which has pressure.

Do not use a steel hammer to disassemble or assemble rim components. Hammering rings or rims with steel hammers may cause small bits of steel to break off from the hammer, the ring, or the rim; these bits could cause bodily harm. Use a lead-, brass-, or rubber-covered steel-headed hammer, or a plastic mallet. The proper procedure for dismounting tube and tubeless tires is illustrated in **Figure 23–28** and **23–29**.

Do not add air to a tire and rim assembly that has been run in a seriously underinflated or flat condition. The rim and/or tire can explode. Instead, deflate the tire completely, remove it from the vehicle, disassemble all tire and rim components, and inspect them for wear and damage.

FIGURE 23–27   Removing tire valve core.

Inspect Parts

Rims and rings must be matched by size and type. These components cannot be interchanged except as provided for in the Multipiece Rim/Wheel Matching Chart. Select the proper tire size and construction to match the manufacturer's rim or wheel rating and size. The diameter of the tire must match the diameter of the rim. Never use any rim or wheel component you cannot positively identify.

Thoroughly inspect all metal surfaces while the tires are being checked, including areas between duals and the inboard side of wheels. Examine for the following:

- Excessive rust or corrosion buildup
- Cracks in metal
- Bent flanges
- Deep tool marks on rings or in gutter areas
- Damaged or missing rim drive plates
- Matched rim parts

Mark defective parts for destruction to ensure that they will no longer be used. Keep in mind that a leak in a tubeless tire assembly might be caused by a cracked rim. Do not put a tube in a tubeless assembly to correct this problem. Cracked rims must be destroyed to avoid accidental use. Do not attempt to weld or otherwise repair cracked, bent, or out-of-shape components.

Mount Tire to Rim

Remove all rust, corrosion, dirt, and other foreign material from all metal surfaces (**Figure 23–30**). This is especially important in the rim gutter and bead seat areas. Check the mating surfaces of side/lock rings in multipiece assemblies.

Paint the rim with fast drying primer to help prevent rust from forming (**Figure 23–31**). Always allow paint to dry before assembling components.

Apply lubricant to bead seat area, tire bead, tire flap, and rim mating surfaces just prior to mounting the tire. Use only those lubricants recommended by the rim and tire manufacturer and never use petroleum, silicon, or water-based lubricants. These can damage the rubber, cause rust buildup, or produce tire-to-rim slipping.

Inflation

Always inflate tires in a safety cage (**Figure 23–32**) or in a portable restraining device (**Figure 23–33**). Check for proper seating of components before removal from the cage or restraint. Failure to follow this procedure when inflating a truck tire could result in serious injury.

FIGURE 23–28 Removing a tire from a flat base rim: (A) positioning tire removing tool, (B) disengaging lock ring from the gutter, (C) removing lock ring, and (D) lifting rim from the tire. (Photo/Image(s) courtesy of Mack Trucks, Inc.)

Use a clip-on air chuck with a remote in-line valve and gauge (**Figure 23–34**). This enables you to stand clear of the tire as it inflates. When a tire is being partially inflated (without restraining) to roundout the tube or seat the side/lock ring, pressure must not exceed 3 psi. Never attempt to correct the seating by hammering, striking, or forcing parts into place while the tire is inflated. Deflate using the remote in-line valve, remove the valve core to make sure that the tire is fully deflated and then remount the tire. When handling and mounting inflated tires, stay out of the trajectory (danger zone) of flying parts whenever possible.

SPOKE WHEEL INSTALLATION

When mounting the rim/wheel on the truck, use the proper size and type of nut, stud, and clamp (spoke wheels only). Spoke wheels use rim studs. Rim studs are threaded on both ends with a nonthreaded section in the middle of the stud. The studs are coated with an anaerobic locking compound.

Rim clamp nuts should be kept tight and checked on a regular basis. Checking alignment of the rim/wheel installation is important because the rims can be drawn out of alignment when improperly

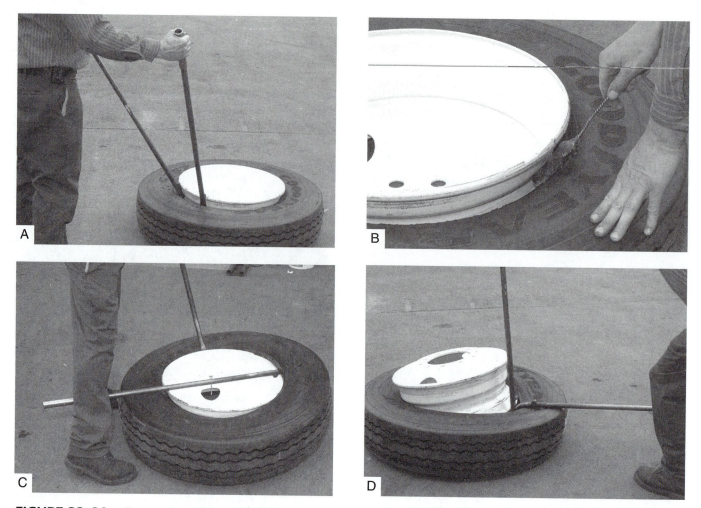

FIGURE 23–29   *Removing a tire from a disc wheel: (A) separating tire bead from wheel; (B) lubricating the tire bead; (C) prying the bead over the wheel; and (D) removing the second bead.*

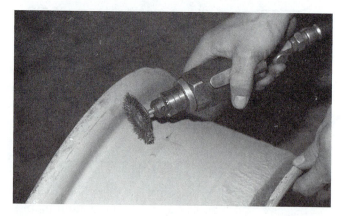

FIGURE 23–30   *Clean all wheel/rim components prior to assembly.*

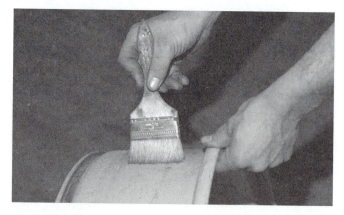

FIGURE 23–31   *Painting the rim.*

tightened. The following are general installation instructions for cast spoke wheels:

1. Slide the inner rear or front tire and rim assembly over the cast spoke wheel and

push it back into position against the tapered mounting surface. Be sure the valve stem faces out, is centered between the two spokes, and clears the disc brake calipers (if applicable).

FIGURE 23–32 Inflating a tire in a safety cage.

FIGURE 23–33 Using a portable lock ring guard.

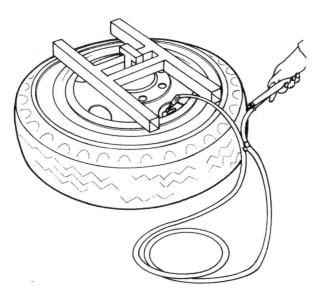

FIGURE 23–34 Using a clip-on chuck with a remote in-line valve and gauge when inflating. (Courtesy of The Budd Company)

2. Slide the spacer ring over the wheel. Check the spacer ring for concentricity by rotating the spacer ring around the cast spoke wheel.
3. Slide the outside rear tire and rim assembly on the wheel, making sure that the valve stem faces inboard and is located in the same relative position as the inner valve stem.
4. Assemble all rim clamps and nuts. Turn the nuts on their studs until each nut is flush with the end of each stud.
5. Turn the top nut 1 until it is snug (**Figure 23–35**).
6. Rotate the wheel and rim until nut 2 is at the top position and snug the nut.
7. Rotate the wheel and rim until nut 3 is at the top position and snug the nut.
8. Rotate the wheel and rim until nuts 4, 5, and 6 are respectively at the top and snug these nuts. Because the entire weight of the tire and rim assembly is on the top spoke, this criss-cross sequence will ensure an even application of force at all points on the rim, keeping the rim in proper alignment.
9. Repeat the sequence of tightening the nuts to the manufacturer's recommended torque.
10. After operating the vehicle approximately 50 miles, check the stud nuts for tightness in the same sequence. Once each week inspect and retorque wheel stud nuts.

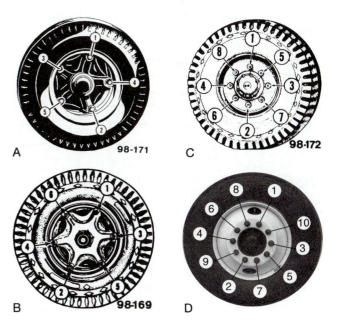

FIGURE 23–35 Wheel nut tightening sequence: (A) six lug, (B) five lug, (C) twelve lug, and (D) ten lug. (Courtesy of Mack Trucks, Inc.)

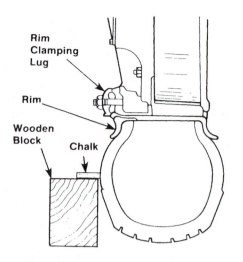

FIGURE 23–36   Checking tire runout using a wood block and chalk. (Photo/Image(s) courtesy of Mack Trucks, Inc.)

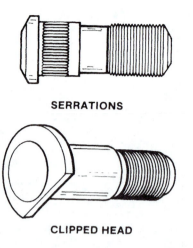

SERRATIONS

CLIPPED HEAD

FIGURE 23–37   Headed wheel studs for disc wheels. (Courtesy of Freightliner Trucks)

Spoke Wheel Runout

Anytime a wheel hub or tire has been reinstalled, the wheel and tire runout must be checked after the wheel stud nuts have been torqued to specifications. To check runout, position a wooden block approximately $1/2$ inch away from the tire as shown in **Figure 23–36**. Slowly rotate the wheel and watch the variation in space between the tire and block. If runout exceeds $1/8$ inch, adjustment is in order.

Position a piece of chalk on the wood block as shown and rotate the wheel so the chalk marks the tire's high spots. The high and low (unmarked) areas show which wheel studs must be loosened and which ones must be tightened.

Loosen the rim clamping nuts that are 180 degrees opposite of the chalk marks slightly and tighten the nuts on the chalk-marked side. Do not overtorque the nuts. Recheck runout and repeat until runout is within $1/8$ inch of spec. If runout cannot be corrected in this way, inspect for part damage or dirt between mating parts.

DISC WHEEL INSTALLATION

Disc wheels are attached to wheel hubs with either threaded studs and nuts or with headed wheel studs. A headed wheel stud has either serrations on the stud body or a flat area on the stud's head to prevent the stud from turning in the wheel hub (**Figure 23–37**). In some disc wheel systems, the end of the stud that faces away from the vehicle is stamped with an "L" or "R" indicating that left- or right-hand threads are used. Studs with an "L" are installed on the driver's side of the vehicle, and right-hand threads are installed on the passenger side. This ensures that the truck's wheel rotation does not loosen the nuts. Other systems use right-hand threads only. Whichever is the case, be sure all hardware is in good condition. Properly install spacer bands on dual spoke rim assemblies.

Disc Wheels with Conical Nuts

The following are general installation instructions for this type of system:

1. Slide the inner/rear or front tire and wheel in position over the studs and push back as far as possible. Use care to avoid damage to threads on the studs and inspect the valve stem to caliper for clearance.

2. Install the outer wheel nut on front wheels and the inner wheel nut on rear dual wheels. Run the nuts on studs until the nuts start to contact the wheel. Rotate the wheel a half turn to allow parts to seat naturally.

3. Draw up the stud nuts alternately following the sequence (criss-cross pattern) illustrated in **Figure 23–38**. Do not fully tighten the nuts at this time. This procedure will allow a uniform seating of nuts and ensure the even face-to-face contact of wheel and hub.

4. Continue tightening the nuts to torque specifications using the same alternating method.

5. Install the outer rear wheel and repeat the preceding method. Be sure that both inner and outer tire valve stems are accessible.

6. After operating the vehicle approximately 50 miles, check the stud nuts for tightness. Some break-in seating may be encountered, and the nuts will loosen. Retighten all nuts to specified torque.

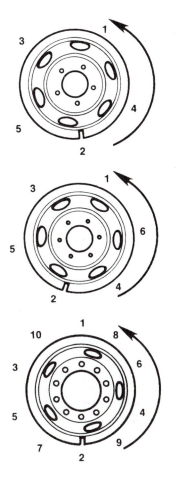

DISC WHEEL TIGHTENING SEQUENCE
(Arrows illustrate 1/2 turn
to seat parts.)

FIGURE 23–38   *Tightening and torquing sequence for disc wheels. (Courtesy of International Truck and Engine Corp.)*

To check and tighten the inner wheel to proper torque, first loosen the outer wheel nuts several turns and tighten the inner nuts, then retighten the outer nuts.

To prevent losing the seating of the outer wheel when checking the inner wheel torque, loosen alternate outer nuts, tighten the inner nuts, and then retighten the outer nuts. Then loosen the remaining outer nuts, tighten inner nuts, and then retighten the outer nuts. Once each week inspect and retorque wheel stud nuts.

Disc Wheels with Flange Nuts

Figure 23–39 illustrates a cross-section view of a disc wheel flange nut installation.

1. Slide the inner rear or front tire and wheel in position over the studs and push back as far

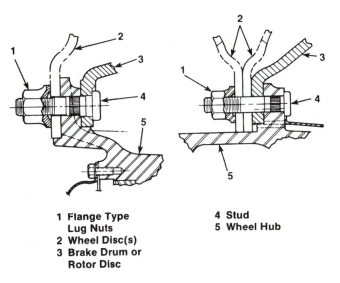

1	**Flange Type Lug Nuts**	4	**Stud**
2	**Wheel Disc(s)**	5	**Wheel Hub**
3	**Brake Drum or Rotor Disc**		

FIGURE 23–39   *Cross section of disc wheel with flange. (Courtesy of International Truck and Engine Corp.)*

as possible. Use care so that the threads on studs are not damaged.

2. Position the outer rear tire and wheel in place over the studs and push back as far as possible. Again use care so that the threads on the studs are not damaged.

3. Run the nuts on the studs until the nuts contact the wheel(s). Rotate the wheel assembly a half turn to permit parts to seat.

4. Draw up the nuts alternately following the (criss-cross) sequence illustrated in **Figure 23–38**. Do not fully tighten nuts at this time. This will allow uniform seating of the nuts and ensure even face-to-face contact of wheel and hub.

5. Continue tightening the nuts to torque specifications using the same alternating sequence.

Tube-Type Radial Tires. Use only rims approved for radial tire usage by the rim manufacturer. Thoroughly clean all rim components, removing all rust and other foreign material. Make certain that rim parts match and are not sprung or broken. Check the inside of the tire before mounting and dry thoroughly if any water is found.

Sparingly lubricate the tire beads and the tube/flat and rim surface with an approved lubricant to minimize the possibility of fluid entering the tire (**Figure 23–40**). Use tubes and flaps that are compatible. Radial tubes will have a permanent red band on the valve stem, below the cap threads, or will have the word "Radial" or the letter "R" molded or stamped either on the valve stem or on a sleeve or ferrule permanently affixed to the valve stem. Radial flaps are also identified by the letter "R" in their designation.

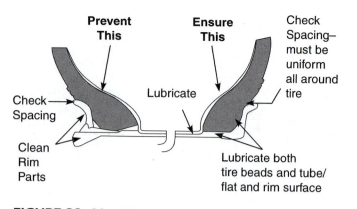

Prevent This

Ensure This

Check Spacing—must be uniform all around tire

Check Spacing

Lubricate

Clean Rim Parts

Lubricate both tire beads and tube/ flat and rim surface

FIGURE 23–40   *Mounting of a tube-type radial tire. (Courtesy of Mack Trucks, Inc.)*

Position the tire assembly in a safety cage and inflate to the recommended operating pressure, deflate completely, and then reinflate to the correct pressure. This will allow the tube, flap, and tire to properly seat.

Visually check the slot and side ring gap (on a two-piece rim) to make sure that the bead is seated. A further check should be made by laying the tire flat and measuring the space between the rim flange and one of the lower sidewall rim line rings. Take measurements around the circumference of the rim flange. If the spacing is uneven, deflate the tire completely and then disassemble, remount, and reinflate.

Tubeless-Type Radial Tires. Be sure to use only rims approved for radial tire usage by the manufacturer. Thoroughly clean the rim, removing all rust and other foreign material. Also, thoroughly lubricate tire beads and rim bead seats with an approved rubber lubricant (see **Figure 23–41**).

Position the tire assembly in a safety cage and inflate to the recommended operating pressure. Due to radial truck tire construction, it may be necessary to use an inflation aid to help seat the beads of tubeless radial tires. The following two types of inflation aids are commonly available:

- **Metal Ring Inflation Aid.** As shown in **Figure 23–42A**, this device uses compressed air to seat the bead.
- **Rubber Ring Inflation Aid.** The ring (**Figure 23–42B**) provides a seal between the tire bead and the rim bead seat, allowing the tire bead to move out and seat.

Check the bead seating by laying the tire flat and measuring the space between the rim flange and one of the three lower sidewall rim line rings (**Figure 23–43**). Take the measurements around the circum-

ference of the rim flange. If the spacing is uneven, deflate the tire completely and then demount the tire, remount, and reinflate.

DUAL TIRE MATCHING

Matching of dual tires is important for several reasons. Tire life is greatly increased when tires are properly matched in tread pattern diameter and circumference. Improperly matched tires can cause costly mechanical problems due to differential carrier failure resulting from constant differential action. Improper traction is also present on mismatched tires and can cause failure of both tires in a very, very short operating time.

The term *mating tires* basically refers to matching tires to the same size. Matching the tread patterns must also be considered in mating. Ideally both duals should be of the same tread design.

Before changing wheels and tires, consider the effect that the change may have on the Gross Vehicle Weight Rating (GVWR) of the vehicle. At the time of vehicle certification, the GVWR is calculated by adding the vehicle's Gross Axle Weight Ratings (GAWR): The GVWR and each of the GAWRs are shown on a certification label (U.S.-purchased tractors) or "Statement of Compliance" label (Canadian-purchased tractors) attached to the left rear door post.

Tire and rim labels certify the tires and rims installed on the vehicle for the given GAWRs. Each GAWR is determined by considering each component of the chassis, including frame suspension, axle, wheels, and tires. The lowest component's capacity is the value used for the rating. Therefore, the tires and rims installed on the vehicle at the time of vehicle manufacture may have a higher load capacity than that certified by the tire and rim label. Tires and rims of the minimum capacity can be installed without changing the load limitations. If tires and rims are installed that have a lower load capacity than that shown on the tire and rim label, then the tires and rims determine the load limitations (the GAWRs and GVWR will be lower).

When pairing tires in a dual assembly, the tire diameters must not differ by more than $1/4$ inch or the tire circumference by more than $3/4$ inch. The total tire circumference of one driving rear axle must match, as nearly as possible, the total tire circumference of the other driving rear axle.

CAUTION: The larger the diameter of the tire, the more likely it will be to overdeflect and overheat. The smaller diameter tire, lacking proper road contact, wears faster and unevenly. Tread or ply separation, tire body breaks, and blowouts can occur from mismatched duals.

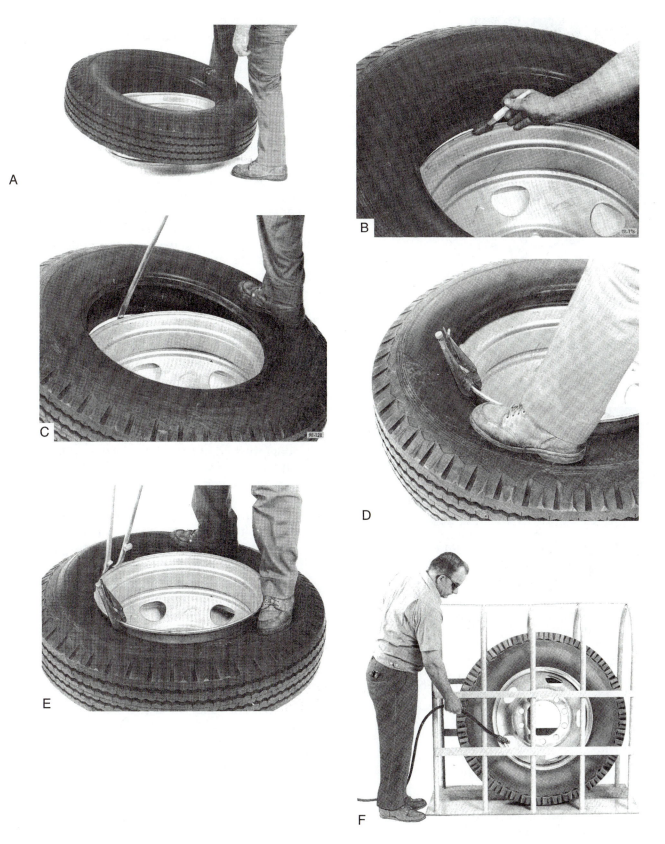

FIGURE 23–41 The procedure for mounting a tubeless tire on a disc wheel is as follows: (A) start the tire on the wheel, (B) lubricate the tire bead, (C) work the tire onto the wheel, (D) start the second bead onto the wheel, (E) work the second bead onto the wheel using two tire tools, and (F) inflate the tire in a safety cage. (Courtesy of Mack Trucks, Inc.)

A   98·146

B   98·134

FIGURE 23–42   *(A) Using a metal ring inflation aid, and (B) installing a rubber ring inflation aid. (Courtesy of Mack Trucks, Inc.)*

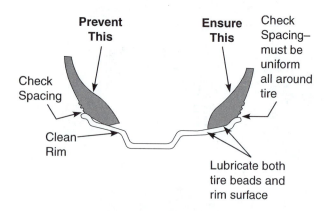

Prevent This — Check Spacing
Ensure This — Check Spacing–must be uniform all around tire
Check Spacing
Clean Rim
Lubricate both tire beads and rim surface

FIGURE 23–43   Mounting of a tubeless-type radial tire. (Courtesy of Mack Trucks, Inc.)

There must be sufficient space between dual tires for air to flow and cool the tires and to prevent them from rubbing against one another. Rims and wheels

of the same size, but of different makes and types, can have different offsets, which would affect dual spacing. If there is sidewall contact between tires or between the inside tire and the chassis, refer to the tire manufacturer's catalog to determine the minimum dual spacing. Refer to the rim or wheel manufacturer's catalog to determine the correct offset.

Installation of a new tire next to a used or worn tire is considered mismatching. It is also critical to use tires of the same construction (bias or radial) on the drive axle.

Ideally, duals are mated so that they are identical in diameter (height), but a diameter variation not exceeding 1/4 inch is generally considered acceptable. If there is any diameter variation between dual tires, place the larger tire on the outside. If two tires are of equal size, but one is slightly more worn than the other, place the less worn tire on the outside. Duals must also be checked for adequate spacing. Make certain they are not "kissing" (contacting each other) especially at the six o'clock position. Always make certain the tires are fully inflated to the specified pressure; otherwise inaccurate measurements will result. Following are several methods of measuring dual sizes:

- **Square Method.** Using a square is the standard method of checking dual diameter matching on the vehicle (**Figure 23–44**). The square leg must be placed parallel to the floor to avoid the tire "bulge." Measure the distance (if any) between the tire tread and the square arm with a ruler. It should not exceed 1/4 inch.
- **Straightedge.** A straightedge can be placed across the four tires of an axle to compare tire diameter. Measurements are taken from the straightedge to the tire tread where gaps are present (**Figure 23–45**). This measurement is doubled to obtain the diameter difference. A taut string can be used in place of the straightedge.

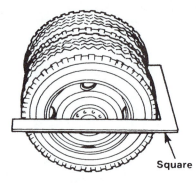

Square

FIGURE 23–44   *Using a square to check dual tire matching. (Courtesy of Freightliner Trucks)*

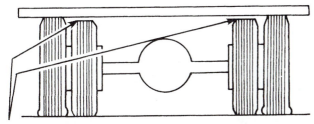

Measure distance between tire tread and straightedge

FIGURE 23-45 *A straightedge positioned across the tires will detect difference in tire size. (Courtesy of Freightliner Trucks)*

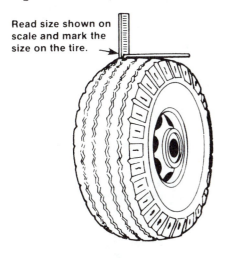

Read size shown on scale and mark the size on the tire.

FIGURE 23-46 *Measuring the size of an unmounted tire using a tire meter. (Courtesy of Freightliner Trucks)*

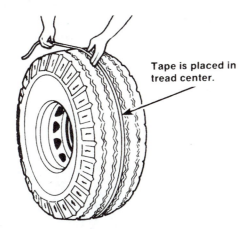

Tape is placed in tread center.

FIGURE 23-47 *Measuring tire circumference with a flexible tape measure. (Courtesy of Freightliner Trucks)*

- **Tire Meter.** A tire meter checks a single tire for size when it is not mounted to the vehicle (**Figure 23-46**).
- **Tape Measure.** A flexible tape measure can be used to check the circumference of an unmounted tire (**Figure 23-47**). Make certain

the tape is in the thread center. A difference of $3/4$ inch in circumference is normally acceptable in mated tires.

DUAL SPACING

Dual spacing for vehicles using disc wheels is determined by the sum of the offset of both wheels used (**Figure 23-48**). Dual spacing for vehicles with rims is determined both by the offset of the demountable rims used and by the width of the spacer band (**Figure 23-49**, **Figure 23-50**).

Three types of spacer bands are available—corrugated, channel, and corrugated-channel. Inspect spacer bands for concentricity to ensure that the

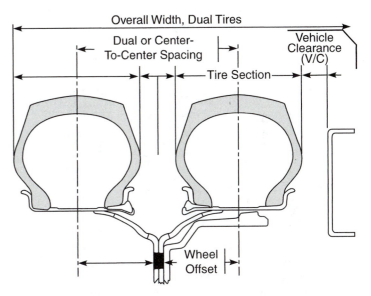

FIGURE 23-48 *Dual spacing with wheel assemblies.*

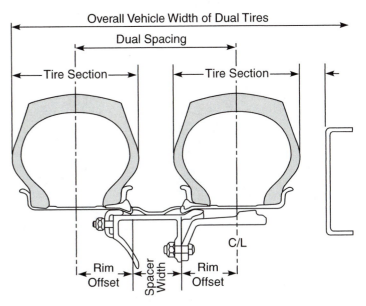

FIGURE 23-49 *Dual spacing with rim assemblies.*

PHOTO SEQUENCE 9

MOUNT A SET OF DUALS TO A CAST SPOKE WHEEL

P9–1 Before disassembling the wheel, check for cracks in the cast spokes, bent and damaged studs, and evidence of the rim rotation on the wheel.

P9–2 Use a wire brush to remove rust from the exposed part of the studs and nuts. Badly rusted studs must be replaced.

P9–3 With an air wrench, back off the nuts about halfway on all the studs. Important! Do not completely remove the wheel nuts.

P9–4 Strike the inside edge of the tire rim, with nuts loosened only, with an 8-lb. hammer until the wedge clamps release; when they release, the tire rim should drop forward. When this happens, the nuts and clamps can be removed, followed by the rims and spacer band.

P9–5 Before reassembling the wheel, check that the 28-degree bevel mounting surfaces on the cast spokes are clean and free of lubricant. The studs should be clean and rust free. Ensure that the wedge clamps are those specified for the wheel.

P9–6 Lift the inside tire and rim assembly onto the wheel and properly position it. Ensure that the valve is centered between two spokes.

P9–7 Now fit the spacer band over the wheel and tap it into position with a rubber mallet.

P9–8 Now fit the outside tire and rim to the wheel. Again, ensure that the valve is centered between two spokes.

P9–9 Install the wedges and nuts over the studs. Use a hammer (tire hammer or 8-lb. sledge) gently on the rim and rotate the wheel to set runout.

PHOTO SEQUENCE 9 (CONTINUED)
MOUNT A SET OF DUALS TO A CAST SPOKE WHEEL

Nut size	Torque specification
⅝"–11	160–200 ft.-lb.
¾"–10	200–260 ft.-lb.

P9–10   Initially torque the nuts in sequence to 50 ft.-lb. Important: do not use an air wrench in place of a torque wrench. Now check wheel runout by rotating the wheel and using an upright hammer head to eyeball the runout. Correct runout if necessary.

P9–11   Torque the wheel nuts in sequence to 80 percent of final torque. Then torque to final torque using the values in the chart at right.

Caution!! Cast spoke wheels are commonly over-torqued. Check studs and nuts for evidence of over-torque and replace. Never exceed the specified torque values.

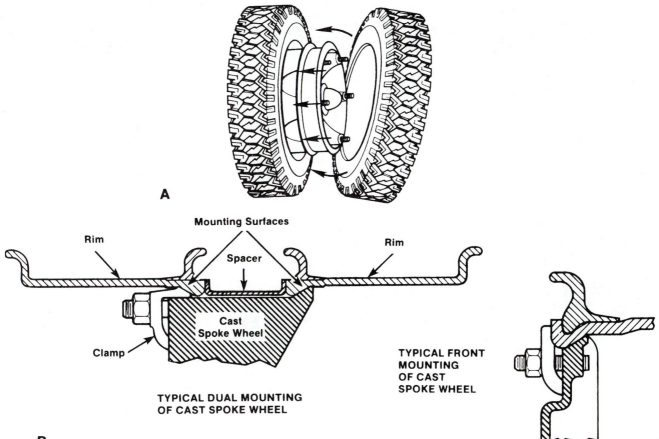

FIGURE 23–50   (A) Position of spacer on a set of duals; (B) sectional view of mounted duals on a cast spoke wheel. (Courtesy of International Truck and Engine Corp.)

PHOTO SEQUENCE 10

INSTALL A SET OF HUB-PILOTED DUALS TO A WHEEL ASSEMBLY

P10–1   Before beginning the disassembly of the wheel, perform a visual inspection. Check that the wheel nuts are properly engaged and look for damaged studs.

P10–2   Remove rust and road dirt from the wheel studs with a wire brush.

P10–3   Use an air gun to remove the wheel nuts.

P10–4   Carefully remove the wheels, ensuring that they are not dragged over the studs.

P10–5   Before reassembling the wheel, clean the hub, wheel, and hub/brake drum mounting faces of rust, dirt, and loose paint. Visually inspect all the studs. Do not paint or apply any other substance on the mounting faces.

P10–6   If an outboard mounted drum is used, make sure that the brakes are released and lift the drum into position, ensuring that it is not sitting on the pilot ledge. Lubricate the fasteners. (Hub-piloted studs must be lightly lubricated to ensure that the correct amount of clamping force is achieved.) Mount the wheels to the hub.

P10–7   Tighten the wheel nuts in sequence in three stages using a torque wrench. First-stage torque value should be 50 ft.-lb. Second-stage torque should be between 50 and 80 percent of the final torque value. Final torque values are listed at right.

P10–8   As a final step, check the wheel runout by rotating the wheel through a full revolution. Do not attempt to correct a runout problem by loosening and retorquing the nuts: disassemble the wheel assembly and locate the problem.

Typical Torque Values for Steel Disc Wheels

Socket size	Thread size	Torque specification
1½" or 3 mm	M22 X 1.5	450–500 ft.-lb.
30 mm	M20 X 1.5	280–330 ft.-lb.

bands have not been distorted or bent. If the spacer band must be replaced due to distortion, misalignment, or corrosion, be sure the replacement band is of equal size. An improperly sized band may alter the overall vehicle width. Vehicle width may be limited by state or federal laws.

WHEEL AND TIRE BALANCING

Wheel balance is the equal distribution of weight in a wheel with the tire mounted (**Figure 23–51**). It is an important factor that affects tire wear and vehicle control.

Front end shimmy and cupping of the tire in random patterns can be caused by wheel assemblies that are out of balance or out-of-round. Even wheel assemblies that are but a few ounces out of balance can make riding uncomfortable and can contribute to accelerated mechanical wear. The centrifugal thrust caused by a wheel assembly that is out of balance will not be reduced by loading the truck. A loaded condition will place additional stress on the wheel, tire, and suspension components, reducing their life expectancy. Although the correct front wheel alignment is necessary for easy steering and maximum tire life (see Chapter 21), the cause of unstable steering can frequently be traced to improper balance of front wheels. When this condition exists, the wheel and tire assembly should be properly balanced. A vulcanized or retreaded tire, or a tire that has a boot in it, may cause an unbalanced condition that cannot be corrected by balancing. In such cases the tire should be replaced before attempting to balance the assembly.

Static Balancing

Radial truck tires are more sensitive to balance and alignment problems than bias ply tires. A tire/wheel assembly that is out of balance or not rolling true can cause uneven tread wear or vibration. In such cases, check for

- **Proper inflation pressure**
- **Proper bead seating.** Improper bead seating can cause vibration, especially on front axles.
- **Balance.** If the tire is out of balance, it can be balanced on the wheel. However, some tire manufacturers recommend vehicle balancing.
- **Proper alignment setting.** Different service conditions may require different settings. Always check with the manufacturer for specific settings if special alignment setting is required.

The following are recommendations for how to obtain proper alignment on a heavy-duty vehicle.

1. **Correct lateral or radial runout.** Runout must be corrected to within $1/8$ inch before attempting the balancing operation. In some cases it may be necessary to remount the tire onto the rim/wheel assembly.

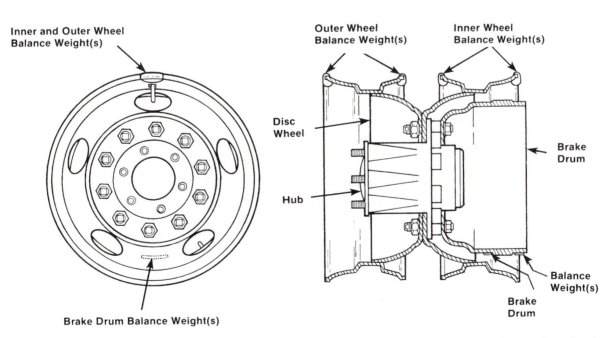

FIGURE 23–51 *Weights attached to the rim flange and brake drum are used to balance the wheel assembly.* (Courtesy of Freightliner Trucks)

2. **Loose wheel bearing and kingpins.** Either, or both, must be corrected to ensure the balancing operation will succeed.

3. **Wheel weights.** A maximum of 18 ounces of wheel weights may be used to balance any one tire. If more weight is required, it is suggested that the tire be removed from the rim/wheel assembly, rotated 180 degrees and remounted. This will, in many cases, bring the assembly within the acceptable limits.

4. **Wheel weight distribution.** Add equal amounts of weights on each side of the rim/wheel when static balancing an assembly. This will avoid throwing the assembly out of balance dynamically; so, if a 16-ounce weight is required, install 8 ounces on each side of the rim/wheel assembly, directly opposite each other.

5. **Preparation of rim/wheel.** Make sure all old balance weights, mud, dirt, and foreign material are removed from the rotating assembly before attempting the balance operation.

6. **Tire inflation.** Be sure tire pressure is correct.

7. **Rim/wheel inspection.** Inspect side ring/lock ring openings on the rim assembly. The gap should not be less than $3/32$ inch or greater than $5/16$ inch. Anything other than this could indicate an improperly seated lock ring assembly.

8. **Correct wheel weights.** Use the proper type of balance weight for the type of rim/wheel involved. There is no point in balancing if the weight will not stay in place because the incorrect weight was used.

Generally speaking, complaints on tire and wheel imbalance usually focus on the front steering axle. Maximum front tire life is experienced when these tires are balanced, when inflation pressure is maintained according to the load, and when front axle alignment is set according to the conditions. Although, in some instances, excessive unbalanced conditions in rear drive wheels can cause ride complaints, the majority of driver complaints will come from unbalanced front tires.

Tires and wheels are affected by two types of unbalance: static and dynamic. Static unbalance will cause the wheel to bounce up and down when rotating. Dynamic unbalance will cause the wheel to wobble or shimmy. A tire/wheel assembly that has been properly static balanced will usually result in a dynamically balanced assembly (although not 100 percent of the time).

There are four methods of balancing tire/wheel assemblies:

1. Spin balancing
2. Off-the-vehicle dynamic balancing
3. Off-the-vehicle static balancing
4. Use of balancing fluid or material introduced inside the mounted tire assembly.

23.6 WHEEL HUBS, BEARINGS, AND SEALS

A wheel hub assembly (**Figure 23–52**) consists of the following major components (also see **Figure 23–53**).

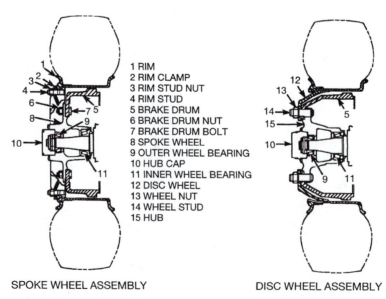

1 RIM
2 RIM CLAMP
3 RIM STUD NUT
4 RIM STUD
5 BRAKE DRUM
6 BRAKE DRUM NUT
7 BRAKE DRUM BOLT
8 SPOKE WHEEL
9 OUTER WHEEL BEARING
10 HUB CAP
11 INNER WHEEL BEARING
12 DISC WHEEL
13 WHEEL NUT
14 WHEEL STUD
15 HUB

SPOKE WHEEL ASSEMBLY   DISC WHEEL ASSEMBLY

FIGURE 23–52 *Cross-section view of spoke and disc wheel assemblies (single-wheel configurations). (Courtesy of Freightliner Trucks)*

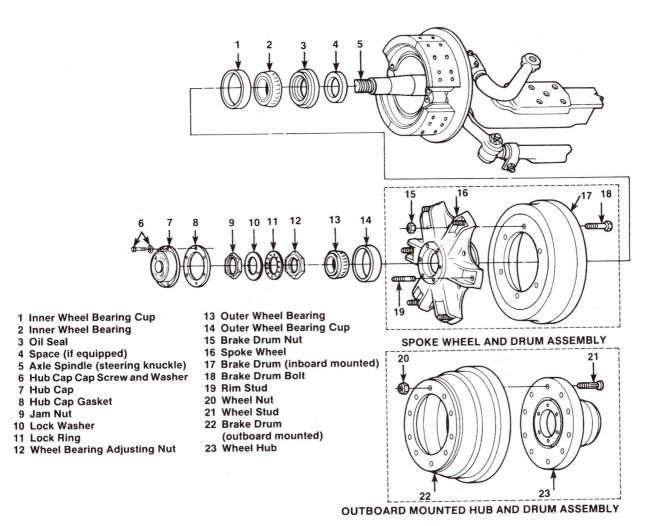

1 Inner Wheel Bearing Cup
2 Inner Wheel Bearing
3 Oil Seal
4 Space (if equipped)
5 Axle Spindle (steering knuckle)
6 Hub Cap Cap Screw and Washer
7 Hub Cap
8 Hub Cap Gasket
9 Jam Nut
10 Lock Washer
11 Lock Ring
12 Wheel Bearing Adjusting Nut

13 Outer Wheel Bearing
14 Outer Wheel Bearing Cup
15 Brake Drum Nut
16 Spoke Wheel
17 Brake Drum (inboard mounted)
18 Brake Drum Bolt
19 Rim Stud
20 Wheel Nut
21 Wheel Stud
22 Brake Drum
   (outboard mounted)
23 Wheel Hub

SPOKE WHEEL AND DRUM ASSEMBLY

OUTBOARD MOUNTED HUB AND DRUM ASSEMBLY

FIGURE 23–53 Typical front axle components (oil lubricated). (Courtesy of Freightliner Trucks)

- **Tapered Wheel Bearings.** A typical tapered wheel bearing assembly consists of a cone, tapered rollers, roller cage, and a separate cup that is press-fit in the hub. All components carry the load with the exception of the cage, which spaces the rollers around the cone. Each hub has a set of inner and outer tapered wheel bearing assemblies. The bearing is locked in place on the axle spindle by an adjusting nut and lock (jam) nut. In other cases, split forged nuts or castellated nuts and cotter pins are used to secure the hub on the axle. Both inner and outer wheel bearing cups and the wheel studs are press-fit in the hub. Spoke wheels combine the wheel and hub into a single unit. On disc wheels, the hub is a separate component in the assembly. On drive axles, the hub is also the interconnecting point for the drive axle shaft and wheels.

- **Seals.** Wheel bearings are lubricated using either grease or oil. The lubricant is held in the hub using special oil or grease seals. There are slight differences in bearing and seal service between grease- and oil-lubricated systems and front and drive axles.

- **Brake Drums.** On a spoke wheel, the brake drum is mounted on the inboard side of the wheel/hub and is held in place with nuts (**Figure 23–53**). Servicing inboard brake drums on spoke wheels involves removing the single or dual wheel/hub and drum as a single assembly. This involves removing the hub nut and disturbing hub components, so bearing and seal service will be required.

On disc wheels, the brake drum is usually mounted on the outboard side of the disc hub (**Figure 23–53**). The drum fits over the wheel studs and is secured between the wheel and hub. This means the wheel and drum can be dismounted without disturbing the hub nut. Outboard drums can be serviced without servicing the bearings and seals.

FRONT AXLE GREASE SEAL/ BEARING SERVICE

Figure 23–53 illustrates the components of a front axle wheel hub assembly. To remove the hub assembly on a disc wheel system, perform the following:

1. Chock the rear tires to prevent vehicle movement and set the parking brake.
2. Raise the front of the truck until the tires clear the ground. Place safety stands under the axle.
3. Back off the slack adjuster to release the brake shoes.
4. Remove the wheel and tire assembly.
5. Remove the brake drum.
6. Remove the hub cap, hub cap gasket, jam or locknut, lockwasher, and lock ring. Exact hardware will vary based on the locking system used to secure the hub.
7. Back off the wheel bearing adjusting nut about two turns or enough to allow the weight of the hub to be lifted from the wheel bearings.
8. Lift the hub until all weight is removed from the wheel bearing. Remove the adjusting nut.
9. Move the hub about 1/2 inch to jar loose the outer wheel bearing. Be careful not to damage the axle threads with the weight of the hub. Do not allow the outer bearing to drop off the end of the axle; carefully lift it off.
10. Remove the hub from the axle spindle.
11. Remove the inner wheel bearing and bearing spacer (if used) from the axle.
12. Remove the seal from the axle if it has not already been removed.

👓 **Shop Talk** _____

If pulling more than one wheel, be sure to keep all of the components of each wheel together and separate from the other wheels.

On spoke wheels, follow previous steps 1 through 3 and then remove the rim and tire assembly from the wheel. Back off the wheel bearing adjusting nut as in previous step 7. Now, using an adjustable wheel support (wheel dolly), raise the wheel and drum assembly until all weight is removed from the bearings. Then remove the adjusting nut. Remove the outer wheel bearing as in step 9 and, using the wheeled support, pull the wheel and drum assembly away from the vehicle, being careful not to damage the axle spindle. Also, do not allow the outer bearing assembly to fall off as the wheel and hub are removed. Next, remove the brake drum nuts and detach the brake drum from the wheel. Finally remove the inner wheel bearing, bearing spacer (if used), and seal.

CLEANING AND INSPECTION

The following is the procedure for cleaning and inspecting the hub cavity and cap:

1. Clean the hub cavity and cap, removing all old lubricant. Inspect the wheel hub mounting flange for wear, warpage, or rough edges. Inspect around the studs for cracks. Inspect the inner surface of the wheel or hub for cracks, dents, or wear. Replace the wheel/ hub if any of these conditions exist.
2. Use a brush to clean the drum and brake mechanism. Wipe the spindle clean.
3. Use a recommended solvent to remove dirt and grease from the bearing and related wheel/ axle parts. Rinse the bearing in clean solvent. Allow the bearings to dry naturally in the air.

CAUTION: Never allow a bearing roller to be spun by compressed air.

4. Inspect the bearing cones and cups. Replace them if they are worn, cracked, pitted, or otherwise damaged.
   To ensure an interference fit, wheel bearing cups are purposely larger than the wheel hub bores they occupy. On aluminum hub bores, heat the hub in the oven to expand the bore so the cups can be driven out. All heating must be even oven heating.

CAUTION: Never use oxyacetylene or similar equipment to heat the bore. This localized heat will weaken the aluminum. If an oven is not available, replace the hub, wheel stud, and bearing assembly as a unit.

Wheel bearing cups on ferrous hubs are removed and installed by driving them out and pressing them in without heating the hub.

5. Dip cleaned bearings in lubricant or coat bearing surfaces with a light grease. Wrap the bearings in waterproof paper and place them in a clean box or carton. Keep bearings covered until ready to install the new seal.

6. Inspect the spindle bearing and seal surface for burrs or roughness. Be careful not to scratch the sealing surfaces when polishing out roughness. Even small marks can permit lubricant to seep out under the sealing lip.

7. Check where the seal lip makes contact. If you can feel a worn groove with your fingernail, there will be leakage, even with a new seal. Replace the bearing spacer if it is grooved or worn.

8. Remove surface nicks, burrs, grooves, and machine marks from the wheel spindle. The shaft end should have no burrs or sharp edges.

9. Be sure that the shaft and bore diameters match those specified for the seal selected and that the seal being installed has not been damaged in any way. Make sure that the new seal faces in the same direction as the original. In general, the lip faces the grease or lubricant to be retained.

REASSEMBLY

Whenreassembling, proceed as follows:

1. If the bearing cups are being replaced, use a mandrel or sleeve to drive the cups into the hub. The bearing cups must be square in the bearing bore (**Figure 23–54**).

 Shop Talk _____

To ease the installation of the bearing cup into the hub, the cup can be cooled and shrunk by placing it in a deep freeze or in an alcohol bath cooled with

dry ice. Aluminum hubs can also be heated in an oven to get the cup to slip into place.

_____

2. If the wheel does not use gear oil to lubricate the bearings, pack the hub cavity between the two bearing cups with an approved wheel bearing grease to the level of the cup's smallest diameter (**Figure 23–55**).

3. Pack grease-type bearing cones, using a pressure packer if possible. If a pressure packer is not available, force the grease into the cavities between the rollers and cage by hand from the large end of the cone. Coat the rollers with grease.

4. Insert the inner bearing cone in the grease-filled hub.

5. Prelubricate the seal by wiping it with the lubricant to be retained. Place the prelubed seal in the hub with the lip facing the bearing cone.

6. Seat the seal using the correct installation tool. Seal drivers should have an outside diameter approximately 0.010 inch smaller than the bore size. The center of the tool should be open so that pressure is applied only at the outer edge of the seal (**Figure 23–56**). Never hammer directly on the surface of the seal. Apply force evenly around the outer edge to avoid cocking the seal.

7. Position the spacer (if used) on the spindle. Align the hole and pin.

8. Apply a light film of lubricant to the spindle.

9. Use a wheel dolly to center the wheel/hub assembly on the spindle. Push the wheel/hub

FIGURE 23–54 *Cutaway view of installing bearing cups in the hub. (Courtesy of Chicago Rawhide)*

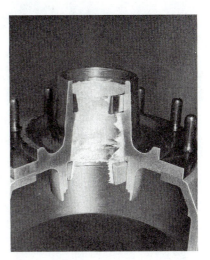

FIGURE 23–55 *Pack the hub cavity with grease as shown in the cutaway view. (Courtesy of Chicago Rawhide)*

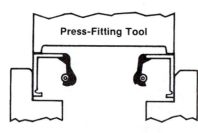

The press-fitting tool should be .010" less than the bore ID.

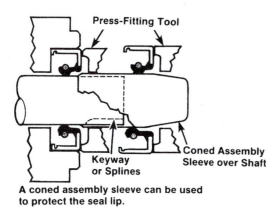

A coned assembly sleeve can be used to protect the seal lip.

FIGURE 23–56   *Press-fitting a grease retaining seal into position. (Courtesy of Chicago Rawhide)*

FIGURE 23–57   *Cutaway view of installing outer bearing cone; grease is removed for clarity. (Courtesy of Chicago Rawhide)*

on far enough so the seal is in safe contact with its riding surface on the bearing spacer or spindle. Install the outer bearing cone (**Figure 23–57**), washer, and adjusting nut in reverse order of removal.

10. Adjust the bearing according to manufacturer's specifications (see the following section). After adjustment, secure the locknut

and locking device. Fill the hub cap with oil or grease, position the new gasket on the hub cap, and install the hub cap.

WHEEL BEARING ADJUSTMENT

Due to an unacceptable number of heavy truck wheel-off incidents in the United States and Canada, some of which have been the result of bearing maladjustments, all the manufacturers of **wheel-end** hardware have approved a single method of wheel bearing adjustment. This method was agreed to through meetings of **The Maintenance Council (TMC)** committee of the American Trucking Association, and the trucking industry has embraced this single standard throughout the continent. Wheel bearing adjustment is a simple but highly critical procedure. It is recommended that technicians learn the procedure, follow it precisely, and use no other method of adjusting wheel bearings. The TMC recommended procedure is reprinted here, word-for-word.

TMC'S RECOMMENDED WHEEL BEARING ADJUSTMENT PROCEDURE

This procedure was developed by TMC's Wheel End Task Force, and it is important to remember that it represents the combined input of manufacturers of wheel end components. **Figure 23–58** identifies the different types of bearing adjustment hardware.

> **Step 1. Bearing Lubrication.**   Lubricate the wheel bearing with clean lubricant of the same type used in the axle sump or hub assembly.
> **Step 2. Initial Adjusting Nut Torque.** Tighten the adjusting nut to a torque of 200 ft.-lb., while rotating the wheel.
> **Step 3. Initial Back Off.**   Back the adjusting nut off one full turn.
> **Step 4. Final Adjusting Nut Torque.** Tighten the adjusting nut to a final torque of 50 ft.-lb. while rotating the wheel.
> **Step 5. Final Back Off.**

Axle Type	Threads Per Inch	Final Back Off
Steer (Single Nut)	12	$1/6$ turn*
	18	$1/4$ turn*
Steer (Double Nut)	14	$1/2$ turn
	18	$1/2$ turn
Drive	12	$1/4$ turn
	16	$1/4$ turn
Trailer	12	$1/4$ turn
	16	$1/4$ turn

*Install cotter pin to lock axle nut in position.

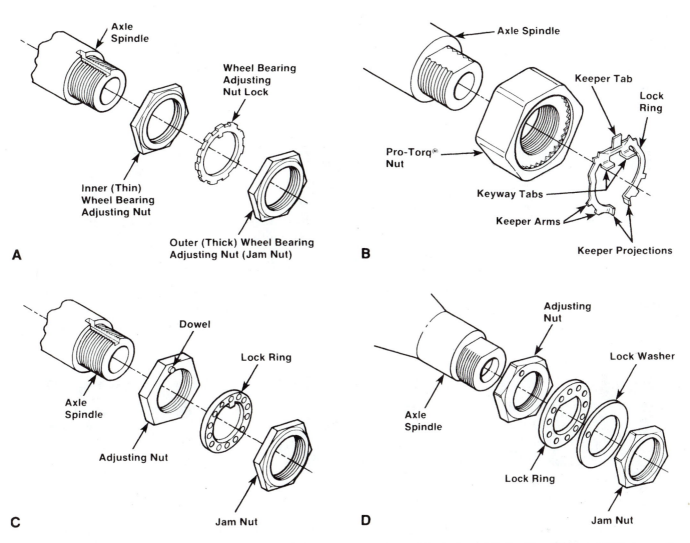

FIGURE 23–58 Bearing setting devices: (A) jam nut and D-shaped lock ring, (B) Pro-Torq™ nut, (C) Eaton axle with adjusting nut and jam nut, and (D) Rockwell axle with adjusting nut and jam nut. (Courtesy of Freightliner Trucks)

Step 6. Jam Nut Torque.

Axle Type	Nut Size	Torque Specifications
Steer (Double Nut)	Less Than 2 5/8"	200–300 ft.-lb.
	2 5/8" And Over	300–400 ft.-lb.
Drive	Dowel Type Washer	300–400 ft.-lb.
	Tang Type Washer	200–275 ft.-lb.
Trailer	Less Than 2 5/8"	200–300 ft.-lb.
	2 5/8" And Over	300–400 ft.-lb.

Step 7. Acceptable End Play. The dial indicator should be attached to the hub or brake drum with its magnetic base. Adjust the dial indicator so that its plunger is against the end of the spindle with its line of action approximately parallel to the axis of the spindle.

Grasp the wheel or hub assembly at the 3 o'clock and 9 o'clock positions. Push and pull the wheel-end assembly in and out while oscillating the wheel approximately 45 degrees. Stop oscillating the hub so that the dial indicator tip is in the same position as it was before oscillation began. Read the bearing end-play as the total indicator movement. Acceptable end-play is 0.001–0.005 inch.

PHOTO SEQUENCE 11

WHEEL-END PROCEDURE: TMC METHOD OF BEARING ADJUSTMENT

The following photo sequence outlines the ATA (American Trucking Association) TMC (The Maintenance Council) recommended wheel end adjustment procedure. This adjustment method was agreed to by ATA member OEMs as a step in the solution to combating increasing wheel-off incidents.

P11–1   Torque the adjusting nut to 200 ft.-lb. to seat the bearing. Ensure that the wheel is rotated during torquing.

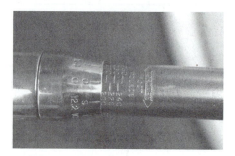

P11–2   Torque wrench specification should read 200 ft.-lb.

P11–3   Now back off the adjusting nut one full turn; this will leave the wheel assembly loose.

P11–4   Now torque the wheel adjusting nut to 50 ft.-lb. while rotating the wheel.

P11–5   Establish endplay. Now back off the wheel adjusting nut. The amount of rotation required to back off the nut will depend on the tpi (threads per inch). For a typical 12-tpi axle spindle, the adjusting nut should be backed off one flat (1/6th) of a turn.

P11–6   Use a jam nut and locking plate on double nut systems. It is critical that the locking nut be torqued to specification. Generally, jam nuts $2^5/_8$ inch or less are torqued to 200–300 ft.-lb. and those larger than $2^5/_8$ inch are torqued to 250–400 ft.-lb. Check OEM specifications.

P11–7   Now verify that endplay exists. Install a dial indicator on the axle spindle and apply a rocking force to the wheel assembly. The endplay reading on the dial indicator must be between 0.001 inch and 0.005 inch. If not within specification, the complete procedure must be repeated.

> *Remember, the consequences of not observing correct wheel-end procedures are wheel-off incidents that can kill.*

DRIVE AXLE GREASE SEALS AND BEARINGS

To remove a drive axle wheel assembly:

1. Jack up the wheel. Support the axle with safety stand.
2. Remove the axle flange nuts and lockwashers. Install pulling screws in axle flange holes, if provided. If not, strike the axle flange in the center sharply with a brass drift and a heavy hammer. It might require several blows to bounce the shaft loose so the tapered washers and axle shaft can be removed.
3. Loosen the bearing lock ring nut setscrew (when used) and remove the locknut. Some drive axles use a lockwasher between the outer nut and the bearing adjusting nut. One or more of the tabs bent over the outer nut and the seal assembly must be bent away from the nut before a socket of the correct size will fit over the nut. Then remove the locknut, lockwasher or ring, and adjusting nut.

All other procedures for wheel removal, inspection, seal installation adjustment, and reassembly are the same as for the front axle.

OIL-LUBRICATED BEARINGS

Beginning in the late 1950s, heavy-duty truck manufacturers introduced oil bath lubricating systems for both front and rear wheels. In the case of a drive axle, oil agitation within the axle housing allows differential oil to flow through the axle spindle tube to the wheel hub cavity (**Figure 23–59**). All other wheels have an isolated wet hub, filled to a specified level with gear lube.

Several different types of oil seals are used on front and rear wheels. For oil lubrication of wheel bearings, rear axle lubricants meeting the requirements of API (American Petroleum Institute) GL-5 are recommended. SAE 90 viscosity grade is normally preferred. For extreme cold environments similar SAE 75W or SAE 80W is preferred; and for extremely warm climates, similar SAE 140 viscosity grade is preferred. Multigrade gear oils like SAE 80W-90, SAE 85W-140 are also satisfactory. The manufacturer of the vehicle usually recommends the type and grade of lubricant to be used.

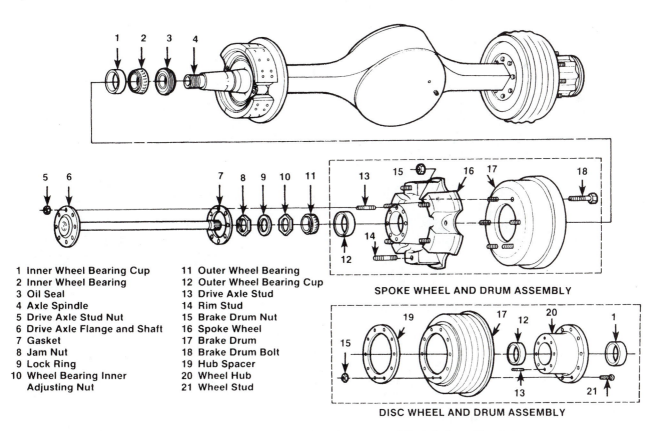

1	Inner Wheel Bearing Cup	11	Outer Wheel Bearing
2	Inner Wheel Bearing	12	Outer Wheel Bearing Cup
3	Oil Seal	13	Drive Axle Stud
4	Axle Spindle	14	Rim Stud
5	Drive Axle Stud Nut	15	Brake Drum Nut
6	Drive Axle Flange and Shaft	16	Spoke Wheel
7	Gasket	17	Brake Drum
8	Jam Nut	18	Brake Drum Bolt
9	Lock Ring	19	Hub Spacer
10	Wheel Bearing Inner Adjusting Nut	20	Wheel Hub
		21	Wheel Stud

SPOKE WHEEL AND DRUM ASSEMBLY

DISC WHEEL AND DRUM ASSEMBLY

FIGURE 23–59   Typical rear axle wheel hub assembly. (Courtesy of Freightliner Trucks)

WARNING: Failure to lubricate any bearing in the specified manner and to maintain proper lubrication can cause a potentially fatal wheel-off incident!

Lip Seal with Wiper Ring

The metal encased lip-type seals shown in **Figure 23–60A** have a wiper ring or wear sleeve mounted on the axle or spindle. The wiper ring provides a smooth contact surface for the lip of the seal, which is also replaced when seals are replaced. The lip of a seal can wear a groove in the axle or spindle at that point where the lip makes contact. The oil-lubricated seal shown in **Figure 23–60B** has an added feature called grit guard. The grit guard is a shielded extension on the wiper ring to prevent grit and road splash from entering the seal.

Unitized Seal

Unitized seals (**Figure 23–61**) in most installations do not require axle wiper rings and minimize wear on the axle spindle as follows. The outer shell of the seal being pressure fit in the wheel hub rotates with the wheel around the sealing element, which is pressure fit on the axle spindle. With the unitized seal during

replacement, the worn surface created by the sealing lip is also replaced, by virtue of a new seal.

Barrier Seal

The rubber-encased seal shown in **Figure 23–62** is a barrier-type seal. The seal effect is between the lips and encased metal ring. This type of seal is installed by hand and should not normally have any lubricant or sealant added to the inner or outer surfaces when installed.

New seals have a lubricant applied to the mounting surfaces to aid in the installation. Do not apply grease to these surfaces. If grease should be applied to the outer surface of the seal, the seal could slip in the hub or on the spindle or axle tube.

Guardian Oil Seal

The rubber-encased lip-type seal with wiper ring shown in **Figure 23–63** is known as the guardian oil seal. It has the grit guard feature similar to the seal shown in **Figure 23–60B** with the advantage of a direct contact between the seal and wiper ring to further guard against the entrance of foreign matter. The wiper ring provides a smooth contact for the lip of the seal. The lip of the seal retains the oil in the bearing cavity. The rubber encasement has grooves around its circumference which retain the seal in place in the hub.

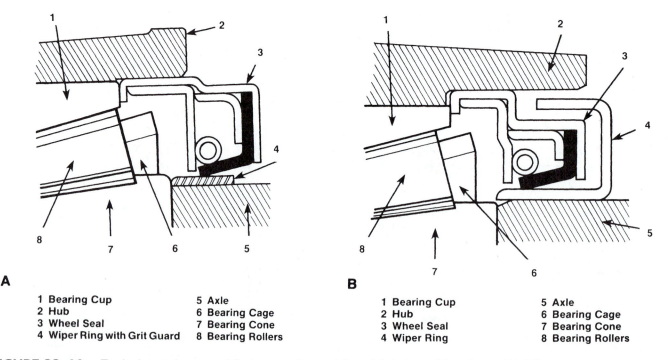

A

1 Bearing Cup	5 Axle
2 Hub	6 Bearing Cage
3 Wheel Seal	7 Bearing Cone
4 Wiper Ring with Grit Guard	8 Bearing Rollers

B

1 Bearing Cup	5 Axle
2 Hub	6 Bearing Cage
3 Wheel Seal	7 Bearing Cone
4 Wiper Ring	8 Bearing Rollers

FIGURE 23–60 Typical metal-encased lip-type seals used for oil-lubricated bearings: (A) Oil-lubricated wheel seal, and (B) wiper ring with grit guard. (Courtesy of International Truck and Engine Corp.)

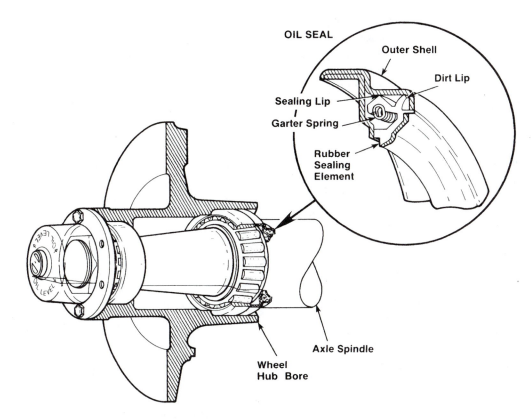

OIL SEAL

Outer Shell

Dirt Lip

Sealing Lip

Garter Spring

Rubber Sealing Element

Axle Spindle

Wheel Hub Bore

FIGURE 23-61 *Example of unitized oil seals. (Courtesy of Freightliner Trucks)*

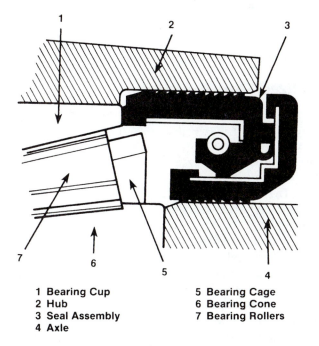

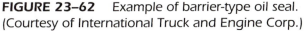

1 Bearing Cup
2 Hub
3 Seal Assembly
4 Axle
5 Bearing Cage
6 Bearing Cone
7 Bearing Rollers

FIGURE 23-62 *Example of barrier-type oil seal. (Courtesy of International Truck and Engine Corp.)*

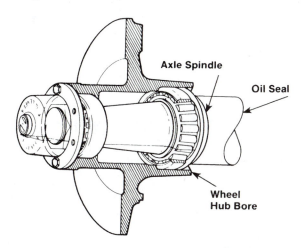

Axle Spindle

Oil Seal

Wheel Hub Bore

FIGURE 23-63 *Example of guardian-type oil seal. (Courtesy of Freightliner Trucks)*

REMOVAL AND INSPECTION

Wheel seals are removed using a brass drift and hammer. Position the drift through the outer opening of the hub and against the bearing. Tap the bearing and seal out through the brake drum side of the hub. Take care so that the seal bore is not damaged.

The wiper or wear ring that is used with an oil-lubricated wheel bearing is removed by using a ball peen hammer and tapping on the ring to expand it. Do not use a chisel to cut the ring because it could damage the machined surface on the axle.

Clean and inspect all components as explained earlier in this chapter under grease-lubricated bearing service. With oil-lubricated systems, inspect for porous or cracked hubs that could be a source of oil leakage. Inspect the hub or wheel bore. Remove any burrs with an emery cloth.

INSTALLING OIL SEALS

Oil seal installation often requires special tools to position and seat the seals. There are also slight differences in installation, depending on seal design.

Wiper Ring Seals

Apply a thin coat of gasket eliminator sealant to the shoulder of the wheel spindle or axle tube. Using the proper installation tool (**Figure 23–64**), place the wiper ring on the spindle or tube. After wiping away excess sealant, tap on the end of the tool to drive the ring firmly onto the shoulder until the installation tool contacts the shoulder. Check the position of the ring to make sure that its edge is parallel with the shoulder.

 Shop Talk

The care with which the ring is installed will affect seal performance. A damaged ring will significantly shorten seal life.

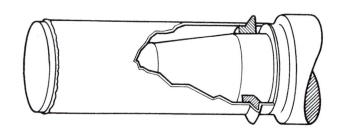

FIGURE 23–64 *Installation tool for wiper ring seals. (Courtesy of International Truck and Engine Corp.)*

Lip Seal with Wiper Ring

Apply a thin coat of sealant to the hub seal bore. This light coat should cover the press fit area. Be sure sealant does not contact the seal lip or contaminate the lube oil.

Lay the wheel on a flat surface with the brake drum facing up. Place the inner wheel bearing into the bearing cup and place the hub seal into the starting position on the hub. Use an installation tool with

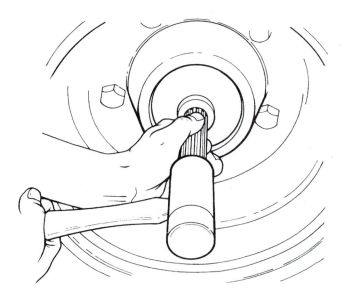

FIGURE 23–65 *Driving the seal into the hub. (Courtesy of International Truck and Engine Corp.)*

a properly sized disc that will apply force to the outer edge of the hub seal and prevent the seal from becoming distorted or damaged. Drive the seal into the hub until it bottoms out in the hub bore (**Figure 23–65**). Do not continue to drive the seal after it is seated because damage to the seal can result.

Unitized Seal

If a seal is being converted to a unitized seal from another type, remove the wear rings if present. Do not use a chisel to remove the wiper rings because it could damage the axle. Instead, use a ball peen hammer and tap the ring lightly on the seal lip surface. Do this in a small area, causing the ring to expand. When expanded, it will slip off the axle housing.

To install a unitized seal in the wheel, first apply a thin coat of gasket eliminator sealant over the press-fit area of the hub bore. Seat the outer face of the seal in the recess of the installation tool adapter (**Figure 23–66**). Insert the centering plug of the tool in the bore of the inner bearing cone (**Figure 23–67**). This center plug prevents cocking of the seal in the bore.

Hold the tool handle firmly and strike the handle with a hammer until the sound of the impact changes: a signal that the seal is seated (**Figure 23–68**). Remove excess sealant. A good check for proper seating is to move the synthetic sealing member with your fingers after installation in the wheel hub. There should be slight in-and-out movement.

PHOTO SEQUENCE 12
INSTALLING WHEEL HUB SEAL

P12–1 Raise the left side of the rear axle with a floor jack.

P12–2 Lower the rear axle onto a safety stand, and remove the floor jack.

P12–3 Remove the rear axle nuts, lockwashers, or split cones, and remove the rear axle shafts.

P12–4 Place a wheel dolly under the left rear tires.

P12–5 Remove the bearing locknut, lockwasher, and adjusting nut.

P12–6 Be sure the wheel, tire, and hub weight is supported on the dolly, and remove the outer wheel bearing. Slide the wheel, tire, and hub assembly straight off the axle spindle.

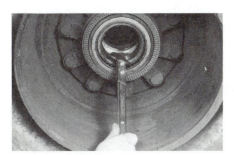

P12–7 Remove the hub seal with the proper puller, and inspect the seal bore in the hub and the seal lip contact surface on the spindle.

P12–8 Lubricate the new seal lip and the seal race on the spindle with the specified rear axle lubricant.

P12–9 Install a coating of non-hardening sealant around the seal case.

PHOTO SEQUENCE 12 (CONTINUED)
INSTALLING WHEEL HUB SEAL

P12–10 *Use a seal driver to install the hub seal. Install a seal protector on the spindle threads, and use the wheel dolly to slide the wheel, tire, and hub assembly straight onto the spindle.*

P12–11 *Install and adjust the adjusting nut. Install the lock-washer and tighten the locknut to the specified torque. Install the axle shaft, lockwasher, or split cone. Install and torque the axle nuts. Be sure the rear hub has the specified lubricant and level.*

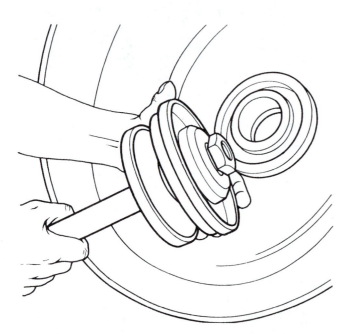

FIGURE 23–66 Seating a unitized seal in the recess of the installation tool. (Courtesy of International Truck and Engine Corp.)

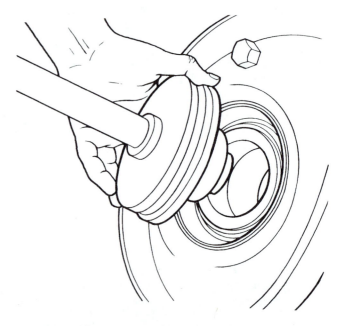

FIGURE 23–67 Centering the plug of the tool in the hub bore. (Courtesy of International Truck and Engine Corp.)

WARNING: Any time a wheel with a unitized seal is removed for any purpose, the seal must be replaced.

Barrier Seal

The barrier-type seal (**Figure 23–62**) is installed by hand, and new seals do not require any special lubricant or sealer the to outer or inner surfaces, except when a new seal appears to be free of lubricant.

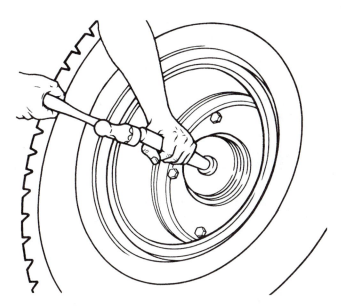

FIGURE 23-68 *Seating the seal with a hammer.* (Courtesy of International Truck and Engine Corp.)

Then it is permissible to wipe the leading edge of inside and outside diameters sparingly with light bearing oil. These seals do not require any special tools for installation.

Position the seal over the hub and use care so that the seal is not cocked when pushed into place. Push the seal into the hub by hand until it is completely bottomed and uniformly seated in the bore. Do not use wheel bearing or chassis grease to install seals, because grease could result in the seal slipping in the hub or on the spindle or axle tube.

If the inner wheel bearing is smaller than the opening in the seal, it is permissible to assemble the bearing on the spindle first and then install the hub assembly with seal. If it is necessary to remove the wheel assembly from the spindle due to hang-ups, recheck the seal for proper installation. In some cases the seal will be separated. The inner half must be removed by hand from the spindle shoulder. Reinsert the inner half in the outer wheel seal by snapping together. Then reinstall the seal assembly.

Guardian Oil Seal

The guardian oil seal grit guard is installed in the same manner as the wiper ring, previously described and shown in **Figure 23-63**. The seal is then positioned on the wiper ring assembly with the lip of the seal toward the bearing cavity of the hub. The hub assembly is then positioned over the axle tube or spindle and seal assembly. This ensures that the seal is in place over the wiper ring (**Figure 23-64**). Do not

install the guardian seal in the hub first because the seal could be pushed into the hub deep enough so that the sealing effect would not be accomplished.

🥽 **Shop Talk** _____

Ask wheel manufacturers what the life expectancy is for their wheels, and the answer received will range from "the life of the truck" to "it depends on the level of maintenance." Both are accurate responses.

PRO-TORQ ADJUSTMENT PROCEDURE

Stemco Pro-Torq wheel nut bearing adjustment is designed to make wheel bearing adjustment safer and more accurate. Ensure that when this procedure is used, it is followed exactly as indicated by the OEM and THC guidelines. See **Figure 23-69**.

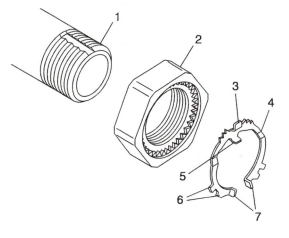

1. Axle Spindle
2. Pro-Torque Nut
3. Keeper Tab
4. Lockring
5. Keyway Tabs
6. Keeper Arms
7. Keeper Projections

FIGURE 23-69 *Axle with Pro-Torq nut and lockring.* (Courtesy of Freightliner Trucks)

CONCLUSION

Wheels do not actually wear out unless abused from overloading, operation over severe road surfaces, or by poor maintenance practices. Most wheel problems are caused by cracked elongated bolt holes usually caused by failing to keep wheel nuts torqued to specification. Failing to tighten inner cap nuts on dual-wheel assemblies is an often overlooked item. The outer nut must be loosened, the inner nut tightened and the outer nut retightened. Loose wheel nuts are by far the most common cause for premature wheel cracks forcing wheel replacement. Properly torqued wheels, checked regularly, are the best insurance against wheel cracking. Rust lines around wheel nuts is an indication that they are loose. Once loose, even slightly, cap nuts will start to gall the chamfer in each wheel hole causing the holes to become egg-shaped as the nut chews out the metal. In worst cases, a severe brake application can snap off wheel studs, causing the vehicle to lose a wheel.

SUMMARY

- Wheels and tires must be properly maintained and serviced.
- Improperly mounted, matched, aligned, or inflated tires can create a dangerous on-road situation.
- Wheel bearings, lubricant seals, and other components in the wheel hub that keep the wheel and tire turning smoothly on the wheel spindle or axle also require regular maintenance and service.
- The rim supports the tire. Wide base wheels can also be referred to as high flotation, super single, wide body, duplex, or jumbo wheels.
- One wide-base wheel and tire replaces traditional dual wheels and tires.
- Compared to steel dual wheels and tires, aluminum wide-base wheels and tires are significantly lighter in weight.
- There are two basic types of tire construction used in heavy-duty applications: bias ply and radial. Radial and bias tires differ in their tread profile, surface contact, and handling characteristics.
- Dual configurations must never use mismatched tires.
- The tire body and belt material can be made of rayon, nylon, polyester, fiberglass, steel, or the newest synthetics—aramid or Kevlar®.
- All tires (new and retread) sold in the United States must have a DOT number cured into the lower sidewall on one side of the tire.
- Low-profile radial truck tires enhance radial design to produce even lower costs per mile.
- Proper tire care and maintenance is second only to fuel mileage in overall cost per mile of truck operation.
- Improper loading will cause the tires on one side of the truck or trailer to carry a greater load than those on the other side.
- Excessive heat produced by running a vehicle at higher-than-rated speeds will shorten tire life.

- Wheel balance is the equal distribution of weight in a wheel with the tire mounted.
- Improper wheel bearing adjustment can result in looseness in the bearings or steering problems.
- Technicians should learn and always use, the TMC method of adjusting bearings.

REVIEW QUESTIONS

1. Which of the following wheel types uses a separate rim that is fastened to the wheel using rim clamps and bolts?
   a. disc wheels
   b. spoke wheels
   c. both a and b
   d. neither a nor b

2. Matching dual tires involves
   a. mounting tires of the same size and tread design at a dual wheel position
   b. mounting worn tires with new tires in a wheel position to achieve a good average tread
   c. using tires of the same construction (bias or radial) on the same axle
   d. a and c only

3. Single piece rims are used in combination with
   a. tubeless tires
   b. tube type tires
   c. both a and b
   d. neither a nor b

4. Multipiece rims using side and lock rings are used in combination with
   a. tubeless tires
   b. tube-type tires
   c. both a and b
   d. neither a nor b

5. The only sure method of properly matching multipiece rim components is by
   a. measuring the component size
   b. locating the component part numbers and checking for matching on a current parts matching chart
   c. using components from a single manufacturer
   d. all of the above

6. When a tire is being partially inflated to round out the tube or seat components, do not exceed
   a. 3 psi
   b. 10 psi
   c. 20 psi
   d. 30 psi

7. On spoke wheels, wheel runout should not exceed
   a. $1/16$ inch
   b. $1/8$ inch
   c. $3/16$ inch
   d. $1/4$ inch

8. On a spoke wheel, the brake drum is
   a. mounted on the inboard side of the wheel/hub; removing it requires removing the hub nut
   b. mounted on the outboard side of the wheel/hub; removing it does not require removing the hub nut
   c. both a and b
   d. neither a nor b

9. Which type of wheel has the hub as part of the wheel itself?
   a. disc wheels
   b. spoke wheels
   c. both a and b
   d. neither a nor b

10. Which type of hub requires heating in an oven to remove the bearing cups?
    a. ferrous
    b. aluminum
    c. both a and b
    d. neither a nor b

11. Oil-lubricated drive axle wheel bearings receive oil from
    a. a separate oil reservoir mounted to the frame
    b. a cavity built into the hub cap
    c. from the ring gear and differential housing of the rear axle
    d. all of the above

12. Which type of oil seal is installed by hand and does not require any special lubricant or sealer to outer or inner surfaces?
    a. wiper ring seals
    b. unitized seals
    c. guardian seals
    d. barrier seals

13. Which of the following methods are used to lock the bearing setting on the axle or spindle?
    a. jam nuts or locknuts
    b. castellated nuts with cotter pins
    c. split forging nuts
    d. all of the above

14. Which of the following lubricants can be used safely to lubricate the bead seat area when mounting tires to rims?

    a. water-based lubricants
    b. silicon-based lubricants
    c. petroleum-based lubricants
    d. none of the above

15. Technician A says that wide base wheels increase fuel efficiency by decreasing weight and rolling resistance. Technician B says that, in hub-piloted systems, a loose inner nut can easily go undetected, eventually pounding out the nut's ball seat. Which is correct?
    a. Technician A
    b. Technician B
    c. both A and B
    d. neither A nor B

16. Which of the following is an advantage of low-profile tires?
    a. better traction
    b. increased tread life
    c. greater vehicle stability
    d. all of the above

17. Technician A always inflates severely underinflated tires in a safety cage. Technician B uses a clip-on air chuck with a remote in-line valve and gauge. Who is correct?
    a. Technician A
    b. Technician B
    c. both A and B
    d. neither A nor B

18. Technician A says that adjustment is in order if wheel runout exceeds $1/2$ inch. Technician B loosens the rim clamping nuts on the chalk marked side and tightens the nuts that are 180 degrees opposite of the chalk. Who is correct?
    a. Technician A
    b. Technician B
    c. both A and B
    d. neither A nor B

19. Which of the following is a method used to measure dual sizes?
    a. square method
    b. tape measure
    c. tire meter
    d. all of the above

20. Which of the following causes a wheel to bounce?
    a. a wheel that is out of balance statically
    b. a wheel that is out of balance dynamically
    c. both a and b
    d. neither a nor b

21. Which type of seal has the grit guard feature of a lip seal with the advantage of a direct contact between the seal and wiper ring?
    a. unitized seal
    b. barrier seal
    c. guardian oil seal
    d. none of the above

22. When using the TMC method of adjusting wheel bearings, end play must be within what range?
    a. 0.001–0.005 inch
    b. 0.001–0.010 inch
    c. 0.005–0.010 inch
    d. 0.005–0.050 inch

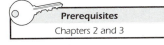

Prerequisites
Chapters 2 and 3

Truck Brake Systems

Objectives

After reading this chapter, you should be able to
- Identify the components of a truck air brake system.
- Explain the operation of a dual-circuit air brake system.
- Understand what is meant by pneumatic and torque imbalance.
- Discuss the effects of the Federal Motor Vehicle Safety Standard No. 121 (FMVSS No. 121) on present-day air brake systems.
- Identify the major components of an air compressor.
- Describe the operation of desiccant and aftercooler air dryers.
- Name the various pneumatic valves and controls used in an air brake system.
- Explain the operation of an air brake chamber.
- Describe the operation of S-cam and wedge-actuated drum brakes.
- Describe the operating principles of manual and automatic slack adjusters.
- List the components and describe the operating principles of an air disc brake system.
- Describe the major components and operation of parking and emergency braking systems.
- Define brake balance.

Key Terms List

aftercooler
air dryer
bobtail proportioning valve
bobtail
brake chambers
brake pads
brake shoes
check valve
compressor
dash control valves
disc brakes
drum brakes
dual-circuit application valve

Federal Motor Vehicle Safety Standard No. 121
   (FMVSS No. 121)
foot valve
gladhands
pop-off valve
purge tank
relay valves
safety pressure relief valve
supply tank
system pressure
tractor protection valve
trailer application valve
treadle valve
wet tank

The brakes on a truck are probably its most vital safety component. It takes a truck with a 450-horsepower engine and a 80,000-lb. load around 90 sec-onds to accelerate to 50 mph. That truck must be brought safely to a stop in less than 5 seconds. If brakes were rated in horsepower, trucks would be required to have brakes rated between 10 and 20 times the horsepower rating of the engine. Brakes have evolved since the earliest wagon brake. A wagon brake consisted of a hand-actuated lever with a friction block of leather or wood on one end. When the wagon brake lever was pulled, the friction block was forced against the rotating outer rim of the wheel. This type of brake became obsolete when the pneumatic tire was introduced on vehicles. Today, truck brakes are either hydraulically actuated or air-actuated. Hydraulic brake systems tend to be used on light-duty trucks that do not usually have to be coupled to trailers. Medium- and heavy-duty trucks use air-actuated brakes almost exclusively.

Hydraulic brake systems use liquid confined in a circuit to transfer mechanical force from the driver's foot to the **brake shoes**. Depressing the brake pedal creates a mechanical force that is transmitted through a pushrod to a piston in the master cylinder. This piston forces the brake fluid through the brake lines to slave pistons in calipers or cylinders located at each wheel assembly. The hydraulic pressure actuating these pistons forces **brake pads** or shoes against a disc or drum that rotates with the wheel assembly. Brake pads and shoes are designed with friction facings. When these friction facings are forced into a rotating disc or drum, the energy of motion (kinetic energy) is converted into heat energy. The friction generated by the contact of brake friction faces with the rotating disc/drum retards or slows its motion . . . and can eventually stop it. When hydraulic brake systems are found on light-duty and a small number of medium-duty trucks, they can be either four-wheel **disc brakes** (more common), or disc brakes on the front wheels and **drum brakes** on the rear wheels. The potential energy of a hydraulic brake system is the mechanical force applied by foot to the brake pedal. In other words, the greater the amount of foot pressure applied to the brake pedal, the greater the amount of retarding or braking effort delivered to the wheels by a hydraulic brake system. Even when hydraulic brakes are power-assisted, brake force at the wheel is directly proportional to the mechanical application pressure at the **foot valve**. A detailed account of hydraulic brakes is provided in Chapter 25.

Air brake operation is similar in some ways to hydraulic brake operation except that compressed air is used to actuate the brakes. The potential energy of an air brake system is in the compressed air itself. The retarding effort delivered to the wheels has nothing to do with mechanical force applied to the brake pedal. The source of the potential energy of an air brake system is the vehicle engine that drives the air **compressor** that discharges to the vehicle air tanks. The brake pedal actuated by the driver meters (precisely measures) some of this com-

pressed air out to **brake chambers**. Brake chambers are used to convert air pressure into mechanical force to actuate foundation brakes. Foundation brakes consist of shoes and drums or rotors and calipers that convert the mechanical force applied to them into friction. In other words, braking effort is achieved in an air brake system in the same way as in a hydraulic brake system. Once again, the force that actuates the foundation brakes is converted to friction, which retards movement. The friction is converted to heat energy, which then must be dissipated (transferred) to the atmosphere.

Air brakes are used almost exclusively in heavy-duty trucks and trailers. They have some very definite advantages over hydraulic brakes in heavy duty highway vehicles, including the following.

- Air is limitless in supply. It must only be cleaned, compressed, stored, and distributed properly in an air brake system. Minor leaks do not result in brake failures.
- The air brake circuit can be expanded easily so trailers can be coupled and uncoupled from the tractor circuit by a person with no mechanical knowledge.
- Besides providing the energy required to brake the vehicle, compressed air is also used as the control medium. That is, it is used to signal the foundation brakes when and with how much force the brakes should be applied in any situation.
- Air brakes are effective even when leakage has reduced their capacity. This means that in the event of air leakage, the system can be designed with sufficient fail-safe devices to bring the vehicle safely to a standstill.

Legislation introduced in 1975 called **Federal Motor Vehicle Safety Standard No. 121 (FMVSS No. 121)** governs all air brake system requirements. It has been modified in small ways to keep up to date with technology, but from the beginning it required all highway vehicles using air brakes to use a dual-circuit application circuit. This means that almost every air brake-equipped vehicle on the highways today has dual-circuit brakes that are FMVSS No. 121-compliant. The dual circuits are named primary and secondary circuits. It is also now a requirement of FMVSS No. 121 that all highway tractors manufactured after March 1, 1997 and trailers manufactured after March 1, 1998 be equipped with ABS. ABS will be explained in Chapter 26. However, it is important to note here that ABS does not significantly alter the basic brake system that will be introduced in this chapter. **Figure 24–1** and **Figure 24–2** are schematics that show the layout of a very basic

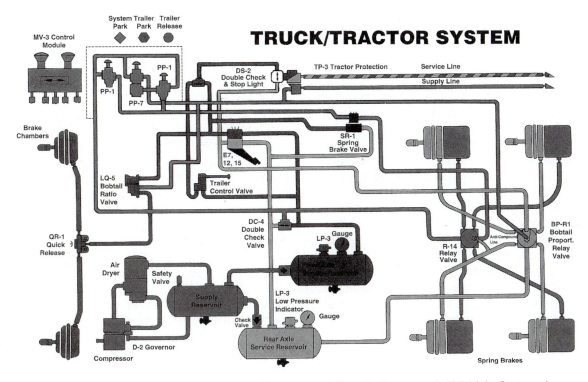

FIGURE 24–1   Truck/tractor air brake system. (Courtesy of Bendix Commercial Vehicle Systems)

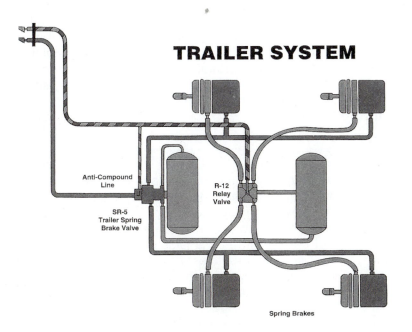

FIGURE 24–2   Trailer air brake system. (Courtesy of Bendix Commercial Vehicle Systems)

air brake system on a tractor and trailer. Neither system would meet today's air brake standards.

AIR BRAKES—A DRIVER'S POINT OF VIEW

If a driver climbs into a parked tractor/trailer combination with no pressure in its air tanks, when the engine is started the air pressure gauges indicating **system pressure** in the primary and secondary cir-

cuits will both read zero. When the ignition key is turned on, a low-pressure warning light and buzzer will alert the driver that the air pressure is low. After the engine is started, the compressor is driven and the air it compresses is discharged to a supply tank, which in turn feeds primary and secondary tanks. In humid and cold climates, an air dryer is usually located between the compressor and the **supply tank**. As pressure builds in the system, pressure gauges indicate the primary and secondary circuit pressure: At 60 psi, the low pressure warning light and buzzer will cease. Pressure will continue to build until governor cut-out pressure is achieved. Governor cut-out pressure in most truck applications will vary between 120 and 130 psi, but we will use 125 psi for this description. At governor cut-out, the compressor is "unloaded"; that is, it continues to rotate but no longer compresses air.

At the moment of governor cut-out, governed air pressure is achieved. Both the primary and secondary dash pressure gauges should indicate 125 psi in the air tanks. If the driver then depresses the red, octagonal Trailer Supply button (older tractors may require that the yellow, diamond-shaped System Park button be actuated before the Trailer Supply button), air will now be charged to the trailer air tanks. When the trailer air system is charged, the vehicle can be moved. The source of the air available to the **dash control valves** is the primary tank, but this is backed up by the secondary circuit and two-way **check valve**.

To move the vehicle, both the System Park and Trailer Supply dash buttons must be depressed. This releases the parking brakes by charging air to the spring brake hold-off circuit on both the tractor and trailer. Any time an air brake-equipped vehicle with properly functioning brakes is moved, air must be in the hold-off circuit, putting the parking brakes into release mode.

While the vehicle is being moved, if the driver applies the foot pedal, a service application of the brakes occurs. When a service application is made, air is delivered from the primary and secondary reservoirs to act on the service chambers. In the commonly used spring brake chambers, which can be used on all but the steering axle brakes, the service chamber is the one closest to the slack adjuster. The air pressure that acts on the service chambers is modulated and varies according to the amount of braking required. Under light-duty braking, there will be low pressure in the service chambers. Under panic braking, system pressure might be delivered to the service chambers. During service braking, air is present in both the service and hold-off chambers in a spring brake unit. Some trucks have a service application gauge. This tells the driver how much pressure is being delivered to the service chambers during service braking. The majority of routine service brake applications involve service application pressures of 20 psi or less.

As air is used by the system, pressure will drop. This can be observed by watching the gauges that indicate pressure in the primary and secondary circuits. When the pressure drops to between 20 and 25 psi below governor cut-out point, the air pressure in the system has to be returned to its governed

pressure. At this value, known as cut-in pressure, an air governor signals the compressor to return to its effective pumping cycle. The compressor will remain in effective cycle until system pressure has returned to 125 psi, at which point the governor will signal the compressor to cut-out mode.

Service brakes perform the braking requirements of a moving vehicle. The braking requirements of a moving vehicle have to be variable. The foot valve is the means of precisely varying service application pressures. Most air brake-equipped vehicles on the road today are required to have a means of mechanically parking. The means used to park air brake-equipped vehicles is usually spring brake assemblies. A driver will bring a vehicle to a standstill using the service brakes; that is, using the foot valve. When the vehicle must be held parked, the driver will exhaust the hold-off air by pulling out the dash System Park valve. This discharges the hold-off air in both the tractor and the trailer circuits and means that the full force of the springs in the spring brake assemblies acts on the slack adjusters in both the tractor and the trailer. The term "dynamiting" is often used by truck drivers to describe exhausting the hold-off air: The result is the mechanical locking up of all the spring brakes on the rig.

This chapter will attempt to explain why and how everything described above happens in an air brake system. The approach is to explain everything by system first, and afterwards to take a look at exactly how the different components and valves in the system operate.

AIR BRAKE SUBSYSTEMS

Truck air brake systems can be divided into a number of subsystems. Some of the components shown in **Figure 24–3** play roles in two or more subsystems. Air brake standards in North America are outlined and updated in federal legislation FMVSS No. 121. This legislation requires a truck air brake system to have separate service and parking brake systems. It also requires all trucks equipped with air brakes to have a dual service application circuit. The term service brakes describes the running brakes on the vehicle applied by the foot valve. The trucking industry is in a process of changing over to electronically controlled antilock braking systems (ABS). This conversion to ABS actually began well ahead of legislation requiring ABS. A complete air brake system on a tractor/trailer combination is made up of the following subsystems:

- **The air supply circuit.** The air supply circuit is responsible for charging the air tanks with clean, dry, filtered air. It consists of an air com-

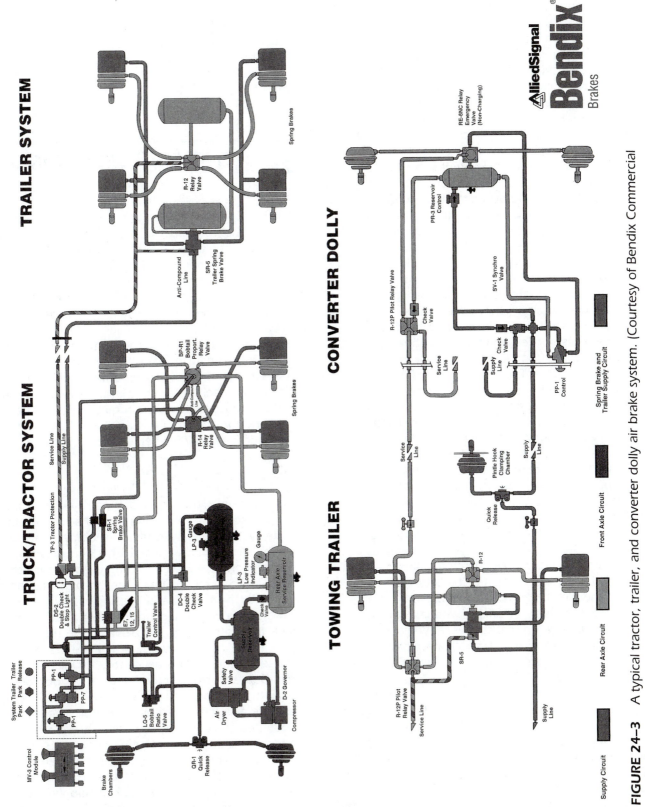

FIGURE 24–3 A typical tractor, trailer, and converter dolly air brake system. (Courtesy of Bendix Commercial Vehicle Systems)

pressor, governor, air dryer/**aftercooler**, supply tank, low-pressure switch, and system safety (pop-off) valve.

- **Primary circuit.** In a tractor/trailer combination, the primary circuit is usually responsible for actuating the tractor rear wheel brakes and actuating the trailer service brakes when applied from the **treadle valve**. It consists of a primary tank, dual circuit application (treadle) valve, pressure gauge, quick release or rear axle **relay valves**, and ABS solenoids. This circuit is sometimes called the rear axle circuit.

- **Secondary circuit.** In a tractor/trailer combination, the secondary circuit is usually responsible for actuating the front axle service brakes and the trailer brakes when actuated by the trailer hand control valve on the tractor. It consists of a secondary tank, dual-circuit application (treadle) valve, pressure gauge, quick release or relay valves, and ABS solenoid valves. This circuit is sometimes called the front axle circuit.

- **Dash control valves and the parking/emergency circuit.** The control circuit consists of 2 or 3 dash-mounted valves that manage the parking and emergency circuits. The parking/emergency circuit consists of spring brake assemblies, **tractor protection valve**, and the interconnecting plumbing.

- **Trailer circuit.** The trailer service and parking/emergency brakes are controlled by the tractor. The tractor and trailer air systems are connected by a pair of hoses and couplers known as **gladhands**. The trailer brake circuit consists of gladhands, air hoses, air tanks, pressure protection valves, and relay valves.

- **Foundation brakes.** The function of foundation brakes is to convert the air pressure supplied by the application circuits (primary and secondary) into the mechanical force required to stop a vehicle. The parking and emergency brakes also use the foundation brakes to effect mechanical application of the brakes. Foundation brakes consist of brake chambers, slack adjusters, S-cams or wedges, shoes, pads, linings, drums, and discs. Vehicles equipped with ABS will additionally have wheel speed sensors mounted on the wheel assemblies.

Note the location of the above dual-circuit air brake system components in **Figure 24–3**. The study of air brake systems will be sequenced in this chapter by first introducing and explaining the operation of the circuits listed above, beginning with the air supply circuit and working through to the foundation brakes. Next, a more detailed description of the criti-

cal air brake components and valves is provided. The student who requires more detail on how a particular component functions while studying one of the systems can refer ahead. For instance, the foot valve plays a role in all service braking whether from the primary or secondary circuits, so to understand all of its functions, the sections on the primary circuit, secondary circuit, and the description of the valve itself should be referred to. Chapter 27 will introduce what is required to maintain, adjust, and repair air brake systems.

24.1 THE AIR SUPPLY CIRCUIT

The air supply circuit on a truck is responsible for charging the air tanks with clean, moisture-free air to be used by the brake system and supply the remaining air requirements of the vehicle. The potential energy of an air brake system is compressed air. The air is compressed by the supply circuit and then distributed and stored in reservoirs (tanks) at pressures typically around 120 psi (827 kPa). Valves controlled by the driver can then deliver compressed air to the brake system components. Trucks today are required to have their parking/emergency brakes fully applied by mechanical force in the event of no system air pressure. So air is required first, to release the parking/emergency brake circuit and then to effect service braking when needed.

AIR COMPRESSORS

The air compressor is driven by the engine. Compressors used on current trucks may be either single or multiple cylinder. The compressor drive may be direct or by means of a controlled clutch. The components of an air compressor are much like those of an engine. The compressor crankshaft is supported by main bearings, which are lubricated with engine oil. Connecting rods are attached to the crank throws and support pistons with rings to seal them in the cylinder bore. When the crankshaft is rotated by the engine accessory drive, the pistons reciprocate in bores machined in the compressor housing. The connecting rod bearings and cylinder walls are also lubricated by engine oil. The engine oil is pumped directly into the compressor crankshaft. The spill and throw-off oil from the bearings drains to a sump. The oil level in the sump is determined by the location of the oil return line. Because the oil pumped into the compressor is allowed to drain back from its sump to the engine crankcase, the

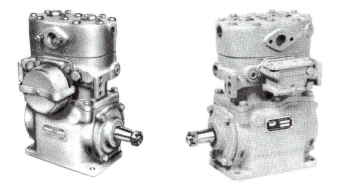

FIGURE 24-4 *Typical air compressors. (Courtesy of Bendix Commercial Vehicle Systems)*

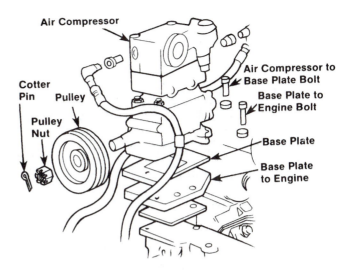

FIGURE 24-6 *Belt-driven air compressor. (Courtesy of Bendix Commercial Vehicle Systems)*

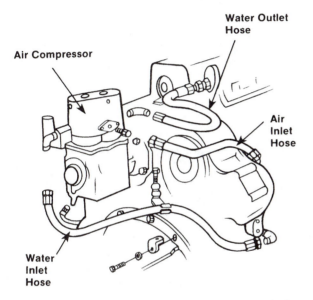

FIGURE 24-5 *Gear-driven, flange-mounted air compressor. (Courtesy of Bendix Commercial Vehicle Systems)*

lubricating oil in the compressor is constantly changing. The cylinder head in the compressor contains valves that are designed to open in one direction only. An inlet valve is designed to open when low pressure is created in the cylinder. A discharge valve is designed to open only when cylinder pressure is greater than the pressure in the discharge line. **Figure 24-4** shows some typical compressors while **Figure 24-5** and **Figure 24-6** show the means of driving them.

Compressor Operation

When the engine is run, it rotates the compressor crankshaft, causing the pistons to reciprocate in the cylinder bores. The internal components of the compressor are shown in **Figure 24-7**. On the downward stroke of the piston, low pressure is created in the cylinder. This low pressure unseats the inlet valve, which allows filtered air to be pulled into the cylinder. When the piston reaches the bottom of its stroke, it reverses. As it is driven upward it pressurizes the air charge. This closes the inlet valve. When the pressure in the cylinder exceeds the pressure on the outlet side of the discharge valve, it opens. When the discharge valve is open, the compressed air in the cylinder is unloaded into the discharge line towards the system supply tank. This operating cycle is shown in **Figure 24-8**.

Compressor Cycles

Most compressors are direct-driven by the vehicle engine. While the compressor is continuously rotated, it may be in either a loaded cycle or unloaded cycle. These cycles are controlled by an air governor. When the compressor is in loaded cycle, it is performing the work of compressing air and delivering it to the discharge circuit. Its operation is therefore exactly that described in the previous paragraph. The governor senses pressure in the supply tank by means of a small gauge signal line. When the governor signals the compressor into unloaded cycle, an air signal is delivered from the governor unloader port to unloader pistons located in the compressor cylinder head. The unloader pistons when actuated hold the inlet valves open. This results in the piston pulling in a charge of air on its downstroke and immediately blowing it back out of the inlet valve on the upstroke. In other words, no compression takes place. The unloaded cycle means that the compressor is being rotated but not performing the work of compressing air. The compressor goes into loaded cycle at governor cut-in.

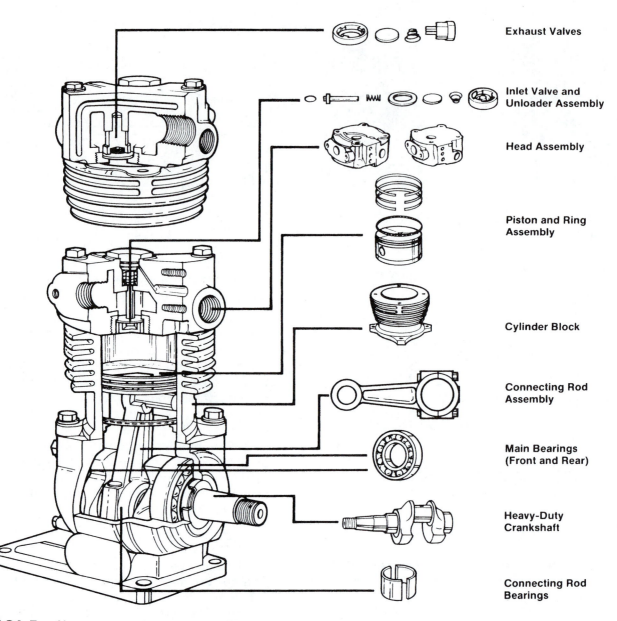

Exhaust Valves

Inlet Valve and Unloader Assembly

Head Assembly

Piston and Ring Assembly

Cylinder Block

Connecting Rod Assembly

Main Bearings (Front and Rear)

Heavy-Duty Crankshaft

Connecting Rod Bearings

FIGURE 24–7 *Air compressor components. (Courtesy of Haldex Brake Systems)*

Governor cut-in occurs when the system pressure has dropped below a specified value and requires recharging. The compressor goes into unloaded cycle at governor cut-out. This is the specified maximum system pressure. **Figure 24–9** shows the unloader mechanism from a typical air compressor.

GOVERNORS

The function of the air governor is to monitor system air pressure and manage the loaded and unloaded cycles of the compressor. The governor assembly can be mounted directly to the air compressor (not generally recommended due to higher temperatures close to the compressor) or remotely, often on the firewall of the engine compartment. **Figure 24–10**

shows a typical mounting location of an air governor.

Air pressure in a truck air system must be carefully managed at what is called system pressure. System pressure is a range of pressures between governed pressure (maximum system pressure) and compressor cut-in pressure (minimum system pressure). Governed pressure is usually set between 120 and 130 psi (827 and 896 kPa), but FMVSS No. 121 specifies that any values between 115 and 135 psi are acceptable. The governor measures system pressure by having a signal line from the supply tank act on the lower sectional area of the governor piston. In opposition to the force acting on this piston is the governor spring. Governor spring tension is adjusted to define the governor cut-out pressure value. When system pressure achieves the specified

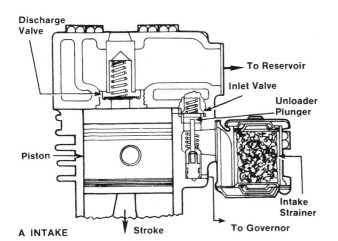

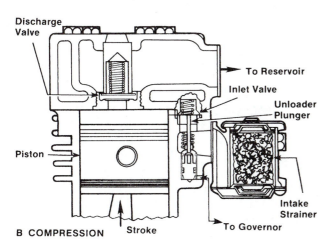

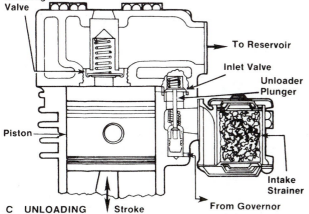

FIGURE 24–8 Air compressor operation. (Courtesy of Haldex Brake Systems)

maximum, an air signal is sent from the governor unloader port to the unloader pistons in the compressor cylinder head. When the unloader pistons are actuated, the inlet valves are held open and the compressor is no longer capable of compressing air. As system air is used by the vehicle air circuits, system pressure drops. When system pressure drops to between 20 and 25 psi below governed pressure, the

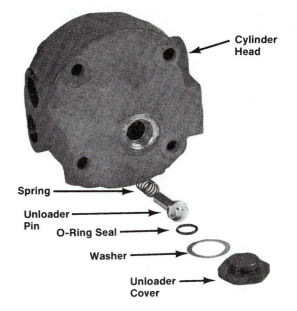

FIGURE 24–9 Unloader mechanism. (Courtesy of Haldex Brake Systems)

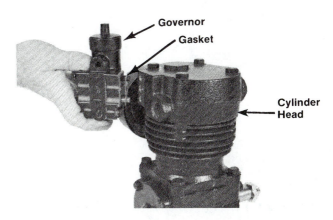

FIGURE 24–10 Governor assembly. (Courtesy of Haldex Brake Systems)

governor spring forces the governor piston to its seated position. This triggers an exhaust valve that dumps the air signal to the compressor unloader pistons, allowing the inlet valves to seat once again. Governor cut-in occurs at the moment the unloader signal to the compressor is exhausted by the governor and the compressor returns to its effective cycle.

System Pressure

The working air pressure on a truck is referred to as system pressure. It should be noted that system pressure is not a specific value on any system; rather, it is any value between the governor cut-in and governor cut-out values. For instance, on an air brake system with governor cut-out set at 125 psi, system pressure would indicate any air pressure value between 125 psi and 100 psi. This definition of system pressure is important when it comes to testing system performance with diagnostic pressure gauges. Most North American-built trucks have governor pressure cut-out set between 120 and 130 psi. A recent ATA TMC (American Trucking Association, The Maintenance Council) recommended practice paper suggested that governor cut-out be set at 125 psi. **Figure 24–11** shows a cutaway view of a typical governor, identifying the connection ports.

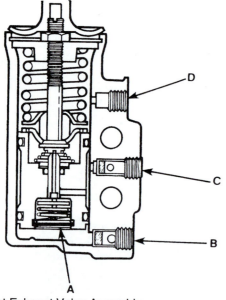

A Inlet Exhaust Valve Assembly
B Exhaust Port
C Unloader Port
D Reservoir Port

FIGURE 24–11 Cutaway view of a governor showing the connection ports. (Courtesy of Bendix Commercial Vehicle Systems)

Shop Talk

FMVSS No. 121 requires that governor cut-out be set within the range of 115 psi to 135 psi. The ATA TMC recommends that governor cut-out be set at 125 psi.

- *It is a legal requirement that governor cut-in occur no more than 25 psi less than governor cut-out pressure.*
- *If the difference between governor cut-out and cut-in is less than 20 psi, the result is too-frequent cycling of the compressor loaded and unloaded cycles.*

AIR DRYER

An **air dryer** is an option on an air brake system. Most trucks on our roads today are equipped with an air dryer, especially those operating in humid or cold conditions. **Figure 24–12** shows the location of an air dryer in the supply circuit. The air compressed by the compressor is atmospheric air, which always contains some percentage of vaporized moisture. This moisture can harm air system valves if it is not removed. The air dryer is located between the compressor discharge and the supply tank. When air is compressed, it is heated to temperatures as high as 300°F (150°C). Air dryers use two methods of separating moisture from the compressed air.

The air dryer functions first as a heat exchanger, that is, an aftercooler, to reduce the air temperature, dropping it to a value cool enough to condense the vaporized moisture it receives. Actually, first-stage

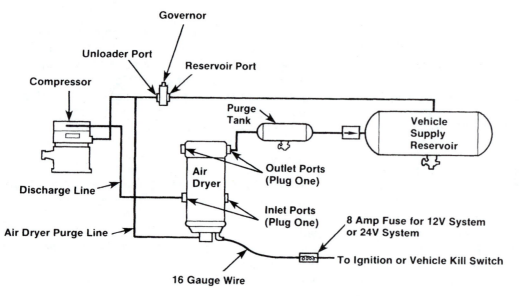

FIGURE 24–12 Typical air dryer installation. (Courtesy of Bendix Commercial Vehicle Systems)

cooling of the hot compressed air occurs in the discharge line that connects the compressor with the dryer. In some older applications, copper lines were used for this purpose. Current steel braided discharge lines still play a role in cooling compressed air. Compressors often use external fins to increase cooling efficiency. When compressed air enters the dryer, it first passes through an oil filter that removes trace quantities of oil that have been discharged by the compressor. Next, the air is passed through a desiccant pack designed to absorb water. Some air dryers use only one of these principles to remove moisture from air, while others use both. Water and oil separated and condensed in the dryer drain to a sump. Dry air exits the dryer and is routed to the supply tank. **Figure 24–13** shows a cutaway view of a typical air dryer.

The water and oil that drain to the sump must be discharged or purged from the dryer. A signal from the unloader port of the governor (the same signal that "unloads" the compressor at governor cut-out) is used to trigger a purge valve in the air dryer sump. Any liquid in the sump is blown out during the dryer purge cycle. Some dryers are equipped with an electric heater to prevent winter freeze-up. A thermostat controls the heater. The dryer shown in **Figure 24–12** is equipped with a heater. It should be noted that air dryers are designed to separate the normal trace quantities of oil discharged by a compressor. A compressor that is pumping oil will rapidly destroy the dryer desiccant packs. Desiccant packs require replacement every so often, largely dependent on how much contaminant they have been exposed to. **Figure 24–14** shows an external view of an air dryer, and **Figure 24–15** shows the operating cycles.

Shop Talk

Air dryers must be fitted with a safety valve. Because the system safety valve is located in the supply tank, if the hose from the dryer to the supply tank were to get plugged (with ice or contaminants) or kinked, the governor would not cut out the compressor effective cycle and the resulting high pressure could explode the air dryer.

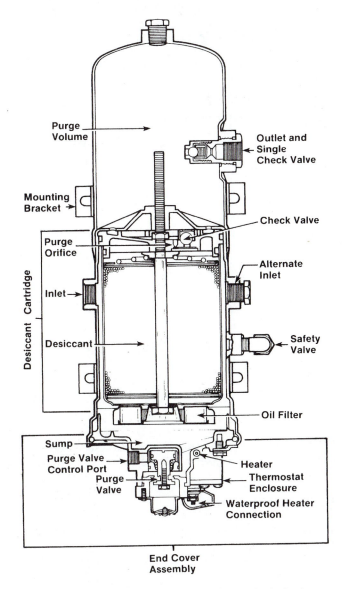

FIGURE 24–13 Cutaway view of a typical air dryer. (Courtesy of Bendix Commercial Vehicle Systems)

FIGURE 24–14 External view of an aftercooler type air dryer. (Courtesy of CR Industries)

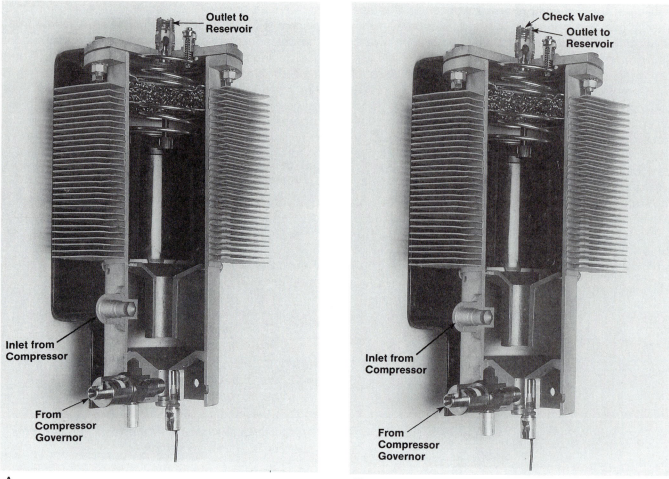

A   **B**

FIGURE 24–15   *Aftercooler operation: (A) airflow pressure cycle, and (B) airflow unload cycle. (Courtesy of CR Industries)*

ALCOHOL EVAPORATORS

Not all of the moisture in the air can be removed. Alcohol evaporators are used in some systems to help prevent freeze-ups. Alcohol vapor is introduced into the system by the alcohol evaporator: This forms a solution with water and lowers its freezing point. Alcohol evaporators should be located downstream from the air dryer. It is important to use only alcohols intended for use in air brake systems. Brake valves contain moving parts and most are lubricated on assembly to last through the component's life. Pure alcohol (methyl hydrate or methyl alcohol) can dry out lubricants in brake valves, causing premature failure. Alcohols sold specifically for use in brake systems contain brake valve lubricant. **Figure 24–16** shows a typical alcohol evaporator unit.

> **WARNING!**   Alcohol evaporators MUST be located downstream from the air dryer in the supply circuit. Alcohol will turn air dryer desiccant into mush if pumped through the system.

FIGURE 24–16   *Alcohol evaporator. (Courtesy of Bendix Commercial Vehicle Systems)*

SUPPLY TANK

The supply tank is the first tank to receive air from the compressor. Therefore, in a typical system, it would be located downstream from the compressor. The supply tank is sometimes referred to as the **wet tank** because in a system with no air dryer, this would be where the air cools and condensing moisture collects. If a compressor discharged oil, this would also collect in the wet tank. Some systems also include a **purge tank** located upstream from the supply tank: This acts to even further reduce the chance of pumping contaminants through the system. The function of the supply tank is to supply all the other tanks in the truck chassis air system. In the brake system, this would include the primary and secondary tanks. **Figure 24–17** shows a typical air tank.

In common with all other tanks on board a truck chassis, the internal walls of the supply tank are coated with an anti-corrosive liner. The anti-corrosive liner protects the steel walls of the air tank from corrosion by moisture. The liner itself, however, is susceptible to the carbon coking that can be caused by a compressor pumping oil. The maximum working pressure of the tank is usually 150 psi: To ensure that this pressure is not exceeded, a **safety pressure relief valve** is fitted. This is often known as a **pop-off valve.** Should something such as a carboned signal hose cause the governor to malfunction and not signal compressor cut–out, the safety valve would trip at 150 psi to prevent damage to the chassis air circuits. **Figure 24–18** shows a typical safety pressure relief valve. Supply tanks are required to have a drain valve. This may be a simple drain cock that must be manually opened and closed, a cable operated manual drain valve, or an automatic moisture ejector. A moisture ejector responds to changes in tank pressure and expels any moisture that has collected in the tank. A moisture ejector is required to have a means of manually opening its dump valve. Some moisture ejectors are equipped with thermostatically actuated heaters that prevent winter freeze-up. **Figure 24–19** shows a manual drain cock and automatic moisture ejector.

FIGURE 24–17   A typical air tank. (Courtesy of Bendix Commercial Vehicle Systems)

FIGURE 24–18   Safety pressure relief valve. (Courtesy of Bendix Commercial Vehicle Systems)

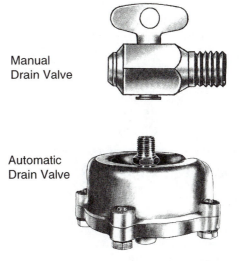

FIGURE 24–19   Manual and automatic reservoir drain valves. (Courtesy of Bendix Commercial Vehicle Systems)

👓 **Shop Talk** _____

Air tanks are pressure vessels. All vehicle air tanks are required to be hydrostatically tested after assembly. They should never be repair-welded. Even if the weld was sound, the heat causes interior anti-corrosive liner to melt and peel away from the tank wall. Replace failed air tanks.

LOW PRESSURE INDICATOR

In an air brake system, if the system air pressure drops too low, the brakes will cease to function properly. A low-pressure indicator is a pressure-sensing electrical switch that closes at any time the pressure drops below its preset value. The low-pressure indicator should be specified to close at 50 percent governor cut-out pressure. In a typical system, that would mean around 60 psi. When the switch closes, a light and a buzzer are actuated.

Air Tanks

FMVSS No. 121 requires that the total volume of compressed air stored on board the tractor dedicated to the brake system be a minimum of 12 times the total brake chamber volume. In dual-circuit brake systems, at least 3 air tanks are required. Most have more. It should be pointed out that what at first sight appears to be a single air tank may in fact be more: Some air tanks are divided by internal bulkheads. One common current design uses a single cylindrical tank divided into supply and secondary chambers by a bulkhead. The supply tank charges the secondary chamber by means of an external line, one-way check valve, and a separate primary tank, also pressure-protected by a one-way check valve. To determine the number of separate chambers there are in an air tank, count the drain cocks and spitter valves. Every air tank is required to have its own drain cock or spitter valve.

24.2 PRIMARY CIRCUIT

On a typical tractor/trailer combination, the primary service circuit is responsible for actuating the brakes over the tractor drive-axle tandem and foot-actuated applications of the trailer service brakes. This is why the primary circuit is sometimes referred to as the rear brake circuit. The secondary circuit is responsible for actuating the tractor front axle brakes and hand valve-actuated applications of the trailer service brakes. This explains why it is sometimes known as the front brake system.

PRIMARY TANK

The primary tank receives air directly from the supply tank. If the primary and secondary tanks are supplied by the supply tank in series, the air must pass through the primary tank first. The supply air enters the primary tank by means of a one-way check valve. This means that compressed air can only travel one way from the supply tank. For instance, if supply tank pressure were to drop below that of the primary tank, air could not drain back. The one-way check valve helps isolate the primary circuit from the remainder of the chassis air system. Should a severe leak take place elsewhere in the system, the primary circuit air supply would be pressure-protected. Like all air tanks, the primary tank must be equipped with a drain cock or automatic moisture ejector.

The primary tank has two main functions. The first is to make a supply of air at system pressure available to the relay valves for direct application of the service brakes. The second is to make a supply of air at system pressure available to the foot valve. The foot valve and primary circuit relay valves are plumbed directly into the primary tank. Primary circuit pressure must be monitored by a dash gauge. A signal line supplies this gauge either directly from the tank or indirectly through the foot valve manifold. The gauge used in a specific truck chassis may be a single unit with separate needles to indicate primary and secondary circuit pressure, or it may be two separate gauges, each identified as primary and sec-

ondary. The primary circuit is generally green so in a single-unit, two-needle gauge, the green needle would indicate the primary circuit pressure.

The primary tank also supplies air to the dash control valve assembly, the push-pull valves or modular dash control valve assembly. This supply is ensured by a two-way check valve, so that in the event of an upstream primary circuit failure, air would be routed to the dash control valves from the secondary tank by shuttling a two-way check valve. In older systems that used a two-way check valve supplied from the primary and secondary tanks, if a line ruptured downstream from the two-way check valve, the air in both circuits would be lost. In the current modular dash control valve assembly, only one circuit is capable of discharging.

Role of the Foot Valve in the Primary Circuit

A detailed description of all foot valve functions is provided later in this chapter and shown in **Figure 24–35**, **Figure 24–36**, and **Figure 24–37**. The primary portion of the foot valve is actuated mechanically. This means that it is directly connected to the treadle. A supply of air from the primary reservoir is made available to the primary supply port at system pressure. Travel of the primary piston will determine how much air is metered out to actuate whatever components are in the primary circuit. Typically, these will be the tractor's tandem drive axles and the trailer's brake system. Also typically, the air metered out by the foot valve is used as a signal to actuate relay valves on both the tractor and the trailer. When the foot valve is actuated, primary circuit air does the following:

- Actuates or signals the brake valves plumbed into the primary circuit
- Actuates the relay piston in the foot valve, that is, actuates the secondary circuit
- Reacts against the applied foot pressure to provide "brakeA feel"

Brake Valves in the Primary Circuit

In most cases, relay valves will be used to actuate the tractor brakes on the primary circuit. These will usually be the rear tractor brakes. The relay valve signal is delivered from the foot valve using a relatively small-gauge signal line. The small-gauge line is used to reduce lag. Signal lag tends to be greatest when the volume of compressed air is greatest. The relay valve is also connected directly to a closer supply of compressed air either by a large gauge air hose or sometimes directly coupled to the tank itself. The relay valve is actuated by the signal pressure

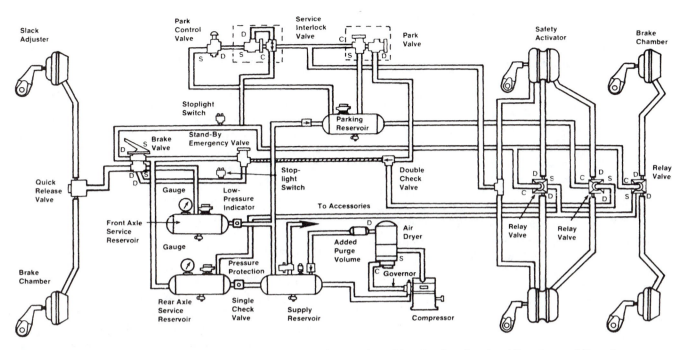

FIGURE 24–20 *Schematic of an older tractor's service and parking brake circuits. (Courtesy of Bendix Commercial Vehicle Systems)*

and delivers a much larger volume of the local supply of air to the service brake chambers it supplies. The signal pressure delivered to the relay valve and the application pressure it delivers to the brake chambers are designed to be identical in most brake systems.

Some systems, mostly much older systems, used quick-release valves in place of the relay valves to route air to the tractor rear service brakes. When a quick-release valve is used to actuate the rear tractor service brakes, the air that acts on the service diaphragms has to be routed through the foot valve. This results in extended lag time. Current FMVSS No. 121 legislation defines maximum actuation and release times for service brakes, so it is unusual to see quick-release valves used on tractor rear service brakes. **Figure 24–20** shows a schematic of an older air brake system using quick-release valves on the steering axle and relay valves on the tandem axles.

Trailer Primary Circuit Brake Signal

When the foot valve is depressed, a signal is routed from the primary portion of the valve to a two-way check valve downstream from the **trailer application valve**. Whenever the trailer application signal from the primary portion of the foot valve exceeds that on the opposite side of the two-way check valve, it shuttles the valve to route the signal to the tractor protection valve. The bias of the two-way check valve is designed to select the highest source pressure and route that to its outlet. Whenever the signal pressure from the foot valve is greater than the signal pressure from the trailer application valve, the trailer service application signal is provided by the primary circuit.

24.3 SECONDARY CIRCUIT

The secondary service circuit is usually responsible for actuating the brakes over the tractor steering axle and trailer application valve-sourced applications of the trailer service brakes. This is why the secondary circuit is sometimes referred to as the front brake circuit.

SECONDARY TANK

The secondary tank can receive air directly from the supply tank, or air may be routed to it indirectly through the primary tank. Air in the secondary tank is pressure-protected by a one-way check valve. This one-way check valve isolates the secondary circuit from the rest of the chassis air system, meaning that loss of compressed air elsewhere in the circuit will not affect the secondary circuit air supply. Like all air tanks, the secondary tank must be equipped with a drain cock or automatic moisture ejector.

The secondary tank has two main functions. The first is to make a supply of air at system pressure available to the relay valves and/or quick-release valves for direct application of the service brakes. The second is to make a supply of air at system pressure available to the relay portion of the foot valve. The foot valve and secondary circuit relay valves are plumbed directly into the secondary tank. Secondary circuit pressure is required to be monitored by a dash gauge. A signal line supplies this gauge either directly from the tank or indirectly through the foot valve manifold. If a single-unit gauge with separate needles to indicate primary and secondary circuit pressures is used, the secondary circuit needle indicator is red. When a two-gauge dash system is used, the secondary circuit gauge is identified as such.

The secondary tank is connected to a two-way check valve supplying the dash control valve assembly. However, under normal operation, the two-way check valve prioritizes the primary circuit feed to the dash valve assembly. This means that secondary circuit air would only be used to supply the dash valve assembly in the event of a primary circuit failure.

Role of the Foot Valve in the Secondary Circuit

Again, make sure you understand the description of the dual-application foot valve earlier in this chapter. When it is operating normally, the secondary or relay portion of the foot valve is actuated pneumatically; that is, by air pressure. Whatever air pressure value is metered by the primary portion of the foot valve to actuate the primary circuit is also metered to actuate the relay portion of the valve. The relay portion of the foot valve is designed to act exactly as a relay valve. This means that it will replicate the signal pressure value from the primary portion of the foot valve, sending out exactly that value using secondary tank source air to actuate the brakes on the secondary circuit.

In the event of total failure of the primary circuit air supply, the foot valve primary piston will travel downward until it mechanically contacts the relay piston. Because the relay piston meters air according to the distance it travels, it will function as usual, metering secondary source air to actuate the secondary circuit brakes. In the event of a total failure of the secondary circuit air, no air will be metered out to actuate the secondary circuit when the relay piston is displaced. Remember, whenever a failure occurs in either the primary or secondary circuits, the driver is alerted to the condition audibly and visually and is required to bring the unit to a standstill and wait for mechanical assistance.

Brake Valves in the Secondary Circuit

Relay valves or quick-release valves may be used to actuate the tractor brakes on the secondary circuit. In most cases, the secondary circuit is responsible for the front service brakes. In many older systems, because the front brakes were located close to the foot valve, they were actuated directly by the relay portion of the foot valve. In other words, the air directed out of the relay portion of the foot valve was the actual air used to charge the front service brake chambers. This air was routed to the chambers by means of a quick-release valve. This provided a short application lag that actually helped ensure balanced application timing, because the air would be required to travel further to reach the rear brakes. Because a quick-release valve was used, release timing lag was exactly what it would be if a relay valve were used.

Most front axle braking is pneumatically proportioned in non-ABS systems whether quick-release or relay valves are used. How the application pressures are proportioned depends on how the truck is being operated. Although proportioning is theoretically unnecessary in ABS systems, in fact it is often still used, because of the redundancy requirement of FMVSS No. 121. The two common proportioning valves used to manage steer axle braking are the ratio valve and the **bobtail proportioning valve**. Both ratio and bobtail proportioning valves are described in some detail later in this chapter. Manually switched front axle limiting valves have not been used for a number of years, but they may sometimes be seen on older vehicles. They operated simply by limiting the application pressure to the front axle brakes by 50 percent whenever the slippery road setting was selected.

When air directed to the front axle service brakes passes through an automatic ratio valve, it is proportioned to limit application pressure under conditions of less severe braking but to permit full front wheel braking under severe braking. Research into the braking dynamics of bobtail tractors and unloaded and loaded tractor/trailer combinations on both dry and slippery pavement has shown that drivers maintain better control under panic and severe braking when full application pressures are delivered to the front axle brakes. For this reason, it has been mandatory since 1984 for all North American highway tractors to be equipped with front axle brakes.

Bobtail proportioning valves alter the brake application proportioning when a tractor is being bobtailed. The brakes over the dual drive axles on a typical tractor have much greater capacity than the front brakes because they are designed to handle the vehicle braking requirements when a loaded trailer is coupled to the tractor. When a tractor is bobtailed, the rear brakes become much too powerful for a vehicle that has most of its weight located close to the front axle. In this instance, it is desirable that the steer axle perform a much larger percentage of the vehicle braking. The braking dynamics of a bobtail tractor are similar to those of a car. When the bobtail tractor is braked, load transfer, suspension dive, and the low percentage of weight over the drive axles throw upward of 60 percent of the braking onto the steer axle. A **bobtail** condition is sensed when there is no air in the trailer air supply line. Bobtail proportioning applies full application air pressure to the front axle for all brake application pressures and proportions the application pressure to the rear axles. That is, it makes the braking of the rear axles less aggressive except under severe braking, when the rear axles also receive full application pressure.

Trailer Secondary Circuit Brake Signal

The trailer application signal from the steering column-mounted trailer application valve (also known as the broker brake, spike, or trolley valve) is sourced from the secondary circuit. A supply of secondary circuit air at system pressure is available at the trailer application valve. When the lever on the valve is actuated, air is metered to the trailer service application signal line. Downstream from this application valve is a two-way check valve. The two source supplies to the two-way check valve are the trailer application valve and the foot valve. The bias of the two-way check valve is designed to select the highest source pressure and route that to its outlet. Whenever the signal pressure from the trailer application valve is greater than the signal pressure from the foot valve, the trailer service application signal is provided by the secondary circuit.

24.4 DASH CONTROL AND THE PARKING/ EMERGENCY CIRCUIT

The parking and emergency circuits in a truck air brake system consist of dash-mounted hand control valves, spring brake chambers, and self-applying, fail-safe features. FMVSS No. 121 legislation requires that all highway vehicles equipped with air brakes have a means of mechanically applying the parking brakes. Also, because air brakes are vulnerable to loss of braking capability when air leakage occurs, a fail-safe mechanically actuated emergency brake system is required. **Figure 24–21** shows a tractor emergency brake schematic.

SPRING BRAKES

Spring brake assemblies are used by most trucks and trailers today to perform service and parking functions. They are dual-chamber brake actuators that are designed to perform both service and the parking/emergency brake function. Service applications are performed by air pressure, whereas the parking/emergency function is performed mechanically; that is, by powerful springs. The two chambers are known as service and hold-off chambers. Each chamber receives its own air supply that is directed to act on a diaphragm. When the service chamber receives air pressure, it performs service braking. The hold-off diaphragm must have air supplied to it in order to release, that is, cage the main spring.

When the chassis system air pressure is zero, there is no air acting on the hold-off diaphragms on the truck, so the main springs are fully applied to the spring chamber pushrod. This means the system is in park mode. When air pressure builds in the system and the parking brakes are released, air is directed to the hold-off chambers and the force of the main springs is "held-off" the brake chamber pushrod. At least 60 psi must be present in all the spring brake hold-off chambers before the vehicle parking brakes can be considered to be released. **Figure 24–22** shows a typical spring brake valve circuit and connections.

The air pressure that acts to hold off the spring brakes is controlled by valves in the tractor dash. The valves that manage the hold-off air pressure usually limit the hold-off pressure to a value below system pressure. This has a way of extending the service life of hold-off diaphragms, and in a true emergency it would result in faster application of the spring brakes. It should be underlined that the effectiveness of parking brakes is always dependent on the brake stroke adjustment at the slack adjusters. In other words, out-of-adjustment service brakes also mean the parking and emergency brakes are out of adjustment. The following summarizes the operation of spring brakes in a tractor/trailer combination:

- When no air is present in the hold-off circuit, the mechanical force of the spring is fully applied to the chamber pushrod and the brake is in park mode.
- To release the parking/emergency brakes, air pressure greater than 60 psi must be present in the hold-off circuit.
- In the event of severe leakage in the hold-off circuit, the spring brakes will begin to apply the moment pressure drops below 60 psi.
- The effectiveness of the parking/emergency brakes depends on the brake stroke adjustment

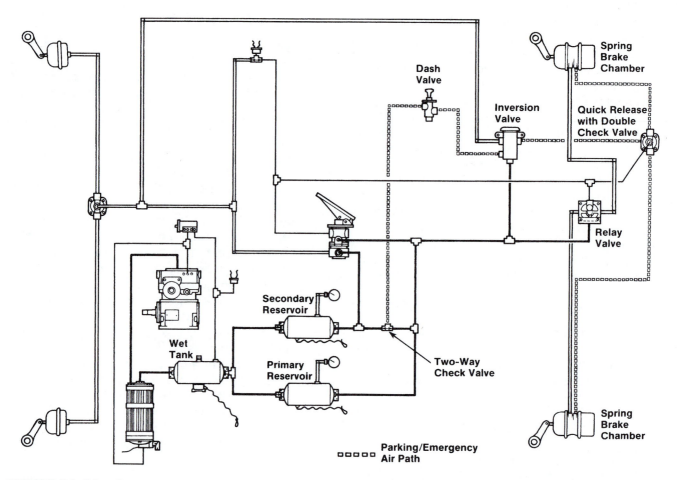

FIGURE 24–21 Tractor emergency brake circuit and components: Note the location of the inversion valve. (Courtesy of Haldex Brake Systems)

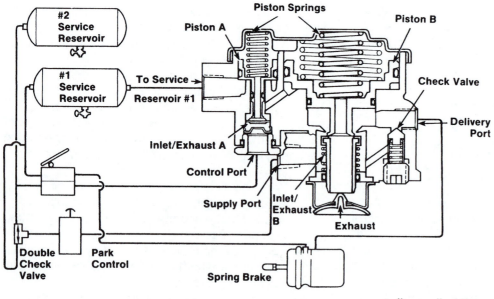

FIGURE 24–22 Spring brake valve circuit and connections.

at the slack adjusters in the same way that service brakes do.

- Because 60 psi air pressure is required to hold off a spring brake, when they are fully mechani-

cally applied they exert about the same amount of force on the foundation brakes as a 60-psi service application.

- Spring brake chambers may not be used on steering axle brakes.

- The spring pressure available in the spring brake assembly can be used for emergency service brake applications when an inversion valve is used: Application signal pressure is used to proportion hold-off air pressure (that is, bleed it down) for a one-time emergency stop.

DASH CONTROL VALVES

Dash control valves are push-pull air valves that enable the driver to release and apply the vehicle parking brakes and supply the trailer with air pressure. Most of these valves are pressure-sensitive, so they will automatically move from the applied to exhaust position should supply pressure drop below a certain minimum. The dash valve assembly receives its air supply from the primary tank, but it is backed up with air from the secondary circuit by means of a two-way check valve. The two-way check valve receives air from two sources, in this case the primary and secondary reservoirs, and is designed to shuttle to output whichever source arrives first or is at the higher pressure. Under normal conditions, primary circuit air is output by the two-way check valve, but in the event of a primary circuit failure, secondary circuit air is routed to the dash control valve assembly. The more recently introduced modular control valve assembly is equipped with a two-way check valve combined with pressure protection for each air pressure source. This means that if an air line failed downstream from the modular control valve, only one of the source air circuits would bleed down.

Two types of dash valve configurations are used: two-valve and three-valve systems. The function, color and shape of each of the three valves is determined in FMVSS No. 121 legislation, so they function exactly the same regardless of which OEM chassis they are in. The shape is important because it means the valves can be identified by touch. The shank size of each valve is different, so it is not possible to install an incorrect knob on a valve without taking extraordinary measures. Every tractor system is required to have the first two valves, the system park and trailer supply valves. The third is optional and, in fact, not often seen in current systems. It is known as the tractor park valve. All three are described below and can be seen in schematic form in several of the air system figures earlier in the chapter.

System Park

This valve is required to have a control knob that is diamond-shaped and yellow. It masters the trailer supply and, if used, the tractor park valves. In older systems, neither of these valves could be actuated until the system park valve had been actuated. When the system park valve is pushed inward, it directs air to the spring brake hold-off chambers in the tractor brake system. The tractor hold-off chambers are charged with air that is routed from the system park valve and directed through quick-release valves before being delivered to the hold-off chambers. When the system park valve is pulled out (exhausted), both the tractor and trailer parking brake systems are put into parking mode—that is, the spring brakes are applied by exhausting all hold-off air. In applications using the current modular control valve assembly, the trailer (trailer supply valve pushed in) may be supplied with air with the tractor brake spring brakes applied, but in these, trailer supply air is immediately cut when the system park valve is exhausted.

When the system park valve is pulled outward, the air supply to the trailer supply valve and the tractor hold-off circuit is exhausted. The trailer immediately goes into park mode. The air in the hold-off circuit up to the quick release valves is exhausted at the system park dash valve and the air in the hold-off chambers is dumped at the quick release valve exhaust. Remember, because the system park masters the trailer supply and, when used, the tractor park valves, the whole tractor/trailer combination is immediately put into park mode.

Trailer Supply

The trailer supply valve is required to be octagonal in shape and red. Unless a modular control valve is used, the trailer supply valve can only be actuated after the system park valve has been actuated. When this happens, air is directed from the valve to the trailer supply port of the tractor protection valve, and from there directed by means of a hose and gladhands to the trailer air circuit. In more recent applications using a modular control valve assembly, the trailer may be charged with air before the system park valve is depressed; however, when both valves are depressed, pulling the system park valve out will also cut the trailer air supply. This immediately results in the trailer supply valve popping out and the application of the trailer parking brakes.

Whenever the trailer supply valve is actuated (pushed in) and the system pressure is above 60 psi, air is directed into the trailer spring brake hold-off chambers, releasing the parking brakes. When the system park valve is pulled out, the trailer supply air is cut, applying the trailer parking brakes. **Figure 24–23** shows a cutaway view of a dash trailer supply valve.

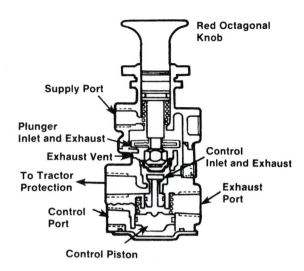

FIGURE 24–23   *Trailer supply valve. (Courtesy of Haldex Brake Systems)*

Tractor Park

The tractor park valve, when used, must be round and blue. It is used in dash three-valve control systems and is mastered by the system park valve. In a dash three-valve system, when the system park valve is depressed, air is made available to the trailer supply and tractor park valves. Both these valves must be depressed before the rig is taken out of park mode. However, to apply the parking brakes in the whole rig, only the system park valve has to be pulled. **Figure 24–24** shows a dash-mounted parking control valve.

 Shop Talk _____

Make sure you understand the function of each dash control valve in the system. Block the wheels and get into a truck with a fully charged air system: Use the controls and observe the effect.

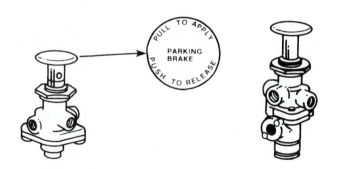

FIGURE 24–24   *Dash-mounted parking control valve. (Courtesy of Bendix Commercial Vehicle Systems)*

24.5   TRAILER CIRCUIT

The trailer service and parking/emergency brake circuits are entirely controlled by the tractor brake system. Although many technicians are employed to work exclusively on trailers, in order to effectively work on trailer brakes, the operation of the tractor brake system must be understood first. See **Figure 24–25**, which shows the trailer brake circuits, but also reference the figures showing the tractor air brake circuits (**Figure 24–3** and **Figure 24–4**). To this point, the means of actuating the trailer service, parking brake, and air supply functions have been mentioned. In this section, they will be reviewed and expanded upon.

TRAILER AIR SUPPLY

The compressed air required by the trailer air circuits is supplied by the tractor. Two hoses couple the tractor air and brake circuits to the trailer. The source of the trailer air supply is the tractor dash control manifold. To send air to the trailer from the tractor, the system park dash valve must first be actuated, followed by the trailer supply valve. This sends air to the tractor protection valve and shuttles it to its on position. The trailer supply air then passes through the tractor to trailer hoses that are coupled to the trailer by means of gladhands. These are explained in more detail a little later in this section. The air in the trailer is stored in reservoirs or tanks. FMVSS No. 121 requires that the volume of air dedicated to the brake system stored on the trailer be at least 8 times the total brake chamber volume.

TRAILER SERVICE SIGNAL

The trailer service signal is sourced from either the tractor foot valve or the trailer application valve. This signal is delivered to the tractor protection valve even when the tractor is being bobtailed. However, the trailer service signal air is prevented from exiting the tractor protection valve when the valve is in its off position. It will be off any time the trailer supply dash control valve is not actuated. As soon as the tractor protection valve is shuttled to the on position, the trailer service brake signal can be transmitted through the tractor protection valve.

GLADHANDS

Two hoses are required to couple the air and brake system of the tractor with that of the trailer. Flexible hoses are required because the tractor and trailer

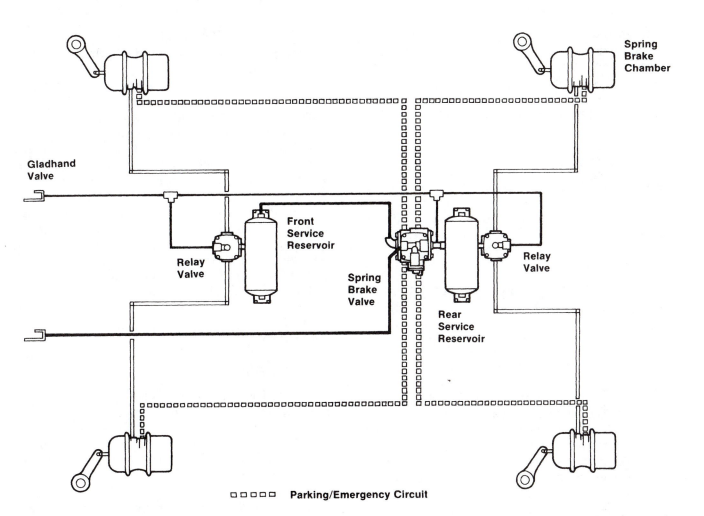

FIGURE 24–25   Tandem axle trailer service and parking/emergency circuits. (Courtesy of Haldex Brake Systems)

articulate at the fifth wheel. The two hoses are usually connected directly to the tractor protection valve. At the end of each hose are gladhands. They are called gladhands because when two are coupled, they have the appearance of a pair of hands in a handshake. The gladhands on the tractor hoses connect with a pair of gladhands on the trailer. Gladhands provide a fast method of coupling the tractor and trailer brake circuits that can be performed by any truck driver without the assistance of a truck technician. All gladhands in North America are universally sized, but some are polarized. Gladhands are sealed by rubber grommets. These deteriorate and require routine replacement—this is the only service required. **Figure 24–26** shows different types of gladhands.

Polarized Gladhands

Polarized gladhands make cross-coupling impossible. When gladhands are polarized, the supply gladhand on the tractor will only engage with the supply gladhand on the trailer, and likewise with the service gladhands.

Aluminum

Cast Iron

FIGURE 24–26   Gladhands. (Courtesy of Bendix Commercial Vehicle Systems)

Quick-Release Gladhands

Quick-release gladhands are sometimes used in the service supply signal to speed up brake release times. When a quick-release gladhand is used in the service signal line to the trailer, it exhausts at the gladhand rather than at the signal source. This can speed up the release of service brakes after an application and reduce drag at the foundation brakes. Brake lining wear, drum life, and release drag are all improved.

Dummy Gladhands

These are located on the tractor. They allow the live gladhands to be connected when the tractor is being bobtailed. Dummy gladhands dead-end the hoses and keep the trailer supply and service lines safely out of the way of rotating driveshafts.

TRAILER SPRING AND SERVICE PRIORITY SYSTEMS

Most trailers manufactured since 1975, that is, those covered by FMVSS No. 121, are required to have a means of mechanically parking the brakes. The means used to achieve this is almost always spring brakes. The trailer parking/emergency brake systems function on the principle of a service brake priority or spring brake priority. Until recently, most trailers were manufactured with service brake priority. A service brake priority system ensured that the service brake reservoir on the trailer would fill with air before the hold-off chambers of the spring brakes were charged to release the parking brakes. This resulted in a delay when hooking up to a trailer whose air tanks were empty or near empty.

A spring brake priority system directs air to the parking/emergency reservoir and the hold-off chambers of the trailer spring brakes. This speeds up shunting, and, in the event of a breakdown, enables a trailer to be moved in the minimum amount of time. Remember that with a spring brake priority system, it is possible to couple to a trailer and move it onto the highway before there is any air in the tanks available for service braking.

TRAILER BRAKE VALVES

Most trailers use multi-functional brake valve single housings that are capable of managing the service and parking/emergency brake requirements, pressure protecting the trailer system, and modulating pressure. The brake system may not be the only pneumatic system on the trailer. Many trailers use air suspensions and air-actuated lift axles. The air supply can also be used for other application-specific functions. In all trailer air systems, the trailer supply line is the means of releasing the parking brakes. This means that in any uncoupled trailer, the parking

brakes should be fully applied. However, the instant the trailer supply is activated, the trailer parking brakes are released.

Generally, when coupling a tractor to a trailer, the tractor should first be mechanically coupled to the trailer at the fifth wheel before applying air to the trailer supply circuit. There are some exceptions to this. One of them is the fact that most jurisdictions in the United States and Canada require drivers in driver testing to observe a specific procedure that requires air lines to be attached first. See the Shop Talk in Chapter 27 for a full explanation of what should be done when coupling a tractor to a trailer. Because trailer parking/emergency brakes are only released when air is present in the trailer supply air line, in the event of a trailer breakaway (trailer separation from the tractor at the fifth wheel, a very unusual occurrence), the trailer supply line would sever, causing immediate application of the trailer parking/emergency brakes.

Park on Air

Until recently, many trailers used park-on-air spring brake systems. This meant that when the trailer was put into park mode with its air tanks fully charged with air, the service brake chambers were charged with air (service brakes applied) and the air remained in the hold-off chambers. In the event of trailer system air leakage, as air bled down, system pressure dropped, as did pressure in the service chambers. However, pressure would also bleed down in the hold-off chambers. As system pressure (and therefore pressure in the hold-off chambers) dropped, spring pressure would gradually take over the parking brake function Most of these park-on-air systems modulated the hold-off pressure at around 90 psi. This resulted in faster parking/emergency response whenever air was exhausted from the trailer supply line, as less compressed air had to be exhausted from the hold-off circuit and extended the life of hold-off pressure.

Park on Spring

All currently manufactured trailers meeting FMVSS No. 121 standards are required to park on spring and have spring brake priority systems. The trailer combination valves used to achieve this incorporate a quick-release valve to direct air to the hold-off diaphragms in the spring chambers. When the trailer supply air is activated, air is directed to the spring brake hold-off chambers, causing them to release almost immediately. When the air pressure exceeds 60 psi in the hold-off circuit, air from the trailer supply line can be directed into the service brake tanks to be made available for service braking. In most cases, the hold-off pressure is regulated below system pressure, usually at a pressure value around 90 psi.

Relay-Emergency Valves

Trailers manufactured before 1975 and those that do not have to comply with FMVSS No. 121 may be equipped with relay-emergency valves. In such cases, the trailer is not equipped with spring brake chambers and has no mechanical park. The brakes used are single-chamber service brakes. The relay-emergency valve functions as a relay valve to effect service braking on a trailer. When the trailer supply air line is charged with air (from the tractor) and a service application signal is received by the relay-emergency valve, it uses air stored in the trailer air tanks to apply the brake chambers. However, when the trailer air supply is cut (trailer supply valve out/trailer breakaway), the trailer brakes go into a park-on-air mode. In park-on-air mode, the relay-emergency valve applies system air pressure to the service chambers. When a trailer is parked for a period of time (or there is an air leak), causing system pressure to bleed down, the result is a parked trailer with no brakes. This means that when coupling a tractor to a trailer, the trailer must be charged with air from the tractor so the brakes can be applied before the fifth wheel and upper coupler are engaged.

> **CAUTION:** When coupling a tractor to a trailer not equipped with spring brake chambers, always connect the gladhands and charge the trailer with air before attempting to couple the fifth wheel.

Trailer Service Brakes

The trailer service brakes are controlled by the tractor. The service brake signal to the trailer may be sourced from either the tractor foot valve or the column-mounted trailer application valve. You should understand exactly how these two application valves operate to manage the trailer service signal, so review the descriptions that appear earlier in this chapter. The service signal is directed to the tractor protection valve. If this is shuttled to its on position, passed through to the trailer service gladhand. Next, the service signal is directed to a trailer relay valve. The relay valve is designed to replicate the service signal pressure, but uses a supply of air stored in the trailer service tanks. This air is sent to the trailer service chambers. This means that if the pressure of the service signal to the relay valve is 20 psi, it will deliver a 20-psi service application using air from the

trailer service tanks. When the service signal pressure drops, the air acting on the trailer spring brake chambers is exhausted at the relay valve. Service signal air is exhausted either at the tractor or, in the event that a quick release gladhand is used, at the gladhand coupler.

24.6 FOUNDATION BRAKES

Up to this point, we have dealt with the components that manage the vehicle braking. The foundation brakes are the components that perform the braking at the end of each axle. They consist of a means of actuating the brake, a mounting assembly, friction material, and a drum or rotor onto which braking force can be applied. Three types of foundation brakes are found on today's highway trucks: the S-cam type, which is by far the most common, disc, and wedge brake assemblies. **Figure 24–27** shows all three types.

S-CAM FOUNDATION BRAKES

In an S-cam brake, the air system is connected to the foundation brake by a slack adjuster. The actuator is a brake chamber. Extending from the brake chamber is pushrod. A clevis is threaded onto the pushrod, and a clevis pin both connects the pushrod to slack adjuster and allows it to pivot. The slack adjuster is splined to the S-camshaft, which is mounted in fixed brackets with bushing type bearings. When force from the brake chamber pushrod is applied to the slack adjuster, the S-camshaft is rotated. The S-camshaft extends into the foundation assembly in the wheel and its S-shaped cams are located between two arc-shaped brake shoes. The foundation assembly is mounted to the axle spider, which is bolted to a flange at the axle end. The brake shoes may either be mounted on the spider with a closed anchor pin that acts as a pivot or open anchor mounted, meaning that they pivot on rollers onto which they are clamped by springs. Opposite the anchor end of the shoe is the actuating roller that rides on the S-cam. The shoes are lined with friction facings on their outer surface. Attached to the wheel assembly and rotating around the brake shoes is a drum. When brake torque is applied to the S-camshaft, the S-shaped cams are driven into actuating rollers on the brake shoes, forcing the shoes into the drum. This action is shown in **Figure 24–28**. When the shoe is forced into the drum, the friction that retards the movement of the drum is created. In

A

B

C

FIGURE 24–27 Foundation brake types: (A) S-cam brake, (B) Wedge brake, (C) Disc brake. (Courtesy of ArvinMeritor Inc.)

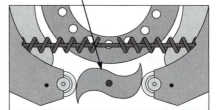

Brakes Applied

Brakes Released

Brake Shoe
Rollers   "S" Cam   Brake Shoes

FIGURE 24–28 S-cam brake operation. (Courtesy of ArvinMeritor Inc.)

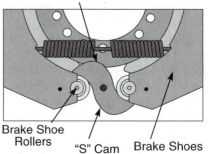

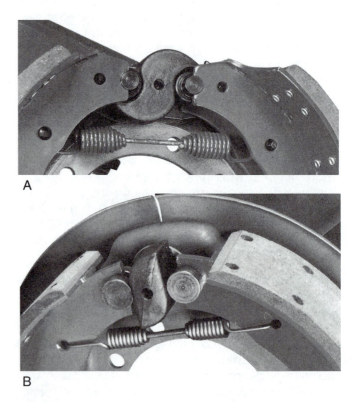

A

B

FIGURE 24–29 Cam geometry: (A) S-cam and rollers, (B) Flat cam and roller. (Courtesy of ArvinMeritor Inc.)

mechanical terms, the energy of motion is converted into friction, which must be dissipated by the drum as heat.

Some cam-actuated brakes use flat cam geometry. This means that the shape of the actuating cam is flat instead of being S-shaped, as shown in **Figure 24–29**.

S-Camshafts

The function of the S-camshaft is to convert the brake torque applied to it by the slack adjuster into the linear force that spreads the brake shoes and drives them into the brake drum. The slack adjuster is splined to the S-camshaft and retained by washers and snaprings. Two different spline configurations, known as coarse spline and fine spline, are used. S-camshafts are supported on the axle housing and on the brake spider by nylon bushings that permit the shaft to rotate to transmit brake torque to the shoes. The bushings are lubricated with chassis grease on a routine basis.

Spiders

The brake spider is bolted or welded to the axle end and provides a mounting for the foundation brake components. It must be tough enough to sustain both the mechanical forces and heat to which it is subjected. The brake shoes are mounted to the spider by means of either open or closed anchors. The anchor is the pivot end of the shoe.

Brake Shoes

Brake shoes are arc-shaped to fit to the inside diameter of the drum that encloses them. They are mounted to the axle spider assembly. Each shoe pivots on its own anchor pin retained in a bore in the brake spider. Today, most shoes are of an open anchor design. This means that the anchor end of the shoe does not enclose the anchor pin that forms the pivot. In most cases, open anchor shoes are retained at the anchor end by a pair of springs that clamp the shoes to each other forcing them against their individual anchors. On the actuation end of the brake shoes, a roller is retained by a clip spring or pin as shown in **Figure 24–30**. A single retraction spring loads the brake shoe rollers onto the S-cam profile. The function of the retraction spring is to hold

the shoes away from the inside surface of the brake drum when the brakes are not being applied. When the brakes are applied, the S-cam rotates, acting on the brake shoe rollers and forcing the brake shoes into the drum. The major advantages of open anchor brake shoes are lower maintenance and faster brake jobs.

Closed anchor shoes operate identically to open anchor shoes. However, on the anchor end of the shoe, a pair of eyes encloses the spider-mounted anchor pin on either side. The disadvantage of the closed anchor brake shoe design was more maintenance time required to perform brake jobs, due mainly to the fact that the anchor pins tended to seize in their spider bores.

Friction faces are attached to the shoes. These are graded by their coefficient of friction rating, which is given a letter code. Coefficient of friction describes the aggressiveness of any friction surface. For instance, sheet ice has lower friction resistance than an asphalt surface, so it can be said to have a lower coefficient of friction. However, the letter codes used on truck shoe friction linings are too general to use as a reliable guide for replacement. Ensure that the OEM recommended friction linings are used when relining shoes. It should also be noted that coefficient of friction alters with temperature, generally decreasing as temperature increases. The friction facing is either riveted or bolted to the shoe. The general practice today is to replace the whole shoe with one with new friction lining attached at each brake job. This reduces the overall time required for the brake job. **Figure 24–31** shoes how a typical S-cam foundation brake is assembled.

Edge Codes

Letters imprinted on the edge of heavy-duty friction facings are used to describe the coefficient of friction. The first letter indicates the cold coefficient of friction, the second the hot coefficient of friction. The further down in alphabetical sequence, the higher the coefficient of friction—that is, the more aggressive the lining. So, a G letter rating indicates a higher coefficient of friction than an F rating. Remember, because of the wide range of coefficient of friction covered by each letter rating, most industry opinion is that this is an inadequate rating of actual brake performance. In fact, the TMC of the ATA has stated that up to a 40 percent variance is possible between shoes with the same nominal ratings. Two letters are always used to rate any lining. For instance, a lining code rated FF would have both cold and hot coefficient of friction ratings within the range 351–450.

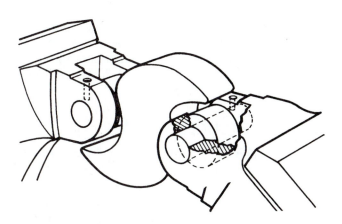

FIGURE 24–30 Brake shoe rollers. (Courtesy of ArvinMeritor Inc.)

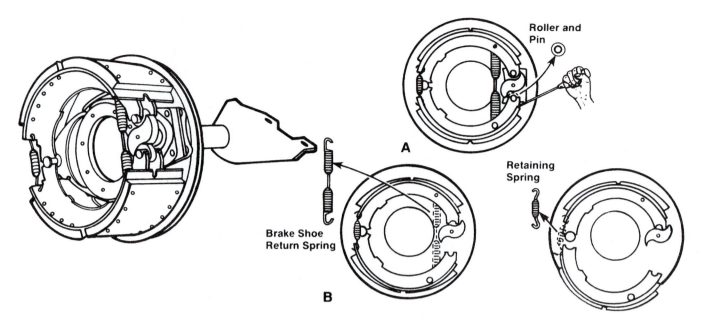

FIGURE 24-31 Typical S-cam brake assembly. (Courtesy of ArvinMeritor Inc.)

Letter	Coefficient of Friction Rating
D	0.150–0.250
E	0.251–0.350
F	0.351–0.450
G	0.451–0.550
H	Over 0.551

Brake Drums

Brake drums are manufactured from cast alloy steels or fabricated steel. The two have different coefficients of friction, so they should never be mixed. Brake drums are manufactured in several different sizes, but 16.5 inches is by far the most common. The friction facings used in today's brake linings, which do not use asbestos, tend to be harder. This means that the friction facings on the shoes last longer but can also be much tougher on brake drums. It has become routine in some applications for the brake drums to be changed at each brake job along with the shoes. Brake drums tend to harden in service. After they show heat checks (small cracks) and heat discoloration (blue shiny areas), they should be replaced rather than machined to oversize. Because drums are relatively low in cost, they are rarely machined after they have been in service. However, because they are prone to out-of-round distortion when they are stored for any length of time, it is good practice to machine drums when new, after fitting them to the wheel assembly, to ensure they are concentric.

When drums are replaced, the correct size and weight class should be observed to ensure the brakes are balanced. Drums of different weights will run at different temperatures. When drums are machined, ensure that the maximum machine and service dimensions are observed. For a 16.5-inch drum, the maximum permitted service dimension is 0.120 inch and the maximum machining dimension is 0.090 inch.

S-cam Foundation Brake Hardware

The S-cam foundation brake hardware includes all the springs, anchor pins, bushings, and rollers on the spider assembly. It is good practice to replace everything at each brake overhaul. The springs used in the foundation brake assemblies must sustain high temperatures, which causes them to lose their tension over time, so it is critically important that these are changed. Cam rollers should be lubricated on the roller pocket race and never on the cam contact face. Lubricating the cam roller pocket race helps avoid flat spots on the cam contact face. The cam contact faces should never be lubricated because the roller is designed to rotate with the S-cam when brake torque is applied to it.

WEDGE BRAKE SYSTEMS

Although not so commonly used today, wedge foundation brakes are used in some applications, especially on the front axles of vocational trucks. In this brake design, the slack adjuster and S-camshaft are replaced by a wedge-and-roller mechanism that spreads the shoe against the drum. They may be single- or double-actuated. In a single-actuated wedge brake system,

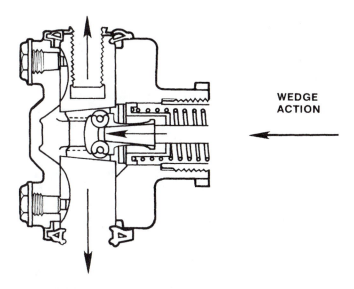

FIGURE 24–32 *Wedge-actuated brake system.*
(Courtesy of ArvinMeritor Inc.)

the shoes are mounted on anchors. A single wedge acts to spread the shoes, forcing them into the drum. Dual-actuated wedge brakes tend to be used in heavy-duty applications. The dual-actuated wedge brake assembly has an actuating wedge located at each end of the shoe. The actuating principle of a wedge brake assembly is shown in **Figure 24–32**.

The brake chambers in a wedge foundation brake system are attached directly to the axle spider. The wedge-and-roller actuation mechanism is enclosed within the actuator and chamber tube. When air is delivered to the service chamber, the actuating wedge is driven between rollers that in turn act on plungers that force the shoes apart. When the service application air is exhausted from the service chambers, retraction springs pull the brake shoes away from the drum. The actuating wedges are usually machined at a 14-degree pitch but may also be 12 or 16 degrees, so they must be carefully checked on replacement. A self-adjusting mechanism is standard on all wedge brakes, and this is contained in the wedge brake actuator cylinder. It consists of a serrated pawl and spring that engages to helical grooves in the actuator plungers. A ratcheting action prevents excessive lining-to-drum clearance occurring as the friction linings wear.

Despite the fact that dual-actuated wedge brakes have higher braking efficiencies than S-cam brakes, they tend to be rarely used today. They are seen by the trucking industry as having a much higher overall maintenance cost. Of note is the tendency of the automatic adjusting mechanism and plungers to seize when units are not frequently used: This necessitates the disassembly of the foundation brake assembly to repair the problem.

AIR DISC BRAKES

Air disc brakes have been around for a couple of decades, but they have never been widely used in the trucking industry. They are probably more popular today than they ever have been in the past. The trucking industry has tended to regard them as having much higher initial, operating, and maintenance costs. Disc brakes are commonly used as bicycle brakes. Their use as the automobile brake of choice has increased, mainly because they provide better control and higher braking efficiencies than drum brakes. **Figure 24–33** shows the operating principle of disc brakes.

The air disc brake assembly consists of a caliper assembly, a rotor, and an actuator mechanism. The actuator mechanism is a brake chamber, usually acting on a slack adjuster to amplify its force and convert the linear force produced by the chamber into brake torque at a power screw or helical ramps. The objective of an air disc brake system is to apply clamping force to a caliper assembly. Running between the friction pads of the caliper is a rotor that is bolted to and therefore rotates with the wheel assembly. When the power screw is actuated, the power piston to which friction faces are attached is forced into the rotor. Because the caliper is floating, when the inboard located piston is forced into the rotor, the outboard friction pads are also forced into the outboard side of the rotor with an equal amount of force. There are also some air actuated disc brakes newly introduced to the market that use brake chambers that act directly on the inboard power pistons.

Air disc brakes provide the advantage of better brake control and higher braking efficiency than S-cam actuated brakes. They also boast a slight overall weight advantage. However, the initial cost tends to be greater, and maintenance costs at this moment in time are also greater. Loaded trucks have

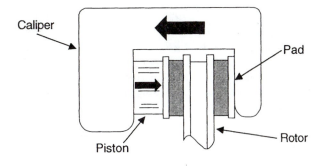

FIGURE 24–33 *As the piston moves out toward the rotor, the caliper moves in toward the rotor, creating balanced clamping force.*

FIGURE 24–34 *Typical air disc brake assembly.* (Courtesy of ArvinMeritor Inc.)

enormous braking requirements. Because the principle of braking a vehicle requires that the energy of movement (kinetic energy) be converted to heat energy, the important factor in determining how long brakes last between service intervals is how well the foundation brakes sustain heat. Air disc brakes on highway trucks have a couple of disadvantages in this respect. First, the total sectional area of the friction faces is much smaller than brake shoes. This means that more force must be applied to them. Next, the physical shape of the rotor is less able than a brake drum to dissipate the heat it absorbs, meaning that it runs hotter and fails earlier due to distortion. A final factor that reduces the number of air-actuated, disc brake-equipped trucks on our highways is the lack of universality. This means less competition for aftermarket parts, higher prices, and greater difficulty in obtaining replacement components. **Figure 24–34** shows a typical air disc foundation brake assembly.

24.7 AIR BRAKE SYSTEM COMPONENTS

This section will describe how different components and valves used in an air brake system function. Most have been previously introduced in the context of the circuit they are located in. For instance, a foot valve plays a role in all service braking, so it has a role in both primary and secondary circuit application pressure management. The description of the foot valve provided in this section outlines all its functions.

FOOT VALVES

The term foot or treadle valve is used to describe the **dual-circuit application valve** required on all current truck brake systems. Understanding how the

foot valve operates is essential because it masters both the primary and secondary service brake circuits on a truck brake system. The foot valve is responsible for service brake applications. Service brake applications cover all the vehicle braking except for parking/emergency braking. During the day-to-day operation of a vehicle, it is necessary to apply the brakes with exactly the right amount of braking force required to safely stop a vehicle. Too little force could result in the vehicle not stopping in time, whereas too much force could lock the wheels. The foot valve is the driver's means of managing the service brakes in an air brake system. It plays the same role as the brake pedal in your car. **Figure 24–35** shows some typical foot valves.

In a hydraulic brake system, the application pressure delivered from the brake pedal is always associated with the amount of mechanical force applied by the foot, even in a power assist system. Because of this relationship with the amount of pressure applied and the braking force, the driver is provided with what is known as "brake feel." The potential energy of an air brake system is compressed air. This means that the mechanical force applied to the brake pedal has no direct relationship with the resulting brake force. Instead, a different means of providing the driver with brake feel is required. In other words, the foot valve in an air brake system must be made to at least feel like a brake pedal in a hydraulic brake system. That is, the pedal should produce increased resistance as braking force increases. The main function of the foot valve is to provide the driver with a means of precisely managing the correct amount of brake force for each stopping requirement.

FIGURE 24–35 *Dual-circuit brake foot pedal control valve: It is also called an application valve or treadle valve.* (Courtesy of Bendix Commercial Vehicle Systems)

The foot valve is divided into two sections. Reference the sectional views of the foot valves in **Figure 24–36** and **Figure 24–37**. The portion closest to the treadle (foot pedal) is known as the primary section and is responsible for primary circuit braking. The portion further from the treadle is known as the relay section and is responsible for secondary circuit braking. The primary section of the foot valve receives a supply of air from the primary tank at system air pressure. The relay section of the foot valve receives a supply of air directly from the secondary tank at system air pressure. When the driver depresses the brake pedal, the supply air from each section of the foot valve is metered (accurately measured) out to apply the brakes on each circuit, primary and secondary. The air from each circuit never comes into direct contact with the other while in the foot valve.

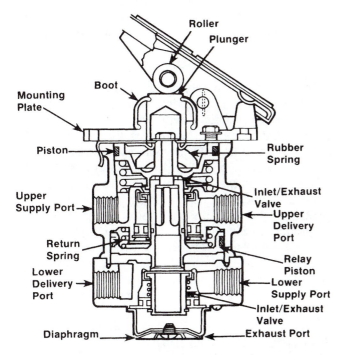

FIGURE 24–36 Cutaway view of a typical dual circuit brake valve. (Courtesy of Bendix Commercial Vehicle Systems)

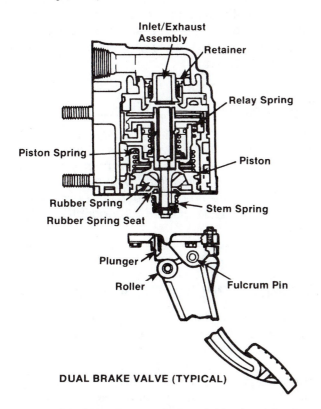

DUAL BRAKE VALVE (TYPICAL)

FIGURE 24–37 Suspended pedal brake valve. (Courtesy of Bendix Commercial Vehicle Systems)

Foot Valve Operation

The primary section of the foot valve is actuated mechanically; that is, by foot pressure. This causes the plunger under the treadle to move the primary piston. Any primary-piston travel off its seat results in the closing of the exhaust valve and then, depending on primary-piston travel, air being metered from the primary supply port out to the primary circuit service brake components. This metered air performs 2 other essential functions. First, it is directed downward to act on the relay (secondary) piston. The relay portion of the foot valve is responsible for actuating the secondary circuit brakes. It is designed to operate exactly like a relay valve. System pressure from the secondary tank is constantly available at its supply port. The air pressure value metered out to actuate the primary circuit brakes is also directed downward to actuate the relay piston. Like most relay valves, the foot valve relay piston is designed so the actuating signal pressure moves the piston to meter out pressure exactly equal to the signal pressure value to the secondary circuit.

Brake feel is an essential feature of an air brake system. Brake feel is provided by directing whatever air pressure value has been metered out to the primary circuit to act against the applied foot pressure. This means that the further the treadle travels the more foot resistance is felt, because the air pressure below it is greater. The idea is to provide air brakes with the feel of hydraulic brakes. In hydraulic brakes, whether power-assisted or not, actual braking force is always proportional to the mechanical force applied to the brake pedal. In air brakes, because the potential energy of the brake system is compressed air, the treadle valve simply meters out the required amount of compressed air to make a stop.

Primary Circuit Application: Normal Operation

Air from the primary reservoir is available to the primary section of the foot valve at system pressure. When the treadle on the foot valve is depressed, air from the primary circuit is metered out to:

1. Actuate the primary circuit brakes
2. Act as a signal to actuate the relay portion of the foot valve
3. Act against the applied foot pressure to provide brake feel

If the treadle on the foot valve is depressed so 20 psi of air is metered out to actuate the primary circuit brakes, 20 psi will also act on the relay piston to actuate the secondary circuit brakes and to act against the applied foot pressure to provide the driver with brake feel. In normal operation, the primary circuit is actuated mechanically.

Secondary Circuit Application: Normal Operation

When the primary circuit is actuated, primary circuit air is directed to enter the relay piston cavity. This air pressure moves the relay piston to first close the secondary exhaust valve and next, to allow air to flow out of the secondary supply port to be metered out to actuate the secondary circuit brakes. This means that in normal operation, the secondary circuit is actuated by air pressure.

Releasing Brakes

When the driver's foot releases the brakes, the foot valve spring causes the primary piston to retract, opening the exhaust valve. Primary signal air, the air acting on the relay piston, and the air acting to provide brake feel are simultaneously exhausted through the primary exhaust valve. When primary circuit air acting on the relay piston is exhausted, the relay piston spring retracts the relay piston, exhausting the secondary circuit signal.

CIRCUIT FAILURES

The main reason for having a dual-circuit service application system is to provide a fail-safe in the event of one circuit failure. When this occurs, either one of the circuits is designed to bring the vehicle to one safe stop. Total failures of one of the two circuits are rare, but not unknown. Physical damage to either of the circuit air tanks would be an example of such a failure. In the event of a failure of either circuit, the driver is trained to bring the vehicle to a standstill

and await mechanical assistance. When such a failure occurs, the driver is alerted to the condition by gauge readings, warning lights, and a buzzer before attempting to stop the vehicle. As a fail-safe for both the primary and secondary circuits, the spring brakes would have to bring the vehicle to a standstill.

Primary Circuit Failure

In the event of a primary circuit failure, the foot valve treadle will depress the primary piston, which will not meter any air because none will be available at the primary supply port. This means that no air will be metered out to actuate the primary circuit brakes, to actuate the relay piston, or to provide brake feel. The driver's foot on the treadle will now travel further with little resistance until the primary piston physically contacted the relay piston. The relay piston will now be actuated mechanically and, according to the distance it travels, meter out air to actuate the secondary circuit brakes. The secondary circuit will operate normally, except that instead of being actuated by air from the primary circuit, it will be actuated mechanically.

Secondary Circuit Failure

In the event of a secondary circuit failure, the primary circuit would operate normally with no loss of brake feel. The air metered into the relay piston cavity would move the relay piston, but no secondary circuit air would be available at the supply port, so none would be metered out to the secondary circuit service brakes.

👓 **Shop Talk** _____

One OEM survey of warranty returns indicates that over half of the foot valves submitted function perfectly. This misdiagnosis results from a failure by the technician to understand how the valve operates. Take some time to understand exactly how this important valve masters the operation of the service brakes.

TRAILER APPLICATION VALVE

The trailer application valve is often known by such names as *broker brake*, *spike*, or *trolley valve*. The valve isolates the trailer service brake function from that of the tractor. It is a hand-operated graduated control valve that is usually mounted on the steering column of the truck. The trailer application valve gets its name *broker brake* because a tractor owner-operator (broker) can use the company-owned trailer brakes to perform a large percentage of the rig's braking, thereby saving the tractor's brakes. Typical trailer application valves are shown in **Figure 24–38**.

In a typical system, the trailer application valve receives a supply of air from the secondary circuit at system pressure. When the application valve is applied, this air is metered out to the tractor protection valve to be used as the trailer service brake signal. This signal could also be delivered from the foot valve, in which case the source air would be the primary circuit. For this reason, a two-way check valve is required. The operation of a two-way check valve is described in some detail later in this section. Essentially, both possible sources of the trailer service brake signal are routed to the two-way check

valve located before the tractor protection valve. The two-way check valve is designed to shuttle and direct the higher source inlet pressure to the outlet. This means that in the event of a service signal from both the foot valve and trailer application valve, the higher pressure signal would be the one routed through the tractor protection valve.

RELAY VALVES

A relay valve is a simple remote-controlled brake valve. Some examples are shown in **Figure 24–39**. The relay valve is designed to have a constant supply of air at system pressure close by. When it receives a remote signal, it uses the local supply to actuate brake chambers. Most relay valves are designed to output the same pressure to the brake chambers as the signal pressure. This means that a 25-psi signal delivered to a relay valve would result in a 25-psi output.

Most current truck air brake systems use relay valves because there is some distance between the application valves (foot valve and trailer spike) and the foundation brakes on a vehicle. This distance varies from the service brakes on the tractor unit to those most far away on the trailer or trailers in the rig. If the air sent to the brake chambers were sourced from the application valve, a considerable "lag" time would result. To speed up the application of brakes, air from the application valve is often used only as a signal. The air used to actuate the vehicle service brakes is contained in a tank mounted as

FIGURE 24–38 Trailer control brake valves. (Courtesy of Haldex Brake Systems)

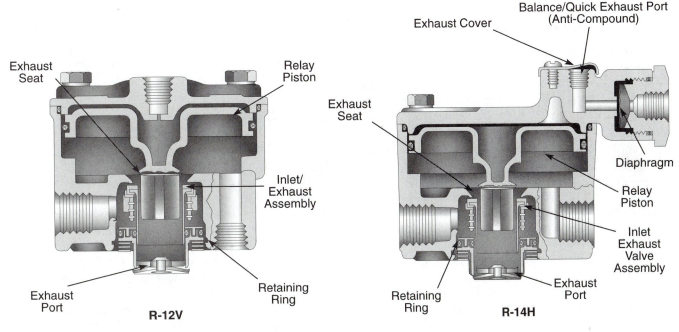

FIGURE 24–39 Relay valves. (Courtesy of Bendix Commercial Vehicle Systems)

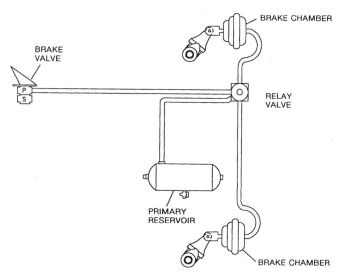

FIGURE 24-40 *Primary circuit connection to a relay valve. (Courtesy of Freightliner Trucks)*

close as possible to the foundation brakes. This is shown in **Figure 24-40**. Relay valves are used to reduce brake application and release times.

A relay valve receives an air signal directly from the application valve. The signal line from the application valve is of a smaller gauge, to minimize the volume of signal air. The relay valve is designed to meter a locally contained supply of air at system pressure, delivering it to the service brake chambers. The signal volume of air is small compared to the volume of air required to actuate a service brake chamber. Because this larger volume of air is stored close to the service brake it supplies, the actuation time is reduced.

Relay Valve Operation

Signal air pressure from the application valve enters the cavity above the relay piston and moves the piston down through its stroke. The exhaust valve seals off the exhaust passage and the inlet valve moves off its seat, permitting system pressure air from the air tank to be metered out to the service brake chambers the relay valve supplies. Relay piston travel depends on the signal air pressure value. The higher the signal pressure, the further the relay piston travels and the higher the service application pressure.

Most relay valves are designed to deliver a service application pressure at exactly the signal pressure. In other words, if the relay valve receives a signal pressure of 20 psi, it will meter exactly 20 psi out to the service chambers.

When the signal air pressure is exhausted at the application valve, the relay piston spring retracts the piston, causing its exhaust seat to move away from the exhaust valve. With the exhaust passage open, air pressure from the service chambers is permitted

to exhaust through the exhaust port of the relay valve.

Relay Valve Crack Pressures

The term *crack pressure* refers to the amount of pressure required to actuate or crack a relay valve so it can begin to meter air. Air is compressible. The larger the volume of confined compressed air, the greater its compressibility. This causes a slight delay between the moment an air signal leaves the application valve and the moment it acts on the relay piston. The delay increases as the distance from the application valve to the relay valve increases. The service brakes in a double trailer combination are actuated by the foot valve in the tractor. The air signal from the tractor foot valve has to travel a considerable distance to actuate the service brakes on the rearmost trailer axle. If all the relay valves on the rig had identical crack pressures, those service brakes closest to the foot valve would be applied before those furthest away, creating a potential danger.

In order to balance and properly time the service brakes on a tractor/trailer combination, relay valves are manufactured with different crack pressures. Generally, those relay valves designed to function furthest away from the tractor foot valve would have the lowest crack pressures. Those on the tractor itself would have slightly higher crack pressures. The objective is that, when the service brakes are applied, the brakes are applied evenly through the whole rig. Valve crack pressures are defined by the strength of the relay piston return spring. Crack pressures range from zero (not common) to values around 4-7 psi. It should be noted that approximately 6 psi of air pressure is required to rotate the S-cams sufficiently to contact the brake shoes against the drum. Properly timed crack pressures on a tractor/trailer combination can help prevent trailer jackknife and other brake balance problems.

QUICK-RELEASE VALVES

Quick-release valves are designed to speed the exhausting of air from an air circuit. They are used in brake and other chassis air systems. They are commonly used to distribute air to the hold-off chambers in spring brake assemblies. In pre-ABS systems, they were used to distribute air to the front axle service brakes. A quick-release valve usually has a single inlet port and two outlet ports. When air pressure is delivered to its inlet port, it is distributed as if it were in a T union. However, when air pressure to the inlet port drops, the air charged to the outlet ports is exhausted at the quick-release valve. As its name suggests, the quick-release valve is designed to speed up the release of air from a circuit by exhaust-

A

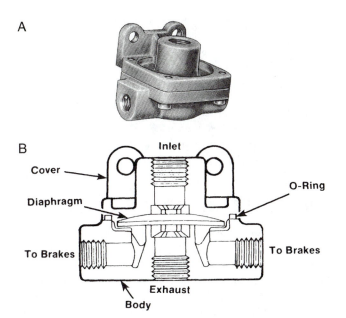

B

FIGURE 24–41 *Typical quick release valve.*
(Courtesy of Bendix Commercial Vehicle Systems)

ing it at the valve rather than having it return to the source valve. **Figure 24–41** shows both external and cutaway views of a quick-release valve.

Several types of quick-release valves are used. They have different external appearances and connection configurations but are functionally the same. Most have a die-cast metal body with an internal diaphragm, spring, and spring seat.

RATIO VALVES

Some years ago, a front wheel limiting valve (also known as a "wet-dry" switch) was used in many air brake systems. Although not currently legal, there are still a few examples of front wheel limiting valves around. This valve was switched from the cab and reduced the application pressure to the front axle service brakes to 50 percent of application pressure when toggled to the *wet road* setting. Trucks meeting the FMVSS No. 121 standard today cannot use a manually switched limiting valve, so when the front axle brakes are required to be ratio-modulated, an automatic ratio valve must be used. **Figure 24–42** shows both external and sectional views of ratio valves.

Automatic ratio valves replaced the front wheel limiting valve. An automatic ratio valve limits front axle braking under light- and medium-duty applications. Typically, the valve functions as follows. The ratio valve will not crack until the application pressure exceeds 10 psi. With application pressures between 10 and 40 psi, the front axle service brakes receive 50 percent of the application pressure. Between 40 and 60 psi, the ratio valve graduates the ratio of air delivered to the front axle service brakes

A

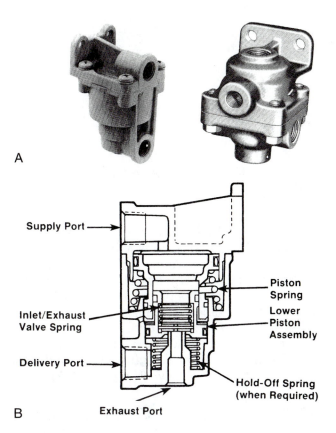

B

FIGURE 24–42 *Ratio valves: (A) typical ratio valves, and (B) sectional view of a ratio valve.*
(Courtesy of Bendix Commercial Vehicle Systems)

to 100 percent: With any application pressure greater than 60 psi, the front brakes receive 100 percent of the application pressure. This means that under severe braking, the front axle brakes receive 100 percent application pressure.

Automatic ratio valves are considered safer than front wheel limiting valves due to the importance of the front axle brakes under any severe braking condition. All research and analysis of panic brake performance on heavy-duty rigs has shown that the driver can maintain better control when the front axle service brakes are provided with 100 percent application pressure. For this reason, some in the trucking industry advocate the elimination of any kind of proportioning of application air pressure to the front brake chambers.

INVERSION VALVES

Inversion valves are used on some current brake systems. They are used to improve braking under emergency conditions. They function by modulating the parking brake hold-off air to provide service braking in a failure condition, resulting in total loss of air pressure in the application circuit source air. When in failure mode, they permit the application signal from the foot valve to proportionally relieve the

FIGURE 24–43   *Inversion valves. (Courtesy of Bendix Commercial Vehicle Systems)*

spring brake hold-off pressure to assist in stopping the vehicle. This takes advantage of the mechanical force of the park spring in the spring brake chamber to help effect an emergency stop. The inversion valve is of the normally-open type. It is closed by air pressure from another source and can only be used in systems that source the parking brake (hold-off) air from an isolated tank. Some examples of inversion valves are shown in **Figure 24–43** and their location in the circuit is shown in **Figure 24–21**.

BOBTAIL PROPORTIONING VALVES

A highway tractor's handling is quite different when the unit is not coupled to a trailer. Operation of an uncoupled tractor is known as bobtailing. A bobtail tractor has a large proportion of its weight loaded over the front axle with a much smaller percentage of the weight over the rear drive axles. Bobtail proportioning valves help balance braking in a bobtailed tractor by increasing the percentage of braking performed by the front axle brakes. A bobtail condition is sensed when there is no air in the trailer supply to the tractor protection valve.

The bobtail proportioning valve is a combination of 2 valves in a single housing. The lower portion contains a standard relay valve that functions exactly as the relay valves already described. The upper portion houses a proportioning valve designed to reduce normal service brake application pressure any time the tractor is not coupled to a trailer. The control port on the valve is connected to the trailer supply air circuit. The result is that during bobtail mode, the front brakes receive full application pressures, whereas the service braking to the rear brakes is proportioned. This condition occurs until severe braking is required: When application pressures are at 90 percent, the rear axle brakes receive full application pressure. The less aggressive braking to the rear axles by a bobtail proportioning valve greatly improves bobtail vehicle handling, especially under slick road conditions. Note the location of the bobtail proportioning valve in **Figure 24–3** at the beginning of the chapter.

SPRING BRAKE CONTROL VALVE

A spring brake control valve is a combination valve that is used to manage both service and parking/emergency brake functions to spring brake chambers. This valve may also incorporate the functions of an inversion valve, regulate spring brake chamber hold-off pressure, and have anti-compounding features. **Figure 24–44** shows some examples of spring brake control valves.

Figure 24–45 shows a schematic of a typical spring brake control valve that includes a sectional view of the valve. Air from the supply line enters at the supply port and depresses control piston A, opening inlet/exhaust port A. This air is also routed to the chamber under pressure protection piston B. When the air pressure builds to approximately 55 psi

FIGURE 24–44   *Spring brake control valves.*

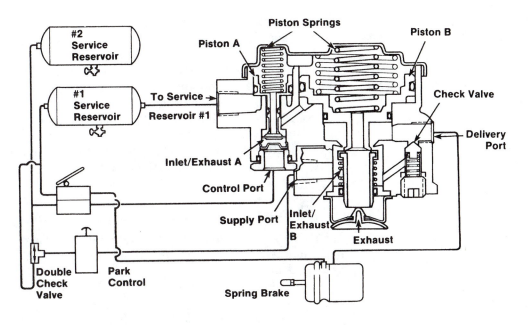

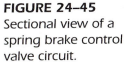

FIGURE 24–45
Sectional view of a spring brake control valve circuit.

beneath piston B, the piston moves up against the spring pressure of the piston and opens the inlet/exhaust port B. This allows the air to flow past a check valve to the number one service reservoir, past the check valve to the delivery port and to the spring brake hold-off chamber. This releases the spring brakes. To apply the parking brake, air is exhausted by the dash park control valve. This dumps air from piston A and piston B, closing intake/exhaust ports and venting hold-off air to the atmosphere exhaust. The moment the spring brake hold-off chamber is exhausted, spring pressure applies the parking brake.

 Shop Talk _____

Current spring brake control valves must prioritize air to the hold-off circuit over air to the air tanks. This means that when air builds up in the air system, it is first delivered to the hold-off circuit, enabling the release of the parking brakes. This enables a vehicle with a disabled air supply system to be moved much more quickly: a much smaller volume of air has to be transferred to the vehicle before release of the parking brakes becomes possible.

CAUTION: Because air is delivered to the trailer hold-off circuit before the trailer tanks are charged, the trailer can be moved before any service braking is possible. Charge trailer tanks before attempting to move the trailer.

TRACTOR PROTECTION VALVES

The tractor protection (TP) valve has two functions. First, it acts as the means of routing and switching the trailer air supply and trailer service signal from the tractor. Second, it functions to protect the tractor air supply in the event of a trailer breakaway or total air loss. When a service brake signal is made from either the foot or trailer application valve, it is routed to the TP valve. This signal will only be transmitted to the trailer if the TP valve is switched on. The TP valve is switched on only when the trailer supply valve in the dash is activated. This prevents loss of air from the trailer service signal line when the tractor is not connected to a trailer. **Figure 24–46** shows a cutaway view of a two-line, TP valve. **Figure 24–47** is a schematic showing the location of a three-line valve in the air brake circuit.

Most current TP valves are known as two-line valves. They have two inlet lines, one from the trailer supply dash control valve or valve assembly and the other from the trailer service signal circuit. The two outlet ports of the TP valve are connected to the two air hoses that enable the tractor air circuit to be coupled to that of the trailer. When no air is present in the trailer supply inlet to the TP valve, it is in an off state. This means that whenever a tractor service application is made, the signal is delivered to its service port, but is prevented from passing beyond. When the tractor is coupled to a trailer and the tractor hoses are connected to the trailer gladhands, the TP valve is switched to its on state when the trailer supply valve is actuated. Trailer supply air passes from the tractor to charge the trailer air circuit. When the TP valve is in its on state, its spool valve shuttles to permit service signal air to be applied from the tractor application circuit to the trailer service signal circuit.

For a TP valve to be in the on state, the air pressure must be above its trigger value. This trigger value is usually 45 psi. In some cases it may be less

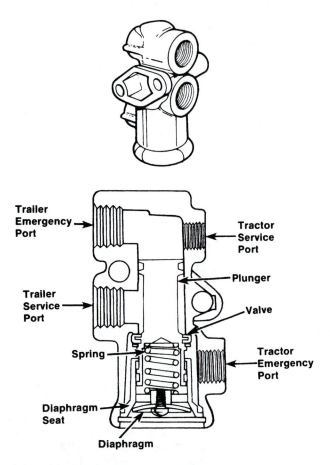

FIGURE 24–46   Two-line tractor protection valve. (Courtesy of Bendix Commercial Vehicle Systems)

(FMVSS No. 121 requires it to be between 20 and 45 psi). It should be noted that when system pressure has dropped to this value, some spring force from the spring brake assemblies is already being applied to the foundation brakes. In the event of a trailer breakaway or massive air loss in the trailer, the TP spool valve will shuttle to the off position after the air pressure drops below the trigger value. After the TP valve shuttles to an off position, the tractor air circuit becomes isolated from the trailer air circuit.

Shop Talk _____

FMVSS No. 121 requires that the TP valve isolate the tractor air supply from that of the trailer between 20 and 45 psi.

As stated before, most current TP valves are two-line systems. Two-line systems are usually controlled by a two-dash-valve manifold. The TP valve trailer supply air is sourced directly from the dash control valve assembly.

In a three-line TP valve, the control system may use a two- or three-dash-valve control system. The three-line TP valve operates much as the two-line version, with the exception that when the valve is switched to its on state, the trailer air supply is sourced directly from one of the tractor air tanks.

CHECK VALVES

A check valve routes air flow in one direction and in most cases prevents any backflow. A couple of different types of check valves are used in air brake systems to perform tasks such as pressure protection and priority routing.

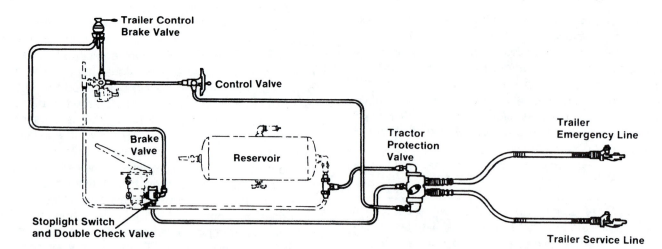

FIGURE 24–47   Three-line tractor protection valve location and plumbing. (Courtesy of Bendix Commercial Vehicle Systems)

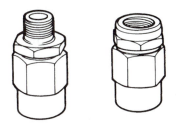

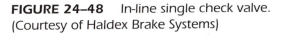

FIGURE 24–48 In-line single check valve. (Courtesy of Haldex Brake Systems)

FIGURE 24–49 Double check valve. (Courtesy of Haldex Brake Systems)

One-Way Check Valve

These devices could not be simpler. They permit air to travel in one direction only. They are often located at the inlet to a tank or subcircuit, in which case they will act as pressure protection devices. This would mean that air can enter the tank or subcircuit, but not exit though the check valve. **Figure 24–48** shows an in-line single check valve.

Two-Way Check Valves

Two-way check valves can be used in several areas of an air brake circuit. They are disc or shuttle valves. The two-way check valve receives air from 2 sources and has a single outlet. The valve core is designed to shuttle so the higher-pressure source air is always delivered to the outlet. In other words, the two-way check valve will transmit the higher-pressure source air to the outlet. It is recommended that these valves be mounted horizontally. **Figure 24–49** shows a typical double check valve.

STOP LAMP SWITCHES

The stop lamp switch is designed to close an electrical circuit when line pressure is sensed. Its function is to illuminate the brake lights on the vehicle. It is rated by the amount of air pressure required to close the electrical circuit. It can be located at various points in the service application circuit.

BRAKE CHAMBERS

Brake chambers convert the compressed air back into the mechanical force required to stop a vehicle. A simple service brake chamber consists of a 2-

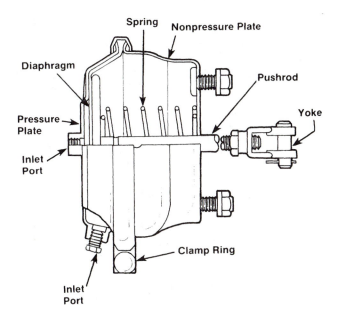

FIGURE 24–50 Air brake chamber.

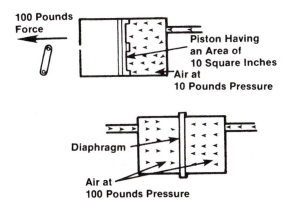

FIGURE 24–51 Compressed air used as potential energy. (Courtesy of Bendix Commercial Vehicle Systems)

piece housing, within which is a diaphragm, a pushrod, and a retraction spring. The two sections of the housing are clamped together so the diaphragm forms an airtight seal. The pushrod is linked to the foundation brakes by means of a slack adjuster. When compressed air acts on the flexible rubber diaphragm, force is applied to the pushrod plate. The amount of linear force is proportional to the amount of air regulated by the driver using the service brake application valve. **Figure 24–50** shows a typical air brake chamber. Note the components. The spring in this assembly is relatively light duty and functions simply to return (retract) the pressure plate after a brake application.

The amount of linear force delivered by a brake chamber can be calculated by knowing the sectional area of the diaphragm and the amount of air pressure applied to it. **Figure 24–51** shows this concept.

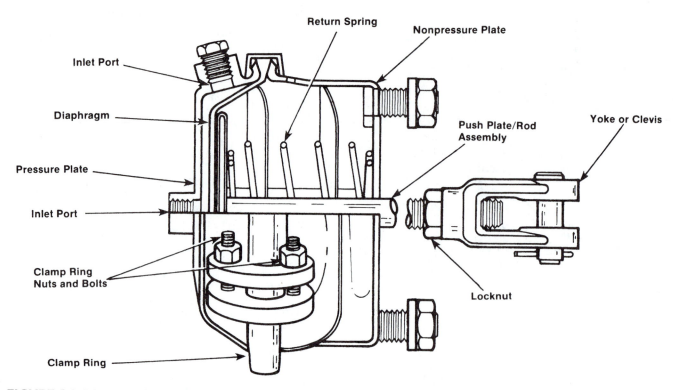

FIGURE 24–52   Cutaway view of a typical air brake chamber.

A common size of diaphragm is 30 square inches. If a 100-psi service application was applied to a 30-square-inch brake chamber, the linear force developed could be calculated as follows:

Linear force = 30 sq. in. × 100 psi = 3,000 lbs

Brake Chamber Stroke

The distance between the pressure and non-pressure plates of the brake chamber defines the limits of its stroke. After the chamber push rod plate is bottomed or stroked out, no more linear force can be applied to the foundation brake regardless of the air pressure. In fact, a stroked-out brake chamber may not apply the brakes at all, resulting in a complete loss of braking. Correct chamber pushrod travel is maintained by properly adjusting the brakes as they wear by means of the slack adjuster, either automatically or manually. **Figure 24–52** shows a sectioned view of a single-pot brake chamber.

Brake Chamber Size

Brake chambers are sized by the effective sectional area of the diaphragm. As of now, this is always measured in square inches. For instance, a brake chamber with a 30-square-inch diaphragm is known as a series 30. Most tractor drive and trailer axles use either series 30 or series 24 diaphragms. Tractor steering axles are more often fitted with series 20, 16, or smaller brake chambers.

Spring Brake Chambers

A spring brake chamber is made up of separate air and mechanical actuators in a single assembly. Each has its own air supply and diaphragm. Under normal operation, the air in each section does not come into contact with the other. The forward portion of the spring brake chamber is a simple air-actuated service brake chamber that functions exactly as the unit described in the previous section. It converts service braking application pressure to mechanical force. Service brake operation of a spring brake chamber functions independently from the parking/emergency function. **Figure 24–51** shows how the brake chamber converts the potential energy of compressed air into mechanical force, and **Figure 24–53** shows how either air pressure or spring force can be used to effect mechanical movement. **Figure 24–54** demonstrates how brake chamber pushrod movement actuates the foundation brakes.

The rear section of the spring brake chamber houses a large compressed spring capable of exerting a force of up to 1,800 lbs. This spring is sometimes known as a *maxi spring* after the manufacturing name of the first spring brakes. The spring brake section of the spring brake chamber

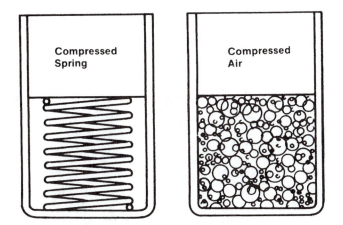

FIGURE 24–53 *Compressed spring—compressed air. (Courtesy of Bendix Commercial Vehicle Systems)*

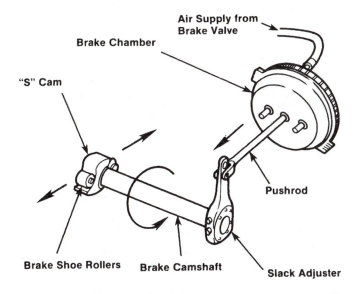
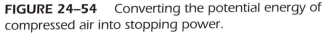

FIGURE 24–54 *Converting the potential energy of compressed air into stopping power.*

has a diaphragm that functions to "hold off" the spring. An air pressure of 60 psi is required to hold off the spring. When hold-off pressure is applied to the hold-off chamber from the park release circuit, the hold-off diaphragm (sometimes known as an "emergency" diaphragm) acts against a cage plate that prevents the spring from loading the intermediate pushrod and service chamber pushrod with spring force. Because the function of this spring is to mechanically apply the brake for parking, hold-off air must be applied to the spring brake chambers on an air brake-equipped vehicle before it can be moved. Whenever no air acts on the hold-off diaphragm, the mechanical force of the spring acts on the intermediate pushrod. The intermediate pushrod is the mechanical connection between the service chamber and the hold-off chamber of the spring brake assem-

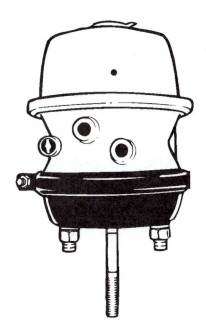

FIGURE 24–55 *Typical spring brake chamber. (Courtesy of the Holland Group Inc.— Anchorlok Div.)*

bly. When there is no air pressure in the hold-off chamber, the force of the spring acts on the intermediate pushrod and is transmitted to the service chamber pushrod and plate. This mechanically applies the brake. **Figure 24–55** shows a typical spring brake assembly, and **Figure 24–56** shows cutaway views of the different types of spring brake chamber used with S-cam and wedge foundation brake assemblies. The operating principle of a spring brake chamber is shown in **Figure 24–57**. Be sure to understand the operation of a spring brake chamber.

> **CAUTION:** The spring force in a spring brake chamber can kill. Never take any risks with spring brake chambers.

Caging Spring Brakes

All spring brakes have a means of mechanically caging them. Caging a spring brake compresses the main spring. A spring brake should be mechanically caged before it is either removed or replaced. Most spring brakes are sold with a cage bolt assembly. Caging of a spring brake consists of inserting the cage bolt lugs into slots in the internal cage plate, rotating the bolt, and then compressing the spring using the cage nut and washer. A caged spring brake should be regarded as armed and dangerous: The procedure for disarming spring brake assemblies is covered in Chapter 27.

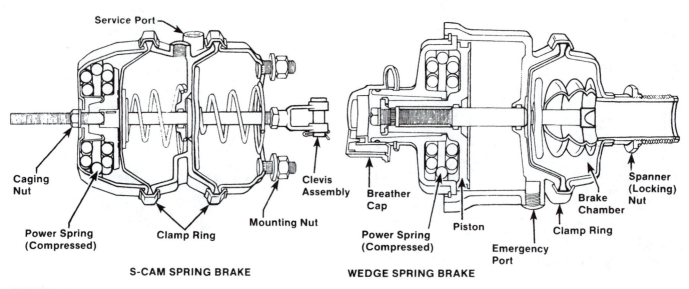

FIGURE 24-56   Spring brake chamber types used with different foundation brakes.

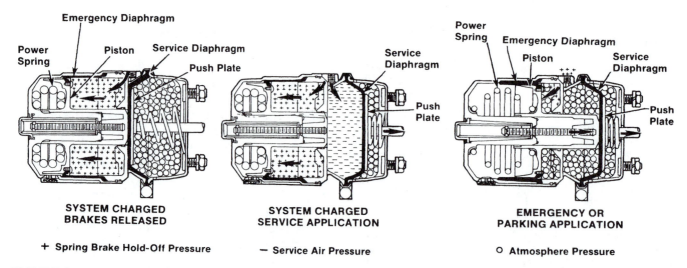

FIGURE 24-57   Spring brake chamber operation.

Summary of spring brake operation:

- No air pressure in either the hold-off or service circuits: brake mechanically applied by the full force of the main spring in the spring brake assembly. This condition is used to mechanically "park" the brake or to effect emergency braking in the event of total loss of air.
- Air pressure in hold-off circuit, none in service chamber: parking brake is released and the vehicle is capable of being moved. To release the parking brakes and move the vehicle, air pressure greater than 60 psi must be present in spring brake hold-off chambers.
- Air pressure in hold-off circuit, service application pressure in service chamber: service brakes applied at the service brake application pressure value. This condition exists at any time the vehicle is mobile and the service brakes are applied.
- No air in hold-off circuit, service application pressure in the service chamber: braking is compounded by having spring and air pressure applied to the foundation brake assembly simultaneously. Compounding of brake application pressures can damage the foundation brake components. Most systems have anti-compounding valves to prevent this condition.
- Spring brakes are mechanically caged with a cage bolt, nut, and washer assembly, often bolted to the side of the spring brake assembly.
- Compressor cut-in must not be lower than 85 psi.

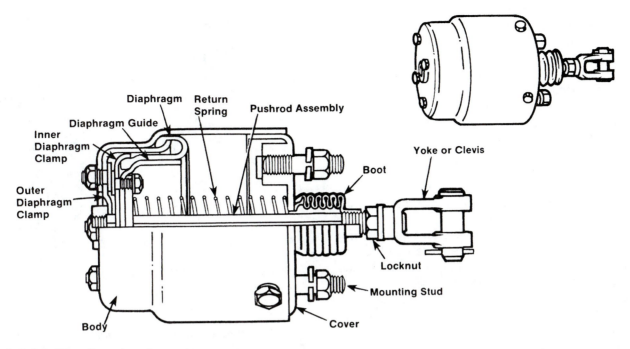

FIGURE 24–58 Rotochamber.

- Compressor build-up time must be less than 3 minutes to get from 50 psi to 90 psi with the engine running at 1,200 rpm.
- There must be sufficient reserve capacity that one full brake application does not drop pressure by more than 18 psi.

CAUTION: Spring brake chambers should be regarded as being potentially lethal. Be sure you read the set-up and disarming instructions in Chapter 27, and always observe the OEM precautions.

Rotochambers

Rotochambers use the same principles as typical brake chambers except that in place of the pancake diaphragm, a rolling-type diaphragm is used. This produces a constant output force throughout the pushrod stroke and provides much longer stroke. Rotochambers are available in a number of sizes and provide a wide range of output forces. They are more likely to be found in off-highway equipment than trucks. **Figure 24–58** shows a typical rotochamber assembly.

Long-Stroke Chambers

Long-stroke brake chambers are designed to produce longer pushrod strokes by increasing the distance between the pressure and non-pressure plates of the chamber. This will typically produce an extra half inch to one inch of stroke travel over the conventional chamber before the adjustment limit. These chambers have a similar appearance to conventional chambers, so they have the following characteristics:

- They use the acronym LS in the serial number
- The airline connection boss is square
- A tag at the chamber clamp bolt states the stroke length
- Cast letter indicates that only long-stroke diaphragms are to be used as replacement

Long-stroke chambers provide an additional safety margin over conventional stroke brake chambers and are often used where the brake configuration makes it difficult to shorten the stroke without creating brake drag. It is important to note that all-long-stroke chambers must meet the same stroke adjustment specifications as non-long-stroke chambers: Using long-stroke chambers does not eliminate the need for routine brake stroke adjustment.

DD3 Safety Actuator

The double diaphragm (DD) 3 safety actuator, like a truck spring brake unit, has three functions:

- Service braking
- Emergency braking
- Park braking

The DD-3 features a mechanical roller locking mechanism that locks the chamber pushrod for parking. Because of this locking roller mechanism, the DD-3 requires the use of special control valves. These valves are used extensively in transit and inter-city buses. They are not often used on highway trucks. DD-3 safety actuators are available in the 24 and 30 series; that is, the diaphragm effective surface areas are 24 and 30 square inches.

SLACK ADJUSTERS

The linear force produced by the brake chamber must be converted into brake torque (twisting force). Rotation of the S-cam forces the shoes into the drum to effect braking. Slack adjusters are used to connect the brake chamber with the S-cam shaft. Look at **Figure 24–54**, which shows the roles of the slack adjuster in the actuation of the brakes. First, it is the mechanical link between the brake chamber and the foundation brake assembly. The slack adjuster is attached to the brake chamber pushrod by means of a clevis and pin and to the S-camshaft by means of internal splines on the slack adjuster and external splines on the S-camshaft. Next to the slack adjuster is a lever. As a lever, it amplifies the force applied to it by the brake chamber. The amount of leverage potential depends on the distance between the clevis pin that connects it to the chamber pushrod and the fulcrum formed at the centerline of the S-cam. This is known as leverage factor. For instance, if the leverage factor of a slack adjuster was 6 inches or 0.5 foot, and a linear force of 3,000 pounds was applied to it by the brake chamber, then the brake torque could be calculated as follows:

Brake torque = Force × Distance
= 3,000 lb. × 0.5 foot = 1,500 lb.-ft.
torque

Figure 24–59 shows a manual and an automatic slack adjuster. The leverage factor of a slack adjuster is critical to the correct operation of a brake system. For this reason, it is important whenever a slack adjuster is changed that the clevis pin be located in the correct slack adjuster hole. A difference of half of an inch can drastically alter the brake torque applied to the foundation brake. This will result in unbalanced braking.

Finally, the slack adjuster is used as the means of removing excess free play that occurs as the brake friction facings wear. All current slack adjusters are required by FMVSS No. 121 to be capable of automatic adjustment. Two general types of automatic slack adjuster are available. The truck technician should also be aware of how a manual slack adjuster

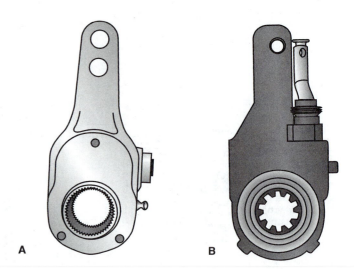

FIGURE 24–59 (A) Manual and (B) Automatic slack adjusters.

operates because there are many still in use today on truck and trailer units manufactured before 1995.

Manual Slack Adjusters

A manual slack adjuster is located on the S-camshaft by means of internal splines on a gear. External teeth on the gear mesh with a worm gear on the adjusting wormshaft. The wormshaft is locked to position by a locking collar that engages to a hex adjusting nut. This locking collar prevents rotation of the wormshaft and sets the adjustment. As friction face wear occurs, slack results; that is, the chamber-actuated pushrod will have to travel farther to apply the foundation brake. This slack is removed by adjusting the brakes. The brake is adjusted by depressing the locking collar and engaging a closed wrench (usually $9/16$-inch) over the hex on the wormshaft. The wrench can now rotate the wormshaft to remove the slack. After the brake has been properly adjusted to obtain the correct amount of pushrod free play, the adjustment is set by re-engaging the locking collar. **Figure 24–60** shows the critical components of a typical manual slack adjuster.

Stroke Sensing Automatic Slack Adjusters

The stroke-sensing automatic slack adjuster responds to increases in pushrod stroke to actuate an internal cone clutch and rotate an adjusting wormshaft. This wormshaft meshes with the external teeth of the gear engaged to the splines of the S-camshaft. This type of slack adjuster works on the assumption that if the applied stroke dimension is correctly maintained, the drum-to-lining clearance will be correct. **Figure 24–61** and **Figure 24–62** show cutaway and external views of an automatic slack adjuster.

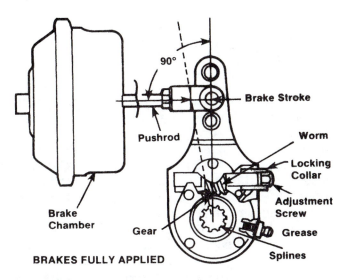

FIGURE 24-60   *Typical manual slack adjuster.*

Clearance Sensing Automatic Slack Adjusters

Clearance-sensing automatic slack adjusters respond only to changes in the return stroke dimen-

FIGURE 24–62   Automatic slack adjuster. (Courtesy of ArvinMeritor Inc.)

sion, so they function almost opposite to stroke-sensing models. Because they "adjust" only when the return stroke dimension increases, if the applied stroke dimension became excessive, no adjustment would occur. Clearance-sensing automatic slack adjusters work on the assumption that if the shoe-to-drum clearance is normal, the stroke should be within the required adjustment limits.

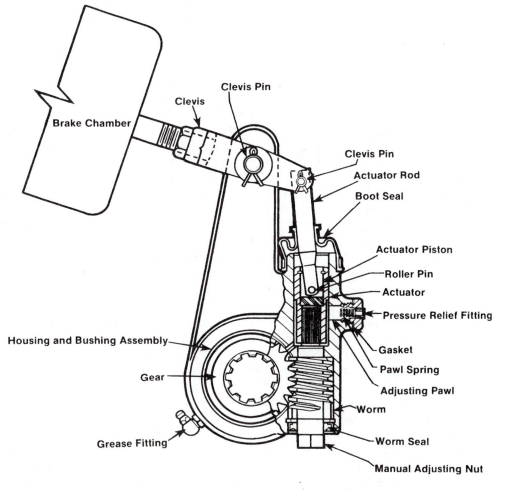

FIGURE 24–61   *Stroke-sensing automatic slack adjuster. (Courtesy of ArvinMeritor Inc.)*

Shop Talk

Slack adjusters without a grease fitting should not be assumed to be functional. Slack adjusters should be lubricated with a low-pressure grease gun.

Types of automatic slack adjusters

Manufacturer	Type	Features
Bendix	Clearance-sensing/apply stroke	Infinitely variable adjustment. Anti-reverse lock. Left and right-hand versions.
Gunnite	Clearance-sensing/apply stroke	Incremental adjustment. Limited correction for over-adjustment.
Haldex	Clearance-sensing/release stroke	Incremental adjustment. Limited correction for over-adjustment.
Rockwell	Stroke-sensing/release stroke	Incremental adjustment. Anti-reverse feature. Front and rear axle versions different.

GAUGES

Air pressure gauges in air brake circuits are used to monitor air pressure in the primary and secondary circuits, and to give the driver an indication of what the application pressure is. Either individual gauges or a two-in-one gauge can be used to display the system pressure in the primary and secondary circuits. When a two-in-one gauge is used, a green needle indicates primary circuit pressure and a red needle indicates secondary circuit pressure. If a loss of pressure occurs in either of the circuits, the driver is alerted to the condition after the pressure drops below 60 psi. The alert is usually visible (gauges/warning lights) and audible (buzzer).

Some systems also have an application pressure gauge. This gauge tells the driver the exact value of a service application. Most application pressure gauges read the primary circuit pressure only, so when the trailer application valve is used, there may be no indicated reading.

24.8 BRAKE SYSTEM BALANCE

Two factors determine the amount of braking force applied to each wheel in a truck air brake system. Because the control and application circuits are managed by compressed air, the first factor of a balanced brake system is pneumatic balance. For a service application to be balanced, the application valve, for example the foot valve, must deliver a signal pressure to relay valves. The relay valves are designed to replicate the signal they receive by sending a local supply of air to service brake chambers they supply. The idea is that during a service brake application, each of the brake chambers on the vehicle is charged with air at exactly the same pressure value. Most brake systems also have some ability to proportion these service application pressures to ensure that the vehicle can be safely stopped with maximum driver control. A good example would be the ratio valve discussed earlier in this chapter. This proportions braking to the front axle brakes according to the service application pressure. Timing is another factor. Because the signal pressure must travel much farther to reach the brakes at the rear of the trailer, unless some accommodation is made, the front brakes on the tractor unit apply well before those on the rear of the trailer. Pneumatic timing means that the brakes on a tractor/trailer combination are applied simultaneously.

Brake torque balance means that the mechanical braking force applied to each wheel assembly is equal. If, for instance, the brakes were out of adjustment on one side of an axle and properly adjusted on the other, when the service brakes were applied it would be possible to lock up the brakes on one side and have no braking force applied to other. Many other factors besides maladjusted brakes could contribute to brake torque imbalance. Mismatched friction facings, glazed linings on one wheel (usually caused by a leaking wheel seal), improper brake chamber pushrod setting, mismatched slack adjuster lengths, glazed drums, and irregularly worn drums could all result in brake torque imbalance.

In the operation of a tractor/trailer combination, each axle and each wheel assembly must bear exactly the proportion of total vehicle load it was designed to carry. Each foundation brake on each wheel must be properly adjusted so it performs its share of the required vehicle braking. Also, the brake applications have to be timed so each application and release of the brakes is practically simultaneous. Anything that changes on the brake or wheel assembly can affect the brake balance. Tire treads and their coefficient of friction (aggressiveness) with the road surface both impact the vehicle braking efficiency. The fitting of low-profile tires can also produce overbraking due to the lower radii of tires.

Although system air pressure is available to the driver for braking the vehicle, only under emergency service applications that would this be required. The majority of service brake applications are made at an application pressure of 20 psi or less. This means that a variance of 4 or 5 psi in the normal pressure required to move the shoes from a retracted position to drum contact (typically 6 psi) on different axles that might be relatively unimportant at a 60-psi application pressure, becomes critical at low application pressures of 15 to 20 psi, where most of the braking takes place.

> **NOTE:** Around 95 percent of braking on a loaded tractor/trailer combination involves brake application pressures of 25 psi or less. 85 percent of braking involves brake application pressures of 15 psi or less.

The key to brake balance is proper maintenance of the brake system. Balanced braking means the braking performance of the vehicle is functioning properly. Vehicle braking requirements are always specified for the maximum load the vehicles is designed to haul. No truck or tractor/trailer combination run at partial or empty loads will brake at optimum efficiency; in most cases, they will be overbraked. In fact, it should be noted that most unloaded trucks are more difficult to handle generally than when they are loaded. Three major factors affect brake balance and thus braking performance:

- **Brake torque balance between tractor and trailer.** To have balanced brakes between tractor and trailer, the brakes on each axle of each unit must produce brake torque in proportion to the load applied up to the capacity of each axle. Brakes are designed and rated for the maximum axle rating such as a 23,000-pound single axle or a 34,000-pound tandem. An empty or partially loaded rig will tend to overbrake. Lightly loaded axles are more likely to lock up. Often this is what causes the black tire streaks on the pavement near intersections. Overbraking causes wheel lockup. The result is flat-spot tires.
- **Air timing balance between tractor and trailer.** Air timing is defined as the time required for signal air to be converted into a brake application. This begins the instant the

driver's foot moves the brake pedal. When a trailer and tractor are coupled in a combination, it is important that the trailer brakes are applied simultaneously with those on the tractor. The brakes on each axle of the entire rig should receive air at the same time and at the same pressure value. To help ensure this, relay valves are often specified with different crack pressures. The crack pressure of a relay valve is the air pressure value required to overcome the relay valve piston return spring. Setting higher crack pressures on the tractor relay valves that are located close to foot valve, and at a lower value on those on the rearmost trailer axles, helps achieve balanced brake applications. For this reason, when replacing relay valves, always properly match the serial numbers. Two valves of identical appearance may actually perform differently. Sequenced crack pressures are most critical in tractor/train combinations such as A and B train doubles: In this type of application, the lowest relay valve crack pressures would be on those valves on the pup trailer and the highest over the tractor axles.

- **Pressure differential between tractor and trailer.** There may be a difference between tractor and trailer air pressure when these units are coupled. Pressure differentials between axles are created by the various valves used in the brake actuation system. Each valve used in a tractor or trailer brake system creates additional potential for pressure drop between input and output pressure. The pressure to the front axle chamber of a tractor normally will have close to the same pressure as the output line of a foot valve. The pressure to the rearmost axle chamber in a combination will have the greatest differential. The differential value is always greatest when service brake application pressures are lowest. Because of the number of valves and length of plumbing required between the foot valve and the rearmost trailer axles, it is difficult to achieve identical application pressures on all axles of a combination.

Balanced brakes extend brake service life and provide the driver with the tools he needs to safely stop a truck during emergency braking. A tractor/trailer jackknife condition is often caused by unbalanced brakes. The condition occurs as often as it does because

- There is a tendency in the trucking industry to maintain tractor brakes to higher overall standards than trailer brakes.

- Compatibility of the tractor brake system with that of the trailer: When the tractor brakes more aggressively than a trailer on a slippery pavement, the chances of jackknife increase.

BRAKE IMBALANCE TERMS

This is a summary of some of the terms used to describe brake imbalance in trucks.

Brake torque imbalance	Occurs when the foundation brakes have variable aggressiveness. Ideally, the brakes on a vehicle should be applied with the same amount of mechanical force and the critical friction surfaces should have the same coefficients of friction.
Jackknife	Jackknife occurs when the rear wheels of the tractor lock, causing the tractor to spin. When this condition occurs, the forward momentum of the trailer during the tractor spin makes the condition difficult to steer out of.
Trailer swing	Trailer swing occurs when the trailer wheels lock causing the trailer to spin around while the tractor remains in a straight-ahead position. An experienced truck driver can often steer out of a trailer swing condition.
Plow-out	The tractor steering wheels lock, making the tractor difficult to steer. Most plow-out is driver error, and experienced drivers can avoid it.
Pneumatic timing	Describes the timing actuation of each brake chamber on the rig. Ideally, all the foundation brakes should be pneumatically actuated at the exactly the same moment, whether they are on the tractor or the rearmost trailer of a truck train.
Pneumatic balance	The ability of the air system on a rig to manage the correct air pressure at every brake chamber on the rig during braking and release.

Consequences of Brake Imbalance

Unbalanced brake torque, timing, and pressure in the brake system of a heavy-duty tractor/trailer combination can cause a number of avoidable problems. Tractor-to-trailer compatibility is also a factor. Properly functioning brake systems on a tractor and a trailer can produce compatibility problems when they are coupled. The effects of brake imbalance can produce one or more of the following conditions:

- Uneven brake lining wear
- Reduced stopping capability
- Brake fade
- Brake drag
- Jackknife
- Wheel hop: suspension damage
- Trailer surge: tractor brakes apply before trailer
- Reduced brake drum life
- Tire wear
- Tire flatspotting

The reason or reasons for brake torque or timing imbalance may not be easy to detect. The causes of pneumatic imbalance include the use of oversized air lines on trailers, air leaks, sticking relay valves, and the use of incompatible air valves. Torque imbalance occurs when some brakes are stronger or weaker than others. There can be a number of causes such as drums machined beyond tolerance, nonuniform adjustment, oil-soaked linings, polished drums and glazed linings, nonuniform drum-to-lining contact, linings with varying friction ratings, incorrect slack adjuster length, nonuniform chamber sizes, or an incorrectly specified axle rating.

Acceptable pneumatic balance usually requires that the air delivered to each axle does not vary more than 2 psi during a 10-to-20 psi application. Pneumatic timing becomes most critical in a multiple trailer combination because the signal air pressure has to travel much further before it arrives at the pup axles.

Maintenance and Wear Conditions That Cause Brake Imbalance

- **Scored, glazed, or "bell-mouthed" brake drums.** Often, brand-new drums deform to an eccentric shape in storage. It is no longer regarded as good practice to machine used drums, but it is good practice to machine new brake drums immediately prior to installation.
- **Linings contaminated by leaking oil seals or excessive grease pumped into the inboard S-cam bushing.** This contamination gets baked into the friction material, and the result is lining glazing.

- **Brake shoe return springs that stretch and lose their temper due to excessive heat.** These springs should be replaced at every brake overhaul.
- **Rollers that have become flat-spotted.** S-cam rollers should be replaced at each brake overhaul.
- **Retrofitting of aftermarket or remanufactured brake valves that can destroy pneumatic balance.** Relay valve crack pressures should be matched when replacing valves.
- **Defective valves.** Valves can slow down or fail when their moving parts become gummed up (water and oil sludge), corroded (water contamination), and carboned due to oil vapor and heat.
- **S-cam profiles that wear significantly.** Although S-cams may not have to be replaced at each brake job, they should be carefully inspected.
- **S-camshaft slop of any kind.** Worn camshaft bushing bearings, bushings, and splines degrade performance. Bushings should be inspected and replaced if necessary at every brake overhaul.
- **Bent spiders.** Bent spiders reduce lining-to-drum contact and the effective leverage of the shoe. Brake spiders tend to bend when seized anchor pins are removed: Use a proper anchor pin puller rather than a sledge hammer and punch to remove seized anchor pins.

MECHANICS OF S-CAM, DISC AND WEDGE BRAKES

The performance of S-cam, disc, and wedge brakes varies significantly, and this can produce problems in a tractor/trailer combination in which the foundation brakes are mixed. For instance, a tractor equipped with S-cam brakes coupled to a trailer equipped with wedge brakes will cause trailer overbrake/tractor underbrake because in moderate to severe braking, double-actuated wedge brakes have superior brake mechanics. A tractor equipped with air disc brakes hauling a trailer with S-cam brakes can cause the opposite problem due to the higher mechanical efficiency of disc brakes over an S-cam actuated drum.

FMVSS No. 121 Requirements

As we said earlier in this chapter, FMVSS No. 121 governs the requirements of air brake systems used on North American highways. Here are a few FMVSS requirements:

- Maximum application time (signal to chamber) for each service chamber must not exceed 0.45 second on a truck or tractor and 0.60 second on a trailer. Application time is defined as the time required to increase pressure from atmospheric to 60 psi.
- Maximum release time for trucks and tractor units is 0.55 second.
- Parking brakes must be able to hold the vehicle loaded to its gross weight stationary on a 20 percent grade.
- Unapplied pressure loss in the supply (wet) tank must not exceed 2 psi for single vehicles and 3 psi for a tractor/trailer combination in one minute.

SUMMARY

- An air dual-circuit brake system is composed of a supply circuit, primary circuit, secondary circuit, parking/emergency control circuit, trailer circuit, and foundation brake assemblies.
- Air compressors are single-stage, reciprocating piston air pumps that are either gear- or belt-driven.
- Air dryers are used to help eliminate moisture and contaminants from the truck's air system.
- Dual-circuit application or foot valves, trailer application valves, bobtail proportioning valves, ratio valves, quick-release valves, relay valves, TP valves, dash control valves, double check valves, and check valves are some of the critical valves of an air brake system.
- The potential energy of compressed air is changed into mechanical force in an air brake system by slack adjusters and brake chambers.
- The most common type of foundation brake assembly used on air brake-equipped trucks is the S-cam type.
- Slack adjusters multiply the force applied to them by the brake chamber into brake torque.
- Brake torque applied to the S-camshafts results in the shoes being forced against the drum.
- Air discs operate by using an air-actuated caliper to squeeze brake pads against both sides of a rotor.
- Wedge brakes use a drum, a pair of shoes, and air-actuated wedges are used to force the shoes against the drum.
- Brake torque balance refers to the ability of a brake system to apply balanced mechanical brake force at all the foundation brake assemblies.
- Pneumatic timing refers to the ability of an air brake system to time the air control and actuation circuits so all the foundation brake components are applied at exactly the same moment.

REVIEW QUESTIONS

1. Which of the following air brake system components functions as a lever?
   a. brake chamber
   b. rotor
   c. caliper
   d. slack adjuster

2. Which of the following components sets the system cut-in and cut-out pressures?
   a. service reservoir
   b. governor
   c. air dryer
   d. supply tank

3. What is the function of a system safety pop-off valve?
   a. releases excess pressure from the compressor
   b. pressure protects air in the supply tank
   c. alerts the driver when air pressure has dropped below system pressure
   d. relieves pressure from the supply tank in the event of governor cut-out failure

4. The function of the foot valve or dual-circuit application valve is to
   a. increase the flow of air to the service reservoirs
   b. reduce the flow of air from the service reservoirs
   c. meter air to the service brake circuit
   d. meter air to the brake hold-off chambers

5. The potential energy of an air brake system is
   a. compressed air
   b. foot pressure
   c. spring pressure
   d. vacuum over atmospheric

6. Which of the following components is a requirement on every air tank?
   a. air dryer
   b. safety valve
   c. drain cock
   d. pressure protection valve

7. What is the minimum total storage volume of air dedicated to brake system requirements of an air brake-equipped trailer?
   a. equal to the total volume of all the brake chambers
   b. two times total chamber volume
   c. four times total chamber volume
   d. eight times total chamber volume

8. The function of an air brake chamber is to
   a. convert the energy of compressed air into mechanical force
   b. multiply the mechanical force applied to the slack adjuster
   c. reduce the air pressure for a service application
   d. increase the air pressure for parking brakes

9. An air brake-equipped truck is traveling on a highway at 60 mph and the driver makes a service brake application. Which of the following should be true in all of the spring brake chambers on the vehicle?
   a. only the hold-off chambers are charged with air
   b. only the service chambers are charged with air
   c. both the service and hold-off chambers are charged with air
   d. neither the service nor the hold-off chambers are charged with air

10. Technician A says automatic slack adjusters can improve brake balance and reduce downtime and maintenance costs. Technician B says manual slack adjusters should not be adjusted after installation. Who is correct?
    a. Technician A
    b. Technician B
    c. both A and B
    d. neither A nor B

11. When an inversion valve becomes functional during a brake failure, what force does it use to brake the vehicle?
    a. air pressure
    b. spring pressure
    c. hydraulic pressure
    d. vacuum

12. Which of the following determines the amount of leverage gained by a slack adjuster?
    a. slack adjuster length
    b. pushrod stroke length
    c. S-cam geometry
    d. number of internal splines

13. Which of the following is an advantage of an open anchor brake shoe?
    a. servo action
    b. anchor pins never have to be changed
    c. faster brake jobs
    d. retraction springs last longer

14. The brake shoes, retraction and fastening hardware are mounted to
    a. a spider
    b. a backing plate
    c. the drum
    d. the axle spindle

15. For a service signal to pass through a TP valve, which of the following must be true?
    a. the service gladhand must be connected to the trailer
    b. the trailer supply dash control valve must be on
    c. there must be air in the primary circuit
    d. there must be air in the secondary circuit

16. What actuates the relay piston in a foot valve on an air brake system when the valve is operating normally?
    a. foot pressure
    b. air pressure
    c. spring pressure
    d. mechanical pressure

17. Which of the following does the most to ensure that the driver has brake feel at the foot valve in a truck air brake system?
    a. spring pressure
    b. metered primary circuit service pressure
    c. metered secondary circuit service pressure
    d. system cut-out pressure

18. Which of the following best describes the term coefficient of friction?
    a. the heat rating of friction materials
    b. the aggressiveness of a friction material
    c. the longevity of friction materials
    d. the melting point of friction materials

19. Technician A says that in a tractor/trailer B-train combination, the crack pressures of the relay valves over the pup axles would be the highest on the rig. Technician B says that quick-release valves on the service gladhands can speed up service brake release times on multi-trailer combinations. Who is right?
    a. Technician A
    b. Technician B
    c. both A and B
    d. neither A nor B

20. A trailer compliant with current FMVSS No. 121 air brake system requirements has no air in its onboard air tanks. When coupled to a tractor air supply, where is the air directed to first?
    a. the service reservoirs
    b. the service brake chambers
    c. the hold-off chambers
    d. the main air reservoir

25

Hydraulic Brakes and Air-Over-Hydraulic Brake Systems

🔑 **Prerequisites**
Chapters 2 and 3

Objectives

After reading this chapter, you should be able to
- Describe the principles of operation of a hydraulic brake system.
- Identify the major components in a truck hydraulic brake system.
- Explain the operation of a hydraulic brake system.
- Describe the operation of drum brakes, both non-servo and servo, disc brakes, and dual brakes in a hydraulic braking system.
- List the major components of a master cylinder.
- Describe the use and operation of wheel cylinders and calipers.
- Name the various hydraulic valves and controls used in an air brake system.
- Explain the operation of a hydraulic power booster.
- List the major components of an air-over-hydraulic braking system.
- Outline some typical maintenance and service procedures performed on hydraulic and air-over-hydraulic brake systems.
- Describe the operation of a typical hydraulic ABS system.

Key Terms List

calipers
combination valve
Federal Motor Vehicle Safety Standard No. 105
   (FMVSS No. 105)
hydraulic power booster
hygroscopic
load proportioning valves (LPVs)
master cylinder

metering valves
primary circuit
pressure differential valve
proportioning valves
secondary circuit
servo brake
slave cylinders
wheel cylinders

When hydraulic brakes are used on trucks in North America, it is mainly on medium- and light-duty applications that are generally not required to haul a trailer. Hydraulic brakes work on the premise that if mechanical force is applied to a liquid in a closed circuit, it can be used to transmit motion or multiply and apply force. A basic hydraulic brake system consists of a **master cylinder** actuated mechanically (by

the driver's boot), hydraulic pipes and hoses that connect the master cylinder with **wheel cylinders** (**slave cylinders**), and foundation brakes that convert the hydraulic pressure into mechanical braking force. Depressing a brake pedal moves a piston in a master cylinder. This piston forces fluid under pressure through brake lines and hoses to wheel cylinders. The force applied at the master cylinder can be amplified at the wheel cylinders if they are of larger diameter than the master cylinder. This concept is shown in **Figure 25–1** and **Figure 25–2**. The amount of braking force delivered at the wheels in a hydraulic brake system is always proportional to the applied force at the master cylinder, even when power-assist devices are used. Most hydraulic brakes use some form of power assist today even in the lightest duty applications. Hydraulic brake standards must comply with **Federal Motor Vehicle Safety Standard No. 105 (FMVSS No. 105)**. This federal legislation performs the same role in hydraulic brake standards as FMVSS No. 121 does with air brakes. FMVSS No. 105 requires that all highway hydraulic brake systems be dual circuit and have the ability to park and perform an emergency stop mechanically. The objective of this section is to provide an introduction to hydraulic brakes, rather than a comprehensive study of them.

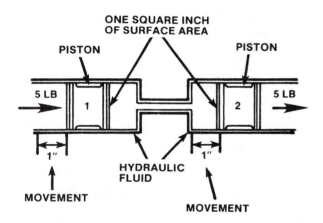

FIGURE 25–1 Force and movement are transmitted equally.

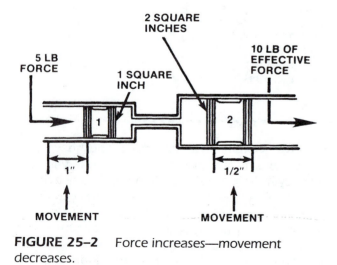

FIGURE 25–2 Force increases—movement decreases.

25.1 HYDRAULIC BRAKE FLUID

Brake fluid standards are set by the Society of Automotive Engineers (SAE) and Department of Transportation (DOT). There are 3 categories of brake fluid known as DOT 3, DOT 4, and DOT 5. DOT 3 and 4 use a polyglycol base. These are **hygroscopic**, meaning that they are designed to absorb moisture that enters the system. This is an advantage for a brake fluid because it prevents the formation of water droplets that could cause localized corrosion, freeze, or boil. DOT 5 is a silicone-based synthetic brake fluid that is not hygroscopic. It has a higher compressibility factor than DOT 3 and 4 fluids and produces a spongy feel. It is not recommended for use in ABS.

 **Shop Talk** _____

Hygroscopic brake fluid can absorb air-born moisture rapidly. If DOT 3 and 4 brake fluids are left uncovered in a container, they can be ruined in as short a period as one hour.

25.2 HYDRAULIC BRAKE SYSTEM COMPONENTS

This section will discuss the components in a typical hydraulic brake circuit, beginning with the hydraulic circuit and finishing with the foundation brakes. **Figure 25–3** shows a simple hydraulic brake circuit in a small truck. **Figure 25–4** shows a schematic of a similar system. Both systems are dual circuit, which is mandatory, and, somewhat typically, they both have a front disc and rear drum configuration.

MASTER CYLINDERS

Since 1967, all highway vehicle hydraulic brake systems have been dual circuit, requiring the use of a tandem master cylinder. A tandem master cylinder has a single bore with two separate pistons and chambers. The two sections or chambers (see **Figure 25–5**) are designed so the fluid from one chamber cannot come into contact with the other. This means that in the event of a total failure in one circuit, the hydraulic action of the other is not affected. A master cylinder is so called because it hydraulically actuates one or more slave cylinders. The force produced by a slave cylinder is always

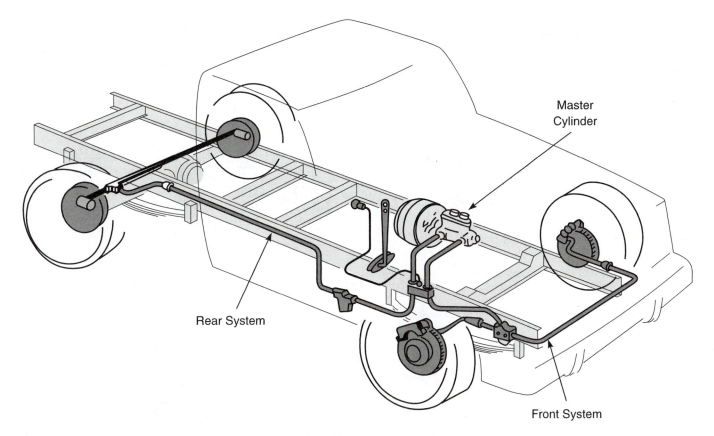

Master
Cylinder

Rear System

Front System

FIGURE 25–3   *Dual circuit brakes with a split disc/drum configuration.*

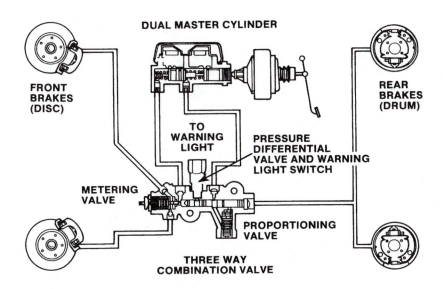

DUAL MASTER CYLINDER

FRONT
BRAKES
(DISC)

REAR
BRAKES
(DRUM)

TO
WARNING
LIGHT

PRESSURE
DIFFERENTIAL
VALVE AND WARNING
LIGHT SWITCH

METERING
VALVE

PROPORTIONING
VALVE

THREE WAY
COMBINATION VALVE

FIGURE 25–4   *Schematic of a dual circuit, split disc/drum hydraulic brake system.*

proportional to the force delivered to the master cylinder. In a brake system, the slave cylinders are known as wheel cylinders.

Each chamber of a tandem master cylinder is connected to a reservoir by means of inlet ports. A dual circuit master cylinder has two internal pistons, one in each of its chambers. When the master cylinder is not actuated, each piston, known as primary and secondary pistons, is held in its retracted position by a spring. When the master cylinder is actuated, the primary piston is pushed past the primary inlet port and pressure is developed in the primary chamber.

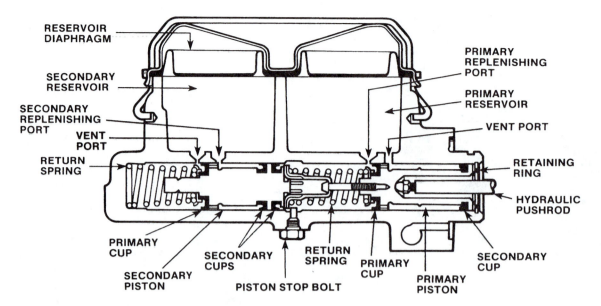

FIGURE 25–5 Cross-section of a typical dual-circuit master cylinder.

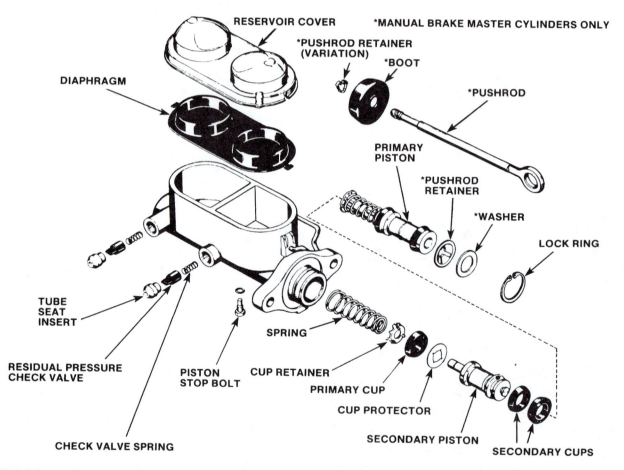

FIGURE 25–6 Exploded view of a typical dual-circuit master cylinder.

This pressure actuates the brakes on the **primary circuit** and, simultaneously, acts on the secondary piston. This moves the secondary piston past its inlet port, applying pressure to the brakes on the **secondary circuit**. When the actuating pressure on the master cylinder is released, compensating ports help vent the brake fluid in the chambers back to the reservoirs. Use **Figure 25–5** and **Figure 25–6** to learn how the components in a master cylinder interact with each other.

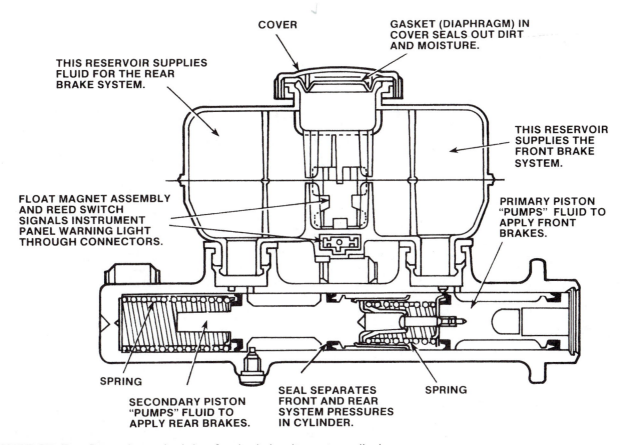

COVER

GASKET (DIAPHRAGM) IN
COVER SEALS OUT DIRT
AND MOISTURE.

THIS RESERVOIR SUPPLIES
FLUID FOR THE REAR
BRAKE SYSTEM.

THIS RESERVOIR
SUPPLIES THE
FRONT BRAKE
SYSTEM.

FLOAT MAGNET ASSEMBLY
AND REED SWITCH
SIGNALS INSTRUMENT
PANEL WARNING LIGHT
THROUGH CONNECTORS.

PRIMARY PISTON
"PUMPS" FLUID TO
APPLY FRONT
BRAKES.

SPRING

SPRING

SECONDARY PISTON
"PUMPS" FLUID TO
APPLY REAR BRAKES.

SEAL SEPARATES
FRONT AND REAR
SYSTEM PRESSURES
IN CYLINDER.

FIGURE 25–7   *Operating principle of a dual circuit master cylinder.*

In the event of a secondary circuit hydraulic failure, the primary circuit will function as normal, but the fluid it sends to act on the secondary piston will produce no effect. A brake warning light would alert the driver that half the hydraulic circuit had failed. If a failure occurs in the primary circuit, the brake pedal would have to travel much further until a stub on the primary piston mechanically contacts the secondary piston. Although the secondary piston is actuated mechanically in this failure mode, the brakes on the secondary circuit should otherwise apply normally. **Figure 25–7** explains the operating principle of a dual-circuit master cylinder.

The hydraulic brake plumbing from the dual master cylinder can be longitudinally split or diagonally split. The longitudinally split system operates the front and rear brakes from each of the tandem circuits. The diagonally split system operates a right front wheel and left rear wheel from one circuit, and the left front wheel and right rear wheel from the other. They are popular on front wheel drive vehicles that have a much greater load transfer to the front axle under severe braking.

Hydraulic Power Boosters

In hydraulic brake systems, the use of a vacuum or **hydraulic power booster** to assist the master cylin-

der in applying the brakes is common. The use of a booster reduces the pedal effort and the pedal travel required to apply the brakes as compared to a non-power assisted system. **Figure 25–8** shows a typical master cylinder with a dash-mounted hydraulic booster. Due to federal regulations requiring that vehicles stop in a shorter distance with less pedal or treadle effort, vacuum boosters are not often seen on

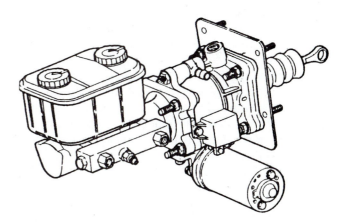

FIGURE 25–8   *Master cylinder with a dash-mounted hydraulic booster. (Courtesy of Bendix Commercial Vehicle Systems)*

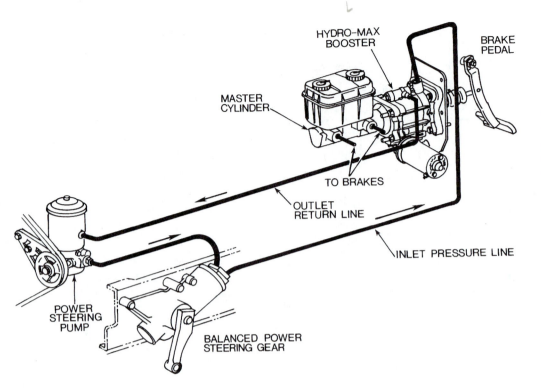

HYDRO-MAX BOOSTER

BRAKE PEDAL

MASTER CYLINDER

TO BRAKES

OUTLET RETURN LINE

INLET PRESSURE LINE

POWER STEERING PUMP

BALANCED POWER STEERING GEAR

FIGURE 25–9
Hydraulic booster powered by the power steering pump. (Courtesy of Bendix Commercial Vehicle Systems)

light- and medium-duty trucks. In other words, most trucks with hydraulic brakes use hydraulic power assist.

A hydraulic booster unit is comprised of an open center valve and reaction feedback mechanism, a large diameter boost power piston, a reserve electric motor pump, and an integral flow switch. The unit is powered by either the truck power steering pump (see **Figure 25–9**) or by a hydraulic pump dedicated to the booster. The reserve electric motor pump (**Figure 25–10**) provides backup power in the event the booster fails. The backup pump is activated by the integral flow switch.

The master cylinder used with the hydraulic booster is usually a dual split system type using separate brake fluid reservoirs, pistons, and output ports for the front and rear brake systems. In some systems, the master cylinder is internally divided into three compartments: primary, front, and rear.

As we have said, the booster can be located in the vehicle hydraulic circuit in one of two ways. First, it can be integrated with the power steering hydraulic pump and steering gear, operating in series with the power steering gear. In this case, the system flow and pressure demands during simultaneous steering and braking are considerable. The fluid flow path is designed to minimize the interaction between the power steering gear and the hydraulic booster.

Second, the unit can be powered by a dedicated hydraulic power source like that shown in **Figure 25–11**. A flexible or rigid pressure line runs from the hydraulic power source to the booster inlet pressure

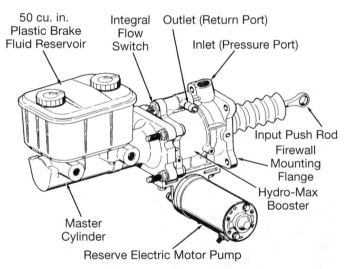

50 cu. in. Plastic Brake Fluid Reservoir

Integral Flow Switch

Outlet (Return Port)

Inlet (Pressure Port)

Input Push Rod

Firewall Mounting Flange

Hydro-Max Booster

Master Cylinder

Reserve Electric Motor Pump

FIGURE 25–10 Hydraulic booster and master cylinder assembly with reserve electric pump.

port. Both maximum inlet port pressure and general principles of operation are similar to steering pump-driven systems.

A typical hydraulic booster interfaces with the following mechanical and electrical components:

- A pressure differential warning valve
- Drum/drum, disc/drum, or disc/disc foundation brakes
- A relay
- The vehicle storage battery(s)

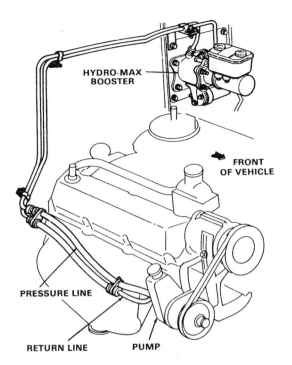

HYDRO-MAX BOOSTER

FRONT OF VEHICLE

PRESSURE LINE

RETURN LINE   PUMP

FIGURE 25–11 Hydraulic booster with a dedicated hydraulic power source. (Courtesy of Bendix Commercial Vehicle Systems)

- A warning light for the pressure differential warning valve
- Electrical connectors and wiring
- A buzzer
- A warning light for the reserve electrical motor
- An optional electronic monitor module

During normal system operation, fluid flow from the hydraulic power source enters the inlet pressure port of the booster, flows through the throttle valve and power piston, through the flow switch, and exits from the outlet (return) port. Force applied to the brake pedal by the vehicle operator is multiplied by the lever ratio of the pedal mechanism to move the input pedal rod of the booster. This movement activates the throttle valve, restricting flow through the power piston applying boosted force to the master cylinder primary piston. A reaction piston inside the power piston subassembly provides the driver with brake feel during an application of the brake pedal. Fluid flow through the flow switch opens the reserve motor pump electrical circuit during normal operation. A separate check valve in the motor pump prevents backflow through the motor pump during normal power applications.

If normal flow from the power steering pump is interrupted, the electric motor provides the power for reserve stops. Upon flow interruption, the integral flow switch closes, energizing a power relay to pro-

vide power to the motor. During reserve operation, fluid is retained within the booster by the inlet port check valve. The motor pump recirculates fluid within the booster assembly with pressure supplied on demand via the throttle valve. In this case, the number of applications is limited only by the electrical capacity of the vehicle. Manual braking is also possible in the event that both systems fail.

Pressure Differential Valve

All dual-circuit hydraulic systems have a **pressure differential valve** or brake warning light system. This is switched by a hydraulically actuated electric switch. Its function is to indicate to the driver when one half of the system is not functioning. It consists of a cylinder inside of which is a spool valve. Each end of the spool valve is subject to pressure from each of the two circuits of the brake system; that is, primary and secondary circuits. When a brake application is made in a dual-circuit hydraulic system, the pressure in the primary and secondary circuits should be the same. If this is the case, the spool valve remains balanced and the brake warning light circuit remains open. However, if no pressure exists in one of the two circuits, the spool valve will be moved off center, contacting a terminal and closing the brake warning light circuit. Closing the brake warning light circuit illuminates the dash brake warning light. On some brake systems, the parking brake will also illuminate the brake warning light circuit when applied.

Metering Valves

Metering valves are incorporated into a hydraulic brake system that uses disc brakes on the front axle and drum brakes on the rear axle. Disc brakes have much lower lag times than drum brakes. Lag time means the time between the first movement of the brake pedal and the moment that braking effort is applied. Metering valves delay the application of the front brakes until pressure has been established in the rear brake circuit. This helps balance the application timing. It also helps reduce the amount of light-duty braking performed by the front wheel brakes.

Proportioning Valves

Proportioning valves are also used in systems using front disc and rear drum brakes. During braking, the vehicle is thrust forward onto the front axle. This is known as load transfer, and the more severe the braking, the greater the load transfer onto the front axle. Load transfer can result in rear wheel drum brake lock-up. **Proportioning valves** prevent pressure delivered to the rear wheel brakes from exceeding a predetermined pressure value. Some trucks use load-sensitive, rear wheel proportioning valves called **load proportioning valves** (**LPVs**). These proportion the pressure applied to the rear wheels according to how much weight is on the rear of the vehicle. The vehicle weight is sensed by mounting the valve on a frame cross-member and using a linkage and lever system attached to the rear axle housing. In this way, as the load increases over the rear axle, the rear brakes can assume a greater proportion of the required vehicle braking. The result is that when the vehicle load is light, the rear wheels are less likely to be over-braked.

Combination Valves

The hydraulic safety switch, metering, and proportioning valves are normally housed in a single valve body called a **combination valve**. This valve is usually located close to the master cylinder assembly. **Figure 25–12** shows how a combination valve is located in a hydraulic brake circuit.

HYDRAULIC DRUM BRAKES

In a typical hydraulic drum brake system, the brakes are actuated by wheel cylinders. The manner in which the brake shoes are mounted on the brake backing plate will determine whether the system is self-energizing or servo-assisted. **Figure 25–13** shows a typical drum brake assembly: Note the location of the components.

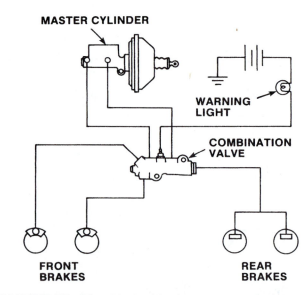

FIGURE 25–12   Hydraulic brake system with a combination valve.

Wheel Cylinders

Hydraulic drum brakes are actuated by a wheel cylinder. The wheel cylinder responds to hydraulic pressure delivered to it by the master cylinder. Wheel cylinders usually contain a pair of pistons, which, when actuated hydraulically, force brake shoes apart and into the drum. Some wheel cylinders are single-actuating and use a single piston to actuate an anchor pin pivoted brake shoe. Cups seal the pistons in the wheel cylinder bore. When the master

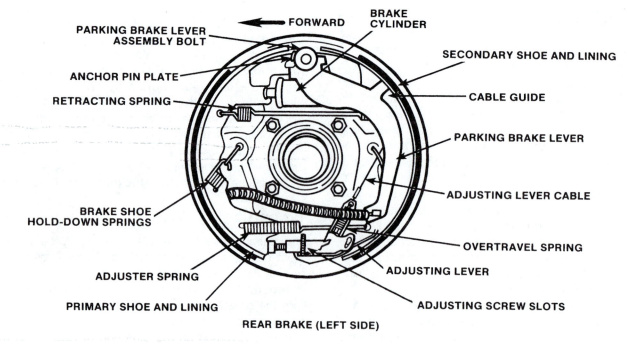

FIGURE 25–13   Typical drum brake components.

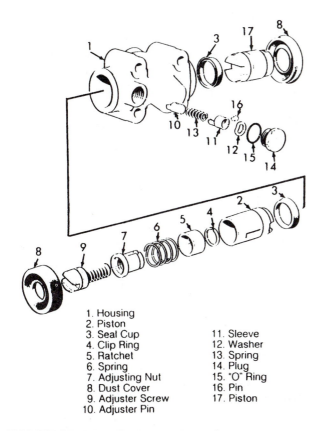

1. Housing
2. Piston
3. Seal Cup
4. Clip Ring
5. Ratchet
6. Spring
7. Adjusting Nut
8. Dust Cover
9. Adjuster Screw
10. Adjuster Pin
11. Sleeve
12. Washer
13. Spring
14. Plug
15. "O" Ring
16. Pin
17. Piston

FIGURE 25–14 Exploded view of a double-acting wheel cylinder. (Courtesy of Volvo Trucks North America, Inc.)

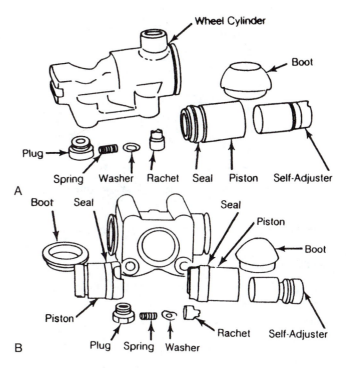

FIGURE 25–15 (A) Exploded view of a single-acting, front wheel cylinder; (B) exploded view of a double-acting, rear wheel cylinder.

cylinder relieves the pressure to the wheel cylinder, the brake shoes retraction springs pull the shoes away from the drum and force the pistons back into their un-applied position in the wheel cylinder. Most wheel cylinders on current hydraulic brake systems are double-acting, meaning each wheel cylinder has a pair of pistons each applied to a different shoe. **Figure 25–14** shows a double-acting wheel cylinder with the components identified. **Figure 25–15** shows both a single-acting front wheel cylinder and a double-acting rear wheel cylinder.

Drum Brakes

Like the air brake, a hydraulic drum brake assembly consists of a cast-iron drum that is bolted to, and rotates with, the vehicle wheel, and a backing plate that is attached to the axle. The shoes, wheel cylinders, automatic adjusters, and linkages are mounted to the fixed backing plate. Additional hardware is required for the parking brakes. The shoes are lined with friction facings, which contact the inside of the drum when the brakes are applied. The shoes are forced outward by the wheel cylinder pistons actuated by hydraulic pressure from the master cylinder. When the shoes are forced into the brake drum, the

kinetic energy (energy of movement) of the rotating drum is transformed by means of friction to heat energy and then dissipated to atmosphere.

When a brake shoe is forced against the drum, the frictional drag acting on its circumference tends to rotate it on its hinge point, the braked anchor. When the rotation of the drum corresponds to outward rotation of the shoe, the frictional drag will act to pull the shoe tighter against the inside of the drum. This is known as self-energizing.

Non-Servo Drum Brakes

In non-servo drum brakes, each shoe is separately anchored at its heel end. Actuating force is applied to the opposite end, the toe. When the brakes are applied with the vehicle in forward motion, the forward or leading shoe is energized. The trailing shoe receives no energization unless the brakes are applied when the truck is in reverse. This reverses the roles of the leading and trailing shoes. For obvious reasons the forward or leading shoes will tend to wear faster, as most of the vehicle braking will be performed when the vehicle is in forward motion. **Figure 25–16** demonstrates the action of a non-servo brake: The brakes are anchored on a pivot and only one of the shoes is energized. Note that their roles are reversed as the direction of rotation of the drum is reversed.

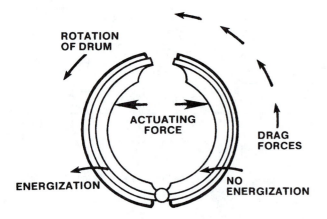

FIGURE 25–16 Non-servo brake action.

Servo-Type Drum Brakes

In many hydraulic brake designs, self-energizing is supplemented by servo action. **Servo brake** action occurs when the action of one shoe is governed by an input from the other. When hydraulic force is applied to the toe ends of the brake shoes, the primary shoe is designed to react first as it has a weaker return spring. This shoe lifts off the anchor (which at this point is only acting as a stop) and is forced against the drum surface. Because the shoe is pivoted at its heel end, the drag forces tend to rotate it with the drum. The secondary shoe is then actuated at its heel via a connecting link by the primary shoe, and it, too, is forced into the drum. In the applied position, the servo brake makes both shoes behave as if they were one continuous shoe. The actuating force pushes on one end while the other is held fast by the anchor. Drag forces cause the whole assembly to be energized. This concept is demonstrated in **Figure 25–17** and **Figure 25–18**. There are

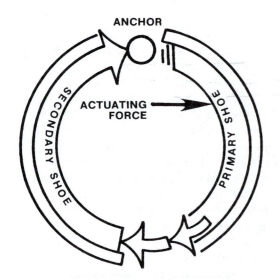

FIGURE 25–17 Servo brake action.

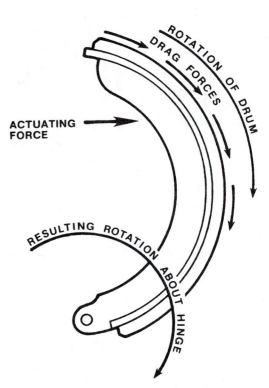

FIGURE 25–18 How a shoe becomes self-energizing.

several different configurations of servo brake. Some hydraulic piston pressure acts against the secondary shoe, but this is incidental.

25.3 AIR-OVER-HYDRAULIC BRAKE SYSTEMS

Some Class 6 and 7 trucks are equipped with an air-over-hydraulic system, especially those manufactured offshore. Such systems combine both air and hydraulic principles to braking. The schematics shown in **Figure 25–19** and **Figure 25–20** show a couple of typical air-over-hydraulic brake systems. There are several variations of air-over-hydraulic brakes used on trucks. In general, the control circuit is managed by air pressure and the application circuit by hydraulics. Depressing the brake pedal moves the application valve plunger. This allows modulated air to flow through quick-release valves and be directed to the booster assemblies. Separate booster assemblies are used for front and rear brakes. Air pressure applied to the booster moves an actuating piston, which creates hydraulic pressure. The actual amount of hydraulic pressure created will depend on the air pressure applied to the diaphragm and its sectional area. This hydraulic pressure is routed to the wheel cylinders, which effect vehicle

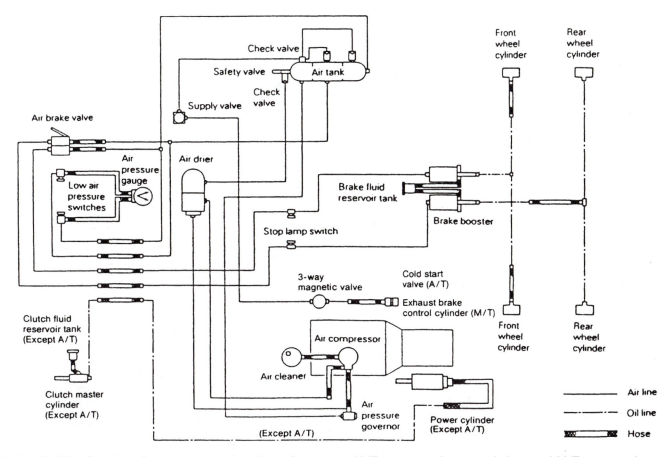

FIGURE 25–19 A typical air-over-hydraulic brake system (A/T—automatic transmission and M/T—manual transmission). (Courtesy of Heavy Duty Trucking)

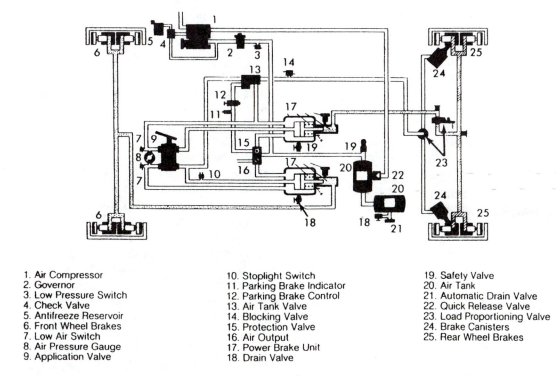

1. Air Compressor	10. Stoplight Switch	19. Safety Valve
2. Governor	11. Parking Brake Indicator	20. Air Tank
3. Low Pressure Switch	12. Parking Brake Control	21. Automatic Drain Valve
4. Check Valve	13. Air Tank Valve	22. Quick Release Valve
5. Antifreeze Reservoir	14. Blocking Valve	23. Load Proportioning Valve
6. Front Wheel Brakes	15. Protection Valve	24. Brake Canisters
7. Low Air Switch	16. Air Output	25. Rear Wheel Brakes
8. Air Pressure Gauge	17. Power Brake Unit	
9. Application Valve	18. Drain Valve	

FIGURE 25–20 Schematic of a different type of air-over-hydraulic brake system. (Courtesy of Volvo Trucks North America, Inc.)

braking in the same manner as in a hydraulic brake system.

Hydraulic pressure forces the pistons of each wheel cylinder outward. Outward movement of the pistons applies brake pressure on brake drum or disc **calipers**. A brake lining wear switch is mounted in each booster to indicate excessive travel of the air piston. A front wheel limiting valve is often used on front axle brakes. This limiting valve proportions fluid pressure during light braking, both to balance the braking and reduce brake lining wear. LPVs can also be used to modulate hydraulic pressure to front and rear brakes based on vehicle loads. The general set-up of the air circuit is identical to that required on a vehicle equipped with air brakes. **Figure 25-21** shows one type of LPV.

All modern air-over-hydraulic systems are required to be dual-circuit systems. As with full air and hydraulic systems, a sudden leak in one portion of a system should not result in a total loss of braking. Because the power for the boost force comes from an engine-driven pump, air-over-hydraulic brakes for medium-duty trucks are required to have fail-safe backup from an electric boost pump. Other fail-safe features include buzzers and lights that warn of low air pressure in the system in much the same manner as an air brake-equipped truck. **Figure 25-22** shows the location of air-over-hydraulic brake components on a truck.

The most common type of parking brake on air-over-hydraulic systems is the spring brake canister type. Its parking brake action is obtained by forcing a wedge assembly between the plungers in the wheel cylinder on each rear wheel. Engagement of the wedge assembly by the spring brake expands the plunger in the wheel cylinder, forcing the brake

Key:
1. Treadle valve
2. Air compressor
3. Parking brake valve
4. Double check valve
5. Valve
   A. Pressure reduction valve (front air circuit)
   B. Quick-release valve (rear air circuit)
6. Rear brake chamber—double spring diaphragm
7. Rear brake air reservoir—with check valve and manual drain valve
8. Front brake air reservoir—with check valve and manual drain valve
9. Wet tank—with automatic drain valve
10. Air pressure regulator valve
11. Master cylinder
12. Caliper (diisc brake assembly)

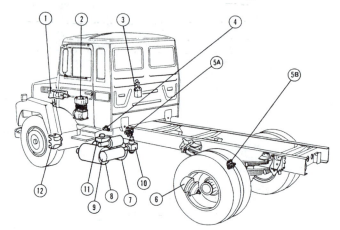

FIGURE 25-22   Location of air-over-hydraulic brake components. (Courtesy of Heavy Duty Trucking)

shoes into the brake drum as shown in **Figure 25-23**. Cable-actuated parking brakes are also found in some air-over-hydraulic systems. The unit is often mounted at the rear of the transmission so it acts on the rear wheels through the driveshaft. The parking brake is incorporated in each rear wheel and is cable controlled from the cab area. For a transmission-located parking brake to function, both rear wheels must be firmly on the ground.

Air-over-hydraulic systems can be used with either drum brakes or disc brakes on all four wheels or mixed disc and drum systems. There are other variations, however. Some systems use hydraulic discs on the front and full air S-cam drums in the rear. A delay valve keeps the hydraulic front brakes from coming on before the slower-acting air brakes.

Some medium-duty trucks mix hydraulic and air brake systems. **Figure 25-24** shows a schematic of a truck with hydraulic discs on the front axle and air-actuated S-cam brakes in the rear.

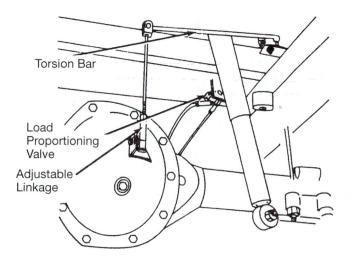

Torsion Bar

Load Proportioning Valve

Adjustable Linkage

FIGURE 25-21   Location of one type of load proportioning valve.

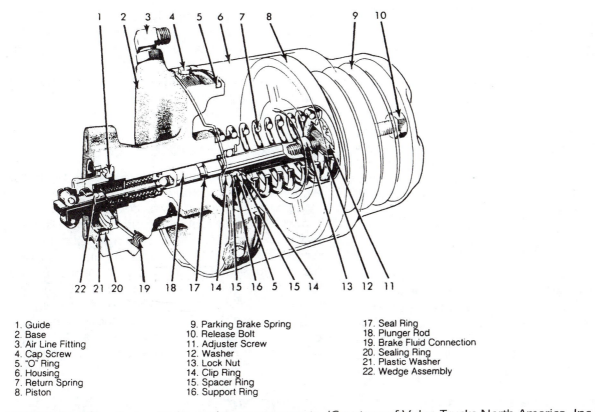

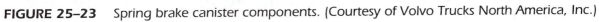

1. Guide	9. Parking Brake Spring	17. Seal Ring
2. Base	10. Release Bolt	18. Plunger Rod
3. Air Line Fitting	11. Adjuster Screw	19. Brake Fluid Connection
4. Cap Screw	12. Washer	20. Sealing Ring
5. "O" Ring	13. Lock Nut	21. Plastic Washer
6. Housing	14. Clip Ring	22. Wedge Assembly
7. Return Spring	15. Spacer Ring	
8. Piston	16. Support Ring	

FIGURE 25–23   Spring brake canister components. (Courtesy of Volvo Trucks North America, Inc.)

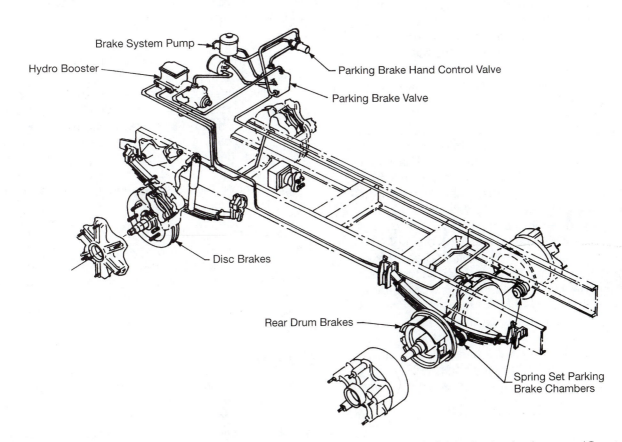

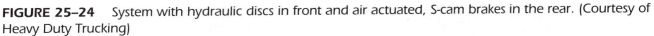

FIGURE 25–24   System with hydraulic discs in front and air actuated, S-cam brakes in the rear. (Courtesy of Heavy Duty Trucking)

25.4 HYDRAULIC BRAKE SERVICE PROCEDURES

The service procedures recommended by the manufacturers of hydraulic brake systems should be observed. There will be differences in each OEM's system maintenance and repair practices. For this reason, it is good practice to always consult the manufacturer service literature when performing repairs. This section describes some general maintenance and repairs that are commonly performed on hydraulic brake systems.

HEAVY-DUTY BRAKE FLUID

Using the manufacturer's specified brake fluid is important to ensure proper operation of the system. The approved heavy-duty brake fluid is designed to retain a proper consistency at all operating temperatures. It is also designed not to damage any of the internal rubber components and to protect the metal parts of the brake system against corrosion and failure.

Some hydraulic brake systems use a non-petroleum base hydraulic brake fluid such as SAE-J-1703 or SAE-J-17021. Other hydraulic systems use petroleum-based brake fluids (mineral oil). It is important to ensure that the correct brake fluid is used in the vehicle brake system and incompatible fluids are not mixed. Use of the wrong brake fluid can damage the cup seals of the wheel cylinder and result in brake failure.

CAUTION: Never use brake fluid from a container that has been used to store any other liquid. Mineral oil, alcohol, antifreeze, cleaning solvents, and water, in trace quantities, can contaminate brake fluid. Contaminated brake fluid will cause piston cups and valves in the master cylinder to swell and deteriorate. One way of checking brake fluid is to place a small quantity of brake fluid drained from the system into a clear glass jar. Separation of the fluid into visible layers is an indication of contamination or mixed types of brake fluid. It is generally regarded as good practice to discard used brake fluid that has been bled from the system. Contaminated fluid usually appears darker. Even brake fluid drained from the bleeding operation may contain dirt particles or other contamination and should not be reused.

Avoid leaving the top off a brake fluid container for any longer than is necessary to pour out fluid and seal the containers tightly. Because of its hygroscopic characteristics, brake fluid can rapidly absorb moisture from air and become saturated. Brake fluids, once opened, have a short shelf life, and it is good practice not to mix brake fluids of the same type if it is not known how old they are. Different types of brake fluid, of course, must never be mixed.

CHANGING BRAKE FLUID

It is a recommended practice to change brake fluid whenever a major brake repair is performed. The system can be flushed with clean brake fluid, isopropyl alcohol or rubbing alcohol. A simple flushing technique is to pour the flushing agent into the master cylinder reservoir and open all bleed screws in the system. The brake pedal is then pumped to force the flushing agent through the system. Ensure that containers are located at each bleed screw nipple to catch the fluid. Replace all the rubber components (seals, cups, hoses, and so on) in the system before adding fresh brake fluid to a flushed, cleaned system.

SERVICING A MASTER CYLINDER

A master cylinder should be removed from a truck using the following procedure:

1. Disconnect the negative battery cable.
2. Disconnect the pressure differential (brake light warning) switch.
3. Disconnect and cap the hydraulic lines from the master cylinder to prevent dirt from getting in them. On some trucks using remote reservoirs, it may be necessary to disconnect the lines connecting the reservoirs with the master cylinder.
4. Remove the master cylinder from the brake booster assembly.

Clean the master cylinder and any other parts to be reused in clean alcohol. Petroleum-based products should not be used to clean the components. Thoroughly inspect all the disassembled components to determine whether they can be reused or not.

Two types of master cylinder are used in truck applications.

Cast Iron

Crocus cloth or an appropriate grit (consult service manual) of cylinder hone should be used to remove lightly pitted, scored, or corroded areas from the bore. Brake fluid can be used as a lubricant while honing lightly. The master cylinder should be replaced if the pitting and scoring cannot be cleaned up. After using crocus cloth or a hone, the master cylinder should be thoroughly washed in clean alcohol or brake fluid to remove all dust and grit. If alcohol is used, air dry the components before reassembly. Mineral solvents should not be used.

Check the clearance between the inside bore wall and the piston (primary piston of a dual circuit master cylinder) and ensure that it is within the OEM specification. The clearance can usually be checked using a thickness gauge. Measuring the bore with a telescoping gauge and micrometer and the piston's outside diameter with a micrometer and subtracting the difference will also produce the clearance dimension. If the clearance is out of specification, the master cylinder should be replaced.

Aluminum Bore

Aluminum is more susceptible to corrosion and physical damage than cast iron. Inspect the bore for scoring, corrosion, and pitting. If the bore is scored or badly pitted and corroded, the assembly should be replaced. An aluminum master cylinder bore can never be cleaned with an abrasive material because this will remove the anodized surface. Anodizing provides a measure of wear and corrosion protection. If this is removed, the result will be rapid failure. Clean the bore with a clean piece of cloth around a wooden dowel and wash thoroughly with alcohol. Some bore discoloration and staining is normal and does not indicate corrosion.

Overhauling a Master Cylinder

Rebuilding aluminum master cylinders may require re-sleeving using the usual machine shop processes. If it is determined that a master cylinder requires reconditioning without re-sleeving, the following general procedure can be used:

1. Remove the snap ring, thrust washer, primary piston, rubber sleeve, and spring.
2. Remove the connection fitting for the rear brake circuit.
3. Remove the sealing ring, valve seat, valve, valve spring, and disc.
4. Remove the stop bolt.
5. Remove the secondary piston with the rubber sleeves and the spring.
6. Remove the connection fitting for the front brake circuit.
7. Remove the sealing ring, valve seat, valve, and valve spring.
8. Clean all parts in brake fluid or denatured alcohol.
9. Before reassembly, lubricate the cylinder bore, pistons, and sleeves with brake fluid.

Use a rebuild kit and assembly fluid or brake fluid to reassemble the master cylinder. Take special care to ensure that new rubber components are not damaged, crimped, or pinched during reassembly. Note how the components fit together when replacing the seals, especially the direction the rubber seals face on the pistons. Some manufacturers' rebuild kits provide a primary piston assembly. The assembly procedure simply reverses the disassembly sequence.

10. When the master cylinder has been reassembled, reinstall and bleed the brakes.

METERING VALVE SERVICE

On systems with front disc and rear drum brakes, inspect the metering valve whenever the brakes are serviced. A trace amount of fluid inside the protection boot does not indicate a defective metering valve, but evidence of a larger amount of fluid indicates wear and the need to replace it. Ensure that the brake lines are correctly connected to the metering valve ports when replacing the valve. Crossed lines will result in the metering valve proportioning oppositely, producing rear brake drag.

🥽 Shop Talk

A pressure bleeder ball must be used to bleed ABS brakes: Use regulated shop air pressure, around 15 psi, to the bleeder ball. When a pressure bleeder is used to bleed a system that includes a metering valve, the valve stem located inside the boot on some valves must either be pushed in or pulled out to open the valve. Never apply excessive pressure that might damage the valve or use a solid block or clamp to hold the valve open. If the valve must be held open, rig the stem with a yield-spring load and take care not to exert more than normal pressure. Manual bleeding using the brake pedal develops sufficient pressure to overcome the metering valve, and the stem does not have to be held open.

PRESSURE DIFFERENTIAL VALVES

The pressure differential valve should re-center automatically on the first application of the brakes after repair work. However, some pressure differential valves may require manual resetting. After repairs have been completed, open a bleeder screw in a portion of the hydraulic circuit that was not worked on. Turn on the ignition to illuminate the dash warning light and then slowly depress the brake pedal.

This should bias the valve piston, causing the brake warning light to go out. If too much pressure is applied during this procedure, the valve piston will go over center, causing the light to flicker out and then come back on. If this occurs, reverse piston movement by closing the open bleed screw and opening a bleed screw in the other hydraulic circuit. Again, slowly depress the pedal to center the valve piston.

PROPORTIONING VALVE

The proportioning valve should also be inspected whenever the brakes are serviced. To check valve operation, install a pair of hydraulic gauges upstream and downstream of the proportioning valve and ensure that the rear brake pressure is proportioned to specification. If this is not the case or the valve is leaking, it must be replaced. Make sure the valve port marked R (it means *rear*) is connected to the rear brake lines.

WHEEL CYLINDERS

Wheel cylinders should be at least externally inspected during any routine brake job. Any evidence of leakage is reason to recondition the unit.

Removing and Installing a Wheel Cylinder

A wheel cylinder should be removed observing the following general procedure:

1. Remove the wheel, drum, and brake shoes.
2. Remove the wheel cylinder mechanical linkages.
3. Disconnect the brake line from the wheel cylinder.
4. Remove the brake cylinder retaining bolts and lockwashers and separate the cylinder from the backing plate.
5. Installation of the wheel cylinder requires reversing the removal sequence.

Disassembly and Reassembly of a Wheel Cylinder

Wheel cylinders are disassembled by removing the dust boot, and then the piston or pistons, cup or cups, depending upon the style of cylinder. Next remove the piston spring. Ensure that the bore is inspected and measured to specification. Cast-iron bores may be resurfaced using a hone, but the piston-to-bore clearance must be measured to specification using the same procedure described previously for master cylinders. Aluminum wheel cylinders may be re-sleeved with a steel insert using the appropriate machine shop processes.

To reassemble the wheel cylinder, lubricate the cylinder bore, pistons, and cups in brake fluid and observe the following general procedure:

1. Install the spring in the wheel cylinder bore.
2. Install a new cup in each end of the cylinder bore with the cup lips facing toward the spring.
3. Insert a piston in either end with the flat side contacting the cup.
4. Install a new rubber boot on each end of the cylinder. Ensure that the boot lip is seated in its groove on the end of the cylinder.
5. Install the bleeder screw in the cylinder body.
6. Mount the wheel cylinder to the backing plate and torque the fasteners.
7. Connect the hydraulic fluid inlet line.
8. Bleed the brakes using the procedure described later in this chapter.

CAUTION: Leaking at the wheel seals after a brake job may occur due to the repositioning of cups onto dirt or sludge.

SERVICING DRUM BRAKES

Whenever a brake drum is removed to inspect the linings or wheel cylinder, make a habit of evaluating the other components in the braking system. Before inspecting drum brakes, release the parking brake. Evidence of grease or oil at the center of the brake assembly indicates that the axle seal needs replacing. If brake fluid drips out of the wheel cylinders when the rubber dust boots are removed, the wheel cylinder needs reconditioning or replacing. Contaminated linings should be replaced. Check for cracked or worn hoses and bent or corroded tubes.

Drum Removal and Inspection

Some brake drums can be replaced without removing the hubs from the vehicle. This is possible on models where the brake drum is mounted outboard from the hub. The drum is fastened to the hub by the wheel studs and nuts. The drum can be removed after removing the tire and wheel assembly. On other models, the hub and drum must be removed from the vehicle.

Drum replacement requires the removal of nuts and bolts securing the hub and drum. The hub and drum can then be separated. Before installing a replacement drum, wash the drum thoroughly with denatured alcohol to remove grease, oil, and other residue. To inspect a drum, proceed as follows:

1. Visually check the drum for cracks, scoring, pitting, or grooves.
2. Check the edge of the drum for chipping and fractures.
3. Drums that are blue, glazed, or heat-checked have been overheated and should not be reused. Severe heat checking in drums will cause rapid lining wear and, even when machined, the drums will fail prematurely. Severe overheating is caused by drums that are machined too thin, improper lining to drum contact, incorrect lining friction ratings, or vehicle overloading. A probable driver complaint would be brake fade.
4. If the brake linings appear to be unevenly worn, check the drum for a barrel-shaped or taper condition (see **Figure 25–25**). Consult the OEM specifications for the maximum drum taper specifications. Also, check the shoe heel-to-toe dimension to ensure that the arc is not spread.
5. Check for drum out-of-round condition using an inside micrometer or drum gauge. Measure the drum at inboard and outboard locations on the machined surface at 4 locations. Compare to the OEM specifications. An out-of-round drum condition produces a pulsating brake pedal condition.
6. Clean the inside of the drum with a water-dampened cloth. New drums may be corrosion-protected with a light grease. This

must be removed with solvent from the friction surface before installing the drum.

Shop Talk

Never paint the outside of the drum. Paint will act as an insulator and slow heat dissipation. The heat absorbed by the drum during braking must be dissipated to the atmosphere around the drum. Overheated brakes cause brake fade!

Brake drums are installed by reversing the order of removal. It is good practice, if reusing brake drums, to install them in their original location. Torque the fasteners to the specification recommended in the service manual.

DISC BRAKE REMOVAL

First, remove the wheel assembly. Next, the caliper assembly must be removed from the brake assembly. Consult **Figure 25–26**. This procedure will depend on the caliper design, but the general procedure is accomplished as follows:

1. Take care when removing the caliper assembly not to damage the bleed screw fitting.
2. Mark the location of all calipers when removing them, for instance, as right-side and left-side, so they can be positioned correctly

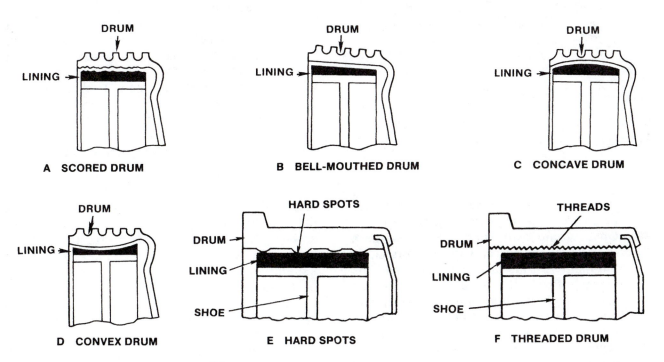

FIGURE 25–25 Drum wear conditions.

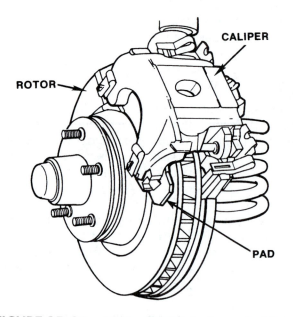

FIGURE 25-26   Hydraulic disc brake assembly.

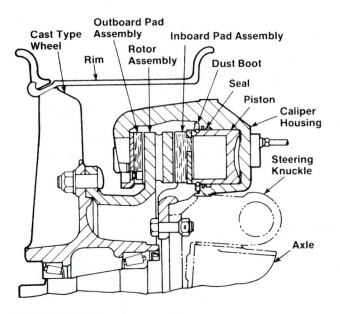

FIGURE 25-27   Cutaway view of a hydraulic disc brake assembly. (Courtesy of International Truck and Engine Corp.)

during installation. **Figure 25-27** identifies disc brake assembly components.

3. On a sliding or floating caliper, install a C-clamp on the caliper, with the solid end of the clamp on the caliper housing and the worm end on the outboard brake pad bracket. Crack the bleed screw on the caliper so fluid is not backed up into the master cylinder. Tighten the clamp until it bottoms the piston in the caliper bore, collect the brake fluid in a container, and then

remove the clamp. Bottoming the piston will create enough clearance for the brake pads to slide over the ridge of rust that accumulates on the edge of the rotor.

4. Disconnect the brake hose from the caliper and cap the brake line and nipple to prevent dirt from entering the hose or caliper assembly. If the object of the service procedure is only to replace the brake pads, the brake hose may not have to be disconnected. If the brake hose is disconnected, cap it with a seal to prevent contaminants from entering.

CAUTION: Many OEMs recommend removing the brake hose when disassembling a disc brake assembly, especially on ABS. Never force the piston inboard without at least opening the bleed screw because contaminants tend to collect in the caliper bore. When the piston is forced inward, this dirt can be forced back to the master cylinder. Also, any sludge in the master cylinder is disturbed. Foreign material half the width of a human hair has been known to render ABS inoperative.

5. Remove the caliper fastening hardware. In front brake systems, this might mean removing the two mounting brackets to the steering knuckle bolts. Support the caliper when removing the second bolt to prevent the caliper from falling.

6. On a sliding caliper, remove the upper bolts, retainer clip, and anti-rattle clips. On a floating caliper, remove the two guide rails/float pins that hold the caliper to the anchor plate. On the older type of fixed caliper, remove the bolts holding it to the steering knuckle. When the fasteners have been removed, separate the caliper from the rotor by prying it upward with heel or pinch bars.

Removing and Installing Brake Pads

Removing and installing a set of brake pads to a disc brake assembly first requires that the caliper be removed using the procedure outlined previously. If the objective is to replace the pads only, the caliper assembly does not have to be hydraulically disconnected. To install a new set of pads to a disc brake assembly, the following procedure should be observed:

1. Disassemble the disc brake foundation.
2. Visually inspect the inside of the caliper for leaks. Clean the caliper and guide rails. Inspect the guide rails for alignment and indi-

cations of corrosion or scoring. Remove any rust buildup using a fine file. Replace the rails if there is any doubt as to their condition.

3. Place a block of wood or metal bar across the face of the caliper pistons. Force the pistons back into the caliper bores. This provides the clearance required for the caliper assembly to slide off the rotor.

4. Lubricate the guide rails/float pins at both ends of the caliper using a silicon base or other approved grease.

5. Install the new pads in the caliper. In some cases, the pads are installed through slots in the lining abutment rails. Install the inboard pad first with the metal side against the pistons. Install the outboard pad with the lining side facing the inboard pad.

6. Slide the caliper assembly over the rotor and onto the guide or support rails. Install the support key for the inboard side and lock it in place with the retaining bolt or screw.

7. Install the lining retaining spring. Typically, the ends are seated in grooves at the caliper rails and the sides extend over the top of the lining backing plates.

Figure 25–28 shows a heavy-duty, dual-piston caliper assembly in an exploded view.

Shop Talk

When axial or radial runout specifications are provided by OEMs, they are expressed in total indicated runout (TIR). TIR is calculated by adding the most positive to the most negative reading through one complete revolution when using a dial indicator. For instance, if, after zeroing the dial indicator, a maximum positive reading of 0.001 inch and maximum negative reading of 0.002 inch occurs through one rotation, perform the following calculation to obtain the TIR:

0.001 inch + 0.002 inch = 0.003 inch

INSPECTING ROTORS

To inspect the rotors, it may or may not be necessary to remove the hub and rotor assembly. This will depend on the configuration used by the OEM. It may also be necessary to separate the rotor from the hub. The following checks are required:

1. Visually inspect the rotor surface for scores, cracks, and heat checks. This check is critical and will determine whether those steps that follow are required. Reject any rotor with evidence of heat checking.

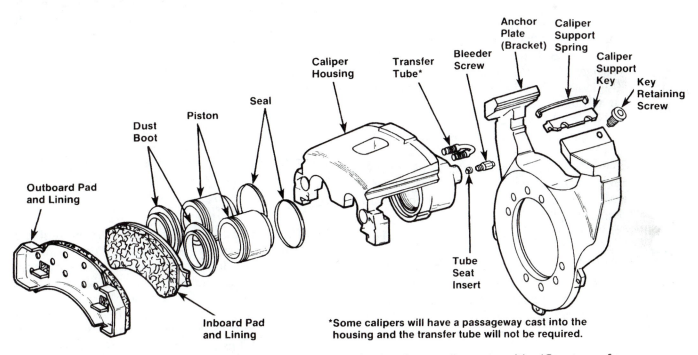

FIGURE 25–28 Components of a hydraulically actuated, two-piston caliper assembly. (Courtesy of International Truck and Engine Corp.)

2. Next, check the rotor or hub and rotor assembly for lateral runout and thickness variation. To check lateral runout,
   - Mount a dial indicator on the steering arm or anchor plate with the indicator plunger contacting the rotor 1 inch from the edge of the rotor.
   - Adjust the wheel bearings so they are just loose enough for the wheel to turn; that is, set them with a little preload.
   - Check the lateral runout on both sides of the rotor. Zero the indicator and measure the total indicated runout (TIR).
   - The lateral runout TIR is always a tight dimension, seldom exceeding 0.015 inch.
3. If the lateral runout exceeds the specification, the rotor will have to be either machined or replaced.
4. Thickness variation should also be measured at 12 equidistant points with a micrometer at about 1 inch from the edge of the rotor.
5. If the thickness measurements vary by more than the specified maximum, usually around 0.002 inch, the rotor should be machined or replaced. This measurement can be made using two dial indicators or a micrometer. If using dial indicators, position them opposite each other and zero the dials. Rotate the rotor and observe the indicator readings for the proper tolerance. Some light scoring or wear is acceptable. If cracks are evident, the hub and rotor assembly should be replaced.
6. With the wheel bearings adjusted to zero endplay, check the radial runout. Rotor radial runout should not exceed the OEM TIR specification, usually around 0.030 inch. Set a dial indicator on the outer edge of the rotor and turn through 360 degrees.
7. Next, inspect the wheel bearings and adjust them using the the OEM-specified procedure.
8. Wear ridges on rotors can cause temporary improper lining contact if the ridges are not removed, so it makes sense to skim rotors at each brake job.

👓 Shop Talk

Excessive rotor runout or wobble increases pedal travel due to caliper piston and can cause pedal pulsation and chatter.

Most rotors fail due to excessive heat. Dishing and warpage are caused by extreme heat. These condi-

tions almost always require the replacement of the rotor. Excessive rotor radial runout can cause noise from caliper housing-to-rotor contact.

Brake rotors have a minimum allowable thickness dimension cast on an unmachined surface of the rotor. This is the replacement thickness. It is illegal to machine a rotor that will not meet the minimum thickness specifications after refinishing. In other words, if a rotor will not "true up" during skimming before the minimum allowable thickness is obtained, it must be replaced.

BRAKE LINES AND HOSES

Brake lines and hoses must be repaired using the OEM specified components. **Figure 25–29** shows the brake line system as it would appear on a typical truck. Note the fasteners that hold the lines properly clamped to the vehicle frame.

CAUTION: Never clamp brake hoses off with Vise-Grips® or "locking pliers." The result is internal damage of the hose and premature failure.

When brake hoses have to be disconnected, ensure that the lock clips are removed and line wrenches used to separate the hose nuts as shown in **Figure 25–30**.

Steel brakes are usually double flare or ISO flare, as shown in **Figure 25–31**: Note the different seats and the fact that they cannot be interchanged. **Figure 25–32** shows how the double flare is made using a double-flaring anvil and cone.

BLEEDING BRAKES

Bleeding brakes is an essential component of servicing hydraulic brakes. The process involves purging air from the hydraulic circuit. Bleeding of brakes may be performed manually, in which case no special equipment is required, or by using a pressure bleeding process. Both methods are outlined here.

Manual Bleeding

Two people or a one-person bleeder kit are required to perform this task. Each wheel cylinder must be bled separately. In most cases, the wheel cylinder furthest from the master cylinder is bled first. This means that the right rear wheel cylinder would be first and then the left rear. After bleeding the rear brakes, the right front brake and then the left front should follow. Ensure that the master cylinder reservoir is topped with brake fluid during the bleeding procedure. The following procedure is used:

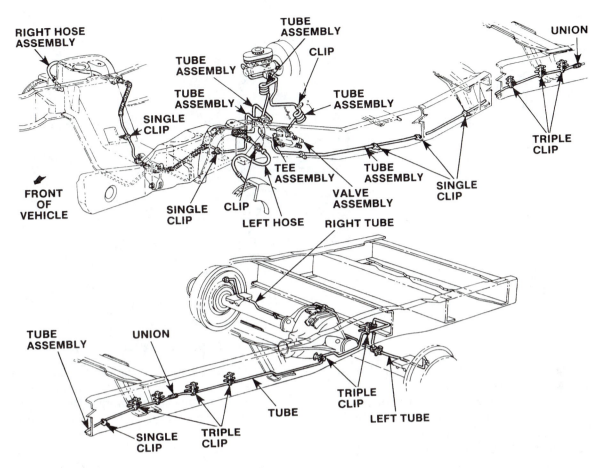

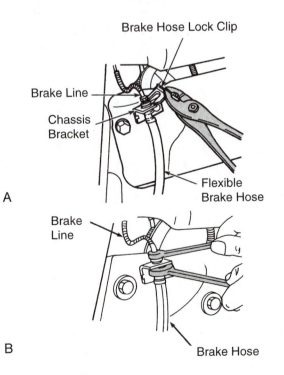

FIGURE 25–29 Typical truck hydraulic brake line system.

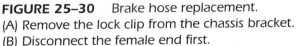

FIGURE 25–30 Brake hose replacement.
(A) Remove the lock clip from the chassis bracket.
(B) Disconnect the female end first.

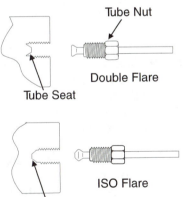

FIGURE 25–31 Two common types of line flares and their seats.

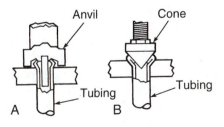

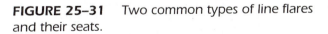

FIGURE 25–32 (A) Anvil folds tubing. (B) Cone performs second fold and doubles seat thickness.

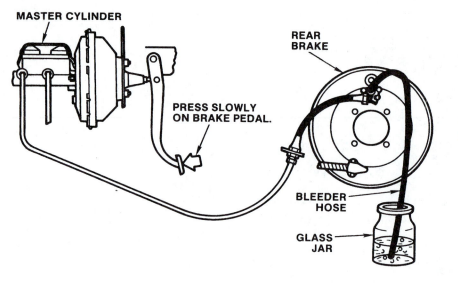

FIGURE 25–33 *Manual brake bleeding technique.*

1. Loosen the master cylinder line nut to the rear brake circuit one complete turn.
2. Push the brake pedal down slowly by hand to the floor of the cab. This will force air trapped in the master cylinder to escape at the fitting.
3. Hold the pedal down and tighten the fitting. Release the brake pedal.
4. Repeat this procedure until bubble-free fluid exits at the fitting and then tighten the line fitting.
5. To bleed a wheel cylinder, loosen the bleed fitting on the wheel cylinder or the caliper with a line wrench. Attach a clear plastic or rubber drain hose to the bleed fitting, ensuring the end of the tube fits snugly around the fitting.
6. Submerge the free end of the tube in a container partially filled with brake fluid as shown in **Figure 25–33**. Loosen the bleed fitting three-quarters of a turn.
7. Push the brake pedal all the way down and hold. Close the bleed fitting and return the pedal to the fully released position.
8. When the fluid exits the bleed tube completely free of air bubbles, close the bleeder fitting and remove the tube.
9. Repeat this procedure at the wheel cylinder or caliper on all the calipers and wheel cylinders beginning at the one furthest from the master cylinder and working towards it.
10. Keep the master cylinder reservoir topped up throughout the bleeding process.
11. When the bleeding operation is complete, top the master cylinder up with brake fluid to the appropriate level in the reservoir.

Pressure Bleeding Brakes

Pressure bleeding equipment must be of the diaphragm type to prevent air, moisture, oil, and other contaminants from entering the hydraulic system. The adapter must limit the pressure to the reservoir to 35 psi.

1. Clean all dirt from the top of the master cylinder and remove the reservoir caps.
2. Install appropriate master cylinder bleeder adapter. A typical pressure bleeder is shown in **Figure 25–34**. Connect the hose from the

FIGURE 25–34 Pressure bleeder. (Courtesy of Wagner Brake, div. of Federal-Mogul Corp.)

bleeder equipment to the bleeder adapter and open the release valve on the bleeder equipment.

3. Using the correct size box or line wrench over the bleeder screw, attach a clear plastic or rubber hose over the screw nipple. Insert the other end of the hose in a glass jar containing enough fluid to cover the end of the hose.

4. Bleed the wheel cylinder or caliper furthest from the master cylinder first, and then the next nearest, and so on until all the valves have been bled.

5. Open the bleeder screw and observe the fluid flow at the end of the hose.

6. Close the bleeder screw as soon as the bubbles stop exiting the hose and the fluid flows in a solid stream.

7. Remove the wrench and hose from the bleeder screw.

8. Fill the master cylinder to the correct level in the reservoir.

Bleeding an Air-Over-Hydraulic System

To bleed a typical air-over-hydraulic system, follow the procedure below:

1. Adjust the air pressure in the air reservoir to 28–43 psi. If the air pressure drops below this range during the bleeding operation, start the engine.

2. When the air pressure is 28–43 psi, stop the engine.

3. Bleed the air from each air bleeder screw in the following order:
   - Front brakes and air booster hydraulic cylinder plug
   - Right front wheel air bleeder screw
   - Left front wheel air bleeder screw
   - Rear brakes and air booster hydraulic cylinder plug
   - Right rear wheel air bleeder screw
   - Left rear wheel air bleeder screw

4. Fill the brake reservoir with fluid. Refill frequently during bleeding so no air enters the brake lines.

5. Attach a vinyl tube to the bleeder screw. Insert the other end of the tube into a bottle partially filled with brake fluid.

6. Depress the brake pedal and at the same time turn the bleeder screw to the left.

7. Tighten the bleeder screw when the pedal reaches the full stroke. Release the pedal.

8. Repeat until no bubbles are observed.

25.5 HYDRAULIC ABS

Full treatment of antilock brake systems (ABS) on air brake systems is provided in Chapter 26. Brief coverage is provided here, focusing on the differences in the system components when ABS is used on a hydraulic brake system. The objective of any antilock brake system is to provide the maximum amount of braking force to each wheel without causing it to lock up. This means that the brake application pressures are modulated or pulsed at high speed. The word *modulate* means to regulate. The pulsing that results in hydraulic ABS may be felt at the brake pedal.

ABS CONFIGURATIONS

When hydraulic ABS is used in truck applications, it is almost always on light-duty, two-axle units, so in the description provided here, a typical two-axle application will be referenced. ABS can be configured in three ways:

- **Single-channel rear wheel.** Only the rear wheels are controlled by ABS. Each rear axle wheel is equipped with a sensor, but hydraulic pressure is modulated to both rear axle wheels. **Figure 25–35** shows a single-channel ABS rear wheel system on a pickup truck.
- **Three-channel.** All four wheels are equipped with wheel speed sensors but each front wheel is modulated individually (two channels) and the rear wheels are modulated as a pair (one channel).
- **Four-channel.** All four wheels are equipped with wheel speed sensors and each wheel on the vehicle is modulated individually.

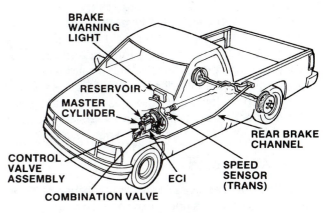

FIGURE 25–35 Typical two-wheel antilock braking system.

Today's ABS systems are self-diagnostic and are designed to revert to normal hydraulic brake operation when either a system or sub-circuit failure is detected. The vehicle operator is alerted when a malfunction occurs by a warning light.

Integrated and Non-Integrated Systems

An integrated ABS system incorporates all the critical system hydraulic actuator components in a single unit at the master cylinder assembly. Integrated systems combine a booster/master cylinder with a valve block. A pump and motor are used to provide pressurized brake fluid to an accumulator. The pump is typically electrically driven and used to maintain pressure in the accumulator at a prescribed value, usually above 2,000 psi. In a non-integrated system, the actuator components, hydraulic boost, master cylinder, and valve block are separate components. Non-integrated ABS system operation is mostly the same in principle as integrated systems. Because these systems vary from manufacturer to manufacturer, always consult the specific OEM service literature before attempting a repair.

ABS COMPONENTS

Hydraulic ABS is a computer-managed system and, as such, requires inputs, control modules and an output circuit to effect the results of processing. Because ABS is designed to default to normal hydraulic braking in the event of a failure, all of the usual hydraulic proportioning built into a normal ABS is still used. This is known as system redundancy. System redundancy is required by federal law in all ABS.

Wheel Speed Sensors

The function of a wheel sensor is to provide the ABS electronic control unit with wheel speed data. A pulse generator is used. A stationary permanent magnet is wound with coiled wire. A rotating toothed ring cuts through the magnetic field of the coil, generating an AC voltage. Both the voltage and frequency increase proportionally with rotational speed. The voltage/frequency is input to the ECU. The wheel speed sensor is located in the wheel hub assembly. Wires connect it to the ECU. **Figure 25–36** shows a couple of different types of wheel speed sensors.

Other ABS Inputs

The ABS ECU receives inputs from other chassis switches and sensors, including the following:

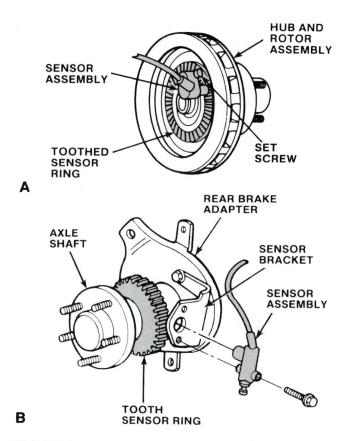

FIGURE 25–36 *Typical ABS wheel sensors: (A) Front, and (B) Rear.*

- **Ignition switch.** Turns system on and off
- **Pressure transducer.** Inputs primary and secondary hydraulic pressure values
- **Differential pressure switch.** Detects primary and secondary pressure differentials
- **Low fluid level switch.** Activates brake warning light and disables ABS function
- **Brake pedal sensor.** A potentiometer-type sensor whose signal is used to compare pedal travel with braking force required
- **Lateral accelerometer.** Inputs cornering speed data. ABS can be managed with respect to load transfer effect.

Electronic Control Unit

Sensor input is delivered to the ECU. Sensor input is received from the wheel speed sensors and the brake pedal sensor. The ECU processes the input information and determines the required outcomes. Essentially, the ECU is programmed to prevent wheel lockup during hard braking. It manages the braking pressure delivered to the wheels by switching solenoids in a control valve assembly. This energizing of

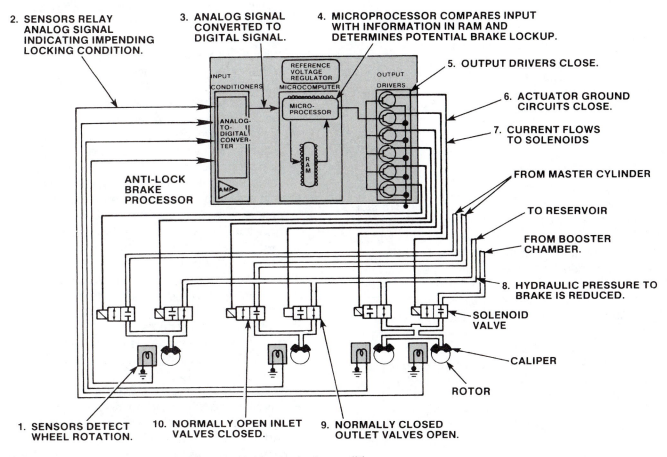

2. SENSORS RELAY ANALOG SIGNAL INDICATING IMPENDING LOCKING CONDITION.

3. ANALOG SIGNAL CONVERTED TO DIGITAL SIGNAL.

4. MICROPROCESSOR COMPARES INPUT WITH INFORMATION IN RAM AND DETERMINES POTENTIAL BRAKE LOCKUP.

5. OUTPUT DRIVERS CLOSE.

6. ACTUATOR GROUND CIRCUITS CLOSE.

7. CURRENT FLOWS TO SOLENOIDS

FROM MASTER CYLINDER

TO RESERVOIR

FROM BOOSTER CHAMBER.

8. HYDRAULIC PRESSURE TO BRAKE IS REDUCED.

SOLENOID VALVE

CALIPER

ROTOR

1. SENSORS DETECT WHEEL ROTATION.

10. NORMALLY OPEN INLET VALVES CLOSED.

9. NORMALLY CLOSED OUTLET VALVES OPEN.

FIGURE 25–37 ABS operation: Potential brake lock condition.

solenoids modulates, that is, alternately relieves (when lockup is sensed) and resumes, the hydraulic braking force to the wheel. **Figure 25–37** shows the role of the control unit in ABS mode.

Modulator Assembly

This unit can also be known as a hydraulic control or modulator/valve assembly. Its function is to effect the results of ECU processing into action. The unit consists of ECU-controlled solenoids and valves. While the ECU receives the same wheel speed data for all wheels on the vehicle, braking is under normal mode; that is, no modulation of pressure to the wheels takes place. However, when the vehicle is being braked and one wheel is slowed proportionally more than the others, hydraulic pressure to the wheel is briefly relieved. The solenoids operate the modulator valves that act to relieve hydraulic pressure to the wheels. The solenoids are controlled or switched by the ECU.

The modulation speed on hydraulic ABS is faster than on air brake systems, due mainly to the incom-

pressibility of hydraulic brake fluid. Modulation frequency can be switched up to 15 times per second on most truck hydraulic ABS. **Figure 25–38** shows a typical ABS hydraulic control unit.

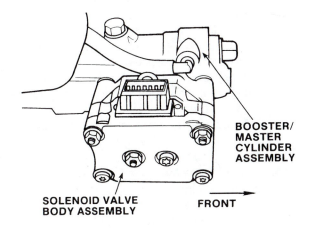

BOOSTER/ MASTER CYLINDER ASSEMBLY

SOLENOID VALVE BODY ASSEMBLY

FRONT

FIGURE 25–38 Typical ABS hydraulic control unit.

Other ABS Outputs

- Brake warning lamp can be illuminated when a hydraulic or ABS malfunction occurs.
- Antilock warning light is illuminated when an ABS malfunction occurs.
- Diagnostic display is used to flash or blink out system codes to aid in troubleshooting the system.

ABS OPERATION

Wheel speed sensors continuously input wheel speed data to the ABS ECU. The system ECU monitors individual wheel speed data and compares it with average wheel speed. During braking, when the ECU senses that a high rate of wheel deceleration is occurring in a wheel, the modulator solenoid for the wheel first functions not to increase hydraulic pressure to the affected wheel. If the wheel continues to decelerate, the ECU signals the modulator solenoid to reduce pressure to the wheel. This results in the wheel accelerating to average wheel speed. When this occurs, hydraulic pressure to the wheel is resumed. This cycling or pulsing can take place up to 15 times per second and performs the pulsing of brakes a skilled driver practices under emergency braking, except at much greater speeds and on a wheel-by-wheel basis.

In the event of a system failure, hydraulic ABS is required to default to normal, that is, non-ABS, hydraulic braking. Should a single wheel speed sensor fail, most systems will alert the driver to the fact that an ABS malfunction has occurred and operate the system in partial ABS mode.

SUMMARY

- Hydraulic brakes tend to be used only on light-duty trucks.
- Air-over hydraulic brakes provide some of the advantages of both air and hydraulic brakes.
- Hydraulic brakes are based on the fact that a liquid does not compress. Pressure applied on one portion of the circuit is transmitted equally to all parts of the circuit.
- In hydraulic braking systems, the mechanical force of the driver stepping on the brake pedal is converted to hydraulic pressure.
- At the wheel, the hydraulic force is once again changed back into mechanical force required to brake the truck.
- Drum brakes can be servo, in which the action of one shoe is governed by input from the other, or non-servo, in which the shoes are separately anchored.

- Hydraulic disc brakes can be of the fixed caliper or floating/sliding caliper type.
- When the brake pedal is depressed, the master cylinder forces brake fluid to the calipers or wheel cylinders, changing mechanical force into hydraulic pressure; the wheel cylinders and calipers change hydraulic pressure back into mechanical force, braking the vehicle.
- Pressure differential valves, metering valves, proportioning valves, combination valves, and load proportioning valves are all operational components of a hydraulic brake system.
- In hydraulic brake systems, a hydraulic power booster is used to assist the master cylinder in applying the brakes. This means that the mechanical force applied by the driver's boot is amplified hydraulically.
- Many Class 6 and 7 trucks are equipped with an air-over-hydraulic brake system. This combines some of the advantages of air and hydraulic brake systems.
- A hydraulic ABS system is designed to modulate hydraulic application pressures to the wheel cylinders to permit maximum braking force without locking the wheels.
- A typical hydraulic ABS consists of wheel speed sensors, an electronic control module, and a modulator assembly.
- Hydraulic ABS is capable of modulation at rates up to 15 times per second.
- All hydraulic ABS systems are required to default to normal hydraulic braking in the event of a system failure.

REVIEW QUESTIONS

1. In hydraulic disc brakes, the pads are forced against the rotors by
   a. shoes
   b. pistons
   c. braked anchors
   d. return springs

2. As the pressure of fluid received by the wheel cylinder of a drum brake system increases, which of the following happens?
   a. wheel cylinder cups and pistons are forced apart
   b. brake shoe retracting springs push the pistons back together
   c. pads are pushed against the rotor
   d. the brake drum locks to the wheel cylinder

3. Which of the following safety devices delays application of front disc brakes in a disc/drum

system until pressure is built up in the hydraulic system?
  a. pressure differential valve
  b. metering valve
  c. proportioning valve
  d. load proportioning valve (LPV)

4. In a typical air-over-hydraulic brake system, depressing the treadle causes which of the following?
  a. hydraulic fluid flows through quick release valves
  b. air pressure forces the pistons of each wheel cylinder outward
  c. an LPV redistributes air pressure to front and rear brakes
  d. air pressure is sent to the booster assemblies

5. Which of the following correctly describes the function of a load proportioning valve (LPV) in an air-over-hydraulic system?
  a. increases the percentage of braking performed by the rear axle(s) as vehicle load increases
  b. decreases the percentage of braking performed by the rear axle(s) as vehicle load increases
  c. changes frame to axle height dimension
  d. compensates for front wheel lockup

6. In a hydraulic brake system, brake fluid transmits pressure applied
  a. equally to all parts of the enclosed circuit
  b. to any surface that will yield
  c. only to slave cylinders in the circuit
  d. to any surface that will not yield

7. In a hydraulic brake system, which of the following best describes what happens when the driver's foot is applied to the treadle?
  a. hydraulic force is converted to mechanical force
  b. mechanical force is converted to hydraulic force
  c. hydraulic force converted to mechanical then back to hydraulic force
  d. mechanical force converted to hydraulic then back to mechanical force

8. What component in a hydraulic brake system helps fluid in the master cylinder return back to the reservoir after a brake application?
  a. compensating port
  b. metering valve
  c. pressure differential valve
  d. check valve

9. Technician A says that in non-servo drum brakes, the forward or leading shoe is energized in normal braking when the vehicle is in forward motion. Technician B says that in non-servo drum brakes, the leading and trailing shoes exchange roles when the vehicle reverses motion. Who is correct?
  a. Technician A
  b. Technician B
  c. both A and B
  d. neither A nor B

10. In hydraulic drum brakes, the shoes are mechanically forced against the drums by
  a. a master cylinder
  b. wheel cylinders
  c. braked anchors
  d. return springs

11. A typical TIR specification for disc rotor lateral runout would more likely be which of the following?
  a. 0.0015 inch
  b. 0.0050 inch
  c. 0.0100 inch
  d. 0.0150 inch

12. In a hydraulic drum brake system, what force pulls the shoes away from the drum when the brakes are released?
  a. hydraulic force at the master cylinder
  b. retraction spring force at the brake shoe assembly
  c. hydraulic force at the wheel cylinder
  d. brake pedal spring force

13. Which of the following devices would play a role in illuminating the brake warning light in the event of a complete hydraulic failure of the front brake circuit?
  a. pressure differential valve
  b. metering valve
  c. proportioning valve
  d. load proportioning valve (LPV)

14. A typical maximum rotor thickness variation specification required by an OEM would likely be
  a. 0.002 inch
  b. 0.005 inch
  c. 0.020 inch
  d. 0.050 inch

15. Technician A says that when replacing disc brake linings, most OEMs recommend that fluid be removed from the master cylinder before compressing the caliper piston with a C-clamp. Technician B says that the caliper bleed screw should be opened before compressing a caliper piston with a C-clamp. Who is correct?
  a. Technician A
  b. Technician B
  c. both A and B
  d. neither A nor B

16. Which of the following best describes the electrical operation of a typical ABS wheel speed sensor?
    a. hall effect sensor
    b. thermistor
    c. potentiometer
    d. pulse generator

17. Which of the following components would best be described as the "brain" of a hydraulic ABS system?
    a. master cylinder
    b. wheel speed sensor
    c. ECU
    d. modulator valve

18. Which of the following would represent the typical maximum modulation speed of a hydraulic ABS system?
    a. 2 times per second
    b. 5 times per second
    c. 15 times per second
    d. 60 times per second

19. Technician A says that an ABS ECU reads the frequency produced by a wheel speed sensor and interprets it as wheel speed. Technician B says that the basis of hydraulic ABS is that hydraulic pressure to the brakes over one wheel is relieved when a lock-up condition is sensed. Who is right?
    a. Technician A
    b. Technician B
    c. both A and B
    d. neither A nor B

20. Technician A says that when hydraulic ABS fails electronically, the system defaults to normal hydraulic braking. Technician B says that a failure of one ABS wheel sensor would result in a vehicle that should not be driven on the highway. Who is right?
    a. Technician A
    b. Technician B
    c. both A and B
    d. neither A nor B

26

ABS and EBS

Learning Objectives

After reading this chapter, you should be able to

- Describe how an antilock brake system (ABS) works to prevent wheel lock-up during braking.
- List the major components of a truck ABS system.
- Describe the input circuit components of an ABS system.
- Outline the role of the ABS module when managing antiskid mode.
- Explain how the ABS module controls the service modulator valves.
- Explain what is meant by the number of channels of an ABS system.
- Describe how trailer ABS is managed.
- Outline the procedure for diagnosing ABS system faults.
- Describe the procedure required to set up and adjust a wheel speed sensor.
- Explain how an electronic brake system (EBS) manages service brake applications.
- Outline the reasons why an EBS system has to meet current FMVSS No. 121 requirements.

Key Terms List

ABS module
chopper wheel
controller
electro-pneumatic
electronic control unit (ECU)
exciter ring

modulator
pulse generator
pulse wheel
redundancy
reluctor wheel
tone wheel
wheel speed sensors

An antilock braking system (ABS) is an electronically managed brake system that monitors and controls wheel speed during braking. An ABS is designed to work with standard air brake systems. An electronic braking system (EBS) is an air brake system in which the control circuit is managed entirely electronically. It is a computerized braking system in which the brake pedal is simply an input to an EBS control module. Although full EBS is used in Europe, when it is used in North America it must be in conjunction with an FMVSS No. 121-compliant pneumatic control system. However, the major truck and bus OEMs

are testing EBS extensively, and efforts are currently underway to modify FMVSS No. 121 to accommodate brake-by-wire. Most of this chapter will address the current ABS systems that are now mandatory on both tractors and trailers. A short introduction to EBS will also be provided. Because ABS is managed by computer, it will help you understand this chapter better if you read Chapter 6 first.

The idea behind ABS is simple. The system monitors wheel speeds at all times and controls brake application pressures when wheel lock is imminent. The main benefits of an ABS system are better vehi-

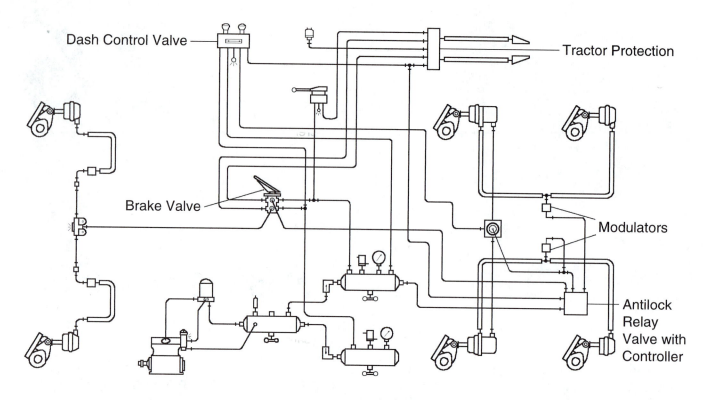

FIGURE 26–1 Typical antilock brake system (ABS) schematic on a highway tractor. (Courtesy of Haldex Brake Systems)

cle stability and control during severe braking. An ABS system permits a less experienced truck driver to maintain much better control during panic stops.

The ABS **electronic control unit** or **ECU** (the system **controller**) receives and processes signals from the **wheel speed sensors** (which signal wheel speed to the ECU). When the ECU detects a wheel lockup, the unit activates the appropriate **modulator** valve (system output, an electrically controlled slave valve), and service air pressure is pulsed. **Figure 26–1** is a schematic of a typical ABS system.

ABS requires no significant changes to the dual-circuit air brake system. Both the control and actuator circuits of the dual-circuit air brake system function pretty much as in a system without ABS. They use compressed air and mostly the same components. In fact, should a malfunction occur in the electronic system, the system is designed to operate exactly as a non-ABS system. Failures tend not to be total. In the case of a typical ABS failure, the affected wheel or axle is ABS-disabled and reverts to normal braking. The other wheels retain ABS function. In the event of any type of ABS failure, a warning lamp alerts the driver as to the status of the system. This warning lamp can also be used to display blink code diagnostics.

26.1 | BASICS OF ABS OPERATION

When compared to the electronic systems required to manage a transmission or an engine, an ABS system is simple. An input circuit is used so ABS can monitor individual wheel speeds under braking. The wheel sensors use a pulse generator principle. This consists of a toothed ring located at the axle ends (on the wheel assembly) or within the axle. The analog voltage output from these sensors is transmitted to an ECU, a microprocessor module that can be located in the vehicle cab or on the frame.

The function of the ECU is to receive input information, process it, and then generate an output by switching something called a modulator. The "brain" of the ECU is a microprocessor. The microprocessor monitors the input information, in this case, wheel speed data, and, based on its programmed software, generates an outcome. Although the process can be called logic processing, the role of ABS is simple, that is, to prevent wheel lockup. Because the ECU is fed with information on exactly how fast each wheel is rotating, at the point where a wheel is about to lock up, the ECU outputs a signal to reduce braking force to the affected wheel. This is achieved by controlling air through a modulator. The modulator is a solenoid-controlled valve capable of dumping air being delivered to the brake chamber. This dumping of air to the brake chamber is momentary and is immediately followed by recharging the brake chamber. The cycling of on/off air to the brake chamber can occur at high speed.

 Shop Talk _____

An ABS system is a simple computer-controlled system. As such, it requires system inputs, processing capability, and outputs. The steps required to produce an outcome are less complex than in an electronically managed transmission, but electronically the process is similar.

SPLIT-COEFFICIENT STOPS

Because in many systems the brakes on each side of the vehicle are individually controlled, ABS permits maximum efficiency stops even when the vehicle is run on a split-coefficient surface. A split-coefficient surface might occur when one side of the truck was running on ice and the other on bare pavement. This type of braking situation is hazardous even for experienced drivers. The ABS module is capable of cycling the brakes on the ice side of the vehicle at a different rate from that on the dry side. This enables it to obtain maximum road traction through the stop.

CAUTION: When driving an ABS-equipped truck, drivers have a tendency to expect too much. Always exercise care when braking, especially under bobtail conditions, even when the vehicle ABS system is fully functional.

ABS FAILURES AND DIAGNOSIS

While the truck is operating, the ABS module continually monitors its own circuitry and performance. If a fault is detected, ABS will alert the driver and shut down either the entire system or a portion of the system. In all North American applications, ABS is required to revert to normal air brake operation when in failure mode. Because the ABS module is capable of self-diagnosis, the technician can then troubleshoot the system using the diagnostics written to the system software.

26.2 TYPES OF ABS

Although all truck ABS systems operate on more or less the same principles, each manufacturer uses different hardware, and most of it is specific to the system. This means that it cannot be interchanged. To make things slightly more complicated, ABS is a rapidly developing technology, which has resulted in each OEM producing different generations of the system. Also, each OEM ABS provides different methods of electronic diagnosis. For example, Bendix uses a light emitting diode (LED) display on the ECU to signal fault codes. Another way of doing this is to flash codes through a dash warning light in the cab. Most current-generation ABS systems allow the use of a handheld electronic service tool (EST) that connects directly to the ECU. The EST may be the same one used to view other chassis electronic systems, although in order to obtain the maximum potential of the tool, a dedicated software cartridge may be required. Some ABS systems permit braking data to be displayed and analyzed on a personal computer (PC) and dedicated software.

26.3 ABS COMPONENTS

A typical truck ABS system requires the following components:

- Wheel speed sensor (system input)
- An **ABS module** or ECU (system processing and switching)
- Brake modulators or ABS valves (system output)
- Interconnecting wiring and connectors (electrically connects the first three sets of components)

Figure 26–2 shows the location of ABS components on the rear drive axles of a highway tractor: The ABS module and the modulator valves are integrated into one unit in some systems.

INPUT CIRCUIT COMPONENTS

As with any computer-managed circuit, the input circuit components are responsible for inputting information into the ECU. An ABS system receives its primary input information from the wheel speed sensors in the ABS circuit. Current ABS systems also connect with other vehicle electronic systems,

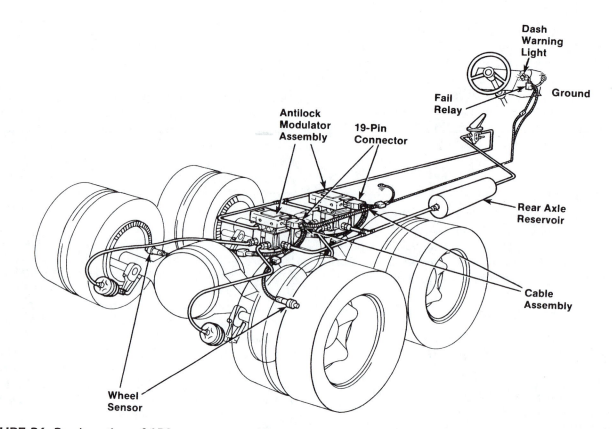

FIGURE 26–2   Location of ABS components. (Courtesy of Bendix Commercial Vehicle Systems)

enabling them to share information that may apply to vehicle braking.

Wheel and Axle Speed Sensors

Wheel and axle speed sensors are electromagnetic devices used to signal wheel speed information to the ABS module. The sensor consists of a toothed **reluctor wheel**. The reluctor wheel can also be known by such names as a **chopper wheel, exciter ring, pulse generator,** or **tone wheel.** However, for the sake of consistency, it will be referred to as a reluctor wheel in this text, and will refer to the entire sensor assembly as a pulse generator. A typical reluctor wheel and sensor assembly is shown in **Figure 26–3.** When the reluctor wheel rotates, it cuts through a permanent magnetic field located in the stationary sensor. This makes its operating principle like that of a generator, and it produces a simple alternating current (AC) signal.

Both the voltage and frequency of this AC signal will rise in exact proportion to wheel speed. This analog voltage signal is sent to the ECU, which converts the pulse to an actual wheel speed so it can be used to manage the air pressure delivered to the brake chambers it controls. On a typical ABS, the frequency varies from 0 (wheel locked or stationary)

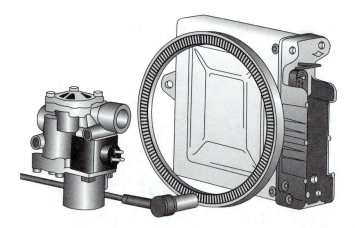

FIGURE 26–3   Major components of a typical ABS system: reluctor wheel, axle sensor, ABS modulator valve, and ECU or ABS module.

to 400 Hz. A graphic representation of the frequency signals produced by a wheel speed sensor is shown in **Figure 26–4.** The ABS module must be capable of rapidly processing the wheel speed data. This data must be converted into a digital format to enable it to be processed by the system ECU.

PHOTO SEQUENCE 13
WHEEL SPEED SENSOR TESTING

P13-1   Block the vehicle wheels.

P13-2   Raise one side of the axle with a floor jack and lower the axle onto a safety stand.

P13-3   Disconnect the wheel speed sensor connector.

P13-4   Connect a pair of AC voltmeter leads to the wheel speed sensor terminals.

P13-5   Rotate the wheel at 30 rpm (1 revolution per 2 seconds) and observe the reading on the AC voltmeter.

P13-6   Compare the meter reading to the specifications in the truck manufacturer's service manual. If the specified AC voltage reading is not obtained, replace the wheel speed sensor.

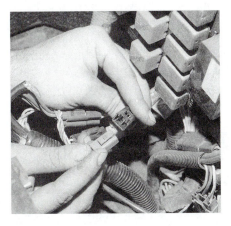

P13-7   Reconnect the wheel speed sensor connector.

P13-8   Raise the axle with a floor jack and remove the safety stand.

P13-9   Lower the wheel onto the shop floor and remove the floor jack.

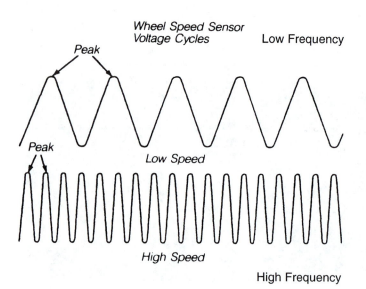

FIGURE 26-4 Frequency signals produced by a wheel speed sensor at low and high speeds. (Courtesy of Bendix Commercial Vehicle Systems)

 Shop Talk

*Systems that use a 100-tooth wheel sensor do so for a reason. The 100 teeth on the **pulse wheel**, when divided into 360 degrees of one full tire revolution, equal 3.6 degrees each. When a radial tire on a loaded-to-specification truck is run down the highway at 60 mph, its dynamic footprint is about 3.6 degrees. One-hundred-tooth ABS reluctor wheels have become the industry standard.*

ELECTRONIC CONTROL UNIT

The ABS module can be called the system controller, or an ECU. It is a simple computer. It receives input signals. Because these input signals in an ABS system are produced by a pulse generator, they must be converted to a digital format before they can be processed. A microprocessor (the brain of the ECU) manages the processing cycle. It is programmed with data that contains the rules and regulations that allow it to produce outcomes. The outcomes required of the ECU in an ABS system are electrical control signals to solenoids in the modulator.

The ABS system ECU continuously monitors input wheel speed information from the sensors in the circuit. Its function is calculation and comparison to information programmed into its memory. When a wheel lockup condition is detected, the ECU outputs an electrical signal to the solenoids that control the air routed to the brake chambers. This function of the ECU is known as switching. The speed at which a computer "thinks" is known as its clock speed. This speed in ABS modules is around 200 times a second. Although this might sound slow compared to a current desktop computer, it is fast enough to manage braking in a truck air brake system. Truck air brake systems are capable of modulating air pressures 3 to 7 times per second, depending on the system. Some ABS modules have two microcomputer circuits, one to act as a primary processor and the second to act as a back-up. This provides for a little additional safety.

Some ABS ECUs contain a diagnostic window and manual reset switch, and they may be integral with the modulator valve assembly (see **Figure 26-5**). Most ABS modules are insulated and carefully protected. When located in the chassis, they function in a harsh environment and are protected from physical damage and electromagnetic radiation. In some cases, the ECU is physically located on the modulator assembly and electrically connected to it by means of sealed pin connectors and a harness.

When a system or subsystem failure is detected by the ABS module, a failure strategy goes into effect. This can mean that entire system or a portion of it defaults to normal non-ABS operation. A warning light alerts the driver that the full range ABS function is no longer available but should also mean that the operation of the service brake system is unaffected. The ability to revert to normal non-ABS air brake operation in the event of a failure is known as **redundancy**.

Many different ABS configurations are used. In one, each vehicle diagonal is controlled by a pair of microprocessors. This arrangement ensures that faults are reliably detected within the microprocessors. Because both microprocessors are fed the same input information when the system is functioning properly, they should produce the same results. Each microprocessor additionally runs constant self-tests and monitors the program execution time. These two monitoring levels function independently of each other and ensure reliable and fast detection of faults. The performance of both the input and output circuit devices is monitored. For example, wheel speed sensors are monitored electrically, enabling open circuits, short circuits, and shunts to be detected even if the vehicle is standing. The air gap between the sensor and the reluctor wheel is also checked through software monitoring. These measures detect not only defects but malfunctions resulting from mechanical vibration in the axle area and electrical interference. In the case of pressure modulating valves, voltage and current are measured in actuated condition and are compared with a permissible range of values. All components are activated and checked each time the ignition is switched on. The ABS warning light is switched off if the system is operational.

ABS Control Logic

We know that the brain of an ECU processing cycle is the central processing unit (CPU). It performs such tasks as data comparison, fetching and carrying data from memory, and arithmetic and logic calculations at very high speeds. We group all of these functions together and collectively call them logic processing. It is much easier to describe logic processing in a simple electronic management system such as ABS than in the more complex management systems required on engines or transmissions. The ABS module monitors the wheel speed data on a continuous basis. This data is logged in the RAM or main memory of the processing cycle. When the vehicle is braked, the rate of deceleration of each monitored wheel is compared with a profile programmed into the system's memory. This essentially tells the CPU how fast a wheel can be decelerated without locking up. So, if under braking, the rate of deceleration identified in the computer's programmed memory is exceeded, the antilock strategy is immediately loaded into the processing cycle. This requires that an output be generated to command the modulator solenoids to momentarily dump service application pressure to the wheel about to lock up. Because the rate of deceleration is immediately changed, wheel speed data signaled to the computer processing cycle indicates that air pressure to the service chamber can be resumed. At that moment the electrical signal from the ECU to the modulator solenoid ceases, allowing full air pressure be sent to the service chamber again.

The ABS module can manage modulator release/apply cycling at very high speeds. In air brake systems, the cycling frequency is slower than in a hydraulic brake circuit, usually up to about 7 times per second. The reason for the slower cycling of air brake systems is due to the compressibility of air when compared to hydraulic brake circuits. The action of ABS occurs in a pulsating manner and simulates pumping of the brakes by the driver, except, of course, that the "pumping" takes place at much higher speeds and with a much higher degree of precision than could be achieved by any driver.

OUTPUT CIRCUIT COMPONENTS

The output circuit is switched by the ABS module. Outputs are the devices that effect the results of the ABS module's processing cycle. The main output device in an ABS system is the brake modulator assembly. A brake modulator assembly combined with an ABS control module is shown in **Figure 26–5**.

Brake Modulators or ABS Valves

The brake modulator or ABS valve assembly is controlled by the ECM. The term used to describe this valve body varies by manufacturer: In this text, the term *modulator* will generally be used. The typical brake modulator assembly contains from one to four

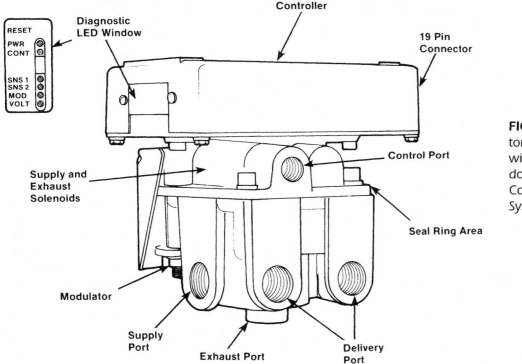

FIGURE 26–5 A modulator controller assembly with a diagnostic LED window. (Courtesy of Bendix Commercial Vehicle Systems)

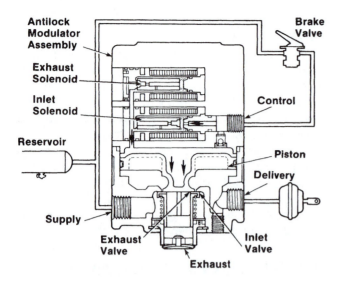

FIGURE 26–6   Modulator valve in balanced position during a normal service application. (Courtesy of Commercial Carrier Journal)

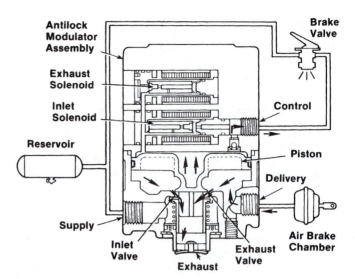

FIGURE 26–7   Modulator valve exhausting during a normal service application. (Courtesy of Commercial Carrier Journal)

solenoid valves. The actual number of solenoid valves used depends on the number of brake circuits that have to be regulated from the unit. Each solenoid is a switched output from the ECM. They are capable of response times as fast as ten milliseconds.

Each modulator valve consists of a solenoid, a piston, and a body housing. During normal (non-antilock) operation, the inlet solenoid is open and the exhaust solenoid remains closed (see **Figure 26–6**). This permits service application air to pass through and be directed to the service chambers. The modulator valve functions as a pilot-controlled relay valve and, when energized, is used to divert pressure to the brake chambers to exhaust. The air-in port is always to the electrical side of the connector.

The modulator valve is located downstream from the relay valve. In practice, it may be integral with the relay valve, but it still functions "downstream" from it. When the system is not managing an antilock stop, the modulator simply routes the air from the relay valve to the service chamber. In this instance, control pressure from the foot valve directs an air signal through to the relay valve, and the relay valve uses reservoir air and delivers it to the service brake chamber, enabling a service brake application in exact proportion to the control pressure. However, when the ABS module determines that a wheel lock-up condition is about to occur based on the input signal from the wheel speed sensor, it switches, or energizes, the solenoids in the modulator to exhaust air pressure routed to the service brake chambers (see **Figure 26–7**). This momentarily relieves the service application pressure on the brake chamber(s).

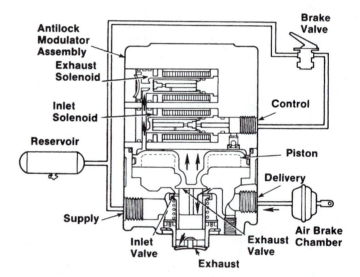

FIGURE 26–8   Modulator valve in ABS mode during a service application. (Courtesy of Commercial Carrier Journal)

When the modulator solenoid and piston are energized in ABS mode, control pressure is exhausted both through check valves located in the solenoid housing and out the exhaust port of the brake valve (see **Figure 26–8**). As the modulator piston moves, the exhaust is opened, allowing the air from the underside of the piston to exhaust through the exhaust port of the modulator.

Study **Figure 26–6**, **Figure 26–7** and **Figure 26–8**, and then take a look at **Figure 26–9**: **Figure 26–9** captures the content of all three figures.

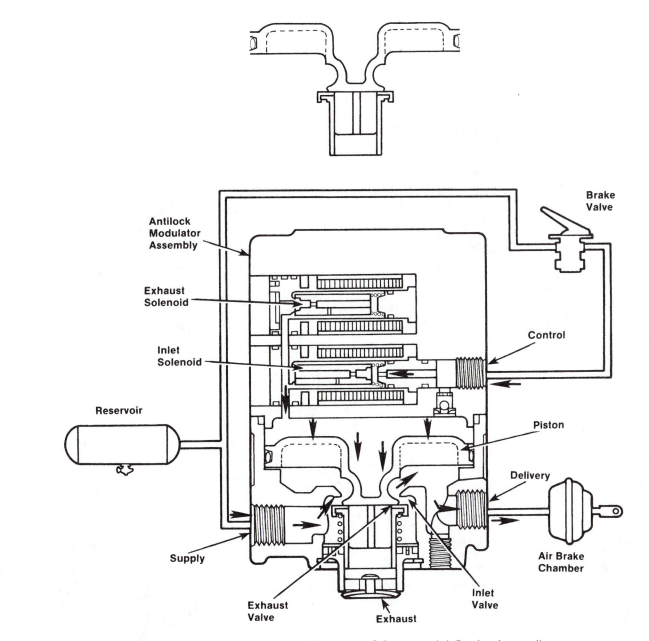

FIGURE 26–9   ABS modulator valve operation. (Courtesy of Commercial Carrier Journal)

A fail relay coil (see **Figure 26–10**) is installed between the chassis power source and the ABS electronic system. The fail light power flows through the normally closed relay contact. Should a failure occur in the system, the fail relay coil will not energize, causing the failure indicator light to illuminate in the dash and alerting the driver.

Some modulator valve assemblies, instead of using solenoids, use motor-driven stepped pistons to regulate braking pressure. They perform the same function as solenoids, basically that of dumping service application air when switched to do so by the ABS module.

To improve severe braking performance, some modulators are designed to provide feedback to the ECU with the objective of smoothing brake pedal feedback. Brake pedal feedback, also known as pulsing, is an undesirable characteristic of ABS. As each new generation of ABS arrives on the marketplace, the response times are improved. The faster the modulator regulates service braking pressure when switched by the ECU, the better is the performance realized from the system. It is important to remember that the cycling performance of an air brake ABS is somewhat slower than a hydraulic ABS due to the compressibility of air.

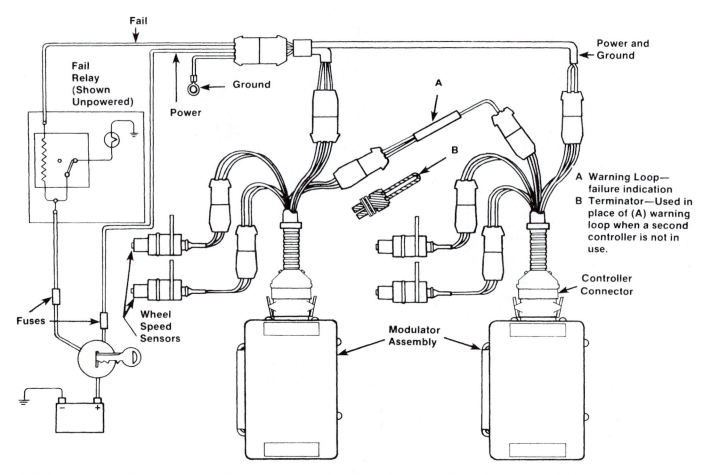

FIGURE 26–10 *Electrical schematic for a typical ABS system. (Courtesy of Bendix Commercial Vehicle Systems)*

Dash Warning Indicator Light

Most antilock brake systems have a warning light. The warning light may be mounted on the dash or above the windshield. This light is used to indicate to the driver that the system is properly operating on start-up, to alert the driver in the event of a malfunction, and to signal diagnostic codes to the technician. The ABS tractor warning light (TRAC ABS) will illuminate when the ignition switch is turned on. A TRAC ABS light is shown located between the tachometer and speedometer in **Figure 26–11**: this means that the driver would find it difficult not to notice when it illuminates. Depending on the OEM, this light will go off when the vehicle starts to move. This may mean that the speed of the vehicle has to reach a preprogrammed value such as 4 mph. If the light stays on a failure in the ABS system is indicated. The light is also designed to illuminate when a malfunction occurs during a trip. Whenever an ABS malfunction occurs, standard full-capability air braking is still available. Most current systems will close down ABS only on the portion of the ABS circuit in which the malfunction has been detected.

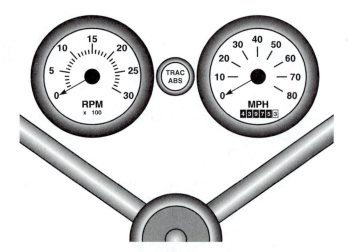

FIGURE 26–11 *ABS tractor (TRAC ABS) warning light.*

In a tractor/trailer combination in which both vehicles are equipped with ABS, there is either a second dash light or a left forward trailer bulkhead light that

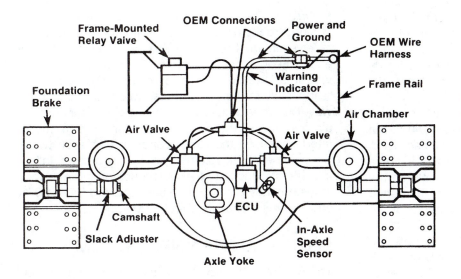

FIGURE 26–12 *Axle-mounted ABS components. (Courtesy of Commercial Carrier Journal)*

can be seen in the driver side mirror that indicates the operating status of the trailer ABS. When the ignition circuit in the tractor is turned on, the trailer ABS (TRLR ABS) light illuminates to indicate that the trailer system is correctly connected to the tractor. Again, when the vehicle begins to move, the TRLR ABS warning light should go out if the trailer ABS module indicates that the trailer ABS circuit is functioning properly.

ATC Indicator Light

On vehicles equipped with automatic traction control (ATC), an ATC light is installed in the dash. This light illuminates when the tractor drive axle wheels spin. The light goes off when the wheel spin stops. ATC operation is described in more detail later in this chapter.

Connectors and Wiring

The wires that carry information (wheel speed data) and electrical power to and from the ABS controller module are harnessed and joined at connections that plug into the components in the system. The wiring harnesses and connectors are weatherproofed and sealed to the connectors. The wire gauge used in the wire harnesses is specific to the task performed.

The antilock system is powered up by the chassis electrical system. The power supply is fuse- or circuit-breaker protected. In a tandem axle system, the electrical supply may be directly to each modulator assembly or can be joined in a common connector and wiring harness assembly. The fail line from the fail relay coil grounds the fail relay whenever the modulator assembly is powered up and in a non-failure mode. An electrical schematic for a typical antilock brake system is illustrated in **Figure 26–10.**

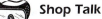

Shop Talk

Always check the vehicle service manual for wire and connector identification. Individual wire identification will differ depending upon the type of connectors in use, the vehicle manufacturer, and the system features in use.

An axle-mounted ABS system is illustrated in **Figure 26–12.** Note that the more vulnerable components are located behind the drive axle carrier.

26.4 ABS CONFIGURATIONS

There are a number of types of ABS systems on trucks today. Just as in automobiles, there is some difference between high performance and basic systems. The type of ABS that will function best on a truck depends on number of axles, axle configuration, axle load, brake circuit distribution, and brake force distribution. For most applications, the steering axle wheels should be included in the ABS loop because the more severe the braking, the greater the role played by the front axle brakes. The term *channel* refers to either an input or output path to the ECM. However, in describing ABS, each manufacturer uses the word to mean slightly different configurations. For instance, a system described as a 6-channel system fitted to a tandem drive straight truck (2 steering wheels/4 drive wheels) would normally consist of 6 inputs, that is, wheel speed data inputs to the ECU, and 6 outputs or modulators. This means that braking on each wheel of the truck would be managed individually. A 6-in/6-out channel system would represent a top of the line system and is seldom used today, mainly because the slightly superior ABS performance achieved does not justify the extra cost.

A more modest system used on truck chassis would be a 4-in/4-out system (known as 4S/4M), but a 6-in/4-out system (known as 6S/4M) is also used. In the 6-in/4-out configuration, there would be a wheel speed sensor on each of the 6 wheels on the vehicle, but only 4 modulators. This means that each modulator would manage a pair of service brake chambers.

Regardless of the number of ABS channels, on most tractors or straight trucks, there is only one ABS module managing the system. Truck ABS systems, whether air or hydraulic, can operate as either split or diagonally arranged systems. A split system means that the modulator valve would manage a pair of service chambers located on one side of the vehicle. This arrangement has a certain advantage when a vehicle is run on a split-coefficient surface, for example, pavement on one side and gravel shoulder on the other. Diagonal arrangements are also used, often on straight trucks, in which case the following is true:

- Diagonal 1 controls the right front wheel and the left rear wheels.
- Diagonal 2 controls the left front wheel and the right rear wheels.

If a system fault occurs in one diagonal, the other diagonal will continue to provide the ABS function. Diagonal configurations of ABS do not perform as well on split-coefficient surfaces.

SIX-CHANNEL SYSTEM

The six-channel system is often described as a 6S/6M system. The S means *sensor* and the M stands for *modulator*. The 6-channel ABS system can be used on straight trucks or tandem drive tractor. A single ECU receives inputs from a wheel speed sensor located on each wheel of the chassis. An ABS modulator for each wheel is located downstream from the relay valve responsible for actuating the service chambers. In this way, the wheel speed of each wheel is monitored and air pressure to the service chambers can be modulated based on the exact conditions at each wheel. Six-channel ABS systems were more common following the introduction of ABS technology into the truck industry. They are not often seen today. **Figure 26–13** shows a full 6-channel (6-in/6-out) system.

SIX/FOUR-CHANNEL SYSTEM

The 6S/4M ABS system locates a wheel speed sensor on each of the 6 wheels of the vehicle, so the speed of each wheel is monitored. However, the system produces only 4 outputs, meaning that 4 of the 6 wheels are managed in pairs. If a wheel lock-up condition is sensed by the ABS module in either of the wheels grouped in a pair, the service application pressure will be modulated to both wheels managed the modulator.

FOUR-CHANNEL SYSTEM

The 4-channel system is often referred to as a 4S/4M system. A wheel speed sensor is located on each of the steering axle wheels and on 2 of the 4 rear wheels. The rear brake wheel speed sensors can be installed on either the forward or rear axle of the tandem, and the actual location usually depends the type of suspension. One sensor must be located on either side, so both sensors have to be on one axle. Extensive field analysis has shown that the simpler and less expensive 4-channel system performs almost as well as a 6-channel system, so they are widely used. Many 4-channel systems incorporate automatic traction control, explained a little later in

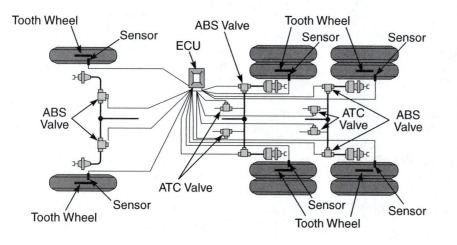

FIGURE 26–13 *Six-channel ABS configuration.*

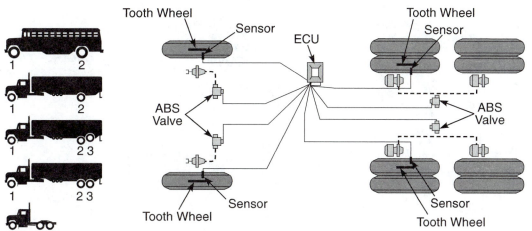

FIGURE 26–14 *Four-channel ABS configuration.*

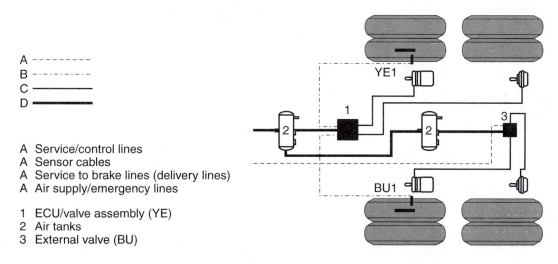

A - - - - - - -
B - · - · - · -
C ————
D ▬▬▬▬

A Service/control lines
A Sensor cables
A Service to brake lines (delivery lines)
A Air supply/emergency lines

1 ECU/valve assembly (YE)
2 Air tanks
3 External valve (BU)

FIGURE 26–15 *Two-channel ABS configuration.*

this chapter. **Figure 26–14** shows a typical 4-channel ABS system.

TWO-CHANNEL SYSTEM

Today, a 2-channel system can be found on trailers or straight trucks. There are limited benefits to an ABS system that only functions over the drive axles on a tractor. However, it is not totally without merit because tractor drive axle wheel lock is the major cause of trailer jackknife. Also, when a tractor is bobtailed, ABS on the drive axles can reduce fishtail, especially where no bobtail proportioning valves are used. **Figure 26–15** shows a typical 2-channel ABS system on a trailer application.

> **CAUTION:** ABS effectiveness is fully dependent on the foundation brake adjustment. Use stroke indicators to check brake adjustment status.

26.5 AUTOMATIC TRACTION CONTROL (ATC) SYSTEMS

An electronic ABS system can also be used to control a wheel spinout condition if the system is set up with traction control. In fact, relatively little extra hardware or expense is required, so ATC has become a popular option. The module that manages the system is now called an ABS/ATC module. Wheel rotational speed is monitored in an ABS system, so when a drive axle wheel begins to spin out, brake pressure is applied to the brakes on that wheel. This transfers driveline torque to the wheels with better traction and arrests the spinout condition. In fact, in most current systems, an ABS/ATC module is multiplexed with an electronically managed

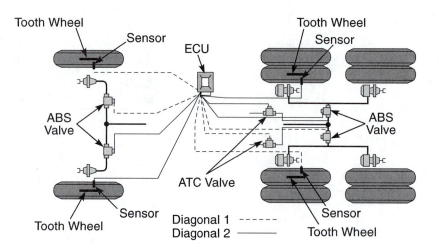

FIGURE 26–16 *Four-channel ABS with an ATC option.*

engine. This means that when a spinout condition is detected by the ABS/ATC module, it can signal the engine electronics to reduce drive torque to the drive axles as part of the traction control strategy.

ATC systems are an option with most 4-channel or 6-channel ABS systems. Just as wheel deceleration can be individually controlled to avoid a skid, acceleration can also be checked to avoid spinning a wheel. When an ATC-equipped vehicle accelerates on a split-coefficient surface, the brakes can be managed to intermittently apply to any drive wheel that is sensed to exceed a preprogrammed acceleration threshold, that is, is about to spin out.

ATC uses two pneumatic solenoid valves and a pair of double check valves. The ATC valve assembly can either be a separate unit or integral with the ABS modulator valve package. Most wheel spinout occurs at relatively low vehicle speeds, usually less than 20 mph. The ATC systems therefore are designed to operate only at vehicle speeds below a certain threshold value. At above 20 mph, an engine derate may take place when a spinout condition is detected. **Figure 26–16** shows a schematic of a 4-channel ABS system with the ATC option.

26.6 TRAILER ABS

All current trailers (built after March 1, 1998) require ABS. The minimum requirement is quite basic, however, and a 2S/1M system on a 4-axle trailer meets the legal requirement for ABS. Any antilock brake system monitors wheel speed and controls the wheel rotational speed under braking. Trailer ABS operates on the same fundamentals already introduced in discussing tractor and straight truck ABS. Again, the various system configurations are described by the number of inputs (wheel speed sensors) and outputs

(modulators) to the ABS module. Some typical configurations used are

2S/1M = 2 wheel speed sensors and a single modulator valve
2S/2M = 2 wheel speed sensors and 2 modulator valves
4S/2M = 4 wheel speed sensors and 2 modulator valves
4S/3M = 4 wheel speed sensors and 3 modulator valves
6S/3M = 6 wheel speed sensors and 3 modulator valves

The operating principles are identical to those on the tractor. Signals from the wheel speed sensors are received and processed by the electronic control unit (ECU). While no lockup condition is sensed, the ABS module allows the service brakes to operate exactly as in a non-ABS system. When wheel lockup is sensed, the ECU begins to manage braking in ABS mode. This means that the ABS relay valve or modulator can cycle the apply/release pressure to the service chamber pressure and prevent the lockup condition. **Figure 26–17** shows some of the more commonly used trailer ABS options.

Trailer ABS has its own ECU that manages antilock braking independently from that of the tractor. Current trailer ABS modules are powered up by the auxiliary pin (pin #7) on an ATA/SAE 7-pin trailer plug (SAE J560 plug). **Figure 26–18** shows a trailer plug and the location of the number 7 auxiliary pin. The function of the 7-pin trailer plug is currently under discussion in the trucking industry due to the desirability of two-way communications between the tractor and trailer and the fact that the auxiliary pin is currently used for functions such as suspension controls and bottom-loading electronics in tankers. Some ABS systems are powered up off the brake light circuit, meaning that the system is only electri-

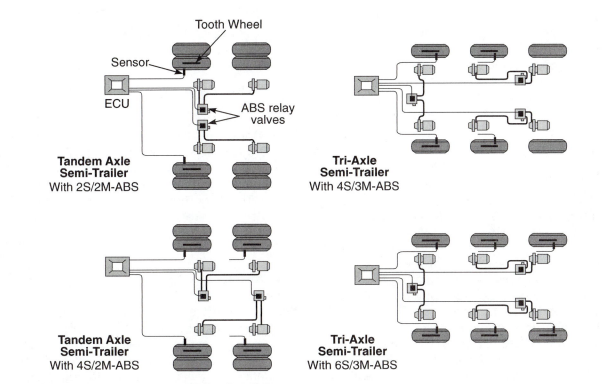

FIGURE 26–17 *Typical trailer ABS configurations.*

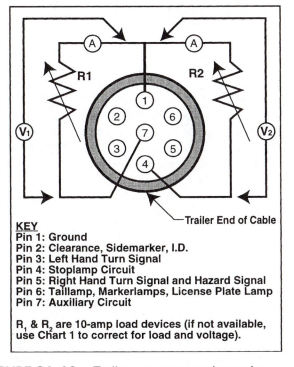

KEY
Pin 1: Ground
Pin 2: Clearance, Sidemarker, I.D.
Pin 3: Left Hand Turn Signal
Pin 4: Stoplamp Circuit
Pin 5: Right Hand Turn Signal and Hazard Signal
Pin 6: Taillamp, Markerlamps, License Plate Lamp
Pin 7: Auxiliary Circuit

R_1 & R_2 are 10-amp load devices (if not available, use Chart 1 to correct for load and voltage).

FIGURE 26–18 *Trailer power test schematic. (Illustration courtesy of The Maintenance Council of the American Trucking Associations, Inc.)*

cally active under braking and only when there is no problem with the brake light circuit. A major disadvantage of this system is that the driver has no way of knowing if there is an ABS malfunction until he

applies the brakes. The processing functions of the ABS module are almost identical to those on a tractor ABS system:

- If an ABS relay valve malfunctions, the wheels controlled by the valve return to standard service braking.
- If a sensor malfunctions in a 2-sensor system, control of that affected side of the trailer returns to standard service braking.
- If one sensor malfunctions in a system that has four sensors on a tandem axle, the ABS will continue to function. In this event, the ECU manages the system using the wheel speed data on the functioning sensor on the same side of the tandem.
- If a problem completely disables the ABS, control returns to standard braking.
- All current trailer ABS use a 100-tooth reluctor wheel.
- Many different manufacturer ABS components are interchangeable, but be sure that components can be cross-referenced before actually doing this. For instance, some Haldex solenoids interchange with ArvinMeritor.

Trailer ABS systems should have a failure warning light mounted on the trailer. Auxiliary-powered systems should also have a second light on the tractor dash. The warning lights indicate the status of the system. The system automatically performs a self-check each time the ECU is activated. The warning light stays on until the self-check is completed. A self-check includes checking the output of the wheel speed sensors. Therefore, the vehicle must be moving at least 4 mph to complete the self-check. Then, if the system is functioning correctly, the warning light will go out. If a fault exists, the light will stay on. The warning light will always go out when the ECU is deactivated.

Shop Talk

Bulb Test: Depress ABS switch for 16 seconds. The ABS light should illuminate even when no codes are present.

When the ECU is activated depends on how the system is powered. A brake light-powered ECU is activated each time the brakes are applied and is deactivated each time the brakes are released. An auxiliary-powered ECU is activated when the ignition is turned on and remains activated until the ignition is turned off. Therefore, how the warning light operates depends on the source of power. **Figure 26–19** shows the tractor power test circuits in an ATA 7 wire trailer receptacle.

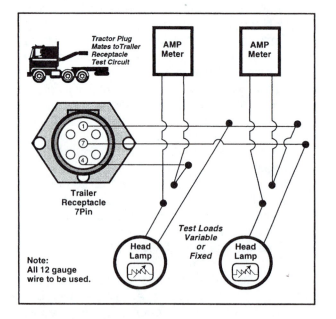

FIGURE 26–19 Tractor power test circuits. (Illustration courtesy of The Maintenance Council of the American Trucking Associations, Inc.)

26.7 ABS DIAGNOSTICS

The electronic control unit or the modulator controller usually provides a quick and easy method of checking ABS failures or faults, as they are called when stored in ECU. At present, there are two diagnostic systems in common use: LED indicators (shown initially in **Figure 26–5** and again in **Figure 26–20**) and blinking light codes (shown in **Figure 26–21**). The following account of ABS roughly follows ArvinMeritor systems, but most other systems are similar to this model.

WARNING: The ABS is an electrical system. When working on ABS systems, the same precautions that must be taken with any other electrical system should be taken (see Chapters 2, 4, and 5). As with any electrical system, there is a danger of electrical shock or sparks that can ignite flammable substances. Always disconnect the battery ground cable before working on the electrical system. This is of special concern when working with flammable or explosive truck or trailer cargos.

LED Indicators

In this system, if the dash warning light remains on after the truck reaches a speed of 4–7 mph, with or without blinking upon the initial start-up of the truck, inspect the ECU for illuminated LEDs. Each LED represents a specific area, as indicated in **Figure 26–20**. The red LEDs illuminate when the ECU senses a fail-

FIGURE 26–20 Typical ECU diagnostic LED window located on a modulator controller assembly. (Courtesy of Bendix Commercial Vehicle Systems)

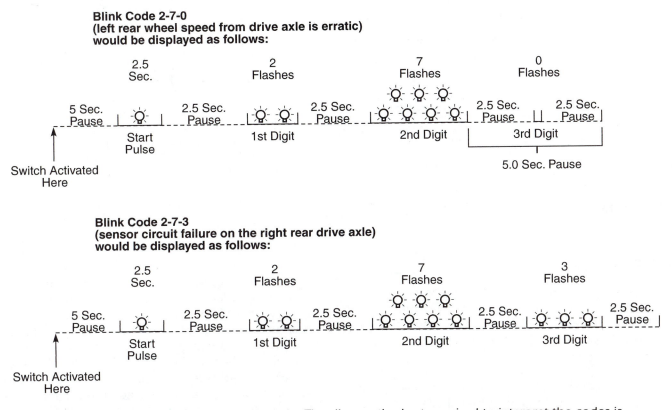

FIGURE 26–21 *Example of a typical blink code. The diagnostic chart required to interpret the codes is located in the service manual.*

ure in the system. When a failure is logged, it is stored in nonvolatile memory. This means that, even if the truck ignition is turned off, the fault code remains.

When tripped, the LEDs are latched into the on position. They can be cleared by activating the reset switch located beside the green LED(1). If failure is indicated, make a note of it for future reference before clearing codes. If the failure indicator LED or LEDs are cleared by the resetting procedure but recur later, an intermittent fault condition could be present. If the LED or LEDs do not clear, the condition is probably still active. Check all wiring and hardware using the OEM troubleshooting guide. Remember always to perform a "wiggle wire" test at each connection. Always perform a road test whenever the brakes have been worked on, making several brake applications before placing the truck back into service.

Blink Codes

When the ABS warning light stays on, blink codes identify the system fault causing the failure. **Figure 26–21** and **Figure 26–22** show how these operate.

The blink code switch and light are usually located on the relay/fuse panel. In an ArvinMeritor system, when the blink code switch is activated, the light flashes are usually either 0.5 second or 2.5 seconds. The pauses between the flashes can also be either 0.5 second or 2.5 seconds. The cycle starts with a 2.5-second flash followed by a three-digit blink code. Blink codes can be interpreted as follows:

- The first digit of the three-digit blink code identifies the ABS configuration installed in the vehicle. For instance, two flashes indicates a 4-channel system.
- The second and third digits identify the specific fault. A fault can only be erased from the ECU after the failure has been repaired. If more than one fault exists, the first fault displayed must be corrected and erased from the ECU memory before the next fault is displayed. When all faults have been corrected and erased from the ECU memory, the blink code 2-0-0 (No Faults—System OK) will be displayed when the blink code switch is activated.

**Blink Code 2-0-0
(no faults — system O.K.)
would be displayed as follows:**

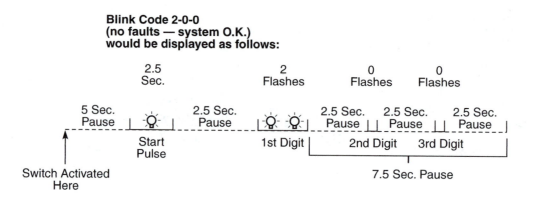

FIGURE 26–22 A typical "no-fault" blink code.

 Shop Talk

Blink codes in most systems do not display faults in the order they were recorded in the ECU memory. Also, if codes are blinked that do not appear in the OEM code chart, the ECU may be defective.

Remember when servicing ABS components that they can be delicate and must be handled with care. If a fault indicates that a component is not functioning properly, it should be removed and replaced as directed in the service manual. Some typical service procedures are covered in the next section.

DIAGNOSING TRAILER ABS

Self-diagnostics are written into the software of all the current systems. Most ABS are designed to perform a self-test on startup. This means that when the ABS circuit is first energized, a clicking noise can be heard from the modulator valve solenoids. There may be some read capability on the ABS module, but in most cases today it is accepted that one type of electronic service tool (EST) will be used to read the system. We will refer to the Haldex-Midland system in describing some of the basics of trailer ABS diagnosis.

ESTs

Haldex recommends the use of their own digital diagnostic unit (DDU), which is supplied with the appropriate wiring and connectors to bus directly to the ABS system. The Haldex DDU is the most user-friendly method of reading Haldex ABS. It performs the following functions:

- Displays active and inactive codes
- Clears fault codes
- Displays valve and sensor configuration data
- Displays auxiliary codes

A ProLink may also be used to read Haldex ABS, providing it is loaded with a Heavy Duty Standard software cartridge and equipped with a Haldex adaptor cable. ProLink does everything the Haldex DDU does, and has the ability to generate a hard copy of any data retained in the ECU. A PC loaded with Haldex software and equipped with a Haldex adapter cable may also be used. The PC has all of the capabilities of the previous instruments but will additionally

- Display odometer readings
- Display an audit trail (archived faults)
- Accept tire data programming
- Provide service reminder data
- Analyze brake performance and display percentage of braking by each wheel

 Shop Talk

Never use a battery charger to perform a dynamic check of a trailer ABS system. ECU damage can result.

26.8 ABS SERVICE PROCEDURE

Technicians are reminded that there is no substitute for the OEM service manual when any maintenance or service procedure must be undertaken on an ABS system. Some typical procedures are outlined here. Again, the ArvinMeritor system is the one referenced in this text: Most of these procedures are similar in other OEM systems.

TIRE SIZE RANGE

Proper ABS/ATC operation with a standard ECU requires that the front and rear tire sizes be within ± 14% of each other. If this tire size range is exceeded without electronically modifying the ECU (PROM), system performance may be affected and the warning lamp can be illuminated. Special ECUs can be ordered if necessary.

Tire size can be calculated using the following equation:

$$\% \ Difference = (rpm \ Steer \div rpm \ Drive) - 1 \times 100$$

rpm = tire revolutions per mile
Steer = steering axle tire
Drive = drive axle tire

Note: Tire rolling radii expressed in revolutions per mile (or kilometer) should be referenced in a tire crossover/reference manual. Measuring and calculating this is not usually sufficiently accurate.

ABS VOLTAGE CHECK

- Voltage should measure between 11 and 14 volts on a 12-volt system (18 and 30 volts for a 24-volt system).
- Remember, the ignition key must be turned on for this test.

Measure voltage at the pins indicated in the table below.

VOLTAGE CHECK PINS

ECU	Connector	Pins
Cab-Mounted	18-Pin	7 and 10
		8 and 11
		9 and 12
Basic	15-Pin	7 and 4
		8 and 9
Frame-Mounted	X1-Grey	1 and 12
		2 and 11

LOCATION OF SENSORS

On the steering axles of most trucks, the sensor is located on the inboard side of the steering knuckle.

On drive axles, the drum assembly sometimes has to be removed to access to the sensor.

CODE INTERPRETATION

Troubleshooting of any ABS system must be performed using the OEM service manual. The following charts may be used to interpret Rockwell WABCO and Haldex-Midland codes.

ArvinMeritor Codes

BLINK CODE IDENTIFICATION

First Digit (Type of Fault)	Second Digit Specific (Specific Location of Fault)
1 No faults	1 No faults
2 ABS modulator valve	1 Right front steer axle (curb side)
3 Too much sensor gap	2 Left front steer axle (driver's side)
4 Sensor short or open	3 Right rear drive axle (curb side)
5 Sensor signal erratic	4 Left rear drive axle (driver's side)
6 Tooth wheel	5 Right rear/additional axle (curb side)*
	6 Left rear/additional axle (driver's side)*
7 System function	1 J1922 or J1939 datalink
	2 ATC valve
	3 Retarder relay (third brake)
	4 ABS warning lamp
	5 ATC configuration
	6 Reserved for future use
8 ECU	1 Low power supply
	2 High power supply
	3 Internal fault
	4 System configuration error
	5 Ground

* Tandem, lift, tag, or pusher axle, depending upon the type of suspension.

Haldex Codes

Haldex codes are also grouped into categories. The following description uses the codes as they would be read by the Haldex DDU.

Codes 00, 07, and 8.8. No faults detected. 00 is dynamic no-fault (moving vehicle) and 07 is static (vehicle parked) no-fault. 8.8 indicates the system is undergoing a self-test sequence, which includes testing the display LEDs. DB indicates that an electrical connection has been made between the chassis and the ECU.

Codes 03, 04, 05, and 06. Indicate a short or open in a wheel speed sensor. The resistance in the indicated sensor should be tested with a DMM. The resistance value should read 980–2,350 ohms.

Codes 13, 14, 15, and 16. Indicate that the output of a moving wheel sensor is insufficient, usually caused by excessive gap between the sensor and the reluctor/exciter wheel. Measure the AC voltage produced while rotating the wheel at 1 revolution every 2 seconds: It should be at least 200 millivolts

(0.2V). If the sensor gap is excessive (sensor dropped out of socket), no fault will be logged because the ECU will assume the vehicle is parked.

Codes 23, 24, 25, and 26. Indicate intermittent loss of wheel sensor input while the vehicle is moving down the road. Causes can be difficult to diagnose, but some possibilities are a broken sensor retaining clip, damaged reluctor wheel, or excessive wheel bearing endplay.

Codes 42 and 43. Occur while the unit is moving and indicate that the wheel speed is slow to recover when ABS releases the brake during an ABS event. Possible causes are a dragging foundation brake, a kinked or otherwise damaged brake hose, or a defective modulator valve.

Codes 62, 63, 68, and 69. Indicate that one of the solenoids or its cable is open or shorted. Disconnect the indicated solenoid and check the resistance across the connectors. Readings across the bottom 2 pins should be 7–9 ohms. Between either bottom pin and the top pin the reading should be between 3.5–4.5 ohms. Remember to check for corrosion at the pins and connectors before condemning a solenoid.

Codes 72, 73, 78, and 79. Indicate a short to ground in a modulator solenoid or its cable. The most likely cause is a damaged cable or solenoid. Disconnect the solenoid connector and check for continuity between each solenoid terminal and chassis ground. Perform tests as outlined in Haldex service literature: This code can also be caused by a defective ECU.

Codes 82, 83, 88, and 89. Indicate a short to battery positive in a modulator solenoid or its cable. The most likely cause is a damaged cable or solenoid. Disconnect the solenoid connector and check for continuity between each solenoid terminal and chassis positive. Perform tests as outlined in Haldex service literature: This code can also be caused by a defective ECU.

Code 90. Indicates ABS voltage is below 8.5 volts. Most likely caused by damaged or corroded terminals or wires. Use voltage drop testing between the trailer connector and the ABS power circuit to locate fault. This fault can also be caused by undersized wiring. If performing the test using a battery powered ABS tester, check the battery voltage.

Code 92. High voltage caused by ABS power supply voltage above 16 volts. Check tractor voltage regulator or battery-powered tester voltage.

Code 80, 93, 99, E0–E9, or EA–EF. Defective ECU. Replace ECU and retest system.

Code CA. Invitation to clear all fault codes. Dynamic faults will not be erased until the vehicle is moved at a speed exceeding 6 mph.

Code CC. A CC code is displayed the third consecutive time a CA is attempted. It is an invitation to clear configuration, so it should be avoided. If a CC is displayed, first power down the system and then power up again.

Sensor Adjustment

1. Push the sensor in until it contacts the tooth wheel.
2. Do not pry or push sensors with sharp objects.
3. Sensors will self-adjust during wheel rotation.

Sensor Signal (Output) Voltage Test

Voltage must be at least 0.200 volt AC at 30 rpm.

1. Turn ignition off.
2. Disconnect ECU (see wiring diagram).
3. Raise wheel from the ground and release brakes.
4. Use a DMM on AC volts mode range set to read up to 1V.
5. Rotate the wheel by hand at 30 rpm (1 revolution per 2 seconds).
6. Measure voltage at the pins indicated in the table below.

SENSOR CHECK PINS			
ECU	Sensor	Connector	Pins
Cab-Mounted	LF	6-Pin	4 and 5
	RF	9-Pin	4 and 5
	LR	15-Pin	5 and 6
	RR	15-Pin	8 and 9
	LR (third Axle)	12-Pin	5 and 6
	RR (third Axle)	12-Pin	8 and 9
Basic	LF	18-Pin	12 and 15
	RF	18-Pin	10 and 13
	LR	18-Pin	11 and 14
	RR	18-Pin	17 and 18
Frame-Mounted	LF	X2—Black	7 and 8
	RF	X2—Black	5 and 6
	LR	X3—Green	1 and 2
	RR	X3—Green	3 and 4
	LR (third Axle)	X4—Brown	3 and 4
	RR (third Axle)	X4—Brown	5 and 6

Sensor Resistance

The sensor circuit resistance must be 700–3,000 ohms. Resistance can be measured at the sensor connector or at the pins on the ECU connector using a DMM. To measure resistance

1. Turn ignition off.
2. Set the DMM to read ohms/resistance.
3. To measure resistance at the sensor connector, disconnect the ECU connector from the ECU.
4. Measure resistance at the pins indicated and check to specification.

Dynamometer Testing Vehicles with ATC

Vehicles equipped with ATC must have the ATC disabled before running the vehicle on a chassis dynamometer. A typical method of disabling ATC is to press and hold the blink code switch for at least three seconds. When the system configuration code appears, ATC has been disabled. The ATC lamp illuminates and stays on while the system is disabled. Some systems require that the ATC fuse be removed: Check with the OEM.

Testing an ABS Modulator Valve

Measure resistance across each valve solenoid coil terminal and ground on the ABS valve to ensure 4.0–8.0 ohms for a 12-volt system (8.0–16.0 ohms for a 24-volt system).

- If the resistance is greater than 8.0 ohms for a 12-volt system (16.0 ohms for a 24-volt system), clean the electrical contacts in the solenoid. Check the resistance again.
- To check the cable and the ABS valve as one unit, measure resistance across the pins on the ECU connector of the harness. Check the diagram of the system you are testing for pin numbers.

ATC Valve

Measure resistance across the two electrical terminals on the ATC valve to ensure 8.0–14.0 ohms for a 12-volt system (16.0–28.0 ohms for a 24-volt system).

- If the resistance is greater than 14.0 ohms for a 12-volt system (28.0 ohms for a 24-volt system), clean the electrical contacts on the solenoid. Check the resistance again.
- To check the cable and ATC valve as one unit, measure resistance across the pins on the ECU

FIGURE 26–23 When working on wheels and axles, the technician should be careful of ABS components such as sensors.

connector of the harness. Check the diagram of the system you are testing for pin numbers.

Front Axle Wheel Speed Sensor Installation

To replace the sensor in the front axle:

1. Connect the sensor cable to the chassis harness. See **Figure 26–23**.
2. Install the fasteners used to hold the sensor cable in place.
3. Apply an OEM-recommended lubricant to the sensor spring clip and sensor.
4. Install the sensor spring clip. Make sure the spring clip tabs are on the inboard side of the vehicle.
5. Push the sensor spring clip into the bushing in the steering knuckle until the clip stops.
6. Push the sensor completely into the sensor spring clip until it contacts the tooth wheel.
7. Remove the blocks and safety stands.

Rear Axle Wheel Speed Sensor Installation

1. Apply an OEM-recommended lubricant to the sensor spring clip and sensor.
2. Install the sensor spring clip. Make sure the spring clip tabs are on the inboard side of the vehicle.
3. Push the sensor spring clip into the mounting block until it stops.

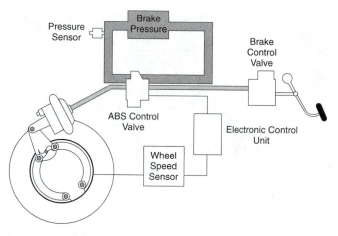

FIGURE 26–24   ABS operation.

4. Push the sensor completely into the sensor spring clip until it contacts the tooth wheel.
5. Insert the sensor cable through the hole in the spider and axle housing flange. Route the cable to the frame rail. Be sure to route the cable in a way that will prevent pinching or chafing and will allow sufficient movement for suspension travel.
6. Connect the sensor cable to the chassis harness.
7. Install the fasteners that hold the sensor cable in place.
8. Install the brake drum on the wheel hub.
9. Complete the installation per the vehicle manufacturer's manual.

Figure 26–24 shows the operating look of an air-actuated, ABS system.

26.9 ELECTRONIC BRAKING SYSTEMS (EBS)

Electronic braking systems (EBS) are currently used on European trucks and US companies Bendix and ArvinMeritor have established themselves as major players in the systems used there. Although Freightliner promised EBS as an option on their Century-class trucks when they were introduced to the market, at the moment of writing, it is not possible to manufacture truck chassis equipped with EBS without also having a FMVSS No. 121-compliant air control circuit as a backup. The problem is the FMVSS No. 121 standards, which assume that any use of EBS would be in addition to the required dual-circuit air brake system with pneumatic controls. Until FMVSS No. 121 addresses EBS issues, any

EBS system introduced in North America can only be in addition to a FMVSS No. 121-compliant dual-circuit air brake system. The European EBS systems cost more than non-EBS systems. They must be sold on the advantages they offer. These advantages include better stopping performance, faster service braking response, driver stability control, and rollover protection. The manufacturers of EBS claim that their systems will result in a lower percentage of brake-related accidents, and early evaluations of European systems would seem to confirm this. At the moment of writing, North American trucks using EBS must also have a completely functional pneumatic control circuit as a backup: This makes EBS a very expensive option.

EBS Advantage

At the present moment, the control circuit in a truck air brake system is pneumatic. By control circuit, we mean all the air signals used to "pilot" or control slave valves such as relay valves in the system. Under ideal circumstances, air pressure signals travel at the speed of sound. Usually the conditions of a pneumatic control circuit are far from ideal, so lag times reduce the efficiency of the control circuit during both application and release. Electrical signals travel at closer to the speed of light. In a true EBS system, the pneumatic control circuit would be eliminated. It would be replaced by a "brake-by-wire" electrical control circuit, which would respond much faster to changes in the command signals from the driver's boot. The actual braking force is still provided by the potential energy of compressed air metered to brake chambers by electrically controlled modulators. **Figure 26–25** shows a circuit schematic of a typical EBS system.

EBS OPERATION

The real advantages of EBS are with control and monitoring. The EBS systems in testing and in the European market have essentially adapted an air brake system to electronic control and monitoring. This means the replacement of the dual circuit pneumatic foot valve with an electrical foot control valve, known as a brake signal transmitter. This electrical foot control valve signals the EBS control module with a driver-initiated brake request. The resulting electrical signal is processed by the EBS ECM, which processes it to switch **electro-pneumatic** valves, known as brake modulators. Brake modulators are located over each axle or axle grouping with brakes and convert the electrical signals they receive from the EBS ECM into an air pressure value, which is delivered to brake chambers.

EBS for 6x4 Tractor

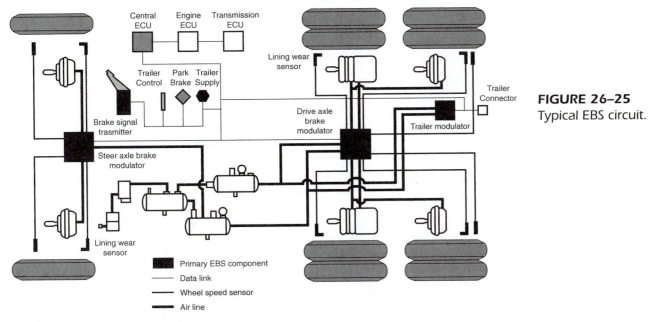

FIGURE 26-25
Typical EBS circuit.

The European experience of EBS has produced benefits that go beyond faster response. EBS can improve air brake system performance in the following areas:

- Balance
- Proportioning
- Antilock

Under normal braking conditions, brake pressures can be balanced to provide the best lining wear. When subjected to heavier braking, the system can be programmed to go into proportioning mode, applying each foundation brake with respect to the load sensed and the wheel rotational speed. Brake-by-wire electronic heavy-duty truck air brakes will be introduced into North America and become the standard means of controlling truck brake systems—it is only a question of time.

ber of axles, axle configuration, axle load, brake circuit distribution, and brake force distribution.
- The ABS is a computer-controlled electrical system.
- LED indicators and blink codes are used to help diagnose ABS failures.
- Electronic brake systems (EBS), also known as electro-pneumatic braking, replace the current FMVSS No. 121-compliant pneumatic control circuit with an electronically managed control circuit.
- EBS provides much faster service braking response and release times because its electrical signals travel at much higher speeds than pneumatic control signals.
- An EBS system controls pressure in the brake service chambers under all braking conditions, unlike ABS, which tends to be a factor only under severe braking conditions.

SUMMARY

- Antilock brake systems (ABS) are designed to help prevent wheel lock-up during severe braking.
- An ABS system uses microcomputer technology to sense and reduce braking force on wheels that are beginning to lockup under braking.
- The components of a typical ABS system are wheel speed sensors, control modules, and brake modulators.
- Several different ABS configurations are used on tractors and trailers, depending on the num-

REVIEW QUESTIONS

1. Which of the following best describes how ABS prevents brake lockup?
   a. senses abnormal braking forces and corrects them
   b. monitors wheel speed and dumps service pressure when lock-up is imminent
   c. decreases service application pressure for all braking
   d. adjusts braking force on every wheel even in light braking situations

2. Which of the following provides the most critical input signal to the ABS module?
   a. slack adjuster travel
   b. wheel speed sensor
   c. modulator valve
   d. engine ECM serial bus

3. Which of the following best describes the type of sensor used to signal wheel speed data to an ABS module?
   a. pulse generator
   b. transducer
   c. hall effect sensor
   d. thermistor

4. Which of the following is the main output device on a truck ABS system?
   a. wheatstone bridge
   b. wheel speed sensor
   c. power supply
   d. modulator assembly

5. Which of the following is the control component in an ABS modulator valve?
   a. engine ECM serial bus
   b. relay valve
   c. solenoid or motor
   d. data display window

6. Which of the following is likely to occur if an ABS system ceases to function?
   a. ABS shuts down and a driver warning light comes on
   b. ABS shifts into default mode
   c. the entire braking system shuts down
   d. LED directs the driver when to apply the brakes

7. Why do some ABS ECUs contain two micro-processor units?
   a. to enable them to process large amounts of information
   b. to enable them to provide continuous diag-nostic information
   c. one microprocessor acts as a fail–safe backup to the other
   d. to double the processing speed

8. Which of the following is true of the ABS tractor warning light?
   a. It illuminates when the ignition key is turned on.
   b. It illuminates when the vehicle is braked.
   c. It goes off when there is an ABS failure.
   d. It goes off only when the vehicle is stationary.

9. When does an ATC light illuminate?
   a. when the interaxle differential engages
   b. whenever a spin-out condition occurs
   c. whenever the ABS system is cycling
   d. when a wheel speed sensor fails

10. Which of the following ABS channel configura-tions tends to be more commonly used in truck air brake systems today?
    a. 2S/2M
    b. 4S/4M
    c. 6S/4M
    d. 6S/6M

11. What is the maximum modulation speed of a typical air brake ABS system?
    a. 2 times per second
    b. 7 times per second
    c. 20 times per second
    d. 200 times per second

12. Technician A says that air brake ABS can be modulated at much higher speeds than hydraulic ABS due to the compressibility of air. Technician B says that a truck air brake ABS control module can "think" (clock speed) at a frequency of 200 times per second. Who is correct?
    a. Technician A
    b. Technician B
    c. both Technicians A and B
    d. neither Technician A nor B

13. Before testing a dual-drive axle truck equipped with ATC with a chassis dynamometer, what should be done?
    a. ATC must be enabled.
    b. ATC must be disabled.
    c. The power divider must be locked out.
    d. The drive shafts on the rear drive should be pulled.

14. An ABS light displays fault codes that are not interpreted in the OEM service manual. Which of the following is the likely problem?
    a. defective wheel speed sensor
    b. defective modulator valve
    c. defective ECM
    d. defective data connector

15. Technician A says that in a true EBS system, the brake pedal does no more than produce an electrical signal to the EBS control module. Technician B says that current FMVSS No. 121 legislation requires that when EBS is used on North American highways, it is in addition to pneumatic controls. Who is right?
    a. Technician A
    b. Technician B
    c. both Technicians A and B
    d. neither Technician A nor B

Prerequisites
Chapters 24 and 26

Air Brake Servicing

Objectives

After reading this chapter, you should be able to

- Understand the safety requirements of working on an air brake system.
- Perform basic maintenance on an air brake system.
- Diagnose common compressor problems.
- Describe the procedure required to service an air dryer.
- Performance test an air dryer.
- Check out the service brakes on a truck.
- Test the emergency and parking brake systems.
- Verify the operation of the trailer brakes.
- Understand the OOS criteria used by safety inspection officers.
- Diagnose some brake valve failures.
- Describe the procedure required to overhaul foundation brakes.
- Determine brake free-stroke and identify when an adjustment is required.

Key Terms List

brake torque balance
Commercial Vehicle Safety Alliance (CVSA)
decelerometer

FMVSS No. 121
out of service (OOS) conditions
pneumatic timing

This section covers some of the most often performed service and repair procedures. As much as possible, describing OEM-specific procedures is avoided, and the approach attempts to be generic. When actually performing service and repair procedures, whether it be in a training center or service garage, technicians should use the OEM service literature specific to the equipment on which they are working. The air brake system is one of the most important safety systems on a truck. Maintenance of the system is simple and can be performed by any trained technician with ordinary shop tools.

27.1 MAINTENANCE AND SAFETY

Although proper maintenance of the brake system is essential to the operation of any highway vehicle, brake work is often the first experience truck technicians have with vehicle systems. Brake work represents a large percentage of the operating maintenance and repairs required on a vehicle and is regarded as requiring fewer skills than other repairs, such as engine work. It is true that brake work can be performed with less training than that required for other chassis systems, but it must always be performed with precision. Quite simply, the consequence of a failed engine repair job is a truck on the side of the road, but the consequence of a failed brake repair job can be loss of life. Never regard brake service and repair work as something that is less important than that required on the other vehicle systems. Use the service literature when performing repairs and make a habit of asking questions when you do not understand something.

The approach in this chapter follows the air brake system through, circuit by circuit, and identifies critical out-of-service criteria, troubleshooting strategies, and some repair practices. Common repair practices required on hydraulic and ABS systems are covered in the chapters dedicated to those subjects.

AIR BRAKE SAFETY CONSIDERATIONS

Before beginning any repairs, it is important to realize that compressed air can be dangerous if not handled with some respect. The potential energy of compressed air can be compared to the potential energy in a compressed coil spring. When this force is released from a storage tank or brake hose, it can cause serious injury. The wheels of the vehicle should always be blocked by some means other than the air brake system. Wheel chocks or wedges are used, and these should be placed on both sides of a wheel not on a steering axle. For many of the tasks required to maintain or service an air brake system, the air will have to be drained from one or more tanks. When draining an air system, wear eye protection and point the release jet of air away from people. Do not disconnect a chassis hose under pressure from its coupler because the hose may whip around as air escapes and cause an injury.

Follow the manufacturer's recommended procedures when working on any air device so as to avoid injury or damage from components that, when released, might be subject to mechanical (spring) or pneumatic propulsion. As system pressure is drained and the emergency brakes are applied, exercise care in the vicinity of brake chambers, pushrods, and the foundation brake assembly. Air tanks closest to the compressor should contain a safety valve specified to trip at the required pop-off pressure, usually 150 psi. When air tanks have to be replaced, do not select old, used reservoirs as replacements. Air tanks should never be repair welded, even as a temporary measure in an emergency, because of the internal coating: this plasticized coating crinkles and separates from the tank wall when heated. The operation of a spring brake chamber must be fully understood (see Chapter 24),

and every truck technician should know that the force exerted by a spring brake can be fatal. When these units have to be removed from a vehicle, they should be caged. Caging requires connecting an external screw to the spring brake chamber cage plate and mechanically compressing the main spring. Note the warning tags on such units and the fact that it is illegal to dispose of a spring brake chamber in a garbage or scrap metal bucket unless it has been disarmed.

GENERAL MAINTENANCE

Maintenance of an air brake system can be broken down into the major areas of pneumatic factors, brake torque balance, and safety considerations.

Pneumatic Factors

A properly functioning or balanced braking system is one in which the braking pressure reaches each actuator at the same time and at the same pressure value. Brake balance is discussed in some detail in Chapter 24, but some of the factors of brake balance are also addressed in this section. Two factors that affect brake balance are application and release times. This is known as **pneumatic timing.** In meeting **FMVSS No. 121** standards for application and release times, vehicle manufacturers carefully select tubing and hose sizes, the valve and fitting flow areas, and valve crack pressures. Air application and release performance is also dependent on the size and volume of the brake chambers and the distance the air must travel to get to them. Original equipment performance is engineered into the vehicle brake system by the manufacturer. The role of the vehicle owner and/or truck technician is to preserve that pneumatic performance. For example, when replacing tubing or hose, the replacement plumbing must match the original size and grade. When replacing valves in the air brake system, ensure that the function and ratings of the replacement valve are comparable with the valve being replaced. Ratings cover factors such as crack pressure and regulated hold-off pressure values. Most valves can be crossed-over to a different brand on replacement, providing the function and ratings exactly match.

👓 **Shop Talk** _____

The brake plumbing can affect pneumatic timing on a truck brake system. When replacing brake hose and fittings, be sure to use the same sizes, or the flow dynamics (movement of air through the system) will be altered. One brake OEM has determined that replacing a straight pipe nipple with a 90-degree elbow is equivalent to adding 7 feet of

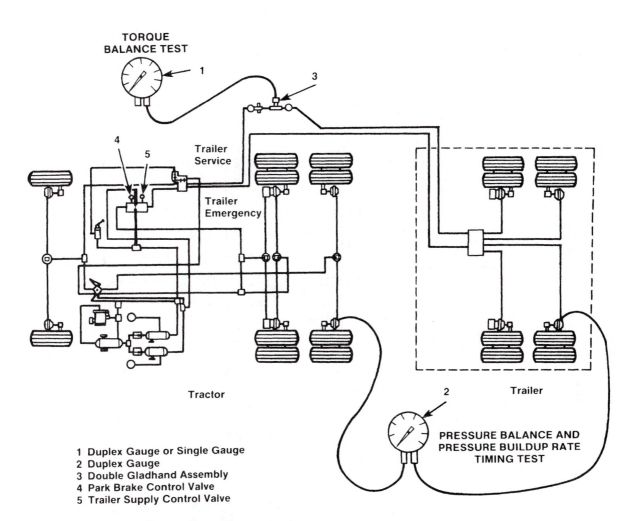

FIGURE 27-1 Single trailer configuration test setup.

1 Duplex Gauge or Single Gauge
2 Duplex Gauge
3 Double Gladhand Assembly
4 Park Brake Control Valve
5 Trailer Supply Control Valve

brake hose into the circuit. Adding a 45-degree elbow is equivalent to adding 3 feet of hose into the circuit. Never alter the brake plumbing, because the increases or decreases in flow restriction can affect pneumatic timing.

Figure 27-1 and **Figure 27-2** show test instrument locations for performing pressure and torque balance tests on air brake-equipped tractor/trailer combinations. The test instruments used are described a little later in this chapter.

Brake Torque Balance

Brake torque balance is when the amount of mechanical braking effort applied to each wheel under braking is balanced. The mechanical components of the typical brake system consist of many levers that must work together to ensure that one wheel is not over-braked and another under-braked. A half-inch difference in slack adjuster length or a shoe that has had its arc spread can make a large

difference in the amount of braking force applied to each wheel assembly. For a brake system to meet the required standards, it must be maintained on a regular basis throughout the life of the vehicle. Because brake failures can have lethal consequences, they are a crucial element of any preventative maintenance program.

Draining Air Tanks

The contaminants that collect in air brake reservoirs consist of water condensed from the air and a small amount of oil from the compressor. This water and oil normally pass into the air tank in the form of vapor because of the heat generated during compression. There is probably no more simple yet more important maintenance than reservoir draining. If water is allowed to collect in any air tank, the storage capacity of the tank is reduced. An air tank with insufficient reserve capacity means that the air pressure drops off too rapidly when air demand on the system is highest. The result can be inadequate braking in high demand situations.

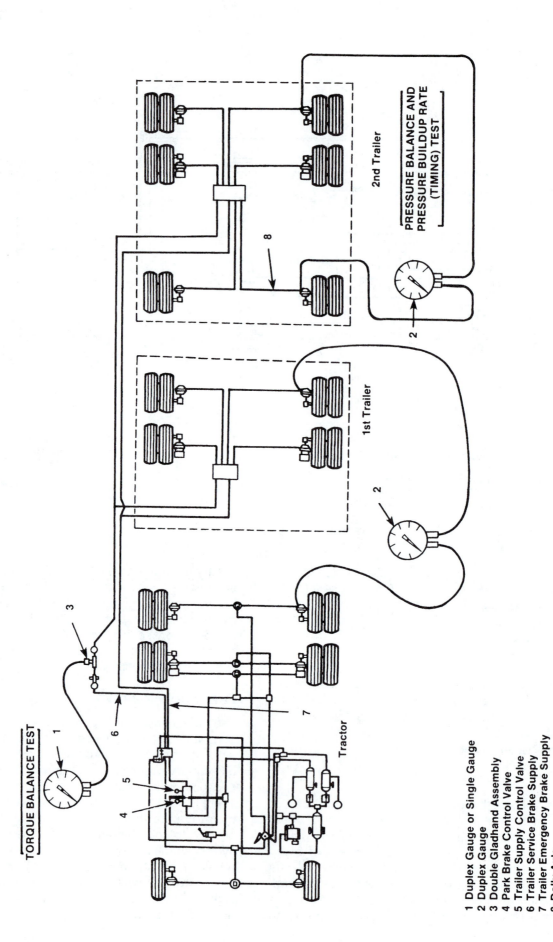

FIGURE 27-2   Multiple trailer configuration test setup.

1 Duplex Gauge or Single Gauge
2 Duplex Gauge
3 Double Gladhand Assembly
4 Park Brake Control Valve
5 Trailer Supply Control Valve
6 Trailer Service Brake Supply
7 Trailer Emergency Brake Supply
8 Dolly Axle

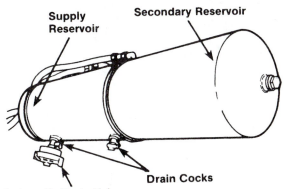

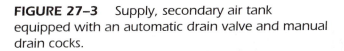

FIGURE 27–3 Supply, secondary air tank equipped with an automatic drain valve and manual drain cocks.

Truck drivers are instructed to drain air tanks daily. The common type of drain valve is a cable-release, spring-loaded drain valve. This enables the driver to pull on the cable for a couple of seconds and to dump the small amount of moisture sitting on the drain valve. Any significant amount of moisture in the tank will exit only after all the compressed air has been dumped. Preferred options to the cable-release dump valve are the automatic drain or spitter valve and the manual drain cock. Least popular with drivers is the manual drain cock because this often requires that the driver climb under the chassis to open it; however, it is the most effective valve for ensuring that the system is completely drained. Automatic drain valves are most popular among drivers because they do not require daily draining. All automatic drain valves (spitter valves) should be checked periodically for proper operation. **Figure 27–3** shows a typical supply, secondary air tank equipped with an automatic drain valve (supply tank) and a manual drain cock (secondary tank).

Many of the most recent air-brake valves contain small orifices and passages that make them more susceptible to contaminants. The prevention of freeze-ups in the system is equally important. The use of air dryers is critical in maintaining a supply of clean, dry air, especially in geographic areas of high humidity. Although alcohol evaporators do nothing to eliminate moisture from an air brake system, they can prevent freeze-ups.

 Shop Talk _____

Trailer hookups. *In just about all jurisdictions in the United States and Canada, driver license testing requires the air lines to be coupled to the trailer*

before backing the tractor under the trailer to couple the fifth wheel. Any observer of day-to-day truck yard coupling procedures would quickly realize that this practice is seldom observed by truck drivers on a daily basis—except when undergoing mandatory state testing. In pre-FMVSS No. 121 system trailers (pre-1975) and in a small number of current special application trailers, an air-actuated, nonmechanical park mode is used. The problem with park-on-air is that in the event of an air leak, the air used to brake the trailer bleeds off leaving the trailer with no brakes. This means that air must be supplied to the trailer to ensure that it is braked before coupling a tractor. The reason drivers can safely engage the fifth wheel to a trailer before the air lines is that most trailers with spring brakes (in compliance with the FMVSS No. 121 requirement for a mechanical park mode) can be assumed to be in park mode and, therefore, the trailer is going to hold. Although attempting to couple air lines to trailer gladhands before the tractor is mechanically coupled to the trailer can place the driver in some danger, driver testing continues to insist that air lines be connected before fifth wheels. The best option is to know the equipment with which you are working. If you have any doubts, couple the air lines first, actuate the trailer brakes, and then back the tractor under to engage the fifth wheel. Also review Chapter 30, Fifth Wheels and Coupling Systems, so that you know how to verify that the fifth wheel is properly engaged.

27.2 TEST EQUIPMENT

The test equipment described here and illustrated in **Figure 27–4**, with the exception of the brake timing unit, is required in conducting pressure balance, pressure buildup rate, and torque balance tests.

- **Double Gladhand Assembly** (with quick connect coupling and shut-off cock). This assembly is used in the service line connection to the trailer for the torque balance test. **Figure 27–5** shows the components that make up this test equipment, which can be assembled from odds and ends found around any truck shop. This equipment is designed to use shop air to simulate trailer supply and service applications in place of a tractor. **Figure 27–6** shows how the unit would be coupled into the circuit for performing tests.

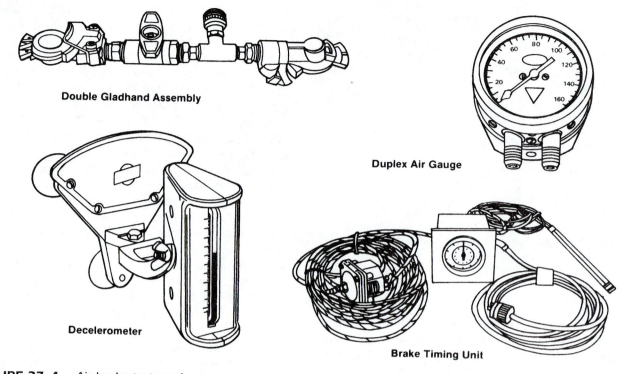

Double Gladhand Assembly

Duplex Air Gauge

Decelerometer

Brake Timing Unit

FIGURE 27–4 Air brake test equipment.

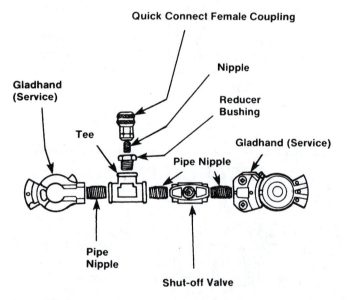

Quick Connect Female Coupling

Nipple

Reducer Bushing

Gladhand (Service)

Gladhand (Service)

Pipe Nipple

Tee

Pipe Nipple

Shut-off Valve

FIGURE 27–5 Double gladhand assembly components.

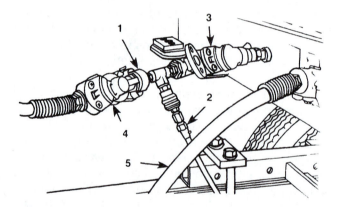

1 **Double Gladhand Assembly**
2 **To Duplex Air Gauge**
3 **Trailer Service Brake Connection**
4 **Tractor Service Brake Connection**
5 **Tractor Emergency Brake Connection**

FIGURE 27–6 Installation of a double gladhand assembly.

- **Duplex Air Gauges.** Two sets are required. Each set includes one duplex gauge and two 25-foot-long air hoses. **Figure 27–7** shows the readings on a duplex air gauge during a pressure balance test, and **Figure 27–8** shows the location of duplex gauges in a brake performance test. These may be used for monitoring application pressures and balance pressures in different parts of the air brake circuit.

- **Additional Air Hoses for Long Single-Trailer and Multiple-Trailer Tests.** Two hoses of sufficient length and with quick connection couplings are required. These hoses are used for making the test connections between the service brake chambers on the various axles of the tractor and trailer(s) and the air gauge(s). **Figure 27–9** shows a test line connection on a service brake hose.

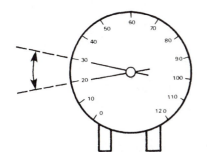

Observe pressure difference between axles (timing lead or lag) during brake application pressures stabilize.

FIGURE 27–7   Gauge readings for a pressure balance check. (Courtesy of International Truck and Engine Corp.)

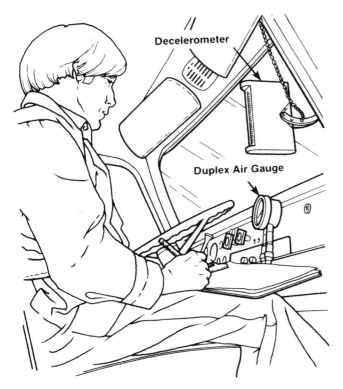

FIGURE 27–8   Installation of test instrumentation in a cab.

- **Decelerometer.** The **decelerometer** measures what is called load transfer dynamic pitch: It provides some indication of brake torque balance. A decelerometer measures the change in angle of a chassis when it is subjected to aggressive braking. This will be dependent on the chassis configuration (bobtail, straight truck, and tractor/trailer combination), the suspension type, payload (loaded or empty), and type of load (static = fixed solid load or dynamic

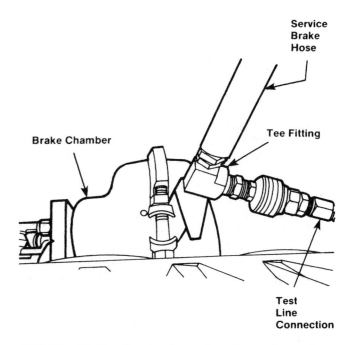

FIGURE 27–9   Connecting a test gauge hose at a service brake chamber.

= liquid or livestock). The location of a decelerometer in its test location in a vehicle cab is shown in **Figure 27–8**.
- **Brake Timing Unit.** This is a sophisticated test instrument that can accurately record brake application and release times. These test units are available from the OEM air brake manufacturers.
- **PC Software.** The read and diagnostic software used by most ABS manufacturers will display detailed brake performance data such as percentage braking by wheel, load transfer effect, adjustment, and fade.

27.3 ASSESSING BRAKE SYSTEM PERFORMANCE

Five tests can be used to verify that the system is functioning properly. These tests are static leakage checks only and do not replace functional, dynamic control and performance tests required to fully check out the air brake system.

Before beginning the testing, perform the following checks:

- Examine all tubing for kinks or dents.
- Examine all hoses for signs of wear, drying out, or overheating.

- Check suspension of all tubing. It should be supported to eliminate vibration.
- Check suspension of all hoses. Position so that the hose will not abrade or be subject to excessive heat.

Test 1. Governor Cutout, Low-Pressure Warning, and Pressure Buildup Test

To perform this test, do the following:

1. Completely drain all air tanks to 0 psi.
2. Start the truck engine and run at a fast idle. The low-pressure warning light in the cab should be on. On some trucks equipped with an antilock braking system (ABS), the warning light will also come on momentarily when the ignition is turned on.
3. The dash warning light should go off at or slightly above 60 psi.
4. Note and observe the pressure build-up time. Pressure should build from 85 to 100 psi within 40 seconds. This is an FMVSS No. 121 requirement, and this test is often used by enforcement officers to verify the performance of the truck air supply circuit because it can be performed quickly.
5. Record the governor cutout pressure value (the air compressor unloads) and check it to specification. This typically will be a value of around 120 to 130 psi.
6. Discharge air from the system by pumping the foot valve and note the governor cut-in value (air compressor effective cycle resumes). The difference between cut-in and cutout pressures is required not to exceed 25 psi. The difference should not be smaller than 20 psi to avoid frequent compressor cycling.

If the air brake system tested does not meet the required specifications, the following troubleshooting sequence can be used to correct the problem:

1. If the low-pressure warning light does not come on
   - Check the wiring.
   - Check the bulb in the warning light.
   - Repair or replace the bulb or low-pressure warning switch(es).
2. If the governor cutout value is higher or lower than specified by the truck manufacturer
   - Adjust the governor using an air pressure gauge of known accuracy. Use **Figure 27–10** and locate the adjusting screw. Back off the locknut and either increase or

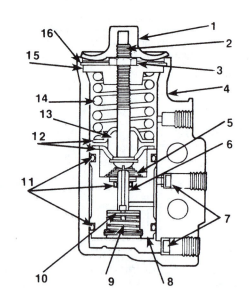

1 Cover	9 Inlet-Exhaust Valve Spring
2 Adjusting Screw	10 Inlet and Exhaust Valve
3 Locknut	11 Grommets
4 Body	12 Lower Spring Seats
5 Exhaust Stem Spring	13 Spring Guide
6 Exhaust Stem	14 Pressure Setting Spring
7 Filters	15 Upper Spring Seat
8 Piston	16 Retaining Ring

FIGURE 27–10 Sectional view of a governor assembly.

decrease the governor spring tension to alter the governor cutout.
   - Check the governor unloader signal.
   - Check the compressor unloader mechanism.
   - Reset or replace the governor.
3. If the low-pressure warning circuit does not activate when the pressure is below 60 psi
   - Check the electrical circuit/sender unit with a digital multimeter (DMM).
   - Repair or replace the faulty low-pressure sender.
4. If buildup time from 85–100 psi exceeds 40 seconds
   - Examine the compressor air supply/strainer and clean or replace as necessary.
   - Check the compressor discharge port and line for excessive carbon buildup. Clean or replace as necessary and then check the air dryer.
   - With the air system charged and the compressor in the unloaded mode, listen at the compressor inlet for leakage. If leakage can be heard, apply a small amount of oil around the unloader pistons to verify. If no leakage is indicated at the inlet, leakage

could be occurring through the compressor discharger valves.

- Check the compressor drive for slippage.

Repeat the test after a defective item has been repaired or replaced to verify proper operation and system integrity (valves, line connections, and so on).

Test 2.　Reservoir Air Supply Leakage Test

To perform the air tank supply leakage test, run the truck engine until the system is at system cutout pressure and then shut the engine off. Then do the following:

1. Allow pressure to stabilize for at least 1 minute.
2. Observe the dash gauge pressures for 2 minutes and note any pressure drop.
   - Pressure drop for a tractor or straight truck: 2 psi drop within 2 minutes is the maximum allowable for either circuit service tank (primary and secondary).
   - Pressure drop for a tractor/trailer combination: 6 psi drop within 2 minutes is the maximum allowable in all the service tanks.
   - Pressure drop for a tractor/trailer train (multiple trailers): no more than 8 psi drop within 2 minutes is allowable in the trailer service tanks.

Shop Talk

Maximum allowable leakage/drop-off rates are often defined by local jurisdictions (state and provincial governments). These regulations may differ from the test values used here. Contact your local transportation enforcement office to identify the specifications used in your area.

To locate a leak in the supply circuit of a truck air system, use a leak detector or soapy solution. The cause is likely to come from one or more of the following:

- Supply lines and fittings (tighten)
- Supply tank
- Safety (pop-off) valve in supply reservoir
- Governor
- Compressor discharge valves
- Air dryer and its fittings
- Alcohol injection components

Test 3.　Manual Parking/Emergency System Test

Before performing an emergency system test, run the engine and idle in the range of 600–900 rpm and make sure that the system pressure is correct (between cut-in and cut-out). Then do the following:

Straight Trucks and Tractors

- Manually operate the park control valve, checking that the parking brakes apply and release promptly as you pull out and push in the control valve button.

Tractor/Trailer Combinations

- Manually operate the Trailer Supply valve (usually red and octagonal) and check that the trailer brakes apply and release promptly as you pull out and push in the control button.
- Manually operate the system park control button (usually yellow and diamond or square shaped) and check that all parking brakes (tractor and trailer) apply and release promptly.

If performance is sluggish in either test, check for

- Dented or kinked lines
- Improperly installed hose fitting
- A defective trailer combination relay emergency valve
- Defective ABS modulator(s)

If the trailer parking brakes do not release and apply when the Trailer Supply valve is cycled, check

- Tractor protection control
- Trailer spring brake valve/combination relay/emergency valve

Figure 27–11 shows a typical dash control module assembly used on a current 2-dash-valve system; note the shape and color coding of the valve knobs. **Figure 27–12** shows a sectional view of the internal components of a typical push-pull dash valve.

Test 4.　Automatic Emergency System Test

This test should be performed with the engine stopped and the air brake system at full system or cutout pressure with the wheels chocked and the parking brakes released.

1. Drain the primary circuit (rear axle) tank in the tractor until the pressure reads 0 psi.

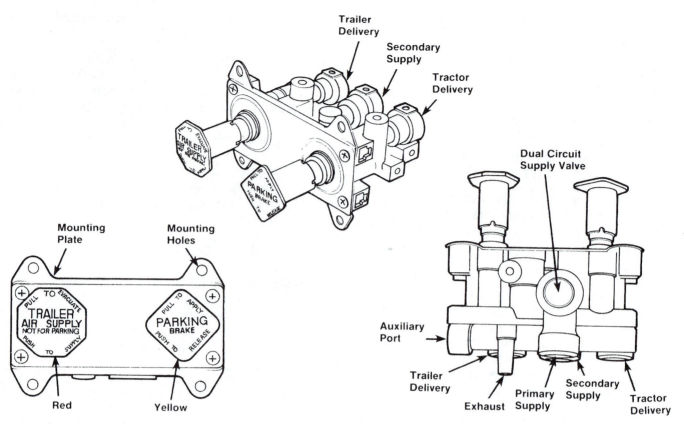

FIGURE 27–11   Tractor dash control valve module. (Courtesy of International Truck and Engine Corp.)

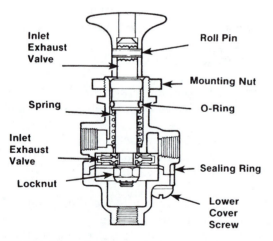

FIGURE 27–12   Cutaway view of a push-pull type control valve. (Courtesy of Bendix Commercial Vehicle Systems)

- Secondary circuit tank should not lose pressure. (This can be read on dash gauge.)
- On tractor/trailer combination vehicles, the trailer air system should remain fully charged.
- Neither the tractor nor trailer parking brakes should apply.

2. With no air pressure in the primary tank, make a brake application.
   - Rear axle brakes should apply and release.
   - On tractor/trailer combinations, the trailer brakes also should apply and release.
   - The stoplights should illuminate.
3. Slowly bleed down the secondary (front axle) tank pressure.
   - The dash System Park push-pull valve should pop out when the secondary circuit pressure drops to between 35 and 45 psi.
   - The TPV should close between 45 and 20 psi, and the trailer supply hose should be exhausted.
   - Trailer parking brakes should be applied immediately after the TPV closes.
4. Close the drain cocks, recharge the system, and this time drain the secondary (rear axle) tank to 0 psi.
   - The secondary (front axle) tank should not lose pressure.
   - On tractor/trailer combinations, the trailer air system should remain charged.
5. With no air pressure in the secondary rear axle reservoir, make a brake application.

- The front axle brakes should apply and release.
- On tractor/trailer combination vehicles, the trailer brakes also should apply and release.

If the vehicle fails to pass the tests outlined, use an OEM brake schematic to check routing and valve location and test the following components for leakage and proper operation:

- Fittings
- Kinked hose or tubing
- Single check valves
- Double check valves
- Tractor protection valve
- Tractor protection control valve
- Parking control valve
- Antilock modulators
- Trailer spring brake control valve
- Inverting relay spring brake control valve

Test 5. Air Brake System Operational Checks

This test outlines a more detailed operational checklist of truck air brake system performance. Perform the following in sequence:

1. Block a set of wheels on the tractor (not the steering axle) and the trailer to prevent the unit from moving when the brakes are released.

2. Inspect the primary and secondary reservoir inlet check valves for correct operation by doing the following:
   - Build air pressure up to system pressure.
   - With the ignition key on, open the drain cock at the supply air reservoir and completely drain the reservoir. The low air pressure buzzer should come on at between 60 and 70 psi.
   - Pressure in both the primary and secondary reservoirs should remain at system air pressure. Use the dash gauges to read this. If loss of air is evident in either system, the one-way inlet check valve could be defective.

3. With the System Park brake dash valve (diamond shaped and yellow) and the Trailer Supply control valve (octagonal and red) in their released positions (pushed in), open the drain cocks in both the primary and secondary tanks.

4. On a coupled tractor/trailer combination, the following should occur when the drain cocks

are opened in the primary and secondary tanks:
   - The Trailer Supply dash valve (octagonal and red) should pop out (trailer park brakes applied) when the circuit tank (primary or secondary) with the highest pressure reaches 40 ± 6 psi. This dash valve could pop immediately if air is depleted rapidly at the trailer supply line.
   - When air pressure in the tank with the higher pressure reaches 30 ± 5 psi, the System Park (diamond shaped and yellow) dash valve may pop out. When the air pressure drops to 25 ± 5 psi, this dash valve must pop out.

5. Close all reservoir drain cocks.

6. Build up air supply in the chassis system to approximately system cutout pressure.

7. With the Trailer Supply dash valve pushed in (trailer parking brakes released), disconnect the trailer emergency gladhands from each other. The Trailer Supply valve should pop out instantly, cutting the supply of air exiting the hose at the gladhand.

8. If the Trailer Supply knob pops out, reconnect the trailer supply hose to the trailer gladhand. Push the System Park in to the released position (in) and pull the Trailer Supply knob out. This will apply the trailer parking brakes and release the tractor brakes.

9. Check the air pressure for leakage by observing air gauges on the instrument panel. Leakage should not be greater than 2 psi in 1 minute.

10. Open the drain cock in the secondary air tank. The drain cock must be opened all the way for quick loss of air. Loss of air can be monitored by observing the dash air gauges. The secondary circuit should show only a loss of air. With the ignition key on, the low-pressure indicator buzzer in the cab should sound at between 60 and 70 psi.

11. Apply the service brakes and observe the slack adjusters and service brake chamber pushrods. The following should occur:
   - The low air warning buzzer should come on.
   - The secondary circuit gauge should indicate zero.
   - All the service brakes throughout the vehicle should apply.
   - Check for special application valves if this test does not confirm that the primary circuit is properly functioning.

12. Close the drain cock in the secondary air reservoir.

13. Build the air supply in both the primary and secondary circuits to system pressure.

14. Fully open the drain cock in the primary reservoir. Observe the secondary circuit air pressure gauge for loss of air; none should occur.

15. Apply the service brakes. What happens here will depend on how the brake system has been optioned.
    - The low air pressure buzzer and warning light should come on.
    - The primary circuit dash gauge should read zero.
    - At minimum, the front service chambers on the tractor and all the service brakes on the trailer should apply. On some systems, all the service chambers will apply.
    - If the brake chamber pushrods move as described, the secondary circuit is functioning properly.

16. Close all drain cocks, build the air pressure to system pressure and return the vehicle to service.

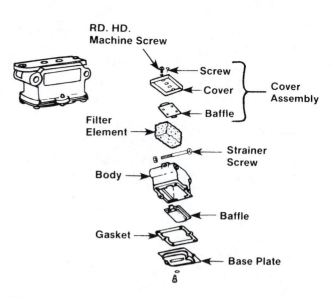

FIGURE 27–13 *Typical air induction system for an air compressor.*

27.4 SUPPLY CIRCUIT SERVICE

The supply circuit of an air brake system includes all the components that compress the air, manage the cycling of the compressor, regulate the system air pressures, remove contaminants from the compressed air, and store it a supply tank. For a full description of the operating principles of the supply circuit components, refer to Chapter 24.

AIR COMPRESSOR SERVICE

Air compressor service should be included in a scheduled maintenance plan to ensure its proper operation and to extend its service life. The OEM maintenance manual should be referred to for specific maintenance instructions. The following would be typical:

- **Every 5,000 miles or monthly.** Service, clean, and replace (as necessary) the air cleaner filter elements. A typical air induction system for a compressor is shown in **Figure 27–13**.
- **Every 25,000 miles or every three months.** Perform the supply circuit operational test outlined earlier in this chapter to verify that the system meets FMVSS No. 121 requirements.
- **Every 50,000 miles or every six months.**
  - Inspect the compressor discharge port, inlet and discharge lines for restrictions, oil, and carboning.

- Check external oil supply (if fitted) and return lines for kinks and flow restrictions.
- Check for noisy compressor operation.
- Check pulley and belt (if equipped) alignment and tension.
- Check compressor mounting bolts and retorque if necessary.

A compressor must have an unrestricted supply of clean filtered air. Part of the preventive maintenance program will be to check the compressor induction system. More frequent maintenance will be required when the vehicle is operated in dusty or dirty environments. Some common methods of providing the compressor with clean filtered air are

- **Polyurethane Sponge Strainer.** This type of strainer element should be cleaned or replaced. Cleaning requires that it be in a commercial solvent or a detergent and water solution. The element should be saturated in clean engine oil and then squeezed dry before replacing it in the strainer. Replace the gasket whenever the air strainer is removed from the compressor intake header.
- **Dry Element/Pleated Paper Air Strainer.** Remove the spring clips from either side of the mounting baffle, separate the cover, and discard the pleated paper filter. Replace the pleated paper filter and remount the cleaned cover, ensuring that the filter is properly positioned. Again, replace the air strainer gasket whenever the air strainer is removed from the compressor intake header.

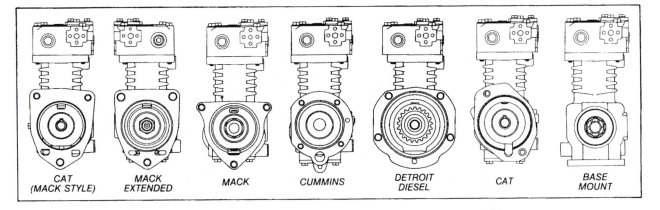

CAT (MACK STYLE)   MACK EXTENDED   MACK   CUMMINS   DETROIT DIESEL   CAT   BASE MOUNT

FIGURE 27-14 Various compressor mountings to meet the engine manufacturer's requirements. (Courtesy of Bendix Commercial Vehicle Systems)

- **Intake Adapter.** Some compressors are fitted with intake adapters that permit the compressor intake to be connected to the engine air intake system. This means that the air delivered to the compressor is filtered by the engine air cleaner system. The delivery of air to the compressor can be upstream or downstream from the turbocharger compressor. When the compressor intake air is ported off downstream from the turbocharger compressor, boosted air is delivered to the compressor. This reduces some of the work that has to be performed by the compressor. When the engine air filter is changed, the compressor intake adapter should be checked. If loose, remove the intake adapter, clean the strainer plate, and replace the intake adapter gasket. Check line connections both at the compressor intake adapter and at the engine or engine air cleaner. Inspect the connecting line for kinks and ruptures and replace it if necessary.

When replacing a compressor assembly, ensure that the mating flange is properly configured to the engine to which it is to be installed. **Figure 27–14** shows some different types of compressor mounting flanges required to mount the compressor to different OEM engines.

Compressor Troubleshooting

Today, compressors are almost never repaired in the field. The reason is that it is usually cheaper for a service facility to stock rebuilt exchange units than the dozens of subcomponents required to rebuild a component such as a compressor. Labor and warranty are also factors. Some exceptions are made: when a persistently occurring malfunction with an OEM product is identified, such as failing valves, it

may be cost effective simply to replace the valves. However, disassembling and reassembling an air compressor during training is a good exercise that will improve the technician's ability to diagnose compressor complaints. **Figure 27–15** shows an exploded view of an air compressor. The technician is usually required to diagnose the cause of failure and, unless the problem is external, to replace the compressor with a rebuilt exchange unit (see **Table 27–1**). Remember, all air compressors will pump trace quantities of the engine oil used to lubricate them. Replacing a compressor diagnosed with trace oil in the discharge is unnecessary; the key is to inspect the air dryer for excessive oil contamination.

AIR DRYER SERVICE

No two vehicles operate under identical conditions, so maintenance and service intervals will vary. Experience is a valuable guide in determining the best maintenance interval for an air dryer. The usual OEM-suggested interval will be somewhere around 25,000 miles linehaul or every 3 months. Service for a typical air dryer, shown in **Figure 27–16**, consists of the following checks:

1. Check for moisture in the air brake system by opening reservoirs, drain cocks, or valves and checking for water. If moisture is present, the desiccant may require replacement. However, the following conditions can also cause water accumulation and should be considered before replacing the desiccant:
   - An outside air source has been used to charge the system. This air may not have been passed through the air dryer.
   - Air usage is higher than normal for a typical highway vehicle. This can be due to accessory air demands or some unusual

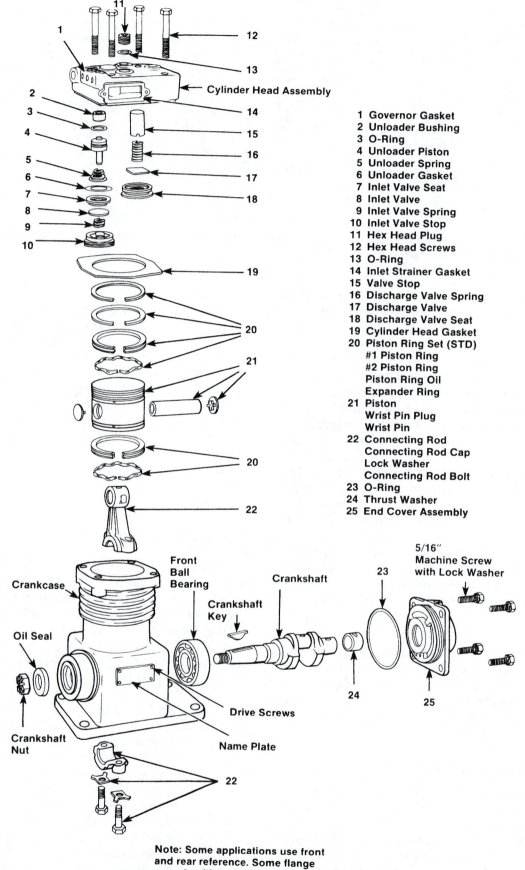

1 Governor Gasket
2 Unloader Bushing
3 O-Ring
4 Unloader Piston
5 Unloader Spring
6 Unloader Gasket
7 Inlet Valve Seat
8 Inlet Valve
9 Inlet Valve Spring
10 Inlet Valve Stop
11 Hex Head Plug
12 Hex Head Screws
13 O-Ring
14 Inlet Strainer Gasket
15 Valve Stop
16 Discharge Valve Spring
17 Discharge Valve
18 Discharge Valve Seat
19 Cylinder Head Gasket
20 Piston Ring Set (STD)
   #1 Piston Ring
   #2 Piston Ring
   Piston Ring Oil
   Expander Ring
21 Piston
   Wrist Pin Plug
   Wrist Pin
22 Connecting Rod
   Connecting Rod Cap
   Lock Washer
   Connecting Rod Bolt
23 O-Ring
24 Thrust Washer
25 End Cover Assembly

Note: Some applications use front and rear reference. Some flange mounts without ball bearings.

FIGURE 27–15 Exploded view of a typical compressor.

	TABLE 27–1: SYMPTOMS OF A TROUBLED COMPRESSOR	
Symptom	**Possible Cause**	**Remedy**
Compressor fails to maintain sufficient pressure or adequate air supply.	1. Restricted intake cleaner	1. Clean/replace element.
	2. Restricted intake tract	2. Clean/repair compressor.
	3. Failure of intake or exhaust valves to seal	3. Repair intake and exhaust valves or replace compressor.
	4. Compressor drive slippage	4. Adjust or replace the compressor drive mechanism.
	5. Failed or maladjusted governor	5. Adjust or replace governor.
	6. Defective cab gauge	6. Replace cab gauge.
	7. Excessive system leakage	7. Locate and repair leaks.
	8. Worn-out compressor	8. Replace compressor.
Compressor operates with excessive noise.	1. Loose drive mechanism	1. Tighten, repair, or replace pulley as necessary.
	2. Restricted cylinder head intake/discharge tracts	2. Repair or replace as necessary.
	3. Worn or burned-out bearings	3. Replace bearings as necessary.
	4. Improper lubrication	4. Service lubrication system as necessary.
	5. Excessive wear	5. Overhaul or replace compressor as necessary.
Compressor not unloading (excessive pressure)	1. Defective unloader pins or seals	1. Replace pins/seals or compressor as necessary.
	2. Defective governor	2. Replace governor.
	3. Restricted reservoir line to governor	3. Repair or replace line as necessary.
	4. Stuck or binding unloader mechanism	4. Repair unloader mechanism or compressor.
	5. Defective gauge	5. Replace gauge.
Compressor pumps excessive oil.	1. Excessive wear	1. Overhaul or replace compressor.
	2. Plugged air cleaner	2. Clean or replace element.
	3. High inlet vacuum (obstructed intake)	3. Service intake circuit.
	4. Restricted oil return line flooding compressor	4. Repair or replace return line.
	5. Excessive oil pressure	5. Service lubrication system.
	6. Defective compressor rear main seal	6. Replace rear main seal or the compressor.
	7. Failed or improperly installed compressor piston rings	7. Remove and reinstall rings or replace compressor.
	8. Back pressure from engine crankcase	8. Check engine crankcase ventilation system/blowby.

air requirement that does not allow the compressor to spend sufficient time in its unloaded cycle. Also check for air system leakage.

- The air dryer is newly installed in a system that previously had no air dryer. The entire air system may be saturated with moisture, and several weeks of operation may be required to dry it out.
- The air dryer is too close to the air compressor, resulting in the air being too hot to condense the moisture.

- In areas where a 30° or more range of temperatures occurs in one day, small amounts of water can accumulate in the air brake system due to condensation. Under these conditions, the presence of small amounts of moisture is normal and should not be considered as an indication that the dryer is not performing properly.

Note: A small amount of oil in the system may be normal and should not, in itself, be considered a reason to replace the desiccant. Desiccant gummed up with oil should

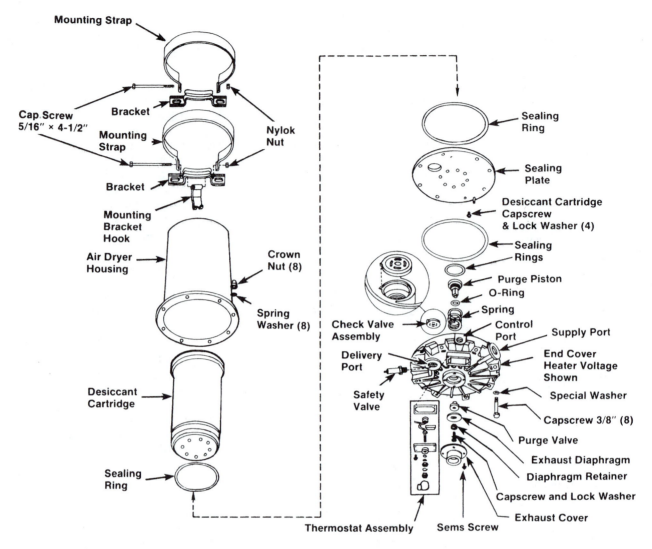

FIGURE 27–16   Exploded view of a typical air dryer.

not be confused with some staining evident in the desiccant bed. Oil-stained desiccant can function adequately.

2. Check mounting bolts for tightness. Retorque to specifications.

A major service event should be conducted every 300,000 miles or 36 months. This requires the air dryer to be disassembled and rebuilt. The overhaul procedure would usually include replacing the desiccant cartridge.

Note: The desiccant change interval may vary from vehicle to vehicle. Although typical desiccant cartridge life is three years, many will perform adequately for a longer period of time. In order to take maximum advantage of desiccant life and ensure that replacement occurs only when necessary, perform an operation and leakage test to determine the need for service.

Air Dryer Operation and Leakage Tests

- Test the outlet port check valve assembly by building the air system pressure to governor cutout and observing a test air gauge installed in the supply tank. Rapid loss of pressure could indicate a failed outlet check valve.

- Locate a failed outlet check valve by bleeding the system down, removing the check valve assembly from the end cover, subjecting air pressure to the unit, and applying a soap solution to the check valve side. Leakage should not exceed a one-inch bubble in 1 second.

- Check for excessive leakage around the dryer purge valve. With the compressor in effective cycle (loaded mode), apply a soap solution to the purge valve housing assembly exhaust port and check that leakage does not exceed a one-inch bubble in 1 second. If it does, service the purge valve housing assembly.

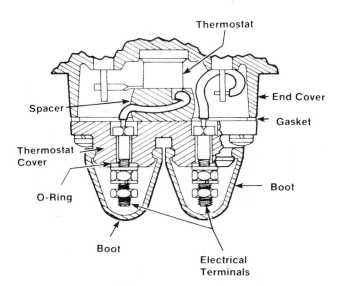
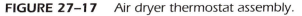

FIGURE 27-17   *Air dryer thermostat assembly.*

- Close all reservoir drain cocks. Build up system pressure to governor cutout and check that the purge valve dumps an audible volume of air. Use the foot valve to pump down the service brakes to reduce system air pressure to governor cut-in. Note that the system once again builds to cutout pressure and is followed by an audible purge at the dryer.
- Check the operation of the safety valve by pulling the exposed stem while the compressor is loaded (compressing air). Air should exhaust while the stem is held, and the valve should reseat when the stem is released.
- Check all lines and fittings leading to and from the air dryer for leakage and integrity.
- Check the operation of the end cover heater and thermostat (see **Figure 27-17**) assembly during cold weather operation as follows:
  - **Electrical supply to the dryer.** With the ignition key on, use a DMM or test light to check for voltage to the heater and thermostat assembly. Unplug the electrical connector at the air dryer and place the DMM test leads on each of the pins of the male connector. If there is no voltage, look for a blown fuse, broken wires, or corrosion in the vehicle wiring harness. Check to see whether a good ground path exists.
  - **Thermostat and Heater Operation.** Turn off the ignition switch and cool the end cover assembly to below 40°. Using an ohmmeter, check the resistance between the electrical pins in the female connector. Compare the readings with the values specified by the OEM. If the resistance is higher than the maximum specified, replace the purge valve housing assembly, which includes the heater

and thermostat. Now heat the end cover assembly to greater than 90° and check the resistance. The resistance should exceed 1,000 ohms. If the resistance values obtained are within the specified limits, the thermostat and heater assembly are operating properly. If the resistance values obtained are outside specifications, replace the purge valve housing assembly.

Compliance Issues Related to the Supply System

Because maintaining a reliable source and supply of compressed air is critical to ensuring that a vehicle can stop as required, a number of regulations addressing potential deficiencies are enforced by most jurisdictions. The regulations place certain degraded brake conditions into more severe categories than other conditions.

Out-of-Service Conditions of the Air Supply System

Out-of-service (OOS) conditions are the most serious vehicle safety defects. The presence of these conditions normally means that a vehicle is detained until repairs are made. The following OOS criteria are defined by the **Commercial Vehicle Safety Alliance (CVSA)** and are used by vehicle safety enforcement officers across North America:

- Low pressure warning device is missing, inoperative, or does not activate at 55 psi and below or half of the governor cutout pressure, whichever is less. (Note that the low-air warning may be audible, an indicator light, or both. At least one must function normally.)
- An air reservoir is separated from its original attachment points.
- A compressor has loose mounting bolts; a cracked, broken, or loose pulley; or cracked or broken mounting brackets, braces, or adapters.

Other Compliance Requirements

When vehicles undergo periodic maintenance inspections or when inspections are carried out at roadside, additional criteria may apply, based on regulations for a particular jurisdiction. These criteria are as follows:

- Compressor buildup time must be less than 3 minutes to get from 50 psi to 90 psi with the engine run at 1,200 rpm.
- Governor cutout pressure must be between 115 and 135 psi.

PHOTO SEQUENCE 14

REBUILDING AN AIR DRYER WITH AN INTEGRAL CARTRIDGE

P14-1 Remove the check valve, delivery port fitting assembly, and the O-ring.

P14-2 Note the position of the purge valve assembly in relation to the end cover so this assembly can be reinstalled in the original position. Remove the three retaining screws from the purge valve assembly and remove this assembly from the end cover.

P14-3 Remove the three O-rings from the exterior of the purge valve. These O-rings may be in the end cover bores.

P14-4 Remove the six cap screws, locknuts, and 12 washers that retain the end cover to the housing and separate the end cover and desiccant cartridge from the housing.

P14-5 Remove the outer housing O-ring from the end cover.

P14-6 Install a strap or chain wrench around the desiccant cartridge about 2 inch (5.08 cm) above the lower end of the cartridge. Rotate the cartridge counterclockwise and remove the cartridge and cartridge O-ring from the end cover.

P14-7 Coat all new O-rings with an appropriate lubricant. Install the three O-rings in their proper locations in the end cover purge valve bore. Install the purge valve assembly in the original position on the end cover with the three retaining bolts.

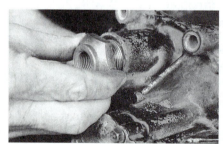

P14-8 Install the delivery port O-ring and fitting in the end cover and tighten this fitting to the specified torque.

P14-9 Place a light coating of lubricant on the bottom of the desiccant cartridge in the O-ring contact area. Screw the desiccant cartridge into the end cover until the cartridge contacts the O-ring.

PHOTO SEQUENCE 14 (CONTINUED)
REBUILDING AN AIR DRYER WITH AN INTEGRAL CARTRIDGE

P14–10 Install a strap or chain wrench 2 in. (5.08 cm) from the lower end of the desiccant cartridge and rotate this cartridge 180–225 degrees clockwise to tighten the cartridge.

P14–11 Install the housing O-ring on the end cover shoulder. Install the housing over the desiccant cartridge and align the holes in the housing and end cover.

P14–12 Install the 6 housing retaining bolts, 12 washers, and nuts. be sure these bolts are in the proper locations. The two longer housing retaining bolts were marked during disassembly. These bolts go through the end cover, housing, and air dryer mounting bracket. Tighten the six housing-to-end cover retaining bolts alternately to the specified torque.

- Governor cut-in must be above 80 psi.
- Sufficient compressed air reserve must be available so that a full brake application does not drop reservoir pressure by more than 18 psi.
- The one-way check valves must be present and functional.

- Apply 2 to 4 drops of oil between the plunger and mounting plate. Never overlubricate!

Every 12 months or 100,000 miles,

- Disassemble the treadle assembly and clean the external components with mineral spirits; replace all rubber components or any worn or damaged part. Check foot valve operation before returning the vehicle to service.

27.5 | SERVICE APPLICATION CIRCUIT

This section deals with service application circuit maintenance and repair requirements.

PREVENTIVE MAINTENANCE OF AIR BRAKE FOOT (TREADLE) VALVES

Figure 27–18 shows a typical foot valve and a sectional view of its internal components. Every 3 months or 25,000 miles

- Visually check for physical damage to the brake valve, such as broken air lines and broken or missing components.
- Clean any accumulated dirt, gravel, or foreign material away from the heel of the treadle, plunger boot, and mounting plate.
- Using light oil, lubricate the treadle roller, roller pin, and hinge pin.
- Check the rubber plunger boot for cracks, holes, or deterioration and replace if necessary. Also, check mounting plate and treadle for integrity.

PERFORMANCE CHECKS FOR FOOT VALVES

The foot valve is the key component in the service brake system of a vehicle. A good PM program will incorporate the following tests.

Function Check

- Test the application pressure of both primary and secondary circuits using accurate test gauges. Depress the treadle to several positions between the fully released and fully applied positions. The application pressure on each test gauge should be equal in each test position and in correct proportion to the movement of the brake pedal; see **Figure 27–19**.
- After a full application is released, the reading on the test gauges drops to zero promptly.

PHOTO SEQUENCE 15
GOVERNOR ADJUSTMENT

P15–1 With the engine running, apply the brake several times to drop system air pressure until the governor cuts in.

P15–2 Now run the engine at 1500 rpm, building system air pressure to cutout. When the governor cuts out observe the pressure on the dash gauge.

P15–3 Determine the specified governor cutout pressure in the truck manufacturer's service manual. If the pressure on the dash gauge does not equal the specified governor pressure, the governor must be adjusted.

P15–4 Shut off the engine and remove the dust boot on the governor.

P15–5 Back off the locknut with a 7/16-inch (11 mm) wrench. Turn the governor adjusting screw counterclockwise to increase the cutout pressure or clockwise to decrease the pressure. When the governor adjusting screw is turned 1/4 turn, the governor cutout pressure is changed approximately 4 psi (27.58 kPa).

P15–6 Start the engine and repeat steps 1 and 2 to verify that the governor cuts out at the specified pressure.

P15–7 With the engine running and the governor cutout pressure available in the supply reservoir, apply the brakes several times. Observe the pressure on the dash gauge when the governor cuts in. The governor cut-in pressure should not be more than 25 psi (172.37 kPa) below the governor cutout pressure. The governor cut-in pressure is not adjustable.

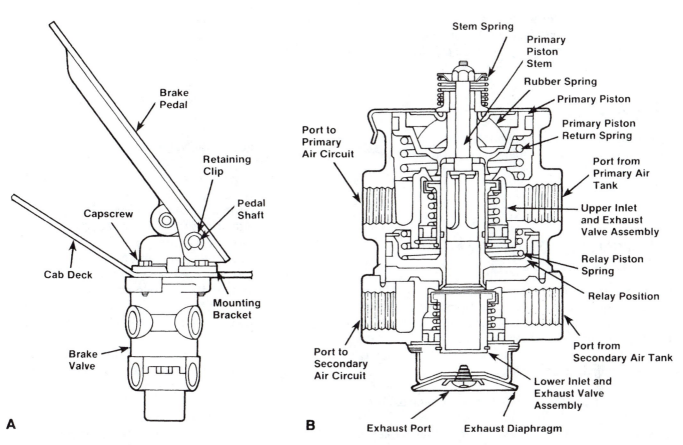

FIGURE 27-18 Typical foot valve. (A) Treadle external components; and (B) cutaway of foot valve.

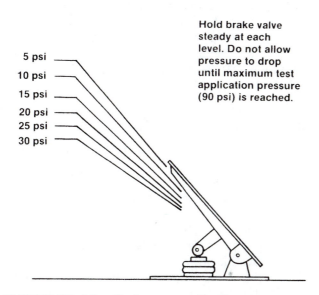

Hold brake valve steady at each level. Do not allow pressure to drop until maximum test application pressure (90 psi) is reached.

5 psi
10 psi
15 psi
20 psi
25 psi
30 psi

FIGURE 27-19 Brake test application pressures. (Courtesy of International Truck and Engine Corp.)

Note that the primary circuit application pressure should normally be about 2 psi greater than the secondary circuit application pressure when both supply tanks are at the same pressure.

Leakage Test

- Make and hold a high-pressure (80 psi) service application.
- Coat the exhaust port and body of the brake valve with a soap solution.
- Leakage permitted is a one-inch bubble in 3 seconds. If the brake valve does not function as described or if leakage is excessive, it is recommended that it be replaced with a new or remanufactured unit.

Removal of a Foot Valve

The removal of a foot valve on some chassis can be labor intensive. Ensure that the valve really requires replacement either because of a performance defect or leakage. It is not smart to make trial-and-error replacements of foot valves.

- Block the vehicle wheels or park the vehicle by mechanical means. Drain all air system reservoirs.
- Identify and disconnect all supply and delivery lines at the brake valve.
- Remove the brake valve and treadle assembly from the vehicle by removing the three cap

screws on the outer bolt flange of the mounting plate. The basic brake valve alone can be removed by removing the three cap screws on the inner bolt flange.

👓 Shop Talk

Carefully label and code every line as it is removed from the foot valve assembly. Failure to do this can considerably lengthen the time required to install the replacement valve.

PREVENTIVE MAINTENANCE OF RELAY VALVES

No two vehicles operate under identical conditions, so once again, the maintenance practices and intervals will vary. Experience is a valuable guide in determining the best maintenance interval for any one particular operation. Typical recommended maintenance practices are as follows.

- Every 3 months or 25,000 miles, check each relay valve for proper operation.
- Every 12 months or 100,000 miles, each relay valve should be performance-tested to ensure that the signal pressure and output application pressures are correct. One OEM recommends disassembling the valve and cleaning the components with mineral spirits. This is a labor-intensive practice that is seldom observed, but when it is, replace all rubber components and other components that are visibly worn or damaged. Check for proper operation before placing vehicle in service.

Operational and Leakage Test

- Block the wheels, fully charge the air brake system, and adjust the brakes.
- Make several service brake applications and check for prompt application and release at each wheel.
- Check for inlet valve and O-ring leakage.
  - Make this check with the service brakes applied when the relay valve is used to control the service brakes.
  - Make this check with the spring brakes released when the relay valve is used to control the spring brakes. Coat the exhaust port

and the area around the retaining ring with a soap solution; a one-inch bubble in 3 seconds is permitted.
- Check for exhaust valve leakage.
  - Make this check with the service brakes fully applied if the valve controls the service brakes.
  - Make this check with the spring brakes fully released if the valve is used to control the spring brakes. Coat the exhaust port with a soap solution; a one-inch bubble in 3 seconds leakage is permitted.
  - Coat the outside of the valve where the cover joins the body to check for seal ring leakage; no leakage is permitted.
- If the valve is used to control the spring brakes, place the park control in the released position and surround the balance port with a soap solution to check the diaphragm and its seat. Leakage equivalent to a 1-inch bubble in 3 seconds is permitted.
  Note: If the anticompound feature is in use, the line attached to the balance port must be disconnected to perform this test.

If the valves do not function as described, or if leakage is excessive, replace the valves with new or remanufactured units. With current ABS systems, each manufacturer has specific methods for diagnosing their components and the technician should always refer to these. Although some components and subcomponents can be cross-referenced between manufacturers, never assume this to be the case. **Figure 27–20** shows a sectional view of a typical antilock relay valve.

Removal of a Relay Valve

- Block and hold vehicle by means other than air brakes.
- Drain air brake system reservoirs.
- If entire valve is to be removed, label the air lines before disconnecting them to facilitate installation.
- Disconnect the air lines from the valve.
- Remove the valve from the reservoir, or if remotely mounted, remove the mounting bolts and then the valve.

CAUTION: Drain all reservoirs before attempting to remove the inlet exhaust valve.

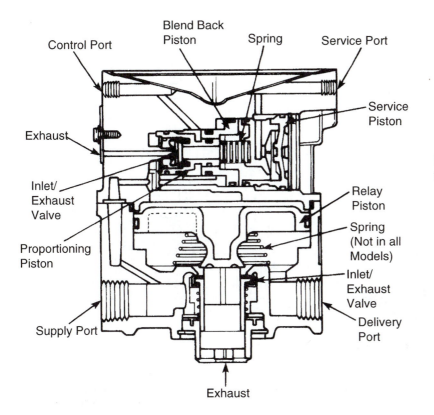

Control Port — Blend Back Piston — Spring — Service Port

Service Piston

Exhaust

Inlet/ Exhaust Valve

Proportioning Piston

Relay Piston

Spring (Not in all Models)

Inlet/ Exhaust Valve

Delivery Port

Supply Port

Exhaust

FIGURE 27-20 Sectional view of a typical antilock relay valve.

TROUBLESHOOTING RELAY VALVES/SERVICE APPLICATIONS

Relay valves can cause a number of brake performance problems. They are vulnerable to external damage and freeze up because of their location on the vehicle axles.

Complaint: Leaks

Determine when the leakage occurs and try to locate the source.
   For static leakage

1. Check for air brake system contamination.
2. Check for backfeed.
   - Spring brakes—emergency diaphragm or piston seal
   - Anticompound—double check or diaphragm in spring brake relay
   - Reduction of signal pressure (park relay)
   - Parking system leakage
   - Park control valve leakage

For dynamic leakage
1. Check for air brake system contamination.
2. Check for reduction of signal pressure (service relay).
   - Total system leakage
   - Brake valve leakage

3. Check for backfeed.
   - Spring brake relay anticompound diaphragm (only occurs with less than full service brake application)

Complaint: Brakes Won't Apply/Release

1. Check the foundation brakes.
   - Worn, damaged, improperly installed or maintained components
2. Check the brake adjustment.
   - Automatic slack malfunction—over/under adjustment
   - Improper manual adjustment—over/under adjustment
3. Make sure that the signal pressure is reaching the relay valve.
4. Check for freeze-ups.
   - Relay piston seized in bore—causing no application pressure or constant application pressure
5. Check for system contamination.
   - Inlet/exhaust sticking in relay
   - Air pressure not reaching relay
6. Check for line restrictions.
   - Pinched, kinked air lines
   - Frozen moisture in lines
7. Make sure that the chamber/slack adjusters are not binding.
   - Incorrect chamber/slack adjuster angle

Complaint: Slow Application/Release

1. Check the brake adjustment.
   - Underadjusted
   - Automatic slack adjusters
2. Check for air brake system contamination.
   - Internal relay valve components sticking
   - See "line restrictions"
3. Check for freezeout.
   - Moisture in air system
   - Mechanical components
4. Check for line restrictions.
   - Pinched, kinked
   - Air system contamination
5. Check the foundation brakes.
   - Worn, defective, or improperly installed components
   - Rust corrosion
6. Check for chambers/slack adjusters binding.
   - Chamber/slack adjuster angle incorrect
   - Bent spider

   Note: The procedure for replacing a service diaphragm in a spring brake unit is covered later on in this chapter.

Trailer Application Valves

These valves are responsible for service applications of the trailer brakes. They meter service signal air to the trailer service application circuit in proportion to the mechanical travel of the valve. **Figure 27–21** shows a sectional view of a trailer hand application valve.

Quick-Release Valves

Quick-release valves are used both in the service application circuit and the hold-off circuits. They distribute as a tee fitting when supplied with air, but exhaust at the valve when the source air ceases. Because they are so simple, diagnosing a malfunction is fairly easy. A common failure is discharge leakage to exhaust while actuated. **Figure 27–22** shows a cutaway view of a quick release valve.

Tractor Protection Valve

The TPV plays an important role in routing service application signal air to the trailer. The full operation of a TPV is explained in Chapter 24. **Figure 27–23** shows a cutaway view of a three-line TPV.

COMPLIANCE ISSUES RELATED TO THE SERVICE BRAKE SYSTEM

Any defect in the service brake application circuit directly affects the ability of the driver to stop the vehicle in the FMVSS No. 121–required stopping time. OOS standards do vary by jurisdiction as does

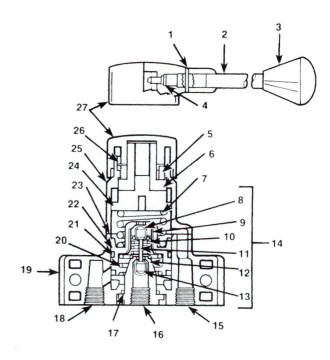

1 Roll Pin	15 Exhaust Port
2 Handle	16 Supply Port
3 Knob	17 Grommet
4 Grommet	18 Delivery Port
5 Adjusting Ring	19 Cover
6 Cam Follower	20 Piston Return Spring
7 Graduating Spring	21 Piston Grommet
8 Exhaust Valve	22 Piston
9 Valve Guide	23 Gasket
10 Spring	24 Cam
11 Stem	25 Body
12 Inlet Valve Seat	26 Adjusting Ring Lock Washer
13 Inlet Valve	27 Head
14 Insert	

FIGURE 27–21 Sectional view of a trailer hand control brake valve.

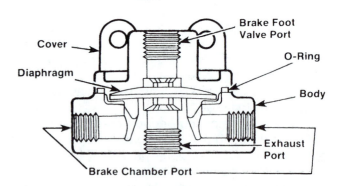

FIGURE 27–22 Cutaway view of a quick release valve. (Courtesy of International Truck and Engine Corp.)

the thoroughness with which they are enforced. OOS regulations grade some service brake defects into more severe categories than others.

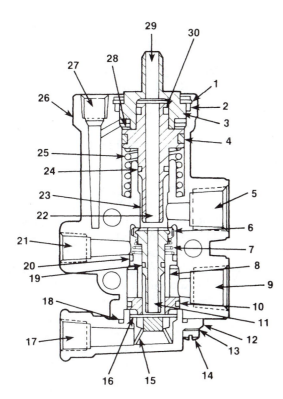

1 Snap Ring	16 Rubber Valve
2 O-Ring Seal	17 Tractor Service Port
3 End Cap	18 O-Ring Seal
4 O-Ring Seal	19 O-Ring Seal
5 Trailer Emergency Port	20 O-Ring Seal
6 Exhaust Valve Assembly	21 Tractor Emergency Port
7 Spring	22 Valve Exhaust Passage
8 Spool	23 Piston
9 Trailer Service Port	24 O-Ring Seal
10 O-Ring Seal	25 Spring
11 Valve Exhaust Passage	26 Body
12 End Cover	27 Control Port
13 Lock Washer	28 Cavity
14 Screw	29 Exhaust Port
15 Spring	30 O-Ring Seal

FIGURE 27–23 _Cutaway view of a three-line tractor protection valve._

OOS Conditions of the Service Brake System

Out-of-service conditions are the most serious vehicle safety defects. The presence of these conditions normally means that a vehicle is detained until repairs are made. Depending on the jurisdiction, a fine also may be levied. The following OOS test criteria defined by the Commercial Vehicle Safety Alliance are used by vehicle safety enforcement officers across North America.

With the governor cut-in, reservoir between 80 and 90 psi, engine idling, and service brakes fully applied, reservoir pressure must be maintained. Reject vehicle if

1. Brake hose bulges or swells under pressure.

2. Audible leak(s) can be heard from any brake hose.
3. Improperly joined or spliced brake hose is discovered.
4. Brake hose is visibly cracked, broken, or crimped.
5. Brake tubing has an audible leak at other than a proper connection.
6. Brake tubing is cracked, damaged by heat, broken, or crimped.
7. Audible air leak is detected at any brake chamber.
8. Mismatched chamber sizes or slack adjuster lengths occur on a steer axle.
9. Any nonmanufactured hole(s) or crack(s) exist in spring brake.

👓 Shop Talk

Only the most serious service brake systems failures are classified as OOS type failures. Many non-OOS faults can still cause brake system underperformance problems. For instance, mismatched chamber sizes on an axle other than the steering axle may not be an OOS infraction, but it is a serious brake performance defect that should be repaired.

27.6 PARKING/EMERGENCY CIRCUIT

FMVSS No. 121 requires that all air brake-equipped vehicles have a means of mechanically applying the parking brakes. In most truck and trailer air brake systems, the parking and emergency circuits are essentially one circuit. This means that they have in common the same actuation hardware and control pneumatics. It cannot be emphasized enough that proper parking/emergency brake performance still requires that the brake adjustment be correct. **Figure 27–24** shows an exploded view of a current two-valve dash control module; these are not normally field repaired. In most cases they are diagnosed as defective and then replaced as a unit.

COMPLIANCE ISSUES RELATED TO THE PARKING/EMERGENCY BRAKE SYSTEM

The regulations are based on FMVSS No. 121 performance requirements and place some brake deficiency conditions into more severe categories than

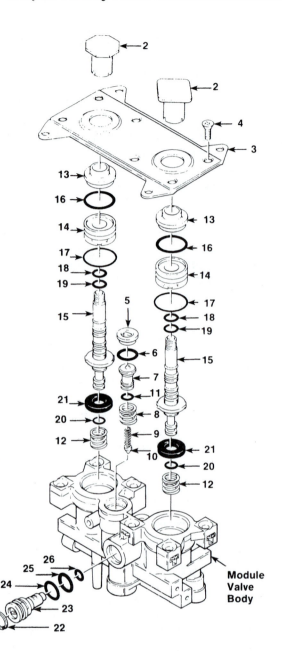

1	Red Button	14	Guide Spool
2	Yellow Button	15	Plunger
3	Cover Plate	16	O-Ring
4	Phillips Head Screws (6)	17	O-Ring
5	Cap	18	O-Ring
6	O-Ring	19	O-Ring
7	Piston	20	O-Ring
8	Spring	21	Exhaust Seal
9	Spring	22	Retaining Ring
10	Check Valve	23	Dual Circuit Valve Assembly
11	O-Ring	24	O-Ring
12	Spring	25	O-Ring
13	Guide Cap	26	O-Ring

FIGURE 27–24   Exploded view of a two-valve dash control module. (Courtesy of Bendix Commercial Vehicle Systems)

others. Again, those classified as OOS conditions are considered to be the most severe.

OOS Conditions of the Parking Emergency Brake System

The following OOS criteria are defined by the Commercial Vehicle Safety Alliance and used by vehicle safety enforcement officers across North America:

1. The parking brakes are applied on a tractor/straight truck and an attempt is made to move the vehicle under engine power. The brakes must prohibit vehicle movement.
2. On a tractor/trailer combination, with the trailer parking brakes applied, they should remain locked when an attempt to move the unit under tractor power is made.
3. Reject if any brake hose bulges or swells under pressure.
4. Reject if there are audible leak(s) at any brake hose.
5. Reject if there are improperly joined or spliced brake hose.
6. Reject if a brake hose is cracked, broken, or crimped.
7. Reject if brake tubing has an audible leak at other than a proper connection.
8. Reject if brake tubing is cracked, damaged by heat, broken, or crimped.
9. Reject if there are audible air leak(s) at any hold-off brake chamber.
10. Reject if there are any nonmanufactured hole(s) or crack(s) in spring brake.

Caging a Spring Brake

This is a common service procedure that is perfectly safe when the potential dangers of spring brakes are realized and the proper procedures are observed. Consult Chapter 24 and ensure that you fully understand the operation of a spring brake assembly.

WARNING: Do not attempt to mechanically release (cage) the spring when the spring brake shows structural damage and/or when safety hooks have been removed. Caging the spring or disassembling the chamber can result in the forceful release of the spring chamber and its contents, which could cause death, severe personal injury, and/or property damage. Remove the complete spring brake chamber and replace with a new unit.

1. Remove the dust plug from the release tool keyhole in the center of the spring brake

chamber. **Figure 27–25** shows a typical spring brake assembly; note the location of the release stud.

2. Remove the release tool from the side pocket of the adapter as shown in **Figure 27–26**. Insert the release tool through the chamber keyhole into the pressure plate as shown in **Figure 27–27**.

3. Turn the release tool one-quarter turn clockwise.

4. Pull on the release tool to seat the release stud cross pin in the cross pin area of the pressure plate.

5. Assemble the release tool washer and nut onto the release tool finger-tight.

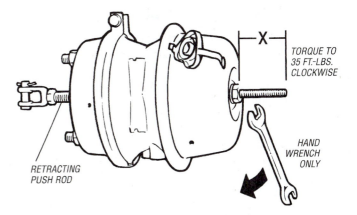

FIGURE 27–27 Insert the release tool lugs into the pressure plate. (Courtesy of the Holland Group Inc.—Anchorlok Div.)

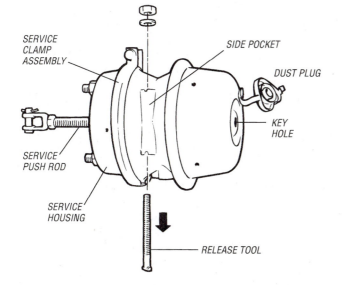

FIGURE 27–25 Spring brake chamber. (Courtesy of the Holland Group Inc.—Anchorlok Div.)

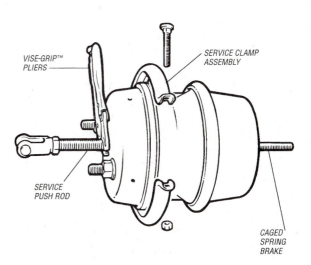

FIGURE 27–28 Caging the main spring. (Courtesy of the Holland Group Inc.—Anchorlok Div.)

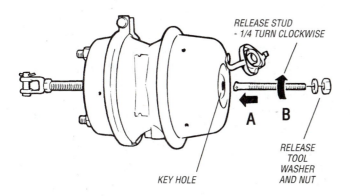

FIGURE 27–26 Release tool removal from spring brake. (Courtesy of the Holland Group Inc.—Anchorlok Div.)

6. Tighten the release tool nut with a hand wrench (do not use an impact wrench) as shown in **Figure 27–28** and make certain that the pushrod is retracting.

7. Do not overtorque the release tool nut. Typical torque values are as follows. Refer to the applicable manufacturer's service manual for specific torque values.
   - S-cam type: 35 lb/ft
   - Wedge type: 27 lb/ft

8. To ensure that the compression spring is fully caged, the release tool length beyond the nut should typically measure:
   - 30-inch chamber: 2.875 inches minimum
   - 24-inch chamber: 2.915 inches minimum

9. Refer to the applicable manufacturer's service manual for specific tool lengths.

Service Diaphragm Replacement

To replace the service diaphragm on a spring brake chamber, perform the following:

1. Manually cage the spring brake chamber as outlined earlier.

> **WARNING:** Always cage the compression spring with the release tool. Never rely on air pressure to keep the spring compressed.

2. To prevent the sudden release of the piggy-back assembly or service pushrod assembly and to facilitate the installation of the new diaphragm, prevent the service pushrod from retracting by clamping it in place with Vise-Grip™ pliers.
3. Remove the service clamp assembly as shown in **Figure 27–29** and discard the old diaphragm.
4. Inspect the service clamp assembly, adapter wall and lip, housing, service return spring, and service pushrod. If any structural damage is noted, replace with new parts.
5. Wipe the surface of the service pushrod plate clean of any oil, grease, or dirt. Check to see that weep holes in the housing are not plugged.
6. Place the new service diaphragm in the adapter and center the housing over the diaphragm and adapter.
7. Make sure that the diaphragm is properly seated between the adapter and housing lip and reassemble the service clamp assembly.

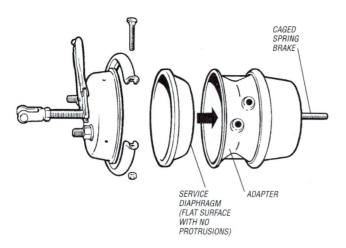

FIGURE 27–29 Service diaphragm replacement. (Courtesy of the Holland Group Inc.—Anchorlok Div.)

Torque the carriage nuts to the manufacturer's specification (typically 18–25 lb/ft). Check carriage bolts and clamp assembly for proper seating around the adapter and housing lip and remove the Vise-Grip™ pliers from the service pushrod.

8. Apply a maximum of 120 psi air pressure to the service port and check the diaphragm seal for leakage by applying a soap-and-water solution to the service clamp area. No leakage is allowed.
9. Uncage the compression spring and reassemble the release tool in the side pocket of the adapter.
10. Replace the dust plug in the release tool keyhole in the center of the chamber.

Changing Position of Mounting Bolts, Clamps, and Air Ports

To change the position of the chamber mounting bolts, clamps, or air ports, do the following:

1. Manually cage the spring brake chamber.
2. If air pressure was used to aid the caging process, exhaust the air pressure.
3. To prevent the sudden release of the housing and to facilitate rotation, prevent the pushrod from retracting by clamping the pushrod with Vise-Grip™ pliers.
4. Loosen the carriage bolts on the clamp and rotate the central housing to locate air ports in the desired position.

Spring Brake Removal

To remove the spring brake from the truck/trailer, do the following:

1. Block the wheels of the truck/trailer to prevent it from moving.
2. Manually release, or cage, the spring brake.
3. After caging the spring brake, release the air pressure in the parking brake system by placing the parking brake control valve in the apply position. Ensure that the service brakes are released.
4. Remove air hoses from the parking brake or service brake chamber. Tag the hoses for proper identification during reinstallation.
5. Loosen the jam nut at the chamber pushrod yoke and disconnect the pushrod yoke from the slack adjuster.
6. Remove the mounting stud nuts and remove the complete spring assembly from the mounting bracket.

CAUTION: The loaded main spring in a spring brake assembly contains a potentially lethal force. Manual caging devices must be used whenever the spring chamber assembly is removed and replaced from the foundation brake mounting plate. Most current spring brake units are sealed with a band clamp that has to be destroyed to separate the spring chamber in order to prevent the unit from being disassembled.

Disarming Spring Brakes

Spring brakes should never be disposed of in scrap metal containers unless they have first been disabled. Disarming a spring brake requires that it is installed in a disarmament chamber. The disarmament chamber is a heavy steel chamber within which the spring brake is placed. After the chamber door has been closed, the chamber has a pair of openings on either side large enough to allow the head of an a oxyacetylene cutting torch to enter. The spring brake chamber band should be cut on either side. In some cases, this will cause the spring chamber to separate (caution: this may occur with explosive force) and release the spring. Other times, a hammer and long chisel will be required to separate the chamber housing. In all cases, great care must be exercised when disarming spring brakes because when the chamber separates a massive amount of force is released.

CAUTION: Never open a disarmament chamber until the main spring has been observed through the inspection windows to be separated from the housing and is under no tension.

Spring Brake Installation

When installing a new spring brake chamber, determine the correct service pushrod length to ensure the proper alignment and efficient operation of the spring brake. To determine the correct pushrod length, measure the B" dimension as shown in **Figure 27–30** and subtract the setup stroke as listed or specified in the manufacturer's installation manual. (Typical stroke values range from $1^3/_8$ inches to $1^3/_4$ inches.) With the spring brake fully caged, B − setup stroke = pushrod length including clevis (**Figure 27–31**).

To mark the pushrod cutoff length, the length of the threaded rod protruding between the clevis legs must not exceed $^1/_8$ inch; this prevents interference with the operation of the slack adjuster as shown in **Figure 27–32**.

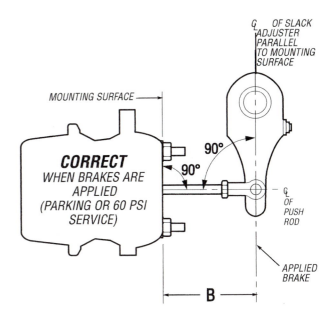

FIGURE 27–30 Determining required pushrod length. (Courtesy of the Holland Group Inc.—Anchorlok Div.)

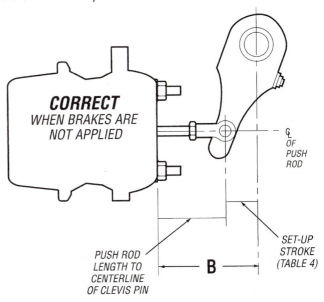

FIGURE 27–31 B dimension − setup stroke = pushrod/clevis length. (Courtesy of the Holland Group Inc.—Anchorlok Div.)

When the proper pushrod length has been marked off, the pushrod can be cut to length with the spring brake fully caged as shown in **Figure 27–33**.

Install the spring brake chamber by performing the following procedures:

1. Install the clevis and jam nut on the pushrod, and install the spring brake to the mounting bracket. (Refer to the manufacturer's installation manual for installation torque.) The clevis must be adjusted so that it has full thread

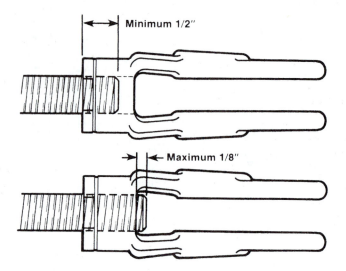

FIGURE 27–32   Installation of clevis onto pushrod. (Courtesy of ArvinMeritor Inc.)

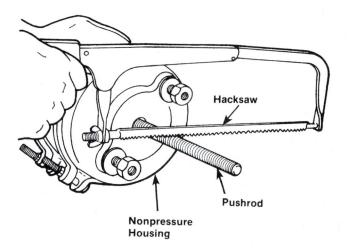

FIGURE 27–33   Cut the pushrod to the proper length with the spring brake fully caged. (Courtesy of International Truck and Engine Corp.)

engagement on the pushrod (from flush to ³/₁₆-inch protrusion).

2. Connect the service and emergency air lines to the proper air ports, and connect the clevis to the slack adjuster.

3. Uncage the compression spring by loosening and removing the release tool nut and washer.

4. Reassemble the release tool and install it into the side pocket of the adapter.

5. Install the dust plug into the release tool keyhole in the center of the chamber.

6. Adjust the slack adjuster to the specified setup stroke. With the brake applied, the following conditions must occur:
   - Pushrod should be 90 degrees to the centerline of slack adjuster.
   - Pushrod should be 90 degrees to the mounting face of the spring brake.

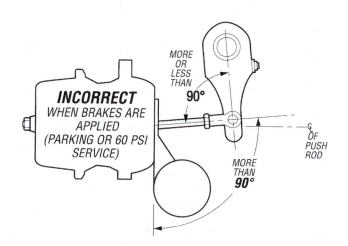

FIGURE 27–34   Pushrod too long. (Courtesy of the Holland Group Inc.—Anchorlok Div.)

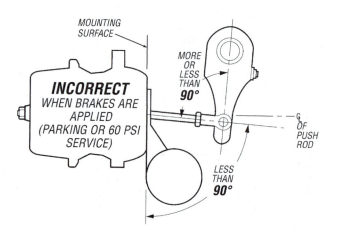

FIGURE 27–35   Pushrod too short. (Courtesy of the Holland Group Inc.—Anchorlok Div.)

7. If the setup results in the condition depicted in **Figure 27–34** or **Figure 27–35**, the spring brake is misaligned and must be corrected by the following:
   - Shorten the pushrod and align the spring brake on the mounting bracket (**Figure 27–34**)
   - Lengthen the pushrod and align the spring brake on the mounting bracket (**Figure 27–35**).

8. If misalignment cannot be corrected, consult with the foundation brake manufacturer to verify the correct mounting bracket position.

9. When the spring brake and pushrod are set as shown in **Figure 27–30**, release the brakes and readjust the slack adjusters to the shortest possible stroke without the brakes dragging.

10. Ensure that the release tool is inside the pocket of the adapter and that the dust plug is installed in the release tool keyhole in the center of the chamber.

27.7 SLACK ADJUSTERS

Remember that it is illegal to fit manual slack adjusters to a system requiring automatic slack adjusters. Slack adjusters are levers; make sure that the specified clevis pin-to-camshaft dimension is always observed. A small difference in slack adjuster effective length can make a big difference in brake torque. Each type of automatic slack adjuster requires a slightly different installation procedure, so the OEM instructions should always be observed. **Figure 27–36** and **Figure 27–37** show cutaway and exploded views of two different types of automatic slack adjusters.

INSTALLATION OF SLACK ADJUSTERS

Installation of a new or rebuilt slack adjuster requires several steps, as outlined in the following sections.

Installation Preparation

1. Check the brake camshaft or powershaft, bushings, and seals for wear and corrosion. If necessary, replace the camshaft, powershaft, bushings, or seals. Make sure that the brake operates smoothly. Turn the camshaft or powershaft by hand. If the camshaft or powershaft does not turn easily, inspect and repair the camshaft, powershaft, bushings, or seals as required.

2. Check the return spring in the air chamber to make sure that the spring has enough tension. Apply the service brake and the spring brake several times. Make sure that the return spring quickly and completely retracts the pushrod. If necessary, replace the return spring or the air chamber.

3. The new slack adjuster must be the same length as the old one. Consult the manufacturer's service manual to determine the correct length of slack adjuster.

4. Make sure that the brake is completely released during the installation and adjustment procedures, except when the directions indicate that the brake must be applied. If the brake is not completely released when the clevis and the slack adjuster are installed and adjusted, the slack adjuster will not adjust the brake correctly. **Figure 27–38** shows a typical slack adjuster installation.

Clevis Installation and Adjustment

Some slack adjuster manufacturers supply installation templates to enable the truck technician to correctly install the adjuster. The templates available can be used with the slack adjuster to check clearances or to ease installation on drum or disc brake

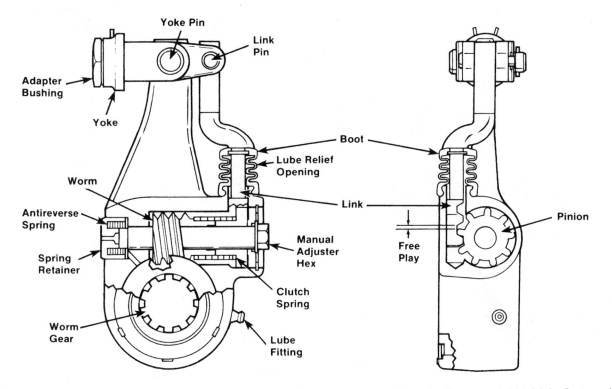

FIGURE 27–36 Cutaway view of a typical slack adjuster. (Courtesy of Bendix Commercial Vehicle Systems)

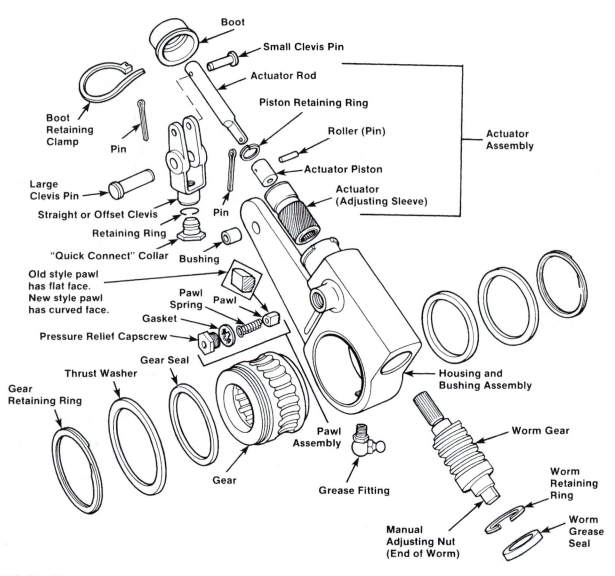

FIGURE 27–37 Exploded view of a slack adjuster. (Courtesy of ArvinMeritor Inc.)

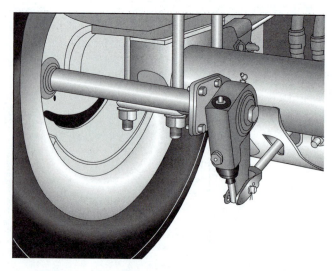

FIGURE 27–38 Typical slack adjuster installation on an S-cam foundation brake.

foundations. A typical installation template is shown in **Figure 27–39**.

Always use the template(s) prescribed by the manufacturer. If templates are not available, refer to the manufacturer's installation manual for specific and detailed installation procedures. The adjustment and installation procedures given here are based on the use of templates.

1. Remove the clevis from the new slack adjuster.
2. Install the new clevis on the pushrod. Do not tighten the jam nut against the clevis.

CAUTION: The clevis must be installed in the correct position on the pushrod, or the slack adjuster will not adjust the brake correctly.

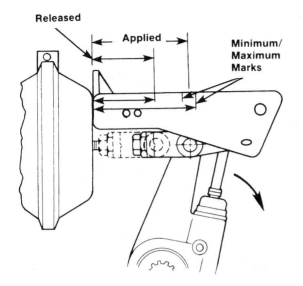

FIGURE 27–39 Measuring free-stroke using a template. (Courtesy of ArvinMeritor Inc.)

3. Measure the length of the slack adjuster with the template as shown in **Figure 27–39**. The marks by the holes in the small end of this template indicate the length of the slack adjuster.
4. Use the template to install the clevis in the correct position. First, put the large clevis pin through the large holes in the template and the clevis. Select the hole in the template that matches the length of the slack adjuster. Hold that hole on the center of the camshaft or powershaft. Look through the slot in the template. The small hole in the clevis must be completely visible. If necessary, adjust the position of the clevis on the pushrod until the small hole in the clevis is completely visible through the slot in the template. At least 1/2 inch of thread engagement must be between the clevis and the pushrod. Also, the pushrod must not extend through the clevis more than 1/8 inch (**Figure 27–32**). If necessary, cut the pushrod or install a new pushrod or a new air chamber.
5. Tighten the jam nut against the clevis to hold the clevis in the correct position. Tighten it to the manufacturer's torque specification.

Installing the Adjuster

To install a new or repaired slack adjuster, do the following:

1. Lubricate the splines on the slack adjuster gear and the splines on the camshaft or the powershaft with anti-seize compound.

Shop Talk

The adjusting pawl assembly can be on either side of the housing or on the front of the housing. Make sure that the pawl assembly can be removed after the slack adjuster is installed. The pawl assembly must sometimes be removed when the slack adjuster is serviced.

2. Install the slack adjuster on the camshaft or powershaft.
3. Install spacing washers and the snap ring until a maximum clearance of 0.060 inch exists between the washer and the snap ring.
4. Remove the pawl assembly from the slack adjuster.

CAUTION: If the pawl is not removed, the teeth will be damaged when the manual adjusting nut is turned.

5. Use a wrench to turn the manual adjusting nut to align the hole in the arm of the slack adjuster with the large hole in the clevis as shown in **Figure 27–40**.
6. Rotate the slack adjuster by hand the same distance as the maximum stroke of the chamber. Make sure that no obstructions will prevent the slack adjuster from rotating when the brakes are applied. Check the

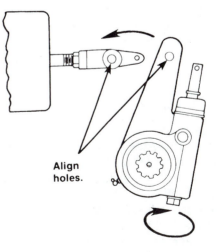

FIGURE 27–40 Use a wrench to turn the adjustment nut to align the arm of the slack adjuster with the large hole in the clevis. (Courtesy of ArvinMeritor Inc.)

OEM specifications to locate the maximum stroke of each size of brake chamber. The adjusted chamber stroke will always be shorter than the maximum stroke.

7. Install both clevis pins through the template, slack adjuster, and clevis. Check again to make sure that the clevis is installed in the correct position. Adjust the clevis if necessary.

8. Remove the template. Apply anti-seize compound to the two clevis pins. Install the clevis pins and install cotter pins to hold the clevis pins in place.

9. Adjust the brakes. Brake adjustment is covered in some detail later in this chapter.

27.8 TRAILER BRAKE SYSTEM

Because the trailer brake system is managed entirely from the tractor unit, it is impossible to develop a proper understanding of the trailer brake system without first understanding the tractor brake system. Trailer service technician specialists must ensure that they understand the principles of the tractor braking system before attempting to work on the trailer.

COMPLIANCE ISSUES RELATED TO THE TRAILER BRAKE SYSTEM

Because maintaining a fully functioning trailer brake system that gives the driver full control of brake applications is critical to ensuring that a vehicle can stop as required, a number of regulations address potential deficiencies and are enforced by most jurisdictions. The regulations place certain degraded brake conditions into more severe categories than other conditions.

OOS Conditions of the Trailer Brake System

OOS conditions are the most serious vehicle safety defects. The presence of these conditions normally means that a vehicle is detained until repairs are made, and sometimes a fine is levied. The following OOS criteria are defined by the Commercial Vehicle Safety Alliance and used by vehicle safety enforcement officers across North America:

1. Discharge trailer supply line, fully apply the tractor service brakes, and ensure that no air discharges from the trailer service line.

2. Fully charge the trailer air system and then disconnect the trailer supply line. The trailer brakes must immediately apply.

3. Audible air leak(s) at brake chamber.

4. Any nonmanufactured hole(s) or crack(s) in spring brake chamber(s).

5. Insecure air reservoir.

6. Brake hose that bulges or swells under pressure.

7. Audible leak(s) at any brake hose.

8. Improperly joined or spliced brake hose.

9. Brake hose that is cracked, broken, or crimped.

10. Brake tubing with an audible leak at other than a proper connection.

11. Brake tubing cracked, damaged by heat, broken or crimped.

Air System Valve Troubleshooting Guide

Table 27–2 identifies some common brake valve failure problems and remedies.

27.9 FOUNDATION BRAKE SERVICE

A brake job refers to the overhaul of the foundation brake assemblies on the tractor and trailer. This is one of the most common service practices in a truck service facility. Although it is thought of as a simple operation, the safety of the entire rig is dependent on it being performed properly.

DISASSEMBLY AND REASSEMBLY OF S-CAM FOUNDATION BRAKE ASSEMBLIES

To disassemble a typical S-cam foundation brake assembly with open anchor shoes, the following procedure should be observed after removing the wheel assembly:

1. Push down on the lower brake shoe. Pull on the cam roller retainer clip to remove the bottom cam roller as shown in **Figure 27–41**.

2. Lift the top brake shoe and pull on the cam roller retainer clip to remove the top cam roller.

3. Lift the lower shoe to release the tension on the brake shoe return spring and remove it.

4. Rotate and drop the lower shoe to release the tension on the brake shoe retainer springs as shown in **Figure 27–42**.

5. Remove the shoe retainer springs and remove both brake shoes.

TABLE 27–2: AIR VALVE TROUBLESHOOTING GUIDE

Valves	Symptom	Remedy
Drain valves—automatic	1. Will not drain 2. Will not drain in cold weather 3. Leaks/malfunctions	1. Repair or replace. 2. Replace with heated units. 3. Repair or replace.
Foot valve	1. Leaks at exhaust with trailer hand valve applied 2. Leaks at exhaust with all brakes released 3. Leaks at exhaust with foot brake applied	1. Check double check valve. Repair or replace. 2. Check anticompound double check valve for backflow. 3. Replace defective foot valve.
Quick-release valves	1. Leaks (when used in service brake system) 2. Leaks (when used in spring brake system)	1. Replace. 2. Replace.
Relay valves	1. Leaks at exhaust port with all brakes released 2. Leaks at exhaust port with service brakes applied	1. Check seal in spring brake for backflow of spring hold-off pressure through service port to open exhaust on valve. Repair or replace. 2. Repair or replace the exhaust valve, which is not seating properly.
System Park control valve (yellow diamond)	1. Leaks at exhaust port 2. Parking brake will not release 3. Parking brake will not apply	1. Replace. 2. Check for full-system pressure delivery through valve. 3. Replace if it will not release pressure.
Trailer Supply valve (red octagonal)	1. Leaks at exhaust port 2. Driver may override automatic trailer brakes when tractor air is below 20 psi. 3. Will not apply trailer immediately when pulled	1. Check for trailer back-leakage. Replace if no back-leakage occurs. 2. Replace if TP system is automatic and 2-line. Repair or replace stoplight switch if TP system is automatic and 3-line. 3. Replace if it will not exhaust. Repair or replace stoplight switch.
Tractor park valve (blue round)	Leaks at exhaust port	Repair or replace.
TPV (nonautomatic type)	1. Will not respond to trailer charge valve 2. Supply line to trailer with slow bleed down does not shut off tractor air and vents between 45 and 20 psi. 3. Leaks at exhaust port or tractor service (back through hand or foot valve) or trailer supply	1. Repair or replace. 2. Check trailer supply valve; it should pop to let TPV vent trailer supply. Repair or replace. 3. Repair or replace.

Prepare an S-cam Foundation Brake for Reassembly

Cleaning

- Use soap and water to clean nonmetal components.
- Some of the heavier components, such as the S-cam, can be cleaned in hot soak tanks.
- Use cleaning solvents to clean the remainder of the components.
- Dry components immediately after cleaning with clean wipers or compressed air.

Corrosion Protection

- If components are assembled immediately after cleaning, lightly lubricate them with grease.
- Avoid getting any grease on friction surfaces such as cam profiles, rollers, and friction faces.
- If components are to be stored after cleaning, apply a corrosion-preventive material. Store components in a special paper or other material that prevents corrosion.

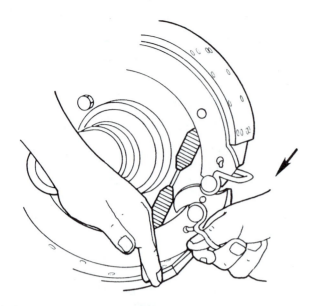

FIGURE 27–41   Push down on the lower brake shoe and pull on the roller retainer clip to remove the lower cam roller. (Courtesy of ArvinMeritor Inc.)

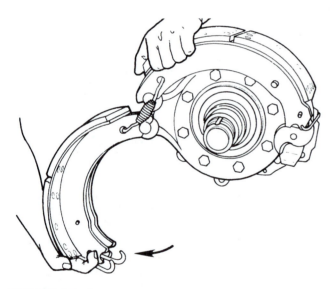

FIGURE 27–42   Rotate the bottom shoe to release the tension on both retaining springs. (Courtesy of ArvinMeritor Inc.)

Inspect Components. It is important to carefully inspect all components before assembly, checking them for wear or damage. Repair or replace them as required.

- Check the spider anchor pin bores for oversizing and cracks. Replace damaged spiders and anchor pin bushings. Tighten spider fasteners to manufacturer's recommended torque as shown in **Figure 27–43**.

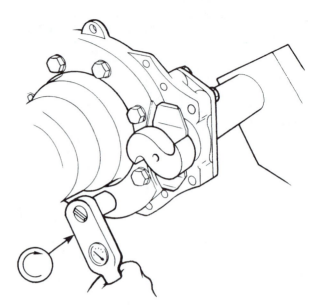

FIGURE 27–43   Tighten all spider fasteners to the manufacturer's specified torque. (Courtesy of ArvinMeritor Inc.)

- Check the camshaft bracket for broken welds, cracks, and correct alignment. Replace a damaged bracket.
- Check anchor pins for corrosion and wear. Replace damaged anchor pins.
- Check brake shoes for rust, expanded rivet holes, broken welds, and correct alignment. Replace a shoe with any of the listed conditions.

CAUTION: On 16.5-inch brake shoe foundation assemblies using 1-inch anchor pins, the spider anchor pin bore must not exceed 1.009 inches (25.63 mm).

- The distance from the center of the anchor pin hole to the center of the roller must not exceed 12.779 inches (32.46 cm). If it does, the shoe is spread and must be replaced; do not attempt to re-arc.
- Check the S-camshaft for cracks, wear, brinnelling, and corrosion. Check the S-cam profile, bearing journals, and splines. Replace damaged camshafts. Consult the OEM service manual to check the S-camshaft and its bushings. The following procedure is typical:
  - Inspect the cam-to-bushing radial free play and axial end play. Radial free play should be tested as shown in **Figure 27–44** and should typically be less than 0.030 inch (0.762 mm).

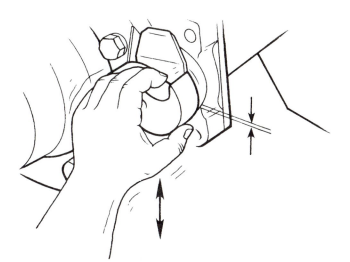

FIGURE 27-44 Testing radial free play of an S-camshaft. (Courtesy of ArvinMeritor Inc.)

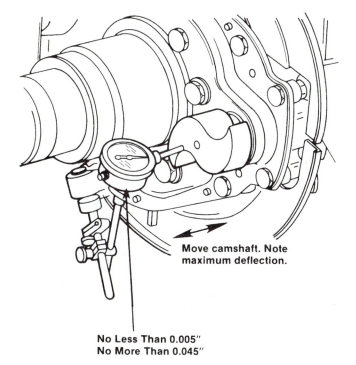

No Less Than 0.005″
No More Than 0.045″

FIGURE 27-45 Measuring S-camshaft endplay. (Courtesy of Roadranger Marketing. One great drive train from two great companies—Eaton and Dana Corporations)

- If radial free play is less than 0.030 inch (0.762 mm), the bushings and seals can be reused.
- If radial free play movement exceeds 0.030 inch (0.762 mm), replace the bushings and seals.
- If axial end play is outside of the specifications window as shown in **Figure 27-45**, remove the snap ring and add/subtract

spacer shims between the slack adjuster and the snap ring to achieve the specified free play, usually between 0.005 inch and 0.045 inch (0.127–1.143 mm).

- Check the drums.
  - Check the brake drums for cracks, severe heat checking, heat spotting, scoring, pitting, and distortion. **Figure 27-46** shows some typical drum conditions. Replace drums as required.
  - It is not recommended to machine used brake drums because it decreases the strength, heat tolerance, and service life of the drum.
  - Measure the inside diameter of the drum in several locations with a drum caliper or internal micrometer. Check for eccentricity, bell-mouthing, and oversizing. Replace the drum if the diameter exceeds any specifications supplied by the drum manufacturer. A typical 16.5-inch drum has a maximum machining tolerance of 0.090 inch and a maximum service diameter of 0.120 inch. This means that the maximum diameter the drum can be machined to and returned to service is 0.090 inch. Any measurement exceeding 0.120 inch means that the drum has to be scrapped.

Shop Talk

Used drums are almost never machined and returned to service today. This is due to the low cost of new replacement drums when compared to the labor required to turn a drum. Additionally, machining used brake drums is technically not a good practice, as the drums have a much shorter in-service life due to much harder friction linings.

- Check dust shields for rust and distortion. Repair or replace damaged shields as necessary.
- Install the drums.
  - Ensure that the correct fastening hardware is used when assembling wheel hardware to brake drums and that the correct torque values are observed.

CAUTION: Cast drums are machined in manufacture. However, when stored, especially for long periods, they can deform and lose their concentricity. It is therefore a good practice to machine brand new drums to ensure that they are perfectly concentric with the wheel. **Figure 27-47** shows a typical heavy-duty drum lathe.

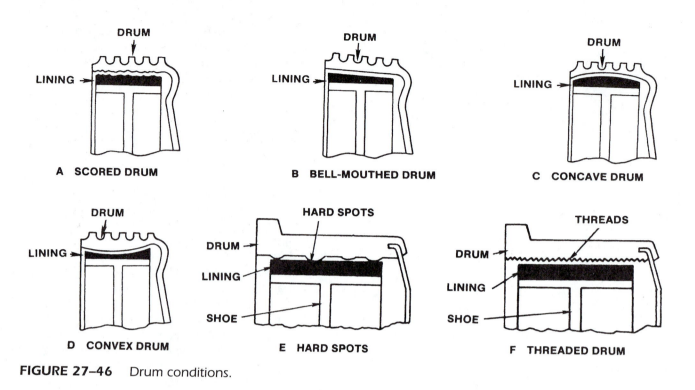

FIGURE 27–46 Drum conditions.

Machine heavy-duty drums with dual wheels weighing up to 1,500 lbs. without outboard support

FIGURE 27–47 Heavy-duty drum lathe. (Courtesy of Hennessy Industries, Inc.)

Linings

If the shoes are to be refaced with new friction linings, ensure that the linings are correctly installed (**Figure 27–48**) and that the correct rivet/torque sequence is observed (**Figure 27–49**).

 Shop Talk

Modern friction facings on brake shoes are much harder than the asbestos base linings used on drums a generation ago. They have a longer service life. The result is that the life of the linings can be the same as the life of the drums in some applications. It is almost impossible to machine/turn down the effects of severe heat checking on a brake drum because the drum becomes so hardened. Cast brake drums are cheap when compared with the labor cost required to turn a drum. Many operations routinely replace brake drums today when the brakes are relined.

Reassembly of an S-cam Foundation Brake

It is recommended to install new camshaft bushings whenever installing a new camshaft.

- Tighten all of the spider bolts to the correct torque.
- Use a seal driver to install new camshaft seals and new bushings in the cast spider and camshaft bracket. If equipped with a stamped spider, install both bushings into the bracket.
- Install the seals with the seal lips toward the slack adjuster. Failure to observe this practice

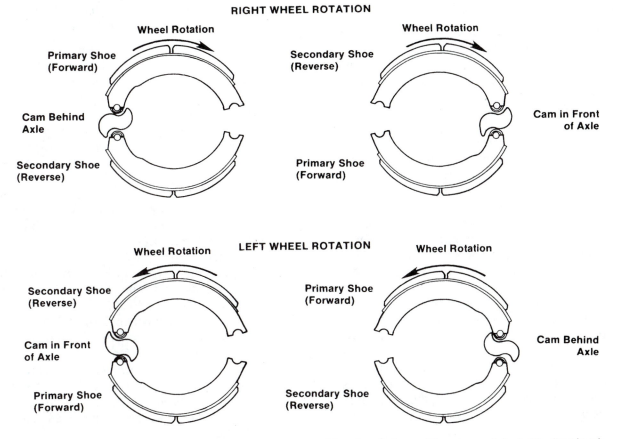

RIGHT WHEEL ROTATION

Wheel Rotation

Primary Shoe
(Forward)

Cam Behind
Axle

Secondary Shoe
(Reverse)

Wheel Rotation

Secondary Shoe
(Reverse)

Cam in Front
of Axle

Primary Shoe
(Forward)

LEFT WHEEL ROTATION

Wheel Rotation

Secondary Shoe
(Reverse)

Cam in Front
of Axle

Primary Shoe
(Forward)

Wheel Rotation

Primary Shoe
(Forward)

Cam Behind
Axle

Secondary Shoe
(Reverse)

FIGURE 27–48 *Correct installation for replacement combination linings. (Courtesy of ArvinMeritor Inc.)*

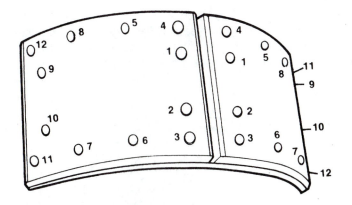

FIGURE 27–49 *Typical rivet installation sequence for installing linings to shoes. (Courtesy of ArvinMeritor Inc.)*

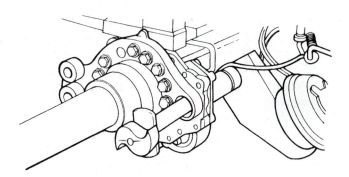

FIGURE 27–50 *Install S-camshaft. (Courtesy of ArvinMeritor Inc.)*

will result in grease being pumped out into the brake linings.
- If the camshaft bracket was removed, install the chamber bracket seal and bracket onto the spider. Tighten the capscrews to the torque specified.

Install the S-camshaft

- Install the cam head thrust washer onto the camshaft. Apply grease to the camshaft bushings and journals.
- Install the camshaft through the spider and bracket as shown in **Figure 27–50** and confirm that the camshaft turns freely by hand.

Installing Manual Slack Adjusters

- Apply the service brake and spring brake several times. Check that the chamber return spring retracts the pushrod quickly and completely. If necessary, replace the service chamber.
- The new slack adjuster should be the same length as the one being replaced.
- If the brake has a spring brake, cage the spring to completely release the brake. No air pressure must remain in the service half of the air chamber.
- Apply anti-seize compound to the slack adjuster and cam splines.
- Install the slack adjuster onto the camshaft.
- If necessary, install spacer washers and the snap ring to a maximum endplay clearance of 0.062 inch (1.57 mm).
- Install the clevis onto the pushrod. Do not tighten the jam nut against the clevis.
- Set the clevis position (turn clockwise or counter-clockwise) so that a 90-degree angle is achieved when it is in the fully applied position. Apply anti-seize compound to the clevis pin(s). Install the clevis pin(s) through the clevis and the slack adjuster.

 Shop Talk

When an axle wheel seal fails, oil saturates the brake friction linings, which rapidly causes them to glaze, lowering their coefficient of friction or aggressiveness. When this occurs, pressure or steam washing removes only the surface oil, so the linings should be replaced. Also replace the linings on the other end of the axle to maintain brake torque balance over the axle.

Installing Automatic Slack Adjusters

A number of different brands of slack adjusters are on the market, and each requires a different setup procedure. Slack adjusters are described in Chapter 24 and earlier in this chapter. Ensure that the OEM instructions are properly observed when fitting automatic slack adjusters to S-camshafts. The consequence of improperly installing automatic slack adjusters is slack adjusters that do not auto-adjust.

 Shop Talk

Take some time to ensure that automatic slack adjusters are properly set up on installation. It takes

much longer to manually adjust automatic slack adjusters than manual slack adjusters.

Installing Brake Shoes

- Place the upper brake shoe into position on the top anchor pin. Hold the lower brake shoe on the lower anchor pin. Install new brake shoe retaining springs, ensuring that the clip eyes face inward.
- Rotate the lower brake shoe forward. Install a new brake shoe return spring with the open end of the spring hooks toward the camshaft as shown in **Figure 27–51**.
- Pull each brake shoe away from the cam to permit enough space to install the cam roller and cam roller retainer. Press the *ears* of the retainer to permit it to fit between the brake shoe webs as shown in **Figure 27–52**.
- Push the cam roller retainer into the brake shoe until its ears lock in the shoe web holes.
- Lubricate the brake components that require lubrication. Avoid lubricating those components that rely on friction to function properly, such as the S-cam profiles, cam roller faces, and friction faces.

 Shop Talk

Avoid lubricating the cam roller face. Lubricate the cam roller bearings. These are designed to grab at their faces and rotate on their bearings in operation.

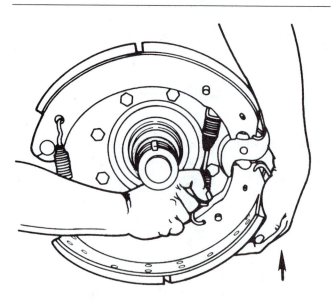

FIGURE 27–51 Rotate the lower brake shoe forward and install the brake shoe return spring. (Courtesy of ArvinMeritor Inc.)

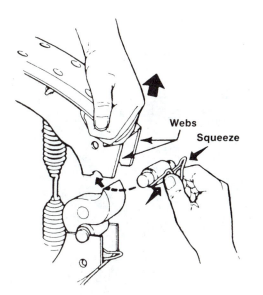

FIGURE 27–52 *Press the ears of the retainer together to permit the roller to fit between the brake shoe webs. (Courtesy of ArvinMeritor Inc.)*

27.10 BRAKE ADJUSTMENT

With the introduction of automatic slack adjusters, the need for frequent brake adjustments on trucks was reduced. However, it is essential that frequent evaluation of the brake adjustment on trucks and trailers be performed both from a performance (driver) and shop inspection point of view. Brake adjustment determines how well the vehicle brake system will function whether it be under service, parking, or emergency braking. The first step in verifying brake performance is to check the pushrod stroke on each chamber throughout the truck or tractor/trailer combination.

 Shop Talk _____

Poorly adjusted brakes affect all vehicle braking, including emergency braking. Never take chances with brake adjustment. Maladjusted brakes can have lethal consequences, and every technician has a stake in the vehicles on which they work.

Checking Pushrod Stroke

Measure pushrod stroke using the following step-by-step procedure:

- Block the vehicle.
- Release parking brakes.
- Place a precise visible reference mark on each pushrod at the face of the brake chamber.
- Establish a reservoir pressure of 90–100 psi by pumping the brake pedal to reduce pressure or running the engine to raise pressure.
- Shut off the engine.
- Make and hold a full service-brake application. (Do not apply the spring or parking brakes.)
- Measure applied stroke from the face of each brake chamber to the reference marks.

The actual means used to mark the brake chamber pushrods limits the accuracy of this method of checking pushrod stroke. This limitation is sometimes used to challenge OOS brake-related charges, and some operators prefer not to leave reference marks on pushrods. However, in practice, having any errors produced through this practice probably works out in favor of carriers more often than not.

ALTERNATE BRAKE ADJUSTMENT INSPECTION METHODS

Several methods may be used for checking brake adjustment with varying degrees of reliability. The following three methods have proven to be consistently reliable.

Visual Stroke Indicators

Well-designed visual stroke indicators confirm correct adjustment at a glance and permit the technician to check brake adjustment without getting under the vehicle or having another person's assistance. Properly fitted visual stroke indicators also help to prevent unnecessary brake re-adjustment.

To determine free-stroke using visual stroke indicators

1. Block wheels.
2. Release parking brakes.
3. Establish reservoir pressure between 90 and 100 psi.
4. Make a full service-brake application and hold.
5. With service brakes applied, use the visual indicators to confirm correct pushrod strokes.

Check Free-stroke

When the foundation brakes are in good condition, checking the length of free-stroke at each brake to determine that brakes are correctly adjusted. Measure the free-stroke length by using a bar or

PHOTO SEQUENCE 16
CHECK FREE-STROKE ON S-CAM FOUNDATION BRAKES

P16–1   Block wheels. If adjusting the brakes on a tractor/trailer combination, block one of the drive axle duals on the tractor and one trailer axle.

P16–2   Release the parking brakes on the rig by depressing both the System Park and Trailer Supply dash valves.

P16–3   Ensure that the system air pressure is above the governor cutout pressure (around 100 psi is ideal) after the brakes have been released.

P16–4   If the rig has visual stroke indicators, get somebody in the cab to make a full service brake application and hold. Inspect each pushrod stroke indicator to confirm the correct pushrod stroke.

P16–5   Correct travel when service brakes are applied.

P16–6   Excessive pushrod travel; brakes require adjustment.

P16–7   If the pushrods on the rig do not have visual stroke indicators, release the parking brakes and then check each brake for pushrod travel using a heel bar.

P16–8   If a pushrod travels between $1/2$ and $3/4$ inch (13–19 mm), the brake has the correct amount of free travel and does not require adjustment.

P16–9   If the pushrod travels more than $3/4$ inch, the brakes require adjustment.

PHOTO SEQUENCE 17

WHEEL-DOWN BRAKE ADJUSTMENT PROCEDURE ON A TRACTOR/TRAILER

P17–1   Block wheels. If adjusting the brakes on a tractor/trailer combination, block one of the drive axle duals on the tractor and one trailer axle.

P17–2   Release the tractor and trailer parking brakes by depressing the System Park and Trailer Supply valves. Ensure that the system air pressure is above the cut-in pressure after the parking brakes have been released.

P17–3   Determine whether the brakes require adjustment by performing the "Check Free-stroke" procedure outlined in Photo Sequence #16.

P17–4   To adjust a manual slack adjuster using the wheel-down method, select the correct-sized combination wrench ($^9/_{16}$ inch) and use the closed end to depress the locking collar and engage the adjusting nut.

P17–5   Now rotate the adjusting nut so that the S-camshaft is rotated in the direction required to actuate the brake. Turn until the brake is fully applied (wrench will not turn further); then back off adjusting nut sufficiently to obtain $^1/_2$-inch free pushrod travel.

P17–6   After the adjusting nut has been backed off (typically, this will require $^1/_3$ to $^1/_2$ of a turn), check for free play by once again levering on the slack adjuster. This should be between $^1/_2$ and $^3/_4$ inch, preferably closer to $^1/_2$ inch.

P17–7   This brake can now be regarded as being properly adjusted. Ensure that the locking collar engages fully over the adjusting screw.

P17–8   Proceed to adjust the remaining slack adjusters on the rig in the same manner.

PHOTO SEQUENCE 18
WHEEL-UP BRAKE ADJUSTMENT PROCEDURE ON A TRACTOR/TRAILER

P18–1 Wheel-up brake adjustments take longer than wheel-down adjustments; however, some fleets prefer this method as it ensures a more accurate adjustment and can help identify other foundation brake problems. First, block the wheels. If adjusting the brakes on a tractor/trailer combination, block one of the drive axle duals on the tractor and one trailer axle. When it comes time to adjust the brakes over the wheels that are blocked, move the blocks to a wheel over which the brakes have already been adjusted.

P18–2 Release the tractor and trailer parking brakes by depressing the System Park and Trailer Supply valves. Ensure that the system air pressure is above the cut-in pressure after the parking brakes have been released.

P18–3 The air pressure must be above the cut-in pressure value.

P18–4 When performing a wheel-up adjustment, the brakes over every wheel on the unit are adjusted regardless of free-stroke. First, use a hydraulic jack to raise one wheel off the floor.

P18–5 Install a safety stand.

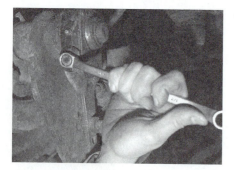

P18–6 To adjust a manual slack adjuster using the wheel-up method, select the correct-sized combination wrench ($^9/_{16}$ inch) and use the closed end to depress the locking collar and to engage the adjusting nut.

P18–7 Now rotate the adjusting nut so that the S-camshaft is rotated in the direction required to actuate the brake. Turn until the brake is fully applied (wrench will not turn further).

P18–8 Now attempt to rotate the wheel.

PHOTO SEQUENCE 18 (CONTINUED)
WHEEL-UP BRAKE ADJUSTMENT PROCEDURE ON A TRACTOR/TRAILER

P18–9   Next, gradually back off the slack adjuster while attempting to rotate the wheel. Back off the slack adjuster until the wheel can be rotated through a full revolution with minimal but definite drag.

P18–10   This brake can now be regarded as being properly adjusted. Ensure that the locking collar engages fully over the adjusting screw.

P18–11   Proceed to adjust the remaining slack adjusters on the rig in the same manner.

lever or by pulling on the adjuster by hand as shown in **Figure 27–53**. Checking the free-stroke confirms the clearance between the brake shoes and the drum. Free-stroke between $1/2$ inch and $3/4$ inch indicates normal running clearance.

To check free-stroke, proceed as follows:

1. Block the vehicle wheels.
2. Release the parking brakes.
3. Extend each pushrod using a heel bar and measure the distance traveled.
4. Each pushrod should move between $1/2$ inch and $3/4$ inch (13–19 mm).

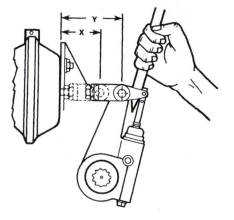

MEASURING "FREE STROKE" WITHOUT TEMPLATE

Drum Brake: Y Minus X Must = 5/8″ – 3/4″
In Service Disc Brake: Y Minus X Must = 3/4″ – 1″
Initial Disc Brake: Y Minus X Must = 7/8″ – 1-1/8″

FIGURE 27–53   Measuring free-stroke without a template. (Courtesy of ArvinMeritor Inc.)

In most cases, when free-stroke is correct, the applied stroke falls within the adjustment limit. However, do not assume that this is true in all cases. Short free-stroke indicates correct brake adjustment only when foundation brakes are in good mechanical condition. Use of this method does not guarantee that applied stroke will fall within the adjustment limit when brakes are in poor condition. Technicians must, therefore, be certain about brake condition to reliably use this method. **Table 27–3** lists the adjustment limits for clamp-type chambers.

27.11 GENERAL BRAKE VALVE TROUBLESHOOTING

This section takes a look at some air brake system complaints and the remedies required to correct them. Remember, however, that it is not entirely comprehensive and is intended only as a guide.

Complaint: Air Leaks

Determine when static leakage occurs:

1. Check for air system contamination such as indications of oil in the air circuit. Remember that trace quantities of oil in the air circuit are normal. However, evidence of oil at the exhaust of the brake system valves and in the dash gauges indicates severe oil contamination, and the compressor should be checked.

TABLE 27–3:   ADJUSTMENT LIMITS FOR CLAMP-TYPE CHAMBERS

Size	Marking	Outside Diameter	Adjustment Limit
6	None	$4^1/_2$ inches (114 mm)	$1^1/_4$ inches (32 mm)
9	None	$5^1/_4$ inches (133 mm)	$1^3/_8$ inches (35 mm)
12	None	$5^{11}/_{16}$ inches (144 mm)	$1^3/_8$ inches (35 mm)
16	None	$6^3/_8$ inches (162 mm)	$1^3/_4$ inches (44 mm)
16L	L stamped in cover, Stroke tag	$6^3/_8$ inches (162 mm)	2 inches (51 mm)
20	None	$6^{25}/_{32}$ inches (172 mm)	$1^3/_4$ inches (44 mm)
20L	L stamped in cover, Stroke tag	$6^{25}/_{32}$ inches (172 mm)	2 inches (51 mm)
24	None	$7^7/_{32}$ inches (183 mm)	$1^3/_4$ inches (44 mm)
24L	L stamped in cover, Stroke tag	$7^7/_{32}$ inches (183 mm)	2 inches (51 mm)
24LS	Square ports, tag, and cover marking	$7^7/_{32}$ inches (183 mm)	$2^1/_2$ inches (64 mm)
30	None	$8^3/_{32}$ inches (205 mm)	2 inches (51 mm)
30	DD3 (Bus/Coach)	$8^1/_8$ inches (206 mm)	$2^1/_4$ inches (57 mm)
30LS	Square ports, tag, and cover marking	$8^3/_{32}$ inches (205 mm)	$2^1/_2$ inches (64 mm)
36	None	9 inches (228 mm)	$2^1/_4$ inches (57 mm)

2. Check for back-feeds from spring brake valves, TPVs, and trailer (if connected) See **Figure 27–54**.

3. Check for leaks at all air line connections and critical valves. Use a soap-and-water solution or an audio air leak detector.

Determine when dynamic leakage occurs:

1. Check for air system leaks with the system at governed air pressure and the engine off.

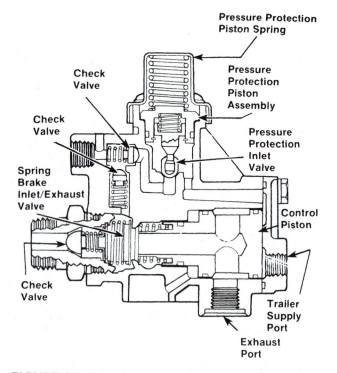

FIGURE 27–54   *Sectional view of a typical spring brake control valve. (Courtesy of International Truck and Engine Corp.)*

Chock the wheels and release the parking brakes. Circle the vehicle and listen for leaks. Apply the trailer service brakes and check for leaks.

2. Have someone apply the tractor service brakes and circle the tractor to check for leaks.

3. Inspect the performance of the double check valve located between the trailer control (hand valve) and foot valve. The actual location of this valve can be difficult to determine; use the OEM pneumatic schematic. If this is the cause, leakage will occur when the hand brake is applied and the foot valve released, or vice versa.

Complaint: Over-sensitive Service Brakes

Check foundation brakes and wheel assembly:

1. Incorrect slack adjuster arm length (check camshaft-to-clevis pin length).
2. Brake chamber clevis connected to wrong hole in slack adjuster (changes leverage).
3. Chamber/slack alignment binding.
4. Inspect friction material.
5. Inspect condition of foundation brake components.
6. Cam rotation counter to wheel rotation. (S-camshafts are directional; it is not unheard of to have them improperly installed.)
7. Inspect tire tread condition and tire mismatch (tread design, flat spotting, standard or low profile).

Check for air system freeze-ups (sticks then breaks free).

1. Inspect relay valve, quick release valve, and other brake valves for proper operation.

2. Check air lines for internal failure causing pressure checking.

Brake valve maintenance items:

1. Plunger/treadle or pedal sticking
   - Treadle or pedal hinge pin not lubricated.
   - Plunger and roller not lubricated.
   - Ensure proper treadle with roller.
   - Plunger cocked in mounting plate.
   - Boot torn, dirt accumulated around plunger.
2. Lubrication
   - Too much (damages rubber spring and/or boot)
   - Too little—see plunger sticking.
3. Installation
   - Body or mounting plate distortion due to uneven floor board or firewall mounting surface.
   - Inspect other air system components for malfunction.

Complaint: Brakes Will Not Fully Apply

1. Brakes out of adjustment: excessive free-stroke.
2. Automatic slack adjuster malfunction (under-adjusting or seized).
3. Air line or component freeze-up.
4. Air line internal failure causing restriction.
5. Relay valve malfunction caused by freeze-up or piston seizure.
6. Defective foundation brake components.
7. Foot valve pedal movement restricted: cab floor obstruction (tools, carpet, gravel, debris).
8. Improper chamber/slack adjuster alignment:
   - Pushrod is binding on chamber nonpressure plate
   - Pushrod is too long/short
9. No air pressure in one brake circuit.
10. Air lines improperly connected to foot valve, relay valve, quick release valve, or spring brake valve.
11. Foot valve plunger/pedal sticking.

Complaint: Brakes Fail to Release Completely

1. Parking brake is applied and will not release.
2. Condition of foundation brake components—springs, shoes, bushings (elliptical/squeezed), or bent spider—causing anchor misalignment.
3. Air lines are pinched, improperly connected.
4. Check foot valve stop adjustment.
5. System contamination causing oil swelling of rubber valve components.

6. Relay valve is contaminated or relay piston is sticking.
7. Quick-release or ratio valve exhaust is restricted.
8. Chamber/slack adjuster mechanically binding.
9. Automatic slack adjuster is malfunctioning—over-adjusting.

27.12 SERVICE PROCEDURES ON AIR DISC BRAKES

These vary somewhat between OEMs as the actuating mechanisms differ. As with any procedure on foundation brakes, the OEM service manual should be consulted. **Figure 27–55** shows an exploded view of one of the more popular disc brake assemblies.

Checking Lining Wear

Lining wear should be inspected by checking the caliper inboard bosses as shown in **Figure 27–56**. Note the lining wear indicator and the wear dimensions.

Lubricating Calipers

Disc brake calipers must be lubricated at required intervals. **Figure 27–57** shows the location of the caliper grease fitting and the caliper pressure relief fittings. Ensure that the caliper is not over-greased.

Checking Rotor Runout

Rotors should be inspected and lateral runout measured each time the brakes are serviced. **Figure 27–58** shows the runout being measured; the runout specifications should be checked in the OEM service specifications.

27.13 BRAKE CERTIFICATION, INSPECTION, AND TESTING

This section addresses the qualifications required of technicians working on brake systems, inspection and testing. Brake systems are one of the most scrutinized systems on a truck, and they are routinely inspected by state and provincial transportation safety enforcement personnel.

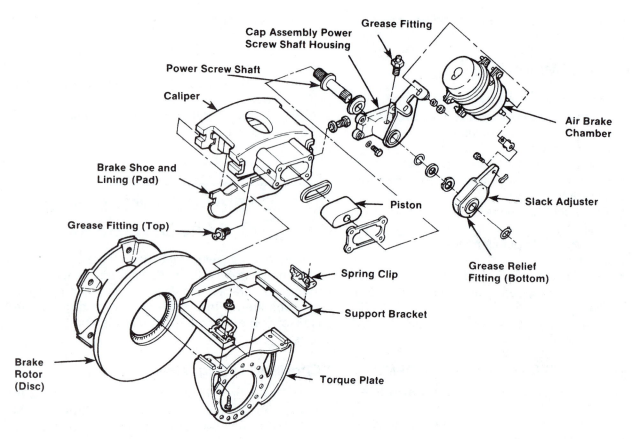

FIGURE 27-55 Exploded view of an air disc brake assembly. (Courtesy of Haldex Brake Systems)

CERTIFICATION

The truck and bus safety Regulatory Reform Act established the minimum training requirements and qualifications for employees responsible for maintaining and inspecting truck brakes and brake systems.

The following fleet personnel need to be certified:

1. Fleet employees responsible for reinspecting, repairing, maintaining, or adjusting the brakes on commercial motor vehicles (CMVs).
2. Those fleet employees who are responsible for the inspection of CMVs to meet periodic inspection requirements.

It is not essential that the person who does the work be certified unless he or she is also responsible for ensuring that the task is done properly. The person who is responsible for the service and inspection under this rule is called a brake inspector.

Training can be the completion of an apprenticeship or training program of a state, a province, a federal agency, or a union-sponsored program, such as

- Training provided by a brake or vehicle manufacturer (or similar commercial program) relating to brake maintenance and inspection tasks

- Experience in performing brake maintenance programs
- Experience in a commercial garage, fleet leasing, or similar facility with brake service and inspection like the assigned inspection and service

Qualification means that individuals understand the brake service or inspection tasks to be done and can do them. They must know and have mastered the methods, procedures, tools, and equipment used and be able to perform the service or inspection based on individual experience, training, or both.

INSPECTION AND TESTING

To check brake systems, it is usually necessary for the inspector to crawl under the vehicle and visually look for defective brake equipment, signs of leaking wheel seals, oily linings, etc., and measure the slack adjuster pushrod travel. Roller brake dynamometers (**Figure 27-59**) can also be used. In minutes, these instruments can gather all sorts of performance data on a vehicle's brakes, including adjustment, lining and drum condition, and brake force under panic braking conditions. They can also spot a dragging brake and check brake timing between the tractor

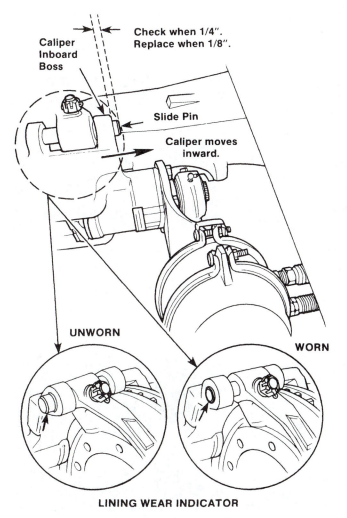

UNWORN   WORN

LINING WEAR INDICATOR

FIGURE 27–56 Checking disc brake lining wear. (Courtesy of Haldex Brake Systems)

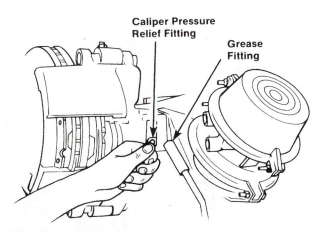

FIGURE 27–57 Lubricating disc brake calipers. (Courtesy of Haldex Brake Systems)

and trailer. In addition to providing better data about brake performance and safety, computer-generated results should help the shop correct brake defects and solve braking problems.

FIGURE 27–58 Checking rotor runout.

FIGURE 27–59 A roller brake dynamometer. (Courtesy of James Winsor, Newport Communications)

Unlike chassis dynamometers that spin wheels at highway speeds and are used to track down engine power complaints, roller brake dynamometers turn the vehicle wheels at less than one mph. It only takes two to three wheel revolutions to get the total braking picture. The dynamometer rollers have enough torque to slowly turn the wheels even when 40 psi or more air pressure is applied by the driver. The dynamometer method of brake testing is used throughout Europe. Because it measures actual forces and is computer-monitored, this type of tester can eliminate human error in determining whether brakes are in or out of adjustment.

PHOTO SEQUENCE 19
MEASURING A BRAKE ROTOR

P19–1 With the front wheel removed and the vehicle chassis supported on safety stands, adjust the front wheel bearings on the side where the rotor is being measured.

P19–2 Mount a dial indicator so the plunger contacts the rotor surface approximately 1 inch (2.54 cm) from the outer edge of the rotor.

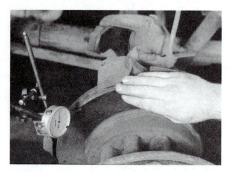

P19–3 Turn the rotor one revolution and observe the runout on the dial indicator.

P19–4 Compare the indicated runout on the dial indicator to the runout specification in the service manual. If runout is more than specified, machine or replace the rotor.

P19–5 Use a micrometer to measure the rotor thickness approximately 1 inch (2.54 cm) from the outer edge of the rotor.

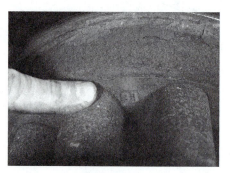

P19–6 Compare the rotor thickness to the minimum discard thickness stamped on the rotor. If the rotor thickness is equal to or less than the discard thickness, replace the rotor.

P19–7 Use the micrometer to measure the rotor thickness at 12 locations around the rotor.

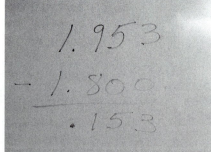

P19–8 Subtract the minimum and maximum micrometer readings to determine the rotor thickness variation.

P19–9 Compare the rotor thickness variation reading to the specifications in the service manual. If the rotor thickness variation exceeds the specifications, machine the rotor.

In a roller brake dynamometer, a computer program plots the brake torque curves. This program calculates the amount of braking force measured in pounds at the tire tread surfaces required to drive the wheels under a gradually increasing brake-apply condition. Data is picked up simultaneously from left and right wheels by sensors on the rollers and is fed to the computer. If brake force is unequal, it will be spotted easily in the printouts and graphs as part of the pass/fail analysis, which compares brake data with base information in the computer. The printout lists pass/fail for each brake on the rig. When this information is put on a graph, two graphed data lines are produced. The lower line represents the minimum standard required to pass the test. The upper line represents the actual performance profile of the brakes being tested. If the actual performance line drops down (indicating low brake force) and crosses the pass line, it means that the brake system has failed the test.

A second method for dynamic brake testing is a plate tester. Unlike the roller tester that can be either portable or a fixed installation, the plate tester is a fixed installation only. It measures brake force as the vehicle rolls over instrumented plates in the floor or pavement. This system measures brake force at each wheel. This technology is used mainly by brake engineers for test work and not for roadside and shop safety inspections.

SUMMARY

- Brake system maintenance and service procedures attempt to ensure that air-brake-equipped vehicles meet FMVSS No. 121 standards while in service.
- Brakes must meet their original standards in terms of pneumatic timing and brake torque balance.
- Pneumatic timing refers to the management of the air in the system to ensure that the air reaches each actuator at the correct time so that the brakes over one wheel are not applied before or after those on another wheel.
- Brake torque balance refers to the requirement that the same amount of mechanical force be delivered to the wheels on each axle and that the brakes over all the axles on a vehicle work together to bring the vehicle to a stop.
- Minimum training requirements and qualifications are necessary for employees who reinspect, repair, maintain, adjust, or periodically inspect the brakes on commercial motor vehicles (CMVs).
- Roller brake dynamometers and plate testers are used in brake inspection testing.

REVIEW QUESTIONS

1. Which of the following is the best practice when instructed to block the wheels on a tractor/trailer combination?
   a. Place a pair of wedged wheel chocks on both sides of one steering axle wheel.
   b. Place a pair of wedged wheel chocks on one tractor steering axle wheel and one trailer axle wheel.
   c. Place a pair of wedged wheel chocks on one tractor drive axle wheel and one trailer axle wheel.
   d. Place a pair of wedged wheel chocks on one trailer axle wheel.

2. What is the required amount of slack adjuster free travel on a properly adjusted, manual S-cam brake?
   a. less than $1/4$ inch
   b. $1/4$ to $1/2$ inch
   c. $1/2$ to $3/4$ inch
   d. $3/4$ to 1 inch

3. Which of the following foundation brake components should be lubricated with chassis grease?
   a. S-cam profile
   b. S-cam roller face
   c. S-cam bushings
   d. brake shoe friction face

4. A $16^{1}/_{2}$-inch brake drum specified to have a machine limit of 0.090 inch over and a maximum dimension of 0.120 inch over, is measured at 0.098 inch over at a brake overhaul. Technician A states that the drum can be turned on a drum lathe, providing that the maximum dimension of 0.120 inch is observed. Technician B says that the drum can be legally returned to service without being turned. Who is right?
   a. Technician A only
   b. Technician B only
   c. both Technician A and B
   d. neither Technician A nor B

5. A tandem axle trailer equipped with S-cam brakes has all 4 slack adjusters originally set with a 6-inch dimension from the clevis pin to camshaft. One slack adjuster is replaced, and because of a different hole arrangement in the new slack adjuster, it is set at $6^{1}/_{2}$ inches. Which of the following is more likely to result?
   a. One of the four wheels will lock up ahead of the others under severe braking.
   b. One of the four wheels will lock up behind the others under severe braking.

c. One of the service brake chambers will require more air pressure than the others.

d. One of the service brake chambers will require less air pressure than the others.

6. When adjusting a manual slack adjuster on an S-cam foundation brake, Technician A states that the lock collar should be retracted and the adjusting nut always rotated clockwise to decrease free travel. Technician B states that free travel can be reduced only when the adjusting nut is rotated counter-clockwise. Who is right?
   a. Technician A only
   b. Technician B only
   c. both Technician A and B
   d. neither Technician A nor B

7. A set of brake linings has become glazed due to oil saturation by a failed wheel seal. Technician A says that pressure washing the foundation brake components should be sufficient to remove the effects of the oil saturation. Technician B states that the best practice is to replace the brake linings on both wheels on the affected axle. Who is right?
   a. Technician A only
   b. Technician B only
   c. both Technician A and B
   d. neither Technician A nor B

8. When adjusting S-cam brakes on a trailer equipped with manual slack adjusters, which of the following must be true?
   a. The tractor parking brakes must be applied.
   b. The trailer service brakes must be applied.
   c. The trailer parking brakes must be released.
   d. The tractor service brakes must be applied.

9. An original equipment 45-degree elbow on a line exiting a relay valve supplying a service chamber is replaced by a 90-degree elbow. Technician A says that this should have no effect on brake performance, because it is the service application pressure that determines braking force. Technician B says that the use of the 90-degree elbow increases flow restriction and can affect pneumatic timing. Who is right?
   a. Technician A only
   b. Technician B only
   c. both Technician A and B
   d. neither Technician A nor B

10. Technician A states that the service diaphragm in a spring brake can be safely changed, providing the spring brake is properly caged with a cage bolt before removing it. Technician B states that it is no longer possible to change the hold-off diaphragms in most spring brake assemblies because the band clamp cannot be unbolted. Who is right?
    a. Technician A only
    b. Technician B only
    c. both Technician A and B
    d. neither Technician A nor B

11. What is the main function of a hold-off diaphragm?
    a. to convert air pressure to mechanical service brake pressure
    b. to prevent compounding of the foundation brakes
    c. to prevent spring brake force from acting on the pushrod
    d. to convert spring pressure into pneumatic pressure

12. When an S-cam foundation brake slack adjuster is in the fully applied position, ideally the slack adjuster-to-pushrod angle should be
    a. less than 90 degrees
    b. exactly 90 degrees
    c. more than 90 degrees
    d. does not matter

13. A tri-axle trailer is equipped with 5 slack adjusters correctly set at an effective length of 6 inches and one incorrectly set at $5^{1}/_{2}$ inches. How much less brake torque will be applied at the brake with the incorrectly set slack adjuster when the service application pressure is 100 psi on a 30-square-inch chamber?
    a. 150 in.-lb.
    b. 1,500 in.-lb.
    c. 15,000 in.-lb.
    d. 15,000 in.-lb.

14. A leaking axle seal has saturated the brake shoe linings and caused them to glaze. Technician A says that the shoes should be immersed in solvent before being returned to service. Technician B says that if the brake shoes on both sides of the axle are not replaced, brake torque balance could be affected. Who is right?
    a. Technician A only
    b. Technician B only
    c. both Technician A and B
    d. neither Technician A nor B

15. Technician A says that it is good practice to machine a new drum before putting it into service because it trues the drum to the wheel. Technician B says that most heat discoloration and checking is easy to remove when brakes are serviced by skimming the drum on a lathe. Who is right?

a. Technician A only
b. Technician B only
c. both Technician A and B
d. neither Technician A nor B

16. When a safety pop-off valve continuously unloads, which of the following should be performed first?
a. replace the governor
b. replace the compressor
c. check unloader air circuit
d. check primary circuit pressure

17. At how much free-stroke travel should a properly adjusted, automatic S-cam brake slack adjuster test?
a. marginal preload
b. zero
c. $1/2$ to $3/4$ inch
d. 1 to $1 1/2$ inch

18. How much S-cam endplay should be set when installing a slack adjuster?
a. none
b. 0.030 inch
c. 0.060 inch
d. 0.120 inch

19. What is the adjustment stroke limit for a 24 Series brake chamber?
a. 1.00 inches
b. 1.75 inches
c. 2.25 inches
d. 3.00 inches

20. What is the adjustment stroke limit for a standard 30 Series brake chamber?
a. 1.50 inches
b. 2.00 inches
c. 2.25 inches
d. 3.00 inches

28

Vehicle Chassis Frame

Prerequisites
Chapters 2 and 3

Objectives

After reading this chapter, you should be able to
- Describe the chassis frame of a heavy-duty truck.
- Define the terms yield strength, section modulus, and resist bend moment (RBM) area.
- List the materials from which frame rails are made and describe the characteristics of each.
- Explain the elements of frame construction.
- Describe how frame damage can occur due to impact and overloading.
- Carry out basic chassis frame alignment checks.
- Explain how the chassis frame, side rails, and crossmembers can be repaired.
- List some guidelines to follow when using frame repair hardware.

Key Terms List

applied moment
area
bending moment
C-channel rail

galvanic corrosion
pounds per square inch (psi)
resist bend moment (RBM)
section modulus
yield strength

The chassis frame is the backbone of all heavy-duty vehicles. The main body of the frame on a highway tractor is shaped like a ladder. Although a ladder's function is far different, its two main components—rails and steps—can be compared to the truck frame rails and crossmembers. The cross sections of the rails resemble a "C" or "I," and as they are increased in size and strength, the "duty rating" of the ladder or truck frame is upgraded. Crossmembers also get stronger as the anticipated workload increases.

The frame supports the cab, engine, transmission, axles, and the various other chassis components. The crossmembers control axial rotation and longitudinal motion of the rails and reduce torsional stress transmitted from one rail to the other. Crossmembers

are also used for vehicle component mounting and protecting the wires and tubing that are routed from one side of the vehicle to the other. The front crossmembers may be of a drop style (**Figure 28–1A**) or a straight-through type (**Figure 28–1B**).

The drop-style crossmember provides for a low center of gravity resulting in better vehicle stability. This design also allows for larger radiators to be used and improves the serviceability of the engine and related components. In addition, the drop crossmembers frequently make excessive cutting or notching of the frame rails unnecessary. In an area to the rear of the cab, on some heavy-duty truck models, a deep section of the frame provides extra strength at points of high stress. However, where lower cost and greater ruggedness are required—for

A

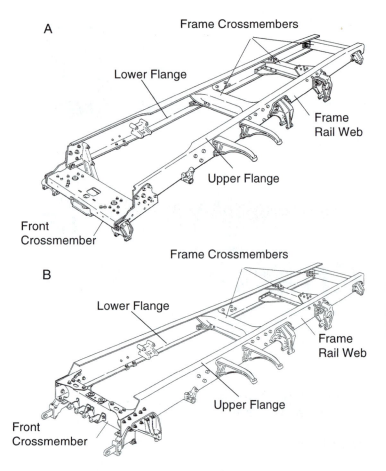

Frame Crossmembers

Lower Flange

Frame Rail Web

Upper Flange

Front Crossmember

B

Frame Crossmembers

Lower Flange

Frame Rail Web

Upper Flange

Front Crossmember

FIGURE 28–1 Typical heavy-duty tractor chassis frames: (A) drop style, and (B) straight-through type. (Courtesy of Freightliner Trucks)

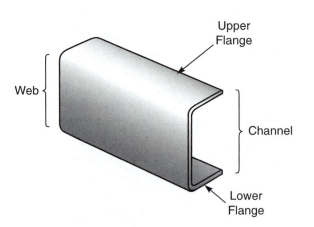

Upper Flange

Web

Channel

Lower Flange

FIGURE 28–2 Typical C-channel rail.

example, off-highway vehicles—straight-through crossmembers are used.

Though I-beams are used in some extra heavy-duty applications like crane carriers, most frame rails used on Class 8 trucks are of the **C-channel rail** type (**Figure 28–2**). Each C-channel has an upper flange, lower flange, and web (the surface between the flanges). The C-channel accepts crossmembers and body mounts well and is generally easy to work with. Each truck manufacturer uses these rails with crossmembers of its own design. Crossmembers may be I-beam, C-section, tubular, boxed, or other shapes and may be bolted or welded together.

All frames are engineered and built to provide the versatility, durability, and performance demanded by the varied uses of the heavy-duty vehicles. These varied uses require that the frames used be rigid enough to resist bending and twisting forces, have sufficient strength to carry anticipated loads, provide resistance to fatigue from repeated flexing, and have the durability to absorb the vibration and shocks encountered in daily operation.

28.1 BASIC FRAME DESIGN

Before undertaking any frame servicing, the technician should understand the following terms and definitions used to describe frame characteristics and strength. Their use will be found throughout this chapter.

- **Yield Strength. Yield strength** is the highest stress a material can sustain without permanent deformation, expressed in **pounds per square inch (psi).** Light and medium trucks may use rails made of steel alloys with a yield strength of 30,000 to 80,000 psi. Class 8 trucks use 110,000-psi, heat-treated steel. A material's yield strength is determined through laboratory testing and the higher the figure, the stronger the material.

- **Section Modulus. Section Modulus** factors the shape and size of the rail but not the material. In other words, it is not related to the frame rail material, but it does help define its flexibility. The deeper the "web" (the vertical portion of the rail) and the wider the "flange" (the horizontal portion attached perpendicular to the web), the greater the rail's section modulus. **Figure 28–3** and the accompanying formula illustrate how section modulus is calculated. Note how the formula emphasizes C-channel height over flange width.

- **Resist Bend Moment (RBM).** This most accurately expresses frame rail strength. **Resist Bend Moment (RBM)** is calculated by multiplying the section modulus of the rail by the yield strength of the material. This term is universally used in evaluating frame rail strength. The higher the RBM, the stronger the frame. This figure

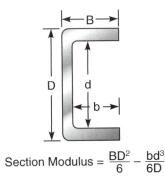

$$\text{Section Modulus} = \frac{BD^2}{6} - \frac{bd^3}{6D}$$

FIGURE 28–3   *Frame section modulus.*

could be in the millions of pounds, and minimum numbers will sometimes be specified by builders of extra heavy-duty equipment, like wreckers, crane carriers, or roll-on/roll-off bodies.

- **Area.** The **area** is the total cross section of frame rail (includes all applicable elements) in square inches.
- **Applied Moment.** The term **applied moment** means that a given load has been placed on a frame at a particular point.
- **Bending Moment.** The term **bending moment** simply means that when a load is applied to the frame, it will be distributed across a given section of the frame material. This load then tends to deflect the frame where the load concentration is greatest. The effect of maximum bending moment on the frame will vary because of different body configurations. For instance, depending on how it is loaded, a straight-body truck will transfer this applied moment in a different manner when compared to a dump truck (**Figure 28–4**).
- **Safety Factor (Margin of Safety).** This is the amount of load that can safely be absorbed by and through the vehicle chassis frame members. This safety factor (SF) is the relationship of the applied moment to the RBM. If the applied moment and the RBM are the same, then the safety factor is classified as 1. This safety factor is necessary because the applied moment is generally calculated based upon a stationary or vehicle static condition.

IDENTIFICATION OF FRAME RAIL MATERIAL

Heavy-duty chassis are usually manufactured with frame rails of either steel or aluminum alloy. Each material must be handled in a specific manner to ensure maximum service life. The frame material must be clearly identified before attempting repair or modification.

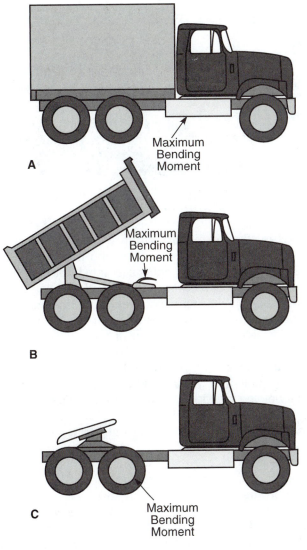

FIGURE 28–4   Frame bending moment location on (A) straight truck, (B) dump truck, and (C) tractor.

🥽 Shop Talk

Although chassis frames are very strong, care must be taken when lifting or moving them, to avoid anything that may scratch, cut, or damage the exposed frame assembly. Cushion all chain hoists or cable slings with a section of heavy hose. If the frame rail is raised with a jack, place a block of wood between the jack and the frame rail.

Steel Frames

Trucks are manufactured with frame rails of mild steel (36,000 psi yield strength), high-strength low-alloy (HSLA) steel (50,000 psi yield strength), or heat-treated steel (110,000 psi yield strength). Each type has different repair procedures.

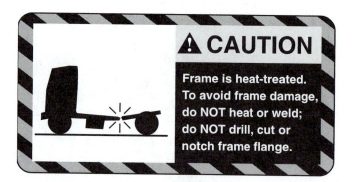

FIGURE 28–5   *Typical heat-treated frame caution label. (Courtesy of Mack Trucks, Inc.)*

Steel frame material can be determined by inspecting the frame or referencing the factory line ticket, service manual, and sales data book.

There are several methods of identifying heat-treated (110,000 psi yield strength) frame rails, the most common of which is a stencil marking on the inside middle or rear section of the rail or a stencil mark on one of the crossmembers. The stencil notes that the rail is heat treated and that rail flanges must not be drilled or welded. This is to caution against the welding of additional brackets or crossmembers or the welding of reinforcements to the rails. This may cause failure due to frame strength reduction at welds and holes.

Material thickness increases as the truck's intended duty becomes more severe. On-highway tractors generally use $1/4$-inch steel; an on/off-road dump truck or mixer chassis may have $5/16$- or $3/8$-inch-thick rails, often with reinforcing members. Multiple frame rails can be used. The depth of the rail also increases with duty severity. Highway tractor rails will usually be $10 1/4$ inches deep, but dump truck rails might be $13 1/2$ inches deep.

Most truck frames carry a decal warning against welding or torching the rails (**Figure 28–5**). This is because high temperatures can adversely affect the steel's heat treatment, weakening the material. Welding or torching is permitted at the ends of the frame rails and for crack repair, however.

During assembly, frame rails are custom drilled to accept specific crossmember locations and the attachment of engine mounts, suspensions, and the like. Hole locations are predetermined, based on the truck's intended application and, therefore, the types of componentry to be attached to the frame. Most builders prefer not to drill extra holes because these weaken the rails.

Aluminum Alloy Frames

It is relatively easy to distinguish aluminum frames from steel frames on the basis of their greater web

and flange thickness and nonmagnetic properties. Frame material can be identified by placing a magnet near the frame. The magnet will be attracted to a steel frame. The aluminum frame will not attract the magnet.

Aluminum alloy frames weigh less than their steel counterparts, but are not as strong as hardened steel frames. The most common aluminum frames are medium strength magnesium-silicon-aluminum alloy (37,000 psi yield strength) and high-strength copper-aluminum alloy (60,000 psi yield strength). Both of these alloys are heat-treated.

Repair procedures for the medium-strength aluminum frame and the high-strength aluminum frame are different. Medium-strength aluminum frames and high-strength aluminum frames cannot be distinguished from one another by appearance. To identify the type of aluminum, consult the factory line ticket or service manual or contact the OEM.

FRAME CONSTRUCTION

Most frames are available with either inside or partial inside channel reinforcements or outside reinforcements. The reinforcements are used to provide a greater RBM than can be obtained by using a single main frame rail.

All crossmembers are designed to connect the frame rails. They provide rigidity and strength, along with sufficient flexibility to withstand the twisting and bending stresses encountered when operating on uneven terrain. A stamped "C" section is a standard type of crossmember. Conventional highway applications use a plate-type crossmember in the first and second positions behind the cab. Stamped "C" section crossmembers have keystone shaped gussets for attaching the crossmembers to the frame. This design resists "weaving" or "out-of-square" frames. Gussets are welded or bolted to the crossmember and then bolted to the frame rail web. Exceptions to this frame rail web attachment may occur at the rearmost crossmember.

All attachments to the frame rails should be made to the frame rail web, and never to the flanges, because stresses are highest in the flange area. It is important that care be taken when attempting to install accessory components such as PTO brackets, spare wheel racks, extra fuel tank brackets, chain racks, and so on, using the frame rails or crossmembers as a foundation. All attachments should be bolted, and either bodybound or flange head bolts should be used in critical locations on the chassis.

As mentioned earlier in the chapter, the common frame on heavy-duty trucks is the single C-channel configuration. Other common channel frame configurations are (**Figure 28–6**):

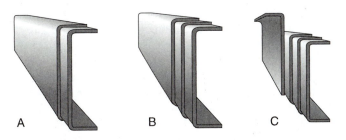

FIGURE 28–6 Two-, three-, and four-element frame rail arrangements.

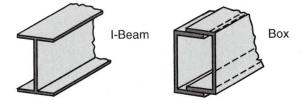

FIGURE 28–7 Two other frame designs: (A) I-beam, and (B) box.

- **Two-Element Rail.** Consists of the main frame rail and a single inside channel frame reinforcement.
- **Three-Element Rail.** Consists of the main frame rail and two frame reinforcements; a single inside channel and a single outside channel frame reinforcement.
- **Four-Element Rail.** Consists of the main frame and three frame reinforcements, a single inside channel, a single outside channel, and a single inverted "L" outside frame reinforcement.

Two frame construction designs—the I-beam and the box—are used in some vehicles (**Figure 28–7**).

Adding reinforcing members to the frame rails increases their RBM value significantly but also adds weight. The weight per foot of rail can range from below 15 pounds to 40 pounds and more, depending on the type of reinforcement you choose. These take various forms. For example, "inserts" may themselves be C-channels, right angles, or inverted or upright Ls clamped within the main rails. Sometimes reinforcing is added to the outside of the rails by using L-channels or nonflanged fishplates. Sometimes both inserts and outside mounted L-channels or fishplates are used. Occasionally triple C-channels are joined together to form super duty rails.

A few builders offer "deep section" or "belly" extensions to the rails where stress is greatest—toward the middle of the frame, between the axles. This is where weight is suspended without direct support, and "bowing" of the frame is possible.

Although it costs more, a deep section frame can add to RBM without the weight penalty incurred by inserts and outside channels or fishplates.

Webs and gussets are added where crossmembers join the main rails, and these are often incorporated in the crossmember's design. In severe duty trucks, extra webbing and gusseting can sometimes extend down much of the frame's length.

Reinforcing channels and plates are usually added at points where stress is concentrated. Examples are beneath a tractor's fifth wheel; at suspension mounting points—especially for add-on tag or pusher axles; at the rear, where heavy lift gates are installed; and where rails are cut or extended to accommodate a shortening or lengthening of the truck's wheelbase.

For mounting of certain truck bodies, subframes consisting of shallow C-channels or hardwood strips are attached to the top of the frame rails. Plates or U-bolts join the subframe to the rails; if U-bolts are used, spacers may be wedged between the rails' top and bottom flanges to keep them from compressing.

CAUTION: Reinforcement of the chassis frame to support additional loading or to concentrate a load, should not be made until it has been fully verified that all other vehicle components, such as the brake system, steering system, suspension system, and so on, can properly and safely carry and support the increased loading.

In any modification of the chassis frame, the addition of holes, reinforcements, welds, clamps, splices, and so on, may cause an increase in local stress in the frame at the point of the modification. These local stress concentrations can significantly affect the life of the chassis frame. The specific effect that the stress concentrator will have on the life of the chassis frame is influenced by its location, the frequency and severity of the loading, and the type of stress concentration. Not observing the repair procedures in the vehicle's service manual may void the manufacturer's warranty.

28.2 FRAME DAMAGE

Damage to the chassis frame generally occurs as a result of impact damage to the vehicle, such as the vehicle being involved in a collision or overturn (**Figure 28–8**). Such damage may often be repaired by straightening and reinforcing the frame, by replacing

FIGURE 28–8   Collision-damaged truck being prepared for straightening. (Courtesy of BeeLine Co., Bettendorf, IA)

the frame sidemembers, or by repairing the damaged area and reinforcing the frame sidemember.

 Shop Talk _____

Frame straightening should be performed only by a qualified frame alignment facility. Under no circumstances should frame alignment be performed by inexperienced or unqualified service personnel. Because impact-damaged frames are corrected by specialty technicians, this type of frame servicing is not covered in this book.

Damage to the chassis frame, such as a crack in the frame sidemember or crossmember which is not associated with impact damage, may be an indication of overloading the vehicle. Overloading can be caused by the following:

- Exceeding either the gross vehicle weight rating (GVWR) or the gross axle weight rating (GAWR) (i.e., loading the frame beyond its design capacity)
- Uneven load distribution
- Improper fifth wheel settings (see Chapter 30)
- Using the vehicle for purposes for which it was not originally designed
- The use of special equipment for which the frame was not designed
- Improper modification of the frame

Frame damage may also be caused by corrosion caused by the contact between dissimilar metals. If aluminum and steel, for example, are allowed to come into direct contact, **galvanic corrosion** can eat away both materials. Aluminum is anodic with respect to steel and will corrode when in the presence of steel. Corrosion of the aluminum frame member will reduce the load-carrying capacity of the frame member and may eventually lead to failure of the frame.

In order to prevent galvanic corrosion, isolation techniques such as nonconductive or barrier type spacers or sealers must be used. It is recommended that a sealer, such as that suggested in the service manual, be painted onto the surface of both the aluminum frame and the steel reinforcement. Steel bolts passing through the aluminum frame and steel reinforcement, as well as the washers under the head of the bolts and nuts, should be sealed also.

28.3   FRAME ALIGNMENT

Both the tractor and trailer frames must be completely straight to ensure the best handling characteristics for over-the-road running as well as city operations. If a frame has been incorrectly straightened or has remained unstraightened following an accident, it is certain to affect the life of the unit.

A bent frame can affect the tracking of a vehicle going down the highway. A bent frame can throw a driveline out of alignment and set up vibrations in the entire vehicle. It can damage the powertrain and cause rapid tire wear. It might also cause steering problems that may result in front wheel misalignment. But most importantly, a bent frame can decrease the control a driver has over a vehicle during an emergency and increase the chances of an accident occurring. If the vehicle itself has been involved in an accident or has been overloaded, it is recommended that the frame be checked for proper alignment following the procedures given in Chapter 21.

Special tools and gauges are available at most heavy-duty truck wheel and alignment shops that permit rapid checking of the frame and component alignment (**Figure 28–9**). The following instructions are given to permit the checking of the frame alignment by an ordinary fleet or small repair shop without these specialized tools and gauges. To make these checks, a plumb bob, a chalk line, and tape or paper will be needed.

FIGURE 28-9   A heavy-duty truck alignment frame. (Courtesy of BeeLine Co., Bettendorf, IA)

Prealignment Checks

When diagnosing misalignment conditions, and before disassembling components, be sure that the following items have been checked and are satisfactory.

- Tire inflation
- Front end alignment
- No visual frame damage or bent axle housings
- Proper wheel and tire balance
- Tires and rims must be of the proper size and type with no mismatching.
- On disc wheel assemblies, the discs should be the same on all wheels.
- Move the vehicle to a level floor and determine that the front wheels are as straight as possible.

Frame alignment can be checked by transferring vehicle reference points to a level floor with a plumb bob. This is done by hanging the plumb bob string through the trunnion and wheel hub centerlines, and from the center of specified frame rail bolts (**Figure 28-10**). Do not run the plumb bob string over the nut,

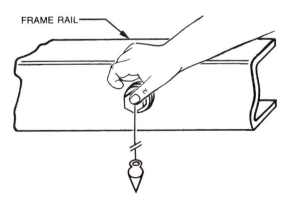

FIGURE 28-10   Method of hanging a plumb bob. (Courtesy of Mack Trucks, Inc.)

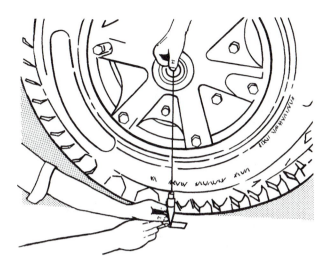

FIGURE 28-11   Marking alignment reference point. (Courtesy of Mack Trucks, Inc.)

as this will give an incorrect reading. The bolt location must also be consistent from side to side.

Use the exact same bolt on each rail. If for any reason a plumb bob cannot be used at the reference points indicated, use the next nearest bolt location on both sides. Place a piece of tape under the plumb bob and accurately mark the location with a pencil (**Figure 28-11**). Reference points must be accurately located and marked if the check is to be dependable; consequently, this is a two-person operation.

FRAME LAYOUT

Three basic types of side rails are found on most heavy-duty vehicles. The first type has straight side rails. Another type has side rails that are equally offset from the vehicle centerline. The third type has side rails that are unequally offset from the vehicle centerline.

When checking the alignment of a frame assembly with an equal offset, the same layout procedures may be used as for a vehicle that has straight frame rails. To check the frame alignment of a frame assembly with an unequal offset, new reference points must be used (**Figure 28-12**). The difference in the side rails will be compensated for by using the lower, outer edge of the front spring front and rear mounting brackets as reference points. The rest of the frame layout may then be checked as listed from Step 4 on. Select the following reference points and project them to the floor with a plumb bob. These points are labeled "M" in **Figures 28-13** to **28-17**. These points must be accurately located and marked.

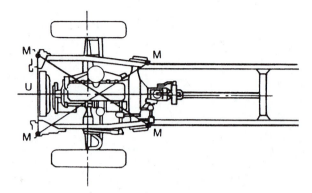

FIGURE 28–12   Unequal offset side rail frame alignment reference points. (Courtesy of Mack Trucks, Inc.)

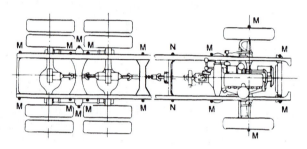

FIGURE 28–13   Straight side rail frame alignment reference points. (Courtesy of Mack Trucks, Inc.)

1. Outer front bolts of front spring front bracket
2. Outer rear bolts of front spring front bracket
3. Outer front bolts of middle crossmember
4. Outer rear bolts of rear crossmember
5. Center of hub cap of front wheels
6. Bogie trunnion spindle centerline
7. Trunnion bracket front and rear mounting bolts

For vehicles with a wheelbase greater than 184 inches, it may become necessary to establish additional reference points. Identify additional reference points as "N." Any additional reference points must be located at a common position on both side rails. Every precaution must be taken to ensure that any new reference points are in exact relationship; otherwise, the alignment check will be adversely affected.

Move the vehicle away from the points marked on the floor. Next connect the points as shown in **Figure 28–14**. Connecting lines can be marked by locating the ends of a chalk line on the floor and snapping it. Then proceed as follows:

- Bisect the line between the points described in step 1, thus locating point "U."
- Bisect the line between the points described in step 4, thus locating point "V."

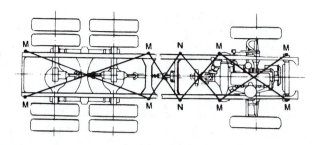

FIGURE 28–14   Connecting alignment reference points. (Courtesy of Mack Trucks, Inc.)

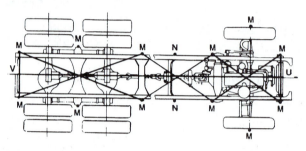

FIGURE 28–15   Locating chassis centerline. (Courtesy of Mack Trucks, Inc.)

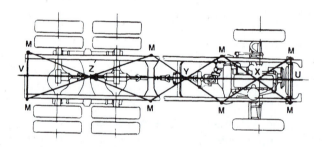

FIGURE 28–16   Checking centerline intersections. (Courtesy of Mack Trucks, Inc.)

- With a chalk line, connect the center points "U" and "V" (**Figure 28–15**).

If the frame is true, the centerline between "U" and "V" will come within 1/4 inch of passing through all intersections "X," "Y," and "Z" (**Figure 28–16**). Variations of greater than a 1/4 inch indicate misalignment in that area. The damaged area may be located by comparing lengths of the pairs of diagonals and noting the variation at the intersection of the centerline.

The lengths of each pair of diagonals should not vary more than 1/4 inch, and they should intersect within 1/4 inch of the centerline (**Figure 28–17**).

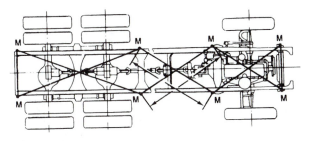

FIGURE 28–17 *Comparing lengths of pair diagonals. (Courtesy of Mack Trucks, Inc.)*

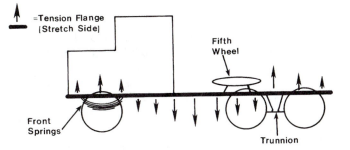

FIGURE 28–18 *Always make reinforcement plates longer than the tension flange edge. (Courtesy of Mack Trucks, Inc.)*

28.4 REPAIR OF FRAME, SIDE RAILS, AND CROSSMEMBERS

Neither truck manufacturers nor the companies that make the frame side rails recommend the repair of these heat-treated items. Replacement is the best procedure. However, side rails and crossmembers are often repaired in the field by owners and operators of the vehicles to minimize downtime and expense.

The effective repair of the frame side rails and crossmembers necessitates some knowledge of factors that cause frame damage in the first place. Damage is primarily caused by high stresses, and usually by a local concentration of high stress. Factors that produce localized stress include abrupt change in section, incorrect bolt patterns, notches, cracks, improper load concentrations, and loose bolts. As weight is applied to the frame of a vehicle, it has a tendency to flex. Where the frame is not directly supported by the suspension, it flexes downward. As this flexing occurs, one frame flange stretches (tension flange) while the other flange is compressed (compression flange). Because frame stress is greatest at the tension flange, reinforcement plates, when used, should be longer on the tension flange edge to give this area additional support **(Figure 28–18)**.

An effective repair must compensate for the factors that caused the original damage and not introduce any factors that would produce new stress concentrations. The following are certain precautions that must be observed when making the following frame repairs.

WELDING REPAIRS

Welding repairs on truck frames should be performed by a qualified welder. The possible exception to this is the repair of minor cracks in the frame.

When making such a repair on steel frames, the shield arc method should be used because the heat generated during welding is localized, and overheating is minimized. Additional advantages are that the finished weld can be ground flush and drilled as necessary.

Recommended methods for crack repair of steel frame members are shielded metal arc welding (SMAW); gas metal arc welding (GMAW), also known as metal inert gas (MIG) welding; gas tungsten arc welding (GTAW), also known as tungsten inert gas (TIG) welding; or flux cored arc welding (FCAW).

Welds on truck frames are common practice, but they should be performed by a skilled welder with some knowledge of how the repair will affect the frame performance. The heat-tempered frame rails used on most highway tractors may be both repair-welded and extended successfully, providing the correct methods are used. The preferred method of repair welding or extending a hardened steel frame is to use a low-hydrogen welding electrode with a wire tensile strength rating similar to the frame rating. An American Welding Society (AWS) E-11018 electrode is ideal. When welding hardened frame rails, try to observe the following:

1. Use the appropriate personal and shop safety equipment during the operation.
2. Multiple passes of a 3.2 mm ($^1/_8$ inch) arc welding electrode with a cool-down period between each pass are preferred to using a MIG technique. MIG welding is much faster. The tendency is to use excess fill during a pass, increasing the chances of overheating the frame material and causing crystallization.
3. Aluminum alloy frames are best welded using a TIG welding process. Ensure that a filler wire compatible with the base material is used. MIG welding can be used, but again, ensure that the base material is not overheated and

that the filler wire is compatible with the base material.

4. When extending both heat-treated alloy steel and aluminum frame rails, ensure that added sections are cut at either a 60- or 45-degree angle. During the welding process, tempering is reduced in the weld area, and an angular weld minimally diminishes RBM over any given section.

5. Properly prepare the weld area. Grind a "V" into the abutting sections. If repairing frame cracks, drill through at one-inch intervals and ensure that the source of the crack is located and drilled through.

6. When the weld has been completed, grind away any profile apparent on either side of the repair so it is flush and preferably not visible. Weld profile increases section modulus through the weld area, and this can form localized stress and create another failure.

> **CAUTION:** Most frame welds fail not due to a deficiency in the welding technique, but to a lack of knowledge of frame repair methods on the part of the welder. A hardened steel frame weld that fails by cracking cleanly through the center of the weld profile often does so because the incorrect filler wire (electrode) has been used. A weld that fails by cracking clean to the sides of the weld profile often does so due to crystallization caused by overheating.

 Shop Talk

Always identify a frame material before attempting to weld it. Most tractors use tempered frame rails with a yield strength of 110,000 psi and require the use of specialty electrodes. Most trailers are fabricated out of mild steel components and can be repair-welded with any general-purpose, mild steel electrodes or filler wires.

Never weld aluminum frames with general-purpose aluminum welding wire: The welds may appear to be sound, but aluminum welding wire is satisfactory only for welding nearly pure aluminums and not the aluminum alloys used in frame rails. When repair welding or extending mild steel frames, the MIG welding process is preferable. Mild steels will not crystallize and because they are not hardened, there is no tempering to remove. This means that angular weld joints are not required through frame sections.

Figure 28–19 shows a typical frame crack. To repair a crack of this type, first "V" out the crack,

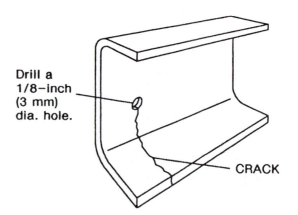

FIGURE 28–19 *Preparing a rail crack for welding. (Courtesy of Freightliner Trucks)*

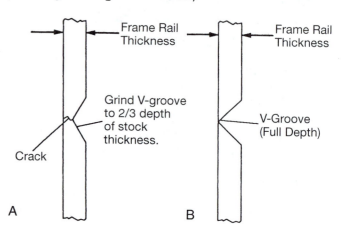

FIGURE 28–20 *Grinding a V-groove. (Courtesy of Freightliner Trucks)*

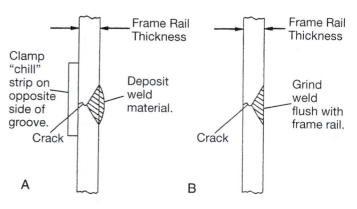

FIGURE 28–21 *Making a one-side weld.*

preferably from the inboard side as shown in **Figure 28–20**. Next, weld in the "V" groove as shown in **Figure 28–21**. Then "V" out the crack from the outboard side as shown in **Figure 28–22** and weld in the "V." If the repair is being performed on a hardened steel frame rail, clamp a chill plate to the opposite side (**Figure 28–22**). This will help prevent overheating and crystallization. Finally, grind the weld profile flush (**Figure 28–22**) so the section modulus is not

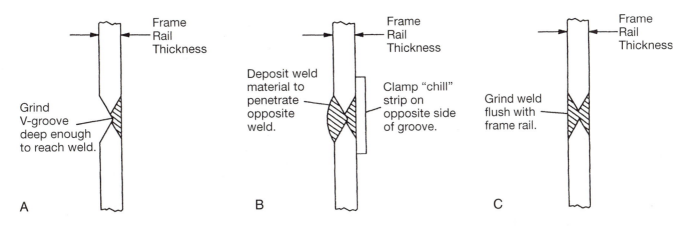

FIGURE 28–22 Grinding, filling, and finishing a two-side weld. (Courtesy of Freightliner Trucks)

increased through the weld section. This repair method should make any reinforcement unnecessary.

 Shop Talk

Avoid reinforcing weld repairs as a matter of procedure. A frame like your backbone is dynamic, that is, designed to flex. Reinforcing increases section modulus by increasing rigidity: This can generate a failure elsewhere in the frame.

Reinforcements must be made with suitable channel stock. The reinforcements should be made from a material matching the strength of the frame. The length of reinforcement is dependent upon the type of service the truck is used in and the frequency and severity of loading. Reinforcements must be bolted rather than welded to the frame rail. Reinforcements may be installed on the outside or the inside of the frame side rail (see **Figure 28–23**). When installing an inner frame reinforcement at a crossmember, it will be necessary to alter the length of the crossmember or replace the one-piece crossmember with a multipiece (bolted) crossmember. Alteration of bolt hole locations in the multipiece crossmember may be required to suit the application.

Should it be necessary to cut the frame, cut the side rail at a 45-degree angle. Angle cutting will distribute the cut load and bolt pattern over a greater area than a cut made at right angles to the frame (**Figure 28–24**). Reinforcement plates must be long enough to extend beyond the critical area so that the ends can be cut on an angle instead of square across the frame section. Avoid any section "gaps" caused by a reinforcement plate stopping short of the ends of adjacent brackets or crossmember gussets (**Figure 28–25**). Also avoid several holes in

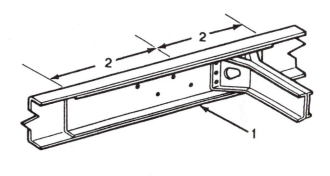

I **REPAIR AREA**
2 **610mm (24.0 INCH) MINIMUM**

FIGURE 28–23 Position of a frame reinforcement. (Courtesy of International Truck and Engine Corp.)

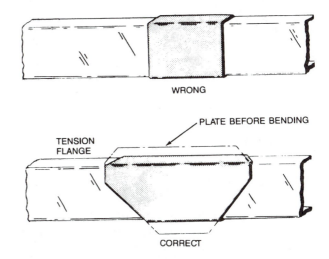

FIGURE 28–24 Reinforcement plate size. (Courtesy of Mack Trucks, Inc.)

direct vertical alignment or holes too close together. A staggered bolt pattern with good spacing and sufficient edge distance is desirable (**Figure 28–26**).

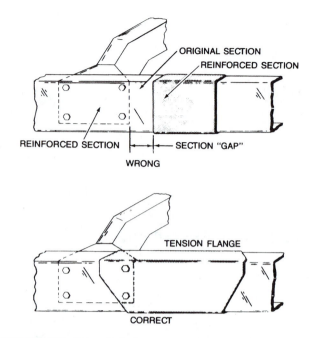

FIGURE 28–25   Reinforcement plate location. (Courtesy of Mack Trucks, Inc.)

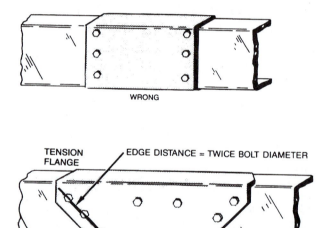

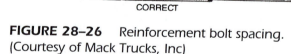

FIGURE 28–26   Reinforcement bolt spacing. (Courtesy of Mack Trucks, Inc)

In addition to their reinforcement recommendations, it is important to remember to never leave a sharp internal angle when cutting reinforcement plates or when modifying members. Sharp angles create high stress zones. It is best to cut material at a 45-degree angle. If this is not possible because of obstructions, try to relieve any high-stress pressures by spreading the load over a curved section (**Figure 28–27**).

Cutting the frame behind the rear axle to alter the wheelbase is usually acceptable to most manufacturers. Consult the manufacturer regarding whether

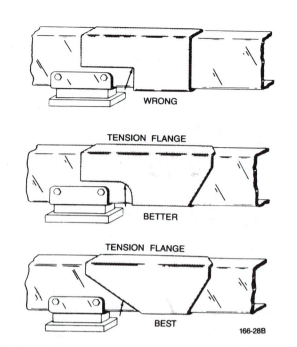

FIGURE 28–27   Reinforcement plate angle. (Courtesy of Mack Trucks, Inc.)

reinforcement is required when extending a frame in this manner.

Shop Talk

When cutting a frame, use a pencil or soapstone to make all lines, points, or other marks. Do not use any scriber or tool that will scratch the surface of the frame rail. Use a machinist's square to project all points from the webs to the upper flanges and to measure inboard from the outside face of the frame rails.

BOLTS AND TORQUE SPECIFICATIONS

Most frames are assembled with Huck™ fasteners, bolts, and nuts. Others are riveted. Bolts must always be used when attaching a reinforcement. Rivets can be replaced by bolts when the frame is repaired and reinforced. Details on working with bolts and rivets can be found in Chapter 3.

In bolted joints, the majority of the load is transferred by the clamping force between the members of the joint. Bolts must be properly torqued to develop and maintain the desired clamping force. Loose or improperly torqued bolts can lead to failure of the joint. The bolts and nuts should be inspected periodically to ensure that proper torque is maintained.

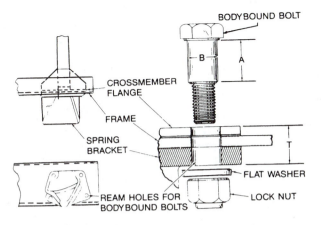

FIGURE 28-28 Determining bodybound bolt size. (Courtesy of Mack Trucks, Inc.)

Bolts and Huck™ fasteners must be always replaced by fasteners that meet the manufacturer's specifications. These may be SAE Grade 5 bodybound (machined with an interference fit shank), SAE Grade 5 or SAE Grade 8. It is poor practice to replace Grade 5 fasteners with Grade 8. Ensure that the nuts also meet the manufacturer's specification. Never consider replacing body-bound bolts with standard shank bolts.

Holes that are enlarged or irregularly worn may be reamed to accept the next larger bolt diameter. For critical areas, bodybound bolts with hardened washers are available and should be used (**Figure 28–28**).

If frame components are aluminum, flange head nuts and bolts should be used, or hardened flat washers must be used next to the aluminum under both the bolt head and nut. If modification or repair requires the replacement of existing bolts with new bolts or bolts of greater length, the old nuts should also be replaced.

DRILLING

Careful consideration should be given to the number, location, and sizes of frame bolt holes in the design of a vehicle. The number, location, and sizes of additional bolt holes put in the frame subsequent to manufacture of the vehicle can adversely affect frame strength.

The drilling of the frame sidemember presents no unusual difficulty. Standard high-speed steel drills of good quality will usually serve, provided they are sharpened properly and not overheated during sharpening or use.

 Shop Talk _____

Cobalt high-speed drills are superior to conventional high-speed drills for drilling hardened frame rails. Drills should be sharpened to give 150 degrees included angle with 7–15 degrees lip clearance.

When drilling the frame, as in any other drilling operation (see Chapter 3), sufficient pressure must be applied to the drill bit to maintain continuous cutting. The drill point should frequently be quenched with cutting oil to help cool the drill. Avoid letting a drill bit turn in the work without cutting. This may result in overheating and eventually damage the drill bit. The drill must be held steady during the drilling operation. Avoid wobble or change of drill angle during drilling.

The following are general guidelines that should be followed when drilling holes in a heavy-duty chassis frame:

- Never drill holes into the flange sections of the frame rails.
- Use existing holes whenever possible.
- Holes should be located as close to the center or neutral axis of the sidemember as possible.
- Maintain a minimum of ³/₄ inch of material between holes.
- There should not be more than three holes located on a vertical line.
- Bolt holes should be no larger than is required for the size of the bolt being used.
- If reinforcements are used, avoid drilling holes closer than 2 inches from the ends of the reinforcement.
- Bolts must be checked periodically to ensure that the proper torque and clamping force is maintained.
- Holes are not permitted in the drop portion of the web of drop center rails.
- The center of the holes must not be closer than 1¹/₂ inches for aluminum rails or 1 inch for steel rails to top or bottom flange face (**Figure 28–29**), nor less than 3 inches apart.
- Bolt holes must not be larger than those existing in the frame, such as for spring bracket bolts.
- On aluminum frame rails, chamfer both sides of all holes 0.02 inch by 45 degrees.

When a pilot hole is used (**Figure 28–30**), it should not be enlarged in successive stages, as rapid wear of drills will occur. A further step to extend drill life is to refrain from breaking through to full size at the bottom of a hole being drilled.

AFTERMARKET MODIFICATIONS

Specialized equipment (such as a dump box, tank, hoist lift, winch, or derrick) is often added to a vehicle

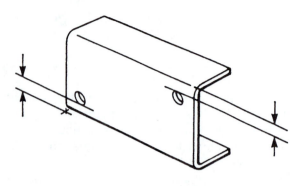

1-1/2" for Aluminum Rails
1" for Steel Rails

FIGURE 28–29   *Proper drilling distances. (Courtesy of Freightliner Trucks)*

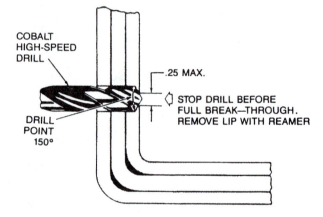

COBALT
HIGH-SPEED
DRILL

.25 MAX.

STOP DRILL BEFORE
FULL BREAK—THROUGH.
REMOVE LIP WITH REAMER

DRILL
POINT
150°

FIGURE 28–30   *Pilot hole drilling. (Courtesy of Mack Trucks, Inc.)*

by the distributor, installer, dealer, or the fleet shop itself. The addition or installation of this equipment on the vehicle can significantly affect the loading of the chassis frame (**Figure 28–31**). In some cases, it may be necessary to reinforce the frame. Care must be exercised to ensure that the gross vehicle weight rating (GVWR) and/or the gross axle weight ratings (GAWR) are not exceeded. Installation of this type of equipment may involve state and federal requirements, which affect vehicle certification for noise emissions, exhaust emissions, brake requirements, lighting systems requirements, etc. The specialized equipment installer is responsible for the safety and durability of their product; in addition, the installer is responsible to ensure that the equipment and its installation comply with all applicable State and Federal Department of Transportation requirements and OSHA regulations.

In any modification of the chassis frame, the addition of holes, reinforcements, welds, clamps, splices, etc., may cause an increase in the local stress in the frame at the point of the modification. This may

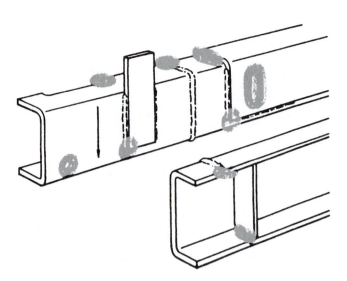

FIGURE 28–31   *Typical side rail stress concentrations. (Courtesy of Freightliner Trucks)*

cause a stress concentration in the frame sidemember(s). These local stress concentrations can significantly affect the life of the chassis frame. The specific effect that the stress concentrator will have on the life of the chassis frame is influenced by the location of the stress concentration, the frequency and severity of the loading, and the type of stress concentration. Any modification of the frame may void the manufacturer's warranty.

The three common methods of attaching specialized equipment to the frame chassis are bolts, U-bolt and clamp attachments, and welding.

Bolt Attachments

When the holes used for fasteners are in the least critical areas possible, bolting is preferred for its versatility and strength. **Figure 28–32** shows a channel section frame side rail loaded as a beam. This type of loading stretches the lower half of the section (produces tension) and compresses the upper half of the section (produces compression), as indicated by the arrows. If the side rail section was without holes, the highest concentrations of stress (tension and compression) would be in the lower and upper flanges of the rail. The uppermost hole is located about one-third of the web depth up from the lower flange. Stress (tension) at the bottom of this hole would be approximately equal to the stress in the lower flange. Any hole located less than this distance (for example, the hole located nearer to the lower flange in the illustration), will have significantly higher stress than will the flange. The reason for the difference in stress among the holes is that the upper hole is located closer to the neutral axis of the rail, or the area where the forces of tension and compression are lower.

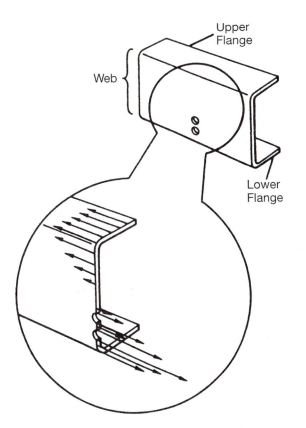

FIGURE 28–32 *A channel section side rail loaded as a beam. (Courtesy of Freightliner Trucks)*

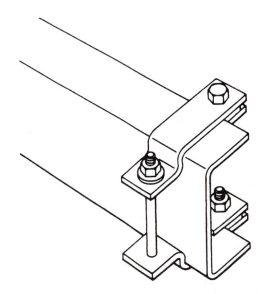

FIGURE 28–33 *Side rail clamp. (Courtesy of Freightliner Trucks)*

The following recommendations must be considered whenever equipment is to be bolted to the chassis frame:

- Use existing holes where possible. When holes must be drilled, they should be located no closer to the top and bottom side rail flanges than the existing holes placed by the factory.
- Avoid drilling holes in any area of the side rail web except the central one-third area. (The web is the surface area between the flanges.) Depending on loading, the top or the bottom of the side rail can be the tension side. Before drilling holes in the side rails, obtain approval from the manufacturer's regional service representative.
- In all cases, avoid drilling holes in the side rail flanges.

U-Bolt and Clamp Attachment

Generally, clamping devices, such as that shown in **Figure 28–33**, are the least time-consuming and least expensive methods of attachment. U-bolts avoid frame drilling, which is especially helpful when the vehicle is equipped with a heat-treated frame. But, because U-bolts or clamps are not locked to the

frame, there is the possibility of heavy equipment moving or being jarred out of place, especially if grease or oil is present on the frame rail. U-bolts and clamps cannot hold high torque loads and can cause local stress concentrations.

The following guidelines must be considered when using U-bolts and clamps:

- Care should be used to block (with spacers) the side rail channel to prevent collapse of the flanges when the U-bolts are tightened. Steel blocks are preferred, because wood may shrink and drop out.
- The blocks used to prevent collapse of the side rail flanges should not interfere with plumbing or wiring routed along the frame rails. In addition, the blocks should not be welded to the side rail flanges.
- Because U-bolts and clamping devices depend on friction and a maintained clamping force for attachment, some bolts should directly connect the attachment to the frame side rail web to prevent the attachment from slipping.
- Do not notch frame side rail flanges in order to "force" a U-bolt fit. If the side rail flanges are too wide, obtain a larger U-bolt or use another method of attachment.

CAUTION: Notching a side rail flange severely weakens the structure and could ultimately result in side rail fracture. With complete side rail fracture, other chassis parts could be damaged with consequent substantial increases in repairs.

Weld Attachment

There are very few cases in which welding on a heat-treated side rail is allowable. Direct welding of the side rail flanges and web should be avoided. To avoid direct welding, equipment bracketry should be welded on a separate reinforcement, and the reinforcement should be bolted to the side rail. It is recommended that before any direct welding to the side rails is performed, the manufacturer be consulted. This step could keep the warranty in effect.

After a straightening job has been completed, the technician should carefully check the repaired vehicle. This check should include an inspection of each rail to make sure that there are no buckles or wrinkles. Rear housings or trailer axles must be set 90 degrees to the centerline of the chassis. The rear end should be checked to make sure it is square with the chassis. Reinforcing sections should fit snugly in the old channels. Finally, a road test can help determine whether a frame was properly straightened.

SUMMARY

- The chassis frame is the backbone of all heavy-duty trucks.
- The frame supports the cab, hood, and power-train components, along with the body and payload.
- The two main components of a ladder-type frame are the two longitudinal members, which are generally referred to as "rails."
- Hardened steel frame rails are formed from high-strength alloy steel, quenched and tempered (heat-treated) to a minimum yield strength of 110,000 psi.
- Most frames are available with either inside or partial inside channel reinforcements or outside reinforcements.
- The reinforcements are used to provide a greater resist bend moment (RBM) than can be obtained by using a single mainframe rail.
- Crossmembers are designed to connect the frame rails. They are made to provide rigidity and strength, along with sufficient flexibility to withstand twisting and bending stresses encountered when operating on uneven terrain.
- All attachments to the frame rails should be made to the frame rail web and never to the flanges, because stresses are highest in the flange area.
- The most common frame rail shape found on heavy-duty Class 8 trucks is the C-channel design. Multiple frame rails are often used.
- Both tractor and trailer frames must be straight to ensure the best handling characteristics for highway as well as city operations.

- A bent frame can decrease the control a driver has over a vehicle during an emergency and increase the chances of an accident occurring.
- Frame reinforcement can be either channel, flat, or angled, but they should be of the same grade and thickness of steel as that of the frame.
- After a straightening job has been completed, the technician should carefully check the repaired vehicle.

REVIEW QUESTIONS

1. Which of the following is a function of the cross-members of the chassis frame?
   a. control axial rotation and longitudinal motion of the rails
   b. protect wires and tubing that are routed from one side of the vehicle to the other
   c. reduce torsional stress transmitted from one rail to another
   d. all of the above

2. The two longitudinal components of the chassis frame are called
   a. crossmembers
   b. frames
   c. rails
   d. channels

3. The highest stress a material can stand without permanent deformation or damage is its
   a. yield strength
   b. resist bend moment
   c. applied moment
   d. bending moment

4. The safety factor of the chassis frame is the reaction of the applied moment to the
   a. resist bend moment
   b. area
   c. applied moment
   d. bending moment

5. Which of the following frame materials has the highest yield strength?
   a. mild steel
   b. high-strength, untempered alloy steel
   c. medium-strength, magnesium-silicon-aluminum alloy
   d. high-strength, copper-aluminum alloy

6. A four-element rail consists of the mainframe and
   a. a single inside channel frame reinforcement and a single outside channel frame reinforcement
   b. two inside and two outside channel frame reinforcement members

c. a single inside channel, a single outside channel, and a single inverted "L" outside frame reinforcement

d. a single inside channel and two outside channel frame reinforcement members

7. Galvanic corrosion is caused by
   a. unevenly distributed loads
   b. two dissimilar metals in contact
   c. improper fifth wheel settings
   d. improper modification of the frame

8. Which of the following effects of a bent frame has the most impact on safety?
   a. damage to the powertrain
   b. vibration in the vehicle
   c. rapid tire wear
   d. decreased driver control

9. To make frame alignment checks, what tool is used to transfer the reference points on the truck to paper or tape on the floor?
   a. a plumb bob
   b. a chalk line
   c. a measuring tape
   d. all of the above

10. Manufacturers recommend that damaged frames, side rails, and crossmembers should be
    a. repaired by welding
    b. repaired by riveting
    c. repaired by bolting
    d. replaced, not repaired

11. Technician A ensures that surface areas to be welded on the frame are clean and free of oil, grease rust, moisture, and paint. Technician B always wears a welder's helmet with the darkest shade of filter lenses while performing gas metal arc welding. Who is correct?
    a. Technician A
    b. Technician B
    c. both A and B
    d. neither A nor B

12. When welding to repair a crack, Technician A widens the crack full length with a grinder. Technician B bevels the crack from one side if welding is to be done from both sides. Who is correct?
    a. Technician A
    b. Technician B
    c. both A and B
    d. neither A nor B

13. Channel stock used for reinforcement must
    a. be welded, rather than bolted, to the frame rail
    b. be made from the same material as the frame rail and matching its strength

c. be installed on only the outside of the frame side rail

d. be installed on only the inside of the frame side rail

14. When cutting a frame, Technician A uses a scriber to make guidelines on the frame rail. Technician B uses a machinist's square to project points from the webs to the upper flanges. Who is correct?
    a. Technician A
    b. Technician B
    c. both A and B
    d. neither A nor B

15. When drilling holes in a heavy-duty chassis frame, the technician should NEVER
    a. locate fewer than three holes on a vertical line
    b. drill holes smaller than those already existing in the frame
    c. drill holes closer than 2 inches from the ends of the reinforcement
    d. chamfer the sides of holes on aluminum frame rails

16. When adding aftermarket modifications to the vehicle, care should be taken to
    a. ensure that the GVWR and the GAWR are not exceeded
    b. make sure all state and federal requirements are met in the installation of the equipment
    c. make sure that modification does not weaken the chassis frame or void the manufacturer's warranty
    d. all of the above

17. In bolting equipment to the chassis frame, Technician A always uses existing holes whenever possible. Technician B never drills holes in the side-rail web. Who is correct?
    a. Technician A
    b. Technician B
    c. both A and B
    d. neither A nor B

18. The main disadvantage of U-bolts and clamps as attaching devices is that
    a. They require a lot of skill to install properly.
    b. They can be jarred out of place.
    c. They are more expensive than other devices.
    d. They require more time to install properly than other devices.

19. Notching a side rail to force a U-bolt fit should not be done because
    a. there will be insufficient friction to hold the U-bolt in place
    b. it requires welding, which might damage the U-bolt

c. it severely weakens the side rail and could cause a fracture

d. installation of the spacer blocks will cause the flanges to collapse when the U-bolt is tightened

20. Which of the following best describes what is required in tractor and trailer frames to ensure the best handling on the road?
    a. flexibility
    b. straightness
    c. strength
    d. all of the above

29

Heavy-Duty Truck Trailers

Objectives

After reading this chapter, you should be able to

- Describe what is meant by semi-trailers and full trailers.
- Identify the various different tractor/trailer and train combinations.
- Describe what is meant by full frame, unibody, and monocoque.
- Explain the various types of hitching mechanisms used and their effects on tractor/trailer or train designation.
- Describe the design characteristics of the dry van, reefer, flatbed, tanker, and other types of highway trailer.
- Explain the operating principles of a reefer trailer.
- Outline some common trailer maintenance practices.

Key Terms List

A-train
B-train
C-train
converter dolly
full trailer
landing gear

lead trailer
monocoque
pup trailer
reefers
semi-trailer
tanker
unibody
van

The term **semi-trailer** refers to any trailer required to be partially supported by another vehicle. A 10-wheel highway tractor is designed to haul a semi-trailer. A common sight on North American highways is the 16-wheel rig as shown in **Figure 29–1**. It is often simply referred to as a tractor/trailer combination. A percentage of the weight of the trailer is supported on the tractor fifth wheel, which acts as both a coupling and weight supporting device. Semi-trailers may be coupled to tractors—or, in the case of highway trains, such as the **B-train**, a fifth wheel on the rear of another trailer. In the B-train configuration, a trac-

tor is coupled to a lead semi-trailer by means of a fifth wheel, and a **pup trailer** is coupled to the **lead trailer**, also by means of a fifth wheel. The B-train is shown in **Figure 29–2**.

The term **full trailer** describes any trailer that fully supports its own weight. In other words, none of its weight is supported by the tow unit. **Figure 29–3** shows a full trailer. An **A-train** is a highway train in which a semi-trailer is coupled to a tractor by a fifth wheel, and a full, pup trailer is coupled to the semi-trailer by means of a pintle and hook assembly. **Figure 29–4** shows a typical A-train double combination. Any

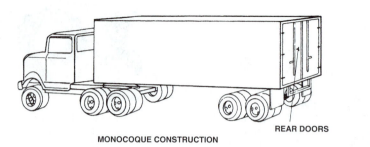

FIGURE 29-1 A tractor and semi-trailer combination. (Courtesy of TTMA)

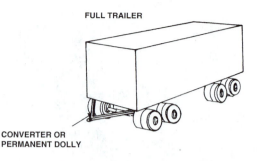

FIGURE 29-3 Full trailer. (Courtesy of TTMA)

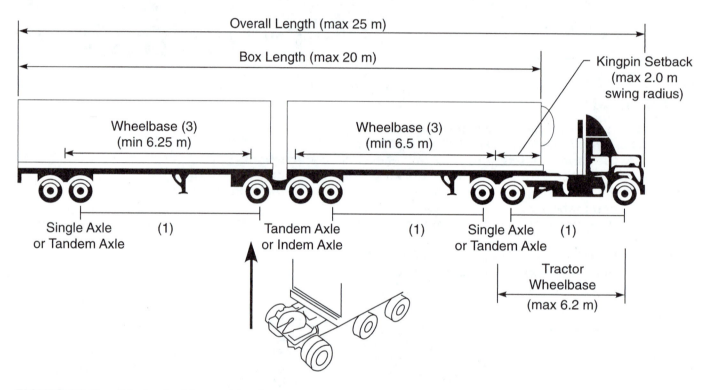

FIGURE 29-2 B-train double combination.

semi-trailer can be converted to a full trailer by coupling it to a **converter dolly.** The converter dolly consists of an axle and frame supporting the fifth wheel. **Figure 29-5** shows a typical converter dolly. The fifth wheel couples to the upper coupler on the pup semi-trailer, and a tow bar with a tow eye is used to couple the converter dolly to a pintle hook on the lead trailer.

Trailers are designed to carry a load. Often they are designed specifically for the load they are to haul. Also, they are usually designed to carry a maximum amount of load while also meeting the federal, state, and provincial regulations of the jurisdictions in which they will be operating. These regulations limit maximum vehicle weight, weight over axle, unit length and

total vehicle length. The maximum allowable width of a tractor/trailer combination in North America is 102 inches. (In most jurisdictions, special permits must be obtained on a per-trip basis if a greater width is required.) Individual semi-trailers are generally limited to 53 feet in length but may be shorter in some jurisdictions. Maximum total and over-axle weights vary by jurisdiction, and the fines for load infractions can be considerable. In fact, most operators use software to figure out whether they are in compliance for each jurisdiction where they are in operation.

A majority of trailers in use on our highways can be described as semi-trailers. The semi-trailer is always equipped with **landing gear** or dolly legs—

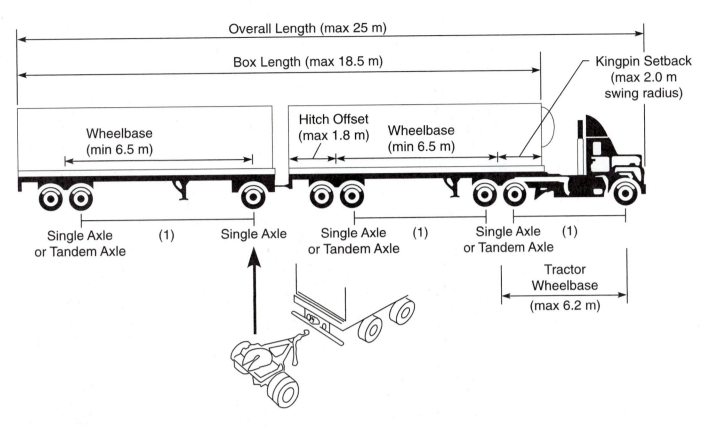

FIGURE 29–4 A-train double combination.

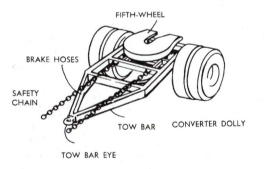

FIGURE 29–5 Converter dolly. (Courtesy of TTMA)

this permits the trailer to be self-supporting when parked and stationary. The semi-trailer has the ability to couple quickly to any tractor with a fifth wheel and an air brake system.

Trailers use a variety of different frames. Some are built on a full frame platform such as the gooseneck platform trailer shown in **Figure 29–6**. This ladder-type frames has many of the same characteristics as the frame on a highway tractor or truck. However, the common construction of a **van** trailer is the

unibody or **monocoque** design. Monocoque design is based on the shell design of a chicken's egg. The chicken's egg is unique in that it can be dropped to the bottom of the deepest ocean without collapsing, a feat that no human-manufactured device has achieved. In the unibody or monocoque frame, all the shell components form the frame. This means that the van floor, roof, bulkhead, doors and sidewalls collectively compose the frame assembly. Damage to any shell component would result in a reduction of frame strength. Think of how fragile the chicken's egg shell is once it has been compromised. **Figure 29–7** shows a typical unibody, semi-trailer van construction.

C-trains are also used on our highways. The C-train is composed of a tractor and two semi-trailers. The lead semi-trailer is coupled to the tractor by the usual fifth wheel coupling. However, the pup trailer is mounted to a special converter dolly with a rigid, double drawbar system that prevents any rotation about the drawbar hitch points. Any rotation that occurs will take place at the converter dolly fifth wheel. This results in a unit that is easier to back up than either A- or B-trains. **Figure 29–8** shows a C-train double configuration.

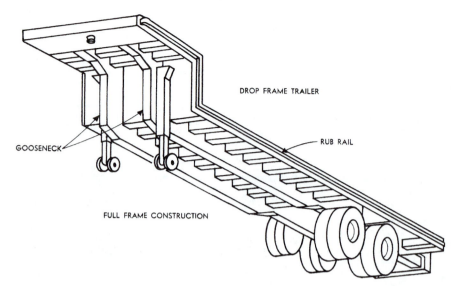

FIGURE 29–6 Gooseneck platform trailer with a ladder frame. (Courtesy of TTMA)

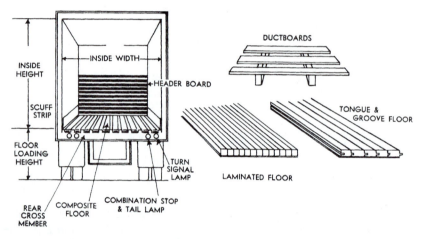

FIGURE 29–7 Unibody, van-type trailer. (Courtesy of TTMA)

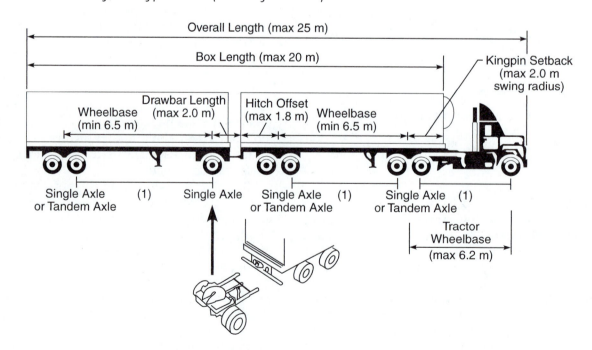

FIGURE 29–8 C-train double configuration.

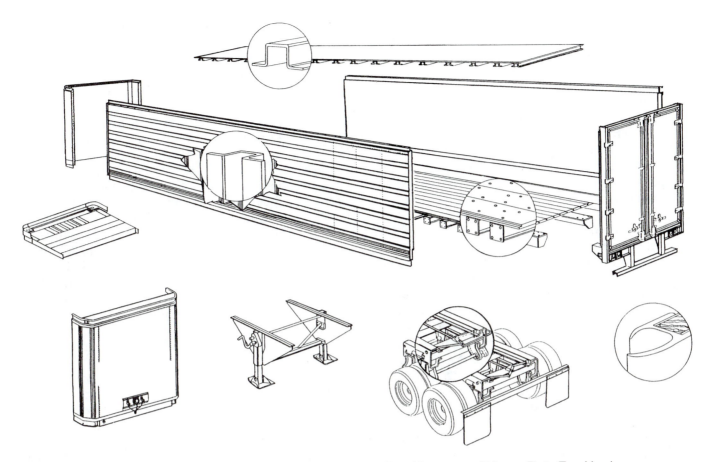

FIGURE 29–9 The frame construction of a heavy-duty trailer. (Courtesy of Heavy Duty Trucking)

29.1 FREIGHT VAN TRAILERS

In the construction of vans, light weight and good aerodynamics are always sought. Despite the need for these two important van trailer characteristics, structural integrity and overall durability are still the two most important features (**Figure 29–9**). The fabrication of the subframe to include steel I-beam crossmember construction has become common in trailer manufacture. Some variables in the spacing of crossmembers can be found, with most designs adding extra support in the rear. While aluminum sheathed plywood-lined sidewalls are common, fiberglass-reinforced plywood (FRP) panels are widely available. FRP is heavier than aluminum and usually more expensive but has high impact strength, good corrosion resistance and a smooth finish.

There are two ways to obtain a light weight without sacrificing strength. One is by using lighter, more expensive materials such as aluminum and high-strength steel in posts and crossmembers (**Figure 29–10**). The other is to give up some cargo space. If the ultimate in interior width and height is not needed, wall posts and floor crossmembers can have a deeper section. Aluminum sheet and post vans can be lighter than FRP vans, but post thickness plus a plywood liner add up to more wall thickness.

FIGURE 29–10 Typical aluminum dry van trailer. (Courtesy of Trailmobile, Inc.)

Why are super-thin trailer walls sometimes important? A 1-inch increase in the width of a 53-foot trailer adds almost 40 cubic feet of space to the interior. Most importantly, it often makes the difference between getting two pallets side by side or not.

Extending trailer length is perhaps the cheapest way to gain cargo space, but legal limits must be met. Fifty-three-foot trailers are still banned in many eastern states, so the trick is to get more interior-space without changing exterior dimensions. On a plywood-lined sheet and post van, one way would be to special-order an FRP front wall designed to meet the DOT strength requirement.

Perhaps the most interesting direction for trailer expansion is down. There is no regulation against lowering the trailer floor. It makes sense to keep the roof at the legal 13 feet 6 inches while stretching downward, because that is where the greatest cargo space can be made. Assuming a completely flat floor for ease of loading, lower limits are determined by tractor fifth wheel height at the front, and trailer floor-to-tire clearance in the rear. Low-profile tires have helped at both ends, with fifth wheel heights coming down from the 48-inch standard to as low as 44 or even 42 inches on tractors with low-profile fifth wheel mounts and lowered frames.

Further gains can also be made without tractor modifications. For fleets with standard-height tractors, trailer cargo area can be increased with a wedge design. Such trailers have floors that slope downward from front to rear, minimizing clearance over the trailer tires and bringing the floor well below standard dock height at the rear. Wedge trailers typically have 110-inch or greater height at the rear door opening and ultra-shallow headers over the door measuring less than $1^1/_2$ inches. Front interior height is usually 108 to $108^1/_2$ inches.

One drawback of lowering the trailer floor is that it will not be the same height as the standard loading dock of 54 inches. Mounting a power hydraulic landing gear assembly at the rear of the trailer has become a common method of dealing with this problem. The driver positions the trailer an inch or so from the dock and flips a switch to raise the floor of the trailer even with the dock. The disadvantages of power hydraulics are cost, weight and complexity. Fortunately, more shippers are adapting to differing trailer heights by installing dock leveler systems, ramps, or reduced-height dock floors, making self-leveling trailers unnecessary.

A real space-consumer in the front of semitrailers is the upper coupler, an assembly about 4 inches thick that anchors the kingpin and forms the load-bearing surface of the fifth wheel. Consisting of top

and bottom plates and internal bracing, the upper coupler assembly needs a certain amount of vertical space for structural strength. But careful design—and use of high-strength steel—can bring its height down, gaining space inside the trailer. Another space-saving technique is leaving the upper plate of the coupler exposed. This means butting the wood floor to the rear of the coupler, which in turn necessitates the use of special crossmembers in the landing gear area. This modification alone can save a cubic foot of space, and is usually available with both standard and low-profile couplers.

A sturdy watertight roof is mandatory to ensure dry cargo. Some trailers are available with one-piece aluminum sheet roofs. They eliminate seams where short sections are riveted together—joints that are prone to leak after a couple of years. Roof bows that protrude down from the ceiling only $1/_2$-inch need to be 6 inches wide for strength; they can be made lighter if allowed to take more interior height. Finally, rear door frames can be made stronger with a deep header that cuts 4 or 6 inches of headroom.

If the van trailer is operated in severe winter conditions, additional roof bows are often added at the rear of the trailer where snow is most likely to collect. Often snow will slide off a building or terminal roof and land on the rear of the trailer when it is backed into a loading door so the roof must sustain this force.

Van Doors

There are basically two types of rear doors: hinged (usually called swingers) and roll-up design (**Figure 29–11A** and **Figure 29–11B**). Swingers come in single and double widths and can be as wide as the body to give a full-width rear door opening. These offer a bit more height and width than roll-ups and are less expensive and easier to maintain, as they have fewer moving parts.

Roll-ups are real time-savers when making deliveries as drivers do not first have to leave their cabs, open the swingers, then re-enter the cab and back into the dock. Often door clearance at loading/unloading areas is very limited. Roll-ups by their nature take up an inch or two of inside height. This may be vitally important with high-cube or stacked loads where every inch counts.

Liners

Van trailers and bodies usually come with some kind of an interior lining to protect both the body and cargo. Plywood frequently is standard as it is the cheapest. But it is also the least durable and doesn't

A

B

FIGURE 29–11 Van doors: (A) swinging doors; and (B) roll-up or overhead door. (Courtesy of American Trailer Industries Inc.)

hold up well to straying lift truck forks. An alternative to plywood are the pressed-wood-fiber products which are lighter, stiffer, and more resistant to water migration than plywood.

Although more expensive, aluminum, steel or plastic composite materials are options, as are scuff plates. An alternative to conventional scuff plates is the bonding of additional layers of fiberglass to an FRP trailer wall when the panel is manufactured. Up to three layers can be specified to any desired height, with resulting cost and weight less than that of conventional scuff liners.

Flooring

Van trailer and body floors are truly one of the critical areas that can lead to high maintenance expense. Floors are commonly hardwood, steel, or aluminum with thicknesses from $1^1/_4$–$1^1/_2$ inches. Some vans contain oak laminate flooring secured with three $^5/_{16}$-inch screws rather than the traditional two screws per board. Thus, it maintains a tighter floor over the years.

Crossmember spacing, especially at the rear of the trailer or body, can cause trouble and high expense from lift truck traffic. A popular procedure is for 6-inch spacing, rather than the more common 12 inches, in the last 4 feet of the trailer/body. With aluminum trailers, some fleets also opt for putting upright posts on 16-inch centers (rather than 24 inches) from the landing gear back.

Other Van-type Trailers

Freight vans are probably the most widely used trailer model in operation. In addition to high-cube and FRP designs, there are several others. **Figure 29–12** illustrates four of the more common designs:

A. **Open Top Van Trailer.** This trailer has no fixed roof, but can be easily covered with a nylon or canvas tarpaulin to allow overhead loading of such items as bar steel, wood chips, refuse, and so on. Cubic capacity is similar to a dry freight van, and when a waterproof tarp is used, vehicles can even double as a dry freight van.

B. **Curtain-Sided Freight Van Trailer.** These trailers feature the easy-loading convenience of flatbeds with all the weather-proof benefits of enclosed vans. The curtains allow for full-side access on both sides as well as the rear.

C. **Electronic/Furniture Van Trailers.** These trailers—originally designed for the electronic industry, but now popular with furniture movers—have a drop floor starting behind the trailer landing gear allowing for greater internal cubic capacity (approximately 3,000 cubic feet) when the maximum drop is used. They usually have one to three side doors, plus the one in the rear. Either swinging or sliding doors may be used.

D. **Doubles.** Using two 28-foot trailer doubles instead of one 48-foot van increases the cubic feet of cargo area by more than 16 percent.

FIGURE 29–12   Other types of cargo vans: (A) open top van; (B) curtain-sided van; (C) drop frame electronic/furniture van; and (D) double van. (Courtesy of American Trailer Industries Inc.)

29.2 REFRIGERATED AND INSULATED VANS

Modern refrigerated trailers or insulated vans are dry freight vans that are completely insulated and have a

FIGURE 29–13   Refrigerated van trailer with the reefer up front. (Courtesy of Carrier Transicold)

refrigerator unit, usually located at the front or nose of the trailer (**Figure 29–13**). These trailer units are called **reefers**. Depending on outside conditions, reefer units must cool and heat to keep cargoes at proper temperatures. To reduce expensive spoilage reefers must be completely reliable, be fuel efficient, and require little maintenance.

The majority of insulated van trailer bodies are constructed of aluminum posts and sheets with 2–4 inches of non-chlorofluorocarbon foam insulation on the inside, covered by an aluminum or fiberglass inner liner (**Figure 29–14**). This arrangement works well in most cases but "hot spots" can be created by the vertical wall posts spaced every 12–24 inches along the trailer's length. A large number of posts can compromise the insulation value of the trailer because metal is such a good heat conductor—creating a "thermal short-circuit, like a teaspoon in a teacup."

To overcome this problem, trailer manufacturers have tried several solutions. For instance, several have used fiberglass posts instead of aluminum. Since fiberglass conducts heat less readily, the 2-inch walls insulate as well as $2^1/_2$-inch walls with metal posts.

Most refrigerated trailers are equipped with floors made of extruded aluminum, although wood floors are available. Aluminum is easy to clean, is not affected by moisture, and can be designed to aid air circulation. Duct-type floors provide channels an inch or two deep for return air, but T-bar floors can flow substantially more air volume. The problem with T-bar floors is their vulnerability to forklift damage (**Figure 29–15**). There is another choice—a hybrid floor duct-T. This combination stands up to abuse better than just a T-bar floor and has more air flow than a duct floor. To help prevent forklift abuse, most trailer manufacturers offer heavier gauge extrusions or reinforced floors with extra aluminum inside the hollow portions. This is in addition to extra floor thickness that is standard in the rear of most trailers, where forklift trucks land off loading ramps. Under the floor is foam insulation, and most trailers have a

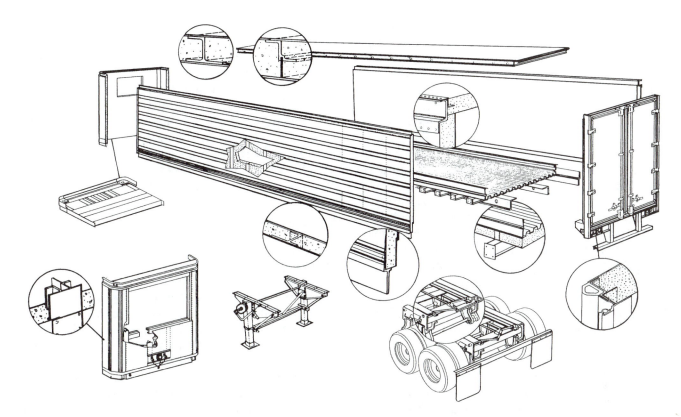

FIGURE 29–14   Construction features of a durable refrigerated van. (Courtesy of Trailmobile, Inc.)

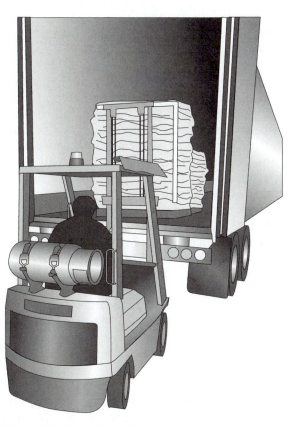

FIGURE 29–15   Forklifts are the number one enemy of trailer floor boards.

FIGURE 29–16   Cutaway view of a reefer unit. (Courtesy of Carrier Transicold)

protective subpan of fiberglass (the lightest choice), aluminum, or plywood. A quarter-inch plywood floor can add over 100 pounds to the trailer's weight.

Reefer Operation

The refrigerator unit functions on some of the same basic principles as the heavy-duty air conditioning system described in Chapter 31. The reefer unit (**Figure 29–16**) has a condenser/radiator and compressor, and is usually driven by its own diesel engine. The reefer unit draws heat out of the air in the cargo compartment. Cooled air is then circulated in the compartment to maintain cargo at the desired temperature. If the reefer must heat the load, the air flow cycle is reversed.

Multiple trailer compartments can be served by a single reefer. This is done by placing a separate evaporator in each compartment, and setting each to run at a different temperature (i.e., 40°F in one and 0°F in another).

The drive engine and refrigeration circuit components must be controlled so the unit can maintain the desired temperature. Many new reefer units contain a microprocessor-based temperature control system (**Figure 29–17**). Separate microprocessor input and logic modules can speed troubleshooting and servicing by providing fault codes and alarm displays.

The capacity of a reefer is its ability to cool and heat, given the volume of cargo it must deal with. Necessary capacity depends first on the type and size of the body or trailer, which in turn is built to haul certain types of products. If a trailer never hauls frozen foods, for example, it is built with less insulation; if deep-frozen products are hauled, the trailer will have more insulation.

Many trucking companies use trailers that will haul anything, so the reefer may be maintaining a temperature of 40°F one day and −20°F the next. Its range capacity, therefore, must be great. Increasing the trailer body size also increases the cargo volume that must be cooled or heated; so, the reefer must be of higher capacity.

Capacity is measured in the number of Btus the reefer can remove at certain desired interior temperatures while the outside (ambient) air is at 100°F. The three interior temperatures commonly required are 35°F, 0°F, and −20°F. Rated Btu capacity runs between 14,000 and 22,000 Btus.

A highway 53-foot high-cube reefer trailer with thin walls (and $1\frac{1}{2}$ inches of insulation) might require a 22,000-Btu reefer unit to maintain lower temperatures. A 45-foot local distribution trailer with $2\frac{1}{2}$–4 inches of sidewall insulation might need a reefer rated at 14,000 to 16,000 Btus.

Transport refrigeration units are built to standards set by the American Refrigeration Institute. Part of the capacity rating equipment is the unit's ability to "pull down" the temperature of the load. This requires efficiency and high capacity.

Reefer Refrigerants

The following refrigerant may be found in trailer reefer systems:

- **R-22.** Unlike CFC-based refrigerant R-12, R-500, and R-502, HCFC refrigerant R-22 is a more ozone-friendly compound. R-22 will be available for use in new equipment until at least 2005 and it will remain available for use in service and replacement until 2020. R-22 is easy to service and recycle, and is compatible with current compressors and oils. It has one of the lowest Global Warming Potential (GWP) ratings of any commercial refrigerant presently on the market. It is also considered a good transition alternative, but it does not do as good a job as high-CFC refrigerants in maintaining box temperatures below 0°F.

- **HFC-134a.** This is the first zero-Ozone Depletion Potential (ODP) refrigerant available in product quantities. It was developed to replace R-12 and is intended for medium and low-capacity application. It does not produce sufficient capacity for frozen cargoes in typical truck and trailer applications. However, HFC-134a does meet the capacity requirements of the refrigerated container industry. It is expected that use of HFC refrigerant R-134a in truck and trailer refrigeration will be limited to chilling applications, or as a component in blended refrigerants.

- **HP-81.** This is a blended refrigerant comprising R-22, R-125, and propane; therefore, it is included in the same phaseout schedule as HCFC refrigerant R-22. Unlike other blends, it exhibits a slightly higher capacity than CFC refrigerant R-502 and has a good performance coefficient.

- **HP-82.** Like HP-81, this is also a blend of refrigerants R-22, R-125, and propane, although in different proportions. Further application testing is necessary before it can be determined whether HP-81 or HP-80 are acceptable alternatives.

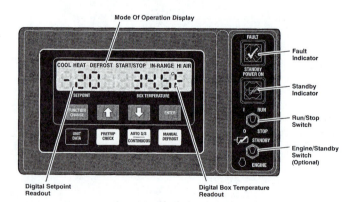

FIGURE 29–17 *Microprocessor control unit of modern trailer refrigeration system. (Courtesy of Carrier Transicold)*

- **HP-62.** HP-62 is a zero-ODP, all-HFC blended refrigerant. It exhibits comparable performance characteristics to R-502 at slightly higher pressures. It also has a higher GWP than R-22. Although full developmental testing on HP-62 is not yet complete, HP-62 may be a potential candidate for long-term replacement of R-502.
- **Propane.** Used extensively in Mexico. Technicians should be aware of explosion hazards when working with propane.

Cryogenic Systems. An alternative to R-refrigerants is cryogenic systems. By using cryogenics to cool air, the system eliminates the use of chlorofluorocarbons and the need for a diesel engine. It usually consists of a stainless steel, double-walled, vacuum-insulated cryogen tank, and a small propane or alcohol tank mounted under the trailer. A heat exchanger, evaporator unit, and various control devices are mounted on the front end of the trailer just as they are in a normal reefer unit. Either liquid nitrogen or liquid carbon dioxide is stored in the tank. The cryogenic liquid passes through an evaporator, turns into a gas, enters a heat exchange coil, and drives a pair of "cryo-motors." The motors power blower fans and drive an alternator to charge the system's battery. After a second pass through the heat exchanger, the gas is vented to the atmosphere. A propane burner supplies energy to the heat exchanger, raising the cryogen's temperature to 800°F.

Cryogenics is not a new technology—systems were built and put into operation in the early 1970s. The system is much simpler than conventional reefer units, weighs less, and has a lower operating cost over the life of the unit. However, the idea has not taken off largely because until now there has not been a reason to change the current systems.

29.3 OTHER TRAILER TYPES AND DESIGNS

In addition to vans, there are several trailer types and designs. One of the most common types is the platform or flat-bed trailer. Available in lengths of 42, 45, 48, 50, and 53 feet, these trailers are used to transport cargo that does not need protection from the elements or is physically too big to fit into a van-type trailer.

There are two basic designs: the straight platform (**Figure 29–18A**) and the lowbed trailer (**Figure 29–18B**). The platforms are designed in all-steel, aluminum, and steel/aluminum, and are constructed as shown in **Figure 29–19**. Lowbeds are manufactured

FIGURE 29–18 (A) Typical straight platform and (B) lowbed trailers. (Courtesy of Dorsey Trailers, Inc.)

in high-tensile and lightweight steel and are capable of carrying much heavier loads (up to 100 tons) than the platform design. The number of axles and tires employed with each design depends on the rated load it is intended to carry.

Dump Trailers

When it is necessary to move large amounts of soil, sand, gravel, stones, or even hot slag from steel mills, the dump trailers shown in **Figure 29–20** can be used. They can be built for dependability in either steel or aluminum in a size to permit maximum payloads within state, provincial, and federal laws. The popular designs are:

A. *Frame-type bathtub dumps* are a rugged design in lengths generally from 21–40 feet. They are engineered for heavy-duty operations.
B. *Slant front bathtub dumps* are available in frame-type and frameless design. They feature an exposed hoist for easy maintenance.
C. *Rock dumps* are engineered to handle the toughest hauling jobs. The illustrated dump trailer is a heavy-duty 100,000 psi yield steel and is available in two sizes: 27-foot, 20.8 cubic yard capacity, and 30-foot, 23 cubic yard capacity.
D. *Bottom hopper dump* semitrailers are generally engineered to haul up to 20 cubic yards of sand, dirt, or aggregates.
E. *Bottom hopper doubles* are usually available in steel or aluminum/steel. Doubles will discharge their loads in 3 to 4 seconds.

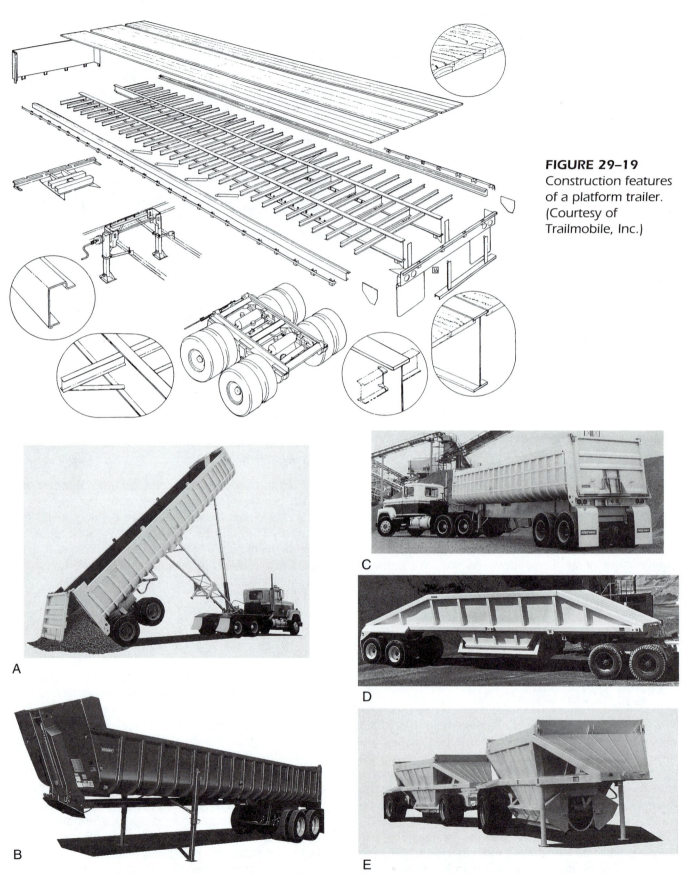

FIGURE 29–19
Construction features
of a platform trailer.
(Courtesy of
Trailmobile, Inc.)

A

B

C

D

E

FIGURE 29–20   Typical dump trailers: (A) frame-type bathtub; (B) slant front bathtub; (C) rock dump;
(D) bottom hopper; and (E) bottom hopper doubles. (Courtesy of American Trailer Industries Inc.)

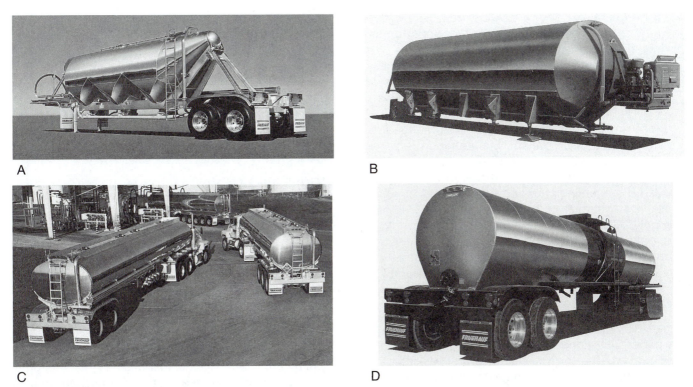

A

B

C

D

FIGURE 29–21 Typical tank trailers: (A) dry bulk; (B) portable bulk storage; (C) petroleum/chemical tank; and (D) hot materials tank. (Courtesy of American Trailer Industries Inc.)

Tanker Trailers

There are two basic designs of **tanker** trailers; liquid (wet) haulers and dry-bulk haulers. They are usually either cylindrical or elliptical in shape. Manufactured in aluminum, steel, and stainless steel, they have various types of internal coatings suitable for the several types of liquid. **Figure 29–21** illustrates four tanker designs that are in common use by the trucking industry:

A. *Hopper pressure dry-bulk trailers* are designed to handle cement as well as a wide variety of dry bulk commodities.
B. *Portable bulk storage units* permit flexible storage and job site mobility. Using the pneumatic discharge system, they keep the contractor's operation running around the clock.
C. *Petroleum/chemical liquid* tankers can haul loads of various petroleum products as well as many chemical items such as acids, milk, vegetable oils, liquid oxygen, butane, and hydrogen. Petroleum tank/trailers must be engineered in full compliance with DOT MC-306-AL specifications.
D. *Hot materials tank trailers* are ideal for hauling asphalt. The insulated steel tanks are designed to easily handle hot liquid materials up to 500°F.

There are several other trailer types designed for specific uses such as log/pole trailers, refuse haulers, livestock trailers, chip haulers, grain and fruit trailers, and auto transporters.

29.4 TRAILER UNDERSIDES

The effectiveness of most undertrailer components depends upon three factors: the proper selection of components, efficient preventive maintenance, and quality repairs. The latter two are the responsibility of the technician.

Tires and Wheels

Trailer tires often have a tread design commonly used for tractor steer axle tires, or they can be trailer-specific. Tire experts claim several benefits for trailer-specific tires.

1. Good mileage and gradual, even wear resulting from the three-groove design and shallow tread depth.
2. Fuel savings are a benefit of the shallower tread, lower rolling resistance and lower weight compared to standard rib tires.

3. Federal excise tax savings are possible due to lighter weight—up to 17 pounds lighter than comparable trailer radials.
4. Excellent retreadability: casings can usually be retreaded and mated with other tires of the same size for either drive or trailer axles.

Most tire designs are generally interchangeable between tractors or trailers. Maintenance requirements are also identical. As described in Chapter 23, the wheel lines of the leading manufacturers are divided by mounting type and tire (tubeless or tube-type), not by trailer or tractor application. Perhaps the only difference is the degree of acceptance of the new technology. For instance, hub-piloted mounts, which have been popular on tractors for several years, are just now being used on trailers.

Brakes

S-cam $16\frac{1}{2}$-inch $\times$ 7-inch brakes are fairly common on trailers. Options may include wedge brakes or disc brakes. Tractor/trailer brakes are basically the same within size, brand, and gross axle weight rating (GAWR). Brake shoes and linings, springs, anchor pins and bushings, rollers and retainers, camshaft hardware, and slack adjusters are similar on both tractors and trailers. The service requirements are also much the same for all brakes. Air brakes are studied in Chapters 24 and 27. ABS is now required on trailers and is discussed in Chapter 26.

 Shop Talk _____

For optimum brake performance and safety, both wheel ends of each axle must have the same type of lining and drum equipment. If the trailer has tandem axles, both axles must also have the same type of wheel equipment. The vehicle's brake-lining thickness must be the same on each brake shoe and on each side of the axle. When brake linings are $\frac{1}{4}$-inch in thickness or less at their thinnest point, they should be replaced.

FIGURE 29–22 illustrates multi-trailer axle and tire arrangement.

Suspension

The suspension system of a trailer depends on the application of the vehicle—type of cargo to be hauled, highest expected gross weight, varying weights carried, turning frequency, degree of stability required, and use (i.e., highway, city, or severe service). A suspension that is too light may bottom out

FIGURE 29–22 *Multi-trailer axle and tire arrangement. (Courtesy of Dorsey Trailers, Inc.)*

and result in impact damage. On the other hand, a rigid suspension can cause fatigue damage in addition to a rough ride for the driver and cargo.

Government regulations dictate many of the design parameters of trailer suspensions—axle load, axle spacing and the number of axles in a group.

In recent years, air-ride suspensions have become more common than steel leaf-spring suspensions in trailers. Other choices—such as solid rubber mount, walking-beam, and single-point suspensions—usually have relatively narrow applications.

Another improvement in trailer suspensions are slide tandems. They increase flexibility, especially in maneuverability and meeting weight laws. Look carefully at the manufacturer's locking pin design. Some are more prone to damage if they do not retract fully when the tandem is released. There should be four locking pins. In addition, look for full-length slide pads on a slider subframe, too.

It is important to remember that there are differences between tractor and trailer suspension system designs. They are not interchangeable. The method of attaching trailer axles differs from that used for any shape of axle on a tractor. There are frame differences, and while a trailer's suspension can handle the same braking forces as a tractor's, it cannot handle engine acceleration force.

 Shop Talk _____

One of the weight-savers now being used by some trailer manufacturers is fiber composite springs (see Chapter 22). They save about 40 pounds per spring, which can amount to 160 pounds per trailer. Composite springs are popular for another reason besides weight saving. While they cost about four times as much as steel, they can last the life of the trailer rather than needing two or three replacements, as with conventional springs.

FIGURE 29–23 *Typical landing gear.*

Landing Gear

There is a wide variety of landing gear found on today's trailers, including telescoping, folding, round leg, square leg, and hydraulic. In addition, landing gear can be mounted inboard or outboard on a trailer (**Figure 29–23**).

Trailer manufacturers and fleet maintenance managers agree that the majority of landing gear damage is caused by driver abuse. Rolling the legs up too far can strip threads in the cranking mechanism. Hitting or bending the legs can knock them out of alignment so they will not crank up or down smoothly. Jamming the high/low range cranking gear can fracture gear teeth.

General maintenance of the landing gear requires lubricating both legs through the grease fittings in the legs at least every six months. At the same time, lubricate the two-speed gears through grease fittings in the gearbox. Other landing gear service repairs are given in the trailer's service manual.

Lighting

Trailer lighting regulations, like those of the truck tractor requirements, are set forth in Federal Motor Vehicle Safety Standard No. 108 (**Figure 29–24**). This standard stipulates where trailer lighting is required.

The main concerns for technicians in servicing the electrical system are the repair or replacement of defective wiring, light housings, and bulbs. Most modern trailers use a no-splice sealed wiring system (**Figure 29–25**) which greatly reduces the chance of defective wiring.

Most trailer wiring is made of multiple strand copper wire covered with polyvinylchloride (PVC) insulation. The best sealed wiring includes protection where the harness connects to the light fixtures.

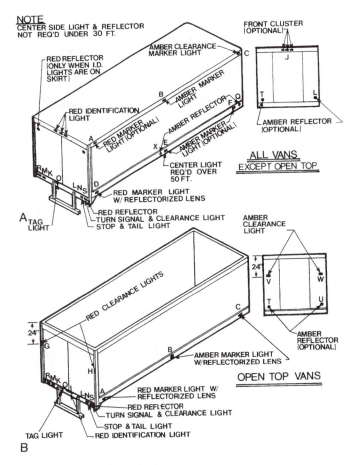

FIGURE 29–24 *Light diagrams for vans: (A) All vans except open top; and (B) open top vans.*

Often this will be some sort of boot-like rubber or plastic. At the very least, it should include heat-shrunk coverings and *never* just plastic electrical tape.

When replacing defective wiring remember the factors which determine wire size are the ampere load and distance to the most distant lamp. Follow the lines or bars to arrive at the proper size of wire for this application. For a 12-volt electrical system, a 60-foot distance combined with an 8-amp load would require a 12-gauge wire size. If the distance were 80 or more feet the technician should use 10-gauge wire. The Maintenance Council (TMC) of the American Trucking Associations (ATA) recommends electrical wire in trailers should never be less than 18-gauge.

In recent years the plug-in type of connector has become more popular because it simplifies replacement of components. Trailer harnesses using plug-in connections are generally "modular," meaning sections of wiring come in set lengths and couple with other sections and lights to make up the complete system. "Male-pin" connectors with two or three pins

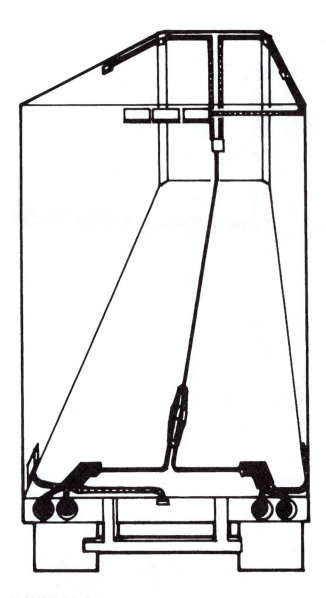

FIGURE 29–25 *In a wiring system such as this, there are no splices for corrosion to attack, and the junction box (a problem spot) has been eliminated. ATA color-coded wires are molded in an abrasion-resistant jacket to form a tough, waterproof cable. Light plugs molded on the cable for positive seal are filled with dielectric grease and have spherical sealing rings to assure positive connections.*

are standard in lighting connectors. Sealed systems use versions of these connectors.

Trailer lamp housings should be made of materials resistant to moisture and salt and sealed against contaminants. To lengthen the life of light bulbs, the fixture should be of a shock-resistant design. Clearance and marker lamps may require "armored" metal brackets surrounding the lens to protect from impacts.

In a basic type of fixture, the housing can be formed of plastic, aluminum, or painted steel. Plastic, nylon or other non-conducting materials are used to insulate the base of the bulb from the housing.

Obviously, a ferrous metal housing will be more likely to rust than the non-ferrous. In addition, the nylon or plastic materials will not corrode, but may be more susceptible to impact damage.

Lenses are usually made of plastic in appropriate colors. But polycarbonate materials should be considered where impacts may crack or break the lens. Even a slight crack breaches the lamp's sealing capability and allows moisture in. This can only lead to corrosion, shorting, and expensive failures.

While sealed beam incandescent bulbs are used for most trailer tail-light systems, the latest type of lamp is the light-emitting diode. This type of lamp uses LED elements instead of incandescent bulbs. As previously mentioned in Chapter 7, LED elements have a greater life expectancy and can be used in clearance, marker, and tail/stop/signal lights.

Rear turn signals should be on a circuit separate from stop lamps. Amber lenses are preferred. Study the tail-light assembly and how it is mounted. Many fleets prefer a simple bolt-on design because they are easily repaired or replaced.

Reflective Taping

The National Highway Safety Administration regulations now require that red and white alternating striped reflective tape, also known as conspicuity tape, must be applied to the full trailer width at the rear of the trailer, near the van base (**Figure 29–26**). The tape height should be as close to 4 feet as possible. But trailer builders have leeway, especially with specialty vehicles like auto transporters. At the rear, white tape should outline the vehicle's extremities, that is, corners or curves.

The regulation applies to all trailers with a gross weight of over 10,000 pounds and a width of over 80 inches. Exempt are pole trailers and converter dollies. Truck tractors or bodies are not covered by the regulation because studies have shown that nearly all collisions, especially of the rear-end variety, involve semi-trailers rather than straight trucks. Headlight glare from approaching vehicles causes reflective tape to glow brightly at oblique as well as right angles. Therefore, it makes the vehicle conspicuous and less likely to be hit by an inattentive driver.

Reflective tape needs only to be applied to a clean surface to adhere. Once installed, all it needs is occasional washing or wiping. The tape uses headlight glare so it needs no extra electrical power or wiring.

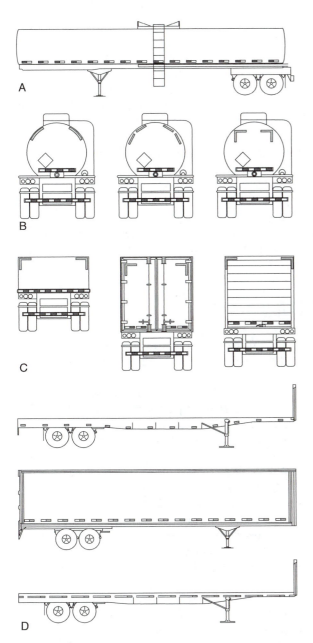

FIGURE 29-26 *The proper location of reflective tape on various trailers: (A) tanker curbside; (B) three alternate rear views; (C) rearview (left to right), platform, swing doors, and roll-up door; (D) sideview layout.*

29.5 TRAILER MAINTENANCE

The maintenance of the various system components—brakes, wheel bearings, tires, suspension, and lighting systems—are discussed in earlier chapters of the book and PM programs are described in

Chapter 4. However, the following trailer-specific maintenance should be performed.

Roof Maintenance

Periodic inspection of the roof will reveal any defects early. Damaged areas should be cleaned and resealed with liquid neoprene or caulking for prevention of roof leaks.

Doors

Door seals should be repaired or replaced with new doors. A high-quality bonding cement can temporarily repair leakage.

Aluminum Panels

To ensure long life and appearance of aluminum panels, a regular cleaning program should be established. The trailer should be washed down with clear water to flush off surface dirt and to bring the metal temperature down to reduce streaking. An acceptable cleaning solution should be applied, either by hand or by brush. This solution should be soluble in hot or cold water, be free-rinsing, and leave no harmful deposits. In both applications, a final complete rinse is required.

Power washing is often necessary when the equipment is infrequently washed, or is subjected to excessive deposits of mud, road tar, oil, etc. Improper power wash solutions may result in permanent metal discoloration and increase its susceptibility to corrosive elements. Also, before power washing, check with local regulations to be sure that it is permissible. For maximum protection, aluminum panels should be waxed to protect the finish, enabling dirt and spot corrosion to be removed more readily. The most economical application requires spraying the surface with a light wax designed for this type of application.

Corrosion Protection

Corrosion is not only a concern for truck and tractor manufacturers, but a concern for trailer OEMs as well. It is primarily caused by environmental elements reacting with metals. Therefore, a good paint job that seals out the environment is the number one defense against corrosion. By far the most important aspect of painting is surface preparation. Paint can not adhere to a surface that is already oxidizing or has an oily film. In some fleet PM shops, shot blasting (like sand blasting, but using steel shot) prepares the surface for paint by completely eliminating any surface oxidation and scale.

The underside of a platform trailer is frequently coated with a proprietary industrial coating often referred to as rust preventive coating (RPC). This coating never completely hardens and is even "self healing" for minor puncture intrusions.

Galvanic corrosion is another problem encountered when dissimilar metals are in contact within the environment. Manufacturers usually avoid that problem by placing vinyl tape between dissimilar metals—like steel crossmembers and aluminum bottom rails on vans and reefers.

Regardless of how much effort the OEMs put into preventing corrosion, the service practices of the fleet also play a role. For instance, corrosion in electrical systems is a serious and common problem. For this to take place, water, primarily road-salt water, must intrude into the wiring system. Sealed wiring harnesses have provided the biggest improvement. There are technicians who continually "spike" harnesses to check for electrical circuit continuity. The hole the spike leaves in the insulation allows moisture to invade a wiring system. In laboratory testing, salt has traveled within a wire in excess of 10 feet because of "wicking."

Fasteners

Use of the frameless stress panel produces a high strength-to-weight ratio. This means that each van sidewall is in itself carrying the load. The size of this section is of vital importance and should be strictly adhered to so maximum body strength can be maintained.

A thorough periodic inspection should reveal any loose, broken, or missing fasteners. All assembly screws and nuts should be tightened and replaced as needed. Rivets should be checked and repaired or replaced as needed. During inspection, a loose rivet, when tapped with a light hammer, will give off a higher pitch (sound) than one that is tight.

The patching of panels should be restricted as much as possible (both in size and number) before complete replacement of damaged panel is undertaken.

Lubrication

Frequent, scheduled lubrication is as important to properly maintaining a trailer as changing the oil is to maintaining a tractor engine. It is best to use the lubricant recommended by the manufacturer. Using the correct amount eliminates the risk of damage from under-lubrication.

For grease-lubricated wheel bearings, a thin layer of grease should be applied to the bearing journals to prevent corrosion. Grease should be forced into the bearing's cavities between the rollers and the cage from the large end of the cone, either by hand packing or by using a pressure packer.

For oil-lubricated wheel bearings, gear oil should be applied to the level indicated on the hub cap. The correct oil level is essential to dynamic bearing lubrication of the wheel-end components. It is a function of hub or wheel design and can differ from model to model. If the fill level is not correct, instead of running in a bath of lube, the bearings may run in a mist, which could contribute to premature bearing failure.

Automatic slack adjusters (ASAs) require a NLGI Number 1 grease or its equivalent, to operate effectively under all conditions and temperatures. ASAs should be lubricated until new grease flows from the pressure relief valve in the pawl capscrew. If the ASA does not have a pressure relief fitting, the pawl assembly must be removed to allow the grease to flow properly.

SUMMARY

- A semi-trailer is any trailer in which a portion of the trailer weight is supported by the tow unit—either a tractor or a lead trailer.
- A full trailer is any trailer in which the trailer weight is self-supported and the unit is towed by a tractor or lead trailer.
- Any semi-trailer can be converted to a full trailer by a converter dolly.
- A train is a combination of a tractor and multiple trailers.
- An A-train double consists of a tractor, semi-lead trailer, and full pup trailer.
- A B-train double consists of a tractor, semi-lead trailer, and semi-pup trailer.
- A C-train double uses a rigid double drawbar and converter dolly to attach the pup trailer.
- Many van trailers use a unibody or monocoque frame.
- A unibody frame is essentially a shell in which all of the shell components play a role in the frame dynamic, including the floor, side panels, bulkhead, doors, and roof.
- The van trailer is the most common truck trailer on North American roads.
- Light weight and good aerodynamics are always sought. Structural integrity and overall durability are still the two most important features.
- There are two ways to obtain light weight without sacrificing strength. One is by using lighter, more expensive materials such as aluminum and high-strength steel in posts and crossmembers. The other is to use wall posts and floor crossmembers that have a deeper section.
- A 1-inch increase in the width of a 53-foot trailer often makes the difference between getting two pallets side by side or not.

- Freight vans are probably the most widely used trailer model in operation.
- Modern refrigerated trailers or insulated vans are dry freight vans that are completely insulated and have a "reefer" (refrigerator) unit usually located at the front or nose of the trailer.
- In addition to vans, there are several trailer types and designs. One of the most common types is the platform or flat-bed trailer.
- The number of axles and tires used with each trailer design depends on the rated load it is intended to carry.
- There are two basic designs of tank trailers: liquid (wet) haulers and dry-bulk haulers.
- Trailer manufacturers and fleet maintenance managers agree that the majority of landing gear damage results from driver abuse.
- The effectiveness of most undertrailer components depends on three factors: the proper selection of components, efficient preventive maintenance, and quality repairs.
- Most wheel and tire designs are generally interchangeable between tractors and trailers.
- An improvement in trailer suspensions are the slide tandems; they increase flexibility, maneuverability, and ability to meet weight laws.
- The maintenance of the various system components will make it possible to obtain maximum life expectancy for the trailer. It is of equal importance that the trailer receive attention in a preventive maintenance program.

REVIEW QUESTIONS

1. What component of van design is a time-saver for making deliveries?
   a. roll-ups
   b. scuff plates
   c. aluminum liners
   d. steel flooring

2. Cargo such as wood chips would be best carried in what kind of a van?
   a. dry freight
   b. reefer
   c. curtain-sided
   d. open top

3. Which of the following materials is NOT a good choice for a reefer interior?
   a. fiberglass
   b. polystyrene
   c. aluminum
   d. rigid foam core

4. The ability to keep cargo cool is a reefer's
   a. efficiency rating
   b. ambient temperature
   c. Btu capacity
   d. refrigeration rating

5. The major difference between the straight platform and lowbed trailers is
   a. the degree of protection from the weather
   b. the height of the load that can be carried
   c. fuel economy
   d. number of axles and tires

6. The two basic designs of tank trailers are
   a. liquid (wet) haulers and dry-bulk haulers
   b. stainless steel and aluminum
   c. cylindrical and elliptical
   d. hopper pressure and portable bulk storage

7. What is the difference between trailer-specific tires and tractor tires?
   a. Trailer tires have a shallower tread.
   b. Trailer tires are lighter in weight.
   c. Trailer tires cannot be retreaded.
   d. both a and b

8. Which of the following is NOT an undercarriage component of a trailer?
   a. tires
   b. refrigerator unit
   c. suspension
   d. landing gear

9. Protection against corrosion in electrical systems is provided by
   a. aluminum junction boxes
   b. spiking wires
   c. modular plug-in connectors
   d. ferrous metal fixture housings

10. Which design is the most widely used trailer model?
    a. freight van
    b. flat bed trailer
    c. dump trailer
    d. tank trailer

11. Most landing gear damage comes from
    a. poor maintenance practices
    b. driver abuse
    c. inadequate design
    d. defective parts

12. The effectiveness of most undertrailer components depends on all BUT which of the following factors?
    a. proper selection
    b. timely and efficient preventive maintenance
    c. quality repairs
    d. efficient loading and unloading practices

13. Trailer maintenance involves which of the following tasks?
    a. frequent, scheduled lubrication
    b. tightening or replacing loose, broken, or missing fasteners
    c. painting or coating surfaces that might corrode
    d. all of the above

14. Which of the following is the usual length limit for a tractor and trailer combination?
    a. 65 feet
    b. 53 feet
    c. 48 feet
    d. 28 feet

15. Which of the following must be used to couple a semi-trailer to a tractor?
    a. converter dolly
    b. fifth wheel
    c. pintle hook
    d. drawbar

16. Which of the following devices will convert a semi-trailer into a full trailer?
    a. lift axle
    b. fifth wheel
    c. pintle hook
    d. converter dolly

17. Which of the following train configurations would be made up of 2 semi-trailers?
    a. any double trailer configuration
    b. A-train
    c. B-train
    d. C-train

18. Which of the following train configurations would have to be equipped with a pintle hook?
    a. any double configuration
    b. A-train
    c. B-train
    d. C-train

19. Which of the following terms would best describe the frame used on a typical van-type trailer?
    a. full frame
    b. ladder frame
    c. truss frame
    d. unibody frame

20. Which type of refrigerant is most commonly used in trailer reefer systems?
    a. R-12
    b. R-22
    c. HFC-134a
    d. HP-82

chapter 30

Fifth Wheels and Coupling Systems

🔑 **Prerequisite**
Chapter 28

Objectives

After reading this chapter, you should be able to

- Describe the various types of fifth wheels.
- Outline the operating principles of the Holland, Fontaine, and ASF fifth wheels.
- Understand the importance of correctly locating the fifth wheel on the tractor.
- Describe the locking principles of each type of fifth wheel.
- Outline the procedure required to couple and uncouple a fifth wheel.
- Service the common types of fifth wheels.
- Describe the procedure required to overhaul a fifth wheel.
- Define high hitch and outline what is required to avoid it.
- Describe the operating principle of a pintle hook and draw bar.
- Outline the function of the kingpin and upper coupler assembly.

Key Terms List

bolster plates
compensating fifth wheel
converter dolly
fully oscillating fifth wheels

kingpin
secondary locks
semi-oscillating fifth wheel
sliding fifth wheel
throat
upper coupler

The fifth wheel is a coupling device. Its purpose is to connect a tractive unit to a towed unit. The tractive unit is normally a tractor, but in the case of multiple trailer train, a fifth wheel also can be found on a lead trailer. The fifth wheel permits articulation between the tractive and the towed units. This articulation, or pivoting, is essential to properly steer a tractor and semi-trailer or tractor and multiple trailer combination on the highway. The fifth wheel must also support some of the weight of a semi-trailer. A full trailer is one in which the entire weight of the trailer is supported on its own axles. Most highway trailers are classified as semi-trailers because the weight on the front portion of the trailer is supported by the tractor; in cases of multiple trailer combinations, the weight is supported by the lead trailer. A full description of tractor/trailer combinations and truck trains is provided in the preceding chapter on trailers.

A fifth wheel consists of a wheel-shaped deck plate usually designed to tilt or oscillate on mounting pins. The assembly is bolted to the frame of the tractive unit. A sector is cut away in the fifth wheel plate (sometimes called a **throat**) to permit a trailer **kingpin** to engage with locking jaws in the center of the fifth wheel. The trailer kingpin is mounted in the trailer **upper coupler** assembly. The kingpin is a hardened steel cylindrical flanged stub, and it protrudes downward from the upper coupler plate. As a tractor is backed into the trailer, the trailer upper coupler first contacts the fifth wheel plate and then the kingpin slides through a sectored recess, the throat, until it contacts the jaw mechanism. The fifth wheel jaw assembly is spring loaded to enclose the kingpin. Most fifth wheels have primary and **secondary locks** to maximize the coupling safety. The flange at the bottom of the kingpin stub prevents a vertical separation from occurring.

Another method of attaching a trailer to either a truck or lead trailer is by using a pintle hook. A drawbar eye engages to the pintle hook assembly. The drawbar extends from the **converter dolly;** in all cases where this coupling method is used, the unit being towed would be classified as a full trailer. A disadvantage of pintle hook and drawbar coupling is that the number of pivot points is increased. A full trailer coupled to a truck by a pintle hook would be capable of articulating both at the pintle hook and at the converter dolly.

 **Shop Talk**

Most truck and trailer coupling mechanisms are simple in design and operating principles. However, because of the extreme consequences of a trailer breakaway, care must be practiced when doing any work on coupling devices. Always consult OEM service literature and ask questions when you are unsure of a procedure.

30.1 TYPES OF FIFTH WHEELS

Most fifth wheels in use in North America are manufactured by one of three companies—Holland, Fontaine, or American Steel Foundries (ASF). Of these, the Holland company has by far the largest market share. Each manufacturer uses distinct locking mechanisms, and some manufacturers use more than one type of lock. The technician should fully understand exactly how each type of fifth wheel lock operates before attempting service and overhaul procedures. Although many different types of special application fifth wheels exist, by far the most common type is the **semi-oscillating fifth wheel.**

SEMI-OSCILLATING

This type of fifth wheel articulates about an axis perpendicular to the vehicle centerline. That is, it rocks

FIGURE 30–1 *Semi-oscillating fifth wheel.* (Courtesy of Holland Hitch Co.)

FIGURE 30–2 Plate mount for a semi-oscillating fifth wheel. (Courtesy of Holland Hitch Co.)

fore and aft. **Figure 30–1** shows a typical semi-oscillating fifth wheel. The fifth wheel plate is either a cast steel or pressed steel plate. The fifth wheel plate has a pair of mounting bosses on either side. The plate is mounted on a pair of saddle brackets fitted on a base plate or frame brackets. Pins are inserted through bosses on the fifth wheel plate and the saddle brackets. These pins are bushed and permit the required articulation. The base plate is either rigidly fitted to brackets that in turn are bolted to the frame rails, as shown in **Figure 30–2**, or to a slide plate that permits fore and aft positioning of the fifth wheel. Most fifth wheels are fitted to the tractor in stationary mounting position. Stationary-mounted fifth wheels are used where the axle loading, trailer kingpin location, and vehicle combination are consistent throughout the fleet for which the tractor is specified.

SLIDING FIFTH WHEEL

A **sliding fifth wheel** assembly in most cases uses a semi-oscillating fifth wheel that can slide fore and aft on the tractor frame. This changes the point at which

FIGURE 30–3   A sliding fifth wheel. (Courtesy of Holland Hitch Co.)

the tractor is loaded with the trailer weight. Sliding fifth wheels provide the tractor with greater flexibility in terms of the types of trailers it hauls. They tend to be fitted to tractors where resale value is critical for this reason. Sliding fifth wheels permit relocation of the point at which the tractor axles are loaded with the weight of the trailer, accommodate trailers with different kingpin settings, and enable the vehicle combination lengths to be varied. **Figure 30–3** shows a sliding fifth wheel.

Sliding fifth wheels must be locked into a set position before they are used to haul a trailer. The fifth wheel base plate mechanically locks to toothed rails on the slide plate. Two methods are used to release the slide locks. A manual release mechanism is used in applications that do not often have to be adjusted. An air release mechanism is used in applications that require frequent adjustment.

A major reason for selecting sliding fifth wheels is driver comfort. Most tractors ride more smoothly when the fifth wheel is located closer to the centerline of the bogie or rear axle. When the axles are not overloaded, the driver can extend the fifth wheel for maximum comfort.

COMPENSATING AND FULLY OSCILLATING

A **compensating fifth wheel** is designed to articulate on two planes. That is, it pivots fore and aft and side-to-side. This type of fifth wheel is used in applications where trailer torque (twist) is critical. In trailers with high rigidity, such as many tankers, trailer torque can be transmitted to a tractor or lead trailer frame and cause handling and premature failure problems. Compensating fifth wheels have made possible the use of aluminum alloy as a frame material in tanker B-trains. Fore and aft pivoting occurs in the same manner as in the common semi-oscillating fifth wheel. Side-to-side pivoting is obtained by hav-

ing trunnion shoes that ride in concave tracks. The Holland Kompensator fifth wheel is used in applications in which the trailer center of gravity is located above the fifth wheel top plate. **Figure 30–4** shows the most common type of compensating fifth wheel, and **Figure 30–5** shows its operating principle. Compensating fifth wheels may have an optional lockout that eliminates side-to-side oscillation when engaged. **Figure 30–6** shows a view of the cradle that enables side-to-side oscillation and the lockout used on a Holland Kompensator fifth wheel.

Fully oscillating fifth wheels are used in applications in which the trailer's center of gravity is located below the fifth wheel top plate, such as in a gooseneck, low-bed trailer. Full oscillation means articulation on all planes, such as would be found in a ball joint and socket.

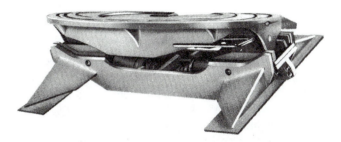

FIGURE 30–4   Holland Kompensator fifth wheel. (Courtesy of Holland Hitch Co.)

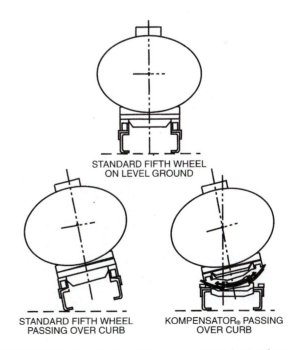

STANDARD FIFTH WHEEL ON LEVEL GROUND

STANDARD FIFTH WHEEL PASSING OVER CURB

KOMPENSATOR® PASSING OVER CURB

FIGURE 30–5   Holland Kompensator operating principle. (Courtesy of Holland Hitch Co.)

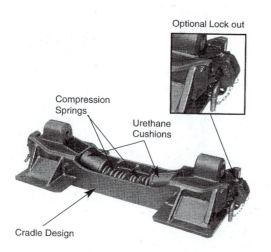

FIGURE 30–6   Holland Kompensator cradle view and lockout mechanism. (Courtesy of Holland Hitch Co.)

SPECIALTY FIFTH WHEELS

A number of special applications exist for fifth wheels. In most cases, the locking mechanics of specialty fifth wheels are no different. A popular specialty fifth wheel is the elevating fifth wheel. Elevating fifth wheels are used on yard shunt tractors to enable trailer hook-ups and movement without retracting the landing gear. They also can be used for unloading operations that require the trailer to be tilted. The fifth wheel is first engaged to the upper coupler and then elevated so that the trailer dolly legs clear ground level by a safe margin. Elevation is achieved either hydraulically or pneumatically. **Figure 30–7** shows a typical air-actuated elevating fifth wheel, and **Figure 30–8** shows a hydraulically actuated version.

Rigid fifth wheels are used in some vocational applications. A rigid fifth wheel is used in some

FIGURE 30–8   Hydraulically actuated elevating fifth wheel. (Courtesy of Holland Hitch Co.)

frameless dump trailers equipped with oscillating **bolster plates**. This results in the articulation taking place at the trailer rather than the fifth wheel. **Figure 30–9** shows a typical rigid fifth wheel.

Turntable or stabilized fifth wheels are used in some converter dolly assemblies. A turntable fifth wheel allows the top plate to rotate with the trailer bolster plate and to use a four-point support on the fifth wheel deck. **Figure 30–10** shows a typical turntable fifth wheel.

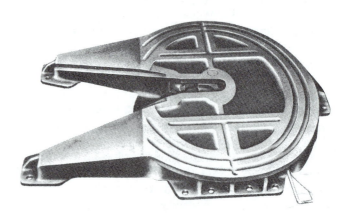

FIGURE 30–9   Rigid-type fifth wheel. (Courtesy of Holland Hitch Co.)

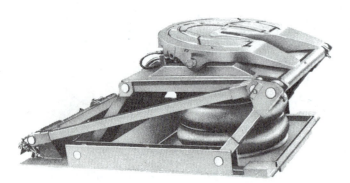

FIGURE 30–7   Air-actuated elevating fifth wheel. (Courtesy of Holland Hitch Co.)

FIGURE 30–10   Turntable-type fifth wheel. (Courtesy of Holland Hitch Co.)

30.2 FIFTH WHEEL HEIGHT AND LOCATION

Most trailers are designed with an intended 47-inch upper coupler (kingpin) plate height. This, plus the fact that any trailers operated on North American roads cannot exceed a height of 13 feet 6 inches, determines the plate height of fifth wheels. When it comes to locating the fifth wheel on the tractor frame, more options exist. These options have to be considered because they will determine how the axles of the tractor are loaded with the weight of the trailer.

The farther forward a fifth wheel is positioned on the tractor frame, the higher the percentage of weight on the tractor front axle. As the fifth wheel is moved rearward, weight is shifted off the front axle and onto the rear tractor axles. On a single-drive axle tractor, the center of the fifth wheel must be located ahead of the axle centerline. On tandem-drive axle tractors, the center of the fifth wheel must be located ahead of the bogie centerline. This is an important consideration when installing sliding fifth wheels; the rearmost possible position should never be behind the centerline of the bogies.

Determining the appropriate location of a fixed fifth wheel is the function of the person responsible for determining the vehicle's specifications. The positioning of sliding fifth wheels has been the source of thousands of axle overload tickets. Dispatchers should call the shots when it comes to instructions for locating a sliding fifth wheel, and this type of decision has to be made while factoring in the trailer weight and permissible weight-over-axle in the jurisdictions in which the unit is to be operated.

Shop Talk

The fifth wheel location on the tractor affects the bridge formula; that is, how the load is distributed over the tractor and the trailer axles. Double-check that the fifth wheel is correctly located when installing. Incorrectly placed fifth wheels can result in weight-over-axle infractions and severely imbalanced dynamic braking.

30.3 PRINCIPLES OF FIFTH WHEEL OPERATION

Three manufacturers of fifth wheels are commonly found on highway trucks in North America. One company, Holland Hitch Co., dominates the market. Holland Hitch Co. uses two distinct types of locking mechanisms in their fifth wheels. The other two companies, ASF (two types) and Fontaine, use different locking mechanisms, so the truck technician should take nothing for granted when working with fifth wheels. Having some understanding of how each type operates helps, and some of the more popular types are introduced in this section.

HOLLAND HITCH A-TYPE

Holland fifth wheels use cast steel top plates. The Holland A-type locking mechanism uses a single swinging lockjaw and plunger. This probably is one of the better known locking mechanisms in the trucking industry, and it is diagrammed in **Figure 30–11** and **Figure 30–12**. When the kingpin enters the

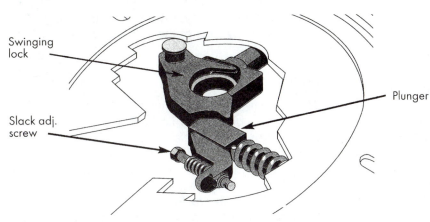

Extra Capacity Type "A" Lock

FIGURE 30–11 Holland A-type locking mechanism. (Courtesy of Holland Hitch Co.)

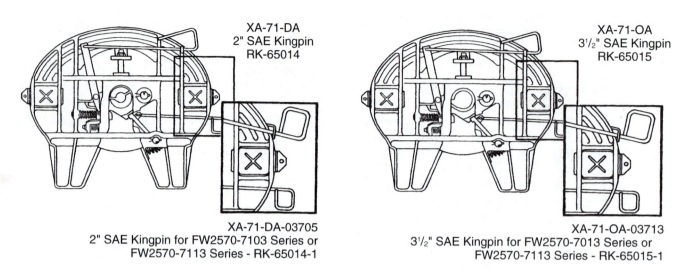

XA-71-DA
2" SAE Kingpin
RK-65014

XA-71-OA
3½" SAE Kingpin
RK-65015

XA-71-DA-03705
2" SAE Kingpin for FW2570-7103 Series or
FW2570-7113 Series - RK-65014-1

XA-71-OA-03713
3½" SAE Kingpin for FW2570-7013 Series or
FW2570-7113 Series - RK-65015-1

FIGURE 30–12 Underside view of a fifth wheel with a Holland A-type lock. (Courtesy of Holland Hitch Co.)

throat of the fifth wheel and exerts pressure on the swinging lock, it pivots, enclosing the kingpin. This action causes a spring-loaded plunger to jam into a concave recess in the swinging lock. Slack caused by mechanical wear can be adjusted by means of an adjusting screw in the throat of the fifth wheel, which acts on the jam plunger.

HOLLAND HITCH B-TYPE

This type of fifth wheel uses a pair of opposed jaws. **Figure 30–13** shows both an exploded and assembled view of a Holland fifth wheel using a B-type locking mechanism, one of the most commonly used in trucking. Consult the references in **Figure 30–14** and **Figure 30–15** that describe the coupling and uncoupling sequences of the fifth wheel described in this section.

Coupling

The Holland B-type locking mechanism uses a pair of pivoting lockjaws as shown in **Figure 30–13**, **Figure 30–14**, and **Figure 30–15**. In the open position, the yoke spring holds the yoke in its unlocked position, and the feet of the yoke contact cam notches on the lock jaws. When the fifth wheel is backed under a trailer, the kingpin enters the throat of the fifth wheel, and when it contacts the inside bore of the lock jaws, it forces them to pivot and close around the kingpin. This action causes the yoke spring to act on the cam assembly, forcing the yoke to wedge itself behind the lock jaw cam notches. At the same time the secondary lock pivots to jam the cam assembly, preventing the yoke from

retracting. As the cam pivots to the locked position, the release handle is pulled inward.

Release

To release the fifth wheel lock, the release handle is pulled outward. This simultaneously swings the secondary lock outward and forces the cam around on its pivot, compressing the yoke spring and pulling the yoke away from its wedged position behind the cam notches on the locking jaws. Now, as the truck is pulled away from the trailer, the trailer kingpin spreads the lockjaws, forcing them into their fully retracted position.

ASF FIFTH WHEELS

ASF fifth wheels use cast steel top plates. The standard ASF locking mechanism consists of sliding front and rear jaws linked by a hinge pin assembly. The front jaw is backed by a rubber block. **Figure 30–16** shows an X-ray view of the ASF lock assembly in the locked position.

Locking

When the fifth wheel is unlocked, the front jaw is on a horizontal plane with its inner bore exposed rearward to the throat. The lower jaw is swung downward, pivoted by the hinge pins. When the fifth wheel is backed under the trailer kingpin, the kingpin contacts the front jaw and forces it to slide toward the rubber block. As the front jaw is driven backward, the rear jaw is pulled up onto a horizontal plane, enclosing the kingpin. In the locked position, the rub-

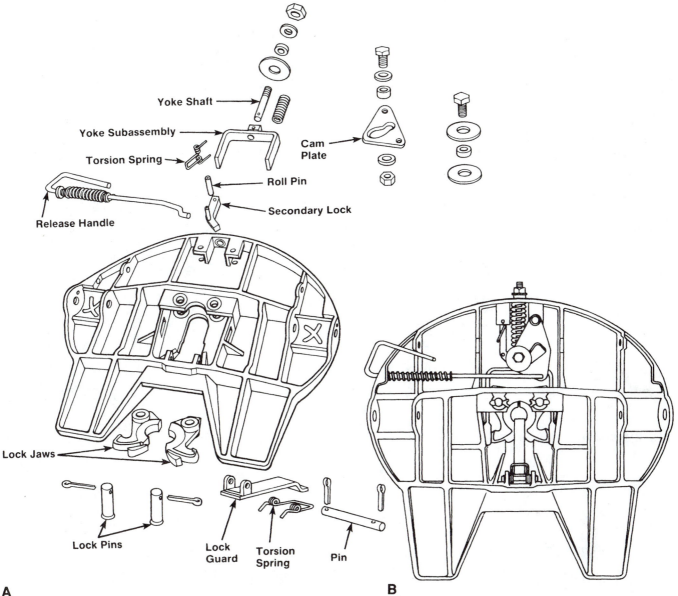

Yoke Shaft

Yoke Subassembly

Torsion Spring

Cam Plate

Roll Pin

Release Handle

Secondary Lock

Lock Jaws

Lock Pins

Lock Guard

Torsion Spring

Pin

A

B

FIGURE 30–13 *Holland B-type locking mechanism: A) Exploded view; and B) assembled. (Courtesy of Holland Hitch Co.)*

ber block is loaded under some compression by the lock arm assembly, which acts on the rear jaw (**Figure 30–17**). When the fifth wheel is engaged, the operating/release handle slides into its locked position (**Figure 30–18**). A safety latch at the operating handle prevents accidental release.

Release

The safety latch on the side of the fifth wheel is raised, allowing the operating/release lever to be forced forward (**Figure 30–19**). This action rotates a cam, permitting the lock arm assembly to swing away and unjamming the rear jaw; a detent on the

cam prevents re-engagement of the lock arm assembly (**Figure 30–20**). Now when the tractor pulls away from the trailer, both sets of jaws slide rearward until the rear jaw swings down, clearing the kingpin, as shown in **Figure 30–21**.

FONTAINE NO-SLACK® II FIFTH WHEEL

Fontaine fifth wheels use pressed steel top plates. This means that where weight is a factor, Fontaine fifth wheels tend to be lighter than their equivalently rated competitor's versions. **Figure 30–22** shows an

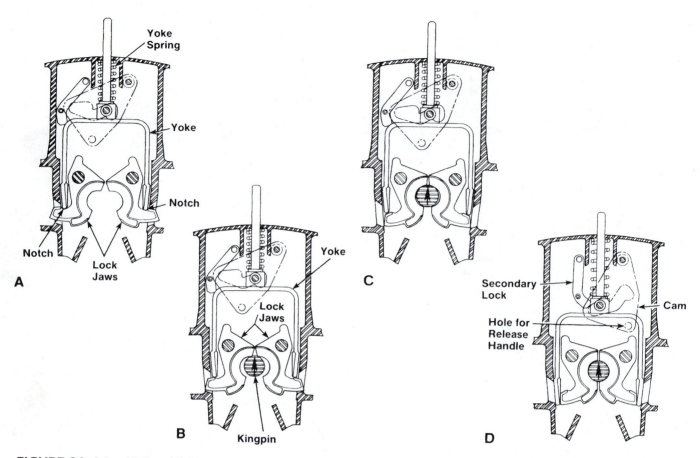

FIGURE 30–14 Holland B-lock coupling. (Courtesy of Holland Hitch Co.)

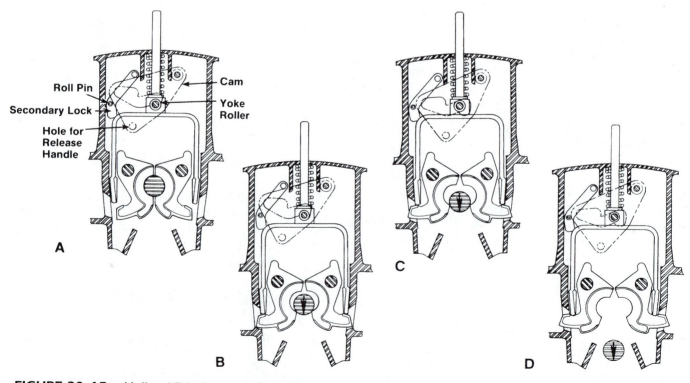

FIGURE 30–15 Holland B-lock uncoupling. (Courtesy of Holland Hitch Co.)

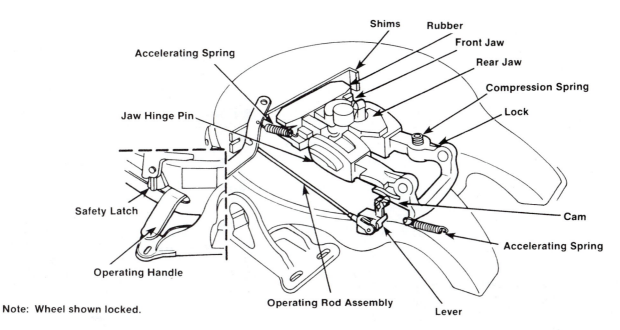
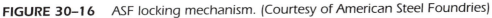

Note: Wheel shown locked.

FIGURE 30-16 ASF locking mechanism. (Courtesy of American Steel Foundries)

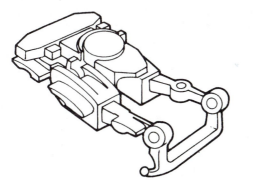

FIGURE 30-17 ASF sliding jaws in locked position. (Courtesy of American Steel Foundries)

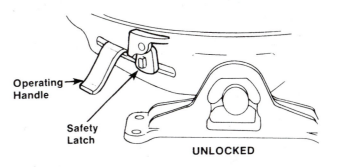

FIGURE 30-19 ASF operating handle in release position. (Courtesy of American Steel Foundries)

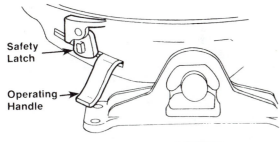

FIGURE 30-18 Side view of an ASF fifth wheel when fully locked. (Courtesy of American Steel Foundries)

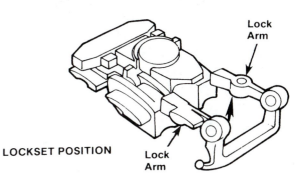
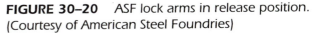

FIGURE 30-20 ASF lock arms in release position. (Courtesy of American Steel Foundries)

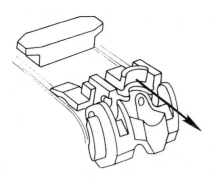

FIGURE 30–21 ASF rear jaw dropped into release position by kingpin. (Courtesy of American Steel Foundries)

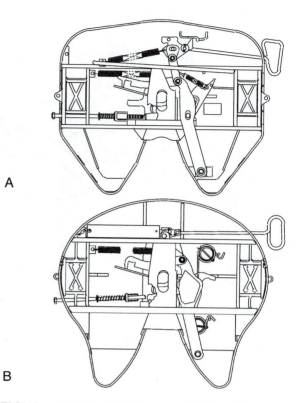

A

B

exploded view of a Fontaine No-Slack® II fifth wheel. Three models are manufactured, standard duty, heavy duty, and extra heavy duty. The standard duty has an improved locking mechanism. Fontaine fifth wheels use a rear jaw that is engaged to the trailer kingpin by a wedge. The function of the wedge is to prevent slack, and the system tends to be automatically adjusting. **Figure 30–23** shows an underview of two types of Fontaine fifth wheels, one of which features an air release.

FIGURE 30–23 (A) Fontaine No-Slack® II; and (B) Fontaine 5092 air release fifth wheel. (Courtesy of Fontaine Fifth Wheel®)

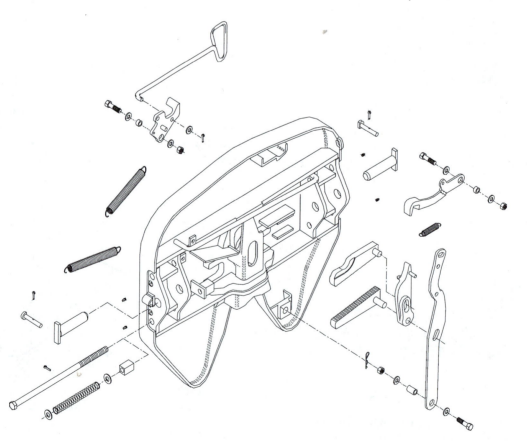

FIGURE 30–22 Exploded view of Fontaine No-Slack® II fifth wheel. (Courtesy of Fontaine Fifth Wheel®)

30.4 FIFTH-WHEEL MAINTENANCE

Each type of fifth wheel requires slightly different maintenance procedures, so it makes sense to consult the manufacturer's recommendations. This section covers some of Holland Hitch's recommended service procedures.

MAINTENANCE INTERVAL

To get the best service life and fifth wheel performance, the first fifth wheel maintenance should be performed within the first 1,000–1,500 miles (1,500–2,500 kms). After this initial service, periodic maintenance at intervals of 30,000 miles (50,000 kms) should ensure trouble-free fifth wheel operation. **Figure 30–24**, **Figure 30–25**, **Figure 30–26**, **Figure 30–27**, and **Figure 30–28** outline some of Holland's recommended maintenance procedures.

MAINTENANCE PROCEDURE

Holland Hitch recommends the following fifth wheel maintenance procedure:

1. Observe fifth wheel operation while coupled to a trailer, checking for slack.
2. Uncouple the fifth wheel from the trailer.
3. Pressure wash or steam clean.
4. Inspect the top plate and brackets for cracks. Replace all damaged, loose, or missing parts.
5. Check the torque on all the fifth wheel mounting bolts.
6. Visually inspect fifth wheel and mounting plate welds.
7. Adjust and check the fifth wheel operation using a lock tester.
8. Lubricate the fifth wheel.

AS-NEEDED LUBRICATION

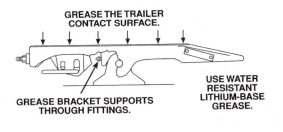

GREASE THE TRAILER CONTACT SURFACE.

GREASE BRACKET SUPPORTS THROUGH FITTINGS.

USE WATER RESISTANT LITHIUM-BASE GREASE.

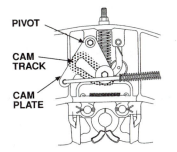

PIVOT

CAM TRACK

CAM PLATE

Lubricate the cam track and pivot with a light oil or diesel oil.

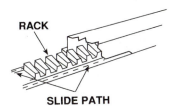

RACK

SLIDE PATH

On sliding fifth wheels, spray a light oil or diesel oil on the rack and slide path.

FIGURE 30–24 Holland's as-needed lubrication. (Courtesy of Holland Hitch Co.)

REQUIRED INSPECTIONS AND ADJUSTMENTS

Perform the following every six months or 60,000 miles, whichever comes first. Thoroughly steam clean all components before inspecting or adjusting.

GENERAL FIFTH WHEEL INSPECTION:

1. Inspect fifth wheel mounting and fifth wheel assembly.

- Check fastener torque
- Replace missing or damaged bolts
- Replace bent, worn or broken parts
  (Use only genuine Holland parts.)

FIGURE 30–25 Holland general fifth wheel inspection. (Courtesy of Holland Hitch Co.)

INSPECTION – LOCKING MECHANISM:

1. Verify operation by opening and closing locks with Holland Kingpin Lock Tester model no. TF-TLN-1000 or TF-TLN-5000.

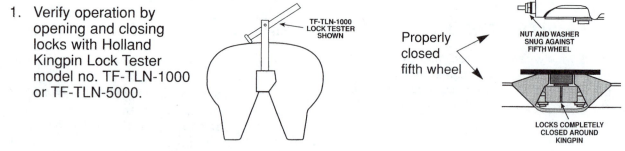

FIGURE 30–26   *Inspection of locking mechanism. (Courtesy of Holland Hitch Co.)*

ADJUSTMENT – LOCKING MECHANISM:

1. Close locks using Holland Lock Tester.

2. Rotate rubber bushing located between the adjustment nut and casting.

3. If the bushing is tight, rotate nut on yoke shank <u>counter-clockwise</u> until bushing is snug, but still can be rotated.

4. Verify proper adjustment by locking and unlocking with the lock tester.

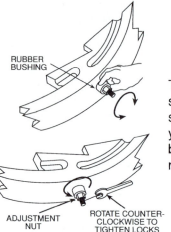

FIGURE 30–27   *Holland adjustment of locking mechanism. (Courtesy of Holland Hitch Co.)*

HOLLAND FIFTH-WHEEL ADJUSTMENT

A-type locking mechanism

1. Using a Holland lock tester, close the fifth wheel.
2. Tighten the 1/2-inch Allen head adjustment screw (clockwise) until tight.
3. Loosen the adjustment screw by turning it counterclockwise 1 1/2 turns.
4. Check the fifth wheel operation by using the lock tester several times.

B-type locking mechanism

1. Close the lock jaws and insert the Holland TF-0237 2-inch diameter plug.
2. Check the fit of the plug and turn the adjusting nut counterclockwise to tighten.
3. When correctly adjusted, the plug is snug in the lock jaws but can be rotated.
4. Check the fifth wheel operation by using the lock tester several times.

Sliding fifth wheel plungers

1. Loosen the locknut and back out the adjustment bolt 1/2 inch.
2. Disengage and engage plungers.
3. Tighten the adjustment bolt (clockwise) until the bolt contacts the slider rack. Turn the bolt an additional half turn.
4. Tighten the locknut securely.
5. Check the plunger operation by operating release mechanism several times.

Routine service lubrication

1. Apply water-resistant, lithium-base grease to the trailer contact surface of the fifth wheel top plate.
2. Apply grease to the bearing surface of the support bracket through the grease fittings on the side of the fifth wheel plate. Do not over-grease as this ends up on the fifth wheel deck plate. The top plate should be tilted back slightly to relieve weight on the bracket while applying the grease.

ADJUSTMENT – FIFTH WHEEL SLIDE MECHANISM:

1. Loosen lock nut and turn adjustment bolt out (counter-clockwise).

2. Disengage and engage the locking plungers. Verify that plungers have seated properly as shown.

3. Now tighten adjustment bolt until it contacts the rack.

4. Turn the adjustment bolt clockwise an additional 1/2 turn then tighten the lock nut securely.

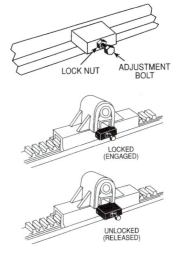

> ⚠️ **CAUTION:** Proper adjustment of the sliding bracket locking plungers must be performed at installation and maintained at regular intervals using the adjustment bolts for both plungers. Proper adjustment is required for proper operation, load transfer, and load distribution.

5. If plungers do not release fully to allow fifth wheel to slide:
   A. Check the air cylinder for proper operation. Replace if necessary.
   B. Check plunger adjustment as explained above.
   C. If a plunger is binding on the plunger pocket, remove the plunger using a Holland TF-TLN-2500 spring compressor. Grind the top edges of the plunger 1/16" as shown. Re-install and adjust the plungers as explained above.

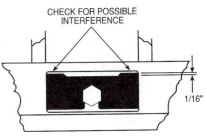

6. If the locking plungers are too loose:
   A. Check plunger adjustment as explained above.
   B. Check plunger springs for proper compression. Replace if necessary.
   C. Check for plunger wear. If necessary, replace as described above.

After inspection and adjustment, relubricate all moving parts with a light, rust resistant oil.

FIGURE 30–28 _Adjustment of fifth wheel slide mechanism. (Courtesy of Holland Hitch Co.)_

---

3. Spray an oil and diesel fuel mixture (80 percent engine oil, 20 percent diesel fuel) on the rack and bracket slide path of sliding fifth wheels.

Lubrication after cleaning

1. Apply a light grease to all moving parts, or spray them with diesel oil.
2. Perform the routine service lubrication procedure.

Fifth-Wheel Service Tools

Figure 30–29 shows some of the basic tools required to test and adjust fifth wheels and kingpins. The kingpin lock testers are available in the SAE standard kingpin dimensions and are used to test fifth wheel locking action in the shop. The C-clamp is used to compress the fifth wheel springs when disassembling and rebuilding fifth wheels. The kingpin gauge or template is used to verify that the kingpin dimensions are to specification; it will test both 2-inch and 3.5-inch kingpins. The kingpin plug shown in **Figure 30–29** is used both to test and reassemble fifth wheels.

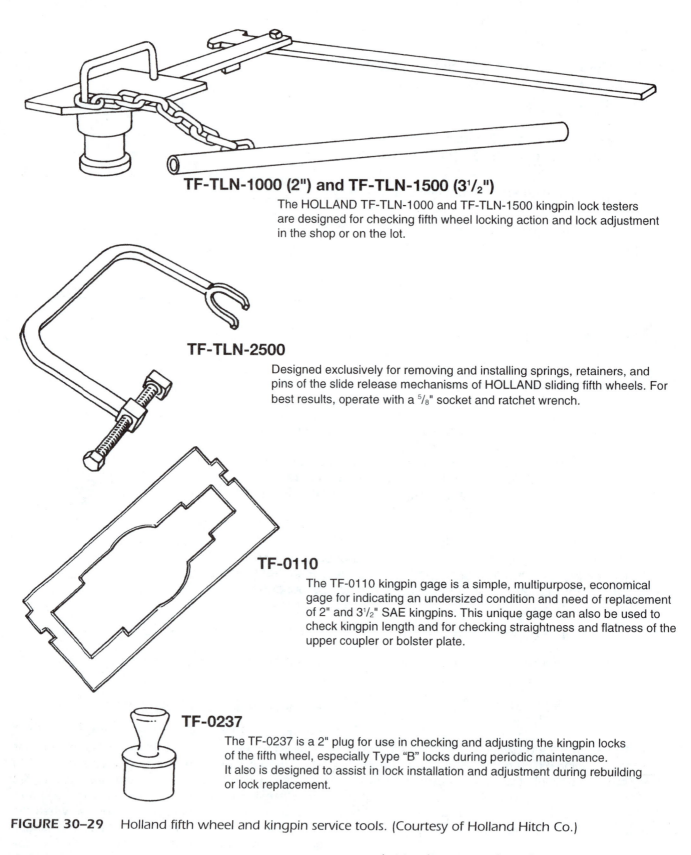

TF-TLN-1000 (2") and TF-TLN-1500 (3¹/₂")

The HOLLAND TF-TLN-1000 and TF-TLN-1500 kingpin lock testers are designed for checking fifth wheel locking action and lock adjustment in the shop or on the lot.

TF-TLN-2500

Designed exclusively for removing and installing springs, retainers, and pins of the slide release mechanisms of HOLLAND sliding fifth wheels. For best results, operate with a ⁵/₈" socket and ratchet wrench.

TF-0110

The TF-0110 kingpin gage is a simple, multipurpose, economical gage for indicating an undersized condition and need of replacement of 2" and 3¹/₂" SAE kingpins. This unique gage can also be used to check kingpin length and for checking straightness and flatness of the upper coupler or bolster plate.

TF-0237

The TF-0237 is a 2" plug for use in checking and adjusting the kingpin locks of the fifth wheel, especially Type "B" locks during periodic maintenance. It also is designed to assist in lock installation and adjustment during rebuilding or lock replacement.

FIGURE 30–29   Holland fifth wheel and kingpin service tools. (Courtesy of Holland Hitch Co.)

Coupling

Each type of fifth wheel uses a different locking mechanism, so while the procedure for coupling each type is more or less the same, the inspection procedure varies. Because the fifth wheel is such a critical safety component on a tractor/trailer combination, be sure to ask to be shown what to look for to ensure that the unit is properly coupled. Properly coupled means that the jaws are fully engaged, the locks are secure, and a high hitch condition does not exist.

1. First ensure that the fifth wheel locks are not engaged and the jaws are in an open position. Also ensure that the fifth wheel top plate is adequately lubricated; lack of fifth wheel lubrication can cause hard and erratic steering.
2. Align the tractor with the trailer so that the fifth wheel throat is aimed at the trailer kingpin.
3. Ensure that the trailer parking brakes are fully applied.
4. Carefully back the tractor up to the trailer. The trailer bolster plate should be slightly lower than the fifth wheel top plate, causing the tractor to lift the trailer. Back up the tractor until solid resistance is felt. All fifth wheels are designed so that when the kingpin fully enters the throat, the jaws engage, and the locks are activated.
5. Next, ensuring that the trailer brakes remain fully applied, nudge the tractor forward. The parking brakes of the trailer should prevent forward movement, and the fifth wheel should remain engaged.
6. Exit the tractor and use a flashlight to verify that the fifth wheel jaws are fully engaged, the locks are applied, and the bolster plate is in full contact with the fifth wheel plate.

Uncoupling

Release levers may pull out linearly or slide axially. If uncoupling a fifth wheel with which you are unfamiliar, ask someone about the correct procedure. The following is a general procedure:

1. Park the tractor/trailer combination and ensure that parking brakes are applied. If the trailer is loaded, ensure that the ground surface can support the landing gear pads.
2. Drop the landing gear. Raise it to relieve some of the trailer weight from the tractor.
3. Disconnect the air and electrical connections from the trailer. Insert the gladhands and trailer plug in the dummy connectors on the tractor and ensure that hoses and the electrical cord are safe from contact with the driveshafts.
4. Release the release lever. Some fifth wheels have secondary locks that must be released before the main release lever is moved. Others have a two-lever release mechanism.

5. Release the tractor brakes and drive the tractor forward, separating the two units.

Shop Talk

Do not attempt to inspect a fifth wheel without removing the grease. The best way to do this is to first scrape off as much as possible, and then apply some solvent before hot pressure washing. Check the turntable plate, throat, and saddle/pivot assembly for cracks. Check the dynamic action of the locking mechanism and the secondary lock integrity. Check all the mounting fasteners and welds.

30.5 MOUNTING FIFTH WHEELS

The method of mounting fifth wheels on the frame is determined by the design of the fifth wheel, the suspension trunnion, and suspension to frame-mounting brackets. The most popular mounts are:

- **Single Angle Outboard Mount.** This type of mount has the fifth wheel deck plate welded or bolted to a base plate. The base plate is welded or bolted to angle iron brackets, to be fitted either side of the tractor frame rails. The angle iron brackets are oriented to fit outboard from the truck main frame rails. Welding should be performed with AWS E-7018 electrodes or MIG equivalent. **Figure 30–30A** shows a single angle outboard mount, probably the most commonly used mounting arrangement in the industry.
- **Double Angle Inboard-Outboard Mount.** This type of mount is used when a fifth wheel assembly is specified with no mounting plate.

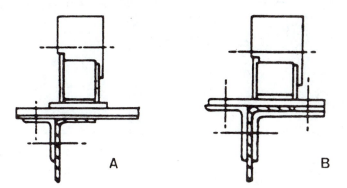

FIGURE 30–30 *(A) Single angle outboard mount; and (B) double angle inboard-outboard mount. (Courtesy of Mack Trucks, Inc.)*

PHOTO SEQUENCE 20
TRAILER COUPLING

Following is a method of coupling a tractor to a spring-brake-equipped trailer.

PS20-1 Inspect the fifth wheel for damaged, worn, or loose components and mountings. Be sure that the fifth wheel jaws are open and the handle is in the unlocked position. The fifth wheel must be properly lubricated and tilted down at the rear.

PS20-2 Check that the trailer parking brakes are applied and block the trailer wheels as a precaution.

PS20-3 Adjust the trailer landing gear so that the trailer bolster plate is just below the fifth wheel.

PS20-4 Slowly back up the tractor, maintaining the alignment between the fifth wheel throat and the trailer kingpin.

PS20-5 When the trailer kingpin enters the fifth wheel throat and the trailer bolster plate rides on the fifth wheel, back up the tractor until there is full trailer resistance.

PS20-6 Connect the air hoses and electrical connectors between the tractor and trailer.

PS20-7 Inspect the trailer bolster plate to be sure that it is supported evenly on top of the fifth wheel. No gap should be visible between the bolster plate and the fifth wheel surface.

PS20-8 Inspect the safety latch to be sure that it swings freely and that the operating handle is behind the safety latch positioned to the rear of its operating slot.

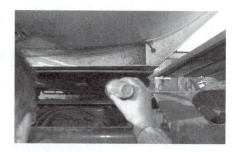

PS20-9 Crawl under the tractor and use a flashlight to verify that the kingpin is properly engaged to the fifth wheel jaws.

PHOTO SEQUENCE 20 (CONTINUED)
TRAILER COUPLING

PS20-10   With the trailer service brakes applied, place the transmission in the lowest gear and partially engage the clutch to create a pulling force between the fifth wheel and the kingpin. The fifth wheel jaws should remain securely locked on the kingpin.

PS20-11   When coupling is completed, apply the tractor and trailer brakes and crank the trailer landing gear to the fully upward position. Remove the blocks from the trailer wheels.

Frame-mounting supports are used on both the inside and outside of the frame rail. A spacer plate or steel spacers are used on the inside frame mounting support to keep the mounting surfaces level. The forward edges of the inside and outside angles must be staggered a minimum of six inches. **Figure 30–30B** shows a double inboard-outboard angle mount used in heavy-duty applications.

The following considerations are based on one manufacturer's recommendations. Before fitting a fifth wheel, check the manufacturer's requirements, which can vary with fifth wheel rating and vehicle application. **Figure 30–31** shows a properly mounted fifth wheel.

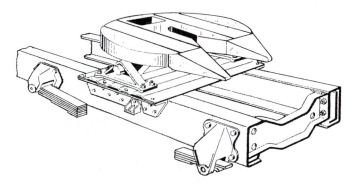

FIGURE 30–31   A properly mounted fifth wheel. (Courtesy of Mack Truck, Inc.)

- Angle iron sections used to mount a fifth wheel must be at least $5/16$-inch ASTM-A36 carbon steel; $3/8$-inch angle iron section also can be used. The angle iron should be at least $3 \times 3\frac{1}{2}$ inches and at minimum 36 inches long. It is good practice to cut a radius on the edges of the mounting angle that contact the tractor frame. Ensure that no bolt hole is located closer than 1 inch to the edge of the angle iron. When recesses are cut into the angle iron to avoid suspension brackets, ensure that there is both sufficient clearance and a radius cut.
- Each section must be secured to the frame rails by at least five $5/8$-inch-diameter, Grade 8 bolts. Space the bolts 8 inches apart and secure them with Grade C lock nuts. Nuts must be torqued to manufacturer's specifications. Mounting angle bolts preferably should be staggered with respect to the horizontal centerline along the length of the mounting angles to avoid weakening the frame. Fifth wheel support brackets are never welded to the truck frame rails.
- For stationary fifth wheel plates, the angle iron bracket must extend at least 18 inches ahead of the fifth wheel's pivot point and at least 12 inches to the rear. For sliding fifth wheels, the angle iron sections must be long enough that these minimum distances are achieved when the fifth wheel is moved.

CAUTION: Installation welding on fifth wheel assemblies should be performed by a qualified welder. Improper welding procedures can result in a trailer separation—the consequences of which can be fatal!

30.6 SLIDING FIFTH WHEELS

Sliding fifth wheels increase the versatility of the tractor, but it is important to understand how tractor weight over axle is affected. Locating a sliding fifth wheel to maximize driver comfort can often result in weight-over-axle fines. The following procedure is typical when adjusting the position of a sliding fifth wheel.

SLIDING PROCEDURE

The fifth wheel should be fully engaged to the trailer for this operation.

1. The tractor and trailer should be parked in a straight line on level ground. Engage the trailer parking brakes.

CAUTION: The trailer must be stationary with its parking brakes applied to prevent damage to the tractor or trailer by uncontrolled sliding of the fifth wheel.

2. Release the sliding locking plungers. For an air slide release (**Figure 30–32**), put the cab control valve into the unlocked position. For a manual slide release, pull the release lever (**Figure 30–33**).
3. Visually check that both plungers have released from the rail teeth. If the locking plungers are jammed in the rack teeth, try lowering the landing gear to relieve pressure on the plungers. This should allow the fifth wheel to slide easier.
4. Release the tractor brakes and drive the tractor forward or backward slowly to position the fifth wheel.
5. After sliding the fifth wheel to the desired setting, engage the slide locking plungers. For an air slide release, put the control valve in the lock position to engage the plungers to the rack teeth. For a manual slide release,

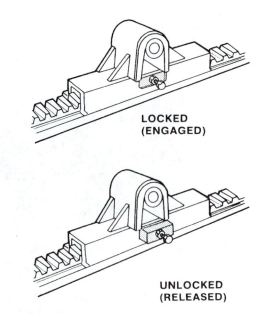

FIGURE 30–32 *Air slide release. (Courtesy of American Steel Foundries)*

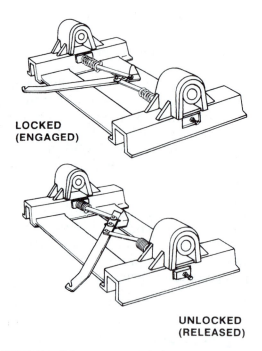

FIGURE 30–33 *Manual slide release. (Courtesy of American Steel Foundries)*

trip the release arm to allow the plungers to engage with the rack teeth.
6. Visually check to see that both plungers are fully engaged to the rail teeth so that no fore and aft movement is possible. Leaving the trailer brakes locked and nudging the tractor slightly might be necessary to engage the plungers in the rack teeth. Raise the landing gear to the fully retracted position.

CAUTION: Do not operate the vehicle if the plungers are not fully engaged and the landing gear fully retracted, because damage to the tractor, trailer, and landing gear can occur.

SLIDING MECHANISM INSPECTION AND ADJUSTMENT

When adjusting the locking plungers

1. Loosen the locknut and turn adjusting bolt out (counterclockwise).
2. Disengage and engage the locking plungers. Check that the plungers are securely seated without binding.
3. Turn adjusting bolt in (clockwise) until it contacts the rack. Turn adjusting bolt an additional one-half turn and then tighten the locking nut securely.

👓 **Shop Talk** _____

On a sliding model fifth wheel, after locating it in the desired position, visually inspect the locking plungers to be sure that they are properly engaged. Also, set the trailer brakes and rock the tractor fore and aft. Do not attempt to shift a sliding fifth wheel while a trailer is in motion!

When locking plungers will not release to permit sliding of the fifth wheel

1. Check the air cylinder for proper operation and replace if necessary.
2. Check plunger adjustment.

CAUTION: Proper adjustment of the sliding bracket locking plungers must be performed at installation and maintained at regular intervals by use of the adjusting bolts provided on both sides. Proper adjustment is required for proper operation and for proper load transfer and distribution.

3. If the adjusted plunger binds on the pocket, grind the top plunger edges 1/16 inch, reinstall, and adjust as just described. Use a spring compressor to remove and reinstall the plunger.

CAUTION: Wear safety glasses when grinding the top plunger edges.

When locking plungers are too loose

1. Check plunger adjustment.
2. Check the plunger springs for proper compression. Replace if necessary.
3. Check for plunger wear and replace if necessary. Use a spring compressor to remove and reinstall the plungers. Adjust the plungers as described.

CAUTION: Rebuilding of fifth wheels is a common shop practice. Although it is a simple procedure, each type of fifth wheel uses distinct locking mechanisms. Always use the manufacturer procedure included in every rebuild kit. A fifth wheel failure can have fatal consequences!

30.7 | KINGPINS AND UPPER COUPLERS

The kingpin stub is mounted on the trailer upper coupler assembly. It is designed to securely engage with the fifth wheel locking mechanism and to permit articulation so that a tractor/trailer or tractor/train combination can turn corners. There are two SAE standard kingpin sizes, 2-inch and 3.5-inch. Kingpins are required to have an SAE-specified Brinell surface hardness that will produce a yield strength of 115,000 psi and a tensile strength of 150,000 psi.

Most kingpins today are welded to the upper coupler assembly, but both bolt-on and removable fasteners are also used. Welding should be performed in accordance with AWS standards. This generally means welding with an AWS E7018 electrode in a single rotational direction with stitched breaks in the weld runs. An E7018 electrode is used because it has a tensile strength in-between the hardened kingpin and the softer upper coupler material.

CAUTION: Always consult the manufacturer for the correct procedure when welding kingpins to upper coupler assemblies.

When kingpins are bolted into the upper coupler assembly, SAE Grade 8 bolts, usually 3/4-inch diameter, should be used. Tapered, removable kingpins are used in applications in which there may be a possibility of damaging an exposed kingpin when the trailer is not coupled to a tractor. Folding gooseneck trailers would be an example. A locking dowel pin both retains and prevents the rotation of removable-type kingpins. **Figure 30–34** shows some examples of different types of kingpins and their required dimensions.

CHECKING KINGPIN DIMENSIONS

Trailer kingpins should be routinely cleaned and inspected for wear, integrity to the upper coupler, and cracks. Most fleets use a template or gauge to check a kingpin for wear; the template slides over the kingpin. A typical kingpin template capable of

2" SAE KINGPIN – KP-T-809 SERIES – 8630 STEEL FORGING – BRINELL 302-363

MODEL	BOLSTER PLT. THICKNESS	HOLES
KP-T-809-CF	.25"	No
KP-T-809-EF	.31"	No
KP-T-809-F	.38"	No
KP-T-809-BF	.50"	No
KP-T-809-C	.25"	8 equally spaced .53" holes on 6.75" diameter for plug welding
KP-T-809-E	.31"	
KP-T-809	.38"	
KP-T-809-B	.50"	

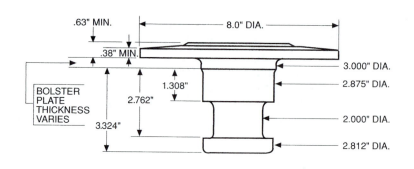

2" SAE KINGPIN – KP-T-880 SERIES – 8630 STEEL FORGING – BRINELL 302-363

MODEL	BOLSTER PLT. THICKNESS	A	B	HOLES
KP-T-880-C	.25"	2.88"	2.12"	No
KP-T-880-E	.31"	2.88"	2.12"	No
KP-T-880	.38"	2.88"	2.12"	No
KP-T-880-B	.50"	2.88"	2.12"	No
KP-T-880-1-C	.25"	2.0"	1.5"	No
KP-T-880-1-E	.31"	2.0"	1.5"	No
KP-T-880-1	.38"	2.0"	1.5"	No
KP-T-880-1-B	.50"	2.0"	1.5"	No

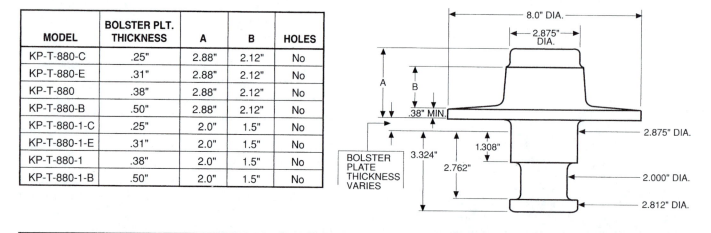

3.5" SAE KINGPIN – KP-T-847 SERIES – 8630 STEEL FORGING – BRINELL 302-363

MODEL	BOLSTER PLT. THICKNESS	HOLES
KP-T-847	.38"	4 equally spaced 1.25" holes on 8.5" diameter for plug welding
KP-T-847-B	.50"	

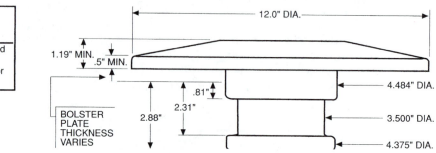

FIGURE 30–34   SAE 2- and 3.5-inch kingpins and specifications. (Courtesy of Holland Hitch Co.)

measuring both SAE 2- and 3.5-inch kingpins is shown in **Figure 30–29**. Standard procedure is to replace a worn or damaged kingpin.

UPPER COUPLERS

Upper coupler assemblies are built into the trailer frame assembly. The lower plate is known as the bolster plate. The kingpin is located through the center of the bolster plate. The bolster plate sits on the tractor fifth wheel plate when coupled and, therefore, provides the weight bearing and pivot surface. Bolster plates may be 0.25-, 0.31-, 0.38-, or 0.50-inch. As the weight of the bolster plate is considerable in itself, the lightest plate with a suitable margin of error is specified.

Fifth wheels and upper couplers are designed to work together to ensure that the trailer load is evenly distributed to the tractor. Upper couplers manufactured from higher tensile strength steels may reduce the upper coupler total weight but also tend to be more flexible. Greater upper coupler flexibility increases the deflection, which can shorten the service life of the assembly. **Figure 30–35** shows both the appearance of a properly loaded fifth wheel and what occurs if the vertical load is concentrated in the center of the fifth wheel assembly. Center loading causes the fifth wheel to act like a beam, flexing with the movement of the vehicle over the road, and can eventually result in failure of the fifth wheel and upper coupler.

An upper coupler that has been distorted into ripples is usually caused by the welding procedure during the upper coupler assembly, especially welding too rapidly on lighter duty bolster plates. The wavy appearance of the bolster plate results in inadequate surface contact to the fifth wheel and high unit pressures where contact does occur. The result is galling to the fifth wheel and rapid wear of the bolster plate. **Figure 30–36** shows a ripple-distorted upper coupler.

A bowed upper coupler is caused by trailer overloading the trailer or by an upper coupler that has been fabricated of gauge materials that are too light. A bowed bolster plate raises the kingpin to a higher than specified height which can make it difficult to release a fifth wheel. This condition can be difficult to detect if the trailer is unloaded as the deflection may take place only when fully loaded. If permanently deformed, a straight edge can locate the problem. **Figure 30–37** shows a bowed upper coupler.

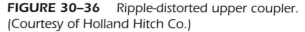

FIGURE 30–36 *Ripple-distorted upper coupler. (Courtesy of Holland Hitch Co.)*

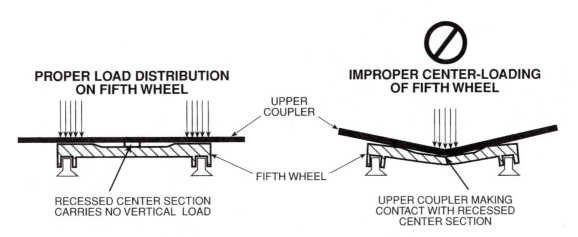

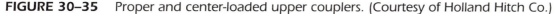

FIGURE 30–35 *Proper and center-loaded upper couplers. (Courtesy of Holland Hitch Co.)*

UPPER COUPLER DESIGN TOO
WEAK TO SUPPORT LOAD.

BOWED UPPER COUPLER

FIFTH WHEEL

FIGURE 30–37 Bowed upper coupler. (Courtesy of Holland Hitch Co.)

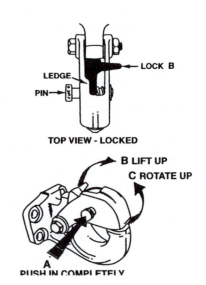

LEDGE

PIN

LOCK B

TOP VIEW - LOCKED

B LIFT UP
C ROTATE UP

A
PUSH IN COMPLETELY

FIGURE 30–38 Typical pintle hook and locking assembly. (Courtesy of Holland Hitch Co.)

30.8 PINTLE HOOKS

A heavy-duty full trailer must be attached to the tractive unit by a pintle hook. The standard design used is the Jeep hook, a name derived from its application on United States military Jeeps. The function of a pintle hook is to engage to the drawbar eye, which itself is bolted to the tongue or drawbar assembly of the trailer. Pintle hooks are available in many different ratings determined by the type of application (off-highway, rough terrain operation requires high strength) and the weights of the trailer to be hauled. Many pintle hooks are mounted to a pneumatic or hydraulic buffer behind the assembly, which is designed to minimize the effect of trailer bumping. **Figure 30–38** shows a typical pintle hook assembly and the means used to lock it.

Drawbars are available in a number of load ratings—again dependent on application—tongue weight, and total trailer weight. Drawbar eyes are normally fastened to the trailer tongue assembly by means of SAE Grade 8 bolts at the shank. Both rectangular and square shanks are used. Two- and three-inch eye diameters are used, with 2.5-inch and 3-inch being the most common. **Figure 30–39** shows a heavy-duty drawbar eye, and **Figure 30–40** shows the method of attaching it to the tongue.

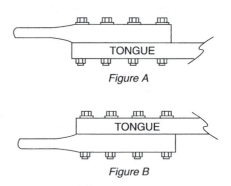

FIGURE 30–39 Heavy-duty drawbar with square shank. (Courtesy of Holland Hitch Co.)

TONGUE

Figure A

TONGUE

Figure B

FIGURE 30–40 Two methods of mounting a drawbar to a tongue. (Courtesy of Holland Hitch Co.)

SUMMARY

- The purpose of the fifth wheel is both to permit the truck/tractor to pivot around the trailer king-pin when turning corners and to support some of the weight of the trailer.

- The semi-oscillating fifth wheel is the most common highway tractor fifth wheel.
- A semi-oscillating fifth wheel articulates or oscillates about an axis perpendicular to the vehicle centerline.
- Compensating fifth wheels are designed to provide both front-to-rear and side-to-side oscilla-

tion between the tractor and semi-trailer. Side-to-side oscillation occurs below the fifth wheel plate surface and lessens the effect of trailer torque or twist acting on the frame of the tractor or lead trailer.

- Fully oscillating fifth wheels articulate on all planes and are used with goose-neck, lowbed trailers.
- Elevating fifth wheels are used on yard shunt tractors. They may be equipped with either air or hydraulic lift mechanisms.
- Many fifth wheels are built with a sliding mechanism that permits longitudinal movement of the fifth wheel on the tractor frame. This allows the point at which the weight of the trailer is supported by the tractor to be altered.
- Relocation of a sliding fifth wheel alters the weight-over-axle distribution on a tractor when coupled to a trailer.
- Fifth wheel mountings may be plate mounted or angle-on-frame mounted.
- The fifth wheel should be inspected every 30,000 miles or three months, whichever comes first.
- Proper lubrication of the fifth wheel includes keeping a water-resistant, lithium-based grease applied to the fifth wheel plate, lubricating all moving parts with a light oil, and applying chassis grease to the grease zerks at the pivot points and the cam.
- Kingpins are used to couple a trailer to the fifth wheel.
- Kingpins come in two SAE standard sizes, 2-inch and 3.5-inch. They are manufactured in special alloy steels and are usually welded to the upper coupler assembly. Bolted and removable kingpins are also used.
- A pintle hook and drawbar assembly is used to couple a full trailer to a truck or lead trailer.

REVIEW QUESTIONS

1. Which type of fifth wheel is most commonly used on highway tractors?
   a. fully oscillating
   b. semi-oscillating
   c. stabilized
   d. elevating

2. Which of the following is an advantage of a sliding fifth wheel?
   a. Weight-over-axle on the tractor can be altered.
   b. Weight-over-axle on the trailer can be altered.
   c. Trailer bridge formula can be altered.
   d. all of the above

3. Which type of fifth wheel permits articulation fore and aft and side to side?
   a. fully oscillating
   b. semi-oscillating
   c. elevating
   d. all of the above

4. How many fifth wheels would a standard B-train combination rig have in total?
   a. 1
   b. 2
   c. 1 and a pintle hook
   d. 2 and a pintle hook

5. Which type of fifth wheel uses a locking mechanism with a pair of opposing lockjaws?
   a. Holland A lock
   b. Holland B lock
   c. Fontaine
   d. all fifth wheels

6. Technician A uses a double angle inboard-outboard mount when mounting a fifth wheel assembly that has no mounting plate. Technician B welds fifth wheel deck plate assemblies to the frame mounting angle iron. Who is correct?
   a. Technician A
   b. Technician B
   c. both A and B
   d. neither A nor B

7. Holland Hitch Co. recommends that a fifth wheel should be inspected every
   a. 50,000 miles
   b. 40,000 miles
   c. 30,000 miles
   d. 20,000 miles

8. When lubricating the fifth wheel, Technician A lubricates all moving parts with a water-resistant, lithium-based grease. Technician B lubricates the rails on a sliding fifth wheel with a mixture of oil and diesel fuel. Who is correct?
   a. Technician A
   b. Technician B
   c. both A and B
   d. neither A nor B

9. When using angle iron to mount fifth wheel brackets to a tractor frame, which of the following methods is recommended?
   a. welding to frame using E-7014 electrodes
   b. welding to frame using E-7018 electrodes
   c. using at least three SAE Grade 5 bolts on either side
   d. using at least five SAE Grade 8 bolts on either side

10. The jaw pair of the fifth wheel lock mechanism should be replaced when wear exceeds
    a. $^3/_8$ inch
    b. $^1/_4$ inch
    c. $^1/_8$ inch
    d. $^1/_{16}$ inch

11. To ensure that the tractor and trailer are properly coupled by the fifth wheel, check by
    a. pulling the tractor forward with the trailer brakes set
    b. measuring clearance between the fifth wheel plate and the trailer upper coupler
    c. listening for an audible snap as the jaws close
    d. observing that the release lever is in the engaged position

12. Which is the most common method of attaching a kingpin to an upper coupler assembly?
    a. SAE Grade 8 fasteners
    b. SAE Grade 5 fasteners
    c. welding
    d. lockpin

13. Trailer weight is transferred to the fifth wheel plate by which of the following?
    a. bolster plate
    b. kingpin
    c. pintle hook
    d. drawbar

14. Which type of fifth wheel is designed to minimize the effect of trailer twist on the tractor frame?
    a. semi-oscillating
    b. rigid
    c. no-tilt
    d. compensating

15. The air release slide mechanism is preferred over the manual release when
    a. slide length is short
    b. slide length is long
    c. frequent adjustment is necessary
    d. safety is a factor

16. Fifth wheel height is defined as
    a. distance from fifth wheel plate to the roof of the trailer when coupled
    b. distance from ground level to the fifth wheel top plate
    c. distance from ground level to the base of the fifth wheel
    d. distance from the base of the fifth wheel to the fifth wheel top plate

31

Heavy-Duty Heating, Ventilation, and Air Conditioning Systems

Objectives

After reading this chapter, you should be able to

- Understand some basic air conditioning theory of heavy-duty truck air conditioning systems.
- Outline the requirements of the Clean Air Act that apply to a heavy-duty truck air conditioning system.
- List the five major components of a heavy-duty air conditioning system and describe how each works in the operation of the system.
- Explain how the thermostatic expansion valve or orifice tube controls the flow of refrigerant to the evaporator.
- Identify the refrigerants used in heavy-duty truck air conditioning systems.
- Describe the function of the main components in a typical heavy-duty air conditioning system.
- Recognize the environmental and personal safety precautions that must be observed when working on air conditioning systems.
- Identify A/C testing and service equipment.
- Test an air conditioning system for refrigerant leaks.
- Outline the procedure required to service a heavy-duty air conditioning system.
- Perform some simple diagnosis of A/C system malfunctions.
- Explain how a truck cab ventilation system operates.
- Describe the role a liquid-cooled heating system plays in a truck cab heating system.
- Describe some types of auxiliary heating and power units.

Key Terms List

accumulator
A/C control module
blower motor/fan assembly
British thermal units (Btus)
Clean Air Act
compressor
condenser
coolant heater
electronic leak detector
evaporator

evaporator thermostat
expansion valve
fuel-fired heater
heater control valve
latent heat
manifold gauge set
ram air
receiver/dryer
retrofitted
refrigerant
sight glass
subcooling

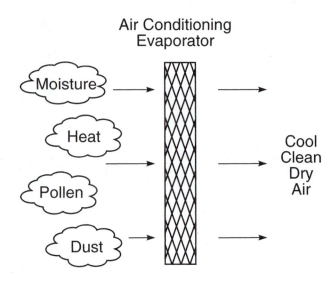

FIGURE 31–1 How heat is removed from the cab air.

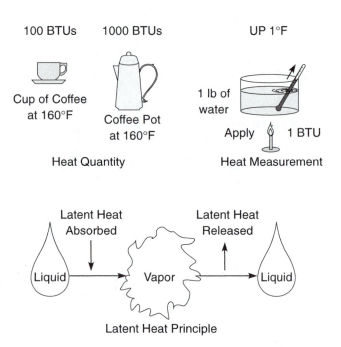

FIGURE 31–2 Principles of heat.

Heating, ventilation, and air-conditioning (HVAC) systems are designed to keep the cab or cab and bunk in a truck at a comfortable temperature. The air conditioning system also helps clean (condition) the air in the cab by removing dust, pollen, smoke, and moisture. This chapter will introduce the basics of air conditioning, followed by some of the repair and diagnostic techniques required of the service technician.

To understand air conditioning theory, it is important for the service technician to have a basic understanding of how heat behaves. The behavior of heat is a branch of physics known> as thermodnamics. An air conditioning system uses some very basic thermodynamic principles to remove heat from the cab of a truck and dissipate or lose it to the atmosphere outside of the truck (**Figure 31–1**). The temperature inside the cab is reduced by removing heat faster than it comes in from the sun, the ventilation system, the engine (through the firewall), and the road pavement.

31.1 BASIC PRINCIPLES OF REFRIGERATION

To understand how an air conditioning system works, we first have to know something about the states of matter, heat flow, and something called **latent heat**.

STATES OF MATTER

The three states of matter are solid, liquid, and vapor. Water can be easily observed in all three of these states. As a solid, water is known as ice; as a

liquid, water, and as a vapor, steam. The temperature of the water will determine which of these three states it is in.

Absolute Heat, Heat Movement, and Measurement

Cold is a relative term. We use it to mean that one substance or atmospheric condition has less heat than another. All matter on Earth contains some heat. Any matter with a complete absence of heat would be at a temperature of −459°F (−273°C). At this temperature, all molecular movement in matter ceases. Heat is a form of energy. Any substance at a temperature higher than absolute zero has at least some heat energy.

Heat always moves away from the heat source. In other words, a warmer substance surrenders heat to cooler substances. This exchange of heat will generally continue until two substances close to each other are at equal temperatures.

Heat is measured in units of energy called **British thermal units (Btus)** or joules. One Btu is the amount of heat energy required to raise the temperature of one pound of water by 1°F at sea level atmospheric pressure. **Figure 31–2** shows some heat principles. One Btu is equivalent to 1,055 joules or 1.055 kj.

Pressure and Heat

The temperature at which a liquid boils depends on the pressure acting on the liquid. Decreasing the pressure lowers the boil point. Increasing pressure raises the boil point. We know that atmospheric pressure is less at high altitudes. This results in the ability to boil water at a lower temperature when sitting on a mountain top. We also know that coolant in a truck cooling system is maintained under pressure. This has the effect of increasing the boil point.

In an air conditioning system, the temperature of **refrigerant,** the cooling medium in the system, is controlled by changing the pressure acting on it. This will also determine the state of the refrigerant. In an A/C system, we are concerned only with the refrigerant in its liquid and vapor states.

Latent Heat

Whenever a substance changes state, it either releases or consumes heat energy. Latent heat is the amount of heat necessary to change a substance from one state to another. When a liquid boils to form a gas, it consumes heat energy. The opposite occurs when a gas condenses to a liquid—when it releases heat. For instance, to change one pound of liquid water into steam at 212°F (100°C) a further 970 Btu must be added simply to enable the change of state. The heat that we add to the water to change its state without increasing its temperature is known as the latent heat of vaporization. When we reverse the change of state, that is, change the water vapor or steam back into liquid, then as the condensation takes place, 970 Btu of heat energy are given up or released. This heat release is known as the latent heat of condensation. Note also that the temperature at which water freezes and at which ice melts are identical. When water freezes, heat energy is released, and when ice melts, heat energy is consumed. **Figure 31–2** demonstrates the principle of latent heat.

The principle of latent heat is the basis of all air conditioning systems. The refrigerants used in air conditioning systems are chosen for their low boil temperatures, capability to change their state readily, and minimal impact on the environment.

SUMMARY OF A/C PRINCIPLES

Remember these air conditioning principles:

- Heat always moves from a warmer area to a cooler area.
- When liquids are heated and evaporate to a vapor state, heat is absorbed.
- When a gas condenses from a vapor to a liquid state, heat is released.

As we go through this chapter, the components of a typical truck air conditioning system will be studied

in some detail. **Figure 31–3** shows the location of these components in a typical highway tractor.

31.2 REFRIGERANT

The function of a refrigerant is to absorb heat from the air in the cab and release it to the atmosphere outside the cab. The refrigerants used in truck A/C systems are industrially manufactured chemicals of some complexity. Until 1995, the common refrigerant used in a truck A/C system was known as R-12. This is classified as a chlorofluorocarbon (CFC). This refrigerant boils at −22°F, and for years, R-12 was considered an ideal mobile vehicle air conditioning refrigerant. However, as a substance of some toxicity and a CFC, its use has become strictly controlled under the guidelines of the Federal **Clean Air Act.**

CFCs, such as R-12 and R-22 (used in trailer reefer systems), are known to deplete the atmosphere's upper ozone protection layer. Ozone in the upper atmosphere acts as a filter that blocks out harmful ultraviolet light. Chemically, ozone is a molecular bonding of 3 oxygen atoms that we know as triatomic oxygen (O_3). When CFCs are released to the atmosphere, they leach oxygen atoms from ozone molecules, reducing it to O_2 (diatomic oxygen), which is the same as the oxygen at ground level atmosphere.

Under the Clean Air Act, R-12 production was to be phased out beginning in 1995. R-12 has been replaced by new ozone-safe refrigerants. Of these CFC-free refrigerants, HFC-134a, a hydrofluorocarbon, has become the choice of truck OEMs for A/C systems. Other refrigerants are being introduced as replacements for R-22 in trailer reefer systems. In Mexico, propane is commonly used as refrigerant in reefer systems, where it functions exactly as it does in the refrigerator in a recreational vehicle. Propane has the advantage of being relatively unharmful if released to the atmosphere, but the disadvantage of being extremely flammable. An ideal refrigerant would be noncombustible, have convenient freeze/boil temperatures, and be completely nontoxic.

HFC-134a is not a recharge option for an R-12 system. There are significant differences. R-12 systems use a mineral oil lubricant that cannot be used with R-134a. In R-12 systems, mineral oil is soluble in the refrigerant. During system operation, the oil, which is in at least partial solution with the refrigerant, is pumped through the system, lubricating the compressor. Mineral oils, when put into a system

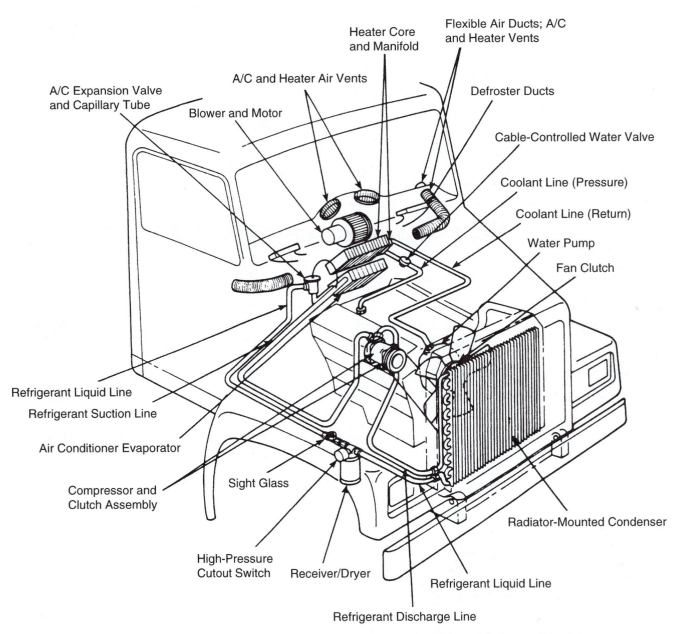

FIGURE 31-3 *Location of components of a typical heavy-duty truck heating and air conditioning system.*

containing R-134a, will not dissolve in the refrigerant and tend to increase in viscosity at the coldest point, the evaporator. Here the mineral oil drops out of circulation, reducing heat transfer and causing the compressor to starve for lubrication. At this point, any moisture in the system will also freeze.

Two new types of lubricants are used with hydrofluorocarbon A/C systems. Polyalkylene glycol (known as PAG) and polyolester (known as POE or ester) are used. Many OEMs specify PAGs in heavy-duty truck A/C units using R-134a. Though the min-

eral oil and PAGs lubricants are quite different, small quantities of mineral oil remaining in an old R-12 system **retrofitted** to R-134a should not normally cause a problem. Ester oil is very hygroscopic (that is, it absorbs moisture) and tends to be used in the aftermarket rather than PAG.

REFRIGERANT CONTAMINATION

In appearance, HFC/R-134a systems look similar to R-12 systems. However, refrigerants should never be mixed in a system, and to emphasize the fact, it has been made illegal. By guidelines issued by the Environmental Protection Agency (EPA), all refrigerants must be approved under the Significant New Alternatives Policy (SNAP). At this moment in time, only HFC-134a meets SNAP guidelines.

Any refrigerant introduced into an HFC-134a system, other than HFC-134a itself, would be considered a contaminant to the system. The correct refrigerant should always be used when servicing A/C systems. Introducing nonapproved refrigerants into a system can cause both performance and/or safety problems. Besides the effect of damaging the service equipment, contaminated refrigerant can cause some confusing problems when troubleshooting a system. For example, with only a 10 percent contamination of R-134a introduced to an R-12 system, the vapor pressure at 100°F would change from approximately 115 psi to 135 psi. At any given temperature, the operating pressure would be much higher than specified, yet evaporator cooling would be inefficient when compared to the system operating with uncontaminated refrigerant. **Figure 31–4** shows a pressure-to-temperature comparison of R-12 and R-134a.

The service equipment required to diagnose and recharge an air conditioning system is specific to the refrigerant, so R-12 equipment should not be used to recover or recharge an R-134a system. In practice, you would have to go out of your way to accomplish this, as R-12 and R-134a systems are equipped with different access service fittings. Older R-12 vacuum pumps will not bring the system to a low enough pressure to boil off moisture, so more powerful vacuum pumps (typically $^3/_4$–1 hp) are used to evacuate R-134a systems.

Refrigerant can be tested before connecting the service equipment. This is done with a refrigerant identifier. These testers identify flammable refrigerants, operate with either R-12 or R-134a, indicate the percentage of the base refrigerant, and indicate that the refrigerant is at an acceptable level (98 percent pure). This type of device helps protect shop equipment and refrigerant supplies. **Figure 31–5** shows a typical refrigerant identifier kit. It should be noted that in Mexico propane is widely used as refrigerant, and if propane is drawn into a recovery station, it will disable it.

 Shop Talk _____

Refrigerant containers for R-12 and R-134a are color coded. R-12 containers remain white in color, and R-134a containers are light blue and clearly marked. In addition, R-134a containers use $^1/_2$-inch 16 acme threads, which cannot be connected to an R-12 gauge set or recovery machine.

REFRIGERANT RECOVERY

The Clean Air Act, passed in 1992, has resulted in major changes in industrial, domestic, and vehicle air conditioning systems.

The Clean Air Act regulates the following:

- Establishes a requirement to recover all ozone-depleting (CFC) refrigerants. This means that they can never be vented to the atmosphere. CFCs must be recovered during the servicing and disposal of air conditioning or refrigeration equipment.
- Requires persons servicing air conditioning and refrigeration equipment to observe practices that reduce refrigerant emissions.
- Establishes equipment, off-site reclaimer, and technician certification programs.
- Includes a sales restriction (that may include licensing) on refrigerants regardless of container size.
- Mandates repair of leaks based on the annual leak rates of equipment.
- Requires that ozone-depleting compounds be removed prior to the disposal of the appliances and that all air conditioning and refrigeration equipment, with the exception of small appliances, be provided with a service fitting that would allow for the recovery of the refrigerant.

A/C CERTIFICATION

To work on vehicle A/C systems, technicians are required to be certified. Certification can be obtained by successfully completing a federally approved technician certification program. Three training programs in particular are recognized by the federal government, the Mobile Air Conditioning Society (MACS), the International Mobile Air Conditioning Association (IMACA), and the National Institute of Automotive Service Excellence (ASE). All along, the EPA has reserved the right to require recertification if and when "environmentally significant changes in equipment for motor vehicle air conditioning systems occur." The EPA also has the authority to revoke certification if a technician fails to demonstrate the ability to properly use the equipment. All recovery/recycling equipment must be Underwriters Laboratory (UL) certified to meet EPA and SAE standards.

Temperature/Pressure Relationship of Refrigerant

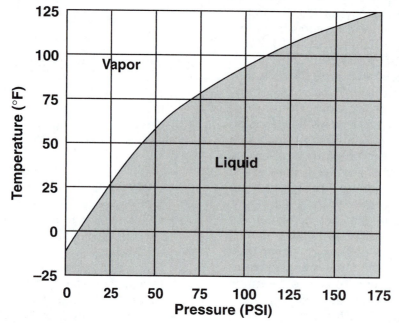

R-12 Temperature/Pressure Chart						
Temp (°F)	Press (PSIG)		Temp (°F)	Press (PSIG)		
16	18.36		100	117.16		
18	19.68		102	120.86		
20	21.04		104	124.63		
22	22.44		106	129.48		
24	23.88		108	133.11		
26	25.36		110	136.41		
28	26.88		112	140.49		
30	28.45		114	144.66		
32	30.06		116	148.91		
34	31.72		118	153.24		
36	33.42		120	157.65		
38	36.07		122	162.15		
40	36.97		124	166.73		
45	41.68		126	171.40		
50	46.70		128	176.16		
55	52.05		130	181.01		
60	57.74		135	193.52		
65	63.78		140	206.62		
70	70.19		145	220.30		
75	76.99		150	234.61		
80	84.17		155	249.54		
85	91.77		160	265.12		
90	99.79		165	281.37		
95	108.25		170	298.30		

R-134a Temperature/Pressure Chart						
Temp (°F)	Press (PSIG)		Temp (°F)	Press (PSIG)		
16	15.33		100	124.27		
18	16.66		102	128.58		
20	18.03		104	132.98		
22	19.45		106	137.48		
24	20.92		108	142.08		
26	22.43		110	146.79		
28	24.00		112	151.59		
30	25.62		114	156.51		
32	27.29		116	161.53		
34	29.01		118	166.66		
36	30.79		120	171.89		
38	32.63		122	177.24		
40	34.53		124	182.70		
45	39.52		126	188.27		
50	44.90		128	193.96		
55	50.69		130	199.76		
60	56.90		135	214.78		
65	63.55		140	230.54		
70	70.67		145	247.08		
75	78.27		150	264.40		
80	88.38		155	282.53		
85	95.01		160	301.49		
90	104.19		165	321.29		
95	113.94		170	341.96		

FIGURE 31–4   Temperature/pressure relationship of refrigerant. (Courtesy of Freightliner Trucks)

FIGURE 31–5 *Refrigerant identifier. (Courtesy of RTI Technologies, Inc.)*

31.3 THE REFRIGERATION CYCLE

All air conditioning systems use a refrigerant medium and cycle it continuously through a closed, pressurized system. The cycle consists of 4 stages that we describe as compression, condensation, expansion, and evaporation. As the refrigerant is pumped through this cycle, heat is removed from a hot area (the cab) and released to the atmosphere outside the vehicle. **Figure 31–6** shows a schematic of a refrigeration cycle. Note that it is divided into a high side and low side. High side and low side are used to describe the pressures of the refrigerant in each side of the system.

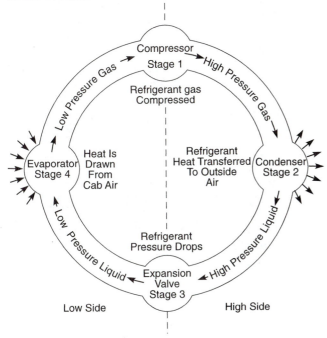

FIGURE 31–6 *The refrigeration cycle.*

COMPRESSOR STAGE

To understand the refrigeration cycle, it is normal to begin at the **compressor.** The function of the compressor is to change a low-pressure refrigerant gas to a high-pressure gas. Some compressors operate on a principle similar to the air compressor used in a truck air system, but other types of pumps are also used. In this case, refrigerant as a low-pressure gas is sucked into the compressor on the downstroke of the piston. The gas is then compressed on the piston upstroke. The process of compressing the refrigerant gas creates heat just as an air compressor heats up the air it compresses. This is known as the heat of compression.

CONDENSER STAGE

High-pressure gas exits the compressor and is forced through the circuit to the **condenser.** The condenser is a heat exchanger. It is usually located in the vehicle air flow so it is subject to the **ram air** effect (that is, the air flow created by vehicle movement) or air movement created by the engine cooling fan. Heat moves from the refrigerant to the outside air. Cooling of the refrigerant causes it to condense. This results in the refrigerant changing to a liquid state. Therefore, cooler, high-pressure liquid exits the condenser.

EXPANSION VALVE STAGE

The high-pressure liquid exiting the condenser is then forced through the system to the **expansion valve.** The expansion valve restricts the flow of the refrigerant, which has the effect of lowering its pressure as it exits the valve. Because it provides the restriction that enables the high pressure to be achieved upstream and reduces pressure downstream, it is the dividing point in the system for high/low pressure. The expansion valve allows the expansion of the liquid refrigerant to form a fine, atomized mist, which will allow it to absorb heat and vaporize more easily. Its action can be compared with that of restricting the outlet of a garden hose, creating a mist of droplets.

EVAPORATOR STAGE

Liquid, low-pressure refrigerant is moved through the system to the **evaporator.** The evaporator is located in the cab of the vehicle. Its function is to remove heat (supplied by the blower) from the cab. This is accomplished when the state of the refrigerant is changed from that of a liquid to a gas. Due to the latent heat of vaporization, when a substance is

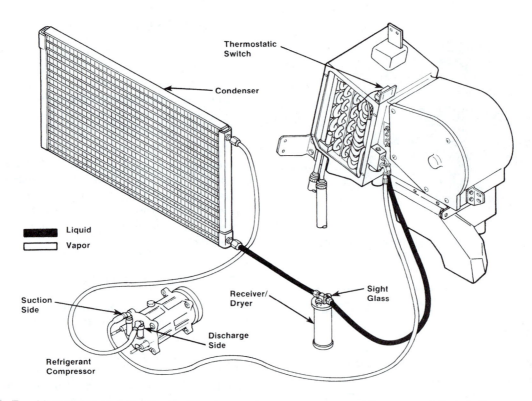

Liquid
Vapor

FIGURE 31-7 Heavy-duty truck air conditioning system components. (Courtesy of Freightliner Trucks)

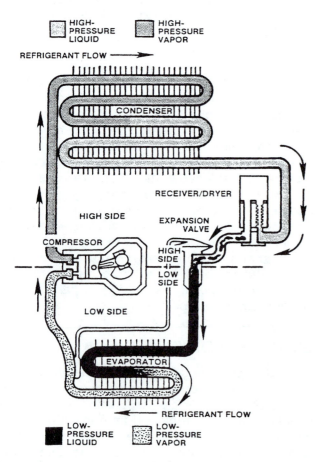

FIGURE 31-8 Refrigerant flow cycle.

changed from a liquid to a gas, it absorbs heat. It is in the evaporator stage that heat is removed from the recirculating cab air. It is this heat that causes the refrigerant in the evaporator to boil. This means that the cooling that takes place in the cab occurs in this stage of the cycle. Low-pressure refrigerant exits the evaporator and is returned to the compressor where the cycle repeats itself.

Figure 31-7 diagrams the components in a typical A/C circuit, and **Figure 31-8** uses schematics to show the same thing. Be sure to note the direction of flow and the state of the refrigerant in each.

31.4 AIR CONDITIONING SYSTEM COMPONENTS

Now we will take a closer look at the components found in a heavy-duty air conditioning system. The major components are

- Compressor
- Condenser
- Receiver/dryer or accumulator
- Expansion valve or orifice tube
- Evaporator

However, to put together an air conditioning system, other components are also required. This section takes a look at both the major and minor components that make an A/C system operate.

THE A/C COMPRESSOR

The compressor is the heart of the heavy-duty air conditioning system. It is one of the components that separates the high-pressure and low-pressure sides of the system (the other is the orifice tube or TXV). It is responsible for the movement of the refrigerant through the circuit. The primary purpose of the A/C compressor is to draw low-pressure vapor from the evaporator and compress it into a higher temperature, high-pressure vapor. This results in the refrigerant having a higher temperature than the surrounding air. The expansion valve located downstream from the compressor discharge provides the restriction required to define the high pressure side. The compressor also circulates the refrigerant in the system by creating a vacuum on its inlet side and pressure on its outlet side. The compressor is located in the engine compartment and is driven by engine power.

Numerous types of compressors are in use today, but they can generally be grouped into the following three types:

Piston-Type Compressors

The piston type of compressor has a crankshaft and its pistons are arranged in an in-line, axial, radial, or V configuration. They operate like a small 2-stroke cycle engine. The pistons are actuated by a crankshaft. An intake stroke takes place on piston downstroke, and a compression stroke takes place on piston upstroke. On the intake stroke, low-pressure refrigerant vapor from the evaporator is drawn into the compressor cylinder through reed valves. These one-way valves are designed to permit gas flow in one direction only and are sealed to prevent reverse flow-by pressure. The inlet reed valve permits refrigerant vapor to be drawn into the compressor cylinder on piston downstroke as shown in **Figure 31–9**. When the piston reverses and is driven upward, the intake reed valves seat, sealing the compressor cylinder. During the upward or compression stroke, the vaporized refrigerant is compressed as shown in **Figure 31–10**. This increases both the pressure and the temperature of the refrigerant gas without adding heat. Outlet or discharge reed valves then open, allowing the high-pressure, high-temperature refrigerant to move to the condenser. The outlet reed valves represent the beginning of the high side of the system. Reed valves are made of spring steel, which can be weakened or broken if improper charging

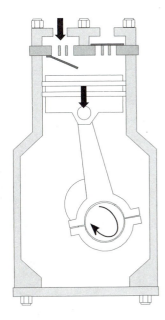

FIGURE 31–9   _Compressor piston downstroke._

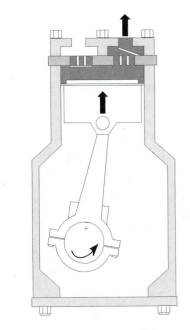

FIGURE 31–10   _Compressor piston upstroke._

procedures are used. **Figure 31–11** shows a cutaway view of a piston type compressor.

Rotary Vane Compressor

This type of compressor consists of a rotor with several vanes driven within a rotor housing. As the compressor shaft rotates, the vanes are moved into the rotor housing by centrifugal force. This forms chambers. As the refrigerant is drawn through suction ports into these chambers, the chamber volume reduces as the rotor turns. **Figure 31–12** shows this

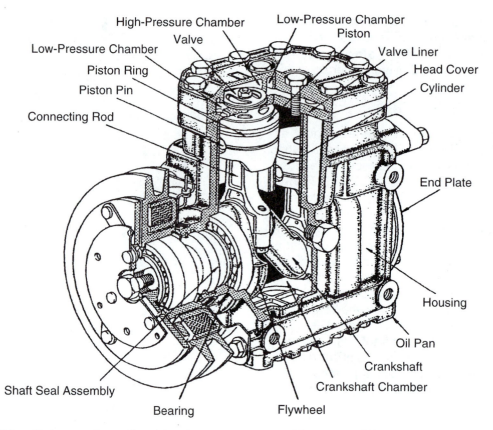

FIGURE 31–11   Cutaway view of a piston-type compressor. (Courtesy of Freightliner Trucks)

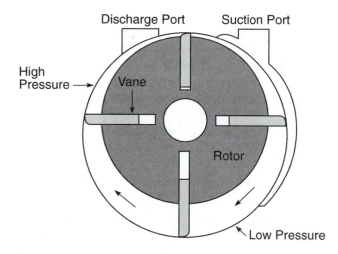

FIGURE 31–12   Operating principle of a rotary vane compressor. (Courtesy of Freightliner Trucks)

principle of operation. The discharge port is located so that compressed gas can be unloaded to be routed to the condenser. No sealing rings are required in a vane-type compressor. The vanes are sealed against the housing by centrifugal force and lubricating oil. The oil sump is located on the discharge side, so the high pressure tends to force it around the vanes into the low-pressure side. This

action ensures continuous lubrication. Because this type of compressor depends on a good oil supply, it is subject to damage if the system charge is lost. A protection device is used to disengage the clutch if pressure drops too low. **Figure 31–13** shows a cutaway view of a rotary vane-type compressor.

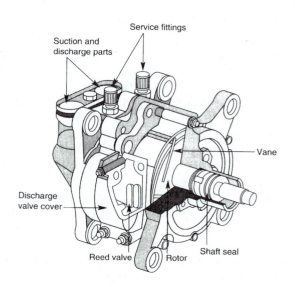

FIGURE 31–13   Cutaway view of a typical rotary vane compressor. (Courtesy of Freightliner Trucks)

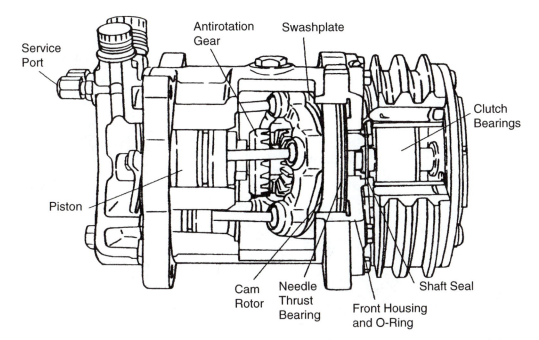

Service
Port

Antirotation
Gear

Swashplate

Clutch
Bearings

Piston

Cam
Rotor

Needle
Thrust
Bearing

Front Housing
and O-Ring

Shaft Seal

FIGURE 31–14 *Cutaway view of a swashplate compressor. (Courtesy of Freightliner Trucks)*

Swashplate-Type Compressors

A swashplate compressor uses a wobble plate principle to actuate the pistons. The cam rotor is actuated eccentrically, permitting a number of pumping cylinders to be contained in a compact space. Compression of the refrigerant gas works in pretty much the same manner as in the piston-type compressor. The essential difference is the absence of a crankshaft. **Figure 31–14** shows a sectional view of a swashplate-type compressor. An advantage of a swashplate-type compressor is that it requires about one half of the power to drive it compared with an equivalent-sized piston compressor.

Compressor Drives

Almost every compressor is equipped with an electromagnetic clutch as part of the compressor pulley assembly. This clutch is designed to engage the pulley to the compressor shaft when the clutch coil is energized. When the clutch coil is energized, engine power is transmitted to the compressor drive shaft, putting the compressor into effective cycle. When the clutch coil is not energized, the compressor pulley freewheels, and the compressor is not rotated. The clutch is therefore engaged by an electromagnetic field and disengaged by springs when the field collapses.

The compressor effective cycles are managed by an electrical circuit, which usually includes a thermostat. In most current truck air conditioning systems, electronics are used to help manage the air conditioning cycles.

Two types of electromagnetic clutches are used. Older A/C systems used a rotating coil-type clutch. The magnetic coil, which engaged or disengaged the compressor to the drive pulley, was mounted within the pulley and rotated with it. Electrical connections for the clutch operation are made through a stationary brush assembly and rotating slip rings, which are part of the field coil assembly.

More recent compressors use stationary-coil-type clutches. These reduce wear and, as a result, increase the service life. The clutch coil does not rotate. When the driver energizes the air conditioning system, the pulley assembly is electromagnetically engaged to the stationary coil on the compressor body, locking it to the clutch. Depending on the system, the magnetic clutch is usually thermostatically controlled and cycles the operation of the compressor, depending on evaporator temperature, to prevent freezing. One OEM uses a constant turning unit on a swashplate compressor and controls pressure with a bypass solenoid. In some system designs the clutch might operate continuously when the system is turned on. When the vehicle is powered by a smaller engine, a solenoid can be mounted between the discharge and inlet pipes, which open during full-throttle application, reducing the parasitic load on the engine. Some truck A/C compressors use a dry reservoir module (DRM) to cut out the compressor to reduce load on the engine at times when full engine power is required. With stationary coil design, service is not usually necessary except for an occasional check on the electrical connections. Some A/C systems, using a two-wire, isolated ground

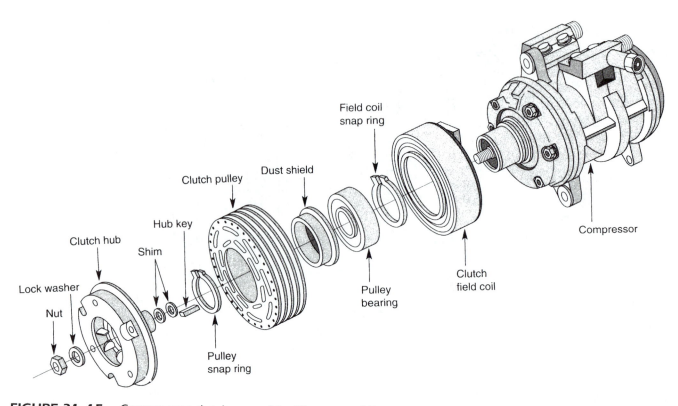

FIGURE 31–15 *Compressor clutch assembly. (Courtesy of Freightliner Trucks)*

circuit, use a diode to suppress voltage spikes. **Figure 31–15** shows the clutch assembly from a typical heavy-duty truck compressor.

CONDENSER

The condenser consists of a coiled tube surrounded by thin cooling fins. This maximizes the heat transfer ability using a minimal amount of space. The condenser is normally mounted just in front of the truck radiator, but if rooftop-mounted, it is located in the center of the cab roof. In either position, it receives plenty of ram air.

The purpose of the condenser is to condense—that is liquefy—the high-pressure, hot refrigerant gas it receives from the compressor. To achieve this, it surrenders its heat. Refrigerant vapor enters the inlet at the top of the condenser, and as the hot vapor passes down through the condenser coils, heat is transferred from the hot refrigerant to the cooler air flowing through the condenser coils and fins. This process causes the refrigerant to change state from a high-pressure hot vapor to a high-pressure warm liquid. The high-pressure warm liquid flows from the outlet at the bottom of the condenser and is routed to the **receiver/dryer** or orifice tube, depending on the system. **Figure 31–16** shows a typical condenser.

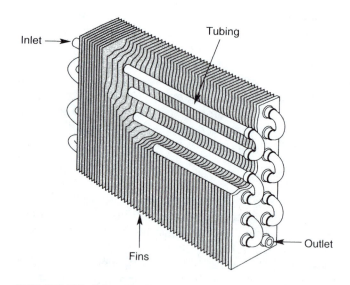

FIGURE 31–16 *Typical condenser. (Courtesy of Freightliner Trucks)*

In an air conditioning system operating under an average heat load, the condenser will have a combination of hot refrigerant vapor (including saturated vapor) in the upper two-thirds of its coils, with the lower third of the coils containing the warm condensed liquid refrigerant.

RECEIVER/DRYER

The receiver/dryer performs three important functions in the refrigeration cycle:

- Filters particles from the liquid refrigerant
- Removes moisture from the liquid refrigerant
- Stores a small portion of the liquid refrigerant until it is needed in the system evaporator(s) to meet high cooling demand

High-pressure liquid refrigerant flows from the condenser to the receiver/dryer. To protect the system from possible moisture-generated acid corrosion, a drying agent, or desiccant (moisture-absorbing material such as silica, aluminum, or silica gel), absorbs and separates any trace moisture contained in the liquid refrigerant. The ability of desiccant to retain moisture is minimal, so anything more than a few drops would present problems. Leaving it uncapped for a period of 5 minutes results in moisture contamination. The cylinder or tank portion of the receiver/dryer then collects and stores liquid refrigerant until system demands require additional refrigerant at the evaporator(s). **Figure 31–17** shows a typical receiver/dryer unit.

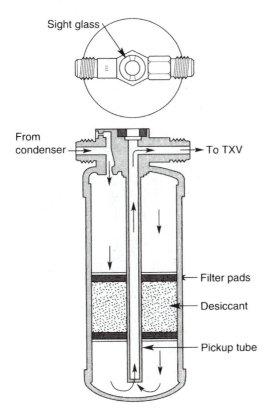

FIGURE 31–17 *Typical receiver/dryer unit.* (Courtesy of Mack Trucks, Inc.)

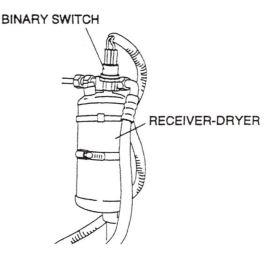

FIGURE 31–18 *Location of a binary switch.* (Courtesy of Freightliner Trucks)

The primary function of the receiver/dryer, however, is the removal of contaminants before they can cycle through the system and cause internal damage. If the compressor or any other system component fails or requires replacement, the receiver/dryer should also be replaced to prevent contamination of the system.

Binary Switch

A binary pressure switch can be located on the receiver/dryer. It provides low- and high-pressure protection for the system. It is called a binary switch because it performs two functions with a single switch. When the high-side pressure falls to a predetermined value below 30 to 15 psi, electrical contacts in the binary switch open. When these contacts open, the compressor clutch will disengage and refrigeration cycle will stop. Electrical contacts in the binary switch also open when the high-side pressure rises above 300 to 350 psi, once again disengaging the compressor clutch and stopping the refrigeration cycle. **Figure 31–18** shows the location of the binary switch on a typical receiver/dryer.

Trinary Switch

Some heavy-duty trucks are equipped with a trinary pressure switch instead of a binary pressure switch. The trinary switch is mounted on the receiver/dryer in a similar fashion to the binary switch. The trinary pressure switch performs three functions to monitor and control pressure inside the A/C system. Two of the functions are the same as on the binary switch system:

1. A low-range pressure function prevents compressor operation when the refrigerant charge has been lost or when ambient temperature is too cold.
2. A mid-range pressure function activates the engine fan clutch as system pressure reaches mid-range. (On some A/C systems, an electrically powered condenser cooling fan is activated instead.) This increases air flow to the condenser and stabilizes or lowers system operating pressures. The switch cycles on and off to maintain operating pressures.
3. A high-range pressure function turns off the compressor if the system pressure is too high.

All three switch functions reset when the proper system pressures are restored.

Pressure Relief Valve

A pressure relief valve is also often located on the receiver/dryer. It provides an added high-pressure relief feature if a failure were to develop in the high-pressure cut-off switch. The pressure relief valve is designed to pop off when the refrigerant pressure exceeds a preset maximum safe pressure value. Pop-off pressure can often be as high as 400 psi.

Fusible Plug

Another protective device is a fusible plug. A fusible plug is usually used in place of a pressure relief valve. The fusible plug will melt when the refrigerant temperature reaches approximately 230°F. When this happens, the refrigerant is discharged from the system. Unlike the pressure relief valve, when the fusible plug melts, the unit must be replaced. The pressure relief valve will close when the excess pressure is exhausted. An example of a fusible plug is shown in **Figure 31–19**.

Accumulator

An **accumulator** performs three functions:

1. Filters particles from the refrigerant
2. Removes moisture from the refrigerant
3. Separates refrigerant into vapor and liquid and allows only vapor to flow to the compressor from the evaporator, preventing compressor damage (A liquid cannot be compressed.)

Two of these functions are similar to those of the receiver/dryer. The important difference is the location of the accumulator. An accumulator is located at the outlet of the evaporator whereas the receiver/dryer is on the inlet side. Systems that use an accumulator operate on what is described as a flooded evaporator system.

The receiver/dryer and the accumulator are often neglected when the air conditioning system is serviced or repaired. Failure to replace either can lead to poor system performance or failure of other system parts. It is recommended service practice to replace these components whenever the system has lost refrigerant or has been open to the atmosphere for any length of time. More importantly, they should be replaced whenever a compressor has malfunctioned and contaminated the system with metal, aluminum, or plastic particles. **Figure 31–20** shows the operating principle of a typical accumulator.

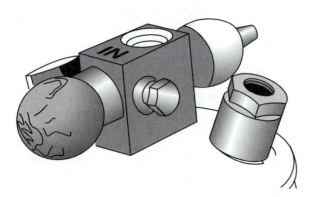

FIGURE 31–19 Fusible plug.

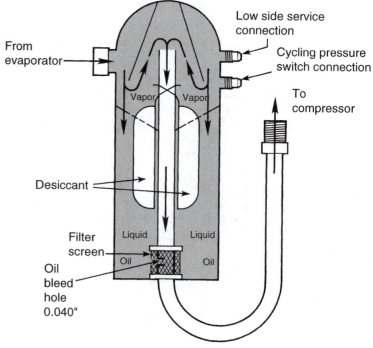

FIGURE 31–20 Typical accumulator. (Courtesy of International Truck and Engine Corp.)

In addition to receiver/dryers and accumulators, some systems use in-line filters and filter dryers. These provide some supplementary drying and filtering capacity. An in-line filter installed upstream from a replacement orifice tube will prevent repetitive plugging and subsequent compressor damage caused by poor lubricant flow through the orifice tube.

THERMOSTATIC EXPANSION VALVE/ORIFICE TUBE

A thermostatic expansion valve or orifice tube is used to divide the high-pressure and low-pressure sides of the refrigerant cycle. The refrigerant flow to the evaporator must be controlled to obtain maximum cooling and to ensure complete evaporation of the liquid refrigerant within the evaporator. Two devices are used to accomplish this—either a thermostatic expansion valve (TXV) or a fixed orifice tube (**Figure 31–21**). One of these devices must be incorporated in the system.

- The TXV or fixed orifice tube both function to provide a pressure drop in the circuit, which changes high-pressure refrigerant liquid into a low-pressure liquid.

- The TXV additionally regulates the amount of refrigerant entering the evaporator to match the cooling demand.

Thermostatic Expansion Valve (TXV)

This valve is mounted downstream from the receiver/dryer before the inlet to the evaporator. It separates the high-pressure side of the system from the low-pressure side. A TXV functions to regulate refrigerant flow to the evaporator. Excess refrigerant flow through the evaporator can cause evaporator flooding and compressor damage due to sludging liquid refrigerant. In operation, the TXV regulates the flow to the evaporator by balancing inlet flow to the outlet temperature.

Both external and internal equalized thermostatic expansion-type valves are used in heavy-duty air conditioning systems. The only difference between the two valves is that an externally equalized TXV uses an equalizer line connected to the evaporator outlet line as a means of sensing evaporator outlet pressure. Internally equalized TXVs sense evaporator inlet pressure through an internal equalizer passage.

Attached to the top of an external pressure-type expansion valve is an externally insulated capillary tube with a temperature measuring bulb, which is clamped to the outlet (suction) pipe of the evapora-

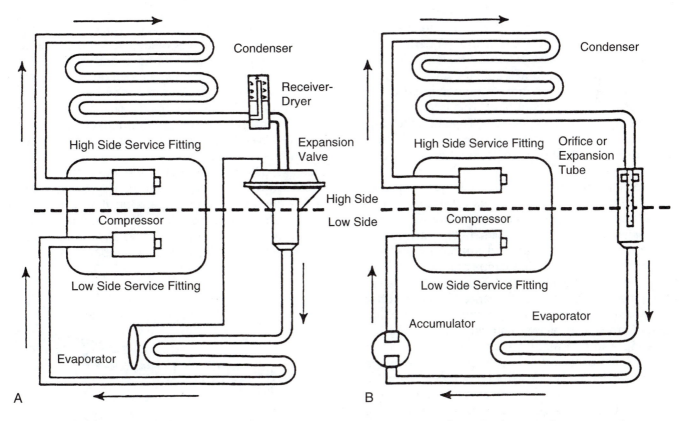

FIGURE 31–21 (A) Expansion valve (receiver/dryer) system; and (B) orifice tube (accumulator system).

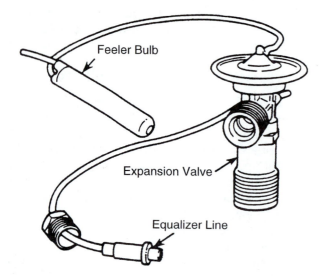

FIGURE 31–22 External pressure-type expansion valve. (Courtesy of International Truck and Engine Corp.)

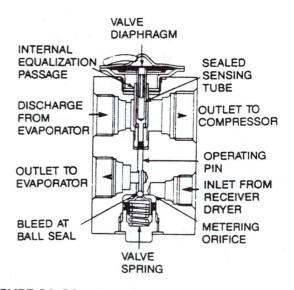

FIGURE 31–23 Block-type, internally equalized-type expansion valve. (Courtesy of Mack Trucks, Inc.)

tor. The sensing bulb and tube are filled with gas (typically the refrigerant used in the system) that expands and contracts according to the temperature surrounding the bulb. The expansion and contraction of the bulb gas exerts pressure on an internal diaphragm, which moves the metering needle valve against the closing spring pressure. The temperature bulb senses the temperature of the refrigerant leaving the evaporator. The externally equalized expansion valve has a second capillary tube attached under the valve diaphragm. It is connected to the evaporator outlet (suction) pipe. This is an equalizer line that senses suction pressure. Together, they regulate the amount of refrigerant entering the evaporator. **Figure 31–22** shows an external-pressure-type expansion valve.

In recent years, the block-type internally equalized TXV has become popular. It is less vulnerable to accidental damage because there is no exposed capillary tube or external connections. Inside the internal-type expansion valve is a sealed sensing tube that responds to changes in the temperature of the vapor flowing out of the evaporator by varying the opening of the metering orifice. As the vapor becomes hotter, the sensing tube expands, and the metering orifice opens further. This allows more liquid refrigerant into the evaporator to meet a higher cooling demand. Likewise, the sensing tube contracts, and the metering orifice becomes smaller as the vapor becomes cooler. This restricts the liquid refrigerant flowing into the evaporator. **Figure 31–23** shows a block-type internally equalized-type expansion valve.

Orifice Expansion Tube

Like the thermostatic expansion valve, the orifice expansion tube is the dividing point between the high- and low-pressure sections of the refrigerant cycle. However, its metering or flow rate control does not depend on comparing evaporator pressure and temperature. It is a fixed orifice, so it simply defines a flow area. This means that the device has no moving parts. The flow rate is determined by the size of the orifice, the pressure difference across the orifice, and by subcooling. **Subcooling** is the term used to describe the additional cooling of the refrigerant in the bottom of the condenser after it has changed from vapor to liquid. The flow rate through the orifice is more sensitive to subcooling than to pressure difference. Some of the latest orifice tubes used with an R-134a system use a spring-loaded orifice check valve; this opens up flow when pressure becomes too high and eliminates the cracking noise notable in some R-134a systems. **Figure 31–24** shows an orifice expansion tube.

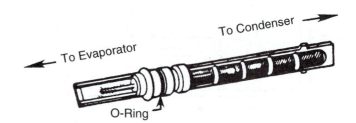

FIGURE 31–24 Orifice expansion tube.

EVAPORATOR

The evaporator, like the condenser, is a heat exchanger. It consists of a refrigerant coil mounted to a series of thin cooling fins. The idea is to provide a maximum amount of heat transfer in a minimum amount of space. The evaporator is located in the cab, frequently beneath the cab's dashboard or instrument panel. The heat transfer function of the evaporator is to draw heat out of the cab air.

The evaporator receives the low-pressure, low-temperature liquid refrigerant from the TXV or orifice tube in the form of an atomized spray. This atomized refrigerant is then evaporated (boiled) by the heat of the inlet air blown through the evaporator. That means its state is changed from a liquid to a gas. Heat from the evaporator core surface area is surrendered due to the latent heat of evaporation.

Actually, it is the heat of the cab air blown through the evaporator that causes the refrigerant to boil. Heat from this air is surrendered to the vaporizing refrigerant, and the result is that the air blown through the evaporator exits much cooler than it entered. In this way, the cab is cooled.

While the process of heat absorption from the air blown through the evaporator core is taking place, any moisture (humidity) in the air condenses on the outside of the evaporator core—as do dust, pollen, and smoke. An air conditioning system is so called because it does more than simply cool air. The condensed water from the cab airflow is drained off through a tube located under the evaporator core, and from there, discharged out of the vehicle. The capability to dehumidify air is a feature of any air-conditioning system that adds to the operator's comfort. It can also be used as a means of controlling fogging of the vehicle windows, and some OEMs engage the A/C compressor in defrost mode. **Figure 31–25** shows a typical evaporator core.

Evaporator cores should be inspected once in a while. Accumulation of dirt over time can plug the cooling fins or restrict the drain. Under high humidity

conditions, the catch basin may overflow onto the cab floor. Plugging of the cooling fins can cause an airflow restriction that can reduce evaporator cooling efficiency, but such failures are not common given the protected location of the evaporator. A simple test is to pour water into the air intake at the windshield and observe whether it drains out onto the ground through the drain pipe.

THERMOSTAT (EVAPORATOR FREEZE PROTECTION DEVICE)

The thermostat is equipped with a temperature probe that is routed to the evaporator core where it is mounted to sense temperature. This switch is often also referred to as a clutch or temperature cycling switch. The temperature probe must be carefully positioned in the evaporator core to ensure satisfactory system performance. A switch in the thermostat acts to cycle the compressor on and off (clutch engaged or disengaged) in response to the preset temperature sensed by the probe in the evaporator core. The thermostatic switch opens to disengage the compressor clutch when it senses that evaporator temperature has dropped to a predetermined low, usually in the region of 34°–37°F. This prevents evaporator freeze-up that could be caused by moisture condensing on the evaporator fins. **Figure 31–26** shows a typical thermostat and the location of the temperature probe.

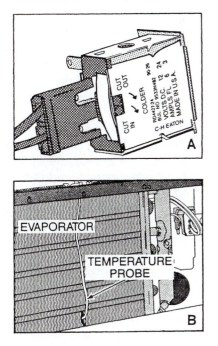

FIGURE 31–26 (A) Typical thermostat and (B) temperature probe. (Courtesy of Mack Trucks, Inc.)

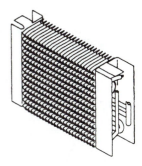

FIGURE 31–25 Typical evaporator core. (Courtesy of Mack Trucks, Inc.)

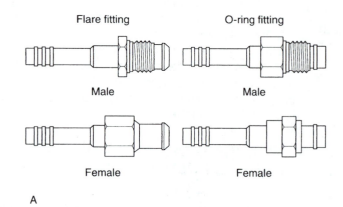
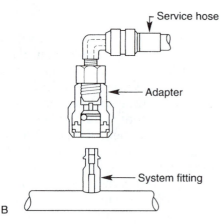

FIGURE 31–27   *Flare, O-ring, and quick fit connectors.*

Although the blower motor and fan continue operation, the refrigeration cycle ceases whenever the compressor is disengaged by its drive clutch. This means that refrigerant is no longer pumped through the system. When this occurs, evaporation continues to take place, but the evaporator temperature is permitted to rise. When sufficient temperature rise has taken place, the thermostatic switch again closes, causing the compressor clutch to engage, and the refrigeration cycle resumes.

REFRIGERATION LINES, HOSES, AND COUPLERS

All the major components of the system have inlet and outlet connections that accommodate either flare or O-ring fittings. The refrigerant lines that connect between these units are made up of special refrigerant hose or tubing with flare or O-ring fittings at each end. The hose or tube end of the fitting is constructed with sealing beads to accommodate a hose or tube clamp connection. Typical flare and O-ring fittings are shown in **Figure 31–27**.

Since the implementation of the Clean Air Act and EPA regulations on refrigerant systems and handling, there are now specified acceptable leakage rates of refrigerant. These rules have changed the design of refrigerant fittings and hose/fitting construction techniques (**Figure 31–28**). All new refrigerant hose is constructed to minimize refrigerant leakage rates. Because R-134a has a smaller molecular structure than the previously used R-12, it can leak through the standard double braid refrigerant hose used for many years in vehicle A/C applications. Current refrigerant hose is constructed with a barrier plastic inner layer in the hose to prevent refrigerant leakage through the hose wall. Examples of this are shown in **Figure 31–29**. These new barrier hoses are compatible with both R-12 and R-134a refrigerants. In order

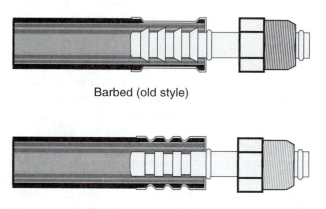

Barbed (old style)

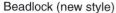

Beadlock (new style)

FIGURE 31–28   *Hose clamp shell crimp fittings will not seal the new style barrier hose.*

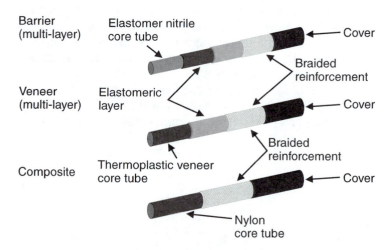

FIGURE 31–29   *New hoses are designed to prevent refrigerant leakage.*

to assemble the new hose, new-style bead lock fittings and special crimping tools are required as shown in **Figure 31–30**.

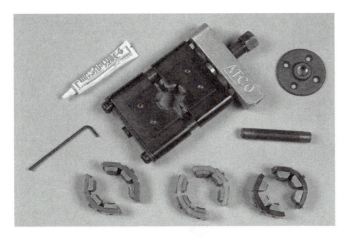

FIGURE 31–30   *Special crimping tools.*

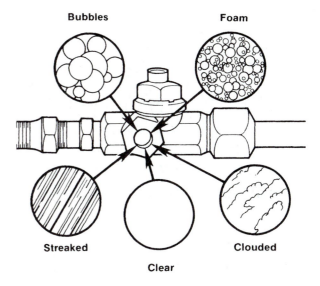

FIGURE 31–31   *Sight glass conditions.*

Refrigerant lines can be classified as follows:

- **Suction Lines.**   These lines, located between the outlet side of the evaporator and the inlet (suction) side of the compressor, carry the low-pressure, low-temperature refrigerant vapor to the compressor. Suction lines can be identified when the system is running, as they are cold to touch. Low-pressure lines are the larger diameter hoses in the system because they have to handle the same volume but at a lower pressure.
- **Discharge Lines.**   Beginning at the discharge outlet on the compressor, the discharge or high-pressure lines connect the compressor to the condenser, the condenser to the receiver/dryer and the receiver/dryer to the inlet side of the expansion valve. Through these lines, the refrigerant travels in its path from a gas state (compressor outlet) to a liquid state (condenser outlet) and from there, to the inlet side of the expansion valve. Discharge lines are warm to the touch and of smaller diameter than suction lines.

SIGHT GLASS

The **sight glass** allows the technician to see the flow of refrigerant in the lines. It can be located either on the receiver/dryer or inline between the receiver/dryer and the expansion valve or tube. Its functions all relate to diagnosing the refrigerant condition.

To perform a sight glass check of the refrigerant, run the vehicle engine, open the windows and doors, and set the A/C controls for maximum cooling with the blower on its highest speed. The system should be run for about five minutes, so make sure that it is in a well-ventilated area or connected to an exhaust

gas extraction system. Always take care when checking the sight glass while the engine is running.

Figure 31–31 shows a typical sight glass and how to interpret some conditions. The following is a guide to interpreting sight glass conditions:

- Oil-streaking indicates that the system is empty.
- Bubbles or foaming indicate that the refrigerant is low. Note that some retrofit systems (to R-134a) can show bubbles on a properly charged and functional system.
- Clouding indicates dessicant breakdown with possible contamination throughout the circuit.
- Clear with no bubbles indicates that the refrigerant level is correct with no contamination or completely empty.
Note: Bubbles normally can be observed when the system is started up.

IN-LINE DRYER

Located between the receiver/dryer and TXV or tube, an in-line dryer absorbs any moisture that gets by the receiver/dryer. It also helps to prevent TXV or tube freeze-up.

MUFFLER

Most heavy-duty truck air-conditioning systems now have a muffler installed in the system. The muffler is usually located on the discharge side of the compressor. It may also be located on the suction side. Its function is to reduce the characteristic pumping noise produced by the compressor. Mufflers should

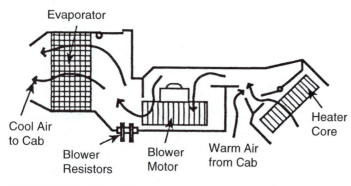

FIGURE 31–32 *Blower motor increases the air flow in the cab.*

be installed into the system correctly, usually with an inlet connection at the top and outlet connection at the bottom. This is done to minimize collection of oil in the unit.

BLOWER MOTOR AND FAN

The **blower motor/fan assembly** is located in the evaporator housing. Its purpose is to blow air through the evaporator and maximize its efficiency as a heat exchanger. The blower draws warm air from the cab, forces it through the coils and fins of the evaporator, and blows the cooled, dehumidified air into the cab. The blower motor is controlled by a fan switch. In most cases, the blower motor/fan assembly also provides warm air to heat the cab in cold weather. The A/C system is designed to cycle only when the blower motor is in the on position; insufficient heat at the evaporator would cause it to freeze in humid conditions. **Figure 31–32** shows the routing of air through the blower motor.

ENGINE COOLING FAN

The engine cooling system plays an important role in the operation of an air conditioner system. The A/C condenser is usually located in front of the engine cooling system radiator. At highway speeds, the ram air through the condenser and radiator usually is sufficient for the operation of the A/C system. However, under high heat and low engine and road speeds the condenser requires additional air, which is delivered by the engine fan.

In some recent truck engine applications, the conventional belt-driven, water-pump-mounted engine coolant fan has been replaced with an electrically driven fan not connected mechanically to the engine coolant pump. The 12-volt, motor-driven fan can be controlled by an engine coolant temperature switch or sensor (thermostat) and/or the air conditioner switch. **Figure 31–33** shows an electrically driven fan.

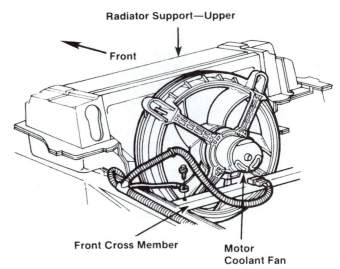

FIGURE 31–33 *Electrically driven fan.*

 **Shop Talk**

Remember that just one drop of water added to refrigerant will initiate corrosion and refrigerant breakdown. Corrosive hydrochloric acid can be produced in the older R-12 system when trace drops of water are added. Also, the smallest amounts of air in the refrigerant system can start chemical reactions that result in system malfunctions.

31.5 SAFETY PRECAUTIONS

Whenever repairs are made to any air-conditioning system, the system must be purged, flushed if contaminated, evacuated, recharged, and leak tested. Refrigerant lines are under some pressure even when the system is not running. This means that they should not be disconnected until the refrigerant has been discharged to a refrigerant recovery unit.

Refrigerants are safe when handled properly. Safety goggles and nonleather gloves should always be worn when testing and servicing A/C systems while discharging, purging, flushing, evacuating, charging, and leak testing the system. When refrigerant contacts leather, the leather can stick to skin.

CAUTION: Refrigerant should never come into contact with skin or eyes. Liquid refrigerant, when exposed to the air, quickly evaporates and will almost instantly freeze skin or eye tissue. Serious injury or blindness could result if you come into contact with liquid refrigerant.

Refrigerant splashed in the eyes should be treated as follows:

1. Flush eyes with cold water to raise the temperature above freezing point.
2. Avoid rubbing eyes.
3. If available, apply an antiseptic mineral oil to the affected area. This will form a protective film over the eyeball to reduce the possibility of infection. Then rinse with a weak boric acid solution.
4. Call an eye specialist or doctor immediately and receive medical treatment as soon as possible.

Refrigerant splashed on the skin should be treated the same as frostbite: Gently pour cool water on the area but do not rub the skin. Keep the skin warm with layers of soft, sterile cloth. Call a doctor right away.

Even though refrigerant does not burn, when it comes into contact with extreme heat or flame, poisonous phosgene gas is created. This gas also is produced when an open flame leak detector is used. Phosgene fumes have an acrid (bitter) smell.

> **CAUTION:** Avoid working in areas where refrigerant may come into contact with an open flame or any burning material, such as a cigarette. When it contacts extreme heat, refrigerant breaks down into poisonous phosgene gas, which, if breathed, causes severe respiratory irritation.

> **CAUTION:** Under current federal Clean Air Act regulations, refrigerants must be recovered and recycled by all users to protect the environment and never released into the atmosphere. Under these regulations, service facilities not having the required recovery and recycling equipment and properly trained and certified personnel are not permitted to perform any refrigeration service work.

Refrigerant Storage

Because of its very low boiling point, refrigerant has to be stored under pressure in containers. These containers should never be exposed to temperatures higher than 125°F. Refrigerant cans should not be left in direct sunlight.

31.6 PERFORMANCE TESTING AN A/C SYSTEM

This section contains a short introduction to air-conditioning system diagnosis. Because there are differences in each OEM system, this is just a general set of guidelines, and the best approach to testing and troubleshooting a system is to use the OEM service manual.

PRELIMINARY CHECKS

Before performance testing an air conditioning system, perform the following checks.

1. Check that the refrigerant compressor drive belt is not damaged and is correctly tensioned. Check the compressor mountings for tightness.
2. Using a thickness gauge, check the compressor clutch clearance to specification.
3. Visually check for broken, burst, or cut refrigerant hoses. Also check for loose fittings on each coupling.
4. Check for road debris buildup on the condenser coil fins. If necessary this can be cleaned using an air pressure nozzle, whisk brush, and soapy water. Avoid bending the cooling fins.
5. Check the sight glass for refrigerant moisture content and state of charge. The air conditioner should be on when checking the sight glass. In fact, it is best if the system has been run for some time.
6. If there is insufficient cab airflow, check that leaves or other debris have not entered the fresh air ducts under the windshield. If debris has entered, it can clog the evaporator fins. Also, ensure that the ducts are connected to the dash louvers and that the air-control flaps are moving properly.

OPERATIONAL TESTING

The following is a guide to conducting a performance test of a truck A/C system.

Cooling Check

Under conditions of extreme heat and high humidity, it will take longer to cool down the cab, especially if the fresh air vents are left open. To make a quick check to determine whether the air conditioner is is cooling sufficiently

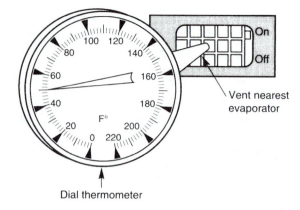

Dial thermometer

FIGURE 31–34 Checking outlet air temperature.

1. Set the engine speed at 1,500 rpm or the OEM-governed rpm level. On some computer-controlled engines, this may be less than 1,500 rpm when the vehicle is stationary.
2. Close the doors and windows.
3. Start the engine. If the unit is in a service shop, install an exhaust extraction pipe.
4. Turn on the air conditioner; set the thermostat and blow controls at maximum. Set the vent ducting to recirculate the air.
5. Ensure that the auxiliary air conditioning, if equipped, is off.
6. Run the system for at least 10 minutes, 15 minutes under the conditions of extreme heat.
7. Insert an A/C thermometer into the louver or air duct closest to the evaporator so that it is fully exposed to the air flow, as shown in **Figure 31–34**.
8. Compare the measured temperature readings to the specifications in the temperature/pressure table in the OEM service manual. Higher than specified temperatures indicate that the system is not cooling enough. Some of the possible causes are identified later in this chapter.

Odor Diagnosis

Under some operating conditions A/C systems may produce a musty odor, usually at start-up in hot weather. This type of odor is usually temporary, and, providing it disappears after a short period of operation, should not be of concern. If it persists, it can be an indication of microbial growth on the evaporator core. This requires the removal of the evaporator core and its thorough cleaning with a disinfectant. Some other odor diagnoses are shown in **Table 31–1**.

TABLE 31–1: ODOR DIAGNOSES

Problem	Possible Cause	Repair
Musty odor	1. External water leaks	1. Seal cab body.
	2. Plugged evaporator drain	2. Unplug drain.
	3. Molds and mildew	3. Clean evaporator.
Coolant odor	1. Anti-freeze solution leaking at heater core or hoses	1. Replace or repair heater core.
		2. Tighten hose clamps or replace heater hose.
Refrigerant leak	1. Refrigerant oil leaking at evaporator core	1. Replace evaporator core.

Refrigerant Check

Look for these four conditions when checking refrigerant in the sight glass:

1. Bubbles normally show only when the system is first started up with R-12 systems. These should disappear within a minute or two. If these bubbles continue to be seen at the sight glass while the system is operating, the refrigerant charge is low and should be recharged. The occasional bubble can be observed during normal A/C operation, particularly when the compressor clutch cycles on from off. Bubbles may appear in the sight glass in an R-134a system when the system is operating normally.
2. A completely clear sight glass means that the system refrigerant charge is correct and the system is operating correctly. However, it can also mean that there is no refrigerant or that it is overcharged.
3. A cloudy sight glass generally means that contamination or broken-down desiccant is being pumped through the system. This is a serious condition that requires a complete flushing of the system and the installation of a new receiver/dryer and expansion valve. If the desiccant cartridge of the receiver/dryer has disintegrated, installation of a new compressor will be required.
4. Oily streaks in the sight glass can mean low refrigerant or too much oil in the system.

Consult **Figure 31–31** to analyze sight glass conditions.

LEAK TESTING A SYSTEM

Testing an A/C system for leaks is an important phase of the troubleshooting process. Over a period of time, the refrigerant charge in all air conditioning systems becomes depleted. In systems that are in good condition, refrigerant losses of up to a half pound per year are considered normal. Higher loss rates indicate the need to locate and repair the leaks. **Figure 31–35** shows some typical potential leak sources.

Leaks are most often found at the compressor hose connections and at the various fittings and joints in the system. Refrigerant can be lost through hose permeation. Leaks can also be traced to pinholes in the evaporator caused by acid etching in R-12 systems. Acids form when water and refrigerant mix. Perform a visual inspection by looking for signs of A/C lubricant leakage and damage or corrosion in the lines, hoses, and components. Suspected leaks can be confirmed by using the leak detection methods that follow:

1. **Soapsuds Solution.** A solution of soap and water can be sprayed onto a suspected fitting or part. If the leak is large enough it will cause bubbles to form in the soapy solution at the point of the leak. This method is used when no other leak detection equipment is available.

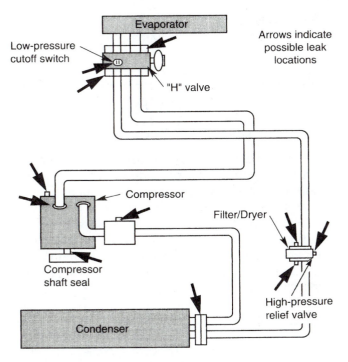

FIGURE 31–35 *Possible refrigerant leak locations.* (Courtesy of Freightliner Trucks)

2. **Fluid Leak Detectors.** Commercial liquid leak detectors perform the same task as the soapsuds solution but can find much smaller leaks. The fluid leak detector is applied to the area to be tested. If a leak is present, the liquid will form clusters of bubbles or large bubbles (depending on the size of the leak) around the source. A fluid leak detector is useful for locating leaks in a confined space in which the sensitivity of an electronic leak detector makes it an impractical instrument.

3. **Flame-Type Propane Leak Detector.** These are not commonly used today as they were designed to test CFC systems. A halide leak detector is less messy than a soapsuds solution and can locate small leaks. It is actually a small torch. When the flame comes in contact with refrigerant, the flame color changes. Propane leak detectors are a potential fire hazard and when refrigerant contacts the flame, poisonous phosgene gas is produced.

4. **Ultraviolet (UV) Leak Detectors.** Sometimes known as black light detectors, UV leak detectors are simple devices capable of identifying small leaks in an A/C system. Systems equipped with this phosphor dye/black light detection system have an identification label on the compressor or receiver/dryer. The phosphor dye may also be introduced to the low side of an A/C system in a specific small quantity. After the phosphor dye is in the system, a standard black light shone over a leak will produce a bright yellow trace at the leak point. This method of leak detection is used for hard-to-find or pinpoint leaks, especially cold or hot leaks. **Figure 31–36** shows a black light detector kit.

5. **Electronic Leak Detectors.** Any shop regularly performing A/C service work should be equipped with an **electronic leak detector.** It is the most sensitive of the leak detection instruments, so sensitive, in fact, that it can make sourcing a leak difficult, especially in a confined space. Electronic leak detectors are small hand held devices with a flexible probe used to seek out refrigerant leaks. When a leak is detected, a flashing light or buzzer signals the technician. The area should be free of oil, grease, and residual refrigerant before beginning the detection process. Suspected leakage areas should be cleaned using soap and water, not a solvent. A detected leak should be an active flow leak, not a residual condition caused by refrigerant trapped

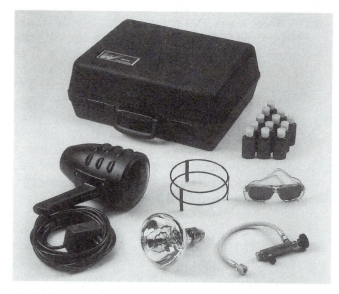

FIGURE 31–36 Black light detection kit. (Courtesy of White Industries, Div. of K-Whit Tools, Inc.)

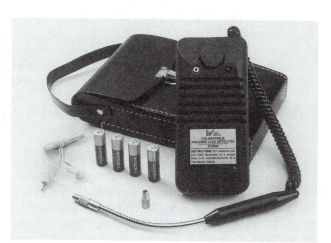

FIGURE 31–37 Handheld electronic leak detector kit. (Courtesy of White Industries, Div. of K-Whit Tools, Inc.)

under an oil film. Electronic leak detectors are usually specific to the refrigerant they are designed to test. Most current electronic leak detectors are of the heated diode type and will test only R-134a refrigerant. The handheld electronic leak detector kit shown in **Figure 31–37** is typical.

CAUTION: Never pressure- or leak-test R-134a service equipment or vehicle air conditioning systems with compressed air. Some mixtures of air and R-134a have been shown to be combustible at elevated pressures.

Shop Talk

Due to their sensitivity, electronic-type detectors must be correctly calibrated before each use to detect the lowest permissible leak rate of the component being checked. Trace the refrigerant system in a continuous path so that no potential leaks are missed. If a leak is found, always continue to test the remainder of the system. At each area checked, the probe should be moved around the location at a rate no more than 1 to 2 inches/second and no more than 1/4 inch from the surface completely around the position. Slower and closer movement of the probe greatly improves the likelihood of finding a leak. It also helps to place the probe lower than a suspected leak because the refrigerant is heavier than air.

ELECTRICAL SYSTEM CHECK

The electrical system should be checked periodically to prevent the truck air-conditioning system from failing unexpectedly. Perform the usual battery checks and follow the wire routing through to the A/C electrical circuit. Because truck air-conditioning systems contain various switches, they can fail completely or intermittently due to corrosion. Check for high resistance at switches and connections using a DMM. Computer managed climate control systems have voltage sensitivity parameters usually set at a low of 11 volts and a high of 16 volts. This helps protect the system against the consequences of low-voltage operation and high-voltage spikes.

31.7 A/C SERVICE EQUIPMENT

Several specially designed pieces of equipment are required to perform test procedures. They include a **manifold gauge set**, service valves, vacuum pumps, a charging station, a charging cylinder, and recovery/recycling systems. Although both R-12 and R-134a use basically the same types of testing equipment, they are not interchangeable. R-134a systems and equipment have metric connectors, but R-12 use English connections.

MANIFOLD GAUGE SETS

A manifold gauge set is used to recycle, recharge, and diagnose A/C systems. The left side gauge of a standard gauge set is blue and known as the low-pressure gauge. The dial is graduated from 1 to 150 psi (with a cushion up to 250 psi) in 1-psi increments, and, in the opposite direction, in inches of mercury (Hg) vacuum from 0 to 30. This gauge is used to check pressure on the low-pressure side of the refrigeration system. The gauge at the right side of the manifold gauge set is red and graduated from 0 to 500 psi in 10-psi graduations. This is the high-side gauge, and it is used for checking pressure on the high-pressure side of the system.

The center manifold fitting is common to both the low and the high side and is for evacuating or adding refrigerant to the system. When this fitting is not being used, it should be capped. A test hose connected to the fitting directly under the low side gauge is used to connect the low side of the test manifold to the low side of the system, and a similar connection is found on the high side. **Figure 31–38** compares the manifold gauge sets used with R-12 and R-134a systems.

The manifold gauge set is designed to access the refrigeration circuit and control refrigerant flow when adding or removing refrigerant. When the manifold test set is connected into the system, pressure is registered on both of the gauges at all times. During testing, both the low- and high-side hand valves should be in the closed position (turned inward until the valve is seated).

Refrigerant will then flow around the valve stem to the high- and low-side gauges to display the operating pressures on each side of the refrigeration circuit. The hand valves isolate the low and high side from the central portion of the manifold. When the gauges are connected to the gauge fittings with the refrigeration system charged, the gauge lines should always

be purged. Purging is done by cracking each valve on the gauge set to allow the pressure of the refrigerant in the refrigeration system to force the air to escape through the center gauge line. Failure to purge lines can result in air or other contaminants entering the refrigeration system. **Figure 31–39** shows the manifold gauge set components.

> **CAUTION:**   Never open the high-side hand valve with the system operating and a refrigerant source at the center hose connection. This will cause refrigerant to exit the A/C system under high pressure into the source container, which could cause it to burst. The only time both hand valves should be open is when evacuating the system.

🥽 Shop Talk

Atmospheric pressure reduces as altitude increases. The gauge pressure must be corrected using an OEM correction chart when checking the system's low-side pressures.

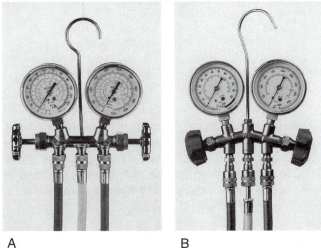

A   B

FIGURE 31–38   *(A) R-12 manifold gauge set and (B) R-134a manifold gauge set. (Courtesy of White Industries, Div. of K-Whit Tools, Inc.)*

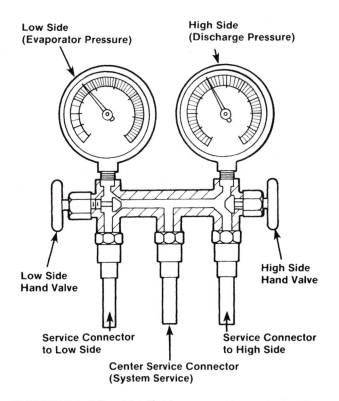

FIGURE 31–39   *Manifold gauge set components. (Courtesy of Mack Trucks, Inc.)*

Service Valves

System service valves incorporate a service gauge port for manifold gauge set connections and are provided on the low and high sides of some air conditioning systems. When making gauge connections, the gauge lines should be purged by cracking the charging valve and bleeding a small amount of refrigerant, followed by immediate connection of the lines. There are two types of service valves used—stem type and Schrader.

Stem-Type Service Valve. The stem-type valve is sometimes used on 2-cylinder reciprocating-piston-type compressors. Access to the high-pressure and low-pressure sides of the system is provided through service valves mounted on the compressor head. The low-pressure valve is mounted at the inlet to the compressor, and the high-pressure valve is mounted at the compressor outlet. Both of these valves can be used to isolate the rest of the air-conditioning system from the compressor when the compressor is being serviced. These valves have a stem under a cap with the hose connection directly opposite. A special wrench is used to open the valve to one of the three following positions:

- **Back seated.** The normal operating position with the valve stem rotated counterclockwise to seat the rear valve face and to seal off the service gauge port.
- **Mid-position.** The test position with the valve stem turned clockwise (inward) $1\frac{1}{2}$ to 2 turns to connect the service gauge port into the system, allowing gauge readings to be taken with the system operating. (A service gauge hose must be connected with the valve completely back seated.)
- **Front seated.** The test position with the valve stem rotated clockwise to seat the front valve face and to isolate the compressor from the system. This position allows the compressor to be serviced without discharging the entire system. The front seated position is for service only and is never used while the A/C is operating.

Note the valve positions as shown in **Figure 31-40**.

Schrader Type. A Schrader valve resembles a tire valve (**Figure 31-41**). They are usually located in the high-pressure line (from compressor to condenser) and in the low-pressure line (from evaporator to compressor) to permit checking of the high side and low side of the system. All test hoses have a

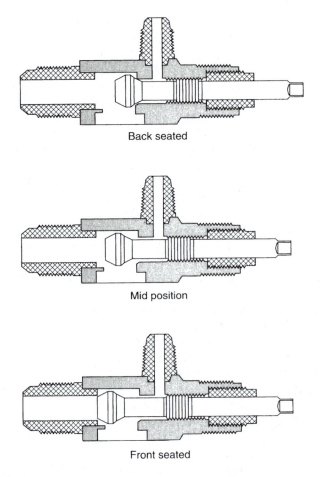
Back seated

Mid position

Front seated

FIGURE 31-40 *Service valve positions.*

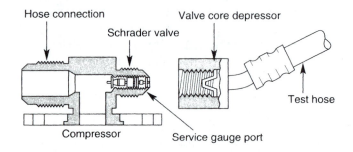

FIGURE 31-41 *Schrader-type service valve. (Courtesy of Mack Trucks, Inc.)*

Schrader valve core depressor in them. As the hose is threaded onto the service port, the pin in the center of the valve is depressed, allowing refrigerant to flow to the manifold gauge set. When the hose is removed, the valve closes and seats automatically.

Shop Talk

In R-12 systems, the Schrader fitting on the high-pressure side is smaller than the low-pressure side, and special adapters are necessary to hook the

high-side service hose into the system. The difference in fitting sizes is to prevent mixing up the high- and low-pressure sides of the system when attaching the gauge set. The different size fittings were introduced to prevent disposable 1-lb. cans of refrigerant from being connected to the high side of the system, causing them to explode. After disconnecting the gauge lines, check the valve areas to be sure the service valves are correctly seated and that the Schrader valves are not leaking. In R-134a systems, the size as well as the type of connection is different, with the larger connection being for the high side.

MANIFOLD SERVICE HOSE SETS

Manifold service hose sets consist of three color-coded (usually blue for the low side, red for the high side, and yellow for the auxiliary) 8-foot-long hose sets (**Figure 31–42**). Separate sets are used for R-134a and R-12 refrigerants. They are high-strength, nylon barrier/low permeation-type sets that meet SAE J2196 standards. They have a burst pressure rating of 2,100 psi and a working pressure rating of 470 psi.

Vacuum Pump

Any air or moisture that is left inside an air conditioning system, even in trace quantities, can reduce the system's efficiency and can eventually lead to major problems, such as compressor failure. Air causes excessive pressures within the system, limiting the refrigerant's capability to change its state from gas to liquid within the refrigeration cycle. Water moisture can cause freeze-up at the expansion valve. This can either restrict or completely block refrigerant flow. These problems result in reduced or no cooling. Additionally, moisture can combine with R-12 refrigerant to form corrosive hydrochloric acid, which can destroy the internal components of the A/C system.

The function of the vacuum pump is to evacuate the A/C system—that is, to remove air and moisture from the system. There are several different types of vacuum pumps, but the rotary vane type is most common in heavy-duty truck A/C work. The idea is to draw all air and moisture out of the system. The system pressure is reduced until any moisture in the system vaporizes and is eliminated, along with the exhausted air. An R-134a system requires at least a $3/4$-hp vacuum pump to create the 29.9 inches of Hg vacuum required to evaluate the system. **Figure 31–43** shows the operation of a vacuum pump in an A/C system, and **Figure 31–44** shows a typical rotary vane-type vacuum pump.

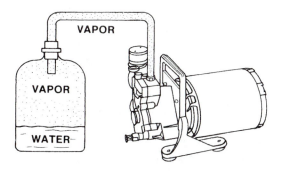

FIGURE 31–43   _A vacuum pump creates and holds a vacuum. (Courtesy of Freightliner Trucks)_

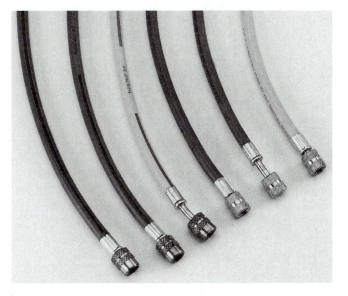

FIGURE 31–42   _Manifold service hose sets. (Courtesy of White Industries, Div. of K-Whit Tools, Inc.)_

FIGURE 31–44   _Rotary vane-type vacuum pump. (Courtesy of Robinair, SPX Corporation)_

Shop Talk

A vacuum pump is unable to remove moisture in liquid state from a system. It lowers the system pressure and, therefore, the boiling point of liquid moisture. It then removes the vaporized moisture. This is why it is necessary to pull a vacuum on a system for an extended period to ensure that no moisture is left in it.

Thermistor Vacuum Gauge

The electronic thermistor vacuum gauge is designed to work with the vacuum pump to measure the last, most critical inch of mercury vacuum during evacuation. It constantly monitors and visually indicates the vacuum level so a technician will know for sure when a system is entirely free of air and moisture. The thermistor vacuum gauge is capable of high accuracy and is calibrated in microns (millionths of a meter).

Refrigerant Management Centers

The refrigerant management center recovers, recycles, and recharges an A/C system. For the service facility specializing in A/C service work, refrigerant management is a requirement with today's refrigeration equipment. The stations are refrigerant-specific, so generally an R-134a station is used only for systems using that refrigerant. Such an arrangement not only ensures clean, single-pass, recycled refrigerant but also provides the accuracy of recharging required by today's "critical charge" A/C systems.

Most refrigerant management centers permit the technician to switch the refrigeration source, allowing the system to be charged from either a recycled or a virgin refrigerant tank. The manifold gauges are built into a refrigerant management center, allowing the center to coupled in series with the vehicle system. Some machines permit the operator to divide the center into a recovery/recycling station and a separate charging station. This provides flexibility for busy shops.

31.8 | A/C SERVICE PROCEDURES

All air conditioning service procedures should be performed referencing the OEM's service literature and their recommended procedure. Some general guidelines are provided here so that you can develop a general understanding of each procedure. System servicing normally comprises

- System recovering and recycling
- System flushing
- Compressor oil level checks
- Evacuation
- System recharging

CHARGING CYLINDER OR CHARGING STATION

A charging cylinder is designed to meter out a specific amount of a refrigerant by weight. Compensation for temperature variations is accomplished by reading the pressure gauge of the cylinder; it is essential that this temperature compensation be performed.

When recharging an air conditioning system, often the pressure in the system reaches a point at which it is equal to the pressure in the cylinder from which the system is being charged. To get more refrigerant into a system to complete the charge, some heat must be applied to the cylinder, or the refrigerant gas can be drawn into the low side with the system running. Some recharging cylinders such as that shown in **Figure 31–45** have heaters that eliminate the problem caused by the equalization of pressure between the cylinder and the system being recharged.

All refrigerant must be recovered from the system before repair or replacement of any component (except for compressors with stem-type service valves). This procedure is accomplished through the

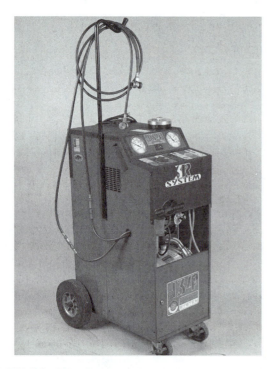

FIGURE 31–45 *Recharging cylinder. (Courtesy of Robinair, SPX Corporation)*

use of a manifold gauge set or a recovery/recycling station, which makes it possible to control the rate of refrigerant recovered, thus minimizing the loss of oil from the system. Two types of recovery stations are shown in **Figure 31–46**, and the components are shown in **Figure 31–47**.

The recovery procedure is performed as follows:

1. Remove the caps from the compressor's suction and discharge service valves (see **Figure 31–48**).
2. Wearing protective goggles and nonleather gloves, couple the manifold gauge set to the valves by performing the following sequence:
   - Close both manifold valves.
   - Connect the manifold's high-side hose to the compressor's discharge service valve.
   - Connect the manifold's low-side hose to the compressor's suction service valve.
   - Connect the manifold's center hose to an approved recovery and recycling station.

3. Follow the recovery and recycling station manufacturer's instructions and recover all of the refrigerant from the system. If the station has a one-pass recovery/recycling operation, the refrigerant that comes out of the A/C system into the machine, through the filtration system, and into the recovery tank is clean, and there is no separate recycling operation required. **Figure 31–49** shows a typical recovery station hookup.
4. Measure oil in the station's oil recovery cup. The compressor will have to be refilled with the same quantity of new refrigerant oil. If the system has become contaminated, all of the compressor oil must be replaced with clean oil after purging and flushing the system. If the system is heavily contaminated with desiccant or grit, replace the compressor, expansion valve, and receiver/dryer.

If the station does not have a one-pass recovery/recycling feature, the recycling can be performed in a typical straight recycling machine as follows:

FIGURE 31–46   Typical refrigerant stations. (Courtesy of White Industries, Div. of K-Whit Tools, Inc.)

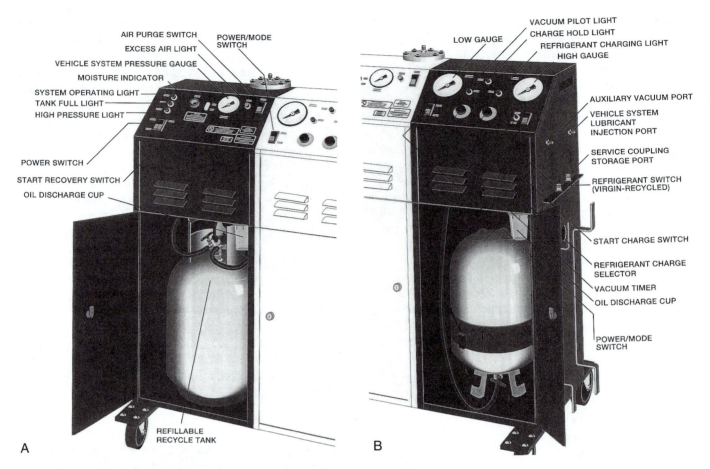

FIGURE 31–47 (A) Components of the recovery and recycle portion of a refrigerant center and (B) recharging portion. (Courtesy of White Industries, Div. of K-Whit Tools, Inc.)

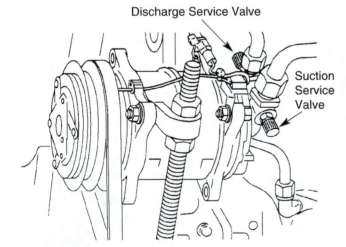

FIGURE 31–48 *Discharge and suction service valves. (Courtesy of Freightliner Trucks)*

1. The liquid valve on the fill tank should be fully opened at this time. Make sure that the hose is properly connected to the port.
2. Push the toggle switch to the recycle position.

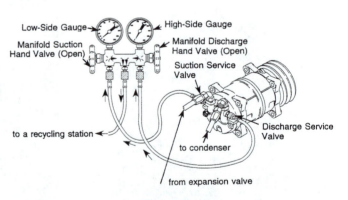
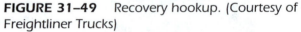

FIGURE 31–49 *Recovery hookup. (Courtesy of Freightliner Trucks)*

3. Momentarily depress the cycle start button (2 to 3 seconds). The system begins its recycling process. Liquid refrigerant should be seen flowing through the moisture indicator glass at this time.
4. Leave the unit in the recycle mode until the moisture indicator turns from yellow to green. This process will take anywhere from

1 to 6 hours, depending on the amount of moisture contamination in the refrigerant.

5. The color sample dot indicates the minimum level of dryness. When the moisture indicator is darker green than the sample, put the rocker switch from the recycle position to the end cycle position. The unit will then shut off completely after a few seconds. It is normal for the machine to vibrate as it works in a vacuum just before shutting off.

6. Close the liquid valve on the tank, disconnect all of the hoses on the tank, and remove the tank from the unit.

7. Apply the magnetic label to the tank to indicate that it has been recycled and is ready to use.

PURGING AND FLUSHING

Purging removes air, moisture, traces of refrigerant, and loose dirt from the A/C circuit, forcing inert, moisture-free gas through the parts. Flushing is more extreme and removes heavy contamination, such as gritty oil, larger particle dirt build-up, metal, or plastics, which cannot be removed by purging with gas. Flushing is performed by forcing liquid refrigerant through the system, allowing it to pick up the contaminants and flush them out.

The decision as to whether to purge or flush a system depends on the extent of contamination. Most routine servicing and recharging does not require the system to be flushed. Flushing with refrigerant requires a recovery and recycling station. It should be performed in the opposite direction of refrigerant flow; in other words, it is back-flushed. Note that some OEM systems should not be flushed due to the type of condenser tubes used. Verify that it is okay to flush a system before doing so.

Purging Procedure

Dry nitrogen gas is usually recommended for purging, but safe, inert, moisture-free gases, such as carbon dioxide or argon, may be substituted. A pressure regulator is required to regulate between 0 to 200 psi. Commercial cylinders of nitrogen are stored at pressures exceeding 2,000 psi; this pressure must be reduced to 200 psi for purging. See **Figure 31–50** and **Figure 31–51**.

1. Recover the refrigerant from the A/C system.

2. Disconnect both ends of the line or component being purged. Seal the rest of the system by capping.

3. Make sure that the nitrogen cylinder control valve, the purging control valve, and the supply line valve are all closed.

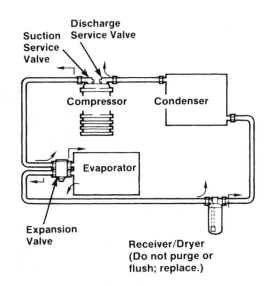

FIGURE 31–50   *Arrows show purge and flush points and the direction of flow. (Courtesy of Freightliner Trucks)*

4. Connect the supply line valve to the outlet end of the component or line.

5. Connect the drain line to the inlet end of the component or line.

6. Place the outlet of the drain line into a recycling system container.

7. Adjust the gas regulator to 200 psi. Open the nitrogen cylinder control valve and the purging control valve. Then slowly open the supply line valve. Check the drain line for gas flow.

8. Let the nitrogen flow at 200 psi for 25 to 30 seconds.

9. Lower the pressure to around 4 psi and let it flow for 1 to 2 minutes. If the line or component was visibly wet, allow the gas to flow until no trace of refrigerant oil or other matter is flowing from the drain tube.

10. Close the nitrogen cylinder control valve and the purging control valve first and then close the supply line valve.

11. Disconnect the supply line valve and the drain line. Tightly cap both ends of the component or line.

Flushing Procedure

Flushing should be performed with the specified refrigerant required in the system. In older systems, flushing agents were used, but these are no longer approved. To flush a typical A/C system, proceed as follows:

1. Recover the refrigerant from the air-conditioning system.

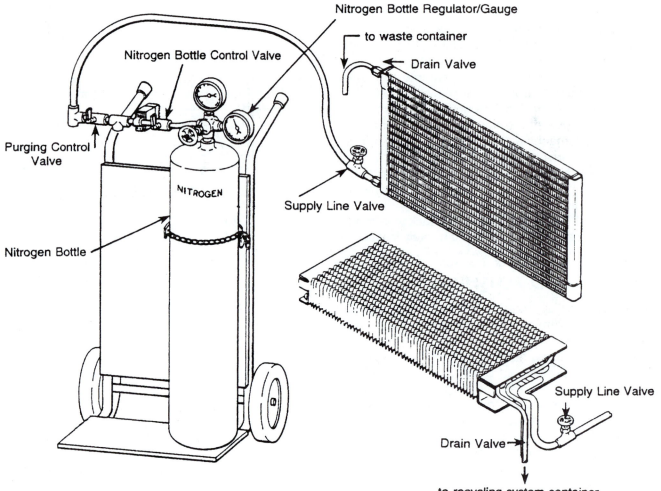

FIGURE 31–51 Purging station. (Courtesy of Freightliner Trucks)

2. The compressor, receiver/dryer, accumulator, or control valves cannot be flushed. To isolate these components from the system, install bypass/interconnect flush fittings. If only one component is to be flushed, only the hoses needed to isolate that component need to be disconnected.

3. Remove the expansion valve or orifice tube. Clean the expansion valve and discard the orifice tube.

4. A system should typically be flushed with at least twice the system capacity to remove all the contaminated lubricant and contaminated matter. Connect the refrigerant (flush) container to the system with an approved refrigerant test hose at the point where the receiver/dryer or accumulator was disconnected. It may be necessary to use a flushing adapter. Connect the recovery equipment to the other position of the receiver/dryer or accumulator lines with the proper adapter and hose. It is permissible to flush in either direction of the normal flow in the circuit.

5. With the connectors coupled as described, invert the refrigerant flush container so that liquid, not gas, is used as a flushing agent. Rotate the valve all the way open.

6. A charging scale, station, or cylinder can be used to ensure that the proper amount of flushing liquid is used. Recycled refrigerant is perfect for system flushing, and new refrigerant is not required.

7. The procedure to flush a specific component is much the same as described except that it is isolated from the circuit. Remember, never vent the refrigerant to atmosphere.

8. After the refrigerant flush has been used, shut off the canister hand valve. Allow the recovery unit to complete its recovery time and cycle out. Shut off the valves and disconnect the equipment and adapters.

9. After the system has been flushed, reconnect or reinstall the compressor. Ensure it has the correct quantity of lubricant. Reinstall a new orifice tube or expansion valve, replacing sealing O-rings or gaskets.

10. Install a new receiver/dryer and/or accumulator and reassemble using new O-rings. Finish by evacuating the system and recharging the system to specification.

CHECKING COMPRESSOR OIL LEVEL

It is unnecessary to routinely check the oil level of the compressor. This is usually only required when one of the following has occurred:

- Ruptured refrigerant hose
- Severe leak at any
- Badly leaking compressor seal
- Collision damage to the system components

When replacing refrigerant oil, it is important to use the specific type and quantity of oil recommended by the compressor manufacturer. **Figure 31–52** shows how oil is drained from the compressor, and **Figure 31–53** shows a compressor being filled with oil. If there is a surplus of oil in the system, too much oil will circulate with the refrigerant, causing the cooling capacity of the A/C system to be reduced. Too little oil will result in poor lubrication of the compressor. When recovering the refrigerant from the system or replacing system parts, add only enough oil to replace that which has been lost with

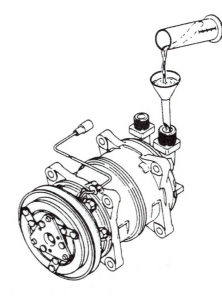

FIGURE 31–53 _Filling compressor with oil._
(Courtesy of Freightliner Trucks)

the refrigerant charge. A rule of thumb is to add 2 ounces of oil for each component replaced.

When handling refrigerant oil

- Oil must be from a container that has not been opened or that has been tightly sealed since its last use. PAG and ester lubricants are hygroscopic (absorb moisture) and are quickly contaminated when left open to the atmosphere.
- The oil should be free of contaminants.
- Never mix any refrigerant lubricants with other types/viscosities of lubricant.
- Tubing, funnels, or other equipment used to transfer the oil should be clean and dry.

During the refrigeration cycle, some of the oil circulates with the refrigerant in the system. When checking the amount of oil in the system, first run the compressor to return the oil to the compressor. The following is a typical procedure for adding oil to a compressor.

1. Turn on the air conditioner (compressor clutch must be engaged) and set the cab blower fan at high speed.
2. Run the compressor for at least 20 minutes between 800 and 1,200 rpm.
3. Shut down the engine and remove the compressor from the engine.
4. Remove the compressor sump drain plug to drain the compressor oil to a clean container. Rotate the drive plate several times by hand to extract the remaining oil through the discharge-side port. About an

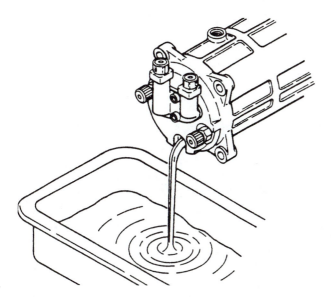

FIGURE 31–52 _Draining oil from compressor._
(Courtesy of Freightliner Trucks)

ounce of oil will remain in the compressor, and this cannot be removed without disassembling it.

5. Measure the drained oil with a measuring cylinder/graduated vial.

6. Visually check the oil for contamination. Unlike engine oil, compressor oil contains no added cleaning agent. Even after extended use, the oil should not become cloudy, as long as there is nothing wrong with the compressor. Inspect the drained oil for any of the following conditions:
   - Cloudiness, or oxidized or moisture-contaminated refrigerant. Severely oxidized refrigerant can cause gumming and clogged filters and passages.
   - Color change to red, which indicates varnishing, the result of exposure to air.
   - Presence of foreign particles (metal powder, etc.) in the oil.

7. Install the drain plug and a new O-ring lightly coated with clean compressor oil. Tighten the plug to the specified torque.

8. Fill the compressor with the specified quantity of oil through the suction-side hose fitting.

EVACUATING THE SYSTEM

The refrigerant in the system must be recovered before evacuation. After discharging the system, the low-pressure gauge hose remains connected to the low-pressure test fitting, and both the high- and low-pressure manifold gauge valves should be closed. **Figure 31–54** shows the evacuator hookup.

Proceed as follows:

1. Connect the high-pressure gauge hose to the high-side service port.

2. Connect the center manifold gauge hose to the vacuum pump.

3. With the manifold gauge set and the vacuum pump properly connected (see **Figure 31–54**), begin evacuation by opening the high- and low-side gauge valves with the vacuum pump operating.

4. Operate the vacuum pump for 20 to 30 minutes after the low-side gauge reaches 28 to 29 inches Hg vacuum for R-12 systems and 29.9 inches Hg for R-134a systems.

5. If the system has been exposed to air and/or moisture (in other words, left open), it should be partially charged and evacuated again.

6. Check the system for leaks by closing off both high- and low-side gauge valves and turning off the vacuum pump. There should be no more than 2 inches vacuum loss in 5 minutes. Greater loss than this indicates a leak in the system.

7. If a leak is indicated, check all the system connections for specified tightness.

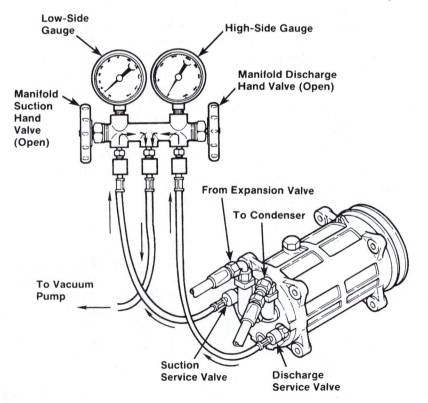

FIGURE 31–54 Evacuation hookup. (Courtesy of Freightliner Trucks)

8. Crack open the refrigerant container valve. Then open the low-pressure valve and allow the refrigerant to charge into the system. Close both the refrigerant container and the low-pressure valve and check the system for leaks. Correct leaks. Repeat the procedure until all leaks are eliminated.

9. Purge or reclaim refrigerant from the system and reevacuate. Do not discharge refrigerant through the vacuum pump.

10. Allow the vacuum pump to run for 20 minutes at a vacuum of 29.9 in. Hg. Close both manifold gauge valves; then turn off the vacuum pump. The system is now ready for recharging.

FULL RECHARGE OF AN A/C SYSTEM

Again, it is always important to follow the manufacturer's instructions. The general procedure is as follows:

1. Ensure that the service valves are closed and then connect the service couplings to the high and low ports of the vehicle A/C system. Open both couplings.

2. Set the station power switch to evacuation recharge.

3. Select the refrigerant source switch to virgin or recycled refrigerant. It is usually recommended that recycled refrigerant be selected.

4. Set the desired charge quantity on the station selector.

5. Press the Start Charge button to begin charging. A charging light will illuminate while charging refrigerant.

6. When charging is complete, the charge selector will read 0, and the charging light will go off.

7. Check High-Low pressure gauge readings and system output temperature to ensure that the system is performing within the OEM's specifications.

8. Close both couplings, disconnect the station from the vehicle, and reinstall the storage plugs.

9. Set the station power switch to Recover and press the Start Recovery button. This will recover the refrigerant vapor left in the hoses.

10. Close the valves on the recycling tank, the valve on the virgin tank, and all hose valves.

TOP-OFF RECHARGE OF AN A/C SYSTEM

Topping-off recharge of an A/C system is not a currently acceptable practice, as it cannot be known how much refrigerant or how much moisture is in a system (**Figure 31–55**). The correct service procedure is to evacuate, purge and recharge the system with the correct weight of refrigerant (**Figure 31–56**).

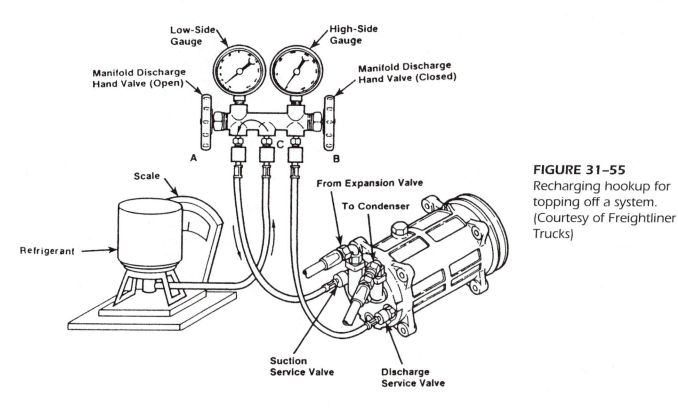

FIGURE 31–55
Recharging hookup for topping off a system. (Courtesy of Freightliner Trucks)

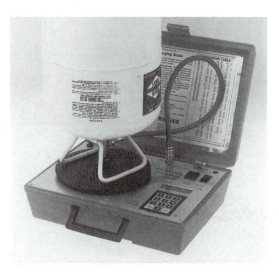

FIGURE 31–56 *Weighing a refrigerant canister. (Courtesy of Robinair, SPX Corporation)*

 Shop Talk

The importance of the correct charge cannot be stressed enough. The efficient operation of the air conditioning system greatly depends on the correct amount of refrigerant in the system. A low charge will result in inadequate cooling under high heat loads, due to a lack of reserve refrigerant, and can cause the clutch cycling switch to cycle faster than normal. An overcharge can cause inadequate cooling because of a high liquid refrigerant level in the condenser. Refrigerant controls will not operate properly, and compressor damage can result. In general, an overcharge of refrigerant will cause higher-than-normal gauge.

CAUTION: Accumulator systems do not have a sight glass. This is because bubbles are always present in the liquid line, even with a full charge. If the refrigerant is added until the bubbles are gone, serious damage and injury can result from overcharging the system.

31.9 COMMON A/C PROBLEMS

This section takes a brief look at some A/C problems and their causes. Remember, this is not a comprehensive list, just a guide to some typical problems and what may cause them. Use the schematic in **Figure 31–57** to locate the system components and

consult **Figure 31–58** for the performance testing hookups. **Figure 31–59** shows a typical A/C electrical schematic. The technician should be sure to consult the electrical schematic of the A/C circuit during system troubleshooting.

NORMAL PRESSURES AND INSUFFICIENT CAB COOLING

- **Coolant control valve at heater core is not completely closed.** When tested, high-side and low-side pressures are normal and the sight glass is clear, but it is simply not cool enough in the cab.
- **Excessive outside air infiltration.** Damaged bunk or door seals or open vent doors on a cab or bunk can make it impossible for the air conditioning system to keep up with the heat load leaking into the cab and bunk.
- **Blend air door/gate in wrong position.** The only time that the air conditioning system and the heating system should be working simultaneously is in demist or defog mode (defrost position).
- **Recirculate/vent door (80/20 door) is jamming in wrong position.** In the max position, there should be a minimum of vent air flow.

HIGH- AND LOW-SIDE PRESSURES ARE LOW

- **System low on refrigerant charge.** High-side and low-side pressures are low; bubbles can be observed in the sight glass; and there is not enough cooling in the cab.
- **Insufficient air flow through evaporator core.** Blower motor speed is low or in the wrong direction; the filter is dirty; dirt is clogging the evaporator fins; or evaporator is frozen due to ice buildup.
- **Moisture in system.** System alternates between normal and poor performance because of moisture freezing at expansion valve and obstructing refrigerant flow. High-side and low-side pressures cycle between normal and low as moisture freezes and thaws at expansion valve.
- **Obstructed expansion valve.** The sight glass is clear, and there is no cooling in the cab.
- **Obstructed orifice tube.** Some but not enough cooling in the passenger compartment of the vehicle.
- **Obstructed receiver/dryer.** The sight glass is clear, and there is not enough cooling in the cab.
- **Expansion valve stuck in closed position.** The sight glass is clear, and there is not enough cooling in the cab.

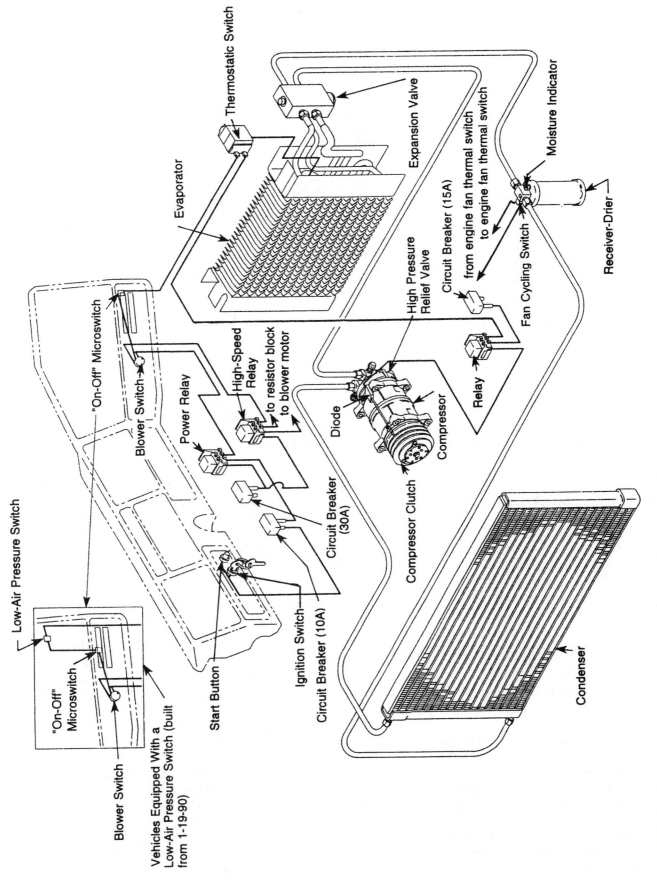

FIGURE 31–57   Truck A/C schematic. (Courtesy of Freightliner Trucks)

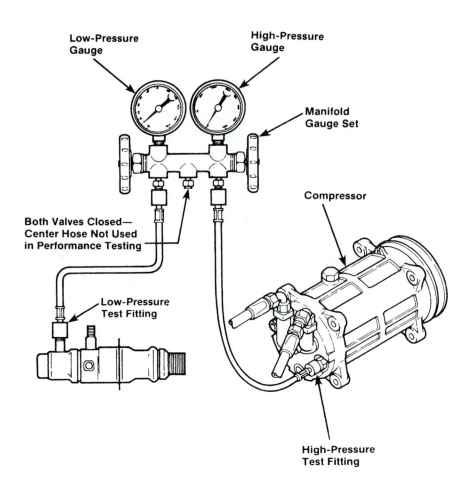

Low-Pressure Gauge

High-Pressure Gauge

Manifold Gauge Set

Compressor

Both Valves Closed— Center Hose Not Used in Performance Testing

Low-Pressure Test Fitting

High-Pressure Test Fitting

FIGURE 31–58   *Performance testing. (Courtesy of Freightliner Trucks)*

- **Defective pressure cycling switch.** Short cycling the on time of the compressor does not allow the high side to build up to normal pressure.
- **Defective evaporator pressure regulating valve.** A defective evaporator pressure regulating valve can restrict the flow of refrigerant, causing low system pressures and not enough cooling in the cab.

HIGH- AND LOW-SIDE PRESSURES ARE TOO HIGH

- **System slightly overcharged with refrigerant.** Sight glass is clear, and cooling in the cab is normal.
- **System heavily overcharged with oil.** Sight glass is clear, and there is not enough cooling in the cab.
- **Large amount of air in system.** Bubbles are in the sight glass, and there is not enough cooling in the cab.
- **Expansion valve stuck in open position.** The sight glass is clear, and there may not be enough cooling in the cab.

- **Poor thermal contact of sensing bulb of TXV with outlet of evaporator.** The sight glass is clear, and there may not be enough cooling in the cab.
- **Insufficient air flow through condenser core.** Sight glass is clear, and there is not enough cooling in the cab.

HIGH-SIDE PRESSURE IS LOW AND LOW-SIDE PRESSURE IS HIGH

- **Damaged compressor.** May be abnormal compressor noise—the sight glass is clear, and there is not enough cooling in the cab.
- **Improperly positioned sensing bulb of TXV.** Poor thermal contact between the sensing bulb and the outlet of the evaporator can cause the TXV to remain in the full open position, providing too little restriction for normal system function.
- **Improperly positioned capillary tube of thermostatic switch.** If the capillary tube of the thermostatic switch is not properly installed (inserted too far) in the evaporator core, the compressor can cut out too soon, not allowing high-side pressure to build or low-side pressure to be pulled down.

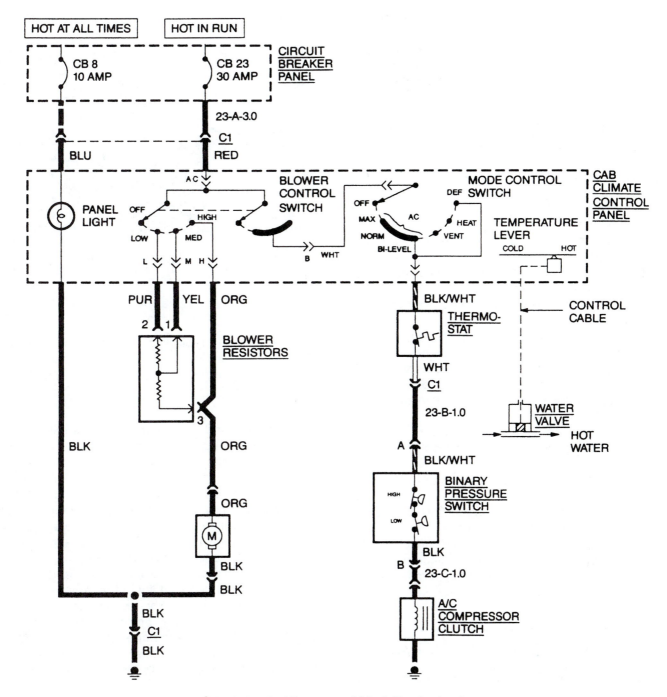

FIGURE 31–59 Typical A/C electrical circuit. (Courtesy of Mack Trucks, Inc.)

31.10 RETROFITTING AN R-134 SYSTEM TO AN R-12 SYSTEM

In 1994 the Mobile Air Conditioning Society (MACS) estimated that 140 million vehicles were using R-12 A/C systems, and they projected that around 40 million would be retrofit to accommodate R-134a refrigerant by 2002. Most of the retrofitting has taken place by now, but while there are R-12 systems on the highways and as R-12 becomes more difficult to acquire, systems will have to be retrofit. Some of the systems are being aftermarket retrofit to blend systems; some have flammable ingredients; and some can compromise system performance. The technician should practice caution when retrofitting or working on A/C systems that are already retrofit.

Personal safety and compliance with federal legislation are both concerns.

Always follow the manufacturer's procedure when retrofitting an A/C system. There are no absolute rules to observe, and the following list is only a set of guidelines:

- Use a refrigerant identifier to ensure that the system contains only R-12 and not a blend.
- Performance test the system, checking for leaks, pressures, and temperatures.
- Use a recycling station to remove all the R-12 from the system.
- Remove the compressor and drain the oil. Record the amount of oil drained from the compressor.
- Remove the TXV or orifice tube. If it is contaminated, flush the condenser.
- Remove the dryer or accumulator; drain and measure the oil in it.
- Use the manufacturer kit instructions and remove all components such as hoses, seals, and gaskets that will have to be replaced.
- Install an R-134a-compatible oil into the system, ensuring that the quantity put in equals the quantity removed.
- Install a new dryer or accumulator.
- Install conversion fittings and a high-pressure cutoff switch.
- Remove the R-12 label(s) and install R-134 conversion labels.
- Connect an R-134a recovery station to the system and evacuate for 30 minutes.
- Charge the system to 80 percent capacity.
- Performance test the system and compare with system performance prior to the retrofit.
- Leak test the system.

31.11 CAB VENTILATING AND HEATING SYSTEMS

The ventilating system on most vehicles is designed to supply outside air to the cab through vents. Several systems are used to route vent air into the cab compartment. The most common method is the flow-through system. In this arrangement, a supply of ram air flows into the truck when it is moving. Outside air can be delivered to the cab either by ram air or ram air helped by the heater/air-conditioner blower.

Most in-cab air movement is accomplished using the ventilation or blower fan whether ram air, the heater, or air conditioning is used to ventilate the cab. The fan is usually of a squirrel cage design (see

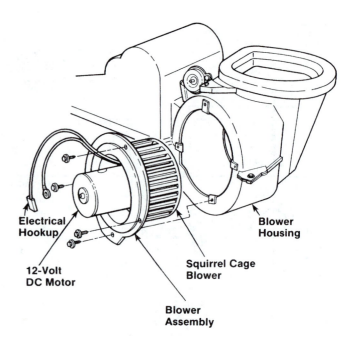

FIGURE 31–60 *Typical squirrel cage fan.* (Reproduced from MOTOR Auto Repair Manual ©1985 by permission of the Hearst Corporation)

Figure 31–60) and is located behind the dashboard. As the squirrel cage blower rotates, it produces suction on the intake and pressure at its outlet. When the fan motor is energized by using the temperature controls on the dashboard, air is moved through the cab compartment. Many trucks are equipped with air-actuated cylinders to open and close vent gates, which determine how air is sourced and routed through the cab and bunk compartment.

Bunk coolers are used in some applications. These use cells fitted with liquid that freezes at relatively high temperatures. The idea is that the cell liquid is chilled to freezing point by the truck's air conditioner while the truck goes down the road, so when air is blown over it after the A/C system has been shut down, the air cools the sleeper berth.

31.12 LIQUID COOLED HEATING SYSTEM

As in any automobile, a truck's cab heating system is built into the engine cooling circuit. Hot engine coolant is routed through a heat exchanger in the cab, and when cool cab air is blown through this heat exchanger, warm air exits. As the hot engine coolant circulates through the heater core, heat is transferred from the coolant to the tubes and fins of the core. Air blown through the core by the blower motor and the fan then picks up the heat from the surfaces of the core and transfers it into the cab compartment of the truck. The circulation of the engine coolant is achieved by the engine coolant pump (**Figure 31–61**). Control doors direct the air into specific areas of the cab compartment to accomplish the heating and defrosting requirements of the vehicle. The doors may be vacuum, cable, or air operated. The motors and cables of the system are controlled by either one or two control levers, which vary the function and temperature of the system.

A typical heater core is designed and constructed very much like a miniature engine cooling radiator. It features inlet and outlet tanks connected by headers to a heat exchanger core. The heater core tank, tubes, and fins can become clogged over time by rust, scale, and mineral deposits circulated by the coolant. When a heater core leaks and must be repaired or replaced, it can be a time-consuming job due to the core's location within the firewall of the vehicle. For this reason, it is good practice to leak test a replacement heater core before installation. Flushing the cooling system and seasonal replace-ment of the coolant can extend the life of all the heat exchangers in the system.

A **heater control valve,** also known as a coolant flow valve, controls the flow of coolant into the heater core from the engine. In a closed position, the valve allows no flow of hot coolant to the heater core, allowing it to remain cool. In a fully open posi-tion, the valve allows hot coolant to circulate through the heater core, providing maximum heating. Settings between these values permit the heating intensity to be controlled.

The thermostat that helps to regulate the engine coolant temperature plays a role in providing the cab with heat. A malfunctioning thermostat can cause the engine to overheat or not reach normal operating temperature, and either of these conditions will impact the cab heater performance.

The combination air conditioner/liquid-cooled heater is a compact unit that heats, cools, and dehu-midifies the cab and defrosts the windshield. The heater and air conditioner cab unit is usually mounted below (**Figure 31–62**), and to the right of the instrument panel. This unit contains the heater core and water flow control valve (**Figure 31–63**) for the heater system, and the TXV, evaporator core,

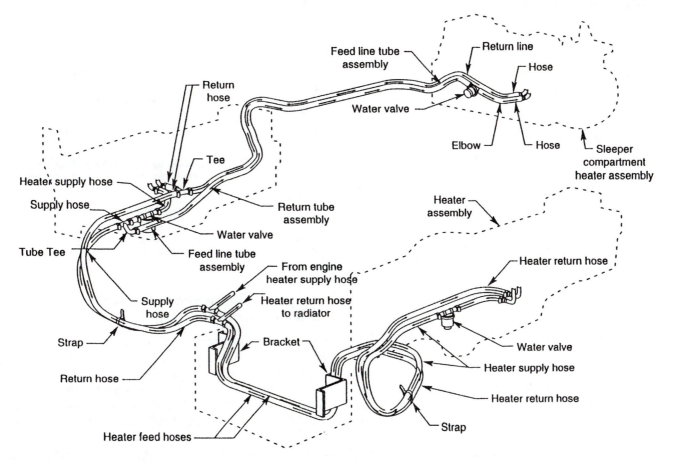

FIGURE 31–61 Coolant hose routing.

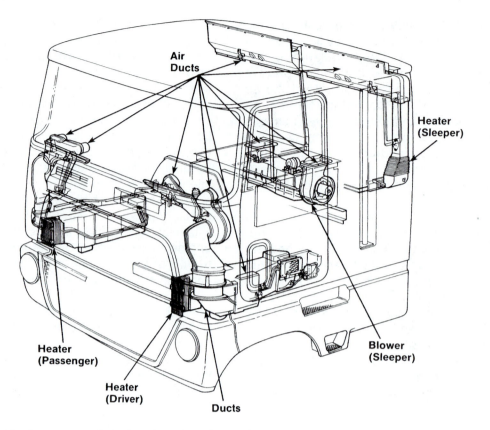

FIGURE 31–62   Heater core and air duct locations in a cab.

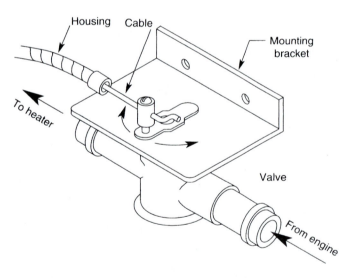

FIGURE 31–63   Cable-controlled coolant valve used to control the flow of coolant through the heater core.

and cold control switch for the air conditioner. A variable speed blower motor is mounted in the top of the unit. The blower motor forces air past both coils and supplies heated or cooled air to the ductwork behind the dash panel. The control panel shown in **Figure 31–64** is typical.

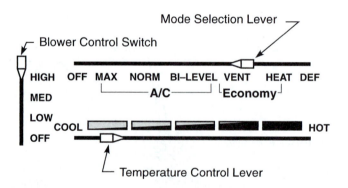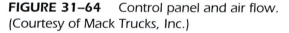

FIGURE 31–64   Control panel and air flow. (Courtesy of Mack Trucks, Inc.)

Sleeper Compartment Heater and Air Conditioner

Sleeper or bunk heater and air conditioner units (or sleeper boxes) include the heater core, evaporator coil, blower, and control valves. These units are usually mounted in the baggage compartment. The sleeper air conditioner is dependent on the cab's air conditioning system sharing the same refrigerant and compressor. The bunk heater is plumbed directly to the engine, usually but not always independently of the cab heater.

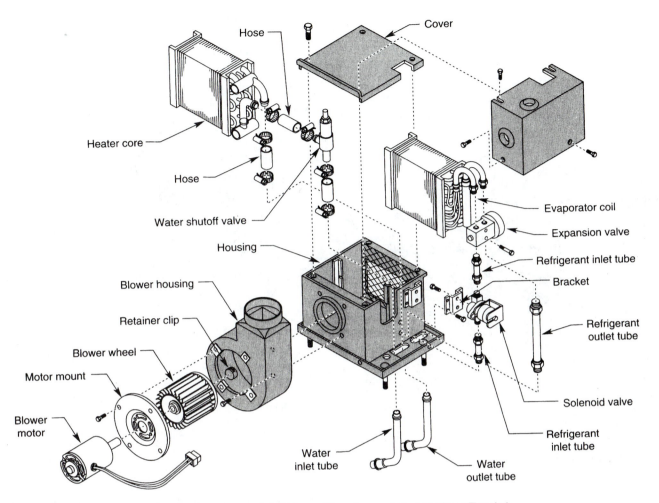

FIGURE 31-65 Typical sleeper heater and A/C box. (Courtesy of Freightliner Trucks)

The operating principles of bunk heaters and A/C evaporators are identical to those of the cab system. The blower forces air through the heat exchangers and from there, to the ducts in the bunk ventilation circuit and to the sleeper. The heater and A/C system schematic shown in **Figure 31-65** is typical of one found in a heavy-duty truck.

SUPPLEMENTARY HEATING DEVICES

Idling engines can greatly shorten engine life. Also, idling engines tend to cool faster than when they are shut down. Many electronically managed engines today have idle defeat programming that results in the engine being shut down when idled for longer than a pre-programmed period so that the engine is not unnecessarily idled. However, in the middle of winter, heat has to be provided when the engine is not running so that the truck driver is not frozen while

asleep in his bunk. This heat can be provided by coolant heating devices and auxiliary power units.

Supplementary Coolant Heater

A supplementary **coolant heater** is a thermostat controlled, diesel fuel heating device whose function is to maintain the engine coolant at a predetermined value. This is usually known as a diesel **fuel-fired heater.** The coolant is maintained at a temperature adequate to provide the truck cab and bunk with sufficient heat during periods of engine shutdown in winter conditions. Most have thermostats operating in conjunction with electronic controls to cycle the system to maintain the required temperature. The unit operates exactly like a kerosene-fired space heater, except that diesel fuel is used as the fuel and the flame acts on a heat exchanger through which coolant is flowed. **Figure 31-66** shows a typical diesel fuel-fired coolant heater.

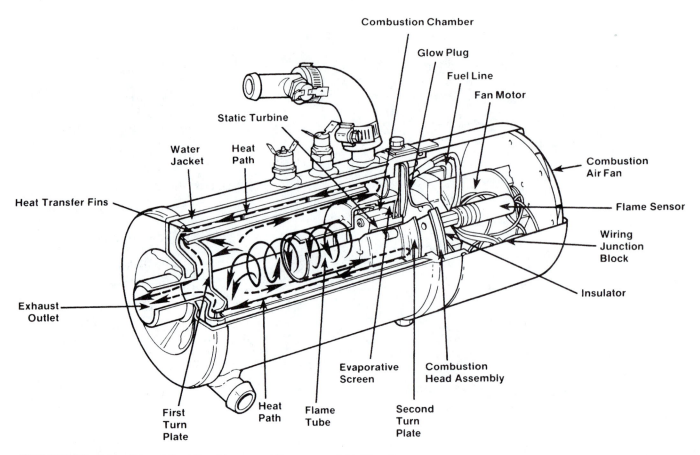

FIGURE 31-66 Diesel fuel-fired heater. (Courtesy of Phillips Temro, Inc.)

Auxiliary Power Units (APUs)

A separate onboard engine and power generator can produce 2,000–4,000 watts or more. These small diesel-fueled units consume one-third to one-sixth of the fuel used by the main engine while idling. This type of power source also makes it possible to cool the cab in hot weather using a 120-volt air conditioner or by driving a separate A/C compressor with the APU's engine. With an APU supplying standard main voltage, all other on-board bunk accessories, from televisions and lighting to microwave ovens, can be used. Electric powered coolant and fuel heaters can also be supplied from an APU system. Some APUs use coolant from their engines to heat the cab, sleeper, and main engine. This involves plumbing the two engines' cooling systems together, and of course, the cab and sleeper heaters. Or, an electric block heater can be installed and run off the APU's power supply. **Figure 31-67** shows a typical APU.

FIGURE 31-67 Typical mounting of an APU. (Courtesy of Heavy Duty Trucking)

31.13 | ELECTRONICALLY MANAGED CLIMATE CONTROL

Some heavy-duty trucks have electronically managed climate control systems that use A/C and heating systems that maintain a specific temperature automatically inside the cab compartment. To maintain a selected temperature, heat sensors send signals to a computer unit that controls the compressor effective cycle, heater valve, blower, and vent door operation. A typical electronic control system might contain input components, such as a coolant temperature sensor, in-truck temperature sensor, outside temperature sensor, high-side temperature switch, low-side temperature switch, low-pressure switch, vehicle speed sensor (to measure ram air effect), sunload sensor, and others. The computer would be used both to control the air conditioning system performance and to help in diagnosing air conditioning system problems. These electronic control systems use the vehicle's SAE/ATA connectors to communicate with other electronic devices on the vehicle and with shop diagnostic tools.

COMPUTER CONTROLLED A/C COMPONENTS

System components include the **A/C control module**, which performs logic processing of the system inputs and switches the outputs. It also includes an input and output circuit. The A/C control module masters A/C operation and tracks diagnostic data. It is usually located somewhere in the engine compartment rather than in the vehicle cab.

System Inputs

The input circuit to the A/C control module necessarily consists of a signal produced from the thermostat and the pressure switches.

- The HVAC thermostat switch is the primary control unit. It is connected in series with the main air conditioning on/off switch and the **evaporator thermostat.** The A/C control module will cycle the air conditioning compressor on and off in response to the thermostat input data. Typically, the thermostat contacts close when the evaporator temperature exceeds 38°F (3°C) on the rise, and open when the temperature decreases below about 32°F (0°C) on the fall. The thermostat is therefore used to turn off the effective cycle of the compressor whenever frost begins to form on the evaporator.

- Pressure switches are critical inputs to the A/C control module. One high-side pressure switch and one low-side pressure switch are used to signal high- and low-side pressure values to the A/C control module. Both switches are installed into the refrigerant lines. These switches monitor air conditioning circuit pressures. Both switches are designed to close at low pressure and open as pressure increases. The high-pressure switch typically opens at 300 psi (2,068 kPa) on the rise and closes at 260 psi (1,793 kPa) on the fall. The low-pressure switch opens at 34 psi (234 kPa) on the rise and closes at 10 psi (68 kPa) on the fall. So that the A/C module can detect an open circuit (broken wire) condition, both switches have a 2,490-ohm resistor in parallel with the contacts. Static pressure in a properly charged air conditioning system will cause the low-pressure switch to be normally open (2,490 ohms resistance) and the high-pressure switch to be normally closed (0 ohms resistance). However, a closed low-pressure switch would usually indicate a partial or total loss of refrigerant. At very low ambient temperatures, the air conditioning system pressures can sometimes drop low enough to close the low-pressure switch, even when the refrigerant charge is adequate. It is important to note that the air conditioning system will not operate when this happens.

A/C Module

The A/C module is a microcomputer. It receives inputs from the A/C input circuit and the vehicle electronic circuits. The inputs are processed along with programmed system instructions to produce outputs. Two types of output are generated. The first category of outputs are those that help optimize system performance and manage the cycling of system components. The other category of outputs are those that protect the system against damage. Truck A/C modules are usually located in the engine compartment rather than in the cab. They are accessible by means of SAE J1587 (using a J1708 data link) or J1939 protocols. The A/C system uses the MID 190. A full description of microprocessors and the processing cycle is provided in Chapter 6.

A/C Module System Outputs

The two primary A/C module outputs are the fan-request signal to the engine ECM and the switch control for the refrigerant compressor clutch.

- Although the engine fan is actually controlled by the engine ECM, the A/C module can provide a fan-request signal input so that other systems, such as air conditioning, may request fan operation. Whether the fan will actually turn on or not depends on the engine processing cycle at the moment the fan-request signal is received.

- The operation of the refrigerant compressor clutch is directly controlled by the A/C module. It is usually a 12-V switched output of the module. The A/C module will switch the compressor clutch as an outcome of its own processing cycle. This means that it will be controlled largely as a result of the thermostat input signal received by the A/C module. The module can discontinue operation of the compressor clutch if it detects a short or open circuit condition. The clutch coil typically has a nominal resistance of 3.5 ohms. The A/C module calculates the actual coil resistance by making current and voltage measurements. If the calculated resistance is less than about 2 ohms or greater than about 12 ohms, the A/C module will de-energize the clutch coil circuit. It will make another attempt to energize the clutch coil after a predetermined period (10–30 seconds). Then if the coil resistance is still out of limits, the circuit will be latched off until the next ignition cycle. When this occurs, a fault code is logged in the A/C module that can be read by means of the ATA connector and the appropriate EST. **Figure 31–68** shows the location of some of the components in an electronically managed A/C system.

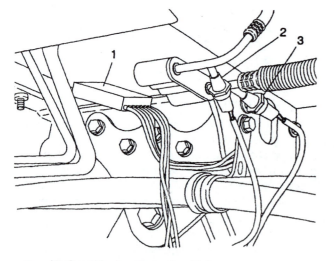

1. Air Conditioning Protection Unit
2. High Pressure Switch
3. Low Pressure Switch

FIGURE 31–68 *Components of a typical electronically managed climate control system. (Courtesy of Freightliner Trucks)*

TABLE 31–2: PARAMETER DEFAULTS	
Missing Parameter	**Default Value**
Ambient air temperature	48°F (9°C)
Road speed	32 mph (51 km/h)
Engine speed	384 rpm
Park brake status	Not set
Air pressure	Okay

Electronically Managed A/C Data Communications

A typical A/C electronic control system is multiplexed into the vehicle electronic system. This allows the A/C electronic system to share hardware with the other vehicle electronic systems. A typical A/C module uses the following data signaled transmitted by other electronic devices to help process the data required to control the truck air conditioning system. In the event of an A/C module becoming disconnected from the databus, the A/C electronics will assume default values for these parameters. This permits the air conditioning system to continue operating. **Table 31–2** lists parameter defaults.

A/C Control Logic

The A/C control module uses the following sequence of logic rules to control the air conditioning system:

1. **Air conditioning start delay.** The air conditioning compressor is not permitted to be driven under effective cycle for the first 15 seconds after the ignition key is switched on and the engine is running at more than 300 rpm.

2. After the ignition has been switched on and the engine has been running at more than 300 rpm for 15 seconds, the air conditioning compressor engages for 15 seconds. At this point, the following conditions must be satisfied:
   - The high- and low-pressure switches do not indicate out-of-limits or fault conditions.
   - No low-air-pressure message is being received from the vehicle databus.
   - Battery voltage sensed by the A/C module is greater than 11 volts and less than 16 volts.
   - There are no logged fault codes that would disable the air conditioning operation. These faults are Low Pressure, Open Clutch Circuit, and Shorted Clutch Circuit.

3. For basic fan operation, the A/C module initiates a fan request to the engine ECM when

the high-side pressure exceeds 300 psig (2,068 kPa), opening the high-pressure switch. The engine ECM should (and normally will) then respond by turning the fan on immediately. When the high-pressure condition ceases to exist—that is, the high-pressure switch closes—the A/C module fan-request to the ECM remains active for a short period and then ceases. However, the actual length of time the fan runs depends on the engine ECM programming.

4. The A/C module switches the air conditioning clutch on and off in response to the thermostat input signal, subject to the following conditions:
   - Data from the databus to the engine indicates that engine speed is greater than 300 rpm.
   - The on/off cycle frequency of the clutch must exceed 15 seconds.
   - The low-pressure switch remains open.
   - If the high-pressure switch opens, the A/C clutch will be de-energized after a short lag.
   - Locally sensed battery voltage exceeds 11.0 volts.
   - The A/C clutch will be disengaged if an over-current or under-current condition in the clutch coil is sensed by the A/C module. If this happens, the A/C control module will wait for a predetermined period of time and then attempt to energize the clutch once again. If the over- or under-current condition still exists, the A/C control module will disable the system until the next ignition cycle.

Trouble Codes

The method of reading the A/C trouble codes is the same as that used for all other truck electronic systems. This process is fully explained in Chapter 8 of this textbook. First, connect an electronic service tool (EST) to the dash ATA connector. Download the entry menu to the EST screen and select the A/C MID (190). Then scroll through the system and select the problem fault mode indicator (FMI). Because of the simplicity of current electronic A/C management systems, locating electronic problems is relatively simple.

SUMMARY

- Heavy-duty heating and air-conditioning systems are designed primarily to keep the cab comfortable despite the outside weather.
- Liquids absorb heat when changed from a liquid state to a gas.
- Gases release heat when changed from a gaseous state to a liquid.
- Heat always moves from a hotter area to a cooler area.
- The temperature at which a liquid changes state to a gas depends on the pressure acting on it.
- Refrigerants are required to have a low boil point and to change quickly from a liquid state to a gas state and back again.
- The basic refrigeration cycle consists of compression, condensation, expansion, and evaporation.
- Compression heats up refrigerant gas. Condensation changes the state of the refrigerant from a gas to a liquid. Expansion reduces the pressure of the liquid refrigerant. Evaporation changes the refrigerant from a liquid state to a gas.
- There are two types of refrigerant lines: suction lines and discharge lines.
- Air conditioning systems are sensitive to moisture and dirt. Clean working conditions are extremely important.
- The electrical system must be checked periodically to prevent the truck air conditioning system from failing unexpectedly.
- Some trucks use electronic climate and A/C controls.

REVIEW QUESTIONS

1. The amount of heat required to raise the temperature of one pound of water by 10°F is
   a. 1 Btu
   b. 10 Btu
   c. 20 Btu
   d. 100 Btu

2. Which of the following is true in a properly functioning refrigeration system?
   a. A liquid changes to a solid and back to a liquid.
   b. A liquid changes to a gas and back to a liquid.
   c. Refrigerant gas reaches a temperature of absolute zero.
   d. Heat flows from a colder object to a warmer one.

3. Which component in an air conditioning system is responsible for removing heat from the circulating refrigerant?
   a. the compressor
   b. the evaporator

c. the receiver/dryer or accumulator
d. the condenser

4. Which of the following best describes the function of the TXV?
   a. It separates the high-pressure side of the system from the low-pressure side.
   b. It regulates refrigerant flow to the evaporator.
   c. It balances the inlet flow to the outlet temperature.
   d. All of the above.

5. The two main refrigerant lines are known as the
   a. high-pressure line and low-pressure line
   b. suction line and discharge line
   c. compressor line and condenser line
   d. inlet line and outlet line

6. Technician A wears safety goggles and rubber gloves while discharging the A/C system. Technician B wears safety goggles and leather gloves when charging and leak testing the system. Who is correct?
   a. Technician A
   b. Technician B
   c. both A and B
   d. neither A nor B

7. When checking an A/C sight glass for the moisture content of the refrigerant, Technician A performs the check with the air conditioner running immediately after starting the engine. Technician B states that some observable bubbles in the sight glass at system start-up are normal. Who is correct?
   a. Technician A
   b. Technician B
   c. both A and B
   d. neither A nor B

8. Which of the following procedures may improve A/C system performance?
   a. cleaning the outer surfaces of the condenser with compressed air and soap
   b. repairing leaks in the evaporator coil
   c. improving the performance of the engine cooling system
   d. all of the above

9. Which method of detecting a leak in the A/C system would be most sensitive when attempting to locate a very small leak?
   a. application of a soapy water solution
   b. using an electronic leak detector
   c. using a flame-type propane leak detector
   d. dye and a black light detector

10. Which piece of equipment is used to meter out a desired amount of specific refrigerant by weight?

a. manifold service hose set
b. Schrader service valve
c. recharging cylinder
d. vacuum pump

11. Purging and flushing an A/C system are alike because both procedures
   a. use dry nitrogen gas
   b. use clean, dry liquid refrigerant
   c. remove contaminants
   d. require the installation of a new receiver/dryer or accumulator

12. Technician A uses refrigerant from an opened container that he knows has been tightly sealed since immediately after its previous use. Technician B states that recycled refrigerant known to be uncontaminated can be reused on recharging an A/C system. Who is correct?
   a. Technician A
   b. Technician B
   c. both A and B
   d. neither A nor B

13. When evacuating an R-12 system, how long should the vacuum pump be run after the low-pressure manifold gauge reaches 28–29 inches Hg vacuum?
   a. 5 minutes
   b. 10 minutes
   c. 20–30 minutes
   d. 40–50 minutes

14. Which of the following is the likely result of a slightly low charge of refrigerant in a truck A/C system?
   a. compressor damage
   b. higher than normal gauge readings
   c. noisy compressor operation
   d. faster than normal clutch switch cycling

15. Which of the following conditions could cause the low-side pressure to read high and the high-side pressure to read low?
   a. system overcharged with refrigerant
   b. insufficient air flow through condenser core
   c. insufficient air flow through evaporator core
   d. damaged compressor

16. Which of the following is a typical input of an electronically managed A/C system?
   a. clutch switch
   b. high-pressure switch
   c. fan-request switch
   d. blower control switch

17. Which of the following components provides the A/C control module with temperature data?

a. thermostat
b. databus
c. ATA connector
d. low temperature switch

18. The expansion valve in an A/C circuit sticks in the open position. What is the likely result when the system is tested?
    a. low high-side pressure
    b. low low-side pressure
    c. high high-side pressure
    d. high low- and high-side pressure

19. Which of the following components is not a heat exchanger?
    a. evaporator
    b. condenser
    c. radiator
    d. compressor

20. Which of the following conditions would be indicated if a sight glass appeared to be cloudy?
    a. desiccant breakdown
    b. system overcharge
    c. low refrigerant
    d. low compressor oil

Glossary

A-Train A combination of two trailers in which a dolly (converter or turntable) is connected by a pintle hook and drawbar. The resulting connection has two pivot points.

Acronym Word made up of the initial letters of other words.

Actuator Device that delivers motion in response to an electrical or hydraulic signal.

Adapter Ring Used by some two-plate clutches when the clutch is installed on a flat flywheel. The adapter ring is bolted between the clutch cover and the flywheel.

Adjustable Pliers Commonly called Channel-locks™, these tools have a multiposition slip joint that allows for multiple jaw opening sizes.

Adjustable Wrench A wrench that can be adjusted to fit bolts of different sizes by rotating a helical adjusting screw.

Aeration Mixing of a gas with a liquid

Aftercooler Removes water and oil from the air by a cooling process. The air leaving an aftercooler is still saturated with water vapor, which will condense when a drop in temperature occurs.

Air Bag An air spring or supplementary restraint device.

Air Brakes Braking system that uses air pressure to actuate the brakes by means of diaphragms, wedges, or cams.

Air Dryer Unit that removes moisture before it enters the brake system.

Air Filter Used to minimize the possibility that moisture-laden air or impurities enter the pneumatic system.

Air-Over-Hydraulic Brakes Brakes using a hydraulic system assisted by an air pressure system.

Air-Over-Hydraulic Intensifier Changes the pneumatic air pressure from the treadle brake valve into hydraulic pressure which actuates the wheel cylinders.

Air Ratchet Wrench This wrench, like the hand ratchet, has a special ability to work in hard-to-reach places. The air wrench looks like an ordinary ratchet but has a large handgrip that contains an air vane motor and drive mechanism.

Air Spring Suspension A single or multiaxle suspension relying on air bags for springs and weight distribution of axles.

Air Timing The time required for the air to be transmitted to or exhausted from each brake, starting the instant the driver moves the brake pedal.

Allen Wrenches Wrenches used to tighten or loosen recessed-hex Allen-head screws.

Alloy Mixture of metallic elements with a base metal.

Amboid Gear This is similar to the hypoid type with one exception: the axis of the drive pinion gear is located above the centerline axis of the ring gear.

Amp (Ampere) The unit for measuring electrical current.

Analog Signal A voltage signal that varies within a given range (from high to low, including all points in between).

Analog-to-Digital Converter (A/D converter) Converts analog voltage signals to a digital format; this is located in a section of the processor called the input signal conditioner.

Analog Volt/Ohmmeter (AVOM) A type of multipurpose, electrical test meter.

Annulus or Ring Gear The largest component of the simple planetary gear set.

Anode Positive electrode.

Anticompounding Valve Prevents simultaneous air and spring forces from being applied to the foundation brakes.

Anticorrosion Agents These agents protect metal surfaces from corrosion.

Antilock Brake System (ABS) A series of sensing devices at each wheel and a microprocessor that control braking action to prevent wheel lockup.

Antilock Relay Valve (ARV) In antilock brake systems, this usually replaces the standard relay valve used to control the rear axle service brakes and performs the standard relay function during tractor/trailer operation.

Antirattle Springs These reduce wear between the intermediate plate and the drive pin in a clutch pack, and smooth clutch release.

Antirust Agent An additive used with lubricating oils; this prevents rusting of metal parts when the engine is not in use.

Applied Moment This term means that a given load has been placed on a frame at a particular point.

Armature The rotating component of the starter motor. It is located between the drive and commutator end frames and the field windings.

Aspect Ratio For any tire, this is calculated by dividing the tire's section height by its section width.

Automatic Slack Adjuster Automatically adjusts the clearance between the brake linings and the brake drum or rotor. The slack adjuster controls the clearance by sensing the length of the pushrod stroke for the air brake chamber.

Autoshift Finger This corresponds to the tab on the end of the gearshift lever in manual systems. The finger engages the shift blocks on the yoke bars.

Auxiliary Filter   Installed in the oil return line between the oil cooler and the transmission. The auxiliary filter prevents debris from being flushed into the transmission and causing a repeat failure. An auxiliary filter must be installed before the vehicle is returned to service.

Axis   Point about which a component rotates.

Axis of Rotation   The centerline around which a gear or part revolves.

Axle   A rod or bar on which wheels turn. It is also a shaft transmitting driving torque to the wheels.

Axle Range Interlock   This feature is designed to prevent axle shifting when the interaxle differential is locked out (or when lockout is engaged). The basic shift system operates the same as the standard shift system to shift the axle and engage/disengage the lockout.

Axle Shims   On vehicles with a leaf spring suspension, these thin wedges can be installed under the leaf springs of single axle vehicles to tilt the axle and correct the U-joint operating angles. Wedges are available in a range of sizes to change pinion angles.

Banjo Housing   Differential carrier housing.

Battery Terminals   These are either two tapered posts or threaded studs on top of the case or two internally threaded connectors on the side.

Bench Grinder   A power tool used to remove metal and sharpen tools.

Bending Moment   This term simply means that when a load is applied to the frame, it is distributed across a given section of the frame material.

Bevel   Slope, slant, or angled edge.

Bias   On a tire, where belts and plies are laid diagonally.

Bit   Binary digit representing one of two values, on or off.

Blade or Spade Fuse   This type of fuse has two male lugs sticking out for insertion in female box connectors.

Block Diagnosis Chart   This chart lists conditions (problem symptoms), causes (problem sources), and remedy (needed repairs) in columns.

Blower Fan   A fan that pushes (or blows) air through a ventilation, heating, or air-conditioning system.

Blowgun   A device attached to the end of an air hose to control and direct air flow while cleaning components.

Bobtail Proportioning Valve   Senses when the tractor is bobtailing and automatically reduces the amount of air pressure that can be applied to the tractor's drive axle(s). This reduces braking force on the drive axles, lessening the chance of a spinout on slippery pavement.

Bobtailing   A tractor running without a trailer.

Bogie   The axle spring suspension arrangement on the rear of a tandem axle tractor or trailer.

Bottoming   Occurs when the teeth of one gear touch the lowest point between teeth of a mating gear.

Box-end Wrench   A wrench designed to fit all the way around a bolt or nut.

Brake Chamber   The brake chamber converts air pressure into mechanical energy by pushing against a plate at the end of a pushrod which is connected to the brake slack adjuster.

Brake Pad   The friction lining and plate assembly that is forced against the rotor to effect braking action in a disc brake.

Brake Shoe   The friction surface-faced component used in the drum brake system to stop the brake drum from rotating.

Breather   Located at the top of the transmission or axle housing. It prevents pressure build-up within the transmission and must be kept clean and the passage open.

Brinelling   Identified as tiny indentations high on the shoulder or in the valley of the bearing raceway caused by improper bearing installation or removal.

British Thermal Unit (Btu)   A measurement of the amount of heat required to raise the temperature of 1 pound of water 1°F.

Bus   Connect in groups—usually refers to electrical bus bars.

Caliper   A disc brake component that changes hydraulic or air pressure into mechanical force to press the brake pads against the rotor, stopping the vehicle. Calipers come in three basic types, fixed, floating, and sliding, and can have one or more pistons.

Cam   An eccentric.

Cam Brakes   Brakes similar in operation and design to the wedge brake, with the exception that an S-type camshaft is used to spread the brake shoes instead of a wedge assembly.

Camber   The attitude of a wheel and tire assembly when viewed from the front of a car. If it leans outward, away from the car at the top, the wheel is said to have positive camber. If it leans inward, it is said to have negative camber.

Capacitor   Electrical charge storage device also known as a condenser.

Cartridge Fuse   This type of fuse is a strip of low melting point metal enclosed in a glass tube. If an excessive current flows through the circuit, the fuse element melts at the narrow portion, opening the circuit and preventing damage.

Caster   Angle formed between the kingpin axis and a vertical axis as viewed from the side of the vehicle.

Caster is considered positive when the top of the kingpin axis is behind the vertical axis.

Cathode Negative electrode.

Cavitation Bubble collapse erosion in a fluid circuit.

Ceramic Fuse A fuse found in some import truck models that has a ceramic insulator with a conductive metal strip along one side.

CFC (chlorofluorocarbon) A compound used in R-12 known to cause damage to the ozone layer.

Charging System Consists of the batteries, alternator, voltage regulator, associated wiring, and the electrical loads of the truck. The purpose of the system is to recharge the battery whenever necessary and to provide the current required to power the electrical truck components.

Check Valve A valve that allows air to flow in one direction only. It is a federal requirement to have a check valve between the wet and dry air tanks.

Chip Semiconductor substrate on which the passive and active elements of an electronic circuit have been photo-infused.

Chisel A bar of hardened steel with a cutting edge ground on one end, driven with a hammer to cut metal.

Clean Air Act These regulations, passed in 1992, have resulted in major changes in heavy-duty truck air conditioning systems.

Clearance Lights Lighting on a commercial vehicle to indicate its dimensions.

Climbing A gear problem caused by excessive wear in gears, bearings, and shafts whereby the gears move sufficiently apart to cause the apex (or point) of the teeth on one gear to climb over the apex of the teeth on another gear with which it is meshed.

Clutch A device for connecting and disconnecting engine torque from the transmission or for a similar purpose in other units.

Clutch Brake Circular disc with friction surfaces mounted on the transmission input shaft between the release bearing and the transmission. Its purpose is to slow or stop the transmission input shaft to allow gears to be engaged without clashing (grinding).

Clutch Pack Consists of an intermediate plate, friction discs, and a pressure plate.

COE Cab-over-engine.

Combination Pliers The most common type of pliers, frequently used in automotive repair. The jaws have both flat and curved surfaces for holding flat or round objects. Also called slip-joint pliers, the combination pliers have many jaw opening sizes.

Combination Wrench Wrench with an open-end jaw on one end and a box-end on the other.

Compressor A mechanical device that increases pressure within a container by pumping air into it.

Computer Electronic processing device with a CPU, data retention, input and output switching.

Condenser A component in an air-conditioning system that cools a refrigerant below its boiling point.

Conductor Any material that permits electrical current to flow easily.

Conduit Channel or pipe used to route fluids, or covering for electrical wires.

Constant Rate Springs Spring assemblies that have a constant rate of deflection.

Controlled Traction A differential that uses a friction plate assembly to transfer drive torque from the truck's slipping wheel to the wheel that has good traction or surface bite.

Converter Dolly An axle, frame, drawbar, and fifth wheel arrangement that converts a semi-trailer into a full trailer.

Coolant Heaters Used to aid engine starting and reduce the wear caused by cold starting.

Coolant Hydrometer An instrument designed to measure coolant specific gravity and determine the amount of antifreeze in the coolant.

Coupling Point The point at which the turbine is turning at the same speed as the impeller.

CPU Central processing unit. The processor or "brain" of the computer processing cycle.

Crankcase Housing within which the crankshaft and other parts of the engine operate.

Cranking Rotating an engine with the objective of starting it.

Cycling Battery charging and discharging. Repeated battery cycling can cause the positive plate material to break away from its grids and fall into the sediment chambers at the base of the battery case.

Dampened Discs Discs that have dampening springs incorporated into the disc hub. When engine torque is first transmitted to the disc, the plate rotates on the hub, compressing the springs. This action absorbs the shocks and torsional vibration caused by today's low rpm, high torque engines.

Dash Control Valves Hand-operated valves located on the dash. They include parking brake valves, tractor protection valves, and differential lock.

Data Electronic information.

Data Links When computers communicate with other electronic devices such as control panels, modules, some sensors, or other computers, they are connected through these circuits.

Deadline To take a vehicle out of service.

Deburring To remove sharp edges from a cut.

Dedicated Contract Carriage Trucking operations set up and run according to a specific shipper's needs. In addition to transportation, they often provide other services such as warehousing and logistics planning.

Detergent Additive Oil additives that keep metal surfaces clean and prevent deposits. These additives suspend particles of carbon and oxidized oil in the oil.

DER Department of Environmental Resources.

Diagnostic Flow Charts Charts that provide a systematic approach to system and component troubleshooting and repair. They are found in service manuals and software, and are vehicle make and model specific.

Diagonal Cutting Pliers Also called cutters, these pliers are used to cut electrical connections, cotter pins, and other automobile wires. Jaws on these pliers have extra hard, cutting edges.

Dial Caliper A versatile measuring instrument. It is capable of taking inside, outside, depth, and step measurements.

Differential A gear assembly that transmits torque from the drive shaft to the wheels and allows two opposite wheels to turn at different speeds for cornering and traction.

Differential Carrier Truck drive axle assembly.

Digital Multimeter (DMM) A multiple mode electrical circuit test instrument.

Digital Signal A signal that has only two values—on or off.

Digital Volt/Ohmmeter (DVOM) A type of test multimeter.

Diode Simple semiconductor device. Formed by joining P-type semiconductor material with N-type semiconductor material. A diode allows current to flow in one direction only.

Direct Drive Gearing of a transmission so that one revolution of the engine produces one revolution of the transmission's output shaft. The final drive ratio of a direct drive transmission would be 1:1.

Disc Brake A type of brake that generates stopping force by the application of pads against a rotating disc (rotor).

Dispatch Sheet Also called a work schedule, this form keeps track of dates when the work is to be completed. Some dispatch sheets follow the job through each step of the servicing process.

Dolly Trailer landing gear or caster-wheel support and hoist device.

Double Reduction Axle Axle that uses two gear sets for greater overall gear reduction and peak torque development. It is favored for severe service applications, such as dump trucks, cement mixers, and other heavy haulers.

Drag Link A connecting rod or link between the steering gear, Pitman arm, and the steering linkage.

Drive or Driving Gear A gear that drives another gear or causes another gear to turn.

Driveline Power transmission components from the flywheel to the wheel end assemblies.

Driven Gear A gear that is driven or forced to turn by a drive gear, a shaft, or some other device.

Driver's Manuals These publications contain information needed by the driver to understand, operate, and care for the vehicle and its components.

Driveshaft An assembly of one or two universal joints connected to a shaft or tube; used to transmit power from the transmission to the differential. Also called the propeller shaft.

Drum Brake Type of brake in which stopping friction is created by the shoes pressing against the interior of the rotating drum.

Dual Hydraulic Braking System Consists of a tandem or double action master cylinder, which is basically two master cylinders usually formed by aligning two separate pistons and fluid reservoirs into a single cylinder.

Eddy Current A small circular current produced inside a metal core in the armature of a starter motor. Eddy currents produce heat and are reduced by using a laminated core.

Electric Retarder Composed of electromagnets mounted in a steel frame. Energizing the retarder causes the electromagnets to exert a dragging force on rotors in the frame which is transmitted directly to the drive shaft.

Electricity The science of electrical energy.

Electromotive Force (EMF) Force that moves electrons between atoms. This force is the pressure that exists between the positive and negative points (the electrical imbalance). It is measured in units called volts.

Electronic Control Unit (ECU) The "brain" of an electronic system.

Electronics The science of electron behavior.

Electrons Negatively charged particles orbiting every atomic nucleus.

End Yoke Connected to the output shaft of the transmission. It transfers engine torque to the driveshaft.

Engine Brake A hydraulically operated device that converts the vehicle's diesel engine into a power-absorbing, retarding mechanism.

Evaporator Air-conditioning system heat exchanger used to heat a refrigerant above its boiling point.

Exhaust Brake A valve installed in the exhaust pipe between the manifold and the muffler. It is a slide mechanism which restricts the exhaust flow.

Extractor A tool that removes screws and bolts broken off below flush.

False Brinelling The polishing of a surface (no damage to the surface).

Farad A unit of capacitance.

Fatigue Failures Usually defined as a progressive destruction of shaft or gear teeth material. This type of failure is usually caused by overloading.

Fault Code A code that is recorded into the computer's memory. Fault codes can be read by a diagnostic data link or flash codes.

Federal Motor Vehicle Safety Standard (FMVSS) Defines the safety standards required on all vehicle systems.

FMVSS 121 Standard that covers air brake systems.

Feeler Gauge A metal strip or blade finished accurately with regard to thickness, and used for measuring the clearance between two parts; such gauges ordinarily come in a set of blades each graduated in thickness by increments of 0.001 inch.

FHWA Federal Highway Administration.

Fiber Composite Springs Springs made of fiberglass, laminated and bonded together by tough polyester resins.

Fifth Wheel A coupling device mounted on a truck and used to connect a semitrailer. It acts as a pivot to allow directional changes of travel between the tractor and semitrailer.

Fixed Value Resistors Designed to have one rating, which should not change. Such resistors are used to control voltage.

Flammable Easily able to catch fire.

Flaring To spread gradually outward.

Flex Plate Sometimes called a flex disc, this is used to mount the torque converter to the crankshaft. The flex plate is positioned between the engine crankshaft and the T/C. Its purpose is to transfer crankshaft rotation to the shell of the torque converter assembly.

Footprint Contact interface between tire and road or the downlink geographic window of a GPS.

Forged Journal Cross (sometimes called a spider or trunnion) Part of a universal joint.

Foundation Brakes The portion of the brake circuit that mechanically effects braking at the wheel.

Franchised Dealership A dealership that has signed a contract with a particular manufacturer to sell and service an exclusive line of vehicles.

Frequency Number of repetitions of a process within a specific time frame.

Fretting A result of vibration that the bearing outer race can pick up the machining pattern.

Friction Plate Assembly This assembly consists of a multiple disc clutch that is designed to slip when a predetermined torque value is reached.

Fully Floating Axles Axle half shafts that transmit only driving torque to the wheels but not bending and torsional loads, as is characteristic of the semi-floating axle.

Fully Oscillating Fifth Wheel A fifth wheel type with fore/aft and side-to-side articulation.

Fuse Link A short length of smaller gauge wire installed in a conductor, usually close to the power source.

Galling Occurs when metal is cropped off or displaced due to friction. Galling is commonly found on trunnion ends.

Gear A wheel with external or internal teeth that serves to transmit or change motion.

Gear Pitch Refers to the number of teeth per given unit of pitch diameter.

Gladhands Connectors between tractor and trailer air lines.

GPS Global positioning satellite. Telecommunications satellite.

Grounded Circuit This is actually a shorted circuit that allows current to return to the battery before it has reached its intended destination.

GVW (Gross Vehicle Weight) The total weight of a fully equipped vehicle and its payload.

Hacksaw A toothed cutting tool used by truck technicians for cutting bolts, angle iron, tubing, etc.

Halogen Light Light with a small quartz/glass bulb that contains filament surrounded by halogen gas. This small, gas-filled bulb is contained within a larger metal reflector and lens element.

Handshake Two-way electronic connection.

Hand Tap A small tool used for hand cutting internal threads.

Hand-Threading Dies Tools used to cut external threads on bolts, rods, and pipes.

Harnesses and Harness Connectors These help to organize the vehicle's electrical system and provide a convenient starting point for tracking and testing many circuits.

Hazardous Materials Substances that are flammable, explosive, or have been shown to produce adverse health effects in people when exposed to the material during normal use.

Head Up Display (HUD) A technology that superimposes data on the driver's normal field of vision. The operator can view the information, which is projected onto the windshield of a conventional tractor or truck. This allows the driver to monitor conditions such as limited road speed without interrupting his normal view of traffic.

Heater Control Valve (sometimes called the water flow valve) This controls the coolant flow into the heater core from the engine.

Heavy-Duty Truck A truck that has a GVW of 26,001 pounds or more.

Hinged Pawl Switch The simplest type of switch, since it either makes or breaks the current in a single conductor or circuit. A toggle switch.

Hydraulic Brake System   A system using fluids under pressure to activate the brakes.

Hydrometer   A tester designed to measure the specific gravity of a liquid.

Hypoid Gears   Gears that intersect at right angles below the centerline axis of the driving gear when meshed. Hypoid gearing uses a modified spiral bevel gear structure that allows several gear teeth to absorb the driving torque and to run quietly.

Impact Sockets   Heavier walled sockets made of hardened steel, designed for use with an impact wrench.

Impact Wrench   A portable handheld reversible air-powered wrench.

In-Line Fuse   An in-series fuse. Used, when necessary, as a protection device for a portion of the circuit even when already protected by a fuse in the fuse box or panel.

In-phase   The in-line relationship between the forward coupling shaft yoke and the driveshaft slip yoke of a two-piece driveline.

Input Retarder   Located between the torque converter housing and the main housing. The unit employs a "paddle wheel" type design with a vaned rotor mounted between stator vanes in the retarder housing.

Installation Templates   Supplied by some truck manufacturers to allow the truck technician to correctly install a slack adjuster.

Insulators   Materials that offer high resistance to the flow of electrons, such as rubber or glass.

Integrated Circuit   A large number of diodes, transistors, and other electronic components, infused on a single piece of semiconductor material and able to perform numerous functions.

Internet   Global network of networks, communications system.

Inversion Valve   An air-operated control valve. Normally in the open position, it is closed by the application of air pressure. The valve is primarily used in emergency and/or parking brake systems.

Jounce   The most compressed position of a spring.

Jump Starting   The procedure used when it becomes necessary to use a booster battery or generator to start a vehicle with a discharged battery.

Jumper Wires   Used to temporarily bypass circuits or circuit components for electrical testing. They consist of a length of wire with an alligator clip at each end. Used to test circuit breakers, relays, lights, and other components.

Jumpout   This occurs when a fully engaged gear and sliding clutch are forced out of engagement.

Kinetic Energy   The energy of motion.

Kingpin   The pin mounted through the center of the trailer upper coupler (bolster plate) that mates with the fifth wheel locks, securing the trailer to the fifth wheel. Also, the linkage pin in a steering knuckle.

Land   Raised surface beside a groove.

Landing Gear   Retractable supports for a semi-trailer to keep the trailer level when the truck/tractor is detached from it.

Lead-Acid Battery   Standard vehicle battery consisting of lead acid cells arranged in series.

Light Beam Alignment System   Alignment system (or a laser alignment system) that uses wheel-mounted instruments to project light beams onto charts and scales to measure toe, caster, and camber, and notes the results of alignment adjustments.

Load Proportioning Valve (LPV)   Used to redistribute hydraulic pressure to front and rear brakes based on vehicle loads. This is a load- or height-sensing valve that senses the vehicle load and proportions the braking between front and rear brakes.

Locking Pliers   Also called Vise-Grips™, these pliers are similar to standard pliers, except that they can be locked closed with a tight grip.

Lockstrap   A manual adjustment mechanism that allows for adjustment of clutch free travel.

Lockup Torque Converter   Eliminates the 10 percent slip that takes place between the impeller and turbine at the coupling stage of operation.

Low-Maintenance Batteries   Conventionally vented lead-acid batteries, requiring normal periodic battery maintenance.

Machinist's Rule   A steel ruler used for measuring. These rulers are available in standard or metric measurements.

Magnetic Base Protractor   Used to determine shaft angles; also called an electronic inclinometer.

Maintenance-Free Batteries   Batteries that do not require the addition of distilled water during normal service life.

Maintenance Manuals   Publications that contain routine maintenance procedures and intervals for vehicle components and systems.

Metering Valve   Used on vehicles equipped with front disc and rear drum brakes. Improves braking balance during light brake applications by preventing application of the front disc brakes until pressure is built up in the hydraulic system.

Micrometer   The common instrument for taking measurements of the outside or inside diameter of a shaft or bore.

Modem   Telecommunications device that translates digital signals to analog for transmission.

Multiple Disc Clutches Transmission clutches that have a drum-shaped housing that can be either a separate casting or part of the transmission housing.

NATEF (National Automotive Technicians Education Foundation) Organization that sets standards and certifies secondary and postsecondary automotive and truck training programs.

National Institute for Automotive Service Excellence (ASE) Institute that has established certification programs for automotive, heavy-duty truck, auto body repair, and engine machine shop technicians, and part specialists.

Needlenose Pliers Tool with long tapered jaws for grasping small parts or for reaching into tight spots. Many needlenose pliers also have wire cutting edges and a wire stripper.

NHTSA National Highway Traffic Safety Administration.

NIOSH National Institute for Occupational Safety and Health.

NLGI National Lubricating Grease Institute.

Nonparallel or "Broken Back" Type In this type of drive shaft installation, the working angles of the U-joints of a given shaft are equal; however the companion flanges and/or yokes are not parallel.

Nose The semi-trailer front.

No-tilt Convertible Fifth A fifth wheel with fore/aft articulation that can be locked out to produce a rigid top plate for applications that have either rigid and/or articulating upper couplers.

OEM Original Equipment Manufacturer.

Ohm Unit of measurement of electrical resistance.

Ohm's Law The basic law of electrical circuits stating that current, resistance, and pressure work together in a mathematical relationship.

Open Circuit An electrical circuit whose path has been interrupted or broken either accidentally (a broken wire) or intentionally (a switch turned off).

Open-End Wrench The jaws of this wrench slide around bolts or nuts where there might be insufficient clearance above or on one side of the nut to accept a box wrench.

Operational Control Valves Used to control the flow of compressed air through the brake system.

OSHA Occupational Safety and Health Administration.

Out-of-Phase A condition of the universal joint which produces an effect somewhat like one man snapping a rope held by a man at the opposite end, resulting in a violent reaction at the opposite end. If both men snap the rope at the same time, the resulting waves cancel each other, a condition that can be described as in-phase.

Output Driver The switching apparatus in an ECM that effects the results of logic processing.

Ovality Something not circular but oval in shape.

Overdrive Gearing of a transmission so that one revolution of the engine produces more than one revolution of the transmission's output shaft.

Overrunning Clutch A clutch mechanism that transmits torque in one direction only.

Oxidation Inhibitor An additive used with lubricating oils; used to keep oil from oxidizing even at very high temperatures.

Packer Garbage truck.

Parallel Circuits Circuits that provide two or more paths for current flow.

Parallel Joint Type In the parallel joint type of driveshaft installation, all companion flanges and/or yokes in the complete driveline are parallel to each other with the working angles of the joints of a given shaft being equal and opposite.

Parts Requisition A form used to order new parts, on which the technician writes the part(s) needed along with the vehicle's VIN or company's identification folder.

PC Personal computer. Used for parameter display, vehicle diagnostics, and data storage.

Phillips Screwdriver A tool whose tip has four radial prongs that fit the four slots in a Phillips-head screw.

Pitman Arm A steering linkage component that connects the steering gear to the drag link.

Pitting Surface irregularities resulting from corrosion.

Planetary Gear Set Gearing modeled after the solar system. A pinion is surrounded by an internal ring gear and planet gears are in mesh between the ring gear and pinion around which all revolve.

Planetary Pinion Gears Small gears fitted into a framework called the planetary carrier.

Pliers Gripping tool with jaws used for working with wires, clips, and pins.

Polarity A state, either positive or negative, of charge differential.

Posi-Drive™ Screwdriver This tool is like a Phillips screwdriver, but with a tip that is flatter and blunter. The squared tip grips the screw's head and slips less than a Phillips screwdriver.

Power The rate of performing work.

Power Synchronizer Speeds up the rotating of the main section gearing for smoother automatic downshifts and slows down the rotation of the main section gearing for smoother automatic upshifts.

Power Train Consists of an engine, clutch, transmission, driveshafts, differential carriers, drive axles, and wheels.

Pressure Differential The difference in pressure between any two points of a system or a component.

Pressure Relief Valve A valve located on the supply tank, usually preset at 150 psi. Limits system pressure if the compressor or governor unloader valve malfunctions.

Printed Circuit Board   Board made of thin sheets of nonconductive plastic onto which conductive metal such as copper has been deposited. Parts of the metal are then etched by acid. The remaining metal lines form the conductors for the various circuits on the board. A printed circuit board can hold many complex circuits in a very small area.

Priority Valve   Ensures that the control system upstream from the valve will have sufficient pressure during shifts to perform its functions.

Programmable Read Only Memory (PROM)   Chip or card that qualifies or enhances ROM data.

Proportioning Valve   Valve used on vehicles equipped with front disc and rear drum brakes. Installed in the lines to the rear drum brakes, and in a split system below the pressure differential valve. By reducing pressure to the rear drum brakes, the valve helps to prevent premature lockup during severe brake application and provides better braking balance.

Protocol   Networking language and rules.

psi   Measurement of pressure in pounds per square inch.

P-type Semiconductors   Positively charged materials. P-type semiconductors are produced by doping a semiconductor crystal.

Pull Circuit   Hydraulic circuit that brings the cab from a fully tilted position, up and over center.

Pull-Type Clutch   Type of clutch that pulls the release bearing toward the transmission for release.

Pulse Generator   Reluctor that generates an AC voltage.

Pump/Impeller Assembly   The input (drive) member of a torque converter that receives torque from the engine.

Punch   Tool for driving out pins, rivets, or shafts; aligning holes; and marking the starting point for drilling a hole. Punches are designated by point diameter and shank shape.

Push Circuit   Hydraulic circuit that raises the cab from the lowered position to the desired tilt position.

Push-Type Clutch   Clutch released by pedal push.

Quick-Release Valve   Device used to exhaust air as close as possible to the service chambers or spring brakes.

Race   Bearing surface runway.

Radial   Radial tires have cord materials running in a direction from the center point of the tire, usually from bead to bead.

RAM   Random access memory. Main memory.

Ram Air   Air forced into the engine or passenger compartment by the force of the vehicle moving forward.

Random-Access Memory (RAM)   The RAM is used during computer operation to store temporary information. RAM is retained electronically and is therefore volatile.

Ratio Valve   Used on the front or steering axle of a heavy-duty truck to proportion the brake application pressure during service braking.

Reactivity   Characteristic of a material that enables it to react with air, heat, water, or other materials.

Read-Only Memory (ROM)   A type of memory used in automotive microcomputers to store information permanently. ROM is magnetically retained.

Recall Bulletins   OEM bulletins that mandate service work or replacement of parts in connection with a recall notice.

Reference Voltage (V-REF)   Voltage supplied to sensors by the ECM; almost always 5V.

Refrigerant   A liquid capable of vaporizing at low temperatures, such as ammonia, HFC-134a, or R-12.

Refrigerant Management Center   Recovers, recycles, and recharges an A/C system.

Relay   An electric switch that allows a small current to control a much larger one. It consists of a control circuit and a power circuit.

Relay/Quick-Release Valve   A valve used to speed the release of air to the service chambers.

Release Bearing   Unit within the clutch used to disengage the clutch; also known as throwout bearing.

Reserve Capacity Rating (RCR)   The ability of a battery to sustain a minimum vehicle electrical load in the event of a charging system failure.

Resist Bend Moment (RBM)   The most meaningful measurement of frame rail strength. The RBM is derived by multiplying the section modulus of the rail by the yield strength of the material. This term is universally used in evaluating frame rail strength.

Resistance   Opposition to current flow in an electrical circuit.

Resource Conservation and Recovery Act (RCRA) This law states that hazardous materials must be properly stored until an approved hazardous waste hauler arrives to take them to the disposal site.

Right-To-Know Law (or the Hazard Communication Regulation)   Passed by the federal government and administered by the Occupational Safety and Health Administration (OSHA). This law requires that any company that uses or produces hazardous chemicals or substances must inform its employees, customers, and vendors of any potential hazards that may exist in the workplace as a result of using the products.

Rigid Discs   Steel clutch plates to which friction linings, or facings, are bonded or riveted.

Rigid Fifth Wheel   A fifth wheel fixed rigidly to a frame. This fifth wheel has no articulation or oscillation. Used in applications where the articulation is provided by other means, such as an articulating upper coupler of a frameless dump.

Rim Component on which a tire is mounted.

Ring Gear Spur gear or outer member of a planetary gearset.

Rivet A headed pin used for uniting two or more pieces by passing the shank through a hole in each piece, and securing it by forming a head on the opposite end.

Roller Clutch Designed with a movable inner race, rollers, accordion (apply) springs, and outer race. Around the inside diameter of the outer race are several cam-shaped pockets. The clutch assembly rollers and accordion springs are located in these pockets.

Rotary Oil Flow Caused by the centrifugal force applied to the fluid as the converter rotates.

Rotation Term used to describe the motion of a gear, shaft, or wheel.

Rotor Part of an alternator that provides the magnetic fields necessary to create a current flow.

rpm Revolutions per minute.

Runout Deviation from specified travel of a rotating object. It is the amount of deviation or wobble a shaft or wheel has as it rotates. Runout is measured with a dial indicator.

Safety Factor (SF) The amount of load which can safely be absorbed by and through the vehicle chassis frame members.

Screw Pitch Gauge A quick and accurate method of measuring the threads per inch.

Secondary Lock The component(s) of a fifth wheel locking mechanism that back up the primary locks. The secondary lock is not required for the fifth wheel to function and can be either manually or automatically applied. On some designs, engagement of the secondary lock can only be accomplished if the primary lock is properly engaged.

Section Height The tread center-to-bead plane on a tire.

Section Width The measurement on a tire from sidewall to sidewall.

Self-Adjusting Clutches Clutches that automatically take up the slack between the pressure plate and clutch disc as wear occurs.

Semiconductor Material or device that can function as either a conductor or an insulator, depending on how its structure is arranged.

Semi-floating Axle In the semi-floating type axle, drive torque from the differential is taken by each axle half-shaft and transferred to the wheels. A single bearing assembly, located at the outer end of the axle, is used to support the axle half-shaft.

Semi-oscillating The standard fifth wheel that oscillates or articulates about an axis perpendicular to the vehicle centerline.

Semi-trailer The standard highway trailer supported at its front end by the tractor fifth wheel.

Sensing Voltage Allows the voltage regulator to sense and monitor the battery voltage value.

Sensor Energy conversion device used to signal a condition or state.

Series Circuit Circuit connected to a voltage source with only one path for electron flow.

Series/Parallel Circuit Circuit designed so that both series and parallel branches exist within the same circuit.

Service Bulletins Publications that provide the latest service tips, field repairs, product improvements, and related information of benefit to service personnel.

Service Manual Book published by the manufacturer that contains service and repair information for vehicle systems and components.

Shift Bar Housing Transmission subassembly that houses shift rails, shift yokes, detent balls and springers, interlock balls, and pin and neutral shaft.

Shift Forks or Yoke Y-shaped components located between the gears on the main shaft that, when actuated, cause the gears to engage or disengage via the sliding clutches. Shift forks are located between low and reverse, first and second, and third and fourth gears in a standard main section.

Shims Gaskets or spacers of a specific thickness.

Shock Absorbers Hydraulic devices used to dampen vehicle spring oscillations by controlling body sway and wheel bounce.

Short Circuit An undesirable connection path between two electrical wires. A short occurs when the insulation is worn between two adjacent wires and the metal in each wire contacts the other.

Single Reduction Axle Any axle assembly that uses only one gear reduction through its differential carrier assembly.

Slave Valve Any air or hydraulic actuator.

Sliding Fifth Wheel A fifth wheel design that enables forward and rearward movement, which affects the weight distribution over the tractor axles and/or overall length of the tractor and trailer.

Slip Rings and Brushes Conduct current to the rotor. Most alternators have two slip rings mounted directly on the rotor shaft; they are insulated from the shaft and from each other. A spring-loaded carbon brush is located on each slip ring to carry the current to and from the rotor windings.

Slipout Occurs when pulling with full power or decelerating with the load pushing. Tapered or worn clutching teeth will try to "walk" apart as the gears rotate, causing the sliding clutch and gear to slip out of engagement.

Slip Yoke Allows the driveshaft to increase in length to accommodate movements of the drive axles.

Socket Wrench A wrench that encloses a bolt or nut and is torqued by a gun or ratchet.

Software   Computer program instructions that manage the processing cycle.

Solenoid   An electromagnet used to perform work. A solenoid is made with one or two coil windings wound around an iron core.

Solid-State Devices   Semiconductor devices, including transistors and diodes.

Solvents   Substances which dissolve other substances.

Spalling   Surface fatigue that occurs when chips, scales, or flakes of metal break off due to fatigue rather than wear.

Specialty Service Shops   Shops that specialize in areas such as engine rebuilding, transmission/axle overhauling, brake work, air conditioning/heating repairs, or electrical work.

Specific Gravity   Scientific measurement of liquid weight based on the ratio of the liquid's mass to an equal volume of distilled water. Commonly used to measure battery state of charge.

Spike   Trailer service brake valve.

Spiral Bevel Gear   Helical gear arrangement that has a drive pinion mesh with a ring gear at the centerline axis of the ring gear. This gearing provides strength and allows for quiet operation.

Splitter Transmission   Standard transmission in which main box gearing has been compounded by an auxiliary section.

Spontaneous Combustion   Process by which a combustible material auto-ignites.

Spring Brake Chamber   A means of providing mechanical parking for an air brake system.

Spring Brake Valve   Limits the hold-off pressure to the spring brake chambers and exhausts hold-off air for parking.

Stall Test   Test performed to determine whether the engine or transmission is at fault.

Stand Pipe   A type of U-joint check valve which prevents reverse flow of the hot liquid lubricant generated during operation. When the universal joint is at rest, one or more of the cross trunnions will be up and without the stand pipe, lubricant would flow out of the trunnion bore.

Starter Circuit   Supplies high current for engine cranking.

Starter Motor   Converts electrical energy from the battery into mechanical energy for cranking the engine.

Starting Safety Switch   Prevents vehicles with automatic transmissions from being started in gear.

Static Balance   Balance at rest. The equal distribution of the weight of the wheel and tire around the axis of rotation so that the wheel assembly has no tendency to rotate by itself regardless of its position.

Stationary Fifth Wheel   A fifth wheel whose location on the tractor frame is fixed once it is installed.

Stator   Located between the torque converter pump/impeller and turbine. It redirects the oil flow from the turbine back into the impeller in the direction of impeller rotation with minimal loss of speed or force.

Stator Assembly   Reaction member or torque multiplier of a torque converter. The stator is supported on a freewheel roller race that is splined to the valve and front support assembly.

Steering Gear   The steering control gearing on a chassis also called a steering box.

Storage Batteries   Batteries that store electrical energy.

Straight Truck   Truck with cargo body mounted on chassis.

Stranded Wires   Wires consisting of a number of small gauge wires twisted together to form a single conductor.

Stud Removers   Stud removers have hardened, knurled, or grooved eccentric jaws that grip the stud tightly when operated—used to remove and install studs.

Sulfation   Occurs when sulfate is allowed to build up on battery plates. Sulfation causes two problems—first, it lowers the specific gravity levels and increases the danger of freezing at low temperatures. Second, in cold weather a sulfated battery may not have the reserve power needed to crank the engine.

Suspension   Means by which axles are attached to the vehicle frame and support the frame and is designed to dampen road shocks through springs (leaf, air, torsion, or other), reducing the forces acting on the frame.

Swaging   To reduce or taper.

Switches   Control on/off conditions and direct the flow of current in a circuit.

System Protection Valves   Protect the brake system against an accidental loss of air pressure, buildup of extreme pressure, or backflow and reverse airflow.

Tachometer   An instrument indicating rotational speed. Tachometers are used to indicate crankshaft rpm.

Tailshaft   Output shaft of the transmission.

Tandem Drive Axle   Combines two axle assemblies through the use of an interaxle differential or power divider and a short shaft that connects the two axles.

Thickness Gauge   Correct term for what is more commonly referred to as feeler gauge.

Three-Speed Differential   Tandem two-speed axle arrangement with the capability of operating two drive axles at different speed ranges simultaneously. The third speed is actually an intermediate speed between the high and low range.

Throw   The number of output circuits in a switch.

Tie-Rod Assembly Transfers the steering motion to the opposite, passenger-side steering knuckle. It links the two steering knuckles and forces them to act in unison. Also called a cross-tube.

Time Guide Used for computing compensation payable by the truck manufacturer for repairs or service work to vehicles under warranty.

Timing Procedure of marking the gear set teeth prior to installation so they can be placed in proper mesh while in the transmission.

Toe A suspension dimension that reflects the difference in the distance between the extreme front and rear of the tire.

Toe-In When the front end of the tire points inward toward the vehicle.

Toe-Out When the front of the tire points outward from the vehicle.

Torque Twisting force.

Torque Converter A fluid coupling using a rotary pump, stators, and a turbine. Torque is transmitted from a driving to a driven member by hydraulic action. It provides varying drive ratios; with speed reduction, it increases torque.

Torque Limiting Clutch Brake Designed to slip when loads of 20–25 pound-feet are reached. This protects the clutch brake from overloading.

Torque Rod Shims Function to rotate the axle pinion to change the U-joint operating angle.

Torque Wrench Wrench with a built-in indicator to measure applied twisting force.

Torsion Bar Suspension Type of suspension system using torsion bars in place of the commonly used steel leaf springs. The typical torsion bar suspension consists of a torsion bar, front crank, rear crank with associated brackets, a shackle pin, and assorted bushings and seals.

Total Pedal Travel The complete distance the clutch pedal must move.

Toxicity A measure of how poisonous a substance is.

Tractor Heavy-duty highway vehicle designed for towing trailers.

Tractor Protection Valve Automatically isolates the tractor air supply from the trailer air supply when tractor system pressure drops to between 25 and 40 psi; also called breakaway valve.

Transfer Case An additional gearbox located between the main transmission and the rear axle. Its function is to transfer torque from the truck's transmission to both front and rear driving axles.

Transistor Three terminal semiconductor devices used to switch and amplify electronic circuits.

Transverse Vibrations Unbalanced driveline vibrations.

Treadle (or foot valve) A dual brake valve that meters air from service reservoirs to the service lines and brake chambers.

Tree Diagnosis Chart Provides a logical sequence for what should be inspected or tested when trying to solve a repair problem.

Trunnions The ends of the universal cross; they are case-hardened, ground surfaces on which the needle bearings ride.

Truth Table Chart showing all outcomes possible from a set of inputs.

Turbine The output (driven) member of a torque converter splined to the forward clutch of the transmission and to the turbine shaft assembly.

Two-Speed Axle Assembly (often referred to as a dual range unit) Two output ratio differential. The driver selects the ratios from controls located in the cab of the truck.

Universal Joints (U-joints) U-joints allow torque to be transmitted to components that are operating at different angles.

Uplink Ground-to-satellite transmission hardware.

Valence Shell Outermost orbital shell in an atom.

Valve Body and Governor Test Stand Special automatic transmission test equipment. The valve body of the transmission is removed from the vehicle and mounted into the test stand. The test stand duplicates vehicle running conditions, so the valve body can be tested and calibrated.

Valve Ring Adjusting Tool Adjustment of automatic transmission valves is performed using this tool.

Variable Pitch Stator Design used in off-highway torque converters.

Vehicle Body Clearance (V/C) Distance from the inside of the inner tire to the spring or other body structures.

Vehicle Retarder Auxiliary braking device used to supplement the service brakes on heavy-duty trucks.

VIN Vehicle Identification Number.

Viscosity Describes oil thickness or flow resistance.

Volt Unit of electrical pressure, electromotive force, or charge differential.

Voltage Generating Sensors Devices which can produce their own input voltage signal.

Voltage Limiter Provides protection by limiting voltage to the gauges and instruments to approximately 5 volts.

Voltage Regulator Controls the voltage in the charging circuit.

VORAD (Vehicle On-board Radar) The VORAD vehicle collision warning system uses Doppler radar to give drivers additional reaction time to respond to potential danger.

Vortex Oil Flow Circular flow that occurs as oil is forced from the impeller to the turbine and then back to the impeller during torque converter operation.

Watt's Law A basic law of electricity used to calculate the power of an electrical circuit expressed in watts. It states that power equals the voltage multiplied by the current (in amperes).

Wattage The measure of electrical power.

Wear Compensator The wear compensator is mounted in the clutch cover and has an actuator arm that fits into a hole in the release sleeve retainer.

Wedge-Actuated Brakes Wedge brakes use air pressure and air brake chambers to force a wedge and roller assembly into an actuator located between adjusting and anchor pistons.

Wheel and Axle Speed Sensors Reluctor-type electromagnetic devices used to obtain vehicle speed information for an antilock controller.

Wheel Balance The equal distribution of weight in a wheel with the tire mounted. An important factor which affects tire wear and vehicle control.

Wheel Hop Wheel or axle oscillation that causes irregular tire-to-road contact.

Wheel Shimmy Steering oscillations either side of the intended steering track.

Windings Electrical coils. The three separate bundles in which wires are grouped in a stator.

Wind-up Axle rotation-caused spring-leaf reactions to braking and acceleration.

Worm Gear Spiraled single-groove gear.

Wrecker A tow truck.

Wrench A tool for twisting and/or holding bolt heads or nuts. The width of the jaw opening determines its size.

Yield Strength The highest stress a material can stand without permanent deformation, expressed in pounds per square inch (psi).

Yoke Fork shaped coupling component.

Yoke Sleeve Kit Installed instead of completely replacing a worn yoke. The sleeve is of heavy walled construction with a hardened steel surface.

Zener Diode Specialty diode designed to conduct with reverse-bias current after a specific voltage value is reached. Often used in electronic voltage regulators.

Zerk A standard grease nipple fitting.

Index